HAESE MATHEMATICS
Specialists in mathematics publishing

Mathematics
for the international student

9
MYP 4

second edition

Michael Haese
Sandra Haese
Mark Humphries
Edward Kemp
Pamela Vollmar

for use with
IB Middle Years
Programme

MATHEMATICS FOR THE INTERNATIONAL STUDENT 9
MYP 4 second edition

Michael Haese	B.Sc.(Hons.), Ph.D.
Sandra Haese	B.Sc.
Mark Humphries	B.Sc.(Hons.)
Edward Kemp	B.Sc., M.A.
Pamela Vollmar	B.Sc.(Hons.), PGCE.

Haese Mathematics
152 Richmond Road, Marleston, SA 5033, AUSTRALIA
Telephone: +61 8 8210 4666, Fax: +61 8 8354 1238
Email: info@haesemathematics.com.au
Web: www.haesemathematics.com.au

National Library of Australia Card Number & ISBN 978-1-921972-49-2

Published by Haese Mathematics.
152 Richmond Road, Marleston, SA 5033, AUSTRALIA

First Edition	2008
Reprinted	2009 (twice), 2010, 2011, 2012
Second Edition	2014

Cartoon artwork by John Martin. Artwork by Brian Houston and Gregory Olesinski.

Cover design by Piotr Poturaj.

Typeset in Australia by Deanne Gallasch and Charlotte Frost. Typeset in Times Roman 10.

Computer software by Adrian Blackburn, Ashvin Narayanan, Tim Lee, Seth Pink, Nicole Szymanczyk, Brett Laishley, and Linden May.

Production work by Gregory Olesinski, Katie Richer, Anna Rijken, and Robert Haese.

Printed in Singapore by Opus Group.

Acknowledgements: While every attempt has been made to trace and acknowledge copyright, the authors and publishers apologise for any accidental infringement where copyright has proved untraceable. They would be pleased to come to a suitable agreement with the rightful owner.

FOREWORD

MYP 4 second edition has been designed and written for the IB Middle Years Program (MYP) Mathematics framework.

This book may also be used as a general textbook at about 9th Grade level in classes where students complete a rigorous course in mathematics. We have developed this book independently of the International Baccalaureate Organization (IBO) in consultation with experienced teachers of IB Mathematics. The text is not endorsed by the IBO.

It is not our intention that each chapter be worked through in full. Teachers must select carefully, according to the abilities and prior knowledge of their students, to make the most efficient use of time and give as thorough coverage of content as possible.

Each chapter begins with an Opening Problem, offering an insight into the application of the mathematics that will be studied in the chapter. Important information and key notes are highlighted, while worked examples provide step-by-step instructions with concise and relevant explanations. Discussions, Activities, Investigations, Puzzles, and Research exercises are used throughout the chapters to develop understanding, problem solving, and reasoning, within an interactive environment.

Two additional chapters are available online:

Chapter 26: Vectors
Chapter 27: Logic

We understand the emphasis that the IB MYP places on the six Global Contexts, and in response there are online links to ideas for projects and investigations to help busy teachers (see p. 6).

Frequent use of the interactive online features should nurture a much deeper understanding and appreciation of mathematical concepts. The inclusion of our Self Tutor software (see p. 4) is intended to help students who have been absent from classes or who experience difficulty understanding the material.

The book contains many problems to cater for a range of student abilities and interests, and efforts have been made to contextualise problems so that students can see the practical applications of the mathematics they are studying.

We welcome your feedback. Email: info@haesemathematics.com.au
Web: www.haesemathematics.com.au

PMH, SHH, MH, EK, PV

ACKNOWLEDGEMENTS

The authors and publishers would like to thank all those teachers who have read proofs and offered advice and encouragement.

ONLINE FEATURES

There are a range of interactive features which are available online.
With the purchase of a new hard copy textbook, you will gain 15 months subscription to our online product. This subscription can be renewed annually for a small fee.

COMPATIBILITY

For iPads, tablets, and other mobile devices, the interactive features may not work. However, the electronic version of the textbook and additional chapters can be viewed online using any of these devices.

REGISTERING

You will need to register to access the online features of this textbook.
Visit www.haesemathematics.com.au/register and follow the instructions. Once you have registered, you can:

- activate your electronic textbook
- use your account to make additional purchases.

To activate your electronic textbook, contact Haese Mathematics. On providing proof of purchase, your electronic textbook will be activated. **It is important that you keep your receipt as proof of purchase.**

For general queries about registering and licence keys:

- Visit our Frequently Asked Questions page: www.haesemathematics.com.au/faq.asp
- Contact Haese Mathematics: info@haesemathematics.com.au

ONLINE VERSION OF THE TEXTBOOK

The entire text of the book can be viewed online, allowing you to leave your textbook at school.

The online text contains the two additional chapters:

- **Chapter 26: Vectors**
- **Chapter 27: Logic**

SELF TUTOR

Self tutor is an exciting feature of this book.

The Self Tutor icon on each worked example denotes an active online link.

Simply 'click' on the Self Tutor (or anywhere in the example box) to access the worked example, with a teacher's voice explaining each step necessary to reach the answer.

Play any line as often as you like. See how the basic processes come alive using movement and colour on the screen.

For example:

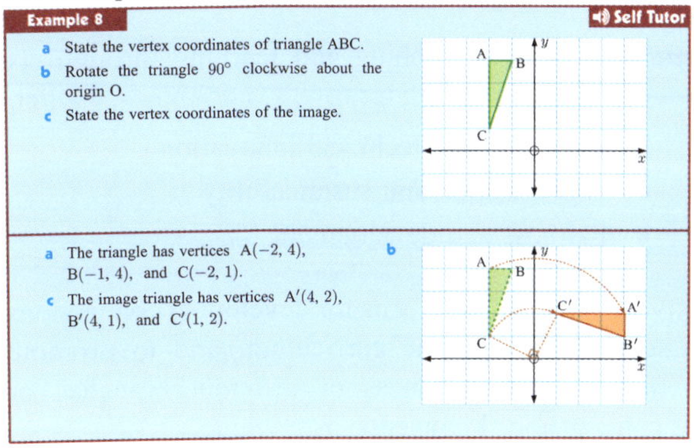
Example 8

a State the vertex coordinates of triangle ABC.

b Rotate the triangle 90° clockwise about the origin O.

c State the vertex coordinates of the image.

a The triangle has vertices $A(-2, 4)$, $B(-1, 4)$, and $C(-2, 1)$.

b

c The image triangle has vertices $A'(4, 2)$, $B'(4, 1)$, and $C'(1, 2)$.

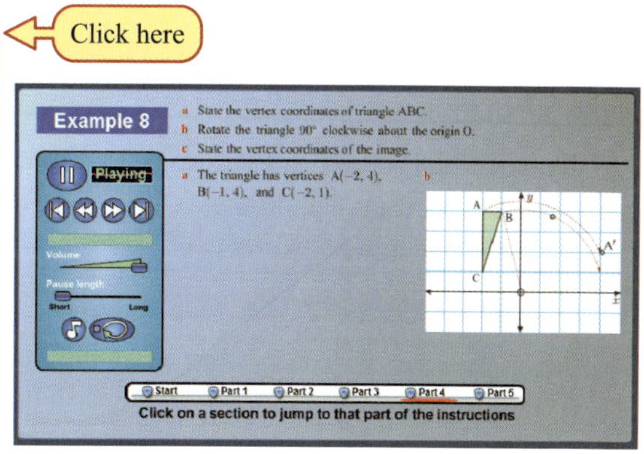

See **Chapter 16**, **Transformation geometry**, p. 334

INTERACTIVE LINKS

Throughout your electronic textbook, you will find interactive links to:

- Graphing software
- Statistics packages
- Geometry packages
- Games
- Demonstrations
- Printable pages

CLICK ON THESE ICONS ONLINE

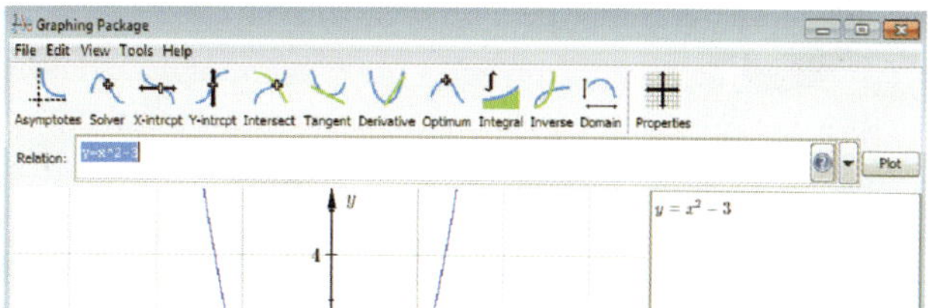

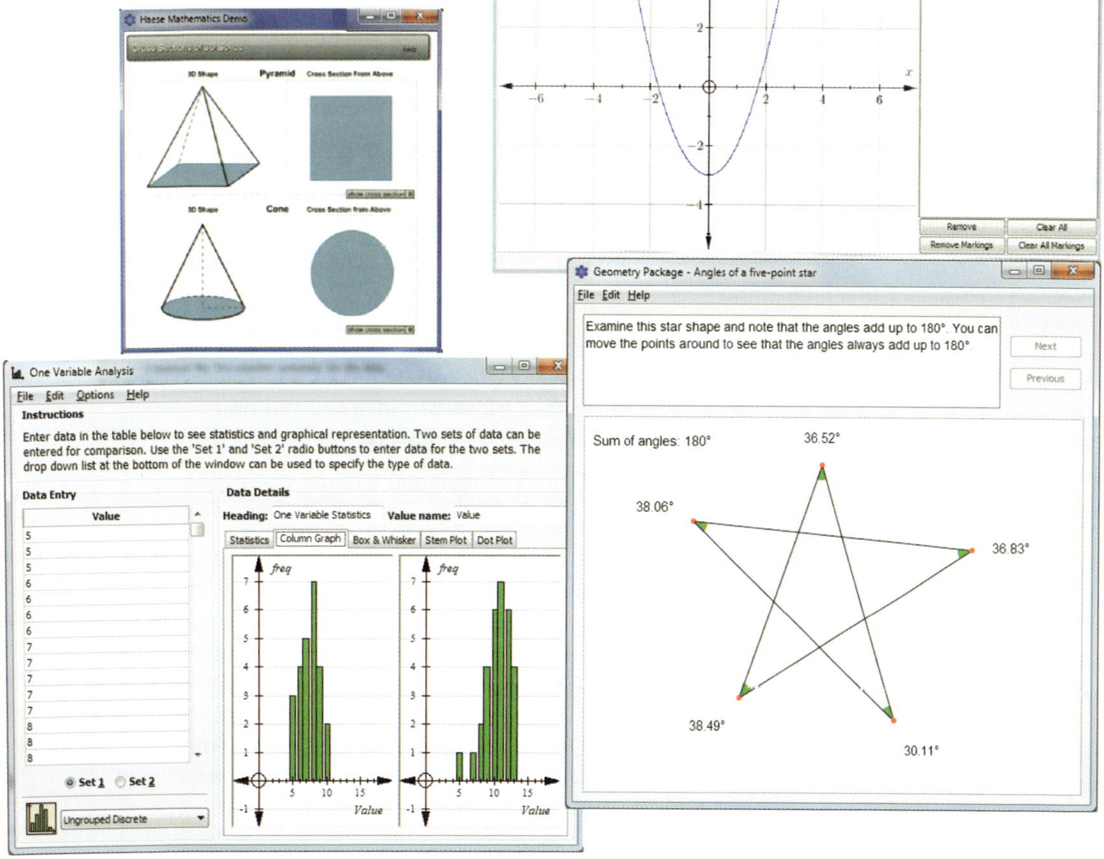

GLOBAL CONTEXTS

The International Baccalaureate Middle Years Programme focuses teaching and learning through six Global Contexts:

- Identities and relationships
- Orientation in space and time
- Personal and cultural expression
- Scientific and technical innovation
- Globalisation and sustainability
- Fairness and development

Click on the heading to access the online link.

The Global Contexts are intended as a focus for developing connections between different subject areas in the curriculum, and to promote an understanding of the interrelatedness of different branches of knowledge and the coherence of knowledge as a whole.

Global context

click here

What is the golden ratio?

Statement of inquiry:	Mathematics can help us to understand beauty in the world around us.
Global context:	Personal and cultural expression
Key concept:	Relationships
Related concepts:	Quantity, Space
Objective:	Knowing and understanding
Approaches to learning:	Thinking, Research

There are six projects in this book, one for each of the Global Contexts:

Each project contains a series of questions, divided into:

- Factual questions (in green)
- Conceptual questions (in blue)
- Debatable questions (in red).

These questions should help guide the unit of work.

The projects are also accompanied by the general descriptor and a task-specific descriptor for each of the relevant assessment criteria, to help teachers assess the unit of work.

TABLE OF CONTENTS

Extension questions and exercises are marked in red

GRAPHICS CALCULATOR INSTRUCTIONS

Graphics calculator instruction booklets are available for the **Casio fx-9860G Plus**, **Casio fx-CG20**, **TI-84 Plus**, and the **TI-*n*spire**. Click on the relevant icon below.

CASIO fx-9860G Plus | CASIO fx-CG20 | TI-84 Plus | TI-*n*spire

When additional calculator help may be needed, specific instructions are available from icons within the text.

GRAPHICS CALCULATOR INSTRUCTIONS

Chapter 1

Algebra

Contents:

OPENING PROBLEM

Holly bought XBC shares for \$2.50 each and NGL shares for \$4.00 each.

Things to think about:

a How much would Holly pay in total for 500 XBC shares and 600 NGL shares?

b Write an expression for the total amount Holly would pay for x XBC shares and $(x+100)$ NGL shares.

c Check your answer to **a** by substituting $x = 500$ into the expression from **b**.

Algebra is a powerful tool used to make problem solving easier. Algebra involves using letters such as x to represent unknown values we want to find, or **variables** which can change depending on the situation.

A ALGEBRAIC NOTATION

When we use algebra to solve problems, the ability to understand algebraic notation is essential.

PRODUCT NOTATION

When we simplify **repeated sums**, we use **product notation**.

For example: $x + x = 2 \times x$ {2 lots of x} and $x + x + x = 3 \times x$ {3 lots of x}
$= 2x$ $= 3x$

In algebra we agree:

- to **leave out** the "×" **signs** between any multiplied quantities
- to write **numerals (numbers) first** in any product
- where products contain two or more letters, we write them in **alphabetical order**.

For example:
- $2a$ is used rather than $2 \times a$ or $a2$
- $2ab$ is used rather than $2ba$.

Example 1 Self Tutor

Simplify using product notation:

a $c + c + c - d \times 7$ **b** $n \times 4 \times m - 3 \times n$ **c** $4 \times (x + y)$

a $c + c + c - d \times 7$
$= 3 \times c - d \times 7$
$= 3c - 7d$

b $n \times 4 \times m - 3 \times n$
$= 4mn - 3n$

c $4 \times (x + y)$
$= 4(x + y)$

INDEX NOTATION

When we simplify **repeated products**, we use **index notation**.

For example: $x \times x = x^2$ and $x \times x \times x = x^3$

Example 2 — Self Tutor

Simplify using index notation:

a $p \times p \times p \times p \times 3$ **b** $5 \times z \times z \times z \times y + y \times y$

a $p \times p \times p \times p \times 3$
$= 3p^4$

b $5 \times z \times z \times z \times y + y \times y$
$= 5yz^3 + y^2$

EXERCISE 1A.1

1 Simplify using product notation:

a $9 \times x$ **b** $f \times 5$ **c** $2 \times a + 3 \times b$
d $a + a - b \times 2$ **e** $a \times c + b + b$ **f** $x - y \times 3 + x$
g $b \times 4 \times a$ **h** $z \times 2y$ **i** $8 \times (a - b)$
j $(c + d) \times 6$ **k** $x \times y \times (z + 5)$ **l** $l \times 3 \times n \times m$

2 Simplify using product notation:

a $k + k + k + k$ **b** $a + a + b + b + b$
c $t + t + t - (u + u)$ **d** $5 + m + m + m$
e $8 - (c + c)$ **f** $9 + g + h + h + h$

3 Simplify using index notation:

a $b \times b \times b \times b$ **b** $4 \times x \times x$
c $y \times y \times y \times 7$ **d** $6 \times q \times q \times q \times p \times p$
e $c \times c \times d \times c \times 2$ **f** $5 \times m \times m + n \times n \times n \times 6$

4 Write in expanded form:

a y^3 **b** $4x^4$ **c** $3a^2b$
d $(6p)^2$ **e** $2g^2 + h^3$ **f** $m^5 - 7n^2$

5 Simplify:

a $9 \times a - a \times b \times 4 \times b$ **b** $t \times t + s \times t \times s$
c $2 \times x \times y \times y \times z - z \times z \times 3$ **d** $a \times a \times a - (a + a)$
e $x + x + x - 8 \times y \times y$ **f** $m \times n \times m \times n \times m - m \times 2$
g $d + d + d - d \times c \times 6$ **h** $s + s + s \times t \times t \times 5 \times t$
i $g \times g \times g \times 4 - (g + g + g)$ **j** $x + x + 5 - x \times y \times y$
k $a \times 3 \times b \times b + b \times a + b$
l $v \times v \times 10 \times w \times v - 6 \times w \times v \times w + 7 \times w$

ALGEBRAIC EXPRESSIONS

An **algebraic expression** is a form which contains numbers, variables, and operation signs.

$2x+3$, $5a-4b$, and $\dfrac{t^2-6}{r}$ are all examples of algebraic expressions.

It is important to be able to convert worded phrases into algebraic expressions, and vice versa.

Example 3 Self Tutor

In words, write the meaning of:

a $x-5$ **b** $a+b$ **c** $3x^2+7$

a $x-5$ is "5 less than x".

b $a+b$ is "the sum of a and b" or "b more than a"

c $3x^2+7$ is "7 more than three times the square of x".

Example 4 Self Tutor

Write the following as algebraic expressions:

a the sum of p and the square of q **b** the square of the sum of p and q

c b less than double a.

a $p+q^2$ **b** $(p+q)^2$ **c** $2a-b$

EXERCISE 1A.2

1 In words, write the meaning of:

a $2a$ **b** pq **c** $\sqrt{m}$ **d** a^2 **e** $a-3$ **f** $b+c$

g $2x+c$ **h** $(2a)^2$ **i** $2a^2$ **j** $a-c^2$ **k** $a+b^2$ **l** $(a+b)^2$

2 Write the following as algebraic expressions:

a the sum of a and c

b the sum of p, q, and r

c the product of a and b

d the sum of r and the square of s

e the square of the sum of r and s

f the sum of the squares of r and s

g the sum of twice a and b

h 5 more than the cube of p

i a less than the square of b

j half the sum of a and b

k the sum of a and a quarter of b

l the square root of the sum of m and n

m the sum of x and its reciprocal

n a quarter of the sum of a and b

o the square root of the sum of the squares of x and y

3 I think of a number, square it, subtract 13, then find the square root of the result. If x is the original number, write an expression for the final result.

4 A jacket costs €50 more than a jumper, and a coat costs €20 more than a jacket. Suppose the cost of a jumper is €x. Write an algebraic expression for the cost of:

a a jacket **b** a coat.

5 Elise is thinking about the price of a holiday. Suppose a one-way flight costs $\$x$. The accommodation each night costs $30 more than each flight. Elise stays for 5 nights. Write an algebraic expression for the:

a total cost of the flights **b** total cost of the hotel

c total cost of the flights and hotel.

ALGEBRAIC EQUATIONS

An **algebraic equation** is a statement that two algebraic expressions are equal.

We separate the algebraic expressions with an **equals** sign.

Example 5 Self Tutor

Write, in sentence form, the meaning of:

a $D = ct$ **b** $A = \frac{b+c}{2}$

a D is equal to the product of c and t.

b A is equal to a half of the sum of b and c,
or, A is the average of b and c.

Example 6 Self Tutor

Write as an equation:
S is equal to the sum of a and the product of g and t.

The product of g and t is gt.

The sum of a and gt is $a + gt$.

So, the equation is $S = a + gt$.

EXERCISE 1A.3

1 Write, in sentence form, the meaning of:

a $L = a + b$ **b** $K = \frac{a+b}{2}$ **c** $M = 3d$

d $N = bc$ **e** $T = bc^2$ **f** $F = ma$

g $K = \sqrt{\frac{n}{t}}$ **h** $c = \sqrt{a^2 + b^2}$ **i** $A = \frac{a+b+c}{3}$

$\frac{n}{t}$ is called the **quotient** of n and t.

2 Write the following as algebraic equations:

- **a** S is the sum of p and r
- **b** D is the difference between a and b, where $b > a$
- **c** A is the average of k and l
- **d** M is the sum of a and its reciprocal
- **e** K is the sum of t and the square of s
- **f** N is the product of g and h
- **g** y is the sum of x and the square of x
- **h** P is the square root of the sum of d and e

B EVALUATING ALGEBRAIC EXPRESSIONS

To **evaluate** an algebraic expression, we **substitute** numerical values for any unknowns, then calculate the result.

Consider the number crunching machine alongside.

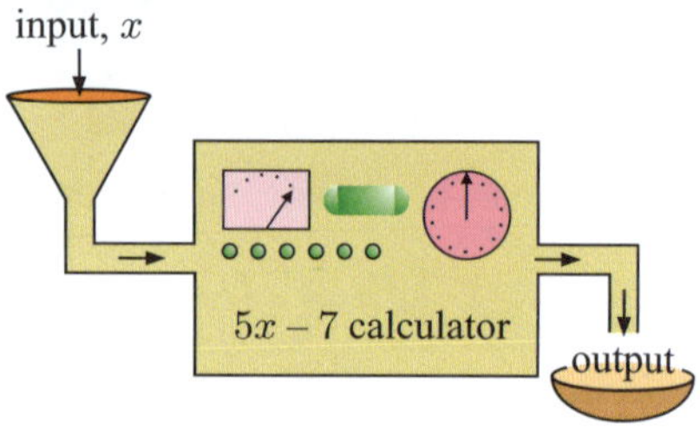

If we place any number x into the machine, it calculates $5x-7$. So, the number is multiplied by 5, and then 7 is subtracted.

For example:

If $x = 2$,
$$\begin{aligned}&5x-7\\&=5\times2-7\\&=10-7\\&=3\end{aligned}$$

If $x=-2$,
$$\begin{aligned}&5x-7\\&=5\times(-2)-7\\&=-10-7\\&=-17\end{aligned}$$

Notice that when we substitute a negative number such as -2, we place it in brackets. This helps us to get the sign of each term correct.

Example 7 **Self Tutor**

If $p=4$, $q=-2$, and $r=3$, find the value of:

a $3q^2-2r$ **b** $2pq+r^3$ **c** $\dfrac{p-2q+2r}{p+r}$

a
$$\begin{aligned}&3q^2-2r\\&=3\times(-2)^2-2\times3\\&=3\times(-2)\times(-2)-6\\&=6\end{aligned}$$

b
$$\begin{aligned}&2pq+r^3\\&=2\times4\times(-2)+3^3\\&=-16+27\\&=11\end{aligned}$$

c
$$\begin{aligned}&\frac{p-2q+2r}{p+r}\\&=\frac{4-2\times(-2)+2\times3}{4+3}\\&=\frac{4+4+6}{4+3}\\&=\frac{14}{7}\\&=2\end{aligned}$$

DISCUSSION

When we substitute a negative value, why should we place brackets around it?

EXERCISE 1B

1 If $k = 3$, $l = 5$, and $m = -2$, find the value of:

a $4k$ **b** $9l$ **c** $2kl$ **d** klm
e $6m + k$ **f** $2k + 4m$ **g** $3m - l$ **h** $8km - 2l$

2 If $p = 4$, $q = -3$, and $r = 2$, evaluate:

a $\frac{p}{q}$ **b** $\frac{p+r}{q}$ **c** $\frac{p+q}{r}$ **d** $\frac{p+4r}{2q}$
e $\frac{p+q+r}{p-r}$ **f** $\frac{8q}{2(p+r)}$ **g** $\frac{3(p+r)}{p+r-2q}$ **h** $\frac{q+5p-r}{r-q}$

3 If $x = -1$, $y = -5$, and $z = 4$, evaluate:

a x^2 **b** z^3 **c** $x^2 + y^2$ **d** $(x + y)^2$
e $x^3 + y^3$ **f** $y^2 - yz$ **g** $(y + z)^3$ **h** $2y^2$

Example 8 Self Tutor

If $p = 4$, $q = -3$, and $r = 2$, evaluate:

a $\sqrt{p-q+r}$ **b** $\sqrt{p+q^2}$

a $\sqrt{p-q+r}$
$= \sqrt{4-(-3)+2}$
$= \sqrt{4+3+2}$
$= \sqrt{9}$
$= 3$

b $\sqrt{p+q^2}$
$= \sqrt{4+(-3)^2}$
$= \sqrt{4+9}$
$= \sqrt{13}$
≈ 3.61

4 If $a = 4$, $b = -1$, and $c = 3$, evaluate:

a $\sqrt{a} + b$ **b** $\sqrt{a+b}$ **c** $\sqrt{a+c}$ **d** $\sqrt{a+bc}$
e $\sqrt{a^2+c^2}$ **f** $\sqrt{b^2+c}$ **g** $\sqrt{a+c-2b}$ **h** $\sqrt{8c+2ab}$

5 Evaluate $\frac{\sqrt{w-2y}}{x^2 z}$ for:

a $w = 13$, $x = 2$, $y = 2$, $z = 4$ **b** $w = -1.5$, $x = -2$, $y = -2.75$, $z = 2.5$

Example 9 **Self Tutor**

A vehicle salesman shows Nigel a red bus and a green bus. The red bus has one more than twice the number of seats as the green bus.

a Let g be the number of seats in the green bus. Write an expression for the number of seats in the red bus.

b If the green bus has 7 seats, find the number of seats in the red bus.

a Let r be the number of seats in the red bus.
Since the red bus has one more than twice the number of seats as the green bus, $r = 2g + 1$.

b If $g = 7$ then $r = 2 \times 7 + 1$
$= 15$

So, if the green bus has 7 seats then the red bus has 15 seats.

6 Reg has an orange tree, a birch tree, and a peach tree in his garden. The birch tree is twice as tall as the orange tree, and the peach tree is 3 metres shorter than the birch tree.

a Let the height of the orange tree be h m. Write an expression for the height of the:

i birch tree **ii** peach tree.

b If the orange tree is 5 metres tall, find the height of the birch tree.

c If the orange tree is 4 metres tall, find the height of the peach tree.

7 There are 24 chocolates in a box. Aislin eats some of the chocolates, then Brindi eats half of the chocolates that are left.

a If Aislin ate x chocolates, find an expression for the number of chocolates that Brindi ate.

b Find the number of chocolates eaten by Brindi if Aislin ate:

i 4 chocolates **ii** 8 chocolates

iii 14 chocolates.

C COLLECTING LIKE TERMS

The **terms** of an expression are algebraic forms separated by + or − signs, the signs being included.
Like terms are terms which contain the same variables (or letters) to the same indices.

For example:
- $3xy$ and $-2xy$ are **like terms**.
- x^2 and $3x$ are **unlike terms** because the indices of x are not the same.

Algebraic expressions can often be simplified by adding or subtracting like terms. We call this **collecting like terms**.

Consider $2a - 3a = \underbrace{a + a}_{2a} \underbrace{- a - a - a}_{-3a} = -a$

Example 10 — Self Tutor

Where possible, simplify by collecting like terms:

a $4x + 3x$ b $5y - 2y$ c $2a - 1 + a$ d $mn - 2mn$ e $a^2 - 4a$

a $4x + 3x = 7x$ {$4x$ and $3x$ are like terms}
b $5y - 2y = 3y$ {$5y$ and $-2y$ are like terms}
c $2a - 1 + a = 3a - 1$ {$2a$ and a are like terms}
d $mn - 2mn = -mn$ {mn and $-2mn$ are like terms}
e $a^2 - 4a$ cannot be simplified since a^2 and $-4a$ are unlike terms.

EXERCISE 1C

1 Simplify, where possible, by collecting like terms:

a $3 + a + 5$ b $2 + 4 + a$ c $b + b + 5$
d $c - 4 + 7$ e $x - 3 + x$ f $4m + m$
g $4m - m$ h $t - 3t$ i $-t - 3t$
j $-t + 3t$ k $a^2 + a$ l $x^2 + x^2 + 2x$
m $2l + 2$ n $a^2 - 3a^2 + 3$ o $2x + 5x - 4$
p $2x + 5x - 4x$ q $3ab + ab$ r $9x^2y - x^2y$

2 Simplify, where possible:

a $6x - 6x$ b $6x - x$ c $6x - 6$
d $2cd + dc$ e $pq - 3pq$ f $3m^2 - m^2$
g $7b^2 - 6 + 2b^2$ h $a + 4 + 3a + 2$ i $a + 4 - 3a - 2$
j $2c + 3d - c + 4d$ k $3a + 8 - a - 2$ l $4x^2 + x - x^2 + 3x$
m $xy + 5 - 6 + 2xy$ n $b^2 - 4b - b^2 - 3$ o $a^2 + 6a + 2a^2 - 2a$
p $4x - y + 5xy$ q $p + q + pq + 2$ r $3x^2 - 8x - x^2 - 3x$

3 Simplify, where possible:

a $3pq + 4pq - qp$ **b** $m^2 - 4 + 2m^2 + 3$ **c** $g + 2 + 3g - 10$

d $-2a + b - 4a - 3b$ **e** $2ab + b + 3b - 2a$ **f** $x^2 + x^3 - x$

g $2x^2 + 4x^3 + 3x^2 - x^3$ **h** $ab^2 + 2ab + a^2b$ **i** $3x^3 - x^2 + 2x^2 - 3x + x^3$

D ALGEBRAIC PRODUCTS

The **product** of two or more factors is the result obtained by multiplying them together.

We use the following steps:

Step 1: Find the product of the **signs**.
Step 2: Find the product of the **numerals** or numbers.
Step 3: Find the product of the **variables** or letters.

For example, the product of the factors $-3x$ and $2x^2$ is:

$$-3x \times 2x^2 = -6x^3$$

$- \times + = -$

$3 \times 2 = 6$

$x \times x^2 = x^3$

Example 11 Self Tutor

Simplify the following products:

a $2x \times -x^2$ **b** $-4ab \times -2b^2$ **c** $5s \times st - 2t \times 3s^2$

a $2x \times -x^2$
$= -2x^3$

b $-4ab \times -2b^2$
$= 8ab^3$

c $5s \times st - 2t \times 3s^2$
$= 5s^2t - 6s^2t$
$= -s^2t$

EXERCISE 1D

1 Simplify the following products:

a $2 \times 5x$ **b** $4x \times 3$ **c** $-2 \times 6x$ **d** $-3x \times 7$

e $3x \times x^2$ **f** $6x \times 3x$ **g** $-4d \times d$ **h** $-9x \times -2x^2$

i $-2x \times x^3$ **j** $-5x \times 3x^2$ **k** $-a \times -a$ **l** $(-2p)^2$

m $3k^2 \times k^2$ **n** $x \times 3x \times 2x^2$ **o** $5t \times -2t^2 \times t$ **p** $-x \times 4x^3 \times -2x$

2 Simplify:

a $3 \times 2x + 4x \times 2$ **b** $5 \times 6a - 7b \times 2b$ **c** $6d^2 + 2d \times 5d$

d $x \times 4y + 3y \times 2x$ **e** $9 \times x^2 - 2x \times 4x$ **f** $5x \times y - 4x \times 3y$

g $6a \times 2b + a \times 3b$ **h** $6x \times y - 5x \times 2y$ **i** $5c \times d - 3c \times 4d$

3 Simplify:

a $7x \times 2y - x^2 \times 3y - 3x \times 3y$

b $5a^2b \times 3ab + 6ab^2 \times 2ab - 4a \times 4a^2b^2$

ACTIVITY — THE ODD ONE OUT

WORKSHEET

Consider the 25 algebraic expressions below.

There are 12 matching pairs of expressions which are equal to each other. For example, the expressions shaded in green are equal.

One of the 25 expressions is 'the odd one out', and does not have a match. Can you find it?

$2x+3y$	$2xy+x^2-xy$	$xy+x^2+3xy-x^2$	$x^3-x^2-x^3$	$x^2y \times 4xy$
$2 \times x \times 3 \times y$	$3x^2-xy+x^2+xy$	x^2+5x	$2 \times y \times x \times 2$	$5-3x-4$
$3+2x^3y^2-2$	$x \times 5 + x \times x$	$(-2x)^2$	$2 \times x + x \times y$	$7x-3x$
$2x+1-5x$	$y \times 5 \times x + xy$	$2x^3 \times 2y^2$	$5x+3y-3x$	$-x+4x$
$x+x+x+x$	$2xy+x-2xy+2x$	$2x \times x - 3x^2$	$2x^2 \times xy^2 + 1$	$xy+x^2$

REVIEW SET 1A

1 Simplify using product notation:

a $8 \times t$ **b** $b \times 7 \times a$ **c** $y+y+y+z+z$

2 Simplify:

a $y+y-y \times y \times y \times y \times 3$

b $a \times a \times b \times a - (b+b)$

c $p \times p \times 5 \times q \times p \times q + p \times q$

d $x \times x + x \times y \times z \times z \times 4$

3 Write in algebraic form:

a "the sum of a and the square root of b"

b "the square root of the sum of a and b"

4 In words, write the meaning of:

a $\sqrt{a}+3$ **b** $\sqrt{a+3}$

5 If $a=-2$, $b=3$, and $c=-5$, find the value of $\dfrac{2(b-c)}{4a}$.

6 Write, in sentence form, the meaning of:

a $T=4s$ **b** $P=\sqrt{\dfrac{x}{y}}$

7 Brothers Lachlan and Simon had \$100 between them to go shopping. Lachlan spent some money, and then Simon spent a quarter of the money that remained.

a If Lachlan spent \$$x$, find an expression for the amount of money that Simon spent.

b Find the amount of money spent by Simon if Lachlan spent:

i \$20 **ii** \$36 **iii** \$60.

8 Simplify, if possible:

a $3+x+5$ **b** $2x-7y+3x+y$ **c** $3a^2+4a-a^2$

d t^2-2t+3 **e** $x^3y+2xy^3+4x^3y$ **f** $x^3+5x+2x^2-5x$

9 Simplify:

a $-4x\times 7$ **b** $5x\times 9x$ **c** $(-3d)^2$

10 Simplify:

a $4\times 4x+3x\times 2$ **b** $6x\times 3y+2x\times xy$ **c** $2a^2\times 7-3a\times 5a$

REVIEW SET 1B

1 In words, write the meaning of:

a $\dfrac{a+b}{2}$ **b** $a+\dfrac{b}{2}$

2 Simplify using index notation:

a $5\times k\times k\times k$ **b** $6\times a\times a-c\times c\times b\times c$

3 Write in algebraic form:

a "3 more than the square of x" **b** "the square of the sum of 3 and x"

4 Daniel is 30 cm taller than Katia. Julian is 5 cm taller than Daniel. Suppose Katia's height is x cm. Write an algebraic expression for the height of:

a Daniel **b** Julian.

5 Write as an algebraic equation:

a K is the sum of m and n

b L is the square root of the product of p and q.

6 If $p=1$, $q=-2$, and $r=-3$, find the value of $\dfrac{4q-p}{r}$.

7 Simplify:

a $10-x^2+6x-3$ **b** $3t^3-t^2+t+2t^2$

8 Simplify:

a $5cd^2\times 10cd$ **b** $2x\times 3y-5x\times x-4y\times x$

9 Simplify:

a $9\times 3z-5z\times 4$ **b** $c^2\times 5d-2cd\times d$ **c** $(4m)^2-2m\times 3mn$

10 I think of a number, subtract 5 from it, then square the result.
Suppose the number is x.

a Write an expression for this statement.

b Find the value of your expression when:

i $x=0$ **ii** $x=4$ **iii** $x=-4$

c Is there an integer value of x that will make this expression negative? Explain your answer.

Chapter 2

Indices

Contents:

OPENING PROBLEM

We have seen previously that $a^n = \underbrace{a \times a \times a \times .... \times a}_{n \text{ times}}$.

This definition of a^n is only meaningful if n is a positive integer.

Things to think about:

a What does it mean if a number is raised to an index which is a *negative* integer? Can you write a^{-3} without a negative index?

b What does it mean if a number is raised to an index which is a fraction?

c **i** Can you use the index laws to show that $(9^{\frac{1}{2}})^2 = 9$?

ii What is the value of $9^{\frac{1}{2}}$?

In this chapter we will study **indices** which are used to represent multiplications. We will see how they are used in **scientific notation**, which helps us write very large and very small numbers more easily.

A EVALUATING INDICES

We often deal with numbers that are repeatedly multiplied together. Mathematicians use **indices** or **exponents** to represent such products more easily.

Indices have many applications in finance, engineering, physics, astronomy, biology, electronics, and computer science.

We have seen that to simplify the product $3 \times 3 \times 3 \times 3$, we can write 3^4.

The 3 is called the **base** of the expression. The 4 is called the **power** or **index** or **exponent**.

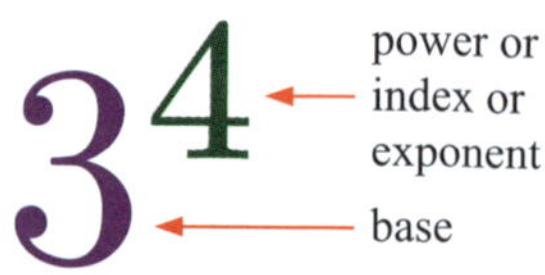

If n is a positive integer, then a^n is the product of n factors of a.

$$\boldsymbol{a^n = \underbrace{a \times a \times a \times a \times \; .... \times a}_{n \text{ factors}}}$$

Example 1 **Self Tutor**

Find the integer equal to: **a** 2^5 **b** $2^3 \times 5^2 \times 7$

a 2^5
$= 2 \times 2 \times 2 \times 2 \times 2$
$= 32$

b $2^3 \times 5^2 \times 7$
$= 2 \times 2 \times 2 \times 5 \times 5 \times 7$
$= 8 \times 25 \times 7$
$= 1400$

EXERCISE 2A.1

1 Find the integer equal to:

a 2^3 **b** 5^2 **c** 2^7 **d** 6^3

e $2 \times 3^2 \times 5^3$ **f** $2^3 \times 3 \times 7^2$ **g** $3^2 \times 5^2 \times 11$ **h** $2^5 \times 5^3 \times 13$

2 Copy and complete the values of these common powers. Try to remember them.

a $2^1 = ...,\ 2^2 = ...,\ 2^3 = ...,\ 2^4 = ...,\ 2^5 = ...,\ 2^6 = ...$

b $3^1 = ...,\ 3^2 = ...,\ 3^3 = ...,\ 3^4 = ...$

c $5^1 = ...,\ 5^2 = ...,\ 5^3 = ...,\ 5^4 = ...$

d $7^1 = ...,\ 7^2 = ...,\ 7^3 = ...$

Example 2 Self Tutor

By dividing continuously by the prime numbers 2, 3, 5, 7, ... write 504 as a product of prime factors in index form.

$$\begin{array}{r|r} 2 & 504 \\ \hline 2 & 252 \\ \hline 2 & 126 \\ \hline 3 & 63 \\ \hline 3 & 21 \\ \hline 7 & 7 \\ \hline & 1 \end{array}$$

$\therefore\ 504 = 2^3 \times 3^2 \times 7$

3 By dividing continuously by the primes 2, 3, 5, 7, ..., write as a product of prime factors in index form:

a 40 **b** 84 **c** 100 **d** 360

e 990 **f** 1152 **g** 2646 **h** 4725

4 Write the following as powers of 2:

a 64 **b** 512 **c** 2048

5 Write the following as powers of 5:

a 125 **b** 3125 **c** 15 625

6 By considering $3^1, 3^2, 3^3, 3^4, 3^5, ...$ and looking for a pattern, find the last digit of 3^{20}.

7 Find integers a and b such that $a^b \times b^a = 800$.

8 **a** The number 30 equals $2 \times 3 \times 5$.

The factors of 30 are 1, 2, 3, 5, 6, 10, 15, and 30.

These factors can be written as:

$2^0 3^0 5^0$ $2^1 3^0 5^0$ $2^0 3^1 5^0$ $2^0 3^0 5^1$

$2^1 3^1 5^0$ $2^1 3^0 5^1$ $2^0 3^1 5^1$ $2^1 3^1 5^1$

Explain why the number of factors of 30 is $2 \times 2 \times 2 = 8$.

b How many factors does the number $540 = 2^2 \times 3^3 \times 5$ have?

c If a, b, and c are non-negative integers, how many factors does the number $2^a 3^b 5^c$ have?

NEGATIVE BASES

So far we have only considered **positive** bases raised to a power.

However, the base can also be negative. To indicate this we need to use brackets.

Notice that $(-2)^2 = -2 \times -2$
$= 4$
whereas $-2^2 = -(2^2)$
$= -1 \times 2 \times 2$
$= -4$

Consider the statements below:

$(-1)^1 = -1$
$(-1)^2 = -1 \times -1 = 1$
$(-1)^3 = -1 \times -1 \times -1 = -1$
$(-1)^4 = -1 \times -1 \times -1 \times -1 = 1$

$(-2)^1 = -2$
$(-2)^2 = -2 \times -2 = 4$
$(-2)^3 = -2 \times -2 \times -2 = -8$
$(-2)^4 = -2 \times -2 \times -2 \times -2 = 16$

From the pattern above it can be seen that:

- a **negative** base raised to an **odd** power is **negative**
- a **negative** base raised to an **even** power is **positive**.

Example 3 — Self Tutor

Evaluate:

a $(-3)^4$ **b** -3^4 **c** $(-3)^5$ **d** $-(-3)^5$

a $(-3)^4$
$= 81$

b -3^4
$= -1 \times 3^4$
$= -81$

c $(-3)^5$
$= -243$

d $-(-3)^5$
$= -1 \times (-3)^5$
$= -1 \times -243$
$= 243$

EXERCISE 2A.2

1 Simplify:

a $(-1)^4$ **b** $(-1)^5$ **c** -1^5 **d** $-(-1)^5$

e -2^4 **f** $(-2)^4$ **g** $-(-2)^4$ **h** $-(-5)^2$

i $-(-5)^3$ **j** -3^3 **k** $(-3)^4$ **l** -7^2

CALCULATOR USE

You can evaluate powers on your calculator. Instructions can be found by clicking on the icon.

GRAPHICS CALCULATOR INSTRUCTIONS

Example 4 **Self Tutor**

Find, using your calculator:

a 7^5 **b** $(-5)^6$ **c** -9^4

Casio fx-CG20

TI-84 Plus

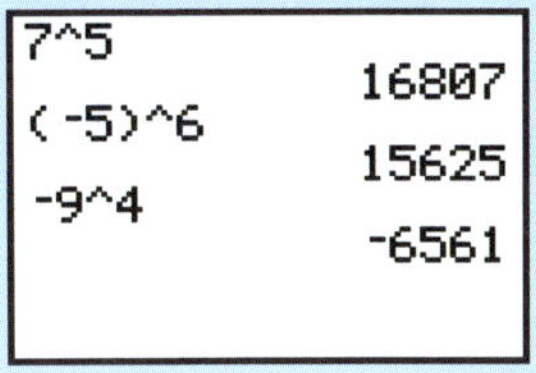

TI-*n*spire

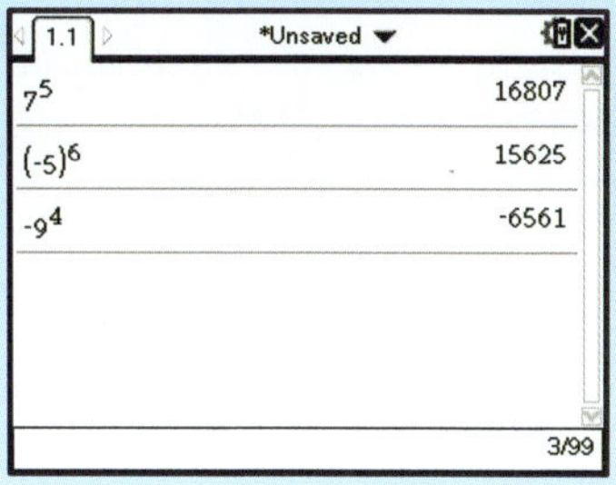

a $7^5 = 16\,807$ **b** $(-5)^6 = 15\,625$ **c** $-9^4 = -6561$

EXERCISE 2A.3

1 Use your calculator to find the value of the following, recording the entire display:

a 2^{12} **b** $(-5)^7$ **c** -3^4 **d** 7^7

e 8^5 **f** $(-9)^3$ **g** -9^3 **h** 1.22^{11}

i -0.972^{15} **j** $(-1.08)^{23}$

2 **a** Use your calculator to find the value of:

i 7^{-1} **ii** $\frac{1}{7^1}$ **iii** 3^{-2} **iv** $\frac{1}{3^2}$

v 4^{-3} **vi** $\frac{1}{4^3}$ **vii** 13^0 **viii** 172^0

b Discuss what happens when a number is raised:

i to a negative power **ii** to the power zero.

B INDEX LAWS

In **Chapter 1** we saw that $a \times a \times a = a^3$.

Using the same principle, the product $a^3 \times a^4$ can be written as $(a \times a \times a) \times (a \times a \times a \times a) = a^7$.

Consider some other expressions and their simplifications:

- $\dfrac{a^5}{a^2} = \dfrac{a \times a \times a \times \cancel{a \times a}^{1}}{\cancel{a \times a}_{1}} = a^3$
- $(a^3)^2 = a \times a \times a \times a \times a \times a = a^6$
- $(a \times b)^2 = a \times b \times a \times b = a \times a \times b \times b = a^2b^2$
- $\left(\dfrac{a}{b}\right)^3 = \dfrac{a}{b} \times \dfrac{a}{b} \times \dfrac{a}{b} = \dfrac{a \times a \times a}{b \times b \times b} = \dfrac{a^3}{b^3}$

These examples can be generalised to the following index laws:

If the bases a and b are both positive, and the indices m and n are integers, then:

$\boldsymbol{a^m \times a^n = a^{m+n}}$ To **multiply** numbers with the **same base**, keep the base and **add** the indices.

$\boldsymbol{\dfrac{a^m}{a^n} = a^{m-n}}$ To **divide** numbers with the same base, keep the base and **subtract** the indices.

$\boldsymbol{(a^m)^n = a^{mn}}$ When **raising** a **power** to a **power**, keep the base and **multiply** the indices.

$\boldsymbol{(ab)^n = a^n b^n}$ The power of a product is the product of the powers.

$\boldsymbol{\left(\dfrac{a}{b}\right)^n = \dfrac{a^n}{b^n}}$ The power of a quotient is the quotient of the powers.

$\boldsymbol{a^0 = 1, \quad a \neq 0}$ Any non-zero number raised to the power of zero is 1.

$\boldsymbol{a^{-n} = \dfrac{1}{a^n}}$ and in particular $\boldsymbol{a^{-1} = \dfrac{1}{a}}$.

Example 5 **Self Tutor**

Simplify using $a^m \times a^n = a^{m+n}$:

a $11^5 \times 11^3$ **b** $a^4 \times a^5$ **c** $x^4 \times x^a$

a $11^5 \times 11^3$
$= 11^{5+3}$
$= 11^8$

b $a^4 \times a^5$
$= a^{4+5}$
$= a^9$

c $x^4 \times x^a$
$= x^{4+a}$
$= x^{a+4}$

EXERCISE 2B

1 Simplify using the index law $a^m \times a^n = a^{m+n}$:

a $5^4 \times 5^2$ **b** $6^5 \times 6^6$ **c** $a \times a^5$ **d** $a^3 \times a^7$

e $b^{12} \times b^2$ **f** $a^2 \times a^n$ **g** $b^m \times b^9$ **h** $p^2 \times p \times p^4$

Example 6 **Self Tutor**

Simplify using $\dfrac{a^m}{a^n} = a^{m-n}$: **a** $\dfrac{7^8}{7^5}$ **b** $\dfrac{b^6}{b^m}$

a $\dfrac{7^8}{7^5} = 7^{8-5}$
$= 7^3$

b $\dfrac{b^6}{b^m} = b^{6-m}$

2 Simplify using the index law $\dfrac{a^m}{a^n} = a^{m-n}$:

a $\dfrac{3^9}{3^2}$ **b** $\dfrac{7^{13}}{7^9}$ **c** $5^7 \div 5^4$ **d** $\dfrac{a^8}{a^3}$

e $\dfrac{b^{18}}{b^{12}}$ **f** $\dfrac{p^n}{p^2}$ **g** $\dfrac{y^5}{y^b}$ **h** $b^{2n} \div b$

Example 7 Self Tutor

Simplify using $(a^m)^n = a^{mn}$:

a $(2^4)^3$ **b** $(x^3)^5$ **c** $(b^7)^m$

a $(2^4)^3$
$= 2^{4\times 3}$
$= 2^{12}$

b $(x^3)^5$
$= x^{3\times 5}$
$= x^{15}$

c $(b^7)^m$
$= b^{7\times m}$
$= b^{7m}$

3 Simplify using the index law $(a^m)^n = a^{mn}$:

a $(2^9)^2$ **b** $(3^7)^3$ **c** $(5^4)^7$ **d** $(a^2)^6$
e $(q^3)^3$ **f** $(d^6)^n$ **g** $(x^y)^8$ **h** $(g^{2a})^5$

Example 8 Self Tutor

Simplify:

a 7^0 **b** x^0 **c** $\frac{y^4}{y^4}$ **d** $2 + 5^0$

a $7^0 = 1$

b $x^0 = 1$ {provided $x \neq 0$}

c $\frac{y^4}{y^4} = y^{4-4}$
$= y^0$
$= 1$ {provided $y \neq 0$}

d $2 + 5^0$
$= 2 + 1$
$= 3$

4 Simplify:

a 4^0 **b** $\left(\frac{2}{3}\right)^0$ **c** 12^0 **d** $2^0 + 2^2$
e a^0 **f** $(13^3)^0$ **g** $3y^0$ **h** $(3y)^0$
i $d^2 \times d^0$ **j** $\frac{c^7}{c^7}$ **k** $2 + 3^0$ **l** $8 - 9^0$

Example 9 Self Tutor

Simplify using the index laws:

a $3x^2 \times 5x^5$ **b** $\frac{20a^9}{4a^6}$ **c** $\frac{b^3 \times b^7}{(b^2)^4}$

a $3x^2 \times 5x^5$
$= 3 \times 5 \times x^2 \times x^5$
$= 15 \times x^{2+5}$
$= 15x^7$

b $\frac{20a^9}{4a^6} = \frac{20}{4} \times a^{9-6}$
$= 5a^3$

c $\frac{b^3 \times b^7}{(b^2)^4} = \frac{b^{3+7}}{b^{2\times 4}}$
$= \frac{b^{10}}{b^8}$
$= b^{10-8}$
$= b^2$

5 Simplify using one or more of the index laws:

a $\dfrac{b^4}{b}$ **b** $3c^2 \times 5c^3$ **c** $\dfrac{a^3b^2}{ab^2}$ **d** $\dfrac{18d^8}{3d^3}$

e $\dfrac{14p^3q^4}{2pq}$ **f** $12st^3 \times 3s^3$ **g** $\dfrac{y^{12}}{(y^2)^5}$ **h** $\dfrac{q^3 \times q^6}{(q^2)^3}$

i $\dfrac{6x^3y^2}{3x^3y}$ **j** $\dfrac{w^2z^4}{(wz^2)^2}$ **k** $\dfrac{a^4 \times a^2}{b(a^2)^3}$ **l** $\dfrac{16p^7q^6}{(2pq^2)^3}$

Example 10 Self Tutor

Express in simplest form with a prime number base:

a 4^4 **b** 9×3^a **c** 49^{x+2}

a $4^4 = (2^2)^4 = 2^{2\times 4} = 2^8$

b $9 \times 3^a = 3^2 \times 3^a = 3^{2+a} = 3^{a+2}$

c $49^{x+2} = (7^2)^{x+2} = 7^{2(x+2)}$

6 Express in simplest form with a prime number base:

a 16 **b** 27 **c** 125 **d** 8^2

e 9^3 **f** $2^a \times 4$ **g** $7^t \div 7$ **h** $8^k \times 2^k$

i $\dfrac{32}{2^x}$ **j** $\dfrac{5^{x+2}}{5^{x-1}}$ **k** $(7^4)^{a-1}$ **l** $3^b \times 3^{2-b}$

m $\dfrac{3^y}{9^x}$ **n** $\dfrac{4^y}{8^x}$ **o** $\dfrac{11^{x+2}}{11^{2-x}}$ **p** $\dfrac{5^t \times 25^t}{125^{t-1}}$

Example 11 Self Tutor

Remove the brackets of:

a $(3a)^2$ **b** $\left(\dfrac{2x}{y}\right)^3$

a $(3a)^2 = 3^2 \times a^2 = 9a^2$

b $\left(\dfrac{2x}{y}\right)^3 = \dfrac{2^3 \times x^3}{y^3} = \dfrac{8x^3}{y^3}$

7 Remove the brackets of:

a $(pq)^3$ **b** $(ab)^5$ **c** $(xy)^4$ **d** $(pqr)^5$

e $(3a)^3$ **f** $(5b)^2$ **g** $(2n)^5$ **h** $(4ab)^3$

i $(9ab)^2$ **j** $\left(\dfrac{x}{y}\right)^4$ **k** $\left(\dfrac{a}{b}\right)^7$ **l** $\left(\dfrac{c}{2d}\right)^6$

m $\left(\dfrac{3x}{y^2}\right)^2$ **n** $\left(\dfrac{w^2}{2v}\right)^2$ **o** $\left(\dfrac{2r}{3s^2}\right)^3$ **p** $\left(\dfrac{3a^2}{ab}\right)^4$

Example 12 — Self Tutor

Express in simplest form, without brackets: **a** $(3a^3b)^4$ **b** $\left(\dfrac{x^2}{2y}\right)^3$

a $(3a^3b)^4$
$= 3^4 \times (a^3)^4 \times b^4$
$= 81 \times a^{3\times4} \times b^4$
$= 81a^{12}b^4$

b $\left(\dfrac{x^2}{2y}\right)^3 = \dfrac{(x^2)^3}{2^3 \times y^3}$
$= \dfrac{x^{2\times3}}{8 \times y^3}$
$= \dfrac{x^6}{8y^3}$

8 Express in simplest form, without brackets:

a $(3a^2)^2$ **b** $\left(\dfrac{2}{p^2q}\right)^4$ **c** $(3cd^4)^3$ **d** $\left(\dfrac{a^4}{3n^7}\right)^2$

e $\left(\dfrac{2x^2}{y^6}\right)^3$ **f** $(2s^3t^4)^5$ **g** $\left(\dfrac{4g^5}{h^2}\right)^2$ **h** $\left(\dfrac{3x^2y^3}{4z}\right)^4$

Example 13 — Self Tutor

Write without brackets or negative indices:
a 2^{-2} **b** $5^0 - 5^{-1}$ **c** $\left(\frac{2}{5}\right)^{-2}$

a 2^{-2}
$= \dfrac{1}{2^2}$
$= \frac{1}{4}$

b $5^0 - 5^{-1}$
$= 1 - \frac{1}{5}$
$= \frac{4}{5}$

c $\left(\frac{2}{5}\right)^{-2}$
$= \left(\frac{5}{2}\right)^2$
$= \frac{25}{4}$
$= 6\frac{1}{4}$

9 Write without brackets or negative indices:

a 2^{-1} **b** 3^{-1} **c** 6^{-1} **d** 8^{-1}
e 3^2 **f** 3^{-2} **g** 2^3 **h** 2^{-3}
i 5^2 **j** 5^{-2} **k** 10^2 **l** 10^{-2}

10 Write without brackets or negative indices:

a $\dfrac{3^2}{3^4}$ **b** $\dfrac{2^{10}}{2^{15}}$ **c** $\left(4^{-1}\right)^2$ **d** $\left(7^2\right)^{-1}$
e $\left(\frac{1}{3}\right)^{-1}$ **f** $\left(\frac{2}{5}\right)^{-1}$ **g** $\left(\frac{4}{3}\right)^{-1}$ **h** $\left(\frac{1}{12}\right)^{-1}$
i $\left(3\frac{1}{2}\right)^{-1}$ **j** $3^0 - 3^{-1}$ **k** $7^{-1} + 7^0$ **l** $2^0 + 2^1 + 2^{-1}$
m $\left(\frac{3}{4}\right)^{-2}$ **n** $5 + 5^{-2}$ **o** $\left(\frac{5}{2}\right)^{-2}$ **p** $\left(2\frac{1}{3}\right)^{-2}$

Example 14 — Self Tutor

Write without brackets or negative indices:

a $(5x)^{-1}$ **b** $5x^{-1}$ **c** $(3b^2)^{-2}$

a $(5x)^{-1}$
$= \dfrac{1}{5x}$

b $5x^{-1}$
$= \dfrac{5}{x}$

c $(3b^2)^{-2}$
$= \dfrac{1}{(3b^2)^2}$
$= \dfrac{1}{3^2 b^4}$
$= \dfrac{1}{9b^4}$

In $5x^{-1}$ the index -1 refers to the x only.

11 Write without brackets or negative indices:

a $(2x)^{-1}$ **b** $2x^{-1}$ **c** $3b^{-1}$ **d** $(3b)^{-1}$

e $(2b)^{-2}$ **f** $\left(\dfrac{2}{b}\right)^{-2}$ **g** $\left(\dfrac{1}{n}\right)^{-2}$ **h** $(3n^{-2})^{-1}$

i $(ab)^{-1}$ **j** ab^{-1} **k** ab^{-2} **l** $(ab)^{-2}$

m $(2ab)^{-1}$ **n** $2(ab)^{-1}$ **o** $2ab^{-1}$ **p** $\dfrac{(ab)^2}{b^{-1}}$

Example 15 — Self Tutor

Write as a power of 2, 3, or 5:

a $\dfrac{1}{8}$ **b** $\dfrac{1}{9^n}$ **c** $\dfrac{25}{5^4}$

a $\dfrac{1}{8}$
$= \dfrac{1}{2^3}$
$= 2^{-3}$

b $\dfrac{1}{9^n}$
$= \dfrac{1}{(3^2)^n}$
$= \dfrac{1}{3^{2n}}$
$= 3^{-2n}$

c $\dfrac{25}{5^4}$
$= \dfrac{5^2}{5^4}$
$= 5^{2-4}$
$= 5^{-2}$

12 Write as a power of 2, 3, or 5:

a $\frac{1}{3}$ **b** $\frac{1}{2}$ **c** $\frac{1}{5}$ **d** $\frac{1}{4}$

e $\frac{1}{27}$ **f** $\frac{1}{25}$ **g** $\dfrac{1}{8^x}$ **h** $\dfrac{1}{16^y}$

i $\dfrac{1}{81^a}$ **j** $\dfrac{9}{3^4}$ **k** 25×5^{-4} **l** $\dfrac{5^{-1}}{5^2}$

m $2 \div 2^{-3}$ **n** 1 **o** 6^{-3} **p** 4×10^2

PUZZLE

The water lily *Nymphaea Mathematicus* doubles its size every day. From the time it was planted until it completely covered a pond, it took 26 days.

How many days did it take to cover half the pond?

DISCUSSION

In **Chapter 1**, we discussed whether we can *always* cancel the common factor of x in $\frac{2x}{x}$.

You should have decided that the cancellation was invalid when $x = 0$.

- Discuss how to simplify $\frac{x}{x}$ using the index laws.
- Use the case where $x = 0$ to explain why 0^0 is undefined.

Global context

Chessboard calculations

Statement of inquiry:	Changing the way we represent numbers can help us perform calculations.
Global context:	Scientific and technical innovation
Key concept:	Form
Related concepts:	Equivalence, Pattern, Representation
Objectives:	Investigating patterns, Communicating
Approaches to learning:	Thinking, Communication

C SCIENTIFIC NOTATION

There are many situations where we need to describe very large and very small numbers. For example:

- There are about 75 000 000 000 000 cells in the human body.
- A glass of tap water contains about 0.000 04 grams of chlorine.

Numbers with so many digits can be hard to comprehend and operate with. We can use **scientific notation** to write these numbers in a way that is easier to understand.

Scientific notation involves writing any given number as *a number between 1 inclusive and 10*, multiplied by a *power of 10*. The result has the form $a \times 10^k$ where $1 \leqslant a < 10$ and k is an integer.

When a number is written in scientific notation:

- If the original number is greater than or equal to 10, then k is **positive**.
- If the original number is less than 1, then k is **negative**.
- If the original number is between 1 and 10, we write the number as it is and multiply it by 10^0, which is 1.

To write a number larger than 10, we start with a number between 1 and 10, then multiply it by a positive power of 10.

For example:

$$27 = 2.7 \times 10$$
$$580 = 5.8 \times 100 = 5.8 \times 10^2$$
$$3040 = 3.04 \times 1000 = 3.04 \times 10^3$$

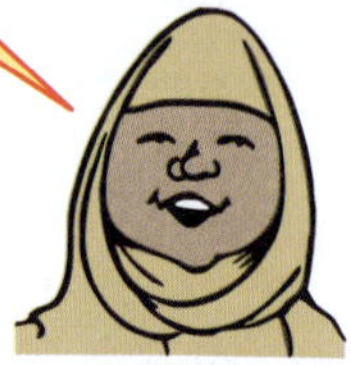

We can express small numbers in the same way. We start with a number between 1 and 10, then *divide* it by a power of 10. This is the same as multiplying by a *negative* power of 10.

For example:

$$0.75 = 7.5 \div 10 = 7.5 \times \frac{1}{10^1} = 7.5 \times 10^{-1}$$

$$0.0006 = 6 \div 10\,000 = 6 \times \frac{1}{10^4} = 6 \times 10^{-4}$$

EXERCISE 2C.1

1 The following values are all equal to 53 000. Which of them is written in scientific notation?

A 53×10^3 **B** 0.53×10^5 **C** 5.3×10^4 **D** $53\,000 \times 10^0$

2 Copy and complete to write the following numbers in scientific notation:

a $376 = 3.76 \times 10^{\cdots}$
b $8000 = ... \times 10^3$
c $0.04 = ... \times 10^{-2}$
d $0.005\,07 = 5.07 \times 10^{\cdots}$
e $9\,040\,000 = ... \times 10^6$
f $0.000\,000\,23 = 2.3 \times 10^{\cdots}$

Example 16 Self Tutor

Express in scientific notation:

a 4 500 000 **b** 0.000 592

a $4\,500\,000 = 4.5 \times 10^6$

b $0.000\,592 = 5.92 \div 10^4 = 5.92 \times \frac{1}{10^4} = 5.92 \times 10^{-4}$

3 Express in scientific notation:

a 425
b 425 000
c 4.25
d 0.425
e 0.000 042 5
f 20.1
g 20 100
h 0.002 01
i 2 010 000
j 2 010 000 000
k 0.000 002 01
l 10

4 Express in scientific notation:

a The distance from the Sun to Mars is about 249 000 000 km.

b A set of kitchen scales are accurate to within 0.0005 kg.

c A bacterium is smaller than 0.005 mm.

d The probability that your six numbers will be selected for Lotto on Monday night is 0.000 000 141 62 .

e The central temperature of the Sun is 15 million degrees Celsius.

f The Great Barrier Reef covers an area of 34 440 000 hectares.

Example 17 — Self Tutor

Write as an ordinary decimal number:

a 3.2×10^2 **b** 5.76×10^{-5}

a 3.2×10^2
$= 3.20 \times 100$
$= 320$

b 5.76×10^{-5}
$= 000005.76 \div 10^5$
$= 0.000\,057\,6$

5 Write as an ordinary decimal number:

a 5×10^4 **b** 3×10^3 **c** 1.8×10^7 **d** 8.1×10^2
e 6.5×10^5 **f** 1.1×10^1 **g** 2.75×10^8 **h** 8×10^6

6 Write as an ordinary decimal number:

a 3×10^{-2} **b** 9×10^{-5} **c** 7×10^{-3} **d** 4.1×10^{-4}
e 8.2×10^{-6} **f** 7.61×10^{-1} **g** 3.25×10^{-7} **h** 2×10^{-8}

7 Express as ordinary decimal numbers:

a The wavelength of blue light is 4.75×10^{-7} m.

b The estimated world population for 2010 was 6.85×10^9 people.

c Physicists in Japan created a model bull which is only 1.2×10^{-5} m long.

d The length of the Earth's equator is approximately 4.01×10^4 km.

e A mosquito weighs about 1.5×10^{-6} kg.

Example 18 — Self Tutor

Which number is larger:

a 3.27×10^4 or 1.56×10^5 **b** 2.6×10^{-4} or 2.79×10^{-4}?

a 1.56×10^5 has a higher power of 10 than 3.27×10^4.
$\therefore\ 1.56 \times 10^5 > 3.27 \times 10^4$.

b Both numbers have the same power of 10.
$2.79 > 2.6$, so $2.79 \times 10^{-4} > 2.6 \times 10^{-4}$.

8 Which is larger:

a 2.2×10^8 or 5.8×10^6

b 7.71×10^{-7} or 3.5×10^{-5}

c 4.9×10^{-3} or 4.9×10^3

d 6.2×10^8 or 6.41×10^8

e 1.006×10^{-6} or 1.01×10^{-6}

f 9×10^{-8} or 0?

9 Electrons and protons are tiny particles present in everything around us.
An electron has mass $0.000\,000\,000\,000\,000\,000\,000\,000\,000\,000\,911$ g.
A proton has mass $0.000\,000\,000\,000\,000\,000\,000\,001\,67$ g.

a Express the mass of an electron in scientific notation.

b Express the mass of a proton in scientific notation.

c Which is heavier, an electron or a proton?

Example 19 **Self Tutor**

Simplify the following, giving your answer in scientific notation:

a $(5 \times 10^4) \times (4 \times 10^5)$ **b** $(8 \times 10^5) \div (2 \times 10^3)$

a
$$\begin{aligned}&(5 \times 10^4) \times (4 \times 10^5)\\ &= 5 \times 4 \times 10^4 \times 10^5\\ &= 20 \times 10^{4+5}\\ &= 2 \times 10^1 \times 10^9\\ &= 2 \times 10^{10}\end{aligned}$$

b
$$\begin{aligned}&(8 \times 10^5) \div (2 \times 10^3)\\ &= \frac{8 \times 10^5}{2 \times 10^3}\\ &= \frac{8}{2} \times 10^{5-3}\\ &= 4 \times 10^2\end{aligned}$$

10 Simplify the following, giving your answer in scientific notation:

a $(2 \times 10^5) \times (3 \times 10^2)$ **b** $(4 \times 10^3) \times (6 \times 10^3)$ **c** $(8 \times 10^5) \times (5 \times 10^6)$

d $(9 \times 10^9)^2$ **e** $(3 \times 10^4)^2$ **f** $(7 \times 10^{-3})^2$

g $(8 \times 10^2) \div (2 \times 10^3)$ **h** $(6 \times 10^7) \div (3 \times 10^4)$

HISTORICAL NOTE — AVOGADRO'S NUMBER

When chemists study how atoms and molecules behave, they need to be able to measure how much of a substance is present. But molecules are very, very small, so it is impossible to count them individually. Instead, the chemist can use the **atomic mass** of each atom in the molecule to calculate its **molecular mass**. Then, given the molecular mass of the substance, if there are that many grams of the substance present, then the number of molecules present is about $6.022\,141 \times 10^{23}$. This special number is called **Avogadro's number**, and corresponds to one **mole** of the substance.

Amedeo Avogadro

Amedeo Avogadro (1776 - 1856) was an Italian physicist and mathematician. He was the first to recognise a relationship between the volume of a gas and the number of atoms it contained. However, he did not name or even calculate the number named after him!

Avogadro's number was first estimated by Austrian Johann Josef Loschmidt in 1865.

French physicist Jean Perrin, who earned a Nobel Prize in Physics, named the number in honour of Avogadro in 1909, for his work in the field of molecular theory.

SCIENTIFIC NOTATION ON A CALCULATOR

Calculators use scientific notation to display very large and very small numbers. However, if we wish, we can tell the calculator to give all its answers in scientific notation. Instructions for writing numbers in scientific notation can be found by clicking on the icon.

GRAPHICS CALCULATOR INSTRUCTIONS

Example 20 Self Tutor

Use your calculator to find:

a $(1.67 \times 10^4) \times (2.3 \times 10^7)$ **b** $(9.49 \times 10^{-3}) \div (2.6 \times 10^6)$

Casio fx-CG20

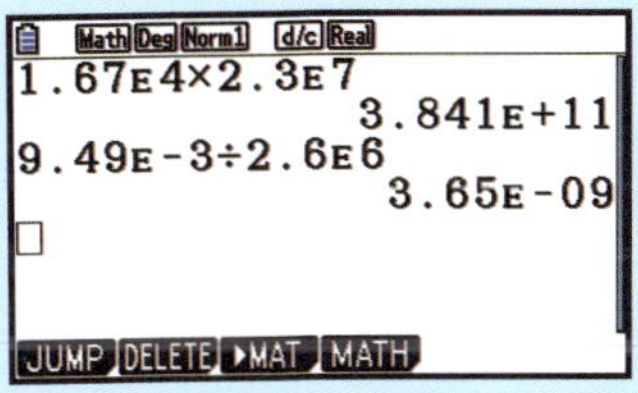

TI-84 Plus

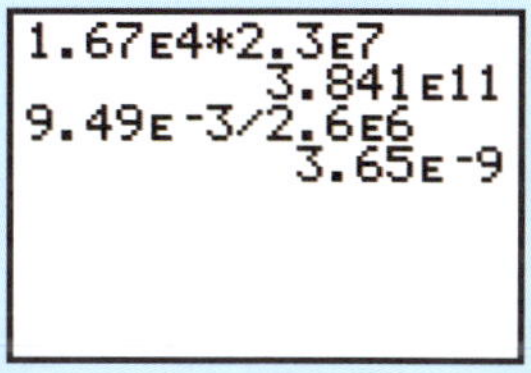

TI-nspire

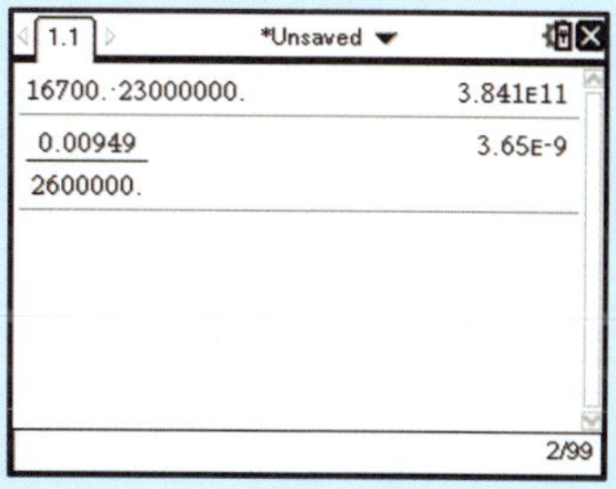

a $(1.67 \times 10^4) \times (2.3 \times 10^7) = 3.841 \times 10^{11}$

b $(9.49 \times 10^{-3}) \div (2.6 \times 10^6) = 3.65 \times 10^{-9}$

EXERCISE 2C.2

1 Calculate the following, giving each answer in scientific notation. The decimal part should be written correct to 2 decimal places.

a $0.0003 \times 0.01 \div 5000$ **b** $375 \times 220 \times 290\,000$

c $876\,000 \times 25\,000$ **d** $800 \times 740 \times 67\,800$

e $0.002\,12 \div 3\,400\,000$ **f** $0.019 \times 0.000\,27 \times 0.02$

2 Find in scientific notation, with the decimal part correct to 2 decimal places:

a $(2.81 \times 10^5) \times (3.4 \times 10^4)$ **b** $(9.81 \times 10^{-4})^2$

c $\dfrac{3.43 \times 10^{-6}}{7 \times 10^7}$ **d** $(8.66 \times 10^{-3}) \times (7.5 \times 10^{-5})$

e $\dfrac{1}{2.5 \times 10^6}$ **f** $(7.59 \times 10^4)^3$

For questions **3** and **4**, give your answers in scientific notation, with the decimal part correct to 2 decimal places. Assume that 1 year $\approx$ 365.25 days.

3 Germany's population is approximately 80 million people. Each German uses an average of 120 litres of water per day. Estimate the total water consumption by Germans:

a in one day **b** in one week **c** in one year.

4 Light travels at a speed of 2.998×10^8 metres per second. How far will light travel in:

a 1 minute **b** 1 day **c** 1 year?

D RATIONAL INDICES

A **rational** number is a number which can be written in the form $\frac{p}{q}$ where p and q are integers and $q \neq 0$. The integers themselves are rational numbers, since for example $5 = \frac{5}{1}$.

The index laws can be applied not just to integer indices, but to rational indices in general. This helps to give meaning to values such as $3^{\frac{1}{2}}$ and $2^{\frac{1}{3}}$.

INVESTIGATION RATIONAL INDICES

This Investigation will help you discover the meaning of numbers raised to rational indices.

What to do:

1 Observe that $5^{\frac{1}{2}} \times 5^{\frac{1}{2}} = (5^{\frac{1}{2}})^2 = 5^{\frac{1}{2} \times 2} = 5^1 = 5$ and $(\sqrt{5})^2 = 5$.

a Copy and complete the following:

i $3^{\frac{1}{2}} \times 3^{\frac{1}{2}} = \ldots\ldots = \ldots\ldots = \ldots\ldots$ **ii** $(\sqrt{3})^2 = \ldots\ldots$

iii $13^{\frac{1}{2}} \times 13^{\frac{1}{2}} = \ldots\ldots = \ldots\ldots = \ldots\ldots$ **iv** $(\sqrt{13})^2 = \ldots\ldots$

b Copy and complete: $a^{\frac{1}{2}} = \ldots\ldots$

2 Observe that $(7^{\frac{1}{3}})^3 = 7^{\frac{1}{3} \times 3} = 7^1 = 7$ and $(\sqrt[3]{7})^3 = 7$.

a Copy and complete the following:

i $(8^{\frac{1}{3}})^3 = \ldots\ldots = \ldots\ldots = \ldots\ldots$ **ii** $(\sqrt[3]{8})^3 = \ldots\ldots$

iii $(27^{\frac{1}{3}})^3 = \ldots\ldots = \ldots\ldots = \ldots\ldots$ **iv** $(\sqrt[3]{27})^3 = \ldots\ldots$

b Copy and complete: $a^{\frac{1}{3}} = \ldots\ldots$

3 Suggest a rule for the general case: $a^{\frac{1}{n}} = \ldots\ldots$

From the **Investigation**, we can conclude that $a^{\frac{1}{2}} = \sqrt{a}$ and $a^{\frac{1}{3}} = \sqrt[3]{a}$.

In general, $a^{\frac{1}{n}} = \sqrt[n]{a}$ where $\sqrt[n]{a}$ is called the ***n*th root of *a***.

Example 21 — Self Tutor

Evaluate:

a $16^{\frac{1}{2}}$ **b** $8^{\frac{1}{3}}$ **c** $16^{-\frac{1}{2}}$ **d** $8^{-\frac{1}{3}}$

a $16^{\frac{1}{2}} = \sqrt{16} = 4$

b $8^{\frac{1}{3}} = \sqrt[3]{8} = 2$

c $16^{-\frac{1}{2}} = \dfrac{1}{16^{\frac{1}{2}}} = \dfrac{1}{\sqrt{16}} = \frac{1}{4}$

d $8^{-\frac{1}{3}} = \dfrac{1}{8^{\frac{1}{3}}} = \dfrac{1}{\sqrt[3]{8}} = \frac{1}{2}$

Example 22 — Self Tutor

Write in index form:

a $\sqrt{3}$ **b** $\sqrt[3]{7}$ **c** $\dfrac{1}{\sqrt[4]{7}}$

a $\sqrt{3} = 3^{\frac{1}{2}}$

b $\sqrt[3]{7} = 7^{\frac{1}{3}}$

c $\dfrac{1}{\sqrt[4]{7}} = \dfrac{1}{7^{\frac{1}{4}}} = 7^{-\frac{1}{4}}$

EXERCISE 2D

1 Evaluate without using a calculator:

a $4^{\frac{1}{2}}$ **b** $4^{-\frac{1}{2}}$ **c** $9^{\frac{1}{2}}$ **d** $9^{-\frac{1}{2}}$

e $36^{\frac{1}{2}}$ **f** $36^{-\frac{1}{2}}$ **g** $27^{\frac{1}{3}}$ **h** $27^{-\frac{1}{3}}$

i $1000^{\frac{1}{3}}$ **j** $1000^{-\frac{1}{3}}$ **k** $125^{\frac{1}{3}}$ **l** $125^{-\frac{1}{3}}$

2 Write in index form:

a $\sqrt{11}$ **b** $\dfrac{1}{\sqrt{11}}$ **c** $\sqrt{12}$ **d** $\dfrac{1}{\sqrt{12}}$

e $\sqrt[3]{26}$ **f** $\dfrac{1}{\sqrt[3]{26}}$ **g** $\sqrt[4]{7}$ **h** $\dfrac{1}{\sqrt[5]{7}}$

Example 23 **Self Tutor**

Use your calculator to find $\sqrt[4]{50}$, rounded to 2 decimal places.

$\sqrt[4]{50} = 50^{\frac{1}{4}}$

Casio fx-CG20

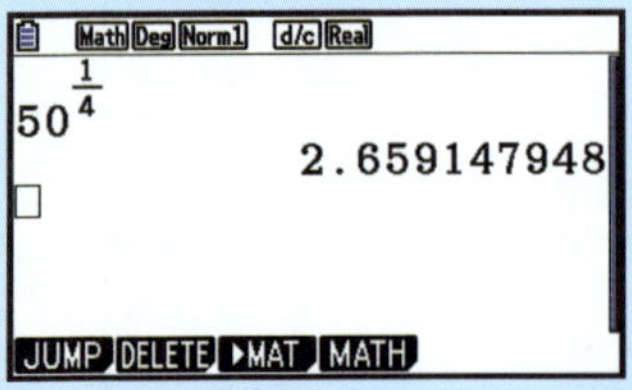

TI-84 Plus

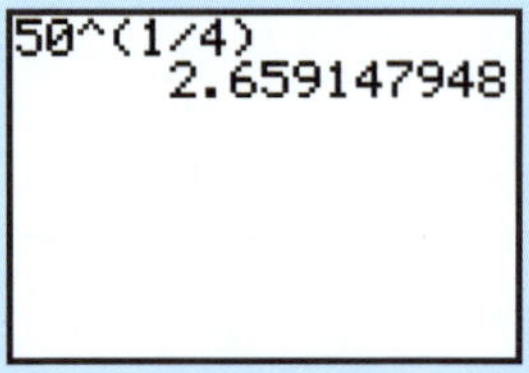

TI-$n$spire

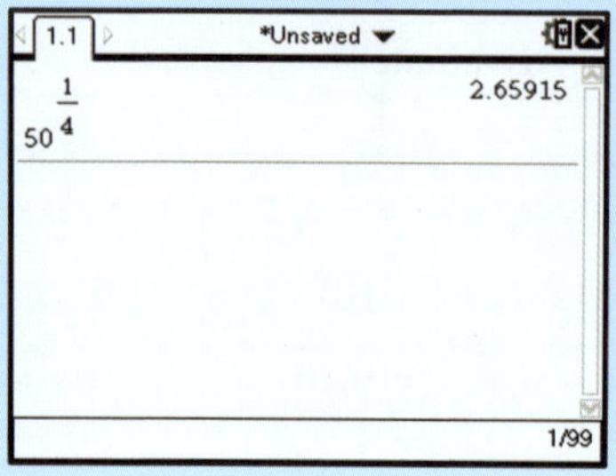

$\sqrt[4]{50} \approx 2.66$

3 Use your calculator to evaluate the following, rounded to 2 decimal places:

a $\sqrt[4]{20}$ **b** $\sqrt[5]{300}$ **c** $\dfrac{1}{\sqrt[4]{80}}$ **d** $\dfrac{1}{\sqrt[6]{12}}$

REVIEW SET 2A

1 Find the integer equal to:

a $2^4 \times 3^2$ **b** $2^2 \times 5^2$ **c** $2 \times 3^3 \times 5$

2 Simplify:

a $(-1)^7$ **b** -3^3 **c** $-(-2)^2$

3 Simplify:

a $q^2 \times q^5$ **b** $\dfrac{a^6}{a^4}$ **c** $(t^{2x})^2$ **d** $(2^4)^0$

4 Simplify the following expressions:

a $(ab)^3$ **b** $\dfrac{3vw^3}{(vw)^2}$ **c** $\left(\dfrac{2a^2}{d^3}\right)^4$

5 Simplify:

a 7^0 **b** $5a^0$ **c** $6 - \left(\frac{1}{4}\right)^0$

6 Write without brackets or negative indices:

a 4^{-2} **b** $\left(4\frac{1}{2}\right)^{-1}$ **c** $6^0 - 6^{-1}$

7 Write each of the following as a power of 2, 3, or 5:

a $\frac{1}{8}$ **b** $\frac{1}{625}$ **c** $\dfrac{1}{(3^t)^5}$

8 Write as an ordinary decimal number:

a 7.71×10^{-3} **b** 4×10^7

9 Evaluate the following, giving your answers in scientific notation with the decimal part correct to 2 decimal places:

a $(1.91 \times 10^{12}) \times (1.6 \times 10^{-16})$ **b** $\dfrac{8.3 \times 10^{18}}{4.9 \times 10^{9}}$ **c** $(2.18 \times 10^{4})^{2}$

10 The diameter of Jupiter is about 142 984 km. Express this in scientific notation, with the decimal part correct to 2 decimal places.

11 Evaluate without using a calculator:

a $81^{\frac{1}{2}}$ **b** $64^{-\frac{1}{3}}$

12 Write in index form:

a $\sqrt[3]{30}$ **b** $\dfrac{1}{\sqrt[4]{11}}$

REVIEW SET 2B

1 Write 1188 as a product of prime factors in index form.

2 Simplify:

a $(c^3)^4$ **b** $3^0 + 3^2$

3 Simplify the following expressions:

a $\dfrac{81g^2h^4}{27g^3h^2}$ **b** $\dfrac{x^{15}}{(x^2)^5}$ **c** $(2a^2bc^3)^3$

4 Express the following as powers of 2:

a $\dfrac{8^2}{4^3}$ **b** $16^k \times 4^k$

5 By considering $9^1, 9^2, 9^3, 9^4, ...$ and looking for a pattern, find the last digit of 9^{99}.

6 Write without brackets or negative indices:

a 3^{-4} **b** $\left(\frac{3}{8}\right)^{-1}$ **c** $2^1 - 2^{-3}$

7 Write without brackets or negative indices:

a $a^{-2}b^3$ **b** $(5a^{-2}b^3)^2$

8 A grain of sand weighs 3.5×10^{-10} kg. Express this as an ordinary decimal number.

9 Evaluate without using a calculator:

a $32^{\frac{1}{5}}$ **b** $100^{-\frac{1}{2}}$

10 Which is larger: 5.4×10^{-6} or 3.9×10^{-4}?

11 Use your calculator to evaluate the following, rounded to 2 decimal places:

a $\sqrt[3]{40}$ **b** $\dfrac{1}{\sqrt[5]{90}}$

12 Thorium is a natural radioactive chemical element. It is named after the Norse god of thunder, Thor.

a The diameter of a Thorium atom is 3.58×10^{-14} m. Write this as an ordinary decimal number.

b Thorium has a half-life of 14.05 billion years. Write this number in scientific notation.

c The *molar mass* of Thorium is 232.04. The mass in grams, of a single atom of Thorium, is this molar mass divided by 6.022×10^{23}.

i Find the mass, in grams, of a single Thorium atom.

ii How many atoms are there in a 10 g block of Thorium?

Chapter 3

Sets and Venn diagrams

Contents:

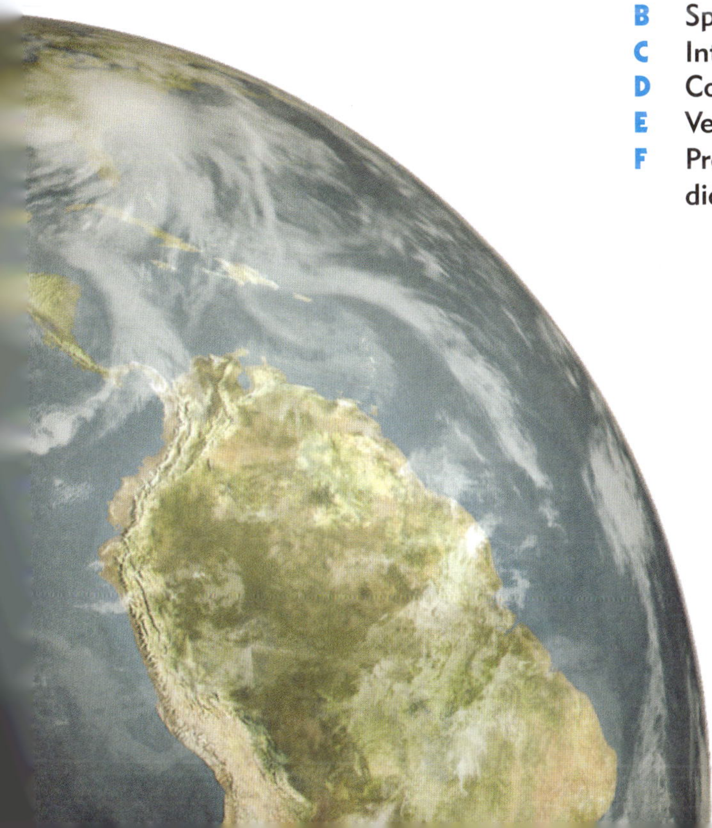

OPENING PROBLEM

50 tourists in Budapest were asked which tourist attractions they had visited. 31 of them had visited Buda Castle, and 23 of them had visited Heroes Square. 10 of the tourists had not visited either place.

Things to think about:

a How can we represent this information using a diagram?

b How many of the tourists had visited:

  i Buda Castle, but not Heroes Square

  ii both of these places?

A SETS

> A **set** is a collection of numbers or objects.
>
> Each object is called an **element** or **member** of the set.

When we record a set, we write its members within curly brackets, with commas between them.

We often use a capital letter to represent a set so that we can refer to it easily.

For example, suppose F is the set of all factors of 12. We write $F = \{1, 2, 3, 4, 6, 12\}$.

SET NOTATION

- $\in$ reads *is an element of* or *is in*
- $\notin$ reads *is not an element of* or *is not in*
- $\{\ \}$ or $\varnothing$ is used to represent an **empty set** which has no elements
- $n(S)$ reads *the number of elements in set S*

For the set $F = \{1, 2, 3, 4, 6, 12\}$, we can write $4 \in F$, $7 \notin F$, and $n(F) = 6$.

SUBSETS

> Set P is a **subset** of set Q if every element of P is also an element of Q. We write $P \subseteq Q$.

For example, suppose $A = \{3, 8\}$ and $B = \{2, 3, 6, 8, 9\}$. Every element of A is also an element of B, so $A \subseteq B$.

UNION AND INTERSECTION

If P and Q are two sets then:

- $P \cap Q$ is the **intersection** of P and Q, and consists of all elements which are in **both** P **and** Q.
- $P \cup Q$ is the **union** of P and Q, and consists of all elements which are in either P **or** Q (or both).

For example, suppose $P = \{1, 2, 4, 5, 6, 8\}$ and $Q = \{0, 2, 3, 5, 6, 7\}$.

- $P \cap Q = \{2, 5, 6\}$ as 2, 5, and 6 are in both sets
- $P \cup Q = \{0, 1, 2, 3, 4, 5, 6, 7, 8\}$ as these are the elements which are in P or Q.

DISJOINT SETS

Two sets are **disjoint** or **mutually exclusive** if they have no elements in common.

If P and Q are disjoint then $P \cap Q = \varnothing$.

Example 1 — Self Tutor

Suppose $A = \{2, 3, 5, 7, 11\}$ and $B = \{1, 3, 4, 6, 7, 8, 9\}$.

a True or false? **i** $4 \in B$ **ii** $4 \notin A$

b List the sets: **i** $A \cap B$ **ii** $A \cup B$

c True or false? **i** $A \cap B \subseteq A$ **ii** $\{3, 6, 10\} \subseteq B$

a **i** 4 is an element of set B, so $4 \in B$ is true.

ii 4 is not an element of set A, so $4 \notin A$ is true.

b **i** $A \cap B = \{3, 7\}$ since 3 and 7 are elements of both sets.

ii Every element which is in either A or B is in the union of A and B.
$\therefore \; A \cup B = \{1, 2, 3, 4, 5, 6, 7, 8, 9, 11\}$

c **i** $A \cap B = \{3, 7\}$ and $A = \{2, 3, 5, 7, 11\}$.
Every element of $A \cap B$ is also in A, so $A \cap B \subseteq A$ is true.

ii $10 \notin B$, so not every element of $\{3, 6, 10\}$ is an element of B.
So, $\{3, 6, 10\} \subseteq B$ is false.

EXERCISE 3A

1 Consider the number plate alongside.

a List the elements of:
  i the set of colours, C
  ii the set of digits, D
  iii the set of letters, L.

b Find: **i** $n(C)$ **ii** $n(D)$ **iii** $n(L)$.

2 **a** List the elements of P, the set of all factors of 45.
b List the elements of Q, the set of letters in the word AUSTRALIA.
c Find $n(P)$ and $n(Q)$.

3 Write in set notation:
a 4 is an element of set P
b 9 is not an element of set D
c $\{2, 3\}$ is a subset of $\{1, 2, 3, 6\}$
d Set A has 6 elements.

4 List all the subsets of: **a** $\{a, b\}$ **b** $\{-1, 0, 2\}$.

5 Find $A \cap B$ and $A \cup B$ for:
a $A = \{a, b, c, d\}$ and $B = \{b, d, f, h\}$
b $A = \{1, 2, 3, 4, 5\}$ and $B = \{6, 7, 8\}$
c $A = \{\text{vanilla, chocolate, strawberry}\}$ and $B = \{\text{vanilla, lemon, chocolate, strawberry, coffee}\}$

6 Suppose $A = \{1, 3, 5, 7\}$ and $B = \{2, 4, 6, 8\}$.
a Find $A \cap B$.
b Are A and B disjoint?

7 True or false? For any two sets A and B:
a $A \cap B \subseteq A$ and $A \cap B \subseteq B$
b $A \cap B \subseteq A \cup B$

8 For each of the following, is $R \subseteq S$?
a $R = \varnothing$ and $S = \{1, 2, 3, 4, 6, 12\}$
b $R = \{1, 2, 3, 4, 5\}$ and $S = \{1, 3, 5\}$
c $R = \{b, c, f, n, p\}$ and $S = \{a, b, c, f, n, q\}$
d $R = \{\text{blue, white}\}$ and $S = \{\text{red, blue, green, white}\}$

9 Let A be the set of prime numbers less than 20, and B be the set of factors of 20.
a List the elements of A.
b List the elements of B.
c Find $n(A)$ and $n(B)$.
d Find $A \cup B$ and $A \cap B$.
e Find $n(A \cap B)$.

10 Consider two sets A and B such that $A \subseteq B$. Find the possible values of x if:
a $A = \{3, 5, 7, x\}$ and $B = \{2, 3, 4, 5, 6, 7, 8\}$
b $A = \{1, 2, x\}$ and $B = \{1, x - 2, x\}$

B SPECIAL NUMBER SETS

Following is a list of some special number sets you should be familiar with:

- $\mathbb{N} = \{0, 1, 2, 3, 4, 5, 6, 7, ...\}$ is the set of all **natural** or **counting numbers.**

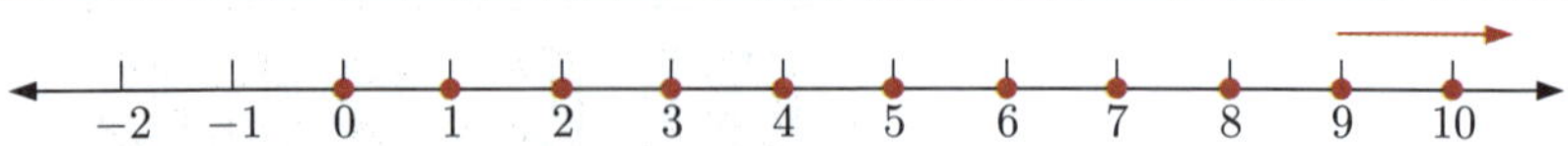

- $\mathbb{Z} = \{0, \pm 1, \pm 2, \pm 3, \pm 4, ...\}$ is the set of all **integers**.

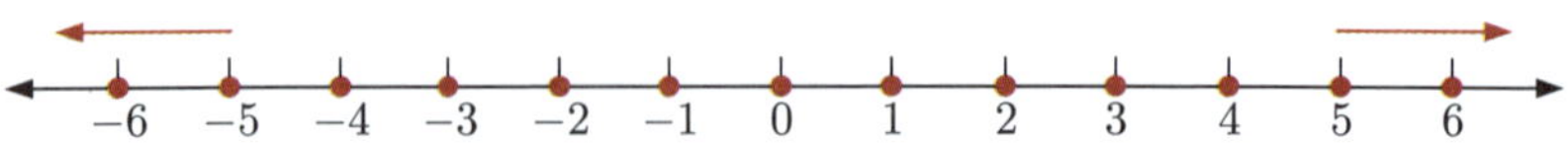

- $\mathbb{Z}^+ = \{1, 2, 3, 4, 5, 6, 7, ...\}$ is the set of all **positive integers**.

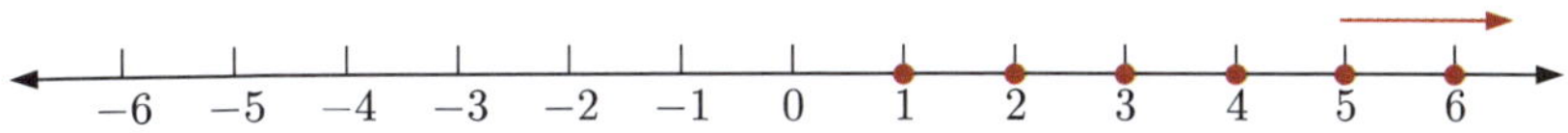

- $\mathbb{Q}$ is the set of all **rational numbers**, or numbers which can be written in the form $\frac{p}{q}$ where p and q are integers, $q \neq 0$.

  For example: $\frac{3}{2}$, $5\ (= \frac{5}{1})$, $0.25\ (= \frac{1}{4})$, and $-\frac{2}{3}$ are all rational numbers.

- $\mathbb{Q}'$ is the set of all **irrational numbers** which cannot be written in rational form.

  For example: $\sqrt{2}$, $\sqrt[3]{7}$, and π all belong to $\mathbb{Q}'$.

- $\mathbb{R}$ is the set of all **real numbers**, which are all numbers which can be placed on the number line.

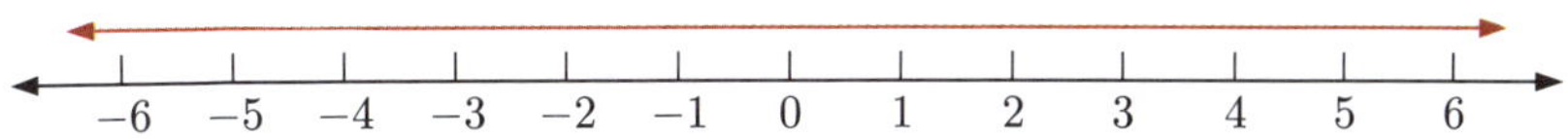

  $\mathbb{R}$ includes all rational and irrational numbers.

All of these special sets have infinitely many elements. We call them **infinite sets**, and say that $n(S) = \infty$.

EXERCISE 3B

1 True or false?

a $5 \in \mathbb{Z}$ **b** $-3 \in \mathbb{N}$ **c** $\sqrt{2} \in \mathbb{R}$ **d** $\frac{3}{7} \in \mathbb{Q}$

e $-8.5 \in \mathbb{Z}^-$ **f** $3\pi \in \mathbb{Q}'$ **g** $6.23 \in \mathbb{R}$ **h** $5\frac{1}{4} \in \mathbb{Q}'$

2 Explain why: **a** $\mathbb{Z}^- \cup \mathbb{Z}^+ \neq \mathbb{Z}$ **b** $n(\mathbb{Z}^- \cap \mathbb{Z}^+) = 0$

3 True or false?

a $\mathbb{N} \subseteq \mathbb{Z}$ **b** $\mathbb{N} \cap \mathbb{Z}^- = \varnothing$ **c** $\mathbb{N} \cup \mathbb{Z}^- = \mathbb{Z}$ **d** $\mathbb{Z}^+ \subseteq \mathbb{Q}$

e $\mathbb{Z} \subseteq \mathbb{Q}$ **f** $n(\mathbb{Z}) = \infty$ **g** $\mathbb{R} \subseteq \mathbb{Q}$ **h** $\mathbb{Q}' \subseteq \mathbb{R}$

4 Find:

a $\mathbb{Q} \cup \mathbb{Q}'$ **b** $\mathbb{Q} \cap \mathbb{Q}'$ **c** $\mathbb{Q}' \cap \mathbb{R}$ **d** $\mathbb{Q}' \cup \mathbb{R}$

DISCUSSION

Which of the following *may* be rational? See if you can provide examples.

- the sum of two irrational numbers
- the difference between two irrational numbers
- the product of two irrational numbers
- an irrational number raised to an irrational power.

C INTERVAL NOTATION

To describe the set of all integers between 3 and 8, we can list the set as $\{4, 5, 6, 7\}$ or illustrate the set as points on a number line. Alternatively, we can write the set using **interval notation** as $\{x \mid 3 < x < 8,\ \ x \in \mathbb{Z}\}$.

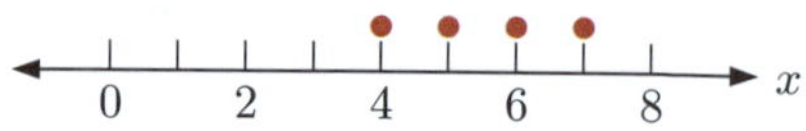

We read this as "the set of all integers x such that x lies between 3 and 8".

The notation $\{x \mid ......\}$ is used to describe "the set of all x such that ......".

Interval notation is very useful if the set contains a large or infinite number of elements and listing them would be time consuming or impossible.

The set of all *real numbers* between 3 and 8 would be written as $\{x \mid 3 < x < 8,\ \ x \in \mathbb{R}\}$.

Unless stated otherwise, we *assume* we are dealing with *real* numbers. Thus, the set can also be written as $\{x \mid 3 < x < 8\}$.

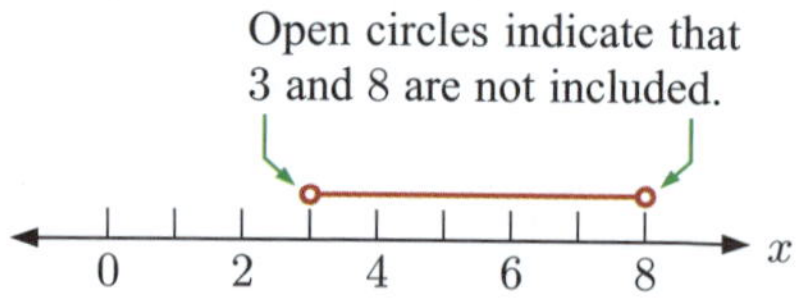

The set of all real numbers between 3 and 8 *inclusive* would be written as $\{x \mid 3 \leqslant x \leqslant 8,\ \ x \in \mathbb{R}\}$ or just $\{x \mid 3 \leqslant x \leqslant 8\}$.

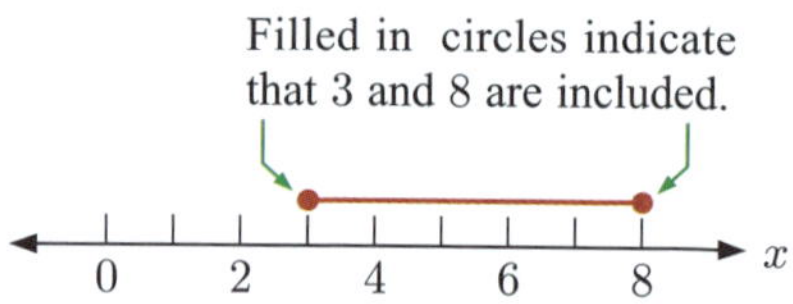

Example 2 Self Tutor

Suppose $A = \{x \mid 0 < x \leqslant 7,\ \ x \in \mathbb{Z}\}$.

a Write down the meaning of the interval notation.
b List the elements of A.
c Find $n(A)$.
d Illustrate A on a number line.

a The set of all integers x such that x lies between 0 and 7, including 7.
b $A = \{1, 2, 3, 4, 5, 6, 7\}$
c $n(A) = 7$
d 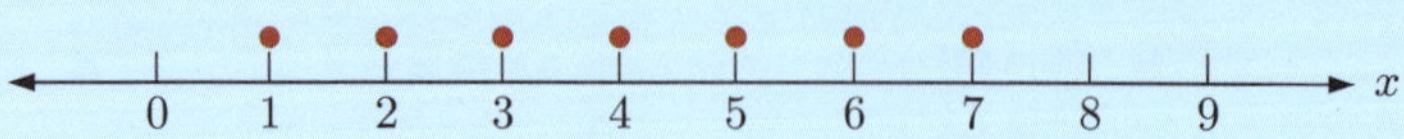

EXERCISE 3C

1 Write using interval notation:

a the set of all integers between 10 and 100
b the set of all real numbers less than 200
c the set of all rational numbers between -1 and 1 inclusive
d 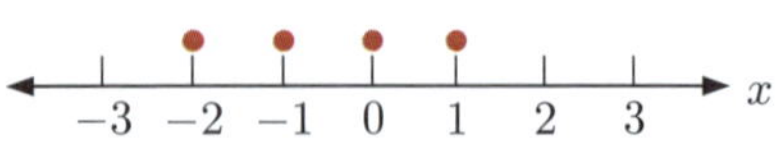
e 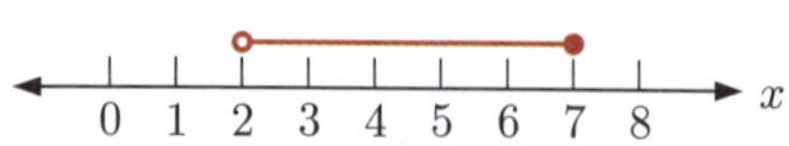

2 Are the following sets finite or infinite?

a $\{x \mid -10 \leqslant x \leqslant 10, \quad x \in \mathbb{Z}\}$

b $\{x \mid x \geqslant 50, \quad x \in \mathbb{N}\}$

c $\{x \mid 4 \leqslant x \leqslant 5, \quad x \in \mathbb{Q}\}$

d $\{x \mid x < -1, \quad x \in \mathbb{R}\}$

3 For each of the following sets B:

i write down the meaning of the interval notation

ii if possible, list the elements of B

iii find $n(B)$

iv illustrate B on a number line.

a $B = \{x \mid -2 < x < 5, \quad x \in \mathbb{Z}\}$

b $B = \{x \mid -6 \leqslant x \leqslant 3, \quad x \in \mathbb{Z}^+\}$

c $B = \{x \mid 1 < x < 2, \quad x \in \mathbb{R}\}$

d $B = \{4 \leqslant x < 7, \quad x \in \mathbb{R}\}$

4 For each of the following, is $C \subseteq D$?

a $C = \{x \mid 0 \leqslant x \leqslant 1000, \quad x \in \mathbb{R}\}$ and $D = \{x \mid x \in \mathbb{R}\}$

b $C = \mathbb{Z}^+$ and $D = \mathbb{Z}$

c $C = \{x \mid 2 \leqslant x \leqslant 5, \quad x \in \mathbb{Z}\}$ and $D = \{x \mid -2 \leqslant x \leqslant 5, \quad x \in \mathbb{R}\}$

d $C = \{x \mid x \in \mathbb{Q}\}$ and $D = \{x \mid x \in \mathbb{Z}\}$

5 Let $P = \{x \mid 10 \leqslant x \leqslant 80, \quad x \in \mathbb{Z}\}$ and $Q = \{x \mid 30 \leqslant x \leqslant 120, \quad x \in \mathbb{Z}\}$.

a Write using interval notation: **i** $P \cap Q$ **ii** $P \cup Q$

b Find:

i $n(P)$ **ii** $n(Q)$ **iii** $n(P \cap Q)$ **iv** $n(P \cup Q)$

DISCUSSION

Can we represent the set of rational numbers $\mathbb{Q}$ on a number line?

D COMPLEMENT OF A SET

UNIVERSAL SETS

Suppose we are only interested in the single digit positive whole numbers $\{1, 2, 3, 4, 5, 6, 7, 8, 9\}$. We would call this set our *universal set*.

The **universal set** under consideration is represented by U.

If we are considering the two-digit positive whole numbers, the universal set would be $U = \{10, 11, 12, 13, ..., 99\}$.

COMPLEMENT OF A SET

The **complement** of A, denoted A', is the set of all elements of U that are **not in** A.

For example:

- If $U = \{1, 2, 3, 4, 5, 6, 7, 8, 9\}$ and $A = \{1, 3, 5, 7, 9\}$ then the *complementary set* of A is $A' = \{2, 4, 6, 8\}$.
- $\mathbb{Q}$ and $\mathbb{Q}'$ are complements in the universal set $\mathbb{R}$.

Notice that:

- $A \cap A' = \varnothing$ {as A and A' have no members in common}
- $A \cup A' = U$ {every element of U is in either A or A'}
- $n(A) + n(A') = n(U)$.

Example 3 — Self Tutor

Suppose $U = \{-3, -2, -1, 0, 1, 2, 3\}$, $A = \{-2, 0, 2, 3\}$, and $B = \{-3, -2, 1, 3\}$.

a List: **i** A' **ii** B' **iii** $A \cap B$ **iv** $A \cup B$ **v** $A \cap B'$ **vi** $A' \cup B'$

b Find $n(A' \cup B')$.

a **i** $A' = \{-3, -1, 1\}$ {all elements of U not in A}

ii $B' = \{-1, 0, 2\}$ {all elements of U not in B}

iii $A \cap B = \{-2, 3\}$ {all elements common to A and B}

iv $A \cup B = \{-3, -2, 0, 1, 2, 3\}$ {all elements in A or B or both}

v $A \cap B' = \{-2, 0, 2, 3\} \cap \{-1, 0, 2\} = \{0, 2\}$

vi $A' \cup B' = \{-3, -1, 1\} \cup \{-1, 0, 2\} = \{-3, -1, 0, 1, 2\}$

b $n(A' \cup B') = 5$ {there are 5 members in this set}

EXERCISE 3D

1 Suppose $U = \{1, 2, 3, 4, 5, 6, 7\}$ and $B = \{2, 4, 6\}$.

a List the set B'.

b Verify that:

i $B \cap B' = \varnothing$ **ii** $B \cup B' = U$ **iii** $n(B) + n(B') = n(U)$

2 Find D', the complement of D, given that:

a $U = \{\text{integers}\}$ and $D = \{0\} \cup \mathbb{Z}^+$

b $U = \mathbb{Z}^+$ and $D = \{\text{odd positive integers}\}$

c $U = \mathbb{Z}$ and $D = \{x \mid x \leqslant 10,\ x \in \mathbb{Z}\}$

d $U = \mathbb{Q}$ and $D = \{x \mid x \leqslant 3 \text{ or } x \geqslant 5,\ x \in \mathbb{Q}\}$

3 Suppose $U = \{0, 1, 2, 3, 4, 5, ..., 24\}$, $F = \{\text{factors of } 24\}$, and $M = \{\text{multiples of } 4\}$.

a List the sets:

i F **ii** M **iii** F' **iv** M'

b List the elements of $F \cap M'$. Explain what this set represents.

4 Suppose $U = \{x \mid 12 \leqslant x \leqslant 20,\ \ x \in \mathbb{Z}\}$, $A = \{12, 14, 16, 18\}$, and $B = \{12, 15, 18\}$.

a List the elements of:

i U **ii** A' **iii** B' **iv** $A \cup B$ **v** $A' \cap B'$

b Show that $(A \cup B)' = A' \cap B'$.

5 Suppose U is the set of the months of the year.
Let $J = \{$months that begin with "J"$\}$ and $K = \{$months containing 31 days$\}$.

a List the elements of:

i J **ii** K **iii** J' **iv** K'

b What does the set K' represent?

c List the elements of $(J \cup K)'$. What does this set represent?

d List the elements of $J' \cap K$. What does this set represent?

6 Suppose $U = \{x \mid 1 \leqslant x < 15,\ \ x \in \mathbb{Z}\}$, $P = \{x \mid 3 \leqslant x \leqslant 6,\ \ x \in \mathbb{Z}\}$, and $Q = \{2, 6, 10, 14\}$.

a List the elements of:

i U **ii** P **iii** P' **iv** Q' **v** $P' \cup Q$ **vi** $P \cap Q'$

b Find:

i $n(P' \cup Q)$ **ii** $n(P \cap Q')$

E VENN DIAGRAMS

A **Venn diagram** consists of a universal set U represented by a rectangle, and subsets within it that are generally represented by circles.

The Venn diagram alongside shows set A within the universal set U.

The **complement** of A is the shaded region outside the circle representing A.

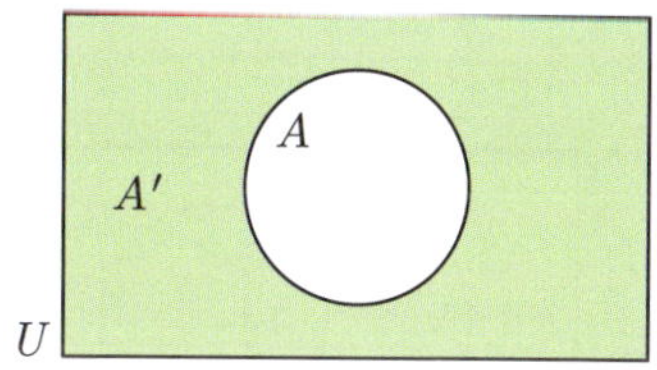

Suppose $U = \{1, 3, 4, 6, 9\}$, $A = \{1, 6, 9\}$, and $A' = \{3, 4\}$.

These sets are represented in the Venn diagram alongside.

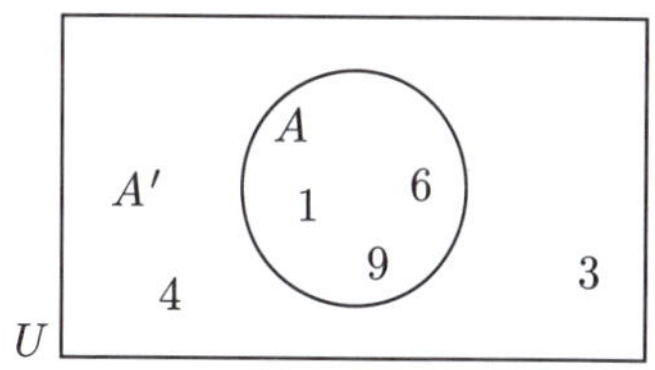

If B is a **subset** of A, then every element of B is also in A.

The circle representing B is placed within the circle representing A.

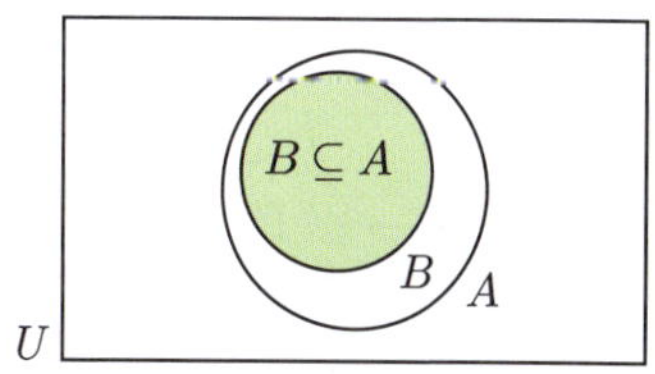

The **intersection** of two sets A and B consists of all elements common to both A and B.

$A \cap B$ is the shaded region where the circles representing A and B overlap.

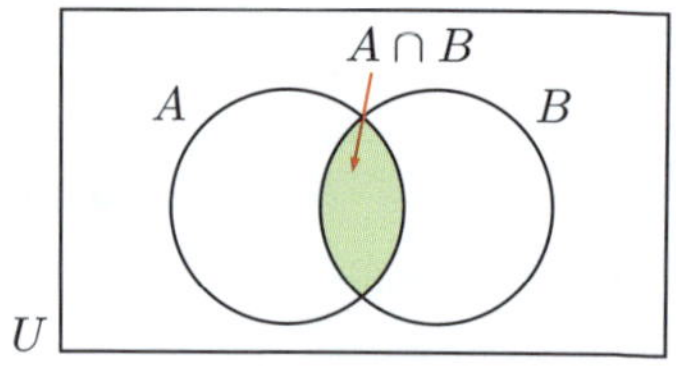

The **union** of two sets A and B consists of all elements in A or B or both.

$A \cup B$ is the shaded region which includes everywhere in either circle.

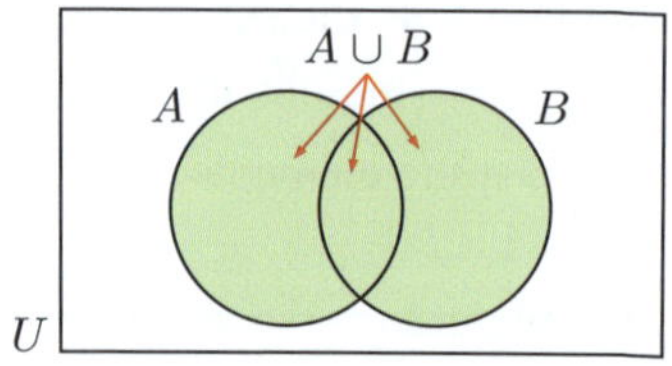

Disjoint or **mutually exclusive** sets do not have common elements.

They are represented by non-overlapping circles.

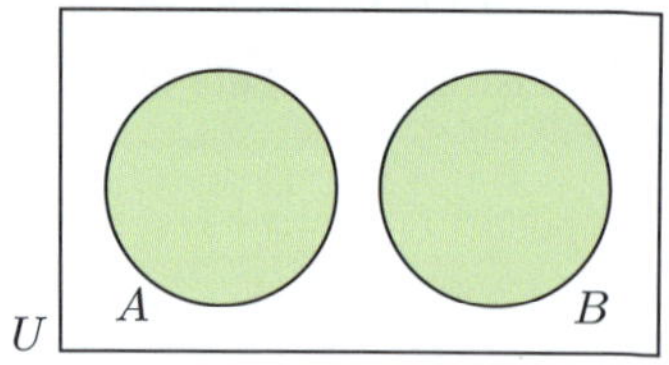

If the sets A and B are **disjoint and exhaustive** then $B = A'$ and $A \cup B = U$.

We can represent this situation without using circles as shown.

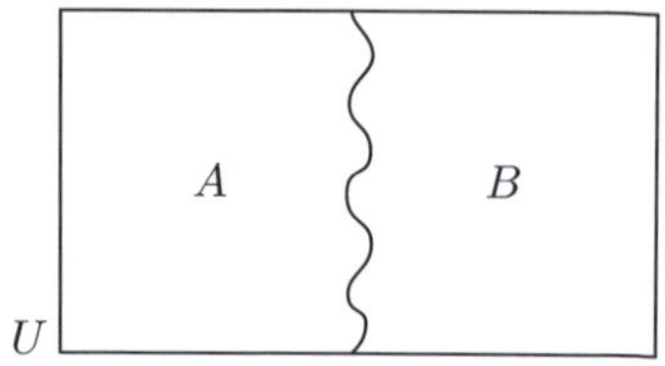

Example 4 — Self Tutor

If we roll an ordinary die, the universal set of the possible outcomes is $U = \{1, 2, 3, 4, 5, 6\}$.
Illustrate on a Venn diagram the sets:

a $A = \{1, 2\}$ and $B = \{1, 3, 4\}$

b $A = \{1, 3, 5\}$ and $B = \{3, 5\}$

c $A = \{2, 4, 6\}$ and $B = \{3, 5\}$

d $A = \{1, 3, 5\}$ and $B = \{2, 4, 6\}$

a $A \cap B = \{1\}$

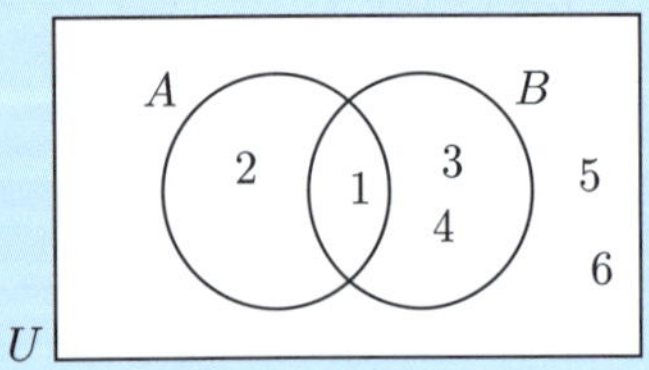

b $A \cap B = \{3, 5\}$, $B \subseteq A$

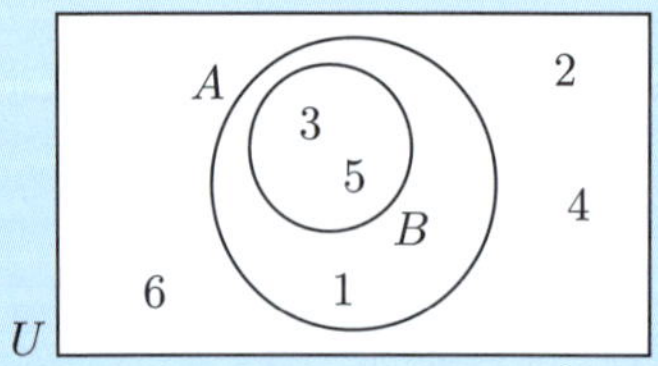

c A and B are disjoint but $A \cup B \neq U$.

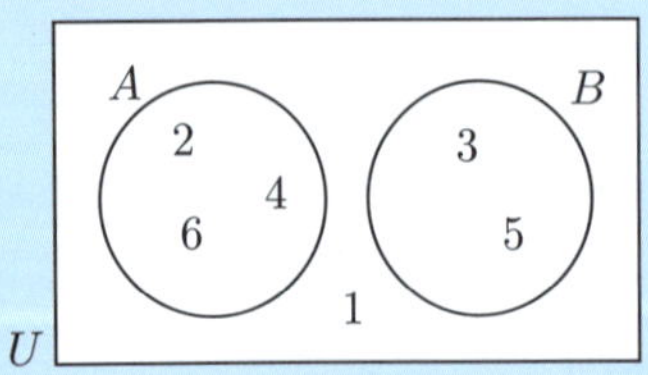

d A and B are disjoint and exhaustive.

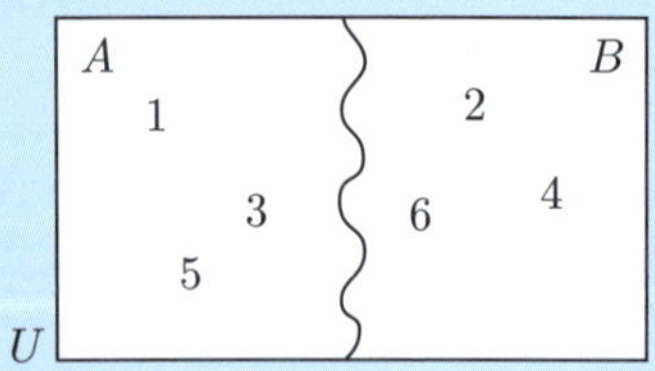

EXERCISE 3E.1

1 For the Venn diagram alongside, list the elements of:

a P **b** Q **c** P'
d Q' **e** $P \cap Q$ **f** $P \cup Q$
g U

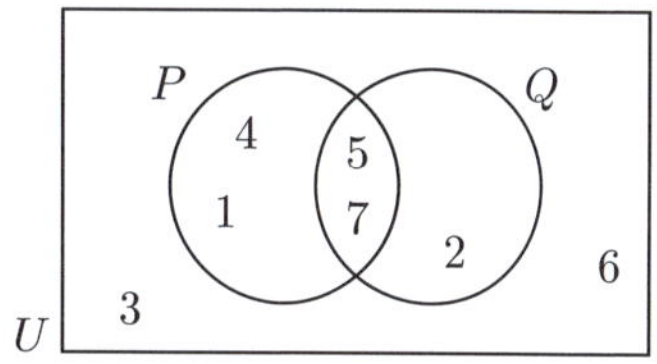

2

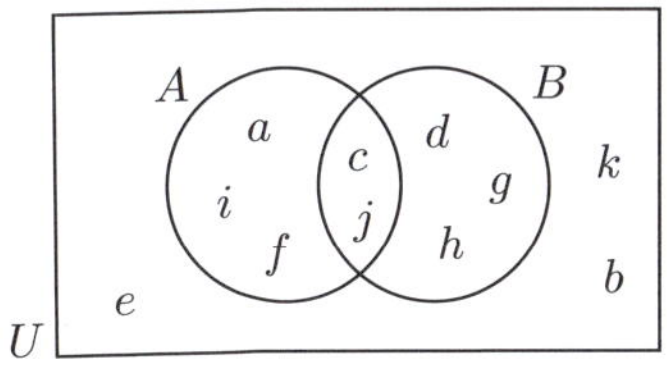

Consider the Venn diagram shown.
List the elements of the set:

a A **b** B **c** A'
d B' **e** $A \cap B$ **f** $A \cup B$

3 Consider the universal set $U = \{0, 1, 2, 3, 4, 5, 6, 7, 8, 9\}$.
Illustrate on a Venn diagram the sets:

a $A = \{1, 2, 4, 8\}$ and $B = \{2, 3, 4, 5\}$
b $A = \{1, 2, 3, 5, 8\}$ and $B = \{4, 6, 7\}$
c $A = \{1, 2, 3\}$ and $B = \{1, 2, 3, 6\}$
d $A = \{0, 1, 3, 7, 9\}$ and $B = \{0, 1, 2, 4\}$

4 Suppose $U = \{1, 2, 3, 4, ..., 12\}$, $A = \{\text{factors of } 12\}$, and $B = \{\text{primes} \leqslant 12\}$.

a List the sets A and B. **b** Find $A \cap B$ and $A \cup B$.
c Represent A and B on a Venn diagram.

5 Suppose $U = \{x \mid x \leqslant 20,\ x \in \mathbb{Z}^+\}$, $R = \{\text{primes less than } 20\}$, and $S = \{\text{composites less than } 20\}$.

a List the sets R and S. **b** Find $R \cap S$ and $R \cup S$.
c Represent R and S on a Venn diagram.

6 A group of students had a Mathematics test and a History test, both out of 20 marks. The results are shown in the table.
Let M be the students who scored 15 or more for the Mathematics test, and H be the students who scored 15 or more for the History test.

a List the elements of M and H.
b Display these sets on a Venn diagram.
c List the elements of $M \cup H$. What does this set represent?
d List the elements of $M \cap H'$. What does this set represent?

Student	*Mathematics*	*History*
Paul	12	15
Tess	11	14
Shaun	16	14
Ken	19	15
Goran	13	12
Pam	13	17
Ling	16	18
Stefan	12	17
Reg	14	16
Claude	10	14
Don	15	17
Vivian	20	13

7 Suppose $U = \{x \mid x \leqslant 25,\quad x \in \mathbb{Z}^+\}$,
$A = \{\text{odd numbers} \leqslant 25\}$,
$B = \{\text{multiples of 5 that are} \leqslant 25\}$, and
$C = \{\text{factors of 50 that are} \leqslant 25\}$.

a Represent A, B, and C on a Venn diagram like the one shown.

b List the elements which are in:

i both A and B

ii A but not C

iii B and C but not A.

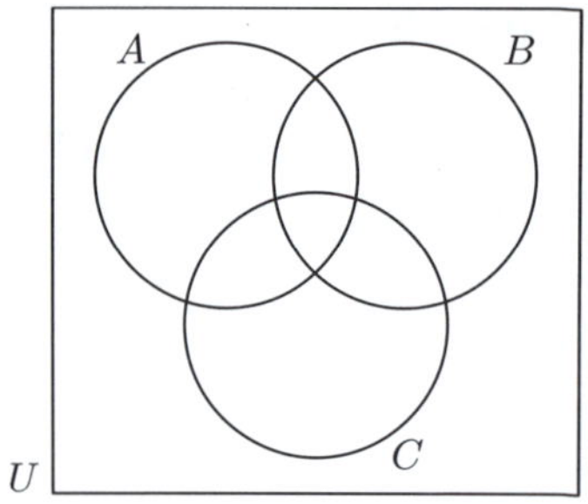

ACTIVITY 1

Click on the icon to practise constructing Venn diagrams.

For cases of two sets you will need to choose the most appropriate layout for the Venn diagram.

VENN DIAGRAM CONSTRUCTION

VENN DIAGRAM REGIONS

We can use shading to show various sets. For example, for two intersecting sets A and B, we have:

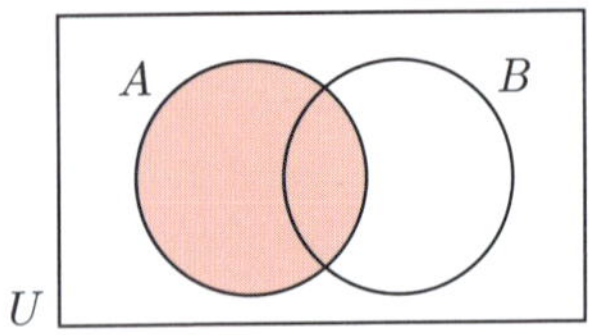

A is shaded

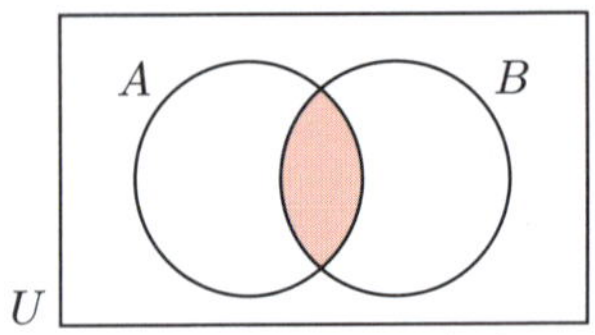

$A \cap B$ is shaded

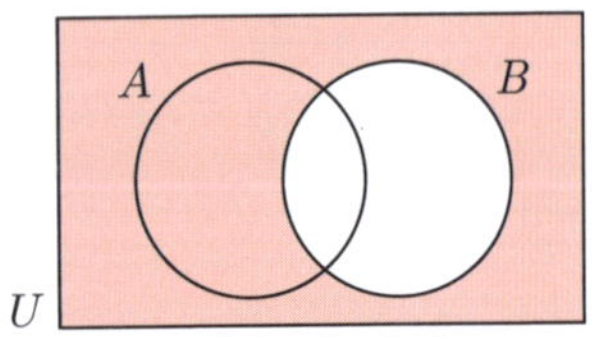

B' is shaded

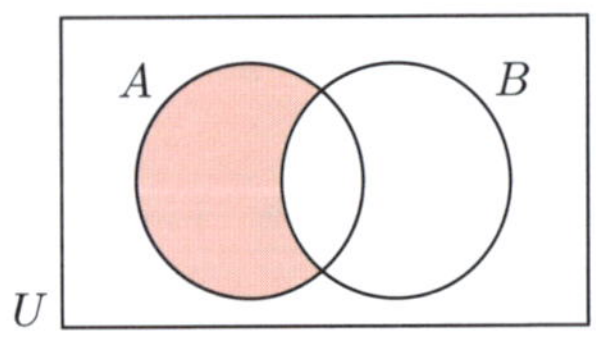

$A \cap B'$ is shaded

Example 5

Self Tutor

On separate Venn diagrams, shade these regions for two intersecting sets A and B:

a $A \cup B$ **b** $A' \cap B$ **c** $(A \cap B)'$

a

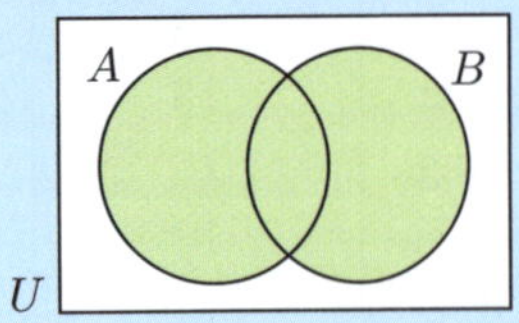

$A \cup B$ means in A, B, or both.

b

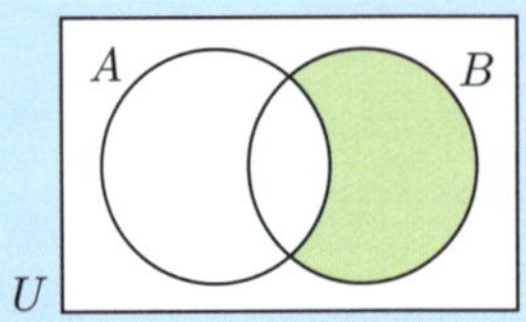

$A' \cap B$ means outside A, intersected with B.

c

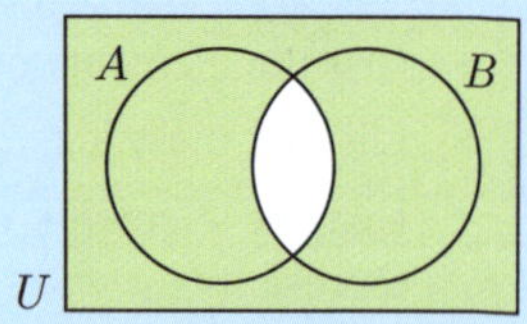

$(A \cap B)'$ means outside the intersection of A and B.

EXERCISE 3E.2

1 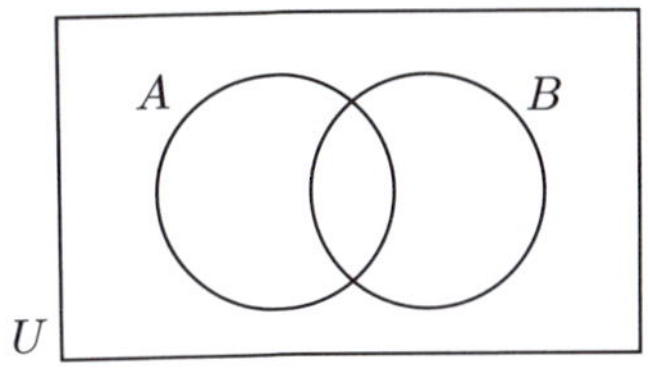

On separate Venn diagrams, shade the region:

- **a** in A
- **b** not in A
- **c** in neither A nor B
- **d** in A and B
- **e** in either A or B, but not both.

2 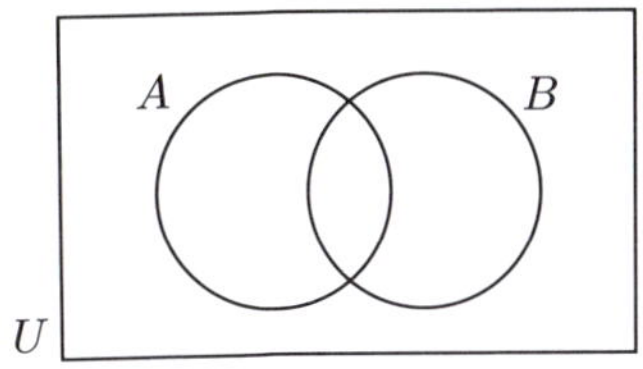

On separate Venn diagrams, shade:

- **a** B
- **b** B'
- **c** $A \cap B'$
- **d** $A \cup B'$
- **e** $(A \cap B)'$
- **f** $(A \cup B)'$

PRINTABLE VENN DIAGRAMS

3 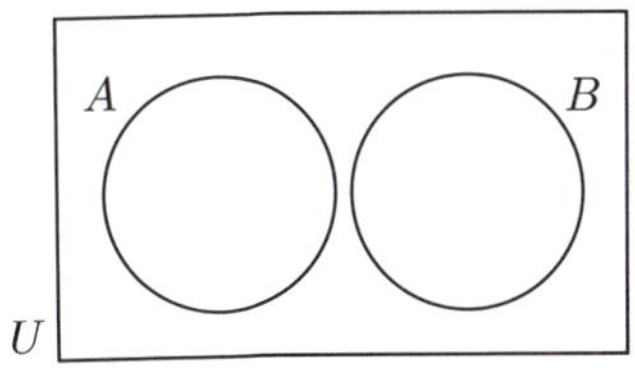

A and B are two disjoint sets. Shade on separate Venn diagrams:

- **a** A
- **b** B
- **c** A'
- **d** B'
- **e** $A \cap B$
- **f** $A \cup B$
- **g** $A' \cap B$
- **h** $A \cup B'$
- **i** $(A \cap B)'$

4 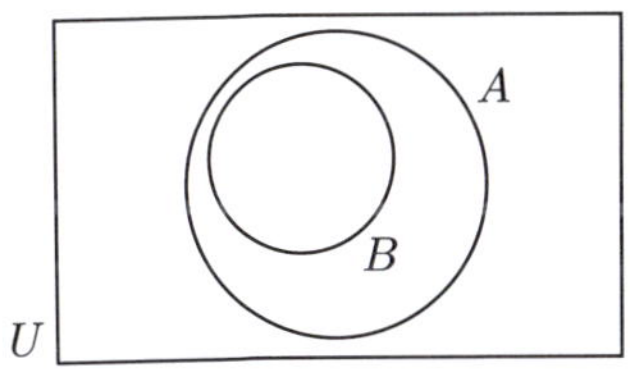

In the given Venn diagram, $B \subseteq A$. Shade on separate Venn diagrams:

- **a** A
- **b** B
- **c** A'
- **d** B'
- **e** $A \cap B$
- **f** $A \cup B$
- **g** $A' \cap B$
- **h** $A \cup B'$
- **i** $(A \cap B)'$

5 On separate Venn diagrams, shade:

- **a** A
- **b** B
- **c** C'
- **d** $A \cup B$
- **e** $B \cap C$
- **f** $A \cap B \cap C$
- **g** $A \cap B' \cap C$
- **h** $(A \cup B)'$
- **i** $A' \cup (B \cap C)$

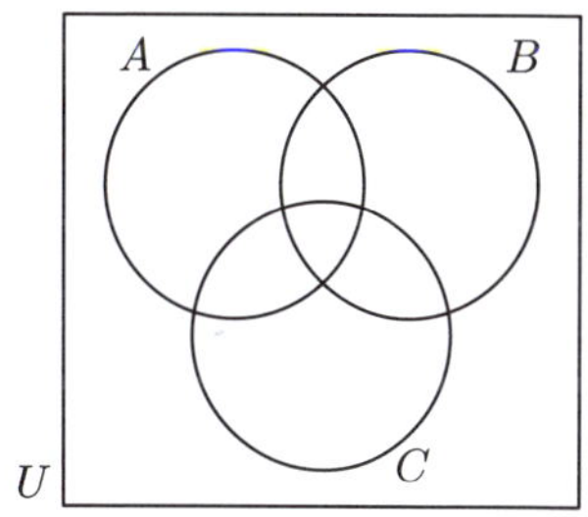

6 In the Venn diagram shown, $U = \mathbb{Z}^+$,
$E = \{\text{positive even integers}\}$,
$O = \{\text{positive odd integers}\}$, and
$P = \{\text{prime numbers}\}$.

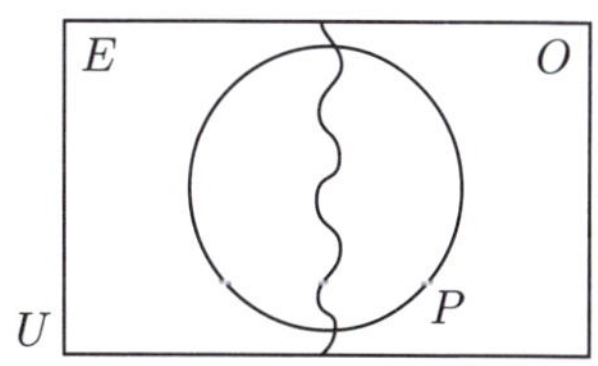

- **a** On separate Venn diagrams, shade the region which represents the set of:
  - **i** odd prime numbers
  - **ii** even composite numbers.
- **b** Explain what the region $E \cap P$ represents. Hence write down $n(E \cap P)$.

ACTIVITY 2

Click on the icon to practise shading regions representing various subsets.
The software includes two and three intersecting sets.

VENN DIAGRAM REGIONS

PUZZLE

Consider the three overlapping sets shown. The universal set contains the smallest positive integers such that each region of the diagram contains exactly one element.

Find the sum of all numbers in the universal set.

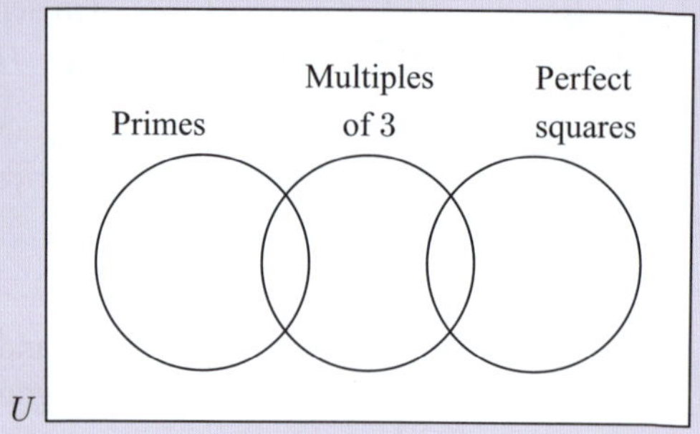

NUMBERS IN REGIONS

For many problems involving Venn diagrams, we are interested in *how many* elements are in each region, but not which elements these are.

To indicate the **number of elements** in a region of a Venn diagram, we write the number in brackets.

Example 6

Self Tutor

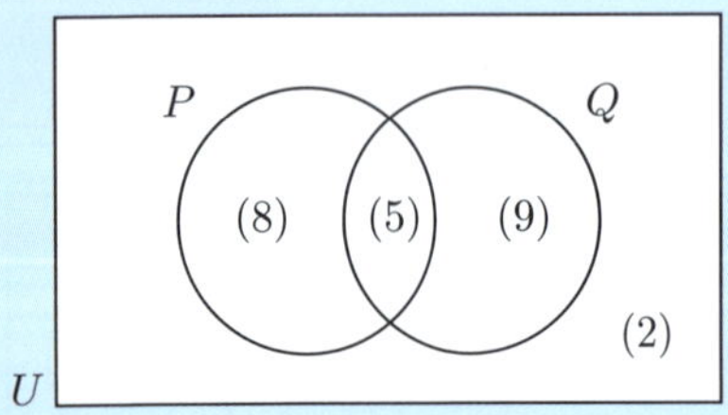

How many elements are there in:

a P **b** Q'

c $P \cup Q$ **d** P, but not Q

e Q, but not P **f** neither P nor Q?

a $n(P) = 8 + 5 = 13$ **b** $n(Q') = 8 + 2 = 10$

c $n(P \cup Q) = 8 + 5 + 9 = 22$ **d** $n(P, \text{ but not } Q) = 8$

e $n(Q, \text{ but not } P) = 9$ **f** $n(\text{neither } P \text{ nor } Q) = 2$

EXERCISE 3E.3

1

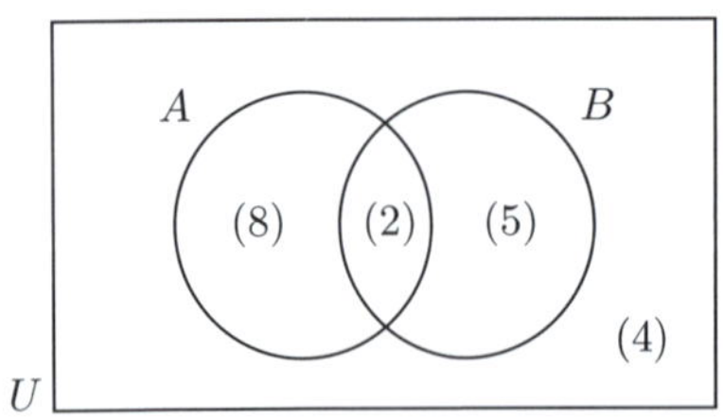

Give the number of elements in:

a B **b** A'

c $A \cup B$ **d** A, but not B

e B, but not A **f** U

2 Give the number of elements in:

- **a** X'
- **b** X and Y
- **c** $X \cup Y$
- **d** $X \cap Y'$
- **e** $Y \cap X'$
- **f** neither X nor Y

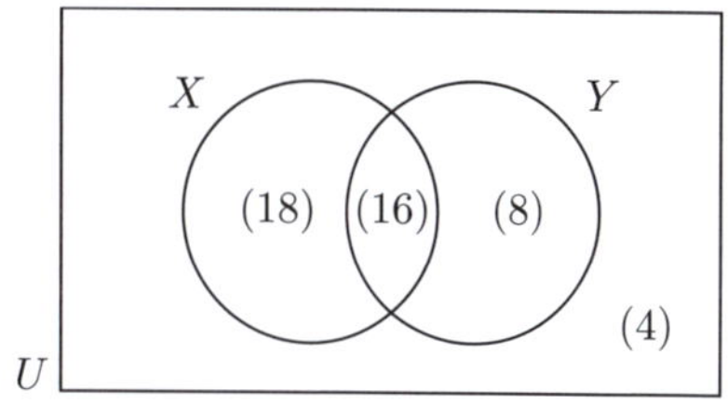

3

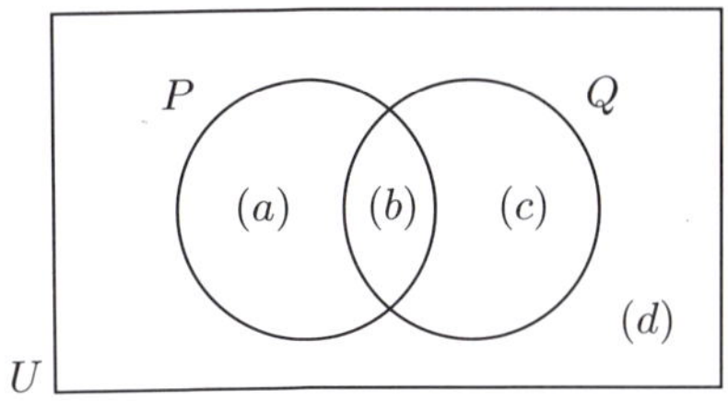

Find, in terms of a, b, c, and d, the number of elements in:

- **a** Q
- **b** P'
- **c** P or Q, but not both
- **d** P and Q
- **e** $(P \cap Q)'$
- **f** $(P \cup Q)'$

4 **a** Find, in terms of a:

- **i** $n(Q)$
- **ii** $n(P \cup Q)$
- **iii** $n(Q')$
- **iv** $n(U)$

b Find a if $n(U) = 43$.

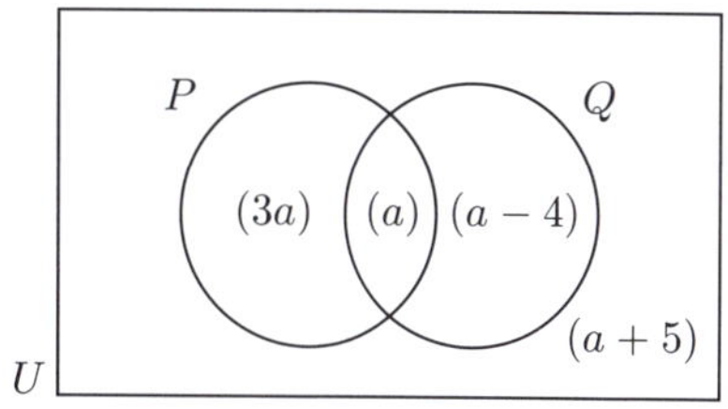

5

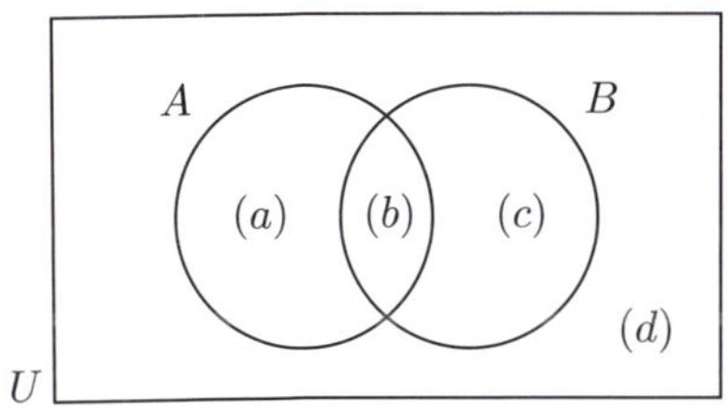

a For the given Venn diagram, find:

- **i** $n(A \cup B)$
- **ii** $n(A) + n(B) - n(A \cap B)$.

b What have you proved in **a**?

Example 7 — Self Tutor

Given $n(U) = 40$, $n(A) = 24$, $n(B) = 27$, and $n(A \cap B) = 13$, find:

- **a** $n(A \cup B)$
- **b** $n(A \cap B')$.

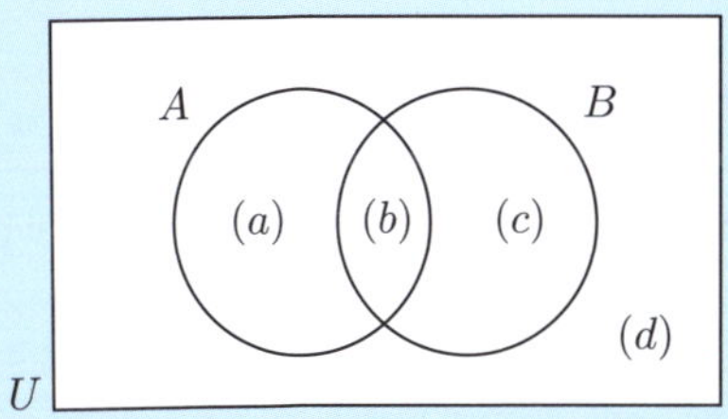

We see that

$$\begin{aligned} b &= 13 && \{\text{as } n(A \cap B) = 13\} \\ a + b &= 24 && \{\text{as } n(A) = 24\} \\ b + c &= 27 && \{\text{as } n(B) = 27\} \\ a + b + c + d &= 40 && \{\text{as } n(U) = 40\} \end{aligned}$$

Using $b = 13$, we find $a = 11$, $c = 14$, and hence $d = 2$.

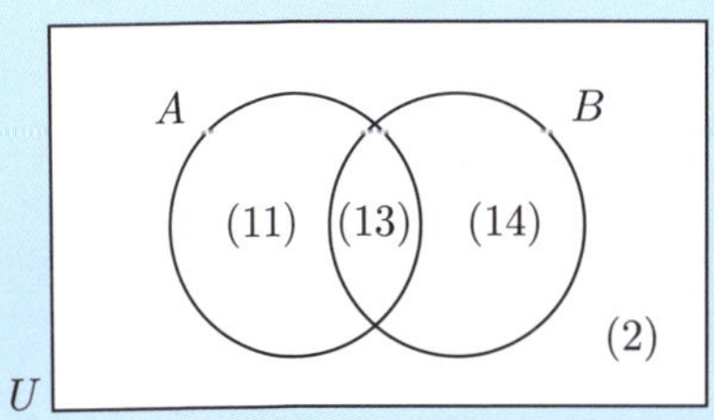

a $n(A \cup B) = a + b + c$
$= 38$

b $n(A \cap B') = a$
$= 11$

6 Suppose $n(U) = 20$, $n(A) = 8$, $n(B) = 9$, and $n(A \cap B) = 2$.

a Copy and complete the Venn diagram alongside.

b Find:

i $n(A \cup B)$ **ii** $n(B \cap A')$

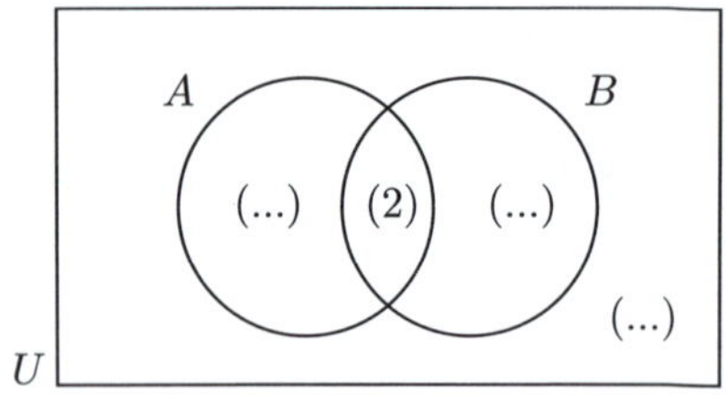

7 Given $n(U) = 35$, $n(N) = 16$, $n(N \cap R) = 4$, and $n(N \cup R) = 31$, find the number of elements in:

a R **b** $(N \cup R)'$ **c** R, but not N.

8

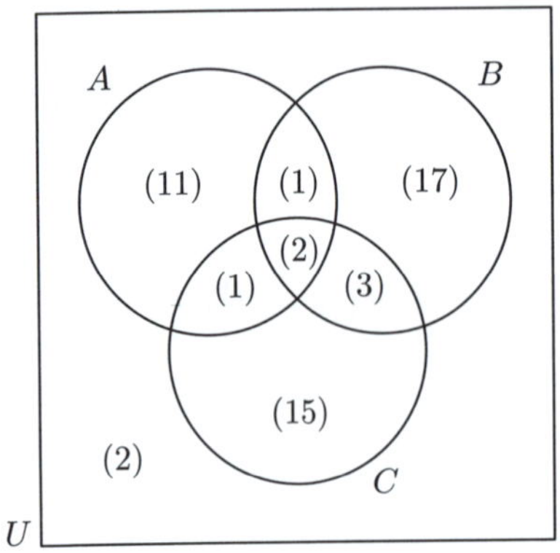

This Venn diagram contains the sets A, B, and C. Find the number of elements in:

a A **b** B

c C **d** A and B

e $A \cup C$ **f** $A \cap B \cap C$

g B and C, but not A

h $(A \cup B) \cap C$

i exactly one of the sets

j at least two of the sets.

F PROBLEM SOLVING WITH VENN DIAGRAMS

By considering the number of elements in different regions, we can use Venn diagrams to solve problems.

Example 8 **Self Tutor**

A tennis club has 42 members. 13 have fair hair, 15 have blue eyes, and 22 have neither fair hair nor blue eyes.

a Show this information on a Venn diagram.

b Find the number of members with:

i fair hair or blue eyes **ii** fair hair but not blue eyes.

---

a Let F represent the set of members with fair hair, and B represent the set of members with blue eyes.

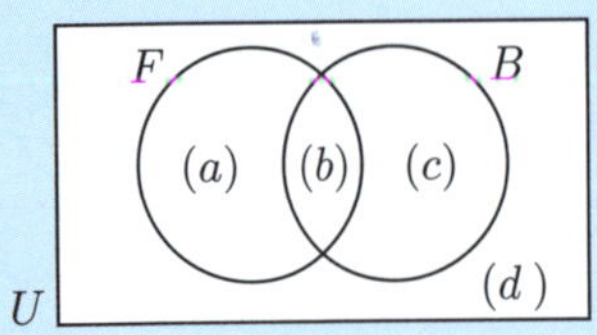

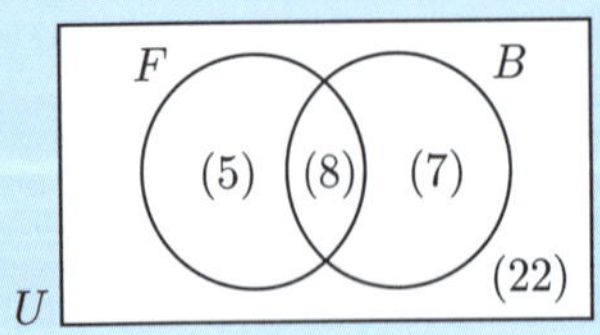

$a + b + c + d = 42$ {42 members in total}

But $d = 22$ {22 have neither}

$\therefore\ a + b + c = 20$

Now $a + b = 13$ {13 have fair hair}

$\therefore\ c = 20 - 13 = 7$

$b + c = 15$ {15 have blue eyes}

$\therefore\ b = 15 - 7 = 8$

and $a = 13 - 8 = 5$

b **i** $n(F \cup B) = 5 + 8 + 7 = 20$

$\therefore$ 20 members have fair hair or blue eyes.

ii $n(F \cap B') = 5$

$\therefore$ 5 members have fair hair but not blue eyes.

EXERCISE 3F

1 A local fair held a baking competition. 29 entrants baked a pie, 38 entrants baked a cake, and 12 entrants baked both. 9 entrants baked neither a pie nor a cake.

a Place the information on a Venn diagram.

b How many people entered the baking competition?

c How many entrants:

i baked a pie but not a cake

ii baked a cake but not a pie

iii did not bake a cake

iv did not bake a pie?

2 Lightpark High has 75 Year 9 students. 24 students play a musical instrument, 40 students play a sport, and 14 students play neither.

a Place the information on a Venn diagram.

b How many students:

i only play a musical instrument **ii** only play a sport

iii play both a musical instrument and sport?

3 Answer the **Opening Problem** on page **44**.

4 The 112 students in a year group in Canada are surveyed to give the origin of cars owned by their family. 84 families own an imported car. 45 families own a Canadian made car. 5 families do not own a car.

a Place the information on a Venn diagram.

b How many families only own an imported car?

c What fraction of the families own both a Canadian and an imported car?

5 41% of students in a Geneva school have been on a holiday to Paris, 46% have been to Rome, and 37% have not been to either Paris or Rome.

a Place the information on a Venn diagram.

b What percentage of students have been on a holiday to:

i both Paris and Rome

ii at least one of Paris or Rome

iii Paris only

iv Paris or Rome but not both?

Example 9 **Self Tutor**

A football team has training on Tuesdays, Wednesdays, and Fridays. Of the 40 players in the squad, 21 attended training on Tuesday, 25 attended training on Wednesday, and 24 attended training on Friday. 11 players attended on Tuesday and Wednesday, 15 attended on Wednesday and Friday, and 14 attended on Tuesday and Friday. 6 players attended all three training sessions.

a Represent this information on a Venn diagram.

b How many players attended training on:

i Wednesday only **ii** Friday, but not Tuesday **iii** exactly two of the days?

---

a Let T represent the players who attended on Tuesday, W represent the players who attended on Wednesday, and F represent the players who attended on Friday.

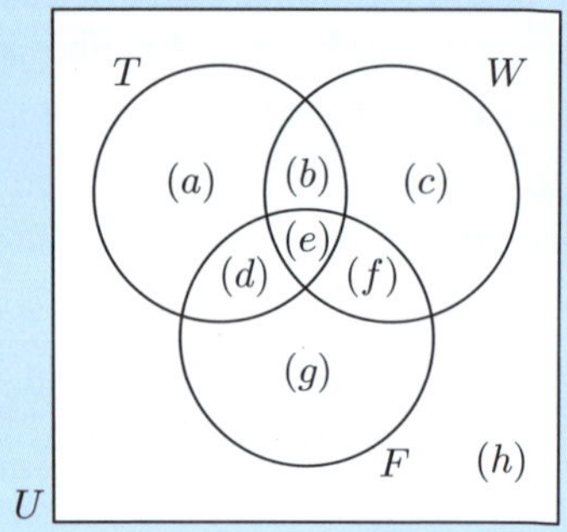

$e = 6$ {6 players attended all three sessions}

Now $b + e = 11$ {Tuesday and Wednesday}
$f + e = 15$ {Wednesday and Friday}
$d + e = 14$ {Tuesday and Friday}

$\therefore$ $b = 5$, $f = 9$, $d = 8$

Also, $a + b + d + e = 21$ {Tuesday}
$\therefore$ $a + 5 + 8 + 6 = 21$ and so $a = 2$
$b + c + e + f = 25$ {Wednesday}
$\therefore$ $5 + c + 6 + 9 = 25$ and so $c = 5$
$d + e + f + g = 24$ {Friday}
$\therefore$ $8 + 6 + 9 + g = 24$ and so $g = 1$

Finally, $a + b + c + d + e + f + g + h = 40$ {total}
$\therefore$ $2 + 5 + 5 + 8 + 6 + 9 + 1 + h = 40$
and so $h = 4$

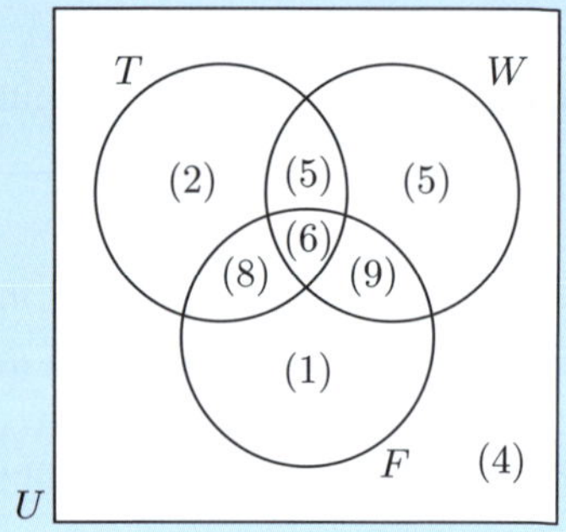

b **i** 5 players attended training on Wednesday only.

ii $9 + 1 = 10$ players attended training on Friday, but not Tuesday.

iii $5 + 9 + 8 = 22$ players attended training on exactly two of the days.

6 88 countries competed in the 2014 Winter Olympics. Of these countries, 21 won gold medals, 22 won silver medals, and 24 won bronze medals.
18 countries won gold and silver medals, 21 countries won silver and bronze medals, and 20 countries won gold and bronze medals.
18 countries won all three types of medal.

a Represent this information on a Venn diagram. State clearly what each set represents.

b How many of the countries won:

i no medals
ii silver and bronze, but not gold medals
iii exactly one type of medal?

7 The records of a men's clothing store indicate that 64% of customers buy shirts, 14% buy ties, and 34% buy jumpers. 7% buy shirts and ties, 20% buy shirts and jumpers, 9% buy ties and jumpers, and 5% buy all three. 19% of customers do not buy any of the three items.

a Illustrate this information on a Venn diagram.

b What percentage of customers buy:

i shirts only **ii** shirts or ties **iii** jumpers but not shirts?

REVIEW SET 3A

1 Consider $A = \{x \mid -2 < x \leqslant 6,\ x \in \mathbb{Z}\}$.

a True or false? **i** $2 \in A$ **ii** $3\frac{1}{2} \notin A$

b Find $n(A)$.

2 Write using interval notation:

a the set of all real numbers between 4 and 8 inclusive

b the set of all rational numbers less than 30.

3 Suppose $U = \mathbb{N}$, $F = \{\text{factors of } 32\}$, and $M = \{\text{multiples of 4 less than } 32\}$.

a List the sets F and M. **b** Is $F \subseteq M$?

c List the sets: **i** $F \cap M$ **ii** $F \cup M$.

d Find $n(F')$.

4 Suppose $U = \{a, b, c, d, e, f, g, h\}$, $A = \{a, b, d, e\}$, and $B = \{a, b, g\}$.

a List the elements of:

i A' **ii** $A \cap B$ **iii** $A \cup B$ **iv** $A' \cap B$ **v** $A \cup B'$

b Illustrate A, B, and U on a Venn diagram.

5 Consider the Venn diagram alongside.

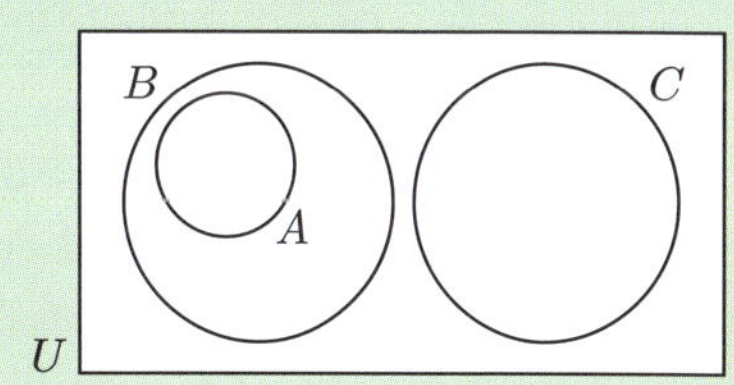

a On separate Venn diagrams, shade the region:

i $A \cap B$ **ii** B' **iii** $A' \cap B$

b True or false?

i $A \cap C = \varnothing$ **ii** $A \subseteq B$ **iii** $B \subseteq C$

6 State the number of elements in:

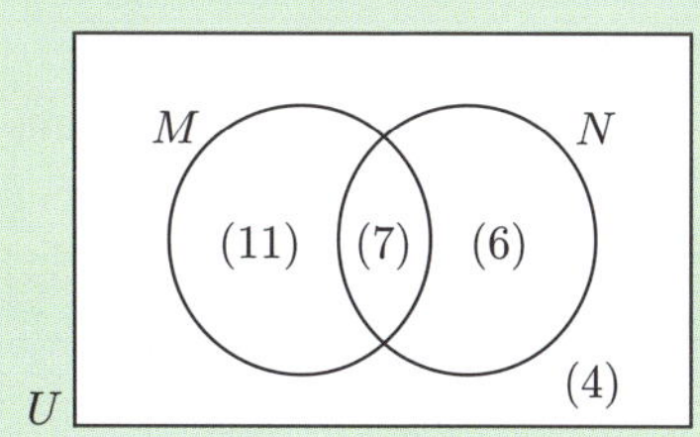

a M' **b** $M \cap N$

c M, but not N.

7 In a class of 27 students, 13 students have brothers, 16 have sisters, and 5 have neither.

a Illustrate this information on a Venn diagram.

b How many students have:

i brothers and sisters **ii** sisters only?

8 **a** Find, in terms of x:

i $n(Z)$ **ii** $n(Y \cup Z)$

b Find x if $n(U) = 21$.

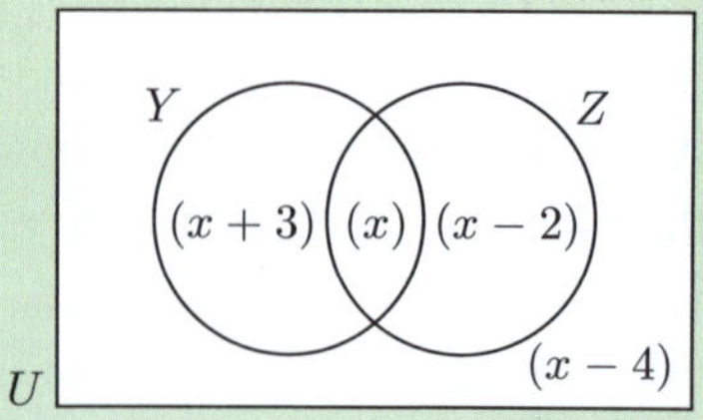

9 A vegan restaurant has 17 dishes on its dinner menu. 15 of the dishes are gluten free, 13 dishes are onion free, and 12 dishes are both gluten and onion free.

a Draw a Venn diagram to represent this information.

b How many dishes:

i are gluten free but not onion free **ii** are onion free but not gluten free

iii contain both gluten and onion?

10 6000 people in Los Angeles auditioned for a reality TV show. 180 were selected for the second round of auditions, and 16 made it to the final.

a Draw a Venn diagram which accurately represents this information.

b How many people:

i made it to the second round, but did not make the final

ii did not make the second round?

c What percentage of the people who auditioned made it to the final?

11 In a game of Bingo, numbers from 1 to 75 are selected at random. So far, 30 different numbers have been selected.

16 of these numbers are multiples of 2, 11 are multiples of 3, and 6 are multiples of 5. 6 numbers are multiples of 2 and 3, 3 numbers are multiples of 3 and 5, and 5 numbers are multiples of 2 and 5. Two of the numbers are multiples of 2, 3, and 5.

Let U be the set of numbers which have been selected,
A be the set of multiples of 2,
B be the set of multiples of 3,
and C be the set of multiples of 5.

a Represent this information on a Venn diagram.

b How many selected numbers are:

i a multiple of 2 or 5 **ii** a multiple of 3, but not 5?

c Determine whether the numbers 35 and 60 have been selected. Explain your reasoning.

REVIEW SET 3B

1 List the elements of set C which is the set of all colours on the flag.

2 Suppose $A = \{a, e, i, o, u\}$ and $B = \{a, b, c, d, e\}$.
Find: **a** $A \cup B$ **b** $A \cap B$

3 Suppose $P = \{\text{square numbers} < 40\}$ and $Q = \{\text{multiples of } 4 < 40\}$.
Find: **a** $P \cap Q$ **b** $n(P \cap Q)$

4 Explain why $\mathbb{N} \subseteq \mathbb{Q}$.

5 Consider the set $\{x \mid x \geqslant -8,\ x \in \mathbb{Z}^-\}$.

- **a** Write down the meaning of the interval notation.
- **b** List the elements of the set.
- **c** Illustrate the set on a number line.

6 For the sets $U = \{\text{positive integers} \leqslant 15\}$, $M = \{\text{multiples of } 5\}$, and $F = \{\text{factors of } 10\}$, list the elements of:

a $M \cap F$ **b** $(M \cap F)'$ **c** $M \cap F'$

7 Suppose $U = \{x \mid x \leqslant 25,\ x \in \mathbb{Z}^+\}$, $A = \{\text{square numbers}\}$, and $B = \{\text{even numbers}\}$.
Represent this information on a Venn diagram.

8 For the Venn diagram shown, give the number of elements in:

a B' **b** $A \cup C$ **c** U.

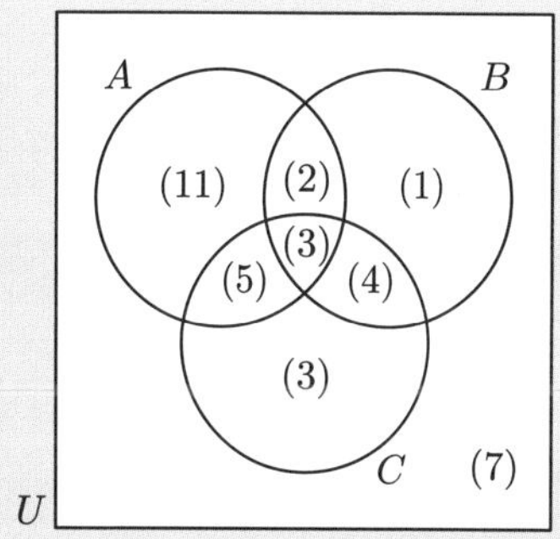

9 A Year 9 class has 25 students. On one day, 7 students forgot their diary, 5 were late, and 2 were late *and* forgot their diary.

- **a** Draw a Venn diagram to represent this information.
- **b** Find the number of students who were on time *and* remembered their diary.

10 In an office of 15 workers, 5 drink tea, 7 drink coffee, and 6 drink neither tea nor coffee.

- **a** Draw a Venn diagram to represent this information.
- **b** Find the number of workers who drink both tea and coffee.

11 Consider $U = \{\text{natural numbers} \leqslant 30\}$, $A = \{\text{multiples of } 5 \leqslant 30\}$, and
$B = \{\text{multiples of } 10 \leqslant 30\}$.

- **a** Describe the set U using interval notation.
- **b** List the elements of: **i** A **ii** B.
- **c** Hence construct a Venn diagram to display sets A and B.
- **d** Find $A \cap B'$.
- **e** How many numbers less than or equal to 30 are multiples of 5 but not 10? Explain your answer.

12 Two rival garden stores have various herbs for sale.
'Pete's Plants' sells lavender, rosemary, parsley, thyme, mint, and basil.
'Garden World' sells basil, coriander, thyme, sage, lemongrass, and chives.

a Draw a Venn diagram representing this information.

b Which plants are sold by:

i both stores

ii 'Pete's Plants' but not 'Garden World'

iii 'Pete's Plants' or 'Garden World', but not both?

c A third store 'Nina's Nursery' opens nearby. It specialises in shrubs, but Nina does stock lavender and rosemary. Draw a new Venn diagram to display the herbs from all three stores.

Chapter 4

Algebraic expansion

Contents:

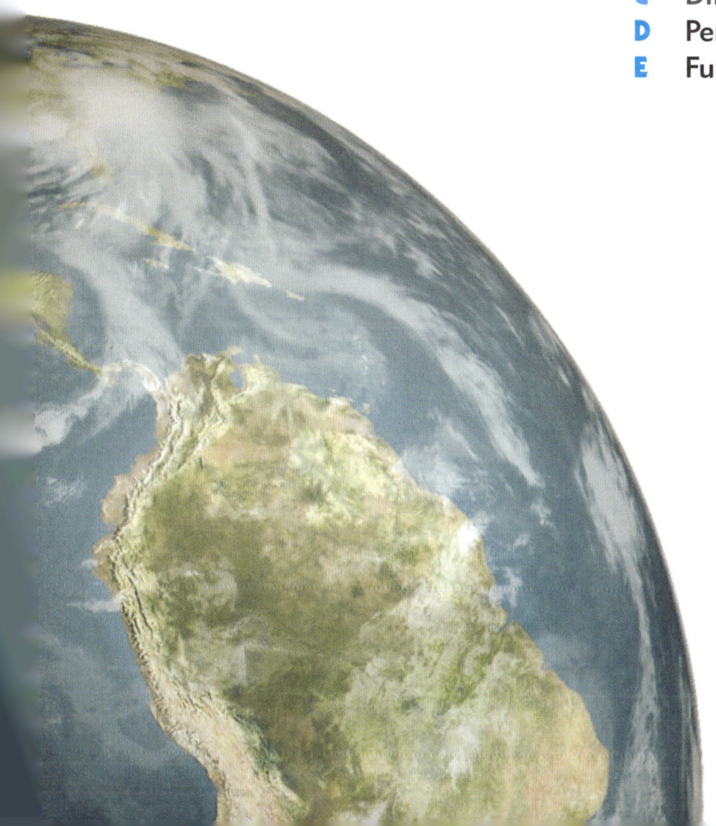

OPENING PROBLEM

Anton thinks that to find the square of the sum of two numbers, you can just square each of the numbers, then add the results.

Things to think about:

a Does $(5+3)^2 = 5^2+3^2$?

b Can you explain why Anton is incorrect?

When we write real life problems in terms of algebra, we often obtain expressions containing brackets. To solve equations containing these expressions, we may need to **expand** the brackets and **simplify** the result. In this chapter we will study the expansion of algebraic expressions.

A THE DISTRIBUTIVE LAW

Consider the expression $2(x+3)$. We say that 2 is the **coefficient** of the expression in the brackets. We can **expand** the brackets using the **distributive law**:

$$\boldsymbol{a(b+c) = ab+ac}$$

The distributive law says that we must multiply the coefficient by each term within the brackets, then add the results.

GEOMETRIC DEMONSTRATION

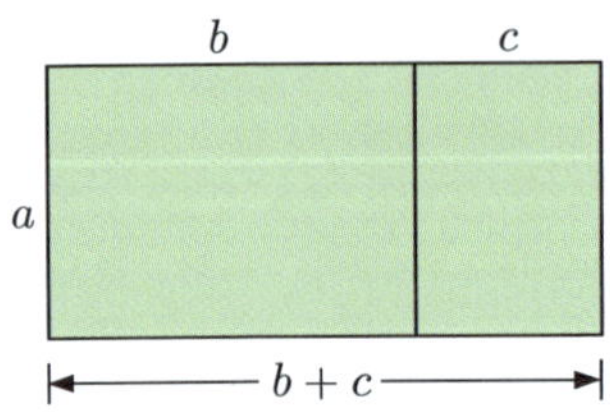

The area of the large rectangle is $a(b+c)$.

However, this area could also be found by adding the areas of the two small rectangles. This is $ab+ac$.

So, $a(b+c) = ab+ac$. {equating areas}

Example 1 Self Tutor

Expand the following:

a $3(4x+1)$ **b** $2x(5-2x)$ **c** $-2x(x-3)$

a
$$\begin{aligned}&3(4x+1)\\&= 3\times 4x + 3\times 1\\&= 12x+3\end{aligned}$$

b
$$\begin{aligned}&2x(5-2x)\\&= 2x(5+-2x)\\&= 2x\times 5 + 2x\times -2x\\&= 10x-4x^2\end{aligned}$$

c
$$\begin{aligned}&-2x(x-3)\\&= -2x(x+-3)\\&= -2x\times x + -2x\times -3\\&= -2x^2+6x\end{aligned}$$

Example 2 — Self Tutor

Expand and simplify:

a $2(3x-1)+3(5-x)$ **b** $x(2x-1)-2x(5-x)$

a
$$\begin{aligned}&2(3x-1)+3(5-x)\\ &=2\times 3x+2\times -1+3\times 5+3\times -x\\ &=6x-2+15-3x\\ &=3x+13\end{aligned}$$

b
$$\begin{aligned}&x(2x-1)-2x(5-x)\\ &=x\times 2x+x\times -1+-2x\times 5+-2x\times -x\\ &=2x^2-x-10x+2x^2\\ &=4x^2-11x\end{aligned}$$

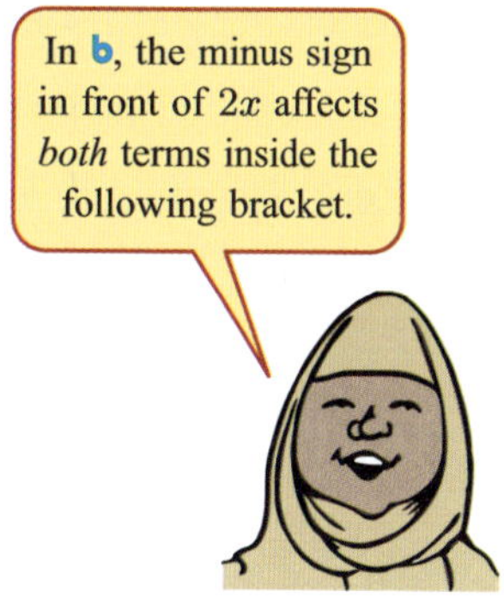

EXERCISE 4A

1 Expand and simplify:

a $2(x+1)$ **b** $5(x+2)$ **c** $4(x-6)$ **d** $7(x-1)$
e $3(2a+b)$ **f** $6(x+3y)$ **g** $8(a-b)$ **h** $4(m^2-n^2)$
i $-3(x+2)$ **j** $-4(2a-1)$ **k** $a(-a+5)$ **l** $3b(b-4)$
m $-(y+2)$ **n** $-5(x-3)$ **o** $-(2-b)$ **p** $-4(a-b)$
q $x(x+y)$ **r** $-x(-x-y)$ **s** $a(2a-1)$ **t** $-3x(7-4x)$

2 Expand and simplify:

a $4(x+y+5)$ **b** $2(3a-5b+1)$ **c** $3x(x-2y+3)$
d $-2(a-3b+5c)$ **e** $6x(x^2+5x-10)$ **f** $4n(n^2-2n-8)$

3 Expand and simplify:

a $2+3(x+1)$ **b** $15-2(x+5)$ **c** $4(x-1)+7$
d $6(x-2)-8x$ **e** $a(a-2)+2a$ **f** $3m(4-m)+m^2$
g $3a(b-a)+4a^2$ **h** $5b-2b(b-2)$ **i** $6a^2-3a(a+4)$

4 Expand and simplify:

a $2(x-3)+3(x+4)$ **b** $4b+(a-b)$ **c** $4b-(a-b)$
d $3(x+2)+5(4-x)$ **e** $6(m-2)-3(2m+1)$ **f** $7n-5(3-2n)$
g $5(y-x)+6(x-y)$ **h** $a(a+2)+5(a-3)$ **i** $x(x+5)-3(x-4)$
j $a^2+a(a+3)$ **k** $-a^2-a(a-1)$ **l** $x(x+y)-y(x+y)$
m $-3(x-6)-(2-x)$ **n** $4(3x-2)-(3x+1)$ **o** $2x(x-5)-3x(2-x)$

B THE PRODUCT $(a+b)(c+d)$

The product $(a+b)(c+d)$ has two **factors**, $(a+b)$ and $(c+d)$.

We can evaluate this product by using the distributive law several times.

$$\begin{aligned}(a+b)(c+d) &= a(c+d)+b(c+d)\\ &= ac+ad+bc+bd\end{aligned}$$

So, $\boldsymbol{(a+b)(c+d) = ac+ad+bc+bd}$

The final result contains four terms:

ac is the product of the **First** terms of each bracket.
ad is the product of the **Outer** terms of each bracket.
bc is the product of the **Inner** terms of each bracket.
bd is the product of the **Last** terms of each bracket.

Example 3 — Self Tutor

Expand and simplify:

a $(x+3)(x+2)$ **b** $(2x+1)(3x-2)$

a $(x+3)(x+2)$

$= x\times x + x\times 2 + 3\times x + 3\times 2$

$= x^2+2x+3x+6$

$= x^2+5x+6$

b $(2x+1)(3x-2)$

$= 2x\times 3x + 2x\times -2 + 1\times 3x + 1\times -2$

$= 6x^2-4x+3x-2$

$= 6x^2-x-2$

EXERCISE 4B

1 Consider the figure alongside.
Give an expression for the area of:

a rectangle 1 **b** rectangle 2
c rectangle 3 **d** rectangle 4
e the overall rectangle.

What can you conclude?

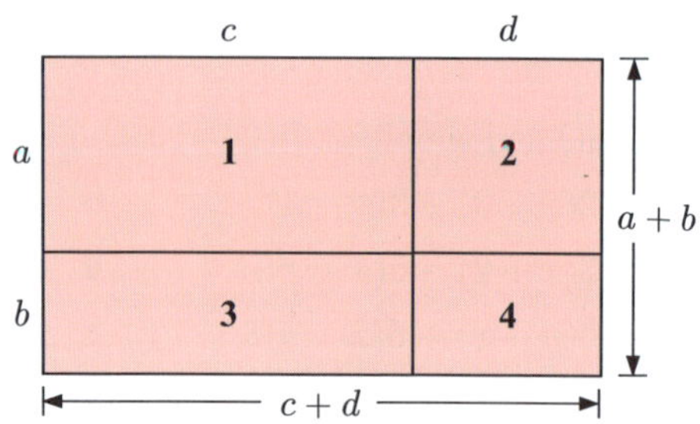

2 Use the rule $(a+b)(c+d) = ac+ad+bc+bd$ to expand and simplify:

a $(x+2)(x+5)$ **b** $(x+4)(x-6)$ **c** $(x-3)(x+7)$
d $(x+5)(x-5)$ **e** $(x-8)(x+2)$ **f** $(3x+2)(4x+1)$

g $(1-3x)(2x+1)$ **h** $(6-x)(2x+5)$ **i** $(4x-3)(1+3x)$

j $(4-x)(4+5x)$ **k** $(8-x)(2x+5)$ **l** $(3x+2)(3x+2)$

Example 4 Self Tutor

Expand and simplify:

a $(x+3)(x-3)$ **b** $(3x-5)(3x+5)$

a $(x+3)(x-3)$
$= x^2 - 3x + 3x - 9$
$= x^2 - 9$

b $(3x-5)(3x+5)$
$= 9x^2 + 15x - 15x - 25$
$= 9x^2 - 25$

3 Expand and simplify:

a $(x+2)(x-2)$ **b** $(a+4)(a-4)$ **c** $(6+x)(6-x)$

d $(3x+1)(3x-1)$ **e** $(5a+2)(5a-2)$ **f** $(7a-3)(7a+3)$

Example 5 Self Tutor

Expand and simplify:

a $(3x+1)^2$ **b** $(2x-3)^2$

a $(3x+1)^2$
$= (3x+1)(3x+1)$
$= 9x^2 + 3x + 3x + 1$
$= 9x^2 + 6x + 1$

b $(2x-3)^2$
$= (2x-3)(2x-3)$
$= 4x^2 - 6x - 6x + 9$
$= 4x^2 - 12x + 9$

4 Expand and simplify:

a $(x+3)^2$ **b** $(x-4)^2$ **c** $(1-x)^2$ **d** $(2+x)^2$

e $(2x+1)^2$ **f** $(2-3x)^2$ **g** $(3+5x)^2$ **h** $(2x-y)^2$

Example 6 Self Tutor

Expand and simplify: $(x+2)(x-3)+5(x+4)$

$(x+2)(x-3)+5(x+4)$
$= x^2 - 3x + 2x - 6 + 5x + 20$
$= x^2 + 4x + 14$

5 Expand and simplify:

a $(x+5)(x+1)+2(x-2)$ **b** $4(x+2)+(x-3)(x+6)$

c $(2a+5)(a-6)+a(a+7)$ **d** $(3x+5)(2x-3)-x(x+3)$

e $(2x+3)(x+2)+(x+6)(x-5)$ **f** $(y+7)(y-4)-(y+1)(y+3)$

g $(k+2)(k-5)-(2k+1)(k-3)$ **h** $(x+2)(x-2)+(x+6)(x-6)$

C DIFFERENCE OF TWO SQUARES

Consider the product $(a+b)(a-b)$.

Using the FOIL rule to expand this product,

$$\begin{aligned}&(a+b)(a-b)\\ &= a^2 - ab + ab - b^2\\ &= a^2 - b^2\end{aligned}$$

Thus, $\boldsymbol{(a+b)(a-b) = a^2 - b^2}$

This is called the **difference of two squares** expansion, because the expression on the right hand side is the difference between the two squares a^2 and b^2.

GEOMETRIC DEMONSTRATION

In the figure alongside,

shaded area = area of large square − area of small square

$= a^2 - b^2$

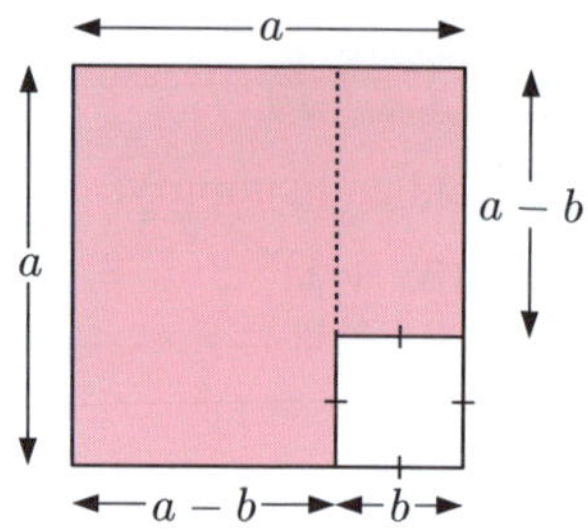

If the rectangle on the right hand side is rotated and placed on top of the remaining shaded area, we form a new rectangle.

$\therefore$ shaded area $= (a+b)(a-b)$

$\therefore$ $(a+b)(a-b) = a^2 - b^2$ {equating areas}

DEMO

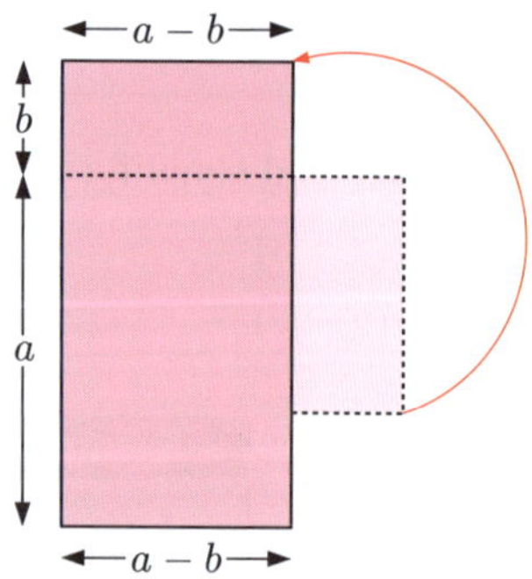

Example 7 Self Tutor

Expand and simplify:

a $(x+5)(x-5)$ **b** $(3-y)(3+y)$

a $\begin{aligned}&(x+5)(x-5)\\ &= x^2 - 5^2\\ &= x^2 - 25\end{aligned}$

b $\begin{aligned}&(3-y)(3+y)\\ &= 3^2 - y^2\\ &= 9 - y^2\end{aligned}$

EXERCISE 4C

1 Expand and simplify using the rule $(a+b)(a-b) = a^2 - b^2$:

a $(x+3)(x-3)$ **b** $(x-3)(x+3)$ **c** $(3+x)(3-x)$ **d** $(3-x)(3+x)$

e $(x+2)(x-2)$ **f** $(2-x)(2+x)$ **g** $(x+6)(x-6)$ **h** $(a+4)(a-4)$

i $(b-1)(b+1)$ **j** $(p-q)(p+q)$ **k** $(5+n)(5-n)$ **l** $(7-y)(7+y)$

Example 8 **Self Tutor**

Expand and simplify:

a $(2x-3)(2x+3)$ **b** $(5-3y)(5+3y)$

a $(2x-3)(2x+3)$
$= (2x)^2 - 3^2$
$= 4x^2 - 9$

b $(5-3y)(5+3y)$
$= 5^2 - (3y)^2$
$= 25 - 9y^2$

2 Expand and simplify:

a $(2x+1)(2x-1)$ **b** $(5x+2)(5x-2)$ **c** $(4a+3)(4a-3)$

d $(3b+5)(3b-5)$ **e** $(4x+1)(4x-1)$ **f** $(1-4x)(1+4x)$

g $(7-2y)(7+2y)$ **h** $(3-2x)(3+2x)$ **i** $(3x+2)(3x-2)$

Example 9 **Self Tutor**

Expand and simplify: $(3x+4y)(3x-4y)$

$(3x+4y)(3x-4y)$
$= (3x)^2 - (4y)^2$
$= 9x^2 - 16y^2$

3 Expand and simplify:

a $(3a+b)(3a-b)$ **b** $(a-3b)(a+3b)$ **c** $(6x+y)(6x-y)$

d $(5x+2y)(5x-2y)$ **e** $(4x+5y)(4x-5y)$ **f** $(2x-7y)(2x+7y)$

INVESTIGATION THE PRODUCT OF THREE CONSECUTIVE INTEGERS

Con was trying to multiply $19 \times 20 \times 21$ without a calculator. Aimee told him to 'cube the middle integer and then subtract the middle integer' to get the answer.

What to do:

1 Find $19 \times 20 \times 21$ using a calculator.

2 Find $20^3 - 20$ using a calculator. Does Aimee's rule work?

3 Check that Aimee's rule works for the following products:

a $4 \times 5 \times 6$ **b** $9 \times 10 \times 11$ **c** $49 \times 50 \times 51$

4 Let the middle integer be x, so the other integers must be $(x-1)$ and $(x+1)$. Use the difference of two squares expansion to help prove Aimee's rule.

D PERFECT SQUARE EXPANSION

$(a+b)^2$ and $(a-b)^2$ are called **perfect squares**.

Now $\quad (a+b)^2 = (a+b)(a+b)$
$= a^2 + ab + ab + b^2 \quad$ {using 'FOIL'}
$= a^2 + 2ab + b^2$

Thus, we can state the perfect square expansion rule:

$$\boldsymbol{(a+b)^2 = a^2 + 2ab + b^2}$$

We can remember this rule as follows:

Step 1: Square the *first term*.

Step 2: Add twice the product of the *first* and *last terms*.

Step 3: Add on the square of the *last term*.

Notice that $\quad (a-b)^2 = (a+(-b))^2$
$= a^2 + 2a(-b) + (-b)^2$
$= a^2 - 2ab + b^2$

Hence: $\quad \boldsymbol{(a-b)^2 = a^2 - 2ab + b^2}$

Once again, we have the square of the first term, twice the product of the first and last terms, and the square of the last term.

Example 10 Self Tutor

Expand and simplify:

a $(x+3)^2$ b $(x-5)^2$

a $\quad (x+3)^2$
$= x^2 + 2 \times x \times 3 + 3^2$
$= x^2 + 6x + 9$

b $\quad (x-5)^2$
$= x^2 + 2 \times x \times (-5) + (-5)^2$
$= x^2 - 10x + 25$

EXERCISE 4D

1 Consider the figure alongside.
Give an expression for the area of:

a square 1 b rectangle 2 c rectangle 3
d square 4 e the overall square.

What can you conclude?

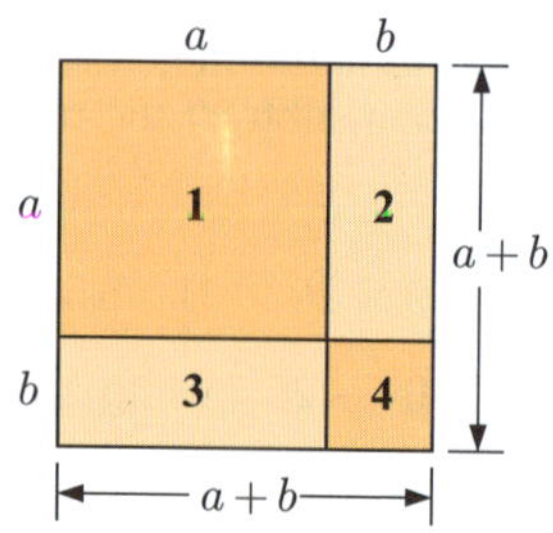

2 Consider the **Opening Problem** on page **66**. Anton thinks that $(5+3)^2 = 5^2 + 3^2$.

a By evaluating each side of the equation, show that Anton is incorrect.

b Check that $(5+3)^2 = 5^2 + 2 \times 5 \times 3 + 3^2$.

3 Use the rule $(a+b)^2 = a^2 + 2ab + b^2$ to expand and simplify:

a $(x+2)^2$ **b** $(x+5)^2$ **c** $(a+4)^2$
d $(y+6)^2$ **e** $(1+b)^2$ **f** $(2+x)^2$

4 Expand and simplify:

a $(x-4)^2$ **b** $(x-2)^2$ **c** $(a-7)^2$
d $(b-8)^2$ **e** $(3-x)^2$ **f** $(5-y)^2$

Example 11 **Self Tutor**

Expand and simplify using the perfect square expansion rule:

a $(5x+1)^2$ **b** $(4-3x)^2$

a $(5x+1)^2$
$= (5x)^2 + 2 \times 5x \times 1 + 1^2$
$= 25x^2 + 10x + 1$

b $(4-3x)^2$
$= 4^2 + 2 \times 4 \times (-3x) + (-3x)^2$
$= 16 - 24x + 9x^2$

5 Expand and simplify:

a $(2x+3)^2$ **b** $(4a-1)^2$ **c** $(3y+5)^2$
d $(3a-4)^2$ **e** $(2x-7)^2$ **f** $(8+3a)^2$
g $(2+5b)^2$ **h** $(6-5x)^2$ **i** $(4-5y)^2$

Example 12 **Self Tutor**

Expand and simplify:

a $(2x^2+3)^2$ **b** $5-(x+2)^2$

a $(2x^2+3)^2$
$= (2x^2)^2 + 2 \times 2x^2 \times 3 + 3^2$
$= 4x^4 + 12x^2 + 9$

b $5-(x+2)^2$
$= 5 - [x^2 + 4x + 4]$
$= 5 - x^2 - 4x - 4$
$= 1 - x^2 - 4x$

6 Expand and simplify:

a $(x^2+3)^2$ **b** $(y^2-2)^2$ **c** $(2a^2+3)^2$
d $(1-4x^2)^2$ **e** $(a^2+b^2)^2$ **f** $(a^2-b^2)^2$

7 Expand and simplify:

a $2x+1-(x+2)^2$ **b** $3x-2+(x+5)^2$
c $(x+4)(x-4)+(x+1)^2$ **d** $(x+4)(x-4)-(x+1)^2$
e $(1-3x)^2+(x-2)(x+3)$ **f** $(3x+2)^2-(x+1)(x-5)$
g $(5x+2)(5x-2)-(x+2)^2$ **h** $(5x+4)(x-3)-(x-1)^2$
i $(3-x)^2+(x+2)^2$ **j** $(3-x)^2-(x-2)^2$

ACTIVITY

Any even number can be written in the form $2a$ where a is an integer.

Any odd number can be written in the form $2a+1$ where a is an integer.

What to do:

Suppose p and q are positive integers, and p^2+q^2 is divisible by 4. Prove that p and q are both even.

Hint: Consider the 4 possible cases, one of which is when p and q are both even.

E FURTHER EXPANSION

In this Section we expand more complicated expressions by repeated use of the expansion laws.

Consider the expansion of $(a+b)(c+d+e)$.

Now $(a+b)(c+d+e)$
$= (a+b)c + (a+b)d + (a+b)e$
$= ac + bc + ad + bd + ae + be$

{Compare with $\square(c+d+e)$
$= \square c + \square d + \square e$}

Notice that there are 6 terms in this expansion. Each term in the first bracket is multiplied by each term in the second bracket.

2 terms in the first bracket × 3 terms in the second bracket ⟶ 6 terms in the expansion.

Example 13 — Self Tutor

Expand and simplify: $(2x+3)(x^2+4x+5)$

$(2x+3)(x^2+4x+5)$

$= 2x^3 + 8x^2 + 10x$ {all terms in 2nd bracket × $2x$}
$\quad + 3x^2 + 12x + 15$ {all terms in 2nd bracket × 3}
$= 2x^3 + 11x^2 + 22x + 15$ {collecting like terms}

Example 14 — Self Tutor

Expand and simplify: $(x+2)^3$

$(x+2)^3 = (x+2)\times(x+2)^2$

$= (x+2)(x^2+4x+4)$ {perfect square expansion}

$= x^3 + 4x^2 + 4x$ {all terms in 2nd bracket × x}
$\quad + 2x^2 + 8x + 8$ {all terms in 2nd bracket × 2}
$= x^3 + 6x^2 + 12x + 8$ {collecting like terms}

EXERCISE 4E

1 Expand and simplify:

a $(x+2)(x^2+x+1)$ **b** $(x+3)(x^2+x-2)$ **c** $(x+4)(x^2-x+2)$
d $(x+6)(x^2-x-1)$ **e** $(3x+1)(x^2+x+5)$ **f** $(5x-2)(x^2-x+4)$
g $(x+3)(2x^2+x-2)$ **h** $(2x-1)(3x^2-x+4)$

2 Expand and simplify:

a $(x+1)^3$ **b** $(x+3)^3$ **c** $(x-1)^3$ **d** $(x-2)^3$
e $(2x+1)^3$ **f** $(2x-3)^3$ **g** $(x+1)^4$ **h** $(x-1)^4$

Example 15 Self Tutor

Expand and simplify:

a $x(x+1)(x+2)$ **b** $(x+1)(x-2)(x+2)$

a $x(x+1)(x+2)$
$= (x^2+x)(x+2)$ {all terms in first bracket $\times\, x$}
$= x^3+2x^2+x^2+2x$ {expanding remaining factors}
$= x^3+3x^2+2x$ {collecting like terms}

b $(x+1)(x-2)(x+2)$
$= (x+1)(x^2-4)$ {difference of two squares}
$= x^3-4x+x^2-4$ {expanding factors}
$= x^3+x^2-4x-4$

3 Expand and simplify:

a $x(x+2)(x+4)$ **b** $x(x-5)(x+3)$ **c** $x(x-4)(x-1)$
d $2x(x+3)(x-2)$ **e** $2x(x+1)(7-x)$ **f** $-x(x+3)(x-6)$
g $-5x(2x-1)(x+3)$ **h** $-x(1-4x)(x-3)$ **i** $3x^2(x+2)^2$

4 Expand and simplify:

a $(x+1)(x+2)(x+4)$ **b** $(x-2)(x+1)(x-3)$ **c** $(x-5)(x+2)(x-1)$
d $(2x-1)(x-4)(x-1)$ **e** $(3x+1)(x+2)(x+3)$ **f** $(4x-1)(4x+1)(x+4)$
g $(1-x)(2-3x)(x+2)$ **h** $(x-5)(1-x)(5x+2)$

5 **a** Expand and simplify:

i $(x-1)(x^2+x+1)$ **ii** $(x-1)(x^3+x^2+x+1)$
iii $(x-1)(x^6+x^5+x^4+x^3+x^2+x+1)$

b Predict the simplification of $(x-1)(x^n+x^{n-1}+x^{n-2}+....+x^2+x+1)$ where $n\in\mathbb{Z}^+$.

6 The difference between the cubes of two consecutive integers is 127. Find the integers.

REVIEW SET 4A

1 Expand and simplify:

a $3(x-4)$ **b** $-5(2x-3)$ **c** $x(1-x)$

2 Expand and simplify:

a $7+4(x-2)$ b $3-2(x+4)$ c $-4(2x-3)-(x+1)$

3 Expand and simplify:

a $(x+4)(x+7)$ b $(x+8)(x-2)$ c $(1-x)(5x+6)$ d $(2a-3)(a-9)$

4 Expand and simplify:

a $(x+3)(x+2)+3(x-4)$ b $(n-3)(n+5)-(n-2)(n+3)$

5 Expand and simplify:

a $(x+4)^2$ b $(x-10)^2$ c $(2x+5)^2$ d $(3-4x)^2$

6 Expand and simplify:

a $(x+9)(x-9)$ b $(3x+2)(3x-2)$ c $x(4+x)(4-x)$

7 Expand and simplify:

a $(x+4)(x^2+x-7)$ b $(4x-3)(x^2-x-5)$

8 Emma receives a large box in the mail. The box is 10 cm longer than it is wide, and the height is 6 cm more than the width.

a If the box has width x cm, write an expression for its i length ii height.

b The area of the base of the box is given by the formula Area = length × width. Write an expression for the area of the base of the box, in terms of x. Expand and simplify your answer.

c The volume of the box is given by the formula Volume = length × width × height. Write an expression for the volume of the box in terms of x. Expand and simplify your answer.

REVIEW SET 4B

1 Expand and simplify:

a $-3(x+4)$ b $x(y+3)$ c $2a(a-5)$

2 Expand and simplify:

a $(x+9)(x-4)$ b $(2x-7)(x+6)$ c $(2a-3)(a-4)$

3 Expand and simplify:

a $(y+3)^2$ b $(3x+2)^2$ c $(4a-b)^2$

4 Expand and simplify:

a $(5-x)(5+x)$ b $(3y+4)(3y-4)$ c $(6a+5b)(6a-5b)$

5 Expand and simplify:

a $-4(x+2)-(x+1)^2$ b $5(y-3)+(y+5)(y-5)$

6 Expand and simplify:

a $(x+4)^2(x-2)$ b $(x+2)(x-3)^2$

7 Expand and simplify:

a $(x^2-3x+4)(2-x)$ b $(x^2+4x-5)(x-4)$

8 a Expand and simplify $(x-3)^3$. b Hence, expand and simplify $(x-3)^4$.

Chapter 5

Radicals

Contents:

OPENING PROBLEM

Jeremy and Wei were each asked to draw a shape with an area of 15 cm^2. Jeremy drew a 5 cm × 3 cm rectangle, while Wei drew a square. The two shapes were placed on top of one another as shown.

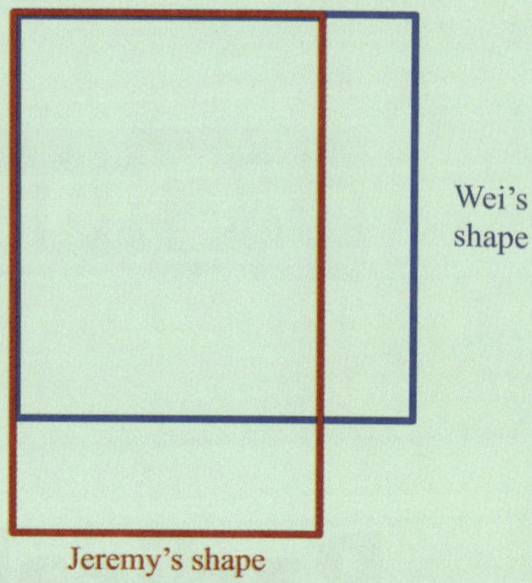

Things to think about:

- **a** Can you explain why the side length of the square is $\sqrt{15}$ cm?
- **b** Can you find the area covered by:
  - **i** Jeremy's shape, but not Wei's shape
  - **ii** both of the shapes?
- **c** Can you find the perimeter of the region formed by the shapes?

A RADICALS AND SURDS

We have previously seen that the set of **real numbers** $\mathbb{R}$ can be divided into the set of **rational numbers** $\mathbb{Q}$, and the set of **irrational numbers** $\mathbb{Q}'$.

A **rational number** is a real number which can be written in the form $\frac{a}{b}$ where a and b are integers, $b \neq 0$.

An **irrational number** is a real number which cannot be written in the form $\frac{a}{b}$ where a and b are integers, $b \neq 0$. The decimal expansion of an irrational number will neither terminate nor recur.

In this chapter we consider **radical numbers**, which may be rational or irrational.

A **radical** is a number that is written using the radical sign $\sqrt{\ }$.

An irrational radical is called a **surd**.

Examples of rational radicals include:

$$\sqrt{9} = 3 = \tfrac{3}{1}$$

$$\sqrt{\tfrac{1}{25}} = \sqrt{\left(\tfrac{1}{5}\right)^2} = \tfrac{1}{5}$$

$$\sqrt{\tfrac{4}{9}} = \sqrt{\left(\tfrac{2}{3}\right)^2} = \tfrac{2}{3}$$

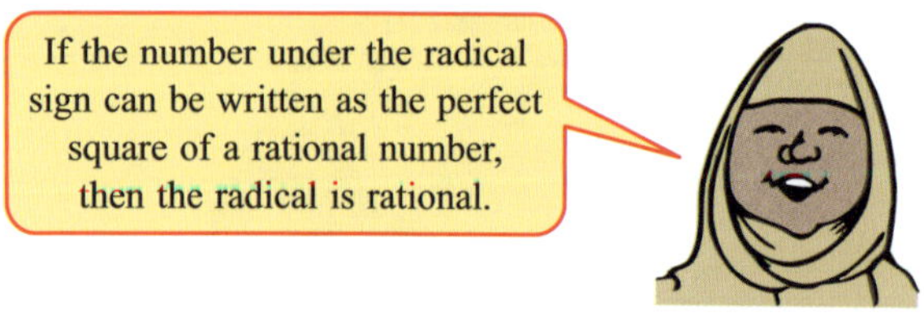

Examples of surds include:

$$\sqrt{2} \approx 1.414\,213\,56\,....$$

$$\sqrt{19} \approx 4.358\,898\,94\,....$$

$$\sqrt{\tfrac{1}{3}} \approx 0.577\,350\,269\,....$$

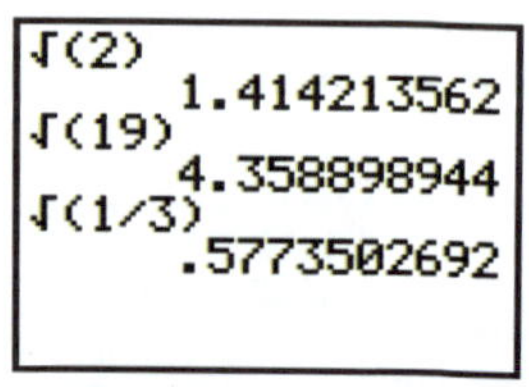

Example 1 — Self Tutor

Determine whether each of the following numbers is a rational radical or a surd. If it is a surd, find its value correct to 4 decimal places.

a $\sqrt{49}$ **b** $\sqrt{53}$

a $\sqrt{49} = 7 = \frac{7}{1}$

$\therefore$ $\sqrt{49}$ is a rational radical.

b $\sqrt{53}$ is not rational, so it is a surd.

```
√(53)
      7.280109889
```

$\sqrt{53} \approx 7.2801$

EXERCISE 5A

1 Determine whether each of the following numbers is a rational radical or a surd. If it is a surd, find its value correct to 4 decimal places.

a $\sqrt{5}$ **b** $\sqrt{36}$ **c** $\sqrt{72}$ **d** $\sqrt{81}$

e $\sqrt{\frac{25}{49}}$ **f** $\sqrt{\frac{1}{10}}$ **g** $\sqrt{\frac{1}{64}}$ **h** $\sqrt{\frac{3}{14}}$

2 Consider the 1000 radicals $\sqrt{1}, \sqrt{2}, \sqrt{3}, ..., \sqrt{1000}$. How many of these radicals are surds?

B SIMPLIFYING RADICALS

In this **Investigation**, we will discover some properties that will help us to simplify radical expressions.

INVESTIGATION 1 PROPERTIES OF RADICALS

Consider the following observations:

- $\sqrt{4 \times 9} = \sqrt{36} = 6$ and $\sqrt{4} \times \sqrt{9} = 2 \times 3 = 6$ so $\sqrt{4} \times \sqrt{9} = \sqrt{4 \times 9}$.
- $\sqrt{\frac{36}{4}} = \sqrt{9} = 3$ and $\frac{\sqrt{36}}{\sqrt{4}} = \frac{6}{2} = 3$ so $\frac{\sqrt{36}}{\sqrt{4}} = \sqrt{\frac{36}{4}}$.

What to do:

Test the following *possible* properties or rules for radicals by substituting different values of a and b where $a \geqslant 0$ and $b \geqslant 0$. Use your calculator to evaluate the results.

1 $\sqrt{a} \times \sqrt{a} = a$ **2** $\sqrt{a} \times \sqrt{b} = \sqrt{ab}$ **3** $\sqrt{\frac{a}{b}} = \frac{\sqrt{a}}{\sqrt{b}}$, $b \neq 0$

4 $\sqrt{a+b} = \sqrt{a} + \sqrt{b}$ **5** $\sqrt{a-b} = \sqrt{a} - \sqrt{b}$

From the **Investigation**, you should have discovered the following properties of radicals:

- $\sqrt{a}\sqrt{a} = (\sqrt{a})^2 = a$
- $\sqrt{a}\sqrt{b} = \sqrt{ab}$
- $\dfrac{\sqrt{a}}{\sqrt{b}} = \sqrt{\dfrac{a}{b}}$

In general, $\sqrt{a+b} \neq \sqrt{a} + \sqrt{b}$ and $\sqrt{a-b} \neq \sqrt{a} - \sqrt{b}$ for $a \geqslant 0$, $b > 0$.

Example 2 Self Tutor

Simplify: **a** $(\sqrt{2})^2$ **b** $(\sqrt{2})^3$ **c** $\left(\dfrac{4}{\sqrt{2}}\right)^2$

a $(\sqrt{2})^2$
$= \sqrt{2} \times \sqrt{2}$
$= 2$

b $(\sqrt{2})^3$
$= \sqrt{2} \times \sqrt{2} \times \sqrt{2}$
$= 2\sqrt{2}$

c $\left(\dfrac{4}{\sqrt{2}}\right)^2$
$= \dfrac{4^2}{(\sqrt{2})^2}$
$= \dfrac{16}{2}$
$= 8$

EXERCISE 5B

1 Simplify:

a $(\sqrt{3})^2$ **b** $(\sqrt{3})^3$ **c** $(\sqrt{3})^5$ **d** $\left(\dfrac{1}{\sqrt{3}}\right)^2$

e $(\sqrt{7})^2$ **f** $(\sqrt{7})^3$ **g** $\left(\dfrac{1}{\sqrt{7}}\right)^2$ **h** $\left(\dfrac{3}{\sqrt{7}}\right)^2$

i $(\sqrt{5})^2$ **j** $(\sqrt{5})^4$ **k** $\left(\dfrac{5}{\sqrt{5}}\right)^2$ **l** $\left(\dfrac{10}{\sqrt{5}}\right)^2$

Example 3 Self Tutor

Simplify: **a** $(3\sqrt{2})^2$ **b** $3\sqrt{3} \times (-2\sqrt{3})$

a $(3\sqrt{2})^2$
$= 3\sqrt{2} \times 3\sqrt{2}$
$= 9 \times 2$
$= 18$

b $3\sqrt{3} \times (-2\sqrt{3})$
$= 3 \times -2 \times \sqrt{3} \times \sqrt{3}$
$= -6 \times 3$
$= -18$

2 Simplify:

a $(2\sqrt{2})^2$ **b** $(4\sqrt{2})^2$ **c** $(2\sqrt{3})^2$

d $(3\sqrt{3})^2$ **e** $(2\sqrt{5})^2$ **f** $(3\sqrt{5})^2$

g $(2\sqrt{7})^2$ **h** $(2\sqrt{10})^2$ **i** $(7\sqrt{10})^2$

3 Simplify:

a $3\sqrt{2} \times 4\sqrt{2}$ **b** $5\sqrt{3} \times 2\sqrt{3}$ **c** $7\sqrt{2} \times 5\sqrt{2}$

d $(-4\sqrt{2})^2$ **e** $(-7\sqrt{3})^2$ **f** $\sqrt{2} \times (-3\sqrt{2})$

g $(-2\sqrt{3}) \times (-5\sqrt{3})$ **h** $(-2\sqrt{7}) \times 3\sqrt{7}$ **i** $\sqrt{11} \times (-2\sqrt{11})$

Example 4 Self Tutor

Write in simplest form:

a $\sqrt{2} \times \sqrt{5}$ **b** $3\sqrt{2} \times 4\sqrt{11}$

a $\sqrt{2} \times \sqrt{5}$
$= \sqrt{2 \times 5}$
$= \sqrt{10}$

b $3\sqrt{2} \times 4\sqrt{11}$
$= 3 \times 4 \times \sqrt{2} \times \sqrt{11}$
$= 12 \times \sqrt{2 \times 11}$
$= 12\sqrt{22}$

4 Simplify:

a $\sqrt{2} \times \sqrt{3}$ **b** $\sqrt{2} \times \sqrt{7}$ **c** $\sqrt{2} \times \sqrt{17}$

d $\sqrt{7} \times \sqrt{3}$ **e** $2\sqrt{2} \times 5\sqrt{3}$ **f** $(4\sqrt{3})^2$

g $5\sqrt{2} \times \sqrt{7}$ **h** $2\sqrt{6} \times 3\sqrt{5}$ **i** $-5\sqrt{2} \times 2\sqrt{7}$

j $(-\sqrt{7}) \times (-2\sqrt{3})$ **k** $(2\sqrt{3})^2 \times 2\sqrt{5}$ **l** $(2\sqrt{2})^3 \times 5\sqrt{3}$

Example 5 Self Tutor

Simplify: **a** $\dfrac{\sqrt{75}}{\sqrt{3}}$ **b** $\sqrt{\dfrac{9}{49}}$

a $\dfrac{\sqrt{75}}{\sqrt{3}}$
$= \sqrt{\dfrac{75}{3}}$
$= \sqrt{25}$
$= 5$

b $\sqrt{\dfrac{9}{49}}$
$= \dfrac{\sqrt{9}}{\sqrt{49}}$
$= \dfrac{3}{7}$

5 Simplify:

a $\dfrac{\sqrt{8}}{\sqrt{2}}$ **b** $\dfrac{\sqrt{3}}{\sqrt{27}}$ **c** $\dfrac{\sqrt{18}}{\sqrt{3}}$ **d** $\dfrac{\sqrt{2}}{\sqrt{50}}$

e $\dfrac{\sqrt{75}}{\sqrt{5}}$ **f** $\dfrac{\sqrt{5}}{\sqrt{75}}$ **g** $\dfrac{\sqrt{18}}{\sqrt{2}}$ **h** $\dfrac{\sqrt{3}}{\sqrt{48}}$

6 Simplify:

a $\sqrt{\frac{1}{16}}$ **b** $\sqrt{\frac{81}{144}}$ **c** $\sqrt{\frac{49}{64}}$ **d** $\sqrt{\frac{121}{9}}$

e $\sqrt{6\frac{1}{4}}$ **f** $\sqrt{1\frac{7}{9}}$ **g** $\sqrt{2\frac{14}{25}}$ **h** $\sqrt{7\frac{1}{9}}$

Example 6 — Self Tutor

Write $\sqrt{32}$ in the form $k\sqrt{2}$.

$\sqrt{32}$
$= \sqrt{16 \times 2}$
$= \sqrt{16} \times \sqrt{2}$ $\{$using $\sqrt{ab} = \sqrt{a} \times \sqrt{b}\}$
$= 4\sqrt{2}$

7 Write in the form $k\sqrt{2}$:

a $\sqrt{8}$ **b** $\sqrt{18}$ **c** $\sqrt{50}$ **d** $\sqrt{98}$

e $\sqrt{200}$ **f** $\sqrt{288}$ **g** $\sqrt{20\,000}$ **h** $\sqrt{\frac{1}{2}}$

8 Write in the form $k\sqrt{3}$:

a $\sqrt{12}$ **b** $\sqrt{27}$ **c** $\sqrt{75}$ **d** $\sqrt{\frac{1}{3}}$

9 Write in the form $k\sqrt{5}$:

a $\sqrt{20}$ **b** $\sqrt{45}$ **c** $\sqrt{125}$ **d** $\sqrt{\frac{1}{5}}$

RESEARCH

- Where did the names *radical* and *surd* come from?
- Why do we use the word *irrational* to describe some numbers?

C SIMPLEST RADICAL FORM

A radical is in **simplest form** when the number under the radical sign is the smallest possible integer.

Example 7 — Self Tutor

Write $\sqrt{8}$ in simplest form.

$\sqrt{8}$
$= \sqrt{4 \times 2}$
$= \sqrt{4} \times \sqrt{2}$
$= 2\sqrt{2}$

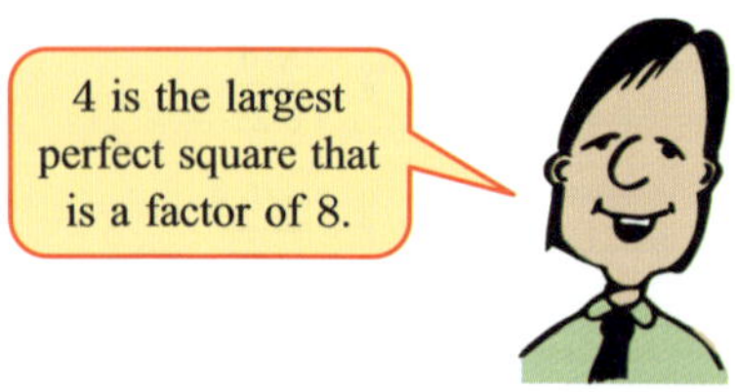

Example 8 — Self Tutor

Write $\sqrt{432}$ in simplest radical form.

$$\begin{aligned} &\sqrt{432} \\ &= \sqrt{2^4 \times 3^3} \\ &= \sqrt{2^4} \times \sqrt{3^3} \\ &= 4 \times 3\sqrt{3} \\ &= 12\sqrt{3} \end{aligned}$$

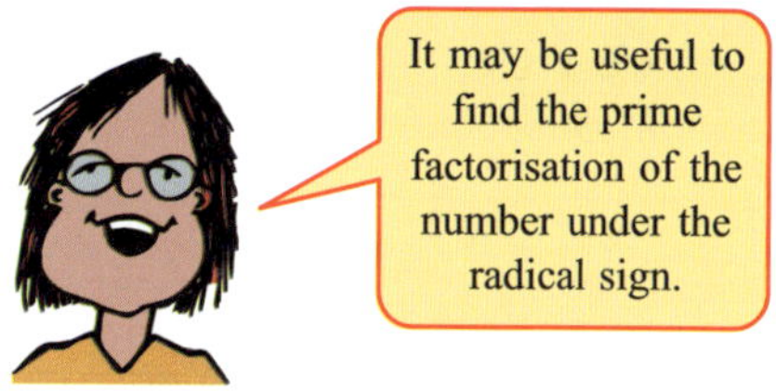

EXERCISE 5C

1 Write in simplest radical form:

a $\sqrt{12}$ **b** $\sqrt{28}$ **c** $\sqrt{54}$ **d** $\sqrt{60}$

e $\sqrt{99}$ **f** $\sqrt{52}$ **g** $\sqrt{40}$ **h** $\sqrt{63}$

i $\sqrt{48}$ **j** $\sqrt{125}$ **k** $\sqrt{147}$ **l** $\sqrt{175}$

m $\sqrt{176}$ **n** $\sqrt{150}$ **o** $\sqrt{275}$ **p** $\sqrt{2000}$

2 Show that $\dfrac{\sqrt{96}}{\sqrt{6}} = 4$:

a by using the rule $\dfrac{\sqrt{a}}{\sqrt{b}} = \sqrt{\dfrac{a}{b}}$

b by first writing $\sqrt{96}$ in simplest form.

D ADDING AND SUBTRACTING RADICALS

We can add and subtract 'like' radicals in the same way as we do 'like' terms in algebra.

For example, $2\sqrt{3} + 4\sqrt{3}$ is 2 'lots' of $\sqrt{3}$, plus 4 'lots' of $\sqrt{3}$, which is 6 'lots' of $\sqrt{3}$.

So, in the same way that $2x + 4x = 6x$, $2\sqrt{3} + 4\sqrt{3} = 6\sqrt{3}$.

Example 9 — Self Tutor

Simplify:

a $3\sqrt{2} + 4\sqrt{2}$ **b** $5\sqrt{3} - 6\sqrt{3}$

a $$\begin{aligned} &3\sqrt{2} + 4\sqrt{2} \\ &= 7\sqrt{2} \end{aligned}$$

{Compare with $3x + 4x = 7x$}

b $$\begin{aligned} &5\sqrt{3} - 6\sqrt{3} \\ &= -1\sqrt{3} \\ &= -\sqrt{3} \end{aligned}$$

{Compare with $5x - 6x = -x$}

EXERCISE 5D

1 Simplify:

a $\sqrt{2}+\sqrt{2}$
b $\sqrt{2}-\sqrt{2}$
c $3\sqrt{2}-2\sqrt{2}$
d $2\sqrt{3}-\sqrt{3}$
e $5\sqrt{7}+2\sqrt{7}$
f $3\sqrt{5}-6\sqrt{5}$
g $6\sqrt{2}-9\sqrt{2}$
h $\sqrt{5}+7\sqrt{5}$
i $3\sqrt{2}+4\sqrt{2}-\sqrt{2}$
j $3\sqrt{2}-5\sqrt{2}-\sqrt{2}$
k $3\sqrt{3}-\sqrt{3}+2\sqrt{3}$
l $3\sqrt{5}+7\sqrt{5}-10$

Example 10 Self Tutor

Simplify: $4\sqrt{5}+3\sqrt{7}-\sqrt{5}+5\sqrt{7}$

'like' radicals

$4\sqrt{5}+3\sqrt{7}-\sqrt{5}+5\sqrt{7}$

'like' radicals

$=3\sqrt{5}+8\sqrt{7}$

2 Simplify:

a $3\sqrt{2}+2\sqrt{3}-\sqrt{2}+5\sqrt{3}$
b $7\sqrt{2}-4\sqrt{3}-2\sqrt{2}+3\sqrt{3}$
c $-6\sqrt{2}-2\sqrt{3}-\sqrt{2}+6\sqrt{3}$
d $2\sqrt{5}+4\sqrt{2}+9\sqrt{5}-9\sqrt{2}$
e $3\sqrt{2}-5\sqrt{7}-\sqrt{2}-5\sqrt{7}$
f $3\sqrt{2}+4\sqrt{11}-\sqrt{2}-\sqrt{11}$
g $6-2\sqrt{2}-\sqrt{2}-5$
h $5\sqrt{3}-5+4\sqrt{3}-8$

3 Show that:

a $\sqrt{12}+\sqrt{27}=\sqrt{75}$
b $\sqrt{80}-\sqrt{5}=\sqrt{45}$

E MULTIPLICATIONS INVOLVING RADICALS

The rules for expanding expressions containing radicals are identical to those for ordinary algebra.

$$a(b+c)=ab+ac$$
$$(a+b)(c+d)=ac+ad+bc+bd$$
$$(a+b)^2=a^2+2ab+b^2$$
$$(a+b)(a-b)=a^2-b^2$$

Example 11 Self Tutor

Simplify:

a $2(2+\sqrt{3})$
b $\sqrt{2}(5-2\sqrt{2})$

a $2(2+\sqrt{3})$
$=2\times 2+2\times\sqrt{3}$
$=4+2\sqrt{3}$

b $\sqrt{2}(5-2\sqrt{2})$
$=\sqrt{2}\times 5+\sqrt{2}\times -2\sqrt{2}$
$=5\sqrt{2}-4$

EXERCISE 5E

1 Expand and simplify:

a $4(3+\sqrt{2})$ **b** $3(\sqrt{2}+\sqrt{3})$ **c** $5(4-\sqrt{7})$ **d** $6(\sqrt{11}-4)$
e $\sqrt{2}(1+\sqrt{2})$ **f** $\sqrt{2}(\sqrt{2}-5)$ **g** $\sqrt{3}(2+2\sqrt{3})$ **h** $\sqrt{3}(\sqrt{3}-\sqrt{2})$
i $\sqrt{5}(6-\sqrt{5})$ **j** $\sqrt{5}(2\sqrt{5}-1)$ **k** $\sqrt{5}(2\sqrt{5}+\sqrt{3})$ **l** $\sqrt{7}(2+\sqrt{7}+\sqrt{2})$

Example 12 **Self Tutor**

Expand and simplify:

a $-\sqrt{3}(2+\sqrt{3})$ **b** $-\sqrt{2}(\sqrt{2}-\sqrt{3})$

a $-\sqrt{3}(2+\sqrt{3})$
$= -\sqrt{3}\times 2 + -\sqrt{3}\times\sqrt{3}$
$= -2\sqrt{3}-3$

b $-\sqrt{2}(\sqrt{2}-\sqrt{3})$
$= -\sqrt{2}\times\sqrt{2} + -\sqrt{2}\times -\sqrt{3}$
$= -2+\sqrt{6}$

2 Expand and simplify:

a $-\sqrt{2}(4+\sqrt{2})$ **b** $-\sqrt{2}(3-\sqrt{2})$ **c** $-\sqrt{2}(\sqrt{2}-\sqrt{7})$
d $-\sqrt{3}(3+\sqrt{3})$ **e** $-\sqrt{3}(5-\sqrt{3})$ **f** $-\sqrt{3}(2\sqrt{3}+\sqrt{5})$
g $-\sqrt{5}(2\sqrt{2}-\sqrt{3})$ **h** $-2\sqrt{2}(\sqrt{2}+\sqrt{3})$ **i** $-5\sqrt{2}(4-2\sqrt{2})$
j $-\sqrt{7}(2\sqrt{7}+4)$ **k** $-\sqrt{11}(2-\sqrt{11})$ **l** $-2\sqrt{3}(1-2\sqrt{2})$

Example 13 **Self Tutor**

Expand and simplify:

a $(2+\sqrt{2})(3+\sqrt{2})$ **b** $(3+\sqrt{5})(1-\sqrt{5})$

a $(2+\sqrt{2})(3+\sqrt{2})$
$= 2\times 3 + 2\times\sqrt{2} + \sqrt{2}\times 3 + \sqrt{2}\times\sqrt{2}$
$= 6+2\sqrt{2}+3\sqrt{2}+2$
$= 8+5\sqrt{2}$

b $(3+\sqrt{5})(1-\sqrt{5})$
$= 3\times 1 - 3\times\sqrt{5} + \sqrt{5}\times 1 - \sqrt{5}\times\sqrt{5}$
$= 3-3\sqrt{5}+\sqrt{5}-5$
$= -2-2\sqrt{5}$

3 Expand and simplify:

a $(1+\sqrt{2})(5+\sqrt{2})$ **b** $(3+\sqrt{2})(3+\sqrt{2})$ **c** $(\sqrt{2}+2)(\sqrt{2}-1)$
d $(4-\sqrt{3})(3+\sqrt{3})$ **e** $(2+\sqrt{3})(2-\sqrt{3})$ **f** $(\sqrt{7}+2)(\sqrt{7}-3)$
g $(\sqrt{11}+\sqrt{2})(\sqrt{11}-\sqrt{2})$ **h** $(\sqrt{2}+1)(3-\sqrt{2})$ **i** $(6-2\sqrt{2})(2+\sqrt{2})$

Example 14 — Self Tutor

Expand and simplify:

a $\left(\sqrt{2}+3\right)^2$ b $\left(\sqrt{5}-\sqrt{3}\right)^2$

a
$$\begin{aligned}&\left(\sqrt{2}+3\right)^2\\&=\left(\sqrt{2}\right)^2+2\sqrt{2}(3)+3^2\\&=2+6\sqrt{2}+9\\&=11+6\sqrt{2}\end{aligned}$$

b
$$\begin{aligned}&\left(\sqrt{5}-\sqrt{3}\right)^2\\&=\left(\sqrt{5}+-\sqrt{3}\right)^2\\&=\left(\sqrt{5}\right)^2+2\sqrt{5}\left(-\sqrt{3}\right)+\left(-\sqrt{3}\right)^2\\&=5-2\sqrt{15}+3\\&=8-2\sqrt{15}\end{aligned}$$

4 Expand and simplify:

a $\left(1+\sqrt{3}\right)^2$ b $\left(\sqrt{2}+5\right)^2$ c $\left(3-\sqrt{2}\right)^2$

d $\left(1+\sqrt{7}\right)^2$ e $\left(\sqrt{3}-\sqrt{2}\right)^2$ f $\left(4-\sqrt{5}\right)^2$

g $\left(\sqrt{3}+\sqrt{5}\right)^2$ h $\left(3-\sqrt{6}\right)^2$ i $\left(\sqrt{6}-\sqrt{3}\right)^2$

j $\left(2\sqrt{2}+3\right)^2$ k $\left(3-2\sqrt{2}\right)^2$ l $\left(3-5\sqrt{2}\right)^2$

Example 15 — Self Tutor

Expand and simplify:

a $\left(4+\sqrt{2}\right)\left(4-\sqrt{2}\right)$ b $\left(2\sqrt{2}+3\right)\left(2\sqrt{2}-3\right)$

a
$$\begin{aligned}&\left(4+\sqrt{2}\right)\left(4-\sqrt{2}\right)\\&=4^2-\left(\sqrt{2}\right)^2\\&=16-2\\&=14\end{aligned}$$

b
$$\begin{aligned}&\left(2\sqrt{2}+3\right)\left(2\sqrt{2}-3\right)\\&=\left(2\sqrt{2}\right)^2-3^2\\&=8-9\\&=-1\end{aligned}$$

5 Expand and simplify:

a $\left(3+\sqrt{2}\right)\left(3-\sqrt{2}\right)$ b $\left(\sqrt{3}-1\right)\left(\sqrt{3}+1\right)$

c $\left(5+\sqrt{3}\right)\left(5-\sqrt{3}\right)$ d $\left(\sqrt{3}-4\right)\left(\sqrt{3}+4\right)$

e $\left(\sqrt{7}-3\right)\left(\sqrt{7}+3\right)$ f $\left(2+\sqrt{2}\right)\left(2-\sqrt{2}\right)$

g $\left(\sqrt{7}-5\right)\left(\sqrt{7}+5\right)$ h $\left(2\sqrt{5}+6\right)\left(2\sqrt{5}-6\right)$

i $\left(3\sqrt{2}+2\right)\left(3\sqrt{2}-2\right)$ j $\left(\sqrt{5}-\sqrt{2}\right)\left(\sqrt{5}+\sqrt{2}\right)$

k $\left(\sqrt{3}-\sqrt{7}\right)\left(\sqrt{3}+\sqrt{7}\right)$ l $\left(2\sqrt{2}+1\right)\left(2\sqrt{2}-1\right)$

F DIVISION BY RADICALS

Numbers like $\frac{6}{\sqrt{2}}$ and $\frac{9}{5+\sqrt{3}}$ involve dividing by a radical.

It is customary to 'simplify' these numbers by rewriting them without the radical in the denominator.

INVESTIGATION 2 DIVISION BY $\sqrt{a}$, $a > 0$

In this Investigation we consider fractions of the form $\frac{b}{\sqrt{a}}$, where a and b are real numbers, $a > 0$.

What to do:

1 **a** Simplify $\frac{6}{\sqrt{2}}$ by writing $6 = 3 \times \sqrt{2} \times \sqrt{2}$.

b Can this method be used to simplify $\frac{7}{\sqrt{2}}$?

2 Consider the fraction $\frac{7}{\sqrt{2}}$.

a If we multiply this fraction by $\frac{\sqrt{2}}{\sqrt{2}}$, are we changing its value?

b Simplify $\frac{7}{\sqrt{2}}$ by multiplying both its numerator and denominator by $\sqrt{2}$.

c The method in **b** is called 'rationalising the denominator'. Will this method work for all fractions of the form $\frac{b}{\sqrt{a}}$ where a and b are real, $a > 0$?

From the **Investigation**, you should have found that:

> For any fraction of the form $\frac{b}{\sqrt{a}}$, $a > 0$, we can remove the radical from the denominator by multiplying by $\frac{\sqrt{a}}{\sqrt{a}}$.

Since $\frac{\sqrt{a}}{\sqrt{a}} = 1$, this does not change the value of the fraction.

Example 16 Self Tutor

Write with an integer denominator:

a $\frac{6}{\sqrt{3}}$ **b** $\frac{35}{\sqrt{7}}$

a $\frac{6}{\sqrt{3}}$

$= \frac{6}{\sqrt{3}} \times \frac{\sqrt{3}}{\sqrt{3}}$

$= \frac{{}^{2}\cancel{6}\sqrt{3}}{\cancel{3}_{1}}$

$= 2\sqrt{3}$

b $\frac{35}{\sqrt{7}}$

$= \frac{35}{\sqrt{7}} \times \frac{\sqrt{7}}{\sqrt{7}}$

$= \frac{{}^{5}\cancel{35}\sqrt{7}}{\cancel{7}_{1}}$

$= 5\sqrt{7}$

EXERCISE 5F.1

1 Write with integer denominator:

a $\frac{1}{\sqrt{3}}$ **b** $\frac{3}{\sqrt{3}}$ **c** $\frac{9}{\sqrt{3}}$ **d** $\frac{11}{\sqrt{3}}$ **e** $\frac{\sqrt{2}}{3\sqrt{3}}$

f $\frac{2}{\sqrt{2}}$ **g** $\frac{6}{\sqrt{2}}$ **h** $\frac{12}{\sqrt{2}}$ **i** $\frac{\sqrt{3}}{\sqrt{2}}$ **j** $\frac{1}{4\sqrt{2}}$

k $\frac{5}{\sqrt{5}}$ **l** $\frac{15}{\sqrt{5}}$ **m** $\frac{-3}{\sqrt{5}}$ **n** $\frac{200}{\sqrt{5}}$ **o** $\frac{1}{3\sqrt{5}}$

p $\frac{7}{\sqrt{7}}$ **q** $\frac{21}{\sqrt{7}}$ **r** $\frac{2}{\sqrt{11}}$ **s** $\frac{26}{\sqrt{13}}$ **t** $\frac{1}{(\sqrt{3})^3}$

2 Write with integer denominator:

a $\sqrt{\frac{1}{5}}$ **b** $\sqrt{\frac{5}{3}}$ **c** $\sqrt{\frac{7}{2}}$ **d** $\sqrt{\frac{3}{10}}$ **e** $\sqrt{\frac{7}{8}}$

RADICAL CONJUGATES

The radical expressions $3+\sqrt{2}$ and $3-\sqrt{2}$ are identical except for opposite signs in the middle. We call these expressions **radical conjugates**.

The **radical conjugate** of $a+\sqrt{b}$ is $a-\sqrt{b}$.

We can use radical conjugates to simplify fractions of the form $\frac{c}{a+\sqrt{b}}$.

INVESTIGATION 3 — RADICAL CONJUGATES

What to do:

1 Expand and simplify:

a $(2+\sqrt{3})(2-\sqrt{3})$ **b** $(\sqrt{3}-1)(\sqrt{3}+1)$

2 What do you notice about your results in **1**?

3 Show that for integers a and b, the following products are also integers:

a $(a+\sqrt{b})(a-\sqrt{b})$ for $b \geqslant 0$ **b** $(\sqrt{a}-b)(\sqrt{a}+b)$ for $a \geqslant 0$.

4 **a** Copy and complete:
To remove the radicals from the denominator of a fraction, we can multiply the denominator by its ......

b What must we then do to the numerator of the fraction to ensure we do not change the value of the fraction?

From the **Investigation** above, you should have found that:

To remove the radicals from the denominator of a fraction, we multiply both the numerator and the denominator by the **radical conjugate** of the denominator.

Example 17 **Self Tutor**

Write $\dfrac{5}{3-\sqrt{2}}$ with an integer denominator.

$$\frac{5}{3-\sqrt{2}} = \left(\frac{5}{3-\sqrt{2}}\right)\left(\frac{3+\sqrt{2}}{3+\sqrt{2}}\right)$$
$$= \frac{5\left(3+\sqrt{2}\right)}{3^2-\left(\sqrt{2}\right)^2}$$
$$= \frac{15+5\sqrt{2}}{7}$$

EXERCISE 5F.2

1 Write with an integer denominator:

a $\dfrac{1}{3+\sqrt{2}}$ **b** $\dfrac{2}{3-\sqrt{2}}$ **c** $\dfrac{1}{2+\sqrt{5}}$ **d** $\dfrac{\sqrt{2}}{2-\sqrt{2}}$

e $\dfrac{10}{\sqrt{6}-1}$ **f** $\dfrac{\sqrt{3}}{\sqrt{7}+2}$ **g** $\dfrac{1+\sqrt{2}}{1-\sqrt{2}}$ **h** $\dfrac{\sqrt{3}}{4-\sqrt{3}}$

i $\dfrac{-2\sqrt{2}}{1-\sqrt{2}}$ **j** $\dfrac{1+\sqrt{5}}{2-\sqrt{5}}$ **k** $\dfrac{\sqrt{3}+2}{\sqrt{3}-1}$ **l** $\dfrac{\sqrt{10}-7}{\sqrt{10}+4}$

Example 18 **Self Tutor**

Write $\dfrac{1}{5+\sqrt{2}}$ in the form $a+b\sqrt{2}$ where $a, b \in \mathbb{Q}$.

$$\frac{1}{5+\sqrt{2}} = \left(\frac{1}{5+\sqrt{2}}\right)\times\left(\frac{5-\sqrt{2}}{5-\sqrt{2}}\right)$$
$$= \frac{5-\sqrt{2}}{25-2}$$
$$= \frac{5-\sqrt{2}}{23}$$
$$= \frac{5}{23}-\frac{1}{23}\sqrt{2}$$

2 Write in the form $a+b\sqrt{2}$ where $a, b \in \mathbb{Q}$:

a $\dfrac{3}{\sqrt{2}-3}$ **b** $\dfrac{4}{2+\sqrt{2}}$ **c** $\dfrac{\sqrt{2}}{\sqrt{2}-5}$ **d** $\dfrac{-2\sqrt{2}}{\sqrt{2}+1}$

3 Write in the form $a+b\sqrt{3}$ where $a, b \in \mathbb{Q}$:

a $\dfrac{4}{1-\sqrt{3}}$ **b** $\dfrac{6}{\sqrt{3}+2}$ **c** $\dfrac{\sqrt{3}}{2-\sqrt{3}}$ **d** $\dfrac{1+2\sqrt{3}}{3+\sqrt{3}}$

4 Write in simplest form:

a $\dfrac{1-\sqrt{3}}{1+\sqrt{3}}+\dfrac{1+\sqrt{3}}{1-\sqrt{3}}$ **b** $\dfrac{\sqrt{2}}{3-\sqrt{2}}-\dfrac{3-\sqrt{2}}{3+\sqrt{2}}$ **c** $\dfrac{2+\sqrt{5}}{3+\sqrt{5}}+\dfrac{2\sqrt{5}}{3-\sqrt{5}}$

5 Write $\sqrt{\dfrac{2+\sqrt{3}}{2-\sqrt{3}}}$ in the form $a+b\sqrt{3}$ where $a, b \in \mathbb{Q}$.

6 **a** Find $\sqrt{3+2\sqrt{2}} \times \sqrt{3-2\sqrt{2}}$ without using your calculator.

b Let $X = \sqrt{3+2\sqrt{2}} - \sqrt{3-2\sqrt{2}}$. Find X^2 without using your calculator.

c Hence, find $\sqrt{3+2\sqrt{2}} - \sqrt{3-2\sqrt{2}}$.

INVESTIGATION 4 APPROXIMATING RADICALS

In this Investigation we will explore a method for approximating radicals. Many calculators use this method for calculating values such as $\sqrt{2}$ and $\sqrt[3]{7}$.

To approximate the value of $\sqrt{2}$, we follow these steps:

Step 1: Make an initial guess of the value of $\sqrt{2}$, and call this guess x.

Step 2: Calculate $\dfrac{2}{x}$.

Step 3: Find the average of x and $\dfrac{2}{x}$. This is your new guess for the value of $\sqrt{2}$.

Step 4: Repeat steps *2* and *3* with the new guess x.

With an initial guess of $x = 1$, the results of this process are shown alongside.

As the process is repeated several times, the values of x get closer to $\sqrt{2} \approx 1.414\,213\,562$.

x	$\dfrac{2}{x}$	*Average of* x *and* $\dfrac{2}{x}$
1	2	1.5
1.5	1.333 333	1.416 667
1.416 667	1.411 765	1.414 216
1.414 216	1.414 211	1.414 214
⋮	⋮	⋮

What to do:

1 Can you see why this method works?

Hint: What can you say about x and $\dfrac{2}{x}$ when x is *exactly* $\sqrt{2}$?

2 Use a similar method to approximate $\sqrt{3}$. Use an initial guess of $x = 2$.

3 Click on the spreadsheet icon. Use the software to approximate other square roots. Experiment with different initial guesses, and investigate the effect this has on the accuracy of the approximations.

SPREADSHEET

4 Adapt this method to approximate cube roots instead of square roots.

REVIEW SET 5A

1 Evaluate the following surds correct to 3 decimal places:

a $\sqrt{7}$ **b** $\sqrt{22}$ **c** $\sqrt{187}$

2 Simplify:

a $(2\sqrt{3})^2$ **b** $\left(\frac{8}{\sqrt{2}}\right)^2$ **c** $3\sqrt{2}\times 2\sqrt{5}$ **d** $\sqrt{12\frac{1}{4}}$

3 Simplify:

a $\frac{\sqrt{15}}{\sqrt{3}}$ **b** $\frac{\sqrt{35}}{\sqrt{7}}$ **c** $\frac{\sqrt{35}}{\sqrt{5}}$ **d** $\frac{\sqrt{2}}{\sqrt{20}}$

4 Simplify:

a $8\sqrt{11}-10\sqrt{11}+5\sqrt{11}$ **b** $\sqrt{7}-2\sqrt{3}+5\sqrt{7}-\sqrt{3}$

5 Answer the **Opening Problem** on page **78**.

6 Write $\sqrt{80}$ in simplest radical form.

7 Expand and simplify:

a $2\left(\sqrt{3}+1\right)$ **b** $\sqrt{2}\left(3-\sqrt{2}\right)$ **c** $\left(1+\sqrt{7}\right)^2$

d $\left(2-\sqrt{5}\right)^2$ **e** $\left(5+\sqrt{6}\right)\left(5-\sqrt{6}\right)$ **f** $\left(3+\sqrt{2}\right)\left(1-\sqrt{2}\right)$

8 Write with an integer denominator:

a $\frac{10}{\sqrt{5}}$ **b** $\frac{\sqrt{3}+2}{\sqrt{3}+1}$ **c** $\frac{1+\sqrt{7}}{1-\sqrt{7}}$

9 Write in the form $a+b\sqrt{5}$, where $a, b\in\mathbb{Q}$:

a $\frac{3}{2-\sqrt{5}}$ **b** $\frac{2\sqrt{5}}{\sqrt{5}+1}$ **c** $\frac{6}{3+\sqrt{5}}$

10 **a** Simplify $\left(3+\sqrt{2}\right)\left(5-3\sqrt{2}\right)$.

b Hence find $\frac{9-4\sqrt{2}}{3+\sqrt{2}}$.

c Check your answer to **b** by simplifying the fraction.

REVIEW SET 5B

1 Simplify:

a $\sqrt{5}\sqrt{2}$ **b** $\frac{\sqrt{8}}{\sqrt{2}}$ **c** $(3\sqrt{6})^2$

2 **a** Simplify $\frac{\sqrt{76}}{\sqrt{4}}$. **b** Evaluate the result in **a** to 4 decimal places.

3 Simplify:

a $\frac{\sqrt{3}}{\sqrt{24}}$ **b** $\sqrt{\frac{1}{49}}$ **c** $\sqrt{11\frac{1}{9}}$

4 Write in simplest radical form: **a** $\sqrt{44}$ **b** $\sqrt{90}$

5 Simplify:

a $5\sqrt{2}-\sqrt{5}+\sqrt{2}+3\sqrt{5}$

b $6-2\sqrt{7}+5\sqrt{7}-10$

6 Expand and simplify:

a $\sqrt{7}\left(\sqrt{2}-1\right)$

b $\left(5-\sqrt{2}\right)^2$

7 Expand and simplify:

a $\left(\sqrt{3}+\sqrt{2}\right)^2$

b $\left(2-\sqrt{5}\right)\left(2+\sqrt{5}\right)$

8 Write with an integer denominator:

a $\dfrac{8}{\sqrt{3}}$

b $\sqrt{\dfrac{7}{13}}$

c $\dfrac{4-\sqrt{5}}{3+\sqrt{5}}$

9 Write in the form $a+b\sqrt{3}$ where $a,\, b \in \mathbb{Q}$:

a $\dfrac{18}{5-\sqrt{3}}$

b $\dfrac{1}{\sqrt{3}-2}$

c $\dfrac{-\sqrt{3}}{3+\sqrt{3}}$

10 **a** Evaluate:

i $\sqrt{20}\times\sqrt{5}$ and $\dfrac{\sqrt{20}}{\sqrt{5}}$

ii $\sqrt{3}\times\sqrt{12}$ and $\dfrac{\sqrt{3}}{\sqrt{12}}$

b Suppose a and b are positive integers, and that $\sqrt{a}\times\sqrt{b}$ is an integer.

Prove that $\dfrac{\sqrt{a}}{\sqrt{b}}$ is rational.

11 Write in simplest form: $\dfrac{1-\sqrt{2}}{3+\sqrt{2}}-\dfrac{2\sqrt{2}}{3-\sqrt{2}}$

Chapter 6

Linear equations and inequalities

Contents:

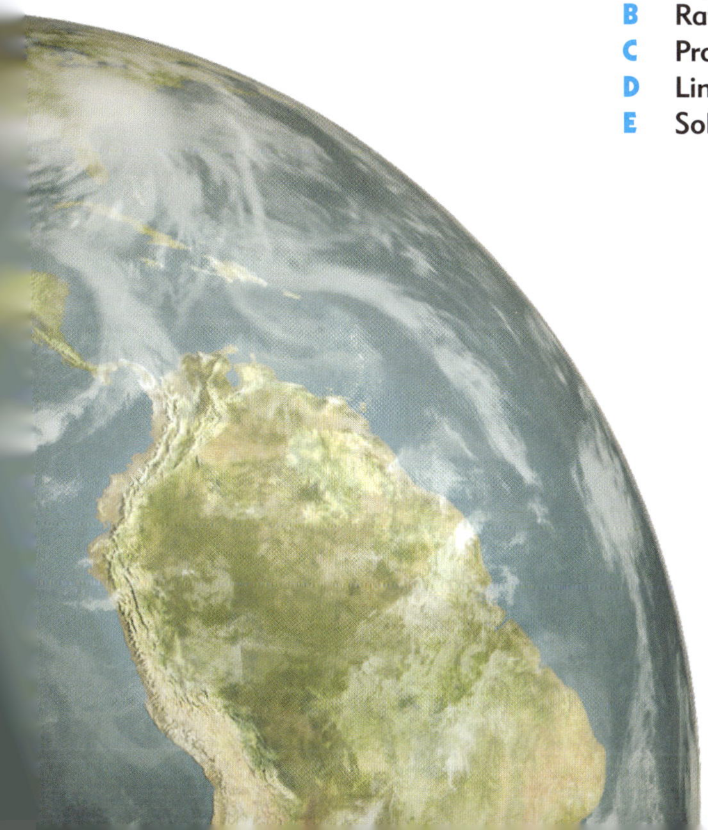

OPENING PROBLEM

Mrs May set her class the following challenge:

Find a fraction whose numerator is 4 more than its denominator, and the value of the fraction is equal to $\frac{1}{3}$.

"That's impossible!" Stan said, "If the numerator is larger than the denominator, the value of the fraction must be greater than 1!"

Things to think about:

a Can you explain why Stan is wrong?

b Can you use algebra to solve the problem?

Many problems in mathematics can be solved by using **equations**.

An **algebraic equation** is a mathematical sentence which indicates that two expressions have the same value. It will always contain the symbol $=$.

We can often convert a worded problem into an algebraic equation, then follow a formal procedure to **solve** the equation and hence the problem.

In other problems, we may be considering when one quantity is greater than or less than another quantity. The problem may then be written as a mathematical **inequality**.

An **algebraic inequality** is a mathematical sentence which compares the values of two expressions. It will contain one of the symbols $>$, $<$, $\geqslant$, or $\leqslant$.

A LINEAR EQUATIONS

Linear equations are equations in which the variable is raised only to the power 1.

All linear equations can be written in the form $ax + b = 0$ where a and b are constants, $a \neq 0$, and x is the variable.

Examples of linear equations include $3x + 6 = 15$, $\frac{x}{5} - 4 = -2$, and $10 - 7.5y = -5$.

ALGEBRAIC SOLUTION TO LINEAR EQUATIONS

In previous years, we have seen how to solve linear equations algebraically:

Step 1: Determine how the expression containing the unknown has been '**built up**'.

Step 2: Perform **inverse** operations on **both sides** of the equation to 'undo' how the expression was 'built up'. In this way we **isolate** the unknown.

Step 3: Check your solution by substitution.

Example 1 **Self Tutor**

Solve for x:

a $4x - 1 = 7$ **b** $5 - 3x = 6$

a
$$4x - 1 = 7$$
$$\therefore\ 4x - 1 + 1 = 7 + 1 \quad \{\text{adding 1 to both sides}\}$$
$$\therefore\ 4x = 8$$
$$\therefore\ \frac{4x}{4} = \frac{8}{4} \quad \{\text{dividing both sides by 4}\}$$
$$\therefore\ x = 2$$

Check: $4 \times 2 - 1 = 8 - 1 = 7$ ✓

b
$$5 - 3x = 6$$
$$\therefore\ 5 - 3x - 5 = 6 - 5 \quad \{\text{subtracting 5 from both sides}\}$$
$$\therefore\ -3x = 1$$
$$\therefore\ \frac{-3x}{-3} = \frac{1}{-3} \quad \{\text{dividing both sides by } -3\}$$
$$\therefore\ x = -\tfrac{1}{3}$$

Check: $5 - 3 \times (-\frac{1}{3}) = 5 + 1 = 6$ ✓

Example 2 **Self Tutor**

Solve for x:

a $\dfrac{x}{5} - 3 = -1$ **b** $\frac{1}{5}(x - 3) = -1$

a
$$\frac{x}{5} - 3 = -1$$
$$\therefore\ \frac{x}{5} - 3 + 3 = -1 + 3 \quad \{\text{adding 3 to both sides}\}$$
$$\therefore\ \frac{x}{5} = 2$$
$$\therefore\ \frac{x}{5} \times 5 = 2 \times 5 \quad \{\text{multiplying both sides by 5}\}$$
$$\therefore\ x = 10$$

Check: $\frac{10}{5} - 3 = 2 - 3 = -1$ ✓

b
$$\tfrac{1}{5}(x - 3) = -1$$
$$\therefore\ \tfrac{1}{5}(x - 3) \times 5 = -1 \times 5 \quad \{\text{multiplying both sides by 5}\}$$
$$\therefore\ x - 3 = -5$$
$$\therefore\ x - 3 + 3 = -5 + 3 \quad \{\text{adding 3 to both sides}\}$$
$$\therefore\ x = -2$$

Check: $\frac{1}{5}((-2) - 3) = \frac{1}{5} \times -5 = -1$ ✓

EXERCISE 6A.1

1 Solve for x:

a $x+7=3$ b $6x=42$ c $x-8=-3$ d $33=-3x$

2 Solve for x:

a $4x+9=21$ b $2x-7=-3$ c $5-3x=-13$ d $-6=4-5x$

e $-5x+3=18$ f $4+7x=-10$ g $12=8x-5$ h $-5+6x=-3$

3 Solve for x:

a $\dfrac{x}{5}=20$ b $\frac{1}{3}x=9$ c $4=\dfrac{x}{-8}$ d $\dfrac{x}{2}+7=-3$

e $\dfrac{x+2}{5}=-3$ f $\frac{1}{4}(x+7)=5$ g $\dfrac{3x-4}{6}=-1$ h $4(2x-3)=28$

i $\dfrac{2x-7}{3}=4$ j $-5(3x+1)=35$ k $\frac{1}{7}(4-x)=-2$ l $\dfrac{1-5x}{3}=-6$

EQUATIONS WITH A REPEATED UNKNOWN

In situations where the unknown appears more than once, we need to **expand** any brackets, **collect like terms**, and then **solve** the equation.

Example 3 Self Tutor

Solve for x: $3(2x-5)-2(x-1)=3$

$$\begin{aligned} 3(2x-5)-2(x-1)&=3 \\ \therefore\ 3\times 2x+3\times(-5)-2\times x-2\times(-1)&=3 && \{\text{expanding brackets}\} \\ \therefore\ 6x-15-2x+2&=3 \\ \therefore\ 4x-13&=3 && \{\text{collecting like terms}\} \\ \therefore\ 4x-13+13&=3+13 && \{\text{adding 13 to both sides}\} \\ \therefore\ 4x&=16 \\ \therefore\ x&=4 && \{\text{dividing both sides by 4}\} \end{aligned}$$

Check: $3(2\times 4-5)-2(4-1)=3\times 3-2\times 3=3$ ✓

If the unknown appears on **both sides** of the equation, we

- **expand** any brackets and **collect like terms**
- move the **unknown to one side** of the equation and the remaining terms to the other side
- **solve** the equation.

Example 4 — Self Tutor

Solve for x:

a $3x - 4 = 2x + 6$

b $4 - 3(2 + x) = x$

a
$$3x - 4 = 2x + 6$$
$$\therefore\ 3x - 4 - 2x = 2x + 6 - 2x \quad \{\text{subtracting } 2x \text{ from both sides}\}$$
$$\therefore\ x - 4 = 6$$
$$\therefore\ x - 4 + 4 = 6 + 4 \quad \{\text{adding 4 to both sides}\}$$
$$\therefore\ x = 10$$

Check: LHS $= 3 \times 10 - 4 = 26$, RHS $= 2 \times 10 + 6 = 26$. ✓

b
$$4 - 3(2 + x) = x$$
$$\therefore\ 4 - 6 - 3x = x \quad \{\text{expanding brackets}\}$$
$$\therefore\ -2 - 3x = x$$
$$\therefore\ -2 - 3x + 3x = x + 3x \quad \{\text{adding } 3x \text{ to both sides}\}$$
$$\therefore\ -2 = 4x$$
$$\therefore\ \frac{-2}{4} = \frac{4x}{4} \quad \{\text{dividing both sides by 4}\}$$
$$\therefore\ -\tfrac{1}{2} = x$$

which means $x = -\frac{1}{2}$

Check: LHS $= 4 - 3(2 + (-\frac{1}{2})) = 4 - 3 \times \frac{3}{2} = 4 - 4\frac{1}{2} = -\frac{1}{2}$, RHS $= -\frac{1}{2}$. ✓

EXERCISE 6A.2

1 Solve for x:

a $2x + 5 + x = 14$

b $3x - 1 + 4x = 27$

c $4(x - 2) - 2x = -18$

d $x - 5(x - 3) = 21$

e $2(x + 4) + 4(x + 3) = 14$

f $4(x - 2) + 5(x + 1) = -30$

g $3(2x + 1) - 2(x - 2) = 23$

h $6(x - 3) + 3(4x - 1) = -3$

i $2(3x - 4) - 3(x - 5) = -11$

j $3(2x - 3) - 4(7 - 2x) = -30$

2 Solve for x:

a $3x - 4 = 2x + 1$

b $2x + 5 = 11 - x$

c $7 - 2x = 3x + 4$

d $-x = 4x + 9$

e $8 - 3x = 4x + 5$

f $3x - 7 = 2 - 3x$

g $5 - x - 2(x + 3) = x + 4$

h $4 - 3(x + 2) = 5x - 1$

i $3 - 4x - (x + 5) = -7$

j $5(2x - 3) - x = x + 1$

k $4(x + 3) = 7 - (2 - 3x)$

l $3(x - 2) + 5 = 7(x - 3)$

m $5x - 2(3x - 1) = 4(x - 6)$

n $6(x - 1) - 5(3x + 4) = 8(x - 3)$

3 Solve for x, and explain your answer:

a $3(2x - 5) + 10 = 6x - 5$

b $6(x - 2) = 3(2x + 1)$

ACTIVITY 1

Click on the icon to practice solving linear equations.

LINEAR EQUATIONS

B RATIONAL EQUATIONS

Rational equations are equations involving fractions. To solve rational equations, we write all the fractions in the equation with the same **lowest common denominator (LCD)**, and then equate the numerators.

For fractions whose denominators involve the variable, the lowest common denominator is found in the same way as for numerical fractions.

For example:

- in $\frac{x}{4} = \frac{x+1}{6}$ the LCD is 12
- in $\frac{5}{3x} = \frac{7}{9}$ the LCD is $9x$
- in $\frac{x-7}{3} = \frac{x}{2x-1}$ the LCD is $3 \times (2x-1) = 3(2x-1)$
- in $\frac{3}{x+1} = \frac{2}{x-2}$ the LCD is $(x+1)(x-2)$

Example 5 Self Tutor

Solve for x: $\frac{x}{2} = \frac{3+x}{5}$

$\frac{x}{2} = \frac{3+x}{5}$ has LCD $= 10$

$\therefore \frac{x}{2} \times \frac{5}{5} = \frac{2}{2} \times \left(\frac{3+x}{5}\right)$ {to create a common denominator}

$\therefore 5x = 2(3+x)$ {equating numerators}

$\therefore 5x = 6 + 2x$ {expanding brackets}

$\therefore 5x - 2x = 6 + 2x - 2x$ {subtracting $2x$ from both sides}

$\therefore 3x = 6$

$\therefore x = 2$ {dividing both sides by 3}

EXERCISE 6B

1 Solve for x:

a $\frac{x}{2} = \frac{5}{9}$ **b** $\frac{x-1}{5} = \frac{x}{10}$ **c** $\frac{x}{3} = \frac{x+1}{2}$

d $\frac{x+2}{5} = \frac{3x-1}{4}$ **e** $\frac{3x}{4} = \frac{6-x}{5}$ **f** $\frac{4x+1}{2} = \frac{3x-2}{5}$

g $\frac{2x-1}{4} = \frac{5x+2}{8}$ **h** $\frac{3x+5}{5} = \frac{7-x}{2}$ **i** $\frac{2x-5}{4} = \frac{3x-1}{-2}$

Example 6 — Self Tutor

Solve for x: $\dfrac{4}{x} = \dfrac{3}{4}$

$\dfrac{4}{x} = \dfrac{3}{4}$ has LCD $= 4x$

$\therefore \dfrac{4}{x} \times \dfrac{4}{4} = \dfrac{3}{4} \times \dfrac{x}{x}$ {to create a common denominator}

$\therefore 16 = 3x$ {equating numerators}

$\therefore x = \frac{16}{3}$ {dividing both sides by 3}

2 Solve for x:

a $\dfrac{3}{x} = \dfrac{2}{7}$ **b** $\dfrac{4}{x} = \dfrac{5}{9}$ **c** $\dfrac{6}{5} = \dfrac{4}{x}$

d $\dfrac{4}{3x} = \dfrac{8}{7}$ **e** $\dfrac{3}{2x} = \dfrac{5}{4}$ **f** $\dfrac{7}{2x} = -\dfrac{1}{8}$

g $\dfrac{9}{2x} = -\dfrac{1}{4}$ **h** $\dfrac{5}{6x} = \dfrac{4}{5x}$

Example 7 — Self Tutor

Solve for x: $\dfrac{2x+1}{3-x} = \dfrac{3}{4}$

$\dfrac{2x+1}{3-x} = \dfrac{3}{4}$ has LCD $= 4(3-x)$

$\therefore \dfrac{4}{4} \times \left(\dfrac{2x+1}{3-x}\right) = \dfrac{3}{4} \times \left(\dfrac{3-x}{3-x}\right)$ {to create a common denominator}

$\therefore 4(2x+1) = 3(3-x)$ {equating numerators}

$\therefore 8x + 4 = 9 - 3x$ {expanding the brackets}

$\therefore 11x + 4 = 9$ {adding $3x$ to both sides}

$\therefore 11x = 5$ {subtracting 4 from both sides}

$\therefore x = \frac{5}{11}$ {dividing both sides by 11}

3 Solve for x:

a $\dfrac{3x+1}{x+2} = \dfrac{2}{3}$ **b** $\dfrac{x+4}{1-2x} = \dfrac{3}{4}$ **c** $\dfrac{2x-3}{4-5x} = -\dfrac{2}{3}$

d $\dfrac{x+5}{2x-1} = \dfrac{4}{5}$ **e** $\dfrac{5x+2}{x-1} = 4$ **f** $\dfrac{3x-1}{2x+1} = -3$

g $\dfrac{6x-1}{3x-2} = -2$ **h** $\dfrac{3x+2}{4-3x} = 2$ **i** $\dfrac{3x-2}{2x-1} = -6$

j $\dfrac{2}{x} = \dfrac{5}{x+3}$ **k** $\dfrac{5}{x-8} = \dfrac{1}{x-2}$ **l** $\dfrac{3}{x-1} = \dfrac{7}{x+2}$

Example 8 **Self Tutor**

Solve for x: $\frac{x}{3} - \frac{1-2x}{6} = -4$

$$\frac{x}{3} - \frac{1-2x}{6} = -4 \quad \text{has LCD} = 6$$

$$\therefore \frac{x}{3} \times \frac{2}{2} - \left(\frac{1-2x}{6}\right) = -4 \times \frac{6}{6} \quad \{\text{to create a common denominator}\}$$

$$\therefore 2x - (1-2x) = -24 \quad \{\text{equating numerators}\}$$

$$\therefore 2x - 1 + 2x = -24 \quad \{\text{expanding brackets}\}$$

$$\therefore 4x - 1 = -24$$

$$\therefore 4x = -23 \quad \{\text{adding 1 to both sides}\}$$

$$\therefore x = -\tfrac{23}{4} \quad \{\text{dividing both sides by 4}\}$$

4 Solve for x:

a $\frac{x}{2} - \frac{x}{4} = 3$ **b** $\frac{x}{3} - 2 = \frac{3x}{4}$ **c** $\frac{x}{6} + \frac{x+2}{3} = -3$

d $\frac{x+1}{2} + \frac{x-3}{5} = -1$ **e** $\frac{3x-1}{4} - \frac{x+2}{8} = 2$ **f** $\frac{x}{6} = 2 - \frac{x+3}{2}$

5 Solve for x:

a $\frac{3x-8}{2} - 1 = \frac{x-4}{8}$ **b** $\frac{x+4}{6} - \frac{x}{2} = \frac{3x-2}{3}$ **c** $\frac{x}{6} - \frac{x+4}{5} = \frac{2}{3}$

d $\frac{x+3}{2} + \frac{x-2}{4} = \frac{x+1}{8}$ **e** $\frac{x-4}{3} - \frac{3x+7}{12} = \frac{x+1}{6}$ **f** $\frac{3x+1}{4} - \frac{1-2x}{6} = \frac{3x+2}{8}$

ACTIVITY 2 SOLVING EQUATIONS

Place the LHS and RHS expressions in the correct boxes so that the resulting equations have the correct solutions shown.

PRINTABLE WORKSHEET

LHS Expressions

$x - 15$	$6(x-3)$	$\frac{x+6}{3}$
$\frac{x}{7} + 3$	$10(x+3)$	$\frac{27}{2x-7}$
$\frac{14}{x+1}$	$8(x-2)$	$\frac{x}{5} + 4$
	$8 - 3x$	

RHS Expressions

-13	7	3	
12	70	32	
-1	4	6	5

LHS		RHS	**Solution**
	=		$x = 1$
	=		$x = 2$
	=		$x = 3$
	=		$x = 4$
	=		$x = 5$
	=		$x = 6$
	=		$x = 7$
	=		$x = 8$
	=		$x = 9$
	=		$x = 10$

C PROBLEM SOLVING

Many problems can be translated into **algebraic equations**. To solve problems using algebra, we follow these steps:

Step 1: Decide on the unknown quantity and allocate it a variable such as x.
Step 2: Translate the problem into an equation.
Step 3: Solve the equation by isolating the variable.
Step 4: Check that the solution satisfies the original problem.
Step 5: Write the answer in sentence form, describing how the solution relates to the original problem.

Example 9 Self Tutor

When a number is trebled and subtracted from 7, the result is -11. Find the number.

Let x be the number, so $3x$ is the number trebled.

When $3x$ is subtracted from 7, we get $7 - 3x$.

$\therefore\ 7 - 3x = -11$

$\therefore\ -3x = -18$ {subtracting 7 from both sides}

$\therefore\ x = 6$ {dividing both sides by -3}

So, the number is 6.

Check: 6 trebled gives 18.
When 18 is subtracted from 7, the result is $7 - 18 = -11$. ✓

Example 10 Self Tutor

What number must be added to both the numerator and the denominator of $\frac{1}{3}$ to obtain a fraction equal to $\frac{7}{8}$?

Let x be the number.

$\therefore\ \dfrac{1+x}{3+x} = \dfrac{7}{8}$ which has LCD $= 8(3+x)$

$\therefore\ \dfrac{8}{8} \times \left(\dfrac{1+x}{3+x}\right) = \dfrac{7}{8} \times \left(\dfrac{3+x}{3+x}\right)$ {to obtain a common denominator}

$\therefore\ 8(1+x) = 7(3+x)$ {equating numerators}

$\therefore\ 8 + 8x = 21 + 7x$ {expanding brackets}

$\therefore\ 8 + x = 21$ {subtracting $7x$ from both sides}

$\therefore\ x = 13$

So, the number is 13.

EXERCISE 6C.1

1 When four times a certain number is subtracted from 23, the result is -5. Find the number.

2 Two less than twice a certain number, is equal to 6 more than four times the number. Find the number.

3 I think of a number. If I divide the sum of 8 and the number by 4, the result is one more than one third of the number. Find the number.

4 A plane flying from Los Angeles to Seattle is carrying 30 rows of passengers, as well as 12 crew members. There are 222 people on board the plane, and every seat is taken. How many passengers are in each row?

5 During a volleyball training session, Keela drank twice as much water as Carol, and Xavier drank 100 mL more water than Keela. Between them they drank 3 litres of water. How much water did Carol drink?

6 At a restaurant, Table A's bill of €72 is split equally between the people at the table. Table B has two more people than Table A, and its bill of €108 is also split equally between the people at the table. Each person at Table A pays the same amount as each person at Table B. How many people were at Table A?

7 What number must be added to both the numerator and the denominator of $\frac{3}{7}$ to obtain a fraction equal to $\frac{7}{9}$?

8 What number must be subtracted from both the numerator and the denominator of $\frac{3}{5}$ to obtain a fraction equal to $\frac{1}{4}$?

9 Write a fraction for which the sum of the numerator and denominator is 20, and the value of the fraction is equal to $\frac{2}{3}$.

10 Consider the **Opening Problem** on page **94**. Solve the challenge set by Mrs May.

DISCUSSION

Consider this problem:

Chiong Min made a batch of dumplings to take to her friend's house. She was putting them in a container when she slipped. Half of the dumplings had to be thrown away. Chiong Min and her friend ate three quarters of what was left. Chiong Min fed two dumplings to her friend's dog, and then there was only one left. How many dumplings did Chiong Min make?

- As a class, write a linear equation to describe this situation.
- Is there a more efficient way to solve this problem?

USING A TABLE

Some problems can be made easier to understand by placing the given information into a table.

Example 11 Self Tutor

Sarah's father is now three times as old as Sarah. In 13 years' time, Sarah's father will be twice as old as Sarah. How old is Sarah now?

Let Sarah's present age be x years, so her father's present age is $3x$ years.

Table of ages:

	Now	*13 years' time*
Sarah	x	$x+13$
Father	$3x$	$3x+13$

So, $3x+13 = 2(x+13)$

$\therefore\ 3x+13 = 2x+26$

$\therefore\ 3x-2x = 26-13$

$\therefore\ x = 13$

$\therefore$ Sarah is currently 13 years old.

Example 12 Self Tutor

Britney has only 2-cent and 5-cent stamps. Their total value is \$1.78, and there are two more 5-cent stamps than there are 2-cent stamps. How many 2-cent stamps are there?

Let x be the number of 2-cent stamps.

$\therefore$ there are $(x+2)$ 5-cent stamps.

Type	*Number*	*Value*
2-cent	x	$2x$ cents
5-cent	$x+2$	$5(x+2)$ cents

$\therefore\ 2x+5(x+2) = 178$ {equating values in cents}

$\therefore\ 2x+5x+10 = 178$

$\therefore\ 7x+10 = 178$

$\therefore\ 7x = 168$

$\therefore\ x = 24$

So, the number of 2-cent stamps is 24.

EXERCISE 6C.2

1. Ellie is now four times as old as her son. In 5 years' time she will be three times as old as her son. How old is Ellie's son now?
2. When George was born, his father was 36 years old. George is now 10% of his father's age. How old is George now?
3. Four years ago, Adrian was one quarter of his brother's age. In two years' time, his age doubled will equal his brother's age. How old is Adrian now?
4. Lana collects 20-cent and 50-cent coins. The total value of her coins is currently \$9.70. If Lana has 4 more 50-cent coins than 20-cent coins, how many of each type does she have?

5 A corner shop has an equal number of 600 mL and 1 litre cartons of milk, and double that number of 2 litre bottles of milk. If there are 84 L of milk in total, how many 600 mL cartons does the shop have?

6 Tickets to a concert cost €20, €35, or €50 each. The number of €20 tickets sold was triple the number of €35 tickets sold. 400 more €50 tickets were sold than €35 tickets. If the tickets sold had a total value of €150 500, how many of each type of ticket was sold?

7 Olivia makes her own fruit and nut mix using dried fruit costing £8 per kilogram, and nuts costing £12 per kilogram. She made a total of 30 kg of fruit and nut mix costing £10.80 per kilogram. How many kilograms of each ingredient did she use?

Example 13 **Self Tutor**

I invest in oil shares which earn me 12% yearly, and in coal mining shares which earn me 10% yearly. If I invest $3000 more in oil shares than in coal mining shares, and my total yearly earnings amount to $910, how much do I invest in each type of share?

Let the amount I invest in coal mining shares be $\$x$.

Type of shares	*Amount invested* ($)	*Interest*	*Earnings* ($)
Coal	x	10%	10% of x
Oil	$(x+3000)$	12%	12% of $(x+3000)$
		Total	910

From the earnings column of the table,

$$10\% \text{ of } x + 12\% \text{ of } (x+3000) = 910$$
$$\therefore\ 0.1x + 0.12(x+3000) = 910$$
$$\therefore\ 0.1x + 0.12x + 360 = 910$$
$$\therefore\ 0.22x + 360 = 910$$
$$\therefore\ 0.22x = 550$$
$$\therefore\ x = \frac{550}{0.22}$$
$$\therefore\ x = 2500$$

$\therefore$ I invest $2500 in coal mining shares and $5500 in oil shares.

8 Canadian Airways shares pay a yearly return of 8%, while BankCorp shares pay 11%. Tom invests $1500 more on BankCorp shares than on Canadian Airways shares, and his total yearly earnings from both investments is $1020. How much did Tom invest in Canadian Airways shares?

9 Inka invested three times as much money in mining shares as she invested in telecommunications shares. Mining shares pay a yearly return of 9%, and telecommunications shares earn 7% yearly. Inka's yearly income from the shares is €816. Find how much money Inka invested in each type of share.

10 Bruce collects stamps. He has six times as many 10-cent stamps as 5-cent stamps, and he has some 20-cent stamps as well. Bruce has 72 stamps with a total value of $\$8.40$. How many of each stamp does he have?

11 Andy has invested in three companies: Baldish, Creweman, and Droners. Baldish shares pay 10% yearly, Creweman shares pay 9% yearly, and Droners shares pay 6% yearly. Andy has invested twice the amount of money in Creweman as in Baldish, and he has invested $\$40\,000$ in total. The yearly return from the share dividends is $\$3325$. How much did Andy invest in each company?

D LINEAR INEQUALITIES

For safety reasons, many rides at theme parks have height restrictions.

For example, it is common to see signs which say:

"You must be at least 120 cm tall to go on this ride."

We could write this requirement as height $\geqslant$ 120 cm or $H \geqslant 120$ cm.

This means that if a person has height $H = 120$, 120.5, 125, or 150 cm, then he or she can go on the ride.

All of these values satisfy the *inequality* $H \geqslant 120$ cm.

An **algebraic inequality** is a mathematical sentence which compares the values of two expressions. It will contain one of the symbols $>$, $<$, $\geqslant$, or $\leqslant$.

Symbol	*Meaning*	*Examples*
$>$	is greater than	$5 > 3$, $a > 7$, $2\frac{1}{2} > x$
$<$	is less than	$3 < 5$, $7 < a$, $x < 2\frac{1}{2}$
$\geqslant$	is greater than or equal to	$5 \geqslant 4$, $b \geqslant 20$, $\frac{4}{5} \geqslant y$
$\leqslant$	is less than or equal to	$4 \leqslant 5$, $20 \leqslant b$, $y \leqslant \frac{4}{5}$

For example, $5x - 2 > 7$ is a *linear inequality* which indicates that the value of the expression $5x - 2$ is greater than 7.

Notice in the examples above that if we interchange the LHS and RHS, then the **inequality sign** needs to be **reversed**.

Example 14 **Self Tutor**

Rewrite the following inequality with the variable on the LHS:

a $8 < x$ **b** $1.5 \geqslant y$

a $8 < x$ is written as $x > 8$ **b** $1.5 \geqslant y$ is written as $y \leqslant 1.5$

SOLUTIONS TO LINEAR INEQUALITIES

While a linear *equation* usually only has one solution, a linear *inequality* is usually true for many values of the variable.

For example:

$x = 2$ indicates that the only value which satisfies the equation is 2

$x > 2$ indicates that any real number greater than 2 will satisfy the inequality, and so there are infinitely many solutions.

We can display the solutions to an inequality using a **number line**.

For example, the solutions to $x > 2$ are shown alongside. The arrow indicates that all values of x to the right of 2 satisfy the inequality. The open circle indicates that 2 is not included.

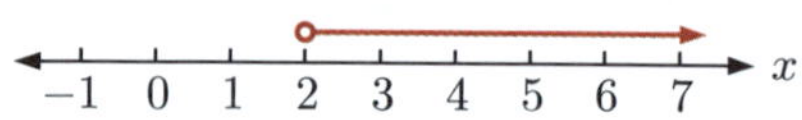

- If the value of the variable may be infinitely large, we use a ray $\rightarrow$ or $\leftarrow$ written above the number line.
- If the sign is $>$ or $<$, the ray has an open circle $\circ\!\!\rightarrow$ or $\leftarrow\!\!\circ$.
  This indicates that the value at the $\circ$ is **not included** in the solution.
- If the sign is $\geqslant$ or $\leqslant$, the ray has a closed circle $\bullet\!\!\rightarrow$ or $\leftarrow\!\!\bullet$.
  This indicates that the value at the $\bullet$ is **included** in the solution.

Example 15 **Self Tutor**

Draw separate number lines to illustrate the following inequalities:

a $a > 1$ **b** $a < -1$ **c** $a \geqslant 1$ **d** $a \leqslant -1$

a

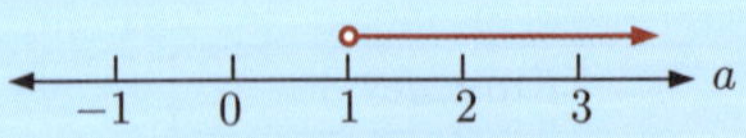

b

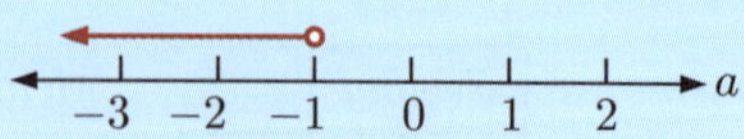

c

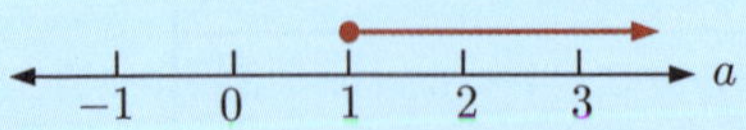

d

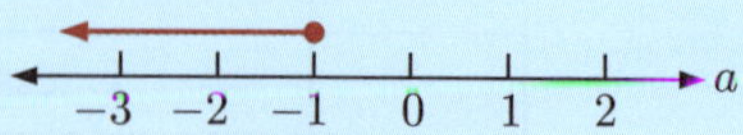

EXERCISE 6D

1 Write a mathematical sentence for:

a the speed S must not exceed 40 km/h
b the age A must be at least 18 years
c a is greater than 3
d b is less than or equal to -3
e d is less than 5
f -20 is greater than or equal to x
g 4 is less than y
h z is greater than or equal to 0.

2 Rewrite the following inequalities with the variable on the LHS:

- **a** $2 < x$
- **b** $5 > b$
- **c** $2\frac{1}{2} \leqslant c$
- **d** $-7 \geqslant d$
- **e** $-19 > a$
- **f** $-3 < p$

3 Draw separate number lines to illustrate the following inequalities:

- **a** $x > 2$
- **b** $a > -3$
- **c** $b \leqslant 2$
- **d** $m \geqslant -1$
- **e** $3 < x$
- **f** $2 > a$
- **g** $x > 3\frac{1}{2}$
- **h** $x \leqslant -1\frac{1}{4}$

4 Given that $a > 5$, decide if the following statements are true or false. Draw a number line to help you if necessary.

- **a** 7 is a possible value for a
- **b** -5.3 is a possible value for a
- **c** 5 is a possible value for a
- **d** 12.2 is *not* a possible value for a
- **e** -100.8 is *not* a possible value for a
- **f** 0 is a possible value for a

5 **a** On the same number line, illustrate the solutions to the inequalities $x > 3$ and $x \leqslant \frac{5}{2}$.

b State the only integer value of x which does not satisfy either of these inequalities.

E SOLVING LINEAR INEQUALITIES

In previous years we have compared a mathematical equation to a balanced set of scales. An inequality is like an **unbalanced** set of scales.

For example, consider the inequality $3x + 2 > 7$. We let a square represent x and a circle represent 1.

To solve a linear inequality, we need to maintain the *imbalance*.

We can carry out the same operation on both sides of the inequality sign, but we need to make sure this will not change its solutions.

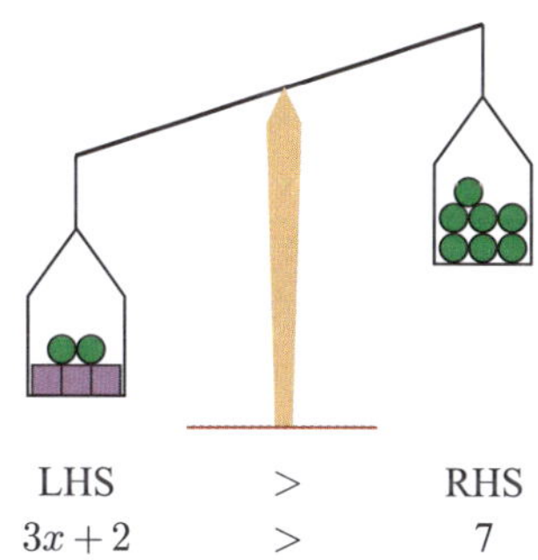

INVESTIGATION OPERATIONS WITH INEQUALITIES

In this Investigation we will begin with inequalities which we know are true, and then perform various operations to both sides of the inequalities. By checking whether the resulting inequalities are true or false, we will establish some rules for operating with inequalities.

PRINTABLE WORKSHEET

Adding or subtracting a positive or negative number

What to do:

1 Copy and complete the following table where T means *true* and F means *false*:

True inequality	*Operation*	*Working*	*Simplify*	*T or F*
$5 > 1$	add 3 to both sides	$5 + 3 > 1 + 3$	$8 > 4$	T
$11 > 7$	add 6 to both sides			
$8 > -3$	add 4 to both sides			
$-7 < -4$	add 5 to both sides			
$-6 < 5$	add 11 to both sides			
$8 > 6$	subtract 2 from both sides			
$9 > 7$	subtract 4 from both sides			
$2 > -1$	subtract 2 from both sides			
$-7 < -5$	subtract 5 from both sides			

2 Does *adding* a number to both sides of an inequality keep the inequality true, or does it make it false?

3 Does *subtracting* a number from both sides of an inequality keep the inequality true, or does it make it false?

Multiplying or dividing by a positive or negative number

What to do:

1 Copy and complete the following table, where T means *true* and F means *false*:

True inequality	*Operation*	*Working*	*Simplify*	*T or F*
$5 > 2$	multiply by 3	$5 \times 3 > 2 \times 3$	$15 > 6$	T
$-3 < -2$	multiply by 4			
$4 > 3$	multiply by (-3)			
$-2 > -5$	multiply by (-4)			
$7 > 3$	multiply by 2			
$-4 < -2$	multiply by 3			
$2 > -6$	multiply by (-2)			
$-5 < -3$	multiply by (-4)			
$6 > 3$	divide by 3			
$-6 > -8$	divide by 2			
$6 > 3$	divide by (-3)	$\frac{6}{-3} > \frac{3}{-3}$	$-2 > -1$	F
$-6 > -8$	divide by (-2)			
$7 < 14$	divide by 7			
$-10 < -5$	divide by 5			
$7 < 14$	divide by (-7)			

2 Does multiplying or dividing both sides of an inequality by a *positive* number keep the inequality true, or does it make it false?

3 Does multiplying or dividing both sides of an inequality by a *negative* number keep the inequality true, or does it make it false?

RULES FOR SOLVING INEQUALITIES

- If we **swap** the LHS and RHS, we **reverse** the inequality sign.
- If we **add to** or **subtract from** both sides, we **keep** the inequality sign.
- If we **multiply** or **divide** both sides by:
  - a **positive** number we **keep** the inequality sign
  - a **negative** number we **reverse** the inequality sign.

Example 16 — Self Tutor

Solve the following inequalities, and show their solutions on separate number lines:

a $a - 4 > 5$ **b** $3b \leqslant 9$ **c** $4 - 2x > 0$

a
$$a - 4 > 5$$
$$\therefore\ a - 4 + 4 > 5 + 4$$
$$\therefore\ a > 9$$

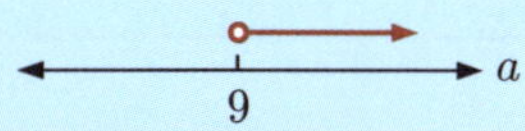

Check: We choose a value of a which satisfies $a > 9$, say $a = 10$.
If $a = 10$, we have $10 - 4 > 5$
$\therefore\ 6 > 5$ which is true. ✓

b
$$3b \leqslant 9$$
$$\therefore\ \frac{3b}{3} \leqslant \frac{9}{3}$$
$$\therefore\ b \leqslant 3$$

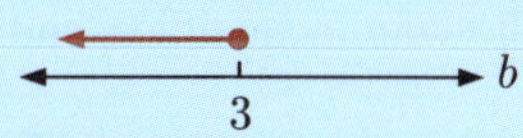

Check: If $b = 2$, we have $3 \times 2 \leqslant 9$
$\therefore\ 6 \leqslant 9$ which is true. ✓

c
$$4 - 2x > 0$$
$$\therefore\ 4 - 4 - 2x > 0 - 4$$
$$\therefore\ -2x > -4$$
$$\therefore\ \frac{-2x}{-2} < \frac{-4}{-2} \quad \{\text{reversing the sign}\}$$
$$\therefore\ x < 2$$

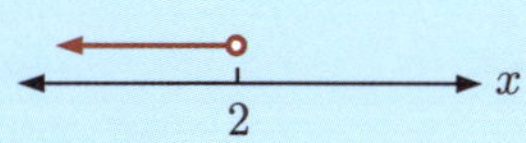

Check: If $x = 0$, we have $4 - 2 \times 0 > 0$
$\therefore\ 4 > 0$ which is true. ✓

EXERCISE 6E

1 Solve the following inequalities, and show their solutions on separate number lines:

a $a + 4 > 6$ **b** $3b \leqslant -9$ **c** $s - 4 < 2$

d $\frac{c}{5} < 2$ **e** $x - 8 \geqslant -3$ **f** $-4b > 16$

g $5 + t > 0$ **h** $5k < -30$ **i** $-\frac{m}{5} \geqslant 12$

2 Solve the following inequalities, and show their solutions on separate number lines:

a $2x+5>11$ **b** $5m-3\leqslant 17$ **c** $3a-2\geqslant 8$

d $4a+9<1$ **e** $7-2b<-3$ **f** $16+7s>2$

g $2a-4\leqslant 0$ **h** $12-5b>-3$ **i** $3b-1\geqslant 0$

j $5n+7<-3$ **k** $18-x>5$ **l** $11-4b\leqslant 4$

Example 17 **Self Tutor**

Solve $\frac{x}{4}-2\geqslant 3$, and show its solution on a number line.

$\frac{x}{4}-2\geqslant 3$

$\therefore \frac{x}{4}-2+2\geqslant 3+2$ {adding 2 to both sides}

$\therefore \frac{x}{4}\geqslant 5$

$\therefore \frac{x}{4}\times 4\geqslant 5\times 4$ {multiplying both sides by 4}

$\therefore x\geqslant 20$

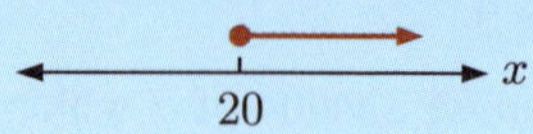

Check: If $x=32$, we have $\frac{32}{4}-2\geqslant 3$

$\therefore 6\geqslant 3$ which is true. ✓

3 Solve the following inequalities, and show their solutions on separate number lines:

a $\frac{x}{3}+1>4$ **b** $\frac{b}{5}-2\leqslant -3$ **c** $\frac{c}{4}+4\geqslant 8$

d $\frac{2x}{3}-2<4$ **e** $\frac{3x}{2}+5\geqslant -2$ **f** $1-\frac{x}{2}<4$

g $2-\frac{x}{4}\geqslant -3$ **h** $3-\frac{2x}{5}<2$ **i** $5-\frac{3x}{4}>-2$

4 Solve the following inequalities, and show their solutions on separate number lines:

a $\frac{a-3}{2}<6$ **b** $\frac{b+5}{4}\geqslant 1$ **c** $\frac{4c+3}{5}\leqslant -1$

d $\frac{4-a}{2}>5$ **e** $\frac{5-3x}{2}\leqslant -6$ **f** $\frac{3-2x}{3}\geqslant -1$

5 Solve the following inequalities by first interchanging the LHS and RHS:

a $5>4+x$ **b** $2\leqslant 6c+14$ **c** $-4>2b$

d $7\leqslant \frac{a}{3}$ **e** $3<\frac{d-2}{4}$ **f** $6\geqslant 3(p+2)$

6 Solve for x, and show the solution on a number line:

a $5x-3>3x+1$ **b** $2x+1\geqslant 4x+7$ **c** $8x+6<3x+1$

d $2x+7>7x+3$ **e** $6x+2\leqslant 3x-7$ **f** $x-11\leqslant 6x-1$

ACTIVITY 3 SIR COOL AND THE INEQUALITY CRUSADE

Click on the icon to help Sir Cool rid the world of inequality!

THE ADVENTURES OF SIR COOL

REVIEW SET 6A

1 Solve for x:

a $4x + 7 = -1$ **b** $\dfrac{5x+4}{3} = -2$

2 Solve for x:

a $2(x-7) - 3(8-x) = 9$ **b** $11 - x + 3(x-2) = x - 4$

3 Four times the result of subtracting 1 from a certain number, is equal to three times the number, add 2. Find the number.

4 Solve for x:

a $\dfrac{x+3}{5} = \dfrac{x-4}{2}$ **b** $\dfrac{4}{3x} = \dfrac{10}{7}$

5 Solve for x:

a $\dfrac{3x+5}{x-2} = -9$ **b** $\dfrac{x+1}{4} + \dfrac{3x}{8} = -2$

6 What number must be subtracted from both the numerator and denominator of $\frac{5}{9}$ to obtain a fraction equal to $\frac{3}{8}$?

7 Three years ago, Evan was one third of his sister's age. In a year's time, Evan's age doubled will match his sister's age. How old is Evan now?

8 Elaine has 88 coins in a jar. They are all 5-cent and 10-cent coins. The total value of the coins is \$7. How many of each type of coin does Elaine have?

9 Simon invests twice the amount in petroleum shares that he does in technology shares. Petroleum shares pay a yearly return of 12%, while technology shares pay 7% per year. Simon's yearly income from these shares is €744. How much money has Simon invested in each type of share?

10 Draw separate number lines to illustrate: **a** $x \geqslant 5$ **b** $3 > x$

11 Solve the following inequalities, and show their solutions on separate number lines:

a $3 - 2x \geqslant -4$ **b** $\dfrac{2x+3}{2} < -3$ **c** $3(x+2) - 1 \leqslant 1 - 4x$

12 **a** Solve the inequality $\dfrac{x}{5} - 3 > -5$ and display the solution on a number line.

b Solve the inequality $8 - 3(x+2) \geqslant 1$ and display the solution on a number line.

c Find the values of x which satisfy *both* inequalities.

REVIEW SET 6B

1 Solve for x:

a $\dfrac{5x+3}{4} = 7$ **b** $7 - 4x = 5x + 6$

2 Solve for x:

a $4-3(x-5)=9x+6$ **b** $6(3x-2)-x=2x+7$

3 At a theme park, Setha went on twice as many rides as Chenda, who went on three more rides than Dara. Together, they went on 25 rides. How many rides did Dara go on?

4 Solve for x:

a $\dfrac{x-3}{4}=\dfrac{2x}{5}$ **b** $\dfrac{3}{5x}=-\dfrac{7}{8}$

5 Solve for x:

a $\dfrac{x+9}{2x-1}=\dfrac{3}{4}$ **b** $\dfrac{5}{x}=\dfrac{7}{x-3}$ **c** $\dfrac{x}{5}-7=\dfrac{2x+3}{4}$

6 Write a fraction for which the sum of the numerator and denominator is 44, and the value of the fraction is $\frac{5}{6}$.

7 When Amanda was born, her mother was 27 years old. Amanda is now one quarter of her mother's age. How old is Amanda now?

8 A supermarket sells tuna in 90 g and 200 g tins. In one day, 112 tins of tuna were sold, with a total mass of 18 kg. How many tins of each type were sold?

9 Edward makes his own lolly bags to give away at a party. He mixes jelly fruits costing \$9 per kilogram and chocolate buttons costing \$12 per kilogram. How many kilograms of each lolly does Edward mix if he produces 15 kg of lolly bags costing \$11.20 per kilogram?

10 Draw separate number lines to illustrate:

a $a \geqslant -2$ **b** $b < -1$ **c** $c > 4\frac{1}{2}$

11 Solve the following inequalities:

a $3-2x \geqslant 0$ **b** $\dfrac{x}{5}-3<-1$ **c** $6(2x+1) \geqslant 6-13x$

12 **a** Solve the equation $\dfrac{3}{5}=\dfrac{2}{x}$.

b Henry used the following method to solve the inequality $\dfrac{3}{5}<\dfrac{2}{x}$:

$$\frac{3}{5}<\frac{2}{x}$$

$$\therefore \frac{3\times x}{5\times x}<\frac{2\times 5}{x\times 5} \quad \{\text{to achieve a common denominator}\}$$

$$\therefore 3x<10 \quad \{\text{equating numerators}\}$$

$$\therefore x<3\tfrac{1}{3} \quad \{\text{dividing both sides by 3}\}$$

i By checking the case when $x=-2$, show that Henry's solution is incorrect.

ii Explain the error which Henry made.

Chapter 7

Measurement

Contents:

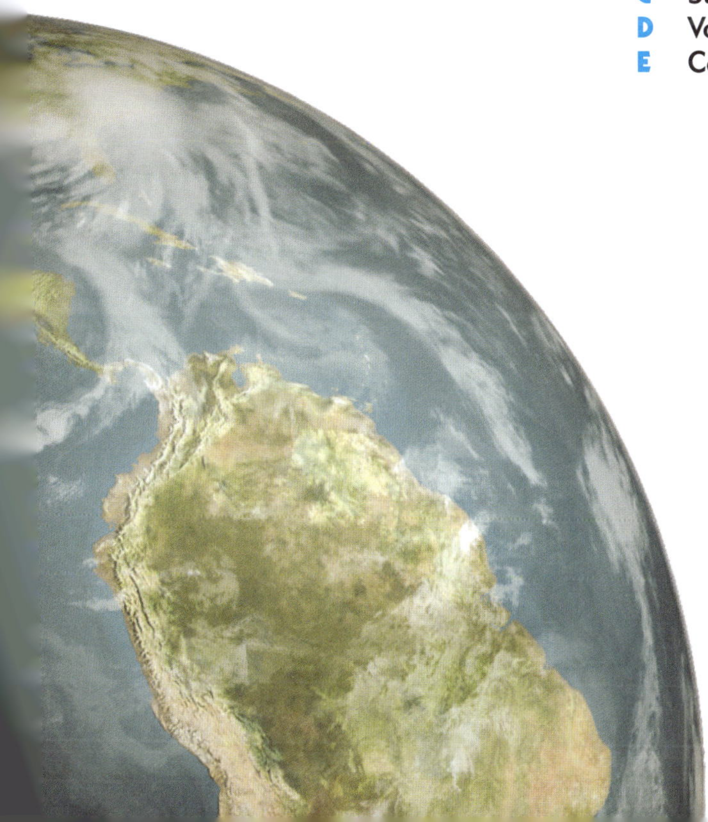

OPENING PROBLEM

Jacob wants to throw a water balloon at his brother. Once the balloon is filled with 1 litre of water, it takes the shape of a sphere.

Things to think about:

- **a** What is the *volume* of the filled balloon?
- **b** What is the *radius* of the water balloon?
- **c** What is the outer *surface area* of the water balloon?

The measurement of **length**, **area**, **volume**, and **capacity** is of great importance.

Constructing a building or a bridge, joining the circuits of a microchip, or rendezvousing in space to repair a satellite, all require the precise use of measurement.

A LENGTH AND PERIMETER

The base unit of length is the **metre** (m). Lengths are also commonly measured in:

- **millimetres** (mm) for example, the length of a bee
- **centimetres** (cm) for example, the width of your desk
- **kilometres** (km) for example, the distance between two cities.

The units of length are related as follows:

1 kilometre (km) = 1000 metres (m)
1 metre (m) = 100 centimetres (cm)
1 centimetre (cm) = 10 millimetres (mm)

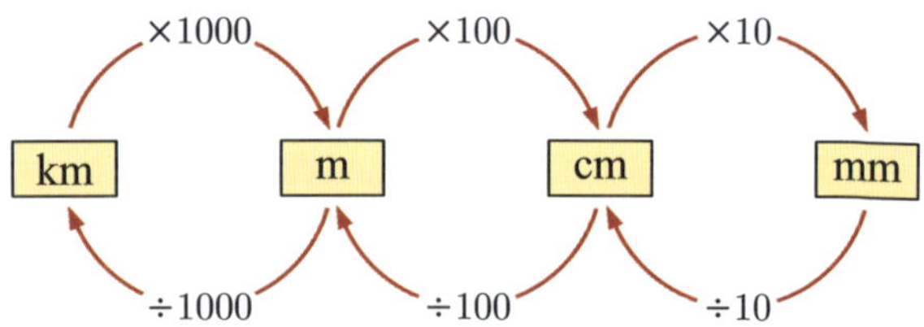

The **perimeter** of a closed figure is the total distance around its **boundary**.

POLYGONS

The perimeter of a **polygon** is the sum of the lengths of its sides.

For example:

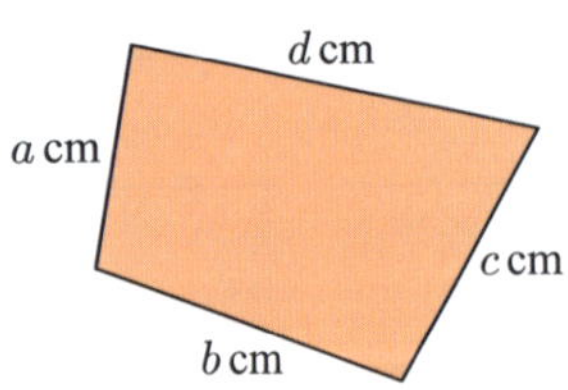

Perimeter $= (a + b + c + d)$ cm

CIRCLES

In the case of the circle, the perimeter is given the special name **circumference**.

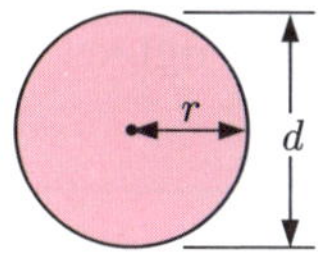

Circumference $C = 2\pi r$ or $C = \pi d$

ARC LENGTH

Consider the red arc shown. The angle at the centre of the circle is 127°.

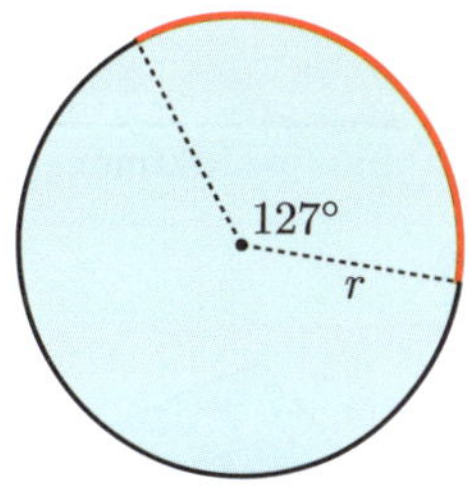

Since there are 360° in the whole circle, the arc makes up $\frac{127}{360}$ of the circle's circumference.

So, the length of the arc is $\left(\frac{127}{360}\right) \times 2\pi r$.

Arc length $l = \left(\frac{\theta}{360}\right) \times 2\pi r$

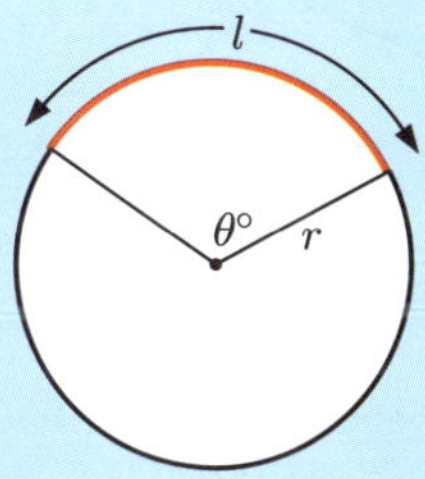

Example 1 **Self Tutor**

Find the perimeter of:

a

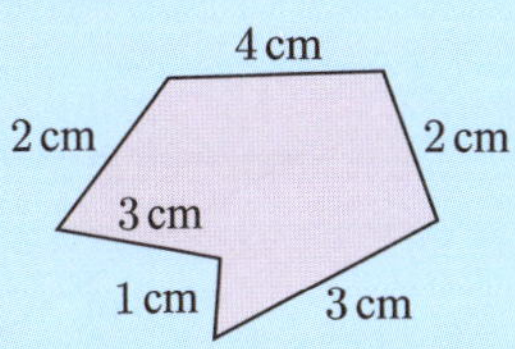

b

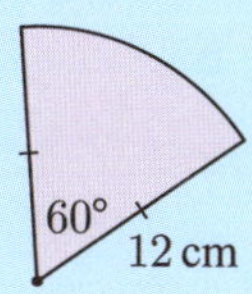

a Perimeter
$= (2 + 3 + 1 + 3 + 2 + 4)$ cm
$= 15$ cm

b Perimeter
$= (12 + 12)$ cm + length of arc
$= 24 \text{ cm} + \left(\frac{60}{360}\right) \times 2 \times \pi \times 12$ cm
≈ 36.6 cm

EXERCISE 7A

1 Find the perimeter of each of the following figures:

a

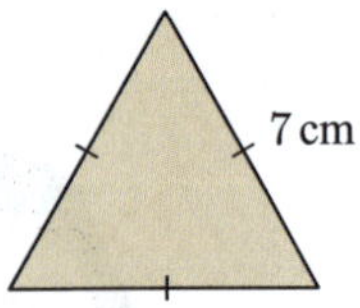

b

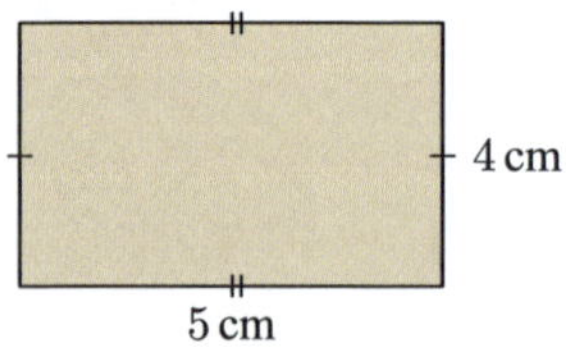

c

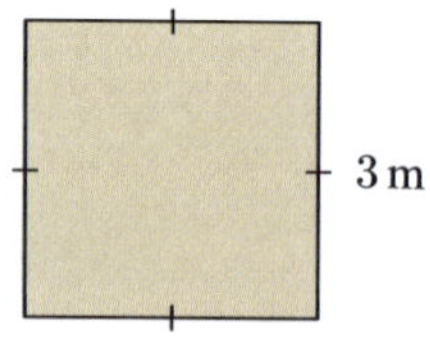

d

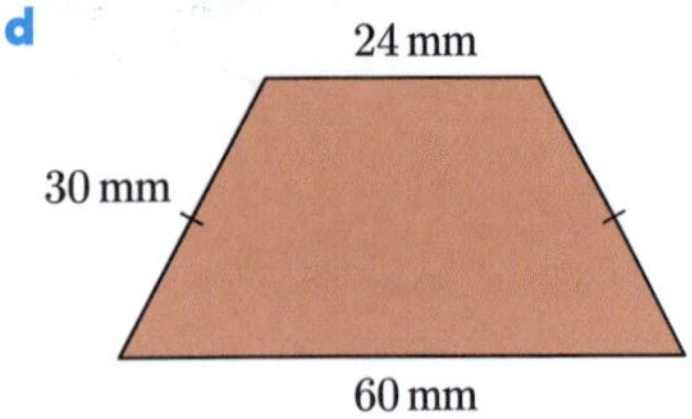

e

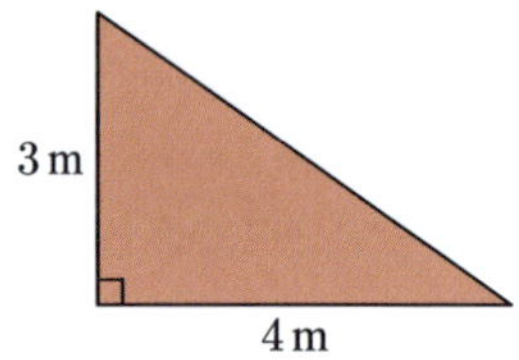

f

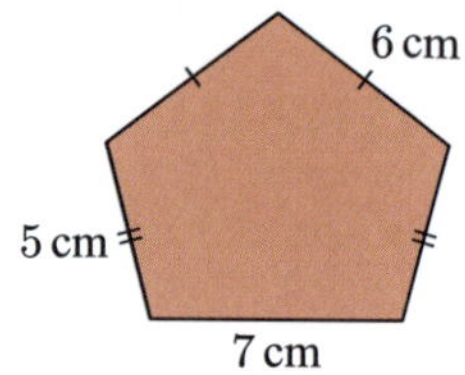

g

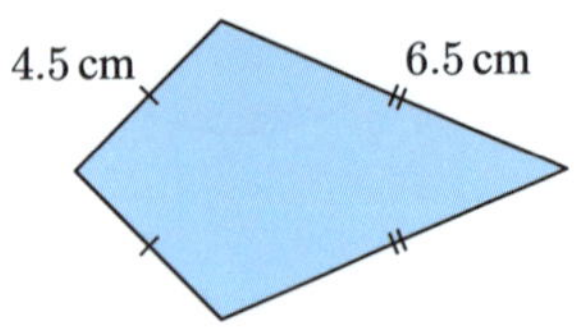

h

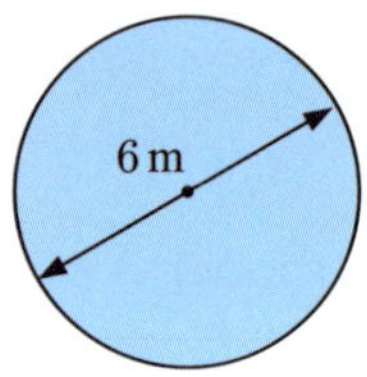

i

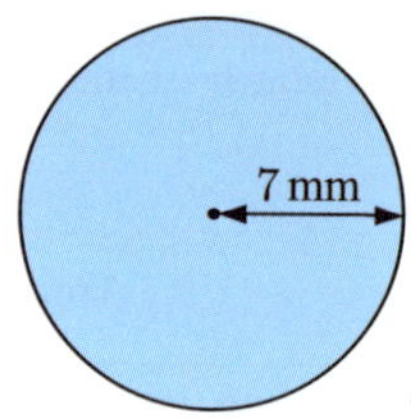

j

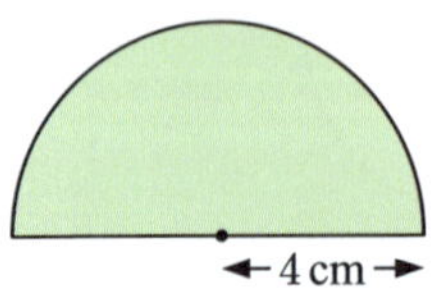

k

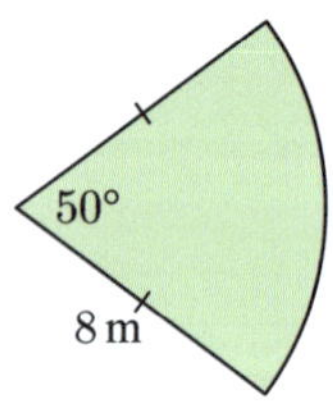

l

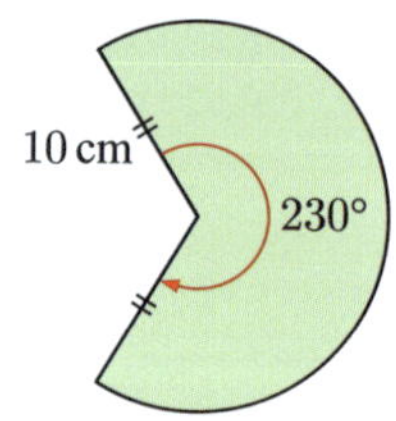

2 Find the perimeter of each of the following shapes:

a

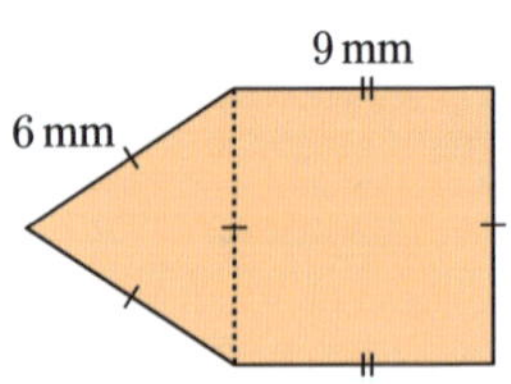

b

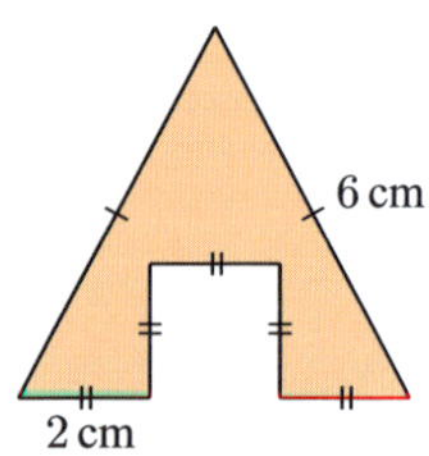

c

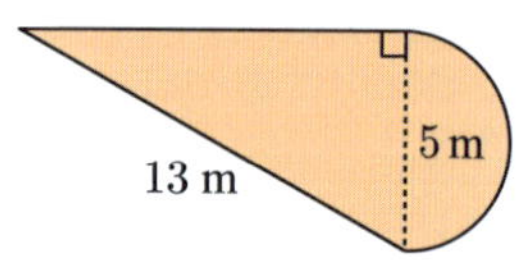

d

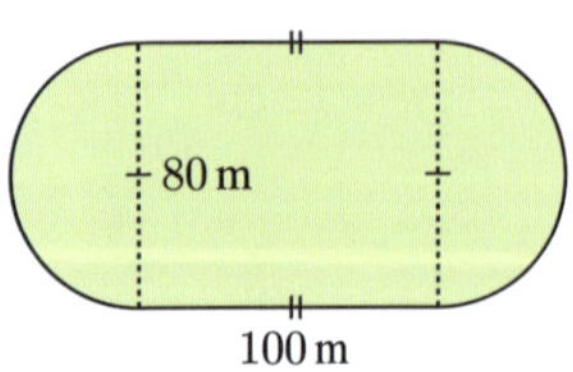

e

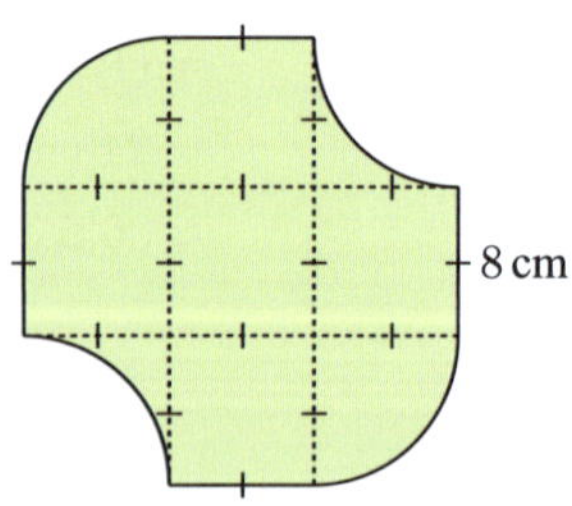

f

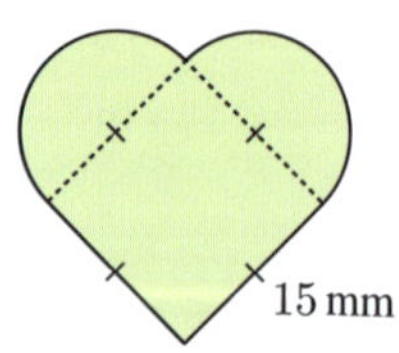

g

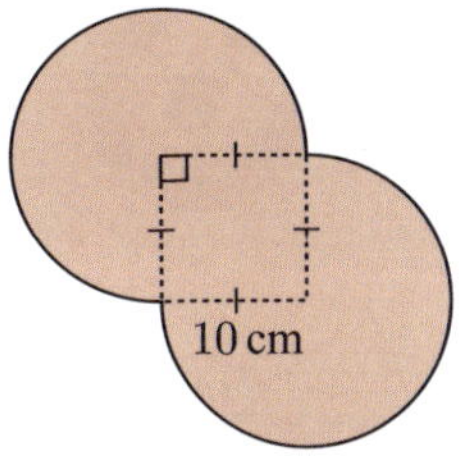

h

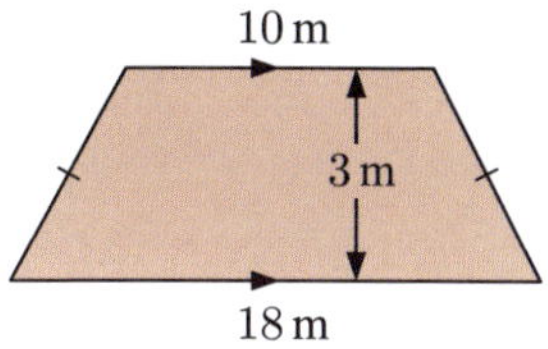

i

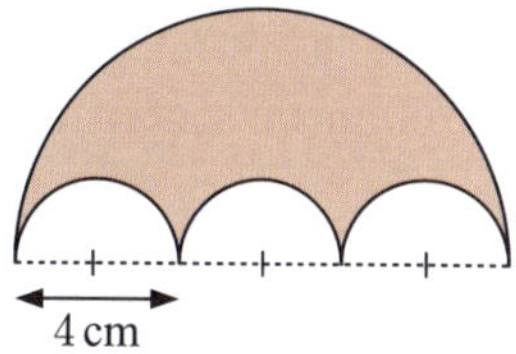

Example 2 **Self Tutor**

Write an expression for the perimeter P of:

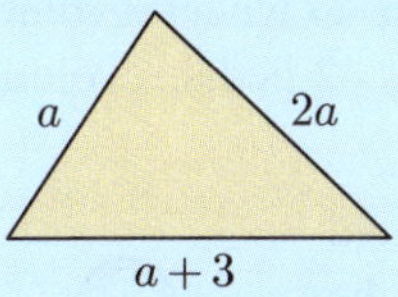

$P = a + 2a + (a + 3)$

$\therefore \; P = 4a + 3$

3 Write an expression for the perimeter P of:

a

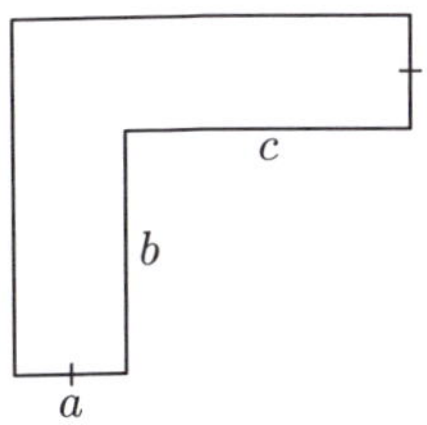

b

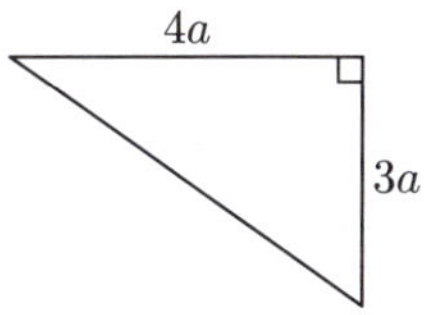

c

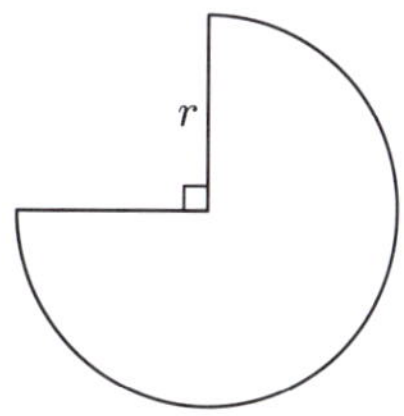

d

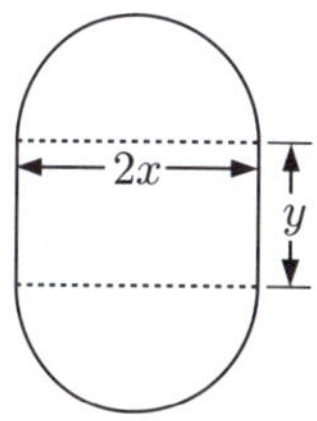

4 An Olympic sized boxing ring has dimensions 6.1 m by 6.1 m, and requires four ropes around each side, as shown. Find:

a the perimeter of the boxing ring

b the total length of the rope required

c the total cost of the rope if it costs \$9.90 per metre.

5 A rectangular room is 3.5 m by 4.8 m. It has a door 80 cm wide, and a window 1.3 m wide. A wallpaper frieze runs around the room just under the cornice, so it goes over the door and window. The frieze is repeated at eye level, and so is broken for the door and window at this height. Find the total length of frieze used in the room.

6 The horses on a merry-go-round complete 20 laps during each ride.

a Find, correct to 1 decimal place, the total distance travelled during a ride by:

i the black horse

ii the white horse.

b How much further does the black horse travel?

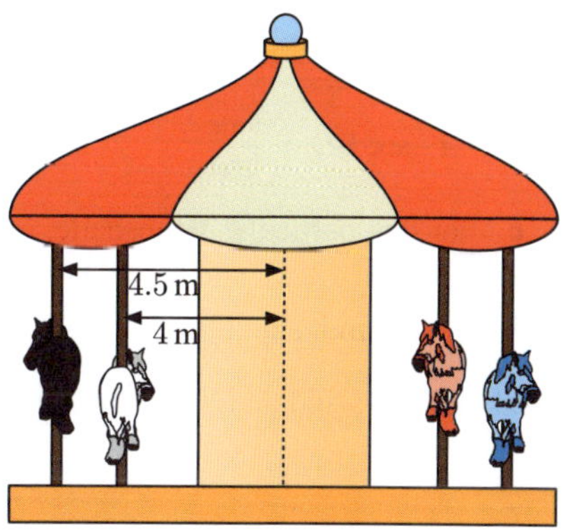

7 A soccer ball has a diameter of 24 cm. How many times must it roll along the ground, to travel the length of a 100 metre soccer field?

8 A lighting company produces conical lampshades from sectors of circles as illustrated. When the lampshades are made, lace is stitched around the circular base. Determine the total cost of the lace for 1500 lampshades if the lace costs €0.75 per metre.

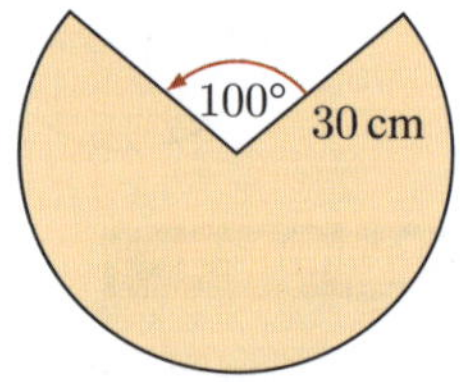

9 A cyclist rides a bicycle with a ratio of pedal revolutions to wheel revolutions of 1 : 6. The diameter of the bicycle wheel is 70 cm, and the cyclist averages 32 pedal revolutions per minute. How long will it take the cyclist to travel 40 km?

10 Three plastic pipes, each of diameter 10 cm, are held together by straps. Find the length of each strap to the nearest centimetre.

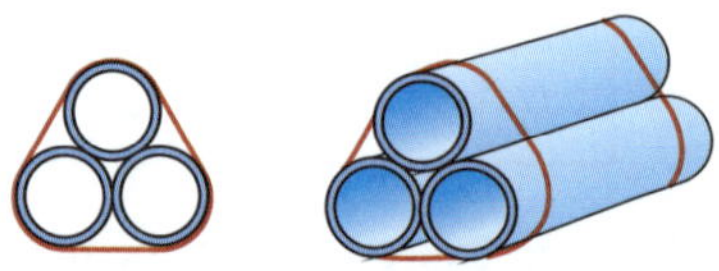

ACTIVITY 1 — CONSTRUCTING A LAMPSHADE

Your company has received an order for lampshades in the shape illustrated. A pattern is required from which the lampshades can be mass-produced.

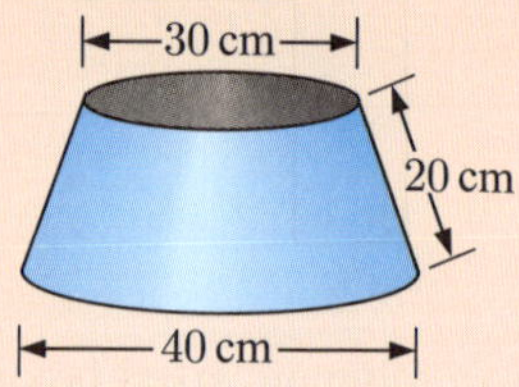

What to do:

1 On a piece of paper, draw two concentric circles. Cut out the *annulus* or washer shape shown.

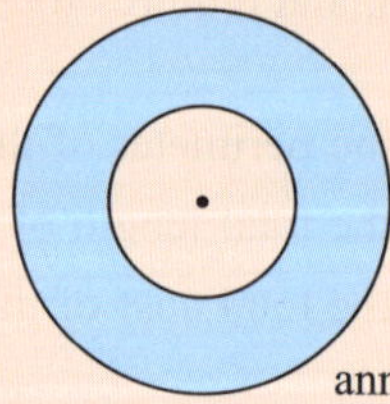

annulus

2 To make the lampshade, you will need to cut the annulus in two places. Suppose the cuts are [AB] and [CD].

Let $B\widehat{O}D$ be $\theta°$ and $OB = OD = x$ cm.

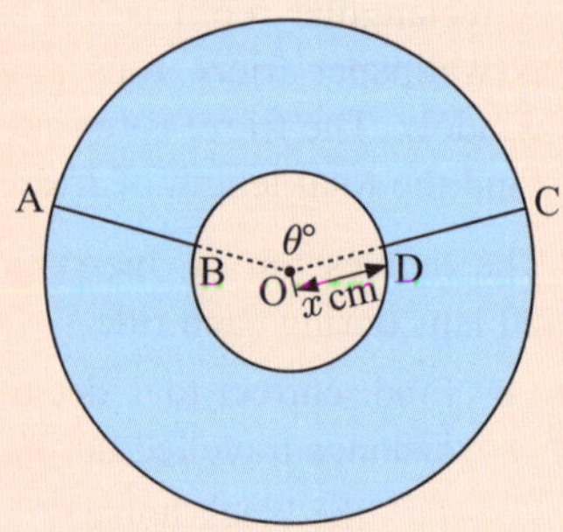

a To make the desired lampshade, what must the length of [CD] be?

b Show that:

i arc BD $= \dfrac{\theta\pi x}{180}$ **ii** arc AC $= \dfrac{\theta\pi(x+20)}{180}$.

c For this particular lampshade, explain why arc BD $= 30\pi$ and arc AC $= 40\pi$.

d Hence find x and θ.

3 Make a half-sized pattern of the lampshade using a large sheet of paper.

B AREA

All around us we see surfaces such as walls, ceilings, paths, and sporting fields. All of these surfaces have boundaries which define their shape.

> The **area** of a region is the amount of **surface** within its boundaries.
>
> The **area** of surface of a closed figure is measured in terms of the number of square units it encloses.

Area can be measured in **square millimetres**, **square centimetres**, **square metres**, and **square kilometres**. There is also another unit called the **hectare** (ha).

$$1\text{ mm}^2 = 1\text{ mm} \times 1\text{ mm}$$
$$1\text{ cm}^2 = 10\text{ mm} \times 10\text{ mm} = 100\text{ mm}^2$$
$$1\text{ m}^2 = 100\text{ cm} \times 100\text{ cm} = 10\,000\text{ cm}^2$$
$$1\text{ ha} = 100\text{ m} \times 100\text{ m} = 10\,000\text{ m}^2$$
$$1\text{ km}^2 = 1000\text{ m} \times 1000\text{ m} = 1\,000\,000\text{ m}^2 \text{ or } 100\text{ ha}$$

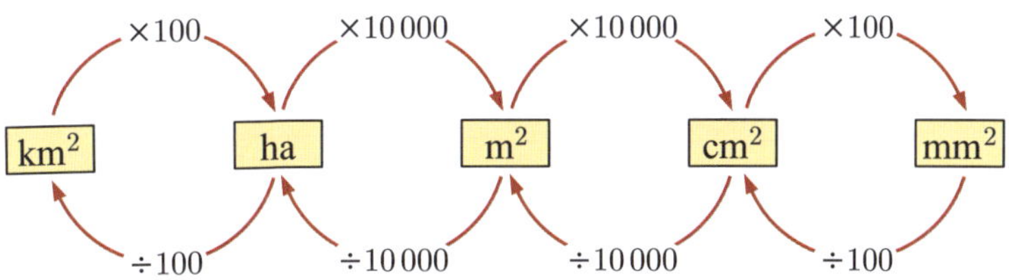

Example 3 — Self Tutor

Convert:

a 4.8 m^2 to cm^2 **b** 4500 m^2 to ha

a 4.8 m^2
$= (4.8 \times 10\,000)\text{ cm}^2$
$= 48\,000\text{ cm}^2$

b 4500 m^2
$= (4500 \div 10\,000)\text{ ha}$
$= 0.45\text{ ha}$

AREA FORMULAE

In previous years we have established the following formulae:

Shape	*Figure*	*Formula*
Rectangle	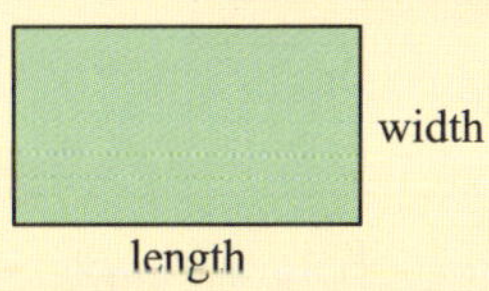	**Area = length × width**
Triangle	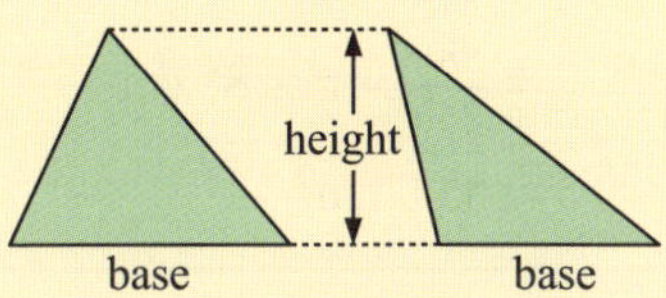	**Area = $\frac{1}{2}$ × base × height**

Shape	*Figure*	*Formula*
Parallelogram	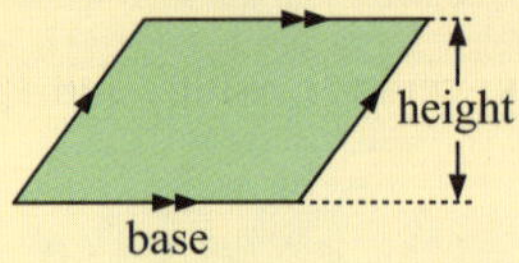	**Area = base × height**
Trapezium	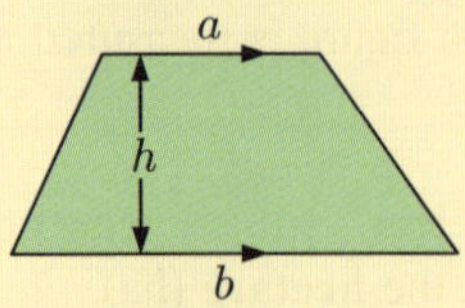	**Area** $= \left(\frac{a+b}{2}\right) \times h$
Circle	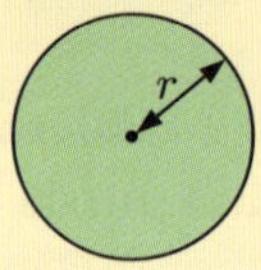	**Area** $= \pi r^2$

SECTORS

The shaded region is known as a **sector** of the circle.

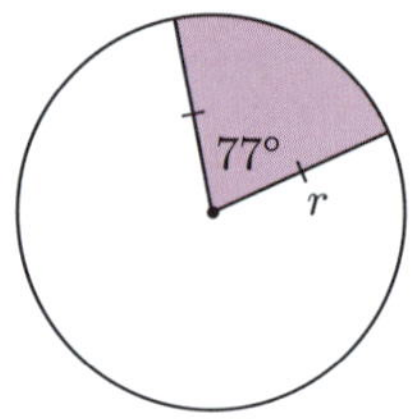

Since there are $360°$ in the whole circle, the shaded region makes up $\frac{77}{360}$ of the whole circle.

So, the area of the shaded region is $\left(\frac{77}{360}\right) \times \pi r^2$.

Area of sector $= \left(\frac{\theta}{360}\right) \times \pi r^2$

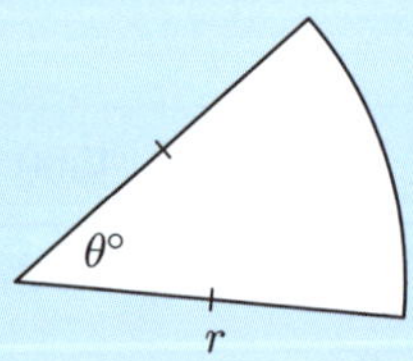

Example 4 **Self Tutor**

Find the area of each of the following figures:

a

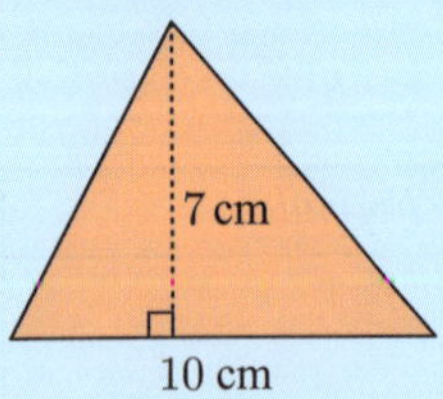

b

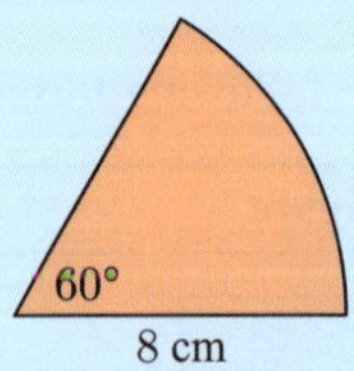

a Area $= \frac{1}{2}(\text{base} \times \text{height})$
$= \frac{1}{2} \times 10 \times 7 \text{ cm}^2$
$= 35 \text{ cm}^2$

b Area $= \frac{\theta}{360} \times \pi r^2 \text{ cm}^2$
$= \frac{60}{360} \times \pi \times 8^2 \text{ cm}^2$
$\approx 33.5 \text{ cm}^2$

Example 5 **Self Tutor**

Find the green shaded area in each of the following figures:

a

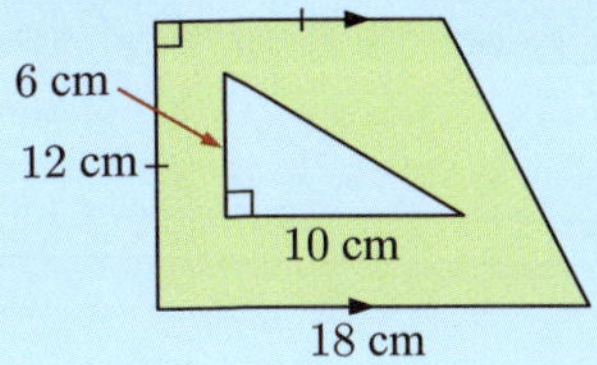

b

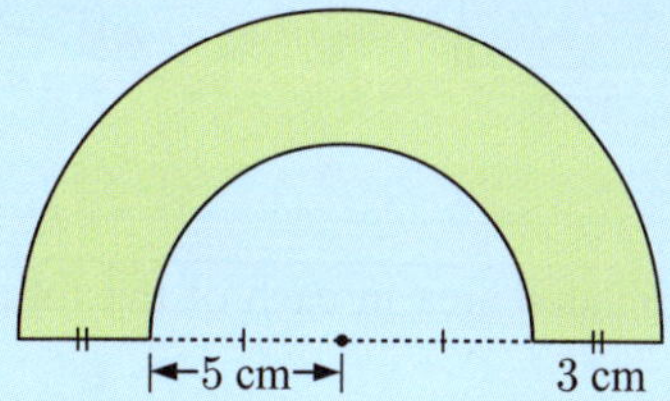

a Area = trapezium area − triangle area

$= \left(\frac{12+18}{2}\right) \times 12\text{ cm}^2 - \frac{1}{2} \times 10 \times 6\text{ cm}^2$

$= (15 \times 12 - 5 \times 6)\text{ cm}^2$

$= 150\text{ cm}^2$

b Area = area of large semi-circle − area of small semi-circle

$= \frac{1}{2} \times (\pi \times 8^2)\text{ cm}^2 - \frac{1}{2} \times (\pi \times 5^2)\text{ cm}^2$

$\approx 61.3\text{ cm}^2$

EXERCISE 7B

1 Convert:

a $38\,400\text{ m}^2$ to ha **b** 25.7 ha to m^2 **c** 3.5 m^2 to cm^2

d $500\,000\text{ cm}^2$ to m^2 **e** 18 cm^2 to mm^2 **f** 460 mm^2 to cm^2

g 9 km^2 to m^2 **h** 35 km^2 to ha **i** 500 ha to km^2

2 Find the area of the shaded region:

a

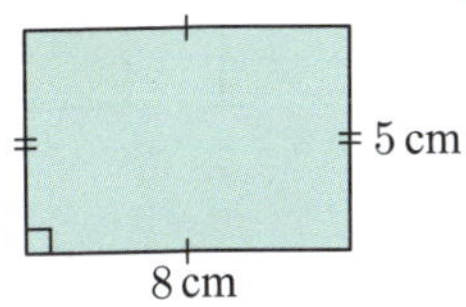

b

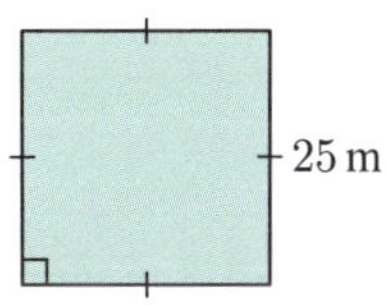

c

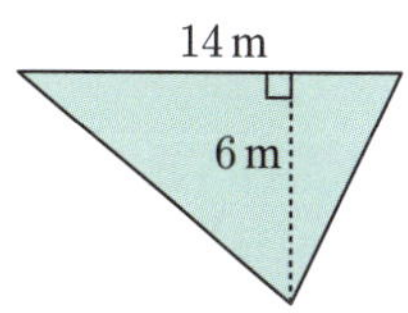

d

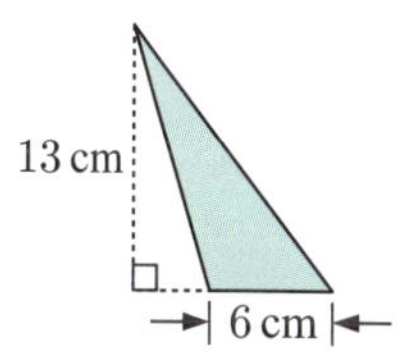

e

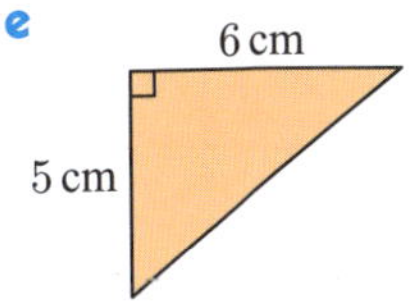

f

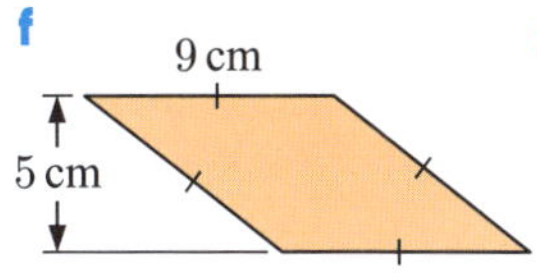

g

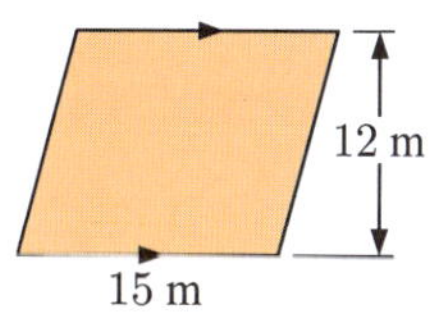

h

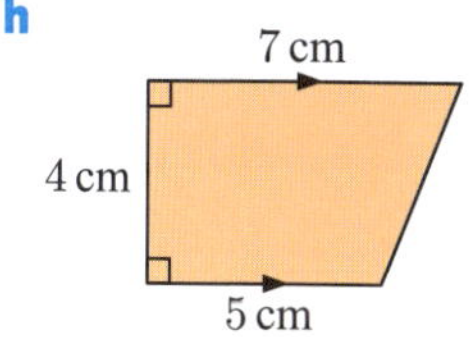

i

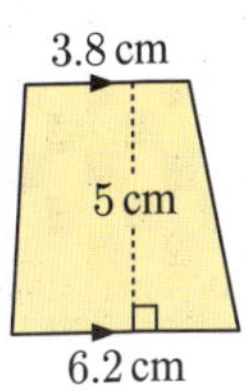

j

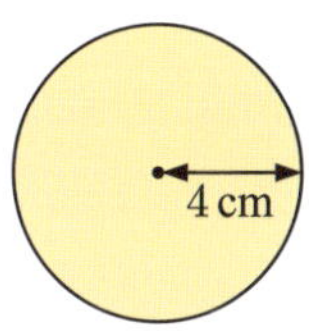

k

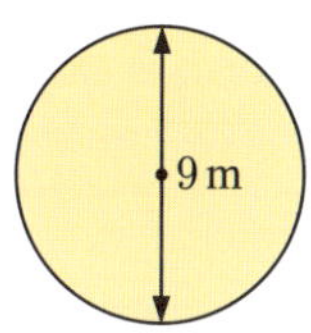

l

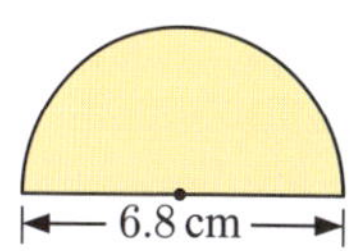

m

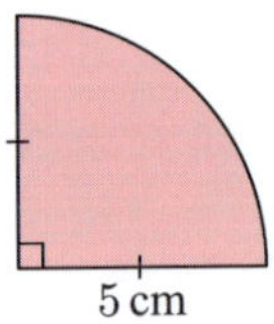

n

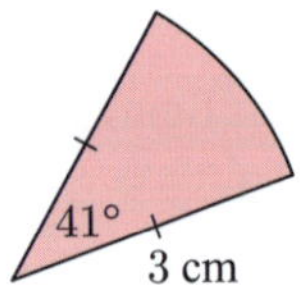

o

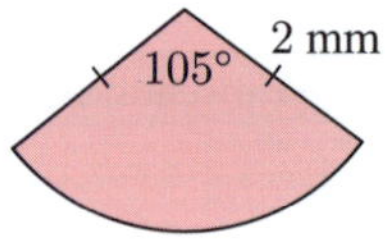

p

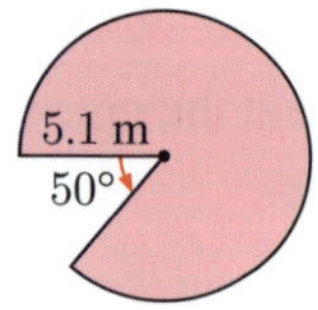

3 Find the shaded area in each of the following figures:

a

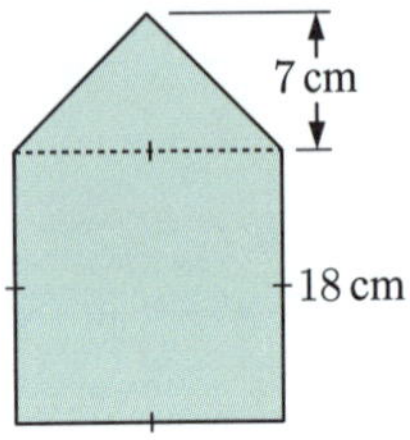

b

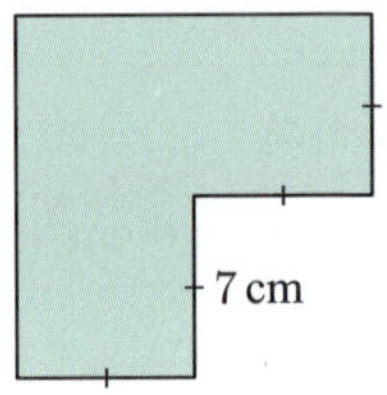

c

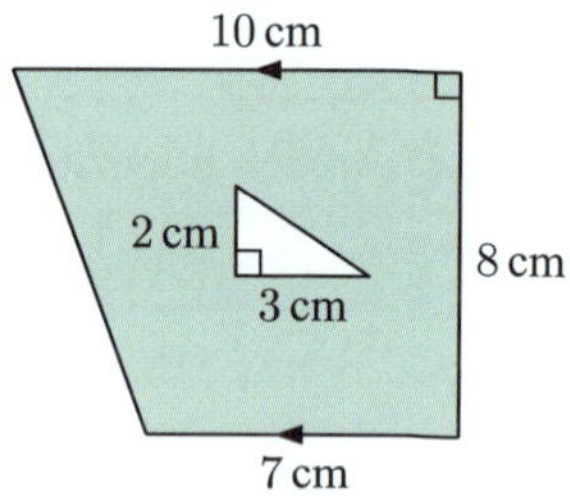

d

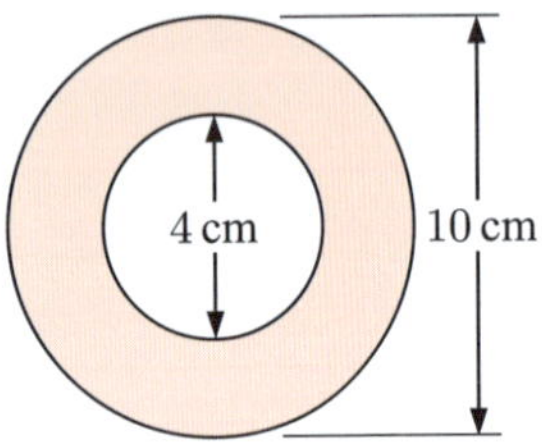

e

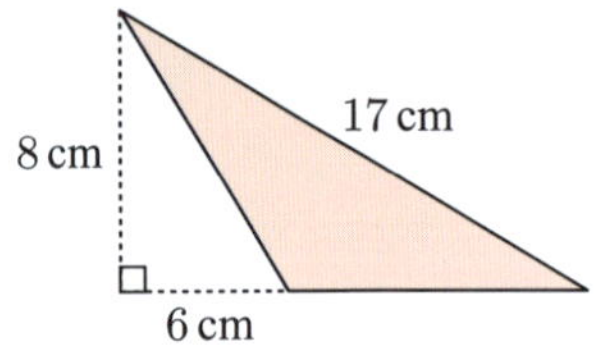

f

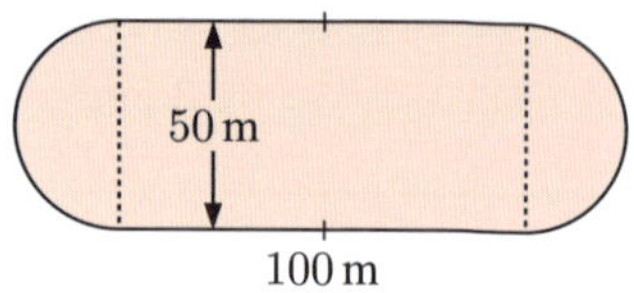

4 A door is in the shape of a rectangle surmounted by a semi-circle. The width of the door is 1.2 m, and the height of the door is 2.5 m. Find the total area of the door.

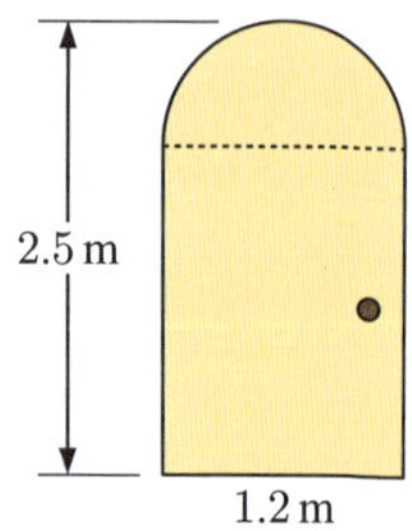

5 A restaurant uses square tables with sides of length 1.3 m, and round tablecloths with diameter 2 m. Determine the percentage of each tablecloth which overhangs its table.

6 A cropduster can carry 480 kg of fertiliser at a time. It is necessary to spread 50 kg of fertiliser per hectare. How many flights are necessary to fertilise a 2.4 km by 900 m rectangular property?

7 A chess board consists of 5 cm squares of blackwood for the black squares, and maple for the white squares. The squares are surrounded by an 8 cm wide blackwood border. Determine the percentage of the board which is made of maple.

8 Connie, Damien, Nick, and Uyen are sharing a pizza.

a Find, correct to 2 decimal places, the area of:

i Connie's slice **ii** Damien's slice

iii Nick's slice **iv** Uyen's slice.

b Find the sum of the areas in **a**. Check your answer by finding the area of the whole pizza.

9 A metal washer has an external diameter of 2 cm and an internal diameter of 1 cm.

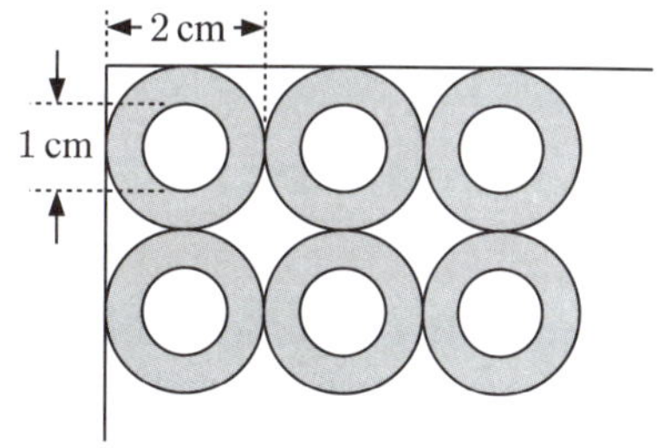

a How many washers can be cut from a sheet of steel 3 m by 1 m?

b If the metal left over was remelted and cast into a sheet of the same thickness and width 1 m, how many additional washers could be cut?

Example 6 Self Tutor

Write a formula for the area A of:

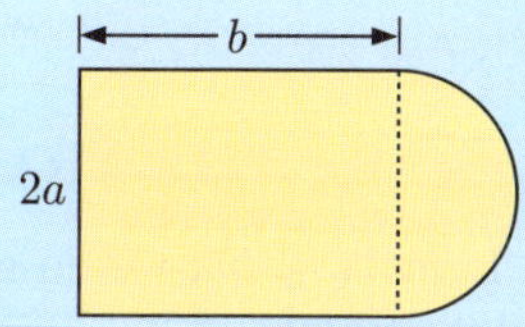

Area = area of rectangle + area of semi-circle

$\therefore\ A = 2a \times b + \frac{1}{2} \times \pi \times a^2$ {the radius of the semi-circle is a units}

$\therefore\ A = \left(2ab + \frac{\pi a^2}{2}\right)$ units^2

10 Write a formula for the area A of each of the following regions:

a

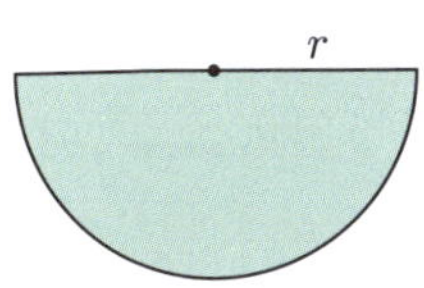

b

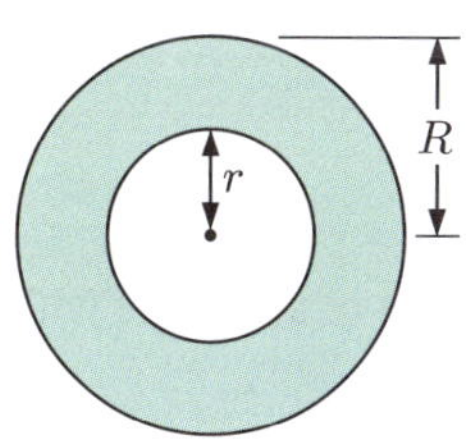

c

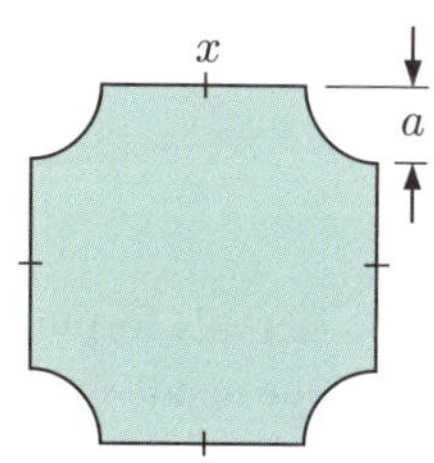

d

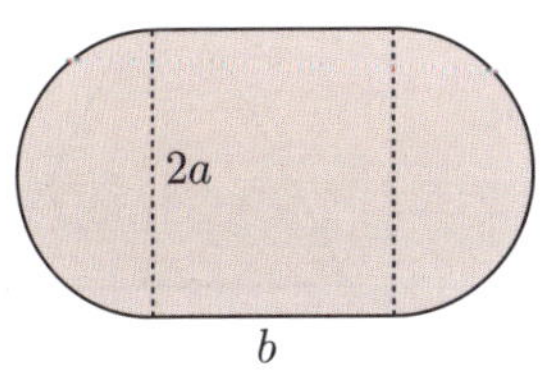

e

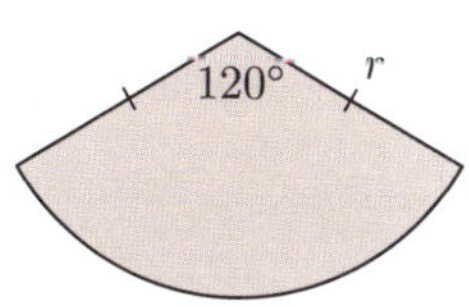

f

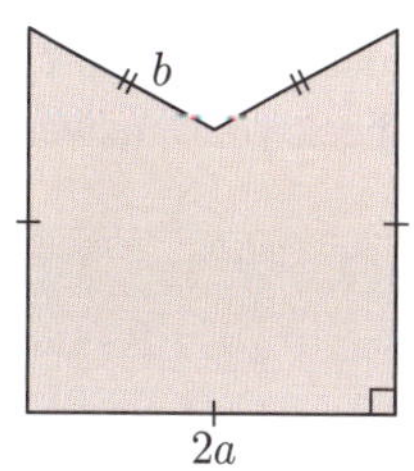

11 Find x:

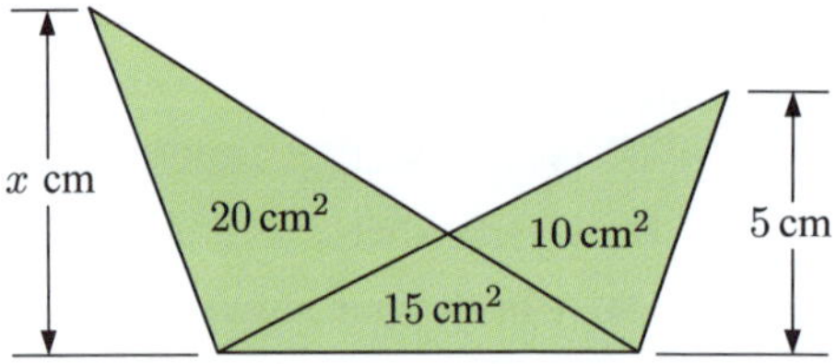

12 PQR is a right angled triangle. 3 semi-circles are constructed on each of its sides as shown.
Prove that the sum of the shaded areas is equal to the area of triangle PQR.

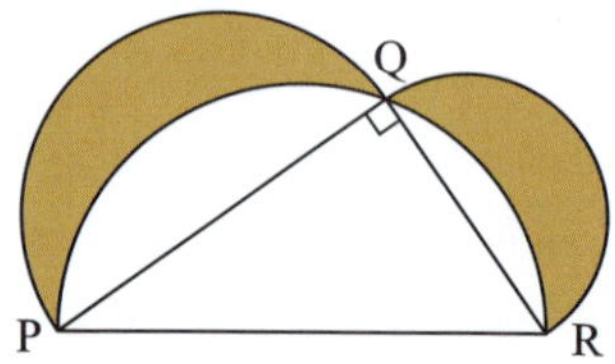

INVESTIGATION 1 — HERON'S FORMULA

Heron or **Hero** of Alexandria was an important geometer and engineer of the first century AD. He invented machines such as the steam turbine, but is most famous for devising a formula for calculating the area of a triangle given the lengths of its three sides:

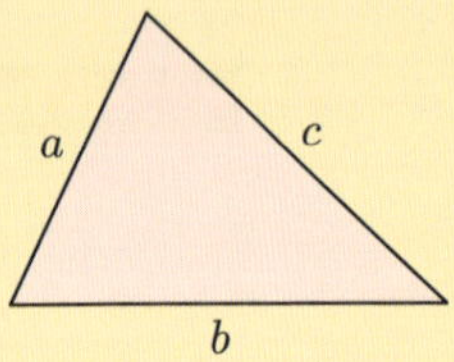

Area $\boldsymbol{A = \sqrt{s(s-a)(s-b)(s-c)}}$

where $\boldsymbol{s = \frac{a+b+c}{2}}$.

s is known as the **semi-perimeter** of the triangle.

The advantage of Heron's formula is that we do not need the perpendicular height of a triangle to find its area. We only need the lengths of its sides.

What to do:

1 **a** Find the area of the triangle shown without using Heron's formula.

b Use Heron's formula to check your answer.

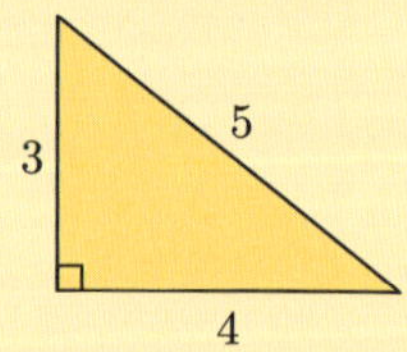

2 Use Heron's formula to find, correct to 1 decimal place, the area of:

a

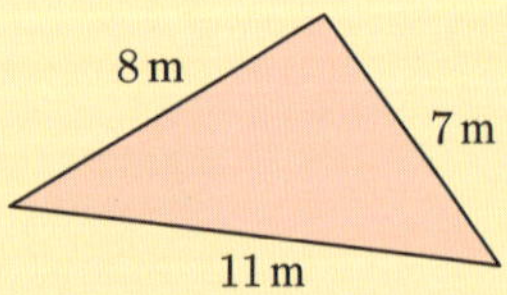

b

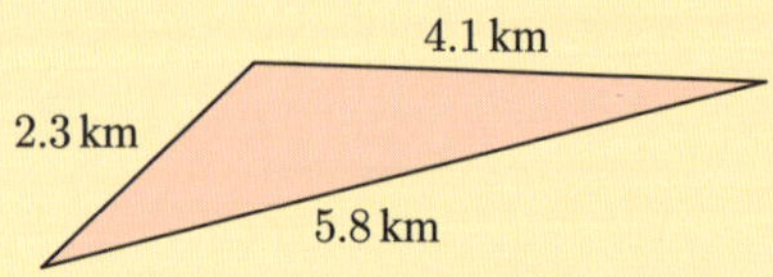

3 Use Heron's formula to find the area of a triangle with side lengths 5 m, 3 m, and 8 m. Comment on your result.

4 Use Heron's formula to find the area of an equilateral triangle with sides of length a cm.
Check your answer using the perpendicular height of the triangle.

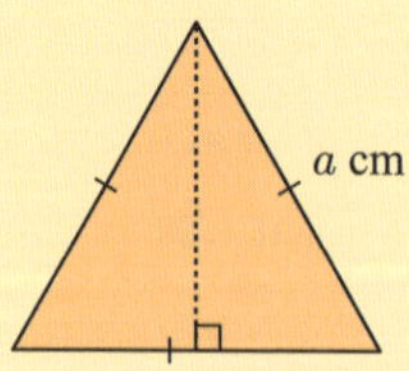

C SURFACE AREA

SOLIDS WITH PLANE FACES

The **surface area** of a three-dimensional solid with plane faces is the sum of the areas of the faces.

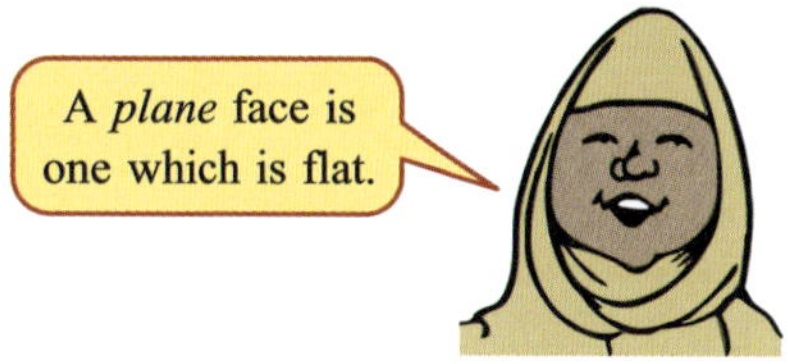

This means that the surface area is the same as the area of the **net** required to make the figure.

For example, the surface area of 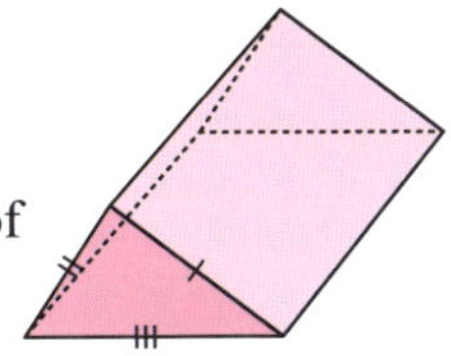 is the area of 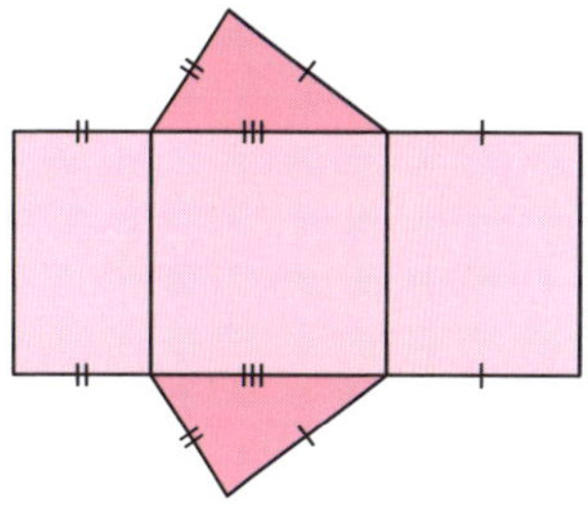

Example 7 — Self Tutor

Find the surface area of the rectangular prism:

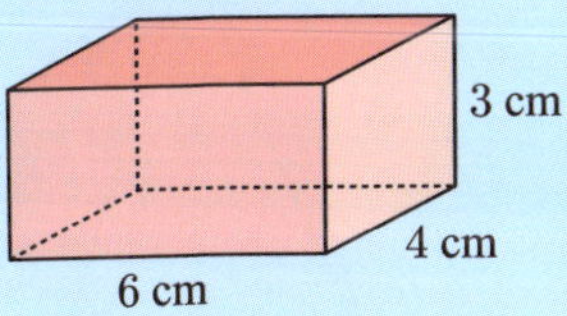

The figure has 2 faces which are $6 \text{ cm} \times 4 \text{ cm}$
2 faces which are $6 \text{ cm} \times 3 \text{ cm}$
and 2 faces which are $3 \text{ cm} \times 4 \text{ cm}$.

Total surface area
$= 2 \times 6 \times 4 \text{ cm}^2 + 2 \times 6 \times 3 \text{ cm}^2 + 2 \times 3 \times 4 \text{ cm}^2$
$= (48 + 36 + 24) \text{ cm}^2$
$= 108 \text{ cm}^2$

EXERCISE 7C.1

1 Find the surface area of each of the following rectangular prisms:

a
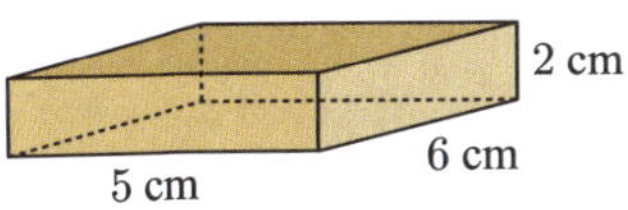

b
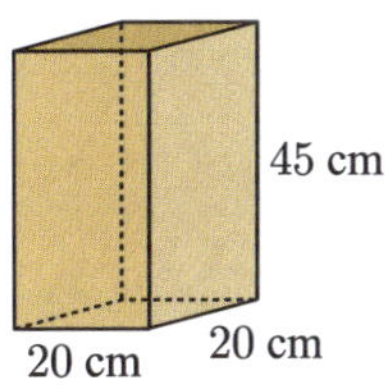

c
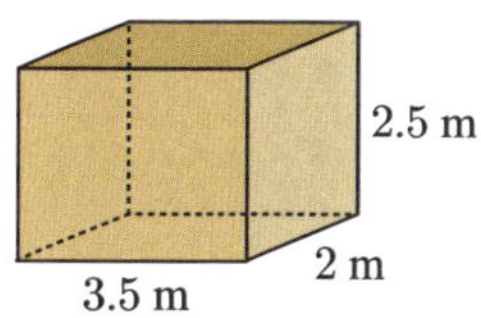

2 Find a formula for the total surface area A of a rectangular box a units long, b units wide, and c units high.

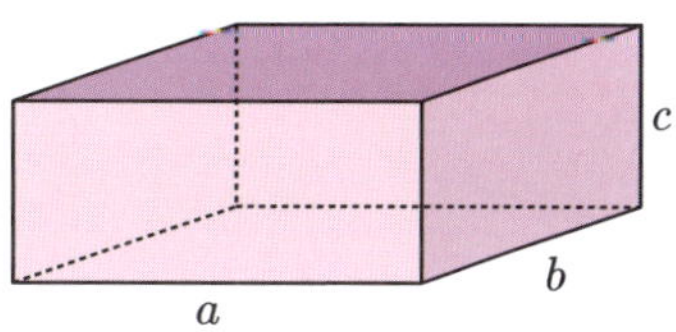

Example 8 — Self Tutor

Find the surface area of the square-based pyramid:

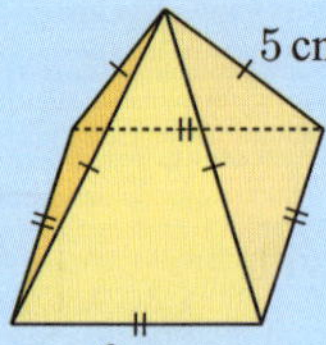

The figure has:

- 1 square base
- 4 triangular faces

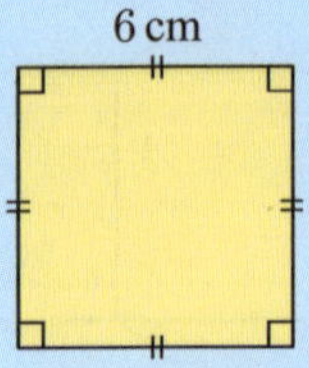

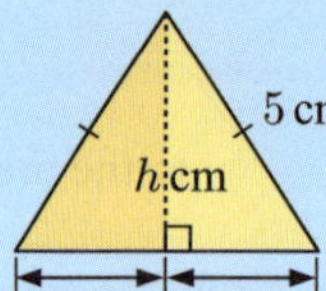

$$h^2 + 3^2 = 5^2 \quad \{\text{Pythagoras}\}$$
$$\therefore\ h^2 + 9 = 25$$
$$\therefore\ h^2 = 16$$
$$\therefore\ h = 4 \quad \{\text{as } h > 0\}$$

Total surface area $= 6 \times 6\ \text{cm}^2 + 4 \times (\frac{1}{2} \times 6 \times 4)\ \text{cm}^2$
$= (36 + 48)\ \text{cm}^2$
$= 84\ \text{cm}^2$

3 Find the surface area of each of the following square-based pyramids:

a

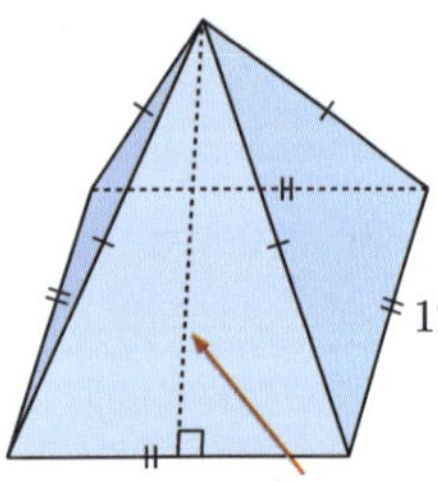

b

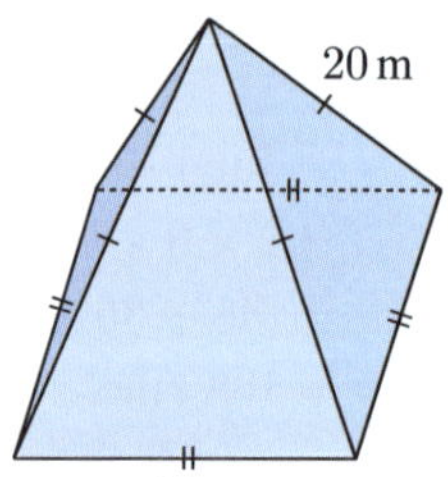

c

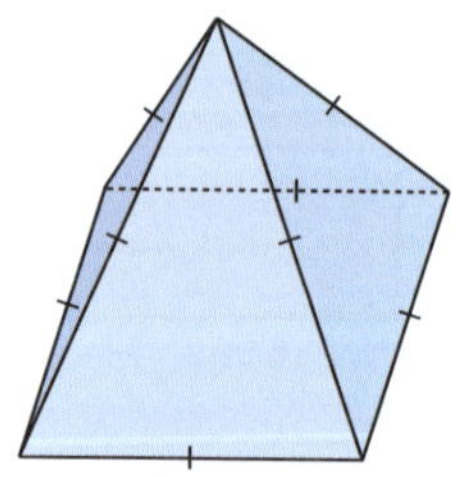

4 Find the surface area of each of the following solids:

a

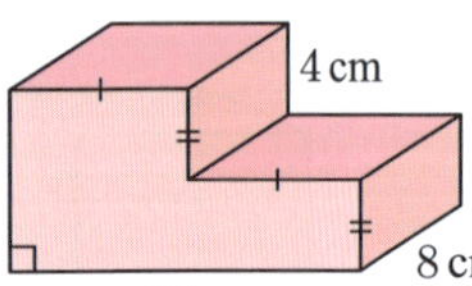

b

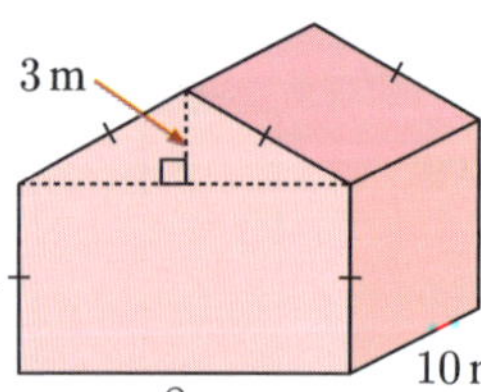

c

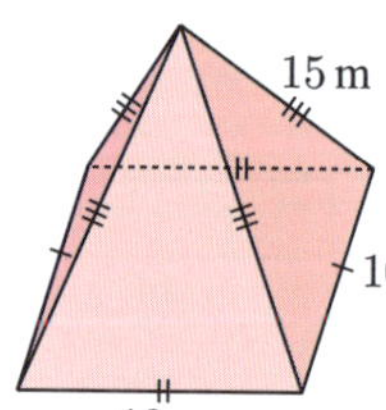

d

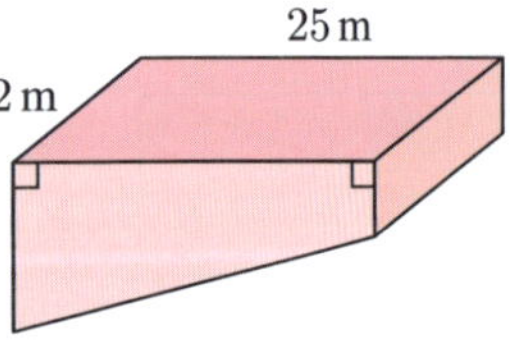

e

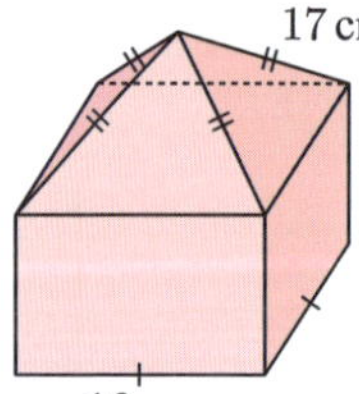

f

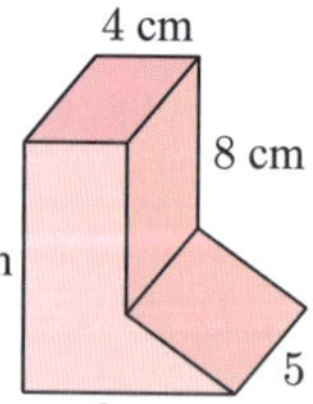

5 Including the floor, find the amount of material required to make this tent.

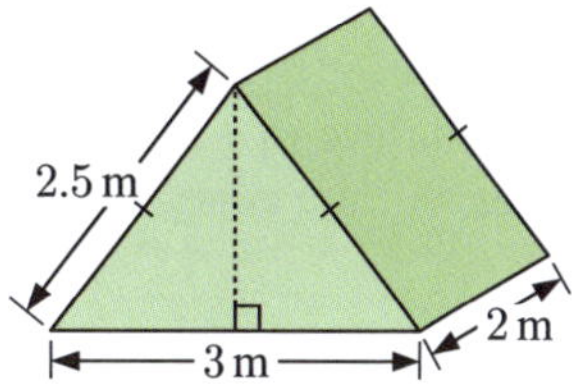

6 All of the edges of an eight-sided die are 2 cm long. Find the surface area of the die.

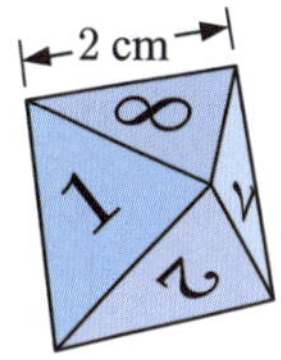

7 An aviary has the shape of a hexagonal prism. Find the area of netting required to cover the aviary, including the floor.

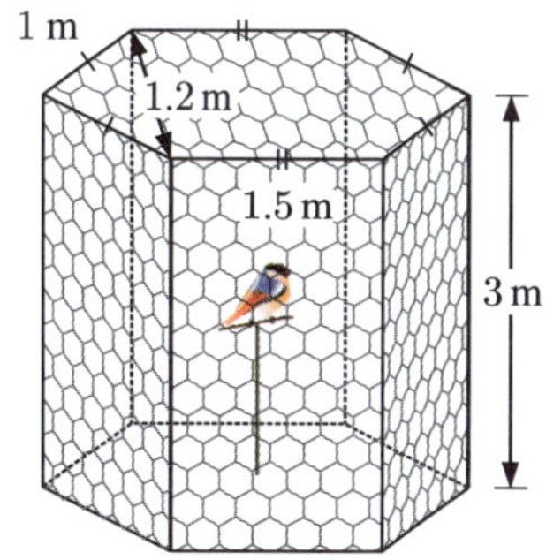

CYLINDERS

The cylinder shown has no top or bottom. If the cylinder is cut, opened out, and flattened, it takes the shape of a rectangle.

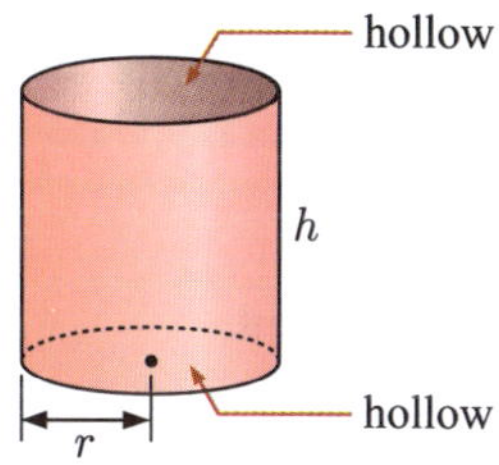

The length of the rectangle is the circumference of the cylinder.

The width of the rectangle is the height of the cylinder.

$\therefore$ for a hollow cylinder, the outer surface area

$$A = \text{area of rectangle}$$
$$= \text{length} \times \text{width}$$
$$= 2\pi rh$$

DEMO

Hollow cylinder (no ends)	**Open can** (one end)	**Solid cylinder** (two ends)
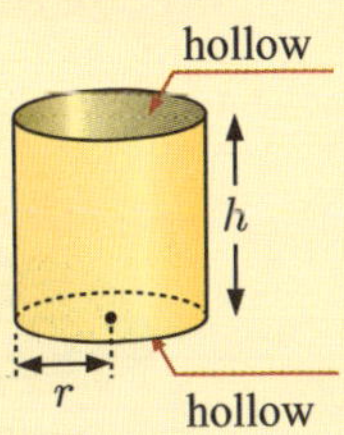	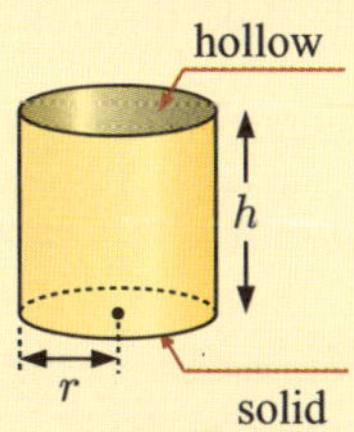	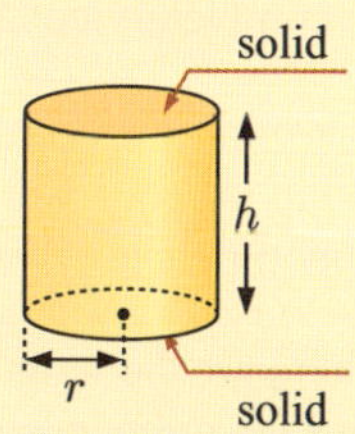
$A = 2\pi rh$	$A = 2\pi rh + \pi r^2$	$A = 2\pi rh + 2\pi r^2$

CONES

The curved surface of a cone is made from a sector of a circle. The radius of the sector is equal to the slant height s of the cone. The arc length AB of the sector is equal to the circumference of the base.

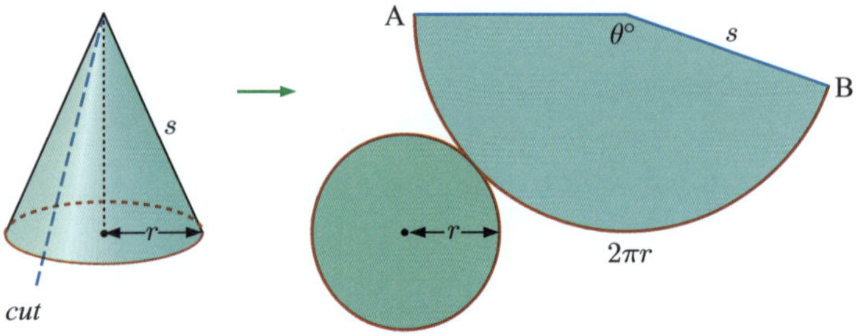

$$\text{arc AB} = 2\pi r$$

$$\therefore \left(\frac{\theta}{360}\right) 2\pi s = 2\pi r$$

$$\therefore \frac{\theta}{360} = \frac{r}{s}$$

$\therefore$ the area of the curved surface

$= $ the area of the sector

$= \left(\frac{\theta}{360}\right) \pi s^2$

$= \frac{r}{s} \times \pi s^2$

$= \pi r s$

Hollow cone (no end)	**Solid cone** (closed end)
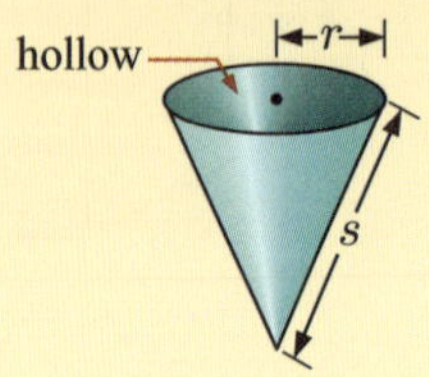 $A = \pi r s$	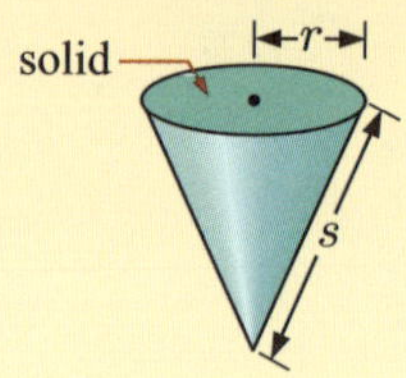 $A = \pi r s + \pi r^2$

SPHERES

INVESTIGATION 2 — SURFACE AREA OF A SPHERE

You will need:

a solid sphere such as an orange or a foam ball, cord, nail, scissors.

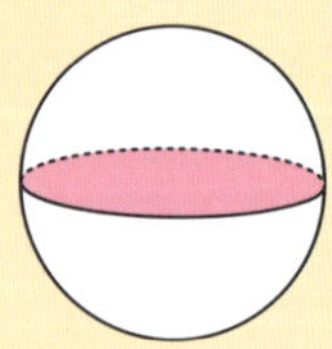

What to do:

1. Cut your sphere in half to obtain two hemispheres.
2. Insert a nail in the centre of the flat circular surface of one hemisphere. Wind a length of cord around the nail in a spiral until the flat surface is completely covered. Cut the length of cord required and put it aside.

3. Now insert the nail in the centre of the curved surface of the hemisphere. Wind the cord around the nail until the curved surface of the hemisphere is completely covered.

4. Compare the lengths of the cord in both cases. Is the area of the curved surface twice the area of the flat circular surface?

From the **Investigation**, you should have found that the curved surface of a hemisphere has twice the area of a circle with the same radius.

So, the area of the curved surface of a hemisphere $= 2\pi r^2$.

This leads to the formula for the surface area of a sphere:

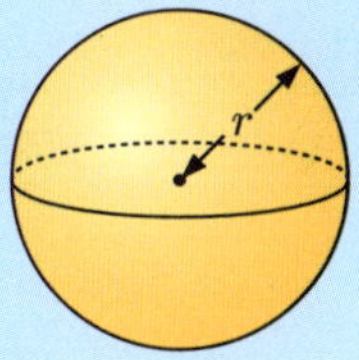

Sphere

$$A = 4\pi r^2$$

The mathematics required to prove this formula is beyond the scope of this course.

Example 9 — Self Tutor

Find the surface area of each of these solid figures:

a

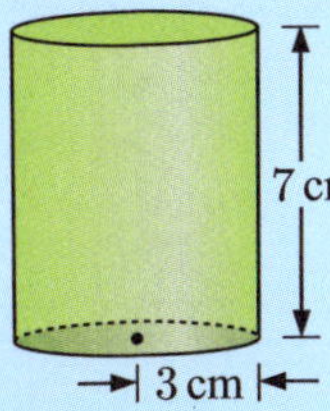

b

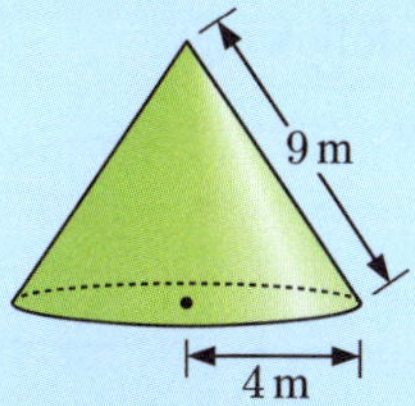

c

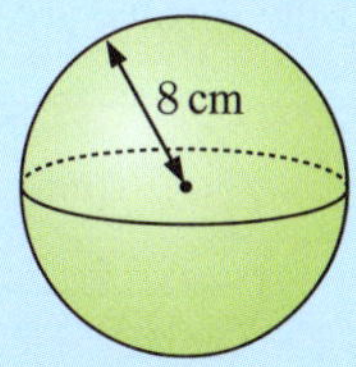

a Surface area $= 2\pi rh + 2\pi r^2$
$= (2 \times \pi \times 3 \times 7 + 2 \times \pi \times 3^2)$ cm^2
≈ 188 cm^2

b Surface area $= \pi rs + \pi r^2$
$= (\pi \times 4 \times 9 + \pi \times 4^2)$ m^2
≈ 163 m^2

c Surface area $= 4\pi r^2$
$= 4 \times \pi \times 8^2$ cm^2
≈ 804 cm^2

EXERCISE 7C.2

1 Find the outer surface area of each of the following cylinders:

a solid

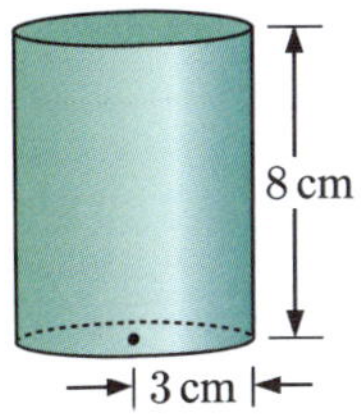

b can (no top)

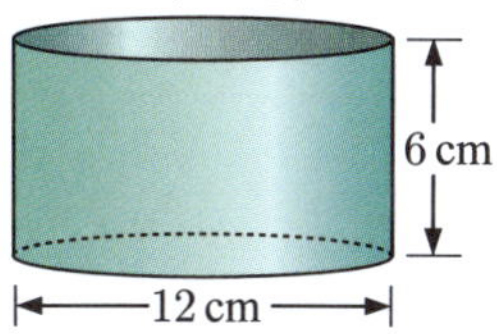

c solid

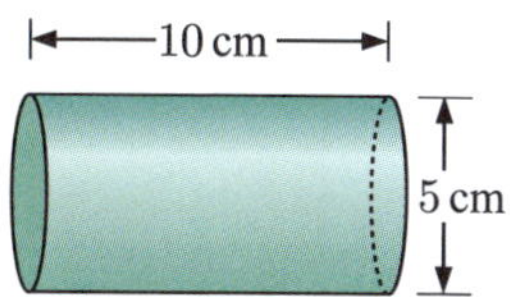

d well (no top)

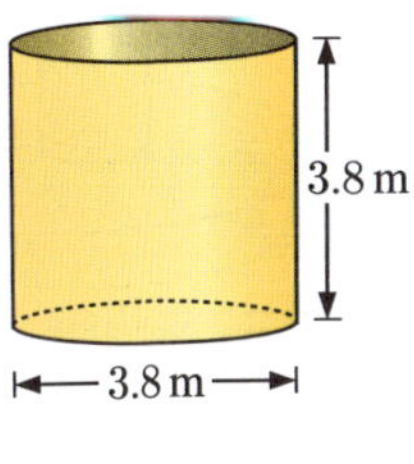

e solid

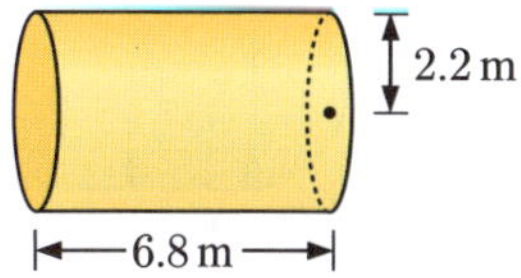

f hollow throughout

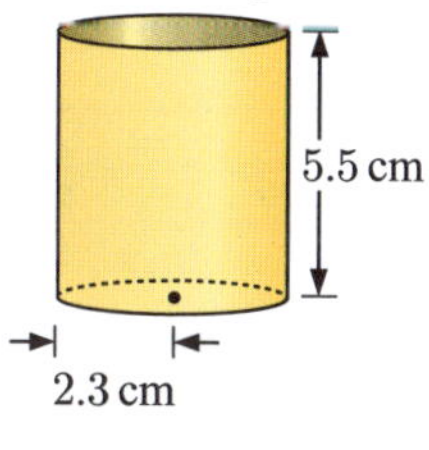

2 Find the outer surface area of each of the following cones:

a solid

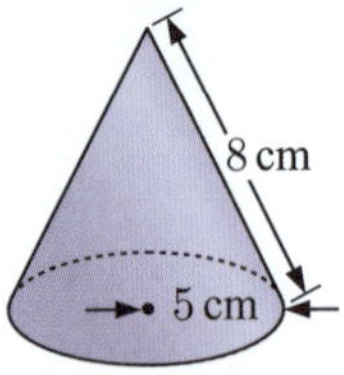

b hollow

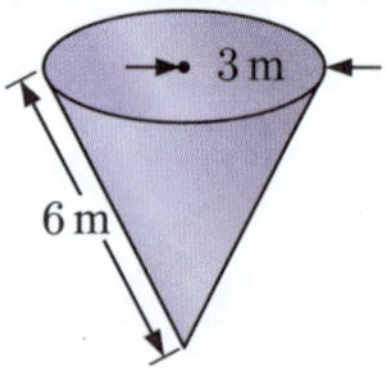

c solid

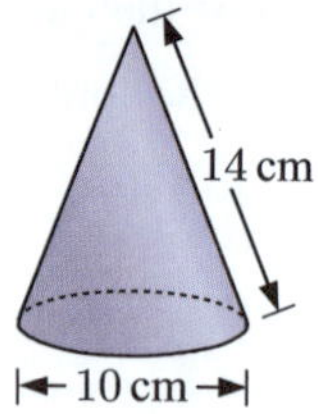

d hollow

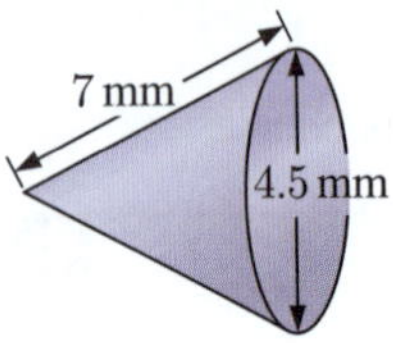

e solid

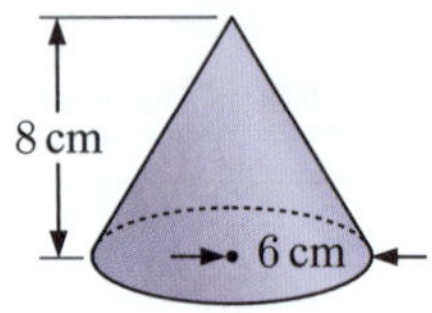

f hollow

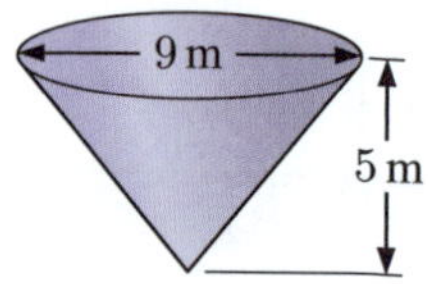

3 Find the surface area of each of the following:

a

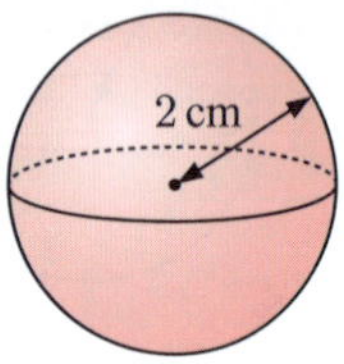

b

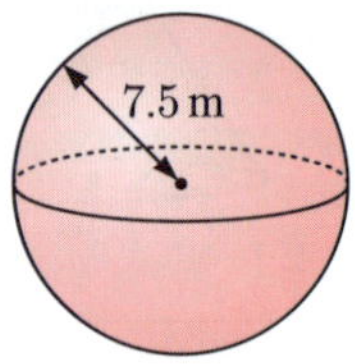

c

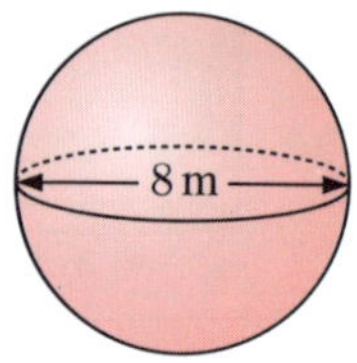

d

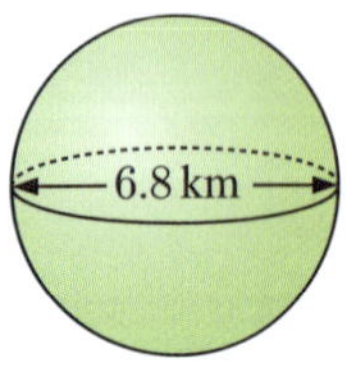

e solid

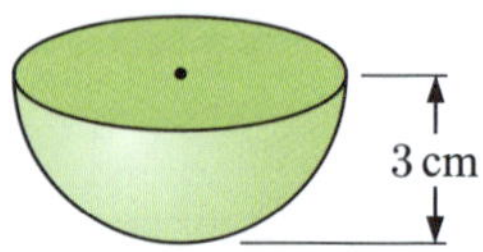

f solid

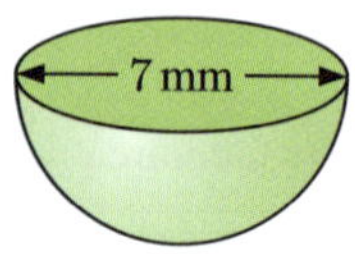

4 Find, correct to 1 decimal place, the surface area of each of the following figures:

a

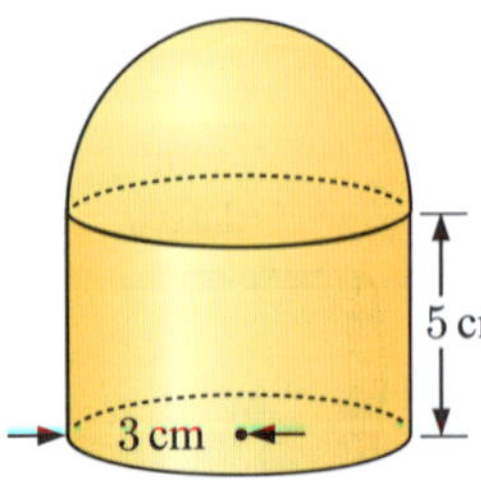

b

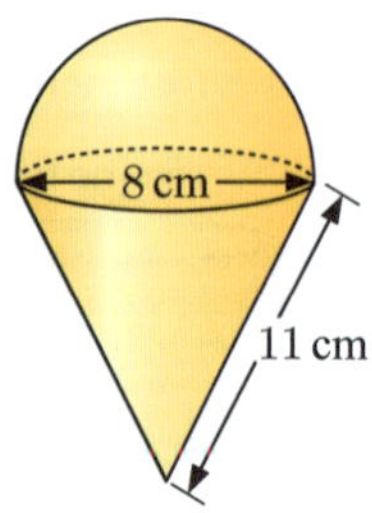

c

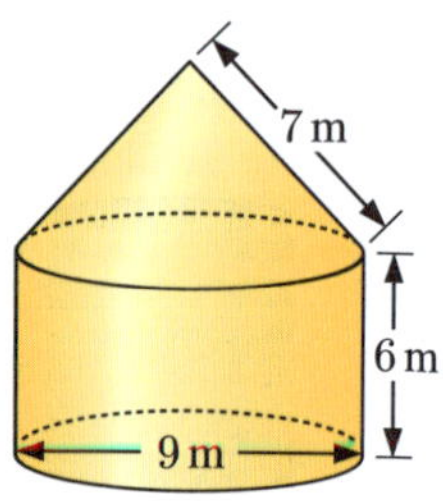

d

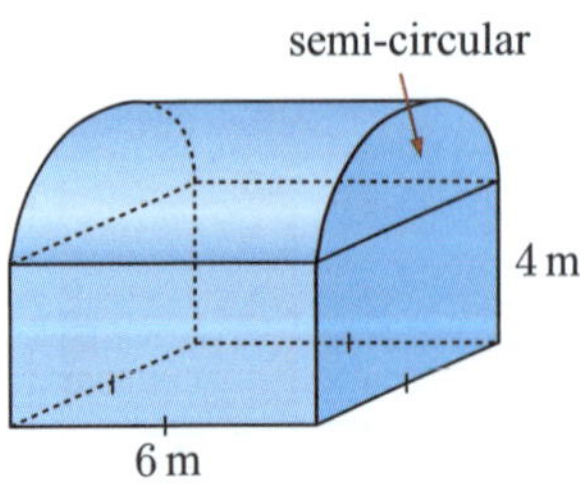

e

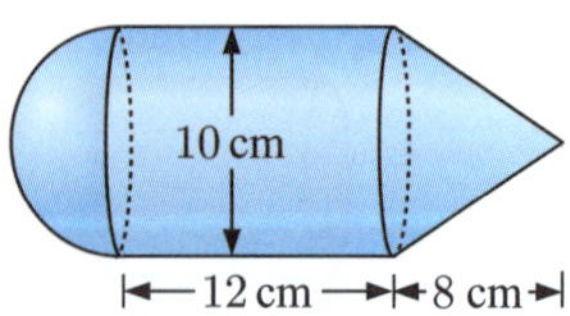

f

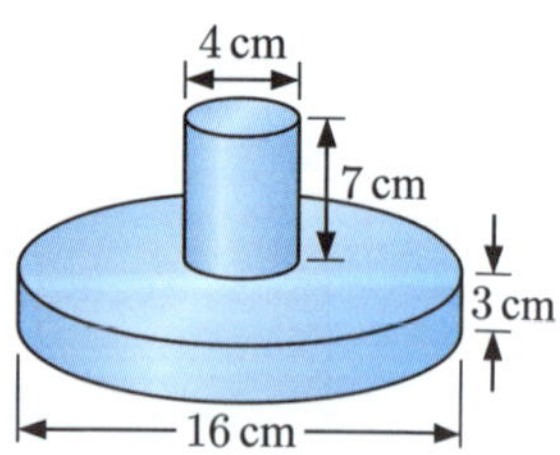

5 A ball bearing has a radius of 1.2 cm. Find the surface area of the ball bearing.

6 A conical piece of filter paper has a base radius of 2 cm, and is 5 cm high.
Find the outer surface area of the filter paper.

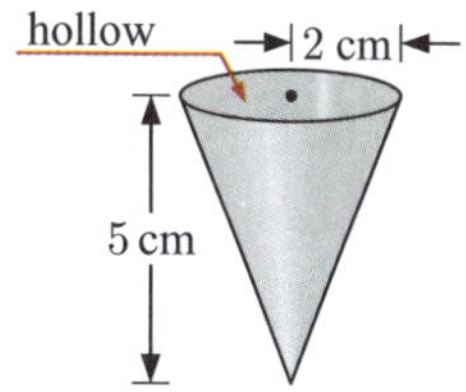

7 How many spheres of 15 cm diameter can be covered by 10 m^2 of material?

8

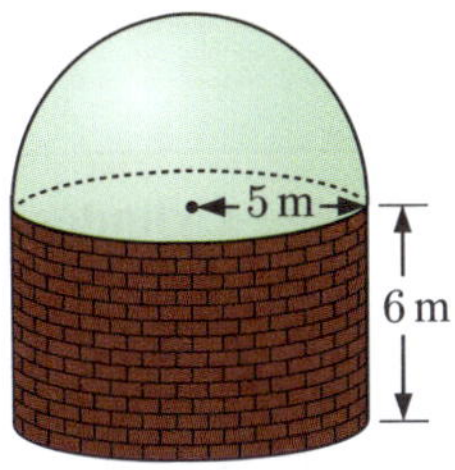

The outside of this observatory needs cleaning. The cost of cleaning the hemispherical glass roof is \$4 per square metre, and the cost of cleaning the cylindrical walls is \$2.50 per square metre. Find, to the nearest dollar, the total cost of cleaning the observatory.

9 A wedge with angle 20° is cut from the centre of a cylindrical cake of radius 13 cm and height 6 cm. Each side of the wedge, excluding the bottom, is to be covered with icing.
Find the total surface area of cake to be iced.

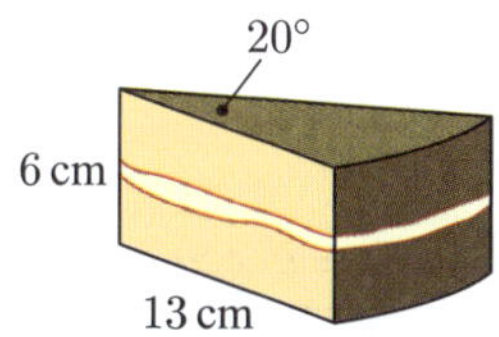

D VOLUME

The **volume** of a solid is the amount of space it occupies.

Volume can be measured in **cubic millimetres**, **cubic centimetres**, or **cubic metres**.

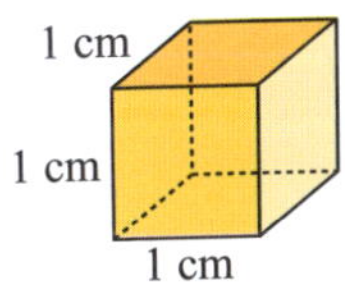

1 cm^3
$= 10 \text{ mm} \times 10 \text{ mm} \times 10 \text{ mm}$
$= 1000 \text{ mm}^3$

Likewise, 1 m^3
$= 100 \text{ cm} \times 100 \text{ cm} \times 100 \text{ cm}$
$= 1\,000\,000 \text{ cm}^3$

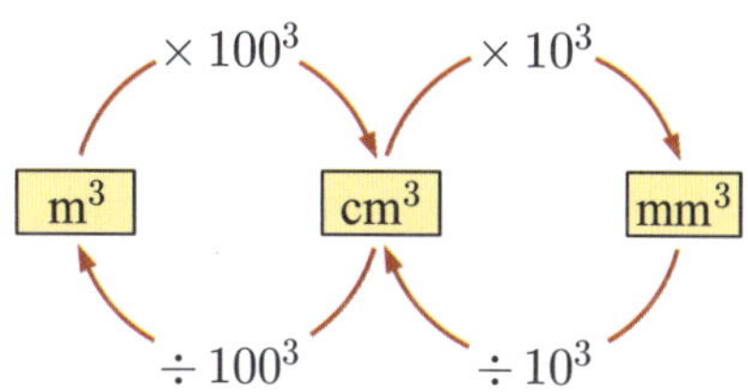

Example 10 Self Tutor

Convert: **a** 5 m^3 to cm^3 **b** $25\,000 \text{ mm}^3$ to cm^3

a 5 m^3
$= (5 \times 100^3) \text{ cm}^3$
$= (5 \times 1\,000\,000) \text{ cm}^3$
$= 5\,000\,000 \text{ cm}^3$

b $25\,000 \text{ mm}^3$
$= (25\,000 \div 10^3) \text{ cm}^3$
$= (25\,000 \div 1000) \text{ cm}^3$
$= 25 \text{ cm}^3$

EXERCISE 7D.1

1 Convert:

a $8.65\ cm^3$ to mm^3
b $86\,000\ mm^3$ to cm^3
c $300\,000\ cm^3$ to m^3
d $124\ cm^3$ to mm^3
e $300\ mm^3$ to cm^3
f $3.7\ m^3$ to cm^3

2 A manufacturer of erasers has $2.4\ m^3$ of rubber. Each eraser has a volume of $8\ cm^3$. How many erasers can be made?

VOLUME FORMULAE

In previous years we have established the following volume formulae:

Rectangular prism	Solid of uniform cross-section	Cylinder
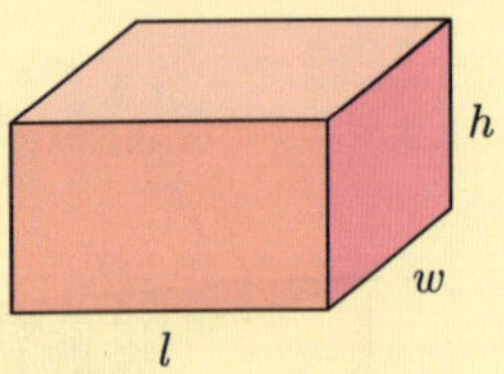	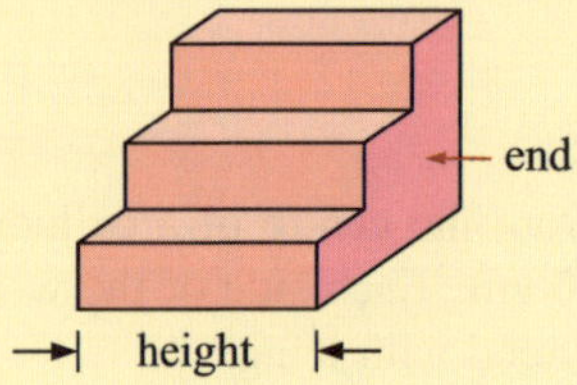	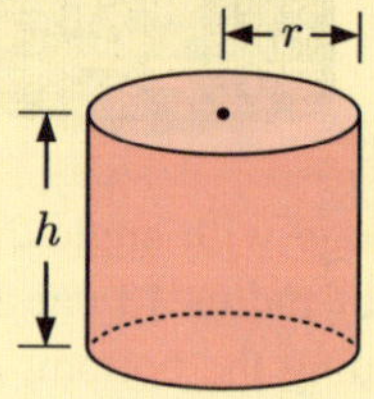
$V = l \times w \times h$	V = **area of end × height**	$V = \pi r^2 h$

Example 11 — Self Tutor

Find the volume of each solid:

a
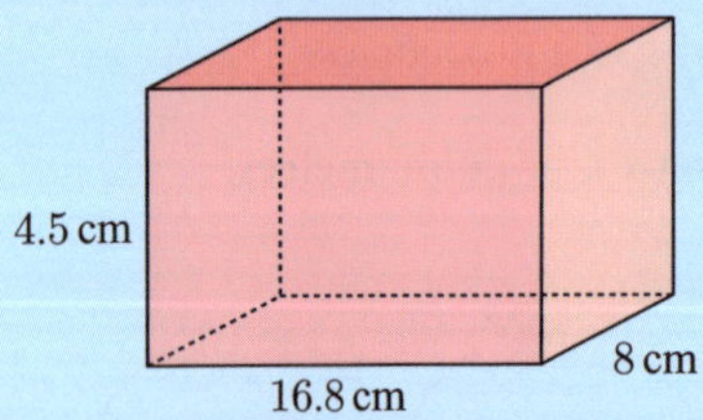

b
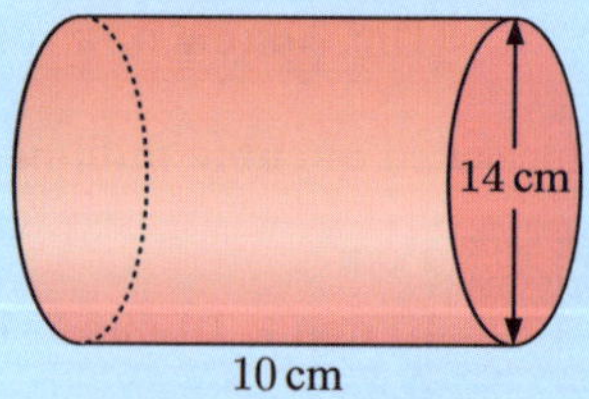

a Volume = length × width × height
$= 16.8 \times 8 \times 4.5\ cm^3$
$\approx 605\ cm^3$

b Volume $= \pi r^2 h$
$= \pi \times 7^2 \times 10\ cm^3$
$\approx 1540\ cm^3$

EXERCISE 7D.2

1 Find the volume of each solid:

a
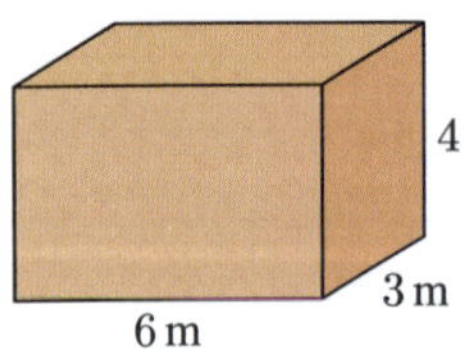

b
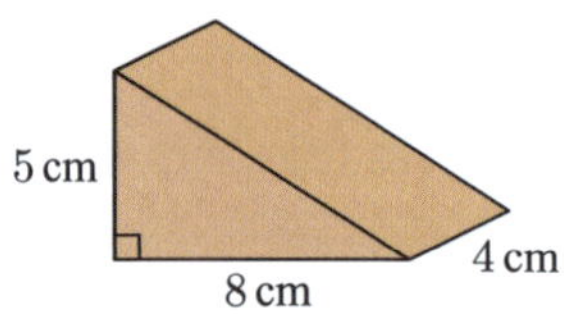

c
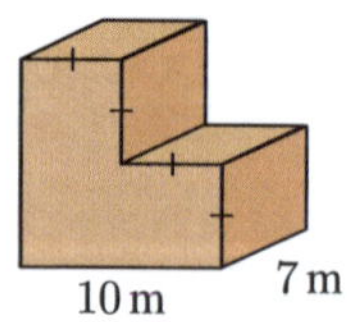

d

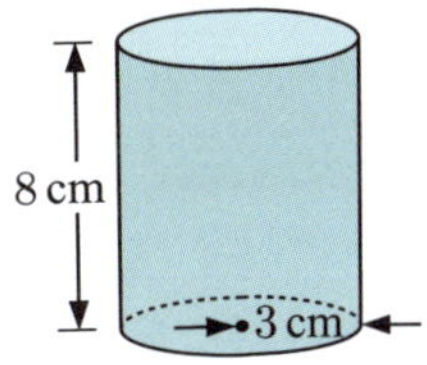

e

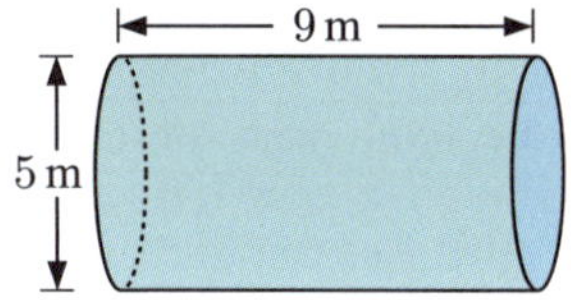

f

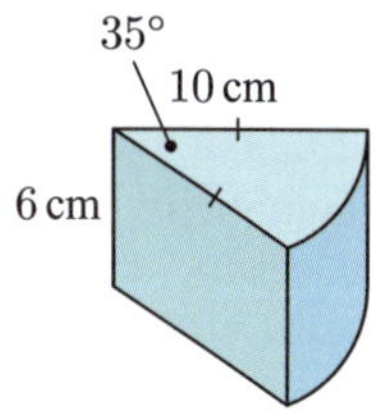

2 Find the volume of this brick.

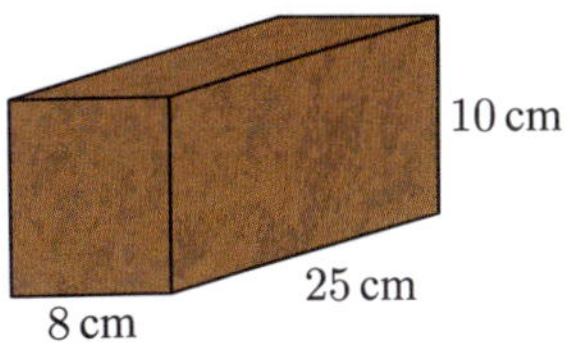

3 A coin is 2.5 cm wide and 2 mm thick. Find the volume of the coin.

4 A concrete contractor is considering his next job. The surface area of concrete to be laid is 85 m^2. The customer has a choice of 8 cm thick concrete or 12 cm thick concrete. Given that the concrete costs $175 per cubic metre, how much extra would the customer have to pay for the thicker concrete?

5 Jenna is erecting a picket fence in front of her house. The fence has 132 pickets, each with the dimensions shown.
Calculate, in m^3, the volume of the pickets in Jenna's fence.

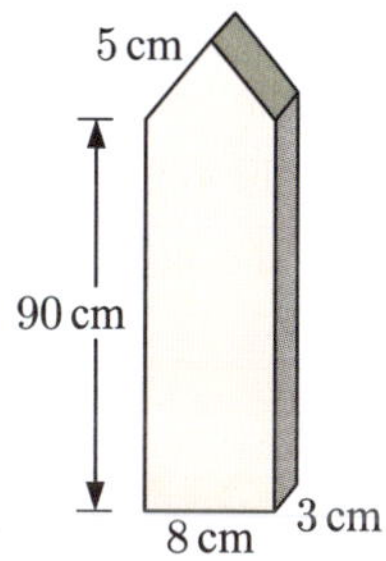

6 A restaurant uses two types of table. Find the volume of wood in each table.

a

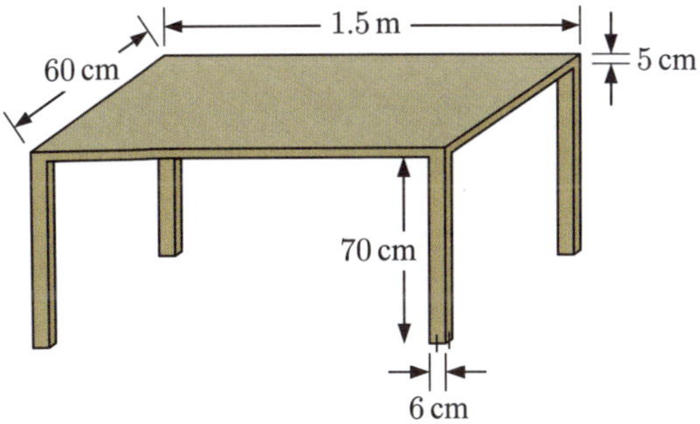

b

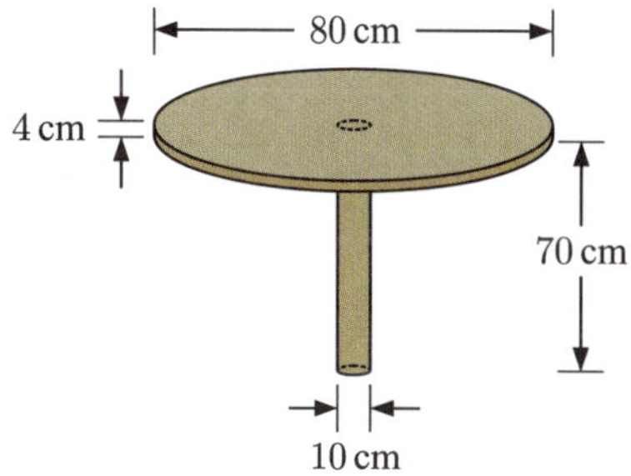

7 Find the total volume of this cat's scratching post.

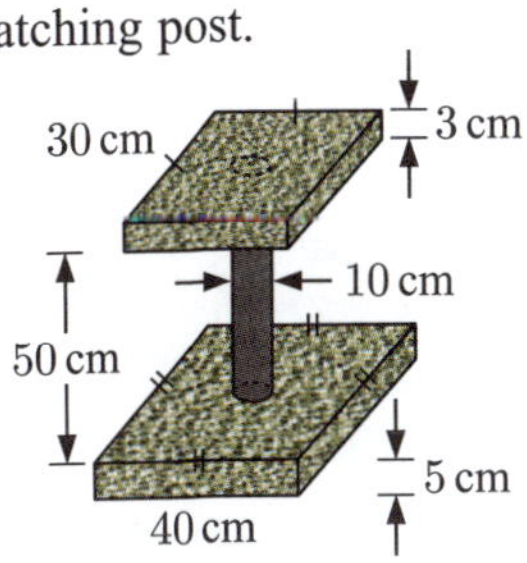

8 Calculate the volume of wood in this model train tunnel.

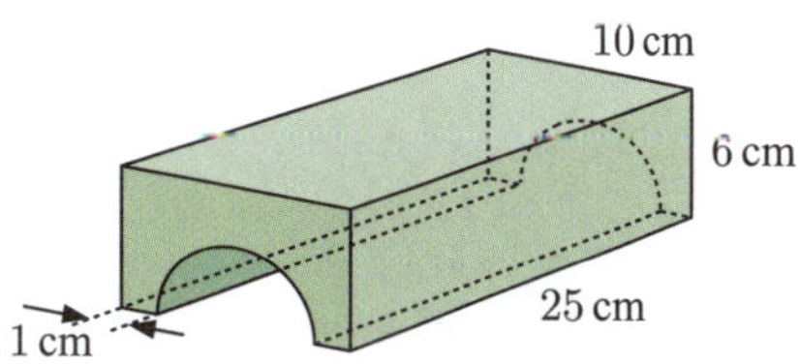

9 **a** Find the volume of the illustrated dumbbell.

b The material used to make the dumbbell weighs 2.523 g per cm^3. Find the total mass of the dumbbell.

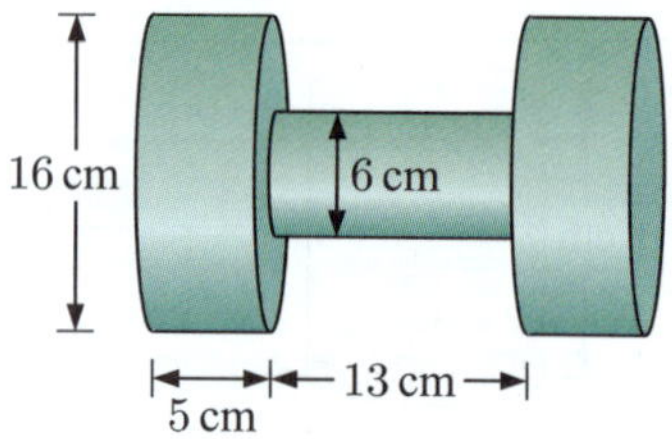

10 A cubic metre of brass is melted down and cast into solid door handles with the shape shown. How many handles can be cast?

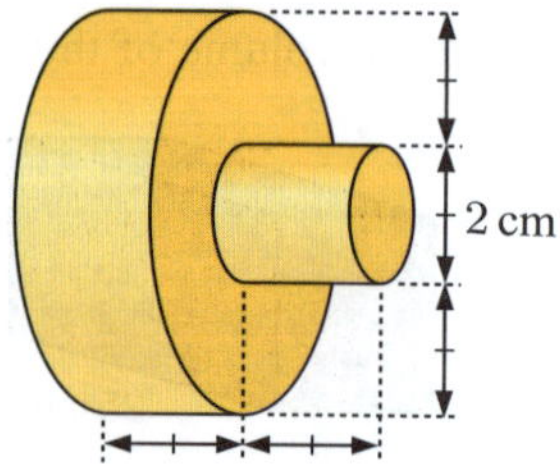

11 A concrete tank has an external diameter of 5 m and an internal height of 3 m. The walls and base of the tank are 20 cm thick. The concrete costs \$142 per m^3.

a Find the volume of concrete required to make:

i the base **ii** the walls.

b Find the total volume of concrete required.

c Find the cost of the concrete required to make the tank.

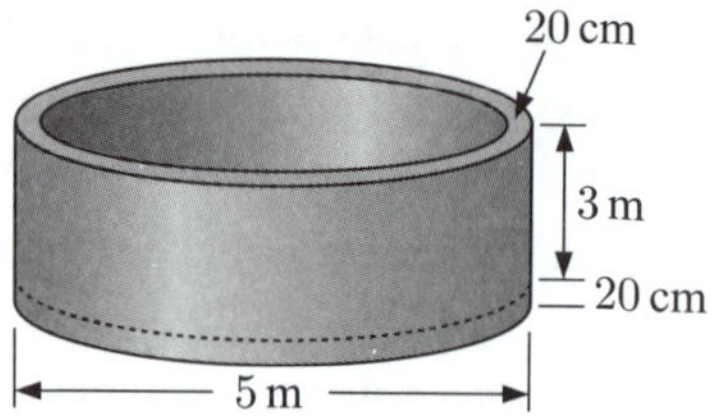

ACTIVITY 2 PACKING BOXES

Maureen has a collection of boxes which need to be stored. She realises that she can save space by packing smaller boxes inside larger ones.

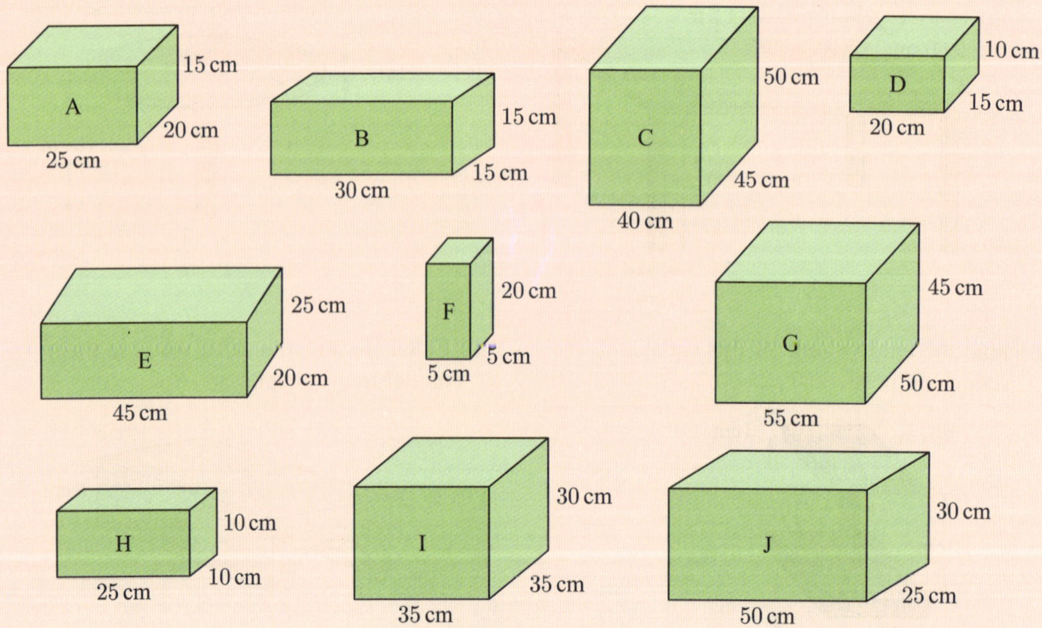

What to do:

1. **a** Show that box B occupies less space than box A.
   b Does this mean that box B can fit completely inside box A? Explain your answer.
2. Given that one box has dimensions a cm $\times$ b cm $\times$ c cm, and another box has dimensions d cm $\times$ e cm $\times$ f cm, how can you determine whether one box will fit completely inside the other?
3. Show how Maureen can pack all of the boxes into two large boxes.

PYRAMIDS AND CONES

Pyramids and cones are known as **tapered solids**. They have a flat base, and come to a point called the **apex**.

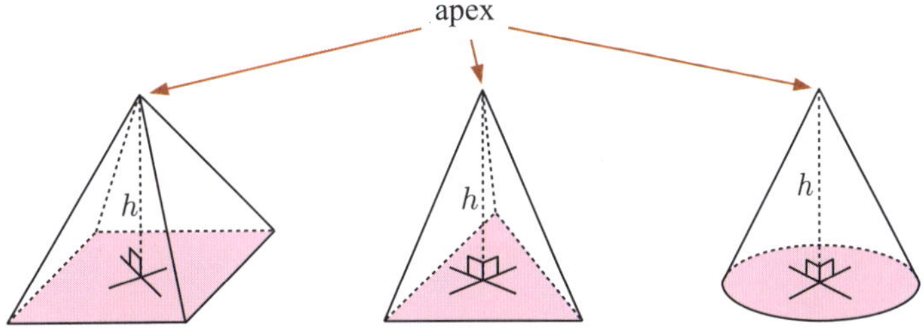

square-based pyramid triangular-based pyramid cone

Tapered solids do **not** have a uniform cross-section. For example, we can see that the perpendicular cross-section of a cone is always a circle, but its radius decreases as we move up the cone.

DEMO

INVESTIGATION 3 VOLUME OF A SQUARE-BASED PYRAMID

In this Investigation we will compare the volume of a square-based pyramid (figure **A**) with the volume of the square-based prism with the same base area and height (figure **B**).

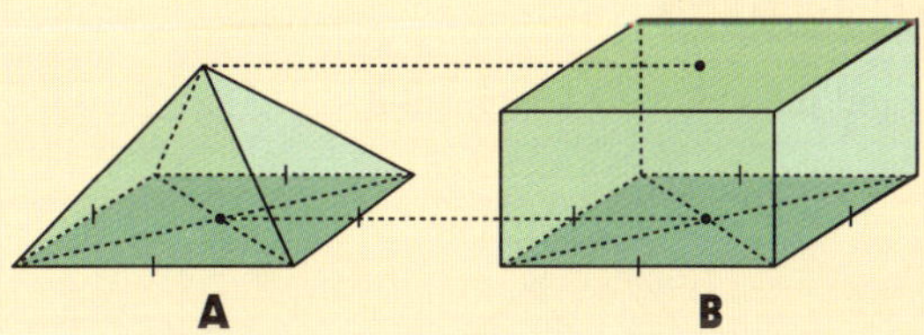

What to do:

1. **a** Click on the icon to obtain a printable sheet containing nets for 6 identical square-based pyramids.
   b Print the nets, cut them out, and construct the pyramids using sticky tape.
   c Arrange the pyramids to form a cube.
   d Explain why the volume of the cube is:
      i six times the volume of figure **A**
      ii twice the volume of figure **B**.
   e Hence write an expression for the volume of figure **A** in terms of the volume of figure **B**.

NETS

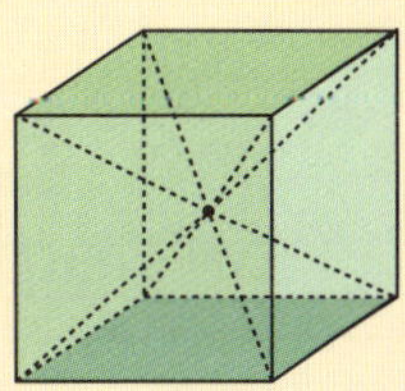

2 We now consider another arrangement of square-based pyramids.

- **a** Click on the icon to obtain a printable sheet containing nets for 3 identical square-based pyramids.
- **b** Print the nets, cut them out, and construct the pyramids using sticky tape.
- **c** Arrange the pyramids to form a square-based prism. Notice that the height of the prism is the same as the height of each pyramid.
- **d** What is the relationship between the volume of the pyramid and the volume of the prism?

NETS

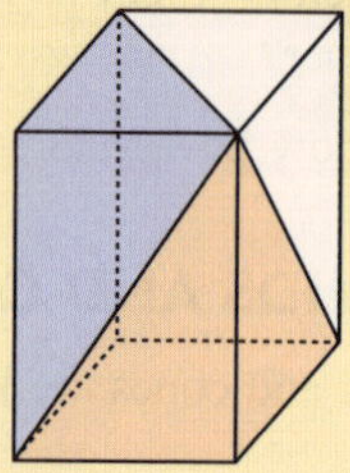

3 Given that the volume of a prism is
V = area of base $\times$ height, predict a formula for the volume of a pyramid.

You should have found that the volume of a square-based pyramid is one third of the volume of a prism with the same base area and height.

Similarly, the volume of a cone is one third of the volume of a cylinder with the same base area and height.

Since the volume of a solid with uniform cross-section is
V = area of base $\times$ height:

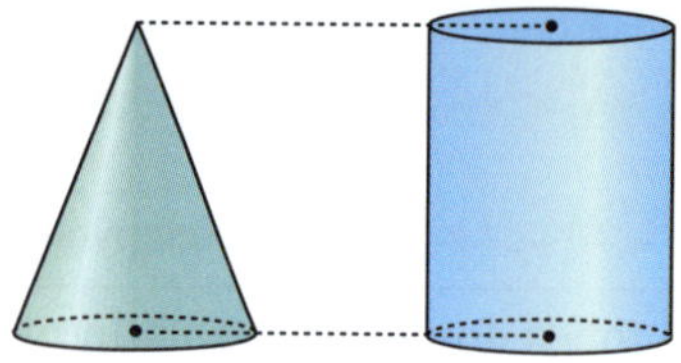

Volume of a tapered solid $= \frac{1}{3} \times$ area of base $\times$ height

Example 12 **Self Tutor**

Find the volume of each solid:

a

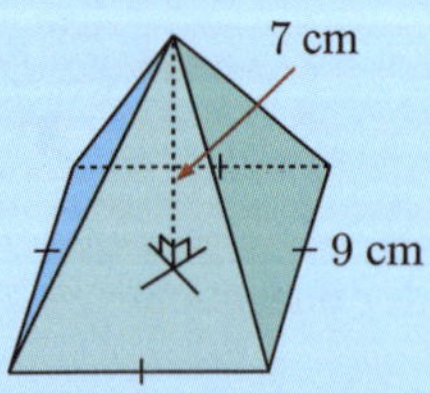

b

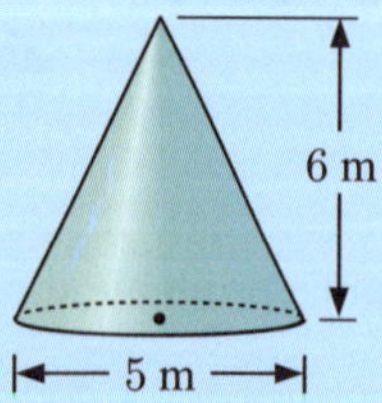

a Volume
$= \frac{1}{3} \times$ area of base $\times$ height
$= \frac{1}{3} \times 9 \times 9 \times 7$
$= 189$ cm^3

b Volume
$= \frac{1}{3} \times$ area of base $\times$ height
$= \frac{1}{3} \times \pi \times 2.5^2 \times 6$
≈ 39.3 m^3

EXERCISE 7D.3

1 Find the volume of each solid:

a

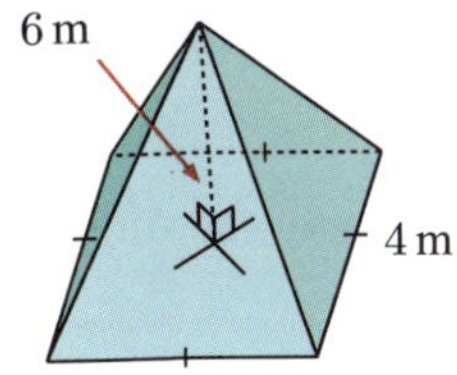

b

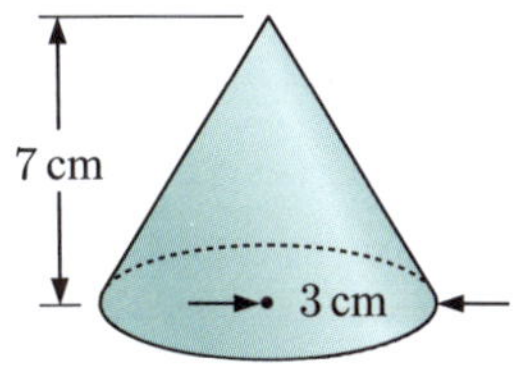

c

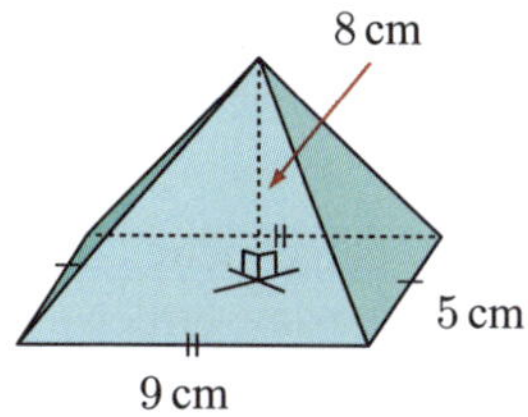

d

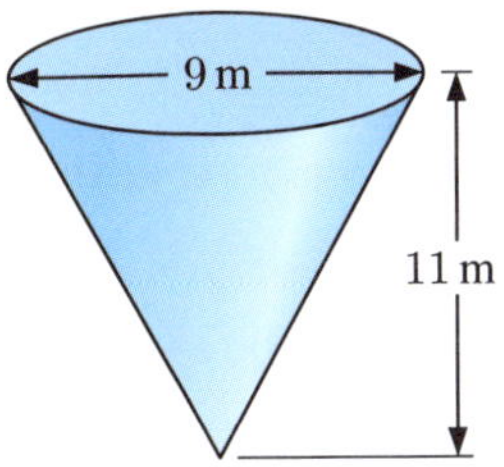

e

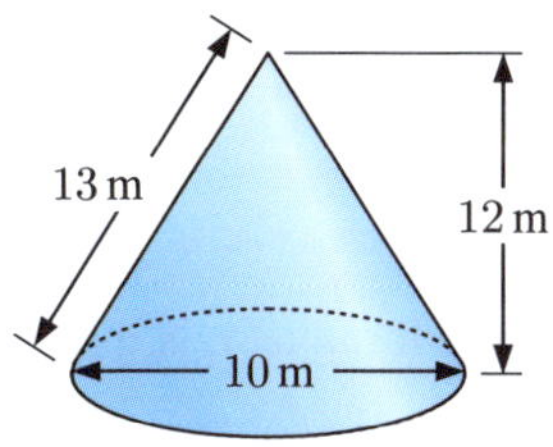

f

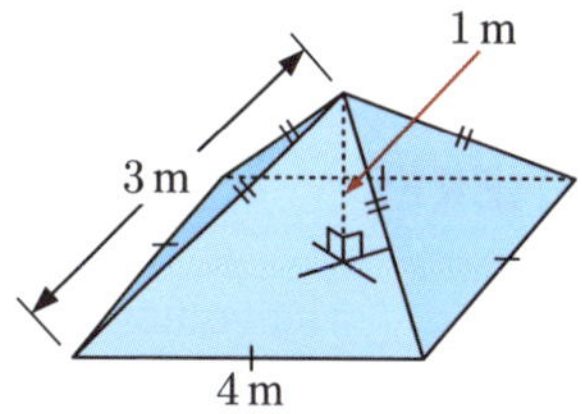

2 Use Pythagoras' theorem to help find the volume of each solid:

a

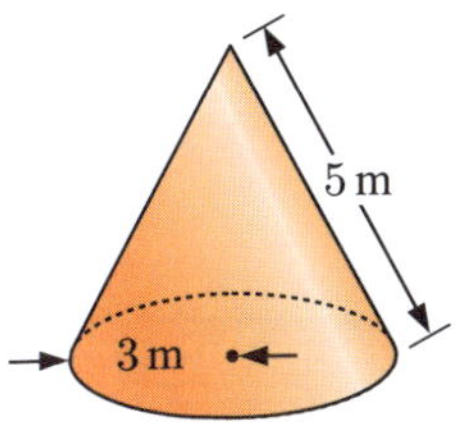

b

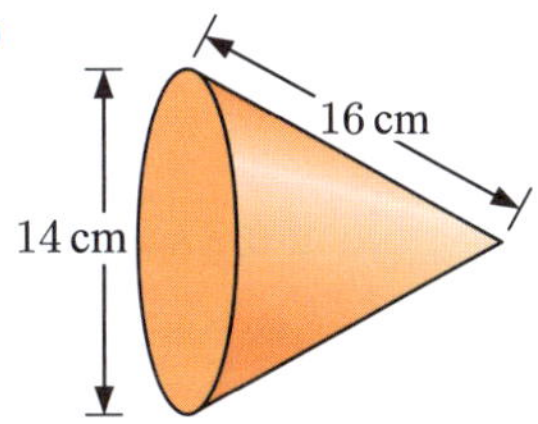

c

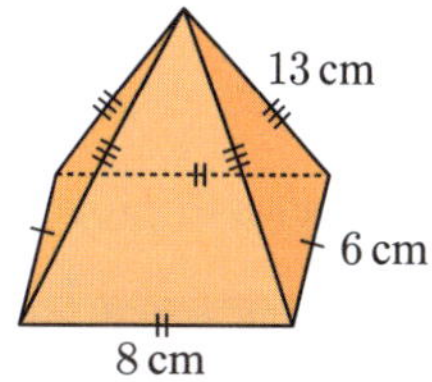

d

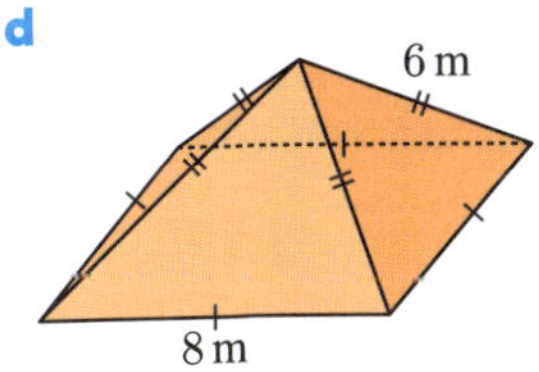

e

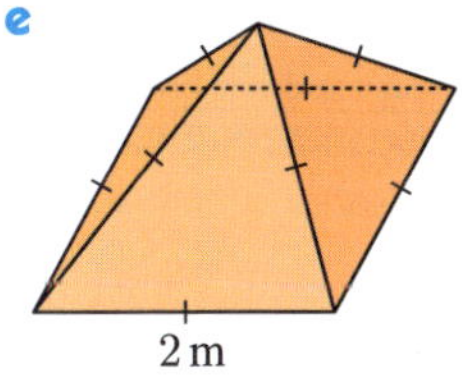

f

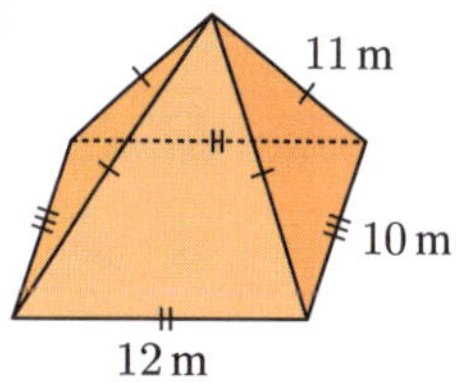

3 Find the volume of each solid:

a

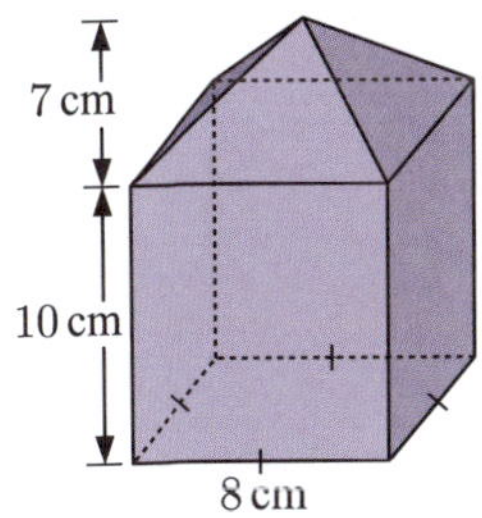

b

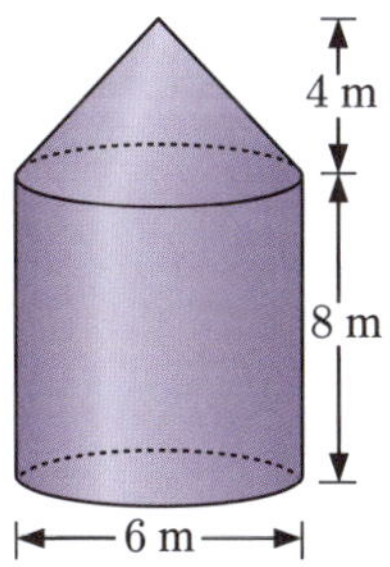

c

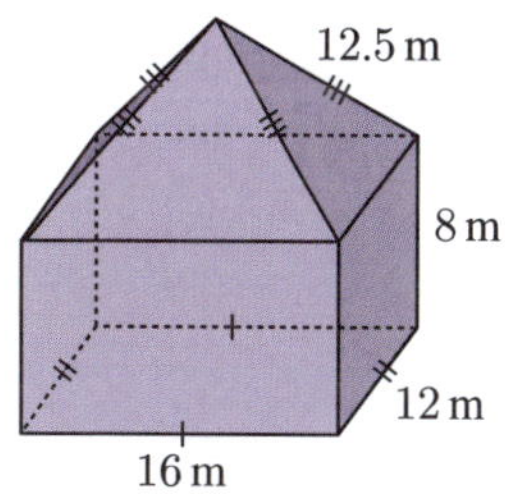

4

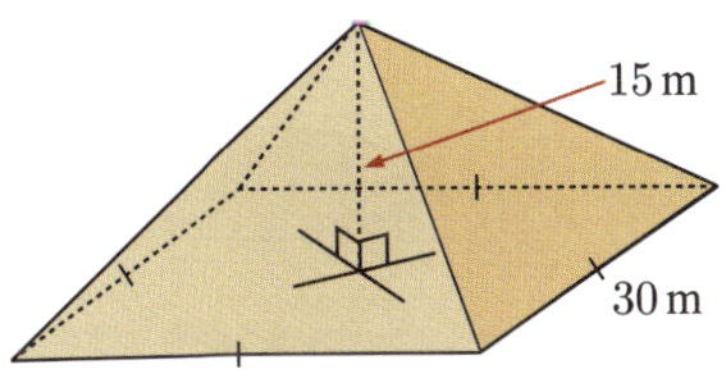

The conservatory for tropical plants at the Botanic gardens is a square-based pyramid with sides 30 metres long and height 15 metres.
Calculate the volume of air in this building.

5 A conical heap of gravel is 1.4 m high and has a diameter of 2.6 m. Find the volume of gravel in the heap.

6 A wax crayon has the dimensions shown. Find the total volume of wax required to make a set of 24 crayons.

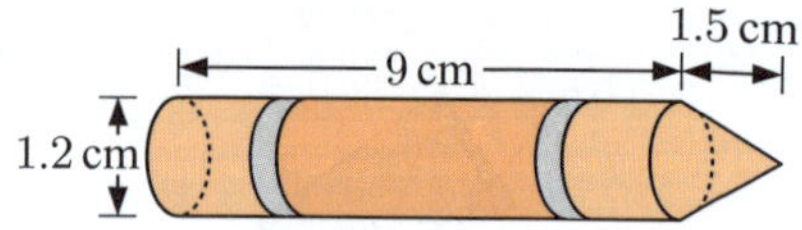

7 0.6 cubic metres of metal is melted down and cast into garden stakes with the dimensions shown. How many garden stakes can be made?

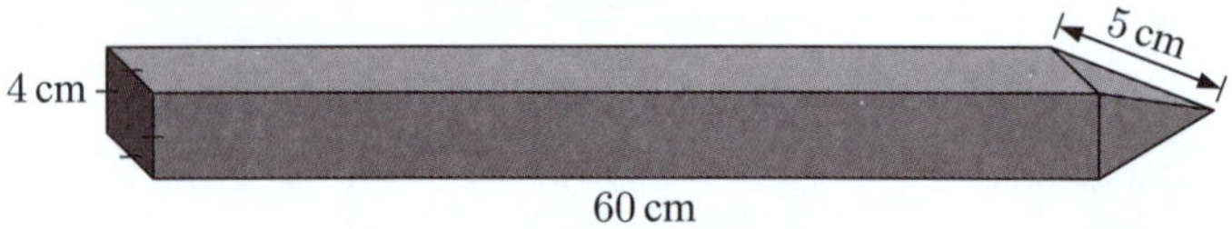

SPHERES

The volume of a sphere with radius r is given by

$$V = \frac{4}{3}\pi r^3$$

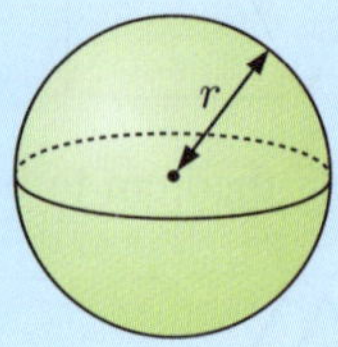

The mathematics to prove this formula is beyond the scope of this course.

Example 13 — Self Tutor

Find the volume of this sphere:

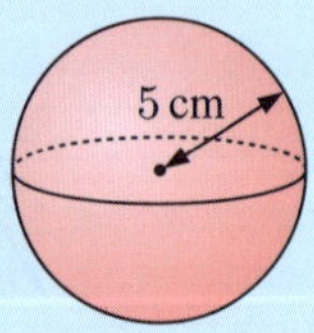

$$\begin{aligned}\text{Volume} &= \tfrac{4}{3}\pi r^3 \\ &= \tfrac{4}{3} \times \pi \times 5^3 \\ &\approx 524 \text{ cm}^3\end{aligned}$$

EXERCISE 7D.4

1 Find the volume of each solid:

a

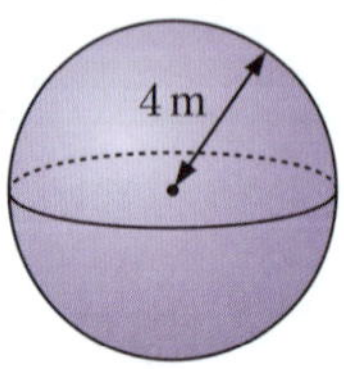

b

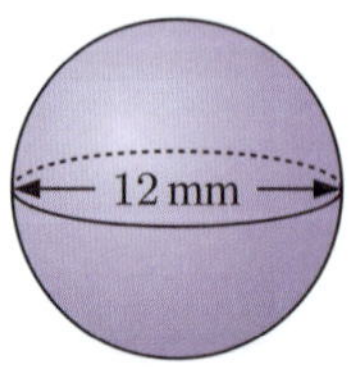

c

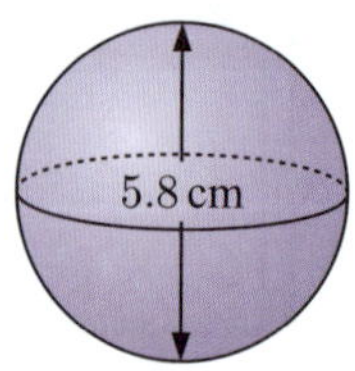

d

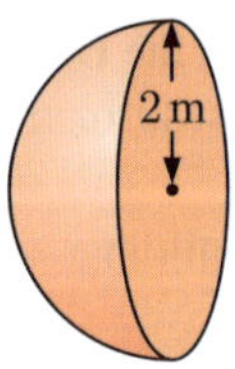

e

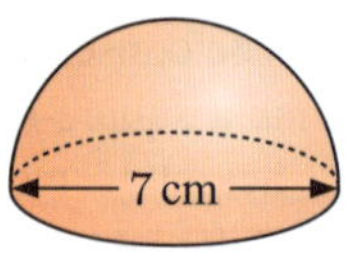

f

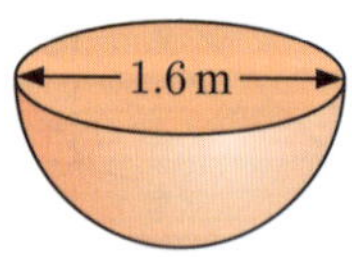

g

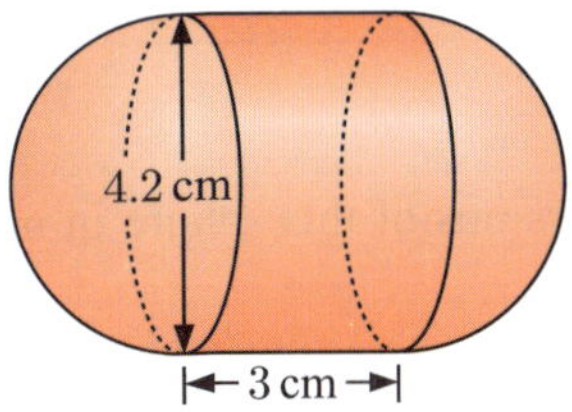

h

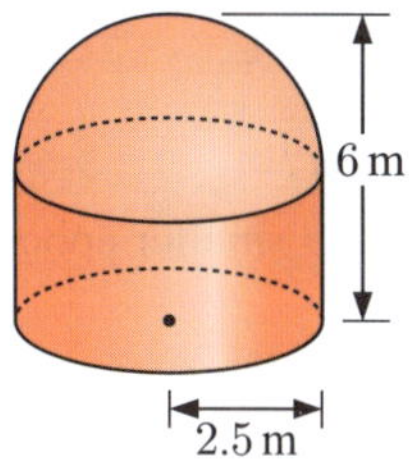

i

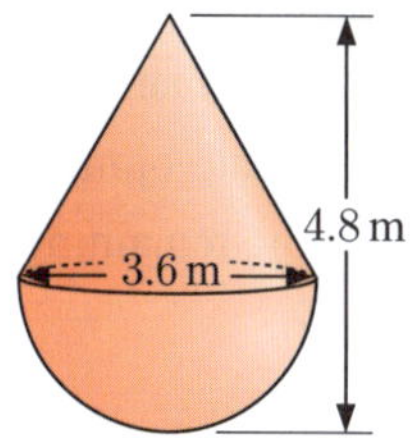

2 A beach ball has a diameter of 1.2 m. Find the volume of air inside the ball.

3 A garden ornament is shaped like a mushroom. The base is a cylinder 8 cm high and 6 cm in diameter. The top of the mushroom is 20 cm in diameter.
What volume of concrete, in m^3, is needed to make 50 mushrooms?

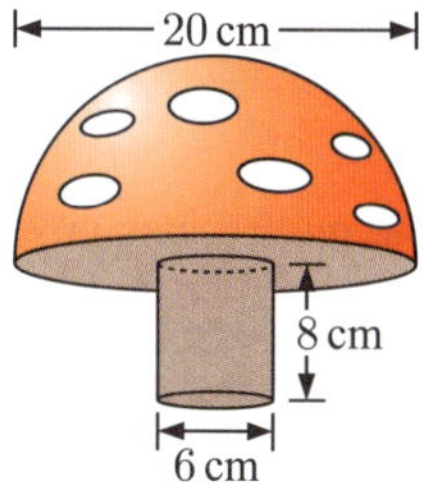

4 Spherical hollow glass baubles have internal diameter 4.9 cm and external diameter 5.0 cm. Find the volume of glass in each bauble.

Example 14 — Self Tutor

A sphere has volume $60\,000\ \text{cm}^3$. Find its radius.

$$\text{Volume} = \tfrac{4}{3}\pi r^3$$
$$\therefore\ \tfrac{4}{3}\pi r^3 = 60\,000$$
$$\therefore\ \tfrac{3}{4} \times \tfrac{4}{3}\pi r^3 = 60\,000 \times \tfrac{3}{4}$$
$$\therefore\ \pi r^3 = 45\,000$$
$$\therefore\ r^3 = \frac{45\,000}{\pi}$$
$$\therefore\ r = \sqrt[3]{\frac{45\,000}{\pi}}$$
$$\therefore\ r \approx 24.3$$

The radius of the sphere is about 24.3 cm.

5 **a** A sphere has volume $500\ \text{cm}^3$. Find its radius.

b A sphere has volume $200\ \text{cm}^3$. Find its:

i radius **ii** surface area.

6 A manufacturer of lead fishing sinkers has $0.237\ \text{m}^3$ of lead. Each sinker is a sphere with volume $5\ \text{cm}^3$.

a How many sinkers can be made?

b Find the radius of each sinker.

c Find the surface area of each sinker.

7 A gold ingot 15 cm by 10 cm by 4 cm is melted down and cast into spheres of radius 1 cm.

- **a** How many spheres can be cast?
- **b** What volume of gold is left over?
- **c** The remaining gold is cast into a smaller sphere. Find the radius of this sphere in millimetres.

E CAPACITY

The **capacity** of a container is the quantity of fluid or gas used to fill it.

The basic unit of capacity is the **litre (L)**.

1 litre = 1000 millilitres (mL)
1 kilolitre (kL) = 1000 litres
1 megalitre (ML) = 1000 kilolitres

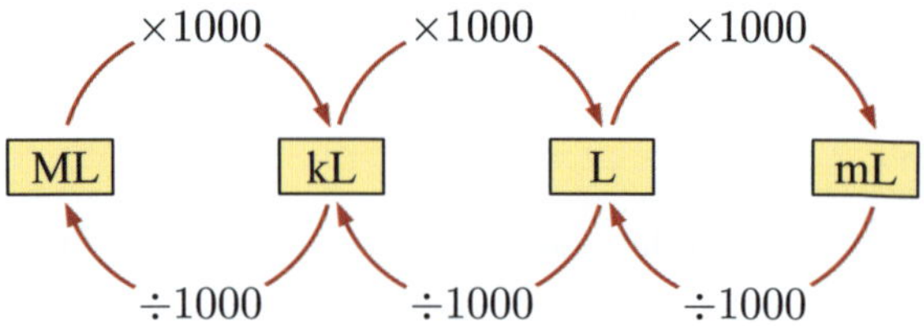

Example 15 — Self Tutor

Convert:

a 4.2 L to mL **b** 36 800 L to kL

a 4.2 L
$= (4.2 \times 1000)$ mL
$= 4200$ mL

b 36 800 L
$= (36\,800 \div 1000)$ kL
$= 36.8$ kL

CONNECTING VOLUME AND CAPACITY

The units for volume and capacity are connected because **1 millilitre (mL)** of fluid fills a container of size 1 cm^3.

We can therefore construct the following table of equivalent units:

$1 \text{ mL} \equiv 1 \text{ cm}^3$
$1 \text{ L} \equiv 1000 \text{ cm}^3$
$1 \text{ kL} = 1000 \text{ L} \equiv 1 \text{ m}^3$

EXERCISE 7E

1 Convert:

- **a** 5.6 L to mL
- **b** 524 kL to L
- **c** 280 mL to L
- **d** 3.5 ML to kL
- **e** 743 000 L to ML
- **f** 0.84 kL to mL

2 Find the capacity of each of the following containers:

a

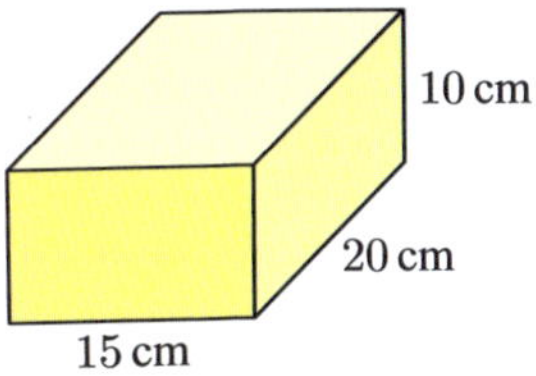

b

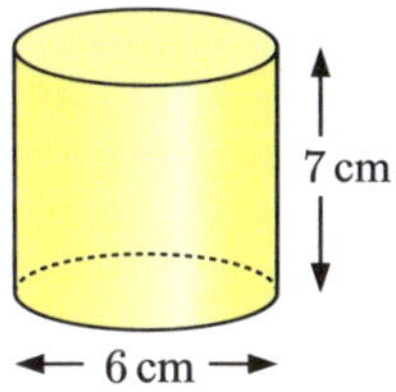

c

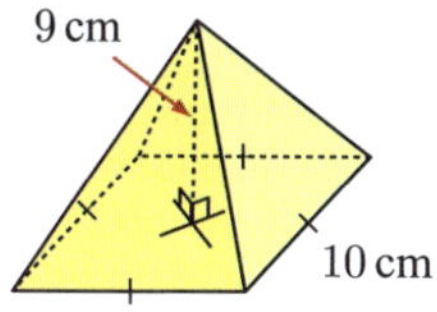

3 A bottle of perfume is spherical with a diameter of 36 mm. Calculate the capacity of the bottle, in mL.

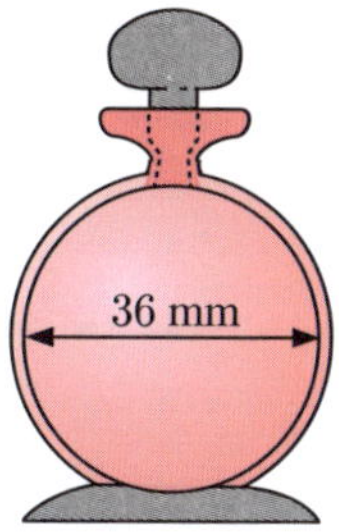

4 The average depth of water in a lake is 1.7 m. It is estimated that the total surface area of the lake is 1.35 ha.

a Convert 1.35 ha to m^2.

b How many kilolitres of water does the lake contain?

5 A car has a rectangular prism petrol tank with dimensions 48 cm by 56 cm by 20 cm.

a Find the capacity of the petrol tank.

b The car consumes petrol at an average rate of 8.7 litres per 100 km. How far could it travel on a full tank of petrol?

6

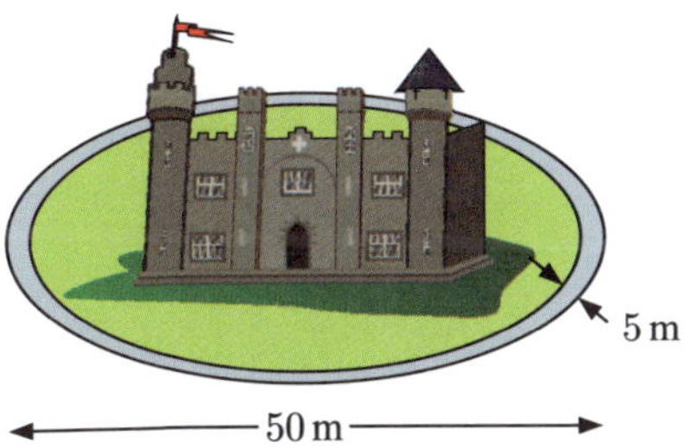

A castle is surrounded by a circular moat which is 5 m wide and 2 m deep. The diameter of the outer edge of the moat is 50 m. Find, in kilolitres, the quantity of water in the moat.

7 How many cylindrical bottles 15 cm high and with 8 cm diameter could be filled from a tank containing 200 litres of olive oil?

8 The inside of a glass vase consists of a cylinder with diameter 8 cm and depth 15 cm, with a hemisphere below it.
Find the quantity of water that the vase can hold.

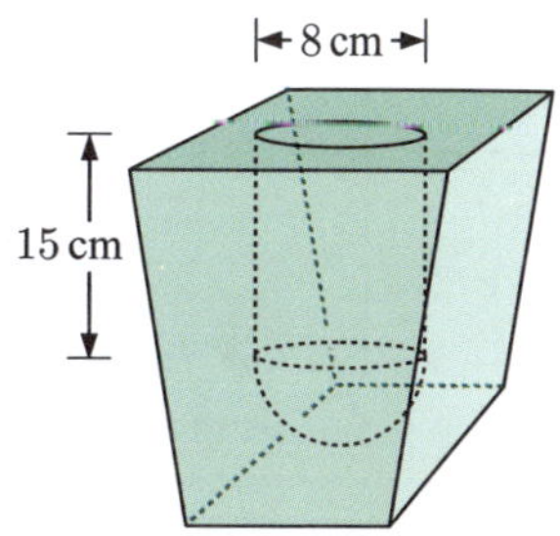

Example 16 **Self Tutor**

14.4 kL of water is pumped into an empty cylindrical tank with base radius 2 m. How high up the tank will the water level rise?

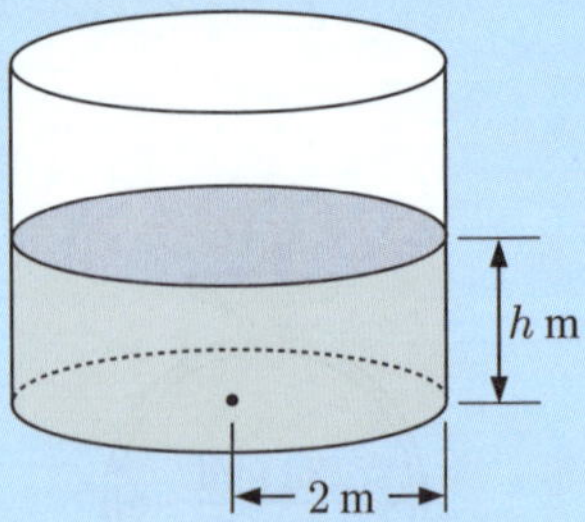

Since $1 \text{ kL} \equiv 1 \text{ m}^3$, the volume of water pumped into the tank is 14.4 m^3.

If the water in the cylindrical tank is h m deep,

its volume $= \pi \times 2^2 \times h$ $\{V = \pi r^2 h\}$

$= 4\pi h \text{ m}^3$

$\therefore 4\pi h = 14.4$ {equating volumes}

$\therefore h = \dfrac{14.4}{4\pi} \approx 1.15$

$\therefore$ the water level will rise to 1.15 m.

9 20 kL of water is pumped into an empty cylindrical tank with base radius 3 m. How high up the tank will the water level rise?

10 Water enters a cylindrical rainwater tank at 80 L per minute. The base diameter of the tank is 2.4 m and the height is 4 m.

a Find the capacity of the full tank.

b How high up the tank will the water level rise after 10 minutes?

11 Consider these three containers:

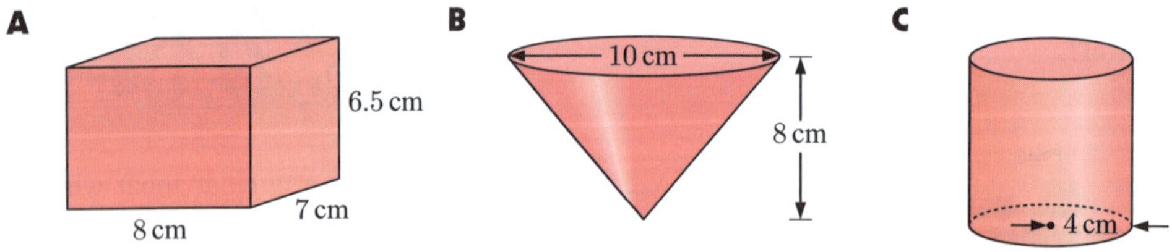

a Find the capacities of containers **A** and **B**.

b Container **A** is filled with oil. Containers **B** and **C** are empty. The oil from container **A** is poured into container **B** until it is full, and the remainder is poured into container **C**. Find the height of the oil in container **C**.

12 Answer the **Opening Problem** on page **114**.

13 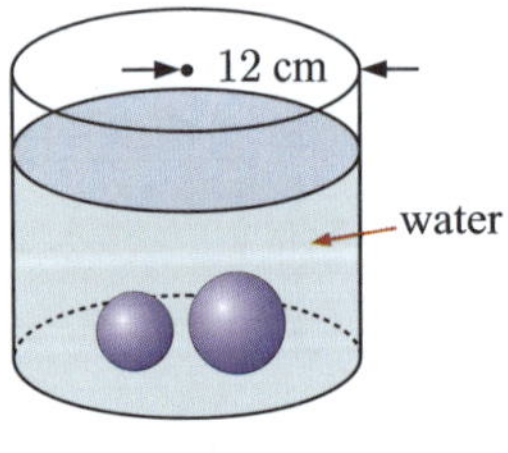

A cylindrical container with radius 12 cm is partly filled with water. Two spheres with radii 3 cm and 4 cm are dropped into the water and are fully submerged. Find the *exact* increase in height of water in the cylinder.

Global context

click here

Rubbish - Where does it go?

Statement of inquiry: Considering finite landfill resources helps us to understand the importance of responsible waste management.

Global context: Fairness and development

Key concept: Form

Related concepts: Measurement, Space

Objective: Applying mathematics in real-life contexts

Approaches to learning: Thinking, Research

REVIEW SET 7A

1 Convert:

a $20\,000\text{ cm}^2$ to m^2 **b** 5 m^3 to cm^3 **c** 0.48 L to cm^3

2 Find the perimeter of:

a

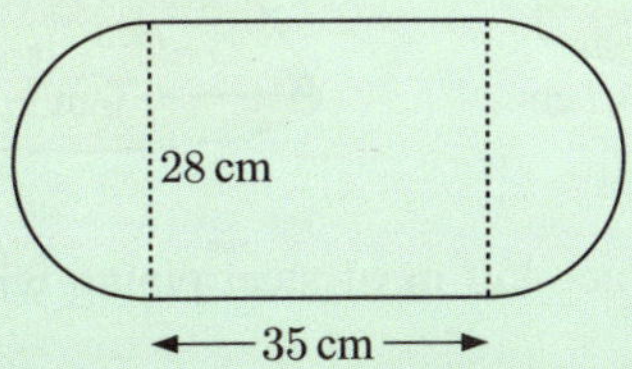

b

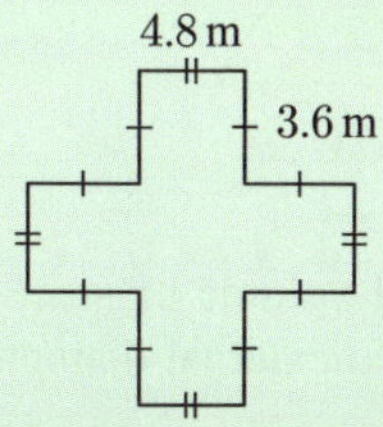

c

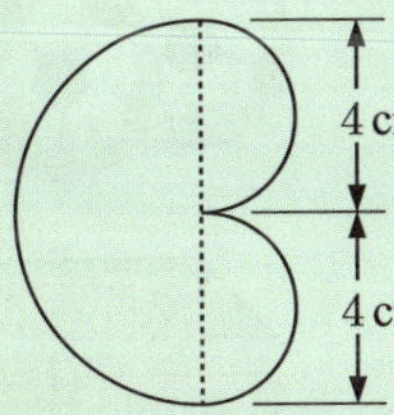

d

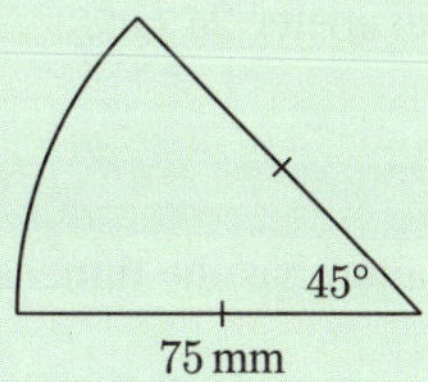

3 A circular playing field has radius 80 metres. The field is surrounded by a fence 10 metres from the edge of the field. Determine the length of the fence.

4 Find the area of:

a a sector with radius 10 cm and angle 120°

b a right angled triangle with base 5 cm and hypotenuse 13 cm

c a circle with diameter 15 cm.

5 Find the surface area of each solid:

a

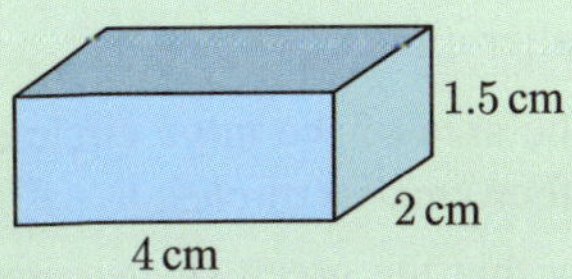

b

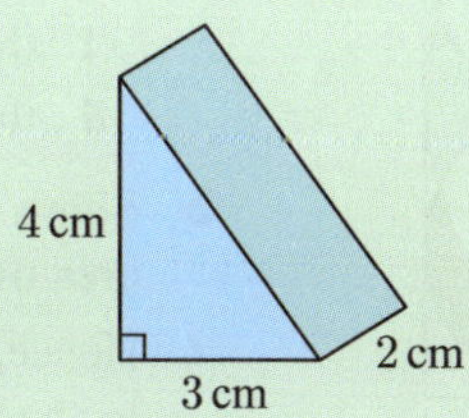

c

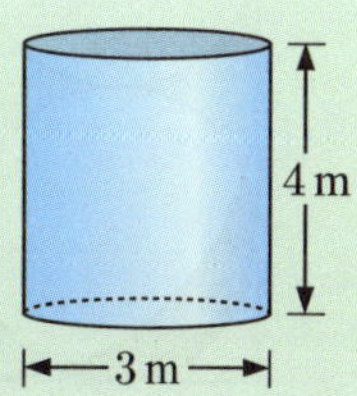

6 A rectangular prism has dimensions $5 \text{ cm} \times 8 \text{ cm} \times 10 \text{ cm}$. Find its:

a volume **b** surface area.

7 Find the total surface area of a solid cone with radius 5 cm and perpendicular height 8 cm.

8 Soup cans have a base diameter of 7 cm and a height of 10 cm.

a Exactly how many such cans can be filled from a vat containing 2000 litres of soup?

b Calculate the total surface area of metal required to make the cans in **a**.

9 A petrol tank has dimensions 24 cm by 40 cm by 60 cm. If the tank can be filled at a rate of 16 litres per minute, how long will it take to fill the tank?

10 Find the volume of the following, rounding your answers to 2 decimal places:

a

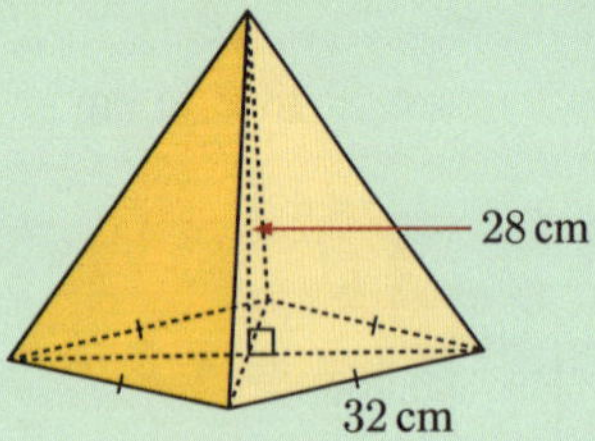

b

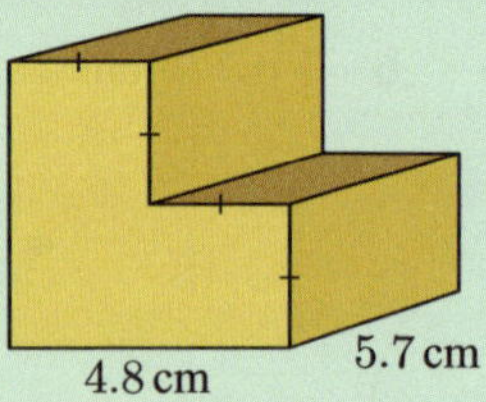

c

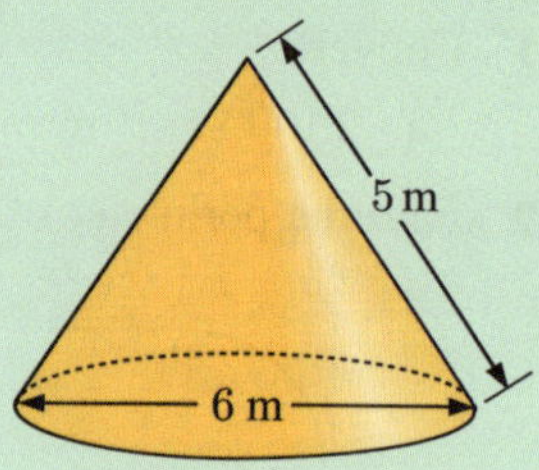

11 Determine the total volume of steel required to construct 450 m of steel piping with internal diameter 1.46 m and external diameter 1.50 m.

12 What area of leather is required to manufacture 10 dozen soccer balls, each with diameter 24 cm?

13 An apple juice carton has the dimensions shown.

a Find:

i the surface area of the carton

ii the amount of juice that the carton holds.

b Megan drinks 50 mL of juice from the carton, then pours the rest into a cylindrical glass with diameter 6 cm. To what height does the juice reach?

14

One large circle and four smaller circles are inscribed as shown inside a square with side length x m.

a Find, in terms of x, the radius of:

i the large circle

ii the smaller circles.

b Show that the area of the large circle equals the area of the four smaller circles.

c Hence, show that the green shaded area is equal to the red shaded area.

REVIEW SET 7B

1 Convert:

a 15 ha to m^2 **b** 3600 cm^3 to mm^3 **c** 4.5 kL to m^3

2 Find the perimeter of:

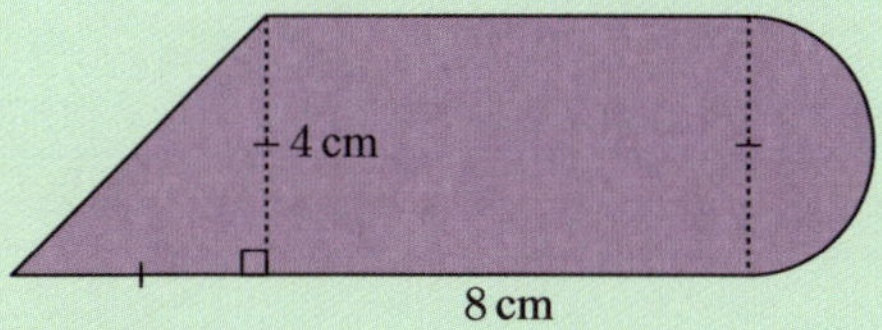

3 Find the perimeter P and the area A of the given figure. Give your answers in simplest form.

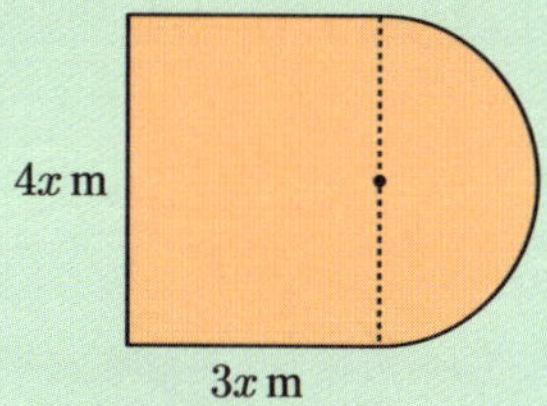

4 Find the shaded area, rounded to 2 decimal places.

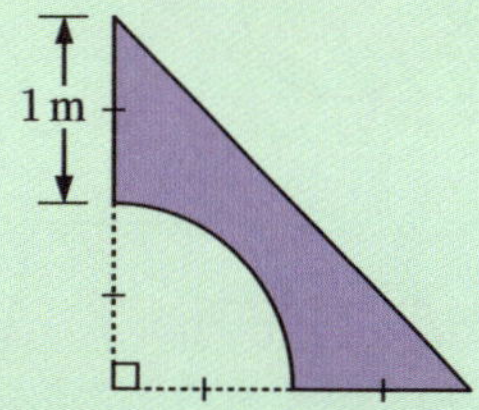

5 A concreting plant manufactures 20 m^3 of concrete. The concrete is used to make cylindrical bollards which are 1.2 m high and have radius 10 cm. How many bollards can be made?

6 Find the surface area of each solid:

a

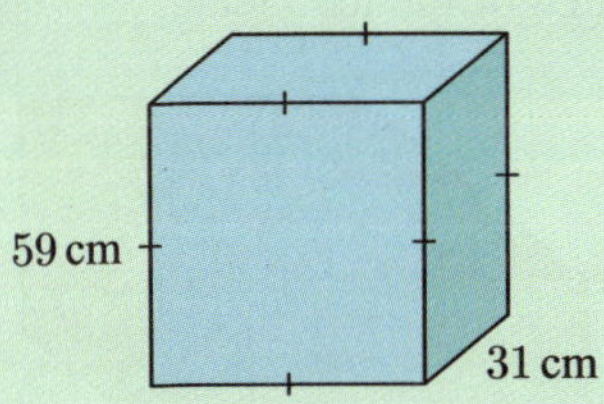

b

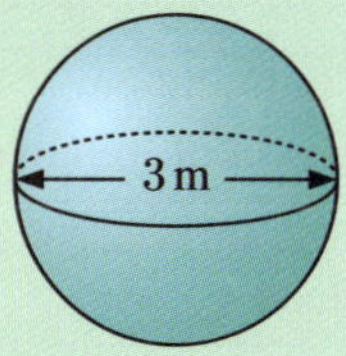

c

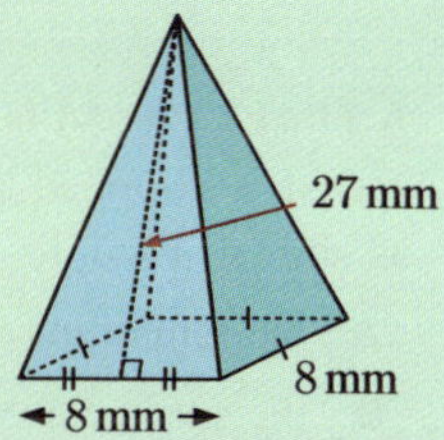

7 Li has a 10 cm × 8 cm × 5 cm block of clay. She wants to mould 50 clay spheres to use as heads for her figurines. Assuming there is no wastage, find the radius of the spheres Li should make.

8 A cylindrical bucket is 20 cm wide and 30 cm high. The bucket is filled with 6 litres of water. How far from the top of the bucket does the water reach?

9 Find the volume of each solid:

a

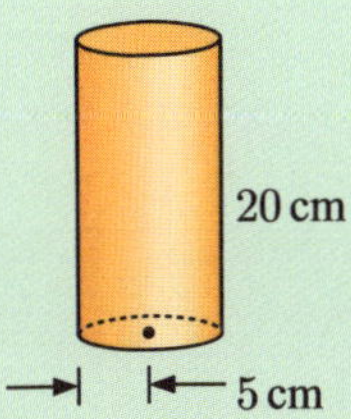

b

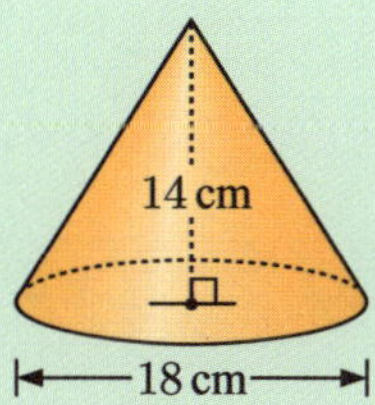

c

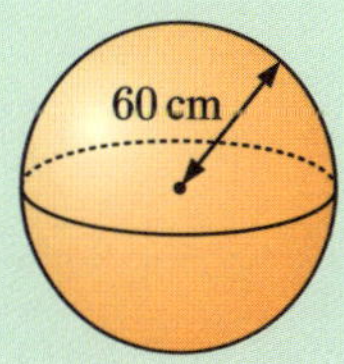

10 A basketball fits exactly into a cube which has sides equal in length to the diameter of a basketball. What percentage of the cube's volume does the basketball occupy?
Hint: Let the ball have radius r cm.

11 Find the **a** volume **b** surface area of the solid alongside.

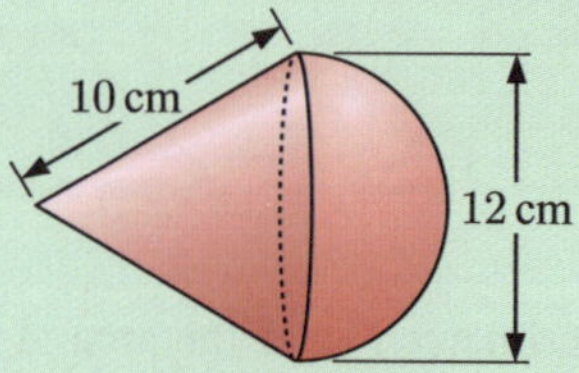

12 A horse's drinking trough has the dimensions shown. How long will it take to fill the trough if water flows into it at 12 litres per minute?

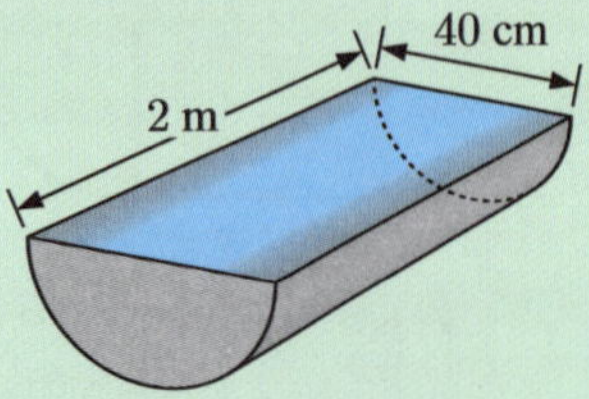

13 An ice block with uniform cross-section has the dimensions shown. Find:

a the volume of the ice block
b the amount of water, in mL, needed to make the ice block
c the number of complete ice blocks that can be made from a 2 L jug of water
d the surface area of the ice block.

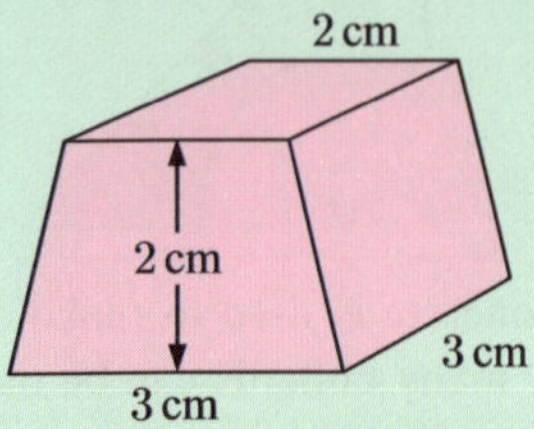

14 A block of wood is 30 cm by 40 cm by 20 cm. Its corner is chopped by an axe, and a section removed as shown. Find, to 1 decimal place:

a the remaining volume of the block
b the surface area of the resulting block.

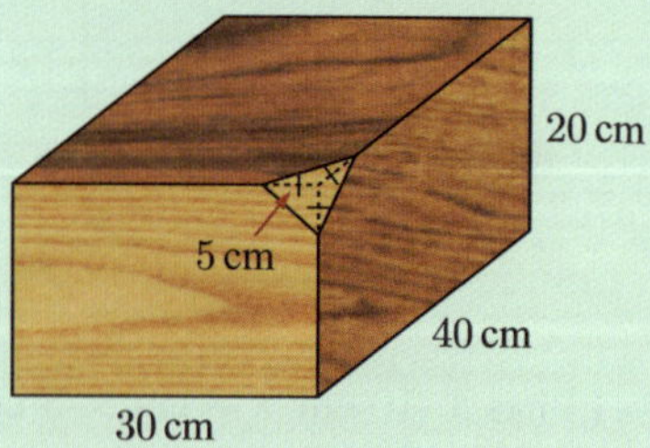

Chapter 8

Coordinate geometry

Contents:

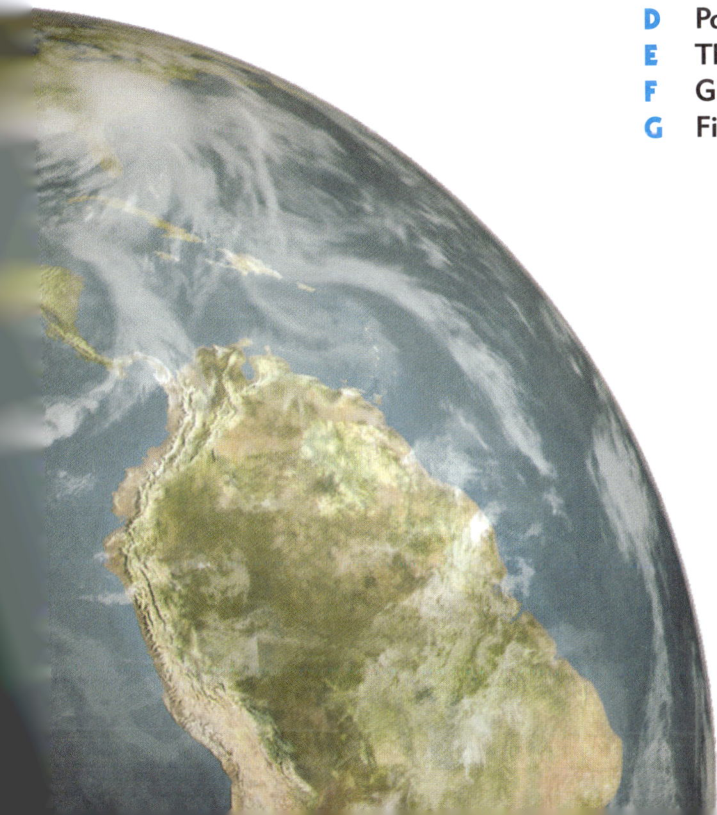

OPENING PROBLEM

The police receive a report that a house at H(1, 2) has been burgled. Police officers from the station closest to the house will attend the scene.

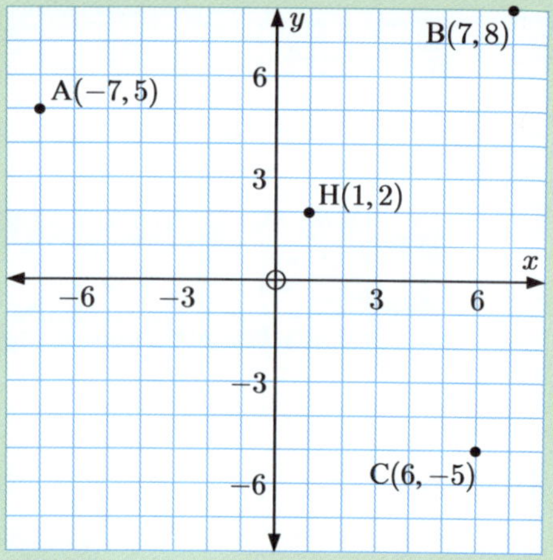

Things to think about:

- **a** Should the house owners expect to see police from station A, B, or C?
- **b** What is the equation of the straight line from the police station to the house?

THE NUMBER PLANE

The position of any point in the **Cartesian** or **number plane** can be described by an **ordered pair** of numbers (x, y).

These numbers tell us the steps we need to take from the origin O to get to the required point.

x is the **horizontal step** from O, and is the **x-coordinate** of the point.

y is the **vertical step** from O, and is the **y-coordinate** of the point.

For example:

- to locate the point $A(4, 2)$, we start at the origin O, move 4 units along the x-axis in the positive direction, and then 2 units in the positive y-direction
- to locate the point $B(-2, 3)$, we start at O, move 2 units in the negative x-direction, and then 3 units in the positive y-direction.

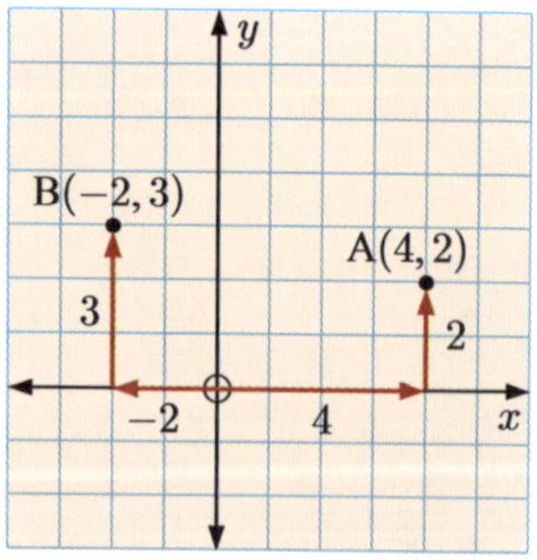

DEMO

A THE DISTANCE BETWEEN TWO POINTS

Suppose we want to find the distance from A(2, 4) to B(7, 1). Let this distance be d units.

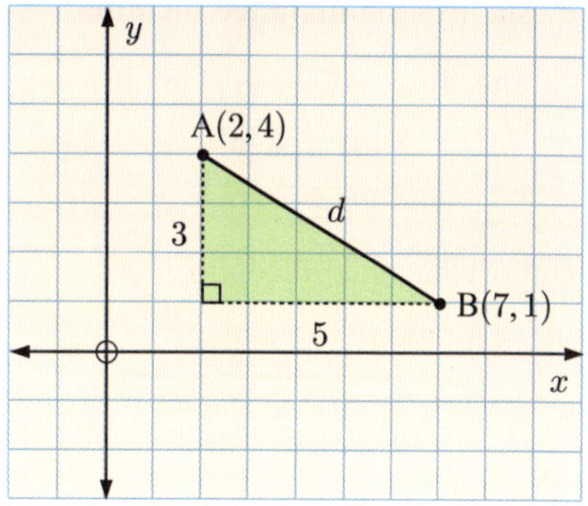

We plot the points on a set of axes, then join [AB] with a straight line. We then construct a right angled triangle, as shown.

DEMO

Notice that $d^2 = 5^2 + 3^2$ {Pythagoras}

$\therefore\ d^2 = 34$

$\therefore\ d = \sqrt{34}$ {as $d > 0$}

The distance from A to B is AB $= \sqrt{34}$ units.

EXERCISE 8A.1

1 If necessary, use Pythagoras' theorem to find the distance between:

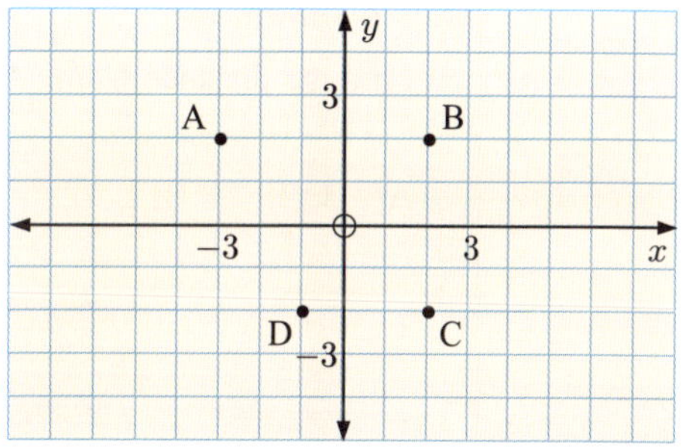

a A and B **b** B and C
c C and D **d** A and C
e B and D **f** O and A.

2 By plotting points and using Pythagoras' theorem, find the distance between:

a O(0, 0) and M(3, 5) **b** C(1, 2) and D(3, 7)
c A(−1, −2) and B(2, 3) **d** A(1, 4) and B(2, −1)
e P(3, 5) and Q(−1, 4) **f** R(−2, 0) and S(0, 3).

3 Two planes are travelling from A(−5, −4) to B(7, 5). The first plane must touch down at C(0, 2) along the way, and the second plane must touch down at D(4, 0) along the way. Each grid unit represents 10 km.

a Plot the points A, B, C, and D on a number plane.
b Find the distance between A and B.
c Find, correct to 1 decimal place, the total distance travelled by:
  i the first plane **ii** the second plane.
d Which plane has the shorter distance to travel?

4 **a** On the same set of axes, plot A(−3, −2), B(5, 2), C(−1, 3), and D(7, −1).
b Draw the lines (AB) and (CD), and find the coordinates of their point of intersection.
c Find the distance from this point of intersection to the closest of the points A, B, C, or D.

THE DISTANCE FORMULA

Instead of graphing points and applying Pythagoras' theorem, we establish the **distance formula**.

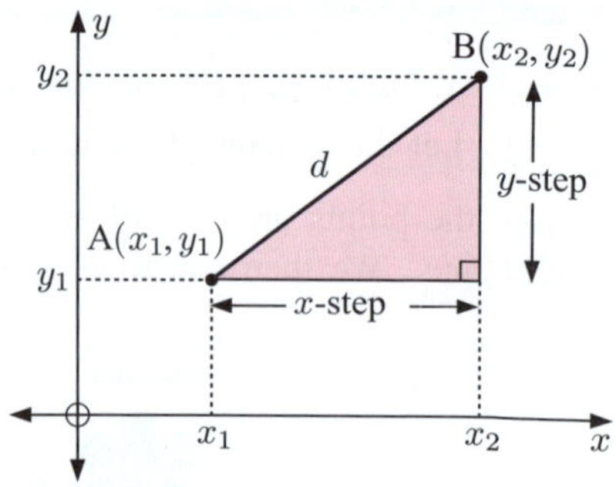

In going from $A(x_1, y_1)$ to $B(x_2, y_2)$,

the x-step $= x_2 - x_1$

and the y-step $= y_2 - y_1$.

Using Pythagoras' theorem,

$$d^2 = (x\text{-step})^2 + (y\text{-step})^2$$

$$\therefore \quad d = \sqrt{(x\text{-step})^2 + (y\text{-step})^2}$$

$$\therefore \quad d = \sqrt{(x_2 - x_1)^2 + (y_2 - y_1)^2}$$

The distance between two points on the number plane, $A(x_1, y_1)$ and $B(x_2, y_2)$, is given by:

$$\mathbf{AB} = \sqrt{(x_2 - x_1)^2 + (y_2 - y_1)^2}$$

Example 1 — Self Tutor

Find the distance from $A(-3, 2)$ to $B(2, 4)$.

$A(-3, 2)$: $x_1 = -3$, $y_1 = 2$ $\quad$ $B(2, 4)$: $x_2 = 2$, $y_2 = 4$

$$\begin{aligned} AB &= \sqrt{(x_2 - x_1)^2 + (y_2 - y_1)^2} \\ &= \sqrt{(2 - -3)^2 + (4 - 2)^2} \\ &= \sqrt{5^2 + 2^2} \\ &= \sqrt{25 + 4} \\ &= \sqrt{29} \text{ units} \end{aligned}$$

We can use the distance formula to help us classify triangles as scalene, isosceles, or equilateral.

We can also test whether a triangle is **right angled**. We find the lengths of its sides, and then use the **converse of Pythagoras' theorem**:

If a triangle has sides of length a, b, and c units, where $a^2 + b^2 = c^2$, then the triangle is right angled. The right angle will be opposite the longest side c.

EXERCISE 8A.2

1 Use the distance formula to find the distance between:

- **a** $O(0, 0)$ and $P(3, -1)$
- **b** $A(2, 1)$ and $B(4, 4)$
- **c** $C(-2, 1)$ and $D(2, 5)$
- **d** $E(4, 3)$ and $F(-1, -1)$
- **e** $G(0, -3)$ and $H(1, 4)$
- **f** $I(0, 2)$ and $J(-3, 0)$
- **g** $P(3, 9)$ and $Q(11, -1)$
- **h** $R(-\sqrt{2}, 3)$ and $S(\sqrt{2}, -1)$.

Example 2 **Self Tutor**

The points P(1, 0), Q(3, 1), and R(−3, 8) form a triangle PQR.

a Use the distance formula to classify the triangle as equilateral, isosceles, or scalene.

b Determine whether the triangle is right angled.

a

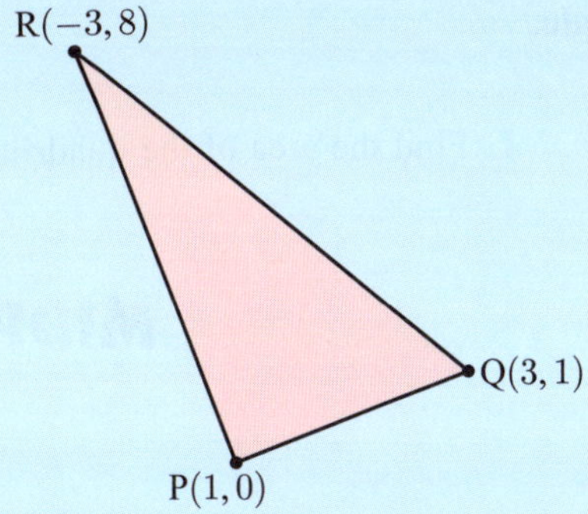

$$PQ = \sqrt{(3-1)^2 + (1-0)^2} = \sqrt{2^2 + 1^2} = \sqrt{5} \text{ units}$$

$$PR = \sqrt{(-3-1)^2 + (8-0)^2} = \sqrt{(-4)^2 + 8^2} = \sqrt{80} \text{ units}$$

$$QR = \sqrt{(-3-3)^2 + (8-1)^2} = \sqrt{(-6)^2 + 7^2} = \sqrt{85} \text{ units}$$

Since none of the side lengths are equal, triangle PQR is scalene.

b The shorter sides are [PQ] and [PR].

Now $PQ^2 + PR^2 = 5 + 80$
$= 85$
$= QR^2$

Using the converse of Pythagoras' theorem, triangle PQR is right angled. The right angle is opposite the longest side, so the right angle is at P.

2 Classify the triangles formed by the following sets of points as scalene, isosceles, or equilateral:

a A(0, −2), B(1, 2), and C(−3, 1)
b X(1, 3), Y(−1, 0), and Z(3, −4)
c H(3, −2), I(1, 4), and J(−3, 0)
d W($2\sqrt{3}$, 0), X(0, 6), and Y($-2\sqrt{3}$, 0).

3 For the following sets of points, determine whether triangle ABC is right angled. If there is a right angle, state the vertex where it occurs.

a A(3, 2), B(1, 5), C(9, 6)
b A(−3, −5), B(0, 4), C(1, −4)
c A(−5, 4), B(1, 6), C(−1, 3)
d A(−7, 5), B(−6, 1), C(2, 3)

4 Find, correct to 1 decimal place, the perimeter of triangle PQR.

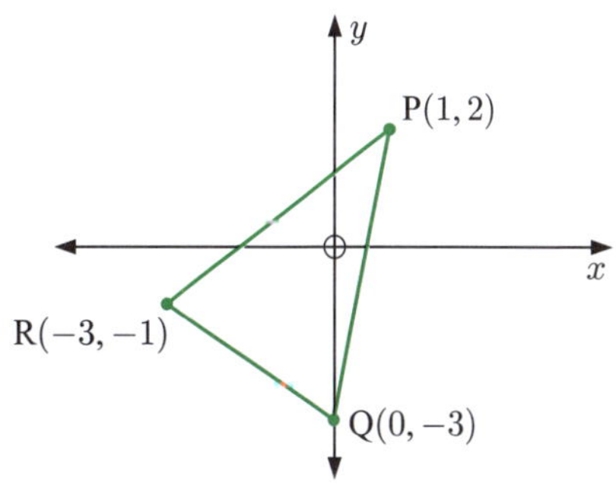

5 Consider triangle LMN with vertices L(−1, 2), M(3, 4), and N(5, 0).

  a Show that the triangle is isosceles.
  b Show that the triangle is right angled at M.
  c Find the size of the angle at L.

6 Quadrilateral WXYZ has vertices W(−3, 2), X(−2, 4), Y(4, 1), and Z(3, −1).

  a Show that WX = YZ and XY = WZ.
  b Find the length of the diagonal [WY].
  c Hence show that the angle at X is a right angle.
  d What type of quadrilateral is WXYZ?
  e Draw the quadrilateral on a number plane.
  f Find the area of the quadrilateral.

B MIDPOINTS

The **midpoint** of line segment [AB] is the point which lies midway between points A and B.

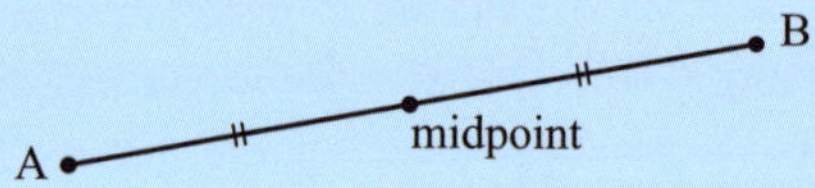

Consider the points A(1, −2) and B(5, 3). From the diagram we see that the midpoint of [AB] is M(3, $\frac{1}{2}$).

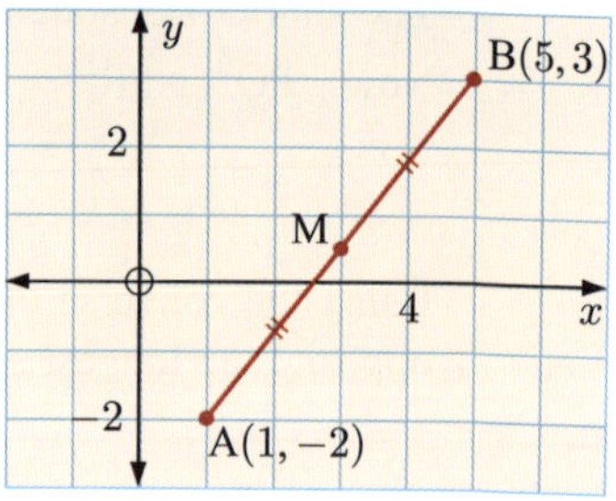

Notice that: the x-coordinate of M $= \dfrac{1+5}{2} = 3$

the y-coordinate of M $= \dfrac{-2+3}{2} = \dfrac{1}{2}$.

The x-coordinate of M is the *average* of the x-coordinates of A and B.
The y-coordinate of M is the *average* of the y-coordinates of A and B.

If A(x_1, y_1) and B(x_2, y_2) are two points, then the **midpoint** of [AB] has coordinates

$$\left(\frac{x_1 + x_2}{2}, \frac{y_1 + y_2}{2}\right).$$

DEMO

Example 3 Self Tutor

Find the midpoint M of [AB] given A(−2, 3) and B(4, 8).

A(−2, 3) B(4, 8)
x_1 y_1 x_2 y_2

The x-coordinate of M
$= \dfrac{x_1 + x_2}{2}$
$= \dfrac{-2+4}{2}$
$= 1$

The y-coordinate of M
$= \dfrac{y_1 + y_2}{2}$
$= \dfrac{3+8}{2}$
$= \frac{11}{2}$

So, the midpoint is M(1, $5\frac{1}{2}$).

EXERCISE 8B

1 **a** Find the midpoint M of [AB].

b Use the distance formula to check that AM = BM.

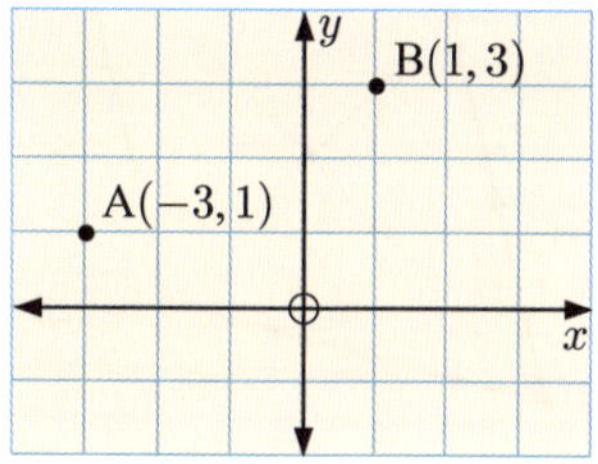

2 Find the midpoint of [AB] given:

a A(5, 2) and B(1, 8)

b A(4, 1) and B(4, −1)

c A(1, 0) and B(−3, 2)

d A(6, 0) and B(0, 3)

e A(0, −1) and B(−3, 5)

f A(−2, 4) and B(4, −2)

g A(−4, 3) and B(9, 5)

h A(−5, 1) and B(−2, 6)

i A(a, 5) and B(2, −1)

j A(a, b) and B($-a$, $3b$).

Example 4 **Self Tutor**

Given that M is the midpoint of [PQ], find the coordinates of Q.

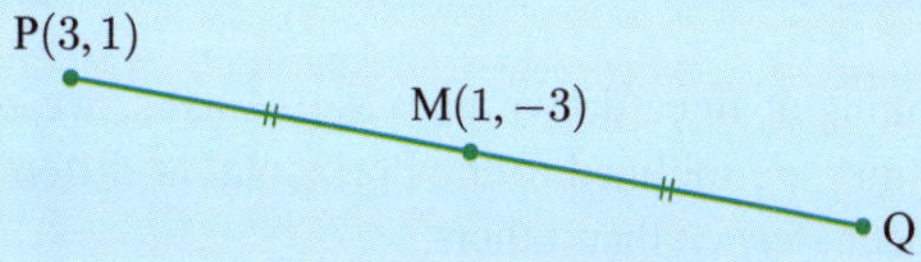

Suppose Q has coordinates (a, b).

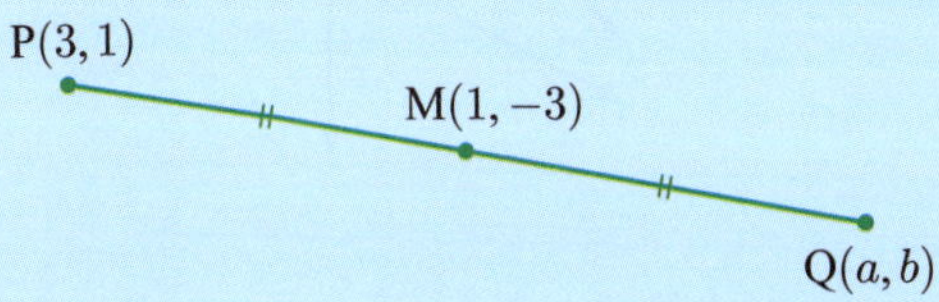

$\therefore$ M is $\left(\dfrac{3+a}{2}, \dfrac{1+b}{2}\right)$

$\therefore$ $\dfrac{3+a}{2} = 1$ and $\dfrac{1+b}{2} = -3$

$\therefore$ $3+a = 2$ and $1+b = -6$

$\therefore$ $a = -1$ and $b = -7$

So, Q is (−1, −7).

3 M is the midpoint of [PQ]. Find the coordinates of Q if:

a P is (−1, 3) and M is (−1, 7)

b P is (−1, 0) and M is (0, −5)

c M is $(2, 1\frac{1}{2})$ and P is (−2, 3)

d M is (−1, 3) and P is $(-\frac{1}{2}, 0)$.

4 T is the midpoint of [AB]. Find the coordinates of A if:

a T is (4, −3) and B is (−2, 3)

b B is (0, 4) and T is (−3, −2).

5 Find the coordinates of:

a B

b D.

6 A circle has diameter [AB]. The centre of the circle is (1, 3), and B has coordinates (4, −1).

a Find the coordinates of A.

b Find the length of the diameter.

7

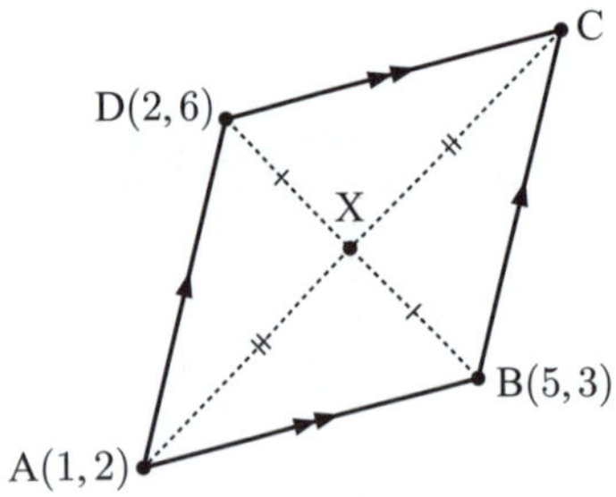

ABCD is a parallelogram. Diagonals [AC] and [BD] bisect each other at X. Find the coordinates of:

a X **b** C.

8 In triangle ABC, M is the midpoint of [AB].
Find the coordinates of N, the midpoint of [CM].

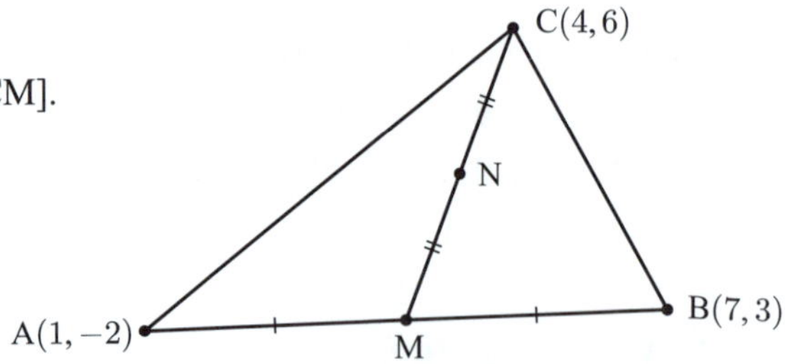

C GRADIENT

When looking at lines drawn on a set of axes, we notice that different lines are inclined to the horizontal at different angles. Some appear *steeper* than others.

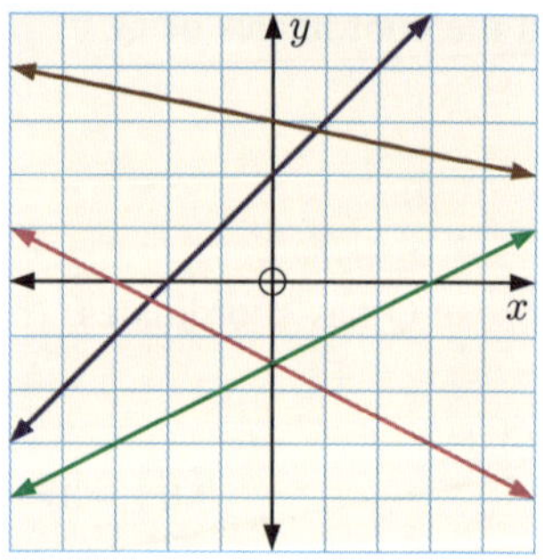

The **gradient** of a line is a measure of its steepness.

If we choose any two points on the line, we can move from one point to the other by making a positive **horizontal step** followed by a **vertical step**.

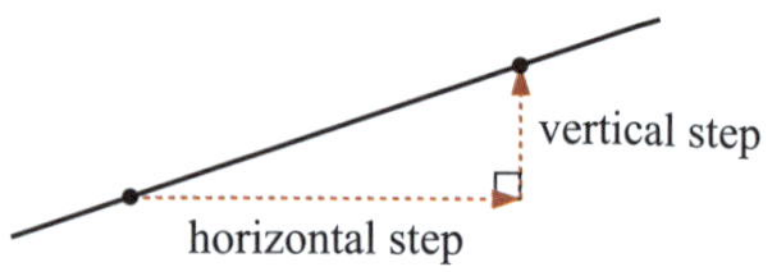

The gradient is calculated by dividing the vertical step by the horizontal step.

GEOMETRY PACKAGE

The **gradient** of a line $= \dfrac{\text{vertical step}}{\text{horizontal step}}$ or $\dfrac{y\text{-step}}{x\text{-step}}$.

If the line is sloping **upwards**, both steps are positive, so the line has a **positive gradient**.

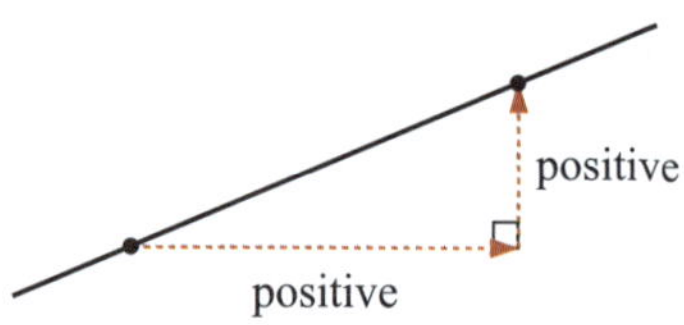

Lines like

are upwards sloping and have positive gradients.

If the line is sloping **downwards**, the horizontal step is positive and the vertical step is negative, so the line has a **negative gradient**.

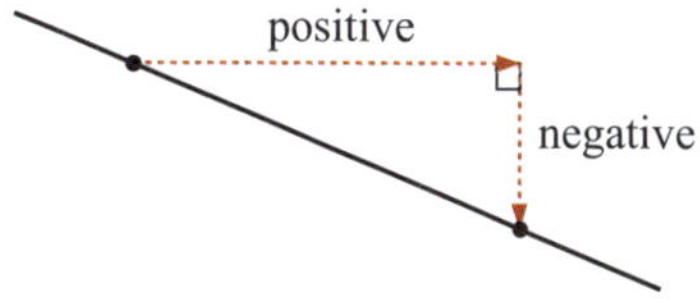

Lines like 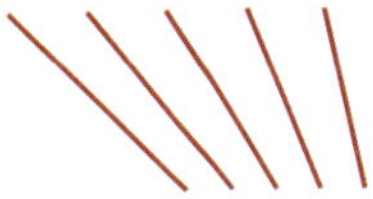 are downwards sloping and have negative gradients.

DISCUSSION

Why do you think we define gradient by $\dfrac{y\text{-step}}{x\text{-step}}$ rather than $\dfrac{x\text{-step}}{y\text{-step}}$?

Example 5

Self Tutor

Find the gradient of each line:

a 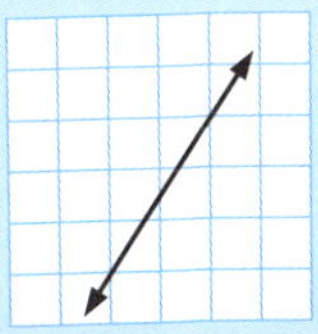

b 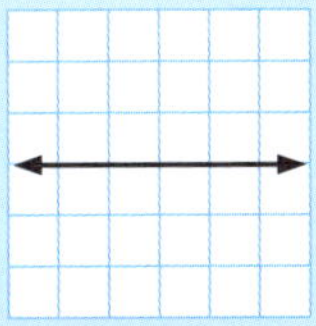

c 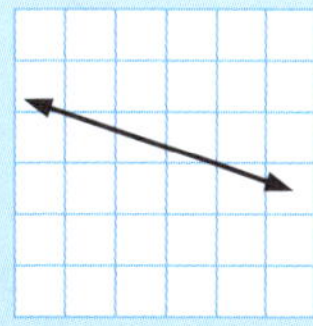

d 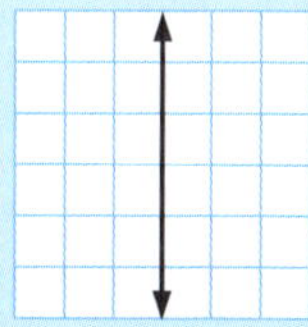

a 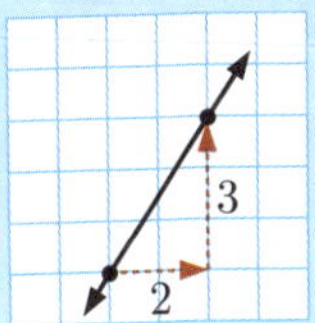

gradient
$= \dfrac{y\text{-step}}{x\text{-step}}$
$= \dfrac{3}{2}$

b 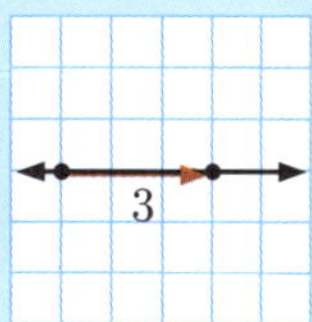

gradient
$= \dfrac{y\text{-step}}{x\text{-step}}$
$= \dfrac{0}{3}$
$= 0$

c 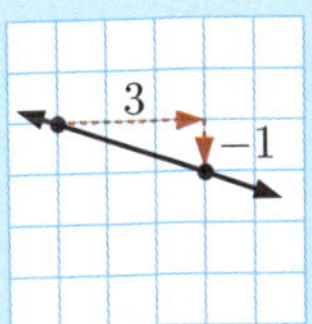

gradient
$= \dfrac{y\text{-step}}{x\text{-step}}$
$= \dfrac{-1}{3}$
$= -\dfrac{1}{3}$

d 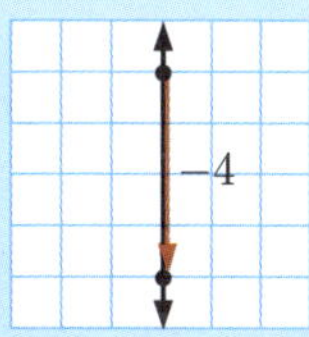

gradient
$= \dfrac{y\text{-step}}{x\text{-step}}$
$= \dfrac{-4}{0}$
which is undefined

Notice that:

- The gradient of all **horizontal** lines is **0**.
- The gradient of all **vertical** lines is **undefined**.

EXERCISE 8C.1

1 Find the gradient of each line:

a 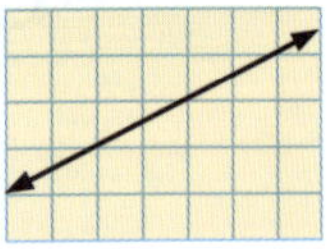

b 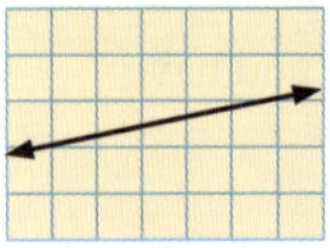

c 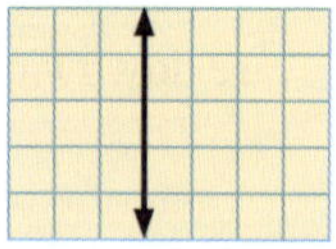

d 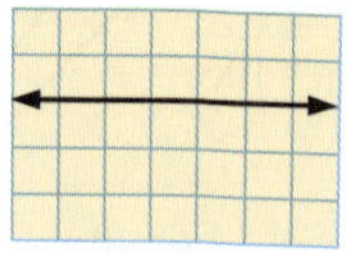

e 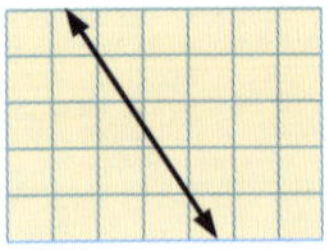

f 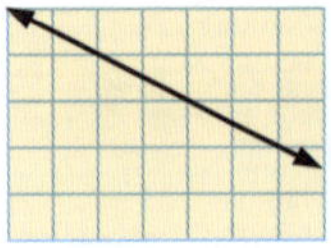

g 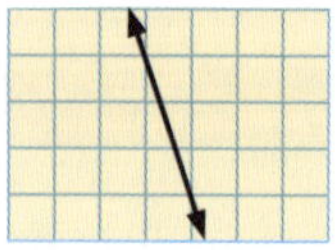

h 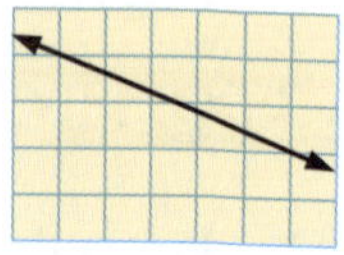

2 Consider the lines:

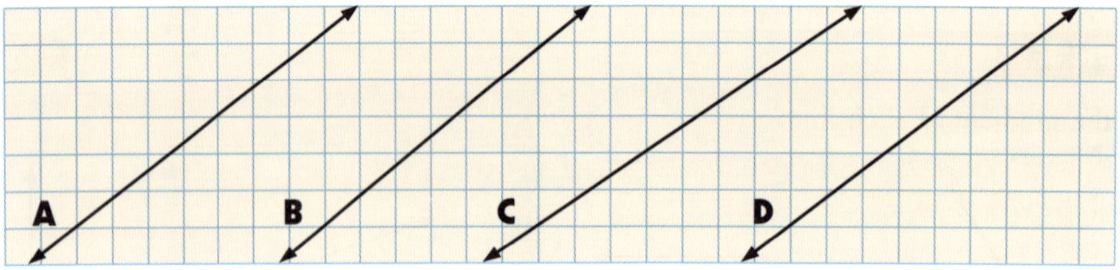

a Which two lines have the same gradient?

b Which line is steepest?

Example 6 — Self Tutor

Draw two lines whose gradients are $\frac{3}{4}$ and -2, through the point $(2, 1)$.

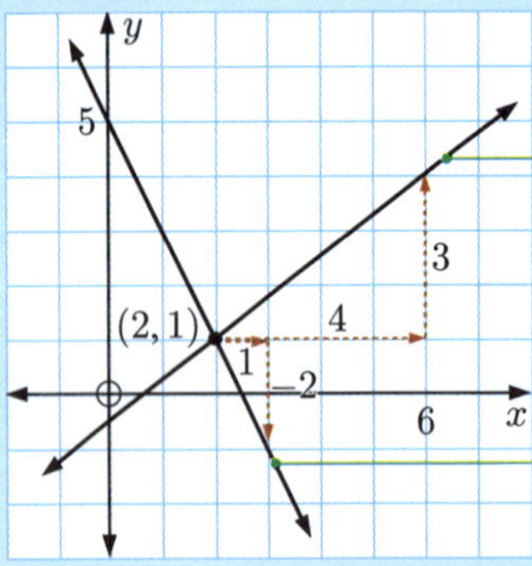

gradient $= \frac{3}{4}$ ← y-step / ← x-step

gradient $= -2$

$= \frac{-2}{1}$ ← y-step / ← x-step

3 On grid paper, draw a line segment with gradient:

a $\frac{1}{2}$ **b** $\frac{2}{3}$ **c** 3 **d** $-\frac{2}{3}$ **e** -2 **f** $-\frac{4}{5}$ **g** 0

4 On the same set of axes, draw lines through $(0, 0)$ with gradients 0, $\frac{1}{3}$, $\frac{3}{4}$, 1, $\frac{3}{2}$, $\frac{5}{3}$, and 5.

5 On the same set of axes, draw lines through $(3, 4)$ with gradients 0, $-\frac{1}{2}$, -1, $-\frac{3}{4}$, and -4.

THE GRADIENT FORMULA

To save us having to plot points on a graph each time, we can use a formula to find the gradient of a line.

If the line passes through $A(x_1, y_1)$ and $B(x_2, y_2)$, the vertical step is $y_2 - y_1$, and the horizontal step is $x_2 - x_1$.

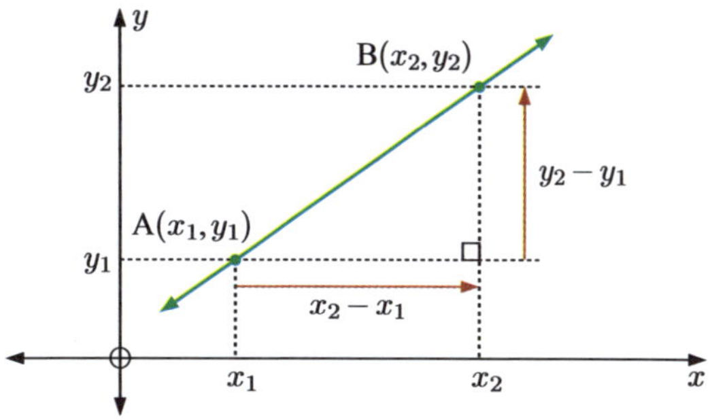

The **gradient** of the line through (x_1, y_1) and (x_2, y_2) is $\dfrac{y_2 - y_1}{x_2 - x_1}$.

Example 7 — Self Tutor

Find the gradient of [PQ] given $P(1, 3)$ and $Q(4, -2)$.

$P(1, 3)$: $x_1 = 1$, $y_1 = 3$ $Q(4, -2)$: $x_2 = 4$, $y_2 = -2$

$$\text{gradient of [PQ]} = \frac{y_2 - y_1}{x_2 - x_1} = \frac{-2-3}{4-1} = \frac{-5}{3} = -\tfrac{5}{3}$$

EXERCISE 8C.2

1 Find the gradient of [PQ] given:

a $P(1, 4)$ and $Q(8, 6)$ **b** $P(2, 7)$ and $Q(6, -9)$ **c** $P(-4, 1)$ and $Q(8, -7)$.

2 Find the gradient of the line segment connecting:

a $(4, 7)$ and $(3, 2)$ **b** $(6, 1)$ and $(7, 5)$ **c** $(6, 3)$ and $(-2, 1)$

d $(0, 0)$ and $(4, -3)$ **e** $(5, -1)$ and $(5, 5)$ **f** $(-4, 3)$ and $(-1, 3)$

g $(-3, -2)$ and $(-1, 5)$ **h** $(0, 2)$ and $(2, -5)$ **i** $(-2, -2)$ and $(-1, 0)$.

3 For the triangle ABC shown, find the gradient of:

a [AB] **b** [BC] **c** [CA].

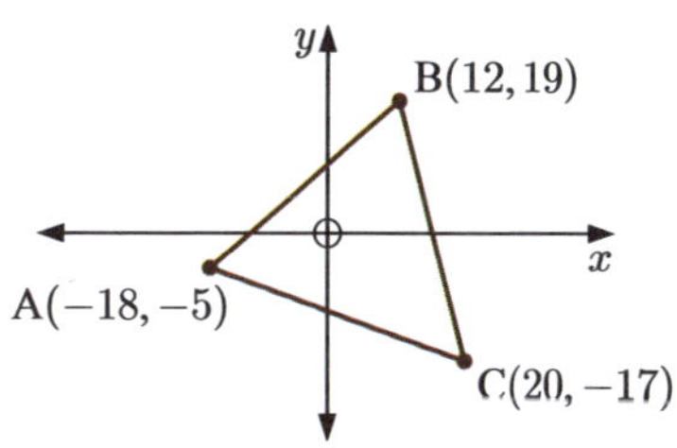

4 Consider the points $A(1, 3)$ and $B(4, 5)$.

a Use the gradient formula to find the gradient of line segment: **i** [AB] **ii** [BA].

b Explain your results in **a**.

5 A quadrilateral ABCD has vertices $A(-4, 2)$, $B(2, 5)$, $C(5, 4)$, and $D(2, -4)$.

a Find the gradient of each side of ABCD.

b Which of the sides slope:

  i upwards **ii** downwards?

c Let M be the midpoint of [AD]. Find the gradient of:

  i [BM] **ii** [CM].

Example 8 **Self Tutor**

The graph alongside shows the distance travelled by a car over time.

a Find the gradient of the line.

b Interpret the gradient found in **a**.

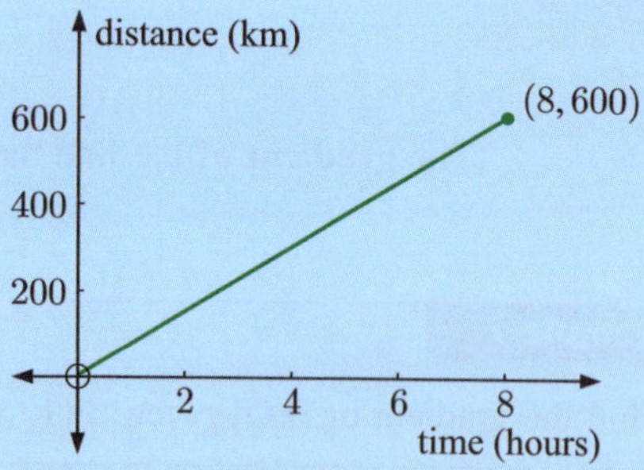

a The line passes through the points $(0, 0)$ and $(8, 600)$.

$$\therefore \text{ gradient} = \frac{600-0}{8-0} = \frac{600}{8} = 75$$

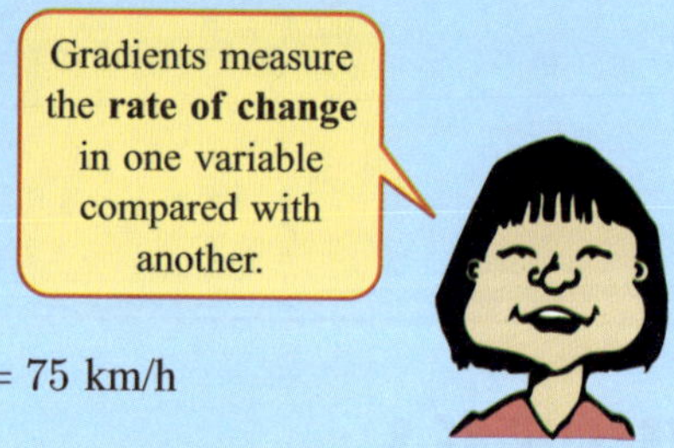

b The car travelled 600 km in 8 hours.

$\therefore$ the average speed of the car $= \dfrac{\text{distance}}{\text{time}} = \dfrac{600 \text{ km}}{8 \text{ hours}} = 75$ km/h

The gradient can be interpreted as the *speed* of the car.

6 The graph alongside indicates the distance run by a sprinter over time.

a Find the gradient of the line.

b Interpret the gradient found in **a**.

c Is the speed of the runner constant? Explain your answer.

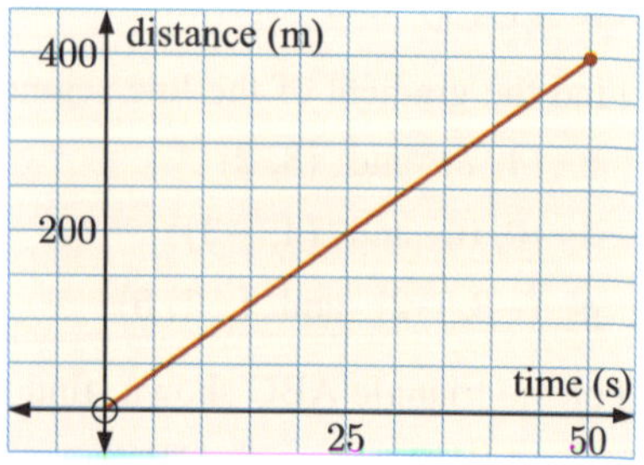

7

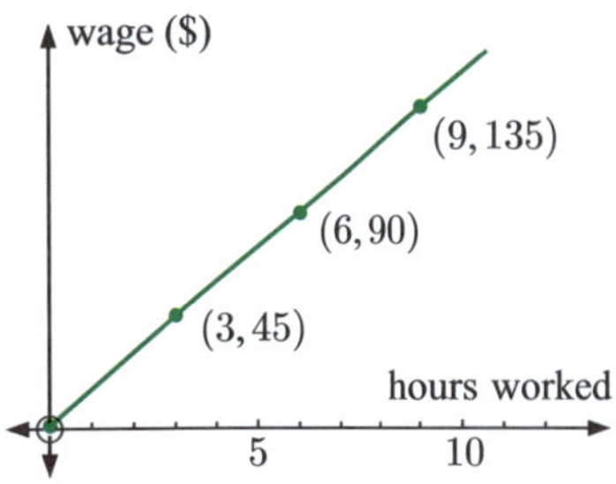

The graph shows the wages paid to sales assistants in a department store.

a Find the gradient of the line.

b Interpret the gradient found in **a**.

c Determine the wage for working:

  i 6 hours **ii** 18 hours.

8 The graph indicates the distance travelled by a truck driver over time during a journey.

  a Determine the average speed from:
    i A to B **ii** C to D.

  b Find the average speed for the whole trip.

  c Determine the time interval(s) over which the average speed was greatest.

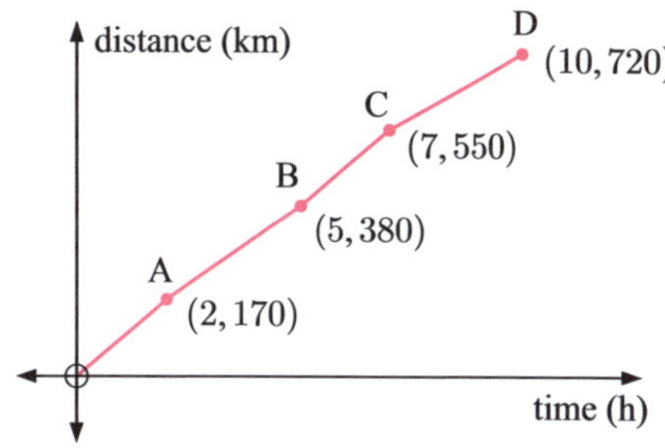

D PARALLEL AND PERPENDICULAR LINES

Two lines are **parallel** if they are side by side with the same distance between them.

Two lines are **perpendicular** if they are at right angles to one another.

INVESTIGATION PARALLEL AND PERPENDICULAR LINES

WORKSHEET

In this Investigation we will discover the relationship between the gradients of parallel and perpendicular lines.

What to do:

1 In the graph alongside, [AB] is parallel to [CD], and [PQ] is parallel to [RS].
Copy and complete:

  a The gradient of [AB] = ...
  The gradient of [CD] = ...

  b The gradient of [PQ] = ...
  The gradient of [RS] = ...

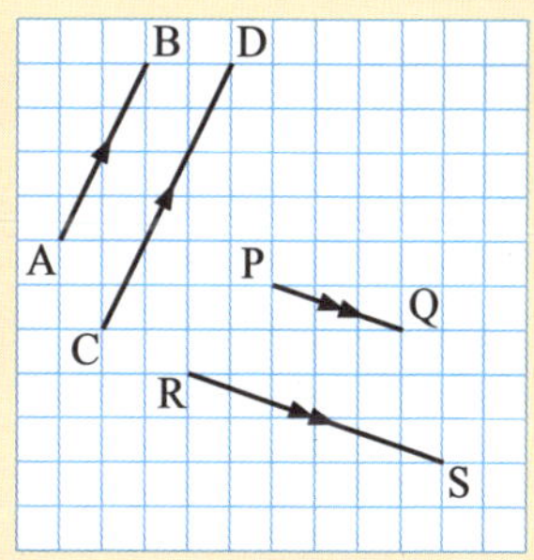

2 In the graph alongside, [AB] is perpendicular to [CD], and [PQ] is perpendicular to [RS].
Copy and complete:

  a The gradient of [AB] = ...
  The gradient of [CD] = ...
  gradient of [AB] × gradient of [CD] = ...

  b The gradient of [PQ] = ...
  The gradient of [RS] = ...
  gradient of [PQ] × gradient of [RS] = ...

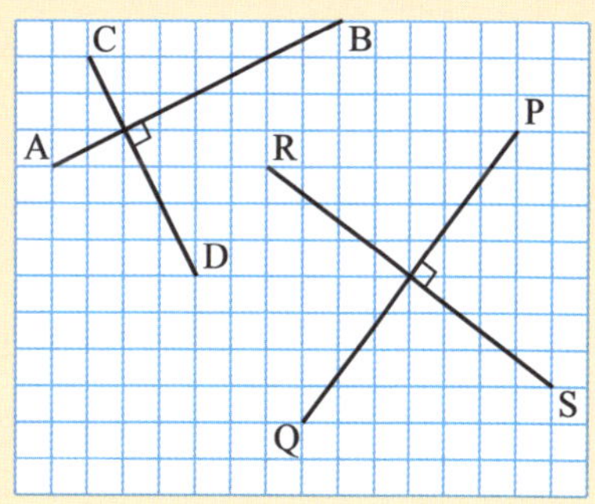

3 Copy and complete:

  a Parallel lines have ......

  b Perpendicular lines have gradients whose product is ......

4 Discuss whether the rules in **3** are valid for horizontal and vertical lines.

From the **Investigation**, you should have discovered that:

For lines which are not horizontal or vertical:

- the lines are **parallel** $\Leftrightarrow$ they have **equal gradient**
- the lines are **perpendicular** $\Leftrightarrow$ their gradients are **negative reciprocals**.

Example 9 — Self Tutor

A line L has gradient $\frac{2}{5}$. Find the gradient of any line which is:

a parallel to L **b** perpendicular to L.

a Parallel lines have equal gradients, so a line parallel to L will also have gradient $\frac{2}{5}$.

b The negative reciprocal of $\frac{2}{5}$ is $-\frac{5}{2}$.

$\therefore$ a line perpendicular to L will have gradient $-\frac{5}{2}$.

Example 10 — Self Tutor

Consider the points A(1, 0), B(2, 4), and C(5, 1).

a Show that triangle ABC is isosceles.

b Find the midpoint M of [BC].

c Use gradients to show that [AM] and [BC] are perpendicular.

d Illustrate **a**, **b**, and **c** on a set of axes.

a AB $= \sqrt{(2-1)^2+(4-0)^2} = \sqrt{1+16} = \sqrt{17}$ units

BC $= \sqrt{(5-2)^2+(1-4)^2} = \sqrt{9+9} = \sqrt{18}$ units

AC $= \sqrt{(5-1)^2+(1-0)^2} = \sqrt{16+1} = \sqrt{17}$ units

Since AB = AC, $\triangle$ABC is isosceles.

b M is $\left(\frac{2+5}{2}, \frac{4+1}{2}\right)$, which is $\left(\frac{7}{2}, \frac{5}{2}\right)$.

c gradient of [AM] $= \dfrac{\frac{5}{2}-0}{\frac{7}{2}-1} = \dfrac{\frac{5}{2}}{\frac{5}{2}} = 1$

gradient of [BC] $= \dfrac{1-4}{5-2} = -\dfrac{3}{3} = -1$

Since the gradients are negative reciprocals, [AM] and [BC] are perpendicular.

d

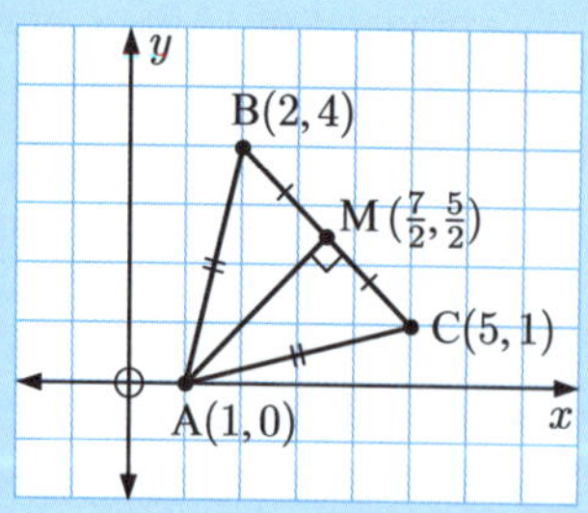

EXERCISE 8D

1 Line M has gradient $\frac{3}{5}$, and line N has gradient 6. Find the gradient of any line which is:

- **a** parallel to M
- **b** perpendicular to M
- **c** parallel to N
- **d** perpendicular to N.

2 Find the gradient of a line which is perpendicular to a line with gradient:

a $\frac{4}{5}$ **b** $\frac{1}{3}$ **c** $-\frac{3}{7}$ **d** 2 **e** -11 **f** 0 **g** undefined **h** $-1\frac{1}{2}$

3 Consider the points A$(-2, 4)$, B$(4, 8)$, C$(7, 5)$, and D$(-11, -7)$.

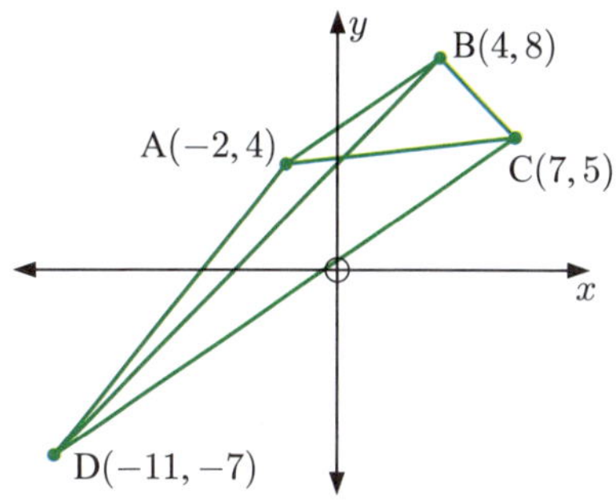

- **a** Find the gradient of:
  - **i** [AB] **ii** [BC] **iii** [CD]
  - **iv** [DA] **v** [AC] **vi** [BD]
- **b** Which of the line segments are:
  - **i** parallel **ii** perpendicular?

4 A$(3, 1)$, B$(2, -4)$, and C$(7, -5)$ are three points in the Cartesian plane.

- **a** Find the gradient of: **i** [AB] **ii** [BC].
- **b** What can be said about [AB] and [BC]?
- **c** Classify triangle ABC.

5 Consider the points A$(-1, 0)$, B$(1, 4)$, and C$(3, 2)$.

- **a** Show that triangle ABC is isosceles.
- **b** Find the midpoint M of [BC].
- **c** Use gradients to show that [AM] and [BC] are perpendicular.
- **d** Illustrate **a**, **b**, and **c** on a set of axes.

6 Triangle ABC has vertices A$(1, 3)$, B$(5, 1)$, and C$(3, -3)$.
M is the midpoint of [AB] and N is the midpoint of [BC]. Show that [MN] is:

- **a** parallel to [AC]
- **b** half the length of [AC].

7 Consider the points A$(7, 7)$, B$(16, 7)$, C$(1, -2)$, and D$(-8, -2)$.

- **a** Show that:
  - **i** [AB] is parallel to [CD]
  - **ii** [AD] is parallel to [BC].
- **b** What kind of quadrilateral is ABCD?

8

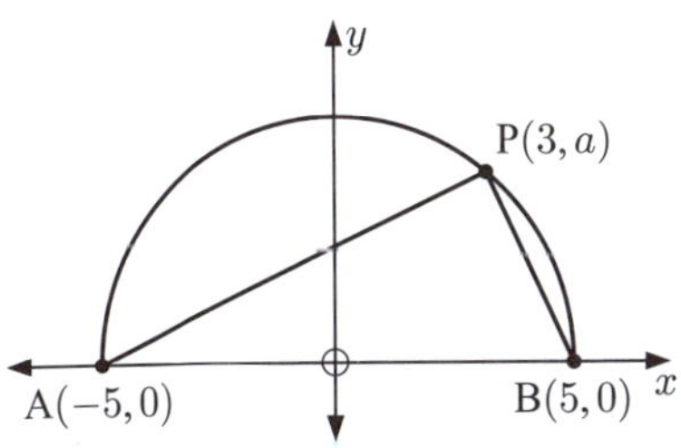

[AB] is the diameter of a semi-circle.
P$(3, a)$ lies on the semi-circle.

- **a** Find a.
- **b** Find the gradient of: **i** [AP] **ii** [BP].
- **c** Show that $\widehat{APB}$ is a right angle.

E THE EQUATION OF A LINE

The **equation of a line** is a rule which connects the x and y-coordinates of **all** points on the line.

The equation of a line is commonly written in either **gradient-intercept form** or in **general form**.

GRADIENT-INTERCEPT FORM

$y = mx + c$ is called the **gradient-intercept form** of an equation of a line.

The line with equation $y = mx + c$ has gradient m and y-intercept c.

For example, the line with equation $y = \frac{1}{2}x + 4$ has gradient $\frac{1}{2}$ and y-intercept 4.

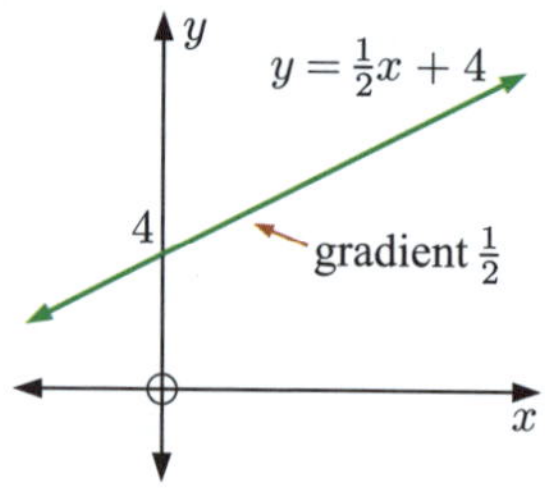

GENERAL FORM

$Ax + By = C$ is called the **general form** of the equation of a line.

For example, the equations $4x + 5y = 3$ and $x - 3y = -4$ are in general form.

Equations in general form are usually written with a positive coefficient of x.

An advantage of the general form is that we can write the equation without fractions.

We can rearrange equations to convert between gradient-intercept form and general form.

Example 11 **Self Tutor**

Write the equation: **a** $y = -\frac{1}{4}x + \frac{3}{4}$ in general form
b $2x + 5y = 7$ in gradient-intercept form.

a $y = -\frac{1}{4}x + \frac{3}{4}$

$\therefore\ 4y = -x + 3$ {multiplying both sides by 4}

$\therefore\ x + 4y = 3$ {adding x to both sides}

b $2x + 5y = 7$

$\therefore\ 5y = -2x + 7$ {subtracting $2x$ from both sides}

$\therefore\ y = -\frac{2}{5}x + \frac{7}{5}$ {dividing both sides by 5}

EXERCISE 8E.1

1 State whether the following equations are in gradient-intercept form or general form:

a $y = 3x - 2$ **b** $5x + 2y = 8$ **c** $x - y = 7$

d $y = -x + 4$ **e** $y = -\frac{1}{5}x + \frac{2}{5}$ **f** $2x - 9y = 8$

2 State the gradient and y-intercept of the line with equation:

a $y = 2x + 5$ **b** $y = -6x - 1$ **c** $y = \frac{3}{5}x - \frac{2}{5}$

d $y = 10 - x$ **e** $y = 1 - 4x$ **f** $y = \frac{5}{4} - \frac{7}{4}x$

g $y = \dfrac{3x + 1}{7}$ **h** $y = \dfrac{4x - 5}{8}$ **i** $y = \dfrac{6 - 7x}{9}$

3 Write the following equations in general form:

a $y = -2x + 11$ **b** $y = 3x - 4$ **c** $y = -\frac{1}{5}x + \frac{6}{5}$

d $y = -\frac{6}{7}x + \frac{5}{7}$ **e** $y = \frac{5}{8}x - \frac{1}{8}$ **f** $y = \frac{4}{9}x + 2$

4 For each of the following lines:

 i write the equation in gradient-intercept form

 ii state the gradient of the line.

a $3x + y = 5$ **b** $2x + 5y = 10$ **c** $7x + 4y = -9$

d $6x - y = 1$ **e** $5x - 11y = 2$ **f** $9x - 2y = -5$

DISCUSSION

What is the gradient of a line with general form $Ax + By = C$?

POINTS ON LINES

A point lies on a line if its coordinates satisfy the equation of the line.

Example 12 — Self Tutor

Determine whether:

a $(3, 7)$ lies on the line with equation $y = 2x + 1$

b $(4, -2)$ lies on the line with equation $5x + 3y = 16$.

a When $x = 3$, we have

$$\begin{aligned} y &= 2(3) + 1 \\ &= 6 + 1 \\ &= 7 \ \checkmark \end{aligned}$$

So, $(3, 7)$ does lie on the line.

b Substituting $x = 4$ and $y = -2$ into the LHS gives

$$\begin{aligned} &5(4) + 3(-2) \\ &= 20 - 6 \\ &= 14 \end{aligned}$$

Since LHS $\neq$ RHS, $(4, -2)$ does *not* lie on the line.

EXERCISE 8E.2

1 Determine whether:

a $(5, 13)$ lies on the line with equation $y = 3x - 2$

b $(-2, -9)$ lies on the line with equation $y = \frac{1}{2}x - 10$

c $(\frac{1}{2}, \frac{3}{2})$ lies on the line with equation $y = 5x - 1$

d $(-\frac{1}{3}, -1)$ lies on the line with equation $y = -\frac{1}{2}x - \frac{7}{6}$.

2 Determine whether:

a $(6, 5)$ lies on the line with equation $3x + y = 23$

b $(-2, 3)$ lies on the line with equation $4x - 3y = 1$

c $(4, \frac{1}{2})$ lies on the line with equation $3x + 8y = 14$

d $(-\frac{2}{3}, \frac{1}{3})$ lies on the line with equation $6x + 9y = -1$.

Example 13 Self Tutor

a Find c given that $(-5, -2)$ lies on the line with equation $y = 4x + c$.

b Find k given that $(4, 7)$ lies on the line with equation $3x - 2y = k$.

a Substituting $x = -5$ and $y = -2$ into the equation gives

$$-2 = 4(-5) + c$$
$$\therefore \ -2 = -20 + c$$
$$\therefore \ c = 18$$

b Substituting $x = 4$ and $y = 7$ into the equation gives

$$3(4) - 2(7) = k$$
$$\therefore \ 12 - 14 = k$$
$$\therefore \ k = -2$$

3 **a** Find c given that $(3, 10)$ lies on the line with equation $y = 2x + c$.

b Find c given that $(9, -8)$ lies on the line with equation $y = -\frac{2}{3}x + c$.

c Find m given that $(-5, \frac{1}{2})$ lies on the line with equation $y = mx + \frac{3}{2}$.

d Find a given that $(a, -\frac{3}{4})$ lies on the line with equation $y = \frac{1}{2}x + \frac{1}{4}$.

4 Find k given that:

a $(5, -1)$ lies on the line with equation $3x + 4y = k$

b $(-2, -7)$ lies on the line with equation $x - 3y = k$

c $(3, k)$ lies on the line with equation $5x + 6y = -9$

d $(k, 0)$ lies on the line with equation $2x + 5y = 20$

e $(0, k)$ lies on the line with equation $3x - 7y = 42$.

5 **a** Show that the point $(5, -2)$ does not lie on the line with equation $y = -2x + 3$.

b Does the point $(5, -2)$ lie above or below this line? Explain your answer.

F GRAPHING LINES FROM EQUATIONS

GRAPHING LINES IN GRADIENT-INTERCEPT FORM

To draw the graph of $y = mx + c$ we:

- Use the y-intercept c to plot the point $(0, c)$.
- Use x and y-steps from the gradient m to locate another point on the line.
- Join the two points and extend the line in either direction.

Example 14 Self Tutor

Draw the graph of $y = -\frac{3}{4}x + 5$.

For $y = -\frac{3}{4}x + 5$:

- the y-intercept is $c = 5$
- the gradient is $m = \frac{-3}{4}$ ← y-step, ← x-step

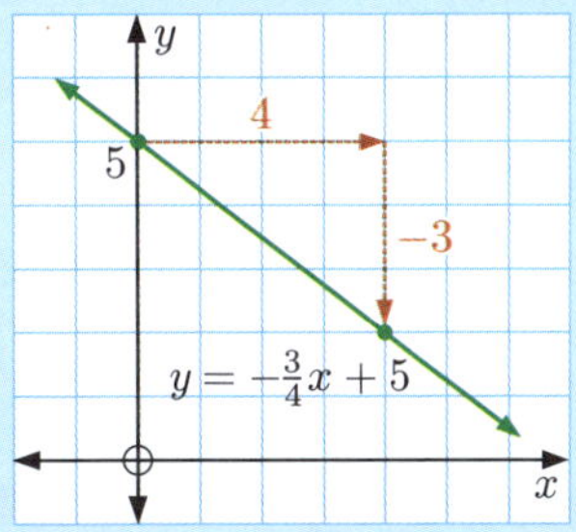

GRAPHING LINES IN GENERAL FORM

To draw the graph of a line in the general form $Ax + By = C$, we:

- Find the y-intercept by letting $x = 0$.
- Find the x-intercept by letting $y = 0$.
- Join the points where the line cuts the axes and extend the line in either direction.

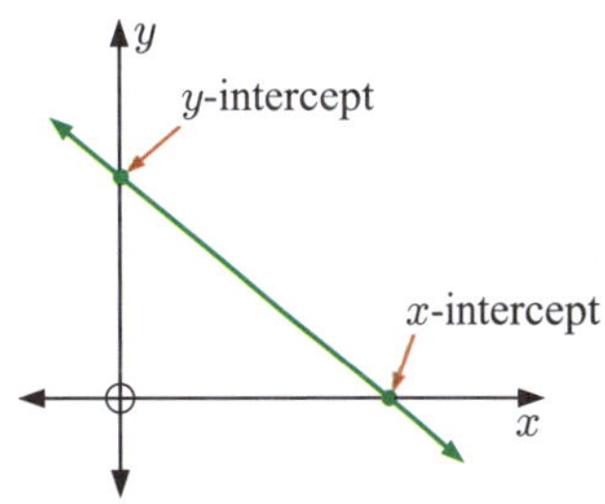

Example 15 Self Tutor

Draw the graph of $4x - 3y = 12$.

When $x = 0$, $\quad -3y = 12$

$\therefore \quad y = -4$

So, the y-intercept is -4.

When $y = 0$, $\quad 4x = 12$

$\therefore \quad x = 3$

So, the x-intercept is 3.

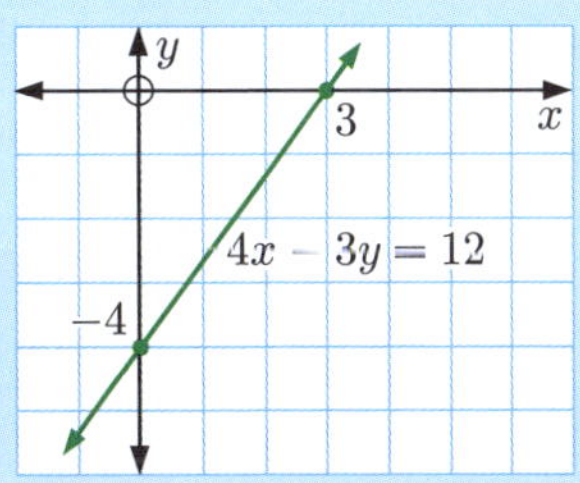

EXERCISE 8F.1

1 Draw the graph of:

a $y = 2x + 1$ **b** $y = 3x - 1$ **c** $y = \frac{2}{3}x$ **d** $y = \frac{4}{3}x - 2$

e $y = -x + 4$ **f** $y = -2x + 2$ **g** $y = -\frac{1}{2}x - 1$ **h** $y = -\frac{2}{5}x - 3$

2 Draw the graph of:

a $x + 3y = 6$ **b** $3x - 2y = 12$ **c** $2x + 5y = 10$

d $4x + 3y = 6$ **e** $x + y = 5$ **f** $x - y = -3$

g $3x - y = -6$ **h** $7x + 2y = 14$ **i** $4x + 9y = -18$

3 Consider the line with equation $y = -\frac{2}{3}x + 4$.

a Find the: **i** gradient **ii** y-intercept.

b Determine whether the following points lie on the line:

i $(-3, 6)$ **ii** $(2, 2)$ **iii** $(8, -\frac{4}{3})$

c Draw the graph of the line, showing your results from **a** and **b**.

4 Consider the line with equation $3x - 5y = 15$.

a Find the: **i** x-intercept **ii** y-intercept.

b Determine whether the following points lie on the line:

i $(-5, -6)$ **ii** $(1, -2)$

c Find c such that $(3, c)$ lies on the line.

d Draw the graph of the line, showing your results from **a**, **b**, and **c**.

VERTICAL AND HORIZONTAL LINES

In previous years, we have seen that:

All **vertical lines** have equations of the form $x = a$ where a is a constant.

All **horizontal lines** have equations of the form $y = c$ where c is a constant.

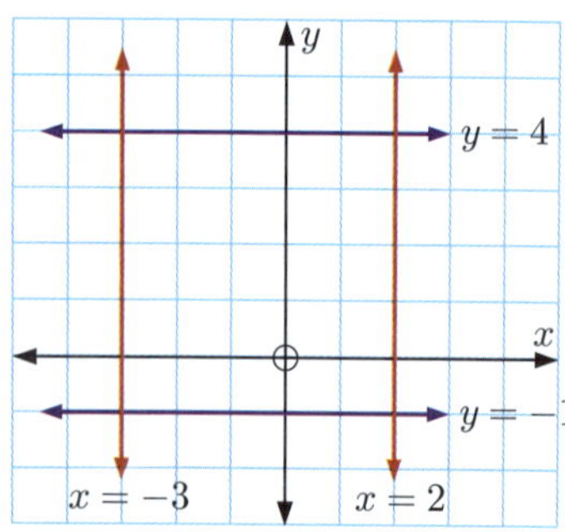

EXERCISE 8F.2

1 Draw the graph of the line with equation:

a $x = 4$ **b** $y = -3$ **c** $x = -6$ **d** $y = -5$

e $x = 0$ **f** $y = \frac{7}{2}$ **g** $y = -\frac{2}{3}$ **h** $x = \frac{5}{4}$

2 Find a such that $(a, 2)$ lies on the line with equation $x = 5$.

3 Find b such that $(b, 3)$ lies on the line with equation $y = 3$.

G FINDING THE EQUATION OF A LINE

To determine the equation of a line, we need to know either:
- its gradient and at least one point which lies on the line, *or*
- two points which lie on the line.

If we know the gradient m and y-intercept c of a line, we can write down the equation of the line immediately as $y = mx + c$.

Example 16 Self Tutor

Find the equation of the line with gradient 5 and y-intercept -3.

The line has gradient $m = 5$ and y-intercept $c = -3$.

So, the equation of the line is $y = 5x - 3$.

If we know the gradient and one point which lies on the line, we can substitute the coordinates of the point into the equation to find c.

Example 17 Self Tutor

A line has gradient $\frac{2}{3}$ and passes through the point $(3, 1)$.
Find the equation of the line.

The gradient $m = \frac{2}{3}$, so the equation has the form $y = \frac{2}{3}x + c$.

Since $(3, 1)$ lies on the line, we substitute $x = 3$, $y = 1$ into the equation.

$\therefore\ 1 = \frac{2}{3}(3) + c$

$\therefore\ 1 = 2 + c$

$\therefore\ c = -1$

So, the equation is $y = \frac{2}{3}x - 1$.

EXERCISE 8G

1 Write down the equation of the line with:

- **a** gradient 3 and y-intercept 4
- **b** gradient 7 and y-intercept -1
- **c** gradient -1 and y-intercept $\frac{1}{3}$
- **d** gradient $\frac{1}{2}$ and y-intercept 0
- **e** gradient 0 and y-intercept 5
- **f** gradient $-\frac{5}{2}$ and y-intercept $-\frac{2}{3}$.

2 Find the equation of the line which has gradient:

- **a** 2 and which passes through $(1, 8)$
- **b** -3 and which passes through $(4, -16)$
- **c** 5 and which passes through $(-3, -1)$
- **d** $\frac{3}{4}$ and which passes through $(8, 12)$
- **e** 0 and which passes through $(-2, 4)$
- **f** $-\frac{1}{2}$ and which passes through $(5, -9)$.

Example 18 — Self Tutor

Find the equation of the illustrated line:

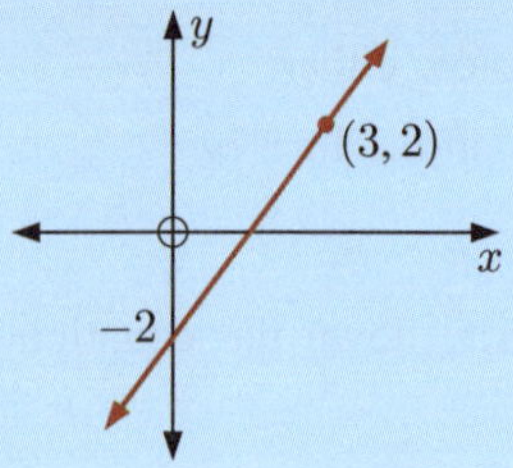

The line passes through $(0, -2)$ and $(3, 2)$.

$\therefore$ the gradient $m = \dfrac{2-(-2)}{3-0} = \frac{4}{3}$

and the y-intercept $c = -2$

$\therefore$ the equation is $y = \frac{4}{3}x - 2$.

3 Find the equations of the illustrated lines:

a

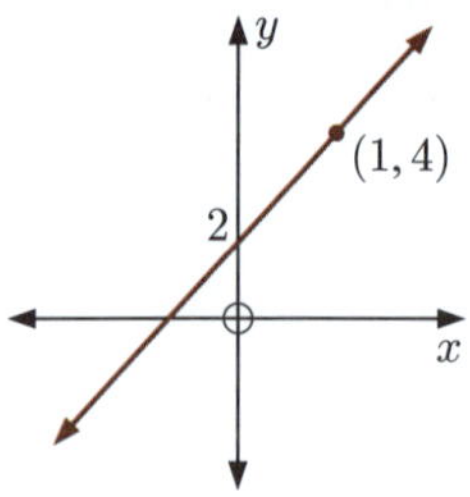

b

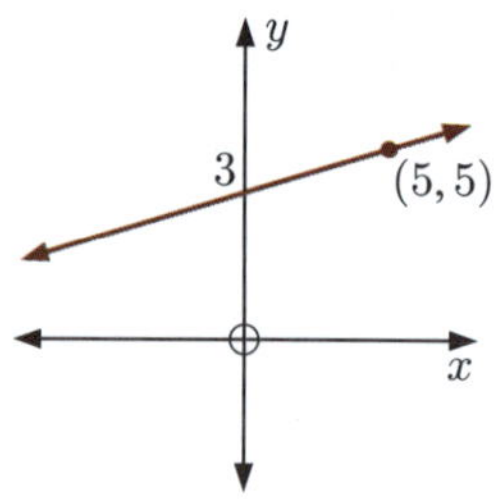

c

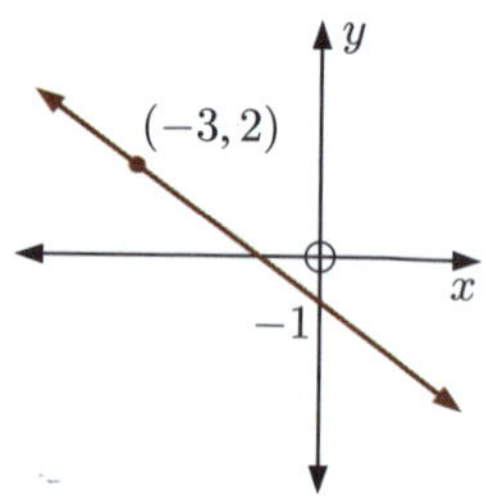

d

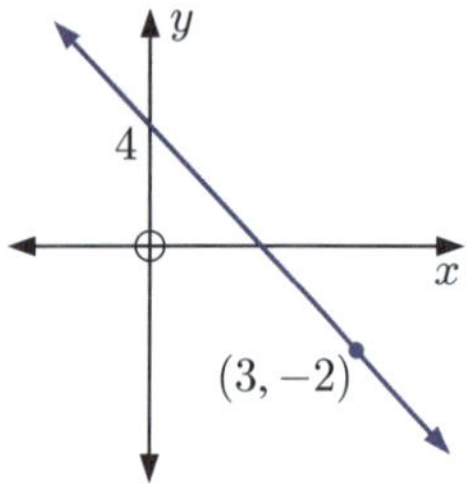

e

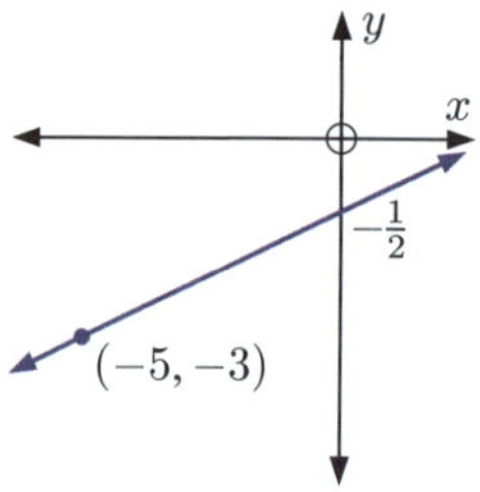

f

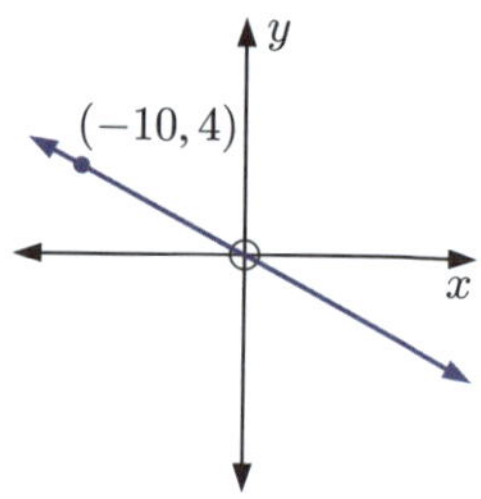

4 Find the equation of the line which has:

a y-intercept -5, and which passes through $(3, 7)$

b y-intercept 7, and which passes through $(5, -8)$

c y-intercept -1, and which passes through $(-8, -4)$.

Example 19 — Self Tutor

Find the equation of the line which passes through $(-2, 1)$ and $(3, 16)$.

The line has gradient $m = \dfrac{16-1}{3-(-2)} = \dfrac{15}{5} = 3$.

$\therefore$ the equation of the line has the form $y = 3x + c$.

Since $(-2, 1)$ lies on the line, we substitute $x = -2$, $y = 1$ into the equation.

$\therefore$ $1 = 3(-2) + c$

$\therefore$ $1 = -6 + c$

$\therefore$ $c = 7$

So, the equation of the line is $y = 3x + 7$.

5 Find the equation of the line which passes through:

a $(2,\ 7)$ and $(5,\ 13)$ **b** $(3,\ 6)$ and $(6,\ 0)$

c $(-3,\ -4)$ and $(6,\ 2)$ **d** $(-4,\ -3)$ and $(3,\ -3)$

e $(-6,\ -14)$ and $(2,\ -2)$ **f** $(-8,\ 11)$ and $(-2,\ 3)$.

6 Answer the **Opening Problem** on page **148**.

7 Find the equations of the illustrated lines:

a

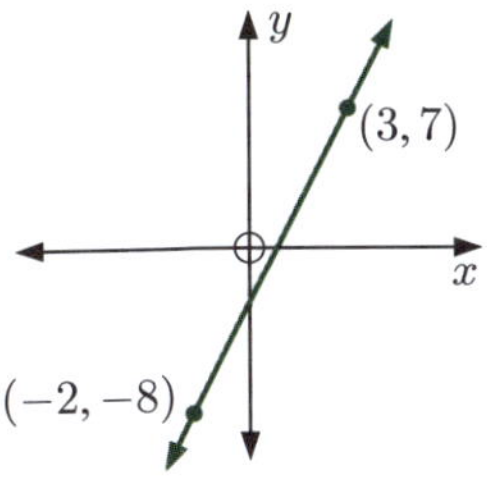

b

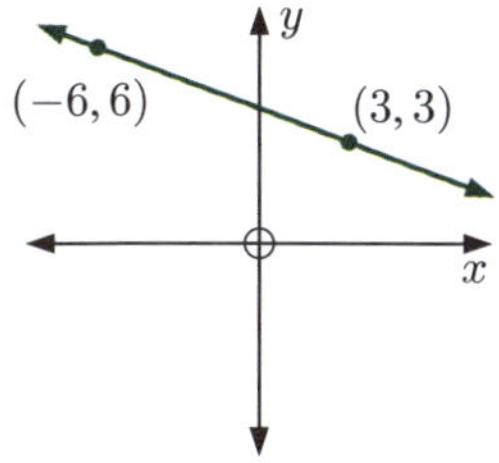

c

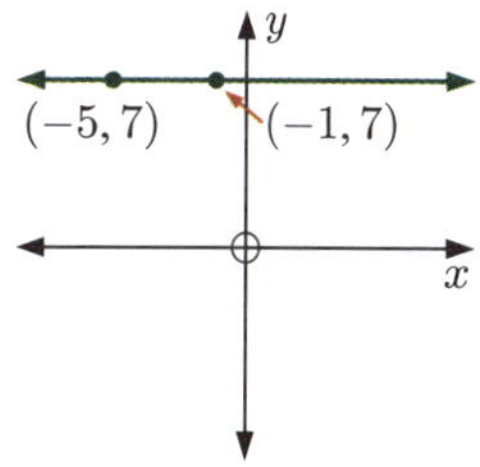

d

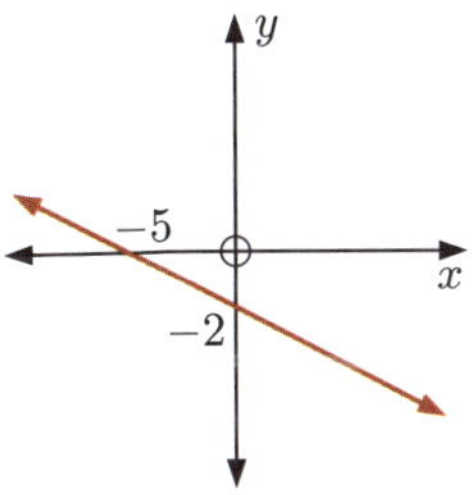

e

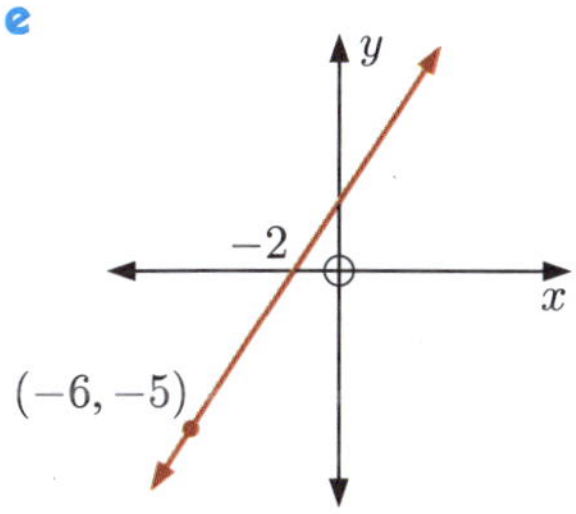

f

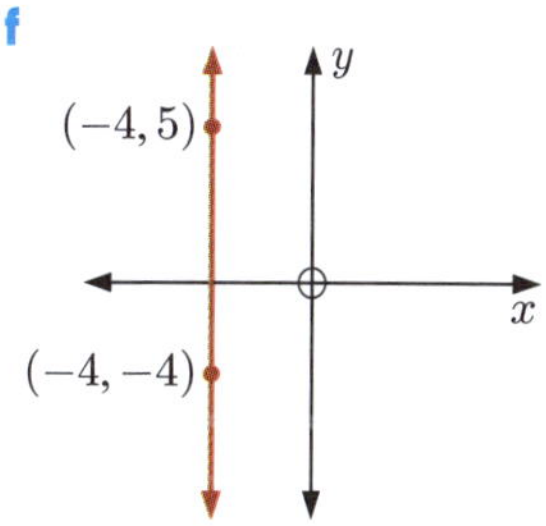

8 Match the following descriptions or graphs of lines with the correct equation:

a the line with gradient -2, passing through the point $(-3,\ 1)$

b the horizontal line with y-intercept -5

c the line with y-intercept 3, passing through the point $(-4,\ 1)$

d the line passing through $(-6,\ -1)$ and $(3,\ 8)$

e the line with x-intercept -4 and y-intercept -8

f

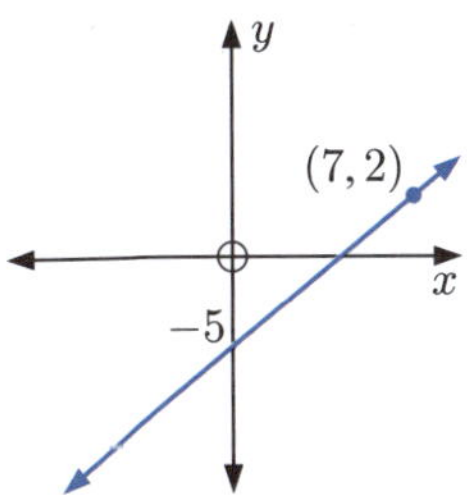

g

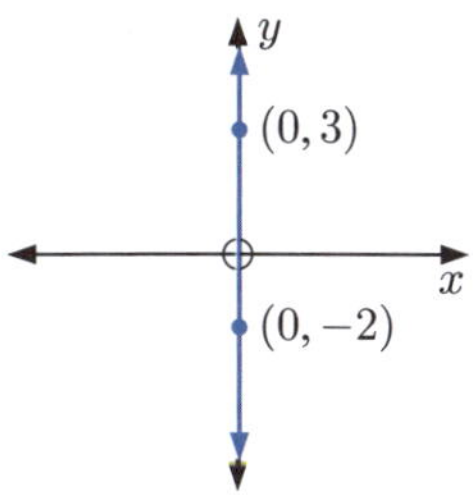

h

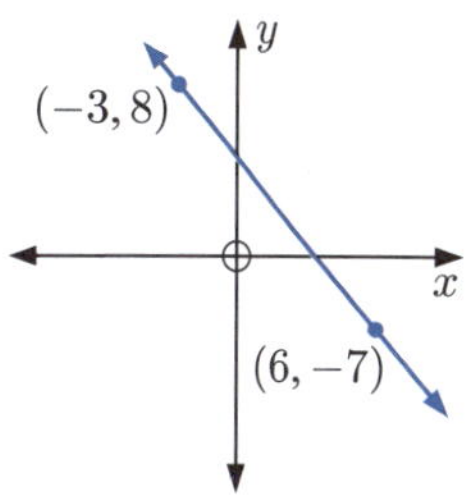

A $y = x - 5$ **B** $y = -\frac{5}{3}x + 3$ **C** $y = -2x - 8$ **D** $y = \frac{1}{2}x + 3$

E $x = 0$ **F** $y = -2x - 5$ **G** $y = -5$ **H** $y = x + 5$

Example 20 **Self Tutor**

Find the equation of the line through $(4, 3)$ which is parallel to the line through $(-5, 1)$ and $(-2, 7)$.

The line through $(-5, 1)$ and $(-2, 7)$ has gradient $\dfrac{7-1}{-2--5} = \dfrac{6}{3} = 2$

$\therefore$ the required line also has gradient 2.

$\therefore$ its equation has the form $y = 2x + c$.

Since $(4, 3)$ lies on the line, we have

$3 = 2(4) + c$

$\therefore\ 3 = 8 + c$

$\therefore\ c = -5$

So, the equation is $y = 2x - 5$.

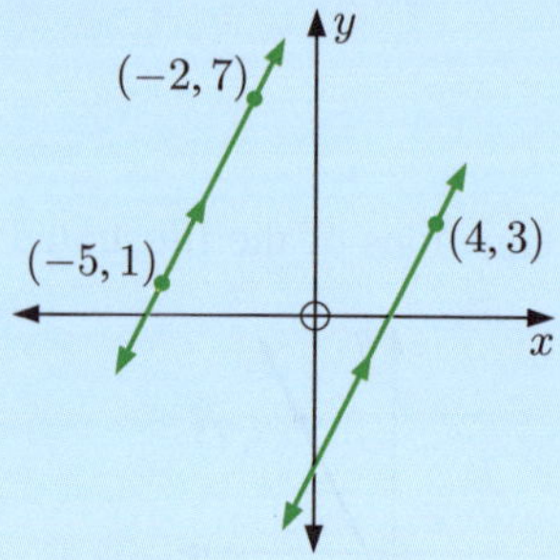

9 Find the equation of the line:

a which is parallel to a line with gradient 3, and passes through $(-1, -2)$

b which is perpendicular to a line with gradient $\frac{1}{2}$, and passes through $(-1, 0)$

c which is perpendicular to a line with gradient -3, and passes through $(4, -1)$

d through $(5, -2)$, and parallel to the line through $(-4, 3)$ and $(5, -6)$

e through $(3, 6)$, and parallel to the line through $(-3, -3)$ and $(1, 7)$

f through $(4, -9)$, and perpendicular to the line through $(5, 4)$ and $(-7, 0)$.

10 Find the equation of each of the lines shown:

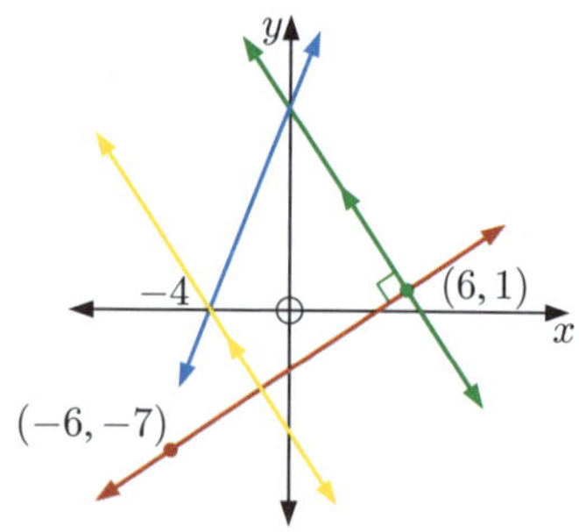

ACTIVITY — MOVING STAIRCASES

In a magical school for mathematics, the staircases are always moving.

Click on the icon to help our young student find his way around the school.

GAME

REVIEW SET 8A

1 Given the points A$(7, -3)$ and B$(-5, 3)$, find:

a the distance between A and B

b the midpoint of [AB].

2 Suppose M is the midpoint of [CD] as shown. Find the coordinates of D.

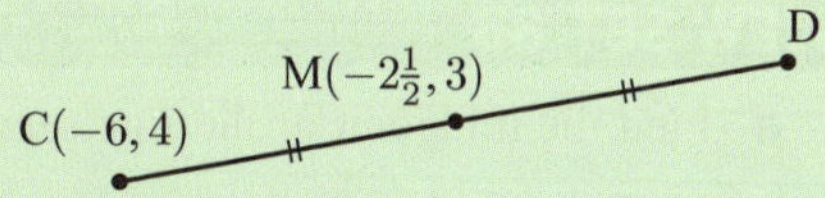

3 Find the gradient of the line segment joining the points:

a $(4, -2)$ and $(6, 3)$ **b** $(-2, 7)$ and $(-2, 5)$.

4 The graph shows the distance travelled by a cyclist over time. Find the average speed of the cyclist:

a from O to A **b** from A to B

c for the whole trip.

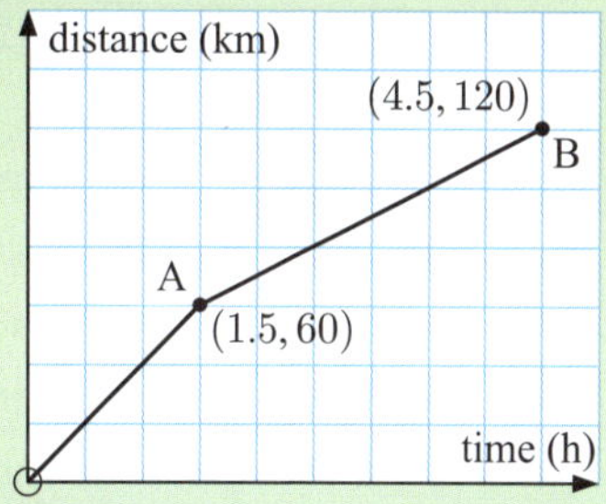

5 **a** Does $(-4, -13)$ lie on the line with equation $y = 2x - 5$?

b Find c given that $(2, 2)$ lies on the line with equation $y = 3x + c$.

6 Line L has gradient $-\frac{1}{4}$, and line M has gradient $\frac{3}{5}$. Find the gradient of any line which is:

a parallel to L **b** parallel to M

c perpendicular to L **d** perpendicular to M.

7 Write the following equations in general form:

a $y = 3x + 7$ **b** $y = -\frac{4}{5}x + \frac{2}{5}$

8 Consider the triangle with vertices $P(2, 1)$, $Q(-4, -3)$, and $R(-2, 7)$.

a Classify the triangle as scalene, isosceles, or equilateral.

b Determine whether PQR has a right angle, and if it does, state where it is.

9 Draw graphs of the lines $y = 7$ and $x = -2$ on the same set of axes.

10 Find the equation of the line:

a with gradient 2 and y-intercept -3

b with gradient $-\frac{1}{2}$ and which passes through $(-6, 5)$

c which passes through $(-4, 6)$ and $(5, -3)$.

11

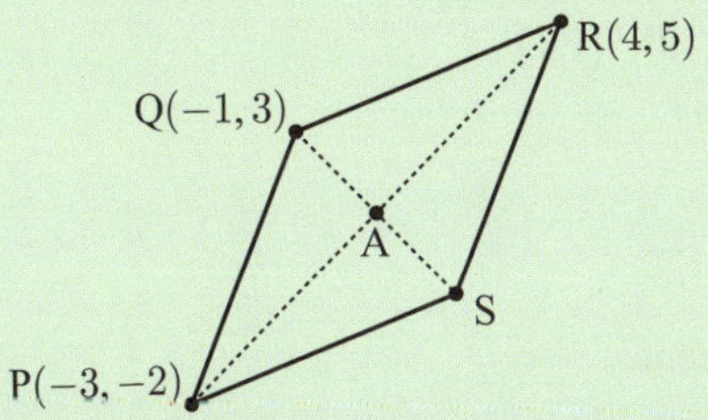

PQRS is a rhombus. Diagonals [QS] and [PR] bisect each other at A. Find:

a the coordinates of A

b the coordinates of S

c the length of each side of the rhombus.

12 Draw the graph of:

a $y = \frac{3}{2}x - 2$ **b** $2x - 3y = 30$ **c** $y = -\frac{1}{4}x + 5$ **d** $4x + 3y = -36$

13 Find the equation of the line through $(1, -7)$, which is perpendicular to the line through $(-3, 2)$ and $(5, 4)$.

REVIEW SET 8B

1 **a** Find the midpoint of the line segment [AB], where A is $(4, -7)$ and B is $(-2, 4)$.

b $M(7, \frac{1}{2})$ is the midpoint of [PQ], where P is $(2, -1)$. Find the coordinates of Q.

2 Find the equation of the line with gradient $\frac{1}{3}$, which passes through the point $(4, 9)$.

3 Find the gradient of the line segment joining $(4, -3)$ and $(2, -6)$.

4 On the same set of axes, draw lines with gradients $\frac{1}{2}$, 1, -2, and 3 which pass through the point $(1, 3)$.

5 Find the equation of the line which has y-intercept 3, and which passes through $(-2, 4)$.

6 The graph alongside indicates the cost of hiring a hall for various lengths of time.

a Find the gradient of the line.

b Interpret the gradient found in **a**.

c Determine the cost of hiring the hall for 9 hours.

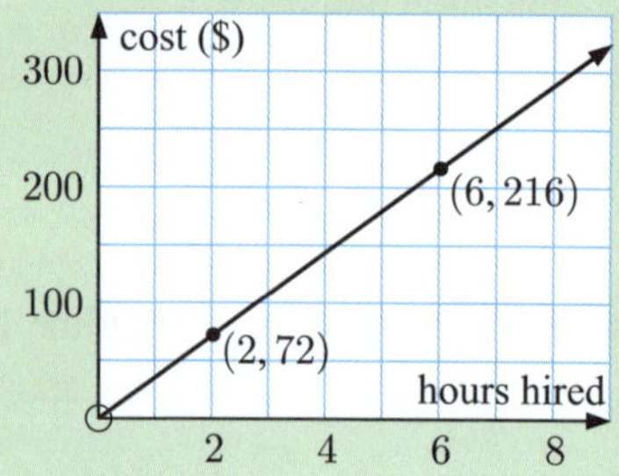

7 A quadrilateral has vertices $A(3, 2)$, $B(9, -2)$, $C(5, -8)$, and $D(-1, -4)$.

a Find the length of: **i** [AB] **ii** [BC] **iii** [CD] **iv** [DA]

b Find the lengths of the diagonals [AC] and [BD].

c Find the midpoints of the diagonals [AC] and [BD].

d Classify the quadrilateral.

8 Find the gradient of a line which is perpendicular to a line with gradient:

a -2 **b** $\frac{3}{4}$ **c** $-3\frac{1}{3}$

9 **a** Write the equation $3x - 4y = 24$ in gradient-intercept form.

b Hence, state the gradient of the line.

10 Draw the graph of the line with equation $y = \frac{2}{3}x - 4$.

11 The point $(k, 5)$ lies on the line with equation $3x - y = -8$. Find k.

12 Consider the points $P(2, 7)$, $Q(7, 4)$, and $R(-4, -3)$.

a Find the gradient of:

i [PQ] **ii** [QR] **iii** [PR]

b Hence, show that triangle PQR is right angled.

c Find the area of triangle PQR.

13 Consider the line with equation $5x + 2y = 20$.

a Find the: **i** x-intercept **ii** y-intercept.

b Determine whether the following points lie on the line:

i $(6, -5)$ **ii** $(2, 4)$

c Find the value of b such that $(1, b)$ lies on the line.

d Draw the graph of the line, showing your results from **a**, **b**, and **c**.

Chapter 9

Quadratic factorisation

Contents:

A **quadratic trinomial** in x is an expression of the form ax^2+bx+c where x is the variable, and a, b, and c are constants, $a \neq 0$.

ax^2	+	bx	+	c
↑		↑		↑
the x^2 term		the x term		the constant term

For example: x^2+5x+6, $4x^2-9$ and $9x^2+6x+1$ are quadratic expressions.

In **Chapter 4** we studied the expansion of algebraic factors, many of which resulted in quadratic trinomials. In this chapter we will consider **factorisation**, which is the reverse process of expansion.

Factorisation is the process of writing an expression as a **product** of its **factors**.

For example:

expansion →

$$(x+2)(x+3) = x^2+5x+6$$

← **factorisation**

Since $x^2+5x+6=(x+2)(x+3)$, we say that $(x+2)$ and $(x+3)$ are the **linear factors** of x^2+5x+6.

You should remember the following expansion rules from **Chapter 4**:

$(x+p)(x+q)=x^2+(p+q)x+pq$	**sum and product** expansion
$(x+a)^2=x^2+2ax+a^2$	**perfect square** expansion
$(x+a)(x-a)=x^2-a^2$	**difference of two squares** expansion

These statements are called **identities** because they are true for all values of the variable x.

Notice that the RHS of each identity is a quadratic expression which has been formed by **expanding** the LHS. So, the LHS of the identities can be obtained by **factorising** the RHS.

A REMOVING COMMON FACTORS

Some quadratic trinomials can be factorised by removing the Highest Common Factor (HCF) of the terms in the expression. In fact, we should always look to remove the HCF before proceeding with any further factorisation.

When we remove a common factor, we write it in front of a set of brackets. The expression within the brackets is found using the reverse of the distributive law.

Example 1 Self Tutor

Factorise by removing a common factor:

a $2x^2+3x$ **b** $-2x^2-6x$

a $2x^2+3x$ has HCF x
$\therefore$ $2x^2+3x=x(2x+3)$

b $-2x^2-6x$ has HCF $-2x$
$\therefore$ $-2x^2-6x=-2x(x+3)$

Example 2 — Self Tutor

Fully factorise by removing a common factor:

a $(x-5)^2 - 2(x-5)$ **b** $(x+2)^2 + 2x + 4$

a
$$\begin{aligned} & (x-5)^2 - 2(x-5) \\ &= (x-5)(x-5) - 2(x-5) \quad \{\text{HCF} = (x-5)\} \\ &= (x-5)[(x-5)-2] \\ &= (x-5)(x-7) \quad \{\text{simplifying}\} \end{aligned}$$

b
$$\begin{aligned} & (x+2)^2 + 2x + 4 \\ &= (x+2)(x+2) + 2(x+2) \quad \{\text{HCF} = (x+2)\} \\ &= (x+2)[(x+2)+2] \\ &= (x+2)(x+4) \end{aligned}$$

EXERCISE 9A

1 Fully factorise by first removing a common factor:

a $3x^2 + 5x$ **b** $2x^2 - 7x$ **c** $3x^2 + 6x$
d $4x^2 - 8x$ **e** $-2x^2 + 9x$ **f** $-3x^2 - 15x$
g $-4x + 8x^2$ **h** $-5x - 10x^2$ **i** $12x - 4x^2$
j $x^3 + x^2 + x$ **k** $2x^3 + 11x^2 + 4x$ **l** $ab + ac + ad$
m $ax^2 + 2ax$ **n** $ab^2 + a^2b$ **o** $ax^3 + ax^2$

2 Fully factorise:

a $3(x+5) + x(x+5)$ **b** $a(b+3) - 5(b+3)$ **c** $x(x+4) + x + 4$
d $x(x+2) + (x+2)(x+5)$ **e** $a(c-d) + b(c-d)$ **f** $y(2+y) - y - 2$
g $ab(x-1) + c(x-1)$ **h** $a(x+2) - x - 2$

3 Fully factorise by removing a common factor:

a $(x+2)^2 - 5(x+2)$ **b** $(x-1)^2 - 3(x-1)$ **c** $(x+1)^2 + 2(x+1)$
d $(x-2)^2 + 3x - 6$ **e** $x + 3 + (x+3)^2$ **f** $(x+4)^2 + 8 + 2x$
g $(x-3)^2 - x + 3$ **h** $(x+4)^2 - 2x - 8$ **i** $(x-4)^2 - 5x + 20$
j $3x + 6 + (x+2)^2$ **k** $(x+1)^3 + (x+1)^2$ **l** $(a+b)^3 + a + b$
m $2(x+1)^2 + x + 1$ **n** $3(x-2)^2 - (x-2)$ **o** $4(a+b)^2 - 2a - 2b$

B DIFFERENCE OF TWO SQUARES FACTORISATION

The **expansion** of $(a+b)(a-b)$ is $a^2 - b^2$, which is the difference between the two squares a^2 and b^2.

Thus, the **factorisation** of $a^2 - b^2$ is:

$$\boldsymbol{a^2 - b^2 = (a+b)(a-b)}$$

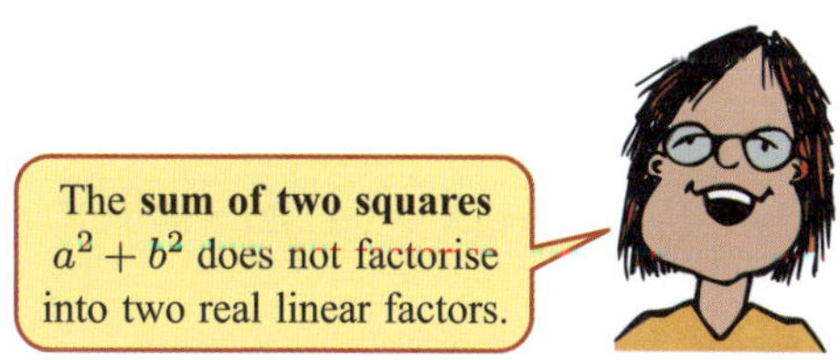

Example 3 Self Tutor

Use the rule $a^2 - b^2 = (a+b)(a-b)$ to factorise fully:

a $9 - x^2$ **b** $4x^2 - 25$

a $9 - x^2$
$= 3^2 - x^2$ {difference of squares}
$= (3+x)(3-x)$

b $4x^2 - 25$
$= (2x)^2 - 5^2$ {difference of squares}
$= (2x+5)(2x-5)$

Example 4 Self Tutor

Fully factorise: **a** $2x^2 - 8$ **b** $-3x^2 + 48$

a $2x^2 - 8$
$= 2(x^2 - 4)$ {HCF is 2}
$= 2(x^2 - 2^2)$ {difference of squares}
$= 2(x+2)(x-2)$

b $-3x^2 + 48$
$= -3(x^2 - 16)$ {HCF is -3}
$= -3(x^2 - 4^2)$ {difference of squares}
$= -3(x+4)(x-4)$

EXERCISE 9B

1 Fully factorise:

a $x^2 - 4$ **b** $4 - x^2$ **c** $x^2 - 81$ **d** $25 - x^2$
e $4x^2 - 1$ **f** $9x^2 - 16$ **g** $4x^2 - 9$ **h** $36 - 49x^2$

2 Fully factorise:

a $3x^2 - 27$ **b** $-2x^2 + 8$ **c** $3k^2 - 75$
d $-5x^2 + 5$ **e** $8t^2 - 18$ **f** $-27x^2 + 75$
g $x^3 - 49x$ **h** $64n^2 - n^4$ **i** $28x^3 - 63x$

Example 5 Self Tutor

Factorise into linear factors:

a $x^2 - 11$ **b** $(x+3)^2 - 5$

a $x^2 - 11$
$= x^2 - (\sqrt{11})^2$
$= (x+\sqrt{11})(x-\sqrt{11})$

b $(x+3)^2 - 5$
$= (x+3)^2 - (\sqrt{5})^2$
$= [(x+3)+\sqrt{5}][(x+3)-\sqrt{5}]$
$= [x+3+\sqrt{5}][x+3-\sqrt{5}]$

3 If possible, factorise into linear factors:

a $x^2 - 3$ **b** $x^2 + 4$ **c** $x^2 - 15$

d $3x^2 - 15$ **e** $(x+1)^2 - 6$ **f** $(x+2)^2 + 6$

g $(x-2)^2 - 7$ **h** $(x+3)^2 - 17$ **i** $(x-4)^2 + 9$

Example 6 Self Tutor

Factorise using the difference of two squares:

a $(3x+2)^2 - 9$ **b** $(x+2)^2 - (x-1)^2$

a
$$\begin{aligned} & (3x+2)^2 - 9 \\ &= (3x+2)^2 - 3^2 \\ &= [(3x+2)+3][(3x+2)-3] \\ &= (3x+5)(3x-1) \end{aligned}$$

b
$$\begin{aligned} & (x+2)^2 - (x-1)^2 \\ &= [(x+2)+(x-1)][(x+2)-(x-1)] \\ &= [x+2+x-1][x+2-x+1] \\ &= [2x+1][3] \\ &= 3(2x+1) \end{aligned}$$

4 Factorise using the difference of two squares:

a $(x+1)^2 - 4$ **b** $(2x+1)^2 - 9$ **c** $(1-x)^2 - 16$

d $(x+3)^2 - 4x^2$ **e** $4x^2 - (x+2)^2$ **f** $9x^2 - (3-x)^2$

g $(2x+1)^2 - (x-2)^2$ **h** $(3x-1)^2 - (x+1)^2$ **i** $16x^2 - (2x+3)^2$

C PERFECT SQUARE FACTORISATION

We know the **expansion** of $(a+b)^2$ is $a^2 + 2ab + b^2$,

so the **factorisation** of $a^2 + 2ab + b^2$ is: $\boldsymbol{a^2 + 2ab + b^2 = (a+b)^2}$

Similarly, the expansion of $(a-b)^2$ is $a^2 - 2ab + b^2$,

so the factorisation of $a^2 - 2ab + b^2$ is: $\boldsymbol{a^2 - 2ab + b^2 = (a-b)^2}$

Example 7 Self Tutor

Fully factorise:

a $x^2 + 10x + 25$ **b** $x^2 - 14x + 49$

a
$$\begin{aligned} & x^2 + 10x + 25 \\ &= x^2 + 2 \times x \times 5 + 5^2 \\ &= (x+5)^2 \end{aligned}$$

b
$$\begin{aligned} & x^2 - 14x + 49 \\ &= x^2 - 2 \times x \times 7 + 7^2 \\ &= (x-7)^2 \end{aligned}$$

EXERCISE 9C

1 Fully factorise:

a x^2+6x+9 b $x^2+8x+16$ c x^2-6x+9
d $x^2-8x+16$ e x^2+2x+1 f $x^2-10x+25$
g $y^2+18y+81$ h $m^2-20m+100$ i $t^2+12t+36$

Example 8 Self Tutor

Fully factorise:

a $9x^2-6x+1$ b $-8x^2-24x-18$

a
$$\begin{aligned}&9x^2-6x+1\\&=(3x)^2-2\times 3x\times 1+1^2\\&=(3x-1)^2\end{aligned}$$

b
$$\begin{aligned}&-8x^2-24x-18\\&=-2(4x^2+12x+9) \quad \{\text{HCF}=-2\}\\&=-2([2x]^2+2\times 2x\times 3+3^2)\\&=-2(2x+3)^2\end{aligned}$$

2 Fully factorise:

a $9x^2+6x+1$ b $4x^2-4x+1$ c $9x^2+12x+4$
d $25x^2-10x+1$ e $16x^2+24x+9$ f $25x^2-20x+4$

3 Fully factorise:

a $-x^2+2x-1$ b $2x^2+8x+8$ c $-3x^2-30x-75$
d $32x^2-16x+2$ e $36x^2+120x+100$ f $-2x^3+28x^2-98x$

DISCUSSION

- What constant term must be added to these expressions to create a perfect square?
  - $x^2-6x+......$
  - $x^2+12x+......$
  - $x^2+16x+......$
- If we know that x^2+bx+c is a perfect square, what is the relationship between b and c?

D FACTORISING EXPRESSIONS WITH FOUR TERMS

Sometimes we can factorise an expression containing four terms by grouping them in two pairs.

For example, $ax^2+2x+2+ax$ can be rewritten as

$$\begin{aligned}&\underbrace{ax^2+ax}+\underbrace{2x+2}\\&=ax(x+1)+2(x+1) \quad \{\text{factorising each pair}\}\\&=(x+1)(ax+2) \quad \{(x+1) \text{ is a common factor}\}\end{aligned}$$

Example 9 **Self Tutor**

Fully factorise:

a $ax + by + bx + ay$ **b** $2x^2 - 15 + 3x - 10x$

a $ax + by + bx + ay$
$= \underbrace{ax + ay} + \underbrace{bx + by}$ {putting terms containing a together}
$= a(x + y) + b(x + y)$ {factorising each pair}
$= (x + y)(a + b)$ {$(x + y)$ is a common factor}

b $2x^2 - 15 + 3x - 10x$
$= \underbrace{2x^2 - 10x} + \underbrace{3x - 15}$ {splitting into two pairs}
$= 2x(x - 5) + 3(x - 5)$ {factorising each pair}
$= (x - 5)(2x + 3)$ {$(x - 5)$ is a common factor}

EXERCISE 9D

1 Fully factorise:

a $bx + cx + by + cy$ **b** $px + 3q + qx + 3p$ **c** $6ax + 3bx + b + 2a$
d $am - bn - an + bm$ **e** $3dr + r - 3ds - s$ **f** $2ac - 5a + 2bc - 5b$

2 Fully factorise:

a $x^2 + 5x + 7x + 35$ **b** $x^2 - 2x - 6x + 12$ **c** $x^2 + 3x + 9 + 3x$
d $x^2 + 8 + 8x + x$ **e** $2x^2 + 3x + 3 + 2x$ **f** $3x^2 + x - 3x - 1$
g $2x^2 + 3x + 10x + 15$ **h** $6x^2 - 3x - 2 + 4x$ **i** $4x^2 + x + 8x + 2$
j $6x^2 + 3x + 10x + 5$ **k** $6x^2 - 4x + 9x - 6$ **l** $4x^2 + x - 8x - 2$
m $3x^2 + 4x + 33x + 44$ **n** $18x^2 + 3x - 12x - 2$ **o** $10x^2 + 4x - 35x - 14$

3 Fully factorise:

a $x^2 - 2x + 1 - a^2$ **b** $x^2 - a^2 + x + a$ **c** $b^2 - x^2 - 4x - 4$
d $c^2 - x^2 + 6x - 9$ **e** $x^2 - y^2 + y - x$ **f** $a^2 + 2ab + a^2 - 4b^2$
g $x^2 + 4x + 4 - m^2$ **h** $x^2 + 2ax + a^2 - b^2$ **i** $x^2 - y^2 - 3x - 3y$

E FACTORISING QUADRATIC TRINOMIALS

Consider the expansion

$$\begin{aligned}(x+1)(x+6) &= x^2 + 6x + x + 1 \times 6 \quad \text{\{using FOIL\}}\\ &= x^2 + (6+1)x + (1 \times 6)\\ &= x^2 + (\textbf{sum} \text{ of 1 and 6})x + (\textbf{product} \text{ of 1 and 6})\\ &= x^2 + 7x + 6\end{aligned}$$

More generally, $(x + a)(x + b) = x^2 + bx + ax + ab$
$= x^2 + (a + b)x + ab$

$$x^2 + px + q = (x + a)(x + b)$$
where a and b are two numbers whose sum is p, and whose product is q.

So, to factorise x^2+7x+6, we need two numbers with a sum of 7 and a product of 6. These numbers are 1 and 6, and so $x^2 + 7x + 6 = (x + 1)(x + 6)$.

We call this the **sum and product method**.

Example 10 Self Tutor

Use the sum and product method to fully factorise:

a $x^2 + 5x + 4$ **b** $x^2 - x - 12$

a We need two numbers with sum 5, and product 4.
The numbers are 1 and 4.
$\therefore \ x^2 + 5x + 4 = (x + 1)(x + 4)$

b We need two numbers with sum -1 and product -12.
The numbers are -4 and 3.
$\therefore \ x^2 - x - 12 = (x - 4)(x + 3)$

EXERCISE 9E

1 Find two numbers which have:

a product 12 and sum 7
b product 15 and sum 8
c product 16 and sum 10
d product 18 and sum 11
e product -36 and sum 9
f product -36 and sum -9
g product -12 and sum -4
h product -30 and sum 13

2 Factorise:

a $x^2 + 4x + 3$
b $x^2 + 11x + 24$
c $x^2 + 10x + 21$
d $x^2 + 15x + 54$
e $x^2 + 9x + 20$
f $x^2 + 8x + 15$
g $x^2 + 10x + 24$
h $x^2 + 9x + 14$
i $x^2 + 6x + 8$

3 Factorise:

a $x^2 - 3x + 2$
b $x^2 - 4x + 3$
c $x^2 - 5x + 6$
d $x^2 - 14x + 33$
e $x^2 - 16x + 39$
f $x^2 - 19x + 48$
g $x^2 - 11x + 28$
h $x^2 - 14x + 24$
i $x^2 - 20x + 36$

4 Factorise:

a $x^2 - 7x - 8$
b $x^2 + 4x - 21$
c $x^2 - x - 2$
d $x^2 - 2x - 8$
e $x^2 + 5x - 24$
f $x^2 - 3x - 10$
g $x^2 + 3x - 54$
h $x^2 + x - 72$
i $x^2 - 4x - 21$
j $x^2 - x - 6$
k $x^2 - 7x - 60$
l $x^2 + 7x - 60$
m $x^2 + 3x - 18$
n $x^2 - 7x - 18$
o $x^2 - 12x + 35$

Example 11 — Self Tutor

Fully factorise by first removing a common factor:

a $3x^2 - 9x + 6$ **b** $-2x^2 + 2x + 12$

a $3x^2 - 9x + 6$
$= 3(x^2 - 3x + 2)$ {HCF = 3}
$= 3(x - 2)(x - 1)$ {sum $= -3$ and product $= 2$
$\therefore$ the numbers are -2 and -1}

b $-2x^2 + 2x + 12$
$= -2(x^2 - x - 6)$ {HCF $= -2$}
$= -2(x - 3)(x + 2)$ {sum $= -1$ and product $= -6$
$\therefore$ the numbers are -3 and 2}

5 Fully factorise by first removing a common factor:

a $2x^2 - 6x - 8$	**b** $3x^2 + 9x - 12$	**c** $5x^2 + 10x - 15$
d $4x^2 + 4x - 80$	**e** $2x^2 - 4x - 30$	**f** $3x^2 + 12x - 63$
g $-2x^2 + 2x + 40$	**h** $-3x^2 + 12x - 12$	**i** $-7x^2 - 21x + 28$
j $-x^2 - 3x - 2$	**k** $-x^2 + 5x - 6$	**l** $-x^2 + 9x - 18$
m $5x^2 + 15x - 50$	**n** $-2x^2 - 8x + 42$	**o** $4x - x^2 + 32$

ACTIVITY — MISCELLANEOUS FACTORISATION

Use the following steps in order to factorise quadratic trinomials:

Step 1: If there is a **common factor**, take it out.

Step 2: Look for a **perfect square** factorisation: $x^2 + 2ax + a^2 = (x + a)^2$ or $x^2 - 2ax + a^2 = (x - a)^2$

Step 3: Look for the **difference of two squares**: $x^2 - a^2 = (x + a)(x - a)$

Step 4: Look for the **sum and product** type: $x^2 + (p + q)x + pq = (x + p)(x + q)$

What to do:

Where possible, fully factorise the following expressions:

a $3x^2 + 9x$	**b** $4x^2 - 1$	**c** $5x^2 - 15$
d $3x - 5x^2$	**e** $x^2 + 3x - 40$	**f** $2x^2 - 32$
g $x^2 + 9$	**h** $x^2 + 10x + 25$	**i** $x^2 - x - 6$
j $x^2 - 16x + 39$	**k** $x^2 - 7x - 60$	**l** $x^2 - 2x - 8$
m $x^2 + 11x + 30$	**n** $x^2 + 6x - 16$	**o** $x^2 - 5x - 24$
p $3x^2 + 6x - 72$	**q** $4x^2 - 8x - 60$	**r** $3x^2 - 42x + 99$
s $-x^2 + 9x - 14$	**t** $-x^2 - 13x - 36$	**u** $-2x^2 - 14x + 36$

F FACTORISING $ax^2 + bx + c$, $a \neq 1$

So far we have considered the factorisation of quadratic trinomials of the form $ax^2 + bx + c$ where:

- $a = 1$, for example $x^2 + 5x + 6 = (x + 3)(x + 2)$
- a is a common factor, for example $2x^2 + 10x + 12 = 2(x^2 + 5x + 6)$
  $= 2(x + 3)(x + 2)$
- we have a perfect square or difference of two squares,
  for example $4x^2 - 9 = (2x)^2 - 3^2$
  $= (2x + 3)(2x - 3)$

Factorising a quadratic trinomial such as $8x^2 + 22x + 15$ appears to be more complicated because it does not fall into any of these categories. We therefore need to develop a method for factorising this type of expression.

FACTORISATION BY 'SPLITTING' THE MIDDLE TERM

Using the FOIL rule, we see that

$$\begin{aligned} &(2x + 3)(4x + 5) \\ &= 8x^2 + 10x + 12x + 15 \\ &= 8x^2 + 22x + 15 \end{aligned}$$

We will now *reverse* the process to factorise the quadratic trinomial $8x^2 + 22x + 15$.

$$\begin{aligned} &8x^2 + 22x + 15 \\ &= 8x^2 + 10x + 12x + 15 && \{\text{'splitting' the middle term}\} \\ &= (8x^2 + 10x) + (12x + 15) && \{\text{grouping in pairs}\} \\ &= 2x(4x + 5) + 3(4x + 5) && \{\text{factorising each pair separately}\} \\ &= (4x + 5)(2x + 3) && \{\text{completing the factorisation}\} \end{aligned}$$

But how do we know how to correctly 'split' the middle term? How do we determine that $22x$ should be written as $+10x + 12x$ rather than $+15x + 7x$ or $+20x + 2x$?

When looking at $8x^2 + 10x + 12x + 15$, we notice that $8 \times 15 = 120$ and $10 \times 12 = 120$.

So, for $8x^2 + 22x + 15$, we need two numbers whose *sum* is 22 and whose *product* is $8 \times 15 = 120$. These numbers are 10 and 12.

Likewise, for $6x^2 + 19x + 15$ we need two numbers with sum 19 and product $6 \times 15 = 90$.

These numbers are 10 and 9, so

$$\begin{aligned} &6x^2 + 19x + 15 \\ &= 6x^2 + 10x + 9x + 15 \\ &= (6x^2 + 10x) + (9x + 15) \\ &= 2x(3x + 5) + 3(3x + 5) \\ &= (3x + 5)(2x + 3) \end{aligned}$$

The order in which the 'split' terms are written does not matter.

The following procedure is used to factorise ax^2+bx+c by 'splitting' the middle term:

Step 1: Find two numbers p and q whose sum is b and whose product is ac.

Step 2: Replace bx by $px+qx$.

Step 3: Complete the factorisation.

Example 12 Self Tutor

Show how to 'split' the middle term of the following so that factorisation can occur:

a $3x^2+7x+2$ **b** $10x^2-23x-5$

a In $3x^2+7x+2$, $ac=3\times 2=6$ and $b=7$.
We need two numbers with a product of 6 and a sum of 7. These are 1 and 6.
So, the 'split' is $7x=x+6x$.

b In $10x^2-23x-5$, $ac=10\times -5=-50$ and $b=-23$.
We need two numbers with a product of -50 and a sum of -23. These are -25 and 2.
So, the 'split' is $-23x=-25x+2x$.

EXERCISE 9F

1 Show how to 'split' the middle term so that factorisation can occur:

a $3x^2+11x+6$ **b** $2x^2+9x+4$ **c** $4x^2+4x-15$

d $5x^2-11x-12$ **e** $3x^2-8x-3$ **f** $12x^2-25x+12$

Example 13 Self Tutor

Factorise by 'splitting' the middle term:

a $6x^2+19x+10$ **b** $3x^2-x-10$

a In $6x^2+19x+10$, $ac=60$ and $b=19$.
We need two numbers with a product of 60 and a sum of 19. These are 4 and 15.

$\therefore$ $6x^2+19x+10$
$=6x^2+4x+15x+10$ {'splitting' the middle term}
$=2x(3x+2)+5(3x+2)$ {factorising in pairs}
$=(3x+2)(2x+5)$ {taking out the common factor}

b In $3x^2-x-10$, $ac=-30$ and $b=-1$.
We need two numbers with a product of -30 and a sum of -1. These are 5 and -6.

$\therefore$ $3x^2-x-10$
$=3x^2+5x-6x-10$ {'splitting' the middle term}
$=x(3x+5)-2(3x+5)$ {factorising in pairs}
$=(3x+5)(x-2)$ {taking out the common factor}

2 Fully factorise:

a $2x^2+7x+6$ **b** $2x^2+15x+25$ **c** $2x^2+11x+15$

d $3x^2+7x+2$ **e** $3x^2+14x+15$ **f** $6x^2+19x+10$

3 Fully factorise:

a $2x^2+5x-3$ **b** $2x^2+x-3$ **c** $3x^2+4x-4$
d $3x^2-8x-3$ **e** $5x^2-8x-4$ **f** $6x^2+7x-3$

4 Fully factorise:

a $2x^2-11x+5$ **b** $2x^2-11x+12$ **c** $2x^2-9x+10$
d $3x^2-14x+8$ **e** $10x^2-9x+2$ **f** $9x^2-15x+4$

5 Fully factorise:

a $12x^2+13x-4$ **b** $6x^2+19x+10$ **c** $8x^2+2x-3$
d $6x^2-19x+10$ **e** $10x^2-13x-3$ **f** $20x^2+31x+12$

6 Fully factorise:

a $4x^2+2x-12$ **b** $18x^2+21x+6$ **c** $8x^2-16x+6$

DISCUSSION

- Can *all* quadratic trinomials be factorised into linear factors?
  For example, consider x^2+x-1 and x^2+x+1.
- Is the factorisation of a quadratic trinomial unique?

REVIEW SET 9A

1 Fully factorise:

a $3x^2+12x$ **b** $15x-3x^2$ **c** $(x+3)^2-4(x+3)$
d $4x^2-9$ **e** $6x^2-24y^2$ **f** x^2-13
g $(x+1)^2-6$ **h** $x^2+8x+16$ **i** $x^2-10x+25$

2 Fully factorise:

a $2x^2+8x+6$ **b** $5x^2-10x+5$ **c** $ax+2a+2b+bx$
d $2cx-2dx+d-c$ **e** $3x^2+2x+8+12x$ **f** $6x^2+9x-2x-3$
g $x^2-8x+24-3x$ **h** $x^2+4x+4-a^2$ **i** $-x^2-3x+28$

3 Fully factorise:

a $2x^2+17x+8$ **b** $2x^2+15x-8$ **c** $2x^2-17x+8$
d $6x^2-11x-10$ **e** $12x^2+5x-2$ **f** $12x^2-8x-15$

REVIEW SET 9B

1 Fully factorise:

a $4x^2-8x$ **b** $16x-8x^2$ **c** $(2x-1)^2+2x-1$
d $9-25x^2$ **e** $18-2a^2$ **f** x^2-23
g $(x+2)^2-3$ **h** $x^2-12x+36$ **i** $2x^2+8x+8$

2 Fully factorise, if possible:

a x^2-9 **b** x^2+9 **c** $9x^2-9$

d $3x^2-9$ **e** $3x^2-9x$ **f** $9x^2-3$

3 Fully factorise:

a $3x^2-6x-9$ **b** $7x^2+28x+28$ **c** $mx+nx-my-ny$

d $3a^2+ab-2b^2-6ab$ **e** $3x+2x^2+8x+12$ **f** $6x+4x^2-2x-3$

g $-x^2+8x-16$ **h** $-2x^2-2x+12$ **i** $-3x^2+9x+12$

4 Fully factorise:

a $3x^2-17x-6$ **b** $3x^2-19x+6$ **c** $3x^2+17x-6$

d $12x^2+7x+1$ **e** $12x^2-23x-2$ **f** $9x^2+12x+4$

HISTORICAL NOTE SRINIVASA RAMANUJAN 1887 - 1920

Ramanujan was born in India in 1887. His parents were poor, but were able to send him to school. He was fascinated by mathematics.

Early attempts to study at University failed because he was required to study other subjects as well as mathematics, and mathematics was the only subject at which he excelled. Also, he was still extremely poor.

Ramanujan taught himself from books and worked at home on mathematical research. He was always meticulous in the recording of his work and his results, but only rarely did he provide proofs for his theories.

Fortunately, he was able to obtain a position at the Madras Port Trust Office, a job that paid a small wage and left him with enough time to continue his research. He was able to take away used wrapping paper on which to write his mathematics. Eventually, Ramanujan obtained a grant from Madras University which enabled him to have access to the library and research facilities, and he was able to undertake his studies and research in a logical way. As a result, Ramanujan was awarded a scholarship to Cambridge University in England. Some of his work revealed amazing discoveries, but it also revealed a lack of background knowledge, and Ramanujan spent most of his time improving his basic knowledge and establishing proofs for some of his discoveries.

The English climate and food did not agree with Ramanujan, but he continued working, publishing 32 important papers between 1914 and 1921 even though he was ill with tuberculosis.

In 1918 Ramanujan was made a Fellow of the Royal Society and was awarded Fellowship of Trinity College. Sadly, he was too ill to accept the position of professor of mathematics at Madras University. He returned to India and died in 1920.

The famous English mathematician Godfrey Hardy wrote of Ramanujan:

"One gift he has which no-one can deny: profound and invincible originality. He would probably have been a greater mathematician if he had been caught and tamed a little in his youth; he would have discovered more that was new, and that, no doubt, of greater importance. On the other hand he would have been less of a Ramanujan, and more of a European professor, and the loss might have been greater than the gain."

PUZZLE — FACTORISATION

Fully factorise the following expressions, and place the letter in the boxes corresponding to the correct solution, to reveal the message.

A	$x^2 - 7x + 10$	**L**	$x^2 + 8x - 48$	**T**	$5x^2 + 20x + 20$
C	$x(x+2) - (x+2)^2$	**N**	$x + x^2 + x^4$	**U**	$(x+2)(x+3) - 2(x+2)$
E	$x^2 - 2x - 3$	**O**	$16x^2 - 25$	**W**	$x^2 - 121$
F	$x^2 - 13x + 22$	**P**	$x^2 - 18x + 81$	**X**	$x^2 + 8x - 20$
G	$-6x^3 + 4x$	**R**	$7x - 21$	**Y**	$x^2 + 3x - 18$
I	$36x^2 - 49y^2$	**S**	$2xy + 6y$		

PRINTABLE PUZZLE

$-2(x+2)$		$(x+6)(x-3)$		$(x+1)(x-3)$	
$(4x+5)(4x-5)$		$(4x+5)(4x-5)$		$(x+10)(x-2)$	
$x(1+x+x^3)$		$(x+1)(x+2)$		$(x-9)^2$	
$2x(2-3x^2)$		$-2(x+2)$		$7(x-3)$	
$7(x-3)$		$(x-2)(x-5)$		$(x+1)(x-3)$	
$(x-2)(x-5)$		$x(1+x+x^3)$		$2y(x+3)$	
$5(x+2)^2$		$(x-2)(x-11)$		$2y(x+3)$	
$(x+1)(x+2)$		$(x-2)(x-5)$		$(6x+7y)(6x-7y)$	
$(x+12)(x-4)$		$-2(x+2)$		$(4x+5)(4x-5)$	
$(x-2)(x-5)$		$5(x+2)^2$		$x(1+x+x^3)$	
$5(x+2)^2$		$(4x+5)(4x-5)$		$2y(x+3)$	
$(6x+7y)(6x-7y)$		$7(x-3)$		$x(1+x+x^3)$	
$(4x+5)(4x-5)$		$(6x+7y)(6x-7y)$		$(4x+5)(4x-5)$	
$x(1+x+x^3)$		$2y(x+3)$		$(x+11)(x-11)$	
$2y(x+3)$		$(x+1)(x-3)$			

Chapter 10

Statistics

Contents:

OPENING PROBLEM

The heights of 1294 Year 10 girls were measured. The results were recorded in classes in the frequency table alongside.

Height h (cm)	*Frequency*
$120 \leqslant h < 130$	1
$130 \leqslant h < 140$	0
$140 \leqslant h < 150$	34
$150 \leqslant h < 160$	139
$160 \leqslant h < 170$	478
$170 \leqslant h < 180$	642
Total	1294

Things to think about:

- **a** Is the data categorical or numerical?
- **b** Why was the data recorded like this?
- **c** Is the data discrete or continuous?
- **d** What does the height $140 \leqslant h < 150$ actually mean?
- **e** How could this data be displayed?
- **f** How can the *shape* of the distribution be described?
- **g** Do we know the exact height of the shortest girl?
- **h** How could we estimate the mean and median of this data?

In today's world vast quantities of information are recorded. We record the populations of countries and how they are distributed, the number of children in families, the number of unemployed people, how much food and manufactured goods are produced, daily temperatures, and much more.

Many groups in our society use the information collected to help them discover facts, make decisions, and predict outcomes. Government departments, businesses, and scientific research bodies all use statistics.

The word **statistics** can be used in three different contexts:

- **Statistics** is a branch of mathematics that is concerned with how information is collected, organised, presented, summarised, and then analysed so that conclusions may be drawn from the information.
- **Statistics** are a collection of facts or data about a group or population.
- The singular **statistic** refers to a quantity calculated from sample data. For example, the mean is a statistic, and the range is a statistic.

HISTORICAL NOTE

The earliest statistical recordings include:

- ancient **Babylonians** recorded their crop yields on clay tablets
- ancient **Egyptian pharaohs** recorded their wealth on stone walls.

More recently:

- **William Playfair** (1759 - 1823) developed the histogram to display data.
- **Florence Nightingale** (1820 - 1910) kept records of the dead and injured during the Crimean War. She developed and used graphs extensively.
- **John Tukey** (1915 - 2000) invented the stem plot in 1972.

Florence Nightingale

A TYPES OF DATA

When we perform a statistical investigation, each fact or piece of information we collect is called a **datum**. The plural of datum is **data**.

The data that we collect is either **categorical** or **numerical** data.

CATEGORICAL DATA

A **categorical variable** describes a particular quality or characteristic. The data is divided into **categories**, and the information collected is called **categorical data**.

The *occupation of workers at a hospital* is a categorical variable. The possible categories include doctor, nurse, secretary, and cleaner.

We have seen in previous years that categorical data can be displayed using a **column graph**, a **bar chart**, or a **pie chart**.

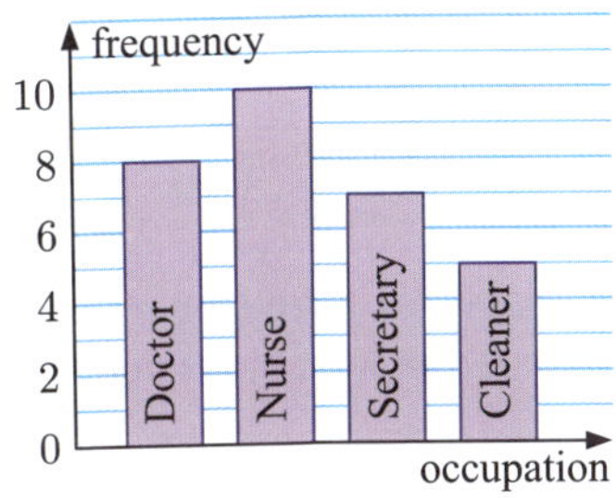

Column graph

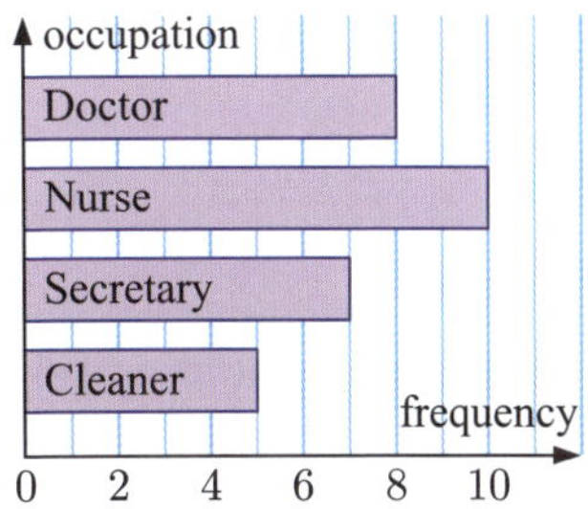

Bar chart

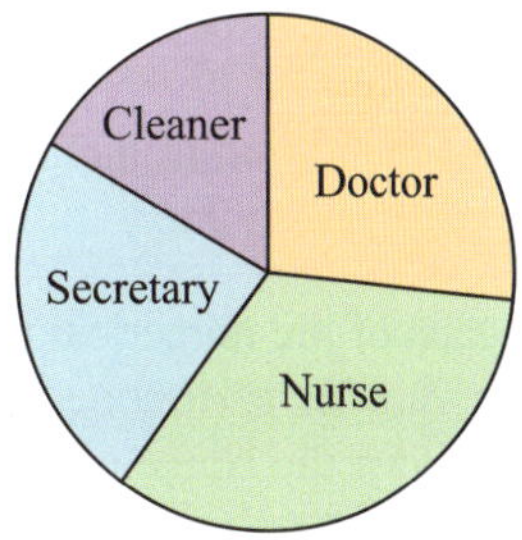

Pie chart

NUMERICAL DATA

A **numerical variable** is one for which data is given in number form. The information collected is called **numerical data**.

Numerical variables can be further classified as **discrete numerical** or **continuous numerical**.

A **discrete numerical variable** can only take distinct values. It is usually the result of **counting**.

The *number of goals scored by a soccer player* is a discrete numerical variable. It can take the values 0, 1, 2, 3, ... It cannot take any value that is not a non-negative integer.

A **continuous numerical variable** can take *any* value within a certain range. It is usually the result of **measuring**.

The *weight of pumpkins* is a continuous numerical variable. It can take any value from about 1 kg to 15 kg.

EXERCISE 10A

1 Classify the following data as categorical, discrete numerical, or continuous numerical:

- **a** the number of pages in a magazine
- **b** the temperature in Brazil
- **c** the manufacturer of a motorcycle
- **d** a person's preferred football code
- **e** the position taken by a netballer
- **f** the width of a person's hand
- **g** the time taken to travel to school
- **h** the area of a block of land
- **i** the number of goals saved by a goalkeeper.

2 A sample of lamp posts was surveyed for the following data. Classify the data as categorical, discrete numerical, or continuous numerical:

- **a** the diameter of the lamp post
- **b** the material from which the lamp post is made
- **c** the location of the lamp post (inner, outer, north, south, east, or west)
- **d** the height of the lamp post
- **e** the time since the last inspection
- **f** the number of inspections since installation
- **g** the condition of the lamp post (very good, good, fair, or unsatisfactory).

3 For each of the numerical variables below, determine whether the variable is discrete or continuous. If the variable is discrete, list the values the variable is likely to take. If the variable is continuous, write down the *range* of values the variable is likely to take.

- **a** The result obtained when an ordinary die is rolled.
- **b** The length of a person's thumb.
- **c** The number of pins knocked down in one roll of ten pin bowling.
- **d** The time it takes to run 100 metres.
- **e** The number of students in a class.

B DISCRETE NUMERICAL DATA

Discrete numerical data with not many different data values can be arranged in a **tally and frequency table**, and displayed in a **column graph** or **dot plot**.

For example, consider the data set:

5 4 5 7 4 5 3 6 5 4
4 5 6 4 6 5 5 6 4 5
5 4 5 5 3 4 6 5 5 4

Number	*Tally*	*Frequency*
3	||	2
4	𝍸 ||||	9
5	𝍸 𝍸 |||	13
6	𝍸	5
7	|	1

Tally and frequency table

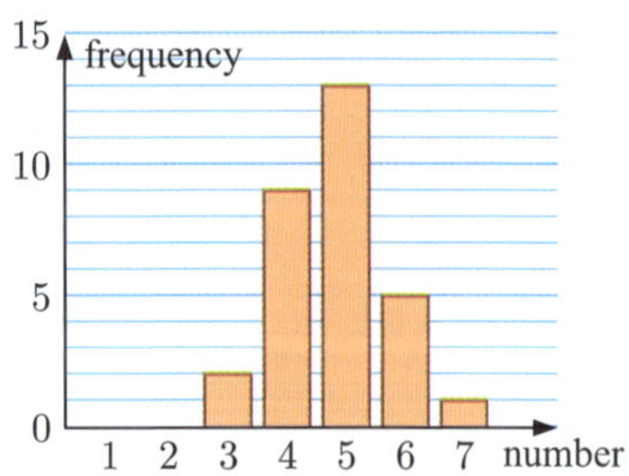

Column graph

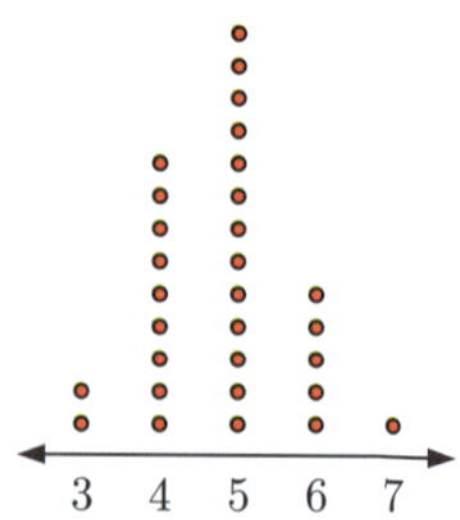

Dot plot

If the data has many different data values, we can display the data in groups using a **stem-and-leaf plot** or **stem plot**. The stem plot displays the frequency of data in each group, but retains the actual data values.

28	47	39	23	17	33
35	26	49	35	9	36
43	44	38	27	32	51
31	11	26	42	24	30

For the data shown, we can construct an unordered or an ordered stem plot.

Unordered stem plot

Stem	*Leaf*
0	9
1	7 1
2	8 3 6 7 6 4
3	9 3 5 5 6 8 2 1 0
4	7 9 3 4 2
5	1

Scale: $5 \mid 1 = 51$

Ordered stem plot

Stem	*Leaf*
0	9
1	1 7
2	3 4 6 6 7 8
3	0 1 2 3 5 5 6 8 9
4	2 3 4 7 9
5	1

Scale: $5 \mid 1 = 51$

You can use the **statistics package** or your **calculator** to display discrete numerical data.

STATISTICS PACKAGE

GRAPHICS CALCULATOR INSTRUCTIONS

Example 1 **Self Tutor**

The scores for a test out of 50 were recorded for 36 students.

25, 36, 38, 49, 23, 46, 47, 15, 28, 38, 34, 9, 30, 24, 27, 27, 42, 16, 28, 31, 24, 46, 25, 31, 37, 35, 32, 39, 43, 40, 50, 47, 29, 36, 35, 33

a Display the data using a stem-and-leaf plot.

b What percentage of students scored 40 or more marks?

a The stems will be 0, 1, 2, 3, 4, 5 to account for numbers from 0 to 50.

Unordered stem plot

Stem	*Leaf*
0	9
1	5 6
2	5 3 8 4 7 7 8 4 5 9
3	6 8 8 4 0 1 1 7 5 2 9 6 5 3
4	9 6 7 2 6 3 0 7
5	0

Scale: $2 \mid 4 = 24$ marks

Ordered stem plot

Stem	*Leaf*
0	9
1	5 6
2	3 4 4 5 5 7 7 8 8 9
3	0 1 1 2 3 4 5 5 6 6 7 8 8 9
4	0 2 3 6 6 7 7 9
5	0

b 9 students scored 40 or more marks.

$\therefore$ the percentage of students scoring 40 or more marks $= \frac{9}{36} \times 100\% = 25\%$.

DESCRIBING THE DISTRIBUTION OF THE DATA SET

Many data sets show **symmetry** or **partial symmetry** about the centre of the distribution. We can see this by drawing a curve over the graph of the data. For example, if we place a curve over the column graph opposite, we see that the curve is symmetric. We say we have a **symmetrical distribution** of data.

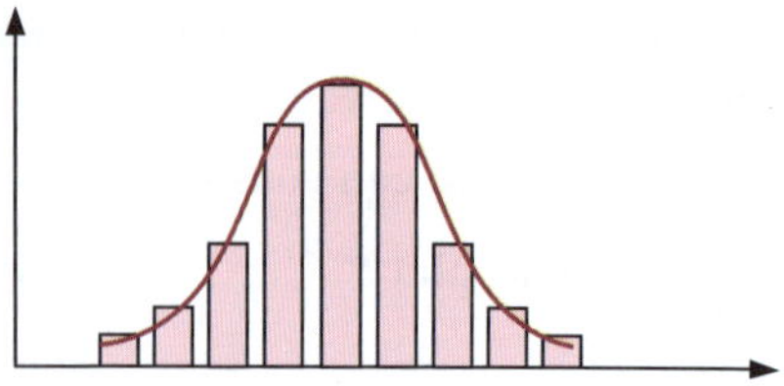

The distribution alongside is said to be **positively skewed** since it is 'stretched' on the right or positive side of the centre.

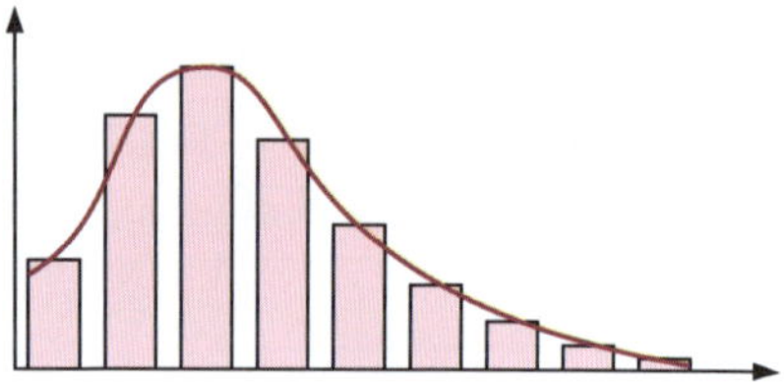

So we have:

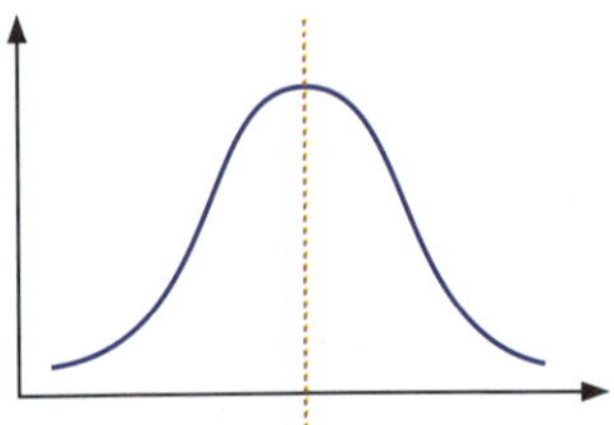

symmetrical distribution

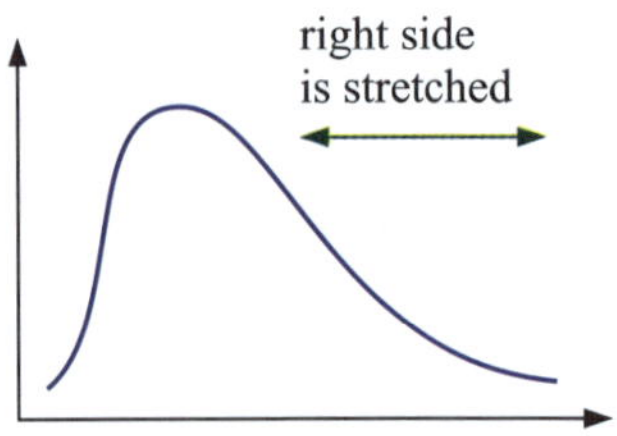

positively skewed distribution

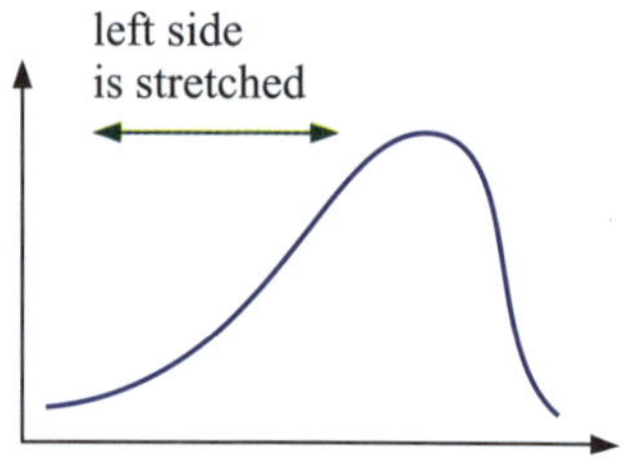

negatively skewed distribution

Example 2 **Self Tutor**

45 people were asked how many times they had visited the doctor this year. The results are given in the frequency table alongside.

a Display this information on a column graph.

b Describe the distribution of the data.

Number of visits	*Frequency*
0	6
1	8
2	9
3	7
4	6
5	4
6	2
7	2
8	1

a **Visits to the doctor**

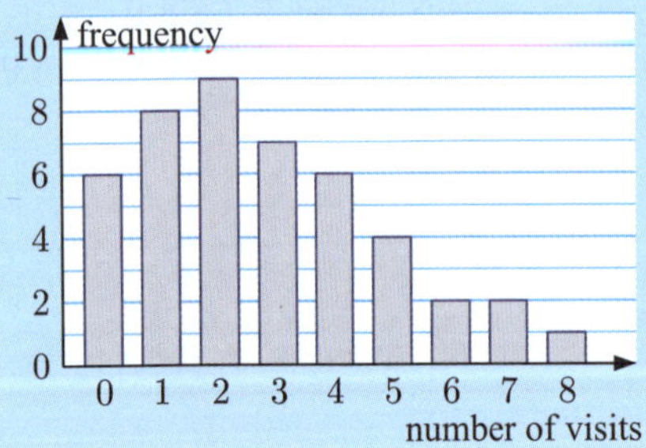

b The right side of the graph is stretched out, so the data is positively skewed.

EXERCISE 10B

1 When Aimee walked through a shopping mall, she recorded the number of customers in each shop. The data is displayed in a dot plot.

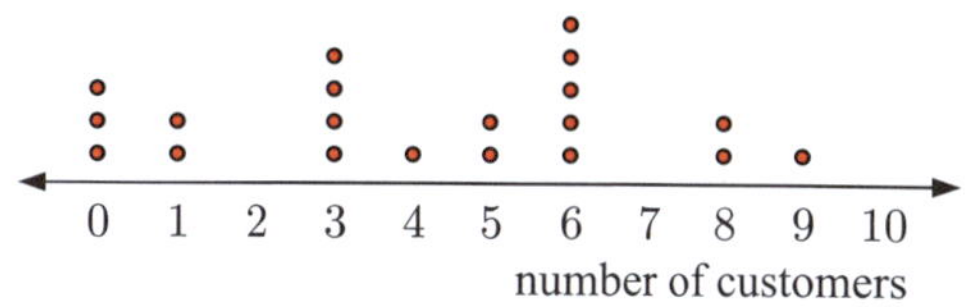

- **a** How many shops did Aimee collect data from?
- **b** How many shops had 3 customers?
- **c** What percentage of shops had:
  - **i** 5 customers
  - **ii** at least 7 customers?

2 The stem-and-leaf plot alongside shows the ten pin bowling scores Ron has obtained this year.

Stem	*Leaf*
7	6 8
8	2 5 8
9	1 1 2 5 9
10	1 2 4
11	0
12	6
13	2 7

Scale: $8 \mid 2 = 82$

- **a** How many games has Ron played?
- **b** Find Ron's:
  - **i** lowest score
  - **ii** highest score.
- **c** Describe the distribution of the data.
- **d** How many times has Ron scored less than 100?
- **e** What percentage of Ron's scores have been 130 or higher?

3 50 boxes of sweets were selected, and the sweets in each box were counted.

- **a** Construct a column graph to display the data.
- **b** Describe the distribution of the data.

Number of sweets	*Frequency*
25	2
26	2
27	4
28	6
29	9
30	16
31	11

4

Number of deliveries	*Frequency*
5	1
6	4
7	8
8	11
9	9
10	5
11	2

Hayden makes deliveries for a whitegoods store. He records the number of deliveries he makes each day for 40 days.

- **a** Draw a dot plot to display the data.
- **b** On how many days has Hayden made at least 10 deliveries?
- **c** Describe the distribution of the data.

5 An ice hockey player scored the following numbers of goals in his last 30 matches:

1 1 3 2 0 0 4 2 2 4 3 1 0 1 0
2 2 5 1 3 7 2 3 1 1 2 1 4 0 3

- **a** Organise the data in a tally and frequency table.
- **b** Construct a column graph to display the data.
- **c** Comment on the distribution of the data.

6 For the following stem plots:

i state the minimum and maximum data values

ii describe the distribution of the data.

a

Stem	*Leaf*
2	7 8
3	0 1 3
4	2 3 5 5 7
5	1 1 3 7 9 9
6	4 3 6 7
7	0 2 8
8	1

Scale: $3 \mid 0 = 30$

b

Stem	*Leaf*
6	3 6 7 8
7	2 2 6 8 9
8	0 5 7 7 8 9 9
9	1 3 4 6 7
10	2 4 5 8
11	0 0 3
12	1 4 7
13	4 4
14	1 3
15	2

Scale: $7 \mid 2 = 72$

c

Stem	*Leaf*
1	4
2	
3	2 5
4	0 1 4
5	1 6 7 9
6	0 2 2 3 5 8
7	1 1 3 4 8 8 9
8	0 2 5 7

Scale: $3 \mid 2 = 3.2$

7 The following marks were scored for a test out of 50 marks:

47 32 32 29 36 39 40 46 43 39 44 18 38 45 35 46 7 44 27 48

a Construct a stem plot to display the data.

b How many students scored 40 or more marks?

c What percentage of the students scored less than 30 marks?

d A score of 25 or more is required to pass the test. What percentage of the students passed?

e Describe the distribution of the data.

8 The number of peanuts in a jar varies slightly from jar to jar. A manufacturer took a sample of sixty jars of peanuts and recorded the number of peanuts in each. The results were:

201	204	213	224	221	193	194	195	178	185	196	210
201	203	207	207	204	192	188	205	207	201	215	201
209	217	189	191	194	194	198	195	204	208	213	224
227	185	198	203	203	213	216	231	182	193	194	203
200	206	210	228	201	196	186	197	199	208	204	189

a Construct a stem plot for the data. The 'stems' are to be split to give a better display of the distribution of numbers. Use the stem 17 for numbers from 170 to 174, 17* for numbers from 175 to 179, and so on.

b What percentage of the jars had 200 peanuts or more?

c What percentage of the jars had less than 190 peanuts?

d Describe the distribution of the data.

e The manufacturer would like at least 95% of his jars to have within 20 peanuts of the stated number which is 200. Is this the case for this sample?

DISCUSSION

As a class, list variables which you would expect to have:

- a symmetric distribution
- a negatively skewed distribution
- a positively skewed distribution.

C CONTINUOUS NUMERICAL DATA

When data is recorded for a continuous numerical variable, there will be many different values. We therefore group the data into **class intervals** which have lower and upper bounds.

We use a **frequency histogram** to display continuous data. This is similar to a column graph, but to account for the continuous nature of the variable, we do not leave gaps between the columns.

The **modal class** is the class with the highest frequency. More of the data lies in this class than in any other class.

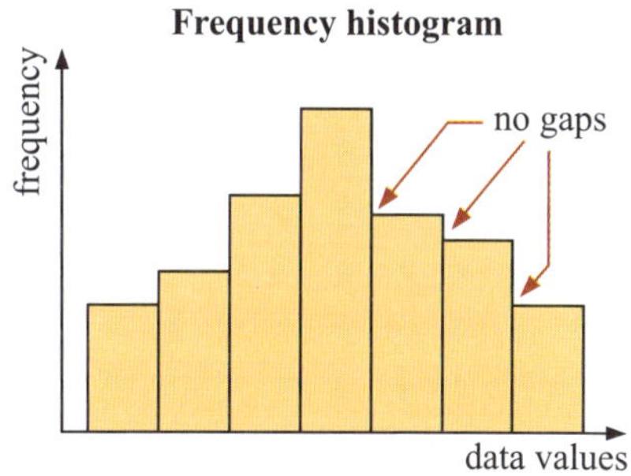

Example 3

Self Tutor

The weight w of 17 pineapples was recorded in kilograms. Each measurement was rounded *down* to the nearest 0.1 kg.

2.1, 3.0, 0.6, 1.5, 1.9, 2.4, 3.2, 4.2, 2.6, 3.1, 1.8, 1.7, 3.9, 2.4, 0.3, 1.5, 1.2

a Organise the data into class intervals.

b Draw a frequency histogram to display the data.

c Find the modal class.

a The lowest weight recorded was 0.3 kg and the highest was 4.2 kg, so we use class intervals of length 1 kg. The class interval $1 \leqslant w < 2$ includes all weights from 1 kg up to, but not including 2 kg.

Weight w (kg)	*Frequency*
$0 \leqslant w < 1$	2
$1 \leqslant w < 2$	6
$2 \leqslant w < 3$	4
$3 \leqslant w < 4$	4
$4 \leqslant w < 5$	1

b

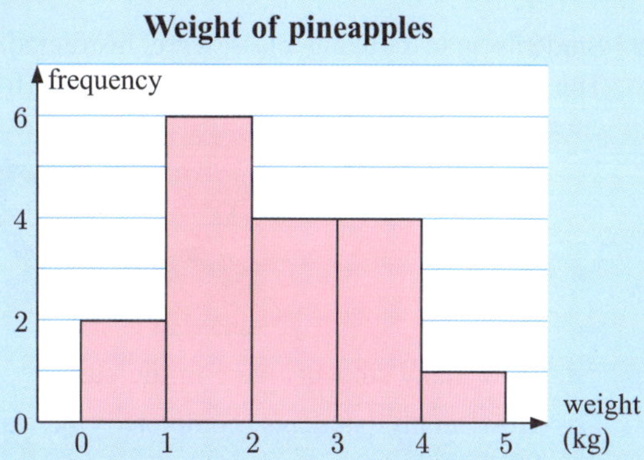

c The modal class is $1 \leqslant w < 2$ kg.

EXERCISE 10C

1 The speeds of vehicles travelling along a section of highway were recorded and displayed in the frequency histogram below.

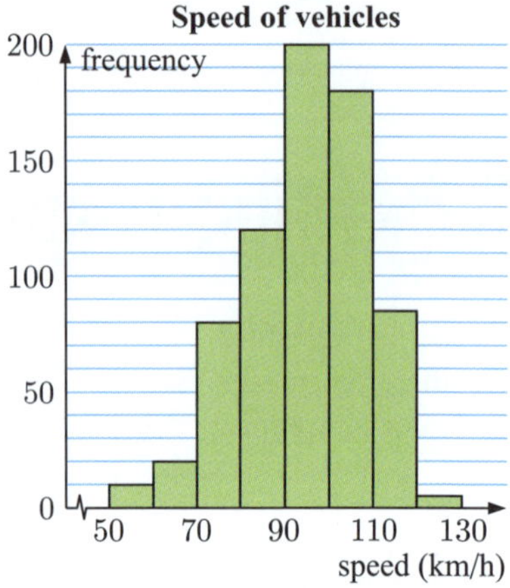

a How many vehicles were included in this survey?

b What percentage of vehicles were travelling at speeds between 100 and 110 km/h?

c What percentage of vehicles were travelling at speeds less than 80 km/h?

d The owners of the vehicles travelling at 110 km/h or more were fined \$252 each. What amount was collected in fines?

2

Height h (cm)	*Frequency*
$170 \leqslant h < 175$	1
$175 \leqslant h < 180$	8
$180 \leqslant h < 185$	9
$185 \leqslant h < 190$	11
$190 \leqslant h < 195$	9
$195 \leqslant h < 200$	3
$200 \leqslant h < 205$	3

A frequency table for the height h of the players in a basketball squad, is given alongside.

a Explain why *height* is a continuous variable.

b Construct a frequency histogram for the data.

c What is the modal class? Explain what this means.

d Describe the distribution of the data.

3 Students in a cooking class were instructed to bake a cake. The table alongside shows the times taken for the students to complete the task.

Time t (min)	*Frequency*
$50 \leqslant t < 60$	2
$60 \leqslant t < 70$	4
$70 \leqslant t < 80$	3
$80 \leqslant t < 90$	5
$90 \leqslant t < 100$	8
$100 \leqslant t < 110$	10
$110 \leqslant t < 120$	13
$120 \leqslant t < 130$	10
$130 \leqslant t < 140$	5

a How many students were in the class?

b How many students completed the task in less than 80 minutes?

c Draw a frequency histogram to display the data.

d Find the modal class.

e Describe the distribution of the data.

f Students taking longer than 2 hours received a penalty. What percentage of students received a penalty?

4 The average monthly water usages w, in kilolitres, of apartments in a building, are:

9.37	8.22	10.14	8.78	11.13	10.03	9.63	7.87	12.46	10.55
8.62	13.09	9.66	7.46	9.14	8.12	8.77	10.43	9.93	11.58
10.11	9.38	11.36	8.53	12.71	10.82	7.92	9.45	8.80	9.62
7.71	8.11	9.70	10.30	7.01	9.12	11.59	8.62		

a Organise the data using the class intervals $7 \leqslant w < 8$, $8 \leqslant w < 9$, ..., $13 \leqslant w < 14$.

b Draw a frequency histogram to display the data.

c Find the modal class.

d Describe the distribution of the data.

e What percentage of apartments used at least 10 kL of water?

D MEASURING THE CENTRE OF A DATA SET

In this course we consider two statistics that are commonly used to measure the **centre** of a data set. These are the **mean** and the **median**.

THE MEAN

The **mean** $\overline{x}$ of a data set is the statistical name for its *arithmetic average*. It can be found by dividing the sum of the data values by the number of data values.

$$\textbf{mean} = \frac{\textbf{the sum of the data values}}{\textbf{the number of data values}}$$

The mean gives us a single number which measures the centre of the data set. It does not necessarily have to equal one of the data values.

Example 4 Self Tutor

12 people were shown a series of television commercials, and then were asked to write down the commercials that they remembered. The data below shows how many commercials each person remembered:

7 10 12 10 8 13 10 6 5 11 6 10

Find the mean number of commercials remembered.

$$\text{Mean} = \frac{7+10+12+10+8+13+10+6+5+11+6+10}{12}$$

$$= \frac{108}{12}$$

$$= 9 \text{ commercials}$$

Example 5 Self Tutor

The table below shows the numbers of aces served by tennis players in their first set of a tournament.

Number of aces	1	2	3	4	5	6
Frequency	4	11	18	13	7	2

Determine the mean number of aces for these sets.

Number of aces	*Frequency*	*Product*
1	4	4
2	11	22
3	18	54
4	13	52
5	7	35
6	2	12
Total	55	179

This is the number of data values. (55)

This is the sum of the data values. (179)

$$\overline{x} = \frac{\text{sum of the data values}}{\text{number of data values}} = \frac{179}{55} \approx 3.25 \text{ aces}$$

THE MEDIAN

The **median** is the *middle value* of an ordered data set.

For an **odd number** of data, the median is one of the data.

For an **even number** of data, the median is the average of the two middle values. The median might not be one of the original data.

If there are n data values, find the value of $\frac{n+1}{2}$.

The median is the $\left(\frac{n+1}{2}\right)$th data value.

For example:

If $n = 23$, $\frac{n+1}{2} = 12$ so the median is the 12th ordered data value.

If $n = 28$, $\frac{n+1}{2} = 14.5$ so the median is the average of the 14th and 15th ordered data values.

Click on the icon for instructions on finding the mean and median using your calculator.

Alternatively, you can use the **statistics package**.

STATISTICS PACKAGE

GRAPHICS CALCULATOR INSTRUCTIONS

Example 6 — Self Tutor

The following sets of data show the number of peas in a randomly selected sample of pods. Find the median for each set.

a 3, 6, 5, 7, 7, 4, 6, 5, 6, 7, 6, 8, 10, 7, 8

b 3, 6, 5, 7, 7, 4, 6, 5, 6, 7, 6, 8, 10, 7, 8, 9

a The ordered data set is:

~~3 4 5 5 6 6 6~~ 6 ~~7 7 7 7 8 8 10~~ {15 data values}

Since $n = 15$, $\frac{n+1}{2} = 8$ $\therefore$ the median is the 8th data value.

$\therefore$ the median $= 6$ peas

b The ordered data set is:

~~3 4 5 5 6 6 6~~ 6 7 ~~7 7 7 8 8 9 10~~ {16 data values}

Since $n = 16$, $\frac{n+1}{2} = 8.5$

$\therefore$ the median is the average of the 8th and 9th data values.

$\therefore$ the median $= \frac{6+7}{2} = 6.5$ peas

Example 7 — Self Tutor

The data in the table below shows the number of people at each table at a restaurant. Find the median of this data.

Number of people	5	6	7	8	9	10	11	12
Frequency	1	0	3	9	12	7	4	2

The total number of data values is $1 + 0 + 3 + 9 + 12 + 7 + 4 + 2 = 38$, so $n = 38$.

$\frac{n+1}{2} = \frac{39}{2} = 19.5$, so the median is the average of the 19th and 20th data values.

Number of people	5	6	7	8	9	10	11	12
Frequency	1	0	3	9	12	7	4	2

13 data values are 8 or less

the 14th to the 25th data values are all 9s

The 19th and 20th data values are both 9s, so the median $= \frac{9+9}{2} = 9$ people.

EXERCISE 10D.1

1 Find the mean of the data set:

a 3, 9, 8, 6, 8, 4, 5, 5

b 17, 23, 14, 37, 25, 30, 22, 27, 31, 36

c 4.5, 6.7, 8.1, 2.4, 6.3, 5.3, 8.4, 4.7, 3.7, 5.8, 6.6, 4.2, 9.3

2 Calculate the median of the data set:

- **a** 21, 23, 24, 25, 29, 31, 34, 37, 41
- **b** 105, 106, 107, 107, 107, 109, 113, 120, 124, 132
- **c** 173, 146, 128, 132, 116, 129, 141, 163, 187, 153, 162, 184

3 The selling prices of the last 10 houses sold in Everton Hills were:

£346 400, £327 600, £411 000, £392 500, £456 400, £332 400,
£348 000, £329 500, £331 400, £362 500

- **a** Calculate the mean and median selling prices of these houses, and comment on the results.
- **b** Which measure would you use if you were:
  - **i** a vendor wanting to sell your house
  - **ii** looking to buy a house in the district?

4 25 children were asked how many Christmas presents they received last year. The results are shown in the table below.

Number of presents	3	4	5	6	7	8
Frequency	2	3	6	8	4	2

Find the mean and median of the data.

5 Packets of chocolate almonds were opened and their contents counted. The table shows the results.

Number in packet	32	33	34	35	36	37	38
Frequency	6	8	9	13	10	3	2

Find the mean and median of the data.

6 Find the mean and median of the data set:

a

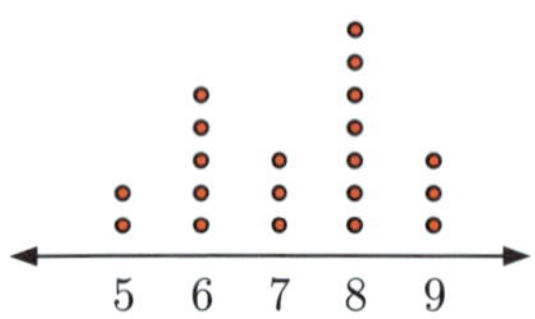

b

Stem	*Leaf*
2	5 8 9
3	1 3 3 6 8
4	0 5 6
5	1 3
6	0 7

Scale: 2 | 5 = 25

7 Consider the three column graphs below:

A

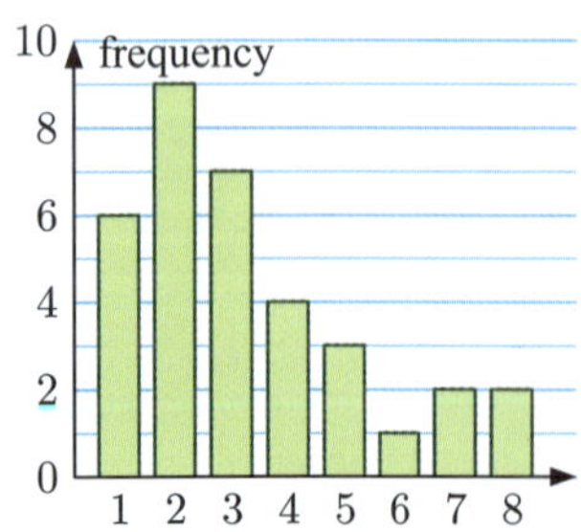

B

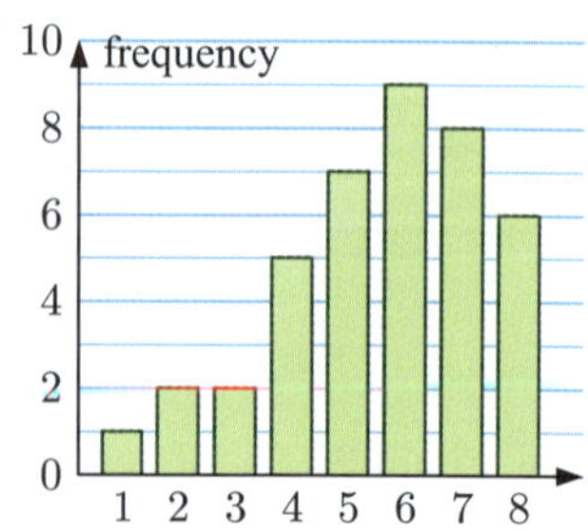

C

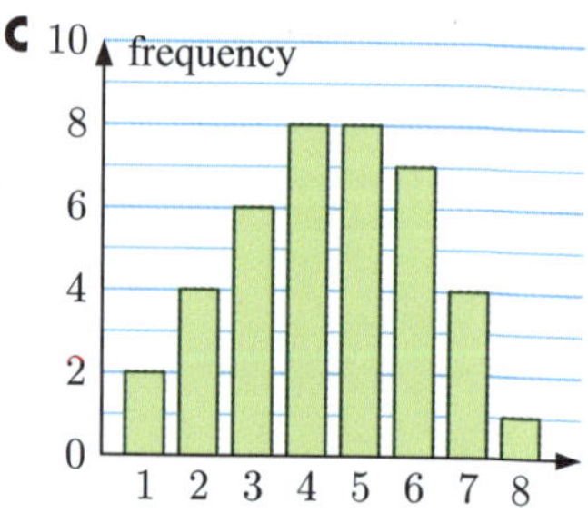

- **a** Describe the distribution of each data set.
- **b** Find the mean and median of each data set.
- **c** Copy and complete:
  - **i** For positively skewed data, the mean is ...... than the median.
  - **ii** For negatively skewed data, the mean is ...... than the median.
  - **iii** For symmetric data, the mean and median are approximately ......

Example 8 **Self Tutor**

Solve the following problems:

a The mean of six scores is 78.5. What is the sum of the scores?

b Find x if 10, 7, 3, 6, and x have a mean of 8.

a $\frac{\text{sum}}{6} = 78.5$

$\therefore$ sum $= 78.5 \times 6$

$= 471$

$\therefore$ the sum of the scores is 471.

b There are 5 scores.

$\therefore \frac{10+7+3+6+x}{5} = 8$

$\therefore \frac{26+x}{5} = 8$

$\therefore 26+x = 40$

$\therefore x = 14$

8 The mean of 12 scores is 8.8. What is the sum of the scores?

9 On an 8 day camping holiday, Lachlan drove an average of 214 km per day. In total, how far did Lachlan drive while on holiday?

10 The mean monthly sales for a CD store are \$216 000. Calculate the total sales for the store for the year.

11 Find x if:

a 2, 2, 6, 4, 7, and x have a mean of 5

b 9, x, 14, 18, x, x, 8, 10, and 4 have a mean of 11.

12 Over a semester, Jamie did 8 science tests. Each test was marked out of 30, and Jamie averaged 25. However, when checking his files, Jamie could only find 7 of the 8 tests. For these he scored 29, 26, 18, 20, 27, 24, and 29. Determine how many marks he scored for the eighth test.

13 Towards the end of the season, a netballer had played 14 matches and had an average of 16.5 goals per game. In the final two matches of the season the netballer threw 21 goals and 24 goals. Find the netballer's new average.

14 A sample of 12 measurements has a mean of 16.5, and a sample of 15 measurements has a mean of 18.6. Find the mean of all 27 measurements.

INVESTIGATION THE EFFECT OF OUTLIERS

In a set of data, an **outlier** or **extreme value** is a value which is much greater than or much less than the other values. In this Investigation we will examine the effect of an outlier on the two measures of central tendency.

What to do:

1 Consider the set of data: 4, 5, 6, 6, 6, 7, 7, 8, 9, 10. Calculate:

a the mean **b** the median.

2 If we add the extreme value 100 to the data set, we now have: 4, 5, 6, 6, 6, 7, 7, 8, 9, 10, 100. Calculate:

a the mean **b** the median.

3 Comment on the effect that the extreme value has on:

a the mean **b** the median.

4 Discuss with your class which measure of central tendency is most affected by the inclusion of an outlier.

DISCUSSION

- Discuss the advantages and disadvantages of the mean and median as measures of the centre of a data set.
- Under what circumstances do you think the mean or the median would be more appropriate to use?

ESTIMATING THE MEAN OF GROUPED DATA

When data is grouped in **class intervals**, the actual data values are not known. It is therefore impossible to calculate the exact mean of the data.

However, we can *estimate* the mean by assuming that each score in an interval takes the value of the interval's **midpoint**.

Example 9 Self Tutor

The speeds of 129 cars were measured at a particular point on a country road. They are shown in the table alongside.

Estimate the mean speed of the drivers.

Speed v (km/h)	*Frequency*
$40 \leqslant v < 50$	1
$50 \leqslant v < 60$	3
$60 \leqslant v < 70$	17
$70 \leqslant v < 80$	39
$80 \leqslant v < 90$	48
$90 \leqslant v < 100$	17
$100 \leqslant v < 110$	4

Speed v (km/h)	*Frequency*	*Interval midpoint*	*Product*
$40 \leqslant v < 50$	1	45	45
$50 \leqslant v < 60$	3	55	165
$60 \leqslant v < 70$	17	65	1105
$70 \leqslant v < 80$	39	75	2925
$80 \leqslant v < 90$	48	85	4080
$90 \leqslant v < 100$	17	95	1615
$100 \leqslant v < 110$	4	105	420
Total	129		10 355

$$\therefore \text{ mean} = \frac{\text{sum of the data values}}{\text{number of data values}} \approx \frac{10\,355}{129} \approx 80.3 \text{ km/h}$$

EXERCISE 10D.2

1 50 adults each kicked a football. The table alongside shows the distances the ball travelled in the air before it bounced.

Distance d (m)	*Frequency*
$20 \leqslant d < 30$	10
$30 \leqslant d < 40$	19
$40 \leqslant d < 50$	14
$50 \leqslant d < 60$	7

a Draw a frequency histogram of the data.

b Is it possible to determine the furthest distance the ball was kicked?

c Extend the table alongside to include an *interval midpoint* and *product* column.

d Estimate the mean distance the ball was kicked.

2 Sixty people were asked: "How many times have you been to the cinema in the last twelve months?". The results are given in the table alongside.

Number of times	*Frequency*
0 - 4	19
5 - 9	24
10 - 14	10
15 - 19	5
20 - 24	2

a Extend the table to include an *interval midpoint* and a *product* column.

b Estimate the mean of the data.

3

Area A (m^2)	*Frequency*
$400 \leqslant A < 500$	5
$500 \leqslant A < 600$	11
$600 \leqslant A < 700$	12
$700 \leqslant A < 800$	10
$800 \leqslant A < 900$	8

The table alongside shows the areas of the blocks of land along a street.

a Find the modal class.

b Estimate the mean land block size.

4 The frequency histogram shows the weights of dogs at a dog show.

a Describe the distribution of the data.

b Construct a frequency table for the data.

c Estimate the mean weight of the dogs at the show.

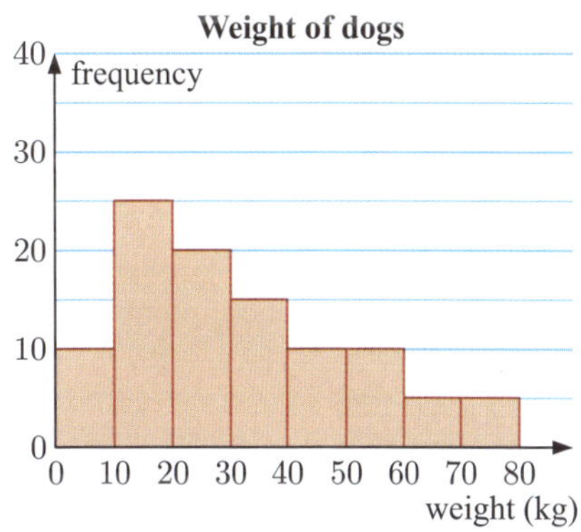

5 The times t that visitors spent in an art gallery were, in minutes:

21.6	35.7	26.2	32.3	22.2	38.0	17.8	31.5
37.5	13.0	29.8	39.1	34.0	12.8	25.4	24.1
33.3	23.3	36.0	14.7	20.1	27.2	32.9	21.6

a Group the data into the class intervals
$10 \leqslant t < 15$, $15 \leqslant t < 20$, ..., $35 \leqslant t < 40$.

b Hence *estimate* the mean time spent in the art gallery, giving your answer correct to 1 decimal place.

c Use the raw data values to find the exact mean. Compare your answer with your estimate in **b**.

E CUMULATIVE DATA

Sometimes it is useful to know the number of scores that lie above or below a particular value. In such situations it is convenient to construct a **cumulative frequency distribution table** and a **cumulative frequency graph**.

The **cumulative frequency** gives a *running total* of the scores up to a particular value.

The cumulative frequency graph can be used to estimate the **median** of grouped data.

Example 10 Self Tutor

The table alongside shows the times 50 people spent on an exercise bike at a gym in one day.

a Construct a cumulative frequency table for the data.

b Represent the data on a cumulative frequency graph.

c Use your graph to estimate how many people used the bike for:

  i less than 12 minutes **ii** more than 23 minutes.

d Estimate the median time spent on the bike.

Time t (min)	*Frequency*
$0 \leqslant t < 5$	5
$5 \leqslant t < 10$	8
$10 \leqslant t < 15$	10
$15 \leqslant t < 20$	12
$20 \leqslant t < 25$	9
$25 \leqslant t < 30$	6

a

Time t (min)	*Frequency*	*Cumulative frequency*
$0 \leqslant t < 5$	5	5
$5 \leqslant t < 10$	8	13
$10 \leqslant t < 15$	10	23
$15 \leqslant t < 20$	12	35
$20 \leqslant t < 25$	9	44
$25 \leqslant t < 30$	6	50

13: This is $5 + 8$

23: $5 + 8 + 10 = 23$ people used the bike for less than 15 minutes.

b At the end of each time interval, we plot the cumulative frequency up to that point. We then draw a curve through these points.

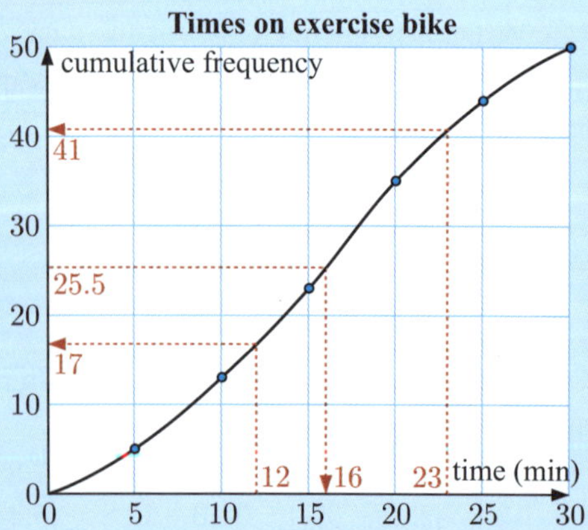

c Reading from the graph, we estimate that:

  i 17 people used the bike for less than 12 minutes

  ii $50 - 41 = 9$ people used the bike for more than 23 minutes.

d The median is the average of the 25th and 26th times.
Reading from the graph, the median ≈ 16 minutes.

EXERCISE 10E

1 This cumulative frequency graph shows the distances jumped by students in a long jump competition.

a How many students were in the competition?

b How many students jumped:

i less than 4 m

ii more than 4.5 m?

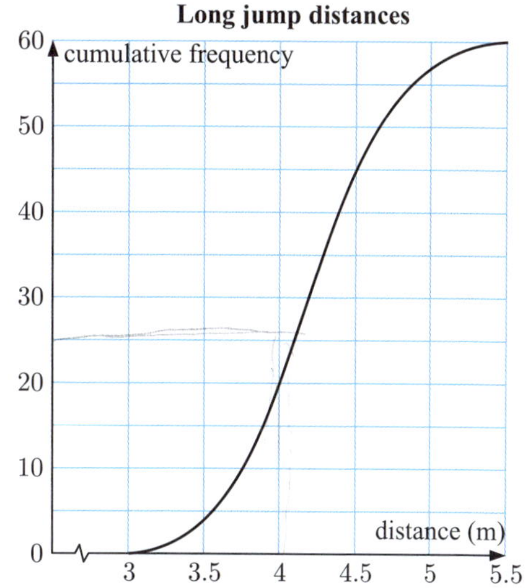

2 The table alongside shows the heights of dancers in a ballet class.

a Construct a cumulative frequency table for the data.

b How many dancers are in the class?

c Represent the data on a cumulative frequency graph.

d Use your graph to estimate the number of dancers who are:

i shorter than 155 cm **ii** taller than 145 cm.

Height h (cm)	*Frequency*
$120 \leqslant h < 130$	5
$130 \leqslant h < 140$	7
$140 \leqslant h < 150$	12
$150 \leqslant h < 160$	10
$160 \leqslant h < 170$	6

3

Speed v (km/h)	*Frequency*
$200 \leqslant v < 220$	12
$220 \leqslant v < 240$	16
$240 \leqslant v < 260$	21
$260 \leqslant v < 280$	18
$280 \leqslant v < 300$	13

An airport authority measured the speeds of planes as they touched down on the runway during a particular day.

a Construct a cumulative frequency table for the data.

b Draw a cumulative frequency graph of the data.

c Estimate the number of planes travelling faster than 230 km/h when they touched down.

d Estimate the median speed.

4 The table alongside shows the weights of competitors in a boxing tournament.

a Estimate the mean weight of the boxers.

b Construct a cumulative frequency table for the data.

c Draw a cumulative frequency graph of the data.

d Boxers weighing between 79 kg and 91 kg are in the cruiserweight class. Estimate the number of boxers in this class.

e Estimate the median weight of the boxers.

Weight w (kg)	*Frequency*
$50 \leqslant w < 60$	14
$60 \leqslant w < 70$	19
$70 \leqslant w < 80$	28
$80 \leqslant w < 90$	23
$90 \leqslant w < 100$	16

DISCUSSION

Which of these cumulative frequency graphs indicates:

a a symmetric distribution

b a positively skewed distribution

c a negatively skewed distribution?

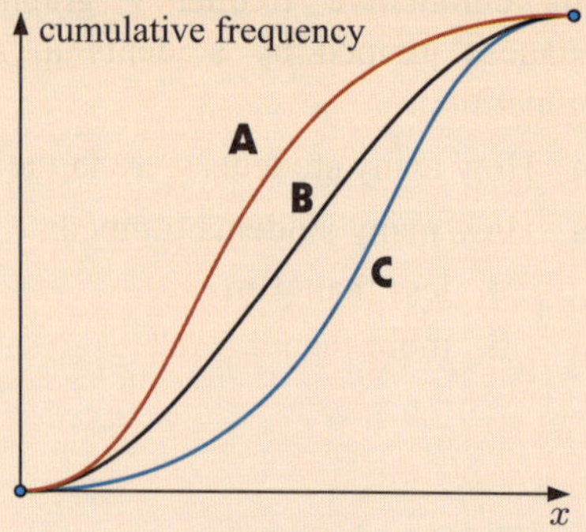

F MEASURING THE SPREAD OF A DATA SET

To accurately describe a data set, we need not only a measure of its centre, but also a measure of its **spread**.

For example, the data sets $1, 4, 5, 5, 6, 7, 8, 9, 9$ and $4, 4, 5, 6, 6, 7, 7, 7, 8$ both have a mean value of 6. However, the first data set is more widely spread than the second data set.

Two commonly used statistics that indicate the spread of a data set are:

- the range
- the interquartile range.

THE RANGE

The **range** is the difference between the **maximum** or largest data value, and the **minimum** or smallest data value.

range = maximum data value − minimum data value

Example 11 Self Tutor

The number of appointments made at a hairdresser each day is shown below.

$$4, 7, 5, 3, 4, 3, 6, 5, 7, 5, 3, 8, 9, 3, 6, 5, 6$$

Find the range of the data.

Searching through the data set, the minimum value $= 3$, and the maximum value $= 9$.

$$\therefore \text{ the range} = 9 - 3 = 6$$

THE INTERQUARTILE RANGE

We have already seen how the median divides an ordered data set into two halves. These halves are divided in half again by the **quartiles**.

The middle value of the *lower* half is called the **lower quartile** or $\mathbf{Q_1}$. One quarter or 25% of the data have values less than or equal to the lower quartile. 75% of the data have values greater than or equal to the lower quartile.

The middle value of the *upper* half is called the **upper quartile** or $\mathbf{Q_3}$. One quarter or 25% of the data have values greater than or equal to the upper quartile. 75% of the data have values less than or equal to the upper quartile.

The data set is thus divided into quarters by the lower quartile $\mathbf{Q_1}$, the median $\mathbf{Q_2}$, and the upper quartile $\mathbf{Q_3}$.

> The **interquartile range** is the range of the middle half of the data.
>
> **interquartile range = upper quartile − lower quartile**
>
> or $\textbf{IQR} = \mathbf{Q_3 - Q_1}$

Example 12 — Self Tutor

For the data set 7, 3, 4, 2, 5, 6, 7, 5, 5, 9, 3, 8, 3, 5, 6, find the:

a median **b** lower and upper quartiles **c** interquartile range.

The ordered data set is: ~~2 3 3 3 4 5 5~~ 5 ~~5 6 6 7 7 8 9~~ {15 data values}

a Since $n = 15$, $\dfrac{n+1}{2} = 8$ $\therefore$ the median is the 8th score, which is 5.

b As the median is a data value, we now ignore it and split the remaining data into two halves:

lower: 2 3 3 **3** 4 5 5 (Q₁ = 3) 5 upper: 5 6 6 **7** 7 8 9 (Q₃ = 7)

Q_1 = median of lower half = 3

Q_3 = median of upper half = 7

c $\text{IQR} = Q_3 - Q_1 = 7 - 3 = 4$

Example 13 — Self Tutor

For the data set 6, 10, 7, 8, 13, 7, 10, 8, 1, 7, 5, 4, 9, 4, 2, 5, 9, 6, 3, 2, find the:

a median **b** lower and upper quartiles **c** interquartile range.

The ordered data set is:

1 2 2 3 4 4 5 5 6 6 7 7 7 8 8 9 9 10 10 13 {20 data values}

a Since $n = 20$, $\dfrac{n+1}{2} = \dfrac{21}{2} = 10.5$

$\therefore$ the median $= \dfrac{\text{10th value} + \text{11th value}}{2} = \dfrac{6+7}{2} = 6.5$

b As the median is not a data value, we split the data into two halves:

lower: 1 2 2 3 **4 4** 5 5 6 6 ($Q_1 = 4$) upper: 7 7 7 8 **8 9** 9 10 10 13 ($Q_3 = 8.5$)

Q_1 = median of lower half = 4

Q_3 = median of upper half = 8.5

c $\text{IQR} = Q_3 - Q_1$

$= 8.5 - 4$

$= 4.5$

Some computer packages (for example, **MS Excel**) calculate quartiles in a different way from this example.

You can use the **statistics package** or your **calculator** to measure the spread of a data set.

STATISTICS PACKAGE

GRAPHICS CALCULATOR INSTRUCTIONS

EXERCISE 10F

1 Find the range of the data set:

a 4, 5, 5, 7, 8, 10, 10, 12, 15, 17, 18, 21, 23

b 39, 28, 50, 63, 37, 46, 61, 33, 25, 47, 29

c 7.5, 13.3, 8.6, 6.5, 9.9, 11.5, 13.4, 6.6, 8.2

2 Kimmi runs a weekly painting class. The numbers of people attending the class over the last 10 weeks have been:

17 14 21 18 19 26 20 22 16 25

Find the range of the data.

3 For the data set 5, 7, 7, 8, 10, 11, 13, 14, 14, 15, 18, find the:

a median **b** lower and upper quartiles **c** interquartile range.

4 The following amounts of money were withdrawn from an ATM on a particular day:

\$100	\$60	\$120	\$90	\$130	\$150	\$200	\$120	\$180
\$70	\$140	\$100	\$50	\$200	\$120	\$80	\$100	\$150

Find the:

a median **b** lower quartile

c upper quartile **d** interquartile range of the data.

5 At a fast-food restaurant, the times taken for 24 customers to receive their orders were recorded in minutes as follows:

1.0	1.6	1.4	2.2	0.9	2.9	2.2	1.0
0.7	1.8	2.8	1.3	0.8	3.3	1.9	2.0
1.3	1.8	1.2	3.4	1.6	2.8	0.8	2.6

a Find the:

i median **ii** lower quartile

iii upper quartile of the data.

b Find the range and interquartile range of the waiting times.

c Copy and complete the following statements:

i "50% of the waiting times were greater than ...... minutes."

ii "75% of the waiting times were less than ...... minutes."

iii "The minimum waiting time was ...... minutes and the maximum waiting time was ...... minutes. The waiting times were spread over ...... minutes."

6 The stem plot shows the heights of bean seedlings two weeks after the seeds germinated. Find the:

a minimum value
b maximum value
c median
d lower quartile
e upper quartile
f range
g interquartile range.

Stem	*Leaf*
6	0 3 8
7	0 1 5 6 7
8	1 1 2 4 4 8 9 9
9	0 4 7 9
10	1

Scale: 7 | 5 = 7.5 cm

7 The table below shows the number of passengers in the cars at a drive-in movie theatre.

Number of passengers	1	2	3	4	5	6
Frequency	2	19	10	17	11	5

a How many cars were at the theatre?
b How many people watched the movie?
c Find the range and interquartile range of the data.

8 The intelligence quotient (IQ) of members of a book club are given below:

118 115 123 108 119 98 111 120 104 109 125 116 101 119 121

a Find the range and interquartile range of the data.
b A person with an IQ of 136 joins the club. Find the new range and interquartile range of the data.
c A person with an IQ of 89 also joins the club. Find the new range and interquartile range of the data.
d Which of the measures of spread is most affected by the introduction of outliers?

G BOX-AND-WHISKER PLOTS

A **box-and-whisker plot** is a visual display of some of the descriptive statistics of a data set. It shows:

- the minimum value (min)
- the lower quartile (Q_1)
- the median (Q_2)
- the upper quartile (Q_3)
- the maximum value (max)

These five numbers form the **five-number summary** of a data set.

For **Example 13**, the five-number summary and corresponding box-and-whisker plot are:

minimum $= 1$
$Q_1 = 4$
median $= 6.5$
$Q_3 = 8.5$
maximum $= 13$

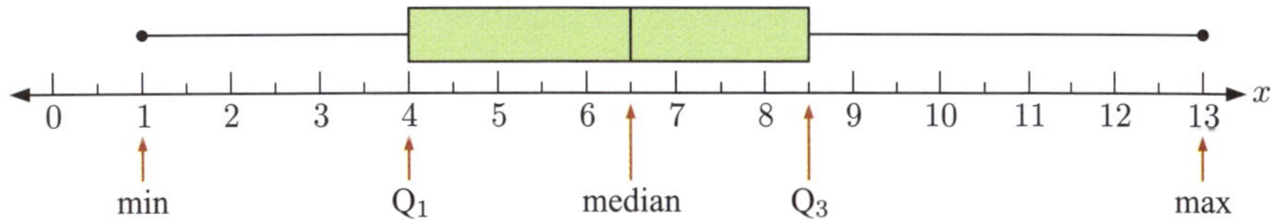

Notice that:
- the rectangular box represents the 'middle' half of the data set
- the lower whisker represents the 25% of the data with smallest values
- the upper whisker represents the 25% of the data with greatest values.

Example 14 **Self Tutor**

Consider the data set: 5 6 7 6 2 8 9 8 4 6 7 4 5 4 3 6 6

a Construct the five-number summary for the data.

b Draw a box-and-whisker plot to display the data.

c Find the: **i** range **ii** interquartile range of the data.

---

a The ordered data set is:

2 3 4 **4 4** 5 5 6 **6** 6 6 6 **7 7** 8 8 9 {17 data values}

$Q_1 = 4$ median $= 6$ $Q_3 = 7$

So, the 5-number summary is: $\begin{cases} \text{min} = 2 & Q_1 = 4 \\ \text{median} = 6 & Q_3 = 7 \\ \text{max} = 9 & \end{cases}$

b

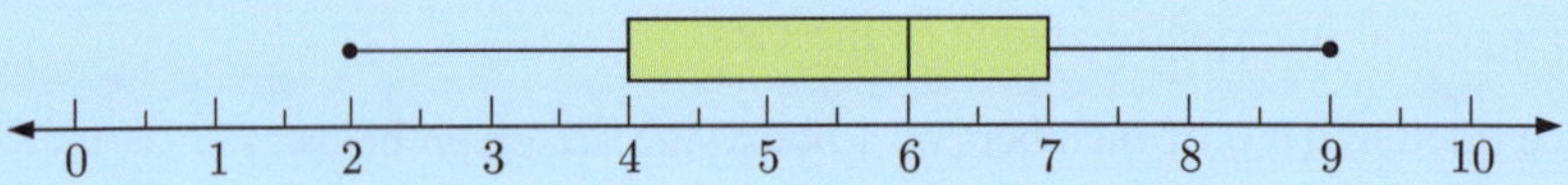

c **i** range = max − min = 9 − 2 = 7 **ii** IQR $= Q_3 - Q_1 = 7 - 4 = 3$

You can use the **statistics package** or your **calculator** to draw box-and-whisker plots.

EXERCISE 10G

1 For the box-and-whisker plot alongside, find the:

a five-number summary

b range

c interquartile range.

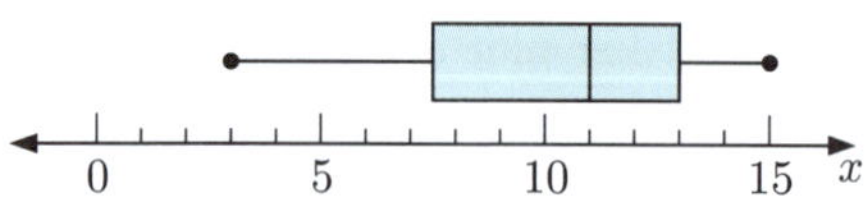

2 A box-and-whisker plot has been drawn to show the distribution of marks for a class in a test out of 100.

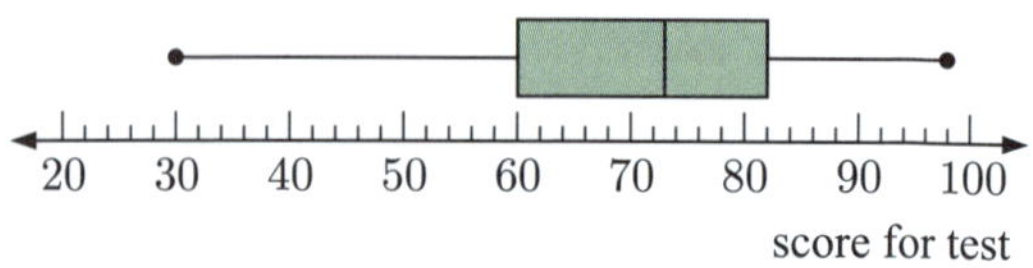

a Find the:

i highest mark **ii** lowest mark.

b What was the median test score for the class?

c What was the range of marks scored for the test?

d What percentage of students scored 60 or more for the test?

e What was the interquartile range for the test?

f If you scored 70 for the test, would you be in the top 50% of students in the class?

3 Consider the data set: 7 13 16 9 12 17 13 10 11

a Construct the five-number summary for the data.

b Draw a box-and-whisker plot to display the data.

c Find the: **i** range **ii** interquartile range of the data.

4 Julie examines a new variety of bean. She counts the number of beans in 33 pods. Her results are:

5 8 10 4 2 12 6 5 7 7 5 5 5 13 9 3 4
4 7 8 9 5 5 4 3 6 6 6 6 9 8 7 6

a Find the median, lower quartile, and upper quartile of the data.

b Draw a box-and-whisker plot of the data.

c Find the range and interquartile range of the data.

5 The data below shows the number of parking tickets handed out by an inspector each day for a month.

21 18 27 25 16 22 23 19 22 24
15 21 22 26 14 18 17 19 21 14
13 19 24 28 23 25 16 15 20 25

a Construct the five-number summary for the data.

b Draw a box-and-whisker plot to display the data.

c The box-and-whisker plot alongside shows the number of parking tickets handed out by the inspector in the previous month. Did the inspector hand out more tickets this month or last month? Explain your answer.

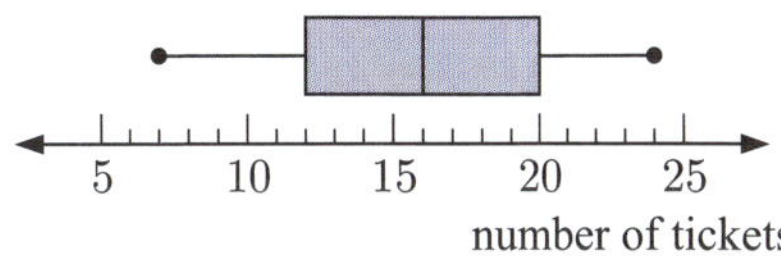

REVIEW SET 10A

1 Classify the following variables as categorical, discrete numerical, or continuous numerical:

a the number of kits in a litter of ferrets

b the flavours of milkshakes sold in a café

c the weights of pumpkins at a greengrocer.

2 The table below summarises the best times of 100 swimmers who swim 50 m.

a Draw a frequency histogram of the data.

b What is the modal class?

c Estimate the mean best time of the swimmers.

d What percentage of swimmers have a best time less than 35 seconds?

Time t (s)	*Frequency*
$25 \leqslant t < 30$	5
$30 \leqslant t < 35$	17
$35 \leqslant t < 40$	34
$40 \leqslant t < 45$	29
$45 \leqslant t < 50$	15

3 The following data set shows the number of points scored by a rugby team in a season of matches:

28, 24, 16, 6, 46, 34, 43, 16, 36, 49, 30, 28, 4, 31, 47, 41, 26, 25, 20, 29, 42

Find the: **a** mean **b** median **c** range **d** interquartile range.

4 A randomly selected sample of small businesses was asked, "How many full-time employees are there in your business?". The results are shown in the bar chart alongside.

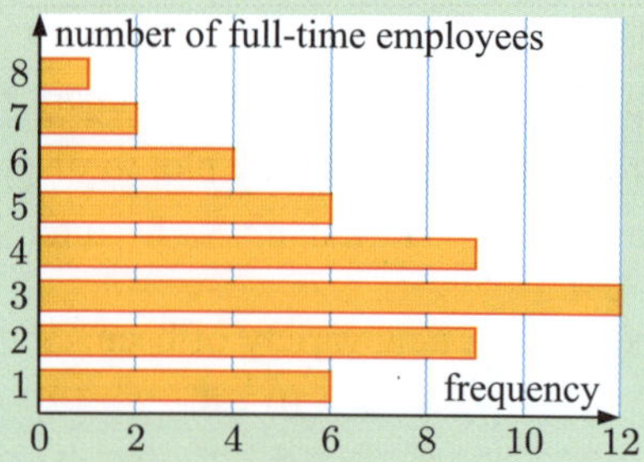

a How many small businesses supplied data for the survey?

b How many of the businesses had only one or two full-time employees?

c What percentage of the businesses had five or more full-time employees?

d Describe the distribution of the data.

e Find the mean of the data.

5 Six scores have mean 8. A seventh score is added, which increases the mean by 1. What is this score?

6 For the data set given, find the:

a minimum value **b** maximum value
c median **d** lower quartile
e upper quartile **f** range
g interquartile range.

Stem	*Leaf*
1	2 4 4 6 8
2	2 3 7 7 7 9 9
3	0 0 2 4 5 8
4	0 2 3

$4 \mid 0 = 40$

7 The given table shows the distribution of scores for a Year 9 spelling test.

Score	*Frequency*
6	2
7	4
8	7
9	12
10	5
Total	30

a For this data, calculate the:

i mean **ii** median **iii** range.

b The average score for Year 9 students across Australia in this spelling test was 6.2. How does this class compare to the national average?

c Describe the distribution of the data.

8 Marion counted the number of cats she saw on the way to school each day for 20 days:

3	2	5	1	1	4	4	0	8	2
0	1	3	7	10	2	4	3	0	3

a Draw a dot plot of the data.

b Find the mean and median of the data.

9 The data below gives the number of letters received by a household each week for 29 weeks:

6 6 6 6 7 7 7 7 7 7 7 8 8 8 8 8 8 8 8 8 9 9 9 9 9 10 10 10

a Find the range of the data set.

b Construct the five-number summary for the data set.

c Construct a box-and-whisker plot for the data set.

10 Consider the data displayed in the cumulative frequency graph alongside.

a How many of the data values are:

i less than 15 **ii** more than 40?

b Estimate the median of the data.

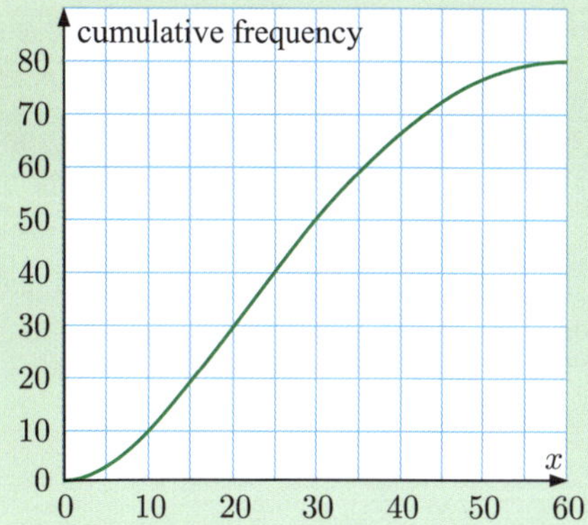

REVIEW SET 10B

1 A class of 20 students was asked "How many children are there in your household?". The following results were collected:

1 2 3 3 2 4 5 4 2 3 8 1 2 1 3 2 1 2 1 2

a What is the variable in the investigation?

b Is the data discrete or continuous? Explain your answer.

c Construct a frequency table to summarise the data.

d Construct a column graph to display the data.

e Describe the distribution of the data.

f Find the median of the data.

2 The numbers of potatoes growing on each of 100 potato plants were recorded and summarised in the table below:

Number of potatoes	0 - 2	3 - 5	6 - 8	9 - 11	12 - 14	15 - 17
Frequency	7	11	25	40	15	2

Estimate the mean number of potatoes per plant.

3 Eight scores have an average of six. When scores of 15 and x are added to the set, the average is increased to 7. Find x.

4 Jenny's golf scores for her last 20 rounds were:

90 106 84 103 112 100 105 81 104 98
107 95 104 108 99 101 106 102 98 101

a Find the median, lower quartile, and upper quartile of the data set.

b Find the interquartile range of the data set and explain what it represents.

5 A microbiologist measured the diameter (in cm) of a number of bacteria colonies 12 hours after seeding. The results were as follows:

0.4, 2.1, 3.4, 3.9, 1.7, 3.7, 0.8, 3.6, 4.1, 0.9, 2.5, 3.1, 1.5, 2.6, 1.3, 3.5

a Construct an ordered stem-and-leaf plot for this data.

b What percentage of the bacteria colonies were greater than 2 cm in diameter?

6 12 of 29 measurements are below 20 cm, and 13 measurements are above 21 cm. Find the median if the other 4 measurements are 20.1 cm, 20.4 cm, 20.7 cm, and 20.9 cm.

7 A squad of netballers each had 20 shots at goal. The number of goals scored by each netballer is shown in the table alongside.

Goals scored	*Frequency*
12	2
13	1
14	3
15	6
16	5
17	4
18	0
19	2
20	1

a Draw a column graph of the data.

b Find the:

i mean **ii** median **iii** range of the data.

8 The table alongside summarises the masses of 50 domestic cats chosen at random.

Mass m (kg)	*Frequency*
$0 \leqslant m < 2$	5
$2 \leqslant m < 4$	18
$4 \leqslant m < 6$	12
$6 \leqslant m < 8$	9
$8 \leqslant m < 10$	5
$10 \leqslant m < 12$	1

a What is the length of each class interval?

b Draw a frequency histogram of the data.

c What is the modal class?

d Estimate the mean mass of a domestic cat.

e Find the percentage of domestic cats with mass 8 kg or more.

9 The data below shows the ages of children at a scouts camp:

10 11 13 12 11 9 10 10 11 12
12 9 11 10 13 12 11 9 11 10
11 14 8 11 12 14 10 13

a Draw a dot plot to display the data.

b Describe the distribution of the data.

c Construct the five-number summary.

d Draw a box-and-whisker plot of the data.

10 The times taken for trampers to climb Ben Nevis were recorded and the results tabled below:

Time t (min)	$175 \leqslant t < 190$	$190 \leqslant t < 205$	$205 \leqslant t < 220$	$220 \leqslant t < 235$	$235 \leqslant t < 250$
Frequency	11	35	74	32	8

a Construct a cumulative frequency table for the data.

b Draw a cumulative frequency graph for the data.

c Estimate the median time for the walk.

d Guides on the walk say that anyone who completes the walk in 3 hours 15 min or less is extremely fit. Estimate the number of extremely fit trampers in this group.

Chapter 11

Algebraic fractions

Contents:

OPENING PROBLEM

Corey has two sisters, Payton and Jade. Payton is the eldest child, so she receives the most pocket money.

Payton receives twice as much as Corey. Jade receives two thirds as much as Corey.

Suppose Corey receives x dollars each week.

Things to think about:

a Can you write an expression, in terms of x, to represent the pocket money that:
  i Payton **ii** Jade receives each week?

b Can you write a single fraction to represent the *total* amount of pocket money given to the children each week?

c What is the *total* amount of pocket money given to the children each week if Corey receives:
  i \$3 **ii** \$15 **iii** \$7.50 each week?

Algebraic fractions are fractions which contain at least one variable.

The variable may be in the numerator, the denominator, or in both of these places.

For example, $\frac{x}{2}$, $\frac{4}{1-a}$, and $\frac{2p-1}{3q}$ are all algebraic fractions.

Algebraic fractions are sometimes called **rational expressions**.

In this chapter we will see how we can simplify, add, subtract, multiply, and divide algebraic fractions in the same way as we do ordinary fractions.

For example:

- $\frac{2}{7}+\frac{4}{7}=\frac{2+4}{7}$ leads to $\frac{a}{c}+\frac{b}{c}=\frac{a+b}{c}$
- $\frac{5}{7}-\frac{2}{7}=\frac{5-2}{7}$ leads to $\frac{a}{c}-\frac{b}{c}=\frac{a-b}{c}$
- $\frac{2}{3}\times\frac{5}{7}=\frac{2\times 5}{3\times 7}$ leads to $\frac{a}{b}\times\frac{c}{d}=\frac{ac}{bd}$

A EVALUATING ALGEBRAIC FRACTIONS

To **evaluate** an algebraic expression, we replace the variables with their known values. We then give our answer in simplest form.

Example 1 — Self Tutor

If $x = 3$, $y = -2$ and $z = 4$, evaluate:

a $\dfrac{y}{z^2}$ **b** $\dfrac{2x - z}{y}$

a $\dfrac{y}{z^2}$
$= \dfrac{(-2)}{4^2}$
$= -\frac{2}{16}$
$= -\frac{1}{8}$

b $\dfrac{2x - z}{y}$
$= \dfrac{2(3) - 4}{(-2)}$
$= \dfrac{6 - 4}{-2}$
$= \dfrac{2}{-2}$
$= -1$

EXERCISE 11A

1 If $a = 2$, $b = 3$, $c = 5$, evaluate:

a $\dfrac{a}{2}$ **b** $\dfrac{b}{6}$ **c** $\dfrac{b+1}{a}$ **d** $\dfrac{a+2b}{c}$

e $\dfrac{c-b}{a}$ **f** $\dfrac{3c-a}{b}$ **g** $\dfrac{a^2}{b-1}$ **h** $\dfrac{ab}{-1-c}$

i $\dfrac{2a^2}{b+1}$ **j** $\dfrac{ac}{b^2+1}$

2 If $x = -1$, $y = 2$, $z = -3$, evaluate:

a $\dfrac{y}{x}$ **b** $\dfrac{z}{x}$ **c** $-\dfrac{z}{y+1}$ **d** $\dfrac{y^2}{-4}$

e $\dfrac{z-y}{x}$ **f** $\dfrac{x-y}{z}$ **g** $\dfrac{y}{x+z}$ **h** $\dfrac{x+y}{x-y}$

i $\dfrac{x+z}{y^2}$ **j** $\dfrac{x^2y}{z-1}$ **k** $\dfrac{x^2+y^2}{10}$ **l** $\dfrac{x^2+z^2}{y}$

m $\dfrac{x-z^2}{y}$ **n** $\dfrac{y^2-z^2}{-2y}$ **o** $\dfrac{3x-2y}{z^2}$

B SIMPLIFYING ALGEBRAIC FRACTIONS

We have seen previously that number fractions can be simplified by cancelling common factors in the numerator and denominator.

For example: $\dfrac{6}{15} = \dfrac{2 \times \cancel{3}^1}{5 \times \cancel{3}_1}$ {cancelling the common factor 3}
$= \dfrac{2}{5}$

If the numerator and denominator of an algebraic fraction are both written in factored form and common factors are found, we can simplify by **cancelling the common factors**.

For example: $\dfrac{2xy}{4x} = \dfrac{{}^{1}\cancel{2} \times {}^{1}\cancel{x} \times y}{{}_{1}\cancel{2} \times 2 \times \cancel{x}_{1}}$ {cancelling the common factor $2x$}

$= \dfrac{y}{2}$

> To simplify algebraic expressions:
> - factorise the numerator and the denominator
> - cancel any common factors
> - simplify the result.

Take care with fractions such as $\dfrac{x+2}{6}$.

The numerator $x+2$ cannot be written as the product of factors other than $1 \times (x+2)$. x and 2 are *terms* of the expression, not factors.

A typical **error** is the cancellation $\dfrac{x+\cancel{2}^{1}}{\cancel{6}_{3}} = \dfrac{x+1}{3}$.

You can check that this cancellation of terms is incorrect by substituting a value for x.

For example, if $x = 2$, $\dfrac{x+2}{6} = \dfrac{2+2}{6} = \dfrac{4}{6} = \dfrac{2}{3}$

whereas $\dfrac{x+1}{3} = \dfrac{2+1}{3} = \dfrac{3}{3} = 1$.

Example 2 — Self Tutor

Simplify: **a** $\dfrac{ad}{2a}$ **b** $\dfrac{a^2b}{ac}$

a $\dfrac{ad}{2a} = \dfrac{{}^{1}\cancel{a} \times d}{2 \times \cancel{a}_{1}}$

$= \dfrac{d}{2}$

b $\dfrac{a^2b}{ac} = \dfrac{{}^{1}\cancel{a} \times a \times b}{{}_{1}\cancel{a} \times c}$

$= \dfrac{ab}{c}$

Example 3 — Self Tutor

Simplify: **a** $\dfrac{3a^4}{-a^2}$ **b** $\dfrac{(3x)^2}{x}$

a $\dfrac{3a^4}{-a^2}$

$= \dfrac{3 \times a \times a \times \cancel{a \times a}^{1}}{-1 \times \cancel{a \times a}_{1}}$

$= -3a^2$

b $\dfrac{(3x)^2}{x}$

$= \dfrac{3x \times 3x}{x}$

$= \dfrac{3 \times x \times 3 \times \cancel{x}^{1}}{{}_{1}\cancel{x}}$

$= 9x$

EXERCISE 11B.1

1 Simplify:

a $\dfrac{20}{2a}$ **b** $\dfrac{12x}{6}$ **c** $\dfrac{3x}{x}$ **d** $\dfrac{4a}{8a}$

e $\dfrac{bc}{b}$ **f** $\dfrac{a}{ab}$ **g** $\dfrac{xy}{xz}$ **h** $\dfrac{a^2}{a}$

i $\dfrac{a}{a^2}$ **j** $\dfrac{p^2}{p^2}$ **k** $\dfrac{2x}{3xy^2}$ **l** $\dfrac{12xy^2}{3x^2}$

2 Simplify:

a $\dfrac{a^2}{-2a}$ **b** $\dfrac{-a}{2a^2}$ **c** $\dfrac{-a^3}{-4a}$ **d** $\dfrac{4a^2}{-a^2}$

e $\dfrac{-6t}{-3t^2}$ **f** $\dfrac{4d^2}{-2d}$ **g** $\dfrac{-ab^2}{2ab}$ **h** $\dfrac{-4ab^2}{6a^2b}$

3 Simplify if possible:

a $\dfrac{(5a)^2}{a}$ **b** $\dfrac{(5a)^2}{5}$ **c** $\dfrac{5a}{(-a)^2}$ **d** $\dfrac{(2a)^2}{-a}$

e $\dfrac{(-3b)^2}{6}$ **f** $\dfrac{(-4c)^2}{8}$ **g** $\dfrac{(3d)^2}{d^2}$ **h** $\dfrac{(e^3)^2}{e^3}$

Example 4 **Self Tutor**

a Can $\dfrac{c+8}{4}$ be simplified by cancelling common factors?

b Write $\dfrac{c+8}{4}$ as a sum of two parts in simplest form.

a There are no common factors, so $\dfrac{c+8}{4}$ cannot be simplfied.

b $\dfrac{c+8}{4} = \dfrac{c}{4} + \dfrac{8}{4}$

$= \dfrac{c}{4} + 2$

4 If possible, simplify by cancelling common factors:

a $\dfrac{m+n}{m}$ **b** $\dfrac{mn}{m}$ **c** $\dfrac{m-n}{n}$ **d** $\dfrac{m+n}{m+n}$ **e** $\dfrac{n}{m+n}$

5 Write each of the following fractions as a sum of two parts in simplest form:

a $\dfrac{a+4}{2}$ **b** $\dfrac{2a+4}{2}$ **c** $\dfrac{2x+3}{2}$ **d** $\dfrac{2x+8}{2}$ **e** $\dfrac{4x+3}{2}$

f $\dfrac{2x^2+4x}{2}$ **g** $\dfrac{2x^2+4x}{4}$ **h** $\dfrac{x^2-6x}{x^2}$ **i** $\dfrac{x^2-6x}{x}$ **j** $\dfrac{x^2-6x}{6}$

6 Check the following simplifications using the given substitutions. If the simplification is false, simplify the algebraic fraction correctly.

a $\dfrac{3x+8}{4} = 3x+2$ (substituting $x = 4$)

b $\dfrac{4x+2}{2} = 2x+1$ (substituting $x = 1$, 2, and 3)

c $\dfrac{4x+3}{2} = 2x+3$ (substituting $x = 3$)

d $\dfrac{2a+b+10}{5} = 2a+b+2$ (substituting $a = 2$ and $b = 1$)

Example 5 **Self Tutor**

Simplify:

a $\dfrac{4(x+8)}{2}$ **b** $\dfrac{3(x-1)^2}{x-1}$

a $\dfrac{4(x+8)}{2}$

$= \dfrac{2 \times \cancel{2}^1 \times (x+8)}{\cancel{2}_1}$

$= 2(x+8)$

b $\dfrac{3(x-1)^2}{x-1}$

$= \dfrac{3(x-1)\cancel{(x-1)}^1}{\cancel{x-1}_1}$

$= 3(x-1)$

7 Simplify:

a $\dfrac{2(a+3)}{2}$ **b** $\dfrac{4(x+2)}{2}$ **c** $\dfrac{6(c+3)}{3}$

d $\dfrac{8(b+5)}{2}$ **e** $\dfrac{2(d-3)}{6}$ **f** $\dfrac{3}{6(z+3)}$

g $\dfrac{4}{2(x+1)}$ **h** $\dfrac{12}{4(2-x)}$ **i** $\dfrac{9}{6(1+x)}$

8 Simplify:

a $\dfrac{10(a+2)}{2(a+2)}$ **b** $\dfrac{3(x+4)}{9(x+4)}$ **c** $\dfrac{12(a-3)}{6(a-3)}$

d $\dfrac{6(x-2)}{8(x-2)}$ **e** $\dfrac{(x+y)(x-y)}{3(x-y)}$ **f** $\dfrac{2xy(x-y)}{6x(x-y)}$

g $\dfrac{5(y+2)^2}{15(y+2)}$ **h** $\dfrac{12(p+q)^2}{8(p+q)}$ **i** $\dfrac{x(x+1)(x+2)}{3x(x+2)}$

DISCUSSION

The algebraic fraction $\dfrac{2x}{x}$ has the common factor x in its numerator and denominator.

Can we *always* cancel the common factor x to say $\dfrac{2x}{x} = 2$?

Is there a particular value of x which would make this cancellation incorrect?

FACTORISATION AND SIMPLIFICATION

It is often necessary to **factorise** either the numerator or denominator before simplification can take place. To do this we use the rules for factorisation that we have seen previously.

Algebraic fraction → **Factorise** the numerator and denominator.
- → If there are common factors, simplify by **cancelling**.
- → If there are no common factors, you **cannot simplify**.

Example 6 — Self Tutor

Simplify:

a $\dfrac{3a-9}{3}$ **b** $\dfrac{4x+6}{3}$

a $\dfrac{3a-9}{3}$

$= \dfrac{{}^{1}\cancel{3}(a-3)}{\cancel{3}_{1}}$

$= a-3$

b $\dfrac{4x+6}{3}$

$= \dfrac{2(2x+3)}{3}$

which cannot be simplified further as there are no common factors.

EXERCISE 11B.2

1 Simplify if possible:

a $\dfrac{2x+10}{2}$ **b** $\dfrac{6x+12}{5}$ **c** $\dfrac{2x-6}{3}$ **d** $\dfrac{2x-6}{2}$

e $\dfrac{12a+4}{2}$ **f** $\dfrac{15a+5}{3}$ **g** $\dfrac{5b-10}{2}$ **h** $\dfrac{5b-10}{5}$

i $\dfrac{3x-15}{3}$ **j** $\dfrac{3x-15}{5}$ **k** $\dfrac{8y+4}{3}$ **l** $\dfrac{8y+4}{4}$

m $\dfrac{4x+16}{2}$ **n** $\dfrac{4x+16}{4}$ **o** $\dfrac{4x+16}{8}$ **p** $\dfrac{x^2+2x}{x}$

Example 7 — Self Tutor

Simplify by factorising: **a** $\dfrac{4x+8}{5x+10}$ **b** $\dfrac{ab-ac}{b-c}$

a $\dfrac{4x+8}{5x+10}$ ← HCF is 4 (numerator), ← HCF is 5 (denominator)

$= \dfrac{4(x+2)}{5(x+2)}$

$= \dfrac{4\cancel{(x+2)}^{1}}{5\cancel{(x+2)}_{1}}$

$= \frac{4}{5}$

b $\dfrac{ab-ac}{b-c}$ ← HCF is a

$= \dfrac{a(b-c)}{1(b-c)}$

$= \dfrac{a\cancel{(b-c)}^{1}}{1\cancel{(b-c)}_{1}}$

$= a$

2 Simplify by factorising:

a $\dfrac{2x+4}{3x+6}$	**b** $\dfrac{2x-10}{3x-15}$	**c** $\dfrac{4x+12}{2x+6}$	**d** $\dfrac{5x-20}{3x-12}$
e $\dfrac{7x-14}{x-2}$	**f** $\dfrac{x+4}{2x+8}$	**g** $\dfrac{xy+3x}{y+3}$	**h** $\dfrac{ax-ay}{x-y}$
i $\dfrac{12-3x}{4-x}$	**j** $\dfrac{ab+bc}{b}$	**k** $\dfrac{ab+bc}{a+c}$	**l** $\dfrac{ab+bc+b}{b}$

Example 8 **Self Tutor**

Simplify by factorising: **a** $\dfrac{x^2-3x}{x}$ **b** $\dfrac{2x^2+4x}{x^2+2x}$

a $\dfrac{x^2-3x}{x}$ ← HCF is x

$= \dfrac{x(x-3)}{x}$

$= \dfrac{^1\cancel{x}(x-3)}{\cancel{x}_1}$

$= x-3$

b $\dfrac{2x^2+4x}{x^2+2x}$ ← HCF is $2x$ / ← HCF is x

$= \dfrac{2x(x+2)}{x(x+2)}$

$= \dfrac{2\overset{1}{\cancel{x}}\cancel{(x+2)}^1}{_1\cancel{x}\cancel{(x+2)}_1}$

$= 2$

3 Simplify by factorising:

a $\dfrac{x^2+3x}{x}$	**b** $\dfrac{4x^2-8x}{4x}$	**c** $\dfrac{3x^2+9x}{x+3}$	**d** $\dfrac{5x^2-10x}{x-2}$
e $\dfrac{4x^2-12x}{x^2-3x}$	**f** $\dfrac{7x^2+14x}{2x^2+4x}$	**g** $\dfrac{4x^2-20x}{2x-10}$	**h** $\dfrac{x^2+2x}{x+2}$
i $\dfrac{x^3-x}{x}$	**j** $\dfrac{x^3-x}{x^2-1}$	**k** $\dfrac{ab^2-cb^2}{a-c}$	**l** $\dfrac{3xy+y^2}{6x+2y}$

Example 9 **Self Tutor**

Simplify:

a $\dfrac{y-x}{2x-2y}$ **b** $\dfrac{x^2y-2xy}{2-x}$

a $\dfrac{y-x}{2x-2y}$

$= \dfrac{-\cancel{(x-y)}^1}{2\cancel{(x-y)}_1}$

$= -\frac{1}{2}$

b $\dfrac{x^2y-2xy}{2-x}$

$= \dfrac{xy\cancel{(x-2)}^1}{-\cancel{(x-2)}_1}$

$= -xy$

4 Simplify if possible:

a $\dfrac{m-n}{n-m}$	**b** $\dfrac{a+b}{b-a}$	**c** $\dfrac{2x-2y}{y-x}$	**d** $\dfrac{4a-4c}{3c-3a}$
e $\dfrac{p-2q}{6q-3p}$	**f** $\dfrac{2a-3b}{12b-8a}$	**g** $\dfrac{1-r}{r^2-r}$	**h** $\dfrac{xy^2-y}{2-2xy}$

Example 10 — Self Tutor

Simplify by factorising and cancelling common factors:

a $\dfrac{x^2-4}{x-2}$ **b** $\dfrac{2a+2b}{a^2-b^2}$

a $\dfrac{x^2-4}{x-2}$

$= \dfrac{(x+2)\cancel{(x-2)}^1}{\cancel{(x-2)}_1}$

$= x+2$

b $\dfrac{2a+2b}{a^2-b^2}$

$= \dfrac{2\cancel{(a+b)}^1}{{}_1\cancel{(a+b)}(a-b)}$

$= \dfrac{2}{a-b}$

5 Simplify:

a $\dfrac{x^2-1}{x-1}$ **b** $\dfrac{a^2-1}{a+1}$ **c** $\dfrac{2+m}{4-m^2}$ **d** $\dfrac{x^2-y^2}{x+y}$

e $\dfrac{x^2-1}{3x+3}$ **f** $\dfrac{x^2-16}{4-x}$ **g** $\dfrac{a^2-b^2}{ab-b^2}$ **h** $\dfrac{x^2-2xy}{2y^2-xy}$

Example 11 — Self Tutor

Simplify: $\dfrac{x^2+x-6}{x+3}$

$\dfrac{x^2+x-6}{x+3}$ ← Two numbers with sum 1 and product -6 are $+3$ and -2.

$= \dfrac{{}^1\cancel{(x+3)}(x-2)}{\cancel{(x+3)}_1}$

$= x-2$

6 Simplify:

a $\dfrac{(x+1)(x-2)}{x-2}$ **b** $\dfrac{(x-y)(x+y)}{y-x}$ **c** $\dfrac{(x-3)(2+x)}{(x+2)(3-x)}$

d $\dfrac{x^2-5x+6}{x-3}$ **e** $\dfrac{x+2}{x^2-3x-10}$ **f** $\dfrac{x^2+6x-7}{x-1}$

g $\dfrac{x^2+8x+12}{x^2+6x+8}$ **h** $\dfrac{x^2-x-6}{x^2-7x+12}$ **i** $\dfrac{x^2-6x+9}{x^2-3x}$

j $\dfrac{1-x^2}{x^2+4x+3}$ **k** $\dfrac{2x^2-18}{x^2+5x+6}$ **l** $\dfrac{x^2+7x+12}{x^2-x-12}$

m $\dfrac{x^2+3x-10}{x^2-10x+16}$ **n** $\dfrac{x^2-8x+16}{4x-x^2}$ **o** $\dfrac{x^2+10x+25}{25-x^2}$

PUZZLE

Click on the icon to load a dynamic puzzle for algebraic fractions.

PUZZLE

C MULTIPLYING ALGEBRAIC FRACTIONS

Variables are used in algebraic fractions to represent unknown numbers. We can treat algebraic fractions in the same way that we treat numerical fractions, since they are in fact *representing* numerical fractions.

To **multiply** two or more fractions, we multiply the numerators to form the new numerator, and we multiply the denominators to form the new denominator.

$$\frac{a}{b} \times \frac{c}{d} = \frac{a \times c}{b \times d} = \frac{ac}{bd}$$

Having multiplied the numerators and denominators, we look to cancel common factors and hence simplify the answer.

Example 12 Self Tutor

Find:

a $\frac{a}{b} \times \frac{b}{2}$ **b** $\frac{n}{3} \times 6$

a $\frac{a}{b} \times \frac{b}{2}$

$= \frac{a \times \cancel{b}^1}{{}_1\cancel{b} \times 2}$

$= \frac{a}{2}$

b $\frac{n}{3} \times 6$

$= \frac{n}{3} \times \frac{6}{1}$

$= \frac{n \times \cancel{6}^2}{{}_1\cancel{3} \times 1}$

$= \frac{2n}{1}$

$= 2n$

EXERCISE 11C

1 Find:

a $\frac{1}{2} \times \frac{x}{2}$ **b** $\frac{b}{3} \times \frac{3}{2}$ **c** $\frac{3}{a} \times \frac{2}{3}$ **d** $\frac{x}{2} \times \frac{x}{3}$

2 Find:

a $\frac{a}{4} \times \frac{2}{a}$ **b** $\frac{x}{2} \times 6$ **c** $\frac{2}{x} \times 6$ **d** $8 \times \frac{n}{2}$

e $\frac{m}{14} \times 2$ **f** $\frac{a}{b} \times \frac{3}{4}$ **g** $\frac{d}{5} \times \frac{5}{d}$ **h** $\frac{c}{d} \times \frac{2d}{c}$

i $\frac{4a}{5} \times \frac{b}{4}$ **j** $\left(\frac{x}{3}\right)^2$ **k** $p \times \frac{1}{p}$ **l** $\frac{x^3}{3} \times \frac{6}{x^2}$

3 Find:

a $\left(\frac{m}{2}\right)^2$ **b** $4 \times \frac{1}{c}$ **c** $\frac{1}{x} \times 8$ **d** $\frac{2}{b} \times \frac{1}{b}$

e $\frac{a}{2} \times \frac{a}{4}$ **f** $\left(\frac{1}{m}\right)^3$ **g** $\frac{2}{a} \times \frac{a}{3}$ **h** $\frac{2}{a} \times \frac{3}{a}$

i $\frac{4a}{3} \times \frac{6}{a}$ **j** $\frac{21}{b} \times \frac{b}{7}$ **k** $\frac{9}{x} \times \frac{x}{18}$ **l** $\frac{a}{b} \times \frac{b}{c} \times \frac{c}{a}$

D DIVIDING ALGEBRAIC FRACTIONS

To **divide** by a fraction we multiply by its reciprocal:

$$\frac{a}{b} \div \frac{c}{d} = \frac{a}{b} \times \frac{d}{c}$$

Example 13 — Self Tutor

Find: **a** $\dfrac{2}{a} \div \dfrac{a}{b}$ **b** $4 \div \dfrac{2}{x}$

a $\dfrac{2}{a} \div \dfrac{a}{b}$

$= \dfrac{2}{a} \times \dfrac{b}{a}$

$= \dfrac{2 \times b}{a \times a}$

$= \dfrac{2b}{a^2}$

b $4 \div \dfrac{2}{x}$

$= \dfrac{4}{1} \times \dfrac{x}{2}$

$= \dfrac{{}^{2}\cancel{4} \times x}{1 \times \cancel{2}_{1}}$

$= \dfrac{2x}{1}$

$= 2x$

EXERCISE 11D

1 Find:

a $\dfrac{a}{b} \div \dfrac{1}{b}$ **b** $\dfrac{a}{b} \div b$ **c** $b \div \dfrac{a}{b}$ **d** $b \div \dfrac{b}{a}$

e $\dfrac{c}{3} \div \dfrac{c}{4}$ **f** $\dfrac{c}{3} \div \dfrac{4}{c}$ **g** $\dfrac{3}{c} \div \dfrac{4}{c}$ **h** $\dfrac{3}{c} \div \dfrac{c}{4}$

i $a \div \dfrac{1}{a}$ **j** $\dfrac{1}{x} \div 3$ **k** $5 \div \dfrac{1}{x}$ **l** $\dfrac{1}{b} \div b$

2 Find:

a $\dfrac{3a^2}{5} \div \dfrac{a}{10}$ **b** $\dfrac{x}{6} \div \dfrac{x^2}{15}$ **c** $\dfrac{a}{b} \div \dfrac{ab}{c}$ **d** $\dfrac{2a}{b} \div \dfrac{a^2}{8}$

e $\dfrac{2x^2}{3} \div \dfrac{2}{x}$ **f** $\dfrac{5}{y} \div \dfrac{y}{2z^2}$ **g** $\dfrac{7a^2}{b} \div \dfrac{3b^2}{a}$ **h** $\dfrac{12x}{y^2} \div \dfrac{8xy}{3}$

E ADDING AND SUBTRACTING ALGEBRAIC FRACTIONS

The rules for the addition and subtraction of algebraic fractions are identical to those used with numerical fractions. We identify the **lowest common denominator (LCD)**, then write both fractions with this denominator.

To find the lowest common denominator, we look for the **lowest common multiple of the denominators**.

For example:
- when finding $\frac{1}{2}+\frac{1}{3}$, the lowest common denominator is 6
- when finding $\frac{2}{3}+\frac{3}{4}$, the lowest common denominator is 12.

The same method is used when there are variables in the denominator.

For example:
- when finding $\frac{2}{3}+\frac{1}{x}$, the lowest common denominator is $3x$
- when finding $\frac{2}{a}+\frac{3}{2a}$, the lowest common denominator is $2a$
- when finding $\frac{3}{4a}+\frac{5}{6b}$, the lowest common denominator is $12ab$.

To **add** two fractions, we write them as equivalent fractions with the *lowest common denominator*, then add the resulting numerators.

$$\frac{a}{c}+\frac{b}{c}=\frac{a+b}{c}$$

To **subtract** one fraction from another, we write them as equivalent fractions with the *lowest common denominator*, then subtract one numerator from the other.

$$\frac{a}{c}-\frac{b}{c}=\frac{a-b}{c}$$

Example 14 Self Tutor

Write as a single fraction:

a $\frac{a}{6}+\frac{2a}{3}$

b $\frac{2x}{3}-\frac{x}{2}$

a $\frac{a}{6}+\frac{2a}{3}$ {LCD = 6}

$=\frac{a}{6}+\frac{2a\times 2}{3\times 2}$

$=\frac{a}{6}+\frac{4a}{6}$

$=\frac{a+4a}{6}$

$=\frac{5a}{6}$

b $\frac{2x}{3}-\frac{x}{2}$ {LCD = 6}

$=\frac{2x\times 2}{3\times 2}-\frac{x\times 3}{2\times 3}$

$=\frac{4x}{6}-\frac{3x}{6}$

$=\frac{4x-3x}{6}$

$=\frac{x}{6}$

EXERCISE 11E

1 Write as a single fraction:

a $\frac{b}{3}+\frac{b}{3}$ **b** $\frac{a}{5}+\frac{4}{5}$ **c** $\frac{2x}{3}+\frac{1}{3}$ **d** $\frac{b}{2}+\frac{b}{3}$

e $\frac{a}{3}-\frac{a}{6}$ **f** $\frac{a}{6}+\frac{x}{12}$ **g** $\frac{5a}{6}-\frac{a}{8}$ **h** $\frac{a}{4}-\frac{b}{3}$

i $\frac{n}{5}+\frac{3n}{2}$ **j** $\frac{p}{7}-\frac{2p}{3}$ **k** $\frac{2k}{3}+\frac{3k}{4}$ **l** $\frac{3s}{8}-\frac{s}{2}$

Example 15 Self Tutor

Write as a single fraction:

a $3+\frac{x}{7}$ **b** $\frac{c}{3}-2$

a $3+\frac{x}{7}=\frac{3}{1}+\frac{x}{7}$ {LCD $=7$}

$=\frac{3\times 7}{1\times 7}+\frac{x}{7}$

$=\frac{21+x}{7}$

b $\frac{c}{3}-2=\frac{c}{3}-\frac{2}{1}$ {LCD $=3$}

$=\frac{c}{3}-\frac{2\times 3}{1\times 3}$

$=\frac{c-6}{3}$

2 Write as a single fraction:

a $\frac{n}{3}+1$ **b** $5+\frac{y}{4}$ **c** $7-\frac{d}{2}$ **d** $\frac{n}{3}+n$

e $x+\frac{x}{2}$ **f** $\frac{m}{3}-2$ **g** $1-\frac{2x}{3}$ **h** $\frac{x}{4}+2x$

3 Answer the **Opening Problem** on page **216**.

Example 16 Self Tutor

Simplify:

a $\frac{3}{x}+\frac{2}{y}$ **b** $\frac{2}{3a}-\frac{1}{2a}$

a $\frac{3}{x}+\frac{2}{y}$ {LCD $=xy$}

$=\frac{3\times y}{x\times y}+\frac{2\times x}{y\times x}$

$=\frac{3y}{xy}+\frac{2x}{xy}$

$=\frac{3y+2x}{xy}$

b $\frac{2}{3a}-\frac{1}{2a}$ {LCD $=6a$}

$=\frac{2\times 2}{3a\times 2}-\frac{1\times 3}{2a\times 3}$

$=\frac{4}{6a}-\frac{3}{6a}$

$=\frac{1}{6a}$

4 Simplify:

a $\frac{2}{x}+\frac{3}{x}$ **b** $\frac{1}{x}+\frac{3}{2x}$ **c** $\frac{1}{2}+\frac{1}{x}$ **d** $\frac{1}{x}-\frac{1}{2}$

5 Simplify:

a $\frac{5}{p}+\frac{3}{q}$ **b** $\frac{7}{r}+\frac{2}{s}$ **c** $\frac{4}{p}+\frac{3}{q}$ **d** $\frac{2t}{s}-\frac{t}{2s}$

e $\frac{a}{b}+\frac{2}{c}$ **f** $\frac{5}{b}-\frac{1}{2b}$ **g** $\frac{a}{b}+\frac{3}{5}$ **h** $\frac{a}{b}+\frac{c}{3b}$

6 Simplify:

a $\frac{3}{a}+1$ **b** $1-\frac{2}{b}$ **c** $c+\frac{1}{c}$ **d** $\frac{2}{d}-3$

e $4+\frac{3}{p}$ **f** $7-\frac{2}{q}$ **g** $8+\frac{1}{3r}$ **h** $4-\frac{3}{2s}$

7 Simplify:

a $\frac{3x}{4}+\frac{x}{3}$ **b** $\frac{5y}{2}-\frac{2y}{3}$ **c** $\frac{5}{2p}+\frac{3}{2p}$ **d** $\frac{7}{2q}-\frac{4}{3q}$

e $\frac{2}{x}+\frac{5}{y}$ **f** $\frac{4}{3x}-\frac{2}{x}$ **g** $\frac{a}{6}+3a$ **h** $3x-\frac{x}{2}$

8 Simplify:

a $\frac{3}{a}+\frac{2}{a^2}$ **b** $\frac{5}{b^2}+\frac{1}{b}$ **c** $\frac{6}{d}-\frac{1}{d^2}$ **d** $\frac{5}{x^2}-\frac{4}{x}$

e $1+\frac{x^2}{2}$ **f** $1+\frac{1}{x^2}$ **g** $x+\frac{2}{x}$ **h** $x+\frac{4}{x^2}$

i $5-\frac{a^2}{4}$ **j** $6-\frac{2}{b^2}$ **k** $x+\frac{x^2}{2}$ **l** $x-\frac{1}{x^2}$

m $4x+\frac{2}{x}$ **n** $6x-\frac{1}{x}$ **o** $\frac{3x}{4}-\frac{1}{x}$ **p** $\frac{1}{2x}+x$

ACTIVITY

What to do:

1 Show that $(a+b+c)^2 = a^2+b^2+c^2+2ab+2bc+2ca$.

2 Given that $a+b+c=3$, $ab+bc+ca=2$, and $abc=-2$, find the values of the following:

a $\frac{1}{a}+\frac{1}{b}+\frac{1}{c}$ **b** $a^2+b^2+c^2$.

REVIEW SET 11A

1 If $a=3$ and $b=-2$, evaluate:

a $\frac{a}{2b}$ **b** $\frac{3b}{a}$ **c** $\frac{a-b}{a+b}$

2 Simplify:

a $\frac{3a^3}{6a}$ **b** $\frac{4x^3y}{8xy^2}$ **c** $\frac{-(2x^2)^2}{(-x)^2}$

3 Write each of the following fractions as a sum of two parts in simplest form:

a $\dfrac{c+6}{2}$ **b** $\dfrac{9x-1}{3}$ **c** $\dfrac{x^2-2x}{2}$

4 Simplify:

a $\dfrac{4x+20}{4}$ **b** $\dfrac{3y+12}{y+4}$ **c** $\dfrac{x-2}{2-x}$

5 Find:

a $\dfrac{m}{6}\times\dfrac{3}{m}$ **b** $\dfrac{k}{4}\times 10$ **c** $\left(\dfrac{5}{p}\right)^2$

6 Simplify:

a $\dfrac{5x^2-25x}{2x-10}$ **b** $\dfrac{x^2-2x-15}{x+3}$ **c** $\dfrac{x^2+3x-10}{2x^2-4x}$

7 Find:

a $\dfrac{x}{8}\div\dfrac{x}{2}$ **b** $\dfrac{3}{t}\div t$ **c** $\dfrac{a^2}{2b}\div\dfrac{3}{ab}$

8 Write as a single fraction:

a $\dfrac{x}{5}+\dfrac{3x}{5}$ **b** $\dfrac{4x}{5}-\dfrac{x}{10}$ **c** $6+\dfrac{n}{2}$

9 Simplify:

a $\dfrac{3}{x}+\dfrac{1}{3x}$ **b** $\dfrac{5}{a}-\dfrac{4}{b}$ **c** $x^2+\dfrac{4}{x}$

10 Simplify:

a $\dfrac{x}{3}+2$ **b** $2-\dfrac{x}{3}$ **c** $\dfrac{3a}{5}\times\dfrac{10}{a^2}$

d $\dfrac{3a}{5}\div\dfrac{10}{a^2}$ **e** $\dfrac{5}{2x}-\dfrac{2}{x}$ **f** $\dfrac{5}{2x}+\dfrac{2}{x^2}$

REVIEW SET 11B

1 If $x=-4$ and $y=2$, evaluate:

a $\dfrac{x}{y}$ **b** $\dfrac{y^2}{x}$ **c** $\dfrac{xy}{y-x}$

2 Simplify:

a $\dfrac{8x^4}{2x}$ **b** $\dfrac{x^2}{5xy^2}$ **c** $\dfrac{(-2x)^2}{x^3}$

3 If possible, simplify by cancelling common factors:

a $\dfrac{a-b}{a-b}$ **b** $\dfrac{a+b}{a}$ **c** $\dfrac{a}{ab}$ **d** $\dfrac{a}{b-a}$

4 Simplify:

a $\dfrac{2(x-3)}{4}$ **b** $\dfrac{6(a-5)}{10(a-5)}$ **c** $\dfrac{18(x+1)^2}{8(x+1)}$

5 Find:

a $\frac{5}{x} \times 3$ **b** $\frac{6a}{5} \times \frac{10}{a}$ **c** $\left(\frac{n}{2}\right)^3$

6 Find:

a $\frac{a}{b} \div \frac{b}{3}$ **b** $12 \div \frac{2}{x}$ **c** $\frac{4c^2}{d} \div \frac{2d}{c}$

7 Simplify:

a $\frac{x^2+4x+4}{x^2-4}$ **b** $\frac{x^2-5x+4}{x^2-x-12}$ **c** $\frac{x^2-14x+45}{x^2-12x+27}$

8 Write as a single fraction:

a $\frac{a}{4} + \frac{a}{7}$ **b** $\frac{k}{5} - 1$ **c** $t - \frac{4}{t}$

9 Simplify:

a $\frac{9}{x} - \frac{3}{2x}$ **b** $\frac{4}{a} + \frac{5}{6}$ **c** $\frac{2x}{3} - \frac{1}{x}$

10 Simplify:

a $\frac{2a}{5} + \frac{a}{2}$ **b** $\frac{8}{x} \div \frac{2}{x^2}$ **c** $\frac{3}{2b} \times \frac{4}{a}$

d $\frac{3a}{5} - a$ **e** $\frac{8}{x} + \frac{2}{x^2}$ **f** $\frac{3}{2b} + \frac{4}{a}$

Chapter 12

Financial mathematics

Contents:

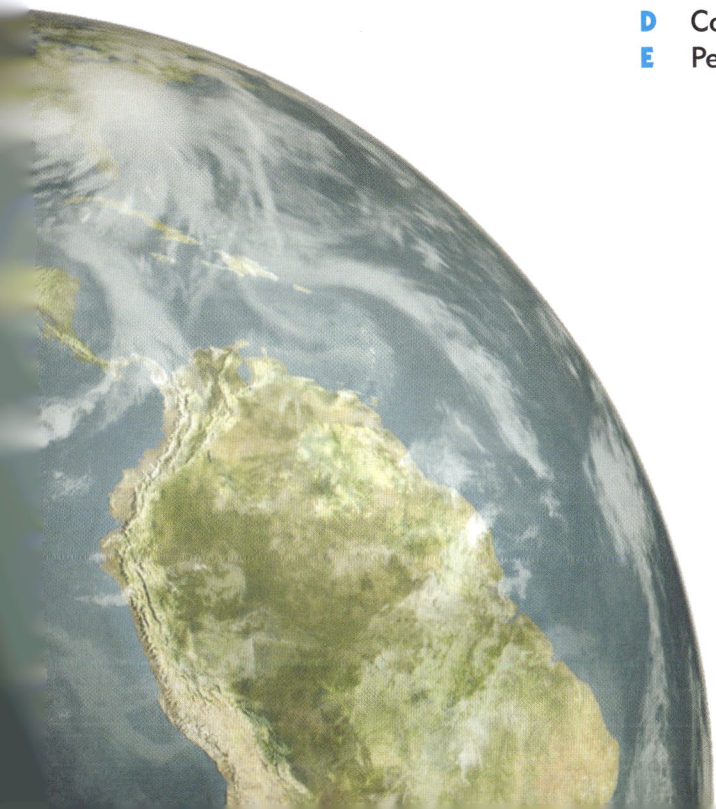

OPENING PROBLEM

Michelle bought an antique clock for \$1500. It appreciated in value by 4% each year.

Michelle sold the clock after 5 years, and invested the money in an account earning 8% per annum compound interest. She left the money in the account for 3 years.

Things to think about:

- **a** How much did Michelle sell the clock for?
- **b** What was the final value of the investment?
- **c** Would Michelle have been better off investing the \$1500 in the account initially, instead of buying the clock?

Money plays a large role in our everyday lives. Performing calculations involving money allows us to make sensible decisions when we run a business, buy or sell items, take out a loan, or invest money.

A BUSINESS CALCULATIONS

Many people choose to run their own business. A successful small business owner needs to have a good understanding of percentages in order to calculate profits, discounts, mark-up, depreciation, and tax.

PERCENTAGE CHANGE USING A MULTIPLIER

Shopkeepers often increase or decrease the price of an item by a percentage. This can be done using a **multiplier**.

For example:

- If a price is increased by 20%, the final price is $100\% + 20\% = 120\%$ of the original price.
  $\therefore$ the multiplier $= 1.2$
- If a price is decreased by 20%, the final price is $100\% - 20\% = 80\%$ of the original price.
  $\therefore$ the multiplier $= 0.8$

The multiplier provides us with a **one-step** method for increasing or decreasing quantities.

new value = multiplier × old value

Example 1 Self Tutor

a Increase \$10 500 by 8%.

b Decrease £120 by 12%.

a new amount
$= 108\%$ of \$10 500 {to increase by 8%, multiply by 108%}
$= 1.08 \times \$10\,500$
$= \$11\,340$

b new amount
$= 88\%$ of £120 {to decrease by 12%, multiply by 88%}
$= 0.88 \times$ £120
$=$ £105.60

We can find a percentage change using: $\text{multiplier} = \dfrac{\text{new value}}{\text{old value}}$

Example 2 — Self Tutor

A house is bought for \$156 000, and is sold six months later for \$175 500. What is the percentage increase on the investment?

$$\begin{aligned}\text{multiplier} &= \frac{\text{new value}}{\text{old value}}\\ &= \frac{175\,500}{156\,000}\\ &= 1.125\\ &= 112.5\% \qquad \therefore \ \text{a } 12.5\% \text{ increase occurred.}\end{aligned}$$

EXERCISE 12A.1

1 Find the multiplier that corresponds to:

a a 10% increase
b a 10% decrease
c a 45% increase
d a 45% decrease
e an 8% decrease
f a 3% increase
g a 100% increase
h a 600% increase
i a 100% decrease

2 For the following multipliers, state the corresponding percentage increase or decrease:

a 1.12
b 1.23
c 0.96
d 0.85
e 1.45
f 0.67
g 2.4
h 0.3

3 a Increase \$50 by 5%.
b Decrease €40 by 30%.
c Decrease £780 by 20%.
d Increase \$68 by 15%.
e Increase \$1600 by 100%.
f Decrease €4500 by 75%.

4 Find the percentage change that occurs when:

a \$80 increases to \$100
b €9000 decreases to €7200
c \$85 reduces to \$82
d £96 increases to £108
e €75 000 increases to €86 000
f RM3500 reduces to RM2400.

5 A restaurant applies a 10% surcharge on Sundays, so the price of each item is 10% higher than normal. Lynette's favourite dish costs \$21.90 during the week. How much will it cost on Sundays?

6 At a hardware store, 3 litre paint cans are 35% cheaper than 5 litre paint cans. Given that the 5 litre cans cost \$40, find the cost of the 3 litre cans.

7 A block of land is bought for €164 000, and sold later for €208 000. Calculate the percentage increase in the land value.

8 Paolo bought a truck for \$47 500. A few months later he found it was no longer large enough for transporting his goods, so he sold it for \$42 000. What was his percentage loss?

MARK-UP

To make a profit, a retailer needs to sell goods at a price *greater* than the cost price which he paid for the goods. Starting with the cost price, the retailer applies a percentage increase called a **mark-up** to calculate the selling price.

Example 3 Self Tutor

Alfred buys shirts for \$20, and marks them up by 60%. At what price does he sell them?

Alfred sells the shirts for $\$20 \times 160\%$
$= \$20 \times 1.6$
$= \$32$

Example 4 Self Tutor

A bag of tomatoes is marked up by 30%, and is sold at \$4.35. Find the cost price of the tomatoes.

selling price $= 130\%$ of cost price
$\therefore$ $\$4.35 = 1.3 \times$ cost price
$\therefore$ cost price $= \dfrac{\$4.35}{1.3} \approx \$3.35$

EXERCISE 12A.2

1 Find the price for which Alfred would sell:

- **a** a shirt costing him \$24, if he applies a 40% mark-up
- **b** a skirt costing him \$50, if he applies a 55% mark-up
- **c** a jacket costing him \$120, if he applies a 35% mark-up
- **d** a suit costing him \$320, if he applies a 30% mark-up
- **e** a pair of trousers costing him \$45, if he applies a 28% mark-up.

2 A baker puts a mark-up of 30% on the bread sold to storekeepers. If it costs the baker, on average, 90 cents per loaf, what price does the baker charge storekeepers?

3 Adriano sells fruit and vegetables. Find the selling price of:

- **a** apples costing him €2.30 per kg, if he marks them up by 60%
- **b** pineapples costing him €0.80 each, if he marks them up by 80%
- **c** cabbages costing him €1.20 each, if he marks them up by 70%.

4 A butcher buys chicken breast fillets for £5 per kg, then adds a 65% mark-up. What price per kilogram will the customer pay?

5 Kane bought a lawnmower for $450 to sell in his shop. If he marks it up by 32%, at what price does he hope to sell it?

6 A jewellery store buys a bracelet for €70, and sells it for €126. What percentage mark-up has been applied?

7 A shop applies a 45% mark-up on fresh fish, and sells it for $25.50 per kg.
Find the cost price of the fish.

8 A litre of milk retails for $1.47 after being marked up by 40%. What was the cost price?

DISCOUNT

To encourage a sale, a retailer may offer a **discount**. This means that the **marked price** is **reduced** or **discounted** by a certain amount or percentage.

Example 5 **Self Tutor**

A dress marked at $150 is discounted by 20% for the end of season sale. Find the sale price of the dress.

$$\begin{aligned}\text{Sale price} &= \$150 \times 80\% \\ &= \$150 \times 0.8 \\ &= \$120\end{aligned}$$

Example 6 **Self Tutor**

A local electrical store offers a 15% discount on all items sold at its summer sale. If a television set is sold for $680, find its marked price.

Since there is a 15% discount,

$$\begin{aligned}\text{selling price} &= 85\% \text{ of marked price} \\ \therefore \quad \$680 &= 0.85 \times \text{marked price} \\ \therefore \quad \text{marked price} &= \frac{\$680}{0.85} = \$800\end{aligned}$$

So, the marked price is $800.

EXERCISE 12A.3

1 Find the sale price of a dress which was originally marked at:

a $80, and is discounted 30%
b €160, and is discounted 25%
c $240, and is discounted 35%
d ¥30 000, and is discounted 12.5%.

2 A CD is marked at $20, and is discounted by 15%. What is its sale price?

3 A refrigerator is marked at $850, and is discounted by 18%. What is its sale price?

4 A bicycle marked at \$560 is to be discounted by 30%. Calculate the sale price.

5 A wok is discounted from \$75 to \$69 in a sale. Find the percentage discount.

6 A vaccuum cleaner has a recommended retail price of 160 dinar. In a clearance sale, it is discounted to 124 dinar. Calculate the percentage discount.

7 A fan was discounted by 20% and sold for \$23.20. What was the marked price of the fan?

8 A pair of bathers sold for \$44 after being discounted by 45%. What was the marked price of the bathers?

TAX ON THE SALES OF ITEMS

In most countries a tax is payable on the sale of items. This tax has various names depending on the country where the items are sold. The two most common names are **value added tax (VAT)** and **goods and services tax (GST)**.

The standard rate of tax varies from one country to another. Examples are shown in the table alongside.

Country	*GST or VAT rate*
Chile	19%
Sweden	25%
Australia	10%
UK	20%
Russia	18%
Singapore	7%

Amount paid by customer = amount kept by provider + tax paid to government

Example 7 Self Tutor

An electrical store in Russia sells a television set for 8500 rubles plus 18% tax. Find:

a the television's selling price

b the tax paid to the government.

a Selling price $= 8500 \times 1.18$ rubles $\{100\% + 18\% = 118\%\}$
$= 10\,030$ rubles

b The amount of tax $= (10\,030 - 8500)$ rubles
$= 1530$ rubles

EXERCISE 12A.4

1 A washing machine has a price of £240 plus 20% VAT.

a What amount is paid by a customer?

b How much is the VAT?

2 A double bed in Canada has a price tag which says \$2165 plus 6% GST.

a What amount is paid by a customer?

b How much is the GST?

3 In Spain, a case of wine sells for €350 plus 20% tax.

a What amount is paid by a customer?

b How much is the tax?

4 In Croatia the sales tax is 22%. A boat is priced at HRK62 500 *including* the tax.

a How much will the seller receive?

b How much is the tax?

CHAIN PERCENTAGE PROBLEMS

Sometimes we need to use a multiplier more than once within a problem.

Example 8 — Self Tutor

Greeting cards are bought for $1.40 by a newsagent. He marks up the price by 50%, then adds on 10% GST. What price does the customer pay?

Final price $= \$1.40 \times 150\% \times 110\%$
$= \$1.40 \times 1.5 \times 1.1$
$= \$2.31$

EXERCISE 12A.5

1 Kirsty buys a skirt for $22. She applies a 55% mark-up, then adds 10% for GST. For what price will Kirsty sell the skirt?

2 A surfboard manufactured for $300 is sold to a retail shop for a 40% profit. The shop then applies a 20% mark-up. What price does the customer pay?

3 A fisherman sells a crayfish to a wholesaler for €35. The wholesaler sells it to a fresh fish shop for 60% profit. However, the fish shop is forced to sell the crayfish at a 25% loss. What price does the customer pay?

4 A rare book is bought by an antiques dealer for $350, and he marks its price up by 100% for sale. One of his regular customers is interested in buying the book, so the dealer offers the customer a 20% discount. What price is offered to the customer?

5 An art supplies store buys canvases for $40, and applies a 40% mark-up. During a sale, a discount is offered of 12.5% for non-members, and 15% for members. Find the difference in the sale price for non-members and members.

B APPRECIATION AND DEPRECIATION

When the value of an investment or an item such as a house increases, we say it **appreciates** in value.

You may have noticed that the prices of everyday goods and services also increase over time. This is known as **inflation**, and is a form of appreciation.

Example 9 — Self Tutor

The inflation rate over the next three years is predicted to be 2%, then 3%, then 3.5%. If an item currently costs \$140, and its cost rises in line with the predicted inflation, find the item's cost in 3 years' time.

Cost in 3 years $= \$140 \times 1.02 \times 1.03 \times 1.035$
$= \$152.23$

Electrical equipment, furniture, vehicles, and machinery all lose value over time. This may be because they become damaged or worn, or because their technology is no longer the latest and best available. We say the items **depreciate** in value.

Example 10 — Self Tutor

A computer is bought for \$1500. It depreciates by 25% in the first year, then 30% in the second year. Find the value of the computer after 2 years.

Value after 2 years $= \$1500 \times 0.75 \times 0.7$
$= \$787.50$

EXERCISE 12B

1 The inflation rate is predicted to be 3%, 3.5%, 4%, and then 5% over the next four years. What would you expect a bicycle costing €500 today to cost in four years' time?

2 Julia's wage was initially \$40 000 per annum. Over a 4 year period it increased by 3%, 8%, 5%, and 10%.

- **a** Find Julia's wage after 4 years.
- **b** Find the overall percentage increase in Julia's wage over this time.

3 Three years ago, an ornate vase was valued at £2000. Over the next three years, it appreciated by 7%, 5%, and 8%. Find the present value of the vase.

4 Sophie bought a car for \$5000. Over the next three years it depreciated by 20%, 15%, and 10%. Find the value of the car after three years.

5 A photocopier is bought for CNY48 000, and depreciates by 15% each year. Find the value of the photocopier after 4 years.

6 Kevin and Don each bought a motorbike for \$8000. Kevin's motorbike depreciated at 22% per year, and Don's motorbike depreciated at 12% per year. Find the difference in value of the motorbikes after 3 years.

Example 11 **Self Tutor**

Over 4 consecutive years, the value of a house increases by 15%, increases by 9%, decreases by 4%, and increases by 18%. What is the overall percentage increase in value over this period?

Let $\$x$ be the original value of the house.

The value after 4 years $= \$x \times 1.15 \times 1.09 \times 0.96 \times 1.18$
$\approx \$x \times 1.4200$

$\therefore$ overall the value of the house increases by 42%.

7 A share fund reported a 9% increase in value for year 1, a 13% decrease in value for year 2, and a 4% increase in value for year 3. What was the overall percentage increase or decrease of the share fund over the 3 years?

8 Jamie bought a piece of cricket memorabilia. Over the next 4 years the piece appreciated by 5%, depreciated by 4%, depreciated by 8%, and appreciated by 6%. Find the overall percentage appreciation or depreciation of the item over the 4 years.

9 An apartment was valued at \$270 000. It appreciated by 5% each year for 3 years, then depreciated by 5% each year for 3 years. Find:

- **a** the value of the apartment after 6 years
- **b** the overall percentage appreciation or depreciation of the apartment.

10 Brenda bought a rare painting. It appreciated by 4% in the first year, and depreciated by 7% in the second year. After 3 years the painting had depreciated by 2% overall. Find the percentage appreciation or depreciation of the painting in the third year.

DISCUSSION

- If an amount is *increased* by 18%, and the resulting amount is then *decreased* by 18%, do we return to the original amount?
- A new model car is 18% more expensive than the model it replaces. What percentage discount must a salesman give in order to sell the new model car at the old model price?

C SIMPLE INTEREST

When a person **borrows** money from a lending institution such as a bank or a finance company, the borrower must repay the loan in full, and also pay an additional **interest** payment. This is a charge for using the institution's money.

Similarly, when money is **invested** in a bank, the bank pays interest on any deposits.

The amount of interest paid is usually given as a *percentage* of the amount which is borrowed or invested.

SIMPLE INTEREST

Simple interest is interest that is calculated each year as a fixed percentage of the **initial amount** of money borrowed or invested.

The fixed percentage is called the **interest rate**, and is usually given **per annum** which means "per year".

Suppose $8000 is invested for 5 years at 10% per annum simple interest.

$$\begin{aligned}\text{The simple interest paid for 1 year} &= 10\% \text{ of } \$8000\\ &= 0.1 \times \$8000\\ &= \$800\\ \text{The simple interest paid for 2 years} &= 10\% \text{ of } \$8000 \times 2\\ &= 0.1 \times \$8000 \times 2\\ &= \$1600\\ &\vdots\\ \text{The simple interest paid for 5 years} &= 10\% \text{ of } \$8000 \times 5\\ &= 0.1 \times \$8000 \times 5\\ &= \$4000\end{aligned}$$

These observations lead to the **simple interest formula**:

Simple interest $I = Pin$

where P is the **principal**, or **initial amount** borrowed or invested,
i is the **flat rate of interest per annum (p.a.)**,
n is the **time** or **duration** of the loan in **years**.

Example 12 **Self Tutor**

Find the simple interest payable on an investment of $20 000 at 12% p.a. over a period of 4 years.

$P = 20\,000$
$i = 12\% = 0.12$
$n = 4$

Now $I = Pin$
$\therefore$ $I = 20\,000 \times 0.12 \times 4$
$\therefore$ $I = 9600$
$\therefore$ the simple interest is $9600.

In some areas of finance, sums of money may be invested over a period of **months** or **days**. However, the interest rate is still normally quoted per annum, so the time period n in the formula must be converted to years.

Example 13 — Self Tutor

Calculate the simple interest payable on an investment of \$15 000 at 8% p.a. over 9 months.

$P = 15\,000$
$i = 8\% = 0.08$
$n = \frac{9}{12} = 0.75$

Now $I = Pin$
$\therefore\ I = 15\,000 \times 0.08 \times 0.75$
$\therefore\ I = 900$
$\therefore$ the simple interest is \$900.

EXERCISE 12C

1 Find the simple interest payable on an investment of:

- **a** \$3000 at 11% p.a. for 2 years
- **b** €5500 at $7\frac{1}{2}\%$ p.a. for 5 years
- **c** £15 000 at 5.75% p.a. for $4\frac{1}{2}$ years
- **d** \$25 000 at 6.2% p.a. for 10 years.

2 Find the simple interest payable on a loan of:

- **a** €2000 at 1.4% p.a. over 9 months
- **b** \$8500 at 8% p.a. over 3 months
- **c** ¥2 750 000 at 7% p.a. over 18 months
- **d** \$60 000 at 3.25% p.a. over 10 months.

Example 14 — Self Tutor

Determine the simple interest payable on an investment of \$100 000 at 15% p.a. from April 28th to July 4th.

From April 28th there are 2 days left in April.

$\therefore$	2 days	left in April	← exclude the first day (April 28)
	31 days	in May	
	30 days	in June	
	4 days	in July	← include the last day (July 4)
	67 days		

$P = 100\,000$
$i = 15\% = 0.15$
$n = \frac{67}{365}$

Now $I = Pin$
$\therefore\ I = 100\,000 \times 0.15 \times \frac{67}{365}$
$\therefore\ I = 2753.42$ {to the nearest cent}

So, the interest payable is \$2753.42.

3 Gemma deposits \$7800 in a special investment account on May 6th. The account pays 8.75% p.a. simple interest. If Gemma withdraws the money on September 9th, how much will her investment have earned over this time?

4 Warren Barr deposited £7500 on July 1st in a special investment account which earns 11% p.a. simple interest. On September 19th he deposited another £4900 in the account. He closed the account on December 20th by withdrawing the total balance.

  a In total, how much interest did Warren's investment earn?

  b Find the total amount Warren withdrew.

Example 15 — Self Tutor

Todd borrows \$6000 for 5 years at 11% p.a. simple interest.

  a Calculate the total amount he must repay.

  b If Todd repays the loan in equal monthly repayments over the 5 years, find the amount he must repay each month.

a $P = 6000$, $i = 11\% = 0.11$, $n = 5$

Now $I = Pin$
$= 6000 \times 0.11 \times 5$
$= 3300$

The total amount to be repaid = initial amount + interest
= \$6000 + \$3300
= \$9300

b Todd must repay the loan over 5 years = 60 months.

$\therefore$ each monthly repayment $= \dfrac{\$9300}{60}$
$= \$155$

5 \$4000 is borrowed under simple interest terms. Find the total amount to be repaid after:

  a 10 years at 7% p.a. **b** 6 years at $9\frac{1}{2}\%$ p.a. **c** 8 months at 13% p.a.

6 Alice borrows \$4700 from a finance company to buy her first car. The rate of simple interest is 17% and she borrows the money over a 5 year period. Find:

  a the total amount Alice must repay the finance company

  b her equal monthly repayments.

7 Cameron borrows €15 000 from a bank to renovate his house. He borrows the money at 9% p.a. simple interest over 8 years. What are his monthly repayments?

8 An electric guitar with all attachments is advertised at RUB 72 000. Ben can afford to pay a deposit of RUB 18 000, but he then has to borrow the remainder at 12% p.a. simple interest over 3 years. What are his monthly repayments?

D COMPOUND INTEREST

A more common method for calculating interest is **compound interest**.

If you leave your money in the bank for a period of time, the interest is automatically added to your account.

With compound interest, any interest that is added to your account will also earn interest in the next time period.

Compound interest is calculated as a percentage of the total amount at the end of the previous compounding period.

In this course we will only consider interest which compounds each year.

Suppose €1000 is placed in an account earning interest at a rate of 10% p.a. The interest is allowed to compound itself for three years. We say it is earning "10% p.a. compound interest".

We can show this in a table:

Year	*Amount at beginning of year*	*Compound interest*	*Amount at end of year*
1	€1000	10% of €1000 = €100	€1000 + €100 = €1100
2	€1100	10% of €1100 = €110	€1100 + €110 = €1210
3	€1210	10% of €1210 = €121	€1210 + €121 = €1331

After 3 years there is a total of €1331 in the account. We have earned €331 in compound interest.

If we construct a similar table for €1000 in an account earning 10% p.a. simple interest for 3 years, we can compare the two different types of interest.

Year	*Amount at beginning of year*	*Simple interest*	*Amount at end of year*
1	€1000	10% of €1000 = €100	€1000 + €100 = €1100
2	€1100	10% of €1000 = €100	€1100 + €100 = €1200
3	€1200	10% of €1000 = €100	€1200 + €100 = €1300

After 3 years there is a total of €1300 in the account, so we have earned €300 in simple interest.

Year	*Compound interest*	*Simple interest*
1	€100	€100
2	€110	€100
3	€121	€100

Example 16 **Self Tutor**

How much interest will I earn if I invest $10 000 for 3 years at:

a 15% p.a. simple interest **b** 15% p.a. compound interest?

a We use the simple interest formula with $P = 10\,000$, $i = 0.15$, $n = 3$.

Now $I = Pin$

$\therefore\ I = 10\,000 \times 0.15 \times 3 = 4500$

The simple interest is $4500.

b

Year	*Initial amount*	*Interest*	*Final amount*
1	$10 000	15% of $10 000 = $1500.00	$11 500.00
2	$11 500	15% of $11 500 = $1725.00	$13 225.00
3	$13 225	15% of $13 225 = $1983.75	$15 208.75

The compound interest = final amount − initial amount
= $15 208.75 − $10 000
= $5208.75

EXERCISE 12D.1

1 $25 000 is invested at 6% p.a. compound interest. Use a table to find:

a the final amount after 3 years **b** the total interest earned in the 3 year period.

2 Calculate the interest earned on an investment of £2000 for 3 years at:

a 5% p.a. simple interest **b** 5% p.a. compound interest.

3 Use a table to determine the interest earned for the following investments:

a 4000 pesos at 7% p.a. compound interest for 2 years

b $500 at 2.5% p.a. compound interest for 3 years

c $11 000 at 5% p.a. compound interest for 4 years.

INVESTIGATION 1 COMPOUND INTEREST SPREADSHEET

A spreadsheet allows us to construct a compound interest table very rapidly.

What to do:

1 Suppose $5000 is invested at 4% p.a. compound interest for 10 years.
Open a new spreadsheet and enter the following:

	A	B	C	D
1	Investment	$5000		
2	Interest rate	0.04		
3				
4	Year	Initial amount	Interest	Final amount
5	1	=B1	=B5*B$2	=B5+C5
6	=A5+1	=D5	=B6*B$2	=B6+C6

SPREADSHEET

Highlight the formulae in row 6. **Fill down** to row 14 for the 10th year of the investment.

a How much interest is paid in: **i** year 1 **ii** year 10?

b How much is in the account after: i 5 years ii 10 years?

2 Suppose \$15 000 is invested at 6% p.a. compound interest for 10 years. Enter 15 000 in B1 and 0.06 in B2.
   a How much interest is paid in: i year 1 ii year 10?
   b How much is in the account after: i 5 years ii 10 years?

3 Use the spreadsheet to find how long it would take for \$8000 invested at 5% p.a. compound interest to double in value.
   Hint: You may need to **fill down** further.

4 Use the spreadsheet to find what compound interest rate is needed for \$12 000 to double in value after 6 years.

THE COMPOUND INTEREST FORMULA

Suppose you invest €1000 in the bank for 3 years, earning 10% p.a. compound interest.

After one year your investment is worth $\$1000 \times 1.1 = \$1100$
After two years your investment is worth $\$1000 \times 1.1 \times 1.1 = \$1000 \times (1.1)^2 = \$1210$
After three years your investment is worth $\$1000 \times (1.1)^2 \times 1.1 = \$1000 \times (1.1)^3 = \$1331$

This suggests that if the money was left in your account for n years, it would amount to $\$1000 \times (1.1)^n$.

This leads us to the **compound interest formula**:

$\boldsymbol{F_v = P_v(1+i)^n}$ where
- F_v is the **future value**
- P_v is the **present value** or initial amount
- i is the **annual interest rate** as a decimal
- $(1+i)$ is the **multiplier**
- n is the **number of years of the investment**.

Notice that the formula for F_v above gives the total future value, which is the original amount plus interest.

To find the **interest only** we use:

$$\textbf{Compound interest} = F_v - P_v$$

Example 17 **Self Tutor**

a What will \$5000 invested at 8% p.a. compound interest amount to after 2 years?
b How much interest is earned?

a $i = 8\% = 0.08$
For 2 years, $n = 2$
$\therefore F_v = P_v(1+i)^n$
$= \$5000 \times (1.08)^2$
$= \$5832$

b Interest earned
$= \$5832 - \$5000$
$= \$832$

EXERCISE 12D.2

1 \$6000 is invested for 4 years at 8% p.a. What will this investment amount to if the interest is:

  a simple interest **b** compound interest?

2 €7500 is borrowed for 10 years at 11% p.a. How much will have to be repaid if the interest is:

  a simple interest **b** compound interest?

3 **a** What will an investment of \$3000 at 9% p.a. compound interest amount to after 5 years?

  b What part of this is interest?

4 How much compound interest is earned by investing 20 000 dirhams for 6 years at 4.5% p.a.?

5 You have \$8000 to invest for 3 years. You have been offered two investment options:

Option 1: Invest at 9% p.a. simple interest. *Option 2*: Invest at 8% p.a. compound interest.

  a Calculate the amount accumulated at the end of the 3 years for both options and decide which option to take.

  b Would you change your decision if you were investing for 5 years?

E PERSONAL LOANS

A common way to borrow money to pay for cars, property, overseas holidays, and education expenses, is to take out a personal loan. Banks, credit unions, and finance companies all offer personal loans with differing terms, conditions, and interest rates.

Interest is calculated on the **reducing balance** of the loan, so the interest reduces as the loan is repaid. The interest each period is added to the balance owed, and the repayment is then deducted.

outstanding balance = previous balance + interest − repayment.

INVESTIGATION 2 LOAN REPAYMENT

This spreadsheet will allow you to investigate how a **reducing balance loan** is paid off.

SPREADSHEET

What to do:

1 Suppose \$2000 is borrowed at a reducing balance interest rate of 12% p.a. Monthly repayments of \$400 are to be made.

  a Click on the icon to load the spreadsheet. Fill row 10 down to the sixth month.

	A	B	C	D	E
1	**Reducing balance loans**				
2					
3	Loan amount	2000			
4	Rate p.a.	12%			
5	Rate per month	=B4/12			
6	Repayment	400			
7					
8	Month	Principal	Interest	Repayment	Balance
9	0	-	-	-	=B3
10	=A9+1	=E9	=B10*B\$5	=B\$6	=B10+C10-D10
11	↓	↓	↓	↓	↓
12			fill down		

b How many repayments are needed to repay the loan?

c Does the final repayment need to be the full \$400? Explain your answer.

2 Erica takes out a loan to buy a car. She borrows €20 000 for 5 years at a rate of 12% p.a. compounded monthly under reducing balance conditions. Determine the closing balance after making monthly repayments of €445 for three months.

3 A loan of £100 000 is taken out for 10 years to buy a share portfolio. Interest is charged at 12% p.a. compounded *quarterly* under reducing balance conditions. Determine the amount still owing after 1 year if quarterly repayments of £4350 are made.

4 Ricky takes out a loan of £23 000 over 5 years at 13.5% p.a. By trial and error, calculate the monthly repayment needed to completely pay off the loan in this time.

LOAN REPAYMENT TABLES

When you apply for a personal loan, you will be advised of the amount you will need to repay each fortnight or each month. This amount is dependent on the loan amount, the duration of the loan, and the interest rate charged.

Banks have **tables of repayments** for various loans. **Loan calculators** are also available on bank websites.

The following **table of monthly repayments** is based on **borrowing 1000 units of currency**.

Table of Monthly Repayments per 1000 units of currency							
Loan term	*Annual interest rate*						
(months)	10.0%	10.5%	11.0%	11.5%	12.0%	12.5%	13.0%
12	87.9159	88.1486	88.3817	88.6151	88.8488	89.0829	89.3173
18	60.0571	60.2876	60.5185	60.7500	60.9820	61.2146	61.4476
24	46.1449	46.3760	46.6078	46.8403	47.0735	47.3073	47.5418
30	37.8114	38.0443	38.2781	38.5127	38.7481	38.9844	39.2215
36	32.2672	32.5024	32.7387	32.9760	33.2143	33.4536	33.6940
42	28.3168	28.5547	28.7939	29.0342	29.2756	29.5183	29.7621
48	25.3626	25.6034	25.8455	26.0890	26.3338	26.5800	26.8275
54	23.0724	23.3162	23.5615	23.8083	24.0566	24.3064	24.5577
60	21.2470	21.4939	21.7424	21.9926	22.2444	22.4979	22.7531

Example 18 — Self Tutor

Erin takes out a personal loan for $16 500 to buy a car. She negotiates a term of 4 years at 11.5% p.a. interest. Calculate the:

a monthly repayments **b** total repayments **c** interest charged.

a From the table, the monthly repayments on each $1000 for 4 years at 11.5% p.a. are $26.0890

$\therefore$ the repayments on $16 500 = $26.0890 × 16.5 {16.5 lots of $1000}

= $430.4685

≈ $430.50 {to next 10 cents}

The monthly repayments are $430.50.

b Total repayments = monthly repayment × number of months

= $430.50 × 48

= $20 664

$\therefore$ $20 664 is repaid in total.

c Interest = total repayments − amount borrowed

= $20 664 − $16 500

= $4164

The total interest charged is $4164.

EXERCISE 12E

1 Ron takes out a personal loan for €15 000 to go overseas. He will repay it over 5 years at 11.5% p.a. Calculate the:

a monthly repayments **b** total repayments **c** interest charged.

2 Jai and Tatiana need £18 000 to fund house renovations. They take out a personal loan over 3 years at 13% p.a. Calculate the:

a monthly repayments **b** total repayments **c** interest charged.

3 Gunter needs €28 000 to buy a boat. His bank offers him a personal loan at 12.5% p.a.
Calculate the total interest Gunter will pay if he repays the loan over:

a 2 years **b** 5 years.

Comment on your answers.

4 Trina wants to borrow $45 000 to purchase some shares. Calculate the total interest charged for the following options:

A Balance Bank offer 11.5% over 5 years

B Cash Credit Union offer 12.5% over 3 years.

Which option would you recommend for Trina?

Global context

click here

Paying off a mortgage

Statement of inquiry:	Decision-making can be improved by using a model to investigate the possible options.
Global context:	Globalisation and sustainability
Key concept:	Relationships
Related concepts:	Change, Model, Quantity
Objectives:	Knowing and understanding, Applying mathematics in real-life contexts
Approaches to learning:	Self-management, Research

REVIEW SET 12A

1 Samantha's water bill is $240. There is a 0.75% surcharge which applies when payments are made by credit card. If Samantha uses her credit card to pay the bill, how much must she pay in total?

2 A toaster is sold to a retailer for €38. The retailer marks it up by 40%, and later discounts it by 15% in a sale. What price does the customer pay?

3 I bought a house for $420 000. Over the next four years it increased in value by 9%, decreased by 3%, increased by 7%, and increased by 12%. Find the value of the house after these four years.

4 A caravan costing £15 000 depreciates at a rate of 16% p.a. Find its value after 5 years.

5 A lounge suite was bought for a basic cost of $1780. It was marked up for a 35% profit, and then 10% GST was added on. Determine the final selling price of the lounge suite.

6 When Luke completed his university degree 3 years ago, he had a student loan of $22 000 to repay. Over the next 3 years, the inflation rate was 2.2%, 1.9%, and 2.6%.

a If Luke's loan rose in line with inflation, find the amount he has to repay now.

b Luke will receive a 10% discount on his loan if he repays the loan in a single payment. How much will he need to pay to repay the loan in a single payment?

7 Find the simple interest payable on a loan of €4000 at 7% p.a. for 5 years.

8 Raj borrows 500 000 rupees from a bank at $8\frac{1}{2}\%$ p.a. simple interest. He will repay the money over a 6 year period, with equal monthly repayments.

a Find the total amount that Raj must repay the bank.

b Find Raj's equal monthly repayments.

9 **a** What will an investment of $8000 at 6% p.a. compound interest amount to after 4 years?

b What part of this is interest?

10 Mizumi takes out a personal loan for ¥600 000 to go on a holiday. Her bank offers her a rate of 11% p.a. over 3 years on a reducing balance basis. Calculate the:

a monthly repayments using the table of monthly repayments on page **247**.

b total repayments

c total interest charged.

REVIEW SET 12B

1 An item of clothing is marked at \$49, but is discounted 15% in a sale. Find the sale price.

2 A share trader buys shares for \$1100, and sells them for \$1240. What percentage profit does she make?

3 The annual rate of inflation is predicted to be 3% next year, then 3.5% in the year after that. If an item currently costs €50, estimate its cost in two years' time.

4 A blender was discounted by 20% and sold for \$64. Find the marked price of the blender.

5 A softball bat was sold for \$120 including 15% VAT.

- **a** How much will the seller receive once the tax is passed on to the government?
- **b** How much is the VAT?

6 Answer the **Opening Problem** on page **232**.

7 Darren borrows \$15 000 from the bank to extend his house. The bank will charge him simple interest of 6.8% p.a. for a 5 year loan. Find the total amount Darren will need to repay.

8 Al borrows 100 000 pesos from the bank to buy solar panels for his house. He borrows the money at 7% p.a. simple interest for 5 years.

- **a** Calculate the interest payable on Al's loan.
- **b** In total, how much money will Al pay back?
- **c** Al wishes to pay back the loan in equal monthly instalments.
  - **i** How many repayments will there be?
  - **ii** What is the size of each repayment?

9 You have €x to invest for 5 years, and have been offered two investment options:

- **A** 6.5% simple interest
- **B** 5.2% compound interest.

Which option should you choose? Explain your answer.

10 A 450 000 rupee car is purchased using a deposit of 150 000 rupees. The balance is repaid over 5 years at $10\frac{1}{2}\%$ per annum on a reducing balance basis. Calculate:

- **a** the amount borrowed
- **b** the size of each monthly repayment
- **c** the total to be repaid
- **d** the total interest paid.

Table of Monthly Repayments per 1000 rupee

Loan term (months)	*Annual interest rate* 10.0%	10.5%	11.0%
12	87.9159	88.1486	88.3817
18	60.0571	60.2876	60.5185
24	46.1449	46.3760	46.6078
30	37.8114	38.0443	38.2781
36	32.2672	32.5024	32.7387
42	28.3168	28.5547	28.7939
48	25.3626	25.6034	25.8455
54	23.0724	23.3162	23.5615
60	21.2470	21.4939	21.7424

Chapter 13

Trigonometry

Contents:

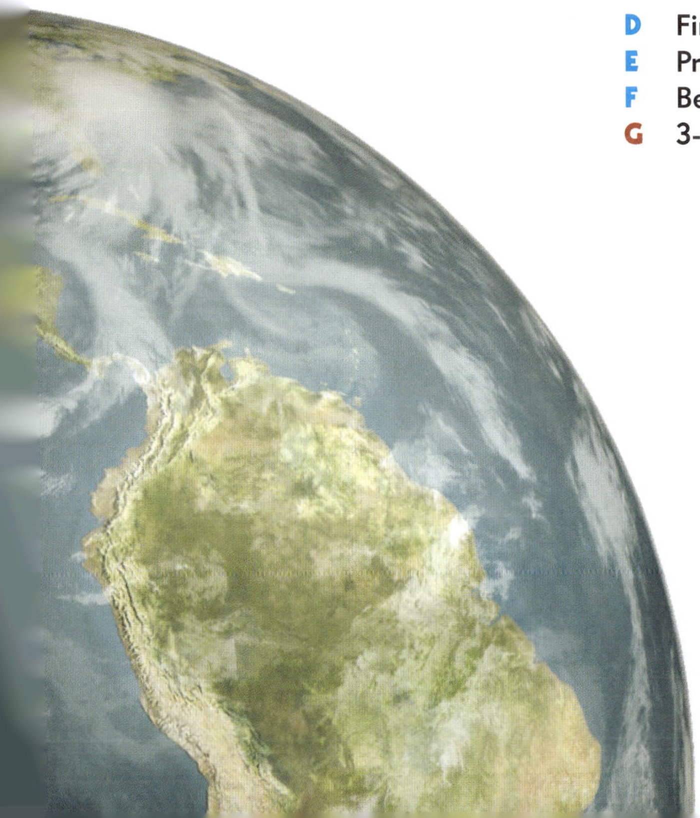

OPENING PROBLEM

A group of students is asked to measure the height of the school gymnasium. It is a symmetric building 24 m wide, as shown.

From a point 32 m from the side wall, the students measure the angles of elevation to the top of the side wall, and to the top of the roof.

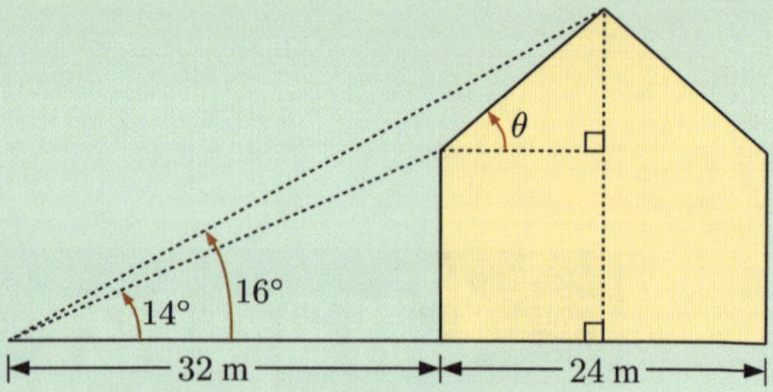

Things to think about:

a How high is:

i the side wall of the gymnasium **ii** the top of the roof?

b Can you find the angle θ which is the pitch of the roof?

Trigonometry is the study of the relationships between lengths and angles of geometrical figures.

Trigonometry is used in many fields, including:

- navigation
- surveying
- engineering
- architecture

A LABELLING RIGHT ANGLED TRIANGLES

In trigonometry, there is a convention for labelling the sides of a right angled triangle.

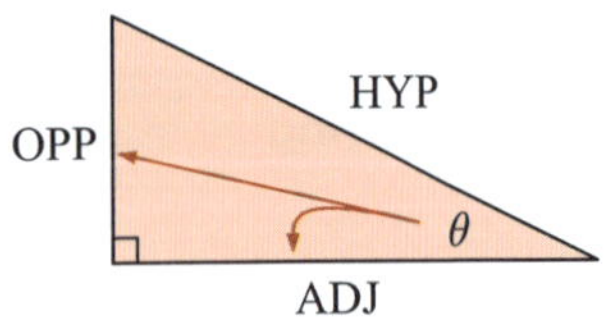

For the right angled triangle with angle θ:

- the **hypotenuse (HYP)** is the longest side
- the **opposite (OPP)** side is opposite θ
- the **adjacent (ADJ)** side is adjacent to θ.

Example 1 **Self Tutor**

For the triangle alongside, name the:

a hypotenuse
b side opposite θ
c side adjacent to θ.

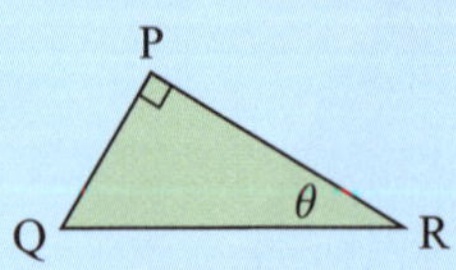

a The hypotenuse is [QR].
b The side opposite θ is [PQ].
c The side adjacent to θ is [PR].

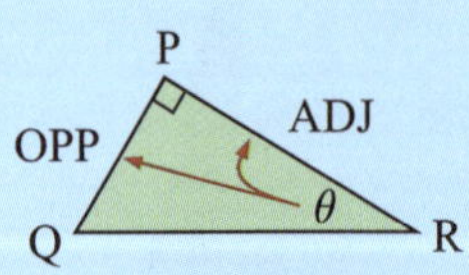

EXERCISE 13A

1 For the triangles given, name the:

i hypotenuse **ii** side opposite angle θ **iii** side adjacent to θ.

a
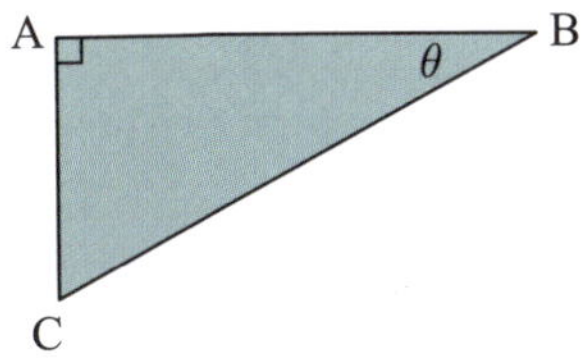

b
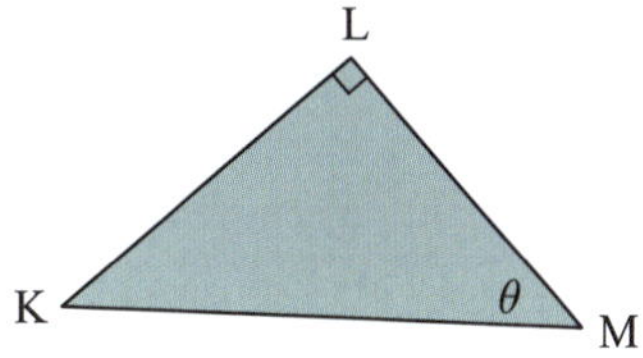

c
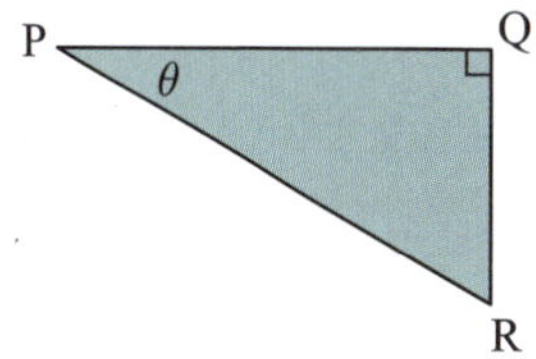

2 For the triangles given, name the:

i hypotenuse **ii** side opposite α **iii** side adjacent to α

iv side opposite β **v** side adjacent to β.

a
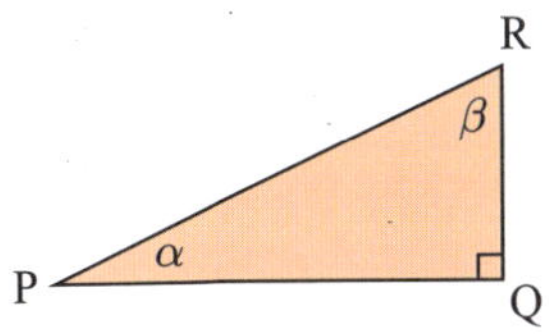

b
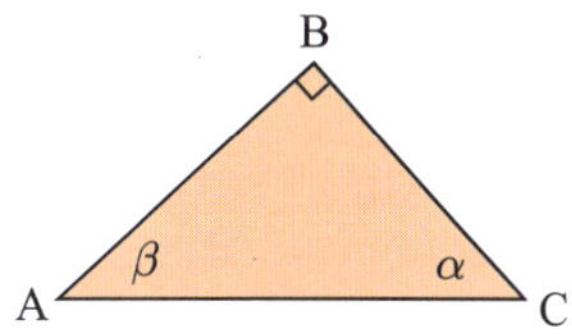

c
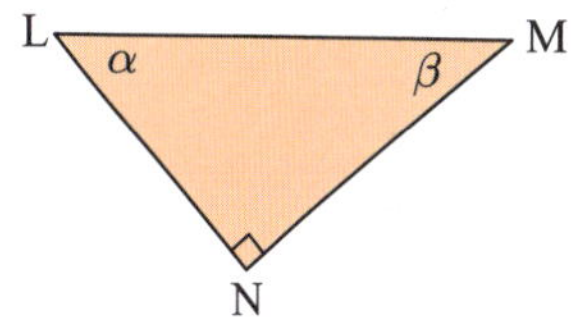

B THE TRIGONOMETRIC RATIOS

We have seen previously that **similar** triangles are equiangular, and that the lengths of their sides are in the same ratio.

In trigonometry, we find there are special ratios between the side lengths that correspond to the angles in the triangle.

INVESTIGATION 1 RATIO OF SIDES OF RIGHT ANGLED TRIANGLES

In this Investigation we consider the ratios $\frac{\text{OPP}}{\text{HYP}}$, $\frac{\text{ADJ}}{\text{HYP}}$, and $\frac{\text{OPP}}{\text{ADJ}}$ in a series of similar right angled triangles.

What to do:

1 For each set of four similar triangles on the next page, complete a table like the one below. Measure each side length to the nearest millimetre, and round each of the ratios to 2 decimal places.

Triangle	HYP	OPP	ADJ	$\frac{\text{OPP}}{\text{HYP}}$	$\frac{\text{ADJ}}{\text{HYP}}$	$\frac{\text{OPP}}{\text{ADJ}}$
1						
2						
3						
4						

PRINTABLE WORKSHEET

a

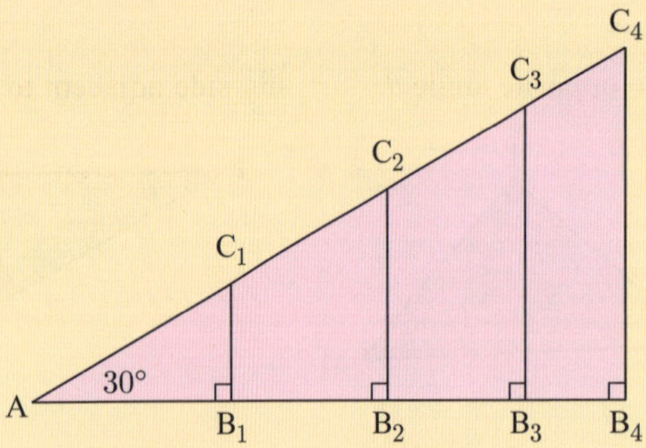

b

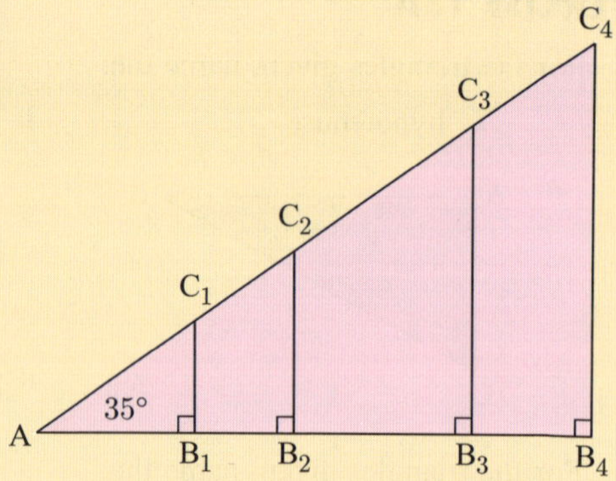

2 Use your calculator to evaluate $\frac{\sqrt{3}}{2}$ and $\frac{1}{\sqrt{3}}$.

3 Summarise your results from **1** and **2**.

From the **Investigation**, you should have discovered that for a right angled triangle with one angle fixed, the ratios $\frac{\text{OPP}}{\text{HYP}}$, $\frac{\text{ADJ}}{\text{HYP}}$, and $\frac{\text{OPP}}{\text{ADJ}}$ are constant.

These ratios have the traditional names **sine**, **cosine**, and **tangent** respectively. We abbreviate them to **sin**, **cos**, and **tan**.

In any right angled triangle with one angle θ, we have:

$$\sin\theta = \frac{\textbf{OPP}}{\textbf{HYP}}, \quad \cos\theta = \frac{\textbf{ADJ}}{\textbf{HYP}}, \quad \tan\theta = \frac{\textbf{OPP}}{\textbf{ADJ}}$$

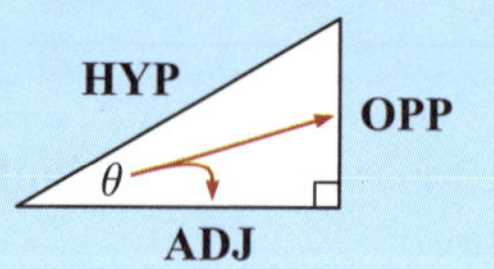

In the previous **Investigation** you should have found that $\sin 30° = 0.5$, $\cos 30° = \frac{\sqrt{3}}{2} \approx 0.87$, and $\tan 30° = \frac{1}{\sqrt{3}} \approx 0.58$.

Trigonometric ratios can be found using a calculator. You should first check that your calculator is in DEGREE mode.

GRAPHICS CALCULATOR INSTRUCTIONS

```
sin 30
                0.5
cos 30
       0.8660254038
tan 30
       0.5773502692
□
JUMP DEL ▸MAT MATH
```

Example 2

Self Tutor

For the given triangle, find $\sin\theta$, $\cos\theta$, and $\tan\theta$.

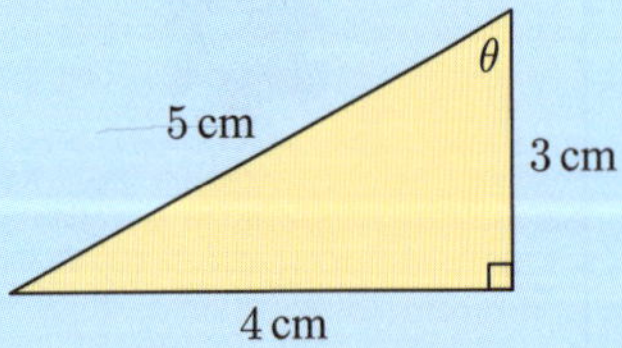

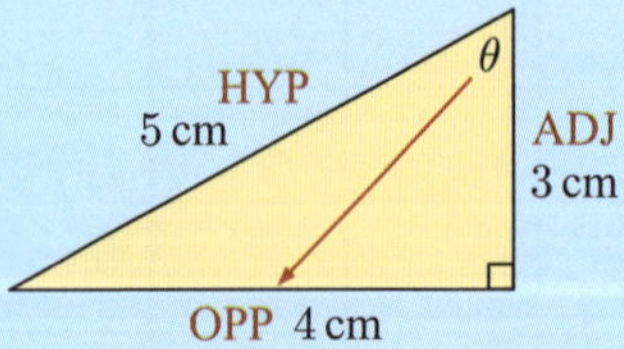

$$\sin\theta = \frac{\text{OPP}}{\text{HYP}} = \frac{4}{5}$$

$$\cos\theta = \frac{\text{ADJ}}{\text{HYP}} = \frac{3}{5}$$

$$\tan\theta = \frac{\text{OPP}}{\text{ADJ}} = \frac{4}{3}$$

EXERCISE 13B

1 For the given triangles, find: **i** $\sin\theta$ **ii** $\cos\theta$ **iii** $\tan\theta$.

a
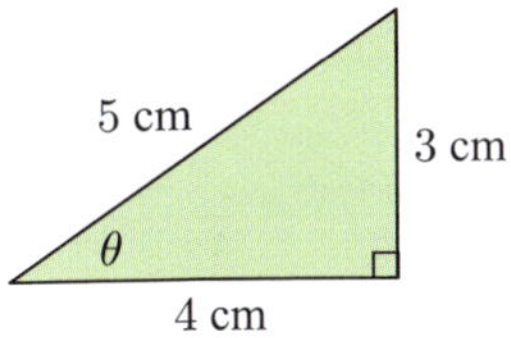

b
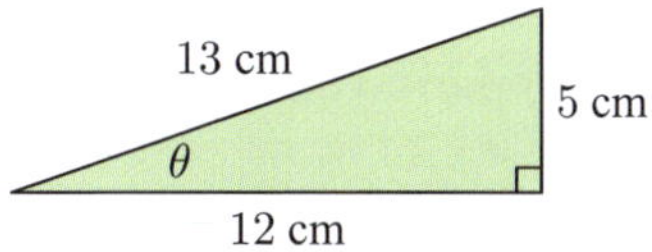

c
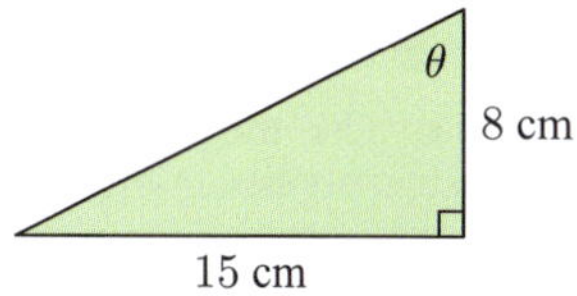

d
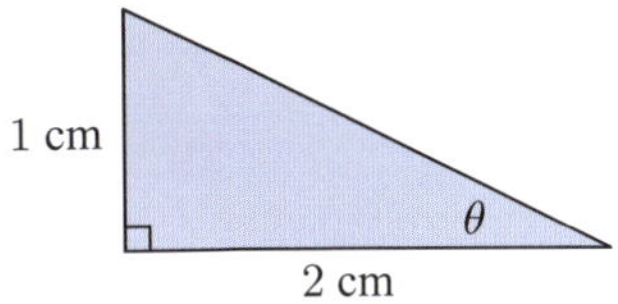

e
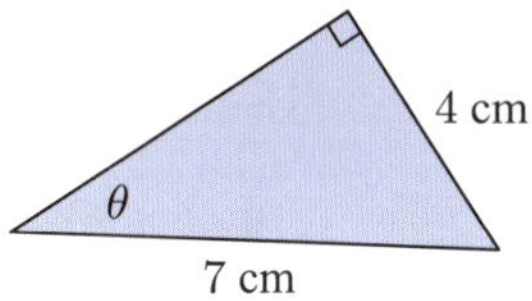

f
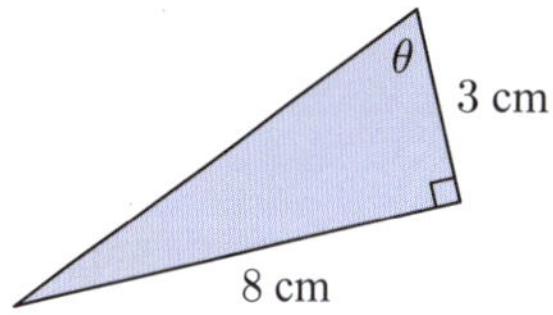

2
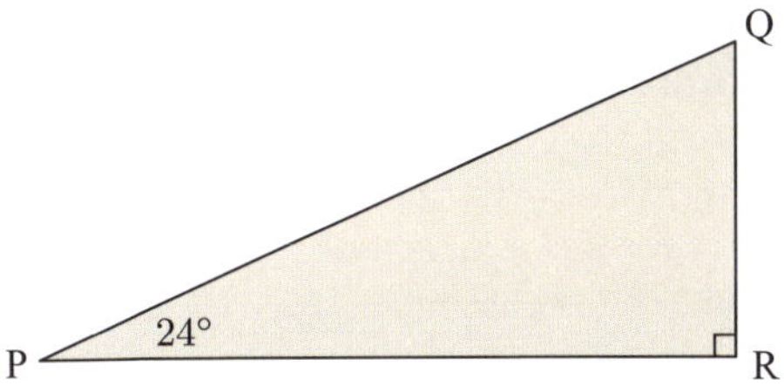

a Use a ruler to find the length of each side of the triangle, to the nearest millimetre.

b Hence, estimate the value of:

i $\sin 24°$ **ii** $\cos 24°$

iii $\tan 24°$

c Check your answers using a calculator.

3 The side lengths in the figure alongside have been rounded to the nearest millimetre.

a Use the diagram to estimate the value of:

i $\sin 36°$ **ii** $\cos 51°$

iii $\tan 54°$ **iv** $\cos 39°$

b Check your answers using a calculator.

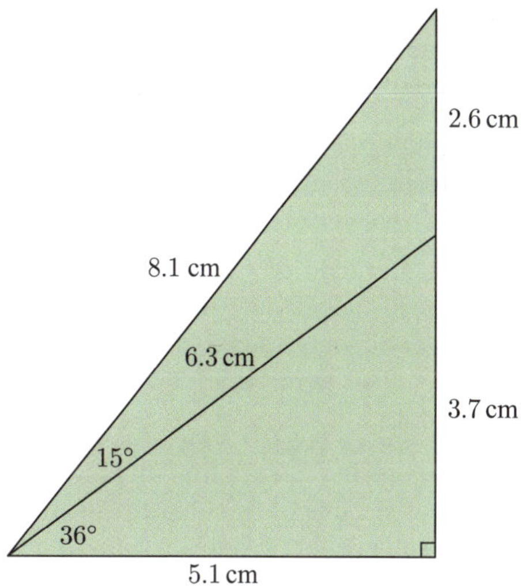

4 $\cos^2\theta$ is shorthand notation for $\cos\theta \times \cos\theta$.

a Use your calculator to find, correct to 4 decimal places:

i $\cos^2 16°$ **ii** $\sin^2 16°$ **iii** $\cos^2 16° + \sin^2 16°$

iv $\cos^2 65°$ **v** $\sin^2 65°$ **vi** $\cos^2 65° + \sin^2 65°$

b For any value of θ, what do you suspect is the value of $\cos^2\theta + \sin^2\theta$?

c Use the figure alongside to prove your answer to **b**.

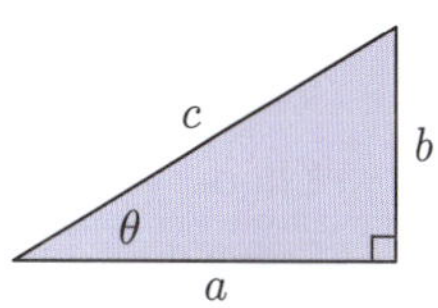

5 **a** Use your calculator to find:

i $\tan 70° \times \tan 20°$ **ii** $\tan 63° \times \tan 27°$ **iii** $\tan 49° \times \tan 41°$

b What do you suspect is the relationship between $\tan\theta$ and $\tan(90° - \theta)$?

c Use the figure alongside to prove your answer to **b**.

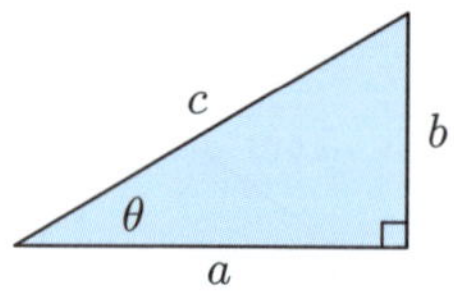

6

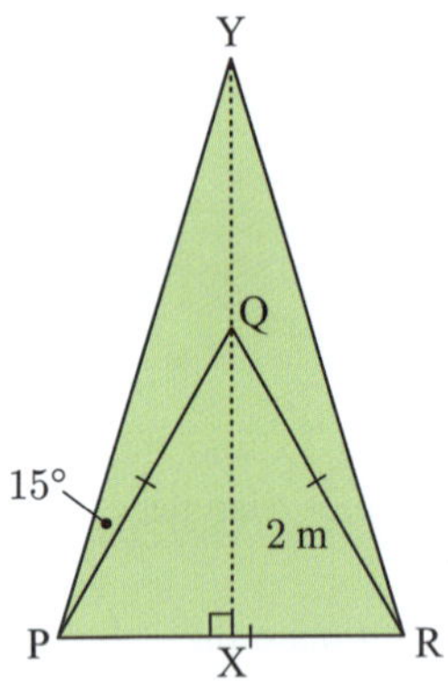

In the diagram alongside, PQR is an equilateral triangle with side lengths 2 m.

Use the diagram to prove that:

a $\cos 60° = \frac{1}{2}$ **b** $\sin 60° = \frac{\sqrt{3}}{2}$

c $\tan 60° = \sqrt{3}$ **d** $\cos 30° = \frac{\sqrt{3}}{2}$

e $\sin 30° = \frac{1}{2}$ **f** $\tan 30° = \frac{1}{\sqrt{3}}$

g $\tan 75° = \sqrt{3} + 2$ **h** $\tan 15° = 2 - \sqrt{3}$

i $\cos^2 75° = \frac{2 - \sqrt{3}}{4}$

HISTORICAL NOTE

The origin of the term "sine" is quite fascinating. **Aryabhata**, a Hindu mathematician who studied trigonometry in the 5th century AD, called the sine-leg of a circle diagram "ardha-jya" which means "half-chord". This was eventually shortened to "jya".

When Arab scholars later translated Aryabhata's work into Arabic, they initially translated "jya" phonetically as "jiba". Since this meant nothing in Arabic, they very shortly began writing the word as "jaib", which has the same letters but means "cove" or "bay".

In 1150, **Gerardo of Cremona** translated this work into Latin. He replaced "jaib" with "sinus", which means "bend" or "curve" but was commonly used in Latin to refer to a bay or gulf on a coastline. The term "sine" that we use today comes from this Latin word.

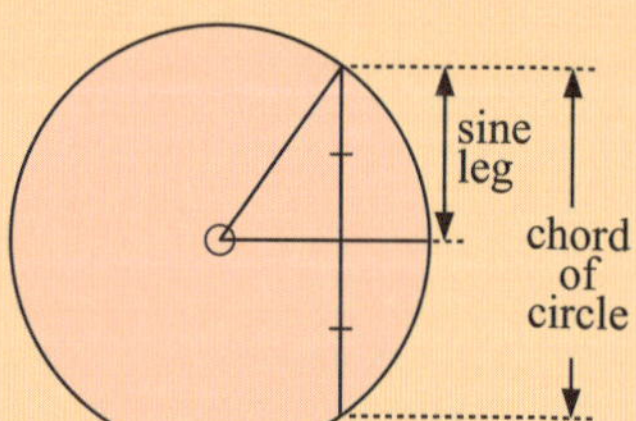

The term "cosine" comes from the fact that the cosine of an angle is equal to the sine of its complement. In 1620, **Edmund Gunter** introduced the abbreviated "co sinus" for "complementary sine".

C FINDING SIDE LENGTHS

Suppose we are given the angles of a right angled triangle, and the length of a side. We can use the trigonometric ratios to find the other side lengths.

Step 1: Redraw the figure and mark on it HYP, OPP, and ADJ relative to a given angle.

Step 2: Choose an appropriate trigonometric ratio, and construct an equation.

Step 3: Solve the equation to find the unknown side length.

Example 3 **Self Tutor**

Find x, giving your answer rounded to 2 decimal places:

a

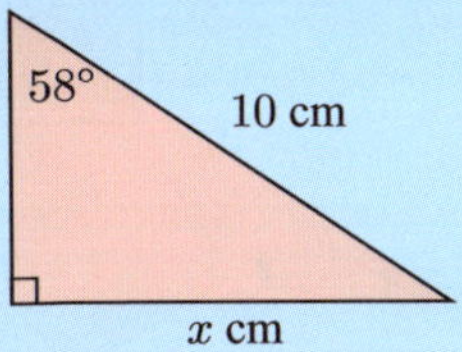

b

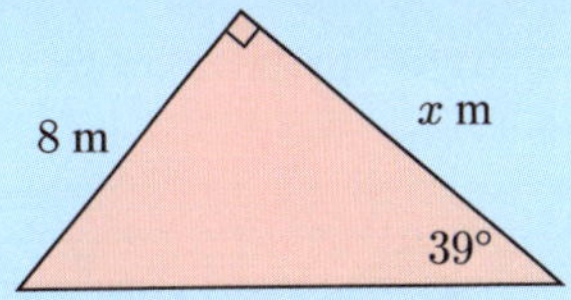

a The relevant sides are OPP and HYP, so we use the *sine* ratio.

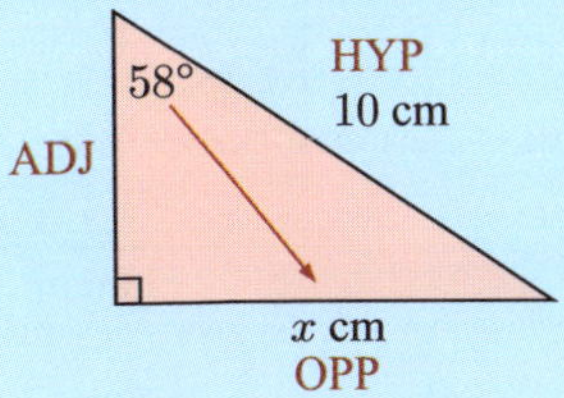

$$\sin 58° = \frac{x}{10} \quad \{\sin\theta = \frac{\text{OPP}}{\text{HYP}}\}$$
$$\therefore \sin 58° \times 10 = x \quad \{\text{multiplying both sides by } 10\}$$
$$\therefore x \approx 8.48 \quad \{\text{calculator}\}$$

b The relevant sides are OPP and ADJ, so we use the *tangent* ratio.

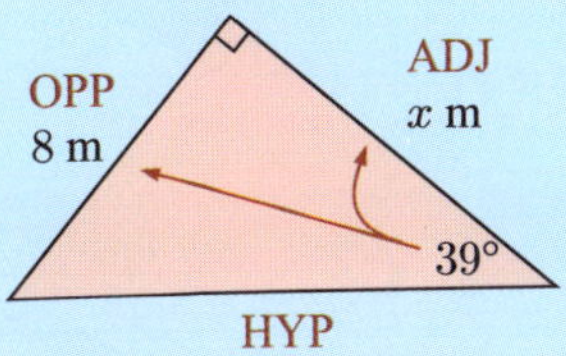

$$\tan 39° = \frac{8}{x} \quad \{\tan\theta = \frac{\text{OPP}}{\text{ADJ}}\}$$
$$\therefore x \times \tan 39° = 8 \quad \{\text{multiplying both sides by } x\}$$
$$\therefore x = \frac{8}{\tan 39°} \quad \{\text{dividing both sides by } \tan 39°\}$$
$$\therefore x \approx 9.88 \quad \{\text{calculator}\}$$

EXERCISE 13C

1 Write down a trigonometric equation connecting the angle and the sides given:

a

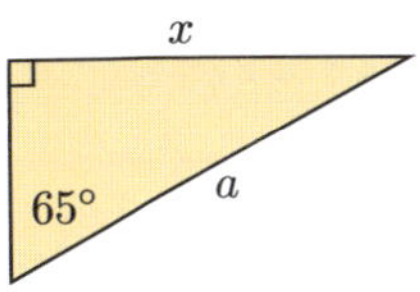

b

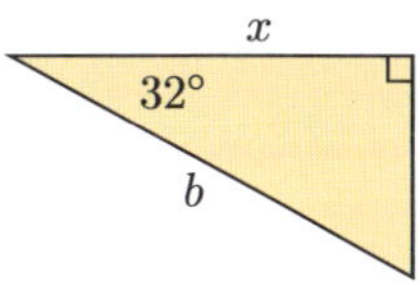

c

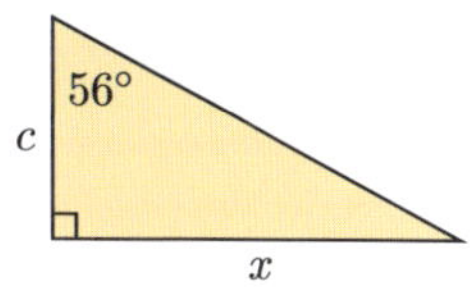

d

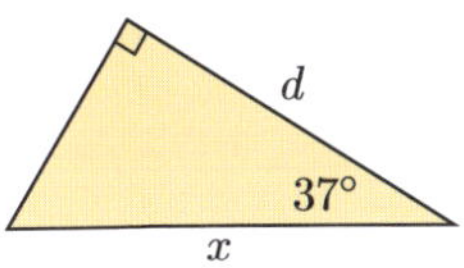

e

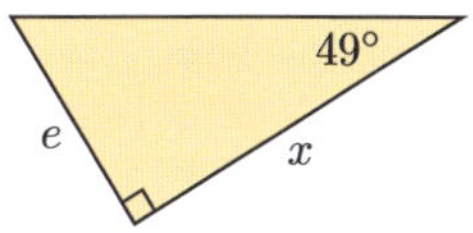

f

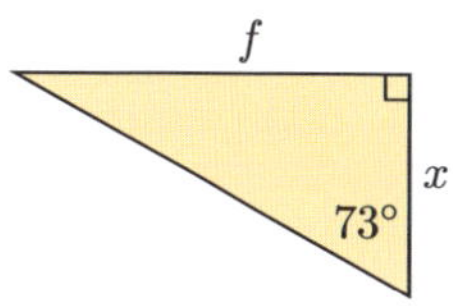

g

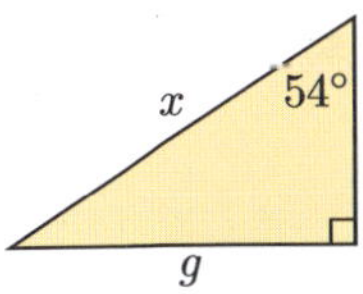

h

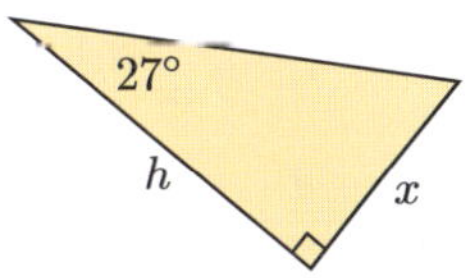

i

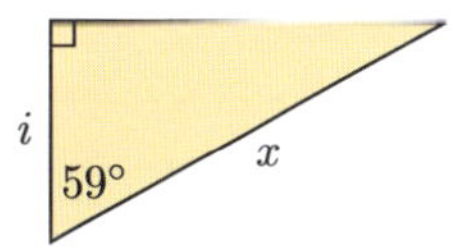

2 Find x, rounding your answer to 2 decimal places:

a
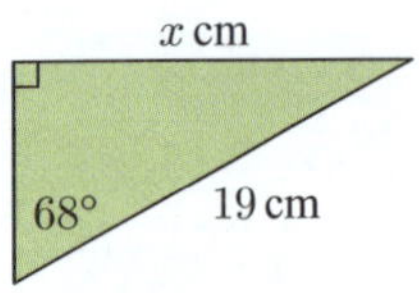

b
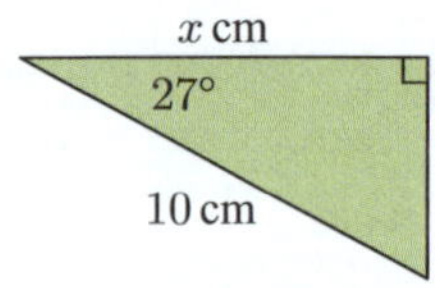

c
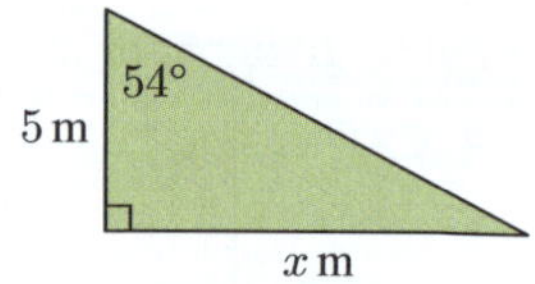

d
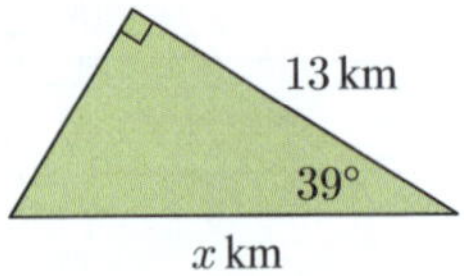

e
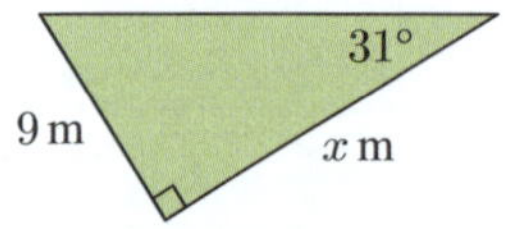

f
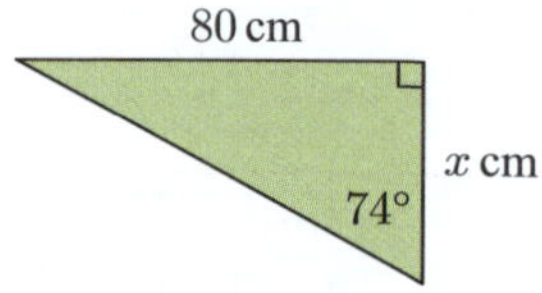

g
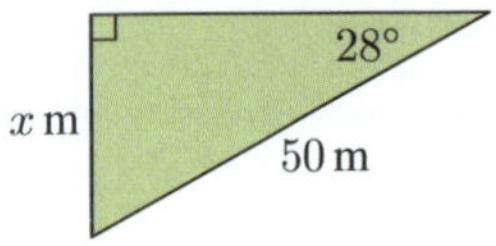

h
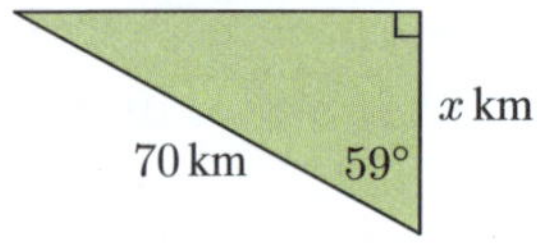

i
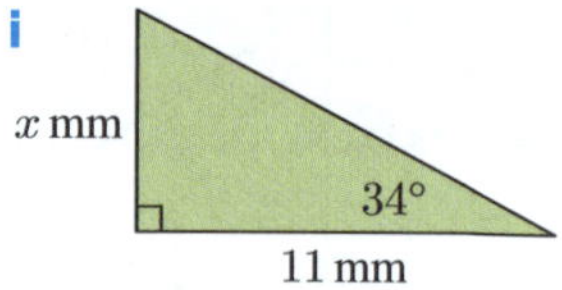

j
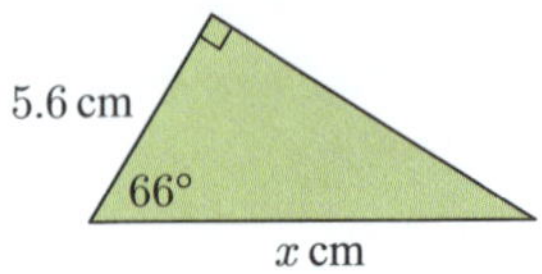

k
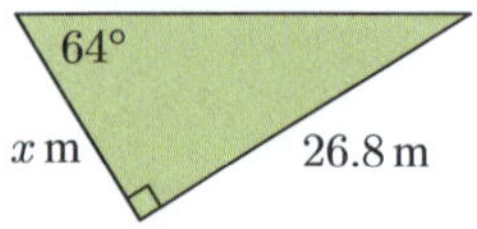

l
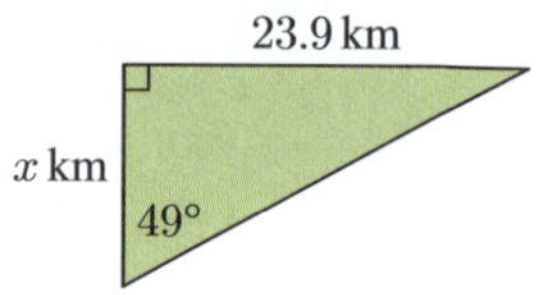

3 Find, to 1 decimal place, *all* unknown angles and sides:

a
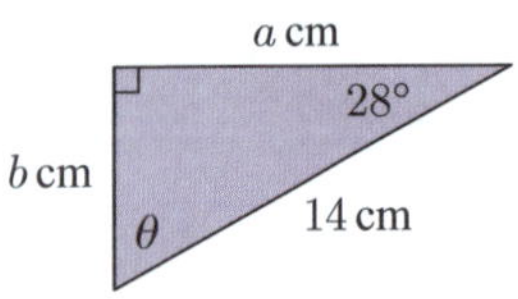

b
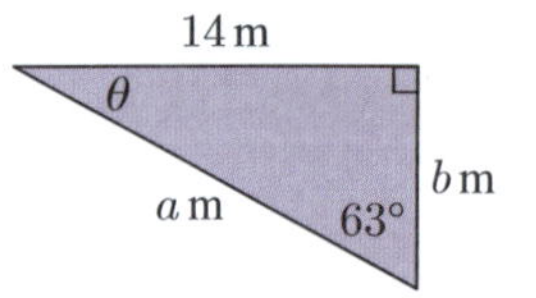

c
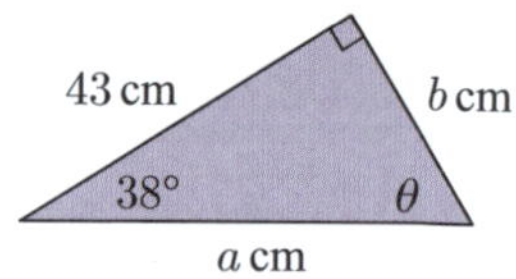

D FINDING ANGLES

If we know two side lengths of a right angled triangle, we can use trigonometry to find the angles.

In the right angled triangle shown, $\cos\theta = \frac{3}{5}$.

To find θ, we need an angle whose cosine is $\frac{3}{5}$.

We say that θ is the **inverse cosine** of $\frac{3}{5}$, and write $\theta = \cos^{-1}\left(\frac{3}{5}\right)$.

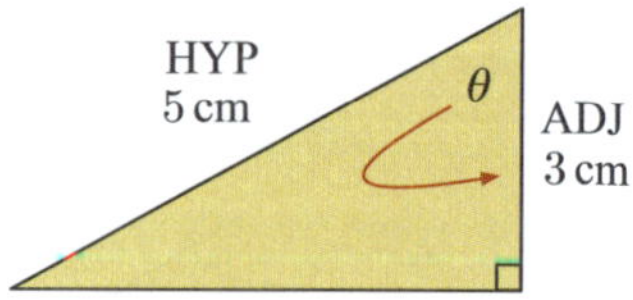

We can use a calculator to evaluate inverse cosines. Click on the icon for instructions.

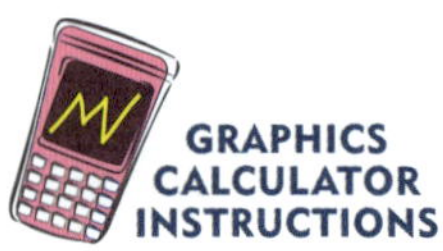

For the right angled triangle with hypotenuse 5 cm and adjacent side 3 cm, $\theta \approx 53.1°$.

We define **inverse sine** and **inverse tangent** in a similar way.

Example 4   **Self Tutor**

Find, to 1 decimal place, the measure of the angle marked θ.

a

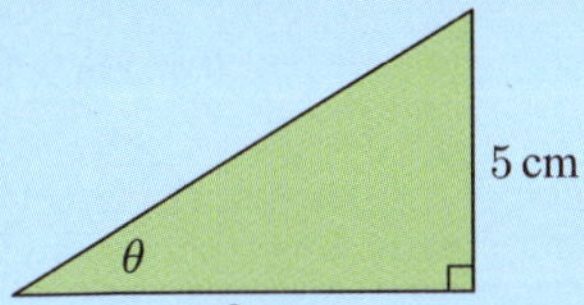

b

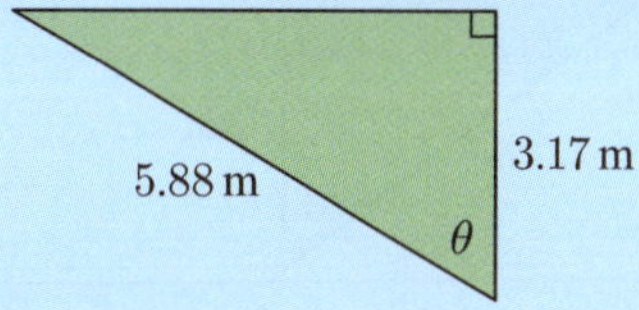

a

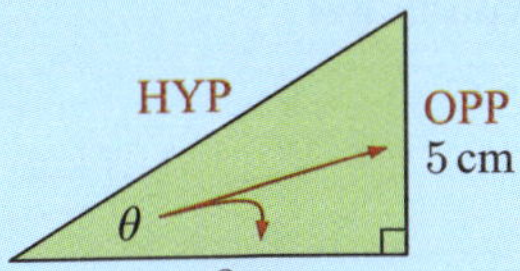

$\tan\theta = \frac{5}{8}$ $\quad \{\tan\theta = \frac{\text{OPP}}{\text{ADJ}}\}$

$\therefore\ \theta = \tan^{-1}\left(\frac{5}{8}\right)$

$\therefore\ \theta \approx 32.0^\circ$ $\quad$ {calculator}

b

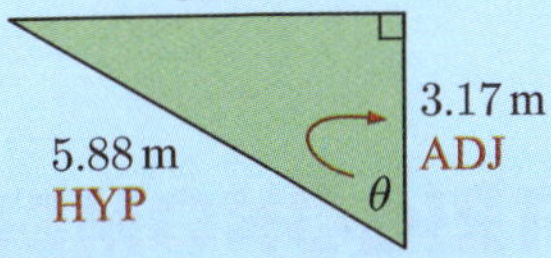

$\cos\theta = \frac{3.17}{5.88}$ $\quad \{\cos\theta = \frac{\text{ADJ}}{\text{HYP}}\}$

$\therefore\ \theta = \cos^{-1}\left(\frac{3.17}{5.88}\right)$

$\therefore\ \theta \approx 57.4^\circ$ $\quad$ {calculator}

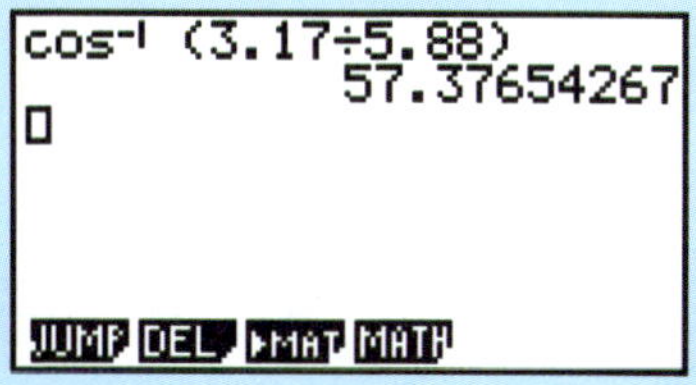

EXERCISE 13D

1 Find, to 1 decimal place, the measure of the angle marked θ.

a

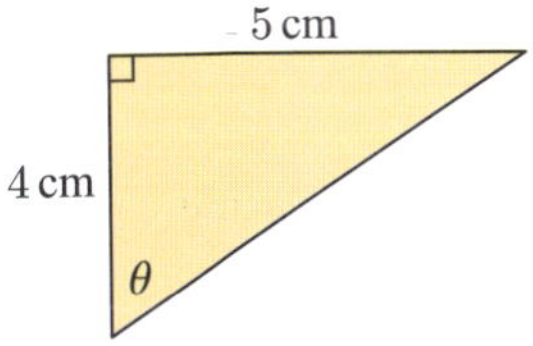

b

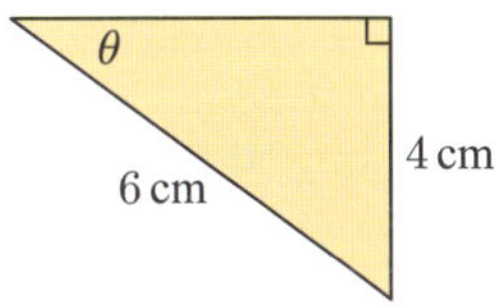

c

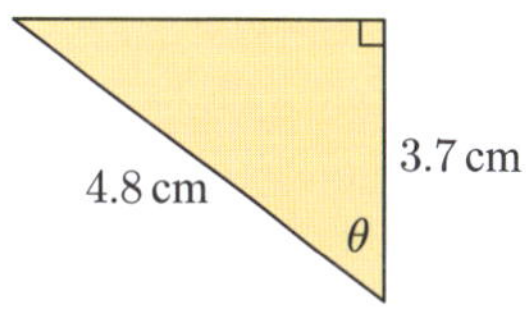

d

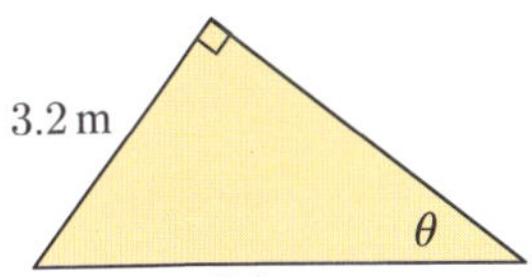

e

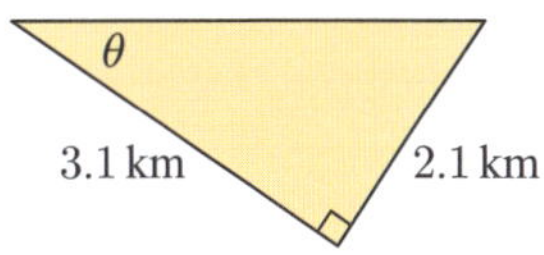

f

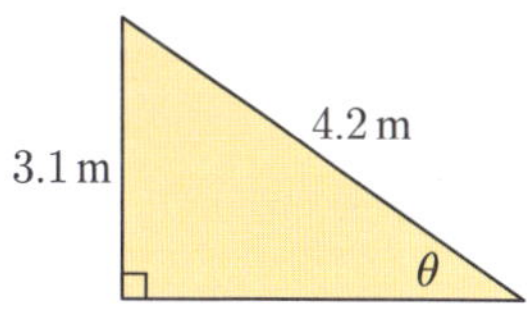

g

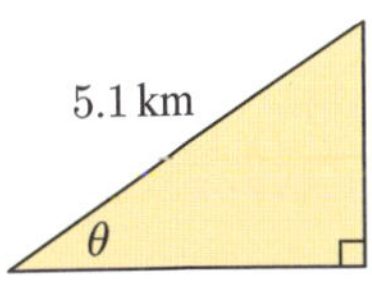

h

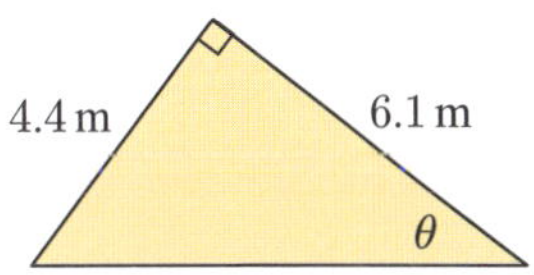

i

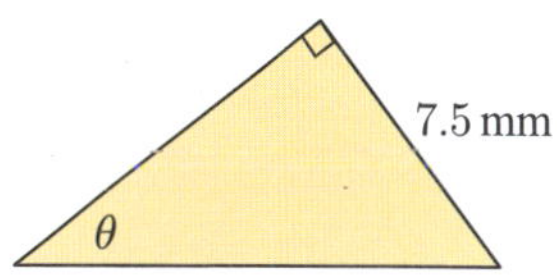

2 Find, to 1 decimal place, all unknown sides and angles in the following triangles. Check your answers for x using Pythagoras' theorem.

a

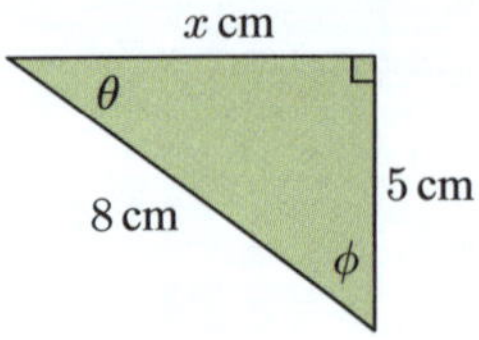

b

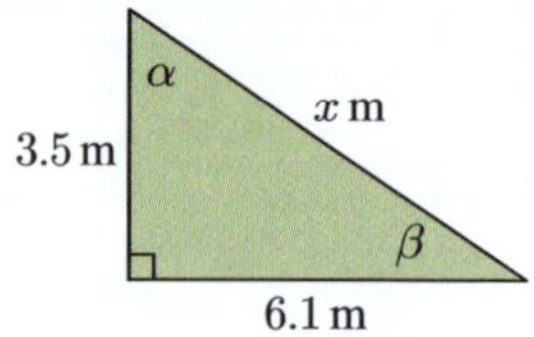

c

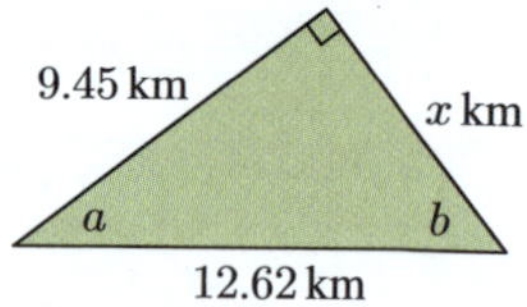

3 Find θ in the following using trigonometry. What conclusions can you draw?

a

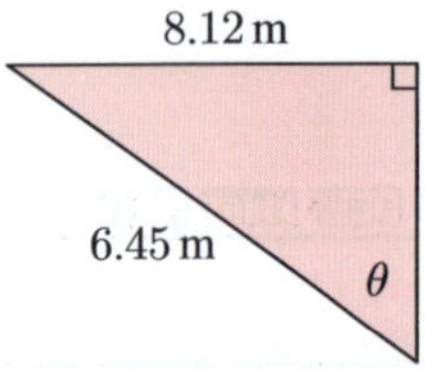

b

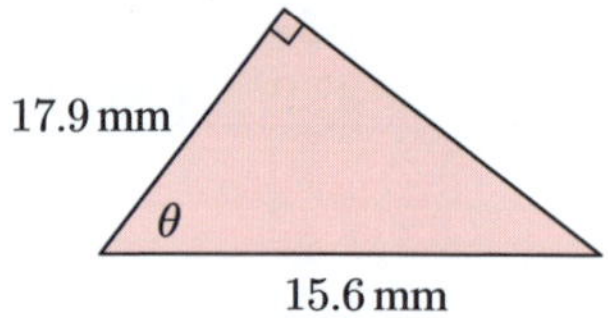

c

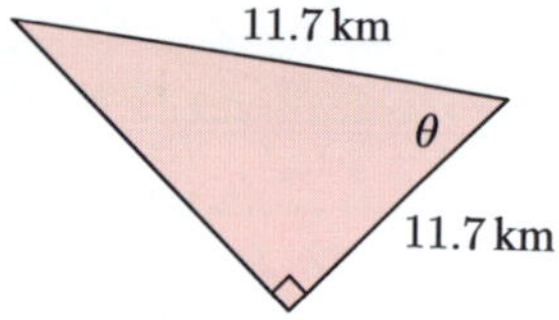

E PROBLEM SOLVING WITH TRIGONOMETRY

The trigonometric ratios can be used to solve a wide variety of problems involving right angled triangles. When solving these problems it is important to follow the steps below:

- Draw a **diagram** to illustrate the situation.
- Mark on the diagram the **unknown** angle or side that needs to be calculated. We often use x for a length and θ for an angle.
- Locate a **right angled triangle** in your diagram.
- Write an **equation** connecting an angle and two sides of the triangle using an appropriate trigonometric ratio.
- **Solve** the equation to find the unknown.
- **Write** your answer in sentence form.

We can use right angled triangle trigonometry to solve problems involving:

- isosceles triangles
- rectangles
- rhombuses
- kites

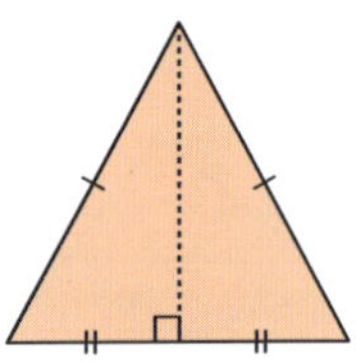

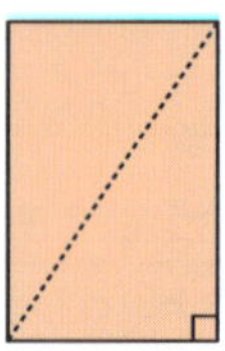

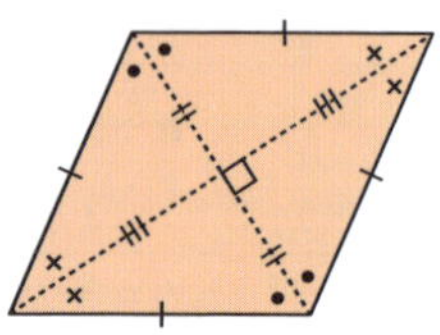

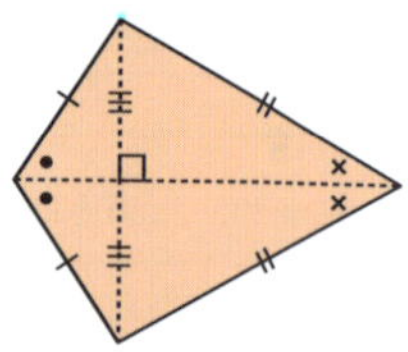

Example 5 **Self Tutor**

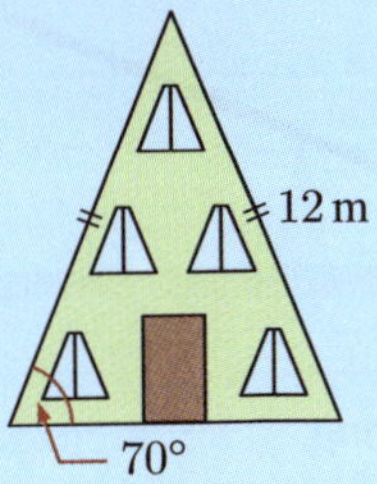

An A-frame house has the shape of an isosceles triangle with base angles of 70°. The oblique walls of the building are 12 m long.

How wide is the building at ground level?

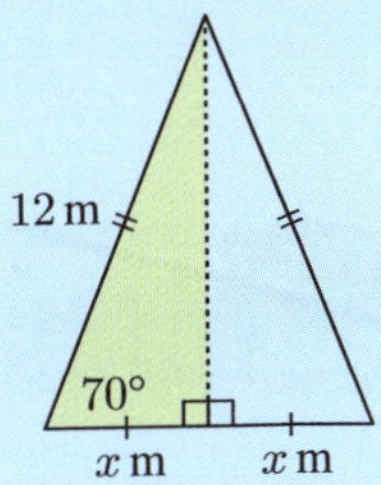

Suppose the building is $2x$ m wide at ground level.

$$\cos 70° = \frac{x}{12} \qquad \left\{\cos\theta = \frac{\text{ADJ}}{\text{HYP}}\right\}$$

$$\therefore \quad 12 \times \cos 70° = x$$

$$\therefore \quad x \approx 4.1042$$

$$\therefore \quad 2x \approx 8.21$$

At ground level, the building is about 8.21 m wide.

Example 6 **Self Tutor**

A staircase 3.6 m in length is needed to access a platform 2.9 m above ground level.

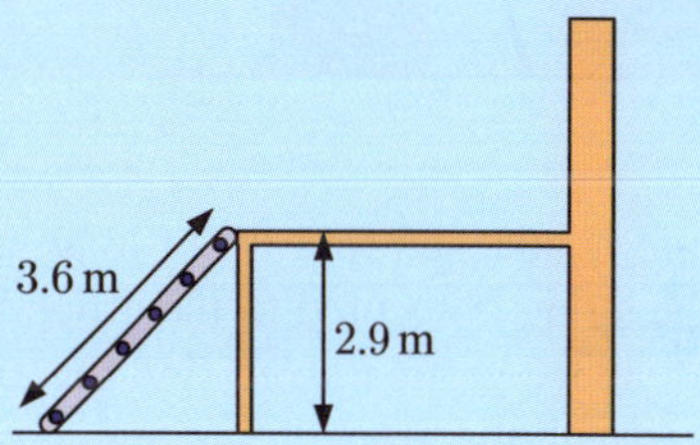

Find, using trigonometry:

a the angle the staircase makes with the ground

b the distance from the foot of the staircase to the post supporting the platform.

a

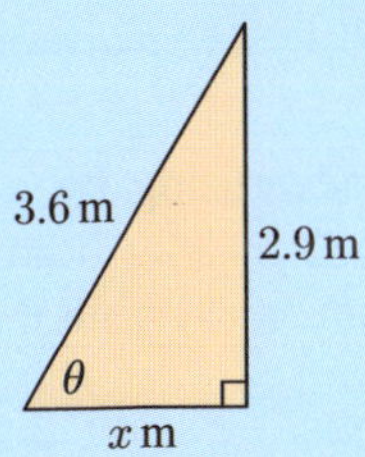

$$\sin\theta = \frac{\text{OPP}}{\text{HYP}} = \frac{2.9}{3.6}$$

$$\therefore \quad \theta = \sin^{-1}\left(\frac{2.9}{3.6}\right)$$

$$\therefore \quad \theta \approx 53.7°$$

The staircase makes an angle of about 53.7° with the ground.

b

$$\cos\theta = \frac{\text{ADJ}}{\text{HYP}}$$

$$\therefore \quad \cos 53.7° = \frac{x}{3.6}$$

$$\therefore \quad 3.6 \times \cos 53.7° = x$$

$$\therefore \quad x \approx 2.13$$

The foot of the staircase is about 2.13 m from the post.

EXERCISE 13E

1 A see-saw has length 5.2 m. When one end is resting on the ground, it makes an angle of 23° with the ground. How far is the other end above ground level?

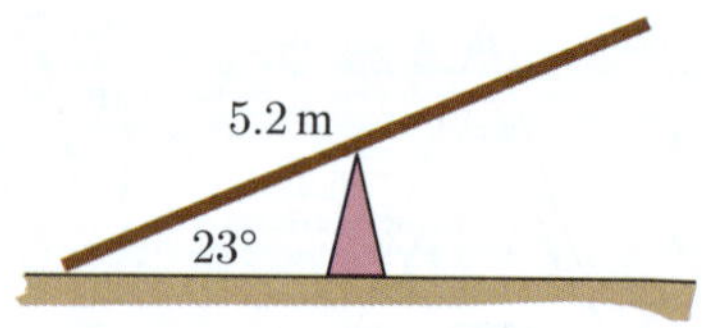

2 The shadow of a tree is 40 m long. The angle from the end of the shadow to the top of the tree is 33°. Find the height of the tree.

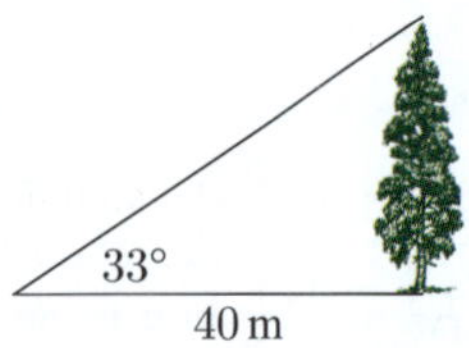

3 A driver travels 2 km up a long steady incline angled 15° to the horizontal. How far has the driver moved horizontally?

4 A ladder makes an angle of 75° with the ground, and rests 4 m up a wall.
Find the length of the ladder.

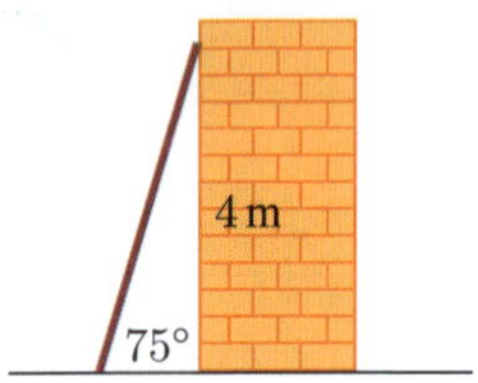

5 A kite is attached to a 30 m long string. The other end of the string is secured to the ground. If the kite is flying 21 m above ground level, find the angle θ that the string makes with the ground.

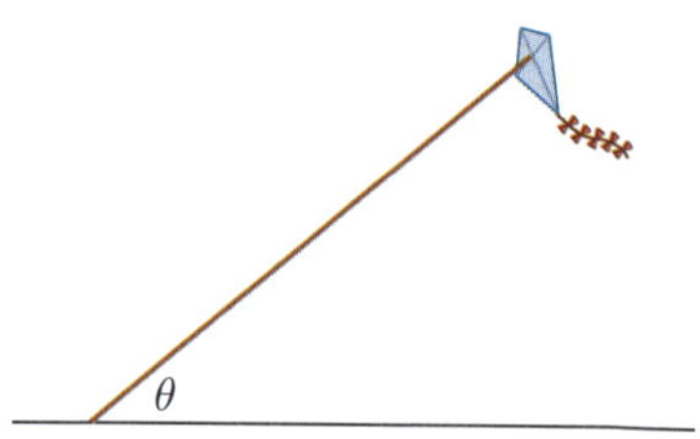

6 A billboard for advertising by a highway is supported by two 5 m long beams. Find the angle that the beams make with the ground.

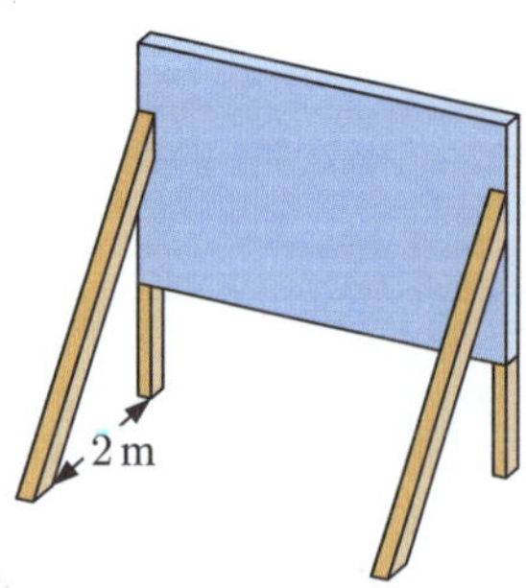

7 A boat has an anchor rope of length 55 m. The boat drifts with the ocean current so that the rope makes an angle of 63° with the surface of the water. Find the depth of the water at the point where the anchor lies on the bottom.

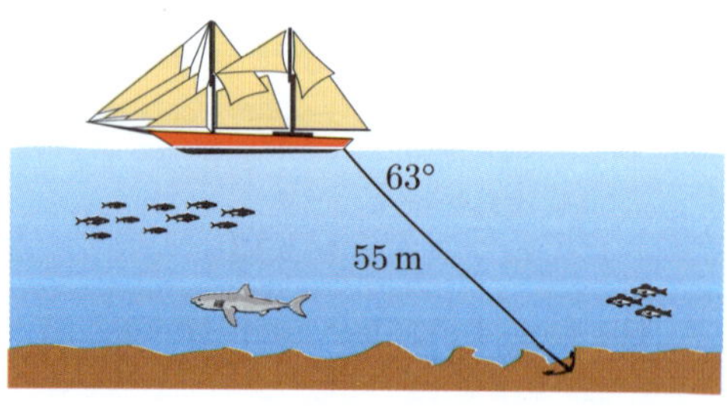

8 A flagpole is supported by four metal braces. Each brace makes an angle of 55° with the ground and meets the pole 6.2 m above ground level. Find the total length of the metal braces.

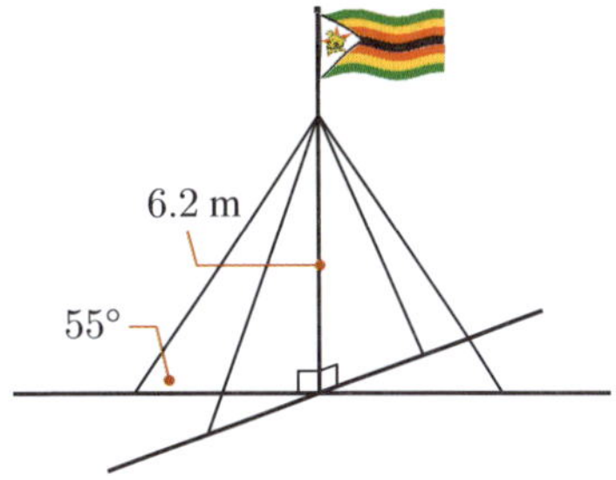

9 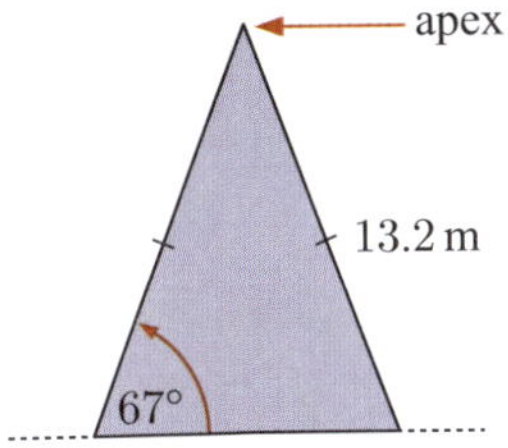

An A-frame house has the shape of an isosceles triangle with base angles 67°. The walls are 13.2 m long. Find:

a the width of the building at ground level

b the height of the apex above the ground.

10 An isosceles triangle has sides 7 cm, 7 cm, and 8 cm in length. Find the measure of the base angles.

11 The metal frame of a rectangular gate has height 1.4 m and length 3.2 m. A diagonal strut is welded in place to support the frame. Find the measure of the angle θ between this diagonal strut and the top of the frame.

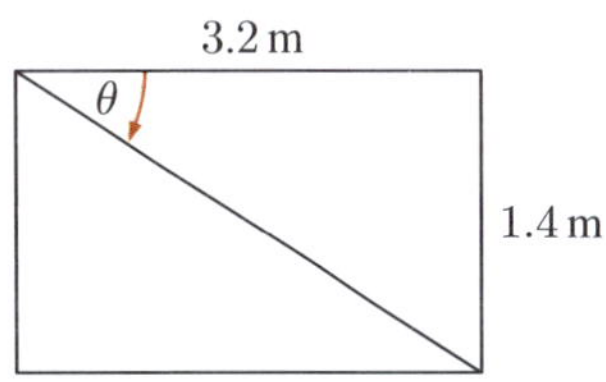

12 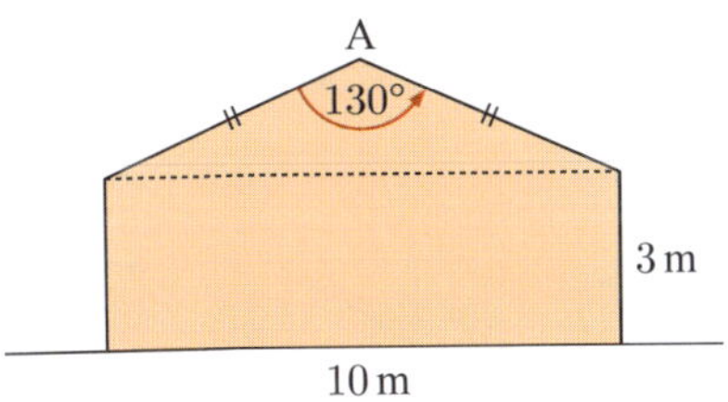

The diagram shows the end wall of a garage. Find:

a the height of the apex A above ground level

b the area of this wall.

13 In the rhombus ABCD alongside, AC = 4.8 m and BD = 6.6 m. Find the measure of $B\widehat{D}C$.

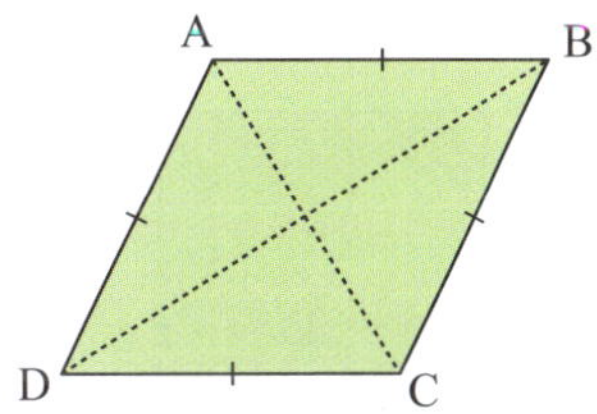

14 Answer the **Opening Problem** on page **252**.

15 

An aeroplane takes off at an angle of 18°. Its average speed in the first 20 seconds of flight is 240 km/h. What is the altitude of the plane at the end of this time?

16 For the kite QRST illustrated, find the measure of $Q\widehat{T}S$.

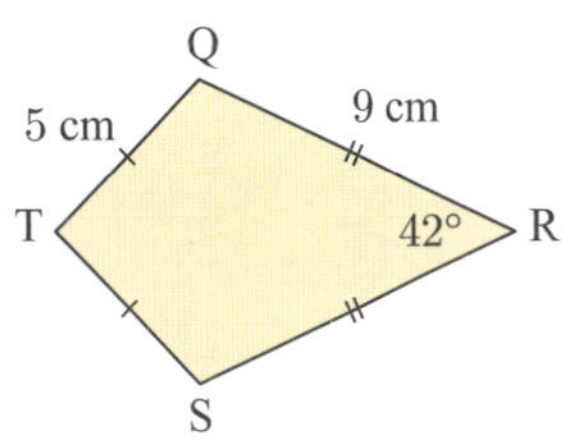

17 A rectangle is 16 cm long and 10 cm wide. Find the acute angle at which the diagonals of the rectangle intersect.

18 A square tiling pattern is created from four identical triangular tiles and a smaller central square tile, as shown. Find:

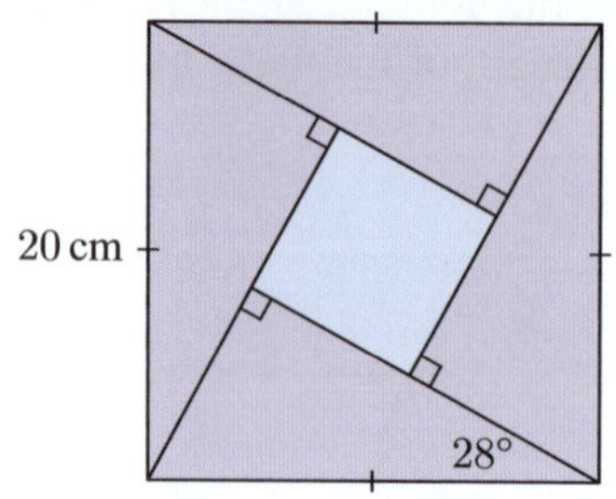

a the perimeter of each triangular tile

b the area of each triangular tile

c the side length of the central square tile

d the percentage of the pattern which is occupied by the central square tile.

19 Julie, Tina, and Brenton have baked a 12 cm by 12 cm square cake. They are trying to work out how to share the cake equally between them.

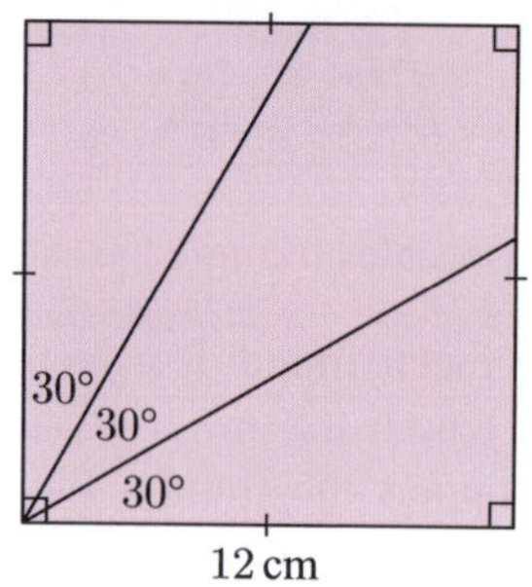

a Brenton suggests cutting the cake at 30° angles from one corner.

i Find the area of each slice.

ii Would each person receive the same amount of cake in this case?

b Find the angles $x°$, $y°$, and $z°$ which should be cut to make three slices of equal size.

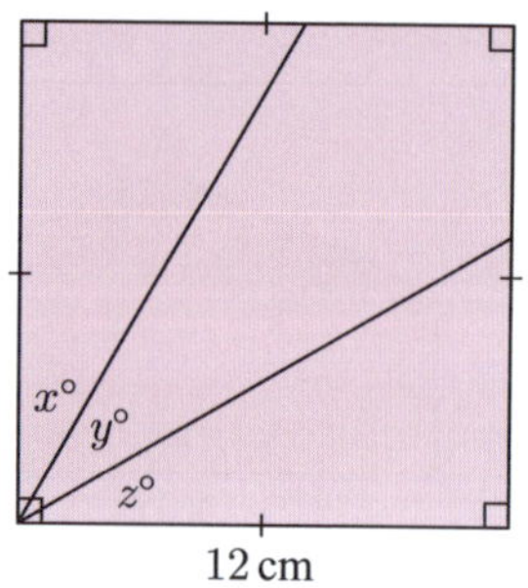

20

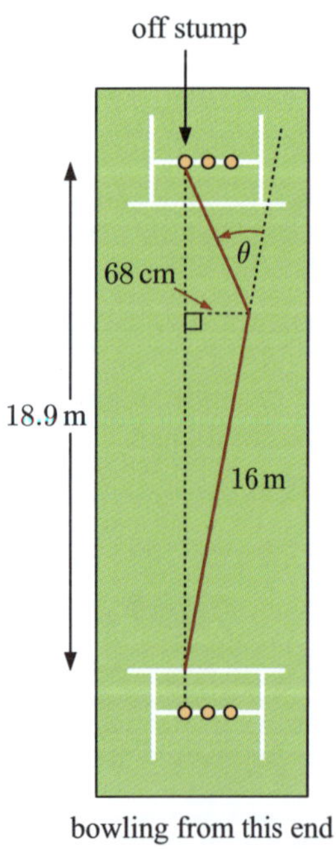

During a cricket match in England, a leg spin bowler delivers the ball from a point 18.9 m from the stumps, and level with the off stump.

The ball travels 16 m before it bounces, and lands 68 cm to the side as shown.

Find the angle θ through which the ball spins in order to hit the off stump.

F BEARINGS

COMPASS BEARINGS

In previous years we have studied the bearings on a compass. We commonly use the **cardinal directions** north, south, east, and west, and also the **ordinal directions** northeast, southeast, southwest, and northwest.

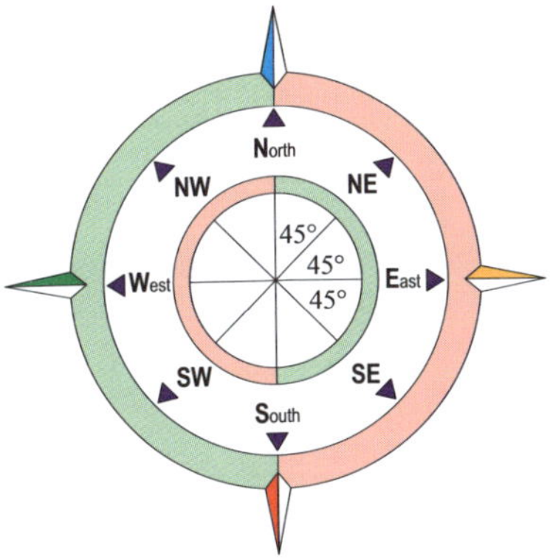

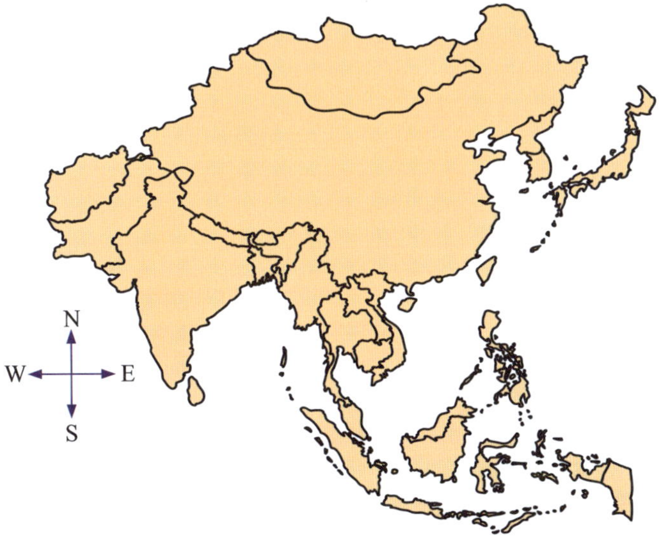

TRUE BEARINGS

We can measure a direction by comparing it with the **true north direction**. We call this a **true bearing**.

Imagine you are standing at point A, facing north. You turn **clockwise** through an angle until you face B. The **bearing of B from A** is the angle through which you have turned.

So, the bearing of B from A is the clockwise measure of the angle between the 'north' line through A, and [AB].

In the diagram alongside, the bearing of B from A is 120° from true north. We write this as 120°T or 120°.

To find the **bearing of A from B**, we place ourselves at point B, face north, then turn clockwise until we face A. The true bearing of A from B is 300°.

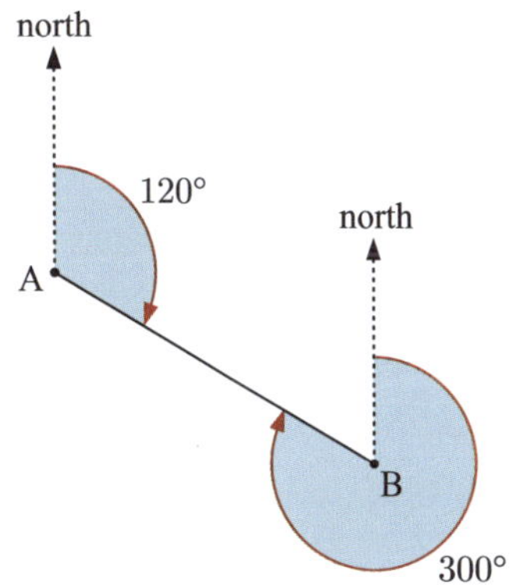

Note:

- A true bearing is always written using three digits. For example, we write 070° rather than 70°.
- The bearing of A from B, and the bearing of B from A, always differ by 180°.
- True north lines are parallel, so we can use angle pair properties to find unknown angles in bearing problems.

EXERCISE 13F

1 Draw diagrams to represent bearings from O of:

a 055° **b** 140° **c** 330° **d** 255°

2 Find the bearing of Q from P if the bearing of P from Q is:

a 124° **b** 068° **c** 244° **d** 321°

3 A, B, and C are checkpoints in an orienteering course. For each of the following, find the bearing of:

i B from A **ii** C from B **iii** B from C

iv C from A **v** A from B **vi** A from C.

a

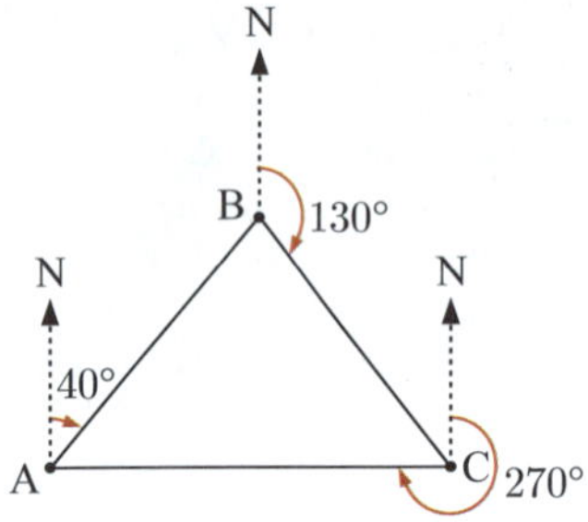

b

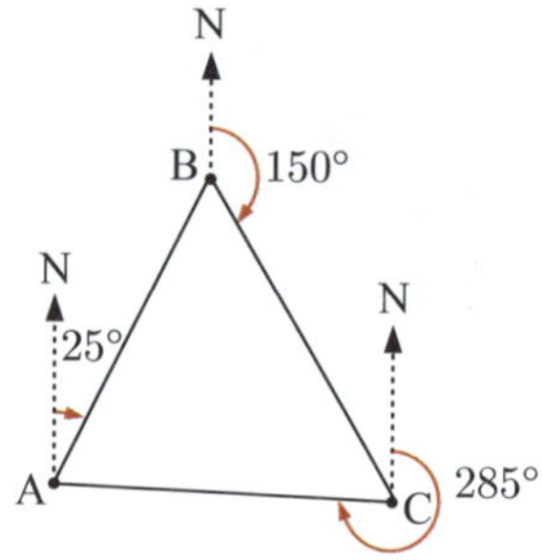

4 For each of the following, find the bearing of:

i B from A **ii** A from B **iii** C from A

iv A from C **v** C from B **vi** B from C.

a

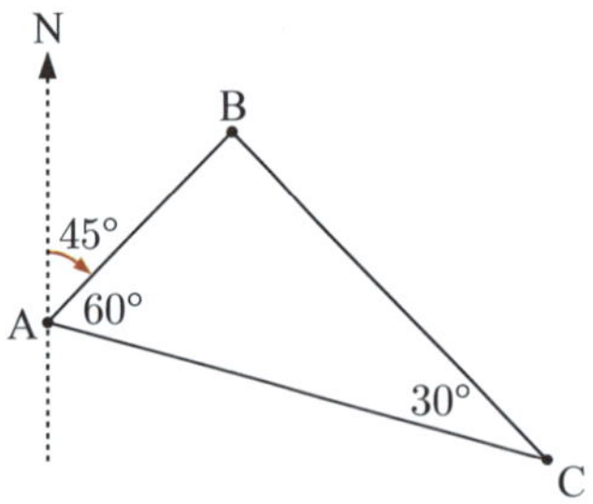

b

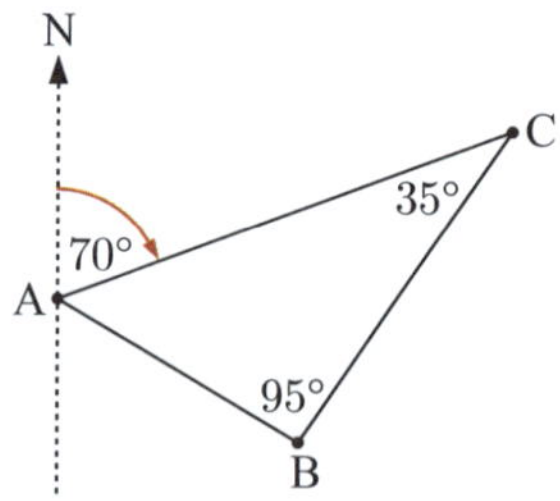

5 The centre of a large city is a square oriented as shown. State the compass bearing and true bearing of:

a B from A **b** C from A **c** D from A

d A from B **e** D from B **f** A from C

g B from C **h** B from D.

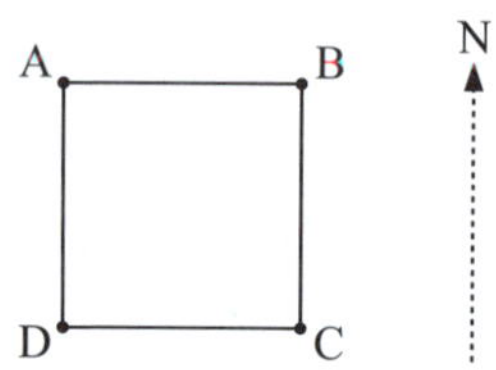

Example 7 Self Tutor

A cyclist rides 21.3 km due west and then 13.8 km due north. Find, to the nearest degree, the bearing of the finishing point from the starting point.

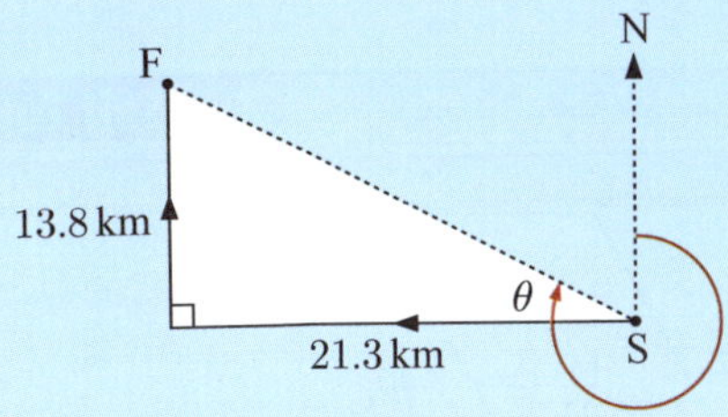

$\tan\theta = \frac{13.8}{21.3}$ $\{\tan\theta = \frac{\text{OPP}}{\text{ADJ}}\}$

$\therefore \theta = \tan^{-1}\left(\frac{13.8}{21.3}\right) \approx 32.9°$

$\therefore 270° + \theta \approx 303°$

So, the bearing of F from S is about 303°.

6 An orienteer is studying the map for a competition. Each grid unit represents 1 km. Find the distance and true bearing from:

a the start to A
b A to B
c B to C
d C to D
e D to the finish.

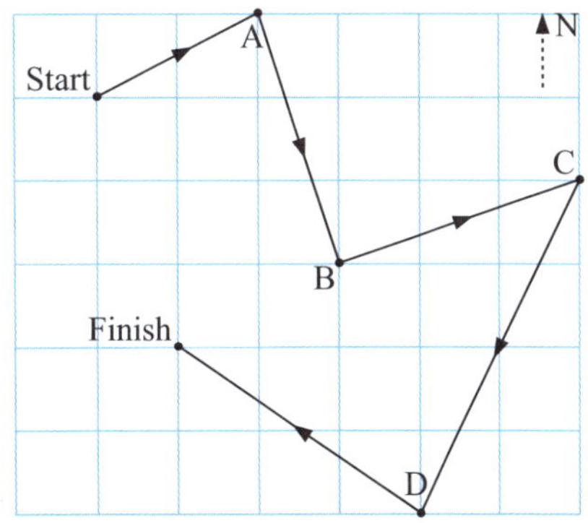

7 A small plane flies between two country airfields. Its destination is 18.4 km south and 27.3 km west of its origin. Find, to the nearest degree, the bearing on which the plane flies.

Example 8 Self Tutor

A ship sails for 180 km on the bearing 238°. How far is the ship west of its starting point?

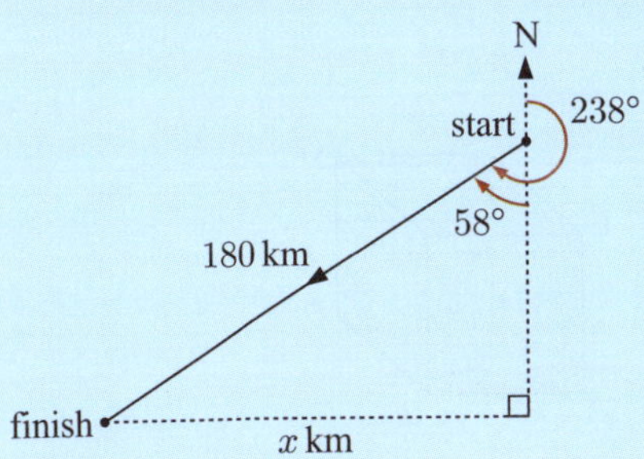

$\sin 58° = \frac{x}{180}$ $\{\sin\theta = \frac{\text{OPP}}{\text{HYP}}\}$

$\therefore x = 180 \times \sin 58°$

$\therefore x \approx 152.6$

The ship is about 153 km west of its starting point.

8 A swimmer swims for 68 m on the bearing 036°.
How far is the swimmer **a** north **b** east of her starting point?

9 A helicopter pilot flies in the direction 147° and lands when he is 12 km south of his starting point. How far did he fly?

10 An aeroplane travels on the bearing 315° until it is 650 km west of its starting point. How far north is it from its starting point?

G 3-DIMENSIONAL PROBLEM SOLVING

Right angled triangles occur frequently in 3-dimensional figures. We can use Pythagoras' theorem and trigonometry to find unknown lengths and angles.

Example 9 Self Tutor

The figure alongside shows a triangular prism.

a Find the length of [AC].

b Find the angle between [AC] and [AF].

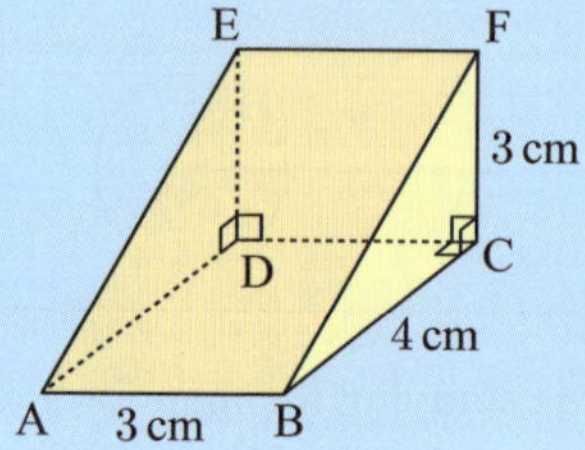

a Consider the base ABCD.
Let [AC] have length x cm.
Using Pythagoras' theorem, $x^2 = 3^2 + 4^2$
$\therefore\ x^2 = 25$
$\therefore\ x = 5$ $\{\text{as } x > 0\}$
$\therefore$ AC = 5 cm

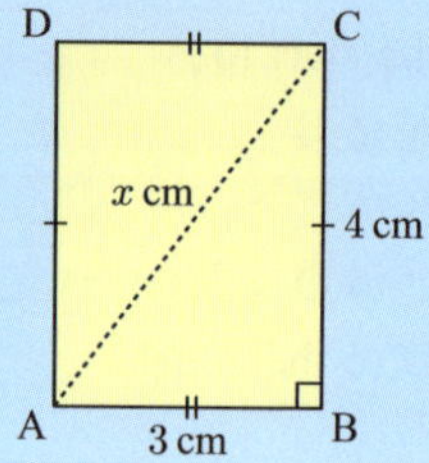

b Consider $\triangle$ACF.
Let the angle between [AC] and [AF] be θ.
Now $\tan\theta = \frac{3}{5}$ $\{\tan\theta = \frac{\text{OPP}}{\text{ADJ}}\}$
$\therefore\ \theta = \tan^{-1}\left(\frac{3}{5}\right)$
$\therefore\ \theta \approx 30.96$
$\therefore$ the angle between [AC] and [AF] is about 31.0°.

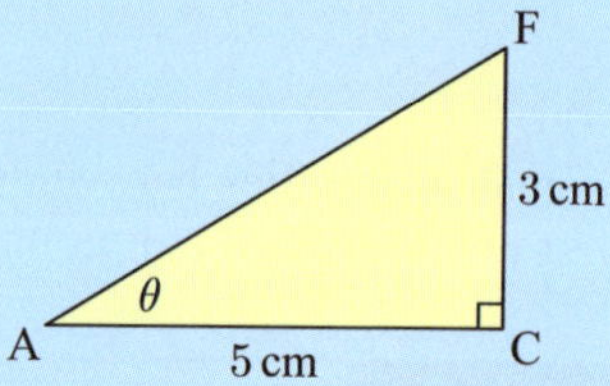

EXERCISE 13G

1 The cube shown has sides of length 1 cm.

a Find the length of [AC].

b Find the angle between [AC] and [AG].

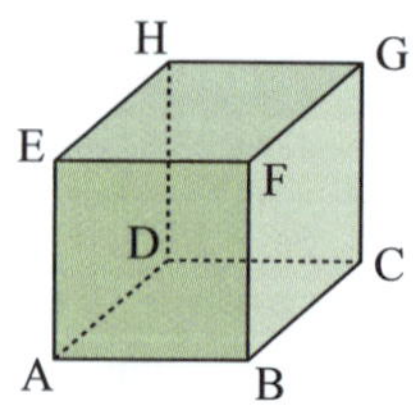

2 Consider the rectangular prism shown.

a Find the length of [BG].

b Find the angle between [CG] and [AG].

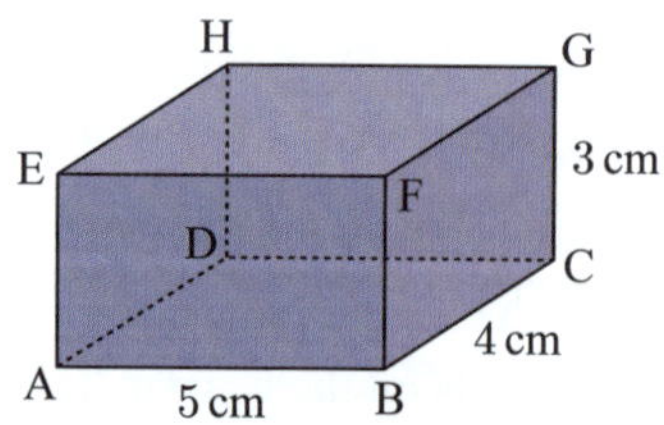

3 Consider the triangular prism shown.

a Find the length of [AE].

b Find the angle between [AC] and [CE].

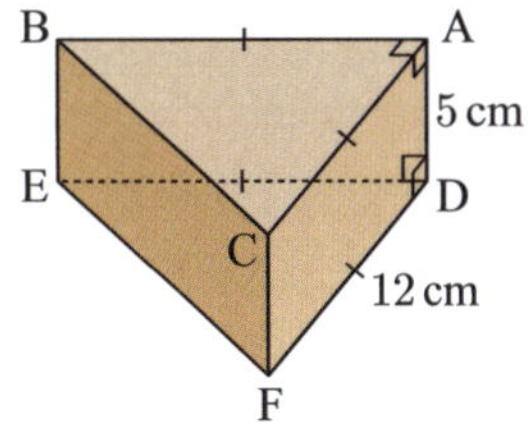

INVESTIGATION 2 THE ANGLE BETWEEN A LINE AND A PLANE

Suppose Jonathon holds his pen at an angle to a piece of paper. If we shine a light from directly above the pen, a shadow will be cast on the paper. This shadow is called the **projection** of the pen onto the paper.

The angle that the pen makes with the paper is the angle between the pen and its projection.

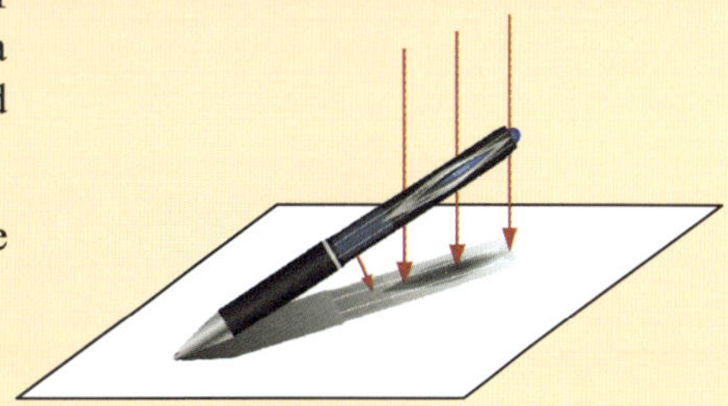

What to do:

1 Jonathon's pen is 16 cm long, and its projection onto the paper is 12 cm long. Find the angle that Jonathon's pen makes with the paper.

DEMO

2 A utility pole is 10 m long, but is not quite vertical to the ground. When the sun is directly above it, the pole casts a 50 cm long shadow on the ground.

a What angle does the pole make with the ground?

b How many degrees is the pole from vertical?

3 **a** Name the projection of each of the following if a light is shone from directly above the figure:

i [BE] **ii** [CF] **iii** [BH]

b Find the angle that each of these line segments makes with the base plane EFGH:

i [DG] **ii** [AH] **iii** [CE]

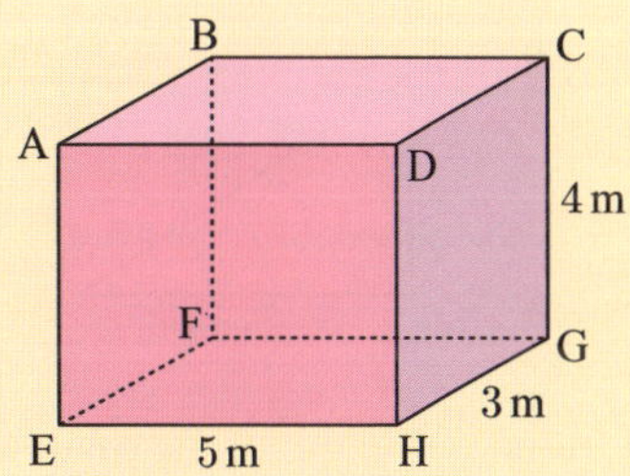

REVIEW SET 13A

1 For the triangle alongside, name the:

a hypotenuse **b** side opposite angle θ.

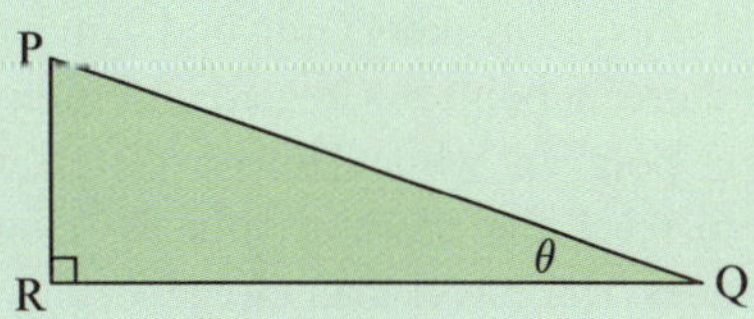

2 The side lengths in the figure alongside have been rounded to the nearest millimetre.

a Use the diagram to estimate the value of:

i $\cos 42°$ **ii** $\sin 64°$

iii $\tan 48°$ **iv** $\sin 26°$

b Check your answers using a calculator.

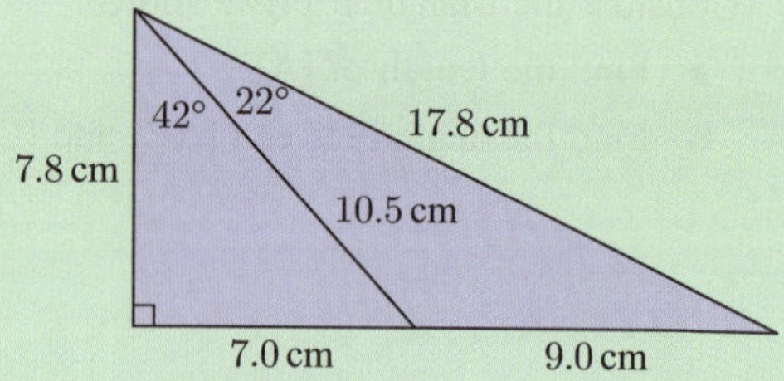

3 Find $\sin\theta$, $\cos\theta$, and $\tan\theta$ for the given triangle.

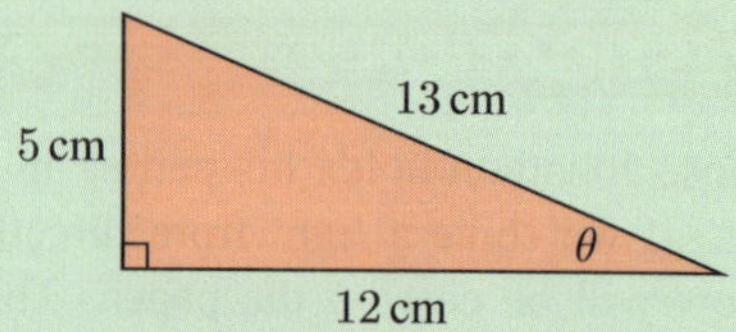

4 Write a trigonometric equation connecting the angle and the sides given:

a

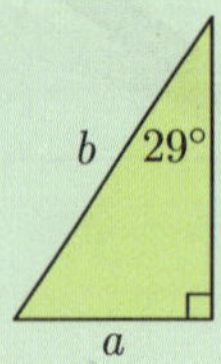

b

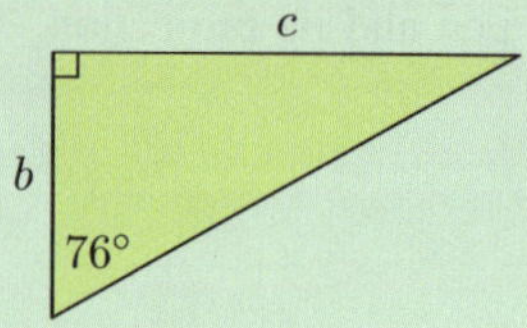

5 Find the value of ϕ:

a

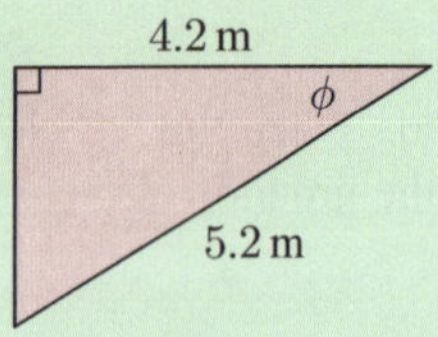

b

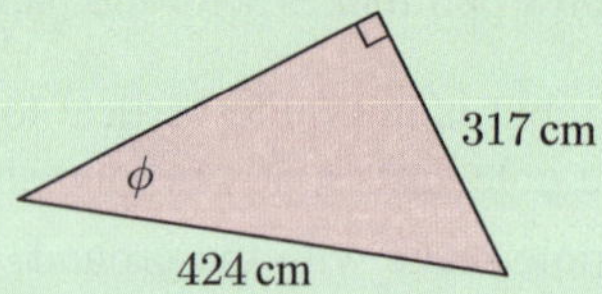

c

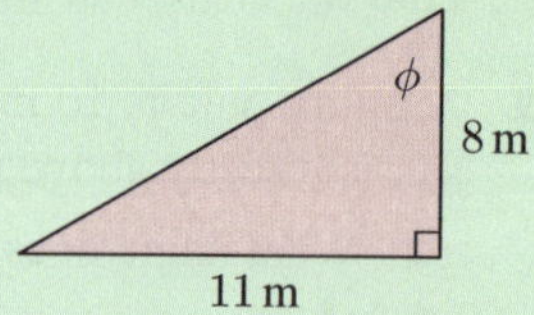

6 Find the value of b:

a

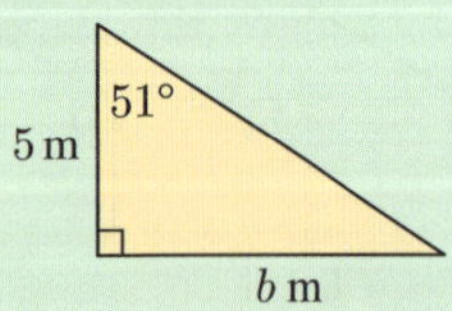

b

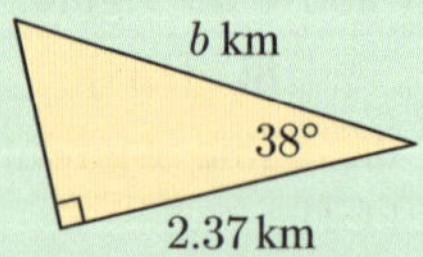

c

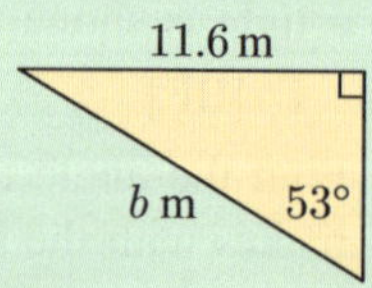

7 Find the length of the shadow cast by a 1.8 m tall boy when the sun is 72° above the horizon.

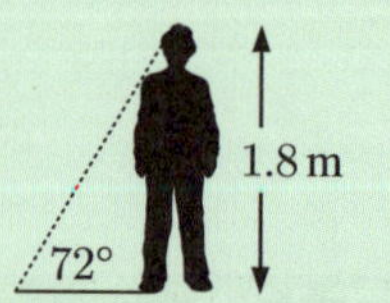

8 Two sea kayakers Greg and Tim start 400 m from shore, at point S. Greg kayaks directly to the shore at point A, while Tim travels at an angle of 32° to reach the shore at point B. How much further does Tim travel than Greg?

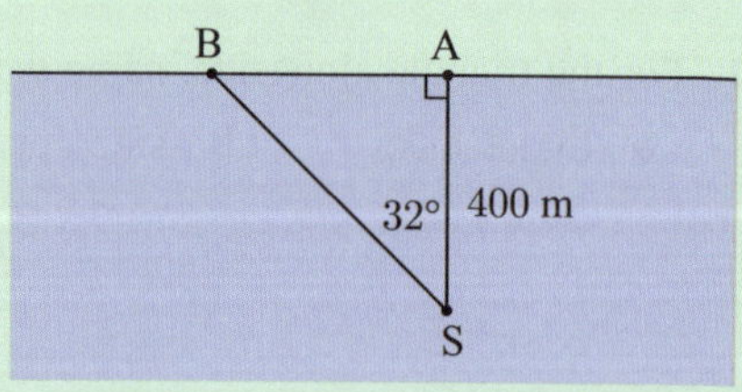

9 A cyclist travels for 5 km up a steady incline. In that time she climbs a vertical distance of 800 m. Find the angle of the incline.

10 An archer standing 30 m away from a target, fires an arrow at the bullseye, but misses to the right by 20 cm. By what angle was the archer off target?

11 During a storm, a flagpole falls and breaks over a fence, as shown. How tall was the flagpole?

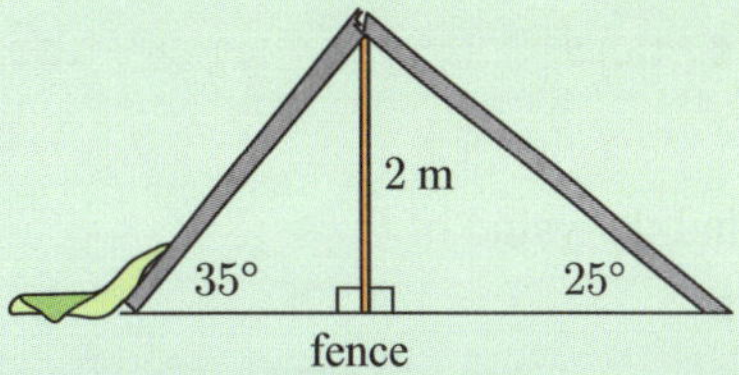

12 A boat sails 50 km on the bearing 054°. How far is the ship north of its starting point?

13 A ship leaves port A and travels 60 km due south. It then sails 40 km due west to an island port B.

a Draw a diagram of the situation.

b How far is B from A?

c The ship wishes to sail back directly to A from B. On what bearing does the captain need to sail?

14

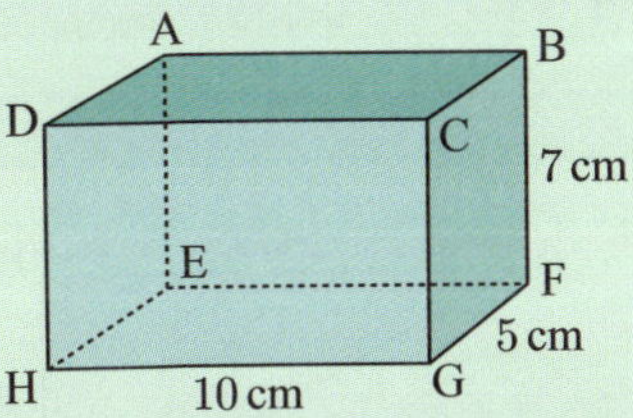

a Find the length of [EG].

b Find the angle between [AG] and [EG].

REVIEW SET 13B

1 State the length of the side:

a opposite to angle θ **b** adjacent to angle θ

c opposite to angle ϕ **d** adjacent to angle ϕ.

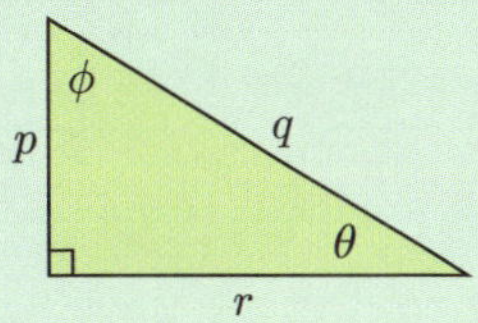

2 **a** Use the figure alongside to find:

i $\cos 30°$ **ii** $\sin 60°$

iii $\cos 60°$ **iv** $\sin 30°$

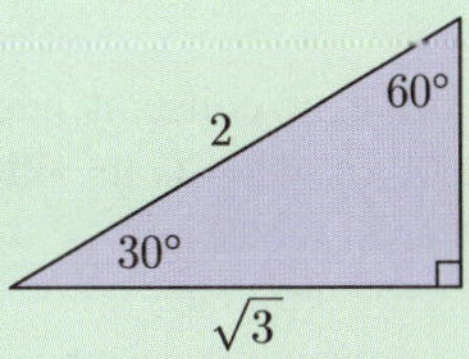

b Use the figure alongside to show that
$\cos\theta = \sin(90° - \theta)$.

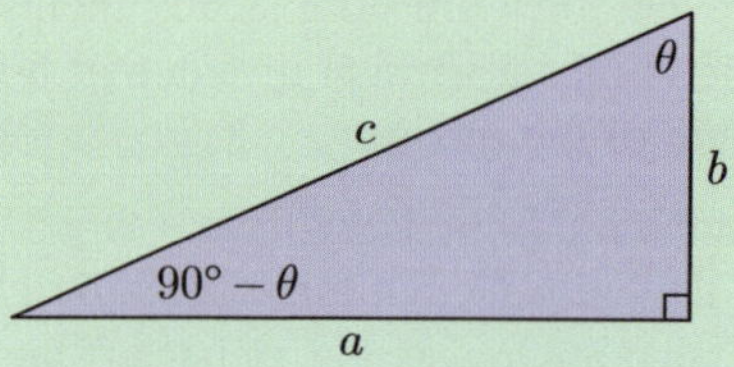

3 Find the exact value of:

a $\sin\theta$

b $\cos\theta$

c $\tan\theta$

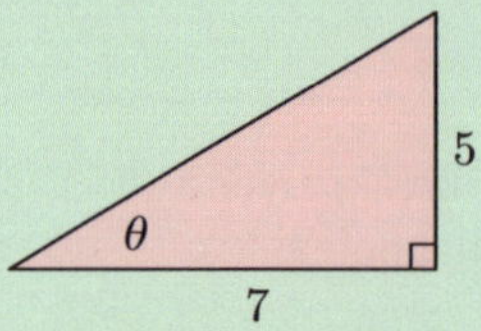

4 Find the value of θ:

a

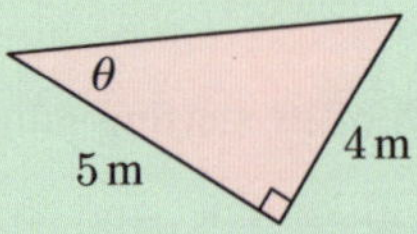

b

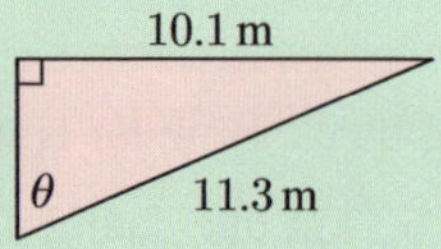

c

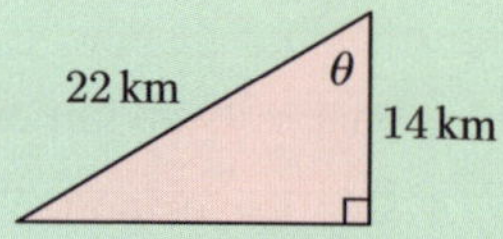

5 Find the value of a:

a

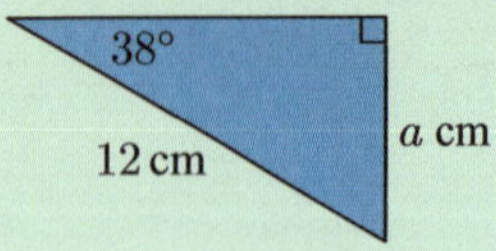

b

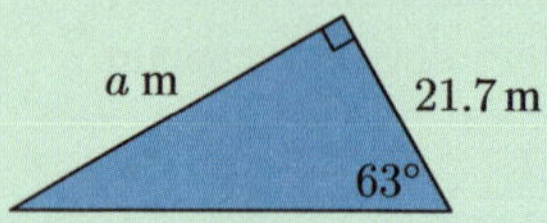

c

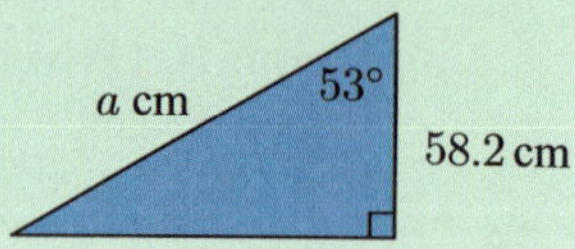

6 **a** Find the exact value of y.

b Name the side opposite θ.

c Find as a fraction:

i $\cos\theta$ **ii** $\sin\theta$ **iii** $\tan\theta$

d Find the unknown angles of the triangle.

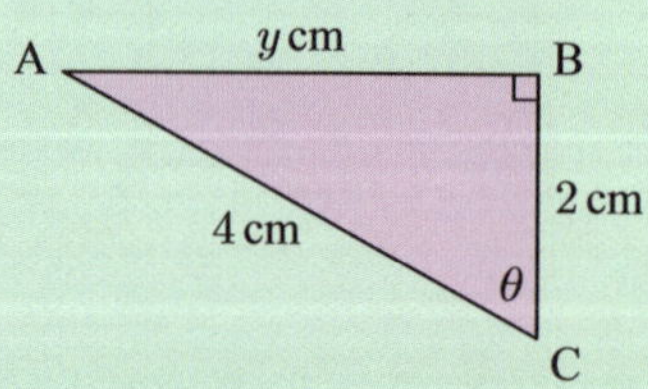

7 Find the pitch θ of the roof alongside.

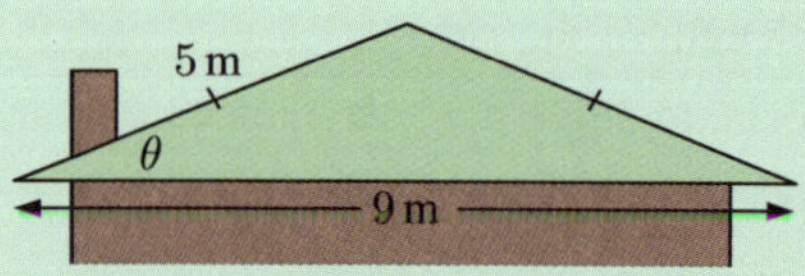

8 An isosceles triangle has equal sides of length 13 cm, and base angles of 40°. Find the length of the base.

9 An aeroplane takes off at a constant angle of 20°. When it has flown 1000 m, what is its altitude? Give your answer to the nearest metre.

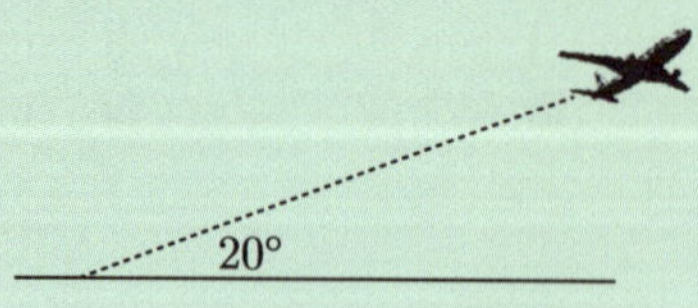

10 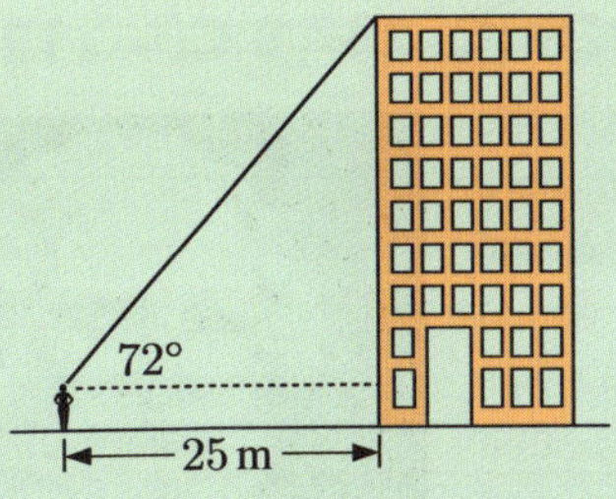

A young girl's eyes are one metre above ground level. She stands 25 m from the base of a tall building. She looks up to the top of the building at an angle of 72°. Find the height of the building.

11 A tree-feller notices that the shadow cast by a tree is 13.2 m when the angle of elevation of the sun is 42°.

The tree is 12 m from the house. If the tree is cut at ground level and falls directly towards the house, will it miss the house?

12 A motorcyclist travels 87 km west and then 63 km north.

a How far is she from her starting point?

b What is her bearing from her starting point?

13 The course for a yachting race is shown alongside. Each grid unit represents 1 km. Find the distance and true bearing from:

a the start to A

b A to B

c B to C

d C to D

e D to the finish.

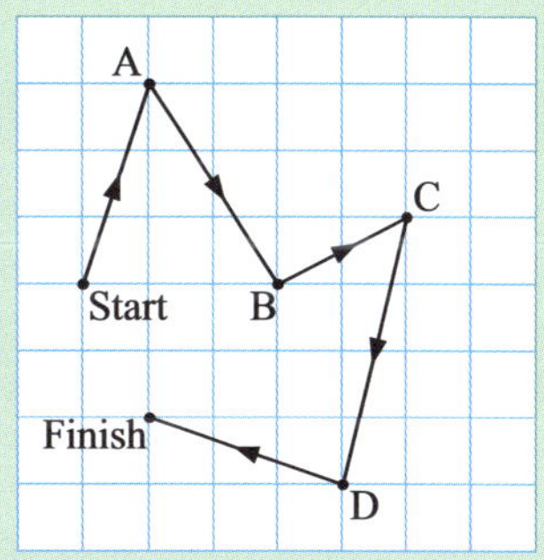

14 Find:

a the size of angle BFC

b the length of [DF]

c the angle between [AF] and [DF].

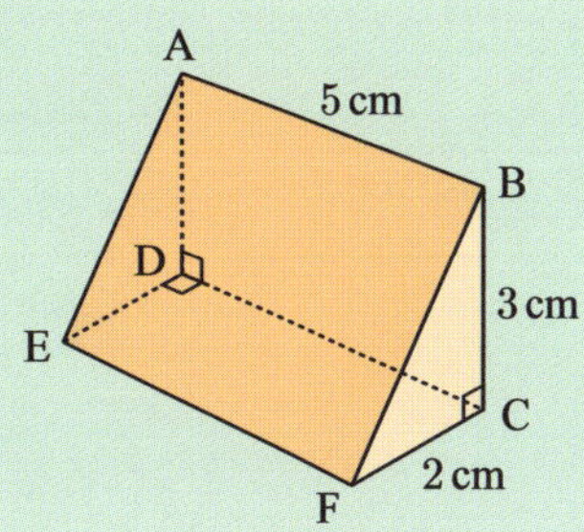

HISTORICAL NOTE ARCHIMEDES (C. 287 B.C. - 212 B.C.)

Archimedes was born about 287 B.C. in Syracuse in Sicily. He was the son of an astronomer, and studied in Alexandria in Egypt. From there Archimedes returned to Syracuse and concentrated on mathematical research. He invented a number of mechanical contraptions which made him famous, but the only one which he recorded in writing was the one that simulated the motion of the planets. There are stories that he designed convex lenses which focused the sun's heat and caused ships to burst into flames when they came within range of the city's walls. Archimedes also pioneered work in mechanics and invented a water screw which is still used for water irrigation in Egypt to this day.

Perhaps the most famous story concerning Archimedes is his discovery of the principle of buoyancy, called Archimedes principle. It is said that Archimedes was stepping into a bath and noticed the water overflowing when he realised that a body immersed in fluid loses as much in weight as the weight of the fluid it displaces. He is said to have run naked into the street shouting "Eureka" as he was so delighted with his discovery.

Archimedes was killed during the Roman capture of Syracuse in 212 B.C. and it is said that he was stabbed by a soldier as he was drawing a mathematical figure in the sand. In accordance with his wishes, his tomb was marked by a sphere inscribed in a cylinder. He is recognised as one of the greatest mathematicians of all time.

Chapter 14

Probability

Contents:

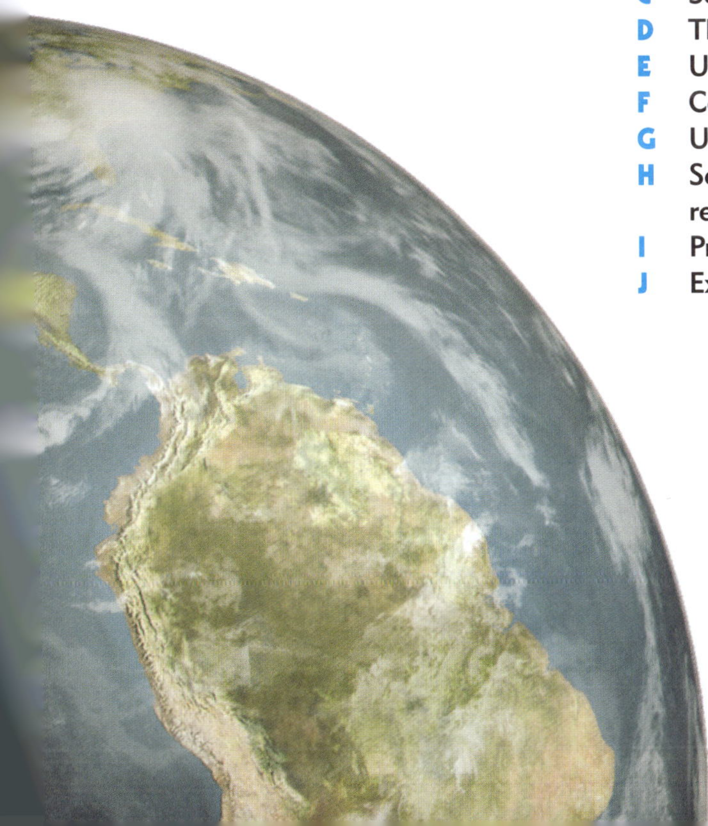

OPENING PROBLEM

Players A, B, C, and D are the final four left in a knock-out tennis tournament. In the semi-finals, A will play B, and C will play D. The winners of these matches will play in the grand final.

The table alongside shows the probabilities of each player beating the others. For example, the shaded square indicates that A has a 40% chance of beating B.

	B	C	D
A	40%	55%	47%
B		42%	54%
C			35%

Things to think about:

- **a** What is the probability that B will beat A?
- **b** How can we represent the possible outcomes of the finals series?
- **c** Which player is most likely to win the grand final?

The study of **probability** deals with the **chance** or **likelihood** of an event happening. For every event we can assign a number which lies between 0 and 1 inclusive.

> An **impossible** event has 0% chance of happening, and is assigned the probability 0.
>
> A **certain** event has 100% chance of happening, and is assigned the probability 1.
>
> All other events between these two extremes can be assigned a probability between 0 and 1.

The number line below shows how we could interpret different probabilities:

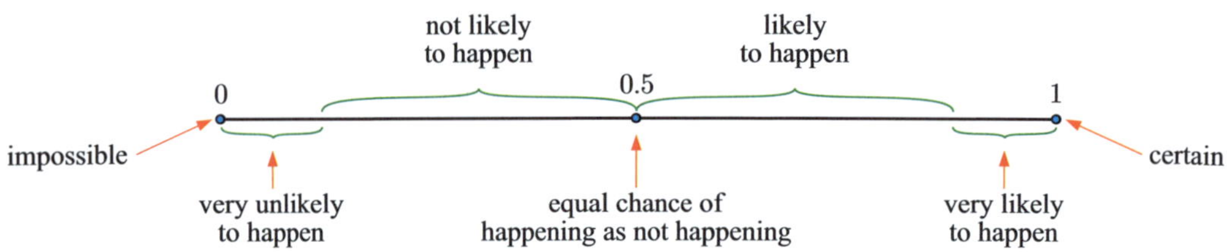

To each event we assign either:

- an **experimental probability** by observing the results of an experiment, or
- a **theoretical probability** by using arguments of symmetry.

The study of chance has important applications in physical and biological sciences, economics, politics, sport, life insurance, quality control, production planning, and many other areas.

Probability theory can be applied to card and dice games to try to increase our chances of success. It may therefore appear that an understanding of probability encourages gambling. However, a better knowledge of probability theory actually helps us to understand why the majority of habitual gamblers lose in the long term.

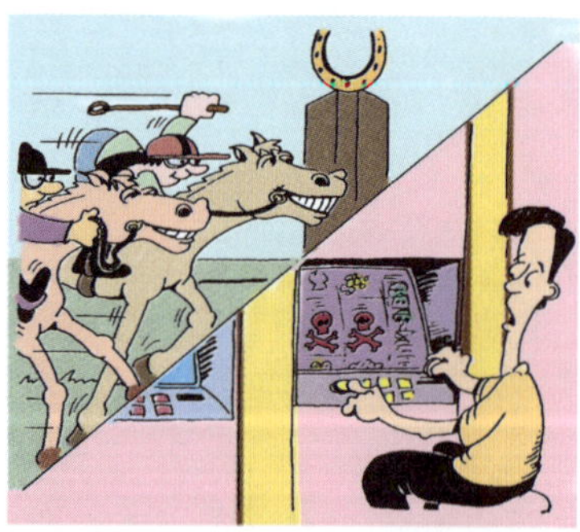

HISTORICAL NOTE

Chevalier de Méré was a French aristocrat and gambler in the 17th century. He wanted to know the answer to this question:

"Should I bet even money on the occurrence of at least one 'double six' when rolling a pair of dice 25 times?"

De Méré's experience of playing dice games convinced him that the answer was yes, but he did not know how to prove it. He therefore asked his friend, the French mathematician **Blaise Pascal**, for help.

In a series of letters between Pascal and fellow mathematician **Pierre de Fermat**, the problem was solved. In the process, they became interested in solving other questions of this kind, and together they laid the foundations of a new branch of mathematics called **theoretical probability**.

Pierre de Fermat

A EXPERIMENTAL PROBABILITY

In experiments involving chance, we use the following terms to describe what we are doing and the results we are obtaining.

- The **number of trials** is the total number of times the experiment is repeated.
- The **outcomes** are the different results possible for one trial of the experiment.
- The **frequency** of a particular outcome is the number of times that this outcome is observed.
- The **relative frequency** of an outcome is the frequency of that outcome expressed as a fraction or percentage of the total number of trials.

$$\textbf{relative frequency} = \frac{\textbf{frequency}}{\textbf{number of trials}}$$

For example, suppose a small plastic cone was tossed into the air 300 times. It fell on its *side* 203 times and on its *base* 97 times. We say that:

- the number of trials is 300
- the possible outcomes are *side* and *base*
- the frequency of *side* is 203, and the frequency of *base* is 97
- the relative frequency of *side* $= \frac{203}{300} \approx 0.677$
- the relative frequency of *base* $= \frac{97}{300} \approx 0.323$.

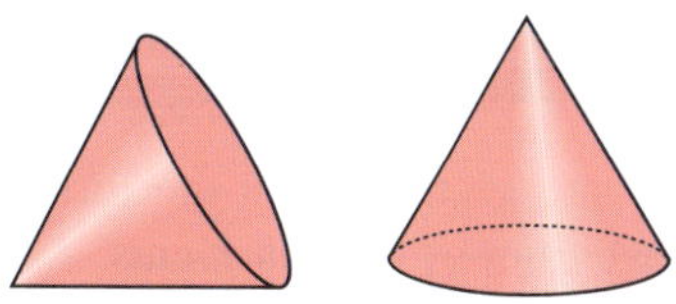

In the absence of any further data, the relative frequency of each outcome is our best estimate of the probability of it occurring.

The **experimental probability** is the **relative frequency** of the outcome.

We write P(*side*) ≈ 0.677, P(*base*) ≈ 0.323.

EXERCISE 14A

1 When a coin is tossed 100 times, it falls heads 47 times. Find the experimental probability that the coin falls:

a heads **b** tails.

2 A batch of 145 paper clips was dropped onto 6 cm by 6 cm squared paper. 113 clips fell completely inside squares, and 32 finished up on a grid line.
Estimate, to 2 decimal places, the probability of a clip falling:

a inside a square **b** on a grid line.

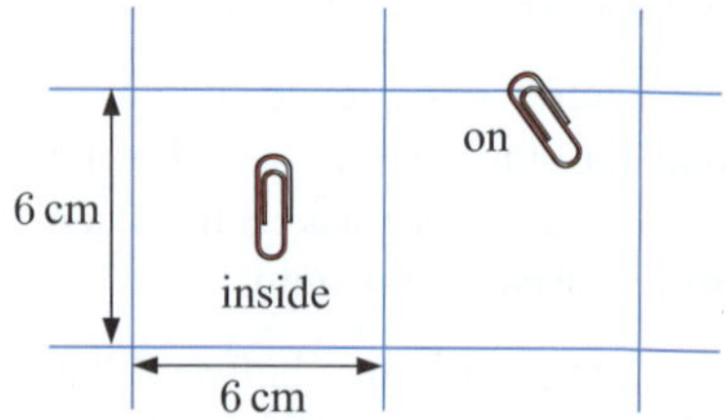

3 Connor visits the same bakery each day for lunch. On 13 out of his last 40 visits, the bakery had sold out of pies. Estimate the probability that the bakery will have sold out of pies on his next visit.

4 During one day at a carnival, 247 people played the coconut shy game. 175 people did not win a prize, 64 people won a minor prize, and the remainder won a major prize. Estimate the probability that the next player will win:

a a major prize **b** any prize.

HISTORICAL NOTE

In the late 17th century, English mathematicians compiled and analysed mortality tables. These tables showed the number of people that died at different ages. From these tables they could estimate the probability that a person would be alive at a future date. This led to the establishment of the first life insurance company in 1699.

B PROBABILITIES FROM TABLED DATA

If we are given data in the form of a table or graph, we can use relative frequencies to estimate the probabilities of various events.

Example 1 Self Tutor

George keeps a record of the number of apples he eats each day.

Estimate the probability that on a randomly selected day, George will eat:

a exactly 1 apple **b** at least 3 apples.

Number of apples	*Frequency*
0	10
1	17
2	12
3	3
4	1

George has recorded data for $10+17+12+3+1=43$ days.

a $P(1 \text{ apple}) \approx \dfrac{\text{frequency}}{\text{number of trials}} \approx \dfrac{17}{43} \approx 0.395$

b $P(\text{at least 3 apples}) \approx \dfrac{\text{frequency}}{\text{number of trials}}$

$\approx \dfrac{3+1}{43}$ {on 3 days he ate 3 apples, on 1 day he ate 4 apples}

≈ 0.093

EXERCISE 14B

1 Pat does a lot of travelling in her car, so she records how often she fills her car with petrol. The table alongside shows the frequencies of the *number of days between refills*.

Days between refills	*Frequency*
1	37
2	81
3	48
4	17
5	6
6	1

Estimate the likelihood that between refills there is:

a exactly a four day gap

b at least a four day gap.

2 José recorded the length of TV commercials in seconds. His results are summarised in the table.

Length	*Frequency*
$0 \leqslant t < 20$	17
$20 \leqslant t < 40$	38
$40 \leqslant t < 60$	19
$t \geqslant 60$	4

Estimate, to 3 decimal places, the probability that a randomly chosen TV commercial will last:

a between 20 and 40 seconds

b at least a minute

c between 20 seconds and a minute.

3

Hours slept	*Frequency*
$5 \leqslant h < 6$	7
$6 \leqslant h < 7$	29
$7 \leqslant h < 8$	46
$8 \leqslant h < 9$	39

This table shows how long Nathan has slept each night recently. Estimate the probability that tonight he will sleep for:

a between 6 and 7 hours

b at least 7 hours

c between 5 and 8 hours.

4 This dot plot shows how many times Kelly went jogging in each of the last 25 weeks.

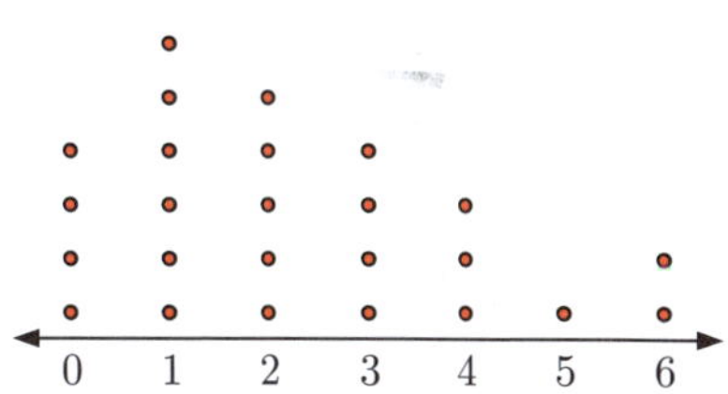

Estimate the probability that next week Kelly:

a will not go jogging

b will go jogging 3 times

c will go jogging at least twice.

5 Paula records the number of phone calls she receives over a period of consecutive days.

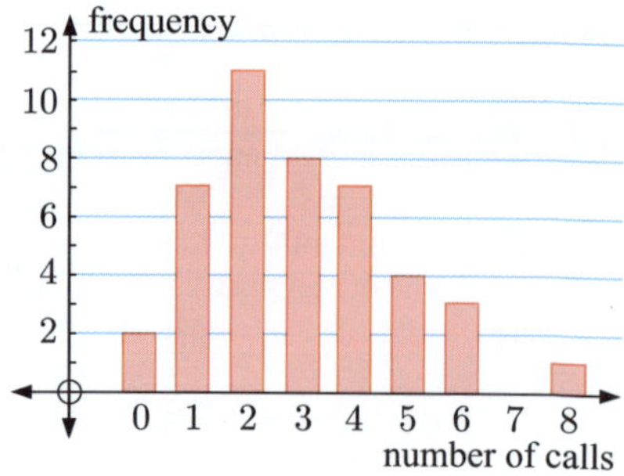

a For how many days did Paula's survey last?

b How many calls did Paula receive over this period?

c Estimate the chance that on a particular day, Paula will receive:

i no phone calls

ii 5 or more phone calls

iii less than 3 phone calls.

C SAMPLE SPACE

The **sample space** of an experiment is the set of its possible outcomes.

We can represent sample spaces in a number of ways, including:

- lists
- grids
- tree diagrams.

LISTING OUTCOMES

The simplest way to represent a sample space is to list the possible outcomes in set notation.

For example, the sample space for spinning this spinner is {red, green, yellow, blue}.

GRIDS

A **grid** or **array** is an efficient way to represent the sample space of an experiment involving two operations. Each point on the grid represents a possible outcome.

Example 2 Self Tutor

Use a 2-dimensional grid to illustrate the possible outcomes when tossing a coin and spinning the spinner shown.

Let H represent a 'head' and T represent a 'tail'.

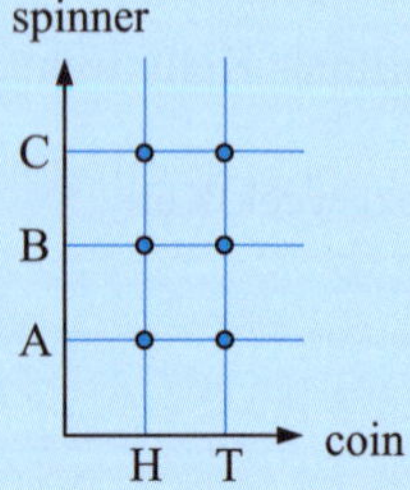

TREE DIAGRAMS

The sample space in **Example 2** can also be represented using a **tree diagram**.

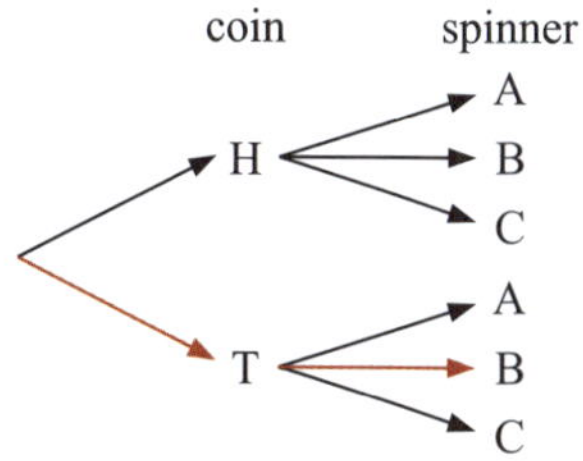

The first set of branches show the possible outcomes when tossing a coin. For each of these outcomes, there are three possible outcomes for spinning the spinner.

The path shown in red represents the outcome TB. The advantage of tree diagrams is that they can be used when more than two operations are involved.

Example 3 — Self Tutor

Illustrate, using a tree diagram, the possible outcomes when drawing three counters from a bag containing many red counters and blue counters.

Let R represent a red counter, and B represent a blue counter.

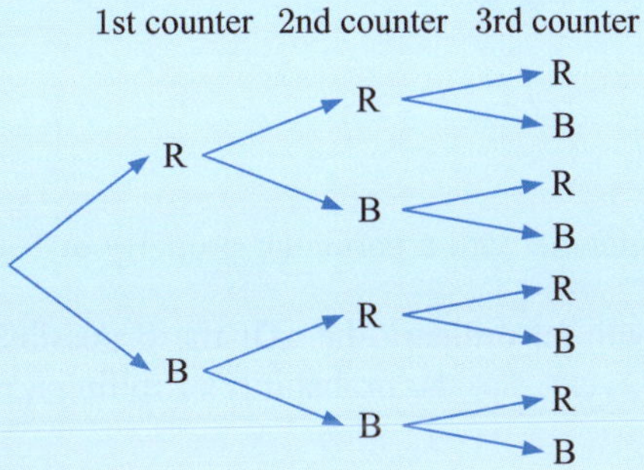

EXERCISE 14C

1 List, in set notation, the sample space for:

- **a** the genders of a 2-child family
- **b** the order in which 3 men can be lined up
- **c** tossing 4 coins simultaneously
- **d** the order in which 4 different rowing teams A, B, C, and D could finish a race.

2 Draw a two-dimensional grid to illustrate the sample space for:

- **a** rolling a die and tossing a coin simultaneously
- **b** rolling a pair of dice
- **c** twirling a square spinner marked A, B, C, D and a triangular spinner marked 1, 2, 3
- **d** selecting one even number and one odd number from the integers 1 to 10.

3 Use a tree diagram to illustrate the sample space for:

- **a** tossing a 10-cent coin and a 20-cent coin simultaneously
- **b** rolling a 6-sided die and tossing a coin simultaneously
- **c** the genders of a 3-child family.

4 Bag A contains red and white marbles and bag B contains blue and yellow marbles. A bag is selected and one marble is taken from it. Draw a tree diagram to illustrate the sample space.

5 Hats A, B, and C each contain pink and white tickets. A hat is selected and then two tickets are taken from it. Draw a tree diagram to illustrate the sample space.

D THEORETICAL PROBABILITY

The sample space when rolling a single die is $\{1, 2, 3, 4, 5, 6\}$.

Since the die is a cube and therefore symmetrical, we expect that each of the six outcomes will be **equally likely** to occur. We say that the **theoretical probability** of any given outcome occurring is 1 in 6, or $\frac{1}{6}$.

For example, $\text{P(rolling a 2)} = \frac{1}{6}$.

If a sample space has n outcomes which are **equally likely** to occur when the experiment is performed once, then each outcome has probability $\frac{1}{n}$ of occurring.

EVENTS

An **event** occurs when we obtain an outcome with a particular property or feature.

Consider the event of *rolling a prime number* with an ordinary die. Of the 6 possible outcomes, the three outcomes 2, 3, and 5 all correspond to this event. So, the probability of rolling a prime number is 3 in 6, or $\frac{3}{6}$.

When the outcomes of an experiment are equally likely, the probability of an event E occurring is given by:

$$\mathbf{P}(E) = \frac{\textbf{number of outcomes corresponding to } E}{\textbf{number of outcomes in the sample space}}$$

Example 4 Self Tutor

A small child has a collection of shapes.

a She chooses one of the shapes at random. Find the probability that it is a triangle.

b The child collects all the circles then chooses one of them at random to give to her mother. Find the probability that the circle she chooses is green.

a $\text{P(triangle)} = \frac{5}{17}$ ← There are 5 triangles. ← There are 17 shapes to choose from.

b $\text{P(the circle is green)} = \frac{2}{7}$ ← 2 of the circles are green. ← There are 7 circles to choose from.

The **complement** of event E is E', and is the event that E does *not* occur.

For any event E, $P(E') = 1 - P(E)$.

EXERCISE 14D

1 A spinner with the numbers 1 to 5 written on equal sectors is spun once. Find the probability of spinning:

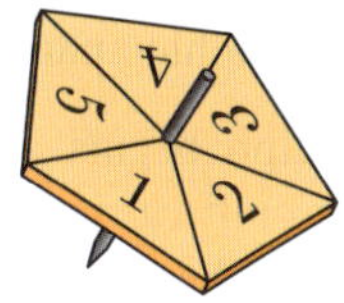

a a 4 b a 1 or a 2 c an odd number.

2 A symmetrical octahedral die has the numbers 1 to 8 marked on its faces. If it is rolled once, determine the probability of getting:

a a 4 b a number less than 5 c a number greater than 8.

3 A bag contains 4 red and 3 green buttons. One button is randomly selected from the bag. Find the probability that the button is:

a red b green c red or green.

4 

A 52 card pack is well shuffled, then one card is dealt from the top of the pack. Find the probability that the card is:

a a Jack b a black card

c a diamond d a diamond or an ace.

5 Find the probability that a person randomly selected in the street has his or her birthday in:

a May b February.

6 Candice has baked 6 blueberry muffins and 12 raspberry muffins. She chooses one to eat at random. Let E be the event that the muffin is blueberry.

a Explain the meaning of the event E'. b Find $P(E)$. c Find $P(E')$.

7 37 people applied for a job at a toy store. 11 were selected for an interview, and from these people, 3 were offered a position at the store.

a Find the probability that a randomly selected applicant was:

i offered a position

ii interviewed, but not offered a position

iii not selected for an interview.

b Find the sum of the probabilities in a. Explain your answer.

c One of the applicants who was interviewed is selected at random. Find the probability that he or she was offered a position.

8 Amy is invited to select a treat from a bag containing 6 strawberry lollies, 3 mints, and 5 chocolates.

a If Amy selects a treat at random, find the probability that she selects:

i a mint ii a mint or a chocolate.

b Amy selects a chocolate, eats it, then passes the bag to Beth. If Beth selects a treat at random, find the probability that she selects:

i a mint **ii** a chocolate.

9 There are 5 different pairs of socks in Vanessa's sock drawer. The light in her room is not working, so she removes 2 socks from her drawer at random. Find the probability that Vanessa has removed a matching pair.

10 Ralph begins a game of 'Battleship' by placing an aircraft carrier on a 10×10 grid. The aircraft carrier must be placed horizontally or vertically, occupying 5 squares. One possible position is shown alongside.

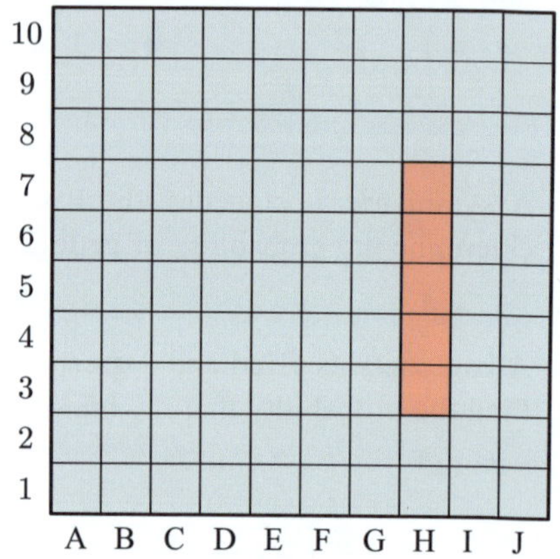

a In how many different positions can the aircraft carrier be placed?

b Ralph selects one of these positions at random. Find the probability that the aircraft carrier:

i is placed horizontally

ii occupies at least one edge or corner square

iii occupies the square B3

iv occupies the square E5.

11 Sean is playing a board game. To win, he needs to roll 10 or more with a pair of dice on his next turn. There are 11 possible outcomes (2, 3, 4, 5, 6, 7, 8, 9, 10, 11, 12), and 3 of these outcomes (10, 11, 12) will result in him winning the game. Sean therefore says the probability that he will win the game is $\frac{3}{11}$. Explain the flaw in Sean's reasoning.

12 A target consists of two concentric circles. The larger circle has diameter 1 m. When a complete novice fires at the target, the probability of hitting the blue circle is the same as the probability of hitting the green region. Find the diameter of the blue circle.

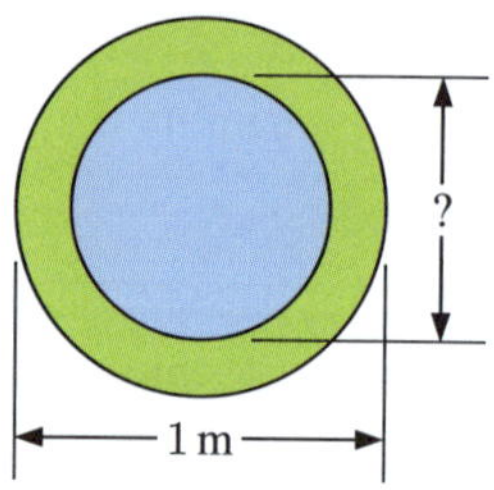

E USING 2-DIMENSIONAL GRIDS

If there are two components to an experiment and the sample space is not too large, a 2-dimensional grid is useful for illustrating the possible outcomes and hence calculating probabilities. Each point on the grid represents a possible outcome, and each outcome is equally likely to occur.

Example 5 Self Tutor

Two square spinners, each with 1, 2, 3, and 4 on their edges, are twirled simultaneously.

a Draw a two-dimensional grid of the possible outcomes.

b Use your grid to determine the probability of getting:

i a 3 with each spinner **ii** a 3 and a 1 **iii** an even result with each spinner.

a

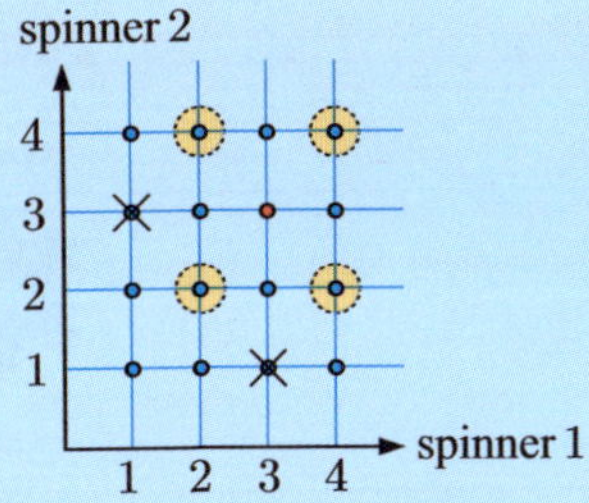

b The sample space has 16 members.

i P(a 3 with each spinner) $= \frac{1}{16}$ {•}

ii P(a 3 and a 1) $= \frac{2}{16} = \frac{1}{8}$ {crossed points}

iii P(an even result with each spinner)

$= \frac{4}{16} = \frac{1}{4}$ {circled points}

EXERCISE 14E

1 Two coins are tossed simultaneously.

a Use a 2-dimensional grid to illustrate the sample space.

b Hence, find the probability of getting:

i two tails **ii** a head and a tail **iii** at least one tail.

2 A coin and a die are tossed simultaneously.

a Draw a 2-dimensional grid to illustrate the sample space.

b Hence, find the probability of getting:

i a tail and a 6 **ii** a tail or a 6

iii neither a 2 nor a 6 **iv** neither a tail nor a 5

v a head and an odd number **vi** a head or an odd number.

3 An ordinary die is tossed and a triangular spinner labelled A, B, and C is spun simultaneously.

a Draw a 2-dimensional grid to illustrate the sample space.

b Hence, find the probability of getting:

i B and 5 **ii** A and a prime number **iii** A or C, and a multiple of 3.

4 Suppose that a pair of dice is rolled. The 36 different possible results are illustrated on the 2-dimensional grid. Use the grid to determine the probability of getting:

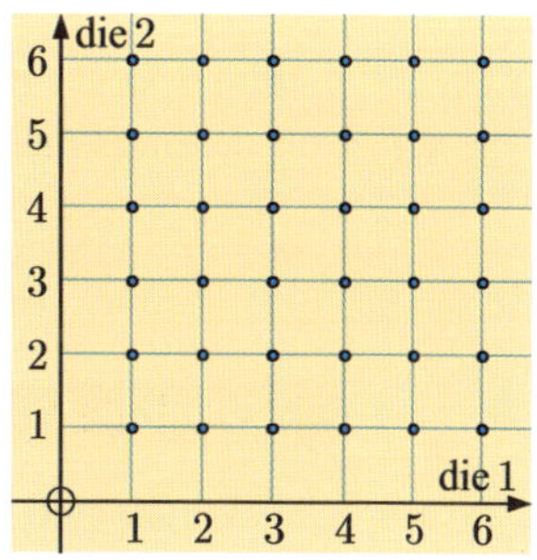

a two 3s **b** a 5 and a 6

c a 5 or a 6 **d** at least one 6

e exactly one 6 **f** no sixes

g a sum of 7 **h** a sum of 7 or 11

i a sum greater than 8 **j** a sum of no more than 8.

5 Two students each select a day of the week at random. Find the probability that:

a a Tuesday and a Friday are selected

b at least one student selects a weekend day

c the selected days start with the same letter

d between them, the selected days contain at least 15 letters.

6 In a role-playing game, Melanie and Neil duel by each rolling an eight-sided die. Neil's character has a high level, so to win the duel Melanie must roll a number at least 2 greater than Neil's roll. Find the probability that Melanie will win the duel.

ACTIVITY

Click on the icon to play a game involving 2-dimensional grids.

GAME

F COMPOUND EVENTS

We will now look more closely at calculating probabilities for **combined events**, which are also called **compound events**.

Consider tossing a coin and rolling a die simultaneously.

We have seen that to determine the probability of getting a head and a 5, we can illustrate the sample space on the two-dimensional grid shown. There are 12 possible outcomes, but only 1 with the property that we want.

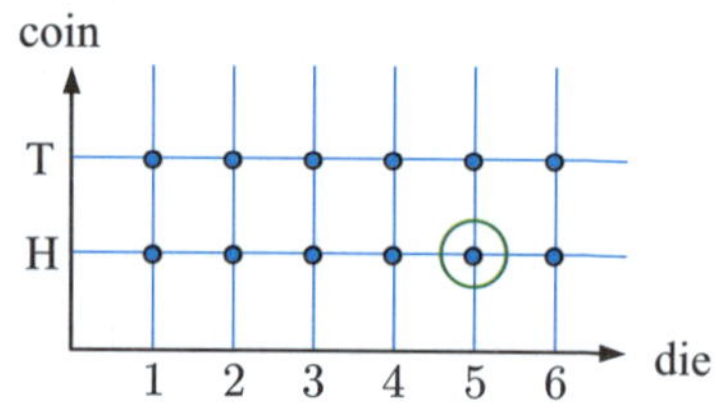

$\therefore$ P(a head and a 5) $= \frac{1}{12}$

However, notice that P(a head) $= \frac{1}{2}$ and P(a 5) $= \frac{1}{6}$.

$\therefore$ P(a head **and** a 5) = P(a head) $\times$ P(a 5).

To find the combined probability, we can **multiply** the separate probabilities.

INDEPENDENT EVENTS

Two events are **independent** if the occurrence of each event does not affect the occurrence of the other.

The two events 'getting a head' and 'rolling a 5' are independent events, since the outcome of the coin toss has no effect on the outcome from the dice roll.

If two events A and B are **independent** then P(A **and** B) = P(A) $\times$ P(B).

Example 6 Self Tutor

A coin and a die are tossed simultaneously. Determine the probability of getting a head and a 3.

The events are independent, since the outcome from the coin does not affect the outcome from the die, and vice versa.

$\therefore$ P(a head and a 3) = P(a head) $\times$ P(a 3)

$= \frac{1}{2} \times \frac{1}{6}$

$= \frac{1}{12}$

EXERCISE 14F.1

1 The spinners alongside are each spun once. Find the probability that each spinner will land on green.

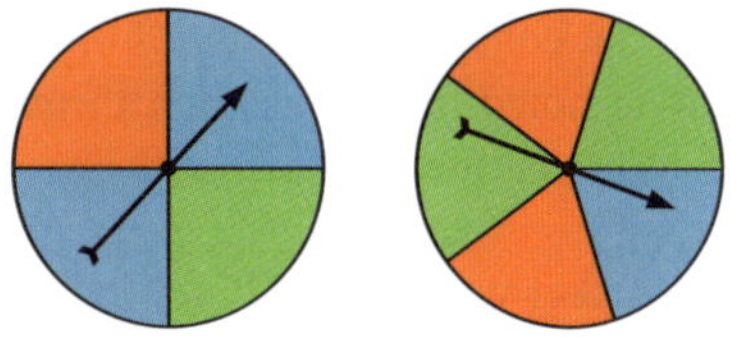

2 A coin and a die are tossed simultaneously. Find the probability of getting a tail and a number larger than 2.

3 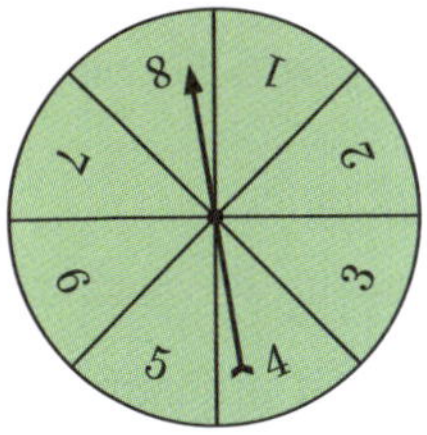

Tony and Ben each spin the spinner alongside. Find the probability that Tony spins a prime number, and Ben spins a multiple of 3.

Example 7 — Self Tutor

Sunil has probability $\frac{4}{5}$ of hitting a target and Monika has probability $\frac{5}{6}$.
If they both fire simultaneously at the target, determine the probability that:

a they both hit

b they both miss.

Let S be the event of Sunil hitting the target and M be the event of Monika hitting it.

a $\text{P(both hit)} = \text{P}(S \text{ and } M)$
$= \text{P}(S) \times \text{P}(M)$
$= \frac{4}{5} \times \frac{5}{6}$
$= \frac{2}{3}$

b $\text{P(both miss)} = \text{P}(S' \text{ and } M')$
$= \text{P}(S') \times \text{P}(M')$
$= \frac{1}{5} \times \frac{1}{6}$
$= \frac{1}{30}$

4 Golfers George and Brett each have a putt 5 metres from the hole. George has probability 0.6 of holing the putt, and Brett has probability 0.3 of holing the putt. Find the probability that:

a they both hole the putt

b they both miss the putt.

5 On any given day at a car dealership, Bronwyn has probability 0.2 of selling a car, and Dwayne has probability 0.15 of selling a car. Find the probability that, on a given day:

a they both sell a car

b neither of them sell a car

c Bronwyn sells a car but Dwayne does not.

6 Each time Morris the dog gets thrown a frisbee, he has probability 0.7 of catching it. If a frisbee is thrown to Morris three times, find the probability that Morris:

a catches it all three times

b misses the first throw, but catches the next two

c catches the first throw, but misses the next two.

DEPENDENT EVENTS

Suppose a hat contains 5 red and 3 blue tickets. One ticket is randomly chosen, its colour is noted, and it is then thrown away. A second ticket is randomly selected. What is the chance that the second ticket is red?

If the first ticket was red, P(second ticket is red) $= \frac{4}{7}$ ← 4 reds remaining / ← 7 to choose from

If the first ticket was blue, P(second ticket is red) $= \frac{5}{7}$ ← 5 reds remaining / ← 7 to choose from

So, the event of the second ticket being red *depends* on what colour the first ticket was.

Two events are **dependent** if they are **not independent**.

The occurrence of one of the events *does* affect the occurrence of the other event.

The rule for finding compound event probabilities for dependent events is different from the rule for independent events.

If A and B are dependent events, then P(A and B) = P(A) × P(B given that A has occurred).

Example 8 Self Tutor

A box contains 4 red and 2 yellow tickets. Two tickets are randomly selected one after the other from the box, the first being removed before the second ticket is chosen. Find the probability that:

- **a** both tickets are red
- **b** the first ticket is red and the second ticket is yellow.

a P(both tickets are red)

= P(first ticket is red and the second ticket is red)

= P(first ticket is red) × P(second ticket is red given that the first ticket is red)

$= \frac{4}{6} \times \frac{3}{5}$ ← 3 red tickets remaining / ← 5 tickets to choose from

$= \frac{12}{30}$

$= \frac{2}{5}$

b P(first ticket is red and the second ticket is yellow)

= P(first ticket is red) × P(second ticket is yellow given that the first ticket is red)

$= \frac{4}{6} \times \frac{2}{5}$ ← 2 yellow tickets remaining / ← 5 tickets to choose from

$= \frac{8}{30}$

$= \frac{4}{15}$

EXERCISE 14F.2

1 A box contains 7 red and 3 green balls. Two balls are randomly selected from the box. The first is *not* replaced in the box before the second is selected. Determine the probability that:

- **a** both balls are red
- **b** the first ball is green and the second is red.

2 Amelie has a bag containing 4 red apples and 6 green ones. She selects one apple at random, eats it, and then takes another, also at random.

a Determine the probability that:

i both apples were red

ii both apples were green

iii the first was red and the second was green

iv the first was green and the second was red.

b Find the sum of the probabilities in **a**. Explain your answer.

3 Marjut has a carton containing 10 cans of soup. 4 cans are tomato and the rest are pumpkin. She selects two cans at random without looking at the labels.

a Find the probability that both cans are:

i tomato soup **ii** pumpkin soup.

b *Hence*, find the probability that Marjut selects one can of each flavour.

4 A lottery has 100 tickets which are placed in a barrel. Two tickets are drawn at random from the barrel to decide the prizes. John has two tickets in the lottery. Determine his probability of winning:

a first and second prize **b** second prize but not first prize.

c none of the prizes.

5 When Jamie plays football, he kicks with his right foot 80% of the time, and his left foot 20% of the time. His right foot kicks are accurate 70% of the time, and his left foot kicks are accurate 40% of the time. Find the probability that Jamie's next kick will be:

a with his right foot and accurate

b with his left foot and inaccurate.

Example 9 Self Tutor

A hat contains 20 tickets with the numbers 1 to 20 printed on them.
If three tickets are drawn from the hat without replacement, determine the probability that they are all prime numbers.

There are 20 numbers, of which 8 are prime. $\{2, 3, 5, 7, 11, 13, 17, 19\}$

$\therefore$ P(3 primes)

= P(1st is prime *and* 2nd is prime *and* 3rd is prime)

$= \frac{8}{20} \times \frac{7}{19} \times \frac{6}{18}$

- 8 primes out of 20 numbers
- 7 primes out of 19 numbers after a successful first draw
- 6 primes out of 18 numbers after two successful draws

$= \frac{14}{285}$

6 A bin contains 12 identically shaped chocolates, 8 of which are strawberry creams. If three chocolates are selected at random from the bin, find the probability that:

- **a** they are all strawberry creams
- **b** none of them are strawberry creams.

7 A bag contains three white and five red marbles. Three marbles are selected simultaneously. Determine the probability that they are all:

- **a** red
- **b** white.

8 A box contains 15 tickets with the numbers 1 to 15 printed on them. If three tickets are drawn from the box without replacement, find the probability that:

- **a** they are all odd
- **b** they are all even
- **c** the first two are odd and the third is even.

G USING TREE DIAGRAMS

We have seen that tree diagrams can be used to illustrate sample spaces and determine probabilities.

Consider the archers Sunil and Monika in **Example 7**. The tree diagram for this information is:

S means Sunil hits the target.
M means Monika hits the target.

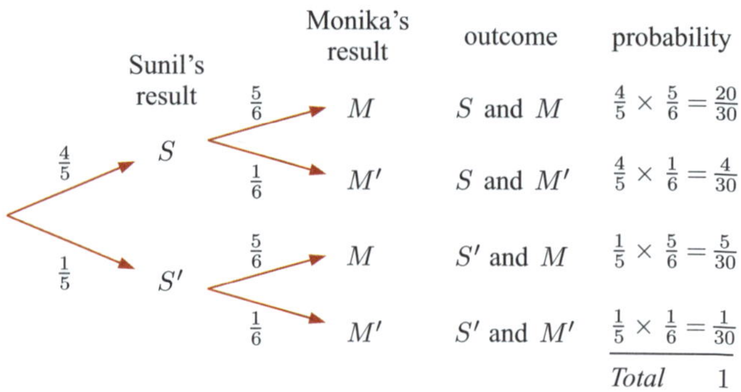

Notice that:

- The probabilities for hitting and missing are marked on the branches.
- There are *four* alternative paths, and each path shows a particular outcome.
- All outcomes are represented.
- The probability of each outcome is obtained by **multiplying** the probabilities along its path.

Example 10 — Self Tutor

Stephano is having computer problems. His desktop computer will only boot up 90% of the time, and his laptop will only boot up 70% of the time. Stephano attempts to boot both machines.

- **a** Draw a tree diagram to illustrate this situation.
- **b** Use the tree diagram to determine the chance that:
  - **i** both will boot up
  - **ii** only the desktop computer boots up.

a $D =$ desktop computer boots up
$L =$ laptop boots up

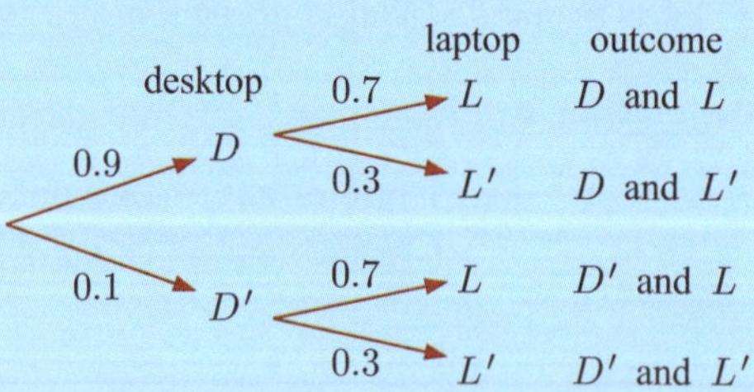

b **i** P(both boot up)
$= \text{P}(D \text{ and } L)$
$= \text{P}(D) \times \text{P}(L)$
$= 0.9 \times 0.7$
$= 0.63$

ii P(desktop boots up but laptop does not)
$= \text{P}(D \text{ and } L')$
$= \text{P}(D) \times \text{P}(L')$
$= 0.9 \times 0.3$
$= 0.27$

If two or more branches meet the description of the event, the probability for each branch is found and the results are **added**.

Example 11 **Self Tutor**

Bag A contains 4 red jelly beans and 1 yellow jelly bean. Bag B contains 2 red and 3 yellow jelly beans. A bag is randomly selected by tossing a coin, and one jelly bean is removed from it. Determine the probability that it is yellow.

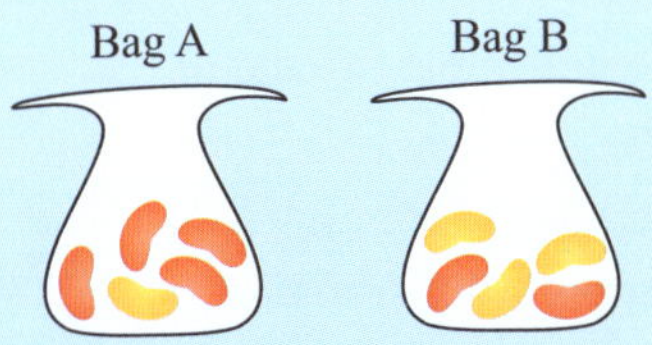

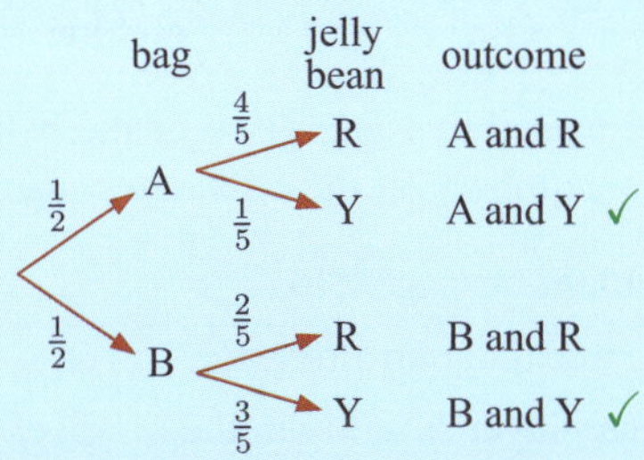

$$\begin{aligned}\text{P(yellow)} &= \text{P(A and Y)} + \text{P(B and Y)} \\ &= \tfrac{1}{2} \times \tfrac{1}{5} + \tfrac{1}{2} \times \tfrac{3}{5} \quad \{\text{branches marked } \checkmark\} \\ &= \tfrac{4}{10} \\ &= \tfrac{2}{5}\end{aligned}$$

EXERCISE 14G

1 Suppose this spinner is spun twice.

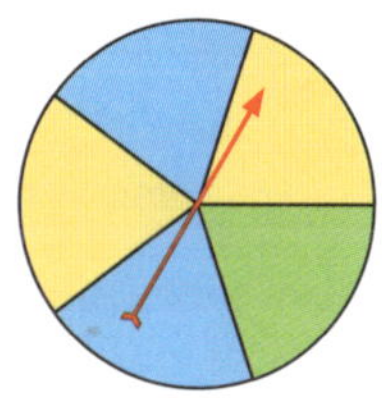

a Draw a tree diagram to illustrate the sample space.

b Determine the probability that:

i blue appears on both spins

ii green appears on both spins

iii different colours appear on the two spins

iv blue appears on *either* spin.

2 The probability of the race track being muddy next week is estimated to be $\frac{1}{4}$. If it is muddy, the horse Rising Tide will start favourite with probability $\frac{2}{5}$ of winning. If it is dry, Rising Tide has a $\frac{1}{20}$ chance of winning.

a Display the sample space of possible results on a tree diagram.

b Determine the probability that Rising Tide will win next week.

3 Tennis star Boris gets his first serve in 72% of the time. If he gets his first serve in, he wins the point 85% of the time. If not, he only wins the point 50% of the time.
Find the probability that Boris will win the next point he serves.

4 Machine A cans 60% of the fruit at a factory. Machine B cans the rest. Machine A spoils 3% of its product, while Machine B spoils 4%. Determine the probability that the next can inspected at this factory will be spoiled.

5 Box A contains 2 black and 3 white blocks. Box B contains 5 black and 1 white block. A box is chosen by the flip of a coin, and one block is taken at random from that box. Determine the probability that the block is white.

6 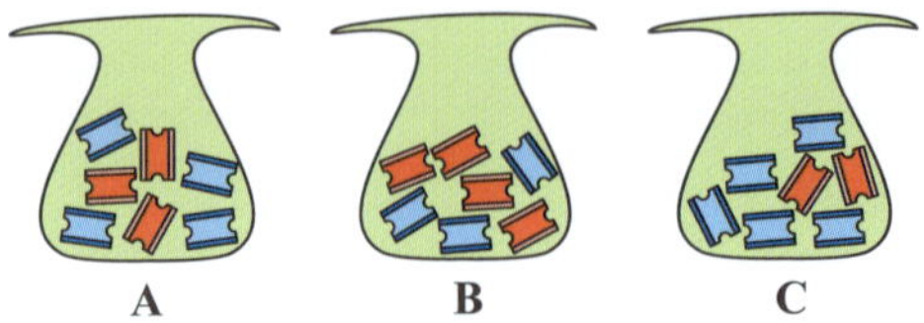

Three bags contain different numbers of blue and red tickets. A bag is selected using a die which has three A, two B, and one C face.
One ticket is selected at random from the chosen bag. Determine the probability that it is:

a blue **b** red.

7 Balls numbered 1 to 10 are placed in a bag. Two of the balls are drawn out at random.
Find the probability that the numbers on the balls are consecutive.

8 Consider the **Opening Problem** on page **276**.

a Copy and complete the tree diagram alongside.

b Find the probability that player A will win the grand final.

c Who is most likely to win the grand final?

d Suppose that instead, A played C and B played D in the semi-finals. Who would now be most likely to win the grand final?

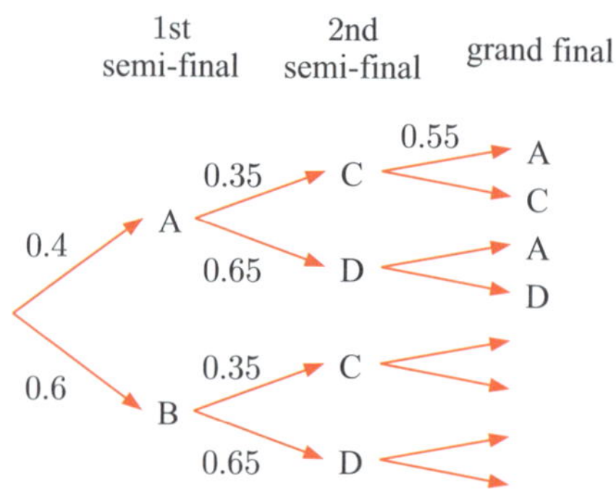

INVESTIGATION THE FINALS SERIES

In many sports competitions, the final is played in a best of 3, 5, or 7 game series, rather than a single match.

If we know each team's probability of winning a single match, we can use a tree diagram to determine each team's probability of winning the series.

What to do:

Suppose team A and team B are to play off in the final. Team A is the better team, and has a 60% chance of winning any given game against team B.

1 If the final consists of a single game, determine the probability that the final will be won by:

a team A **b** team B.

2 Suppose the final consists of a best of 3 game series. The series will continue until a team has won 2 games. The tree diagram below displays the possible outcomes.

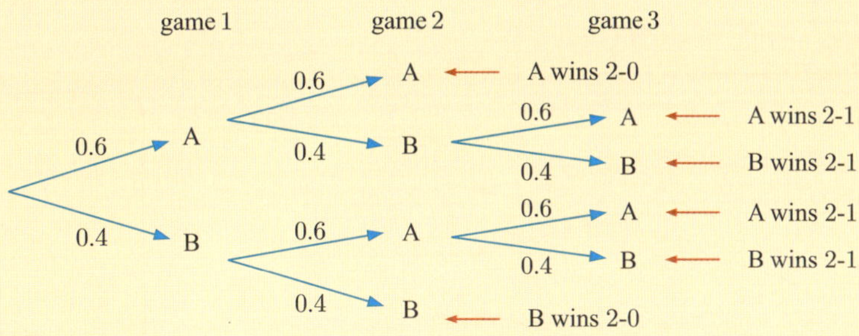

Determine the probability that the finals series will be won by:

a team A **b** team B.

3 Now suppose the final consists of a best of 5 game series. The series will continue until a team has won 3 games.

a Draw a tree diagram to display the possible outcomes.

b Determine the probability that the finals series will be won by:

i team A **ii** team B.

c Determine the probability that the finals series will last:

i 3 games **ii** 4 games **iii** 5 games.

4 Use a tree diagram to determine each team's probability of winning a best of 7 game series.

5 As the series gets longer, would you expect the better team to be more likely or less likely to win the series? Do your results in the Investigation support this?

H SAMPLING WITH AND WITHOUT REPLACEMENT

Sampling is the process of selecting one object from a large group and inspecting it for some particular feature. The object is then either:

- put back, which we call **sampling with replacement**, or
- put to one side, which we call **sampling without replacement**.

Sometimes the inspection process makes it impossible to return the object to the large group. In these cases we have no option but to sample without replacement.

For example:

- Is the chocolate hard or soft-centred? Bite it or squeeze it to see.
- Does the egg contain one or two yolks? Break it open and see.
- Is the object correctly made? Pull it apart to see.

Example 12 **Self Tutor**

A bin contains 4 blue and 5 green marbles. A marble is selected at random from the bin, its colour is noted, and it is then *replaced*. A second marble is then drawn and its colour is noted. Determine the probability that:

a both marbles are blue **b** the first marble is blue and the second is green

c there is one marble of each colour.

Tree diagram:

B = blue

G = green

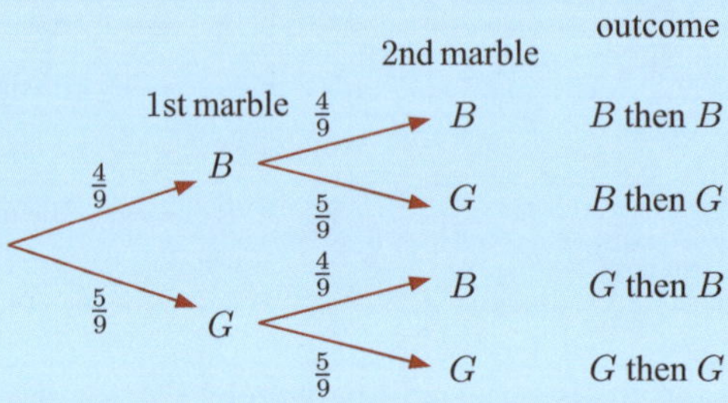

a P(both blue)
$= \text{P}(B \text{ then } B)$
$= \frac{4}{9} \times \frac{4}{9}$
$= \frac{16}{81}$

b P(first is blue and second is green)
$= \text{P}(B \text{ then } G)$
$= \frac{4}{9} \times \frac{5}{9}$
$= \frac{20}{81}$

c P(one of each colour)
$= \text{P}(B \text{ then } G \text{ or } G \text{ then } B)$
$= \text{P}(B \text{ then } G) + \text{P}(G \text{ then } B)$
$= \frac{4}{9} \times \frac{5}{9} + \frac{5}{9} \times \frac{4}{9}$
$= \frac{40}{81}$

EXERCISE 14H

1 A box contains 6 red and 3 yellow tickets. Two tickets are drawn at random, the first being *replaced* before the second is drawn.

a Draw a tree diagram to represent the sample space.

b Hence, determine the probability that:

i both tickets are red **ii** both tickets are yellow

iii the first ticket is red and the second is yellow

iv one ticket is red and the other is yellow.

2 7 tickets numbered 1, 2, 3, 4, 5, 6, and 7 are placed in a hat. Two of the tickets are taken from the hat at random *without replacement*. Determine the probability that:

a both are odd **b** both are even

c the first is even and the second is odd **d** one is even and the other is odd.

3 Jessica has a bag of 9 acid drops which are identical in shape. 5 are raspberry flavoured and 4 are orange flavoured. She selects one acid drop at random, eats it, and then takes another, also at random.

a Determine the probability that:

i both acid drops were orange flavoured **ii** both were raspberry flavoured iii the first was raspberry and the second was orange

iv the first was orange and the second was raspberry.

b Explain why the sum of your answers in **a** is 1.

4 A cook selects an egg at random from a carton containing 7 ordinary eggs and 5 double-yolk eggs. She cracks the egg into a bowl and sees whether it has two yolks or not. She then selects another egg at random from the carton and checks it.

Let S represent a single-yolk egg, and D represent a double-yolk egg.

a Draw a tree diagram to illustrate this process.

b Find the probability that both eggs had:

i two yolks ii only one yolk.

5 In a particular board game there are nine tiles. Five tiles are green, and the remainder are brown. The tiles start face down on the table so they all look the same.

a A player picks a tile at random. Determine the probability that it is:

i green ii brown.

b Suppose a player has to pick two tiles, replacing the first and shuffling them all before the second is selected. Use a tree diagram to help determine the probability that:

i both tiles are green ii both tiles are brown

iii tile 1 is brown and tile 2 is green iv one tile is brown and the other is green.

6 A sporting club runs a raffle in which 200 tickets are sold. There are two winning tickets which are drawn at random, in succession, without replacement.

Adam buys 8 tickets in the raffle. Determine the probability that he:

a wins first prize **b** does not win first prize

c wins both prizes **d** wins neither prize

e wins second prize *given that* he does not win first prize.

7 In the manufacturing of transformers for electric lights, there is a 3% chance that any transformer made is faulty. Benny leaves for a job installing 2 new lights. He carries 3 transformers in his van.

a When selecting a transformer, is this sampling with or without replacement?

b Draw a tree diagram to illustrate the sampling process.

c Find the probability that Benny will not be able to finish the job with the transformers in his van.

I PROBABILITIES FROM VENN DIAGRAMS

In **Chapter 3** we studied **Venn diagrams**. We saw that they consist of a rectangle which represents the universal set, and circles within it which represent subsets. In probability questions, the circles in a Venn diagram are used to represent particular events.

Example 13 — Self Tutor

A wheel with sectors numbered 1 to 20 is spun once. Let A be the event of spinning an odd number, and B be the event of spinning a number greater than 13.

a Represent A and B on a Venn diagram.

b Find the probability that the number spun is odd *and* greater than 13.

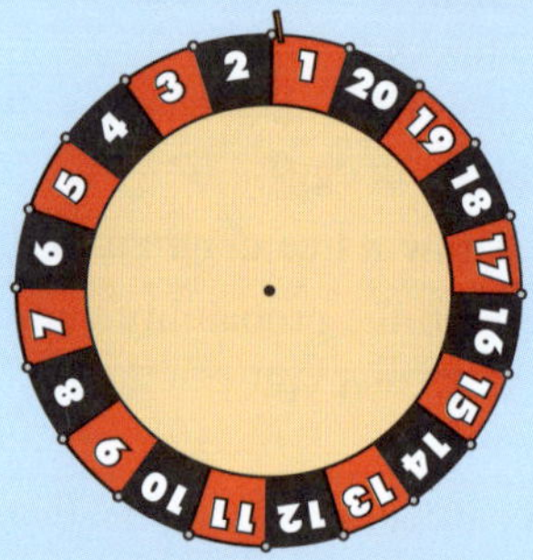

a $A = \{1, 3, 5, 7, 9, 11, 13, 15, 17, 19\}$,
$B = \{14, 15, 16, 17, 18, 19, 20\}$
$A \cap B = \{15, 17, 19\}$

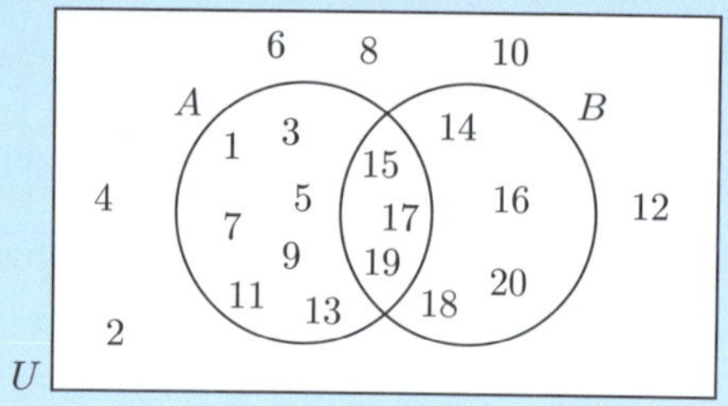

b 3 of the 20 elements are in $A \cap B$
$\therefore$ P(odd *and* greater than 13) $= \frac{3}{20}$

Example 14 — Self Tutor

In a class of 30 students, 19 study Physics, 17 study Chemistry, and 15 study both of these subjects.

a Display this information on a Venn diagram.

b Determine the probability that a randomly selected student studies exactly one of the subjects.

c Given that a student does not study Physics, determine the probability that the student studies Chemistry.

a

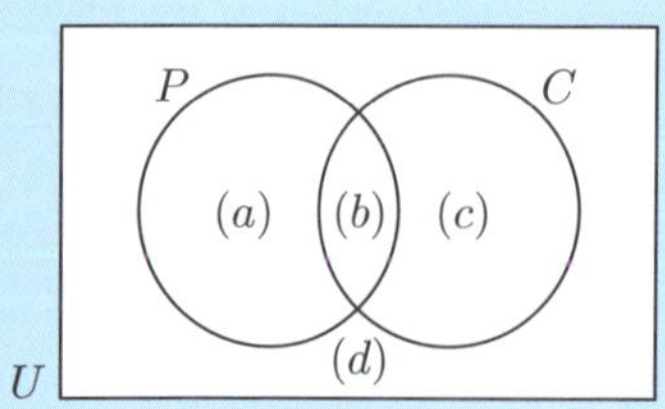

Let P represent students studying Physics and C represent students studying Chemistry.

Now $a + b = 19$ {19 study Physics}
$b + c = 17$ {17 study Chemistry}
$b = 15$ {15 study both}
$a + b + c + d = 30$ {there are 30 in the class}
$\therefore$ $b = 15$, $a = 4$, $c = 2$, $d = 9$.

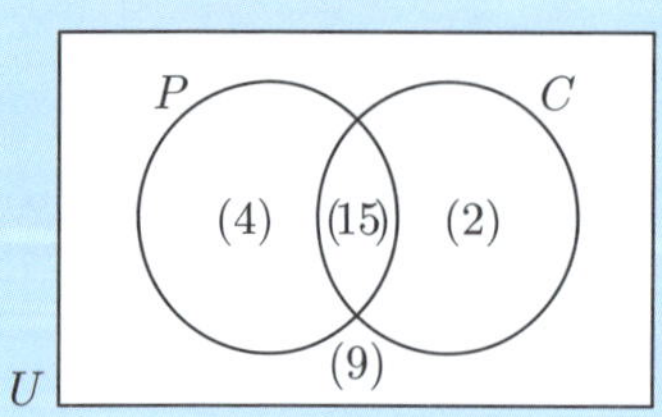

b P(exactly one of P and C) $= \dfrac{4+2}{30}$

$= \frac{6}{30}$

$= \frac{1}{5}$

c Of the $9+2=11$ students who do not study Physics, 2 study Chemistry.

$\therefore$ P(a student who does not study Physics, studies Chemistry) $= \frac{2}{11}$

EXERCISE 14I

1 The numbers from 1 to 25 are placed in a bag, and one number is selected at random. Let A represent selecting a multiple of 3, and B represent selecting a multiple of 4.

- **a** Represent A and B on a Venn diagram.
- **b** Find the probability of selecting:
  - **i** a multiple of 3
  - **ii** a multiple of 3 *and* 4
  - **iii** a multiple of 4, but not a multiple of 3
  - **iv** a multiple of 3 or 4, but not both.

2 A letter of the alphabet is chosen at random. Use a Venn diagram to find the probability that the letter is in:

- **a** both STATISTICS and PROBABILITY
- **b** STATISTICS, but not PROBABILITY
- **c** STATISTICS or PROBABILITY
- **d** neither word.

3 50 married men were asked whether they gave their wife flowers or chocolates for her last birthday. 31 gave chocolates, 12 gave flowers, and 5 gave both chocolates and flowers. If one of the married men was chosen at random, determine the probability that he gave his wife:

- **a** chocolates or flowers
- **b** chocolates but not flowers
- **c** neither chocolates nor flowers.

4 In a class of 40 students, 19 play tennis, 20 play netball, and 8 play neither of these sports. A student is randomly chosen from the class. Determine the probability that the student:

- **a** plays tennis
- **b** does not play netball
- **c** plays at least one of the two sports
- **d** plays exactly one of the sports
- **e** plays netball, given that they do not play tennis.

5 The medical records for a class of 30 children show whether they have previously had measles or mumps. 24 have had measles, 12 have had measles *and* mumps, and 26 have had measles *or* mumps. If one child from the class is selected at random, determine the probability that he or she has had:

- **a** mumps
- **b** mumps but not measles
- **c** neither mumps nor measles.

6 Neville checked his letter box each day for 80 days. He received letters on 40 days, bills on 14 days, and junk mail on 41 days. He received letters *and* bills on 9 days, and bills *and* junk mail on 7 days. He received all three items on 4 days, and no items on 14 days.

  a Display this information on a Venn diagram.

  b Find the probability that, on a randomly chosen day, Neville received:

    i junk mail only
    ii letters or junk mail, but not both
    iii bills and letters, but not junk mail
    iv bills, given that he did not receive junk mail
    v junk mail or bills, given that he received letters.

J EXPECTATION

When a chance experiment is performed many times, we cannot predict *exactly* what will happen. However, we can use probability to calculate how many times we *expect* a particular event to occur.

For example, when a fair coin is tossed 100 times, we would *expect* half of the results to be heads. We therefore expect $100 \times \frac{1}{2} = 50$ heads.

Similarly, if a normal six-sided die is rolled 300 times, we would expect a '4' to be rolled $\frac{1}{6}$ of the time. We therefore expect $300 \times \frac{1}{6} = 50$ results to be a '4'.

If there are n trials of an experiment, and the probability of an event occurring in each trial is p, then the **expectation** of the occurrence of that event is np.

Example 15 — Self Tutor

When an archer fires at a target, there is probability $\frac{2}{5}$ that he hits the bullseye.

In a competition, the archer is required to fire 40 arrows. How many times would you expect him to hit the bullseye?

$p = \text{P(bullseye)} = \frac{2}{5}$ and $n = 40$

$\therefore$ the expected number of bullseyes is $np = 40 \times \frac{2}{5}$
$= 16$

EXERCISE 14J

1 A goalkeeper has probability $\frac{3}{10}$ of saving a penalty attempt. How many goals would he expect to save out of 90 penalty shots?

2 During the snow season there is a $\frac{3}{7}$ probability of snow falling on any particular day. If Dan skis for five weeks, on how many days could he expect to see snow falling?

3 A hat contains three yellow discs and four green discs. A disc is drawn from the hat and its colour recorded. The disc is then returned to the hat and the procedure is repeated 350 times. On how many occasions would you expect a green disc to be drawn?

4 Two dice are rolled simultaneously 180 times. On how many occasions would you expect to get a double?

5 In a random survey of a district, people were asked whether they will vote for politician A, B, or C. The results are shown alongside.

A	B	C
165	87	48

a How many people took part in the survey?

b Estimate the probability that a randomly chosen voter in the district will vote for:

i A **ii** B **iii** C.

c There are 7500 people in the district. How many of these would you expect to vote for:

i A **ii** B **iii** C?

6 The pair of spinners alongside are spun 600 times. How many times would you expect the sum of the spins to equal 5?

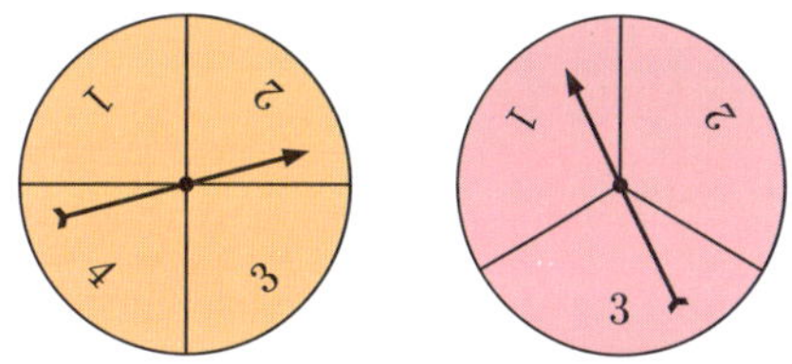

7 200 people are each given a bag containing 2 red balls and 3 green balls. Each person draws two balls from their bag without replacement. How many people would you expect to draw:

a two red balls **b** two green balls?

8 Each hockey game, the probabilities of Gemma, Tiffany, and Melanie scoring a goal are 0.4, 0.15, and 0.2 respectively. Over a 40 game season, in how many games would you expect:

a Gemma to score a goal

b Melanie, but not Tiffany to score a goal

c none of them to score a goal

d exactly one of them to score a goal?

Global context

click here

Genes

Statement of inquiry:	Knowledge of probability can help us understand how the world works.
Global context:	Identities and relationships
Key concept:	Relationships
Related concepts:	Justification, Representation, System
Objectives:	Knowing and understanding, Applying mathematics in real-life contexts
Approaches to learning:	Thinking, Research

REVIEW SET 14A

1 Over a 35 day period, Lorna records the number of emails she receives. The graph of her data is shown alongside.

Estimate the probability that on any day Lorna will receive:

- **a** no emails
- **b** at least 3 emails
- **c** between 1 and 5 emails inclusive.

Number of emails

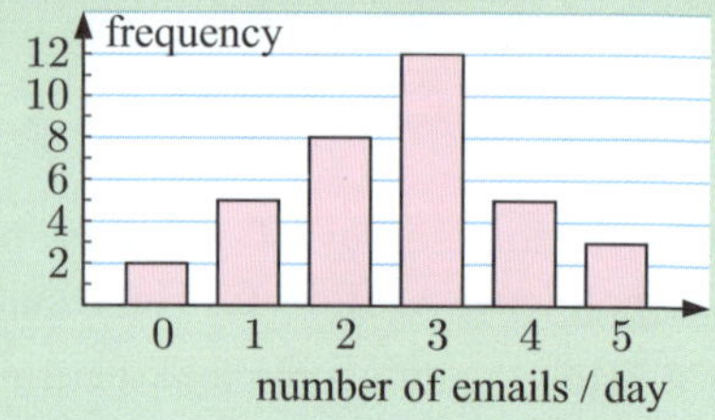

2 A letter of the alphabet is chosen at random. Find the probability that the letter is:

- **a** M or N
- **b** contained in the word CHOCOLATE
- **c** after the letter S in the alphabet.

3 A bag of mixed lollies contains 10 mints and 6 chocolate caramels. A second bag contains 8 mints and 8 chocolate caramels. A bag is randomly chosen by tossing a coin, and a lolly is then taken from it.

- **a** Construct a tree diagram to show the sample space.
- **b** Hence determine the probability that the lolly selected will be:
  - **i** a chocolate caramel
  - **ii** a mint.

4 Marisa and Caleb each roll a die. Find the probability that Marisa rolls a number greater than 1, and Caleb rolls a 3 or a 4.

5 Alec Smart and Joe Slow sit for an examination in Chemistry. Alec has a 95% chance of passing and Joe has a 25% chance of passing. Determine the probability that:

- **a** both pass
- **b** both fail
- **c** Joe passes *and* Alec fails
- **d** Alec passes *and* Joe fails.

6 At the local girls' school, 65% of the students play netball, 60% play tennis, and 20% play neither sport.

- **a** Display this information on a Venn diagram.
- **b** Hence, determine the likelihood that a randomly chosen student plays:
  - **i** netball
  - **ii** netball but not tennis
  - **iii** at least one of these two sports
  - **iv** exactly one of these two sports.

7 In a test there are 3 multiple choice questions, each with 4 choices. A student randomly guesses the answers to each question.

- **a** Draw a tree diagram to show all possible ways the student can answer the questions, and the probability for each outcome.
- **b** Find the probability that the student gets:
  - **i** the first question right, but the second and third questions wrong
  - **ii** exactly 1 of the 3 multiple choice questions correct.

8 From past experience, a surfer has probability 0.83 of catching a wave. In one week she tries to catch 75 waves. How many waves do you expect her to catch?

9 A hat contains 12 tickets with the numbers 1 to 12 printed on them. If two tickets are drawn from the hat without replacement, find the probability that they are both prime numbers.

10 Each morning when Harold has a shower, there is a 90% chance that the hot water is working. The probability that Harold has a long shower is 80% if the hot water is working, and 10% if the hot water is not working.

a Find the probability that, on any given day, Harold will have a long shower.

b During a 365 day year, how many long showers would you expect Harold to have?

11 The probability of a delayed flight on a foggy day is $\frac{9}{10}$. When it is not foggy the probability of a delayed flight is $\frac{1}{12}$. The probability of a foggy day is $\frac{1}{20}$.

a Construct a tree diagram to show this information.

b Hence, find the probability of:

i a foggy day and a delayed flight

ii a delayed flight **iii** a flight which is not delayed.

c Comment on your answers to **b ii** and **iii**.

12 A class consists of 25 students. 15 have blue eyes, 9 have fair hair, and 3 have both blue eyes and fair hair.

a Represent this information on a Venn diagram.

b Hence, find the probability that a randomly selected student from the class:

i has neither blue eyes nor fair hair **ii** has blue eyes, but not fair hair

iii has fair hair, given that he or she has blue eyes

iv does not have fair hair, given that he or she does not have blue eyes.

REVIEW SET 14B

1 What is meant by saying that two events are *independent*?

2 When a box of drawing pins was dropped onto the floor, 49 pins landed on their backs and 32 landed on their sides. Estimate, to 2 decimal places, the probability of a drawing pin landing:

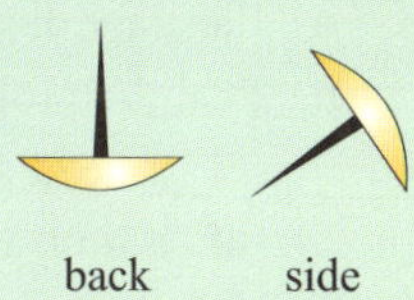

a on its back **b** on its side.

3 93 people arriving at the beach are asked their age. The results are shown opposite. Assuming they give honest replies, estimate the probability that a randomly selected person on the beach will be aged:

a 30 or more **b** between 10 and 30.

Age	*Frequency*
0 - 9	17
10 - 19	25
20 - 29	26
30 - 39	20
40+	5

4 Illustrate on a 2-dimensional grid, the possible outcomes when a coin and a pentagonal spinner with sides labelled A, B, C, D, and E are tossed and spun simultaneously.

5 A shared garden area has 12 patches owned by 12 different people. The patches are separated by fences as shown. If a patch is selected at random, find the probability that it has:

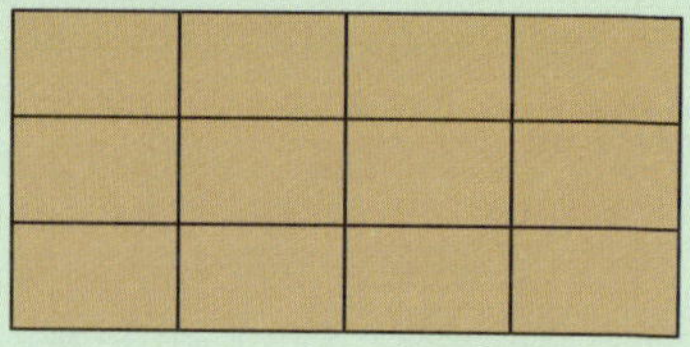

a two shared fences **b** three shared fences.

6 Bag X contains three white and two red marbles. Bag Y contains one white and three red marbles. A bag is randomly chosen and two marbles are drawn from it.

a Illustrate the given information on a tree diagram.

b Find the probability of drawing two marbles of the same colour.

7 Two fair six-sided dice are rolled simultaneously. Determine the probability that the result is a 'double', which means both dice show the same number.

8 In an office of 15 employees, 10 drive a car to work, 6 use public transport, and 3 use both.

a Represent this information on a Venn diagram.

b Find the probability that a randomly selected employee uses neither a car nor public transport to get to work.

9 A waterpolo player has probability $\frac{1}{4}$ of scoring a goal each time she shoots. If she has 24 shots at goal, how many goals would you expect her to score?

10 The letters A, B, C, D, ..., N are put in a hat.

a If one of the letters is chosen at random, determine the probability of drawing a vowel (A, E, I, O, or U).

b Two letters are drawn *with* replacement. Copy and complete the tree diagram, including all probabilities. Hence, determine the probability of drawing:

i a vowel and a consonant

ii at least one vowel.

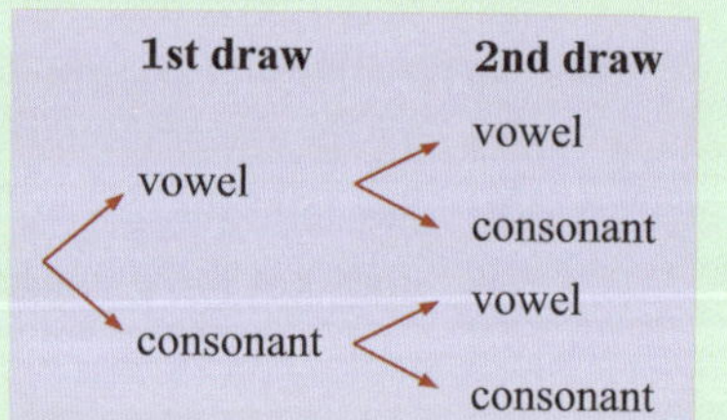

11 A coin is tossed and a die is rolled simultaneously.

a Illustrate the sample space on a grid.

b Find the probability of getting:

i a head and an even number **ii** a head and a non-3

iii a 5 or a 6 **iv** a head or an even number.

c If the experiment is performed *twice*, find the probability of getting two heads and two numbers whose sum is 7.

12 Two dice are rolled simultaneously.

a Illustrate the possible outcomes on a 2-dimensional grid.

b Determine the probability of getting:

i a double 5 **ii** at least one 4

iii a sum greater than 9 **iv** a sum of 5 or 6.

c If the dice were rolled 900 times, on how many occasions would you expect the sum of the rolls to be prime?

Chapter 15

Formulae

Contents:

OPENING PROBLEM

A solid cylinder has base radius r and height h. The surface area of the cylinder is given by the formula $S = 2\pi r^2 + 2\pi rh$.

If we know the base radius and height of a cylinder, we can use this formula to calculate its surface area.

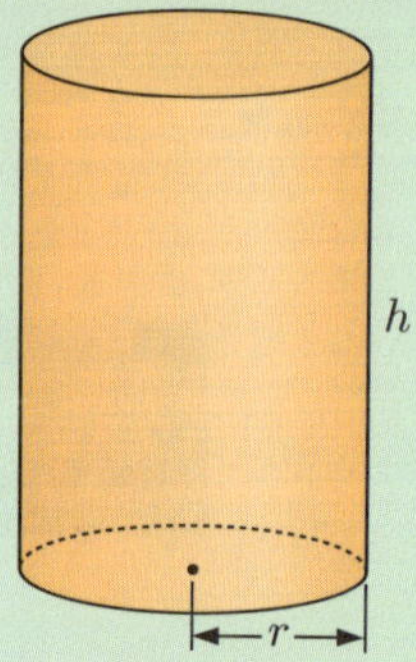

Things to think about:

a Can you use the formula to find the surface area of a solid cylinder with radius 3 cm and height 8 cm?

b Can you rewrite the formula so that the height of the cylinder h is given in terms of its radius r and surface area S?

c Why might it be useful to rearrange the formula in this way?

A **formula** is an equation which connects two or more variables.

For example, the formula for simple interest $I = Pin$ connects the variables *interest* (I), *principal* (P), *interest rate* (i), and *duration* (n) of a loan.

In this case the formula gives the interest I in terms of the other variables. We say that I is the **subject** of the formula.

We can **rearrange** the formula to make one of the other variables the subject.

For example, the rate of interest for a simple interest loan is $i = \frac{I}{Pn}$.

The plural of formula is **formulae** or **formulas**.

A FORMULA CONSTRUCTION

Formulae are often constructed by generalising numerical observations. To construct a formula, we start with a specific numerical situation. Once we understand this specific situation, we then **generalise** the result.

For example, the perimeter of the rectangle shown is given by

$$P = 3 + 6 + 3 + 6 \text{ metres}$$
$$\therefore \quad P = (2 \times 3) + (2 \times 6) \text{ metres}$$

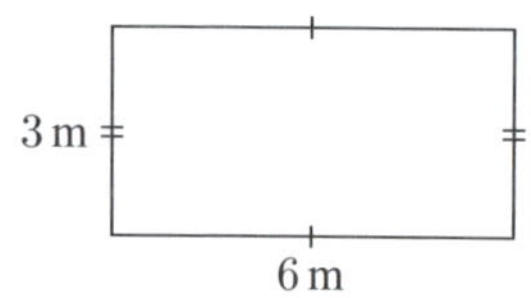

$\therefore$ P is double the width plus double the length of the rectangle.

We *generalise* the result by writing the formula $P = 2a + 2b$ or $P = 2(a + b)$.

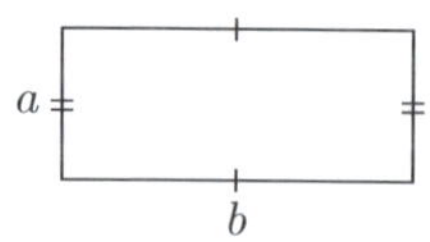

Example 1 **Self Tutor**

Write a formula for the total cost $\$C$ of a taxi trip, which includes:

a a fixed charge of \$3, plus \$0.55 per km for 12 km
b a fixed charge of \$3, plus \$0.55 per km for k km
c a fixed charge of \$3, plus $\$d$ per km for k km
d a fixed charge of $\$F$, plus $\$d$ per km for k km.

a $C = 3 + (0.55 \times 12)$

b $C = 3 + (0.55 \times k)$
$\therefore\ C = 3 + 0.55k$

c $C = 3 + d \times k$
$\therefore\ C = 3 + dk$

d $C = F + dk$

EXERCISE 15A

1 A mobile gardener charges a \$30 call-out fee, plus \$40 per hour thereafter. Write a formula for the gardener's total fee $\$F$ for a service lasting:
a 2 hours **b** h hours.

2 Write a formula for the amount €A in a savings account, given monthly deposits of:
a €250 over 6 months **b** €250 over m months **c** €D over m months.

3 Find a formula for the total mass M kg of:
a a 20 kg wheelbarrow containing 12 bricks weighing 5 kg each
b a 20 kg wheelbarrow containing b bricks weighing 5 kg each
c a 20 kg wheelbarrow containing b bricks weighing n kg each
d a w kg wheelbarrow containing b bricks weighing n kg each.

4 Write a formula for the amount $\$A$ in Chris' wallet if initially he had:
a \$100, and he bought 6 books costing \$12 each
b \$100, and he bought b books costing \$12 each
c \$100, and he bought b books costing $\$p$ each
d $\$d$, and he bought b books costing $\$p$ each.

5 Write a formula for the amount W litres of water in a tank, if initially the tank held:
a 3500 litres, and 10 litres per minute was let out for 200 minutes
b 3500 litres, and r litres per minute was let out for 200 minutes
c 3500 litres, and r litres per minute was let out for m minutes
d L litres, and r litres per minute was let out for m minutes.

6 Helen works at a garden centre. When it is rainy, she works a 5-hour day. When it is fine, the centre is busier, so she works an 8-hour day. Write a formula for the total number of hours H that Helen works if:
a she works for 6 days, 2 of which are rainy
b she works for 7 days, r of which are rainy
c she works for d days, r of which are rainy.

7 An art gallery has a series of paintings to be hung along a wall. The paintings are to be hung such that there is an equal gap between each painting, as well as at either end of the wall.

Find a formula for the length of the gap G metres between each painting, if:

a 3 paintings, each 2 metres wide, are to be hung along a 14 metre wall

b 3 paintings, each w metres wide, are to be hung along a 14 metre wall

c 3 paintings, each w metres wide, are to be hung along a d metre wall

d n paintings, each w metres wide, are to be hung along a d metre wall.

8 Write down a formula for the perimeter P mm of:

a

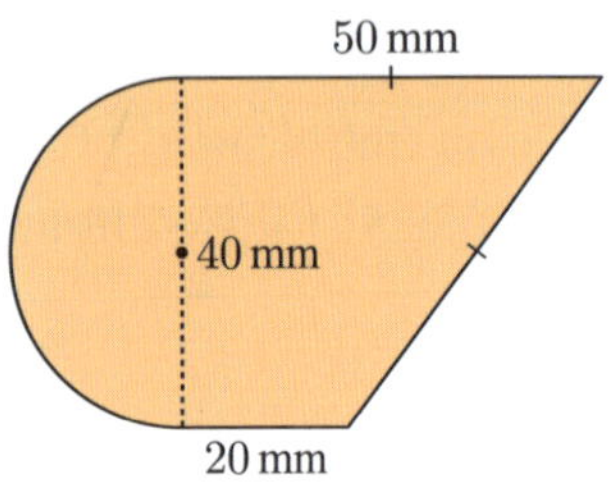

b

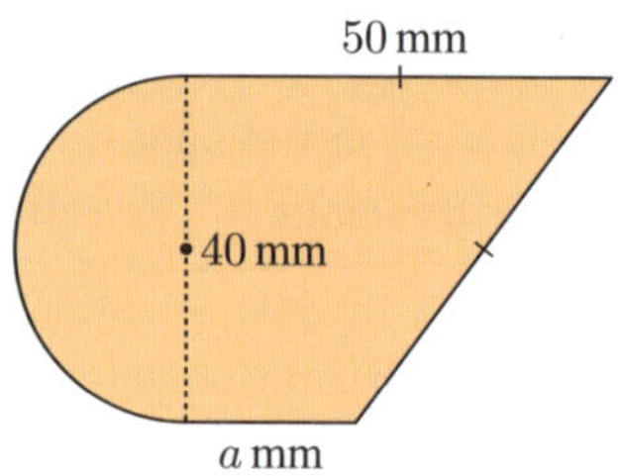

c

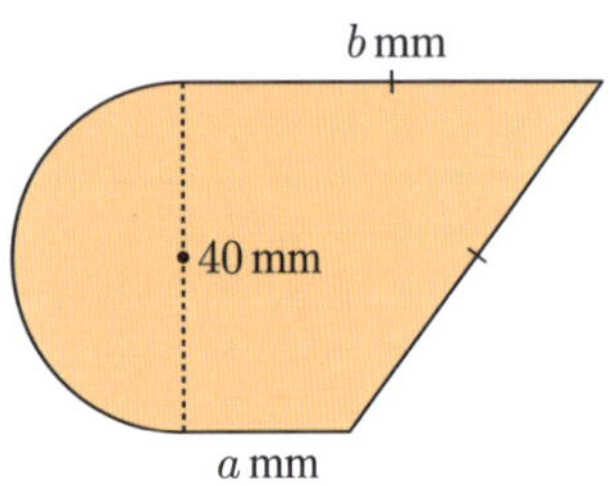

d

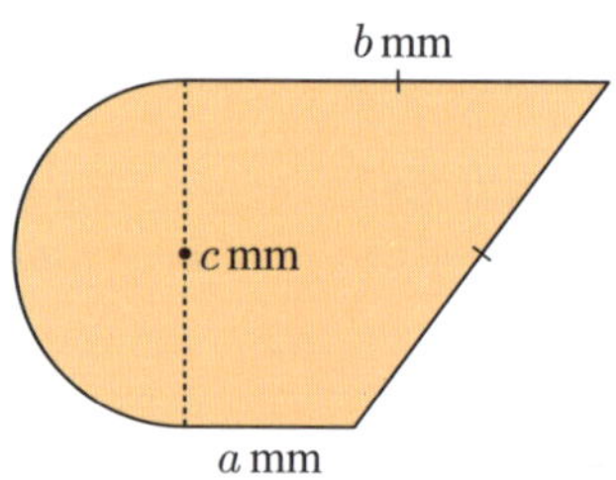

9 Write down a formula for the area A cm^2 of:

a

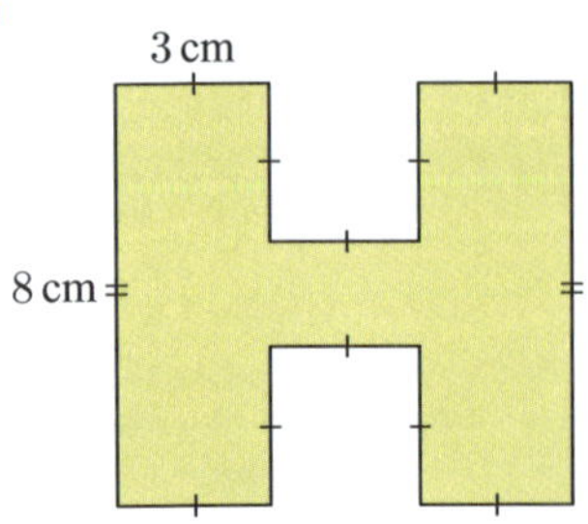

b

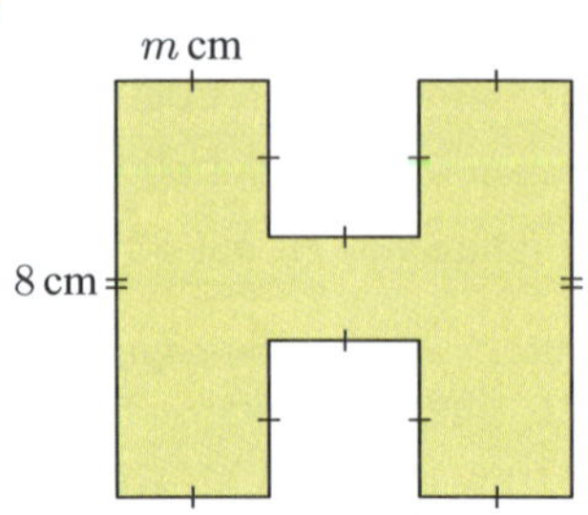

c

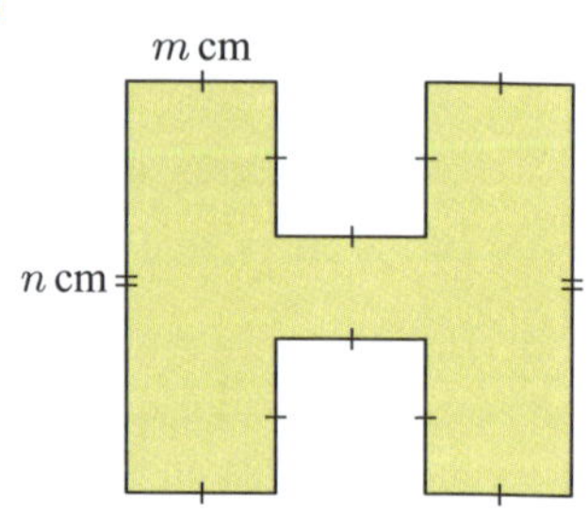

10 Write down a formula for the volume V m^3 of:

a

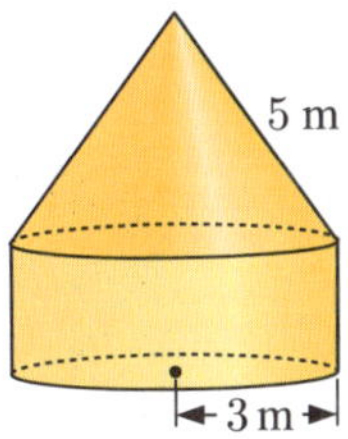

b

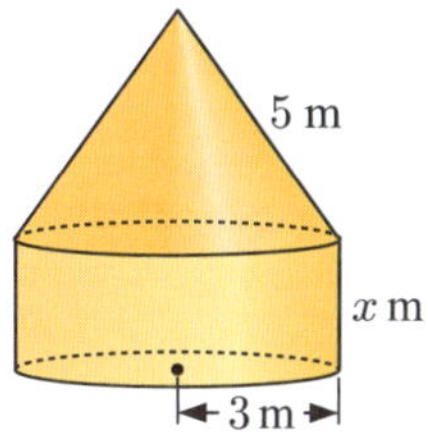

c

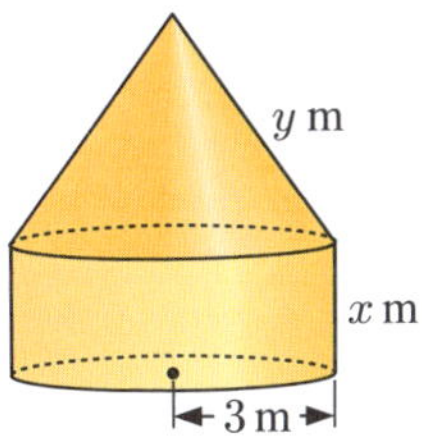

d

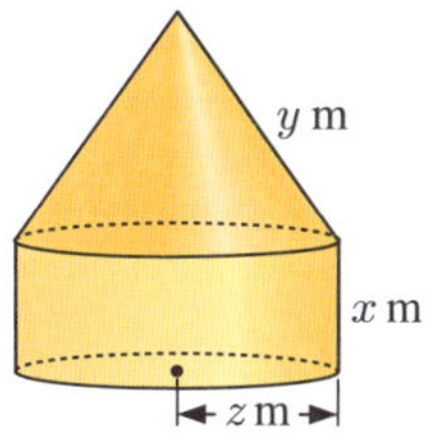

The volume of a cone with base radius r and height h is given by $V = \frac{1}{3}\pi r^2 h$

B SUBSTITUTING INTO FORMULAE

A formula is normally written with one variable by itself on the left hand side of the equation. We call this variable the **subject** of the formula. The other variables are contained in an expression on the right hand side.

To find the value of any of the variables, we need to know the values of all the other variables in the formula. We follow this procedure:

Step 1: Write down the formula.
Step 2: State the values of the known variables.
Step 3: Substitute the known values into the formula to form a one variable equation.
Step 4: Solve the equation for the unknown variable.

Example 2 **Self Tutor**

Consider the formula $v = u + gt$.

a Find v when $u = 15$, $g = 9.8$, and $t = 10$.
b Find t when $v = 45$, $u = 20$, and $g = 9.8$.

a $v = u + gt$ where $u = 15$, $g = 9.8$, and $t = 10$

$\therefore\ v = 15 + 9.8 \times 10$

$\therefore\ v = 15 + 98$

$\therefore\ v = 113$

b $v = u + gt$ where $v = 45$, $u = 20$, and $g = 9.8$

$\therefore\ 45 = 20 + 9.8t$

$\therefore\ 25 = 9.8t$

$\therefore\ \dfrac{25}{9.8} = t$

$\therefore\ t \approx 2.55$

EXERCISE 15B

1 Consider the formula $F = g + 5h$.

- **a** Find F when $g = 2$ and $h = 3$.
- **b** Find g when $F = 32$ and $h = 5$.
- **c** Find h when $F = 21$ and $g = 9$.

2 The simple interest earned by an investment is given by $I = Pin$ where P is the principal, i is the interest rate expressed as a decimal, and n is the duration of the investment in years.

- **a** Find the simple interest earned by an investment of \$10 000 at 8% p.a. for 5 years.
- **b** Find the rate of interest required for an investment of €8000 to earn €1848 in simple interest over $3\frac{1}{2}$ years.

3 The momentum P of an object is given by $P = mv$ kg m/s, where m is the mass of the object in kg, and v is the speed of the object in m/s.

- **a** Find the momentum of:
  - **i** a 1500 kg car travelling at 15 m/s
  - **ii** a 3 g bullet travelling at 500 m/s.
- **b** Sidney hit a 170 g hockey puck with his hockey stick. The puck travelled with momentum 6.8 kg m/s. At what speed was it travelling?

Example 3 **Self Tutor**

The volume of a cylinder with base radius r and height h is given by $V = \pi r^2 h$. Find:

- **a** the volume of a cylinder with base radius 2 cm and height 5 cm
- **b** the base radius of a cylinder with volume 10 m^3 and height 3 m.

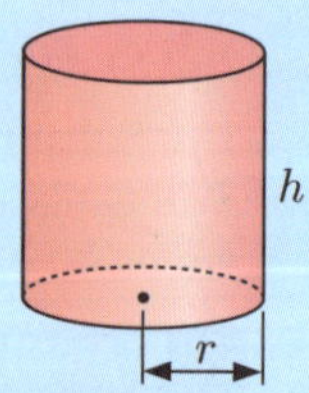

a $V = \pi r^2 h$ where $r = 2$ and $h = 5$

$\therefore V = \pi \times 2^2 \times 5$

$\therefore V \approx 62.8$

The volume is approximately 62.8 cm^3.

b $V = \pi r^2 h$ where $V = 10$ and $h = 3$

$\therefore 10 = \pi \times r^2 \times 3$

$\therefore r^2 = \dfrac{10}{3\pi}$

$\therefore r = \sqrt{\dfrac{10}{3\pi}}$ {r must be positive}

$\therefore r \approx 1.03$

The radius is approximately 1.03 m.

4 The volume of a cone with base radius r and height h is given by $V = \frac{1}{3}\pi r^2 h$. Find:

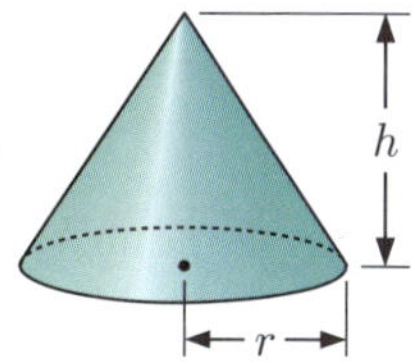

- **a** the volume of a cone with base radius 5 cm and height 8 cm
- **b** the height of a cone with volume 200 cm^3 and base radius 3.5 cm
- **c** the base radius of a cone with volume 30 m^3 and height 6.2 m.

5 Fulvio's formula for the price of his pizzas is $P = F + \frac{r^2}{32}$ euros, where r is the radius of the pizza in cm, and F is the profit in euros. Find:

- **a** the price of a 12 cm radius pizza which makes €3 profit
- **b** the profit for a 15 cm radius pizza which sells for €9.90
- **c** the radius of a €19.90 pizza which makes €7.50 profit.

6 Heron's formula for the area of a triangle with side lengths a, b, and c is $A = \sqrt{s(s-a)(s-b)(s-c)}$ where $s = \frac{a+b+c}{2}$. Use Heron's formula to find the area of:

- **a** a triangular garden plot with side lengths 6 m, 7 m, and 8 m
- **b** an isosceles triangle with side lengths 3 m, 3 m, and 2 m.

7 The formula $D \approx 4\sqrt{h}$ kilometres allows us to estimate the distance to the horizon when we are h metres above our surroundings. Estimate:

- **a** the distance to the horizon for a person at the top of a cliff 100 metres above sea level
- **b** the height of a tower if a person standing at the top can see for 44 km to the horizon.

ACTIVITY 1 — FALLING OBJECTS

When an object is dropped, the distance it falls in the first t seconds is approximately given by $D = 4.9t^2$ metres.

You will need: a tennis ball, stopwatch, tape measure.

What to do:

1. Working in pairs, find a place in your school where you can drop a tennis ball from a reasonable height. For example, you could use a balcony or the top of a staircase.
2. While one member of the pair drops the tennis ball, the other person uses the stopwatch to time how long it takes for the ball to hit the ground. To improve the accuracy of your experiment, drop the ball several times and average your results.

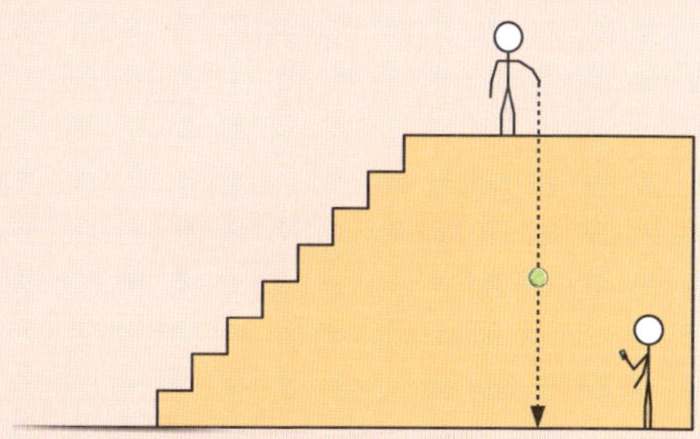

3. Use the formula $D = 4.9t^2$ to calculate the height from which the ball was dropped.
4. Use your tape measure to check your answer. Suggest reasons for any inaccuracies in your results.

C REARRANGING FORMULAE

Consider the formula $M = 3x + k$. M is the subject of the formula, because M is expressed in terms of the other variables.

A formula can be **rearranged** to make a different variable the subject. We do this using **inverse operations** to isolate the variable, just as we do when solving equations.

Example 4 Self Tutor

Make x the subject of $M = 3x + k$.

$$M = 3x + k$$
$$\therefore\ 3x + k = M$$
$$\therefore\ 3x = M - k \quad \{\text{subtracting } k \text{ from both sides}\}$$
$$\therefore\ x = \frac{M-k}{3} \quad \{\text{dividing both sides by } 3\}$$

EXERCISE 15C.1

1 Make x the subject of:

a $y = 5x$
b $y = x - 3$
c $y = 3x + 2$
d $y = \frac{1}{2}x - 1$
e $y = 3 - x$
f $y = 2x$

2 Make x the subject of:

a $B = 7x + n$
b $T = 3xy$
c $G = x - 5p$
d $K = l - 9x$
e $P = wx + yz$
f $C = 4xy^2 - z$

3 Make y the subject of:

a $2x + 3y = 9$
b $3x - y = 4$
c $4x + 2y = 3$

4 The ideal gas law is the equation of state for an ideal gas. If an amount n of gas has pressure P, volume V, and temperature T, then $PV = nRT$ where R is the universal gas constant. Rearrange this formula to make T the subject.

Example 5 Self Tutor

Make g the subject of $n = \frac{a}{g}$.

$$n = \frac{a}{g}$$
$$\therefore\ ng = a \quad \{\text{multiplying both sides by } g\}$$
$$\therefore\ g = \frac{a}{n} \quad \{\text{dividing both sides by } n\}$$

5 Make a the subject of:

a $x = \dfrac{a}{y}$ **b** $K = \dfrac{c}{a}$ **c** $M = \dfrac{abc}{n}$

d $T = \dfrac{a+s}{p}$ **e** $N = \dfrac{b+d}{2a}$ **f** $F = \dfrac{3a-d}{2g}$

6 The density D of an object with mass M and volume V is given by $D = \dfrac{M}{V}$. Rearrange this formula to make:

a M the subject **b** V the subject.

7 The perimeter P of a pentagon with three sides of length a and two sides of length b, is given by $P = 3a + 2b$. Rearrange this formula to make:

a a the subject **b** b the subject.

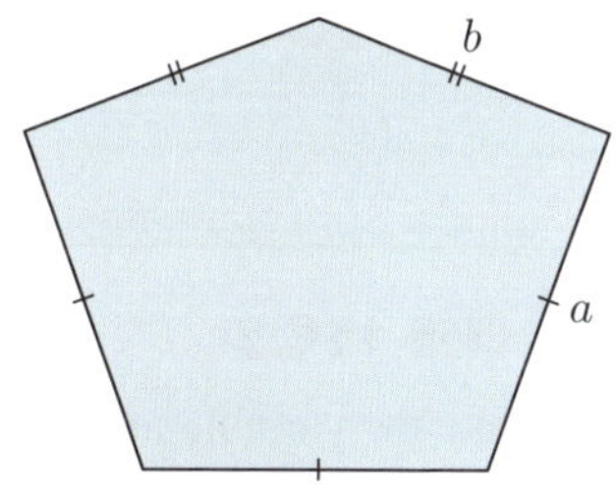

Example 6 **Self Tutor**

Make x the subject of $P = \dfrac{t}{x^2}$, where $x > 0$.

$$P = \frac{t}{x^2}$$

$\therefore\ x^2 P = t$ {multiplying both sides by x^2}

$\therefore\ x^2 = \dfrac{t}{P}$ {dividing both sides by P}

$\therefore\ x = \sqrt{\dfrac{t}{P}}$ {as $x > 0$}

8 Make x the subject of the following, where $x > 0$:

a $D = x^2 + 1$ **b** $K = 5x^2$ **c** $F = \dfrac{x^2}{y}$

d $N = \dfrac{7z}{x^2}$ **e** $C = 2x^2 + y^2$ **f** $T = 3x^2 z - w$

9 The *body-mass index* B of a person with mass M kg and height H metres, is given by $B = \dfrac{M}{H^2}$. Rearrange this formula to make:

a M the subject **b** H the subject.

10 The surface area of a sphere of radius r is given by $S = 4\pi r^2$. Rearrange this formula to make r the subject.

11 The distance s travelled by an object with initial speed u, final speed v, and acceleration a is given by $s = \dfrac{v^2 - u^2}{2a}$. Rearrange this formula to make u the subject.

12 Make x the subject of:

a $xy = 2x - 3$ **b** $\dfrac{3x}{x-2} = y$ **c** $y = \dfrac{3}{\sqrt{5-x}}$

FORMULAE INVOLVING FRACTIONS ON BOTH SIDES

When a formula contains a fraction on both sides, we multiply both sides by the **lowest common denominator**.

Example 7 — Self Tutor

Make c the subject of the formula $\frac{b}{a} = \frac{2}{c}$.

The lowest common denominator of $\frac{b}{a}$ and $\frac{2}{c}$ is ac.

$\frac{b}{a} = \frac{2}{c}$

$\therefore\ bc = 2a$ {multiplying both sides by ac}

$\therefore\ c = \frac{2a}{b}$ {dividing both sides by b}

EXERCISE 15C.2

1 Make the variable in brackets the subject of the formula:

a $\frac{2}{q} = \frac{p}{5}$ $\{q\}$ **b** $\frac{w}{x} = \frac{y}{z}$ $\{y\}$ **c** $\frac{e}{f} = \frac{g}{h}$ $\{h\}$

d $\frac{e}{f} = \frac{g}{h}$ $\{f\}$ **e** $\frac{s}{t} = \frac{t}{r}$ $\{r\}$ **f** $\frac{k}{2m} = \frac{n}{k^2}$ $\{m\}$

2 The heights and bases of the triangles alongside are connected by the formula $\frac{H}{B} = \frac{h}{b}$. Rearrange this formula to make:

a H the subject **b** B the subject

c h the subject **d** b the subject.

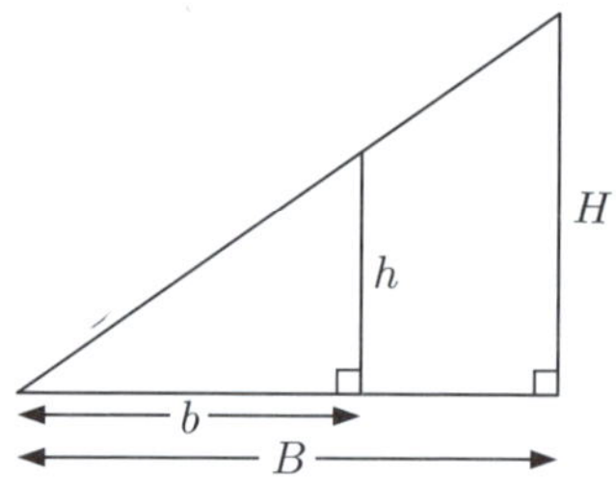

3 Make x the subject of:

a $\frac{x+2y}{x-y} = \frac{5}{7}$ **b** $\frac{4x}{2x+y^3} = \frac{2}{xy+1}$.

4 Make u the subject of $\frac{1}{f} = \frac{1}{u} + \frac{1}{v}$.

D REARRANGEMENT AND SUBSTITUTION

To understand the usefulness of rearranging formulae, suppose we know the areas of several circles and want to find the radius of each circle.

Instead of substituting the area A into the formula $A = \pi r^2$ and solving for r each time, it is quicker to first **rearrange** the formula to make r the subject, and then **substitute** for the particular cases.

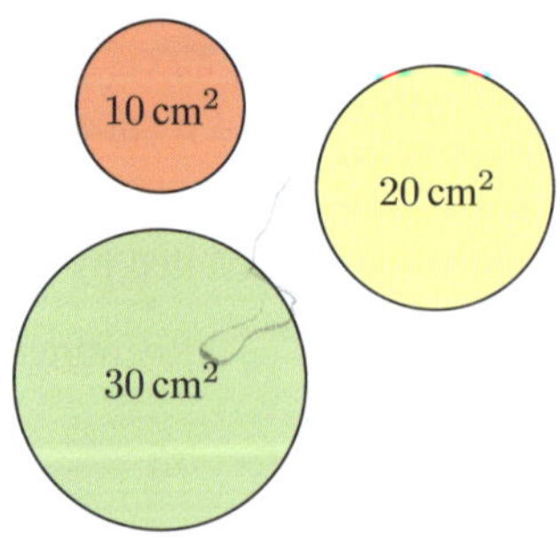

Example 8 Self Tutor

The area of a circle of radius r is given by $A = \pi r^2$.

a Rearrange this formula to make r the subject.

b Hence find the radius of a circle with area:

i 10 cm^2 **ii** 20 cm^2 **iii** 30 cm^2.

a $A = \pi r^2$

$\therefore r^2 = \dfrac{A}{\pi}$ {dividing both sides by π}

$\therefore r = \sqrt{\dfrac{A}{\pi}}$ {as the radius r is positive}

b **i** When $A = 10$, $r = \sqrt{\dfrac{10}{\pi}} \approx 1.78$ $\therefore$ the radius is approximately 1.78 cm.

ii When $A = 20$, $r = \sqrt{\dfrac{20}{\pi}} \approx 2.52$ $\therefore$ the radius is approximately 2.52 cm.

iii When $A = 30$, $r = \sqrt{\dfrac{30}{\pi}} \approx 3.09$ $\therefore$ the radius is approximately 3.09 cm.

EXERCISE 15D

1 **a** Make t the subject of the formula $M = \dfrac{4t}{s^3}$.

b Hence, find the value of t when:

i $M = 12$, $s = 3$ **ii** $M = 15$, $s = 2$ **iii** $M = 2$, $s = 5$.

2 The circumference C of a circle with radius r, is given by the formula $C = 2\pi r$.

a Make r the subject of the formula.

b Find the radius of a circle with circumference:

i 20 cm **ii** 45 cm **iii** 62.3 cm.

3 The area of a triangle with base length b and height h, is given by the formula $A = \frac{1}{2}bh$.

a Rearrange this formula to make h the subject.

b Find the height of a triangle with:

i area 30 cm^2 and base 6 cm

ii area 32 m^2 and base 8 m

iii area 19.5 mm^2 and base 5 mm.

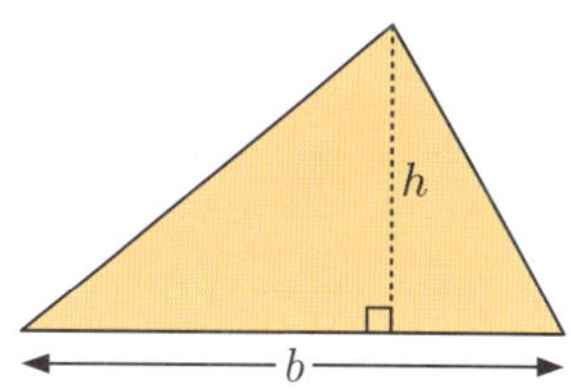

4 When a car travels a distance of d kilometres in the time t hours, the average speed s for the journey is given by the formula $s = \dfrac{d}{t}$ km/h.

a Make d the subject of the formula. Hence find the distance covered by a car travelling at an average speed of:

i 60 km/h for 3 hours

ii 80 km/h for $1\frac{1}{2}$ hours

iii 95 km/h for 1 h 20 min.

b Make t the subject of the formula. Hence find the time required for a car to travel:

i 180 km at an average speed of 60 km/h

ii 140 km at an average speed of 84 km/h

iii 220 km at an average speed of 100 km/h.

5 A right angled triangle has a hypotenuse of length c, and its other sides have lengths a and b. The side lengths are connected by the formula $c^2 = a^2 + b^2$.

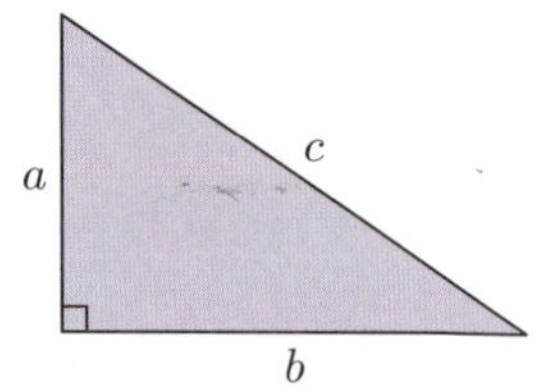

a Make a the subject of the formula.

b Hence find a in the following triangles:

i

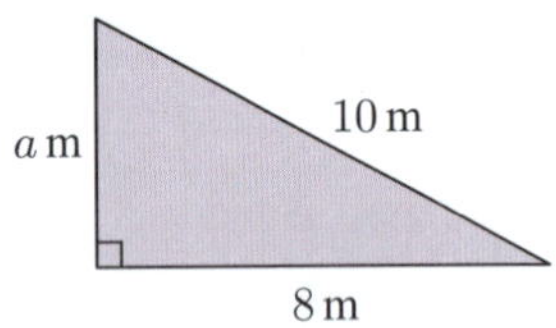

ii

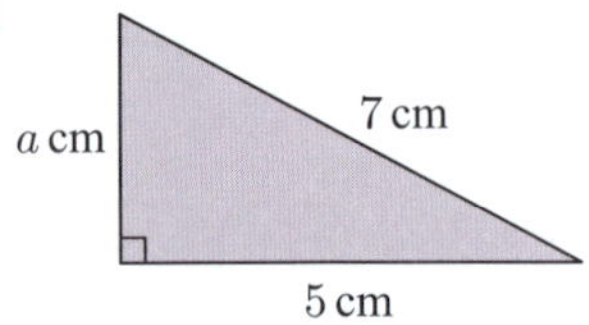

iii

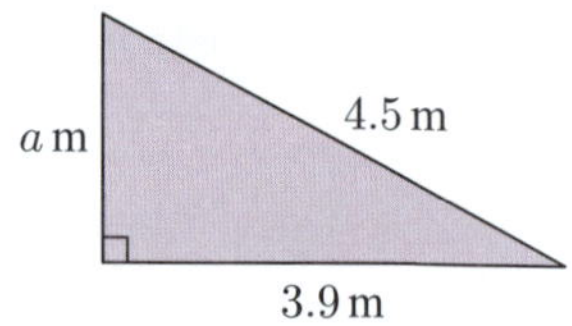

6 The volume of a square-based box with base length x and height h, is given by the formula $V = x^2h$.

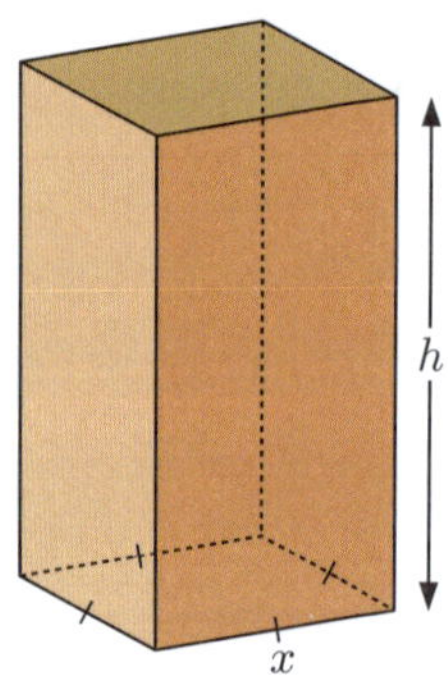

a Rearrange the formula to make h the subject. Hence find the height of a box with:

i volume 250 cm^3 and base length 5 cm

ii volume 800 m^3 and base length 2.5 m.

b Rearrange the formula to make x the subject. Hence find the base length of a box with:

i volume 960 cm^3 and height 15 cm

ii volume 600 cm^3 and height 7 cm.

7 **a** Suppose the figure alongside has perimeter P m. Write a formula for P.

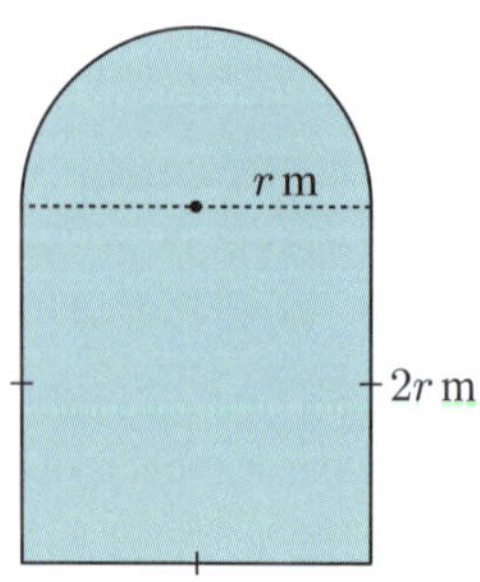

b Rearrange the formula to make r the subject.

c Find r given that the figure has perimeter:

i 5 m **ii** 8 m **iii** 10 m.

8 **a** Suppose the figure alongside has area A cm^2. Write a formula for A.

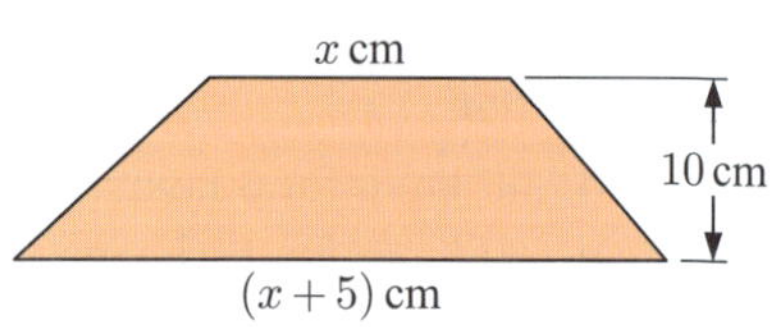

b Rearrange the formula to make x the subject.

c Find x given that the figure has area:

i 75 cm^2 **ii** 125 cm^2 **iii** 200 cm^2.

GAME

This game can be played by 2 or more players.

SCORESHEET

You will need: 3 dice, printed scoresheet

What to do:

The players take turns to roll the dice. The 3 numbers rolled are used as values for a, b, and c to substitute into the formula $D = \frac{a^2 + 2b}{c}$.

The player may use the numbers, in any order, to obtain a value of D which is an integer from 1 to 10. The player ticks this value off, on his or her scoresheet.

For example, if the player rolls 2, 5, and 6, the value $D = 8$ can be obtained by substituting $a = 6$, $b = 2$, and $c = 5$. The player ticks off the value 8.

The first player to obtain all 10 values of D is the winner.

REVIEW SET 15A

1 A trough is initially empty. Write a formula for the amount of water V litres in the trough if:

- **a** 6 buckets of water each containing 8 litres, are poured into it
- **b** n buckets of water, each containing 8 litres, are poured into it
- **c** n buckets of water, each containing l litres, are poured into it.

2 Write a formula for the total cost $\$C$ of packing parcels of books for despatch if there is a fixed charge of:

- **a** \$2 per parcel plus \$1.20 for one book
- **b** \$2 per parcel plus \$1.20 per book for 5 books
- **c** $\$p$ per parcel plus \$1.20 per book for b books
- **d** $\$p$ per parcel plus $\$x$ per book for b books.

3 When painting a house, Hendrik and Gemma charge a fixed amount of $\$A$ for setting up their equipment, plus $\$V$ per hour of time spent painting. Find the total cost $\$M$ if they spend:

- **a** 50 hours painting
- **b** h hours painting.

4 Suppose $M = a + \frac{b^2}{c}$ where $b > 0$.

- **a** Find M when $a = 5$, $b = 8$, and $c = -2$.
- **b** Find b when $M = 210$, $a = 70$, and $c = 14$.

5 Consider an object moving with initial velocity u m/s and acceleration a m/s^2. Its velocity after t seconds is given by the formula $v = u + at$ m/s.

- **a** A ball is dropped with initial velocity 1 m/s and acceleration 9.8 m/s^2. Find its velocity after 5 seconds.
- **b** A motorcycle initially travelling at 5 m/s, accelerates at 3.125 m/s^2. How long will it take for the motorcycle to reach a speed of 20 m/s?

6 Make x the subject of the formula:

a $mx + c = y$ **b** $ax = \dfrac{2}{x}$

7 A cylinder is capped with a cone as shown.

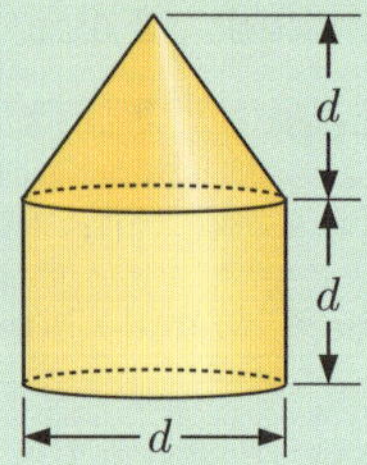

a Write a formula for the total volume V of the solid, in terms of d.

b Find the total volume of the solid if $d = 3$ m.

c If the total volume is 20 m^3, find the length d.

8 Make y the subject of:

a $3x + 4y = 10$ **b** $5x - 3y = 8$ **c** $\dfrac{3}{y} = \dfrac{a}{b}$

9 Consider the figure alongside with area A m.

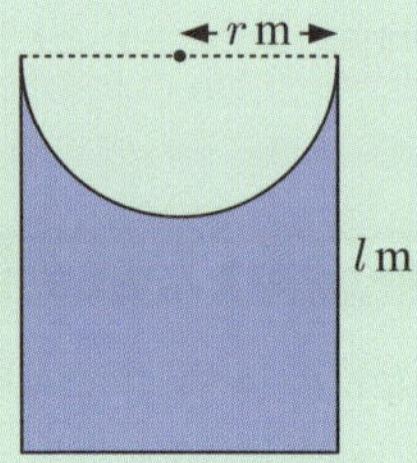

a Write a formula for A.

b Make l the subject of the formula.

c Find l given that the area is 40 m^2 and $r = 3$.

10 Answer the questions in the **Opening Problem** on page **304**.

11 The illustrated silo is made up of a hollow cylinder topped by a hemisphere (half sphere).

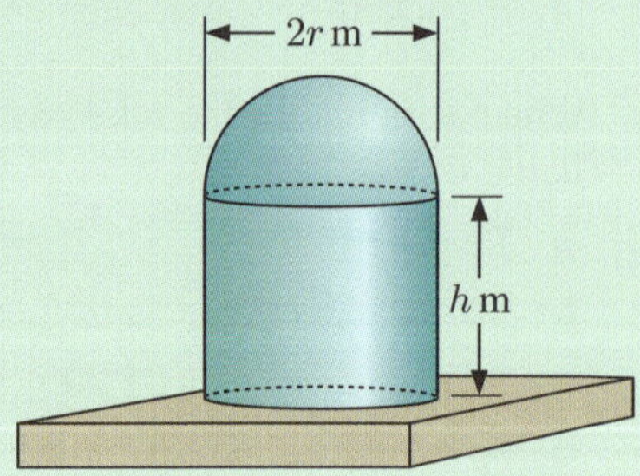

a Find the radius of the cylinder and hemisphere.

b Write a formula for the surface area of the hemisphere.

c Write a formula for the area of the curved surface of the cylinder.

d Hence find the total surface area of the silo, assuming it has no bottom.

REVIEW SET 15B

1 A gardener charges a fixed amount for travel and then an amount per hour for the time he works. Find a formula for the total charge C dollars if:

a the fixed amount is \$40, and he works for 3 hours at \$30 per hour

b the fixed amount is \$40, and he works for n hours at \$30 per hour

c the fixed amount is $\$F$, and he works for n hours at \$30 per hour

d the fixed amount is $\$F$, and he works for n hours at $\$a$ per hour.

2 The period of a pendulum is the time for one complete swing. It can be estimated using the formula $T = \frac{1}{5}\sqrt{l}$ seconds, where l is the length of the pendulum in cm. Find the period of a pendulum with length 74 cm.

3 Suppose $A = \frac{b+c}{2}$. Find c when $b = 5$ and $A = 6$.

4

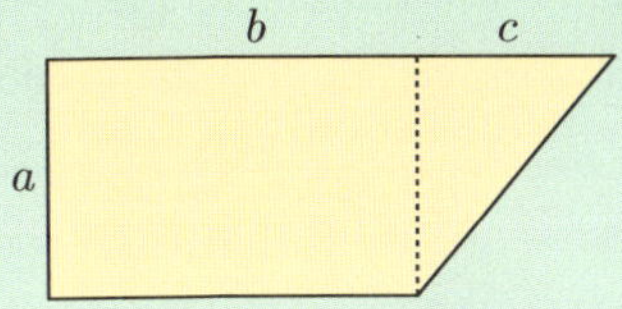

The diagram consists of a rectangle and a right angled triangle. Find a formula for calculating its:

a area A **b** perimeter P.

5 The equation of a straight line is $2x - y = 4$.

a Rearrange this equation into the form $y = mx + c$, where m and c are constants.

b State the gradient of the line, m.

c State the y-intercept of the line, c.

6 Make x the subject of the formula: **a** $y = 3q - 2x$ **b** $3y = \frac{4}{x}$

7 The volume of a cylinder is given by the formula $V = \pi r^2 h$, where r is its radius and h is its height. Find:

a the volume of a cylinder with height 23 cm and base radius 10 cm

b the height of a cylinder with base radius 12 m and volume 426.4 m^3.

8 Make b the subject of:

a $2a + 3b = c$ **b** $a - \frac{4}{b} = c$ **c** $\frac{3}{b} = \frac{b}{e}$ **d** $\frac{a}{c} = \frac{4}{b}$

9 The area of a trapezium is given by $A = \left(\frac{a+b}{2}\right) \times h$ where a and b are the lengths of the parallel sides, and h is the height.
Find a if the area of the trapezium is 144 cm^2.

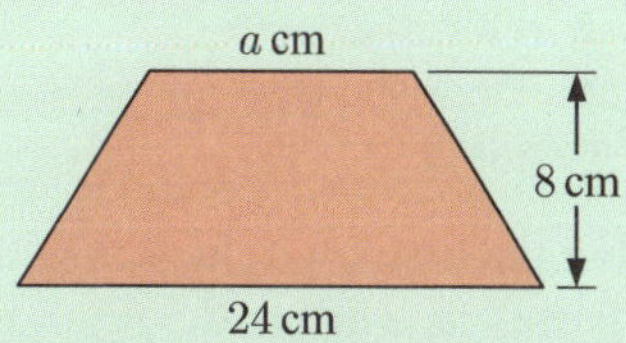

10 **a** Make a the subject of the formula $d = \frac{1}{2}at^2$.

b Hence, find the value of a when $t = 2$ and $d = 14$.

11 **a** 'Spaceship Earth' is a spherical building at a theme park. Its diameter is 50 m. Find the surface area of 'Spaceship Earth' to the nearest m^2.

b Rearrange the formula $A = 4\pi r^2$ to make r the subject.

c A spherical gas tank has surface area 50 m^2. Find its radius correct to two decimal places.

ACTIVITY 2 — PREDICTING FORMULAE

When we observe a pattern in a few simple cases, it is tempting to predict that the pattern is true in general.

For example:

$$1 = 1^2$$
$$1+3 = 4 = 2^2$$
$$1+3+5 = 9 = 3^2$$
$$1+3+5+7 = 16 = 4^2$$

Assuming this pattern continues, we predict that "the sum of the first n odd integers is n^2".

How many cases would have to be checked if we wanted to know for certain that the prediction is true? Could this ever be achieved?

What to do:

1 One way to prove the result on the sum of odd integers is to consider

$$S = 1 + 3 + 5 + 7 + \dots + 2n-3 + 2n-1$$
and $$S = 2n-1 + 2n-3 + \dots \dots + 3 + 1$$

a Add these equations vertically to complete the proof.

b Can you now be certain that the result is true for all positive integers n?

2

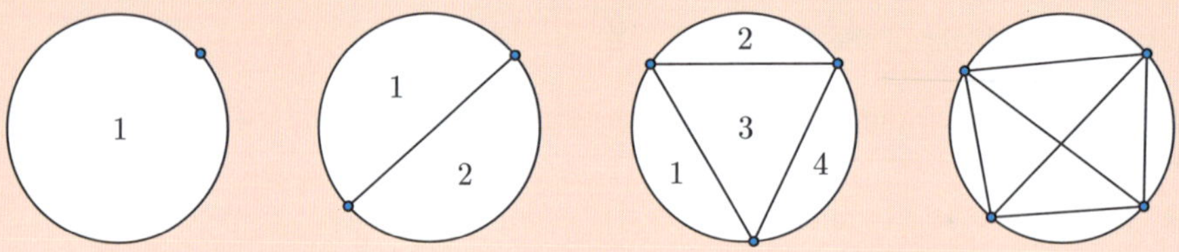

$n = 1$ $n = 2$ $n = 3$ $n = 4$

The above figures show the first four cases of n points being placed on a circle's circumference, and each point joined with a line to every other point. Let R_n be the number of regions formed. So, $R_1 = 1$, $R_2 = 2$, and $R_3 = 4$.

a Find R_4.

b Hence, predict a formula for R_n.

c Draw the case for $n = 5$, and hence find R_5. What does this case say about your formula for R_n?

3 For the first few values of n, the table below shows the sum of the first n prime numbers and 2, as well as whether the result is prime or composite:

n	*Sum*	*Value*	*Prime / Composite*
1	$2+2$	4	composite
2	$2+3+2$	7	prime
3	$2+3+5+2$	12	composite
4	$2+3+5+7+2$	19	prime
5	$2+3+5+7+11+2$	30	composite
6	$2+3+5+7+11+13+2$	43	prime

a Continue this table up to $n = 10$.

b Use the word 'prime' or 'composite' to complete these predictions:

i if n is odd, the value is ......

ii if n is even, the value is ......

c For each prediction in **b**, either prove it true, or find a counter-example to prove it false.

Chapter 16

Transformation geometry

Contents:

OPENING PROBLEM

How can we describe the transformation of the green semi-circle to produce the following images?

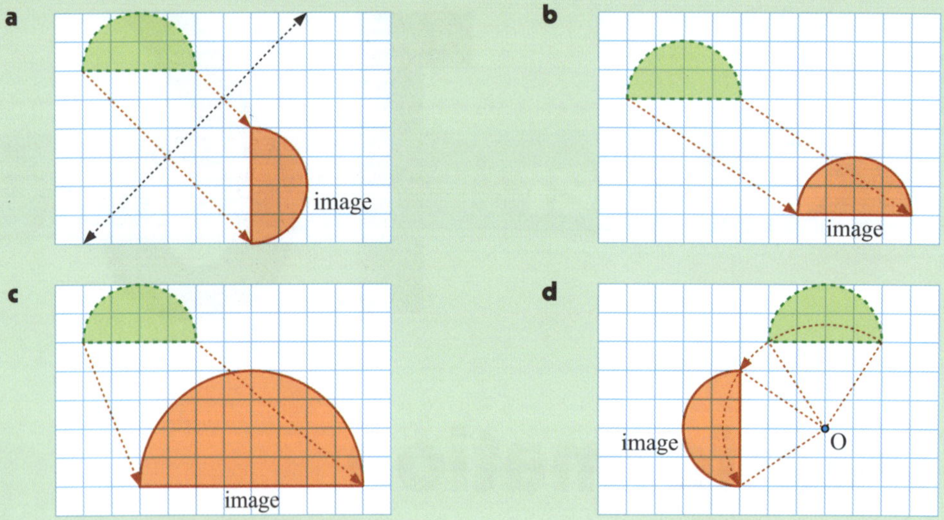

TRANSFORMATIONS

A change in the size, shape, orientation, or position of a figure is called a **transformation**.

Reflections, rotations, translations and enlargements are all examples of transformations. We can describe these transformations mathematically using **transformation geometry**.

In transformation geometry, figures are transformed in size, shape, orientation, or position, according to certain rules.

The original figure is called the **object** and the new figure is called the **image**.

HISTORICAL NOTE

Sir D'Arcy Wentworth Thompson (1860 - 1948) was a mathematical biologist from Scotland, most noted for his book *On Growth and Form*. He emphasised the similarities between some plants and animals which develop because a particular characteristic is physically beneficial to both organisms.

In the chapter *The Comparison of Related Forms*, Thompson described how differences in the forms of related animals could be described using mathematical transformations. For example, the pictures below show how a salmon can be transformed into a schnapper by using a curved grid.

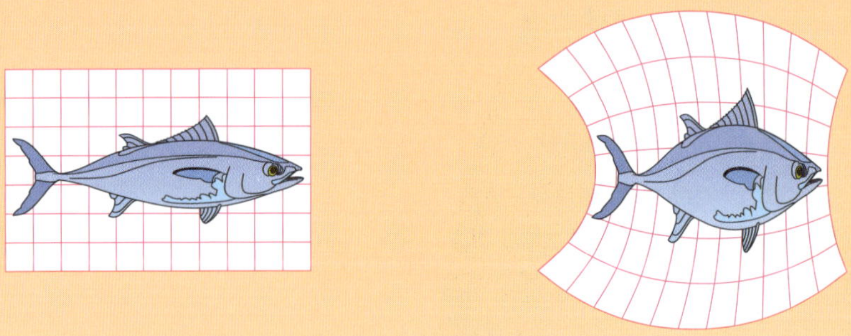

The transformations that we will consider in this chapter are:

- **translations**, where every point moves a fixed distance in a given direction

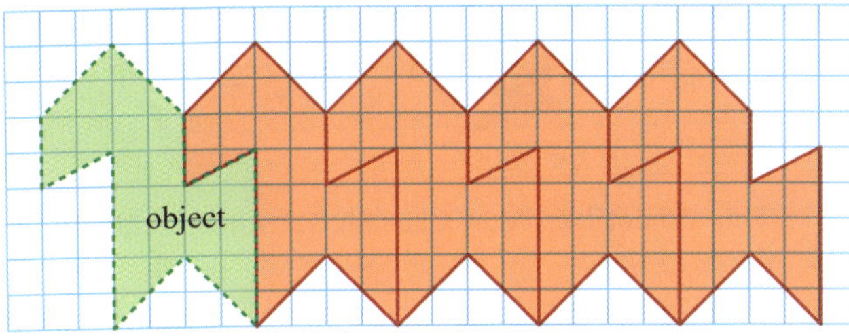

slide the original 4 units to the right to find each new image

- **reflections** or mirror images

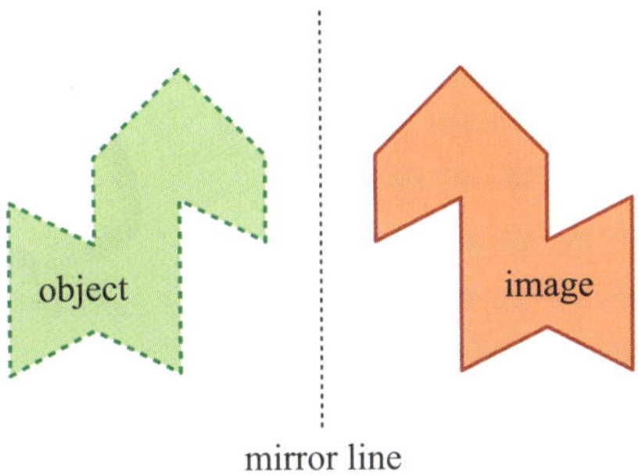

- **rotations** about a point through a given angle

DEMO

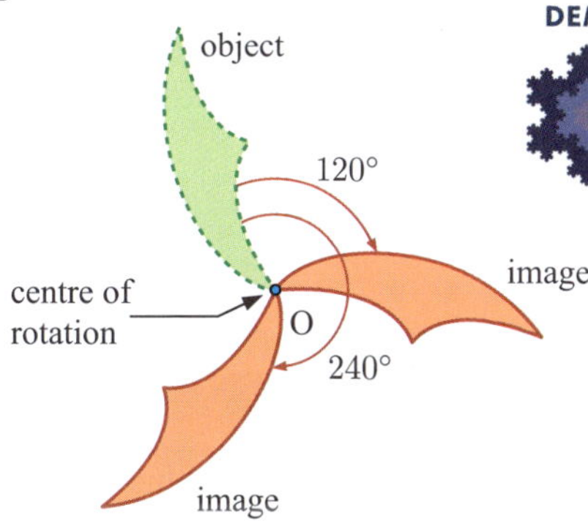

- **enlargements** and **reductions**, where objects are transformed into larger or smaller objects of the same shape.

DEMO

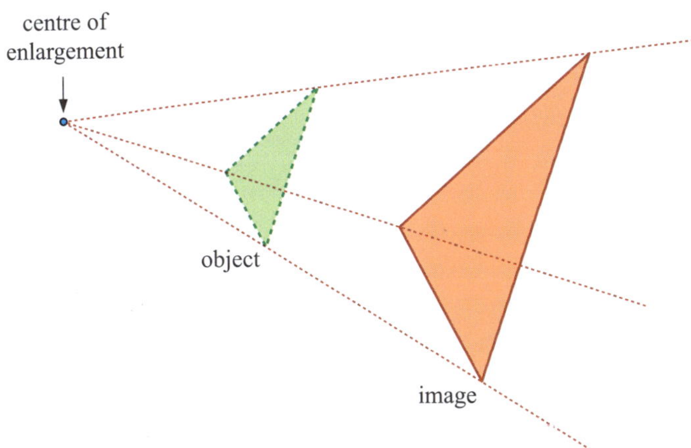

A TRANSLATIONS

A **translation** is a transformation in which every point on the figure moves a fixed distance in a given direction.

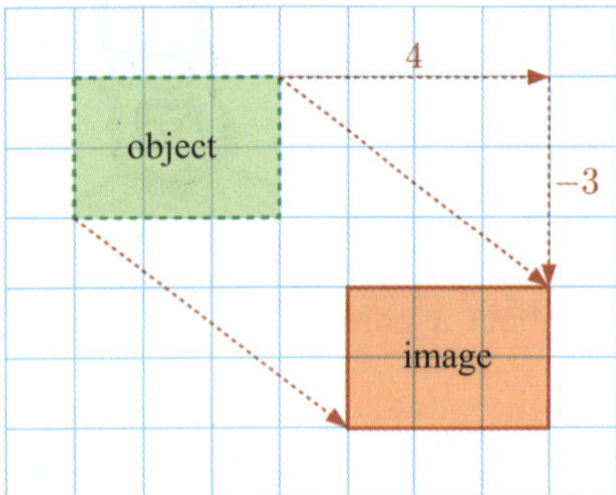

This object has been translated 4 units right and 3 units down.

We can describe the translation using the **translation vector** $\begin{pmatrix} 4 \\ -3 \end{pmatrix}$ ← horizontal movement, ← vertical movement

We can use Pythagoras' theorem to find the distance that the object moves.

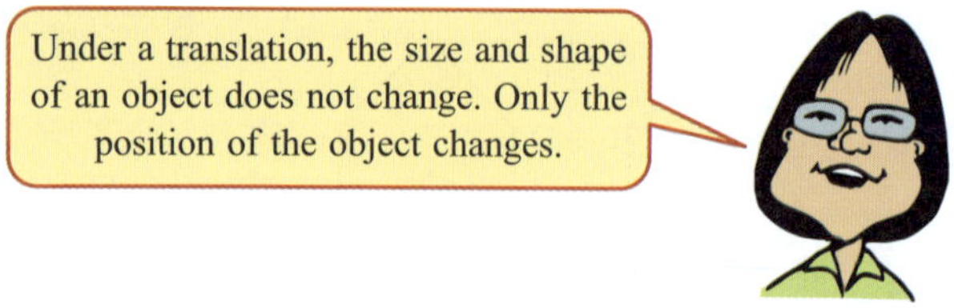

Example 1 — Self Tutor

a Translate this object by the translation vector $\begin{pmatrix} -5 \\ 2 \end{pmatrix}$.

b How far has the object moved?

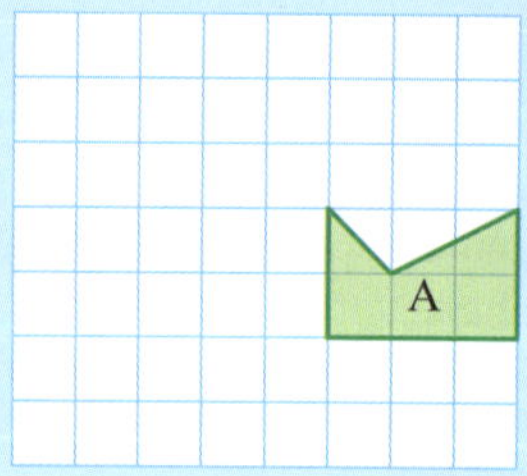

a The translation vector $\begin{pmatrix} -5 \\ 2 \end{pmatrix}$ means that the object moves 5 units left and 2 units up.

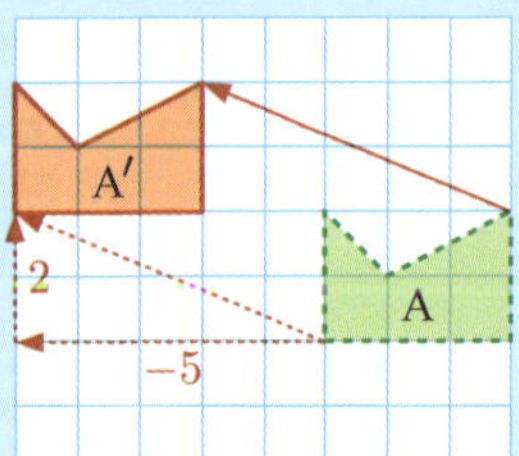

b Let the distance be x units.

Using Pythagoras' theorem, $x^2 = 5^2 + 2^2$

$\therefore\ x^2 = 29$

$\therefore\ x = \sqrt{29}$ {as $x > 0$}

$\therefore$ the object has moved $\sqrt{29}$ units.

We can also translate objects plotted on a **Cartesian plane**.

Example 2 Self Tutor

a Translate the quadrilateral ABCD 2 units right and 4 units up.

b Use a translation vector to describe this translation.

c State the vertex coordinates of the image quadrilateral.

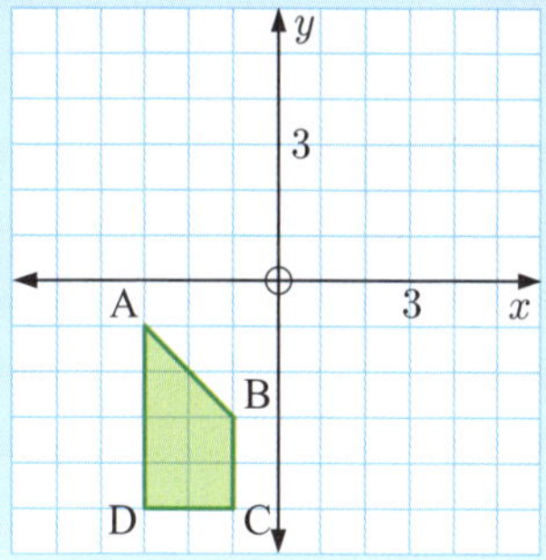

a

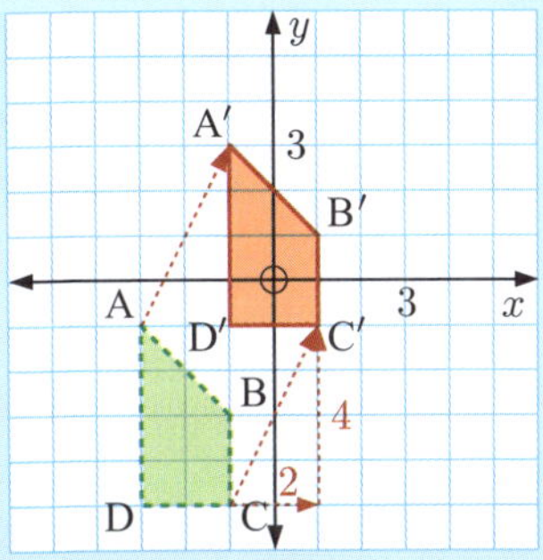

b The translation vector is $\begin{pmatrix} 2 \\ 4 \end{pmatrix}$.

c The vertices of the image quadrilateral are A′(−1, 3), B′(1, 1), C′(1, −1), and D′(−1, −1).

Example 3 Self Tutor

Is this transformation a translation?

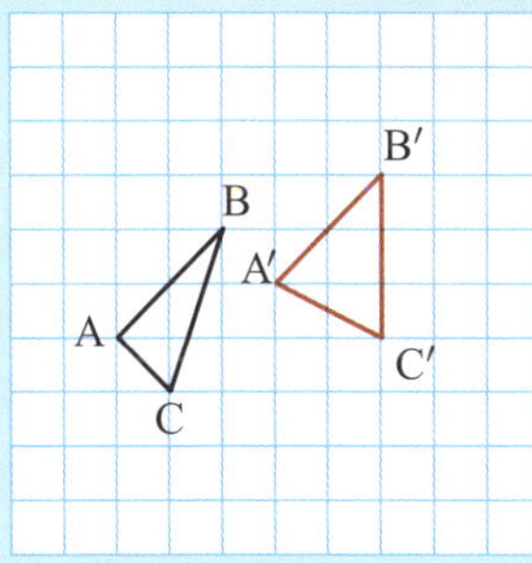

In a translation, every point on the figure moves the same distance in the same direction.

In this case, A has moved 3 units right and 1 unit up, while C has moved 4 units right and 1 unit up.

This transformation is **not** a translation.

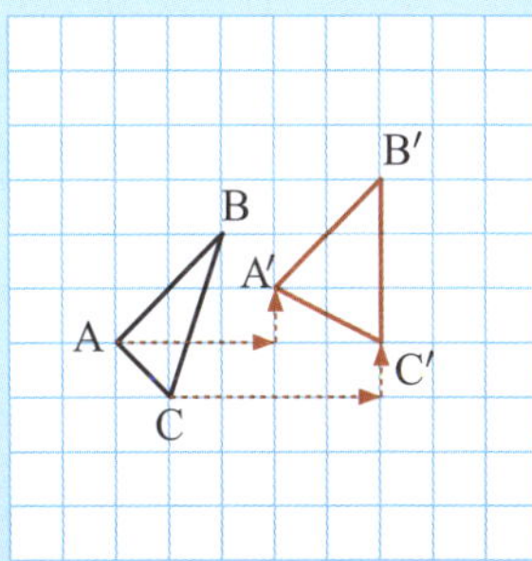

EXERCISE 16A

1 Translate the given figures in the direction indicated. In each case, describe the translation using a translation vector.

PRINTABLE DIAGRAMS

a
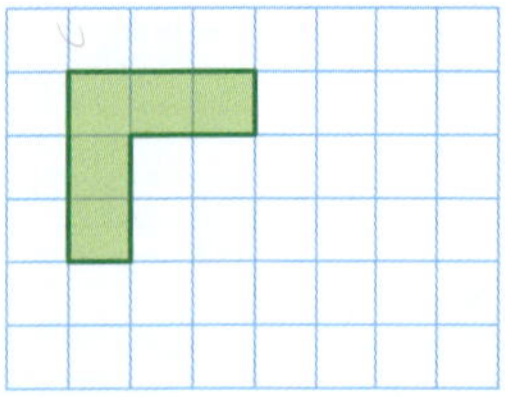
3 units right,
2 units down.

b
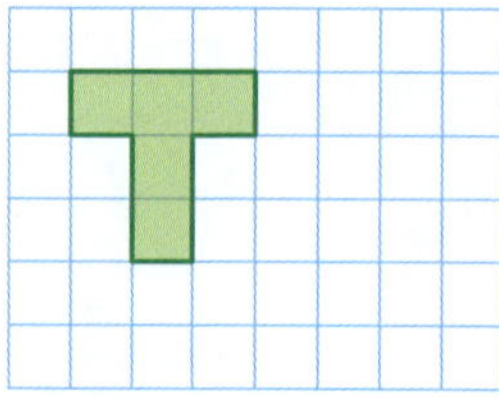
4 units right.

c
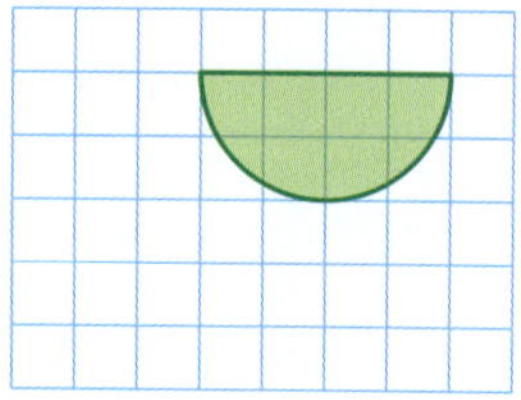
3 units down.

d
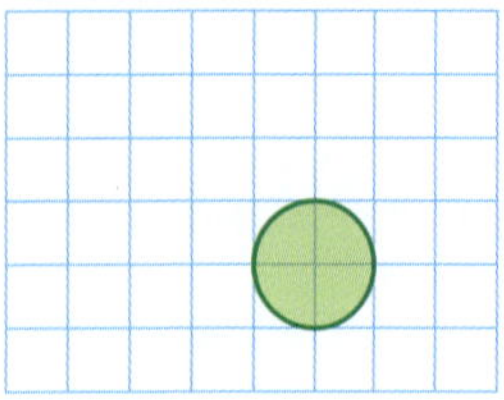
3 units left,
1 unit up.

e
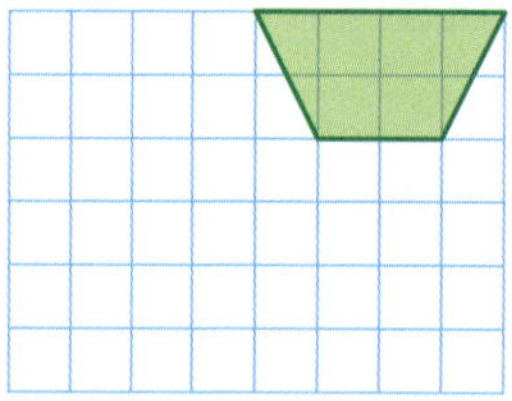
1 unit left,
4 units down.

f
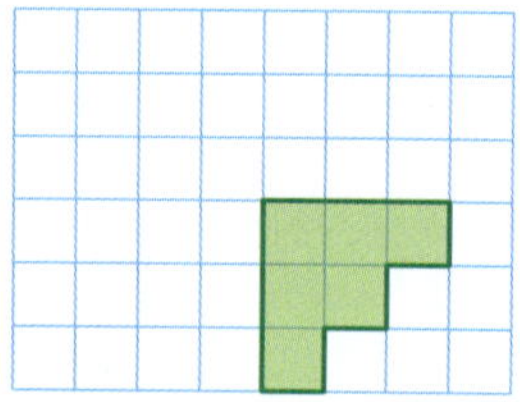
3 units left,
3 units up.

2 Describe the translation from A to A′ in words and using a translation vector.

a
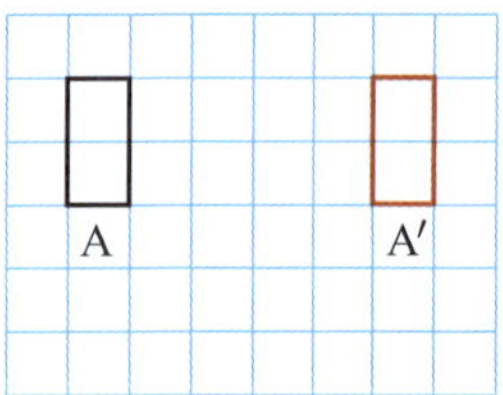

b
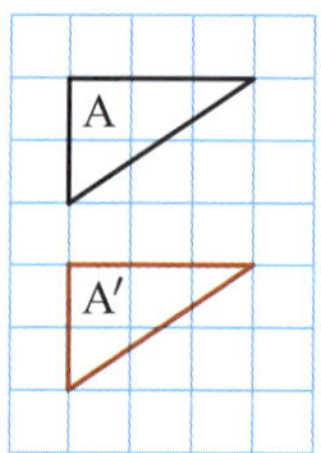

c
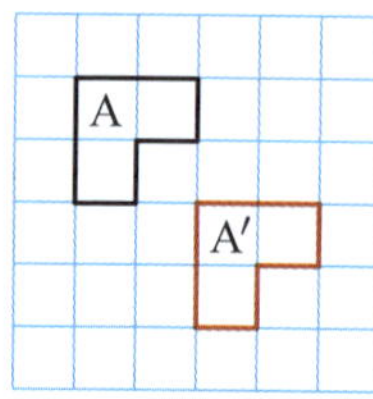

d
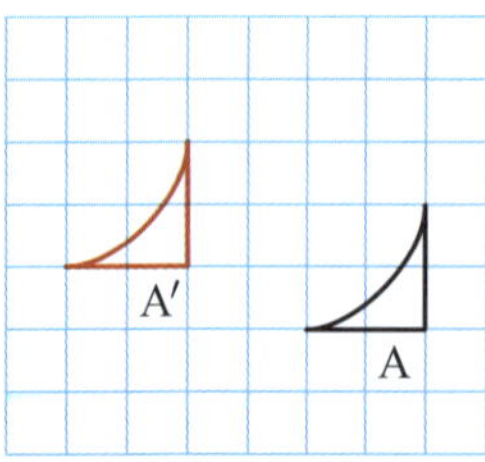

e
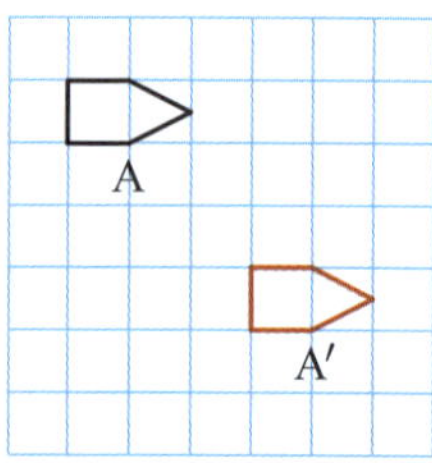

f
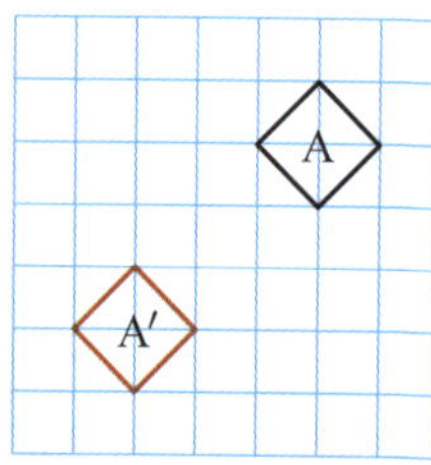

3 An object A is translated $\begin{pmatrix} 5 \\ -3 \end{pmatrix}$ to A′.

a How far has the object moved?

b Describe the translation from A′ back to A.

4 Translate each figure in the direction given, and state the vertex coordinates of the image:

a

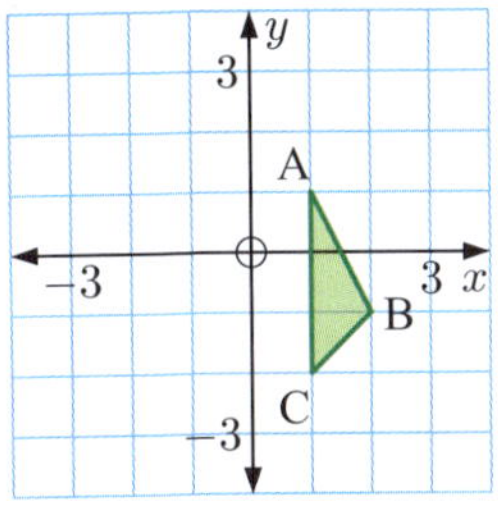

Translate 4 units left.

b

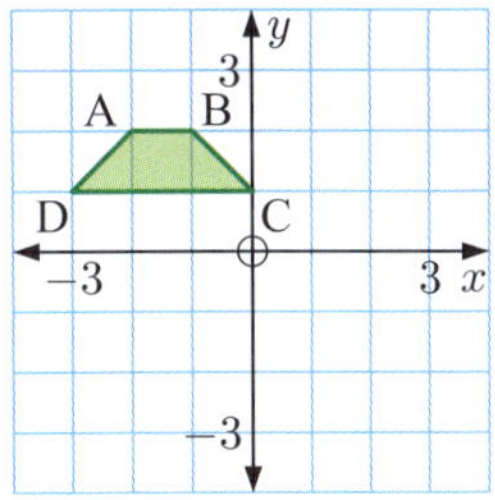

Translate 2 units right, then 1 unit down.

c

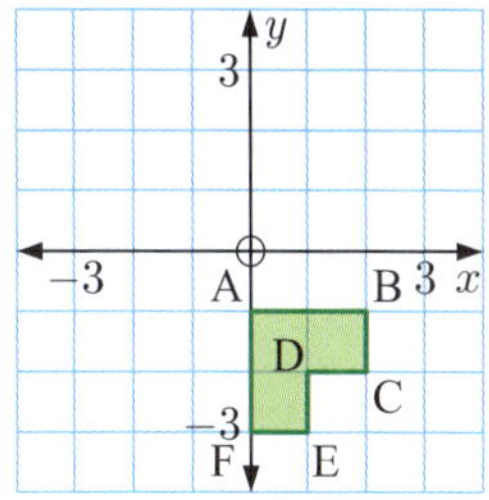

Translate 3 units left, then 4 units up.

5 The figure A has been translated to B, then B has been translated to C.

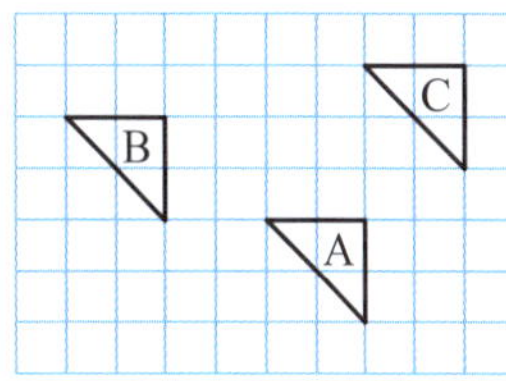

a Give the translation vector from A to B.

b Give the translation vector from B to C.

c What translation vector would move A directly to C?

6 Are the following transformations translations? If so, describe the translation.

a

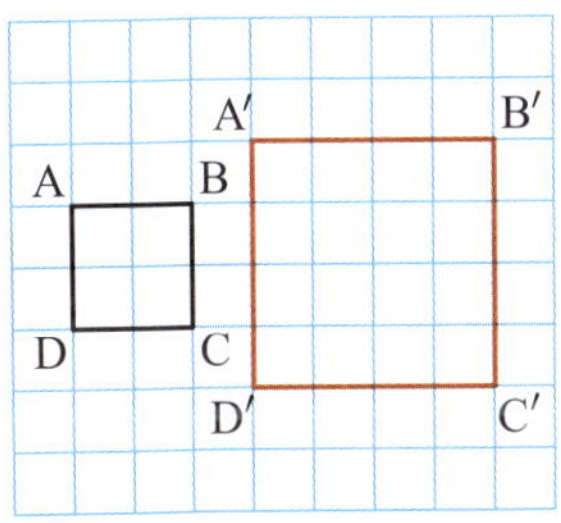

b

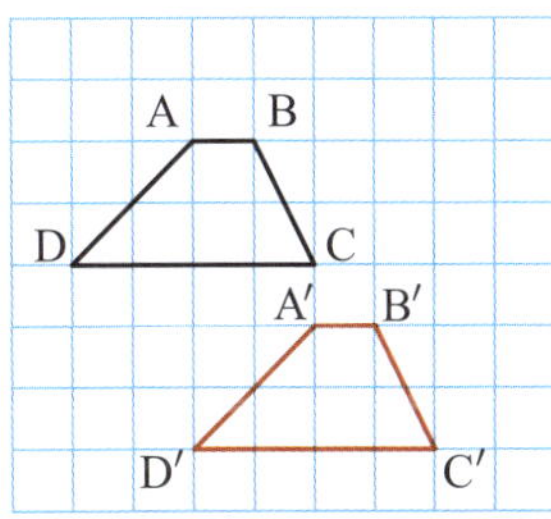

c

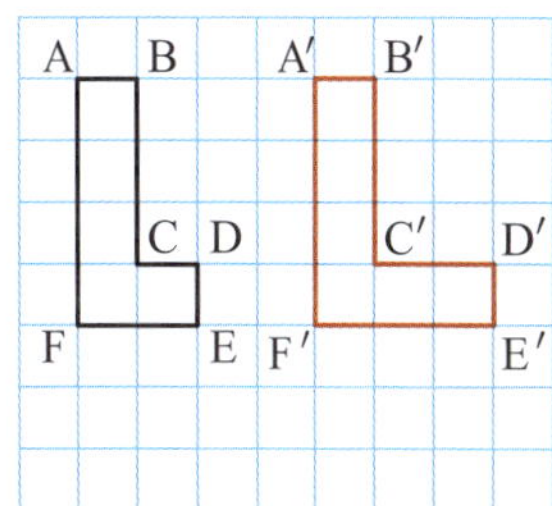

7 Consider the figures alongside.

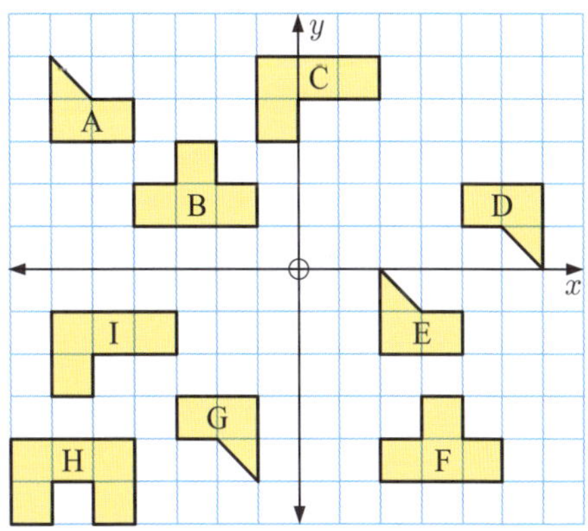

a Which of these figures is a translation of figure C? Describe the translation from figure C to this figure.

b Which of these figures is a translation of figure G? Describe the translation from figure G to this figure.

c Which of these figures cannot be translated to any other figure?

8 Draw triangle ABC where A is $(-1, 3)$, B is $(4, 1)$, and C is $(0, -2)$.

a Translate the figure with translation vector $\begin{pmatrix} 4 \\ -2 \end{pmatrix}$.

b State the coordinates of the image vertices A′, B′, and C′.

c Find the gradient of [AA′], [BB′], and [CC′].

d By what distance has every point moved?

B REFLECTIONS

We encounter reflections every day as we look in a bathroom mirror, peer into a pond of water, or glance at the traffic in a car rear-view mirror.

The object alongside has been **reflected** in the **mirror line** to form its image. In this case we might call it the **mirror image**.

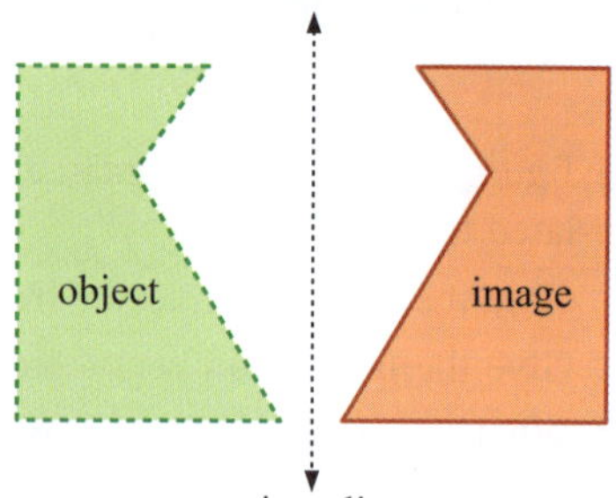

INVESTIGATION 1 REFLECTIONS

PRINTABLE FIGURES

You will need: A mirror, paper, pencil, ruler.

What to do:

1 Make two copies of the figures shown below:

a

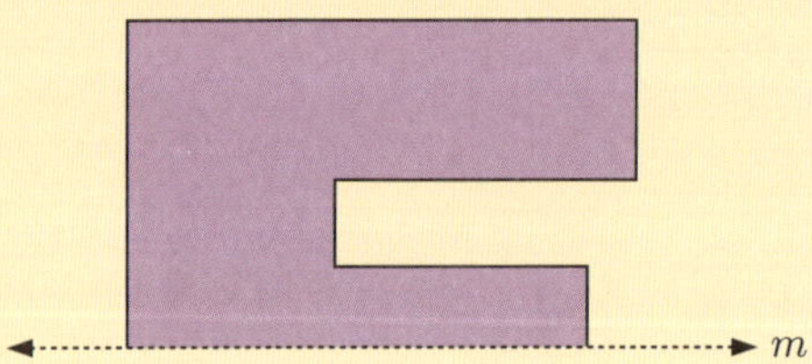

b

c

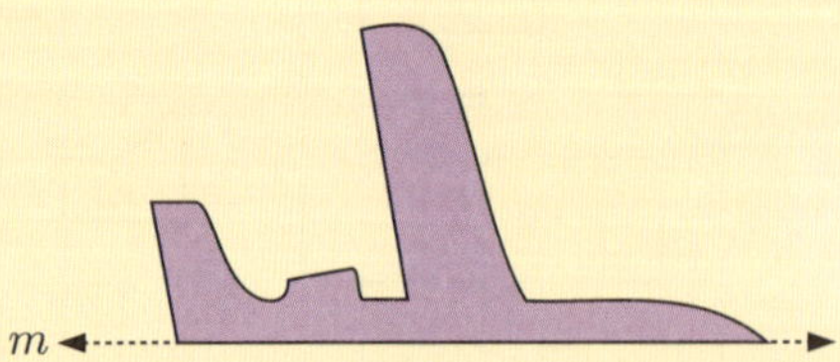

d

2 Put the mirror along the mirror line m on one copy. Draw the reflection as accurately as you can on the second copy.

3 Cut out the second copy with its reflection and fold it along the mirror line. You should find that the two parts of the figure can be folded exactly onto one another along the mirror line.

INVESTIGATION 2 PROPERTIES OF REFLECTION

The diagram shows a triangle and its reflection in a mirror line.

PRINTABLE DIAGRAM

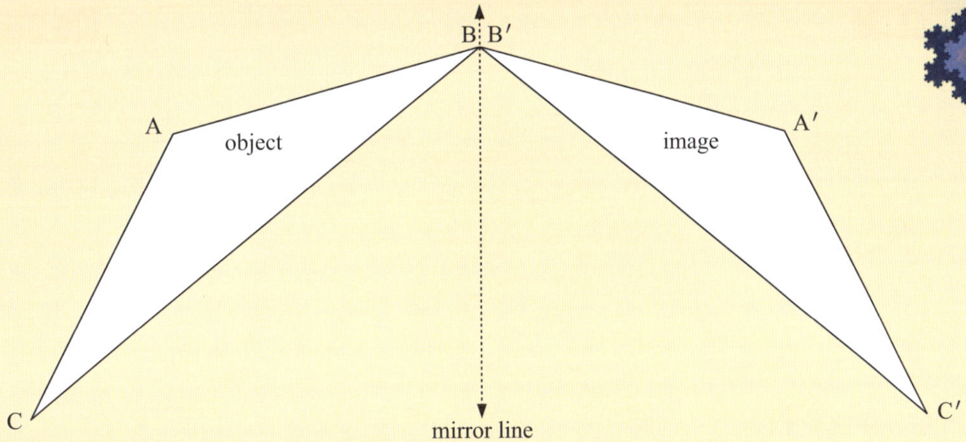

What to do:

1. Copy or print the diagram.
2. Join point A to its image point A′ on the reflection, and join C to C′.
3. Measure the distance from A to the mirror line and the distance from A′ to the mirror line. What do you notice?
4. Measure the angle that [AA′] makes with the mirror line.
5. Repeat **3** and **4** using the point C and its image point C′.
6. Measure the lengths of the sides of triangle ABC and the lengths of the sides of the image triangle A′B′C′. What do you notice?
7. Measure the angles of triangle ABC and the angles of triangle A′B′C′. What do you notice?
8. What do you notice about the image of a point that is on the mirror line?

From the **Investigation**, you should have found:

For any reflection:

- the image is as far behind the mirror line as the object is in front of it
- the line joining any image point to the corresponding point on the object is at right angles to the mirror line
- all lengths and angles are the same size in the image as they were in the object
- points on the mirror line are **invariant**, which means they do not move.

The mirror line is the **perpendicular bisector** of the line segment from any point on the object to its corresponding point on the image.

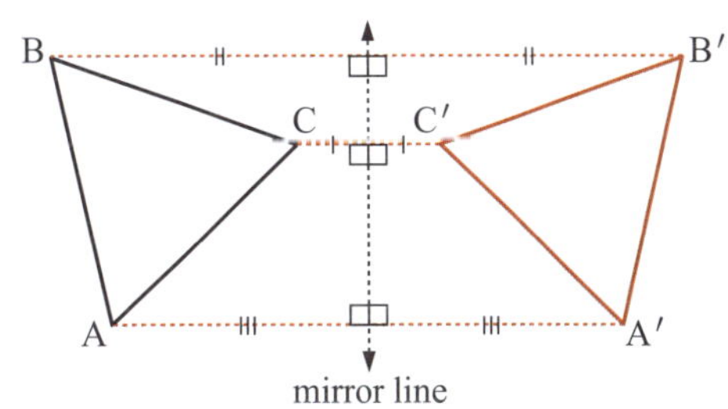

We can use these facts to help us draw reflections. It is easy to work on grid paper because we can count squares and see right angles.

Example 4 **Self Tutor**

Reflect each figure in the given mirror line:

a

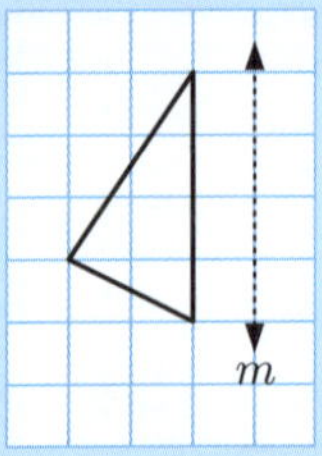

b

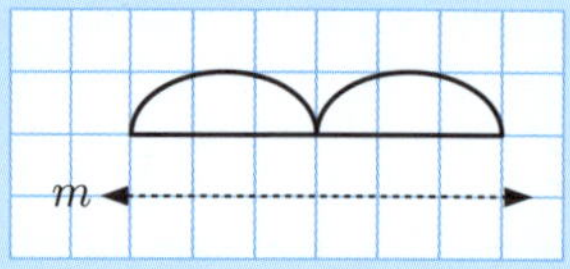

c

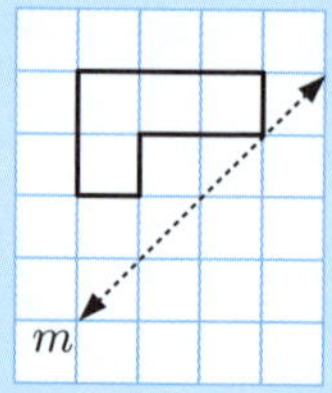

a

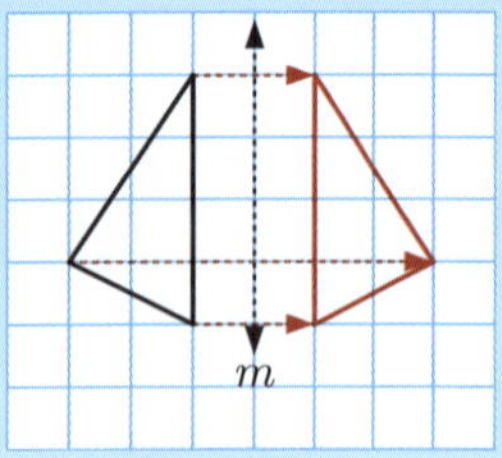

b

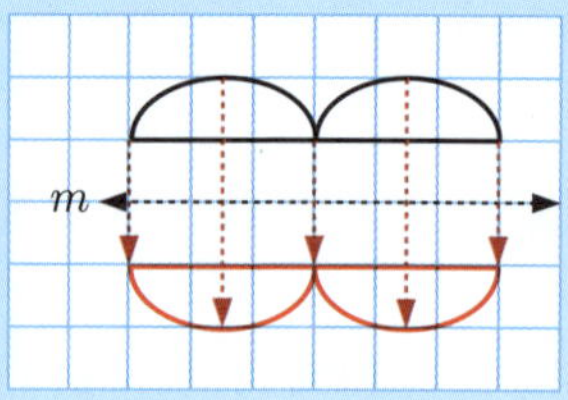

c

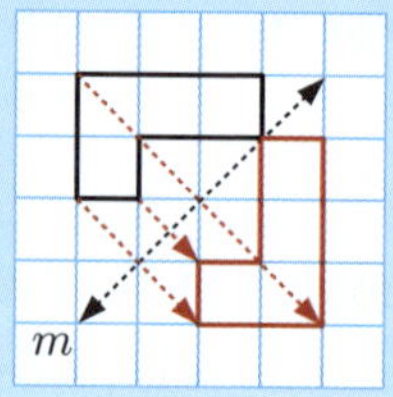

EXERCISE 16B.1

1 Copy the following figures onto grid paper and reflect them in the given mirror lines:

PRINTABLE FIGURES

a

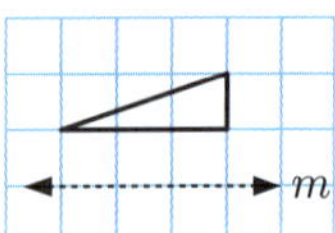

b

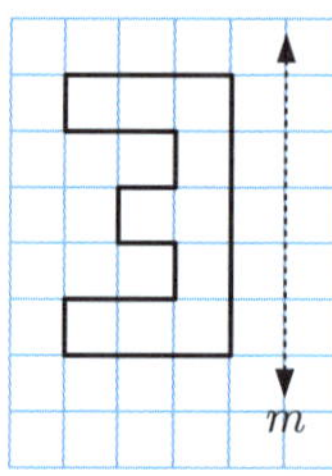

c

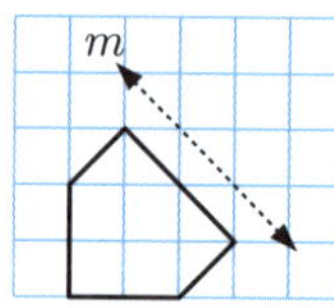

d

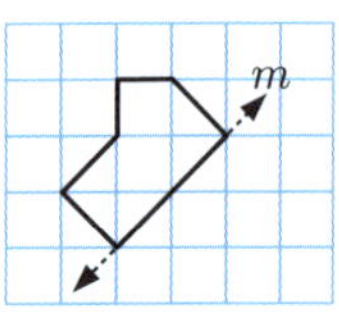

e

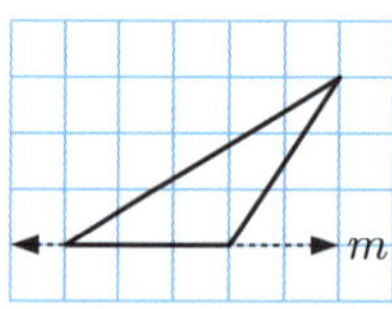

f

2 Which of the following transformations represent reflections?

a

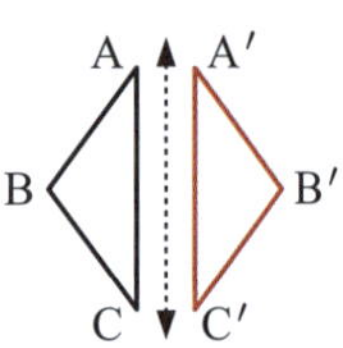

b

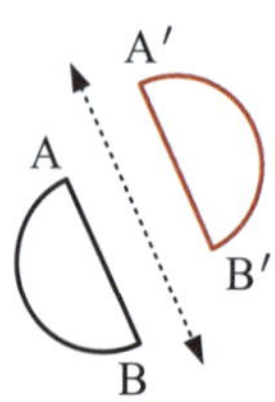

c

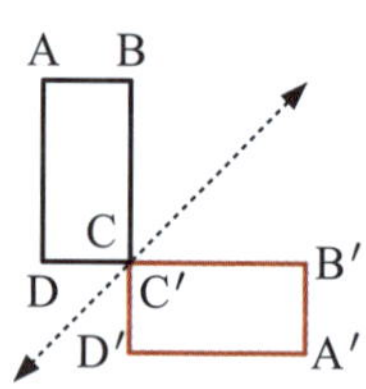

d

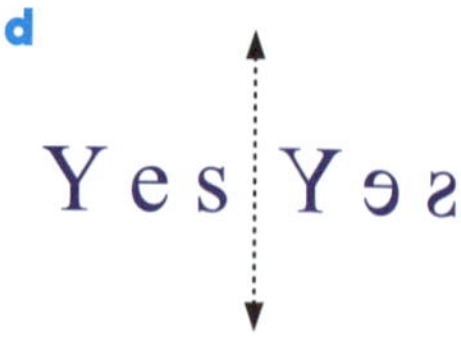

Example 5 **Self Tutor**

Reflect this figure in the y-axis.

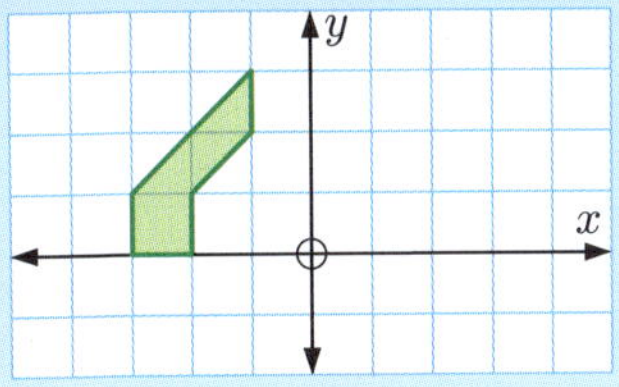

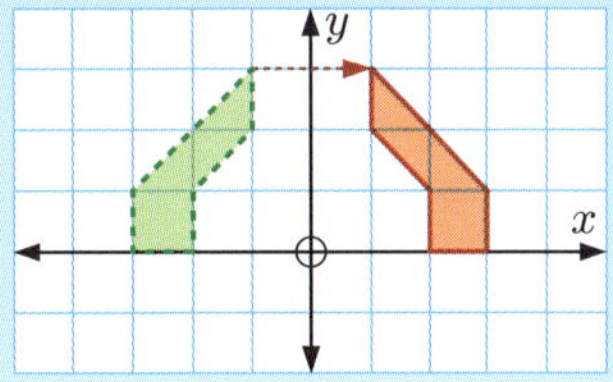

3 Reflect each figure in the x-axis:

a

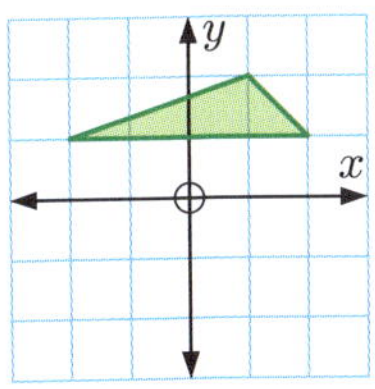

b

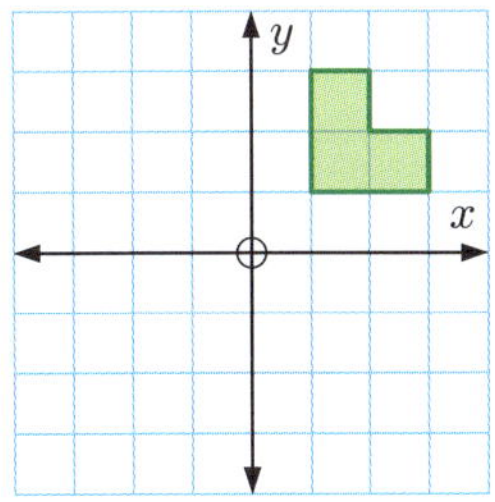

c

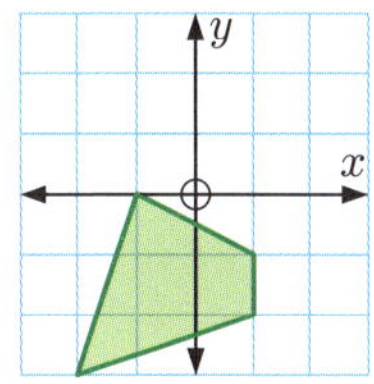

4 Reflect each figure in the y-axis:

a

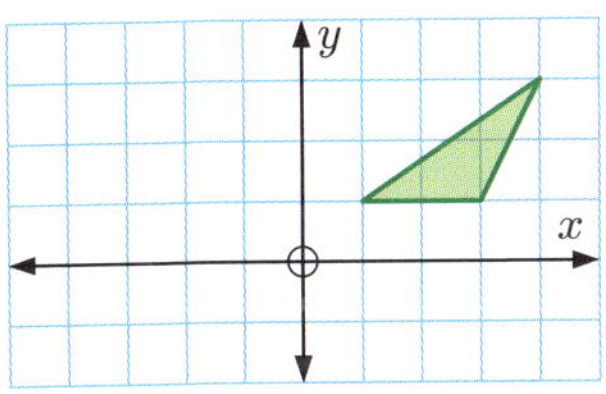

b

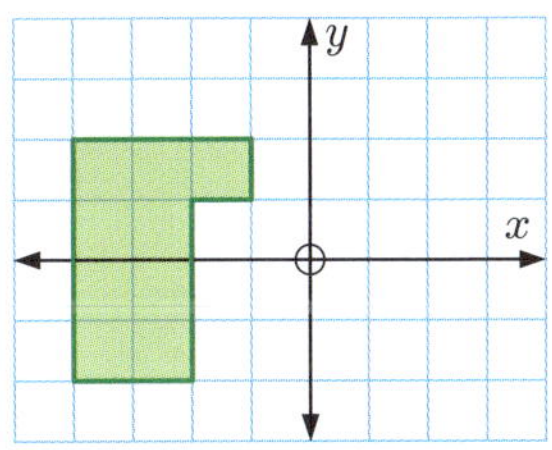

c

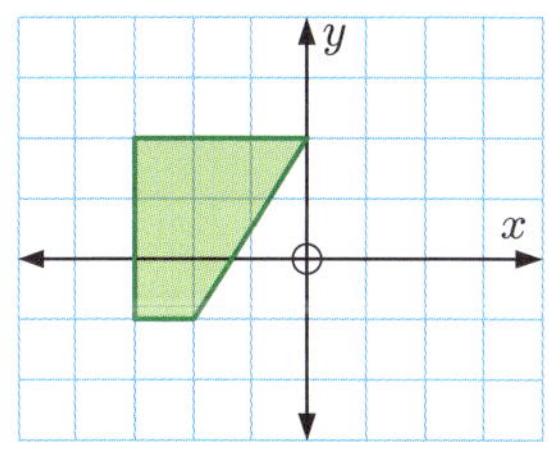

5 **a** A quadrilateral has vertices A(1, 5), B(4, 3), C(4, 1), and D(1, 1).

  i Plot the quadrilateral ABCD on a Cartesian plane.

  ii Reflect ABCD in the x-axis, and state the vertex coordinates of the image.

  iii Reflect ABCD in the y-axis, and state the vertex coordinates of the image.

  iv Reflect ABCD in the line $y = x$, and state the vertex coordinates of the image.

b Copy and complete:

  i When the point (a, b) is reflected in the x-axis, the image has coordinates (...., ...).

  ii When the point (a, b) is reflected in the y-axis, the image has coordinates (...., ...).

  iii When the point (a, b) is reflected in the line $y = x$, the image has coordinates (...., ...).

6 For each diagram, determine whether A′ is a reflection of A in the x-axis:

a

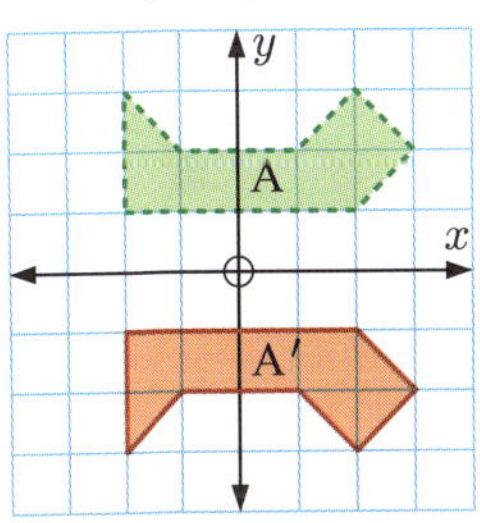

b

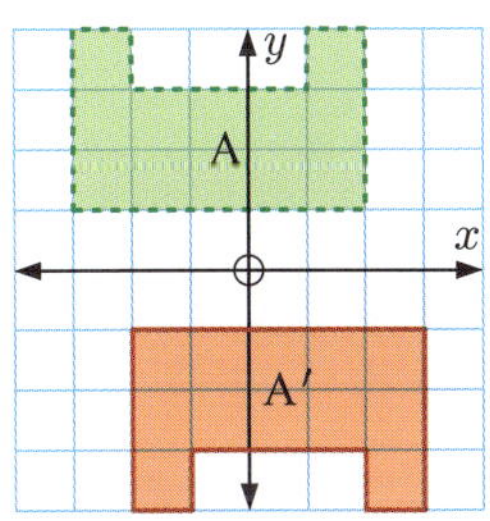

c

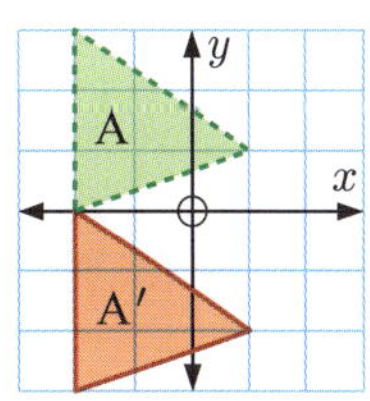

7 **a** Which two of the figures are reflections of each other?

b Which of the axes is the mirror line for this reflection?

c Which two of the figures are translations of each other?

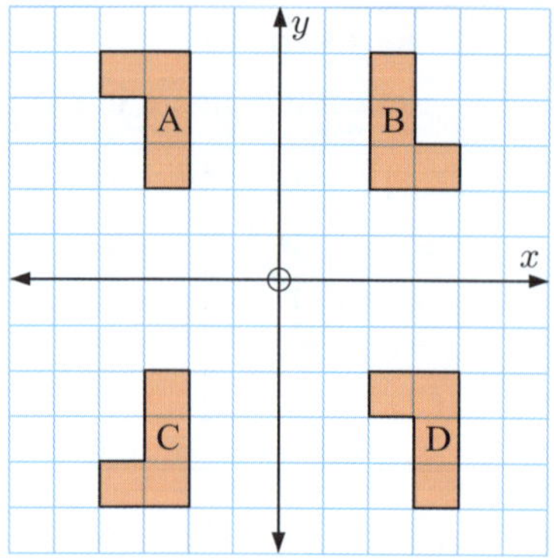

8 A service track must go from A to the pipeline and then to B by the shortest route.

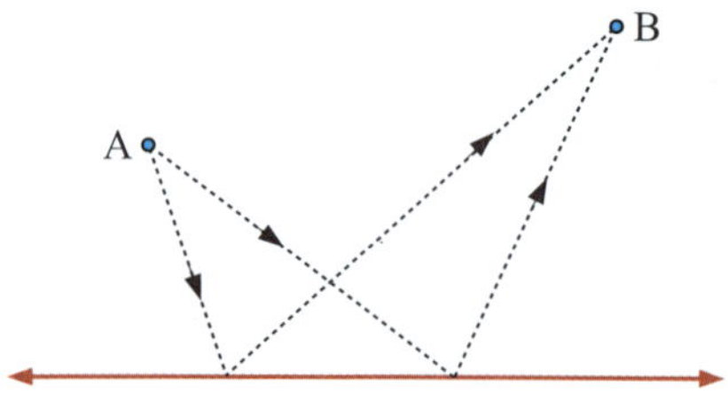

Simran's solution is as follows:

Step 1: Use the pipeline as a mirror line to reflect B.

Step 2: Join [AB′], and let [AB′] meet the pipeline at X.

Step 3: Join [BX].

Step 4: AX + XB is the shortest distance.

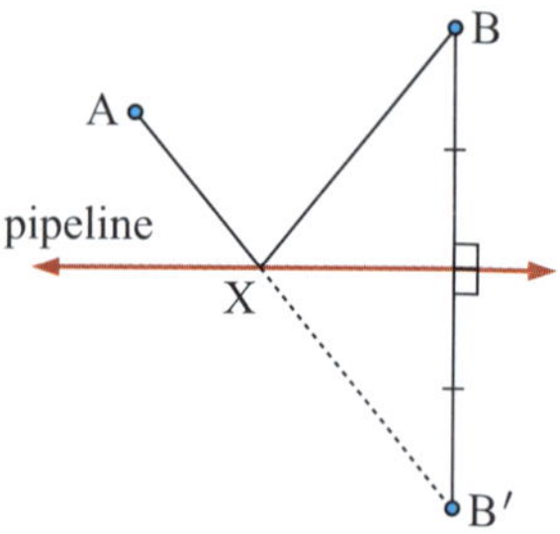

a What can be said about triangle BXB′? Explain your answer.

b Explain why Simran's solution is correct.

Hint: Let Y be any point on the pipeline other than X. Join [AY] and [YB′].

LINE SYMMETRY

A figure has a **line of symmetry** if it can be reflected in that line so that each half of the figure is reflected onto the other half of the figure.

For example, an isosceles triangle has one line of symmetry. It is the line from the apex to the midpoint of the base of the triangle.

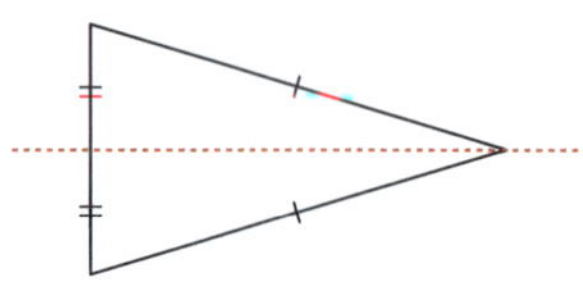

A square has 4 axes of symmetry.

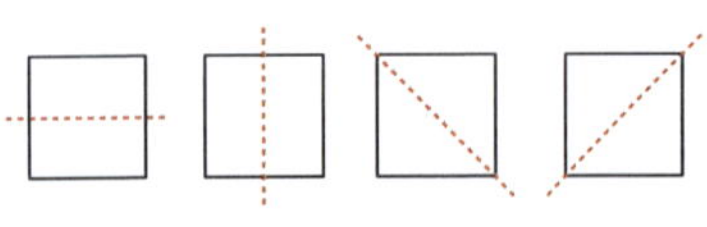

Example 6 **Self Tutor**

For each figure, draw all lines of symmetry:

a

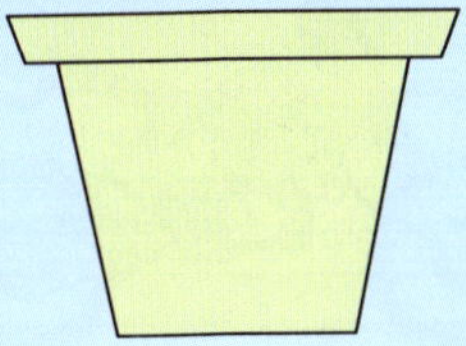

b

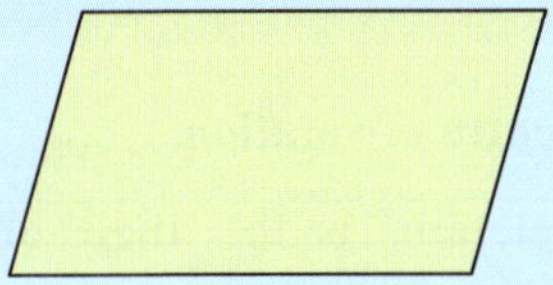

c

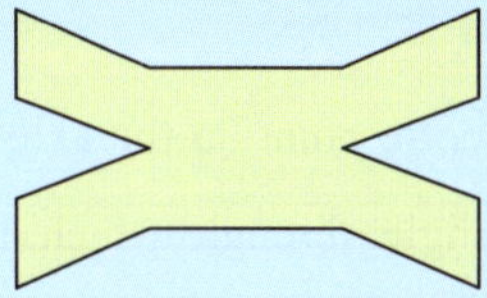

a

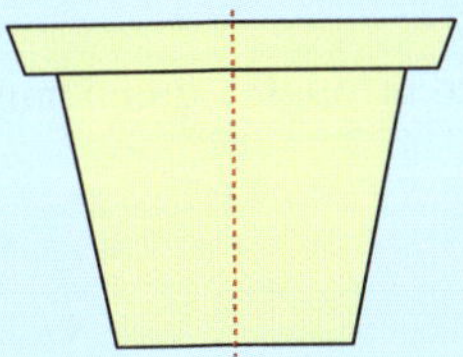

1 line of symmetry

b

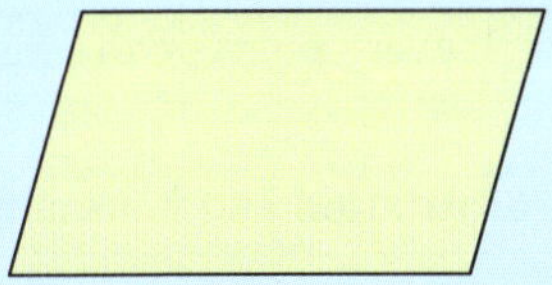

no lines of symmetry

c

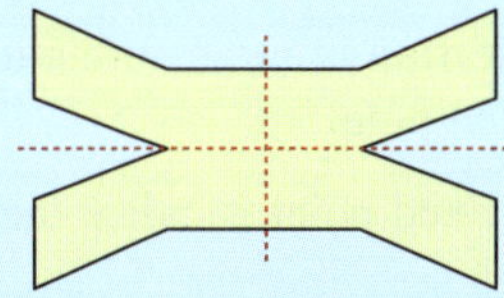

2 lines of symmetry

EXERCISE 16B.2

1 For each figure, draw all lines of symmetry:

a

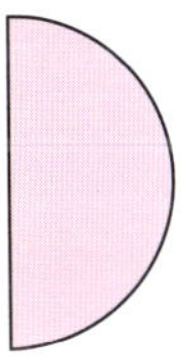

b

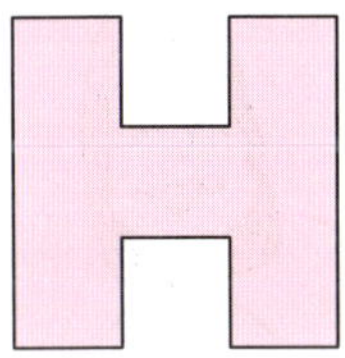

c

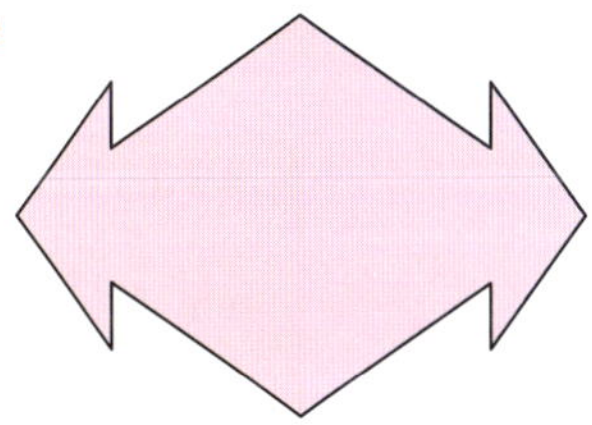

d

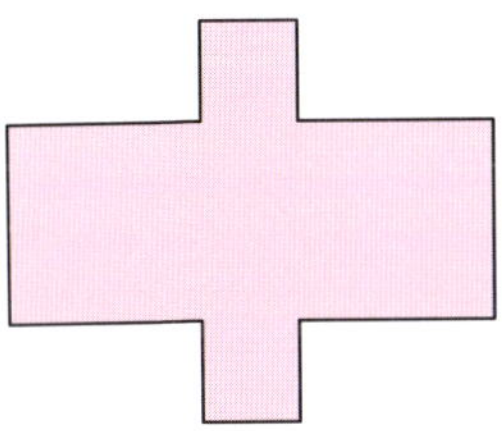

e

f

2 Draw each of the following figures, and draw all lines of symmetry. Record the number of lines of symmetry in each case.

a a square
b an equilateral triangle
c a parallelogram
d a rectangle
e a rhombus
f a regular pentagon
g a circle
h a kite
i an isosceles triangle

C ROTATIONS

When a wheel moves about its axle, we say that the wheel *rotates*.

The centre point on the axle is the **centre of rotation**.

The angle through which the wheel turns is the **angle of rotation**.

Other examples of rotation are the movement of the hands of a clock, and opening and closing a door.

> A **rotation** is a transformation in which every point on the figure is turned through a given angle about a fixed point.
>
> The fixed point is called the **centre of rotation** and is usually labelled O.

For example, the object alongside has been rotated 90° anticlockwise about O.

DEMO

Click on the icon to view the rotation.

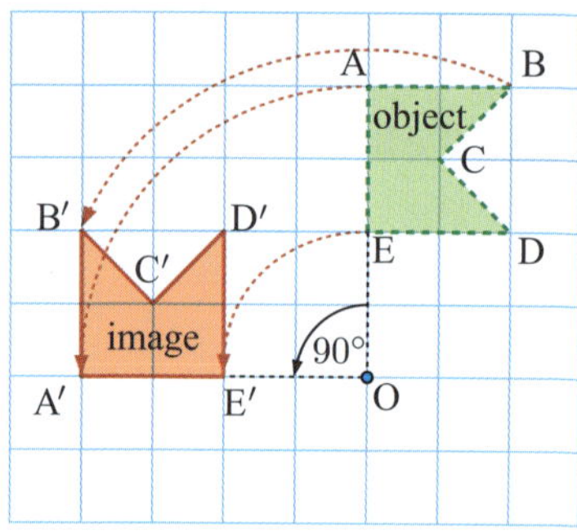

When a point is rotated about O, that point and its image are the same distance from O.

$OA = OA'$, $OB = OB'$, and so on.

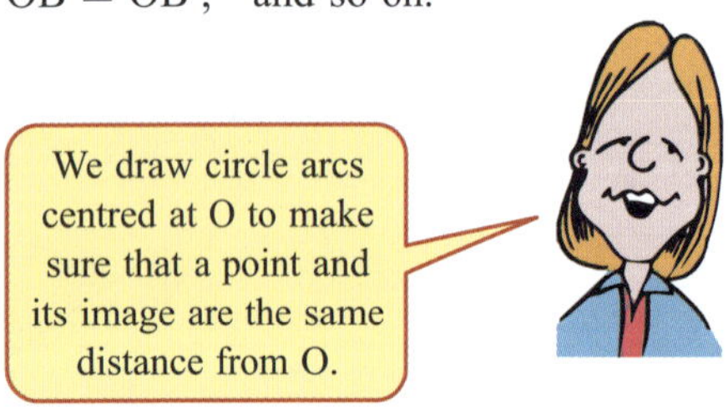

To completely describe a rotation we need to know:

- the **centre** of the rotation
- the **direction** of the rotation (clockwise or anticlockwise)
- the **angle** of the rotation.

Example 7 Self Tutor

Rotate each figure about O through the angle indicated:

a

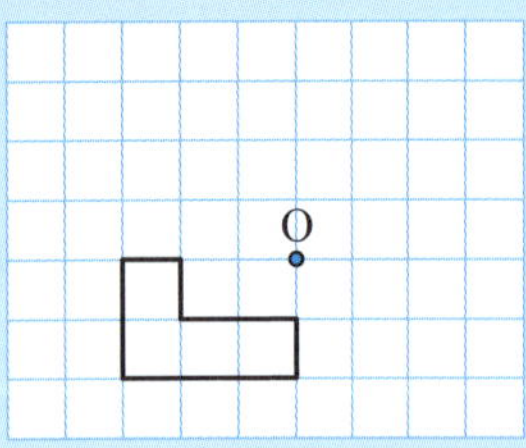

180° clockwise

b

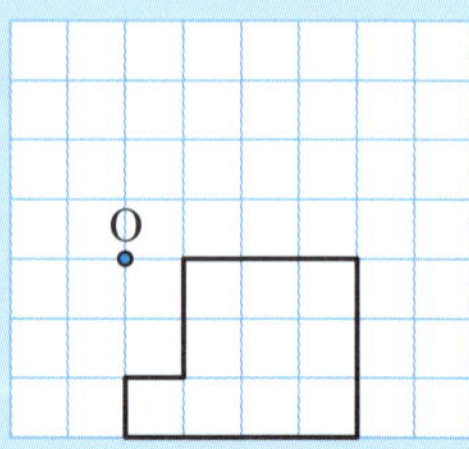

90° anticlockwise

c

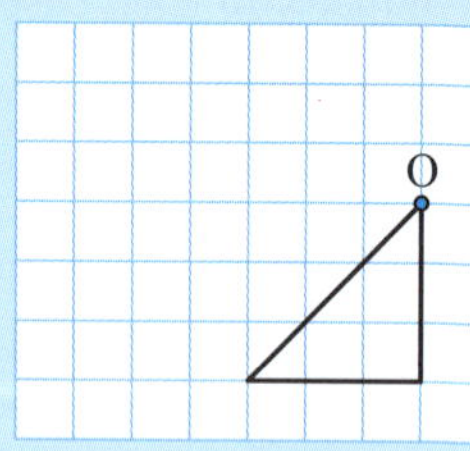

90° clockwise

a
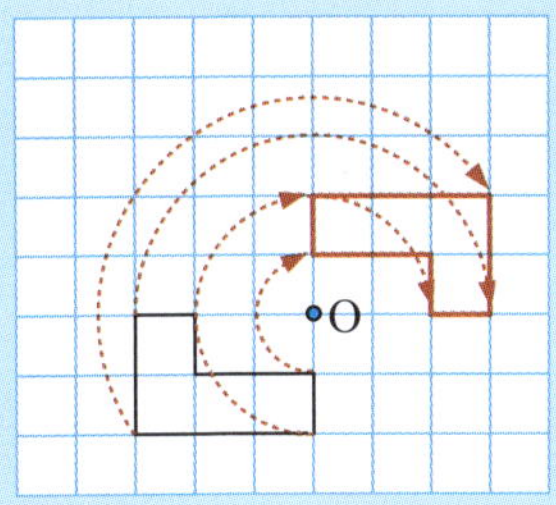

b
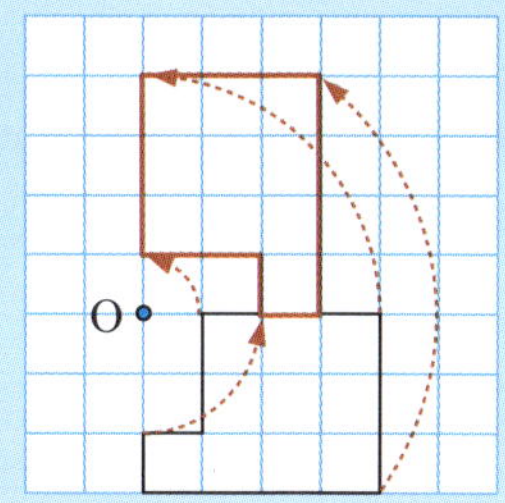

c
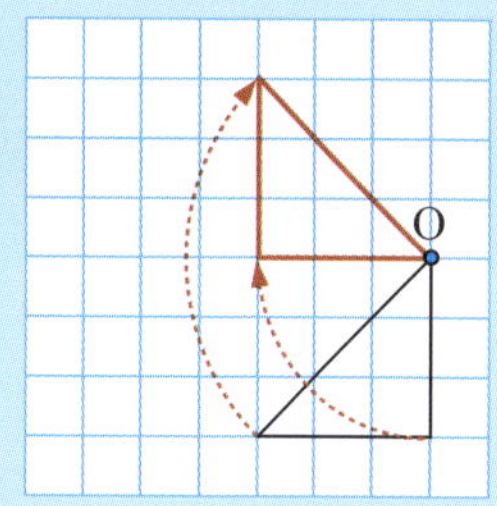

EXERCISE 16C

1 Rotate each figure about O through the angle given:

a
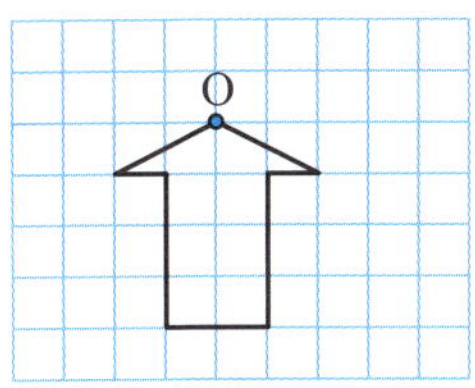

180°

b
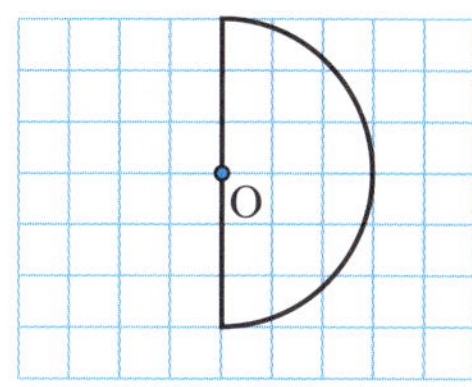

180°

c
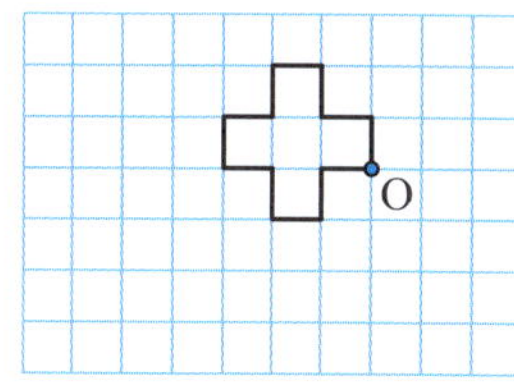

90° anticlockwise

d
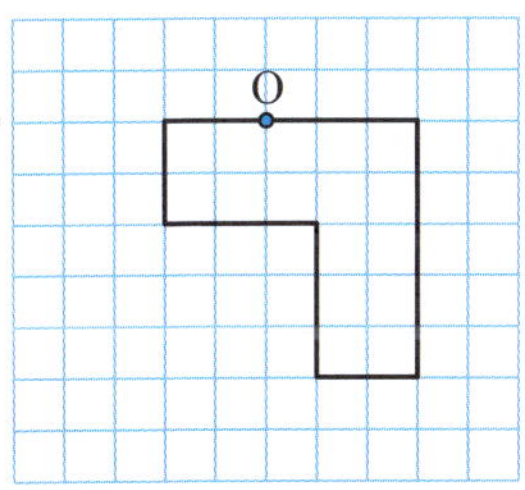

90° clockwise

e
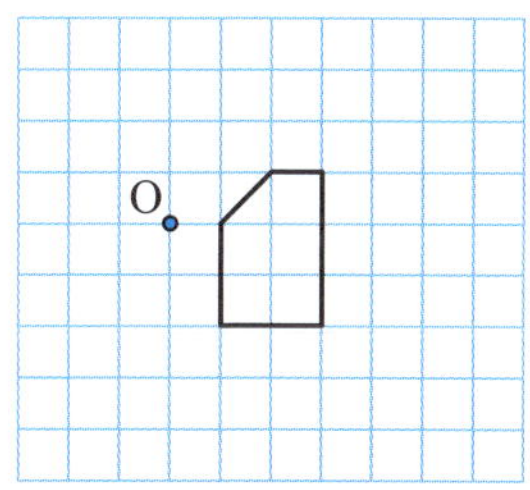

90° clockwise

f
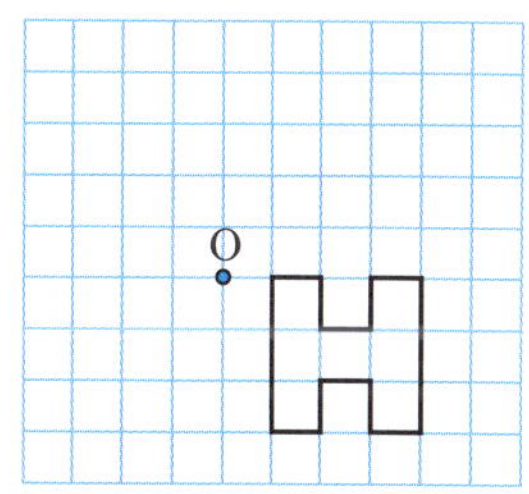

90° anticlockwise

2 Which of the following letters of the alphabet either remain the same letter or become a different letter under a rotation? Do not include full rotations of 360°.

A B C D E F G H I J K L M
N O P Q R S T U V W X Y Z

3 Derek is sitting in chair A. Which chair will he move to if he rotates anticlockwise about O through an angle of:

a 90° **b** 270° **c** 180°?

4 A figure is rotated 75° clockwise about a point O. Describe a rotation which will return the figure to its original position.

5 **a** Which of the figures alongside is a rotation of A about the origin?

b Determine the angle of rotation from figure A to this figure.

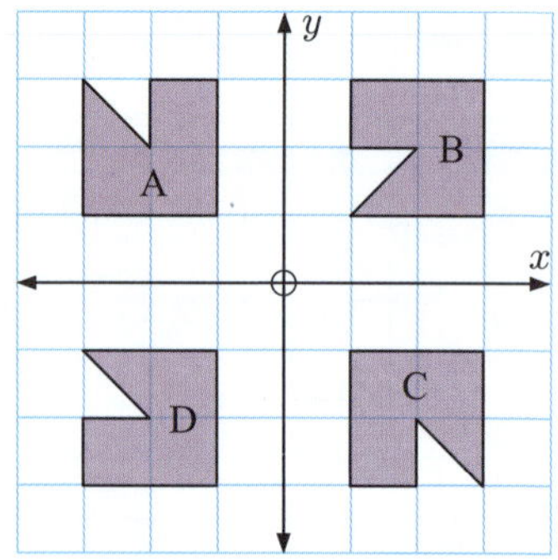

Example 8 **Self Tutor**

a State the vertex coordinates of triangle ABC.

b Rotate the triangle 90° clockwise about the origin O.

c State the vertex coordinates of the image.

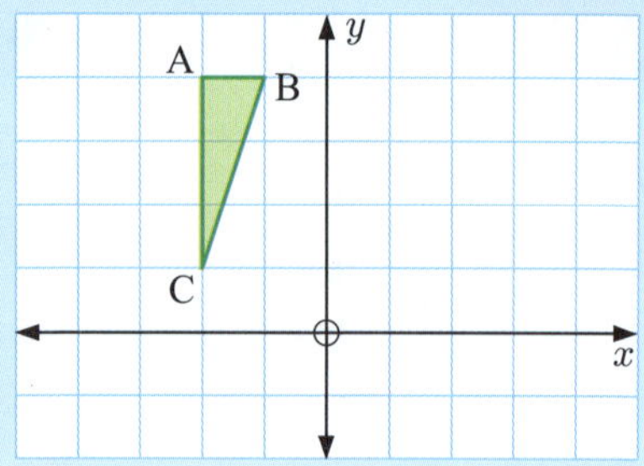

a The triangle has vertices A$(-2, 4)$, B$(-1, 4)$, and C$(-2, 1)$.

c The image triangle has vertices A$'(4, 2)$, B$'(4, 1)$, and C$'(1, 2)$.

b

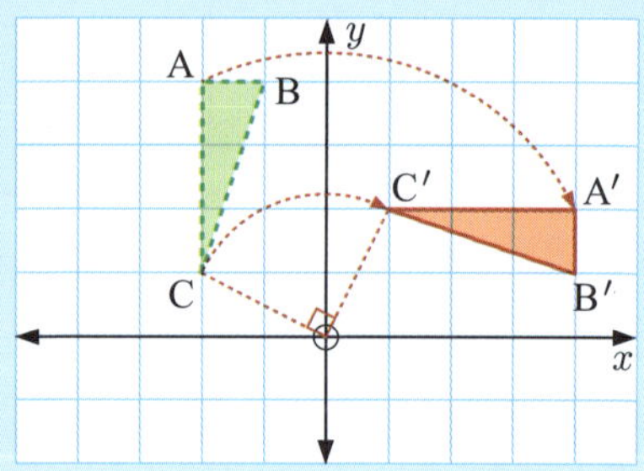

6 **a** State the vertex coordinates of triangle PQR.

b Rotate the triangle 90° anticlockwise about the origin O.

c State the vertex coordinates of the image.

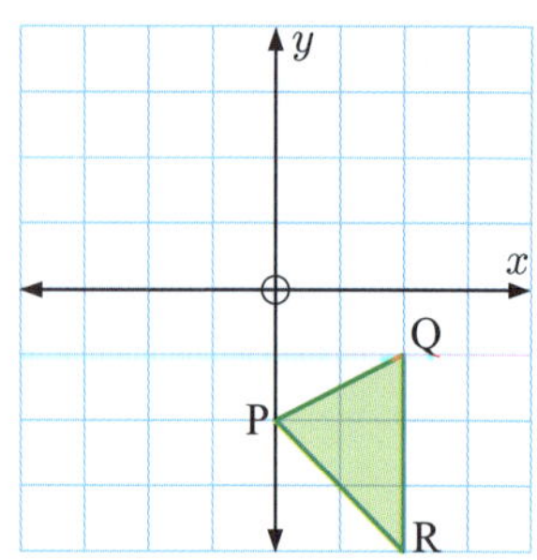

7 A quadrilateral has vertices A$(-4, -3)$, B$(-1, -3)$, C$(-1, -4)$, and D$(-4, -4)$.

a Plot ABCD on a Cartesian plane.

b Rotate ABCD 90° clockwise about the origin O.

c State the vertex coordinates of the image.

8 **a** Find the image of each point under a 90° anticlockwise rotation about O:
 i $(4, 1)$ **ii** $(5, -2)$ **iii** $(0, -5)$ **iv** $(-1, -5)$ **v** $(-3, 4)$

b Hence predict the image of (x, y) under a 90° anticlockwise rotation about O.

9 **a** Find the image of each point under a 180° rotation about O:
 i $(-3, 5)$ **ii** $(3, 0)$ **iii** $(1, 4)$ **iv** $(-3, -2)$ **v** $(4, -2)$

b Hence predict the image of (x, y) under a 180° rotation about O.

10 **a** Find the image of each point under a 90° clockwise rotation about O:
 i $(0, -2)$ **ii** $(4, 2)$ **iii** $(-1, -5)$ **iv** $(4, -2)$ **v** $(-3, 2)$

b Hence predict the image of (x, y) under a 90° clockwise rotation about O.

11 O is the midpoint of [AC] and $AB \neq BC$.
Redraw the figure and rotate it through 180° about O.

a Name the resulting figure which is made up of the original and its image.

b List the geometrical properties of the resulting figure. Give reasons for your answer.

c Describe the case where $A\widehat{B}C$ is a right angle.

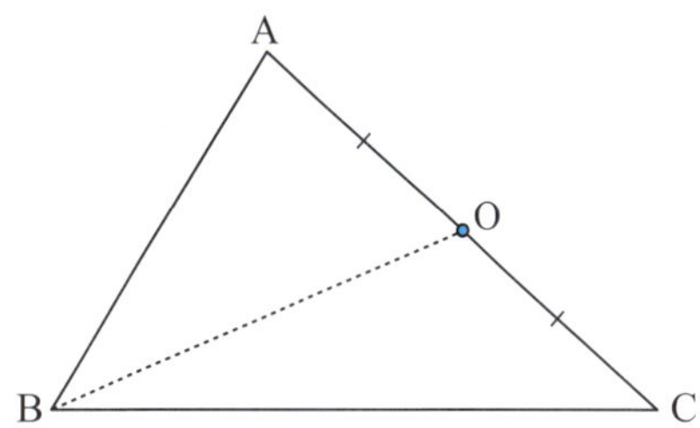

D ENLARGEMENTS AND REDUCTIONS

We are all familiar with **enlargements** in the form of photographs, zoom tools in computer software, or looking through a microscope. For an enlargement the image is larger than the original.

Plans and maps are examples of **reductions**. In these cases the image is smaller than the original.

Most photocopiers can make images either smaller or larger than the original.

The **scale factor** of an enlargement (or reduction) is the ratio by which the side lengths of the object are enlarged (or reduced).

Consider the diagrams below. They are clearly not the same *size*, but they do have the same *shape*. Their side lengths are in the same proportions.

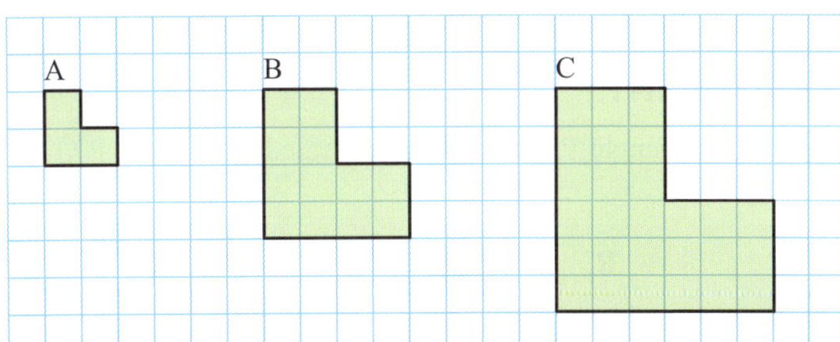

B is an enlargement of A, and is a reduction of C.

When A is enlarged to B, the side lengths are all doubled. The scale factor $k = 2$.

When C is reduced to B, the side lengths are all multiplied by $\frac{2}{3}$. The scale factor $k = \frac{2}{3}$.

If the scale factor is greater than 1, an **enlargement** occurs.

If the scale factor is less than 1, a **reduction** occurs.

DEMO

EXERCISE 16D.1

1 In the following diagrams, A has been enlarged to A′. Find the scale factor in each case.

a
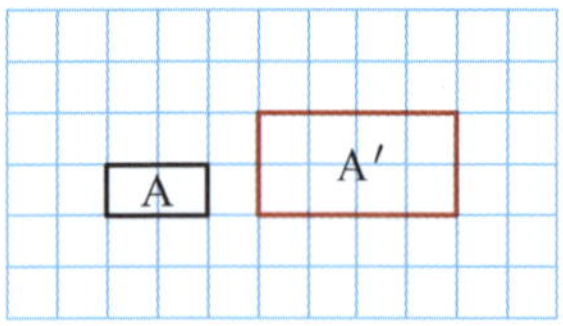

b
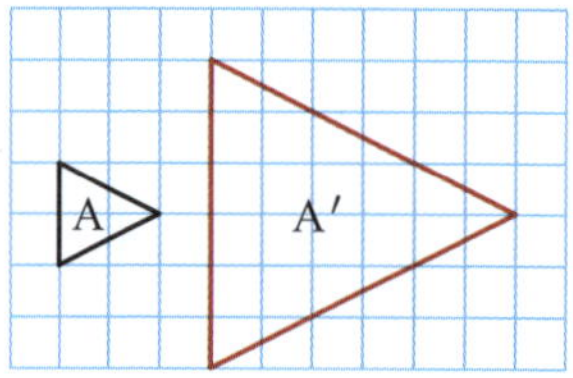

c
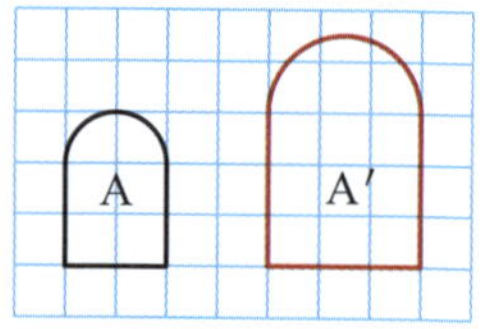

2 In the following diagrams, B′ is a reduction of B. Find the scale factor in each case.

a
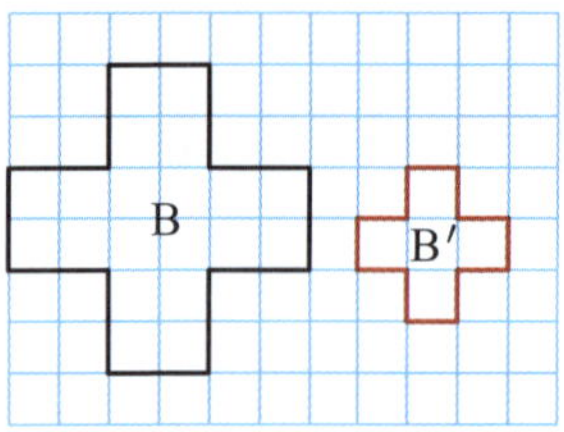

b
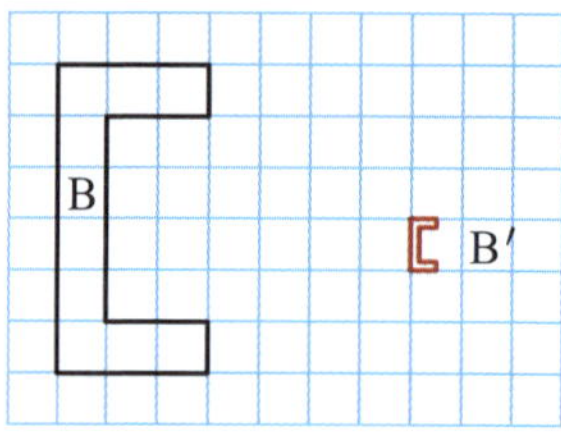

c
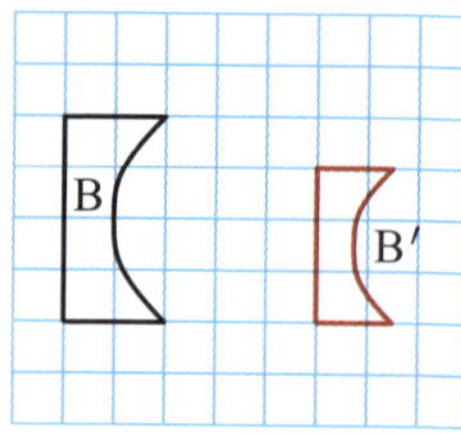

3
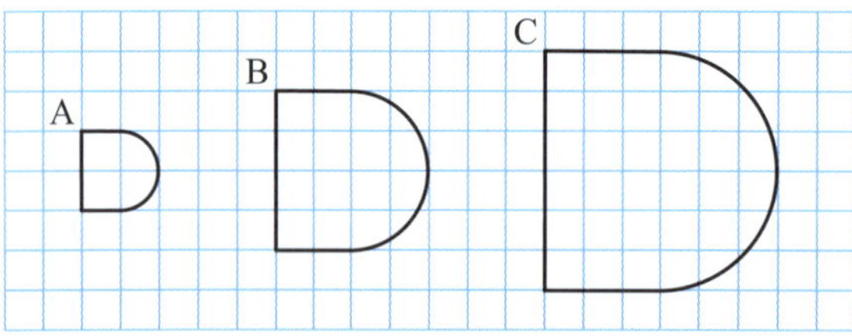

Describe the transformation from:

a A to B **b** B to A
c A to C **d** C to A
e B to C **f** C to B.

CENTRE OF ENLARGEMENT

In each of the diagrams below, triangle PQR has been enlarged to triangle P′Q′R′ using a scale factor of 3.

A
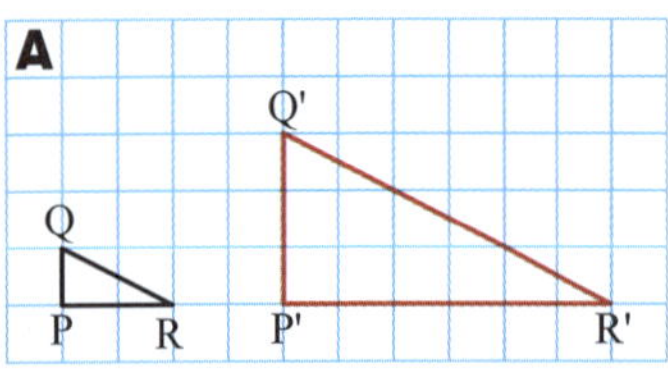

B
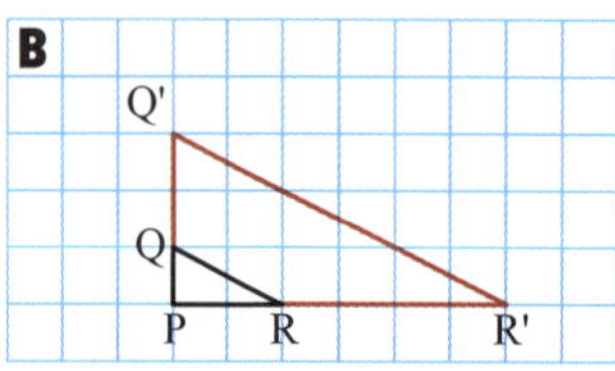

C
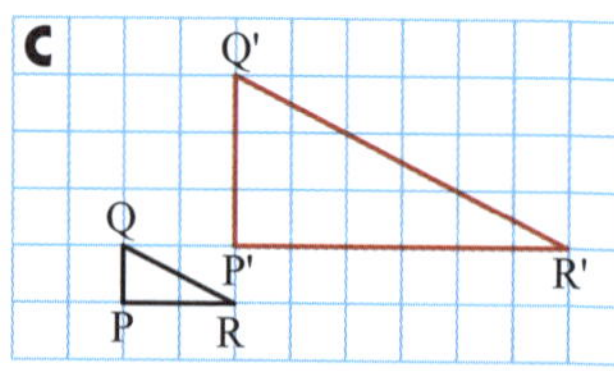

DISCUSSION

1 What is the difference between the enlargements on the previous page?

2 What does 'enlargement with scale factor 3' tell you?

3 What additional instructions do you need to be able to draw an enlargement in the correct position?

A

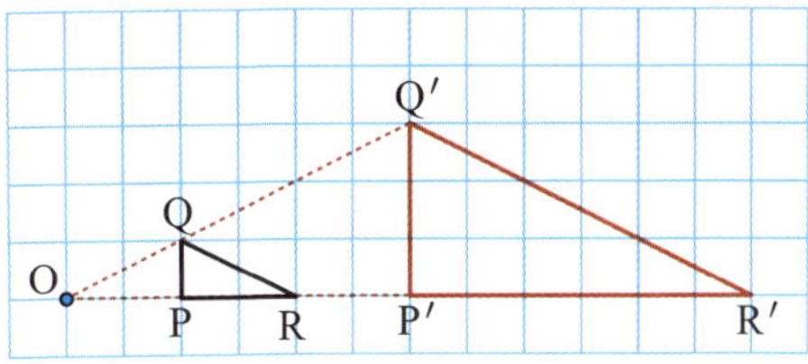

C

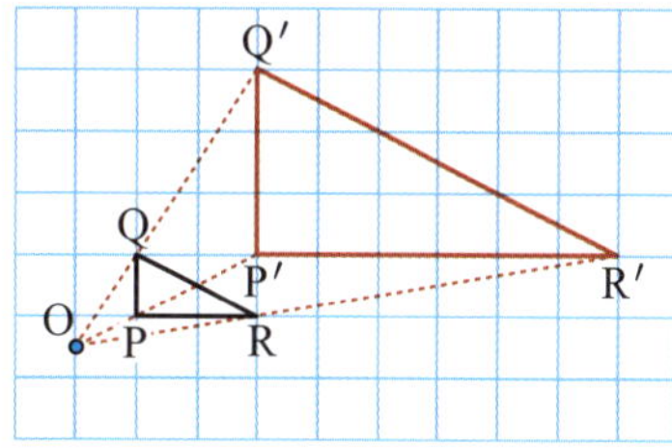

Consider again the diagrams **A** and **C** above. Suppose we draw a line through each vertex of the object and its corresponding point on the image. We see that the lines always meet at a point. We label this point O and call it the **centre of enlargement**.

Notice on each diagram that $OP' = 3 \times OP$, $OQ' = 3 \times OQ$, and $OR' = 3 \times OR$. This corresponds to the scale factor $k = 3$.

DISCUSSION

What is the centre of enlargement for diagram **B**?

Example 9 Self Tutor

Find the image of each figure for the centre of enlargement O and scale factor given:

a scale factor 2

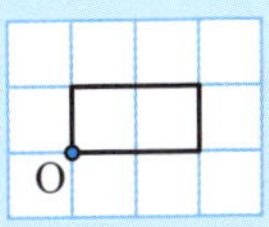

b scale factor 3

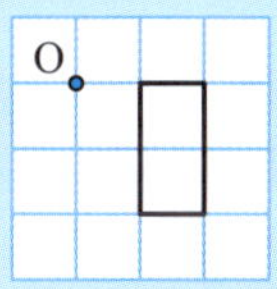

c scale factor $\frac{1}{2}$

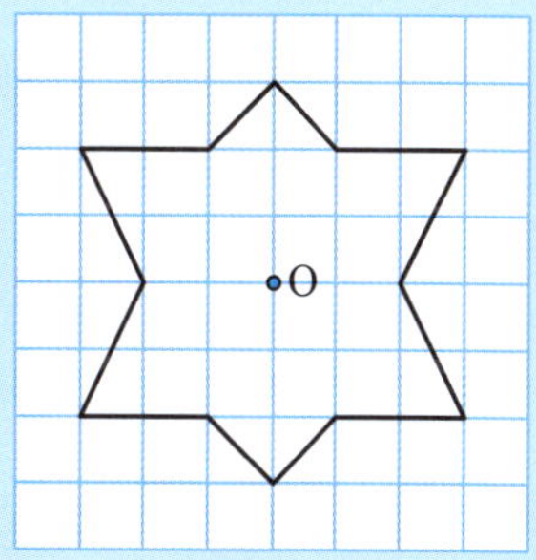

a

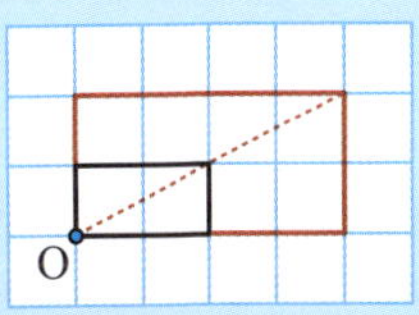

b

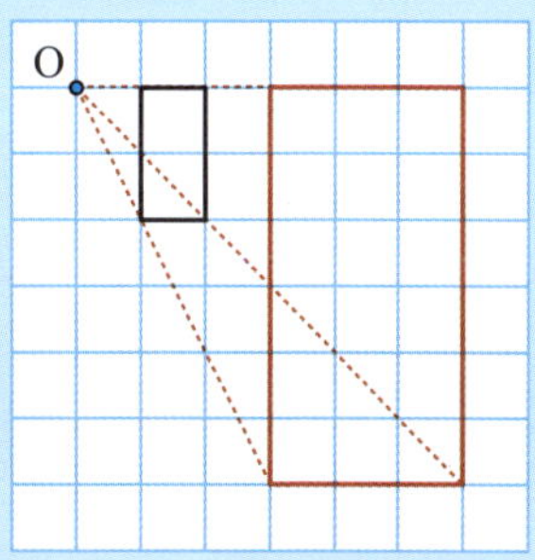

c

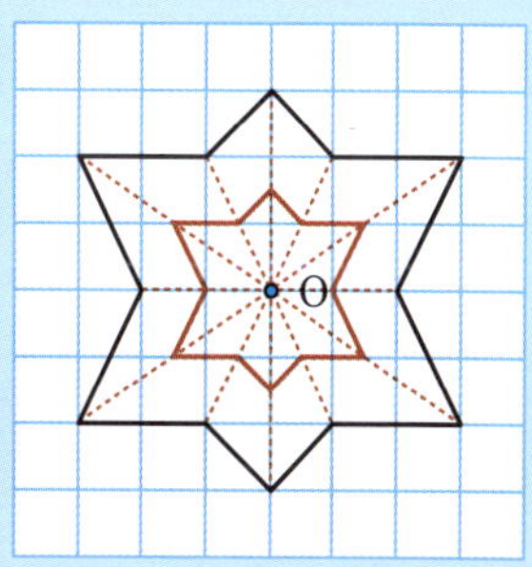

EXERCISE 16D.2

PRINTABLE DIAGRAMS

1 Copy the following diagrams. Locate the centre of enlargement in each case.

a

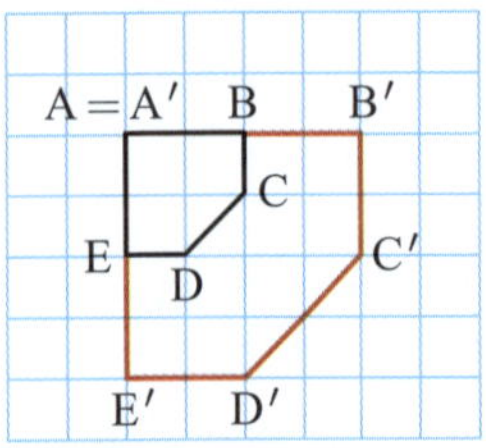

b

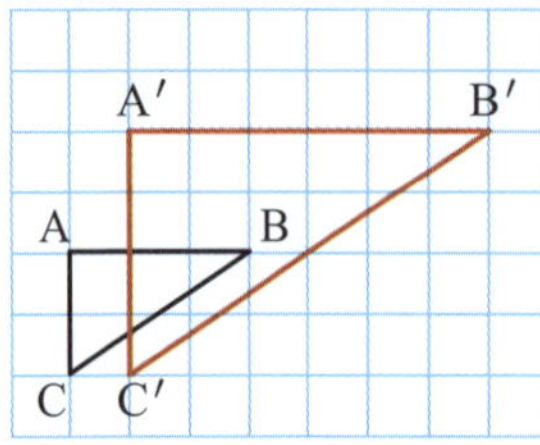

c

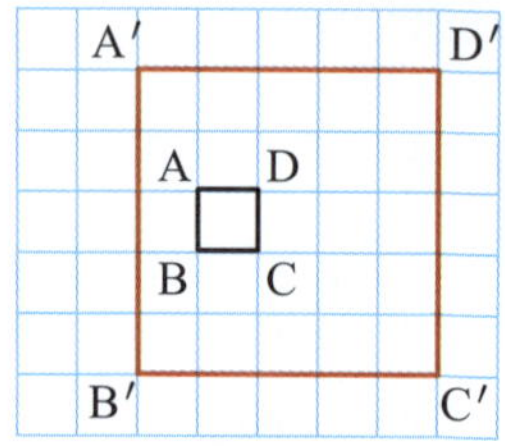

2 Find the image of each figure for the centre of enlargement O and scale factor given:

a scale factor 2

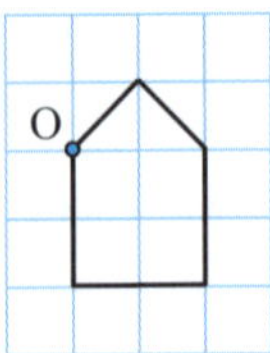

b scale factor 3

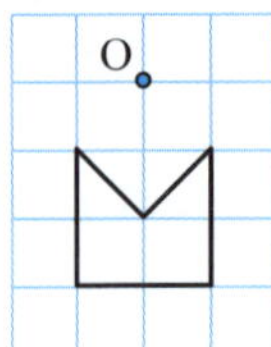

c scale factor $\frac{1}{2}$

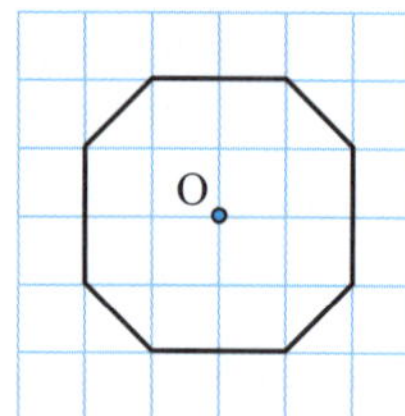

d scale factor 2

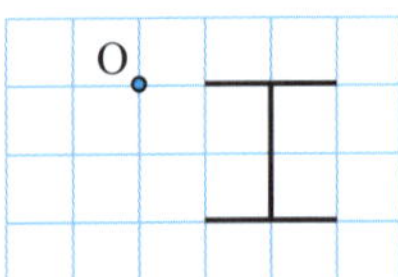

e scale factor $\frac{1}{3}$

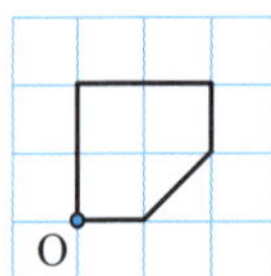

f scale factor 3

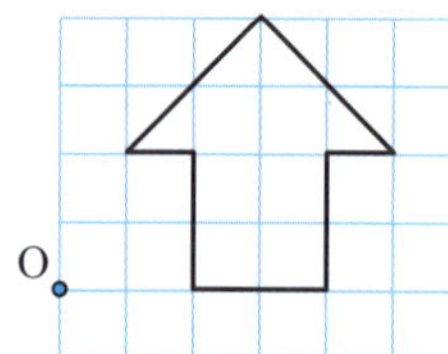

3 **a** Find the image of each figure for the centre of enlargement O and scale factor given:

i scale factor 2

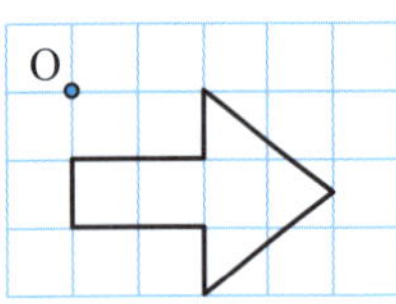

ii scale factor $\frac{1}{2}$

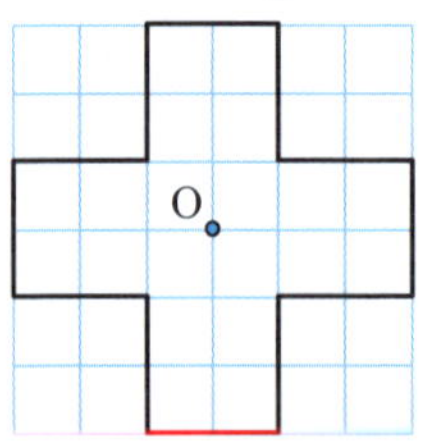

iii scale factor 3

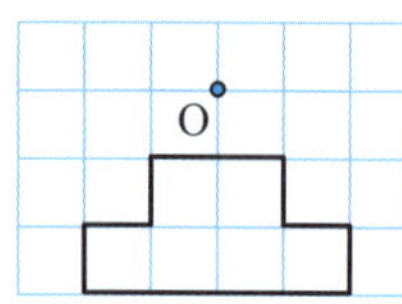

b Copy and complete the table below for the diagrams in **a**.

Part	*Area of object*	*Area of image*	$\frac{\text{Area of image}}{\text{Area of object}}$	*Scale factor, k*	k^2
i					
ii					
iii					

c Comment on your results in **b**.

4 Jason made a $\frac{1}{100}$ scale model of a yacht. The area of the mainsail of his scale model was 56 cm^2. What was the actual sail area, in m^2, of the full sized yacht?

5 A 10 cm by 8 cm photograph is to be enlarged so that the longer side will be 25 cm.

- **a** Find the scale factor k for the enlargement.
- **b** Find the length of the shorter side on the enlargement.
- **c** The cost of printing a photo is proportional to its area. If the smaller photograph costs \$3.00 to print, find the cost of printing the larger photograph.

6 A gardener has a rectangular garden bed 5 m by 8 m. He wants to build a new garden bed which is an enlargement of the old one, with double the area. What dimensions should the new garden bed have?

ACTIVITY — TESSELLATIONS

A **tessellation** is an arrangement of figures with the same shape, without gaps, in a pattern which could be extended indefinitely. Rotations and reflection may be used to make the figures fit together.

Some examples of tesselations made from polygons are:

What to do:

1 Draw tessellations using the following shapes:

a
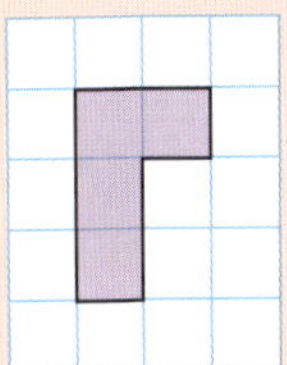

b
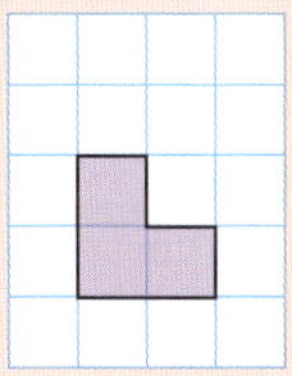

c
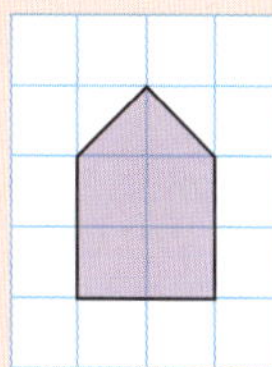

2 What transformation is needed to form a tessellation of these elephants?

3 Which *regular* polygons tessellate?

4 Copy or print this diagram of a popular paving brick.

a Show how it has been created by altering the sides of rectangle ABCD.

b Draw several of these pavers to show how they tessellate.

c Apart from their visual appeal, what advantage do these pavers have over rectangular ones?

d The manufacturer has decided that it is time to market different pavers. He has hired you as a designer. Your job is to design two new pavers, both based on a rectangle.

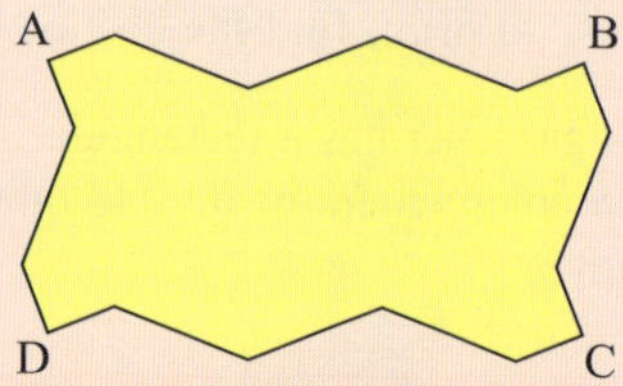

PRINTABLE DIAGRAM

5 **Maurits Cornelis Escher** (1898 - 1972) was a Dutch graphic artist who created many clever and fascinating tessellations. The pattern alongside resembles one of his well-known drawings.

a To make the 'bird', follow these steps:

Step 1: Begin with an equilateral triangle.

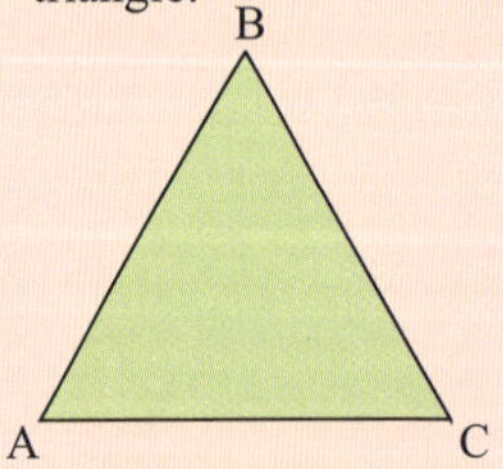

Step 2: Change side [AB] as shown.

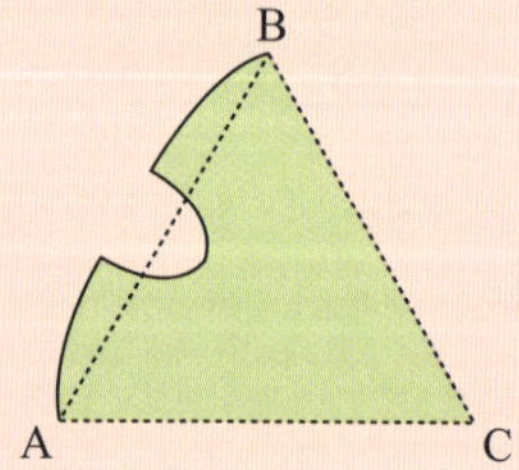

Step 3: Rotate the change 60° clockwise about vertex A to side [AC].

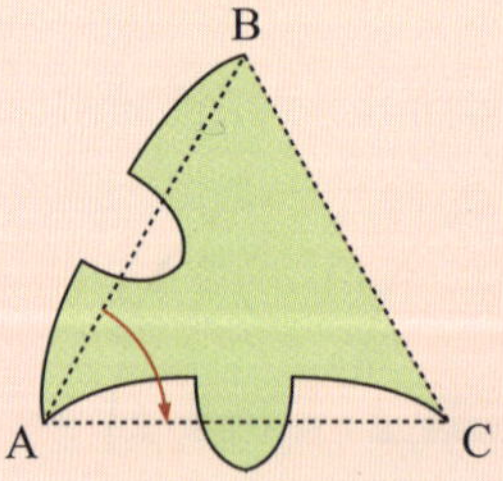

Step 4: Change side [BC] as shown, where M is the midpoint of [BC].

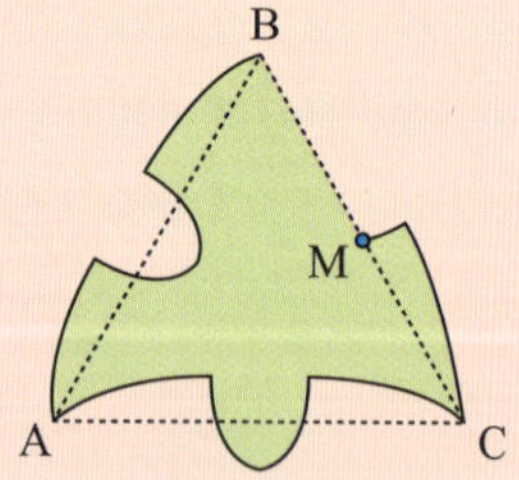

Step 5: Rotate the change 180° about the midpoint M.

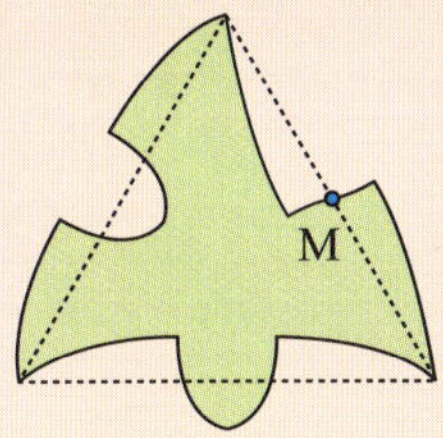

Step 6: Add the final detail.

b Photocopy your 'bird' many times and carefully cut each one out with scissors.

c Show how six of these shapes tessellate by rotating them about the point A.

d Continue the tessellation to cover a wider area.

REVIEW SET 16A

1 Translate the given figures in the direction indicated:

a

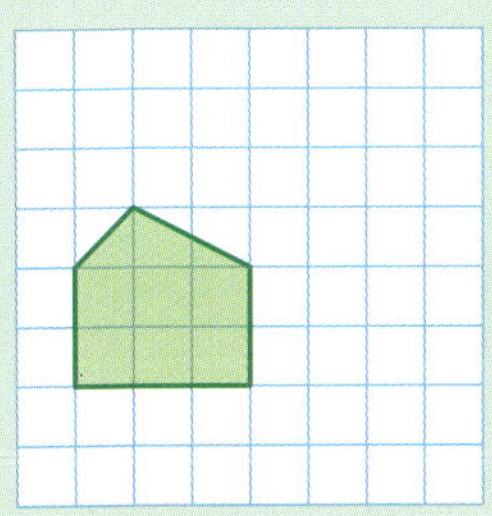

3 units right

b

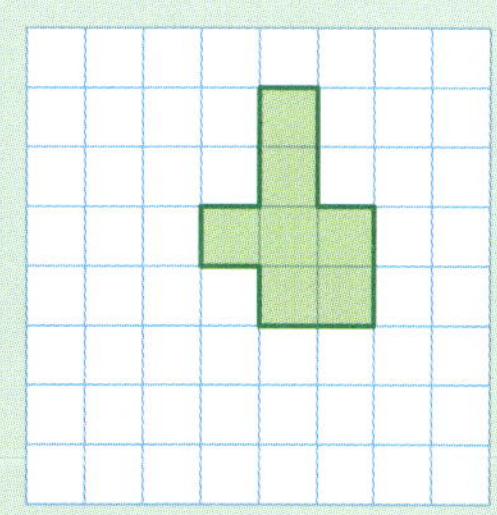

2 units left and 2 units down

PRINTABLE DIAGRAMS

2 Figure A has been translated to B, then B has been translated to C.

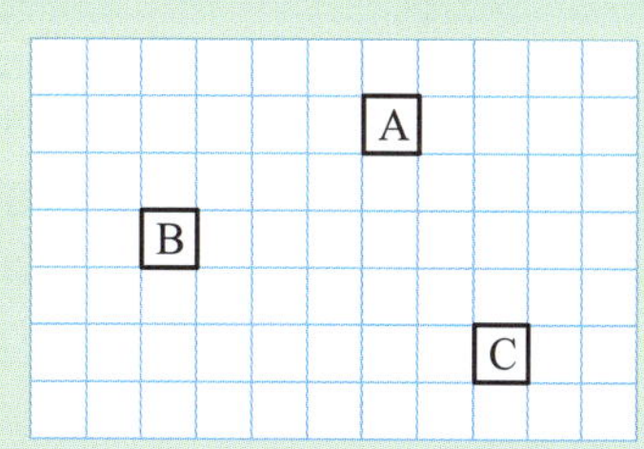

a Give the translation vector from A to B.

b Give the translation vector from B to C.

c What translation vector would move A directly to C?

3 Complete the table alongside with the images of the given points under a reflection in the y-axis.

	P	P′
a	(1, 5)	
b	(−2, 4)	
c	(−4, −3)	
d	(3, 2)	

4 Copy the figure and reflect it in the mirror line shown.

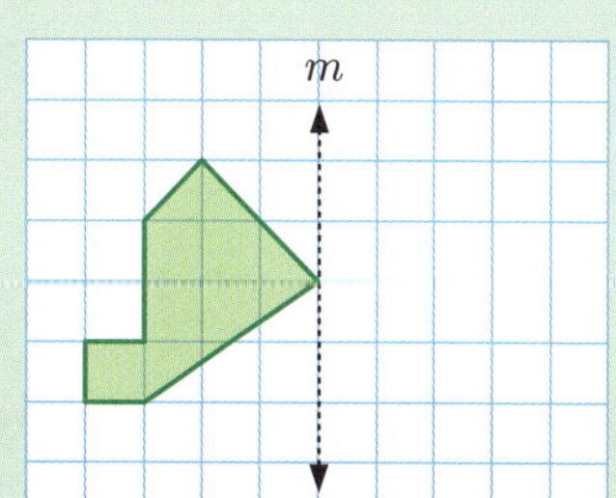

5 Describe the following transformations:

a

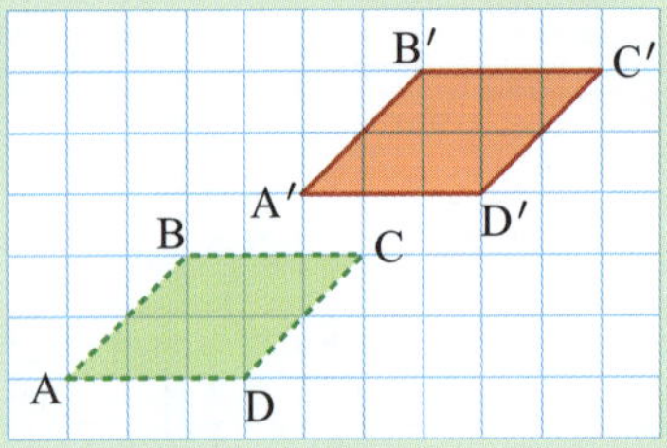

b

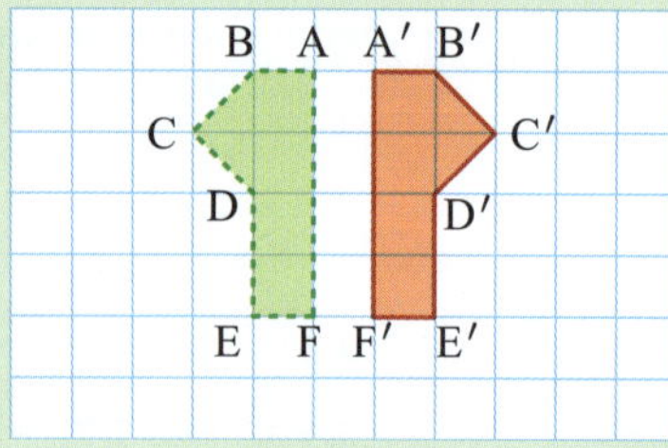

6 Draw any rectangle and all of its lines of symmetry.

7 Rotate each figure about O through the angle indicated:

a

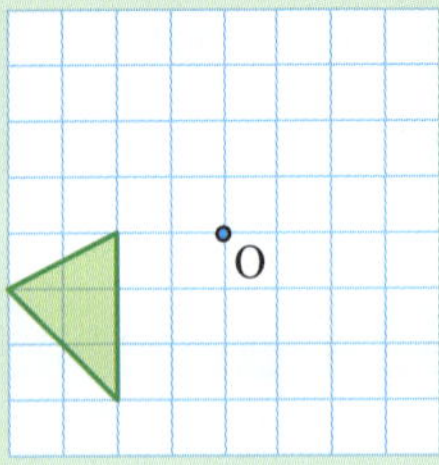

180°

b

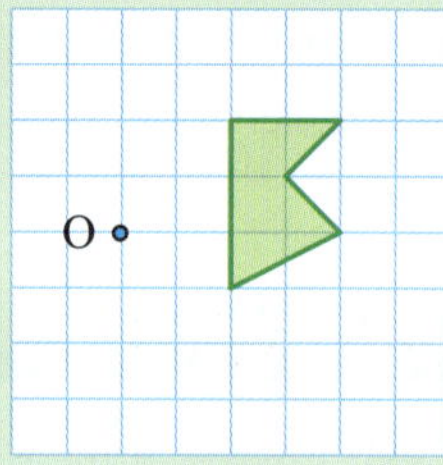

270° anticlockwise

8 **a** Which of the figures alongside is a rotation of A about the origin?

b Determine the angle of rotation from figure A to this figure.

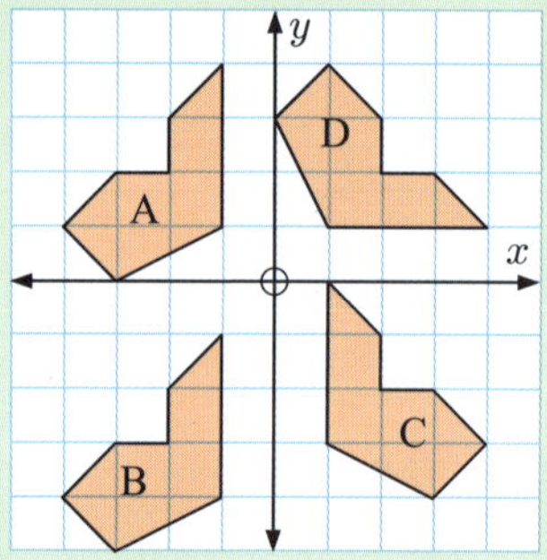

9 A figure is enlarged with a scale factor of 2. What happens to the area of the figure?

10 Find the image of each figure for the centre of enlargement O and scale factor given:

a scale factor 3

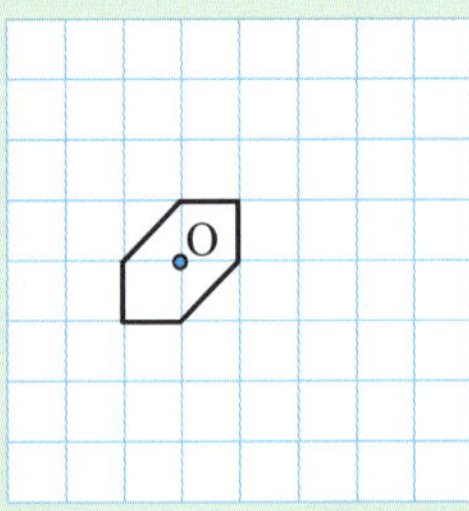

b scale factor $\frac{1}{3}$

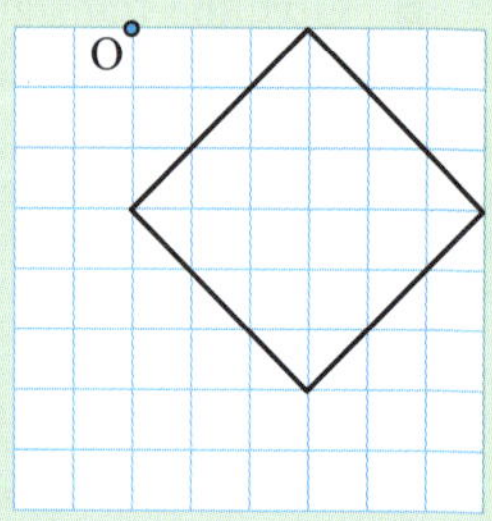

11 Copy the figure alongside. On separate diagrams carry out the following transformations:

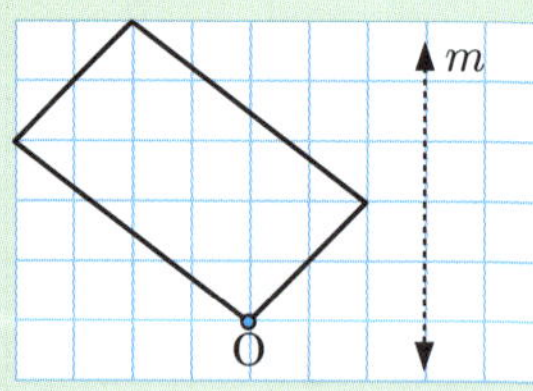

a a translation of $\begin{pmatrix} 3 \\ -2 \end{pmatrix}$

b a rotation about O, 90° clockwise

c a reflection in the mirror line m

d a reduction with scale factor $\frac{1}{2}$ and centre of enlargement O.

12 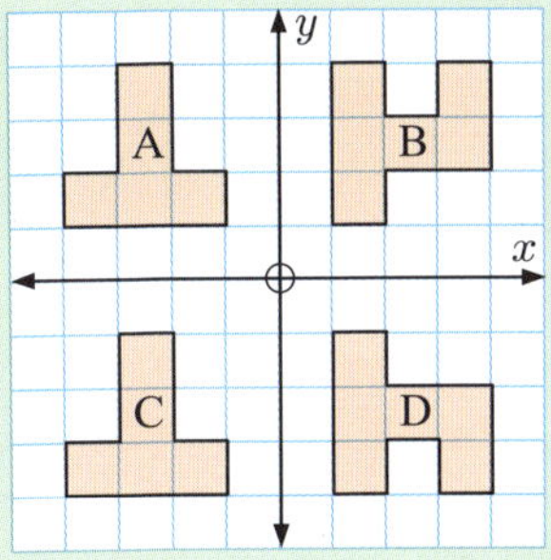

a Which two of the figures alongside are translations of each other?

b Which two of the figures are reflections of each other?

c Which of the axes is the mirror line for this reflection?

REVIEW SET 16B

1 Figure A has been translated to give the image B.

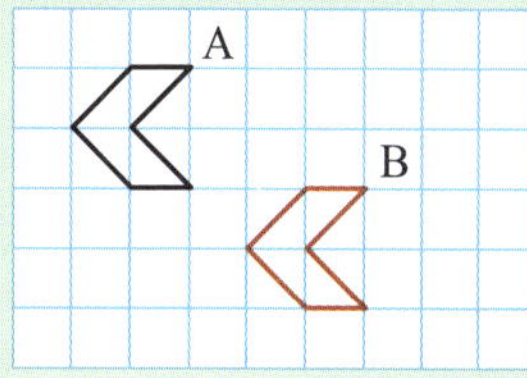

a State the translation vector.

b How far has the object moved?

c Describe the transformation needed to return B to A.

2 Draw triangle ABC where A is $(3, 3)$, B is $(-3, -2)$, and C is $(3, -2)$.

PRINTABLE DIAGRAMS

a Translate the figure using the translation vector $\begin{pmatrix} 3 \\ 2 \end{pmatrix}$.

b State the coordinates of the image vertices A′, B′, C′.

3 Complete the table alongside with the images of the given points under a reflection in the x-axis.

	P	P′
a	$(1, 4)$	
b	$(-2, 3)$	
c	$(-3, -1)$	
d	$(4, 2)$	

4 Reflect each figure in the axis indicated:

a

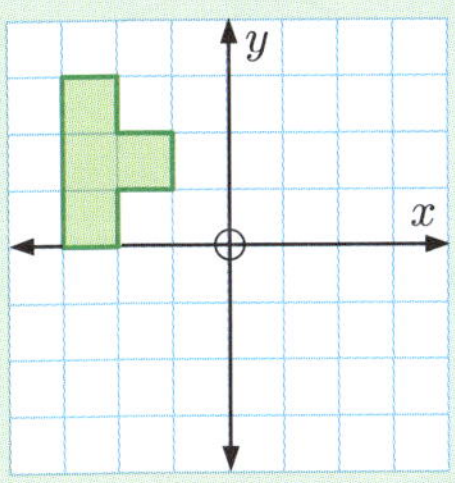

x-axis

b

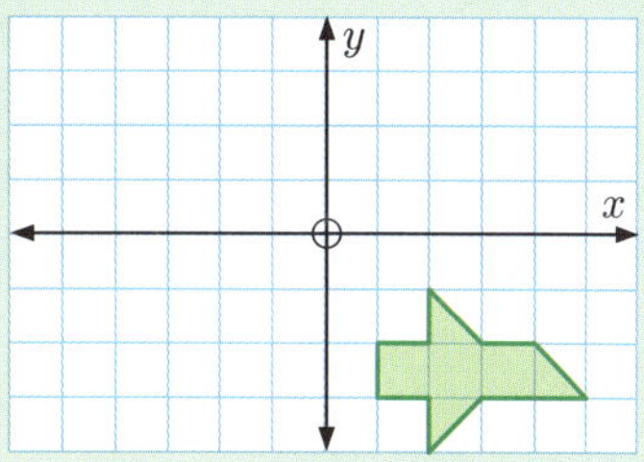

y-axis

5 A triangle has vertices A$(2, 1)$, B$(4, 3)$, and C$(3, 0)$.

a Plot triangle ABC on a Cartesian plane.

b Reflect $\triangle$ABC in the x-axis, and state the vertex coordinates of the image.

c Reflect $\triangle$ABC in the y-axis, and state the vertex coordinates of the image.

6 Find the image of each point under a $90°$ anticlockwise rotation about O:

a $(0, 2)$ **b** $(-5, 0)$ **c** $(1, -3)$ **d** $(-2, -4)$ **e** $(-3, 2)$

7 Copy the following figures and draw in all lines of symmetry:

a 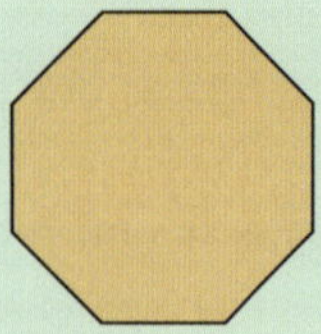 **b**  **c** 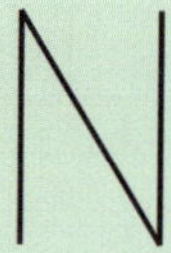

8 **a** State the vertex coordinates of the quadrilateral PQRS.

b Rotate the quadrilateral 90° anticlockwise about the origin O.

c State the vertex coordinates of the image.

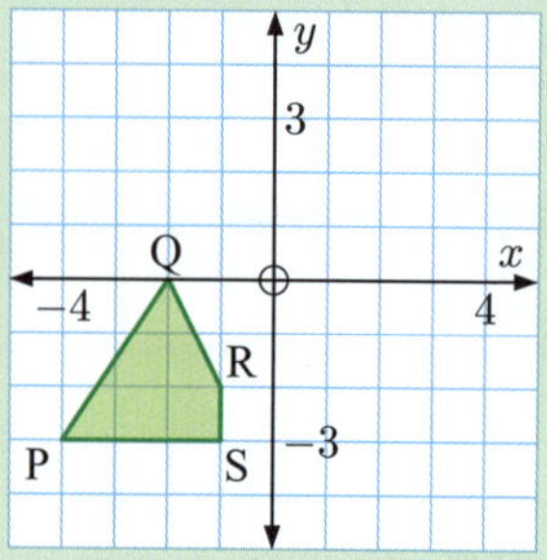

9 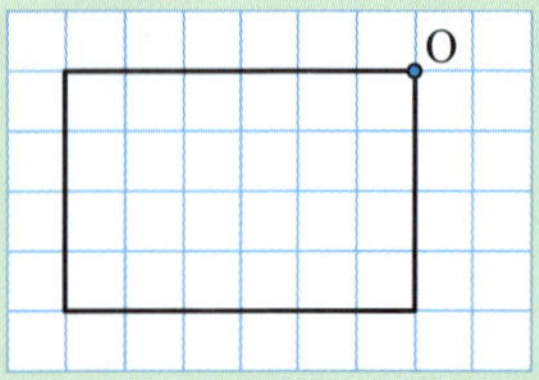

Reduce this figure with scale factor $\frac{1}{2}$ and centre O.

10 Copy the given diagram and locate the centre of enlargement.

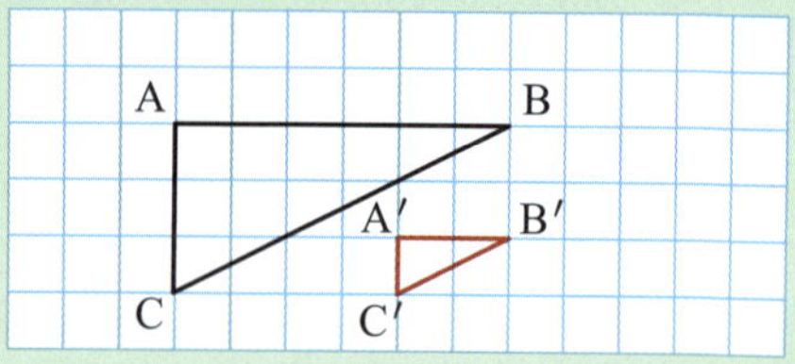

11 Copy the figure alongside. On separate diagrams carry out the following transformations:

a a translation of $\begin{pmatrix} -3 \\ 4 \end{pmatrix}$

b a rotation about O, 180° clockwise

c a reflection in the mirror line m

d an enlargement with scale factor 2 and centre of enlargement O.

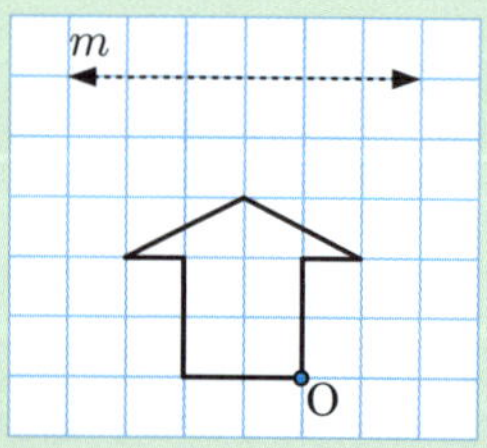

12 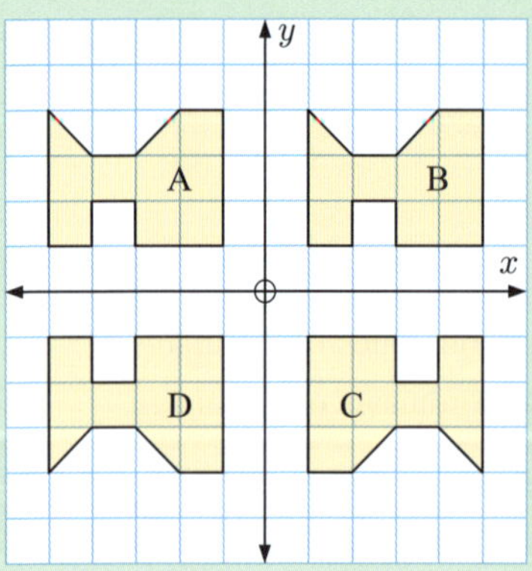

a Which of the figures alongside is a *translation* of A?

b Which figure is a *reflection* of A?

c Which figure is a *rotation* of A?

d Describe the transformation from C to D.

Chapter 17

Comparing numerical data

Contents:

OPENING PROBLEM

At the local soft drink factory, there are two machines which place screw caps on the plastic bottles. The quality control officer Hugh suspects that machine A causes fewer faulty screw caps than machine B. Each day 1000 bottles are randomly selected from the machines, and their caps are tested for faults. Hugh records the number of faulty caps each day for 50 days.

The results are:

Machine A

4 7 5 6 6   5 6 6 7 6   6 5 5 8 6   6 5 7 4 8   9 6 7 5 6
6 8 6 5 9   6 8 7 5 6   6 4 6 5 5   7 5 6 4 6   5 6 6 5 6

Machine B

5 9 9 7 3   9 8 10 8 6   7 7 9 8 5   9 9 4 11 9   6 9 7 9 8
9 7 9 8 8   9 4 10 8 6   9 7 3 8 7   5 9 8 7 6   8 4 9 10 9

Things to think about:

- **a** Can you clearly state the problem that the quality controller Hugh wants to solve?
- **b** How has Hugh tried to make a fair comparison?
- **c** How can we display these data sets in a way that makes them easy to compare?
- **d** How can Hugh best indicate the most typical number of faulty caps for each machine?
- **e** How can Hugh indicate the spread of the data for each machine?
- **f** Can Hugh make a satisfactory conclusion?

Analysing a single set of data can be useful, but sometimes to fully understand a situation we need to compare two data sets. We can do this by comparing statistics such as measures of **centre** and **spread**, and also **graphically**.

A COMPARING THE MEASURES OF CENTRE AND SPREAD

To compare the statistics of two data sets, we use:

- the **mean** or **median** of each data set to compare their centres
- the **range** or **interquartile range** of each data set to compare their spreads.

Example 1 **Self Tutor**

Julie and Andrea both play goalshooter for their respective netball teams.

Julie played 11 games and scored 15, 19, 22, 20, 24, 19, 21, 23, 25, 16, and 23 goals.

Andrea played 10 games and scored 16, 10, 21, 14, 24, 8, 11, 13, 17, and 20 goals.

- **a** Find the mean number of goals scored for each player.
- **b** Which player scored the higher number of goals per game?
- **c** Find the range of scores for each player.
- **d** Which player had the greater spread of scores?

a Julie's mean $= \frac{15+19+22+20+24+19+21+23+25+16+23}{11}$

$= \frac{227}{11}$

≈ 20.6 goals

Andrea's mean $= \frac{16+10+21+14+24+8+11+13+17+20}{10}$

$= \frac{154}{10}$

$= 15.4$ goals

b On average, Julie scored the higher number of goals per game.

c Julie's range $= 25 - 15 = 10$ goals

Andrea's range $= 24 - 8 = 16$ goals

d Andrea had the greater spread of scores.

EXERCISE 17A

1 A local council counted the number of cars which drove along Park Street and High Street each day over a number of months. The measures of centre and spread for the data are shown in the table alongside.

	Park Street	*High Street*
Mean	543	619
Median	562	603
Range	120	93
IQR	55	44

a Which street is used by more cars per day?

b Which street has the greater spread in the number of cars that use it each day?

2 The points scored by two basketball teams over a 12 match series are:

Team A: 91, 76, 104, 88, 73, 55, 121, 98, 102, 91, 114, 82

Team B: 87, 104, 112, 82, 64, 48, 99, 119, 112, 77, 89, 108

Which team had the higher mean score?

3 A school in Toronto has two Year 9 classes. The data below shows the number of days the students in each class were away sick during the school year.

Class 9A

0	5	3	4	0	13
3	6	7	2	3	1
5	5	9	11	4	9
0	6	4	0	8	

Class 9B

8	5	9	0	2	6
4	7	12	8	3	9
10	17	9	2	0	13
21	8	14	11	6	3

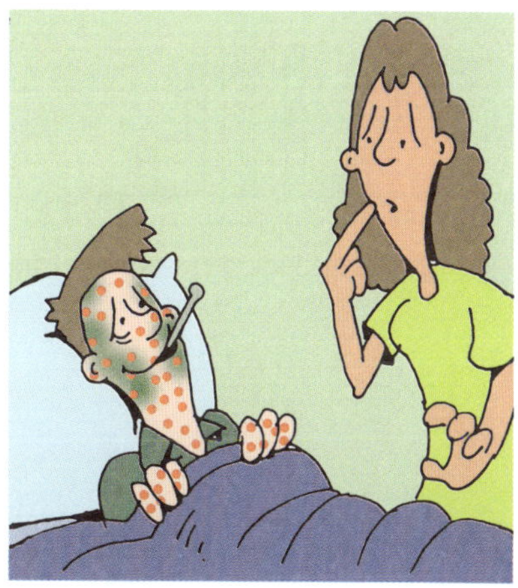

a How many students from each class did not have any sick days?

b Calculate the median of each data set.

c Calculate the interquartile range of each data set.

d In which class did the students on average take more sick days?

e In which class was there a greater spread in the number of sick days?

4 The chess points ratings for members of two chess clubs are given below:

The Rookies:	1176	1208	1322	1105	1098	1301	1111	1225	1076
	1099	1233	1176	1153	1258	1358	1236	1083	1184
Best Mates:	1208	1053	1133	1181	1313	1214	1102	1107	1057
	1157	1410	1030	972	1095	1020	1046	1152	1216

a Calculate the mean and median rating for the members of each club.

b Calculate the range and interquartile range for each club.

c Which club generally has the higher rated players?

d Which club has the greater spread in player ratings?

5 Suzanne wanted to know whether Sydney is generally warmer or cooler than Paris. She recorded the daily maximum temperature for each city during the month of July, in °C:

Sydney:	16.6	19.3	20.7	20.4	17.1	17.6	17.6	17.0	17.0	16.4	17.9
	18.9	12.3	15.6	13.2	16.3	16.1	20.3	14.7	18.5	13.3	
	13.9	15.6	15.4	19.1	18.4	17.3	20.4	20.0	20.2	21.4	
Paris:	24.0	23.1	23.8	26.3	30.5	23.9	23.6	23.8	21.8	23.9	27.0
	24.3	18.9	23.1	26.2	20.1	20.2	20.0	16.1	18.9	21.7	
	21.5	21.1	22.7	24.1	23.0	19.2	22.3	23.8	25.4	30.3	

a Calculate the mean and median temperature for each city.

b Would you conclude that Sydney is generally warmer or cooler than Paris? Discuss your answer.

B COMPARING DATA SETS GRAPHICALLY

Two data sets can be compared **graphically** using:

- a side-by-side column graph

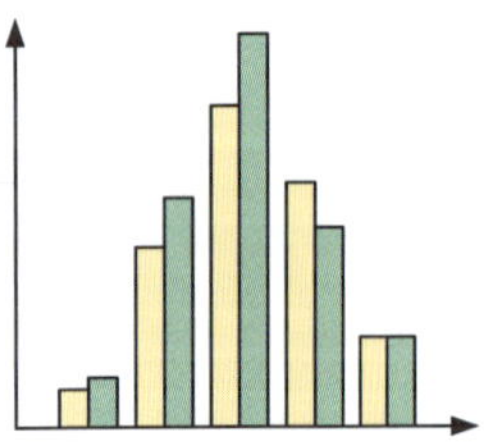

- a back-to-back bar chart

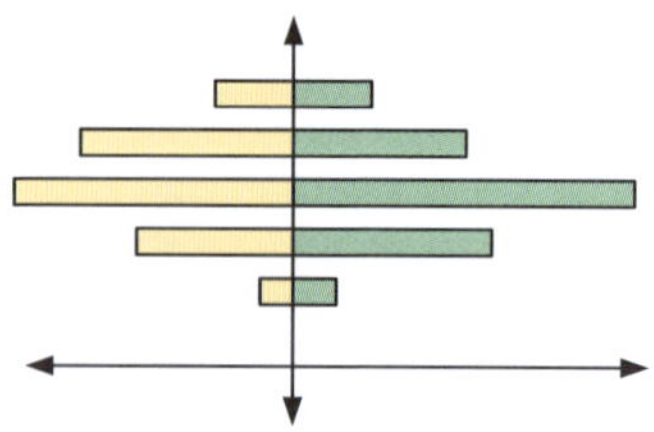

- a back-to-back stem-and-leaf plot.

	12	6
4	11	2 3 7
7 1	10	6 7 8 8
9 8 7 6 6 5 3 3 2 0	9	0 3 4 5 5 7 9
7 5 3 2 1	8	2 2 4 5 8
2 0	7	0 5 6
	6	1

Scale: 6 | 1 means 6.1

Graphing the two data sets together makes it easier to visually compare the distribution of each data set.

Example 2 **Self Tutor**

Sam and Kate are each given a weekly spelling test out of 20 marks. The results over 35 weeks are displayed in a side-by-side column graph.

a How many times did each student score full marks?

b Describe the distribution of each data set.

c Find the range of the marks for each student.

d Who generally scored higher marks?

e Who had the greater spread of marks?

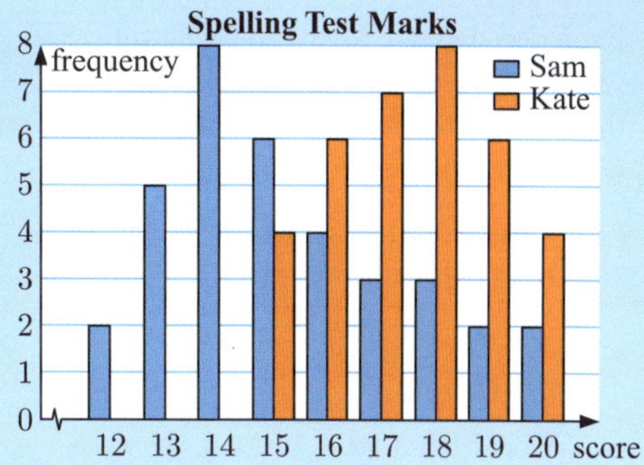

a Sam scored full marks 2 times. Kate scored full marks 4 times.

b Sam's data is positively skewed. Kate's data is approximately symmetrical.

c The range of Sam's marks is $20 - 12 = 8$ marks.
The range of Kate's marks is $20 - 15 = 5$ marks.

d From the column graph, Kate has more scores at the higher end than Sam.
Kate generally scored higher.

e Comparing the ranges, Sam had the greater spread of marks.

EXERCISE 17B

1 Sophie randomly selected 50 Grade 9 classes from California and Oregon, and recorded the number of students in each class. The results are displayed in the back-to-back bar chart below.

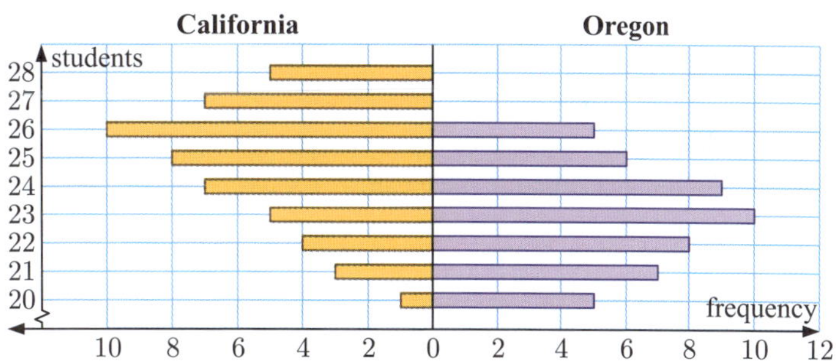

a How many of the selected classes from each state contain at least 25 students?

b Describe the distribution of each data set.

c Copy and complete: "The selected classes from ...... generally have larger class sizes."

2 The ages of members from two youth groups are displayed in the side-by-side column graph opposite.

a How many members from each group are aged 14 years or younger?

b Describe the distribution of each data set.

c Which youth group has the greater range in the ages of its members?

d Which youth group generally has the younger members?

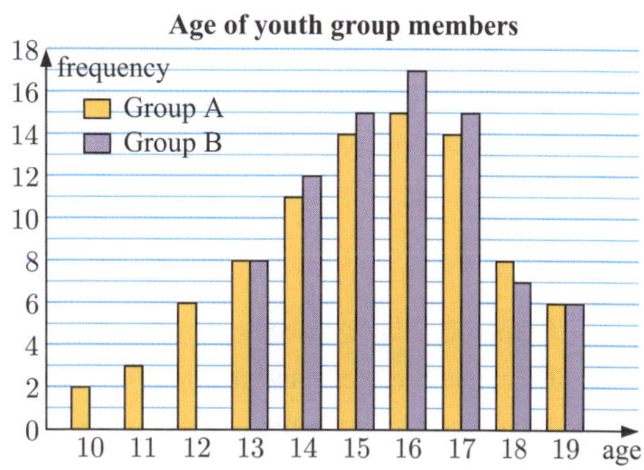

Example 3 **Self Tutor**

Pedro operates a Mexican restaurant. He is about to launch a new advertising campaign, and he wants to be able to measure its effect. He records the number of customers in his restaurant for thirty days before he starts advertising, and then for thirty days after he starts advertising. He collects the following data:

Before advertising:	74	82	63	77	89	91	74	71	57	62	83	79	75	62	59
	84	93	78	68	94	73	77	84	66	72	61	56	76	88	64
After advertising:	93	98	108	84	89	76	92	69	93	85	101	78	67	99	90
	88	70	93	104	108	95	80	103	79	92	98	84	106	98	101

a Construct a back-to-back stem-and-leaf plot for this data.

b Find the mean of each data set.

c Discuss the effect of the advertising.

a

Before advertising		After advertising
9 7 6	5	
8 6 4 3 2 2 1	6	7 9
9 8 7 7 6 5 4 4 3 2 1	7	0 6 8 9
9 8 4 4 3 2	8	0 4 4 5 8 9
4 3 1	9	0 2 2 3 3 3 5 8 8 8 9
	10	1 1 3 4 6 8 8

$6 \mid 7 = 67$ customers

b **Before advertising:** $\text{mean} = \dfrac{74 + 82 + 63 + ... + 88 + 64}{30} = 74.4$ customers

After advertising: $\text{mean} = \dfrac{93 + 98 + 108 + ... + 98 + 101}{30} = 90.7$ customers

c The number of customers has *increased* by approximately 16 customers per night as a result of the advertising.

3 To determine if a weight-loss program is effective, a group of 25 men trialled the program for three months. Their weights in kilograms before starting the program were:

95	104	93	86	82	111	100	125	117
121	119	97	120	104	132	111	98	83
123	79	101	135	114	99	122		

Their weights after completing the program were:

79	92	84	78	68	99	85	106	99
107	102	83	110	93	125	96	79	74
113	68	89	122	95	83	106		

a Construct a back-to-back stem-and-leaf plot for the data.

b Calculate the mean and median for each data set.

c Would you say that the weight-loss program is effective? Explain your answer.

4 A psychologist wants to know whether single men or women own more pets. She surveys 50 men and 50 women, each of whom live by themselves, and records how many pets they own. The results are shown alongside.

Single men

Pets	*Frequency*
0	15
1	20
2	10
3	3
4	2
5	0

Single women

Pets	*Frequency*
0	7
1	10
2	15
3	10
4	5
5	3

a Display the data on a back-to-back bar chart.

b Calculate the mean and median of each data set.

c What conclusion could the psychologist make?

5 Vasilios owns a vending machine which dispenses drinks and snacks. He records the number of drinks and snacks sold each week.

Drinks: 32 57 44 36 40 28 23 31 22 32
27 33 24 42 25 64 39 60 37 48
30 61 72 67 47 36 52 49 41 52
75 48 39 56 81 55 74 50 84 68

Snacks: 44 58 42 50 53 67 49 51 37 28
35 47 60 62 73 56 66 42 23 31
26 37 39 43 70 73 52 74 48 67
53 46 61 77 68 40 31 38 26 50

a Construct a back-to-back stem-and-leaf plot to display the data.

b Describe the distribution of each data set.

c Find the mean and median of each data set.

d Find the range and IQR of each data set.

e Does Vasilios generally sell more drinks or snacks?

f Is the range of sales greater for drinks or snacks?

6 Randomly selected 14 year olds from New Zealand and Japan were asked how many subjects they studied at school. The results were:

New Zealand: 6 7 6 5 6 7 7 5 6 7 8 7 6 7 6 8 5 7 6 5 7 6 6
5 7 4 6 7 6 5 8 7 6 7 7 7 6 5 7 4 5 7 8 6 7

Japan: 7 6 6 8 5 7 6 8 7 5 6 6 7 9 6 7 5 8 7 6 5 8 7
5 8 7 6 8 9 8 7 6 8 5 7 8 6 5 7 6 9 8 5 7 7

a Draw a side-by-side column graph to illustrate the data.

b Find the mean and median for each data set.

c Find the range of each data set.

d Copy and complete: "The selected students from ...... generally studied more subjects."

7 The heights of players from two rugby squads are displayed on a back-to-back histogram.

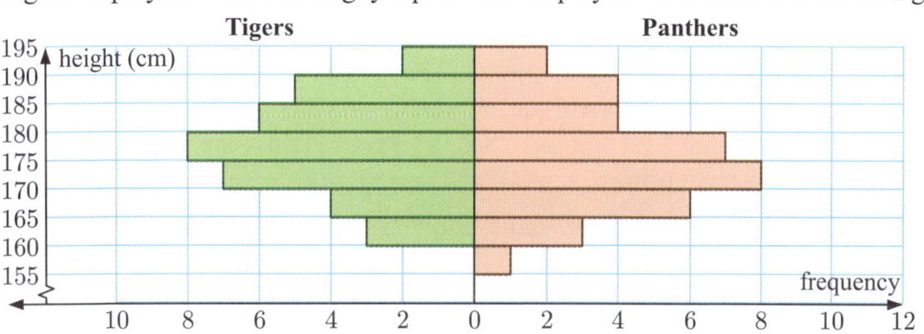

a How many players from each squad are at least 180 cm tall?

b Find the modal class for each squad.

c Describe the distribution of each data set.

d Estimate the mean of each data set.

e Which team appears to have the greater average height? Discuss the reliability of your answer.

C PARALLEL BOX-AND-WHISKER PLOTS

In **Chapter 10**, we used a **box-and-whisker plot** to display some of the descriptive statistics of a data set.

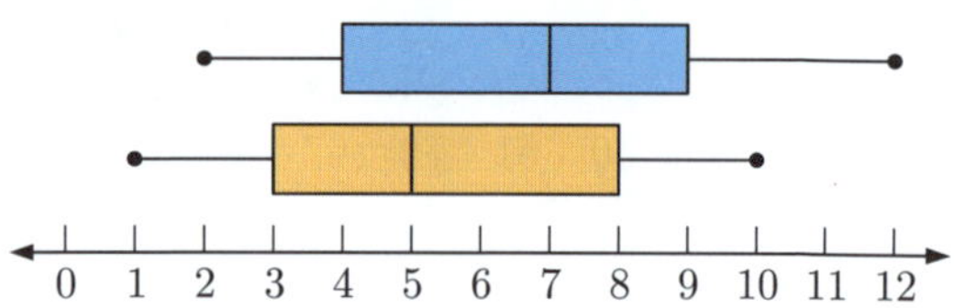

If we want to compare two data sets, we can draw a box-and-whisker plot for each data set on the same scale. This is know as a **parallel box-and-whisker plot**.

When comparing data sets which are not of equal size, a parallel box-and-whisker plot is often more useful than the graphs we have studied so far.

Example 4 **Self Tutor**

Mrs Freeman has a Year 8 and a Year 12 music class. She asks each student in her classes how many times they have practised their musical instrument in the last month. The responses are:

Year 8 class: 5 2 10 8 3 16 6 10 9 9 3 11 5 7 6

Year 12 class: 13 19 15 10 14 11 15 13 14 16 15 18 12

a Draw a parallel box-and-whisker plot for the data sets.

b Which class generally has practised their musical instruments more often?

c Which class is more varied in their number of practices?

a We first construct the five-number summary for each data set.

Year 8: min $= 2$ $Q_1 = 5$ median $= 7$ $Q_3 = 10$ max $= 16$

Year 12: min $= 10$ $Q_1 = 12.5$ median $= 14$ $Q_3 = 15.5$ max $= 19$

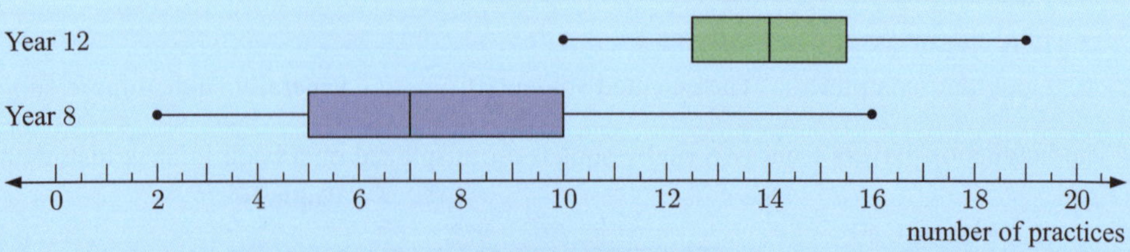

b Looking at the parallel box-and-whisker plot, the values for the Year 12 class are generally higher than those for the Year 8 class. So, the Year 12 class has generally practised their musical instruments more often.

c The values for the Year 8 class have greater spread, so the Year 8 class was more varied in their number of practices.

You can use your **graphics calculator** or the **statistics package** to draw parallel box-and-whisker plots.

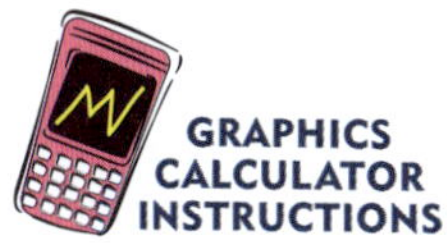

EXERCISE 17C

1 For each of the following parallel box-and-whisker plots, determine which data set:

i generally has larger values **ii** has a greater spread of values.

Give reasons for your answers.

a

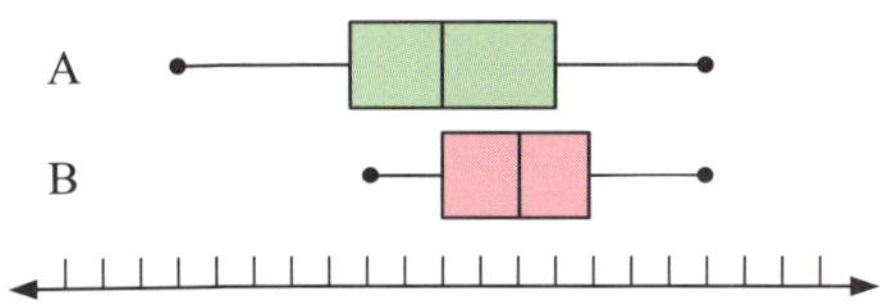

b

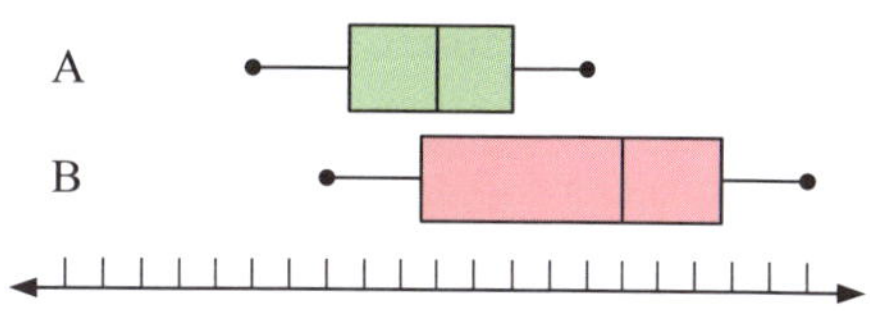

2 This parallel box-and-whisker plot shows the numbers of students on school buses A and B over a one month period.

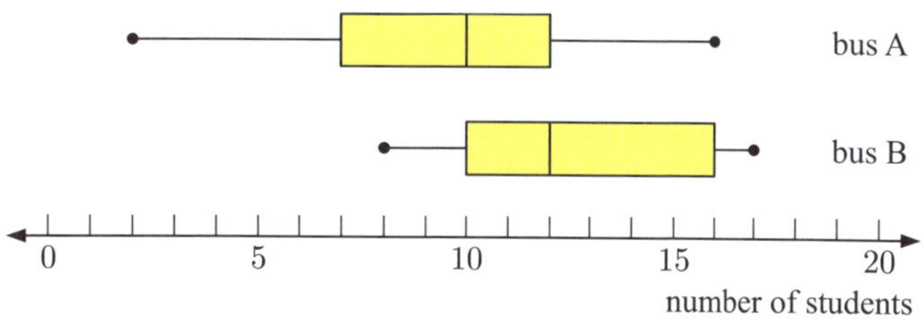

a On what percentage of trips were there 10 or more students on:

i bus A **ii** bus B?

b Which bus generally has the higher number of students?

3 Maria and Sophie play in the same softball team. They are fierce but friendly rivals when it comes to scoring the most runs. The parallel box-and-whisker plot below shows the numbers of runs scored by each player during a season.

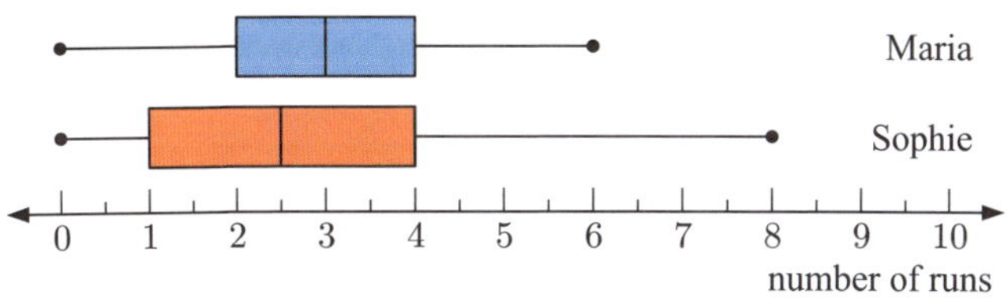

a Find the five-number summary for each player.

b For each player, find the:

i range **ii** interquartile range.

c Compare the performances of the players.

4 The data below shows the number of test matches played by players from the Zimbabwean and Indian cricket teams.

Zimbabwe:	19	4	4	11	11	8	25	4	2	19	22	12	8
India:	81	19	6	6	90	23	6	4	17	3	20	53	9

a Find the five-number summary for each data set.

b Draw a parallel box-and-whisker plot for the data sets.

c Which team's players have generally played more games?

d Compare the spread of the data sets.

5 Frank makes batteries. He claims that his brand lasts *twice as long* as those manufactured by a competitor. Thirty of each brand of battery were randomly selected and tested. The results, to the nearest hour, were:

Frank's	135	139	141	145	150	163	135	128	145	143	140	149	136	146	129
batteries:	139	142	132	146	146	135	137	144	146	145	128	151	136	154	152
Competitor's	87	89	76	88	86	90	88	91	84	94	90	97	92	93	87
batteries:	89	87	92	94	87	88	93	83	89	88	90	85	95	87	96

a Find the five-number summary for each data set.

b Draw a parallel box-and-whisker plot to display the data sets.

c Find the mean of each data set.

d Does Frank's claim appear to be justified? Explain your answer.

6 Consider the **Opening Problem** on page **346**.

a What is the problem that Hugh wants to solve?

b Find the mean for each data set.

c Find the range and interquartile range of each data set.

d Draw a parallel box-and-whisker plot to display the data sets.

e Use your findings to compare the machines.

INVESTIGATION — COMPARING DATA

In this Investigation, you must use statistics to answer a question which is of interest to you.

What to do:

1 Decide on a question to investigate, which involves comparing two sets of data. It should be a question for which data is readily available.

2 Use the internet or your library to find data relevant to your question.

3 Display the data using an appropriate graph.

4 Calculate the relevant statistics for each data set.

5 Draw a conclusion from your results, and discuss the accuracy of what you have found.

Global context

click here

The printing press

Statement of inquiry:	Analysing data helps us to identify important events in our history.
Global context:	Orientation in space and time
Key concept:	Relationships
Related concepts:	Change, Model, Representation
Objectives:	Communicating, Applying mathematics in real-life contexts
Approaches to learning:	Communication, Research

REVIEW SET 17A

1 The heights of the volleyball players from the two squads playing in the finals were recorded in centimetres. The results were:

Team A:	190	188	176	191	205	187	193	177
	180	175	186	183	179	181	192	
Team B:	185	184	172	200	183	188	186	192
	179	189	194	196	180	182	181	

Which team had the greater mean height?

2 A primary school recorded the number of students who were injured on the monkey bars and the swings each week over 25 weeks. The results are displayed in this side-by-side column graph.

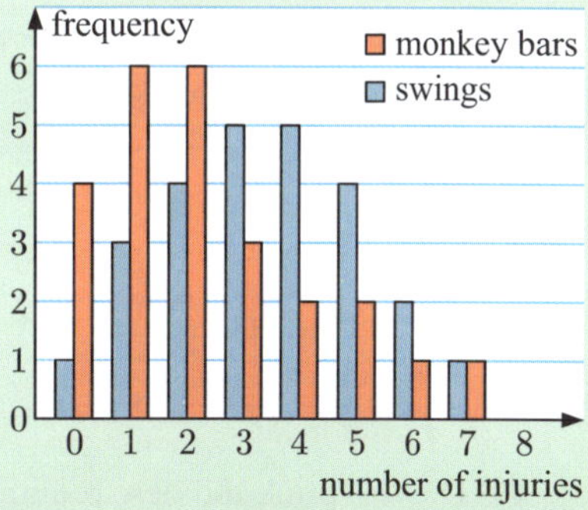

a Describe the distribution of each data set.

b On how many weeks were there no injuries on the:

i monkey bars **ii** swings?

c Which piece of equipment is more dangerous? Explain your answer.

3 Taxi drivers Peter and John each claim that he is the more successful driver. They agree to randomly select 25 days on which they work, and record their fare totals for the day. To the nearest dollar, the data collected are:

Peter:	194	99	188	208	95	168	205	196	233	183	155	190	147
	116	132	153	205	191	182	118	140	270	93	154	223	
John:	260	152	127	163	180	161	110	153	139	110	147	162	223
	139	142	161	97	116	129	215	241	160	159	174	158	

a Draw a back-to-back stem-and-leaf plot for this data.

b Find the mean and median fare total for each driver.

c Find the range and interquartile range for each driver.

d Which driver is more successful? Give reasons for your answer.

4 For each of the following parallel box-and-whisker plots, determine which data set:

i generally has larger values **ii** has a greater spread of values.

Give reasons for your answers.

a

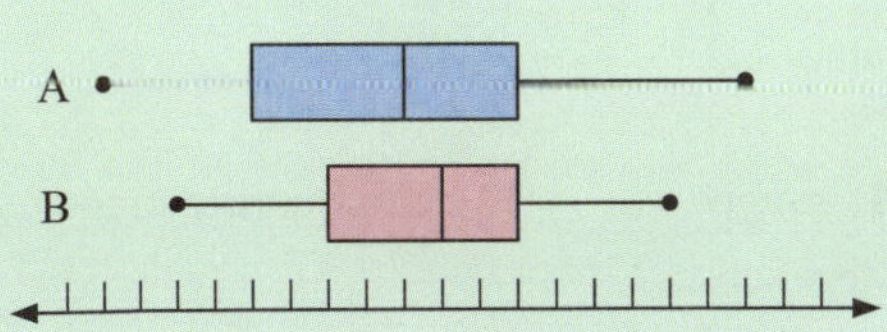

b

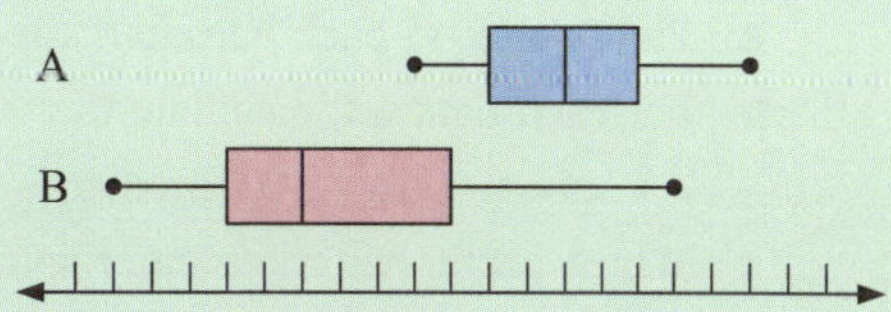

5 Zac and Imran are football goalkeepers. The lengths of a number of their goal kicks were measured in metres:

Zac:	46	43	49	41	43	43	44	47	45	44	43	37	46	38	44
	46	45	42	43	44	42	42	46	40	41	41	40	38	44	42
Imran:	49	45	43	33	45	50	42	46	46	47	45	42	45	44	48
	35	45	38	54	45	40	51	46	49	46	44	44	45	46	47

a Draw a parallel box-and-whisker plot to display these data sets.

b Who generally kicks the ball further? Explain your answer.

c Which goalkeeper has greater variability in the length of their kicks?

REVIEW SET 17B

1 A speed reading tutor wants to test the effectiveness of a new program on a group of readers. He records their times before the program begins, and then again after completing the program. The results in seconds are:

Before:	20.4	20.8	21.3	21.5	21.7	22.6	25.1	25.8	26.8	28.1
	28.6	28.6	29.3	29.8	31.7	33.7	34.6	37.6	39.0	39.1
After:	20.4	20.5	21.0	21.5	21.8	22.2	22.4	25.8	26.9	28.3
	28.4	29.6	30.2	31.2	31.2	32.7	33.5	34.6	35.4	42.7

a For each data set, find:

i the mean **ii** the median **iii** the IQR.

b Do you think the new program has been effective? Explain your answer.

2 Over the course of a 35 game season, basketballers Rhys and Evan kept a record of the number of rebounds they made each game. The results were:

Rhys

8	6	4	9	6	10	9	9	8
7	5	7	8	7	4	7	8	5
8	9	6	3	9	10	8	7	10
9	7	8	8	10	9	6	8	

Evan

7	7	6	8	6	7	6	6	8
6	5	7	7	8	5	6	6	5
7	9	7	6	6	5	8	7	7
8	5	7	7	6	8	7	8	

a Construct a side-by-side column graph for the data.

b Which player generally collected more rebounds?

3 A researcher believes that boys are not as good at mathematics as girls. To examine this claim, 32 Year 9 boys and 32 Year 9 girls were selected at random. They were given a test containing 20 questions worth one mark each. The results were:

Boys:	18	10	11	13	12	14	7	15	14	13	16	12	16	17	15	13
	18	14	12	15	17	11	17	16	15	9	17	12	15	14	13	14
Girls:	8	14	16	19	12	14	15	13	10	16	15	15	14	18	17	17
	11	14	14	15	17	15	18	16	14	9	13	12	15	13	11	13

a Plot a back-to-back bar chart for this data.

b For each data set, calculate the:

i mean **ii** median **iii** range **iv** interquartile range.

c Which group had the greater spread of test scores?

d Which group generally performed better in the test?

4 This parallel box-and-whisker plot compares times taken by athletes to run 1000 m at the start and at the end of the athletic season. The times are shown in seconds.

a For each set of data, find the:

- **i** fastest time
- **ii** slowest time
- **iii** median time
- **iv** range
- **v** IQR.

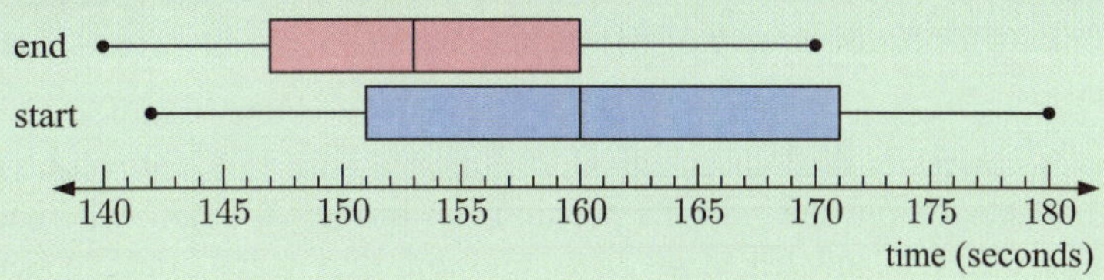

b What percentage of athletes took 160 seconds or more to run the 1000 m:

i at the start of the season **ii** at the end of the season?

c Was there an improvement in times during the season? Explain your answer.

d At which recording time was the variability greatest?

5 The data below shows the number of people who went on the merry-go-round and the bouncy castle each day of a local carnival.

Merry-go-round:	63	70	59	77	82	65	61	71	79	62
Bouncy castle:	46	72	85	98	63	87	32	79	85	71

a Find the five-number summary for each data set.

b Draw a parallel box-and-whisker plot to display the data.

c Which attraction appears to be more popular? Explain your answer.

d Compare the spread of the two data sets.

HISTORICAL NOTE — CAROLINE HERSCHEL

Caroline Lucretia Herschel was born in 1750 in Hanover, Germany. Her father was an army musician, and her brother William became famous for his discovery of the planet Uranus.

Caroline was a determined and independent young woman, and after her father died she planned to pursue a career as a governess. However, before she gained a position, her brother William, who had moved to England, asked her to go and join him as his housekeeper.

While living with William, Caroline studied a variety of subjects and commenced music lessons in singing and violin.

She was also fascinated with her brother's work in astronomy, and became his assistant. Between them, they built and operated the largest telescopes of the time. William supervised the construction of the telescopes which they built and sold to help support themselves, while Caroline summarised and collated the results of their observations. Caroline began to achieve fame in her own right as an astronomer. She taught herself geometry and logarithms to assist with the difficult calculations involved.

After William's death, Caroline produced a number of books on astronomy from the notes they had collated over the years. Her contribution to astronomy was so significant that she was awarded honours by the kings of both Denmark and Prussia, and was awarded membership of the Royal Irish Academy and the Royal Astronomical Society. She died in 1848, at the age of 97.

Chapter 18

Quadratic equations

Contents:

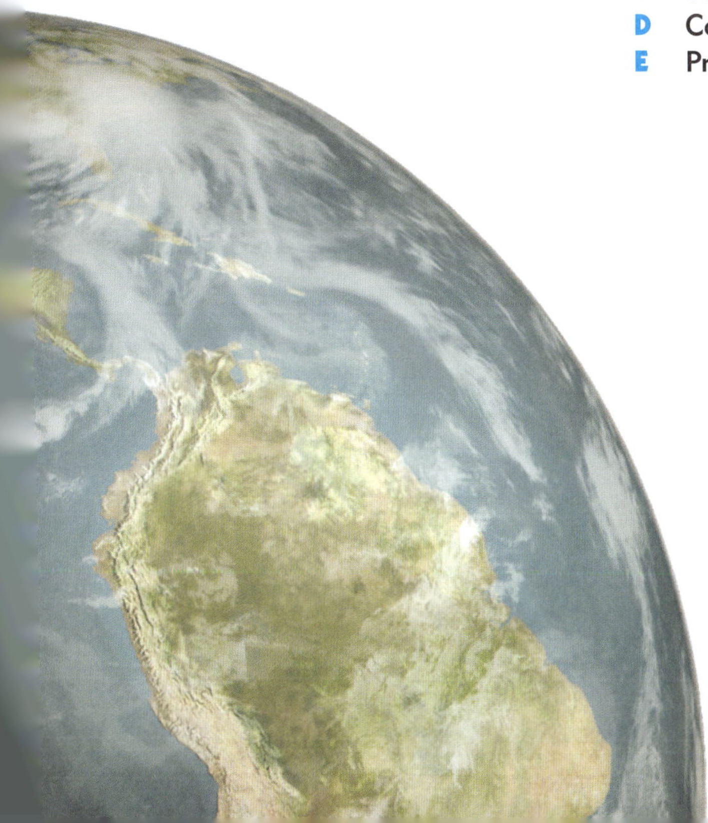

OPENING PROBLEM

Natalie is cutting a rectangular dish of lasagne for her sons Simon and Todd. Simon likes to have a square slice. Todd, who is older, eats twice as much as Simon.

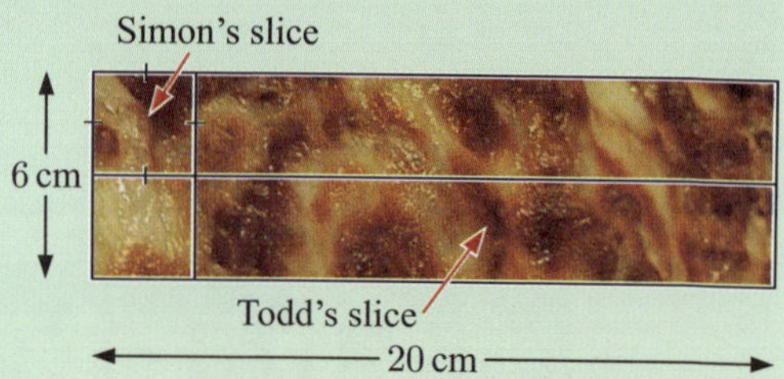

Things to think about:

- **a** Suppose Simon's slice is $3\text{ cm} \times 3\text{ cm}$. Will Todd's slice be twice the size of Simon's?
- **b** Suppose Simon's slice is a square of length x cm. Can you write an expression in terms of x for the dimensions of Todd's slice?
- **c** Can you find the dimensions of Simon's slice so that Todd's slice is twice the size of Simon's?

In the past we have solved many equations of the form $ax + b = 0$, $a \neq 0$. These are called **linear equations**, and have *only one* solution.

For example, $2x + 5 = 0$ is a linear equation which has the solution $x = -\frac{5}{2}$.

A **quadratic equation** is an equation which can be written in the form $ax^2 + bx + c = 0$, where a, b, and c are constants, $a \neq 0$.

A quadratic equation may have *two*, *one*, or *zero* solutions.

Here are some simple quadratic equations which show the truth of this statement:

Equation	$ax^2 + bx + c = 0$ *form*	*Solutions*	
$x^2 - 1 = 0$	$x^2 + 0x - 1 = 0$	$x = 1$ or $x = -1$	{**two solutions**}
$(x-3)^2 = 0$	$x^2 - 6x + 9 = 0$	$x = 3$	{**one solution**}
$x^2 + 1 = 0$	$x^2 + 0x + 1 = 0$	none as x^2 is always $\geqslant 0$	{**zero solutions**}

We can verify that the solutions to a quadratic equation are correct by substituting them back into the equation.

For example, the quadratic equation $x^2 + 5x + 6 = 0$ has solutions $x = -2$ and $x = -3$.

If $x = -2$,
$$\begin{aligned} & x^2 + 5x + 6 \\ &= (-2)^2 + 5 \times (-2) + 6 \\ &= 4 - 10 + 6 \\ &= 0 \end{aligned}$$

If $x = -3$,
$$\begin{aligned} & x^2 + 5x + 6 \\ &= (-3)^2 + 5 \times (-3) + 6 \\ &= 9 - 15 + 6 \\ &= 0 \end{aligned}$$

A EQUATIONS OF THE FORM $x^2 = k$

Consider the equation $x^2 = 2$.

If $x = \sqrt{2}$, then $x^2 = \left(\sqrt{2}\right)^2$
$= 2$

If $x = -\sqrt{2}$, then $x^2 = \left(-\sqrt{2}\right)^2$
$= \left(-\sqrt{2}\right) \times \left(-\sqrt{2}\right)$
$= 2$

So, the solutions of $x^2 = 2$ are $x = \pm\sqrt{2}$.

$$\text{If } x^2 = k \text{ then } \begin{cases} x = \pm\sqrt{k} & \text{if } k > 0 \\ x = 0 & \text{if } k = 0 \\ \text{there are \textbf{no real solutions}} & \text{if } k < 0. \end{cases}$$

Example 1 **Self Tutor**

Solve for x:

a $x^2 + 3 = 6$ **b** $3 - 2x^2 = 7$

a $x^2 + 3 = 6$
$\therefore x^2 = 3$ {subtracting 3 from both sides}
$\therefore x = \pm\sqrt{3}$

b $3 - 2x^2 = 7$
$\therefore -2x^2 = 4$ {subtracting 3 from both sides}
$\therefore x^2 = -2$ {dividing both sides by -2}
which has no real solutions as x^2 cannot be < 0.

This principle can be extended to other perfect squares.

For example, if $(x-1)^2 = k$ where $k > 0$
then $x - 1 = \pm\sqrt{k}$.

We can therefore solve equations of the form $(x \pm a)^2 = k$ without having to expand the LHS.

Example 2 **Self Tutor**

Solve for x:

a $(x-2)^2 = 25$ **b** $(x+1)^2 = 7$

a $(x-2)^2 = 25$
$\therefore x - 2 = \pm\sqrt{25}$
$\therefore x - 2 = \pm 5$
$\therefore x = 2 \pm 5$
$\therefore x = 7$ or -3

b $(x+1)^2 = 7$
$\therefore x + 1 = \pm\sqrt{7}$
$\therefore x = -1 \pm \sqrt{7}$

EXERCISE 18A

1 Solve for x:

a $x^2 = 16$ **b** $x^2 = 19$ **c** $2x^2 = 18$

d $12x^2 = 72$ **e** $3x^2 = -12$ **f** $4x^2 = 0$

g $2x^2 + 1 = 51$ **h** $1 - 3x^2 = 10$ **i** $13 = 7 + 2x^2$

2 Solve for x:

a $(x-2)^2 = 9$ **b** $(x+4)^2 = 25$ **c** $(x+3)^2 = -1$

d $(x-4)^2 = 2$ **e** $(x+3)^2 = -7$ **f** $(x-2)^2 = 0$

g $(x+5)^2 = 3$ **h** $(x-6)^2 = 2$ **i** $(x+8)^2 = 13$

j $(3x+5)^2 = 0$ **k** $(5x+1)^2 = 4$ **l** $4(2x-1)^2 = 8$

B THE NULL FACTOR LAW

For quadratic equations which are not of the form $x^2 = k$, we need an alternative method of solution.

If a quadratic equation is given in **factorised form** then we can use the **Null Factor law**.

The **Null Factor law** states that:

When the product of two or more numbers is zero, then *at least one* of them must be zero.

So, if $ab = 0$ then $a = 0$ or $b = 0$.

Example 3 Self Tutor

Solve for x using the Null Factor law:

a $2x(x-4) = 0$ **b** $(x+3)(2x-5) = 0$

a $2x(x-4) = 0$

$\therefore$ $2x = 0$ or $x - 4 = 0$

$\therefore$ $x = 0$ or 4

b $(x+3)(2x-5) = 0$

$\therefore$ $x + 3 = 0$ or $2x - 5 = 0$

$\therefore$ $x = -3$ or $2x = 5$

$\therefore$ $x = -3$ or $\frac{5}{2}$

EXERCISE 18B

1 What can be deduced from the following equations by applying the Null Factor law?

a $pq = 0$ **b** $2a = 0$ **c** $-4p = 0$ **d** $2xy = 0$

e $-3mn = 0$ **f** $\dfrac{ac}{3} = 0$ **g** $wx^2 = 0$ **h** $wxyz = 0$

2 Solve for x using the Null Factor law:

a $x(x+3)=0$	**b** $2x(x-5)=0$	**c** $(x-1)(x-3)=0$
d $4x(2-x)=0$	**e** $-3x(2x+1)=0$	**f** $5(x+2)(2x-1)=0$
g $(2x+3)(2x+1)=0$	**h** $11(x+2)(x-7)=0$	**i** $-6(x-5)(3x+2)=0$
j $2(x-3)^2=0$	**k** $-4(2x+1)^2=0$	**l** $x^2=0$
m $2x(3x-5)=0$	**n** $-3x(6-5x)=0$	**o** $(3x+1)(2x-5)=0$
p $8(3x+1)^2=0$	**q** $-5(3-4x)=0$	**r** $-6x(4-3x)=0$

C SOLUTION BY FACTORISATION

To solve quadratic equations which are not in factorised form, we use the following procedure:

Step 1: If necessary, rearrange the equation so one side is **zero**.
Step 2: **Fully factorise** the other side (usually the LHS).
Step 3: Use the **Null Factor law**: if $ab=0$ then $a=0$ or $b=0$.
Step 4: **Solve** the resulting linear equations.

To factorise the quadratic, we use the techniques we learnt in **Chapter 9**. We first take out any common factors, then recognise the type of factorisation required. We look for:

- **Difference of two squares** $x^2-a^2=(x+a)(x-a)$
- **Perfect squares** $x^2+2ax+a^2=(x+a)^2$ or $x^2-2ax+a^2=(x-a)^2$
- **Sum and product method** $x^2+bx+c=(x+p)(x+q)$ where $p+q=b$ and $pq=c$
- **Splitting the x-term** $ax^2+bx+c,\quad a\neq 0$
  - Find numbers p and q whose sum is b and whose product is ac.
  - Replace bx by $px+qx$.
  - Complete the factorisation.

Example 4 **Self Tutor**

Solve for x: $x^2=7x$

$$\begin{aligned} x^2 &= 7x \\ \therefore\ x^2-7x &= 0 \quad \{\text{making the RHS} = 0\} \\ \therefore\ x(x-7) &= 0 \quad \{\text{factorising the LHS}\} \\ \therefore\ x=0 \text{ or } x-7 &= 0 \quad \{\text{Null Factor law}\} \\ \therefore\ x &= 0 \text{ or } 7 \end{aligned}$$

EXERCISE 18C

1 Solve for x:

a $x^2+12x=0$	**b** $3x^2+9x=0$	**c** $4x^2=16x$
d $3x^2=21x$	**e** $2x^2+7x=0$	**f** $2x^2=18x$
g $3x^2=7x$	**h** $4x^2=9x$	**i** $0=3x^2+8x$

Example 5 **Self Tutor**

Solve for x: $x^2 + 2x = 8$

$x^2 + 2x = 8$

$\therefore\ x^2 + 2x - 8 = 0$ {making the RHS = 0}

$\therefore\ (x+4)(x-2) = 0$ {the numbers 4 and -2 have sum 2 and product -8}

$\therefore\ x + 4 = 0$ or $x - 2 = 0$ {Null Factor law}

$\therefore\ x = -4$ or 2

Check: If $x = -4$, then $x^2 + 2x = (-4)^2 + 2(-4) = 16 - 8 = 8$ ✓

If $x = 2$, then $x^2 + 2x = 2^2 + 2(2) = 4 + 4 = 8$ ✓

2 Solve for x:

a $x^2 + 6x + 8 = 0$ **b** $x^2 + 11x + 24 = 0$ **c** $x^2 + 4x + 4 = 0$

d $x^2 - 49 = 0$ **e** $x^2 + 2x = 48$ **f** $x^2 + 25 = 10x$

g $x^2 = 2x + 15$ **h** $x^2 = 5x + 14$ **i** $x^2 = 7x - 12$

j $x^2 + 60 = 17x$ **k** $x^2 = 9x + 22$ **l** $x^2 = 3x + 18$

3 Solve for x:

a $-x^2 - 5x + 24 = 0$ **b** $2x^2 + 16x + 24 = 0$ **c** $5x^2 - 5x - 100 = 0$

d $2x^2 + 8x = 42$ **e** $3x^2 - 18 = 15x$ **f** $7x^2 + 7 = 14x$

g $4x^2 + 72 = 36x$ **h** $5x^2 + 30x = -45$ **i** $60 - 2x^2 = 26x$

Example 6 **Self Tutor**

Solve for x: $2x^2 = 3x - 1$

$2x^2 = 3x - 1$

$\therefore\ 2x^2 - 3x + 1 = 0$ {making the RHS = 0}

In $2x^2 - 3x + 1$, $ac = 2$ and $b = -3$.

We need two numbers with a sum of -3 and a product of 2. These are -2 and -1.

$\therefore\ 2x^2 - 2x - x + 1 = 0$ {'splitting' the middle term}

$\therefore\ 2x(x-1) - 1(x-1) = 0$ {factorising in pairs}

$\therefore\ (x-1)(2x-1) = 0$ {$(x-1)$ is a common factor}

$\therefore\ x - 1 = 0$ or $2x - 1 = 0$ {Null Factor law}

$\therefore\ x = 1$ or $\frac{1}{2}$

4 Solve for x:

a $2x^2 + 11x + 5 = 0$ **b** $5x^2 - 21x + 4 = 0$ **c** $3x^2 - 7x + 2 = 0$

d $3x^2 + 10x = 8$ **e** $2x^2 - 13x = 7$ **f** $7x^2 = 11x + 6$

g $3x^2 + 10 = 17x$ **h** $2x^2 + 20 = 13x$ **i** $3x^2 + x = 10$

j $2x^2 = 13x + 15$ **k** $2x^2 - 7x = 15$ **l** $3x^2 = 11x + 4$

5 Solve for x:

a $x(x+2)+3(x-1)=11$
b $x(3+x)+3=31$
c $(x+2)(x-7)=8x$
d $3x(x+2)=9$
e $2x(x-1)-3(x+2)=-3$
f $3x(x+4)=x-10$

6 Solve for x:

a $5x^2+8x=0$
b $x^2-10x=39$
c $15x^2+17x-4=0$
d $3x^2-24x+48=0$
e $x(8-x)+20=-13$
f $2x^2=11x$
g $2x^2-88=-14x$
h $5x^2+4=12x$
i $54-3x^2=21x$
j $(x+5)(x-1)=8x$
k $(8x+1)(x-2)=13-x$
l $6x^2+13x=x^2+6$

D COMPLETING THE SQUARE

Factorising quadratic expressions can be very difficult, and we have generally only dealt with cases where the factorisation involves integers.

Quadratic equations such as $x^2+6x-1=0$ have solutions which are *irrational*. They cannot be written as a fraction in the form $\frac{p}{q}$ where p, q are integers.

Quadratic equations like this cannot readily be solved by factorisation, so instead we use a method called **completing the square**. This involves use of the perfect square factorisations $x^2+2ax+a^2=(x+a)^2$ or $x^2-2ax+a^2=(x-a)^2$.

Consider the following method for solving $x^2+6x-1=0$:

$$\begin{aligned} x^2+6x-1&=0 \\ \therefore\ x^2+6x&=1 \\ \therefore\ x^2+6x+9&=1+9 \\ \therefore\ (x+3)^2&=10 \\ \therefore\ x+3&=\pm\sqrt{10} \\ \therefore\ x&=-3\pm\sqrt{10} \end{aligned}$$

We add 9 to both sides to 'complete' a perfect square on the LHS.

The hardest part of this process is knowing that we need to add 9 to both sides to complete the square.

From our previous study of perfect squares, we observe that:

$(x+4)^2=x^2+8x+4^2$ (half of 8)

$(x-5)^2=x^2-10x+(-5)^2$ (half of -10)

$(x+6)^2=x^2+12x+6^2$ (half of 12)

So, the number we must add to both sides is found by **halving the coefficient of x, then squaring this value**.

In the above example, the coefficient of x is 6, so half the coefficient of x is 3. We therefore add $3^2=9$ to both sides of the equation.

Example 7 **Self Tutor**

For each equation:

- **i** Find the number which must be added to both sides of the equation to create a perfect square on the LHS.
- **ii** Write the equation in the form $(x+p)^2 = k$.

a $x^2 + 2x = 6$ **b** $x^2 - 8x = -3$

a **i** In $x^2 + 2x = 6$, half the coefficient of x is $\frac{2}{2} = 1$.
So, we add 1^2 to both sides.

ii The equation becomes $x^2 + 2x + 1^2 = 6 + 1^2$

$\therefore\ (x+1)^2 = 6 + 1$

$\therefore\ (x+1)^2 = 7$

b **i** In $x^2 - 8x = -3$, half the coefficient of x is $\frac{-8}{2} = -4$.
So, we add $(-4)^2 = 4^2$ to both sides.

ii The equation becomes $x^2 - 8x + 4^2 = -3 + 4^2$

$\therefore\ (x-4)^2 = -3 + 16$

$\therefore\ (x-4)^2 = 13$

Example 8 **Self Tutor**

Solve for x by completing the square. Leave your answers in simplest radical form:

a $x^2 + 4x - 4 = 0$ **b** $x^2 - x - 7 = 0$

a

$x^2 + 4x - 4 = 0$

$\therefore\ x^2 + 4x = 4$ {moving the constant term to the RHS}

$\therefore\ x^2 + 4x + 2^2 = 4 + 2^2$ {adding $(\frac{4}{2})^2 = 2^2$ to both sides}

$\therefore\ (x+2)^2 = 8$

$\therefore\ x + 2 = \pm\sqrt{8}$

$\therefore\ x = -2 \pm 2\sqrt{2}$

b

$x^2 - x - 7 = 0$

$\therefore\ x^2 - x = 7$ {moving the constant term to the RHS}

$\therefore\ x^2 - x + (\frac{1}{2})^2 = 7 + (\frac{1}{2})^2$ {adding $(\frac{-1}{2})^2 = (\frac{1}{2})^2$ to both sides}

$\therefore\ (x - \frac{1}{2})^2 = 7 + \frac{1}{4}$

$\therefore\ (x - \frac{1}{2})^2 = \frac{29}{4}$

$\therefore\ x - \frac{1}{2} = \pm\sqrt{\frac{29}{4}}$

$\therefore\ x = \frac{1}{2} \pm \frac{\sqrt{29}}{2}$

$\therefore\ x = \dfrac{1 \pm \sqrt{29}}{2}$

EXERCISE 18D

1 What number must be added to the following to create a perfect square?

a $x^2 + 4x$ **b** $x^2 - 4x$ **c** $x^2 - 6x$ **d** $x^2 + 10x$
e $x^2 - 10x$ **f** $x^2 + 16x$ **g** $x^2 - 18x$ **h** $x^2 - 20x$
i $x^2 + 3x$ **j** $x^2 - 3x$ **k** $x^2 + 5x$ **l** $x^2 - 5x$

2 Write each equation in the form $(x + p)^2 = k$:

a $x^2 + 2x = 2$ **b** $x^2 - 2x = 4$ **c** $x^2 + 6x = -4$
d $x^2 - 6x = 7$ **e** $x^2 + 10x = -23$ **f** $x^2 + 8x = -13$
g $x^2 + 6x = 2$ **h** $x^2 + 3x = -1$ **i** $x^2 - 7x = 5$

3 Solve for x by completing the square. Leave your answers in simplest radical form.

a $x^2 - 4x - 2 = 0$ **b** $x^2 + 2x - 6 = 0$ **c** $x^2 - 2x - 2 = 0$
d $x^2 + 4x - 1 = 0$ **e** $x^2 + 6x - 11 = 0$ **f** $x^2 - 6x + 1 = 0$
g $x^2 + 8x + 5 = 0$ **h** $x^2 + x - 3 = 0$ **i** $x^2 - 3x - 5 = 0$

Example 9 **Self Tutor**

If possible, solve for x by completing the square: $x^2 - 2x + 7 = 0$

$x^2 - 2x + 7 = 0$
$\therefore\ x^2 - 2x = -7$ {moving the constant term to the RHS}
$\therefore\ x^2 - 2x + 1^2 = -7 + 1^2$ {adding $(\frac{-2}{2})^2 = 1^2$ to both sides}
$\therefore\ (x - 1)^2 = -6$

which has no real solutions as $(x - 1)^2$ cannot be < 0.

If $(x - a)^2 = k$ where $k < 0$, then there are no real solutions.

4 If possible, solve for x by completing the square:

a $x^2 - 2x + 3 = 0$ **b** $x^2 + 4x + 1 = 0$ **c** $x^2 - 4x + 5 = 0$
d $x^2 + 6x - 2 = 0$ **e** $x^2 - x - 4 = 0$ **f** $x^2 + 5x + 8 = 0$

5 **a** If you were asked to solve $2x^2 + 6x - 1 = 0$ by completing the square, explain why you would start by dividing both sides by 2.

b Solve by completing the square:

i $2x^2 + 8x + 1 = 0$ **ii** $3x^2 - 6x + 2 = 0$ **iii** $5x^2 + 10x + 1 = 0$

INVESTIGATION — THE QUADRATIC FORMULA

What to do:

1 Use factorisation techniques or completing the square to solve:

a $x^2 + 3x - 10 = 0$ **b** $2x^2 - 7x - 4 = 0$ **c** $x^2 - 4x - 3 = 0$
d $4x^2 - 12x + 9 = 0$ **e** $x^2 + 6x + 11 = 0$ **f** $4x^2 - 8x + 1 = 0$

2 The **quadratic formula** for solving $ax^2 + bx + c = 0$ is $x = \dfrac{-b \pm \sqrt{b^2 - 4ac}}{2a}$.

Check that this formula gives the correct answers for each equation in **1**.

3 What is the significance of $b^2 - 4ac$ in determining whether the quadratic equation $ax^2 + bx + c = 0$ has real solutions?

4 Derive the quadratic formula by completing the square on $ax^2 + bx + c = 0$.
Hint: Start by dividing each term by a.

5 Use the quadratic formula to solve:

a $x^2 - 3x + 1 = 0$	**b** $x^2 - 7x + 2 = 0$	**c** $x^2 + 2x - 5 = 0$
d $x^2 - 4x + 6 = 0$	**e** $x^2 + 10x + 25 = 0$	**f** $x^2 - 4x - 5 = 0$
g $x^2 - 10x + 22 = 0$	**h** $2x^2 - 4x - 7 = 0$	**i** $x^2 + x + 2 = 0$
j $4x^2 - 12x + 9 = 0$	**k** $9x^2 - 12x + 11 = 0$	**l** $5x^2 - 6x + 1 = 0$

E PROBLEM SOLVING

When practical problems are converted into algebraic form, a quadratic equation may result. To solve these problems, follow these steps:

Step 1: Carefully **read the question** until you understand the problem. A **sketch** may be useful.

Step 2: Decide on the **unknown** quantity and label it with a variable such as x.

Step 3: Use the information given to write an **equation**.

Step 4: **Solve** the equation.

Step 5: **Check** that any solutions satisfy the equation and are reasonable in the context of the problem.

Step 6: Write your answer to the question in **sentence form**.

Example 10 — Self Tutor

The difference between a number and its square is 20. Find the number.

Let the number be x, so its square is x^2.

$$x^2 - x = 20$$
$$\therefore\ x^2 - x - 20 = 0 \quad \{\text{rearranging}\}$$
$$\therefore\ (x-5)(x+4) = 0 \quad \{\text{factorising}\}$$
$$\therefore\ x = 5 \text{ or } x = -4$$

So, the number is -4 or 5.

Check: The square of -4 is 16, and the difference between -4 and 16 is 20. ✓
The square of 5 is 25, and the difference between 5 and 25 is 20. ✓

Example 11 **Self Tutor**

Use Pythagoras' theorem to find x:

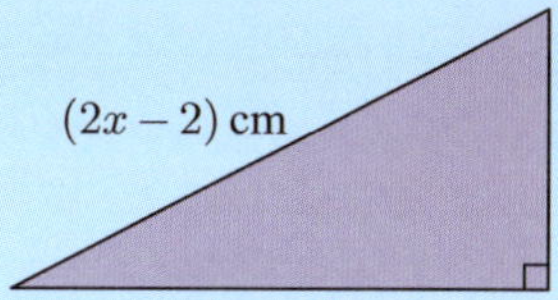

Using Pythagoras' theorem, $(x+2)^2 + x^2 = (2x-2)^2$

$\therefore\ x^2 + 4x + 4 + x^2 = 4x^2 - 8x + 4$

$\therefore\ 2x^2 + 4x = 4x^2 - 8x$

$\therefore\ 2x^2 - 12x = 0$

$\therefore\ 2x(x-6) = 0$

$\therefore\ x = 0$ or 6

But $x > 0$ as lengths are positive quantities. $\therefore\ x = 6$

Example 12 **Self Tutor**

A rectangular housing block has perimeter 64 m and area 220 m^2. Find the dimensions of the block.

Suppose the block is x m long and y m wide.

$\therefore\ 2x + 2y = 64$ {perimeter = 64 m}

$\therefore\ x + y = 32$

$\therefore\ y = 32 - x$

The area is 220 m^2, so $xy = 220$

$\therefore\ x(32-x) = 220$

$\therefore\ 32x - x^2 = 220$

$\therefore\ x^2 - 32x + 220 = 0$

$\therefore\ (x-10)(x-22) = 0$

$\therefore\ x = 10$ or 22

When $x = 10$, $y = 22$, and when $x = 22$, $y = 10$.

So, the block is 10 m × 22 m.

EXERCISE 18E

1. When 15 is added to the square of a number, the result is eight times the original number. Find the number.

2. Two numbers differ by 13, and the sum of their squares is 125. Find the numbers.

3. When a number is squared and then multiplied by 6, the result is one more than the original number. Find the number.

4 Use Pythagoras' theorem to find x:

a

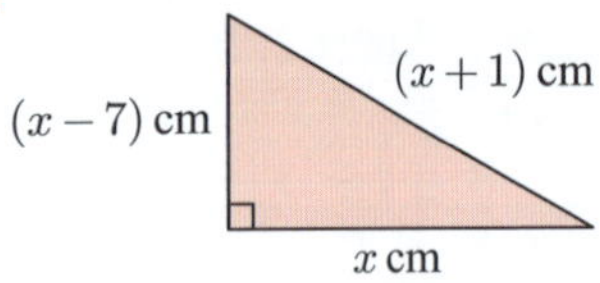

b

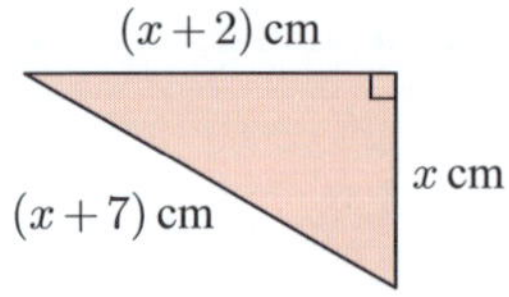

c

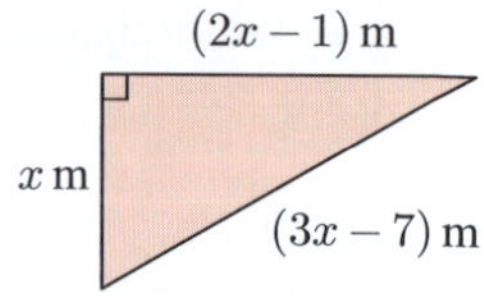

5 A right angled triangle has sides which are 2 cm and 7 cm shorter than its hypotenuse. Find the length of the hypotenuse:

a exactly **b** to the nearest millimetre.

6 A rectangular field has perimeter 600 m and area 21 600 m^2. Find the dimensions of the field.

7

A 52 m long fence is constructed on three sides of a property with area 240 m^2.
The fourth side facing the road is left open.
Find the dimensions of the property.

8 A rectangular chicken yard was built against an existing barn wall. 30 m of fencing was used to enclose 108 m^2. Find the dimensions of the yard.

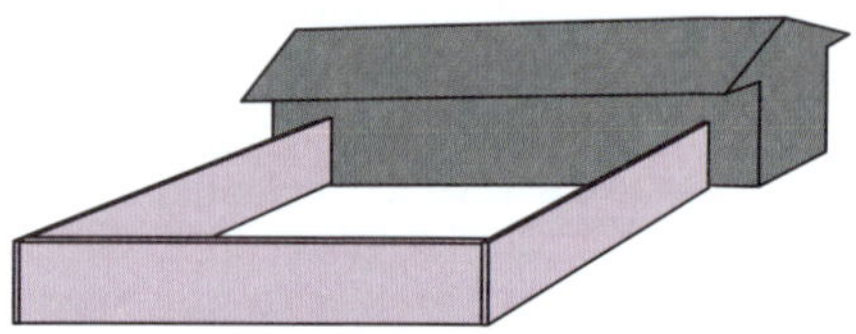

9 A gardener plants 90 seedlings. The number of seedlings in each row is 3 more than twice the number of rows. How many rows did the gardener plant?

10 Answer the **Opening Problem** on page **360**.

11 A rectangular lawn measures 10 m by 6 m. One circuit around the outer edge of the lawn will mow $\frac{1}{5}$ of the lawn. Find the width of the mower, to the nearest centimetre.

Global context

click here

What is the golden ratio?

Statement of inquiry:	Mathematics can help us to understand beauty in the world around us.
Global context:	Personal and cultural expression
Key concept:	Relationships
Related concepts:	Quantity, Space
Objective:	Knowing and understanding
Approaches to learning:	Thinking, Research

REVIEW SET 18A

1 Solve for x:

a $3x^2 = 48$ **b** $65 - 2x^2 = 15$

2 Solve for x:

a $(x+1)^2 = 36$ **b** $(x-3)^2 = -16$ **c** $(3x-1)^2 = 5$

3 What can be deduced from the following equations by applying the Null Factor law?

a $mn = 0$ **b** $-7p = 0$ **c** $ab^2 = 0$

4 Solve for x:

a $2x^2 + 8x = 0$ **b** $x^2 + 5x - 66 = 0$ **c** $x^2 = 13x - 36$

5 Solve for x:

a $2x^2 - 2x - 24 = 0$ **b** $3x^2 - 15 = 12x$ **c** $2x^2 + 3x - 9 = 0$

6 What number must be added to the following to create a perfect square?

a $x^2 - 8x$ **b** $x^2 + 12x$ **c** $x^2 - 7x$

7 Solve for x:

a $x(x+3) + 2(x-4) = 6$ **b** $3x(x+2) - 10(2x-1) = 2$

8 Solve for x by completing the square:

a $x^2 - 2x - 10 = 0$ **b** $x^2 + 5x - 2 = 0$

9 When the square of a number is increased by 10, the result is seven times the original number. Find the number.

10

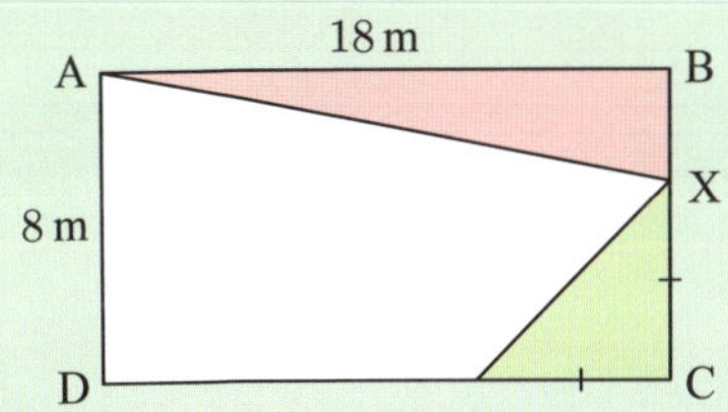

In the figure alongside, the two shaded triangles have equal area. Find the length of [BX].

REVIEW SET 18B

1 Solve for x:

a $2x^2 = 22$ **b** $5 - 4x^2 = 45$

2 Solve for x:

a $(x+2)^2 = 17$ **b** $(2x+1)^2 = 0$ **c** $(4x-3)^2 = -4$

3 Solve for x using the Null Factor law:

a $x(x+3) = 0$ **b** $-5(x-1)(3-x) = 0$ **c** $2(3x-1)(2x+7) = 0$

4 Solve for x:

a $2x^2 = 9x$ **b** $x^2 + 18 = 11x$ **c** $2x^2 - 8x = 42$

5 Write in the form $(x+p)^2 = k$: $x^2 - 6x = 10$

6 Solve for x:

a $x^2 + x - 30 = 0$ **b** $60x - 4x^2 = 200$ **c** $4x^2 - 15 = 17x$

7 Use Pythagoras' theorem to find x:

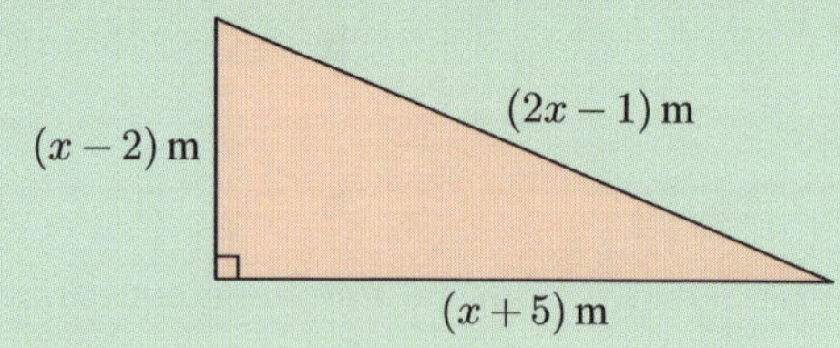

8 If possible, solve for x by completing the square:

a $x^2 - 10x - 2 = 0$ **b** $x^2 - 5x + 7 = 0$

9 Solve for x:

a $x(x+7) - 4(x+10) = 68$ **b** $4x(x+1) - 6(2x+3) = 3$

10 Two flagpoles are 6 m high and 13 m apart. Wires supporting the flagpoles are connected to a hook on the ground at X. The wires are perpendicular to each other.
Find the distance AX between the hook and the nearer flagpole.

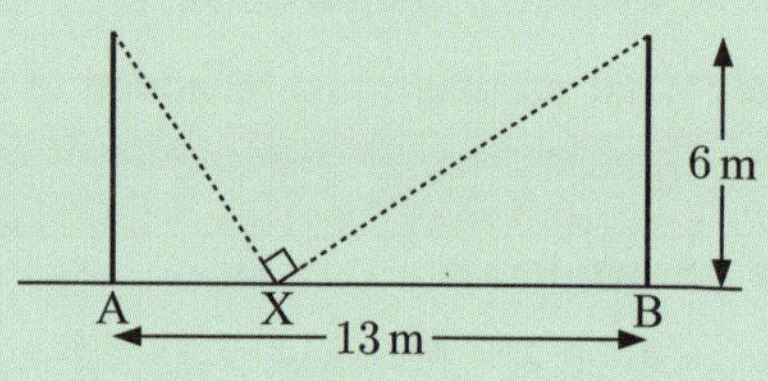

Chapter 19

Simultaneous equations

Contents:

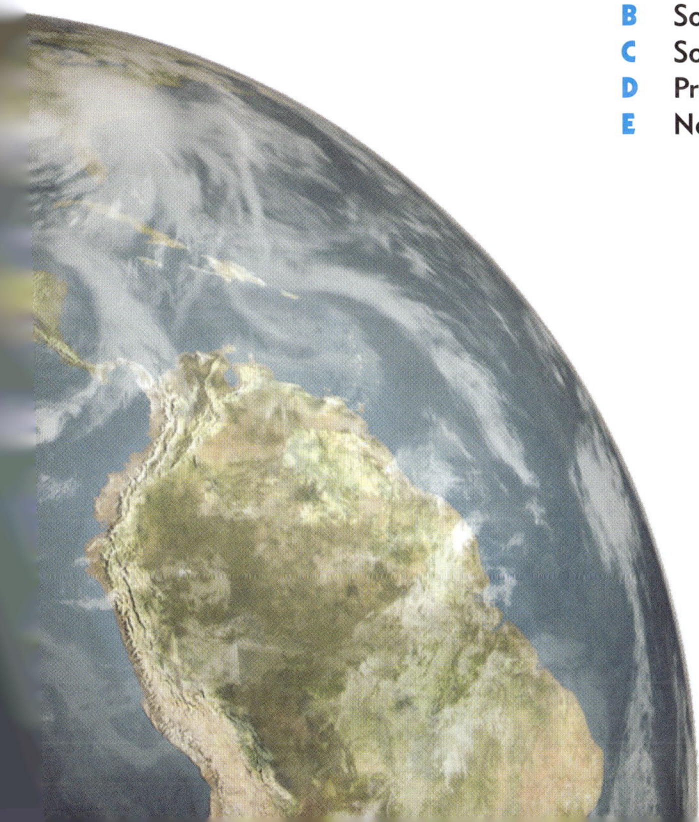

OPENING PROBLEM

Ewen wants to buy a pie, but only has \$3.50. He sees his friend André leaving the canteen with a pie, and asks him how much it cost.

"I'm not sure," said André, "I bought a pie and a sandwich. They cost me \$7 altogether."

Ewen saw another friend Samuel with a pie, and asked him how much it cost.

"Well, I bought 2 pies and 3 sandwiches, and they cost me \$17 altogether."

Things to think about:

a Can Ewen determine the cost of a pie using:

  i André's information only **ii** Samuel's information only?

b Suppose a pie costs \$$x$, and a sandwich costs \$$y$.
Can you explain why: **i** $x+y=7$ **ii** $2x+3y=17$?

c Can you find values for x and y which make *both* equations true?

d Does Ewen have enough money to buy a pie?

To answer the **Opening Problem**, we need values for x and y which satisfy both $x+y=7$ and $2x+3y=17$ *at the same time.*

We say that $\begin{cases} x+y=7 \\ 2x+3y=17 \end{cases}$ is a set of **simultaneous equations**.

Notice that if $x=4$ and $y=3$, then $x+y=4+3=7$,
and $2x+3y=2(4)+3(3)=17$.

So, $x=4$, $y=3$ is a **solution** to the set of simultaneous equations $\begin{cases} x+y=7 \\ 2x+3y=17. \end{cases}$

A GRAPHICAL SOLUTION

One way to solve simultaneous equations is to draw the graph of both equations on the same set of axes. The **point of intersection** corresponds to a solution to the simultaneous equations.

A set of two linear simultaneous equations can have either one, zero, or infinitely many solutions.

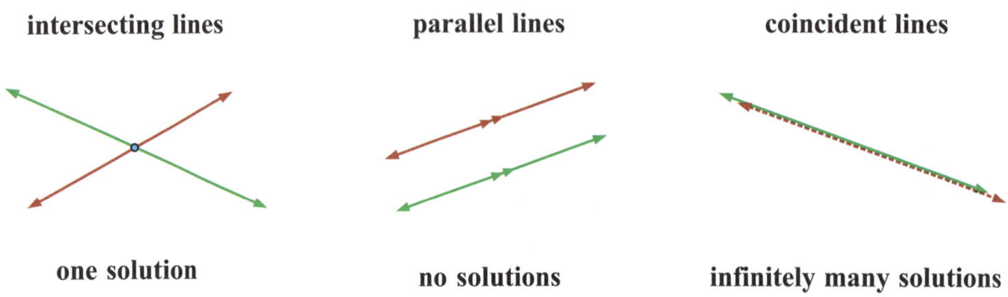

Example 1 **Self Tutor**

Solve the following simultaneous equations graphically: $\begin{cases} y = 2x - 4 \\ 2x + 3y = 12 \end{cases}$

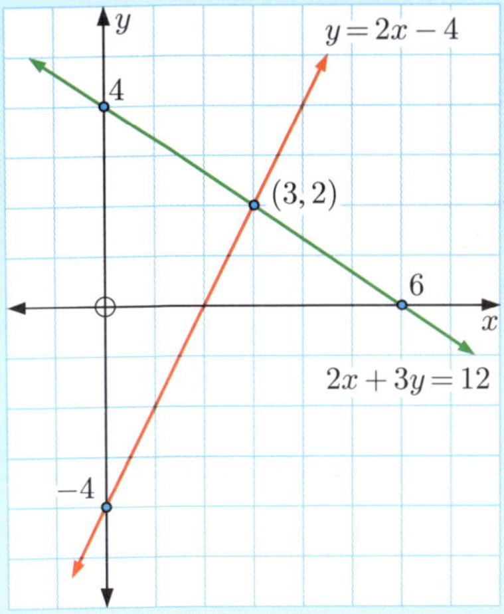

We draw the graphs of $y = 2x - 4$ and $2x + 3y = 12$ on the same set of axes.

The graphs meet at the point $(3, 2)$.

$\therefore$ the solution is $x = 3$, $y = 2$.

Check:

Substituting these values into:

- $y = 2x - 4$ gives $2 = 2(3) - 4$ ✓
- $2x + 3y = 12$ gives $2(3) + 3(2) = 12$ ✓

EXERCISE 19A.1

1 Solve the following simultaneous equations graphically:

a $\begin{cases} y = 4x - 1 \\ y = 2x - 3 \end{cases}$ **b** $\begin{cases} y = 3x \\ y = -2x + 5 \end{cases}$ **c** $\begin{cases} y = x + 2 \\ y = -3x - 6 \end{cases}$

2 Solve the following simultaneous equations graphically:

a $\begin{cases} y = 2x - 8 \\ 2x + 5y = 20 \end{cases}$ **b** $\begin{cases} 4x + y = 8 \\ 2x - 3y = 18 \end{cases}$ **c** $\begin{cases} 3x - y = -6 \\ 3x + 4y = -36 \end{cases}$

3 Try to solve the following simultaneous equations graphically. State the number of solutions in each case.

a $\begin{cases} y = 4x + 1 \\ y = 4x - 3 \end{cases}$ **b** $\begin{cases} x - \frac{1}{2}y = -3 \\ y = 2x + 6 \end{cases}$ **c** $\begin{cases} 5x + 3y = 2 \\ 5x + 3y = -7 \end{cases}$

USING TECHNOLOGY

If the solutions to a set of simultaneous equations are not integers, it may be difficult to find them on a graph drawn by hand. In these situations, we can use technology to graph the equations and find the solution.

You can use your **graphics calculator** or the **graphing package** to graph the equations. When using your graphics calculator, any equations in the general form $Ax + By = C$ will need to be rearranged to make y the subject.

GRAPHICS CALCULATOR INSTRUCTIONS

GRAPHING PACKAGE

Example 2 Self Tutor

Use technology to solve the following simultaneous equations: $\begin{cases} y = 4x + 13 \\ x + y = 7 \end{cases}$

We rearrange the second equation, so the system is now: $\begin{cases} y = 4x + 13 \\ y = 7 - x \end{cases}$

Casio fx-CG20	TI-84 Plus	TI-*n*spire
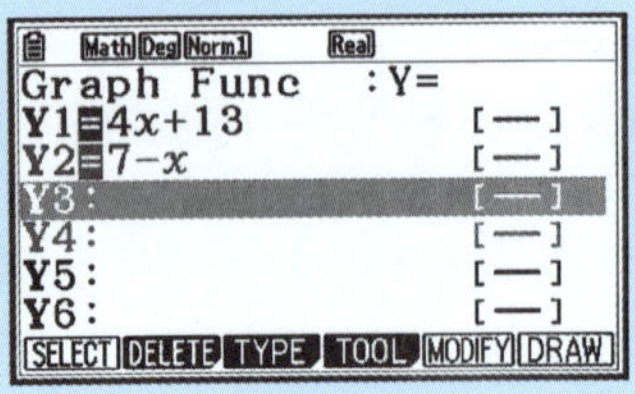	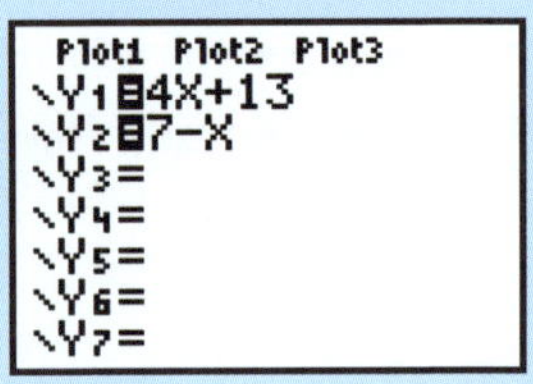	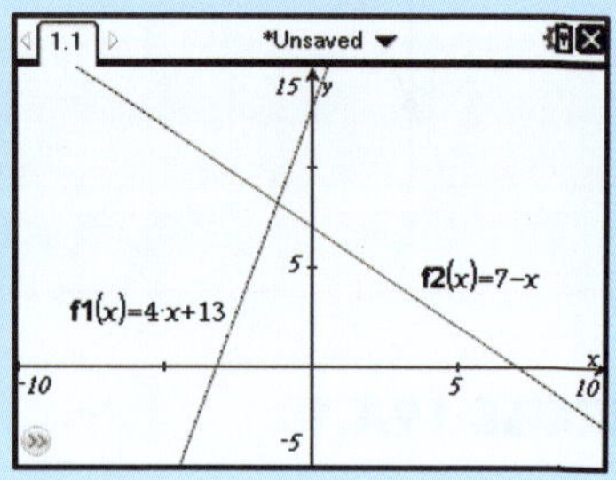
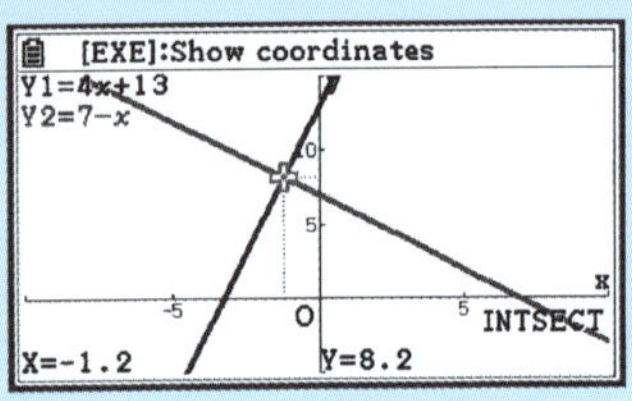	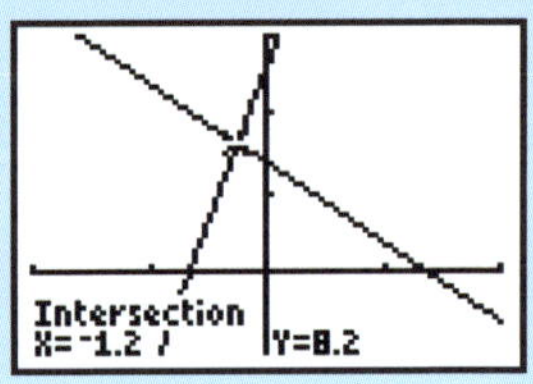	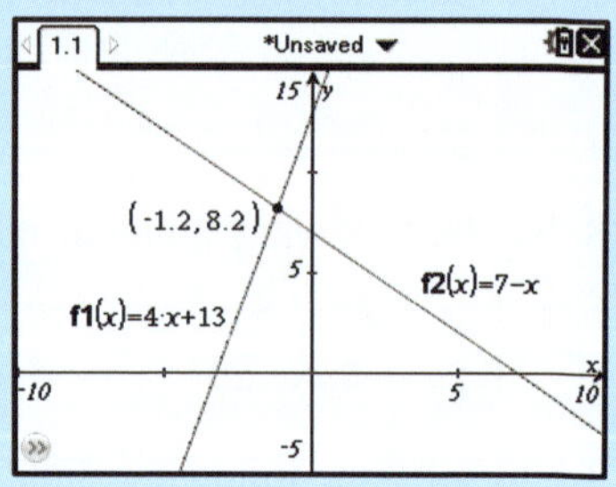

The solution is $x = -1.2$, $y = 8.2$.

EXERCISE 19A.2

1 Use technology to solve the following simultaneous equations:

a $\begin{cases} y = 5x - 2 \\ y = x + 1 \end{cases}$ **b** $\begin{cases} y = 3x - 6 \\ y = 1 - 2x \end{cases}$ **c** $\begin{cases} y = 3x - 4 \\ x + y = -5 \end{cases}$

d $\begin{cases} x + y = 6 \\ 6x + y = 15 \end{cases}$ **e** $\begin{cases} x + 2y = -8 \\ 7x + 3y = -5 \end{cases}$ **f** $\begin{cases} 3x - 4y = -2 \\ 2x + 3y = -10 \end{cases}$

2 Try to solve the following simultaneous equations using technology. State the number of solutions in each case.

a $\begin{cases} y = 1.5x + 2 \\ 3x - 2y = -4 \end{cases}$ **b** $\begin{cases} 4x + 5y = -3 \\ y = -0.8x + 1 \end{cases}$ **c** $\begin{cases} x - 4y = 7 \\ 2x - 8y = -1 \end{cases}$

B SOLUTION BY SUBSTITUTION

We will now consider some **algebraic** methods for solving simultaneous equations.

The method of **solution by substitution** is used when at least one equation is given with either x or y as the **subject** of the formula. We substitute an expression for this variable into the other equation.

Example 3 — Self Tutor

Solve simultaneously by substitution: $\begin{cases} y = x + 5 \\ 3x - y = 1 \end{cases}$

$y = x + 5$ .... (1)
$3x - y = 1$ .... (2)

Substituting (1) into (2) gives $3x - (x + 5) = 1$
$\therefore 3x - x - 5 = 1$
$\therefore 2x = 6$
$\therefore x = 3$

Substituting $x = 3$ into (1) gives $y = 3 + 5$
$\therefore y = 8$

The solution is $x = 3$, $y = 8$.

Check: In (1): $8 = 3 + 5$ ✓
In (2): $3(3) - 8 = 9 - 8 = 1$ ✓

EXERCISE 19B

1 Solve simultaneously by substitution:

a $\begin{cases} y = x - 3 \\ 2x + 3y = 26 \end{cases}$
b $\begin{cases} y = 3x - 2 \\ 5x - 2y = 5 \end{cases}$
c $\begin{cases} 7x + 4y = -7 \\ y = 8 - 5x \end{cases}$
d $\begin{cases} y = 2x - 12 \\ y = 13 - 3x \end{cases}$
e $\begin{cases} y = 3x + 4 \\ 5x + 3y = 5 \end{cases}$
f $\begin{cases} 5x - 6y = 2 \\ y = 3 - 7x \end{cases}$

2 Solve simultaneously by substitution:

a $\begin{cases} x = y + 9 \\ 5x + 2y = 10 \end{cases}$
b $\begin{cases} 7x - 5y = 1 \\ x = 2y + 4 \end{cases}$
c $\begin{cases} x = 2y - 7 \\ x = 13 - 6y \end{cases}$
d $\begin{cases} y = 2x - 2 \\ x = 5y + 7 \end{cases}$
e $\begin{cases} 2x + 4y = 1 \\ x = 10y - 7 \end{cases}$
f $\begin{cases} x = -4y - 3 \\ 9x + 5y = 2 \end{cases}$

3 Solve simultaneously by substitution:

a $\begin{cases} y = \frac{1}{2}x - 2 \\ 3x - 8y = 11 \end{cases}$
b $\begin{cases} x = -\frac{2}{3}y \\ 6x + 7y = 6 \end{cases}$
c $\begin{cases} 5x - 12y = 18 \\ y = \frac{1}{4}x - 1 \end{cases}$
d $\begin{cases} y = -\frac{1}{2}x - 1 \\ 4x + 3y = -2 \end{cases}$
e $\begin{cases} 2x - 3y = 3 \\ x = \frac{11}{5}y + 2 \end{cases}$
f $\begin{cases} 14x + 15y = 25 \\ y = \frac{2}{5}x - 3 \end{cases}$

4 **a** Try to solve by substitution: $\begin{cases} y = 3x + 1 \\ 9x - 3y = 5 \end{cases}$

b What is the simultaneous solution for the equations in **a**?

5 **a** Try to solve by substitution: $\begin{cases} y = 3x + 1 \\ 6x - 2y = -2 \end{cases}$

b How many simultaneous solutions do the equations in **a** have?

6 Consider the simultaneous equations: $\begin{cases} 4x + 5y = -4 & \text{.... (1)} \\ 3x + 2y = -5 & \text{.... (2)} \end{cases}$

a Rearrange equation (1) so that y is the subject.

b Hence, solve the simultaneous equations by substitution.

C SOLUTION BY ELIMINATION

If both equations are presented in the general form $Ax + By = C$, then solution by substitution is tedious. We instead use the method of **elimination**.

In this method, we make the coefficients of x (or y) the **same size** but **opposite in sign**. We then **add** the equations, which has the effect of **eliminating** one of the variables.

Example 4 Self Tutor

Solve simultaneously by elimination: $\begin{cases} 4x + 3y = 2 \\ x - 3y = 8 \end{cases}$

The coefficients of y are the same size but opposite in sign.

We **add** the LHSs and the RHSs to get an equation which contains x only.

$$\begin{aligned} 4x + 3y &= 2 \quad \text{.... (1)} \\ x - 3y &= 8 \quad \text{.... (2)} \\ \hline \text{Adding,} \quad 5x \quad &= 10 \\ \therefore \ x &= 2 \end{aligned}$$

Substituting $x = 2$ into (1) gives $4(2) + 3y = 2$

$\therefore \ 8 + 3y = 2$

$\therefore \ 3y = -6$

$\therefore \ y = -2$

The solution is $x = 2$, $y = -2$.

Check: In (2): $(2) - 3(-2) = 2 + 6 = 8$ ✓

In problems where the coefficients of x (or y) are **not** the **same size** or **opposite in sign**, we must first **multiply** one or both equations by a constant.

Example 5 **Self Tutor**

Solve simultaneously by elimination: $\begin{cases} 3x + 2y = 7 \\ 2x - 5y = 11 \end{cases}$

$3x + 2y = 7$ .... (1) $\qquad 2x - 5y = 11$ .... (2)

To make the coefficients of y the same size but opposite in sign, we multiply (1) by 5 and (2) by 2.

$$\therefore \quad 15x + 10y = 35 \quad \{(1) \times 5\}$$
$$4x - 10y = 22 \quad \{(2) \times 2\}$$

Adding, $\quad 19x = 57$

$$\therefore \quad x = 3$$

Substituting $x = 3$ into (1) gives $3(3) + 2y = 7$

$$\therefore \quad 9 + 2y = 7$$
$$\therefore \quad 2y = -2$$
$$\therefore \quad y = -1$$

The solution is $x = 3, \quad y = -1$.

Check: In (2): $2(3) - 5(-1) = 6 + 5 = 11$ ✓

We can choose to eliminate either x or y, so we choose the variable which is easiest to eliminate. In **Example 5** we could have eliminated x by multiplying (1) by 2 and (2) by -3.

EXERCISE 19C

1 What equation results when the following are added vertically?

- **a** $3x + 4y = 6$
 $8x - 4y = 5$
- **b** $2x - y = 7$
 $-2x + 5y = 5$
- **c** $7x - 3y = 2$
 $2x + 3y = 7$
- **d** $6x - 11y = 12$
 $3x + 11y = -6$
- **e** $-7x + 2y = 5$
 $7x - 3y = 6$
- **f** $2x - 3y = -7$
 $-2x - 8y = -4$

2 Solve simultaneously by elimination:

- **a** $\begin{cases} 5x - y = 4 \\ 2x + y = 10 \end{cases}$
- **b** $\begin{cases} 3x - 2y = 7 \\ 3x + 2y = -1 \end{cases}$
- **c** $\begin{cases} -5x - 3y = 14 \\ 5x + 8y = -29 \end{cases}$
- **d** $\begin{cases} 4x + 3y = -11 \\ -4x - 2y = 6 \end{cases}$
- **e** $\begin{cases} 2x - 5y = 14 \\ 4x + 5y = -2 \end{cases}$
- **f** $\begin{cases} -6x - y = 23 \\ 6x + 5y = -13 \end{cases}$

3 Give the equation that results when both sides of the equation:

- **a** $2x + 5y = 1$ are multiplied by 5
- **b** $3x - y = 4$ are multiplied by -1
- **c** $x - 7y = 8$ are multiplied by 3
- **d** $5x + 4y = 9$ are multiplied by -2
- **e** $-3x - 2y = 2$ are multiplied by 6
- **f** $4x - 2y = 3$ are multiplied by -4.

4 Solve simultaneously by elimination:

- **a** $\begin{cases} 2x + y = 8 \\ x - 3y = 11 \end{cases}$
- **b** $\begin{cases} 3x + 2y = 7 \\ x + 3y = 7 \end{cases}$
- **c** $\begin{cases} 5x - 2y = 17 \\ 3x - y = 9 \end{cases}$

d $\begin{cases} 2x + 5y = -14 \\ -6x + 2y = -9 \end{cases}$ **e** $\begin{cases} 7x + 3y = 5 \\ 5x - 6y = 9 \end{cases}$ **f** $\begin{cases} 4x + 9y = 24 \\ 12x - 7y = -30 \end{cases}$

5 Solve simultaneously by elimination:

a $\begin{cases} 2x + 3y = 13 \\ 3x + 2y = 17 \end{cases}$ **b** $\begin{cases} 4x - 3y = 1 \\ 2x + 5y = 7 \end{cases}$ **c** $\begin{cases} 2x + 5y = 14 \\ 5x - 3y + 27 = 0 \end{cases}$

d $\begin{cases} 7x + 2y = 20 \\ 13x + 3y = 34 \end{cases}$ **e** $\begin{cases} 3x + 7y = 5 \\ 5x + 11y = 10 \end{cases}$ **f** $\begin{cases} 5x - 7y - 9 = 0 \\ 4x - 5y - 5 = 0 \end{cases}$

6 Try to solve by elimination:

a $\begin{cases} 2x - y = 3 \\ 4x - 2y = 6 \end{cases}$ **b** $\begin{cases} 3x + 4y = 6 \\ 6x + 8y = 7 \end{cases}$

Comment on your results.

D PROBLEM SOLVING

Many problems can be described mathematically by a **pair of linear equations**. The **Opening Problem** is one example.

Once the equations are formed, they can then be solved simultaneously, and we can then answer the original problem.

Step 1: Decide on two unknowns such as x and y. Do not forget the units.
Step 2: Write down **two** equations connecting x and y.
Step 3: Solve the equations simultaneously.
Step 4: Check your solutions with the original data given.
Step 5: Give your answer in sentence form.

The form of the original equations will help you decide whether to use substitution or elimination.

Example 6 **Self Tutor**

Two numbers have a sum of 45 and a difference of 13. Find the numbers.

Let x and y be the unknown numbers, where $x > y$.

From the information given,

$x + y = 45$ .... (1) {'sum' means add}

and $x - y = 13$ .... (2) {'difference' means subtract}

Adding, $2x = 58$

$\therefore\ x = 29$

Substituting $x = 29$ into (1) gives $29 + y = 45$

$\therefore\ y = 16$

The numbers are 29 and 16.

Check: In (1): $29 + 16 = 45$ ✓ In (2): $29 - 16 = 13$ ✓

EXERCISE 19D

1 The sum of two numbers is 72, and their difference is 40. Find the numbers.

2 Two numbers have a sum of 30. Half of their difference is 7. Find the numbers.

3 The larger of two numbers is three times the smaller number, and their difference is 34. Find the two numbers.

Example 7 **Self Tutor**

When shopping in Jamaica, 5 coconuts and 14 bananas cost \$8.70, and 8 coconuts and 9 bananas cost \$9.90.

Find the cost of each coconut and each banana.

Let each coconut cost x cents, and each banana cost y cents.

$$\therefore \ 5x + 14y = 870 \quad .... \ (1)$$
$$\text{and} \quad 8x + 9y = 990 \quad .... \ (2)$$

To eliminate x, we multiply (1) by 8 and (2) by -5.

$$\begin{aligned} \therefore \quad 40x + 112y &= 6960 \qquad \{(1) \times 8\} \\ -40x - 45y &= -4950 \quad \{(2) \times -5\} \\ \hline \text{Adding,} \qquad 67y &= 2010 \\ \therefore \quad y &= 30 \end{aligned}$$

Substituting $y = 30$ in (2) gives
$$\begin{aligned} 8x + 9(30) &= 990 \\ \therefore \ 8x + 270 &= 990 \\ \therefore \ 8x &= 720 \\ \therefore \ x &= 90 \end{aligned}$$

Check: In (1): $5 \times 90 + 14 \times 30 = 450 + 420 = 870$ ✓
In (2): $8 \times 90 + 9 \times 30 = 720 + 270 = 990$ ✓

The coconuts cost 90 cents each, and the bananas cost 30 cents each.

4 Three pieces of fish and two serves of chips cost a total of £8.10. Five pieces of fish and three serves of chips cost a total of £13.25. Find the cost of each piece of fish and each serve of chips.

5 Seven cups of coffee and four muffins cost a total of €25.30. Two cups of coffee and three muffins cost a total of €9.55. Find the cost of each item.

Example 8 **Self Tutor**

I have only five cent and ten cent coins in my pocket. I have 24 coins altogether, and their total value is \$1.55. How many of each type of coin do I have?

Let x be the number of five cent coins, and y be the number of ten cent coins.

$\therefore\ x + y = 24$ .... (1) {the total number of coins}

and $5x + 10y = 155$ .... (2) {the total value of the coins in cents}

To eliminate x, we multiply (1) by -5.

$$\therefore\ -5x - 5y = -120 \quad \{(1) \times -5\}$$
$$5x + 10y = 155 \quad \{(2)\}$$

Adding, $5y = 35$

$\therefore\ y = 7$

Substituting $y = 7$ into (1) gives $x + 7 = 24$

$\therefore\ x = 17$

Check: In (1): $17 + 7 = 24$ ✓

In (2): $5 \times 17 + 10 \times 7 = 85 + 70 = 155$ ✓

I have 17 five cent coins and 7 ten cent coins in my pocket.

6 Margaret saves fifty cent and ten cent coins. She has 56 of these coins, and their total value is \$17.60. How many of each coin type does she have?

7 Anthony and Michelle have €65.25 between them. Anthony's amount is two thirds of Michelle's amount. How much money does each have?

8 Milk is sold in either 600 mL or 1 L cartons. A supermarket manager ordered 79.8 litres of milk and received 93 cartons.
How many of each size carton did the manager receive?

9 The triangle alongside is equilateral.
Find a and b.

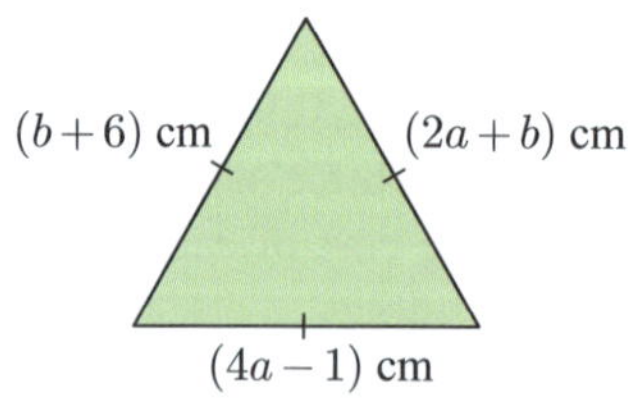

10 A rectangle has perimeter 56 cm. If 4 cm is taken from the length and added to the width, the rectangle becomes a square. Find the dimensions of the original rectangle.

11 A car park charges a certain rate per hour before 3 pm, and a higher rate per hour after 3 pm.
Parking from 11 am to 5 pm costs \$18, and parking from 9 am to 7 pm costs \$31.
Find the rates charged before and after 3 pm.

12 An art store sells small and large painting sets. The small sets contain 2 brushes and 3 tubes of paint, and the large sets contain 5 brushes and 8 tubes of paint. Leanne bought some of each of these sets to use as gifts. In total she received 37 brushes and 58 tubes of paint. How many of each set did she buy?

INVESTIGATION — LARGER SYSTEMS OF EQUATIONS

Simultaneous equations like $\begin{cases} 2x + 3y = 11 \\ 3x - 2y = 10 \end{cases}$ are called 2 by 2 systems of equations, because there are 2 equations involving 2 unknowns.

In this Investigation we examine a method for solving 3 by 3 systems of equations.

Consider the 3 by 3 system of equations $\begin{cases} 2x + 3y - z = 5 \\ -x - 3y + 4z = 2 \\ x + 6y - 5z = 7. \end{cases}$

To solve this system of equations, we write the first two equations with the z variable on the RHS, and find x and y in terms of z:

$$\begin{aligned} \therefore \quad 2x + 3y &= 5 + z \quad \text{.... (1)} \\ -x - 3y &= 2 - 4z \quad \text{.... (2)} \\ \text{Adding,} \quad x \quad &= 7 - 3z \quad \text{.... (3)} \end{aligned}$$

Substituting $x = 7 - 3z$ into (1) gives

$$\begin{aligned} 2(7 - 3z) + 3y &= 5 + z \\ \therefore \quad 14 - 6z + 3y &= 5 + z \\ \therefore \quad 3y &= 7z - 9 \\ \therefore \quad y &= \frac{7z - 9}{3} \quad \text{.... (4)} \end{aligned}$$

We then substitute these values into the original third equation to solve for z:

$$\begin{aligned} (7 - 3z) + 6\left(\frac{7z - 9}{3}\right) - 5z &= 7 \\ \therefore \quad 7 - 3z + 14z - 18 - 5z &= 7 \\ \therefore \quad 6z &= 18 \\ \therefore \quad z &= 3 \end{aligned}$$

Substituting into (3) and (4), $x = 7 - 3(3) = -2$, and $y = \frac{7(3) - 9}{3} = 4$.

The solution is $x = -2$, $y = 4$, $z = 3$.

What to do:

1 Solve simultaneously:

a $\begin{cases} x - y + z = 9 \\ 3x + y - z = 7 \\ 4x - 2y + 5z = 35 \end{cases}$

b $\begin{cases} 2x - y - z = 3 \\ 3x + 2y - 2z = 13 \\ 5x + y - 3z = 18 \end{cases}$

2 At a greengrocer:

- 3 pineapples, 2 cabbages, and 6 bananas cost a total of \$27
- 2 pineapples, 3 cabbages, and 4 bananas cost a total of \$23
- 4 pineapples, 1 cabbage, and 8 bananas cost a total of \$31.

Find the cost of each item.

3 Outline a method which could be used to solve a 4 by 4 system of equations.

Test your method by trying to solve: $\begin{cases} a+b+2c+d=14 \\ 2a-b-c+d=8 \\ a+2b+c-d=15 \\ 3a-2b+3c+2d=24 \end{cases}$

E NON-LINEAR SIMULTANEOUS EQUATIONS

So far we have only considered simultaneous equations which are linear. This means that both equations could be written in the form $Ax+By=C$, where A, B, and C are constants.

In this Section we consider some non-linear simultaneous equations.

Depending on the form of the equations, we may be able to solve them by substitution. Otherwise, we can solve them graphically.

Example 9 **Self Tutor**

Solve simultaneously by substitution: $\begin{cases} y=x^2+4x+2 \\ 5x+y=12 \end{cases}$

$y=x^2+4x+2$ .... (1)
$5x+y=12$ .... (2)

Substituting (1) into (2) gives $5x+(x^2+4x+2)=12$

$$\therefore \ x^2+9x-10=0$$
$$\therefore \ (x-1)(x+10)=0$$
$$\therefore \ x=1 \text{ or } -10$$

Substituting $x=1$ into (1) gives $y=(1)^2+4(1)+2=7$
Substituting $x=-10$ into (1) gives $y=(-10)^2+4(-10)+2=62$

The solutions are $x=1$, $y=7$ or $x=-10$, $y=62$.

Example 10 **Self Tutor**

Solve simultaneously by substitution: $\begin{cases} xy=12 \\ y=2x+2 \end{cases}$

$xy=12$ .... (1)
$y=2x+2$ .... (2)

Substituting (2) into (1) gives $x(2x+2) = 12$

$$\therefore \; 2x^2 + 2x = 12$$
$$\therefore \; 2(x^2 + x - 6) = 0$$
$$\therefore \; 2(x+3)(x-2) = 0$$
$$\therefore \; x = -3 \text{ or } 2$$

Substituting $x = -3$ into (2) gives $y = 2(-3) + 2 = -4$
Substituting $x = 2$ into (2) gives $y = 2(2) + 2 = 6$

The solutions are $x = -3, \; y = -4$ and $x = 2, \; y = 6$.

Example 11 **Self Tutor**

Solve graphically: $\begin{cases} y = x^3 - 3x \\ y = -2x \end{cases}$

Casio fx-CG20	**TI-84 Plus**	**TI-*n*spire**

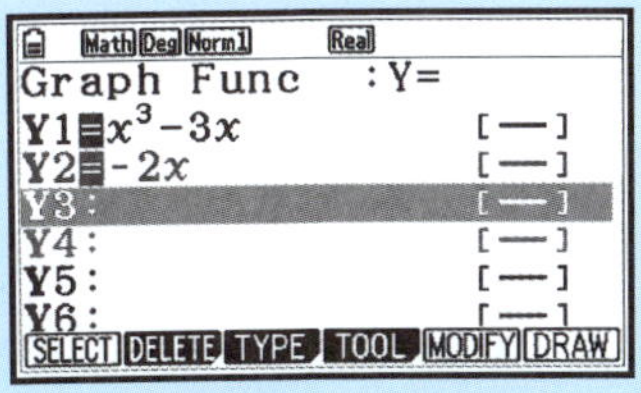

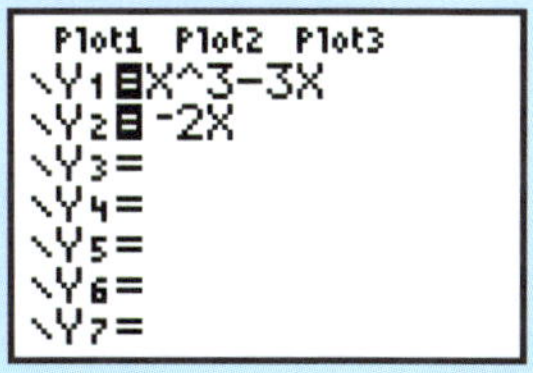

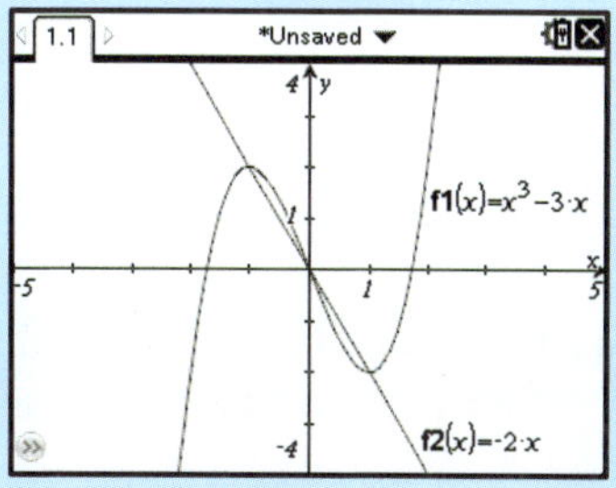

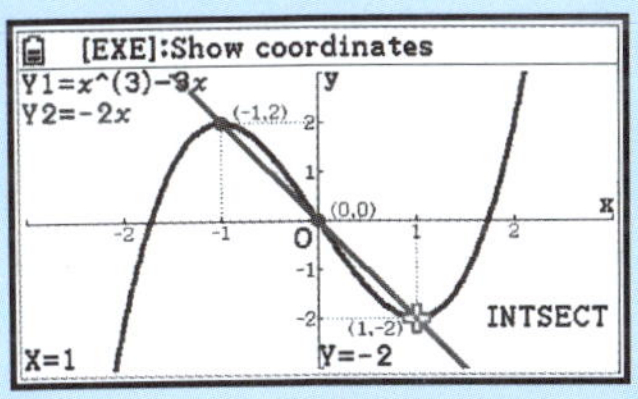

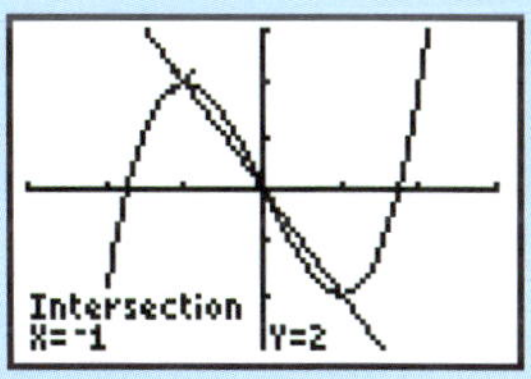

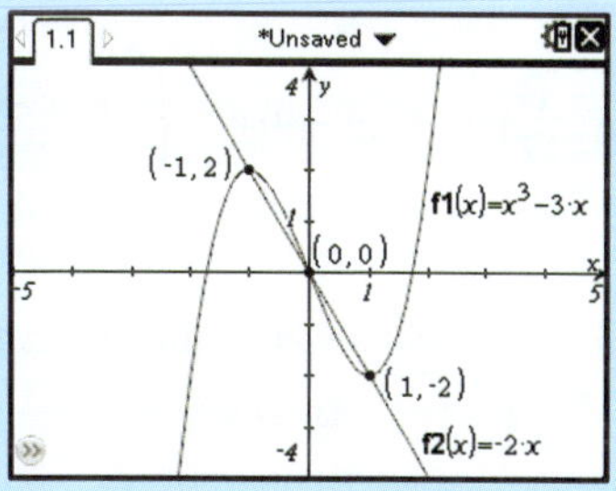

The solutions are $x = -1, \; y = 2$ and $x = 0, \; y = 0$ and $x = 1, \; y = -2$.

EXERCISE 19E

1 Solve simultaneously by substitution:

a $\begin{cases} y = x^2 - 2x + 5 \\ 3x + y = 11 \end{cases}$ **b** $\begin{cases} 2x + y = 17 \\ y = x^2 + 5x - 1 \end{cases}$ **c** $\begin{cases} y = 2x^2 - x + 3 \\ 5x + y = 9 \end{cases}$

d $\begin{cases} y = 3x^2 - 5x - 4 \\ 4x - y = -26 \end{cases}$ **e** $\begin{cases} y = 2x^2 + x - 5 \\ y = x^2 + 5x + 7 \end{cases}$ **f** $\begin{cases} y = x^2 - 7x - 11 \\ y = 3x^2 + 5x + 7 \end{cases}$

2 Solve simultaneously by substitution:

a $\begin{cases} xy = 20 \\ y = x + 1 \end{cases}$ **b** $\begin{cases} y = 16 - 2x \\ xy = 32 \end{cases}$ **c** $\begin{cases} xy = 5 \\ y = 6x + 7 \end{cases}$

3 Use substitution and the quadratic formula to try to solve:

a $\begin{cases} y = x^2 - 4x + 10 \\ 3x + 2y = 7 \end{cases}$ **b** $\begin{cases} xy = 6 \\ y = x + 4 \end{cases}$ **c** $\begin{cases} y = 5x - 7 \\ xy = -4 \end{cases}$

4 Show algebraically that it is impossible for two real numbers x and y to have a sum of 10 and a product of 30.

5 Solve graphically:

a $\begin{cases} y = x^3 - x^2 + x \\ y = x \end{cases}$ **b** $\begin{cases} y = \dfrac{1}{x} \\ y = \dfrac{1}{2-x} \end{cases}$ **c** $\begin{cases} y = x^3 + x \\ y = \dfrac{4x}{x^2+1} \end{cases}$

REVIEW SET 19A

1 Solve graphically without using technology:

a $\begin{cases} y = 5x + 2 \\ y = 2x - 7 \end{cases}$ **b** $\begin{cases} 2x - 3y = 18 \\ x + y = 4 \end{cases}$

2 Use technology to solve simultaneously:

a $\begin{cases} y = 3x - 4 \\ y = 9 - x \end{cases}$ **b** $\begin{cases} 3x - 4y = 3 \\ 7x - 11y = 8 \end{cases}$

3 Solve simultaneously by substitution:

a $\begin{cases} x = 8 - 3y \\ 2x - 7y = -10 \end{cases}$ **b** $\begin{cases} y = 5x + 3 \\ 4x + y = 1 \end{cases}$

4 **a** Try to solve $\begin{cases} 4x - 2y = 3 \\ y = 2x - 5 \end{cases}$ by substitution.

b How many solutions do the simultaneous equations in **a** have?

5 What equation results when the following are added vertically?

a $4x - 3y = 8$
$2x + 3y = -14$

b $5x - y = 7$
$-5x - 3y = 33$

6 Solve simultaneously by elimination:

a $\begin{cases} -2x + 7y = 13 \\ 2x - 3y = -9 \end{cases}$ **b** $\begin{cases} 4x - 6y = 21 \\ 3x + 5y = -8 \end{cases}$

7 Solve simultaneously:

a $\begin{cases} 3x - 2y = 18 \\ y = x - 7 \end{cases}$ **b** $\begin{cases} 4x + 5y = 2 \\ 5x - 11y = 14 \end{cases}$

8 6 adult tickets and 5 student tickets for the theatre cost \$168. 2 adult tickets and 3 student tickets cost \$72. Find the cost of each type of ticket.

9 **a** On the same set of axes, draw the graphs of:

i $y = 2x - 5$ **ii** $x - 4y = -8$ **iii** $y = -x - 5$

b Hence, solve the following simultaneous equations:

i $\begin{cases} y = 2x - 5 \\ x - 4y = -8 \end{cases}$ **ii** $\begin{cases} y = 2x - 5 \\ y = -x - 5 \end{cases}$ **iii** $\begin{cases} x - 4y = -8 \\ y = -x - 5 \end{cases}$

10 Carmel has some fifty cent coins and some twenty cent coins in her purse. She has 21 coins in total, and their total value is \$8.40. How many of each type of coin does Carmel have?

11 Solve simultaneously by substitution: $\begin{cases} y = x^2 - 4x - 5 \\ 2x + y = 10 \end{cases}$

12 Solve graphically using technology: $\begin{cases} y = x - \dfrac{4}{x} \\ y = -x^3 + 2x^2 \end{cases}$

REVIEW SET 19B

1 Solve graphically without using technology: $\begin{cases} y = 3x - 8 \\ y = -2x + 7 \end{cases}$

2 Solve by substitution: $\begin{cases} y = 4x - 7 \\ x = 9 - 2y \end{cases}$

3 Try to solve $\begin{cases} y = \frac{3}{5}x - 1 \\ 3x - 5y = 5 \end{cases}$ using technology.

State how many solutions this set of linear equations has.

4 State the equation that results when both sides of $2x - 3y = -4$ are multiplied by -3.

5 Solve by elimination: $\begin{cases} 13x + 6y = 1 \\ 7x + 4y - 5 = 0 \end{cases}$

6 Solve simultaneously:

a $\begin{cases} 6x - 5y = 2 \\ y = 3x + 2 \end{cases}$ **b** $\begin{cases} 5x - 7y = 2 \\ -3x + 4y = 5 \end{cases}$

7 **a** Try to solve $\begin{cases} y = 4x - 2 \\ 8x - 2y = 4 \end{cases}$ by substitution.

b How many solutions do the simultaneous equations in **a** have?

8 In an egg and spoon race, children must transport eggs from one end of a course to the other. Points are awarded for each egg successfully transported, and points are deducted for each egg which drops and breaks.

Jenny transported 5 eggs and dropped 2 eggs, for a score of 11 points. Heath transported 8 eggs and dropped 5 eggs, for a score of 5 points.

How many points are deducted for dropping an egg?

9 **a** At a school fete, Nicole buys 5 books and 2 CDs for \$50, and Sharyn buys 2 books and 3 CDs for \$31. Find the price of:

i a book **ii** a CD.

b Later in the day, the prices at the fete are dropped. Nicole's and Sharyn's purchases would now cost \$29 and \$16 respectively. Find the new price of:

i a book **ii** a CD.

10 A sports store sells packs of 3 tennis balls for \$5, and packs of 8 tennis balls for \$12. In one day the store received \$151 in sales of tennis balls. If the store sold 97 tennis balls in total, how many packs of 3 balls did they sell?

11 Solve simultaneously by substitution: $\begin{cases} xy = 36 \\ y = 2x + 6 \end{cases}$

12 Jason paints signs on buildings. He can paint letters which contain all straight edges quite easily, but letters which contain a curved edge take longer.

It takes Jason 25 minutes to paint the word CHEMIST, and 43 minutes to paint the word SUPERMARKET.

a On average, how long does Jason take to paint each:

i straight edged letter

ii curved edged letter?

b How long would it take Jason to paint the word ACCOUNTANT?

Chapter 20

Congruence and similarity

Contents:

OPENING PROBLEM

Jacob makes conical tanks for storing grain. His tanks have different *sizes*, but all have the same *shape*. His smallest tank has a height of 2 m and a capacity of 7 kL.

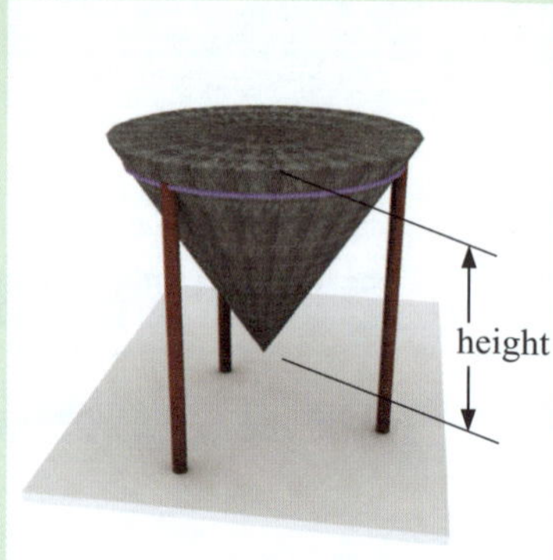

Things to think about:

a What do we mean by saying two objects are different in *size* but have the same *shape*?

b Jacob wants to manufacture a tank which is 4 m high.

  i How many times higher is the new tank than the original tank?

  ii What will the capacity of the new tank be?

A CONGRUENCE OF FIGURES

Two objects are **congruent** if they are identical in every respect except for position. The objects have the same size *and* the same shape.

EXERCISE 20A

1 Which of the following figures are congruent?

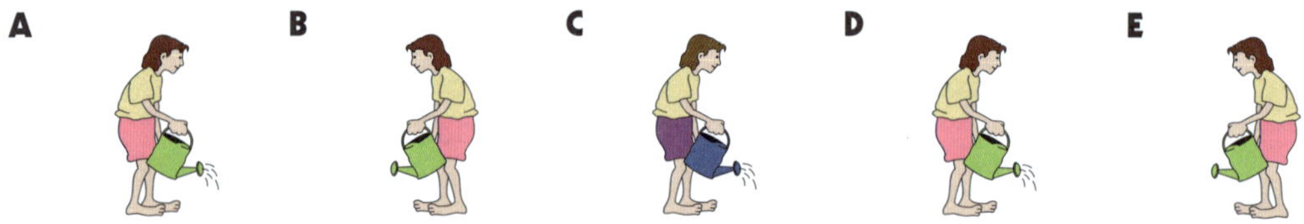

2 Which of the following figures are congruent?

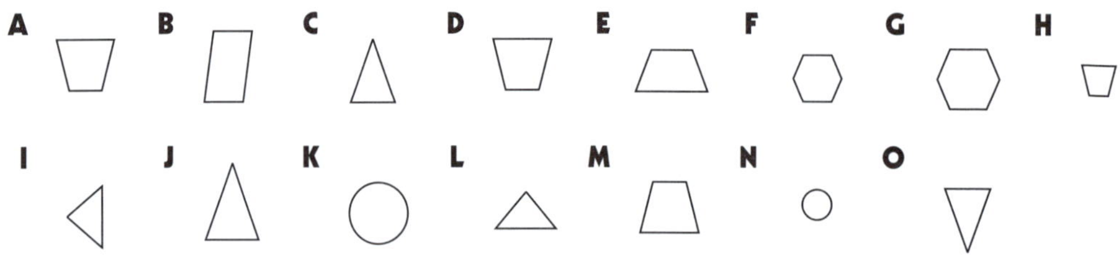

3 For each pair of congruent geometric figures:

  i identify the side in the second figure corresponding to the side [AD] in the first figure

  ii identify the angle in the second figure corresponding to $A\widehat{B}C$ in the first figure.

a

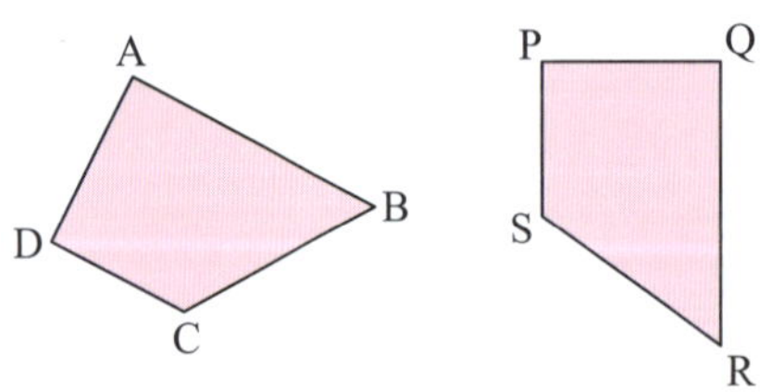

b

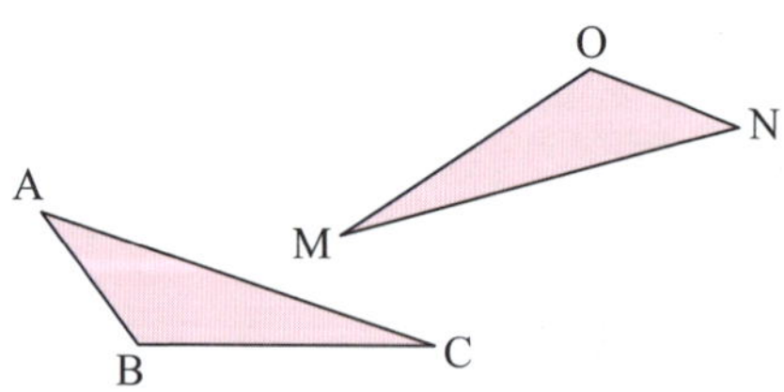

c

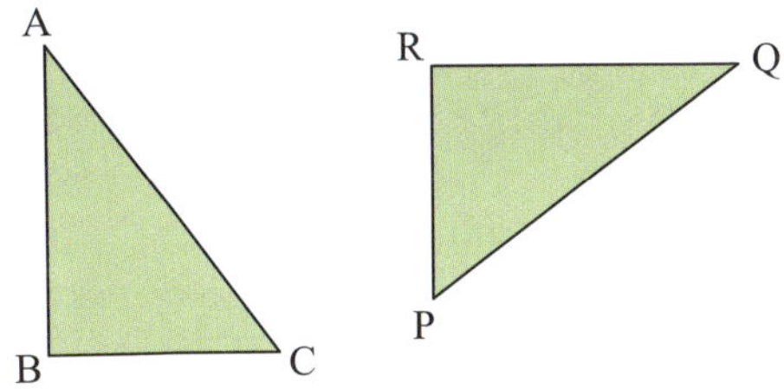

d

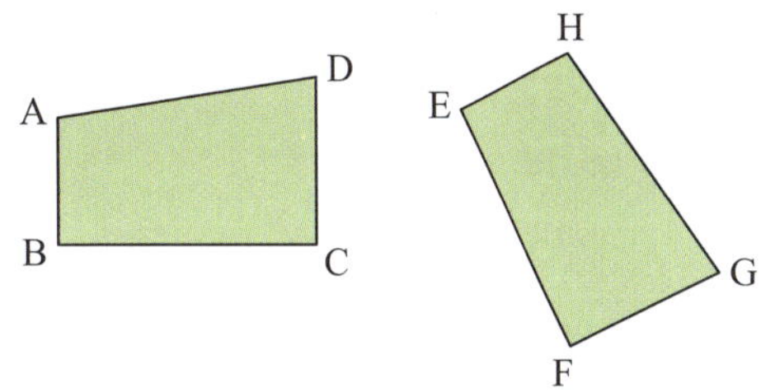

B CONGRUENT TRIANGLES

Two triangles are **congruent** if they are identical in every respect except for position. The triangles have the same shape and size.

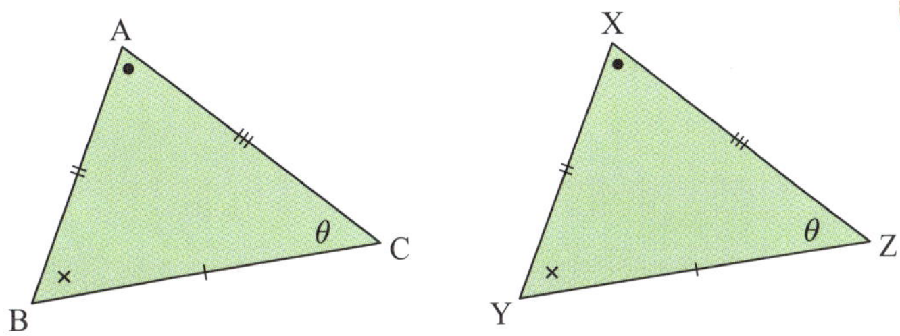

Triangles ABC and XYZ are congruent.

We write $\triangle ABC \cong \triangle XYZ$, which reads "triangle ABC is congruent to triangle XYZ".

Notice that we list the vertices that are in corresponding positions in the same order. We therefore say triangle ABC is congruent to triangle XYZ rather than triangle YXZ or triangle ZYX.

If we are given sufficient information about a triangle, there will be only one way in which it can be drawn. Any two triangles which have this information in common must be **congruent**.

INVESTIGATION CONSTRUCTING TRIANGLES

If several people are asked to accurately draw triangle ABC in which $AB = 3$ cm and $BC = 2$ cm, many different shaped triangles will probably be drawn. Three possible answers are drawn below:

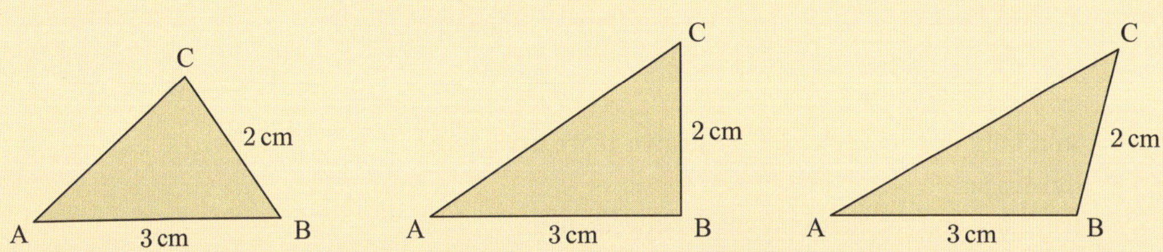

The information given is insufficient to draw a triangle of one particular shape.

In this Investigation we consider what information is 'sufficient' to uniquely describe a triangle. All triangles drawn using this information will be congruent.

PRINTABLE WORKSHEET

What to do:

For each of the following cases, draw a triangle from the information given.

Decide whether all triangles drawn using the given information will be congruent.

1 Given three sides
Draw a triangle ABC with $AB = 4$ cm, $BC = 5$ cm, and $AC = 6$ cm.

2 Given three angles
Draw a triangle ABC with $A\widehat{B}C = 50°$, $B\widehat{C}A = 60°$, and $C\widehat{A}B = 70°$.

3 Given two sides and the included angle between them
Draw a triangle ABC with $AB = 6$ cm, $BC = 5$ cm, and $A\widehat{B}C = 45°$.

4 Given two sides and a non-included angle
Draw a triangle ABC with $AB = 6$ cm, $BC = 5$ cm, and $C\widehat{A}B = 45°$.

5 Given two angles and a side
Draw a triangle ABC with $A\widehat{B}C = 70°$, $B\widehat{C}A = 30°$, and $BC = 7$ cm.

6 Given a right angle, a hypotenuse, and a side
Draw a triangle ABC with $A\widehat{B}C = 90°$, $AC = 10$ cm, and $AB = 8$ cm.

From the **Investigation**, you should have discovered that:

- If we are given three angles, more than one triangle can be drawn.
  For example, these triangles have identical angles, but triangle **B** is larger than triangle **A**.

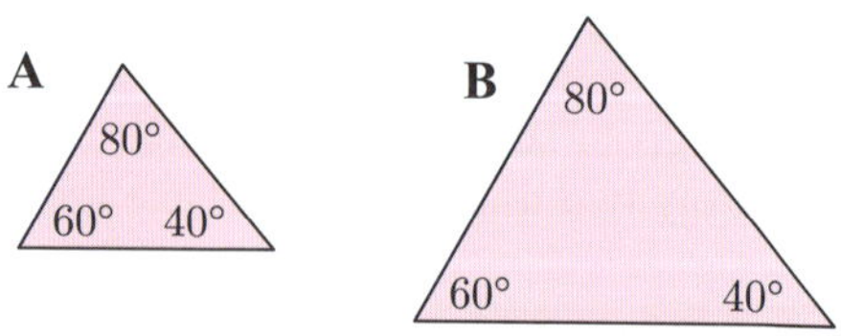

- If we are given two sides and a non-included angle, more than one triangle can be drawn.
  For example, triangles **C** and **D** have two equal sides and the same non-included angle, but they are *not* congruent.

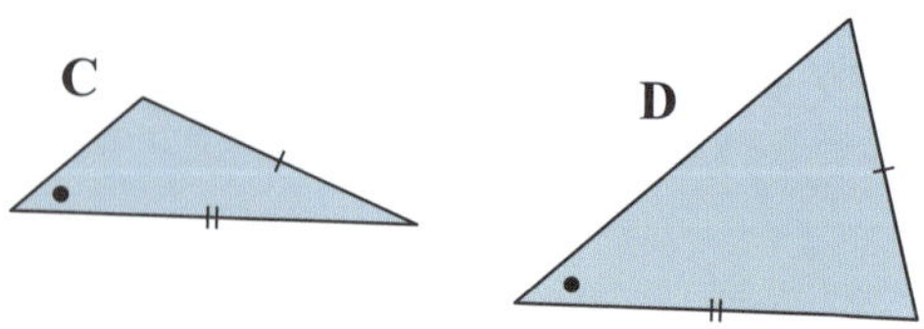

- **One and only one triangle can be drawn if we are given:**
  - three sides
  - two sides and the included angle between them
  - two angles and a side
  - a right angle, the hypotenuse, and one other side.

There are therefore four acceptable tests for the **congruence** of two triangles.

Two triangles are congruent if one of the following is true:

- All corresponding sides are equal in length. **(SSS)**

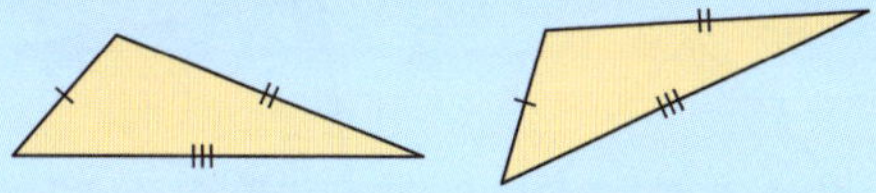

- Two sides and the **included** angle are equal. **(SAS)**

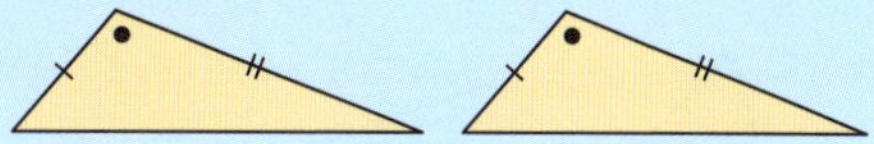

- Two angles and a **corresponding side** are equal. **(AAcorS)**

- For right angled triangles, the hypotenuses and one pair of sides are equal. **(RHS)**

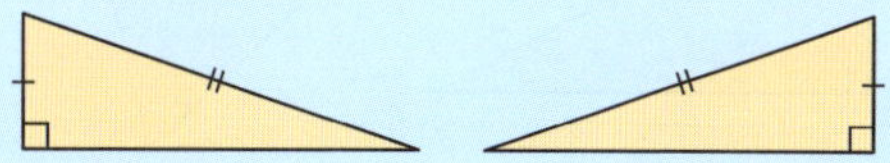

The information we are given will help us decide which test to use to prove that two triangles are congruent.

Example 1 | Self Tutor

Determine whether the following pairs of triangles are congruent. Give reasons for your answers.

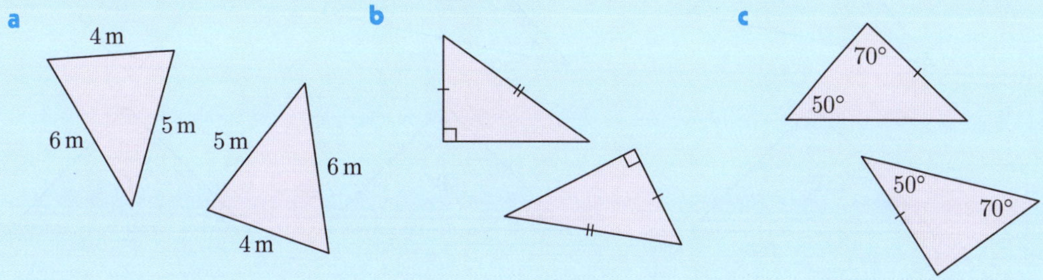

a The triangles are congruent. {SSS}

b The triangles are congruent. {RHS}

c The third angle of each triangle is 60°. {angle sum of triangle is 180°}

$\therefore$ the triangles are not isosceles.

We cannot use the rule AAcorS since the equal sides are not in corresponding positions. One is opposite the 50° angle, the other is opposite the 70° angle.

$\therefore$ the triangles are not congruent.

EXERCISE 20B

1 Determine whether the following pairs of triangles are congruent. Give reasons for your answers.

a

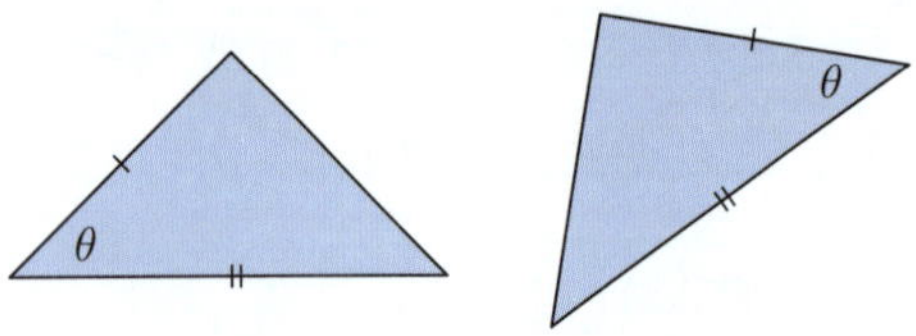

b

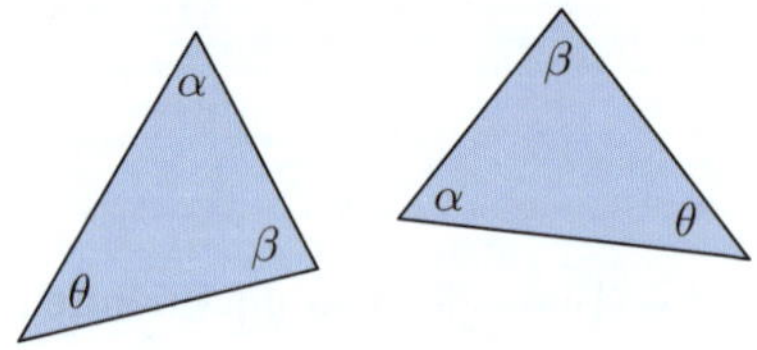

c

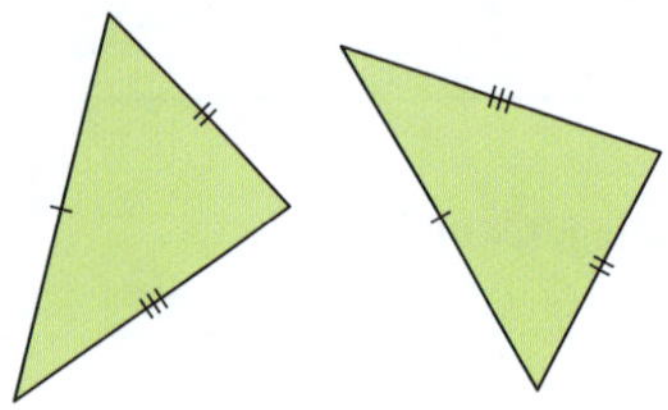

d

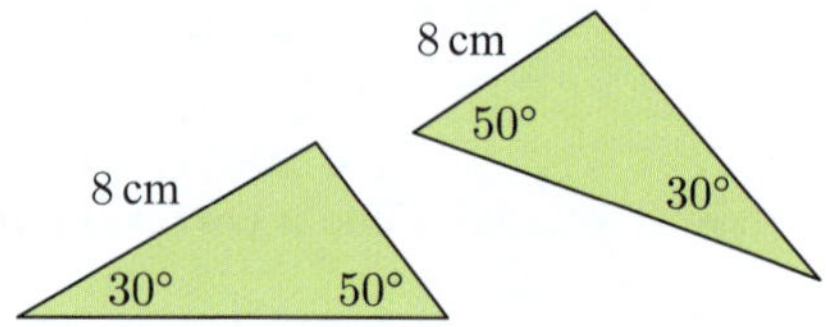

e

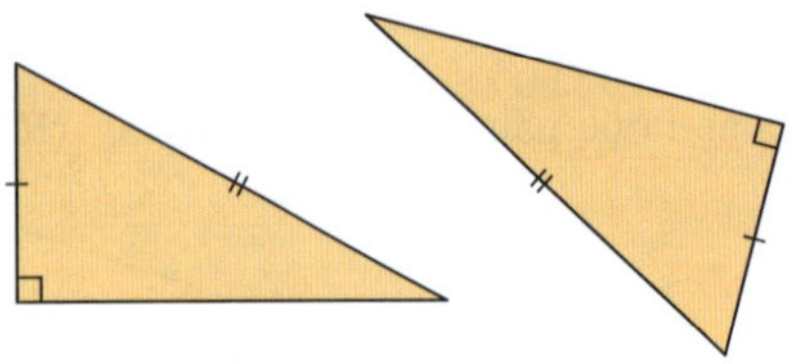

f

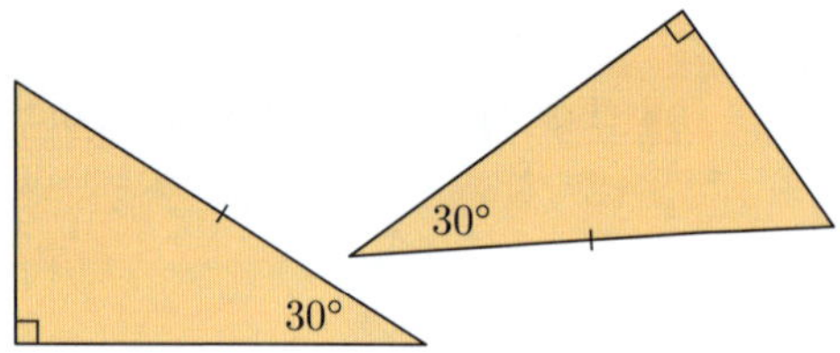

2 In each set of three triangles, two are congruent. The diagrams are *not* drawn to scale, but the information marked on them is correct.
State which pair of triangles is congruent, and give a reason (SSS, SAS, AAcorS, or RHS).

a

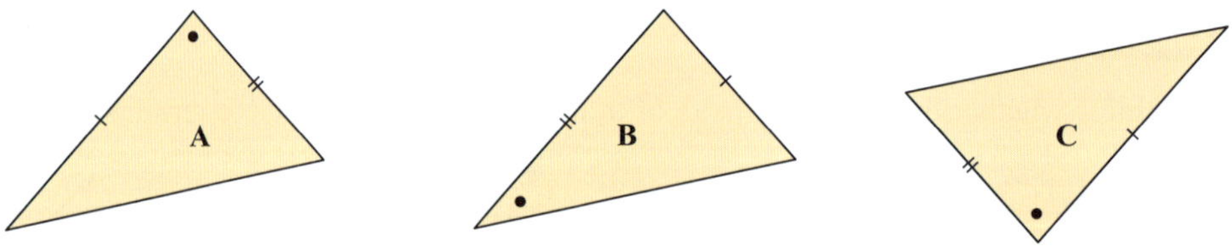

b

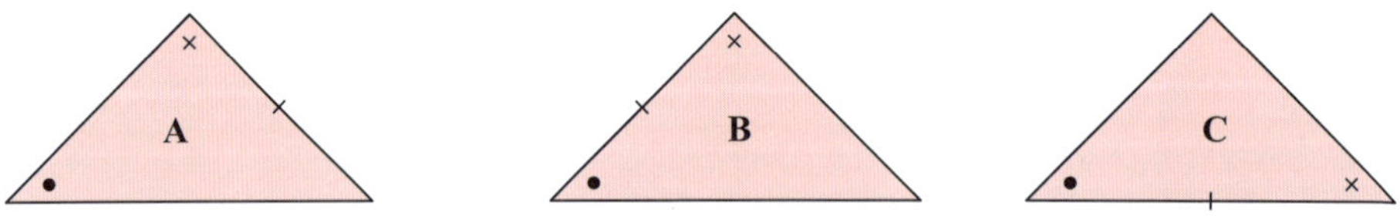

c

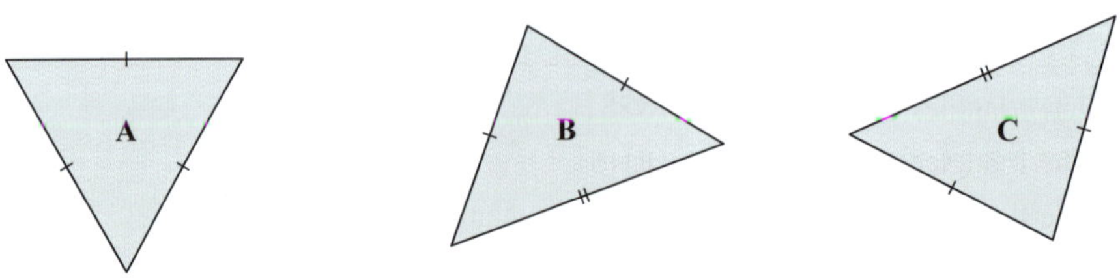

d

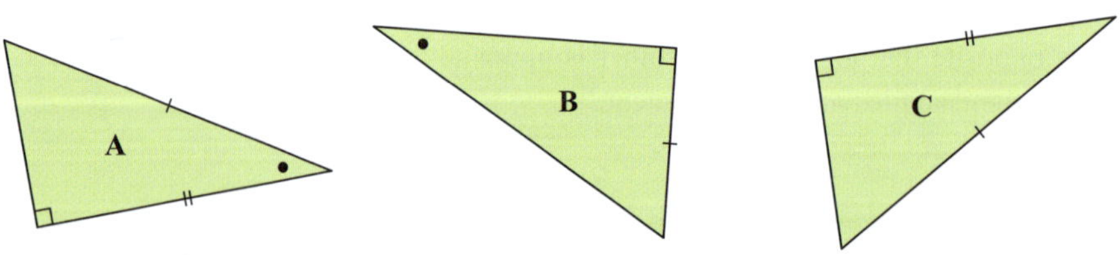

C PROOF USING CONGRUENCE

If we can show that two triangles are congruent, we can deduce that the remaining corresponding sides and angles of the triangles are equal.

Congruence is a useful tool in geometrical proofs.

When we identify congruent triangles in labelled diagrams, we:

- list the corresponding vertices in the same order
- use the symbol $\cong$ to mean 'is congruent to'.

Example 2 **Self Tutor**

Triangle ABC is isosceles with $AC = BC$. [BC] and [AC] are produced to E and D respectively so that $CE = CD$.

Prove that $AE = BD$.

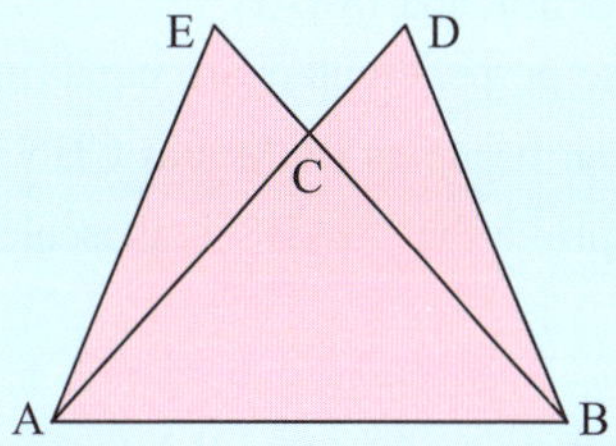

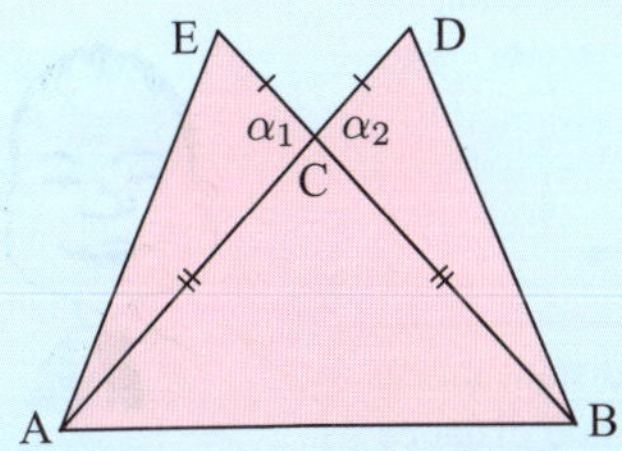

In triangles ACE and BCD:

- $AC = BC$ {given}
- $\alpha_1 = \alpha_2$ {vertically opposite angles}
- $CE = CD$ {given}

So, $\triangle ACE \cong \triangle BCD$. {SAS}

Equating corresponding sides, $AE = BD$.

EXERCISE 20C

1

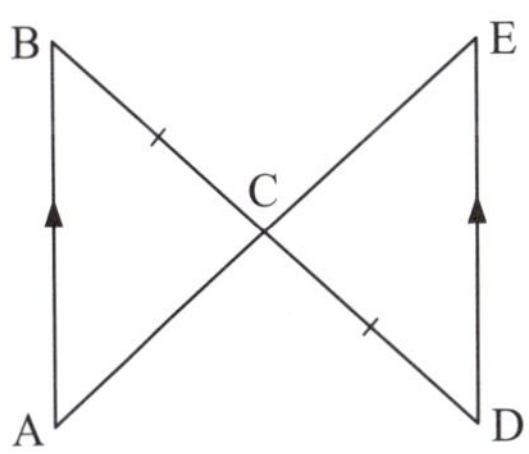

a Explain why triangles ABC and EDC are congruent.

b If $AC = 5$ cm, find the length of [EC].

2 [AB] and [CD] are both diameters of the circle. Show that $AC = BD$.

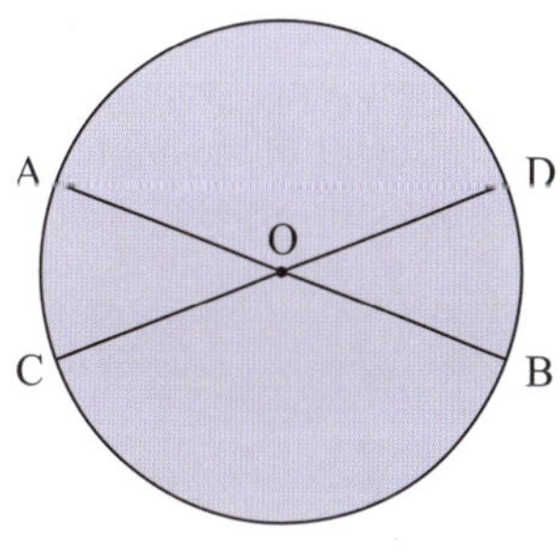

3 Two diagonals are drawn from vertex P of a regular pentagon. Show that these diagonals are equal in length.

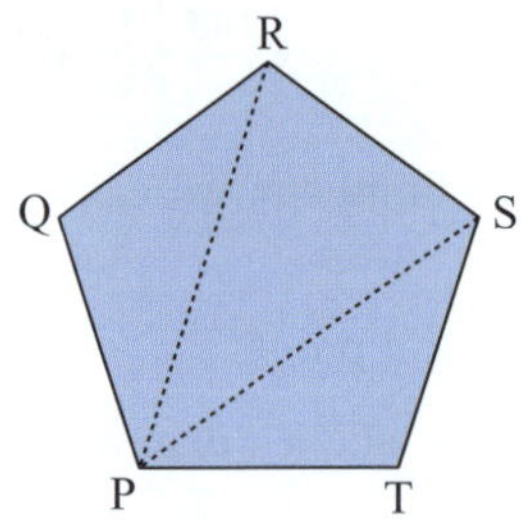

4 In $\triangle$ABC, [BM] is drawn perpendicular to [AC], and [CN] is drawn perpendicular to [AB]. Suppose these perpendiculars are equal in length.

  a Prove that triangles BCM and CBN are congruent.

  b Hence, prove that $\triangle$ABC is isosceles.

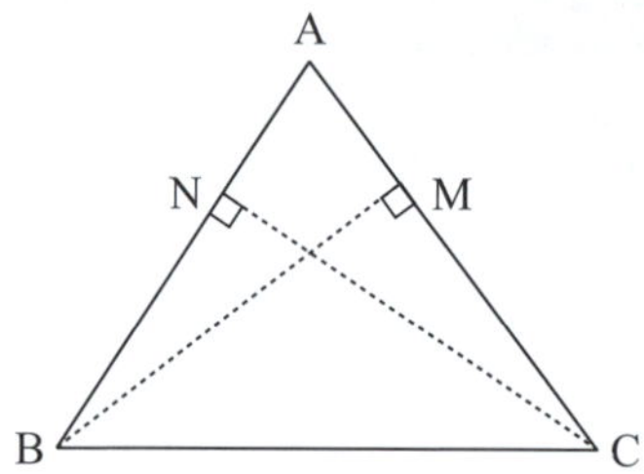

5 Given two points A and B, the **perpendicular bisector** of [AB] is the line which bisects [AB] and is perpendicular to [AB]. Use congruent triangles to show that any point on the perpendicular bisector of [AB] is equidistant from A and B.

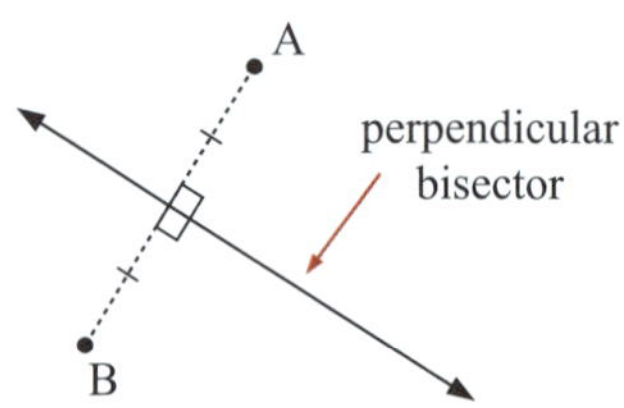

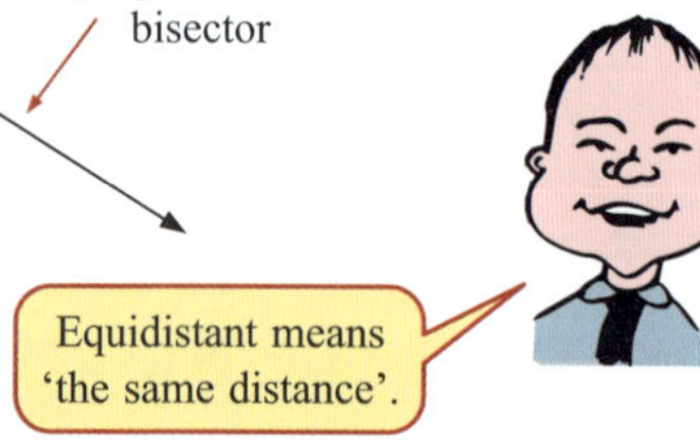

6 An **isosceles trapezium** is a trapezium where the non-parallel sides are equal in length.

  a By drawing two lines perpendicular to [DC] through A and B respectively, show that $A\widehat{D}C = B\widehat{C}D$.

  b Hence, show that the diagonals of an isosceles trapezium are equal in length.

ACTIVITY 1 — ARE ALL TRIANGLES ISOSCELES?

Charles Dodgson, better known as Lewis Caroll, the author of *Alice in Wonderland*, used congruence to devise the following "proof" that all triangles are isosceles:

Step 1: Draw any triangle ABC. Draw the angle bisector of the angle at A, and the perpendicular bisector of [BC]. Let these lines meet at X.

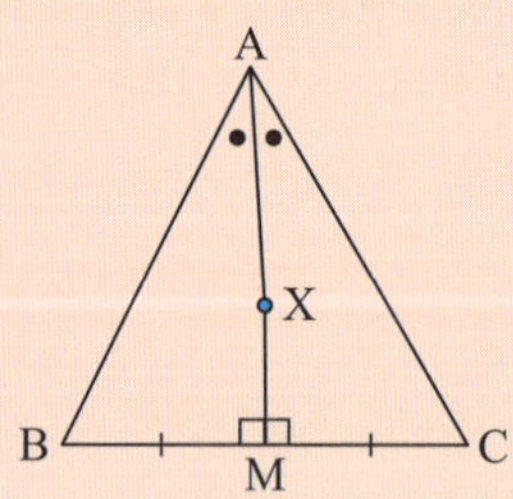

Step 2: Join [BX] and [CX]. Draw [XP] perpendicular to [AB], and [XQ] perpendicular to [AC].

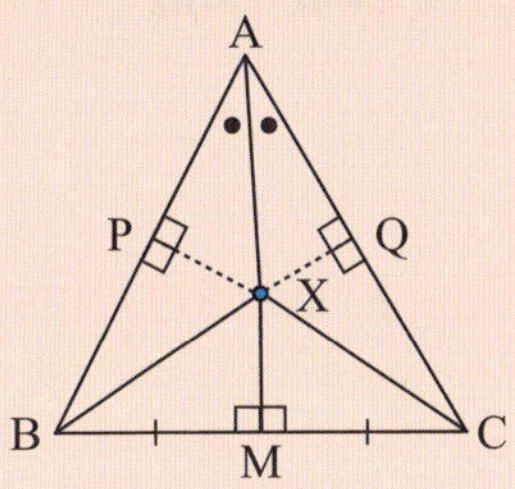

Step 3: $\triangle BXM \cong \triangle CXM$ {SAS}

$\therefore$ BX = CX

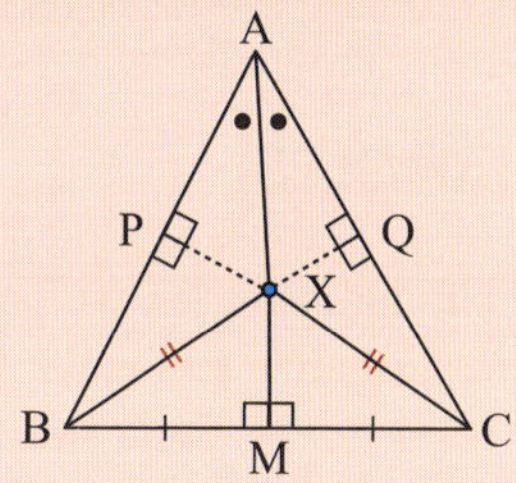

Step 4: $\triangle APX \cong \triangle AQX$ {AAcorS}

$\therefore$ AP = AQ .... (1)

and PX = QX

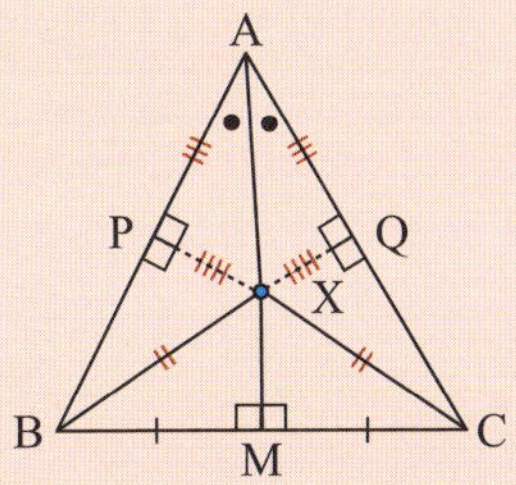

Step 5: $\triangle XPB \cong \triangle XQC$ {RHS}

$\therefore$ PB = QC .... (2)

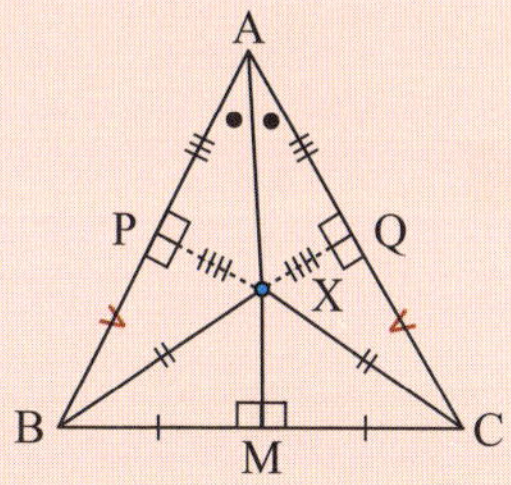

Combining (1) and (2) gives AB = AC, so $\triangle ABC$ is isosceles.

What to do:

Lewis Caroll's proof is clearly flawed, since we know that not all triangles are isosceles! Your task is to identify the problem with this "proof".

D SIMILARITY

Two figures are **similar** if one is an enlargement of the other. The objects have the same shape, but are different in size.

The figures alongside are similar, as A′B′C′D′ is an enlargement of ABCD with scale factor 2.

Notice that

$$\frac{A'B'}{AB} = \frac{B'C'}{BC} = \frac{C'D'}{CD} = \frac{D'A'}{DA} = 2,$$

so the corresponding side lengths are in the **same ratio**.

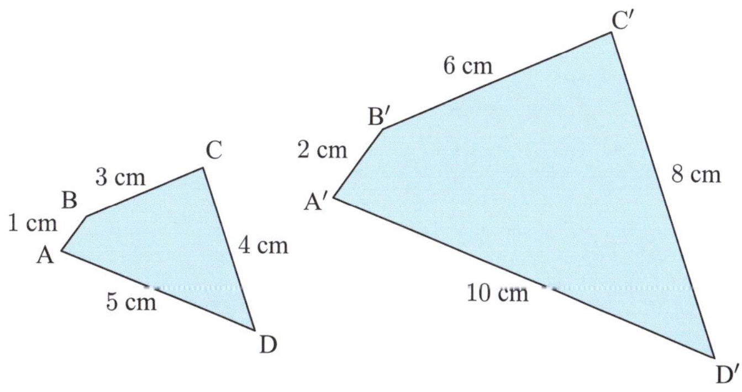

When a figure is enlarged or reduced, the size of its angles does not change. The figures are therefore **equiangular**.

Two figures are **similar** if:

- the figures are **equiangular** *and*
- the corresponding side lengths are in the **same ratio**.

SIMILAR TRIANGLES

When testing for similar *triangles*, we only need to show that *one* of the above conditions is true. This is because if two triangles are equiangular then their side lengths *must* be in the same ratio, and vice versa.

Two triangles are similar if *either*:

- they are equiangular *or* • their side lengths are in the same ratio.

If we can show that two of the angles in one triangle are equal in size to two of the angles in another triangle, then the remaining angles must also be equal, since the angles in each triangle sum to 180°.

Example 3 **Self Tutor**

Show that the following figures possess similar triangles:

a

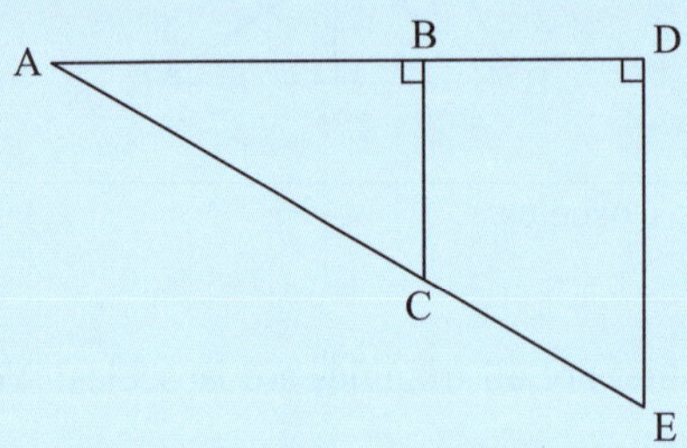

b

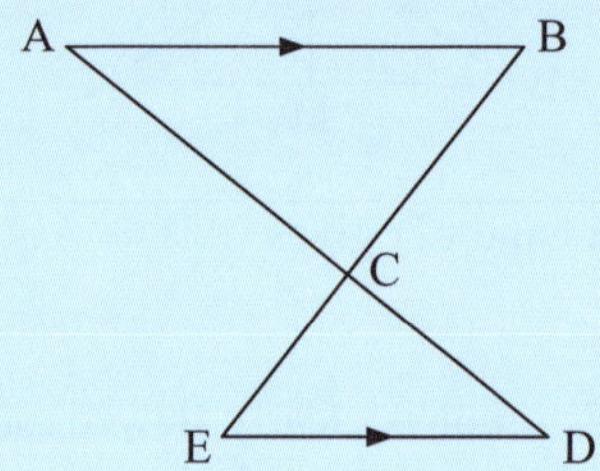

a

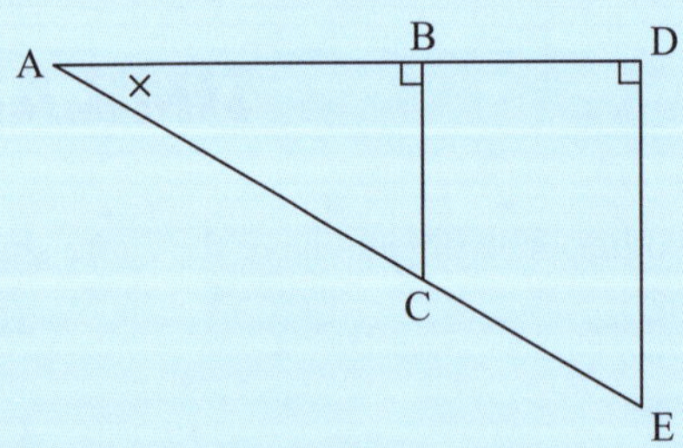

$\triangle$s ABC and ADE are equiangular as:

- $A\widehat{B}C = A\widehat{D}E = 90°$ {given}
- $\widehat{A}$ is common to both triangles

$\therefore$ the triangles are similar.

b

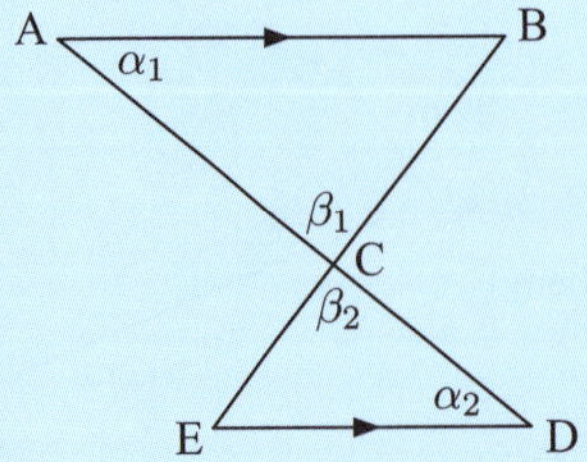

$\triangle$s ABC and DEC are equiangular as:

- $\alpha_1 = \alpha_2$ {alternate angles}
- $\beta_1 = \beta_2$ {vertically opposite angles}

$\therefore$ the triangles are similar.

EXERCISE 20D.1

1 Decide whether the following pairs of triangles are similar:

a

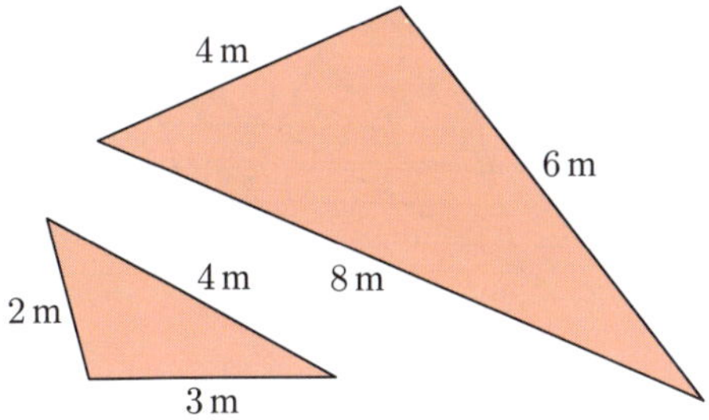

b

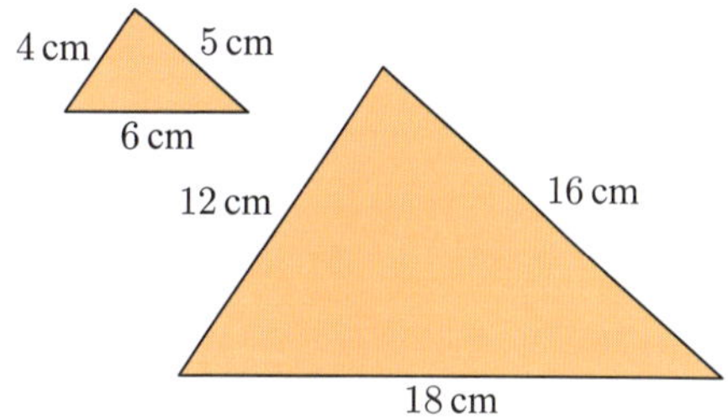

c

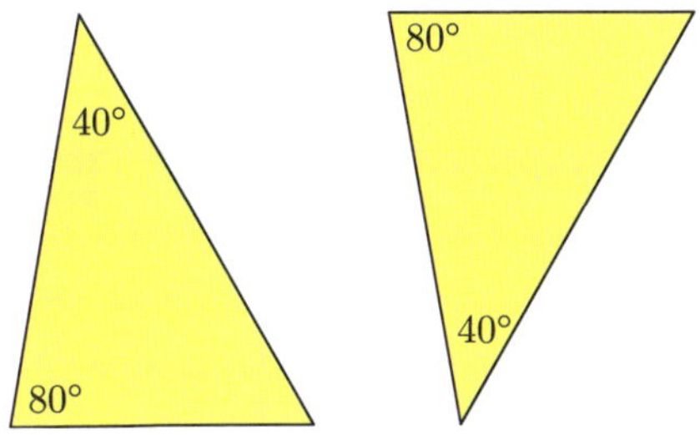

d

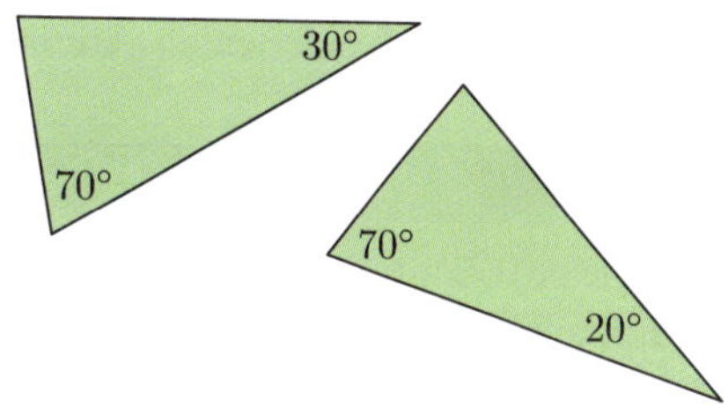

e

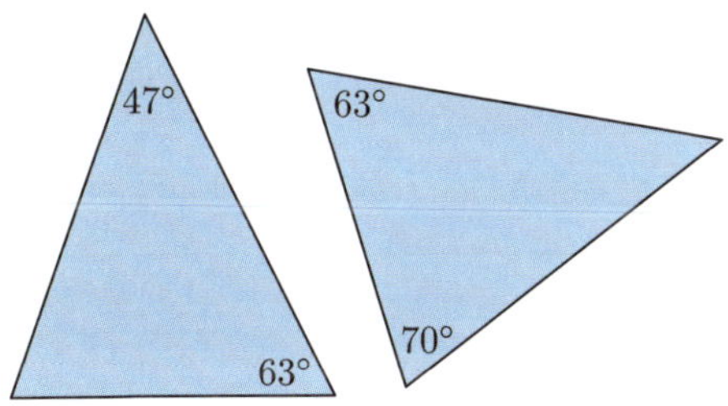

f

2 Show that the following figures posses similar triangles:

a

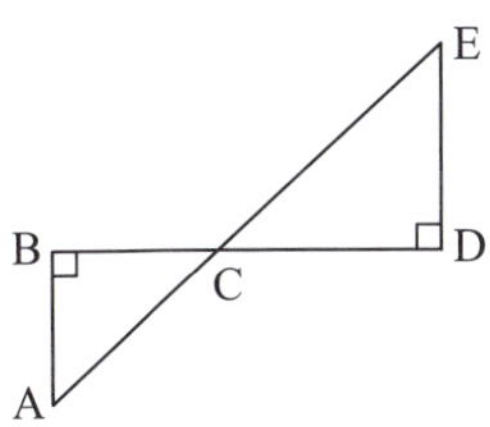

b

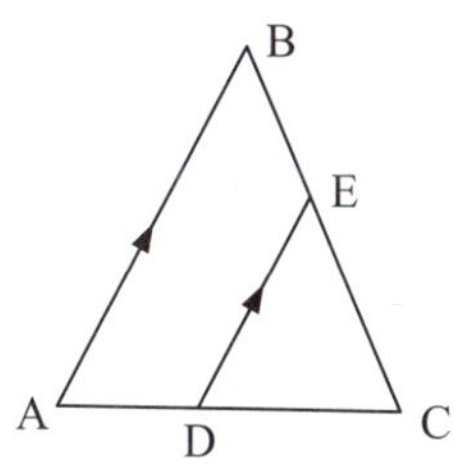

c

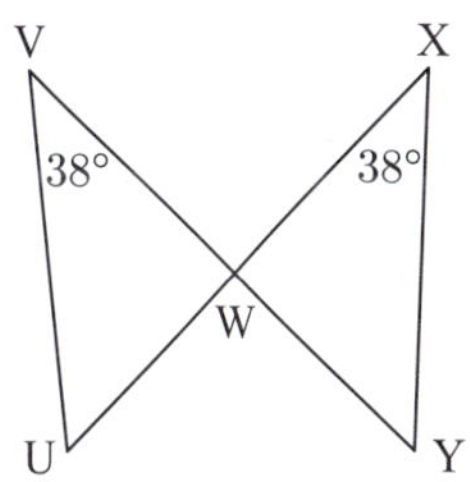

d

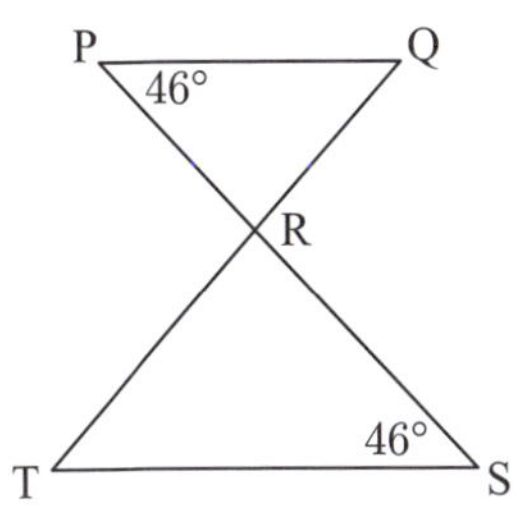

e

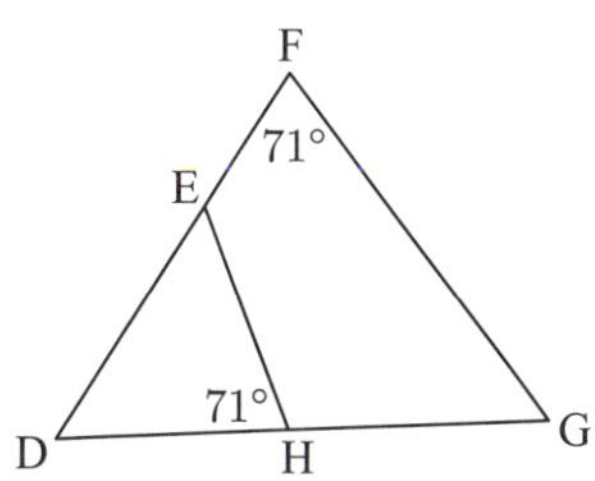

f

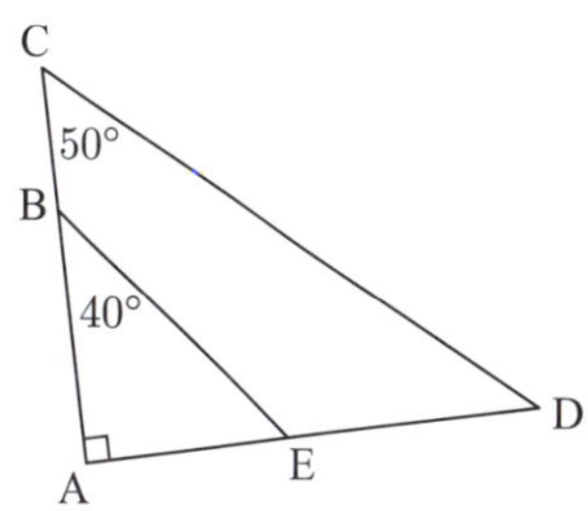

3 Congruent triangles ABC and CDE are placed next to each other so that ACE is a straight line.
Show that the shaded triangles are similar.

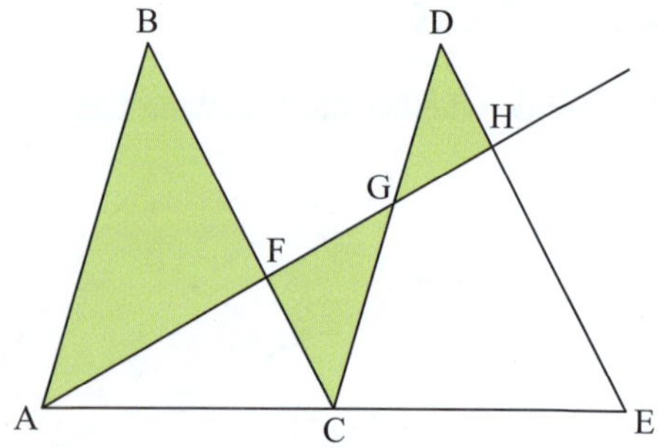

FINDING SIDE LENGTHS

We can use similar triangles to find unknown side lengths. When doing so, we must be careful to correctly identify which side lengths are in the same ratio.

Example 4 Self Tutor

Triangle A′B′C′ is similar to triangle ABC.
Find the value of x.

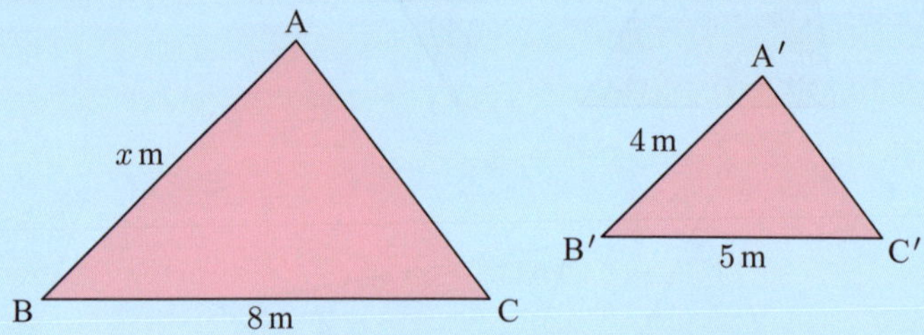

The corresponding sides are in the same ratio.

$$\therefore \quad \frac{AB}{A'B'} = \frac{BC}{B'C'}$$

$$\therefore \quad \frac{x}{4} = \frac{8}{5}$$

$$\therefore \quad x = \frac{32}{5} = 6.4$$

EXERCISE 20D.2

1 Find x given that triangle A′B′C′ is similar to triangle ABC:

a

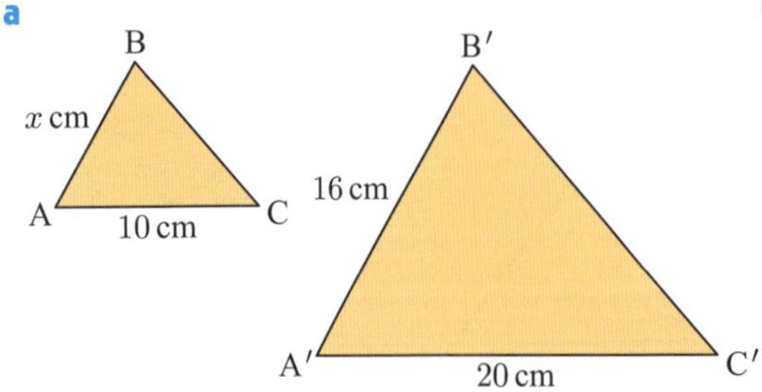

b

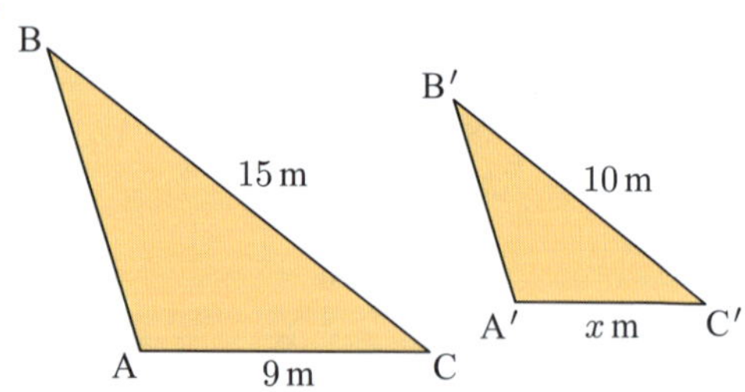

c

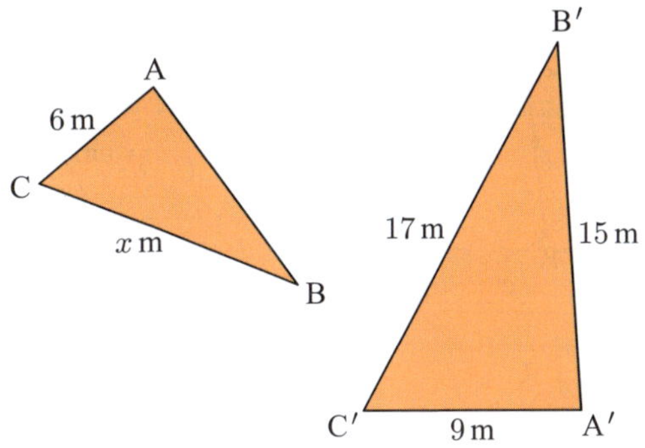

d

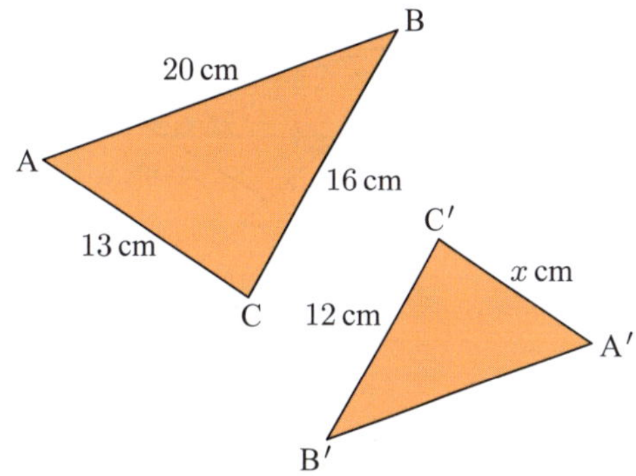

Example 5 Self Tutor

In the figure alongside, BD = 20 cm.

Establish that a pair of triangles is similar, and hence find x.

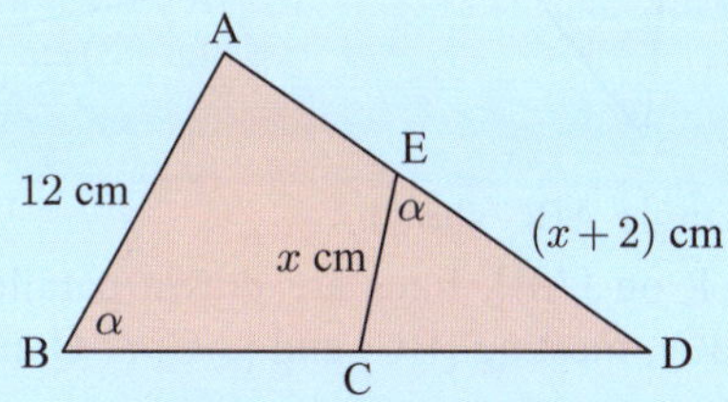

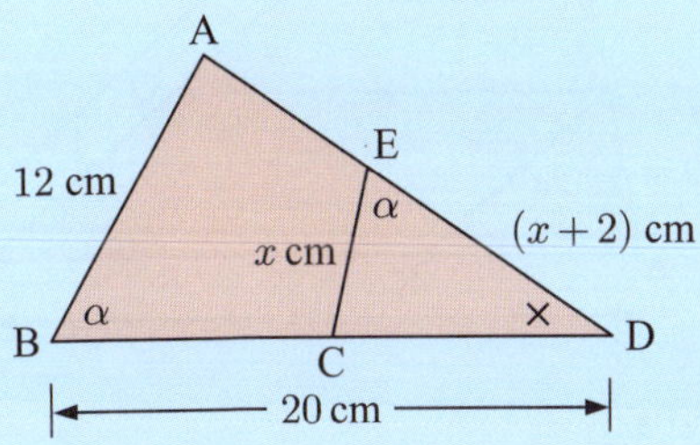

$\triangle$s CED and ABD are equiangular since:

- $C\widehat{E}D = A\widehat{B}D$ {given}
- $\widehat{D}$ is common

$\therefore$ $\triangle$s CED and ABD are similar.

Corresponding sides must be in the same ratio.

$$\therefore \frac{ED}{BD} = \frac{CE}{AB}$$

$$\therefore \frac{x+2}{20} = \frac{x}{12}$$

$$\therefore 12(x+2) = 20x$$

$$\therefore 12x + 24 = 20x$$

$$\therefore 24 = 8x$$

$$\therefore x = 3$$

2 In each of the following figures, establish that a pair of triangles is similar, and hence find x.

a

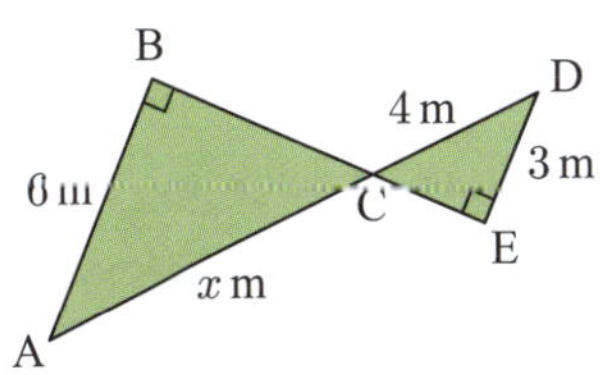

b

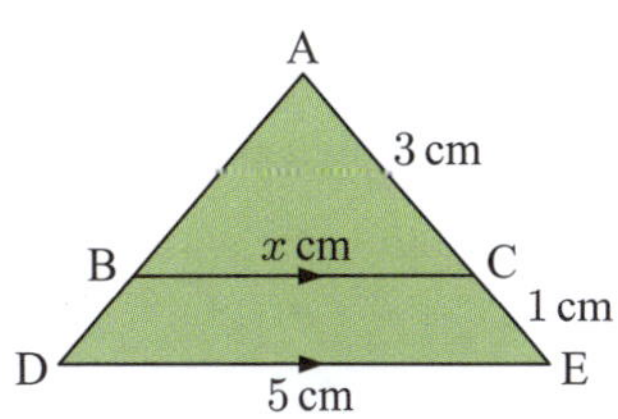

c

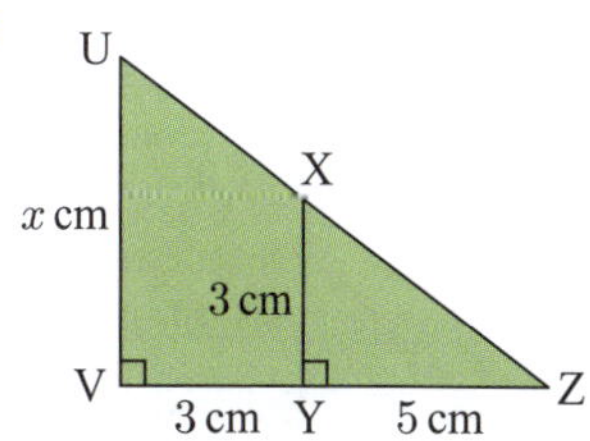

d 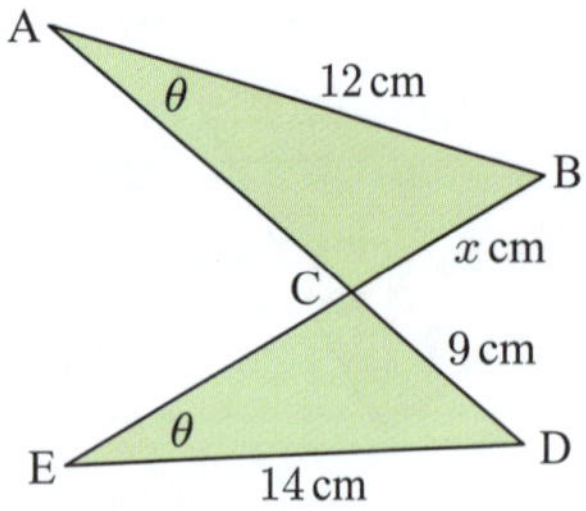

e 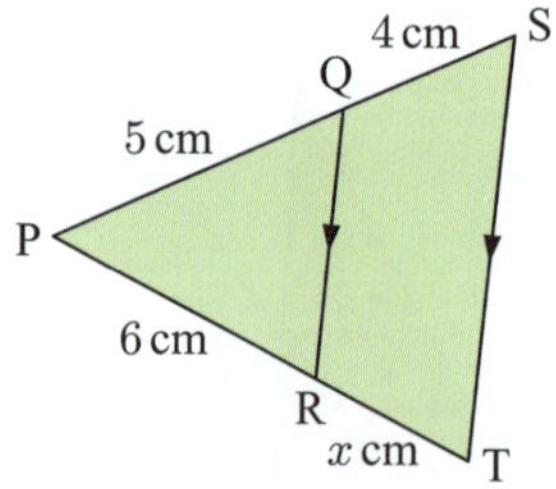

f 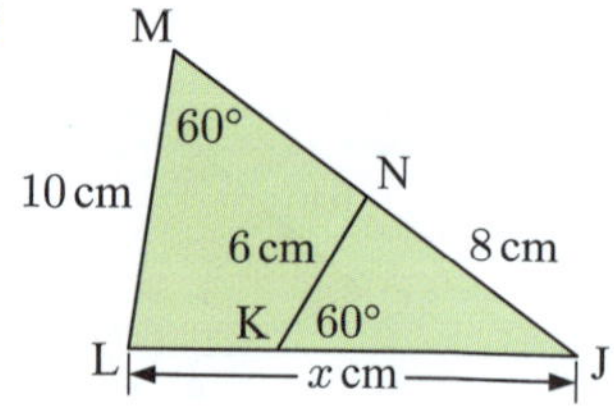

g 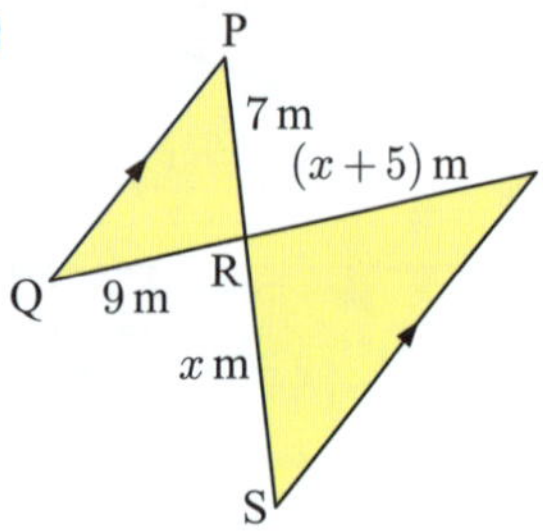

h 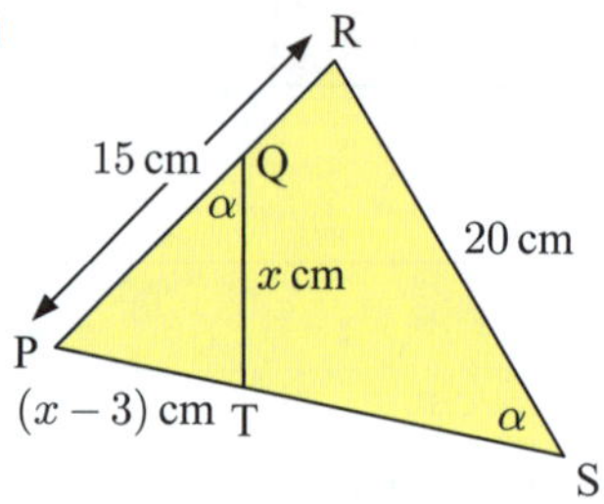

i 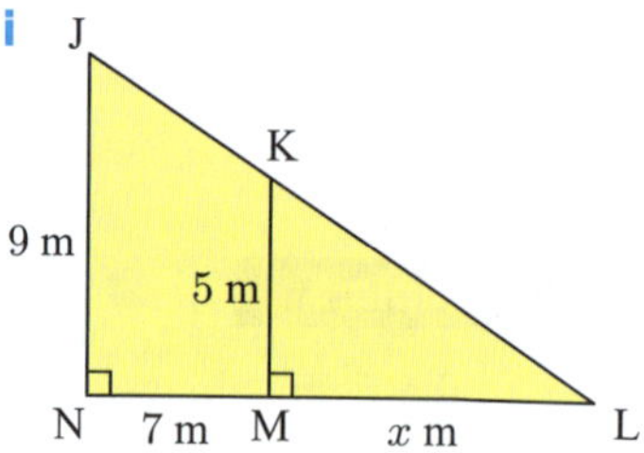

3 Consider triangle ABC shown.

From point P on [AB], lines are drawn parallel to [AC] and [BC], meeting [BC] and [AC] at Q and R respectively.

Find the perimeter of parallelogram PQCR.

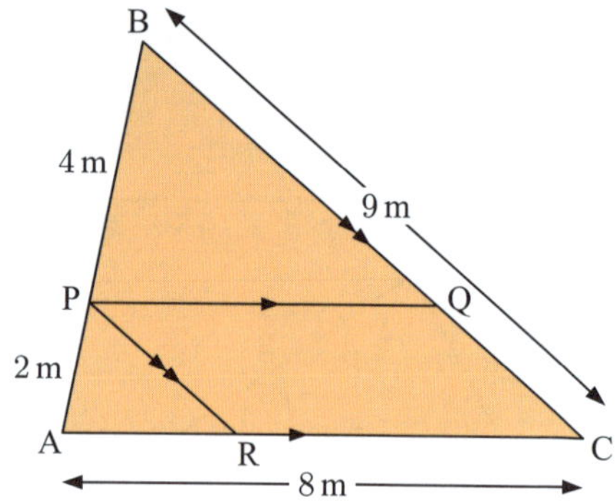

4 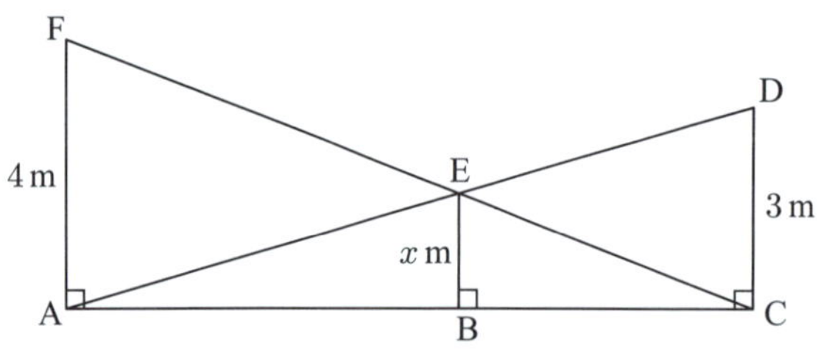

Find x.

E PROBLEM SOLVING

To solve problems involving similar triangles, we follow these steps:

Step 1: Sketch the situation, showing all of the given information.

Step 2: Use a variable such as x to represent the unknown quantity to be found.

Step 3: Establish that a pair of triangles are similar, and hence set up an equation involving the variable.

Step 4: Solve the equation.

Step 5: Answer the question in a sentence.

Example 6 | Self Tutor

One sunny morning, Phan and Khuyen compare the length of the shadow of their school hall with that of a 1 metre ruler. Their measurements are shown in the diagram.

Use the information to find the height of the hall.

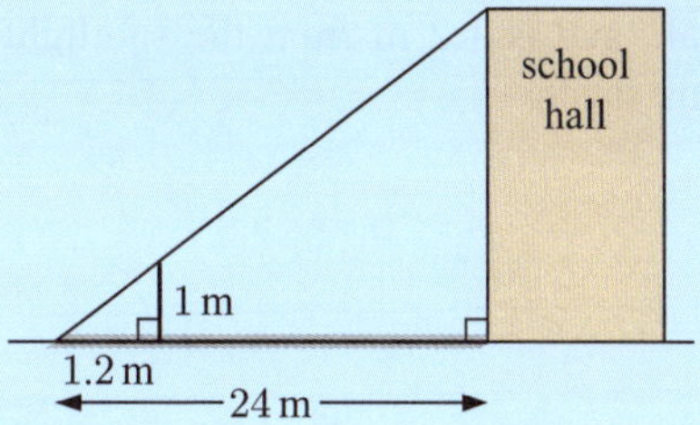

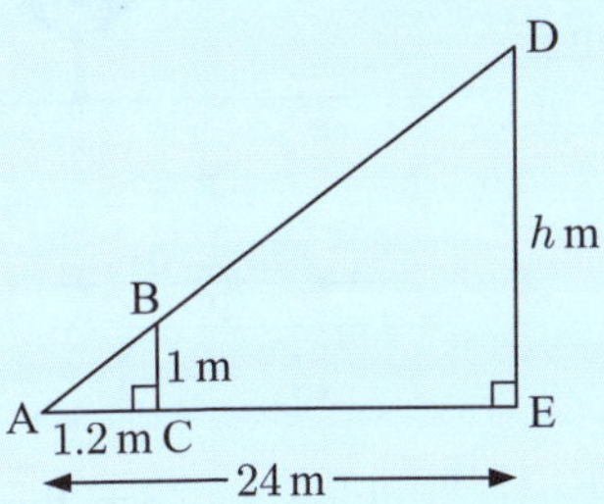

Let the height of the hall be h m.

$\triangle$s ABC and ADE are equiangular as:

- $A\widehat{C}B = A\widehat{E}D = 90°$
- $\widehat{A}$ is common to both triangles

$\therefore$ $\triangle$s ABC and ADE are similar.

Corresponding sides must be in the same ratio.

$$\therefore \frac{DE}{BC} = \frac{AE}{AC}$$

$$\therefore \frac{h}{1} = \frac{24}{1.2}$$

$$\therefore h = 20$$

So, the hall is 20 m high.

EXERCISE 20E

1

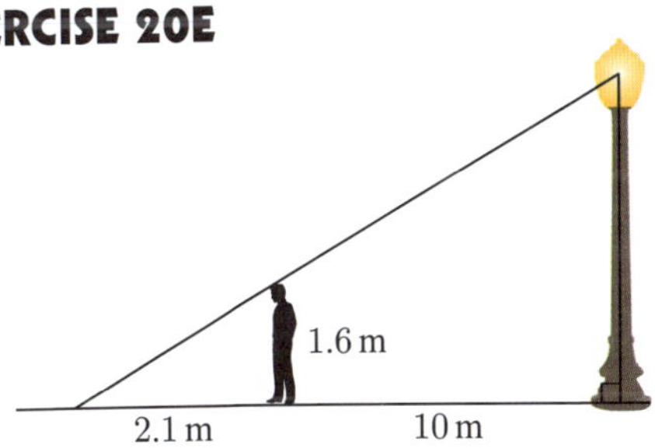

When a 1.6 m tall person stands 10 m from the base of an electric light pole, the shadow of the person is 2.1 m long.

Find the height of the globe above ground level.

2 A road has a constant slope of 2 in 11. This means that if you travel 11 m horizontally along the road, you will rise 2 m.

  a Over what horizontal distance will the road rise 300 m?

  b What is the length of the road [AB] to the nearest metre?

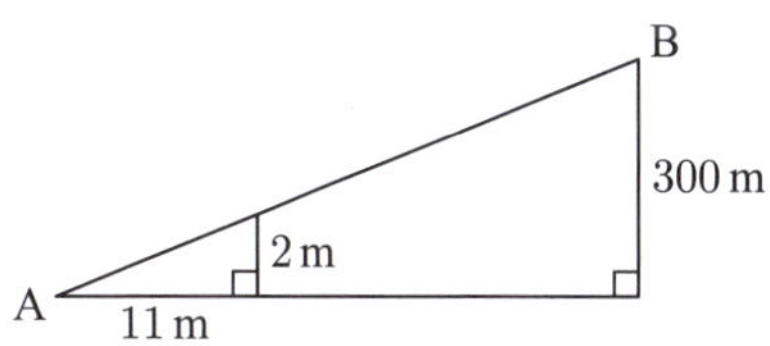

3

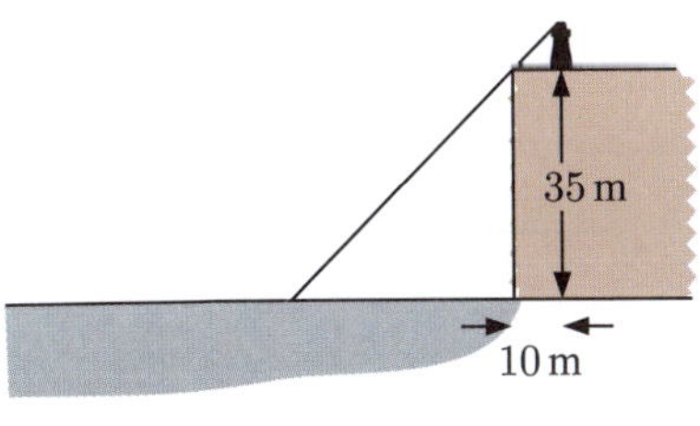

A lighthouse is situated 10 m back from the edge of a 35 m high cliff. The beacon is 8 m above the base of the lighthouse. How far from the cliff does the shadow extend?

4 Curtis walks in front of an in-ground spotlight shining on a wall 5 m away. Curtis is 156 cm tall, and is 1.2 m from the spotlight. How tall is his shadow?

5 Enrico is standing at point A on one side of a river, directly opposite the tree T. 10 m along the bank is a post P. Enrico walks to P and then a further 6 m along the bank to C. He then walks 8 m directly away from the river until he reaches D, which is in line with P and T.

a Show that triangles CDP and ATP are similar.

b How wide is the river at A?

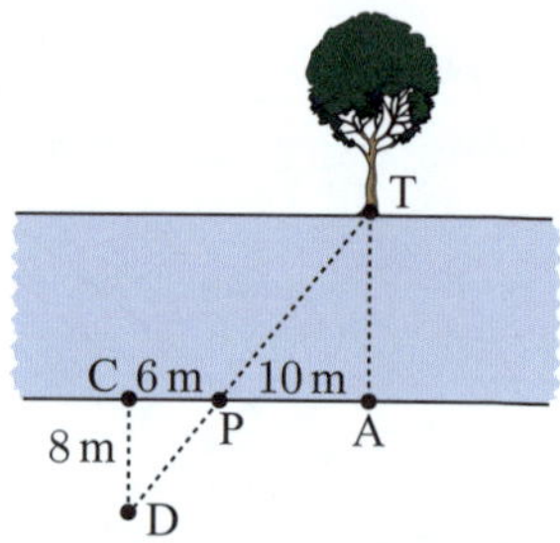

6

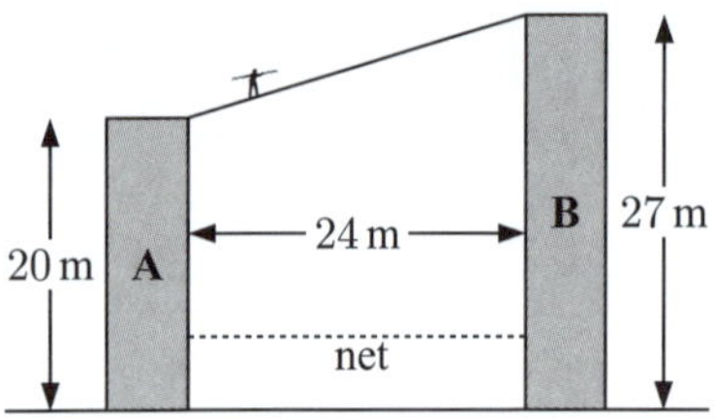

A tightrope walker is walking between the roofs of two buildings A and B. The buildings are 24 m apart. There is a safety net 5 m above ground level.

a Find the length of the tightrope.

b After travelling 15 m, the tightrope walker falls off. How far will he fall before landing in the net?

7 Squash player Kenny is about to play a shot at K. He wants to rebound the ball off the wall at P, so that it hits the corner at Q. Locate the position of P by finding the length PR.

You may assume that the ball leaves the wall at the same angle with which it entered, so $K\widehat{P}R = Q\widehat{P}S$.

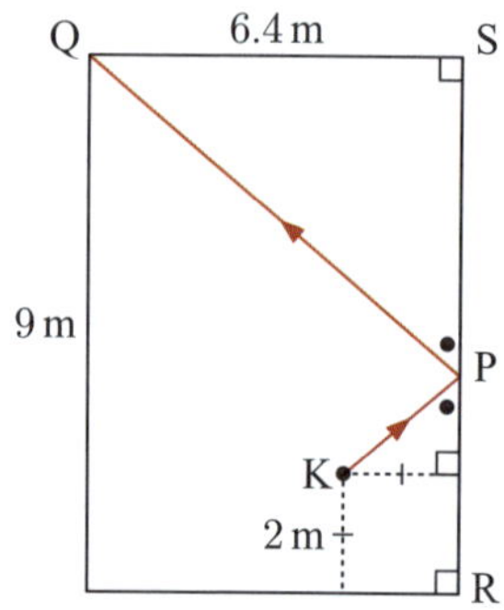

8 Alison and Brianna are in the basement of a department store.

Alison steps onto an escalator to the ground floor, and Brianna steps onto an escalator to the first floor. The escalators' paths cross at P. When Alison reaches P, she has completed $\frac{11}{20}$ of her escalator trip.

a What fraction of Brianna's escalator trip has been completed at P?

b Find the ratio of the height from the basement to the ground floor, compared with the height from the ground floor to the first floor.

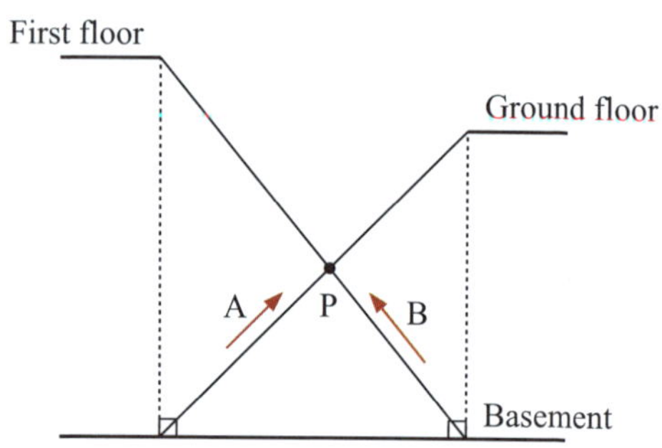

F AREAS OF SIMILAR OBJECTS

We have seen that if two objects are similar, then there is a relationship between the lengths of their sides.

For objects which are 2-dimensional, there is also a relationship between the **areas** of the figures.

For example:

- The two circles shown are similar. Circle **B** is an enlargement of circle **A** with scale factor k.

  Area of $\mathbf{B} = \pi(kr)^2$
  $= k^2 \times \pi r^2$
  $= k^2 \times$ area of **A**

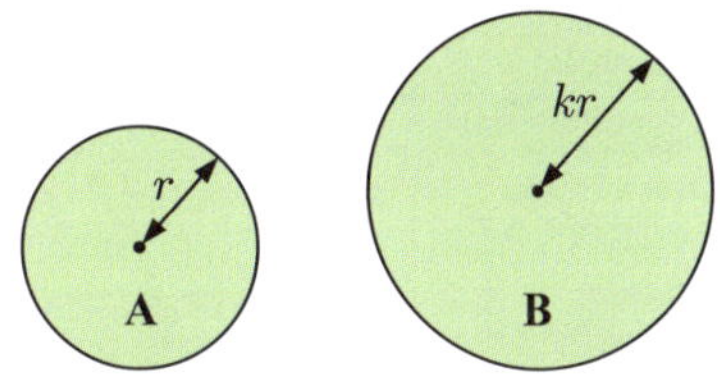

- The two rectangles shown are also similar. Rectangle **B** is an enlargement of rectangle **A** with scale factor k.

  Area of $\mathbf{B} = ka \times kb$
  $= k^2 \times ab$
  $= k^2 \times$ area of **A**

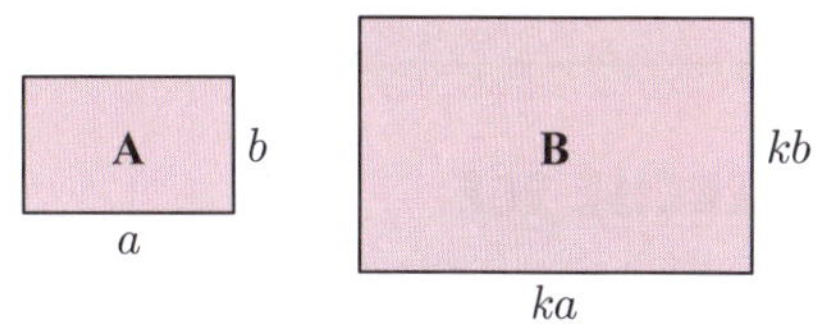

Using examples like this we can conclude that:

> If a figure is enlarged with scale factor k to produce a similar figure, then
> the new area $= k^2 \times$ the old area.

Example 7 Self Tutor

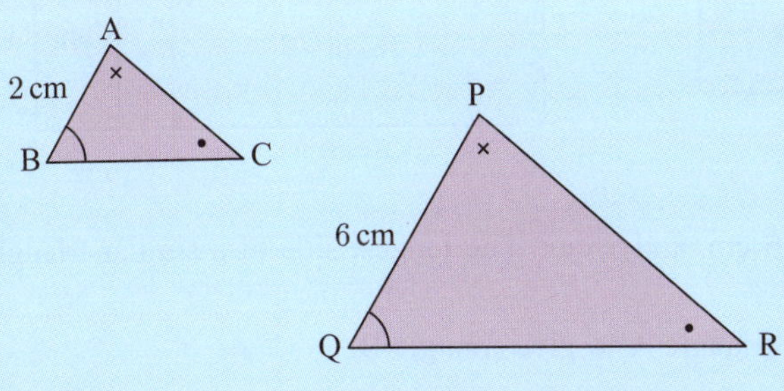

Triangles ABC and PQR are equiangular and therefore similar.

AB = 2 cm and PQ = 6 cm.

The area of $\triangle$ABC is 5 cm^2.

Find the area of $\triangle$PQR.

Suppose we enlarge $\triangle$ABC with scale factor k to give $\triangle$PQR.

$$\therefore \ k = \frac{6}{2} = 3$$

Now area $\triangle PQR = k^2 \times$ area $\triangle ABC$
$= 9 \times 5$ cm^2
$= 45$ cm^2

Example 8 — Self Tutor

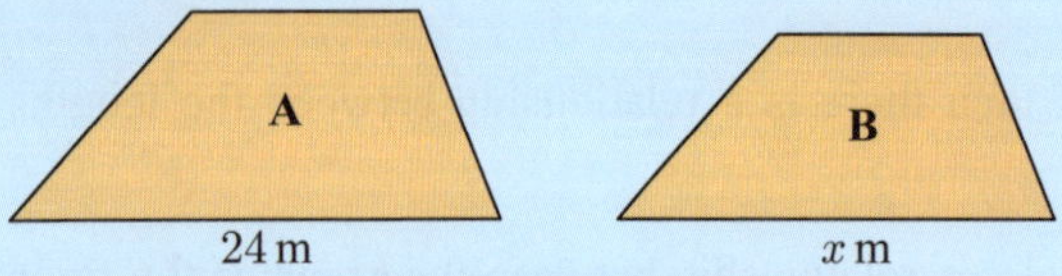

Figures **A** and **B** have areas 160 m^2 and 90 m^2 respectively.

Given that the figures are similar, find x.

Figure **A** is reduced with scale factor k to give figure **B**.

Area of **B** $= k^2 \times$ area of **A**

$\therefore\ 90 = k^2 \times 160$

$\therefore\ \frac{9}{16} = k^2$

$\therefore\ k = \frac{3}{4}$ $\quad \{k > 0\}$

Now $x = \frac{3}{4} \times 24$ $\quad$ {sides in the same ratio}

$\therefore\ x = 18$

EXERCISE 20F

1 For each pair of *similar* shapes, find the unknown length or area:

a

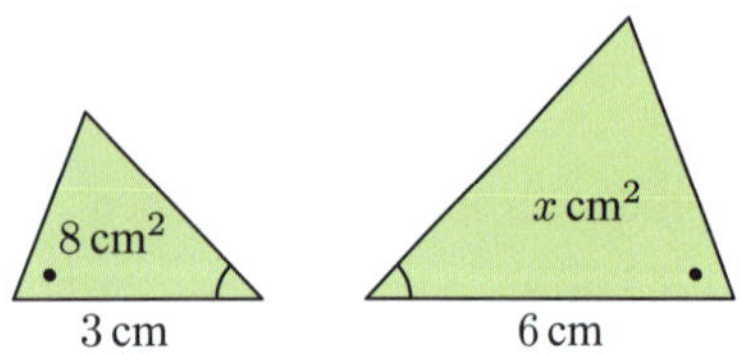

b

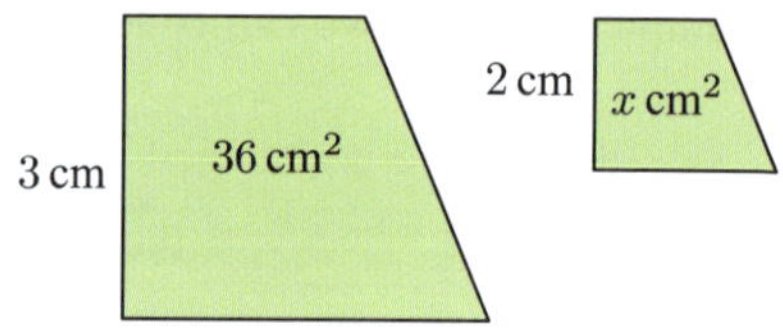

c

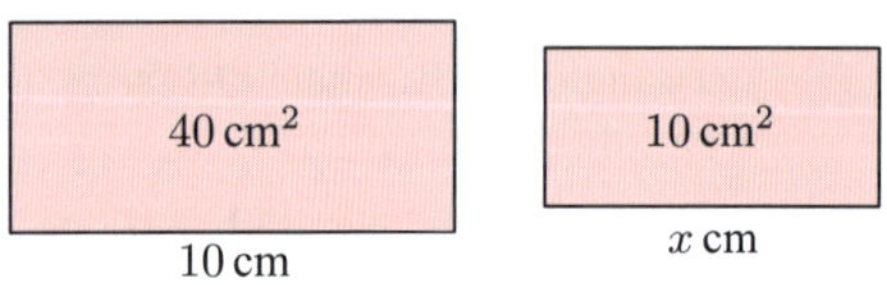

d

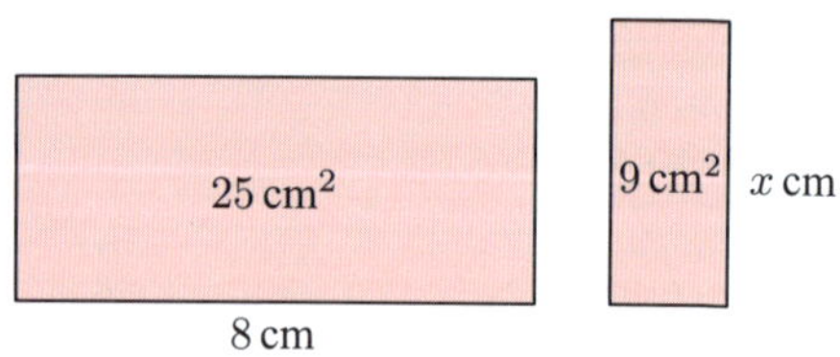

2 The side lengths of triangle A are 5 cm, 6 cm, and 7 cm. The longest side of a similar triangle B is 28 cm.

- **a** Find the scale factor when enlarging triangle A to give triangle B.
- **b** Find the length of the other two sides of triangle B.
- **c** How many times larger is the area of triangle B than the area of triangle A?

3

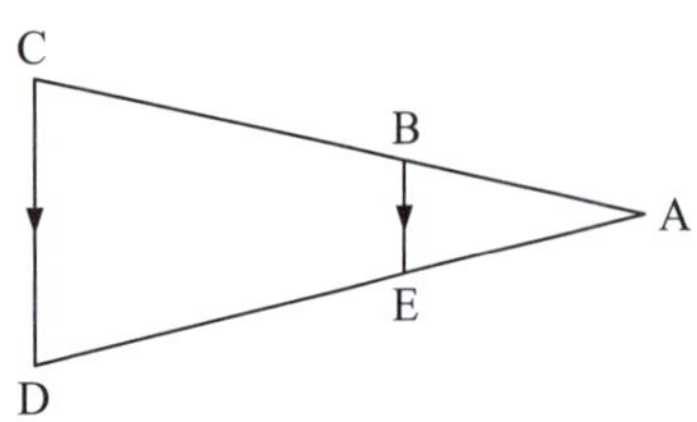

In the given figure, DE $= 6$ cm, AE $= 4$ cm, and $\triangle$ABE has area 16 cm^2.

- **a** Show that $\triangle$ABE and $\triangle$ACD are similar.
- **b** Find the scale factor to enlarge $\triangle$ABE into $\triangle$ACD.
- **c** Find the area of:
  - **i** $\triangle$ACD
  - **ii** BCDE.

4 In the given figure, AE = 3 cm, $\triangle$ABE has area 5 cm^2, and BEDC has area 6 cm^2.
Find the length of [ED] rounded to 3 significant figures.

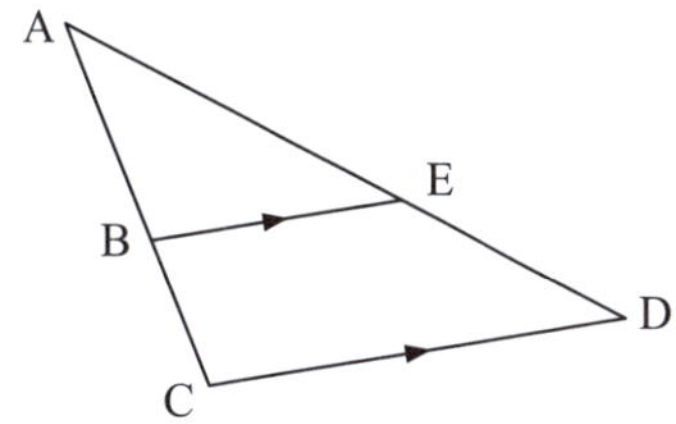

5 Figures **A** and **B** are similar. The dots shown are equally spaced at intervals of 1 unit.
If figure **A** has area 1.6 $units^2$, find the area of figure **B**.

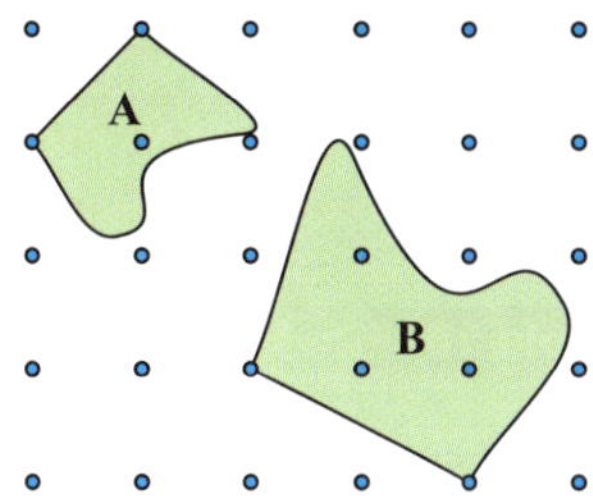

G VOLUMES OF SIMILAR OBJECTS

For 3-dimensional objects which are similar, there is also a relationship between the **volumes** of the figures.

For example:

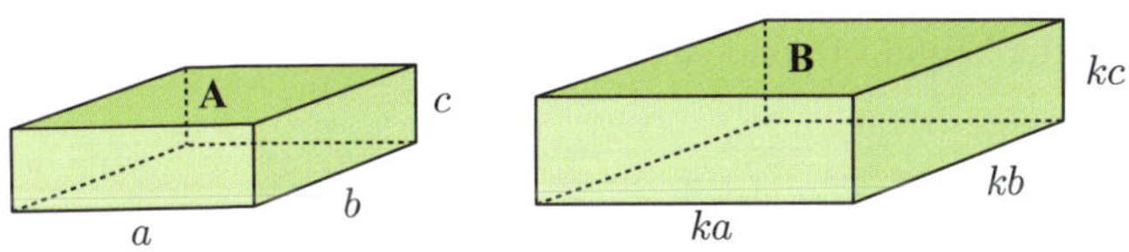

Volume of box **B**
$= ka \times kb \times kc$
$= k^3 \times abc$
$= k^3 \times$ volume of box **A**

If a 3-dimensional object is enlarged with scale factor k to produce a similar object, then the new volume $= k^3 \times$ the old volume.

Example 9 **Self Tutor**

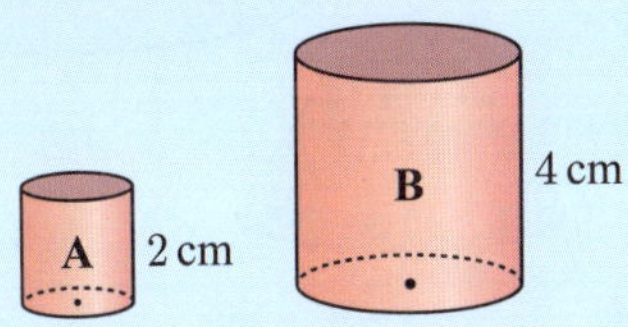

Cylinders **A** and **B** are similar with heights 2 cm and 4 cm respectively.

Cylinder **A** has volume 10 cm^3.

Find the volume of cylinder **B**.

Suppose cylinder **A** is enlarged with scale factor k to give cylinder **B**.

$\therefore \quad k = \frac{4}{2} = 2$

Volume of **B** $= k^3 \times$ volume of **A**

$\therefore$ the volume of cylinder **B** $= 8 \times 10$ cm^3

$= 80$ cm^3

EXERCISE 20G

1 Consider the following *similar* solids. Find the unknown volume:

a

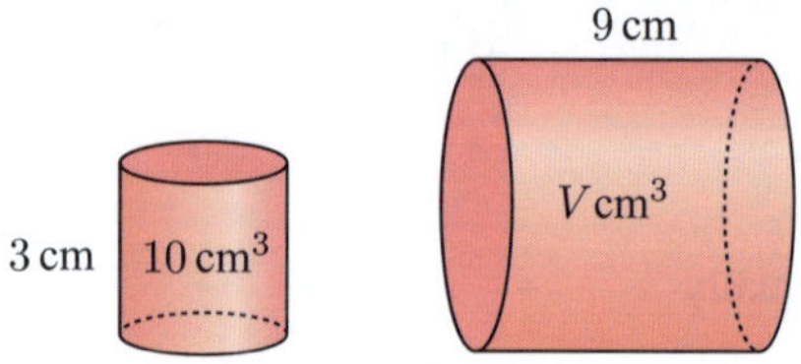

b

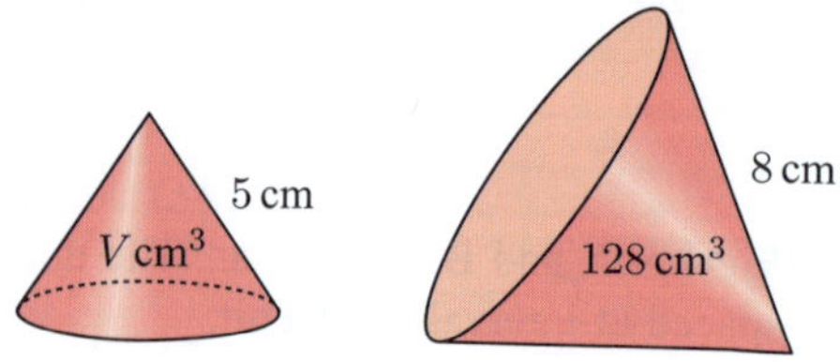

Example 10 — Self Tutor

A and **B** are similar cones with volumes 54 cm^3 and 250 cm^3 respectively.

If cone **A** is 6 cm high, find the height of cone **B**.

Suppose cone **A** is enlarged with scale factor k to give cone **B**.

Volume of cone **B** $= k^3 \times$ volume of cone **A**

$\therefore \quad 250 = k^3 \times 54$

$\therefore \quad \frac{125}{27} = k^3$

$\therefore \quad k = \frac{5}{3}$

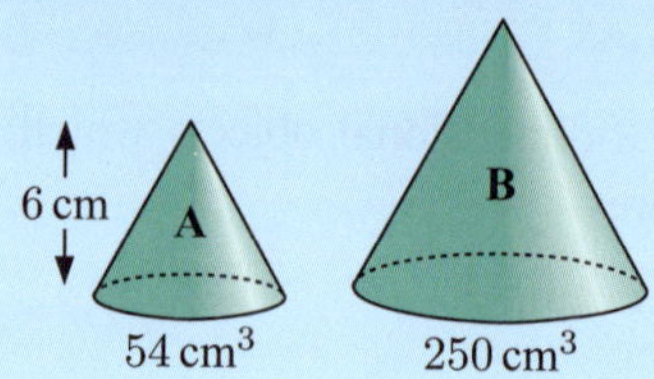

So, the height of cone **B** $= \frac{5}{3} \times$ the height of cone **A**

$= \frac{5}{3} \times 6$ cm

$= 10$ cm

2 For each pair of *similar* solids, find the unknown length:

a

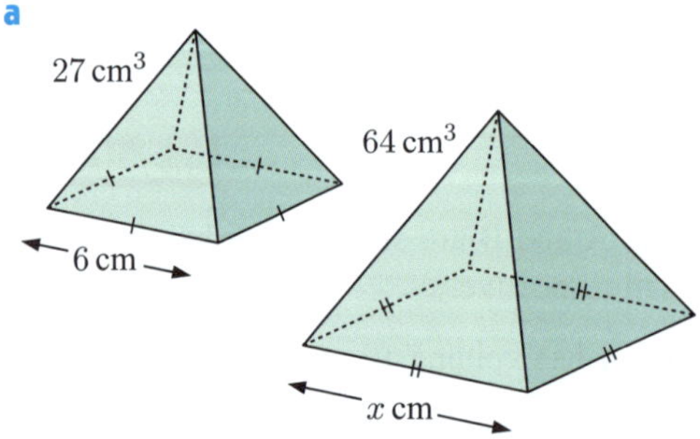

b

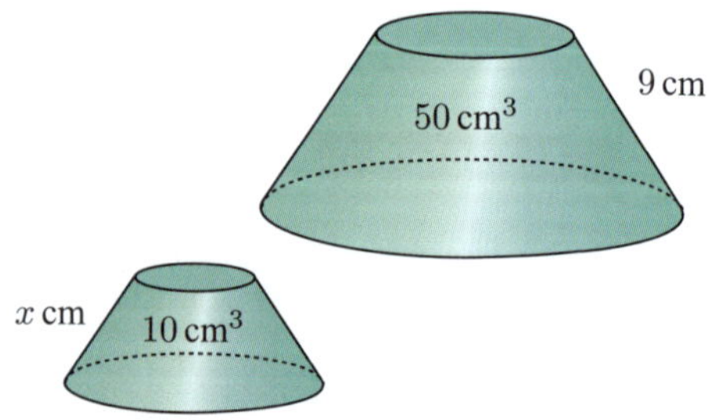

3 Two similar pyramids have volumes 7500 cm^3 and 480 cm^3. If the larger pyramid is 30 cm high, find the height of the smaller pyramid.

4 A pet store sells small and large fish tanks. The small fish tank is 30 cm by 60 cm by 50 cm high. The large fish tank is similar to the small tank, and has capacity 150 litres. Find, rounded to 3 significant figures, the scale factor for the enlargement.

5 Two similar statues have heights 3.5 m and 4.9 m.

a If the volume of the smaller statue is 1.5 m^3, find the volume of the larger statue.

b If the surface area of the larger statue is 7.35 m^2, find the surface area of the smaller statue.

6 Two cylindrical containers have capacities 875 mL and 2240 mL respectively.
Thin straws are placed in the cylinders as shown. They have length 15 cm and 24 cm respectively.
Are the cylinders similar? Give evidence to support your answer.

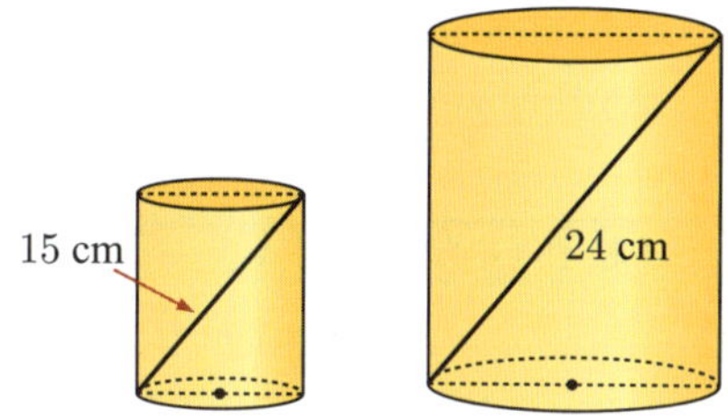

ACTIVITY 2 — MEASURING INACCESSIBLE DISTANCES

We can use similar triangles to find distances which cannot be measured directly.

What to do:

1. Working in pairs, find two landmarks A and B (these could be posts, trees, or bins) such that you cannot measure the distance between A and B directly. For example, there may be a fence or a river between them.

2. Mark a line from A which is perpendicular to [AB] using a length of rope or a piece of string. Mark two points C and D on this line.

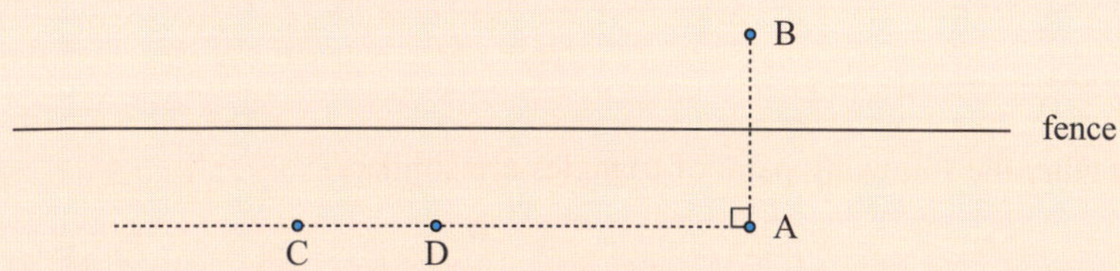

3. Stand at point C, with your partner standing at point D. Your partner now walks perpendicular to [AD] until you see they are directly between yourself and landmark B. Mark this point E.

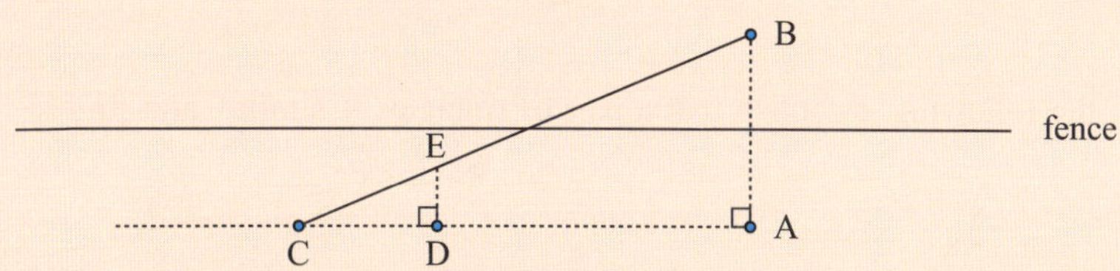

4. Measure the lengths of [CD], [DE], and [AC].
5. Use similar triangles to find the distance between A and B.
6. Choose two different points for C and D, and repeat steps **3** to **5**. This time swap roles with your partner.

7 Discuss the following questions with your partner and with your class:

a Did you get the same result for the distance between A and B when you chose different points for C and D?
Think of some sources of error in this experiment which may explain any differences you obtained.

b How could you check in **3** whether the person at D indeed walked perpendicular to [AD]?

REVIEW SET 20A

1 For the pair of congruent figures alongside:

a identify the side in the second figure corresponding to [AB] in the first figure

b identify the angle in the second figure corresponding to $D\widehat{C}B$ in the first figure.

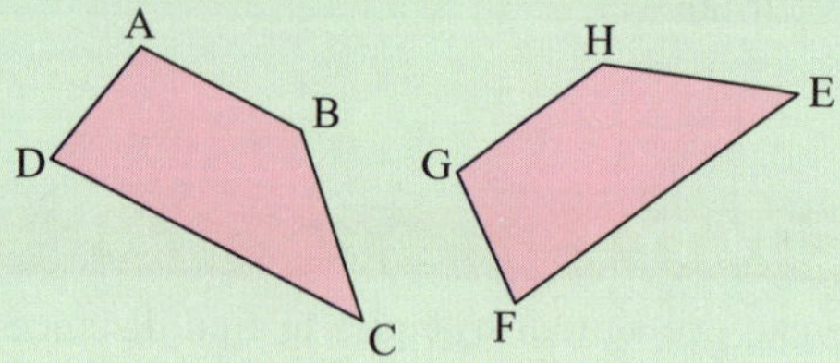

2 State whether these pairs of triangles are congruent, giving reasons for your answers.

a

b

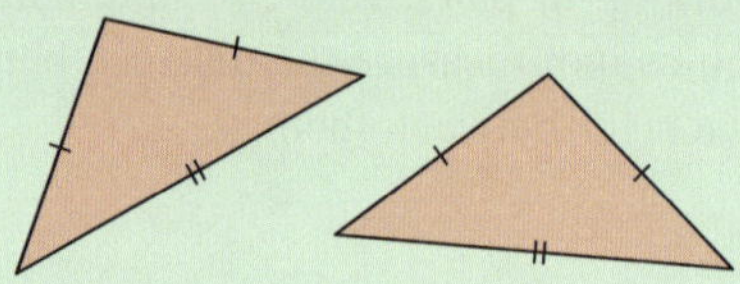

3

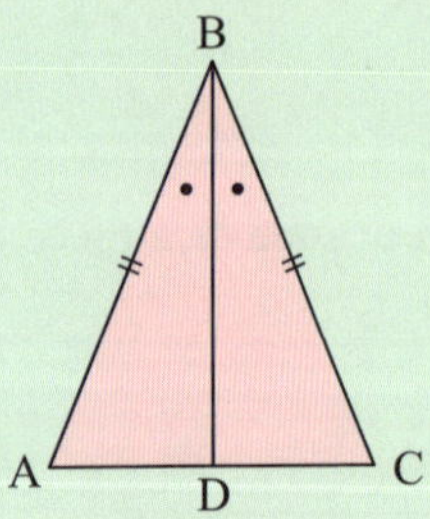

Triangle ABC is isosceles with $BA = BC$.
[BD] is the angle bisector of $A\widehat{B}C$.

a Show that triangles DAB and DCB are congruent.

b Hence explain why [BD] is perpendicular to [AC].

4 Decide whether the following pairs of triangles are similar:

a

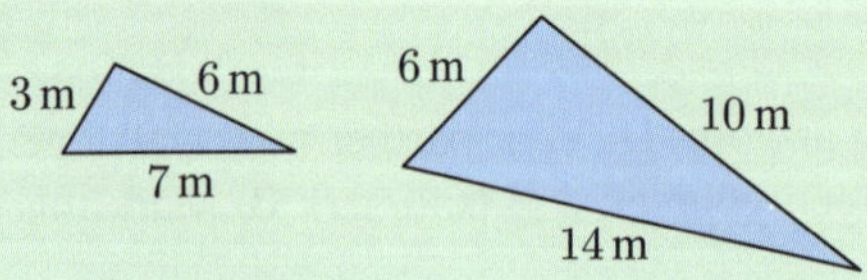

b

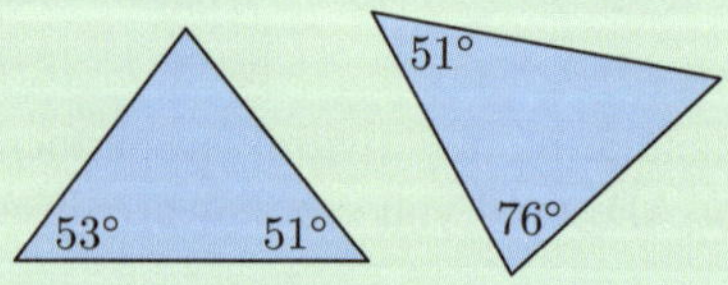

5 In each of the following, establish that a pair of triangles is similar, and find x:

a

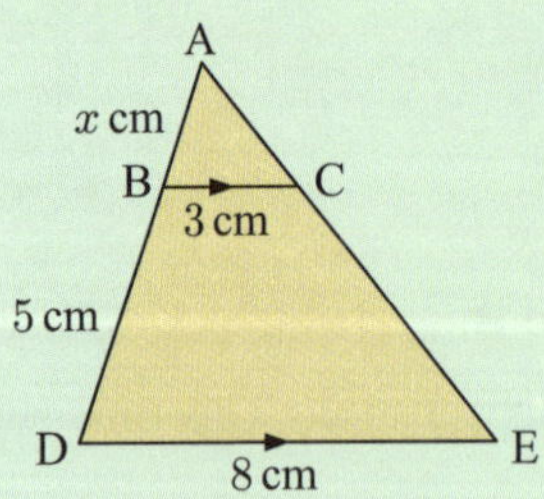

b

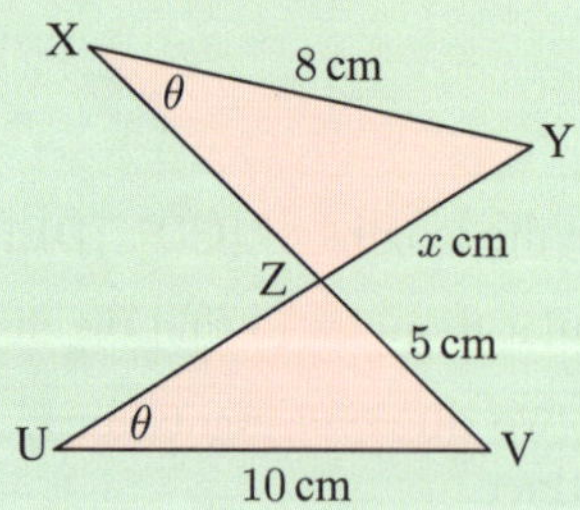

6 Find the length of [AC].
Your answer should not involve x.

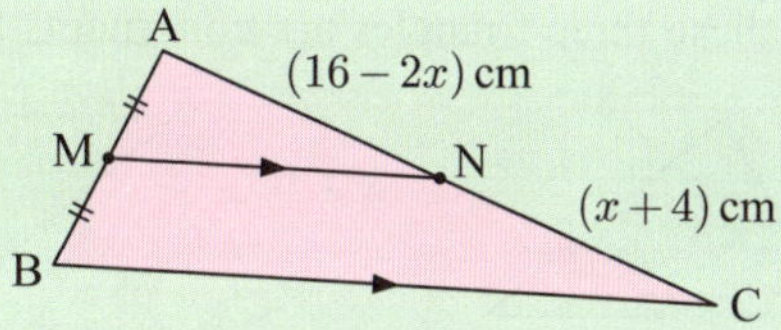

7

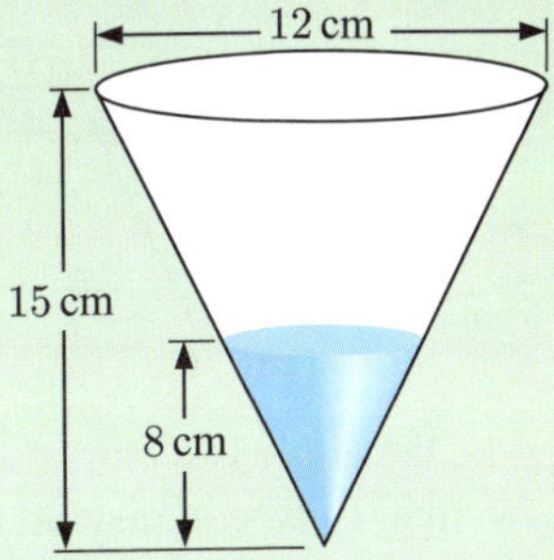

A conical flask has height 15 cm and base diameter 12 cm. Water is poured into the flask to a depth of 8 cm.

a Show that triangles ABC and MNC are similar.

b Hence show that $x = 3.2$.

c Find the diameter of the surface of the water.

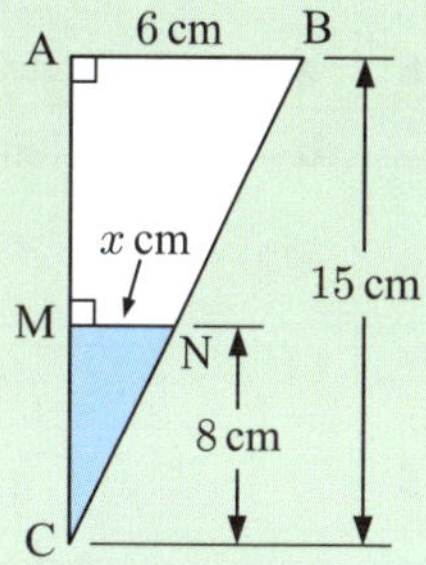

8 Find x for these similar figures:

a

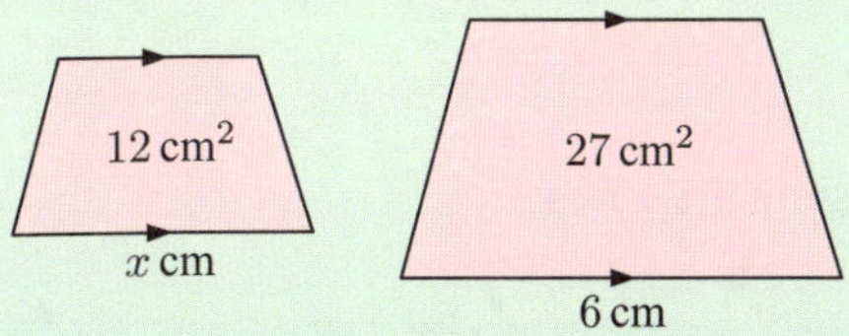

b

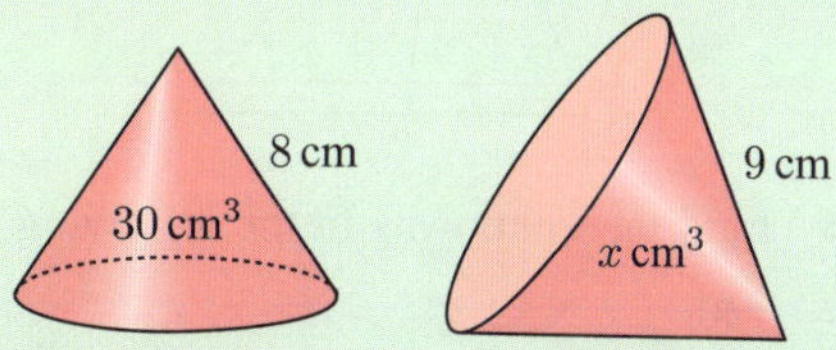

9 Parallelograms ABCD and EFGH are similar. Find:

a x **b** y

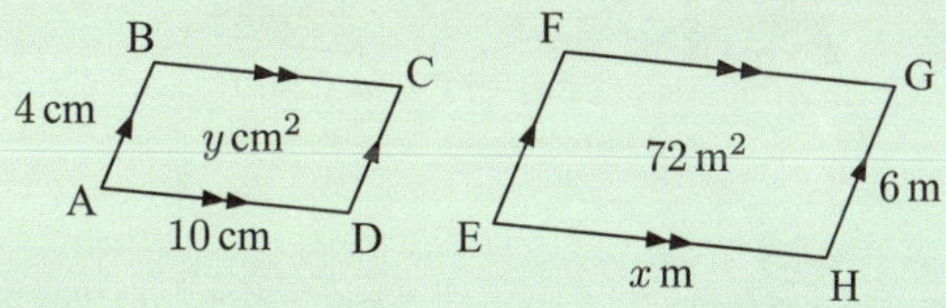

10 The surface areas of two similar cylinders are 250 cm^2 and 360 cm^2 respectively. Find the volume of the larger cylinder if the volume of the smaller cylinder is 375 cm^3.

REVIEW SET 20B

1 State whether these pairs of triangles are congruent, giving reasons for your answers:

a

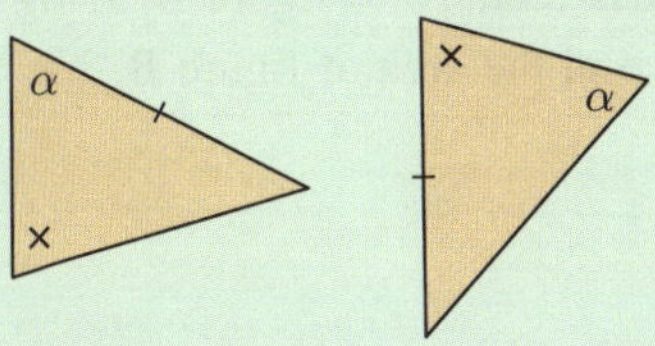

b

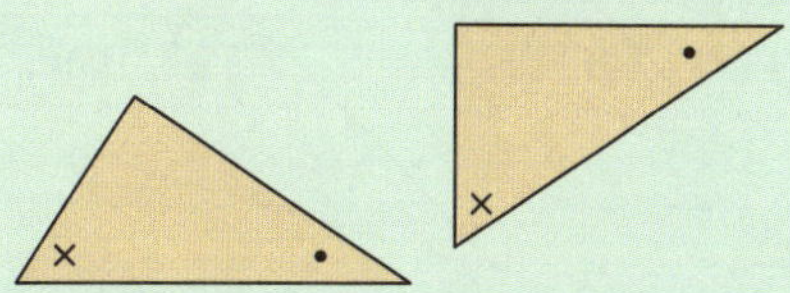

2 Find x given that triangle P′Q′R′ is similar to triangle PQR.

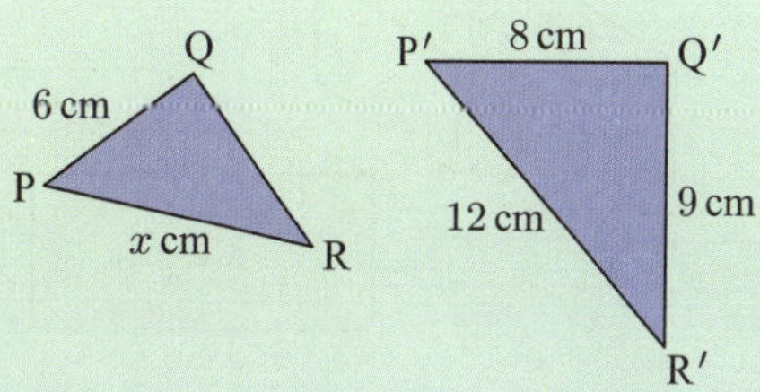

3 Two of these three triangles are congruent. State which pair is congruent, and give a reason.

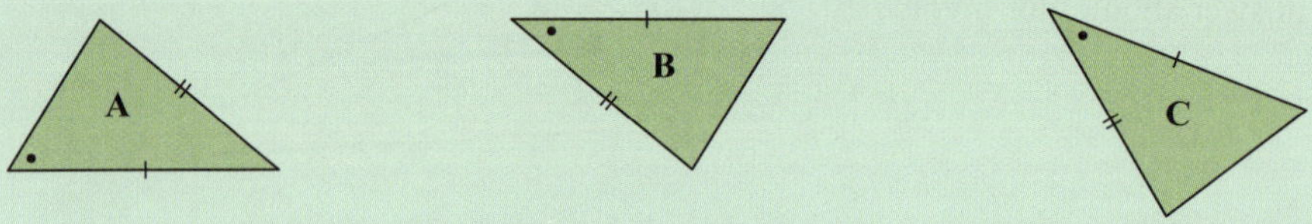

4 **a** Explain why triangles ABE and ACD are similar.

b Find the length of [CD] given that BE = 4 cm.

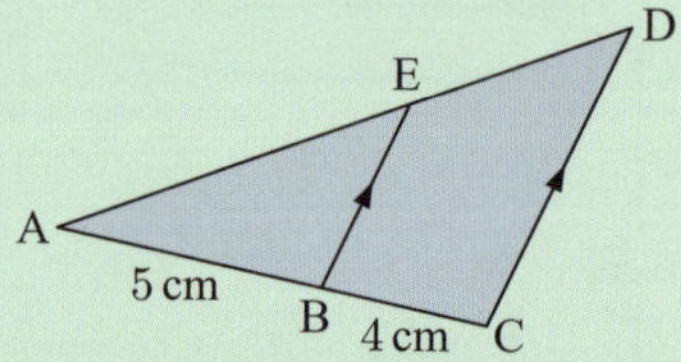

5 In triangle ABC alongside, the angle bisector of $A\widehat{B}C$ meets [AC] at P. Show that P is equidistant from [AB] and [CB].

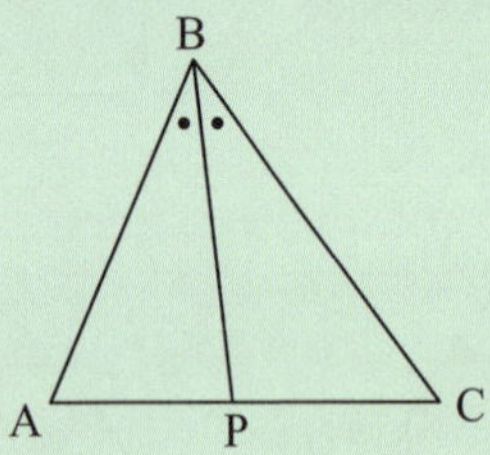

6 Find the unknowns in the following pairs of similar figures:

a

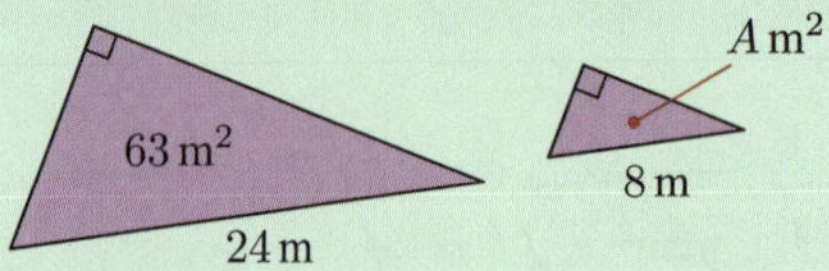

b

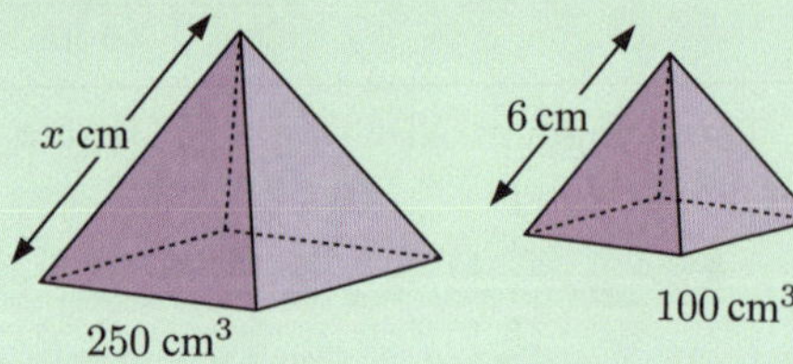

7 Figure ABCD is a rectangle. Show that the three triangles in this figure are all similar to each other.

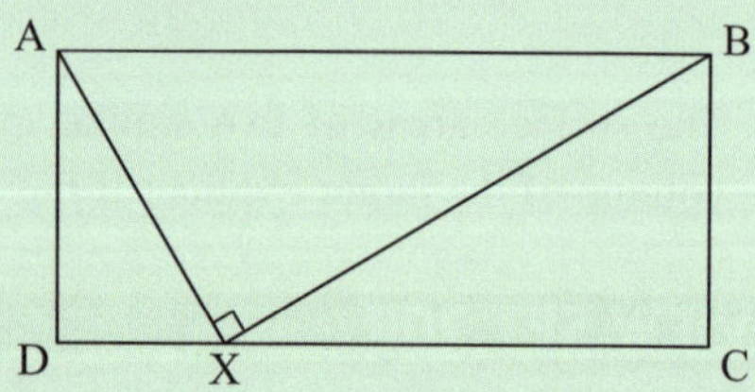

8 Figure **A** is reduced to figure **B**. If figure **A** has area 15 cm^2, find the area of figure **B**.

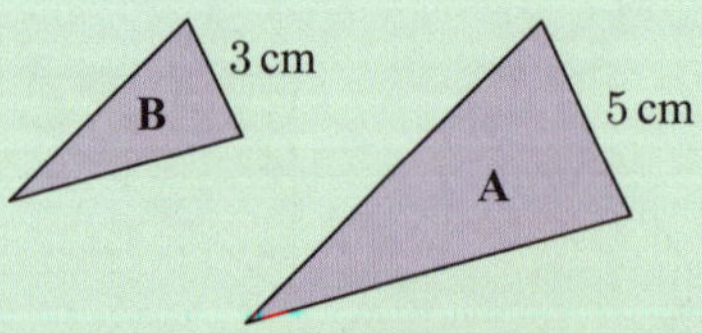

9 When placed in the appropriate orientation, it is clear that two blocks of cheese are similar.

The surface areas which are uppermost are 20 cm^2 and 45 cm^2.

a Find x.

b Find the ratio of the volumes of the blocks of cheese.

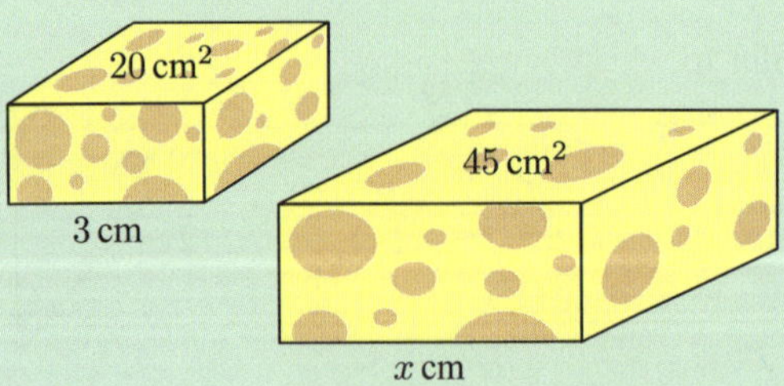

10 The dimensions of a tennis court are shown in the diagram.
Samantha hits a shot from the base line corner at S. The ball passes over her service line at T such that UT = 1.92 m.
The ball then travels over the net and lands on the opposite baseline [AD].

a Find the lengths of [SU] and [BC].

b A ball landing on the baseline is "in" if it lands between points B and C.
Assuming the ball continues along the same trajectory, will it land "in"?

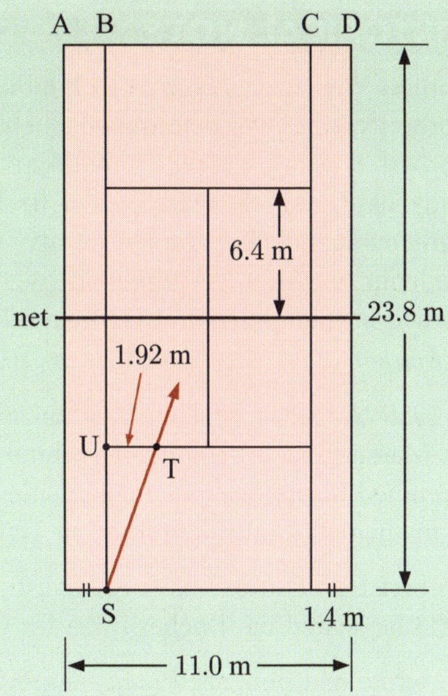

HISTORICAL NOTE — NORBERT WIENER 1894 - 1964

Norbert Wiener was born in Columbia, Missouri, USA in 1894. Norbert's father was determined that his son should succeed, and Norbert, who was obviously highly intelligent, was expected to study hard. At 18 years of age he achieved his doctorate in mathematics at Harvard University. Being too young for an academic position, he travelled overseas on scholarship, and studied in England and in Germany. His first paper, *Messenger in Mathematics*, was published in 1913.

In 1919 Norbert accepted a position as mathematics instructor at the Massachusetts Institute of Technology (MIT). In time he was promoted to the chair of mathematics, and remained professor of mathematics until he retired in 1960.

Norbert travelled extensively, investigating the latest research and findings by mathematicians. In 1933 he shared the Bocher Prize for mathematicians.

He wrote and published many important mathematical papers, including *Differential Space* (1923), *Generalised Harmonic Analysis* (1930), and *Tauberian Theorems* (1932).

During World War 2, Norbert's work on gunfire control and range finding led him to discover the theory on predicting stationary time series, which in turn led to his most famous work on the concept of cybernetics. His famous book *Cybernetics or Control and Communication in the Animal and the Machine* was first published in 1948.

In 1964 Norbert was awarded the National Medal of Science. His reputation as a teacher helped to make MIT one of the world's finest centres for learning.

Chapter 21

Quadratic functions

Contents:

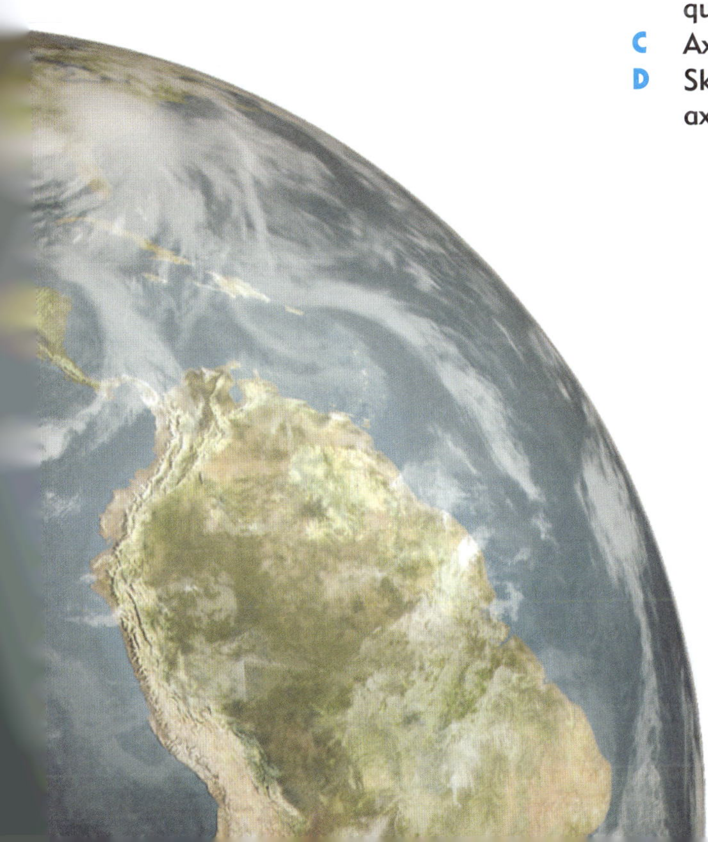

In this chapter, we study the relationship between two variables which are linked by a **quadratic function** of the form $y = ax^2 + bx + c, \quad a \neq 0$.

OPENING PROBLEM

Stunt motorcyclist Marvin is attempting a world record jump of 110 metres.

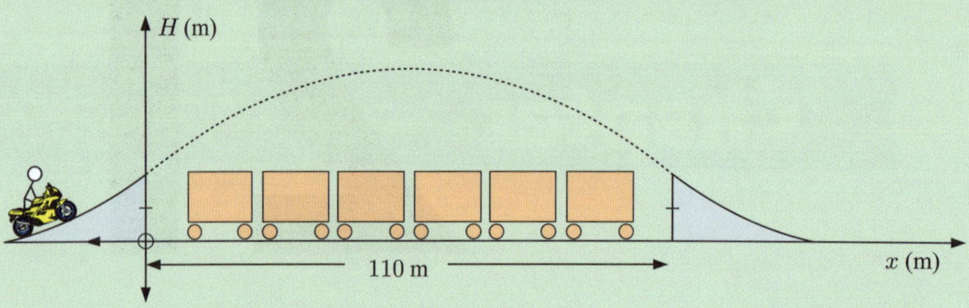

His height above ground at the point where he has travelled x metres horizontally from the take-off ramp, is given by $H = -0.009x^2 + x + 6$ metres, for $0 \leqslant x \leqslant 110$.

Things to think about:

- **a** What type of function is this?
- **b** How high is Marvin above the ground:
  - **i** the instant he leaves the take-off ramp
  - **ii** after travelling 80 metres horizontally?
- **c** Will Marvin safely reach the landing ramp?

A QUADRATIC FUNCTIONS

A **quadratic function** is a relationship between two variables which can be written in the form $\boldsymbol{y = ax^2 + bx + c}$, where x and y are the variables, and a, b, and c are constants, $a \neq 0$.

The constants a, b, and c are called the **coefficients** of the quadratic function.

For example:

- $y = 3x^2 - 2x + 1$ is a quadratic function with $a = 3, \quad b = -2, \quad c = 1$
- $y = 2x^2 + 3x$ is a quadratic function with $a = 2, \quad b = 3, \quad c = 0$.

Example 1 — Self Tutor

Explain whether each function is quadratic:

a $y = -x^2 + 3x - 4$ **b** $y = 2x^2 + 1$ **c** $y = x^3 + 2x + 7$

a Yes, with $a = -1, \quad b = 3, \quad c = -4$.
b Yes, with $a = 2, \quad b = 0, \quad c = 1$.
c No, as it has an x^3 term rather than an x^2 term.

FINDING y GIVEN x

For any given value of x, the corresponding value of y can be found by **substituting** into the function.

Example 2 — Self Tutor

Suppose $y = 2x^2 - 3x + 4$. Find the value of y when:

a $x = 2$ **b** $x = -1$

a When $x = 2$,
$$\begin{aligned} y &= 2(2)^2 - 3(2) + 4 \\ &= 8 - 6 + 4 \\ &= 6 \end{aligned}$$

b When $x = -1$,
$$\begin{aligned} y &= 2(-1)^2 - 3(-1) + 4 \\ &= 2 \times 1 + 3 + 4 \\ &= 2 + 3 + 4 \\ &= 9 \end{aligned}$$

EXERCISE 21A.1

1 State whether each function is quadratic. If it is, give values for a, b, and c. If it is not, explain why.

a $y = 2x^2 + x + 4$ **b** $y = 3x + 8$ **c** $y = -x^2 - x - 1$

d $y = x^2$ **e** $y = \frac{1}{2}x^2 - 6$ **f** $y = -2x + 5x^3$

g $y = -3x^2$ **h** $y = -x^{-2} - 4x$ **i** $2y + 4x^2 + 5 = 0$

2 Suppose $y = x^2 + 2x - 5$. Find y when:

a $x = 3$ **b** $x = 0$ **c** $x = -2$

3 Suppose $y = 3x^2 - x + 4$. Find y when:

a $x = 0$ **b** $x = 2$ **c** $x = -3$

4 Suppose $y = -2x^2 - 4x + 7$. Find y when:

a $x = 1$ **b** $x = -3$ **c** $x = \frac{1}{2}$

GRAPHS OF QUADRATIC FUNCTIONS

The graph of a quadratic function is called a **parabola**.

RESEARCH — **PARABOLIC CURVES**

We use parabolas almost every day without realising it. The parabola is the shape used in the mirror surfaces of motor vehicle headlights, torches, and satellite dishes. Parabolas are also used as arches in buildings and spans of bridges.

What to do:

1 Find out what is meant by the **focus** and **directrix** of a parabola.

2 Where does the name **parabola** come from?

3 Research the use of parabolic curves. Your work should include sketches, properties of the parabola, and reasons for using this shape rather than other shapes.

4 Research the parabolic "Whispering Wall" in the Barossa Valley, Australia.

The simplest quadratic function is $y = x^2$. Its graph can be drawn from a table of values.

x	-3	-2	-1	0	1	2	3
y	9	4	1	0	1	4	9

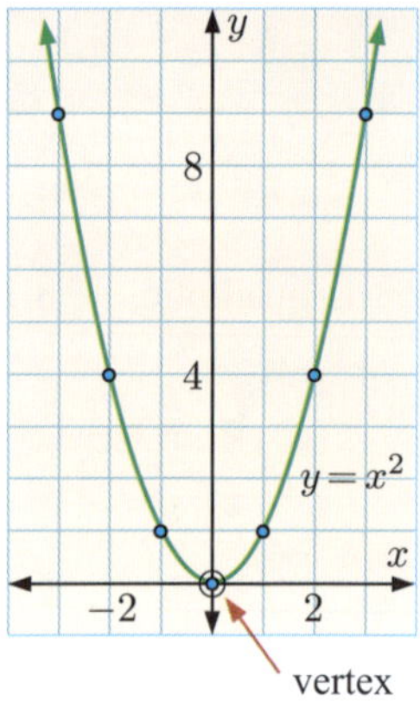

Notice that:

- The curve is a **parabola** and it opens upwards.
- There are no negative y values, since $x^2 \geqslant 0$ for all x. The curve does not go below the x-axis.
- The curve is **symmetrical** about the y-axis.
  For example: when $x = -3$, $y = (-3)^2 = 9$
  when $x = 3$, $y = 3^2 = 9$.
- The curve has a **turning point** or **vertex** at $(0, 0)$.

Example 3 **Self Tutor**

Draw the graph of $y = x^2 - 2x - 1$ from a table of values from $x = -3$ to $x = 3$.

When $x = -3$, $y = (-3)^2 - 2(-3) - 1$
$= 9 + 6 - 1$
$= 14$

When $x = -2$, $y = (-2)^2 - 2(-2) - 1$
$= 4 + 4 - 1$
$= 7$

When $x = -1$, $y = (-1)^2 - 2(-1) - 1$
$= 1 + 2 - 1$
$= 2$

When $x = 0$, $y = (0)^2 - 2(0) - 1$
$= -1$

When $x = 1$, $y = 1^2 - 2(1) - 1$
$= 1 - 2 - 1$
$= -2$

When $x = 2$, $y = 2^2 - 2(2) - 1$
$= 4 - 4 - 1$
$= -1$

When $x = 3$, $y = 3^2 - 2(3) - 1$
$= 9 - 6 - 1$
$= 2$

The tabled values are:

x	-3	-2	-1	0	1	2	3
y	14	7	2	-1	-2	-1	2

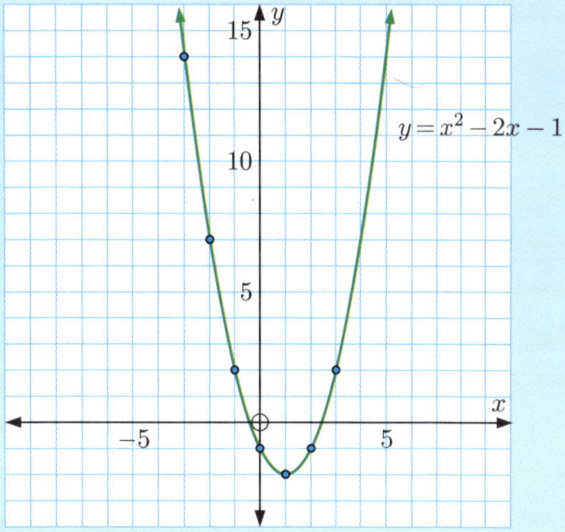

Given the coordinates of a point, we can use **substitution** to test if the point lies on the graph of a function. We substitute the x-coordinate of the point into the function, and see if the resulting value matches the y-coordinate of the point.

Example 4 **Self Tutor**

Do A(2, −8) and B(−3, 4) lie on the graph of the quadratic function $y = -x^2 - 3x + 2$?

When $x = 2$,
$$y = -(2)^2 - 3(2) + 2$$
$$= -4 - 6 + 2$$
$$= -8$$

The coordinates of A satisfy the function, so A lies on its graph.

When $x = -3$,
$$y = -(-3)^2 - 3(-3) + 2$$
$$= -9 + 9 + 2$$
$$= 2 \quad \text{which is} \neq 4$$

The coordinates of B do not satisfy the function, so B does not lie on its graph.

EXERCISE 21A.2

1 Use a table of values from $x = -3$ to $x = 3$ to draw the graph of:

a $y = x^2$ **b** $y = x^2 - 2x$ **c** $y = x^2 - 2x - 2$

d $y = 2x^2$ **e** $y = 2x^2 + 1$ **f** $y = \frac{1}{2}x^2 - 2$

g $y = -x^2$ **h** $y = -x^2 + x$ **i** $y = -x^2 + 2x + 1$

Use technology to check your answers.

GRAPHICS CALCULATOR INSTRUCTIONS

GRAPHING PACKAGE

2 Do A(3, 10) and B(−2, 13) lie on the graph of the quadratic function $y = 2x^2 - 3x - 1$?

3 Do P(0, 2), Q(2, 7), and R(−1, −4) lie on the graph of the quadratic function $y = -x^2 + 5x + 2$?

4 For the quadratic function $y = ax^2 + bx + c$, which of the coefficients a, b, or c affects:

a whether the parabola opens upwards  or downwards 

b whether the parabola is thinner  or wider 

c where the parabola cuts the y-axis?

5 **a** The point $C(2, k)$ lies on $y = x^2 + 3x - 7$. Find k.

b The point $D(-3, d)$ lies on $y = 4x - 3x^2$. Find d.

c The point $E(5, m)$ lies on $y = -x^2 + 6x - 12$. Find m.

d The point $F(-\frac{1}{2}, n)$ lies on $y = 6x^2 + x + 3$. Find n.

6 Consider the quadratic function $y = -2x^2 + 4x + 1$.

a Determine whether $(2, 1)$ and $(-2, -13)$ lie on its graph.

b The points $(0, k)$ and $(3, m)$ lie on the graph of the quadratic function. Find the values of k and m.

c Sketch the graph of the quadratic, including the information from **a** and **b**.

FINDING x GIVEN y

When we substitute a value for y, we are left with a **quadratic equation** which we need to solve for x. Since the equation is quadratic, there may be 0, 1, or 2 possible values for x for any given value of y.

We can observe this on the graph of the parabola, because if we draw a horizontal line with a particular value of y, it will meet the graph in either 0, 1, or 2 places.

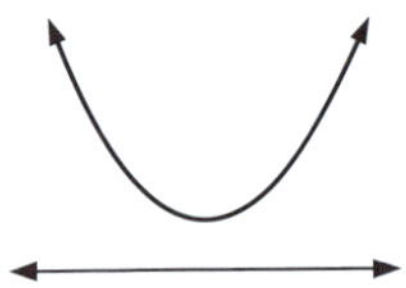
0 solutions

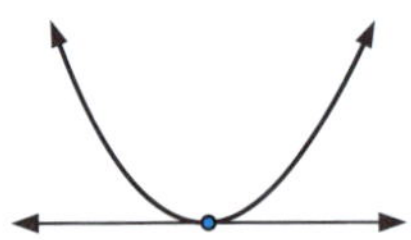
1 solution

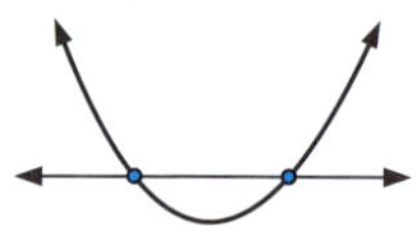
2 solutions

Example 5 — Self Tutor

Suppose $y = x^2 + 4x - 5$. Find the value(s) of x for which:

a $y = 0$

b $y = -9$

a When $y = 0$,
$$x^2 + 4x - 5 = 0$$
$$\therefore (x + 5)(x - 1) = 0$$
$$\therefore x = -5 \text{ or } x = 1$$
There are two solutions, $x = -5$ or 1.

b When $y = -9$,
$$x^2 + 4x - 5 = -9$$
$$\therefore x^2 + 4x + 4 = 0$$
$$\therefore (x + 2)^2 = 0$$
$$\therefore x = -2$$
There is only one solution, $x = -2$.

EXERCISE 21A.3

1 Suppose $y = x^2 + 2x + 3$. Find the value(s) of x for which:

a $y = 3$ **b** $y = 2$ **c** $y = 18$

2 Suppose $y = x^2 - 4x + 1$. Find the value(s) of x for which:

a $y = 6$ **b** $y = 1$ **c** $y = -3$

3 Consider the quadratic function $y = 3x^2 + 15x - 2$.

a Find the value of y when: **i** $x = 3$ **ii** $x = -2$

b Find the value(s) of x for which: **i** $y = -20$ **ii** $y = 40$

4 **a** The point A$(m, 7)$ lies on $y = x^2 + 3x - 3$. Find m.

b The point B$(n, 29)$ lies on $y = 2x^2 + 20x + 7$. Find n.

5 A cannonball is fired into the air. Its height is given by $H = 60t - 5t^2$ metres, where t is the time in seconds after its release.

a Find the height of the cannonball after:

i 1 second **ii** 3 seconds.

b Find the time(s) at which the cannonball is:

i 100 m **ii** 160 m above the ground.

c How long does it take for the cannonball to hit the ground?

6

Bjorn makes wooden display cabinets. The profit when he makes x cabinets per day is given by $P = 320x - 20x^2 - 320$ dollars.

a Calculate Bjorn's profit if he makes:

i 0 **ii** 4 **iii** 10 cabinets per day.

b How many cabinets per day must Bjorn make to receive a $460 profit?

7 Answer the **Opening Problem** on page **416**.

8 The graph of $y = x^2 - 2x - 3$ is shown alongside.

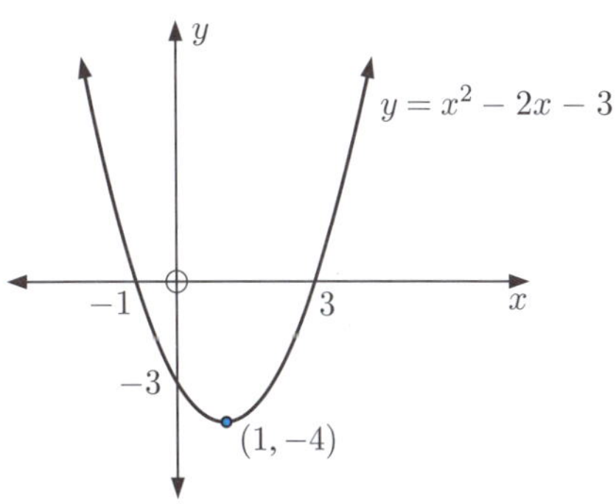

a How many points on the graph are there with y-coordinate:

i 1 **ii** -4 **iii** -6?

b Suppose $x^2 - 2x - 3 = k$. For what values of k does the equation have:

i two solutions **ii** one solution

iii no solutions?

B TRANSFORMATIONS TO SKETCH QUADRATICS

In this Section we will graph quadratic functions without finding a table of values. We do this by transforming the graph of $y = x^2$.

INVESTIGATION 1 GRAPHS OF QUADRATIC FUNCTIONS

In this Investigation you can use the **graphing package** or your **graphics calculator** to draw the graphs.

GRAPHICS CALCULATOR INSTRUCTIONS

GRAPHING PACKAGE

Part 1: Graphs of the form $y = x^2 + k$

What to do:

1 Graph the two functions on the same set of axes, and state the coordinates of the vertex of each function.

a $y = x^2$ and $y = x^2 + 1$

b $y = x^2$ and $y = x^2 - 1$

c $y = x^2$ and $y = x^2 + 3$

d $y = x^2$ and $y = x^2 - 3$

2 What effect does the value of k have on:

a the position of the graph

b the shape of the graph?

3 What transformation is needed to graph $y = x^2 + k$ from $y = x^2$?

Part 2: Graphs of the form $y = (x - h)^2$

What to do:

1 Graph the two functions on the same set of axes, and state the coordinates of the vertex of each function.

a $y = x^2$ and $y = (x - 1)^2$

b $y = x^2$ and $y = (x + 1)^2$

c $y = x^2$ and $y = (x - 3)^2$

d $y = x^2$ and $y = (x + 3)^2$

2 What effect does the value of h have on:

a the position of the graph

b the shape of the graph?

3 What transformation is needed to graph $y = (x - h)^2$ from $y = x^2$?

Part 3: Graphs of the form $y = (x - h)^2 + k$

What to do:

1 Graph the two functions on the same set of axes, and state the coordinates of the vertex of each function.

a $y = x^2$ and $y = (x - 1)^2 + 2$

b $y = x^2$ and $y = (x + 2)^2 - 3$

c $y = x^2$ and $y = (x - 3)^2 - 1$

d $y = x^2$ and $y = (x + 4)^2 + 5$

2 Copy and complete:

- The graph of $y = (x - h)^2 + k$ has the same shape as the graph of ......
- The graph of $y = (x - h)^2 + k$ is found by translating the graph of $y = x^2$ ...... units horizontally and ...... units vertically.
- The vertex of the graph of $y = (x - h)^2 + k$ is at (......, ...).

Part 4: Graphs of the form $y = ax^2$, $a \neq 0$

What to do:

1 Graph the two functions on the same set of axes.

a $y = x^2$ and $y = 2x^2$ **b** $y = x^2$ and $y = 3x^2$ **c** $y = x^2$ and $y = -x^2$

d $y = x^2$ and $y = \frac{1}{2}x^2$ **e** $y = x^2$ and $y = -2x^2$ **f** $y = x^2$ and $y = -3x^2$

2 What effect does a have on:

a the shape of the graph **b** the direction in which the graph opens?

From the **Investigation** you should have discovered:

- Graphs of the form $y = x^2 + k$ have the same shape as the graph of $y = x^2$.
  The graph of $y = x^2$ is translated k units vertically to give the graph of $y = x^2 + k$.
  - If $k > 0$, the graph moves upwards.
  - If $k < 0$, the graph moves downwards.
- Graphs of the form $y = (x-h)^2$ have the same shape as the graph of $y = x^2$.
  The graph of $y = x^2$ is translated h units horizontally to give the graph of $y = (x-h)^2$.
  - If $h > 0$, the graph moves to the right.
  - If $h < 0$, the graph moves to the left.
- Graphs of the form $y = (x-h)^2 + k$ have the same shape as the graph of $y = x^2$.
  The graph of $y = x^2$ is translated h units horizontally and k units vertically to give the graph of $y = (x-h)^2 + k$. The **vertex** is at (h, k).
- If $a > 0$, $y = ax^2$ opens upwards 

  If $a < 0$, $y = ax^2$ opens downwards 

  If $a < -1$ or $a > 1$, then $y = ax^2$ is 'thinner' than $y = x^2$.
  If $-1 < a < 1$, $a \neq 0$, then $y = ax^2$ is 'wider' than $y = x^2$.

Example 6 **Self Tutor**

Sketch each of the following functions on the same set of axes as $y = x^2$.

a $y = x^2 + 2$ **b** $y = (x-3)^2$

a We draw $y = x^2$, then translate it 2 units upwards.

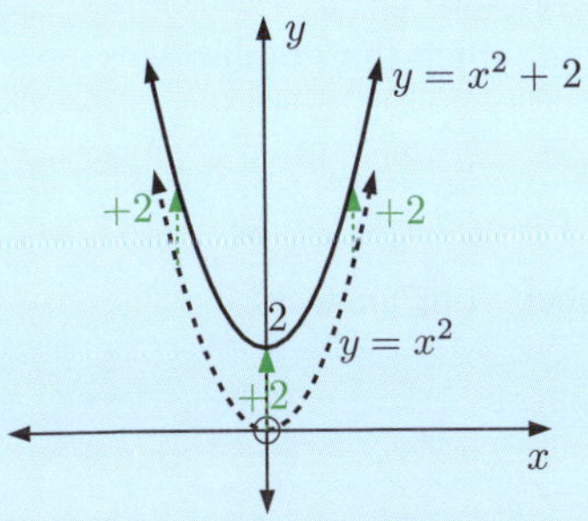

b We draw $y = x^2$, then translate it 3 units to the right.

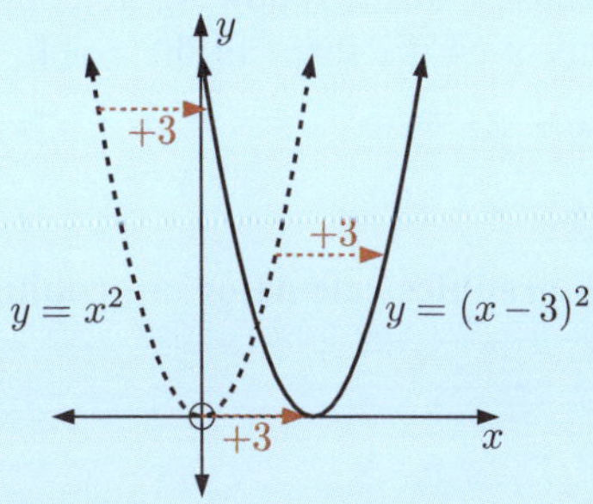

EXERCISE 21B.1

1 Sketch each of the following functions on the same set of axes as $y = x^2$. Use a separate set of axes for each part.

GRAPHING PACKAGE

a $y = x^2 + 1$ **b** $y = x^2 + 3$ **c** $y = x^2 - 1$
d $y = x^2 - 4$ **e** $y = x^2 + 6$ **f** $y = x^2 - 5$

Use a **graphics calculator** or **graphing package** to check your answers.

2 Sketch each of the following functions on the same set of axes as $y = x^2$. Use a separate set of axes for each part.

a $y = (x-1)^2$ **b** $y = (x+2)^2$ **c** $y = (x-4)^2$
d $y = (x+3)^2$ **e** $y = (x+4)^2$ **f** $y = (x-5)^2$

Use a **graphics calculator** or **graphing package** to check your answers.

3 State the coordinates of the vertex of the quadratic function:

a $y = x^2 + 2$ **b** $y = (x+1)^2$ **c** $y = x^2 - 3$ **d** $y = (x-6)^2$

Example 7 Self Tutor

Sketch each of the following functions on the same set of axes as $y = x^2$.

a $y = 2x^2$ **b** $y = -2x^2$

a $y = 2x^2$ is 'thinner' than $y = x^2$.

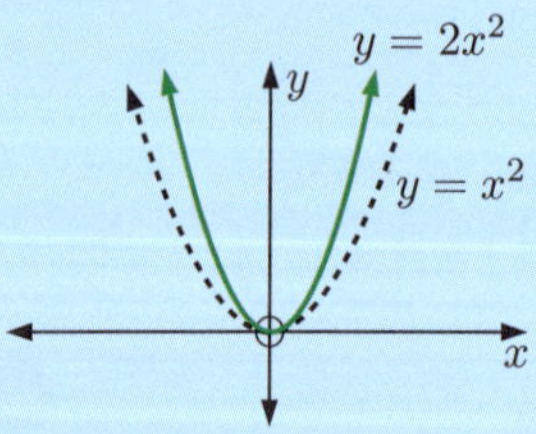

b $y = -2x^2$ has the same shape as $y = 2x^2$, but opens downwards.

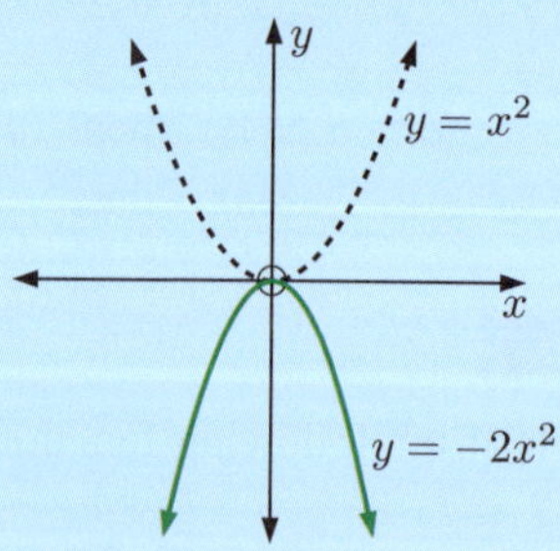

4 Sketch each of the following functions on the same set of axes as $y = x^2$. Comment on the shape of the graph, and the direction in which the graph opens.

a $y = 3x^2$ **b** $y = -3x^2$ **c** $y = \frac{1}{2}x^2$
d $y = -\frac{1}{2}x^2$ **e** $y = -x^2$ **f** $y = \frac{1}{5}x^2$

Use a **graphics calculator** or **graphing package** to check your answers.

Example 8 **Self Tutor**

Sketch each of the following functions on the same set of axes as $y = x^2$. In each case state the coordinates of the vertex.

a $y = (x-3)^2 - 2$ **b** $y = (x+1)^2 + 4$

a We draw $y = x^2$, then translate it 3 units to the right and 2 units downwards.

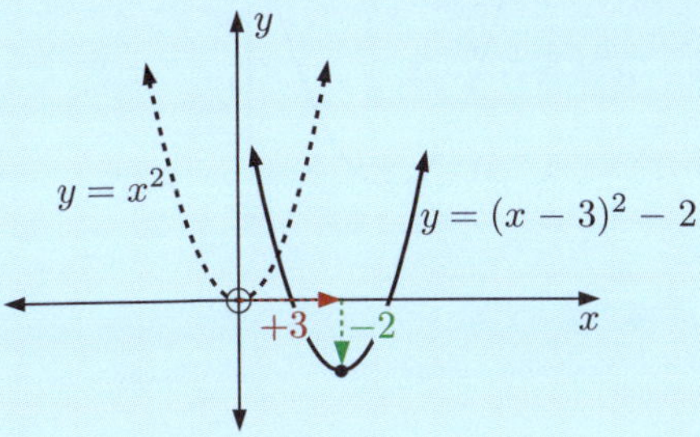

The vertex is at $(3, -2)$.

b We draw $y = x^2$, then translate it 1 unit to the left and 4 units upwards.

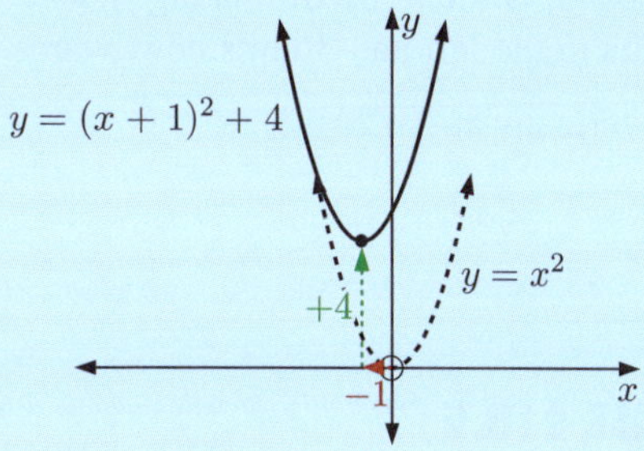

The vertex is at $(-1, 4)$.

5 Sketch each of the following functions on the same set of axes as $y = x^2$. In each case state the coordinates of the vertex.

a $y = (x-2)^2 + 1$ **b** $y = (x+3)^2 + 4$ **c** $y = (x-1)^2 - 3$

d $y = (x+4)^2 - 2$ **e** $y = (x-4)^2 - 4$ **f** $y = (x+5)^2 + 1$

COMPLETING THE SQUARE

In **Chapter 18** we used the process of completing the square to help solve quadratic equations which did not factorise.

This process can be used to convert quadratic functions such as $y = x^2 - 2x + 5$ into the form $y = (x-h)^2 + k$, and hence make them easier to graph.

Consider $y = x^2 - 2x + 5$

$\therefore\ y = \underbrace{x^2 - 2x + 1^2} + 5 - 1^2$

$\therefore\ y = (x-1)^2 + 4$

So, the graph of $y = x^2 - 2x + 5$ can be found by translating the graph of $y = x^2$ 1 unit to the right and 4 units upwards.

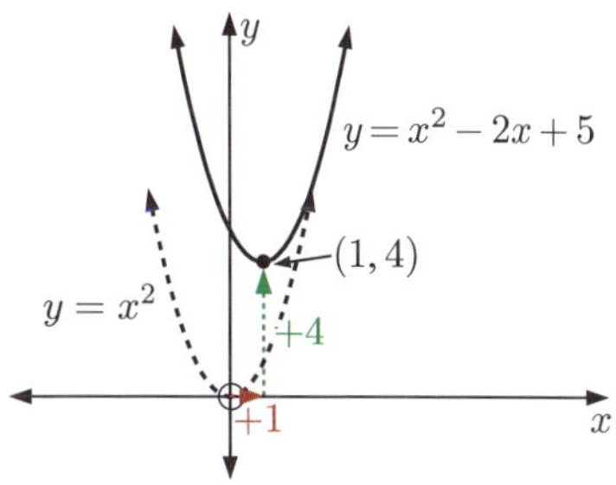

Example 9 **Self Tutor**

Write $y = x^2 + 4x + 1$ in the form $y = (x - h)^2 + k$. Hence, sketch the graph of $y = x^2 + 4x + 1$, stating the coordinates of the vertex.

$$y = x^2 + 4x + 1$$
$$\therefore\ y = x^2 + 4x + 2^2 + 1 - 2^2$$
$$\therefore\ y = (x + 2)^2 - 3$$

The graph is found by translating $y = x^2$ 2 units to the left and 3 units downwards.

The vertex is at $(-2, -3)$.

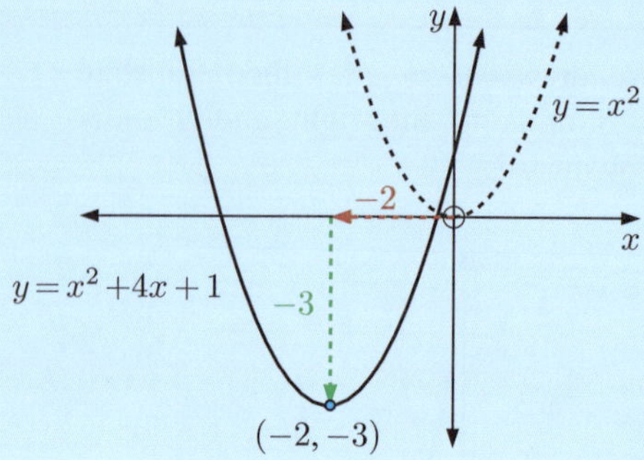

EXERCISE 21B.2

1 Write each of the following quadratic functions in the form $y = (x - h)^2 + k$. Hence, sketch the function, stating the coordinates of its vertex.

a $y = x^2 + 2x + 2$ **b** $y = x^2 - 4x + 2$ **c** $y = x^2 + 2x - 3$

d $y = x^2 - 6x + 1$ **e** $y = x^2 + 2x$ **f** $y = x^2 - 4x$

g $y = x^2 + x - 1$ **h** $y = x^2 - 3x + 2$ **i** $y = x^2 + 5x - 5$

GRAPHING PACKAGE

Use a **graphics calculator** or **graphing package** to check your answer.

C AXES INTERCEPTS

Suppose we are given a function and its graph.

- An ***x*-intercept** is a value of x where the graph meets the x-axis.
  x-intercepts are found by letting y be 0 in the equation of the function.
- A ***y*-intercept** is a value of y where the graph meets the y-axis.
  y-intercepts are found by letting x be 0 in the equation of the function.

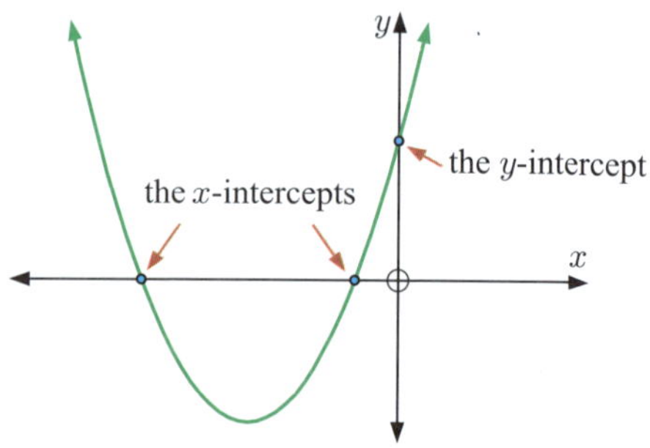

INVESTIGATION 2 AXES INTERCEPTS

In this Investigation we will discover the connection between a quadratic function and the axes intercepts of its graph.

What to do:

GRAPHING PACKAGE

1 For each of the following quadratic functions, use a **graphing package** or **graphics calculator** to:

i draw the graph **ii** find the y-intercept.

a $y = x^2 - 3x + 2$ **b** $y = -x^2 + x + 5$ **c** $y = x^2 - 3x$
d $y = 2x^2 - 4x - 7$ **e** $y = -3x^2 - x + 4$ **f** $y = 4x^2 - 6$

2 Using your observations from **1**, state the y-intercept of a quadratic function in the form $y = ax^2 + bx + c$.

3 For each of the following quadratic functions, use a **graphing package** or **graphics calculator** to:

i draw the graph **ii** find any x-intercepts.

a $y = (x-1)(x-3)$ **b** $y = (x-2)(x-7)$ **c** $y = -(x+4)(x-5)$
d $y = 2(x+3)(x-3)$ **e** $y = -3x(x+5)$ **f** $y = 5(x-2)(x-6)$
g $y = (x-4)^2$ **h** $y = -(x+6)^2$ **i** $y = 3(x-2)^2$

4 Using your observations in **3**:

a State the x-intercepts of a quadratic function of the form $y = a(x-\alpha)(x-\beta)$.

b Comment on the x-intercepts of quadratic functions in the form $y = a(x-\alpha)^2$.

THE y-INTERCEPT

For a quadratic function of the form $y = ax^2 + bx + c$, the y-intercept is the constant term c.

Proof: Consider the function $y = ax^2 + bx + c$.
If we let $x = 0$ then $y = a(0)^2 + b(0) + c = c$
$\therefore$ the y-intercept is c.

THE x-INTERCEPTS

For a quadratic function of the form $y = a(x-\alpha)(x-\beta)$, the x-intercepts are α and β.

Proof: Consider the function $y = a(x-\alpha)(x-\beta)$.
If we let $y = 0$ then $a(x-\alpha)(x-\beta) = 0$
$\therefore$ $x - \alpha = 0$ or $x - \beta = 0$
$\therefore$ $x = \alpha$ or β {Null Factor law}
$\therefore$ the x-intercepts are α and β.

x-intercepts are therefore easy to find when the quadratic is in **factorised** form.

Example 10 Self Tutor

Find the x-intercepts of:

a $y = (x+1)(x-2)$ **b** $y = -(x-2)^2$

a When $y = 0$, $(x+1)(x-2) = 0$
$\therefore$ $x = -1$ or $x = 2$
$\therefore$ the x-intercepts are -1 and 2.

b When $y = 0$, $-(x-2)^2 = 0$
$\therefore$ $x = 2$
$\therefore$ the x-intercept is 2.

FACTORISING TO FIND x-INTERCEPTS

For any quadratic function of the form $y = ax^2 + bx + c$, the x-intercepts can be found by solving the equation $ax^2 + bx + c = 0$.

Since the quadratic equation may have *two solutions*, *one solution*, or *no solutions*, the graph of the corresponding quadratic function will have *two x-intercepts*, *one x-intercept*, or *no x-intercepts*.

We can use factorisation to solve the equation and hence find the x-intercepts.

Example 11 — Self Tutor

Find the x-intercept(s) of the quadratic functions:

a $y = x^2 + 4x + 4$ **b** $y = -x^2 + 2x + 3$

a When $y = 0$, $x^2 + 4x + 4 = 0$

$\therefore (x+2)^2 = 0$

$\therefore x = -2$

$\therefore$ the x-intercept is -2.

b When $y = 0$, $-x^2 + 2x + 3 = 0$

$\therefore x^2 - 2x - 3 = 0$

$\therefore (x-3)(x+1) = 0$

$\therefore x = 3$ or $x = -1$

$\therefore$ the x-intercepts are -1 and 3.

EXERCISE 21C

1 State the y-intercept of each of the following functions:

a $y = x^2 + x + 1$ **b** $y = x^2 - 2x + 3$ **c** $y = 3x^2 + 4x - 2$

d $y = -x^2 - x + 4$ **e** $y = -2x^2 + x + 6$ **f** $y = 4x^2 - 2x + 1$

g $y = 4 + x^2$ **h** $y = 3 + x - 5x^2$ **i** $y = 2x - x^2 + 7$

2 Find the y-intercept of each of the following functions:

a $y = (x+1)(x-2)$ **b** $y = (x-1)(x+3)$ **c** $y = -(x+2)(x-1)$

d $y = x(x-3)$ **e** $y = -(x-4)(x+1)$ **f** $y = -(x-5)(x-2)$

3 For each of the following functions, find the x-intercepts:

a $y = (x-1)(x+3)$ **b** $y = -(x+2)(x-3)$ **c** $y = (x-3)(x+7)$

d $y = 3(x-2)(x-5)$ **e** $y = (x+5)^2$ **f** $y = -2(x-7)^2$

4 For each of the following functions, find the x-intercepts:

a $y = x^2 - 4$ **b** $y = x^2 - 9$ **c** $y = x^2 - 7x + 12$

d $y = x^2 - x - 6$ **e** $y = 3x - x^2$ **f** $y = -x^2 + 7x - 10$

g $y = x^2 - 6x + 9$ **h** $y = 2x^2 + 4x + 2$ **i** $y = -x^2 - 5x + 6$

j $y = 3x^2 + 12x - 15$ **k** $y = -x^2 + 6x + 16$ **l** $y = -2x^2 - 8x + 42$

m $y = 2x^2 + 7x - 4$ **n** $y = 3x^2 - 5x - 2$ **o** $y = 4x^2 - 8x - 5$

Check your answers using technology.

GRAPHING PACKAGE

D SKETCHING QUADRATICS FROM THEIR AXES INTERCEPTS

Once we have found the axes intercepts of a quadratic function, we can use them to sketch the graph of the function.

Example 12 — Self Tutor

Consider the quadratic function $y = -(x+1)(x-3)$.

a Find the x-intercepts. **b** Find the y-intercept. **c** Hence, sketch the function.

$y = -(x+1)(x-3)$

a When $y = 0$, $-(x+1)(x-3) = 0$

$\therefore x = -1$ or $x = 3$

$\therefore$ the x-intercepts are -1 and 3.

b When $x = 0$, $y = -(1)(-3) = 3$

$\therefore$ the y-intercept is 3.

c

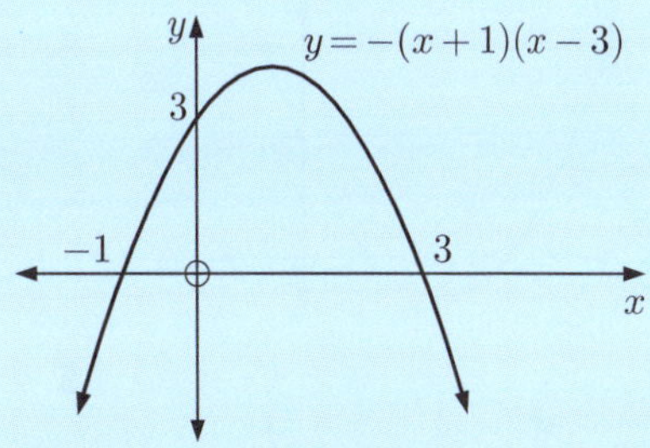

Example 13 — Self Tutor

Consider the quadratic function $y = (x-2)^2$.

a Find the x-intercepts. **b** Find the y-intercept. **c** Hence, sketch the function.

$y = (x-2)^2$

a When $y = 0$, $(x-2)^2 = 0$

$\therefore x = 2$

$\therefore$ the x-intercept is 2.

b When $x = 0$, $y = (-2)^2 = 4$

$\therefore$ the y-intercept is 4.

c

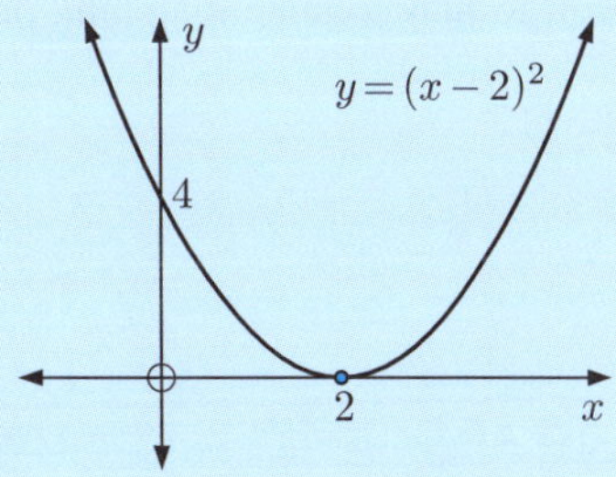

EXERCISE 21D

1 Sketch a graph of the quadratic function which has:

- **a** x-intercepts -2 and 3, and y-intercept -5
- **b** x-intercepts -4 and 1, and y-intercept 6
- **c** x-intercepts 2 and 3, and y-intercept 4
- **d** x-intercept -2, and y-intercept 7
- **e** x-intercepts -8 and -3, and y-intercept -10
- **f** x-intercept 4, and y-intercept -12.

2 For each of the following quadratic functions:

i find the x-intercepts **ii** find the y-intercept **iii** sketch the function.

a $y=(x-1)(x-3)$ **b** $y=-(x-1)(x-5)$ **c** $y=(x+2)(x-4)$

d $y=(x-1)^2$ **e** $y=3(x+2)(x-3)$ **f** $y=-2(x+4)(x-2)$

g $y=-\frac{1}{3}(x+3)^2$ **h** $y=x^2+3x-10$ **i** $y=-x^2-4x-4$

j $y=x^2+7x+6$ **k** $y=2x^2-20x+50$ **l** $y=3x^2+4x-4$

3 Match each of the following quadratic functions with its graph:

a $y=(x+5)(x-3)$ **b** $y=-(x+3)^2$ **c** $y=x(x-4)$

d $y=2x^2+14x+20$ **e** $y=-x^2-2x+15$ **f** $y=3x^2-18x+15$

A

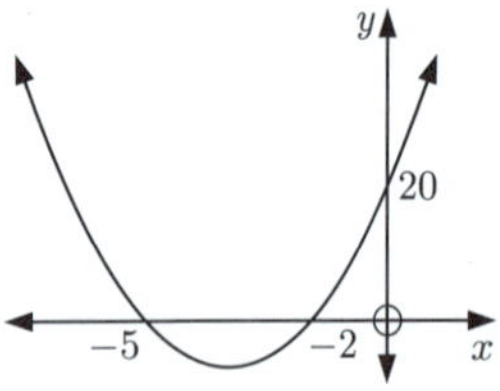

B

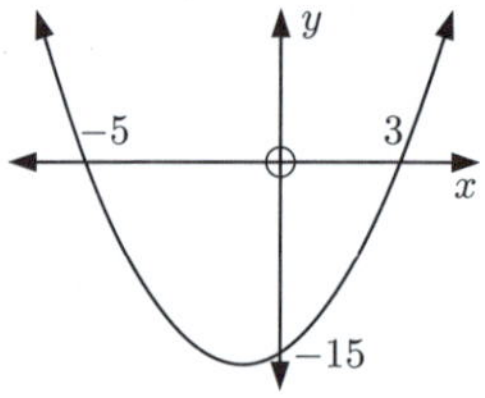

C

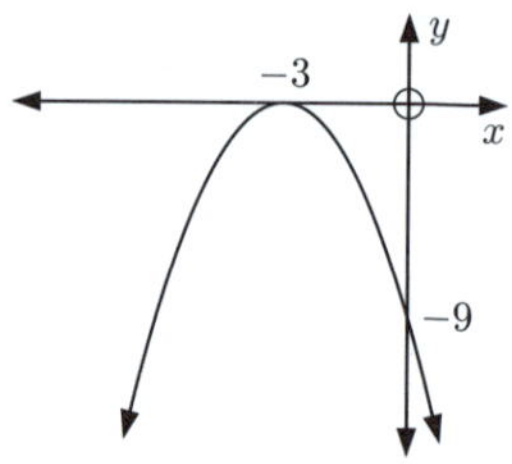

D

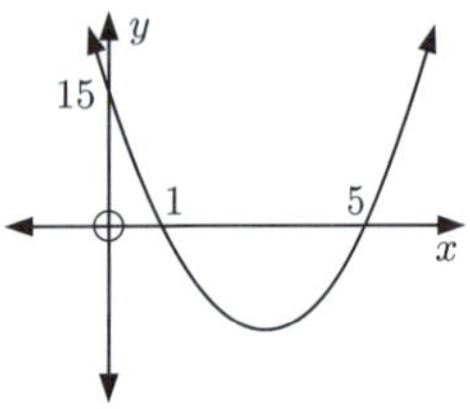

E

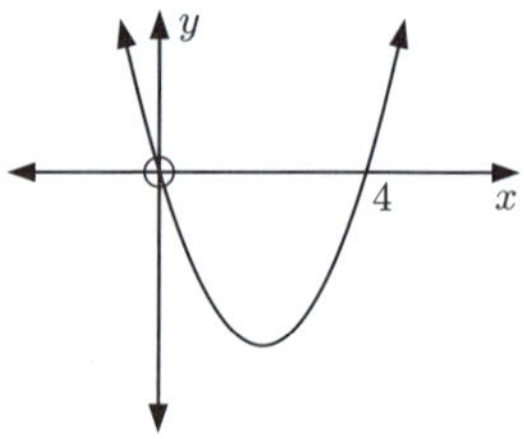

F

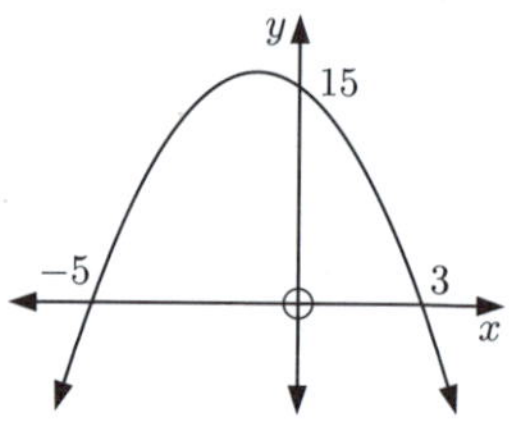

ACTIVITY — MATCHING QUADRATICS

Click on the icon to play a game matching properties of quadratics to their graphs.

QUADRATICS GAME

REVIEW SET 21A

1 The point $(t, -2)$ lies on the graph of the function $y=x^2-3x-2$. Find t.

2 Consider the quadratic function $y=4x^2-3x+2$. Find the value of y when:

a $x=-2$ **b** $x=1$ **c** $x=3$

3 A manufacturer of cricket bats finds that the profit for making x bats per day is given by the formula $P=120x-4x^2$ dollars.

a Calculate the profit when:

i 7 bats are made **ii** 20 bats are made.

b How many bats per day need to be made in order to:

i break even **ii** make a \$500 profit?

4 Using a table of values from $x = -3$ to $x = 3$, draw the graph of $y = x^2 - 2x + 3$.

5 Sketch each of the following functions on the same set of axes as $y = x^2$. In each case state the coordinates of the vertex.

a $y = x^2 - 2$ **b** $y = (x+1)^2$

6 Sketch each of the following functions on the same set of axes as $y = x^2$. Comment on the shape of the graph, and the direction in which the graph opens.

a $y = 4x^2$ **b** $y = -\frac{1}{3}x^2$

7 **a** Write $y = x^2 - 4x + 3$ in the form $y = (x-h)^2 + k$.

b Hence, sketch the graph of $y = x^2 - 4x + 3$, stating the coordinates of the vertex.

8 Find the x-intercepts of:

a $y = (x+2)(x-3)$ **b** $y = -x^2 + x + 12$

9 Find the y-intercept of each of the following functions:

a $y = -2(x-5)(x+1)$ **b** $(2x-3)^2$

10 Consider the quadratic function $y = (x+3)(x-2)$.

a Find the x-intercepts. **b** Find the y-intercept.

c Hence, sketch the graph of the function.

REVIEW SET 21B

1 The point $(a, 3)$ lies on the graph of the function $y = x^2 + 3x + 5$. Find a.

2 Suppose $y = x^2 - 4x + 14$. Find:

a y when $x = 3$ **b** the value(s) of x for which $y = 19$.

3 Using a table of values from $x = -3$ to $x = 3$, draw the graph of $y = -x^2 - x + 2$.

4 Sketch each of the following functions on the same set of axes as $y = x^2$. In each case state the coordinates of the vertex.

a $y = (x-3)^2$ **b** $y = x^2 - 6$ **c** $y = -5x^2$

5 A ball is hit directly upwards. Its height above the ground is given by $H = 20t - 5t^2$ m, where t is the time in seconds after the ball is hit. Find the time at which the ball is:

a 15 m **b** 20 m above the ground.

6 Sketch each of the following functions on the same set of axes as $y = x^2$. In each case state the coordinates of the vertex.

a $y = (x+2)^2 + 5$ **b** $y = (x-5)^2 - 3$

7 Find the y-intercept of the quadratic function:

a $y = -2x(3-x)$ **b** $y = (x-7)(x+3)$

8 **a** Write $y = x^2 - x + 3$ in the form $y = (x-h)^2 + k$.

b Hence, sketch the graph of $y = x^2 - x + 3$, stating the coordinates of the vertex.

9 Find the x-intercepts of:

a $y = -(x+3)(x-6)$

b $y = 2x^2 + x - 6$

10 Consider the quadratic function $y = 6x^2 - 11x - 10$.

a Find the y-intercept.

b Find the x-intercepts.

c Hence, sketch the graph of the function.

Chapter 22

Deductive geometry

Contents:

OPENING PROBLEM

GEOMETRY PACKAGE

Kelly draws many different figures with the 'star' shape alongside. The figures vary in shape and size.

For each figure she draws, Kelly measures the five angles of the star with a protractor. She adds them together and always gets an answer of approximately 180°.

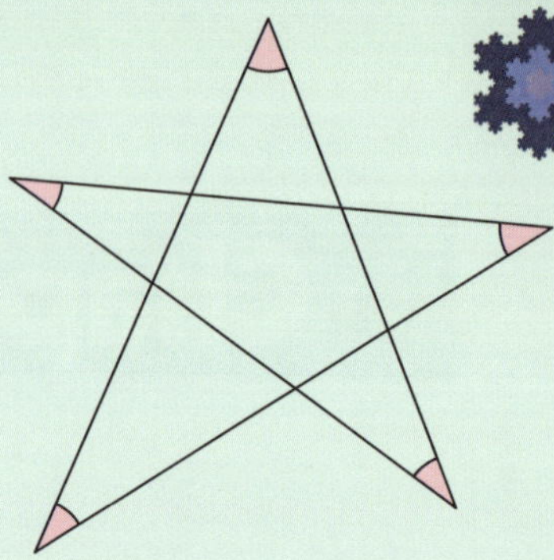

Kelly therefore *conjectures* that they actually do add to 180°.

Things to think about:

- **a** What do we mean by a conjecture?
- **b** Can you prove that Kelly's conjecture is true?

Simple geometrical figures often have very interesting and useful properties. In this chapter we will discover some of these properties and we will prove that they are true.

HISTORICAL NOTE

Around 300 BC, **Euclid of Alexandria** developed his theory of **Euclidean geometry**.

His thirteen books, collectively called the *Elements*, included definitions, theorems, and proofs based on self-evident truths called **axioms**.

They formed the basis of all geometry until the 19th century, at which time mathematicians such as **Gauss**, **Lobachevski**, and **Bolyai** challenged one of the Euclidean axioms.

This challenge led to major breakthroughs in other branches of mathematics, and enabled **Albert Einstein** to produce his *Theory of Relativity* early last century.

Albert Einstein

ACTIVITY — PAPPUS' THEOREM

In this Activity you will discover Pappus' theorem, named after **Pappus of Alexandria** (290 - 350).

What to do:

GEOMETRY PACKAGE

1. Draw two long straight lines on a sheet of paper. They do not have to be parallel, but do not let them intersect on the page.
2. Mark any three points A, B, and C on one line, and three points D, E, and F on the other line.

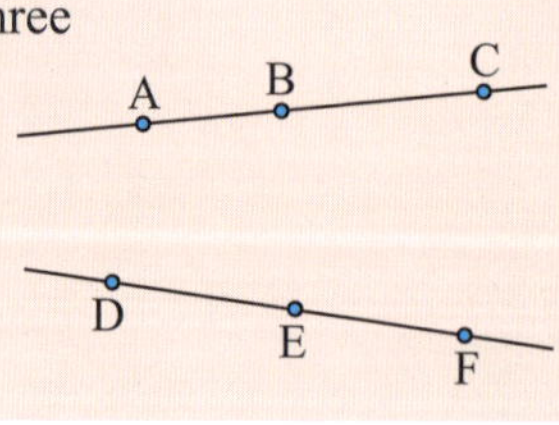

3 **a** Draw lines (AE) and (BD), and mark their intersection point X.

b Draw lines (AF) and (CD), and mark their intersection point Y.

c Draw lines (BF) and (CE), and mark their intersection point Z.

4 What do you notice about the points X, Y, and Z?

5 Repeat these steps with another pair of straight lines. Comment on your observations. Does it matter if your chosen lines intersect?

From previous years you should be familiar with the following geometric facts:

Name	*Statement*	*Figure*
Angles at a point	The sum of the sizes of the angles at a point is 360°.	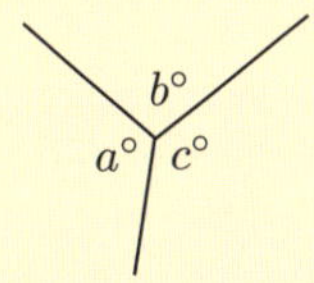 $a+b+c=360$
Angles on a line	The sum of the sizes of the angles on a line is 180°.	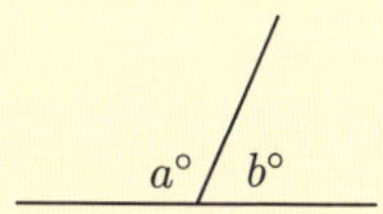 $a+b=180$
Vertically opposite angles	Vertically opposite angles are equal.	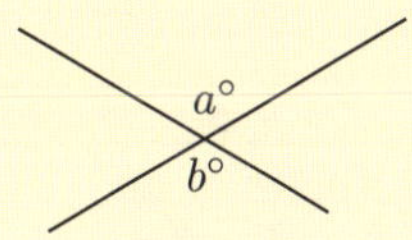 $a=b$
Corresponding angles	When two **parallel** lines are cut by a third line, the angles in corresponding positions are equal.	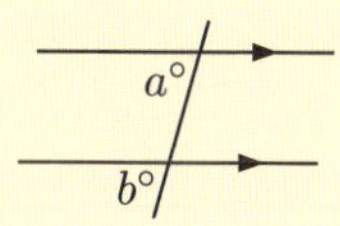 $a=b$
Alternate angles	When two **parallel** lines are cut by a third line, the angles in alternate positions are equal.	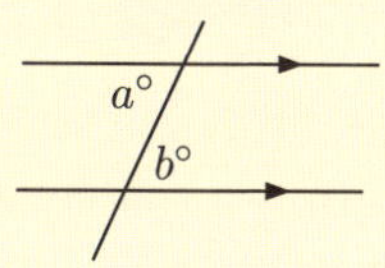 $a=b$
Co-interior or Allied angles	When two **parallel** lines are cut by a third line, the angles in co-interior positions are *supplementary*. This means they add to 180°.	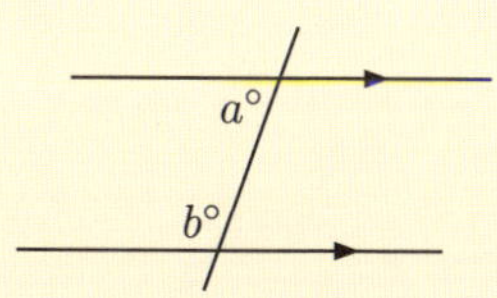 $a+b=180$

Name	*Statement*	*Figure*
Angles of a triangle	The sum of the sizes of the interior angles of a triangle is 180°.	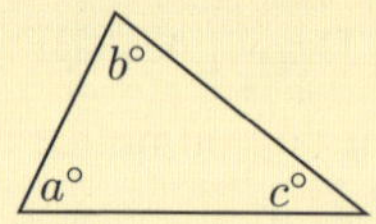 $a+b+c=180$
Exterior angle of a triangle	The size of an exterior angle of a triangle is equal to the sum of the interior opposite angles.	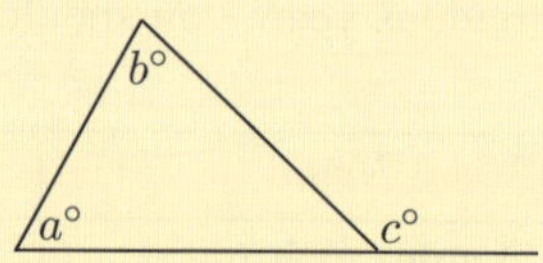 $c=a+b$
Isosceles triangle	In an isosceles triangle: • base angles are equal • the line joining the apex to the midpoint of the base is perpendicular to the base and bisects the angle at the apex.	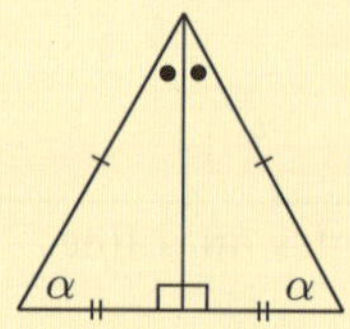
Equal angles of a triangle	If a triangle has two equal angles then the triangle is isosceles.	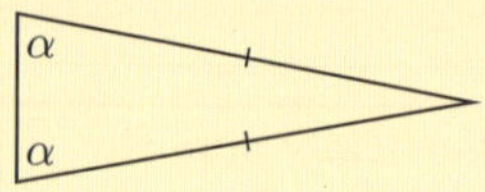
Angles of an n-sided polygon	The sum of the sizes of the interior angles of an n-sided polygon is $(n-2)\times 180°$.	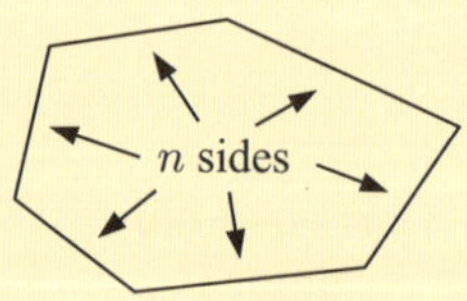 sum of angles $=(n-2)\times 180°$

A DEDUCTIVE GEOMETRY

In geometrical figures where one or more angles or lengths are unknown, we can use **theorems** to find the value of the unknowns.

When using a theorem, we should state the theorem's name as a reason to justify a particular step in the argument.

Example 1 — Self Tutor

Find the value of x:

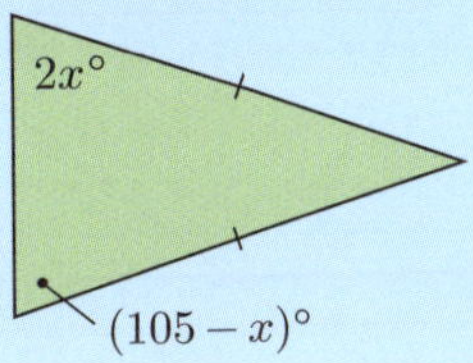

Two sides of the triangle have equal lengths, so the triangle is isosceles.

$\therefore \quad 2x = 105 - x$ {isosceles triangle}

$\therefore \quad 3x = 105$

$\therefore \quad x = 35$

Sometimes we need to draw an additional line or lines so that we can solve a problem. These lines are called **constructions**.

Example 2 — Self Tutor

Find the value of a:

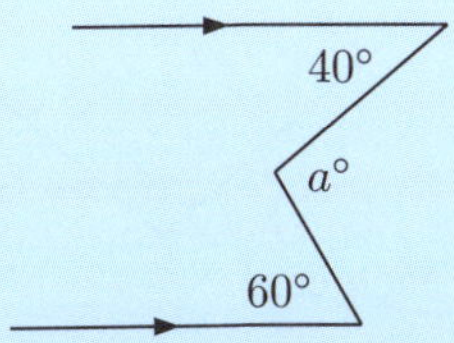

We label the figure and draw a construction line [CX].

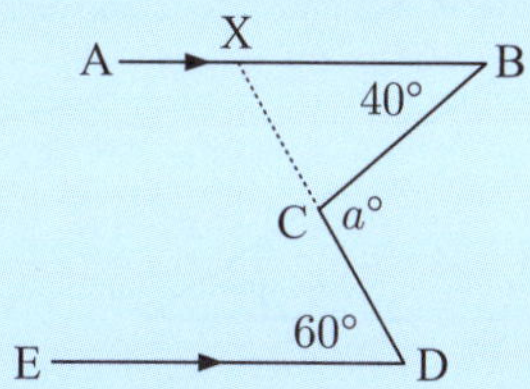

$B\widehat{X}D = X\widehat{D}E$ {alternate angles}

$\therefore \quad B\widehat{X}D = 60°$

$\therefore \quad a = 60 + 40$ {exterior angle of a $\triangle$}

$\therefore \quad a = 100$

GEOMETRIC PROOF

In **deductive geometry**, we use logical reasoning, and theorems we already know to be true, to prove that certain observations about geometrical figures are correct.

Example 3 — Self Tutor

Prove that the angle bisectors of angles ABD and CBD form a right angle.

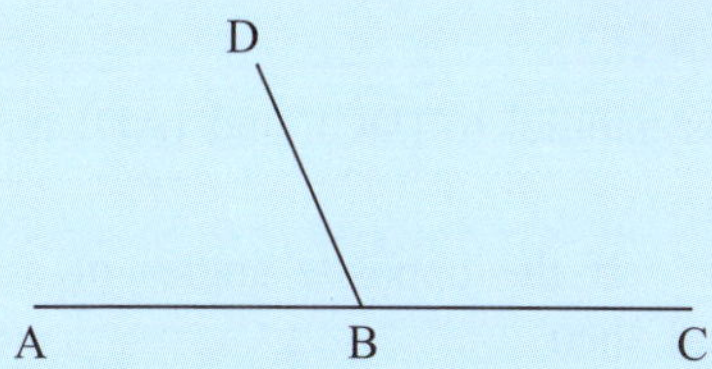

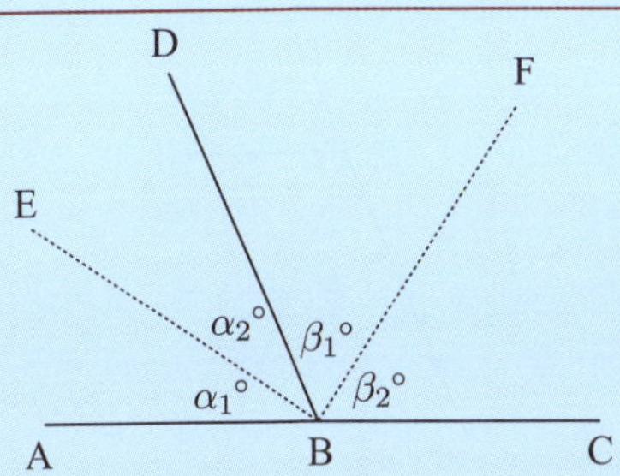

We draw angle bisectors [BE] and [BF].

$\therefore \quad \alpha_1 = \alpha_2$ and $\beta_1 = \beta_2$

But $2\alpha + 2\beta = 180$ {angles on a line}

$\therefore \quad \alpha + \beta = 90$

$\therefore \quad E\widehat{B}F = 90°$

EXERCISE 22A

1 Find the values of the unknowns:

a

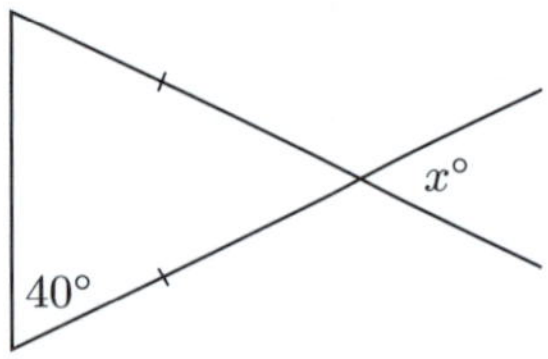

b

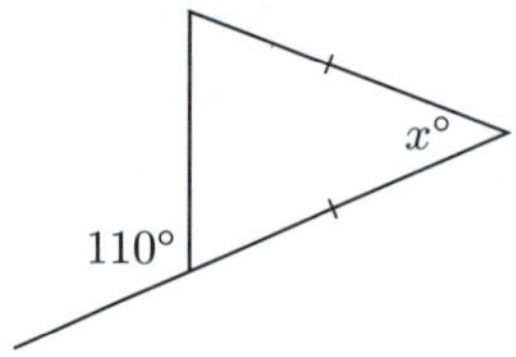

c

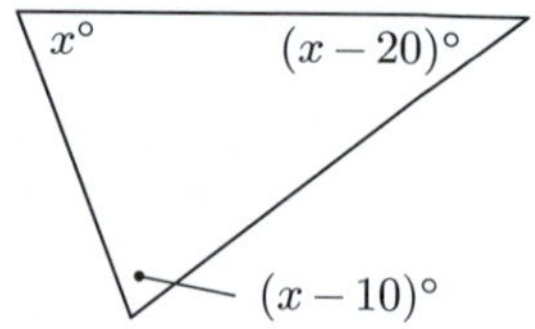

d

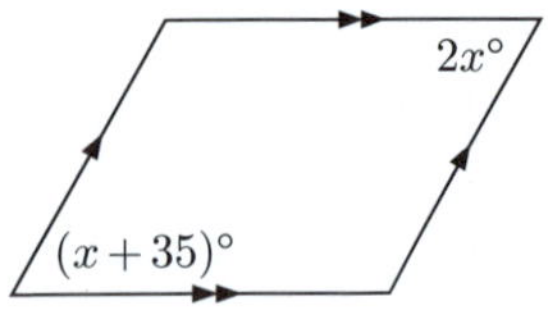

e

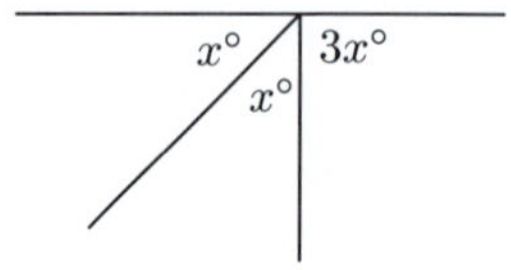

f

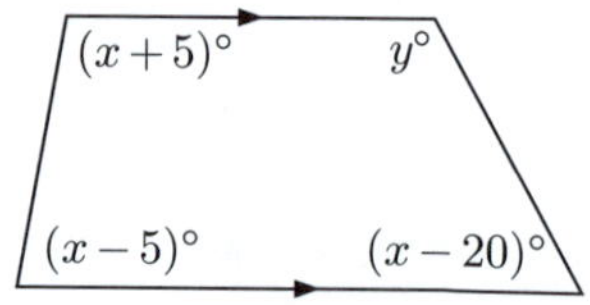

g

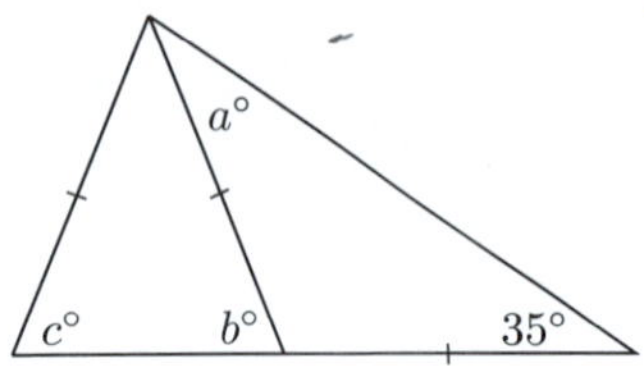

h

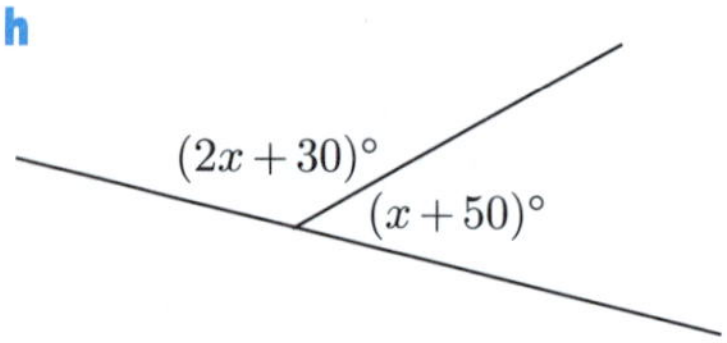

i

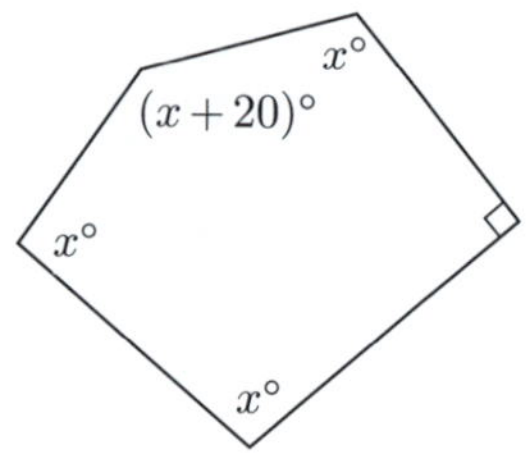

2 Find a in each figure. You will need to construct an additional line in each case.

a

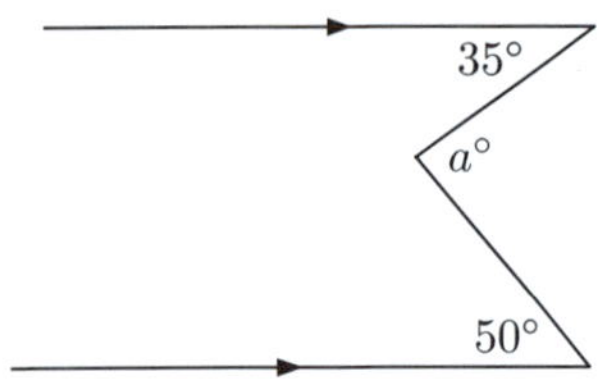

b

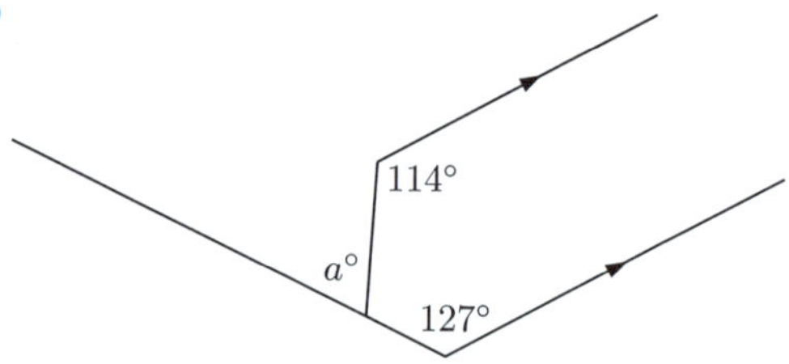

3 The opposite angles of this quadrilateral are equal in size.

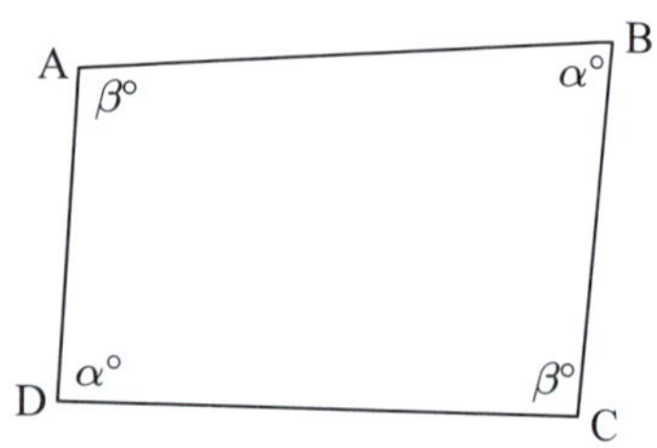

a Find the value of $2\alpha + 2\beta$, giving reasons for your answer.

b Find the value of $\alpha + \beta$.

c Explain why [AB] is parallel to [DC], and [AD] is parallel to [BC].

d Copy and complete: "If the opposite angles of a quadrilateral are equal, then ......"

4 ABC is an isosceles triangle in which AB = AC. [BA] is extended to D, and [AE] is drawn parallel to [BC]. Let $A\widehat{B}C = x°$.

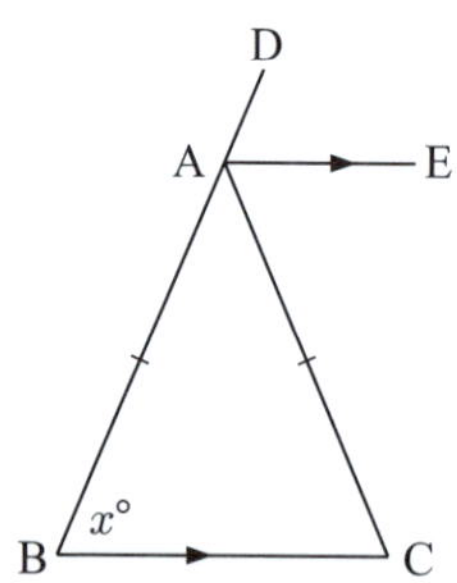

a Find, in terms of x:

i $A\widehat{C}B$ **ii** $C\widehat{A}E$ **iii** $D\widehat{A}E$

b Copy and complete:

"The line segment [AE] ...... $C\widehat{A}D$."

5 In the given figure, [AB] and [ED] are parallel. By adding a construction line, show that $B\widehat{C}D = (a+b)^\circ$.

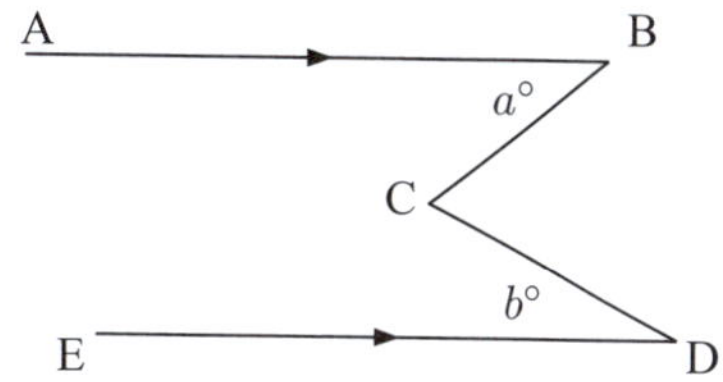

6 ABCD is a rhombus. M is the point on [AD] such that [CM] bisects $A\widehat{C}D$.
Show that $D\widehat{M}C$ is 3 times larger than $A\widehat{C}M$.

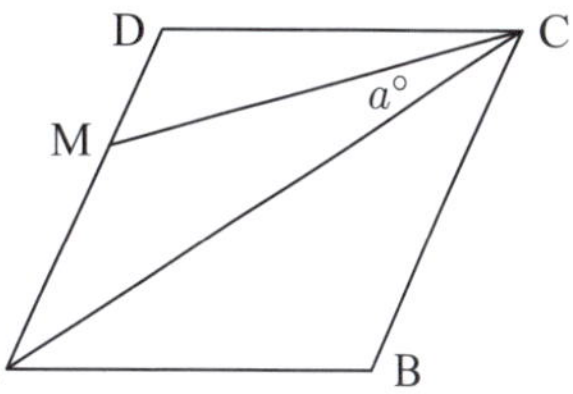

7 [AB] is the diameter of a semi-circle with centre O. P is any point on the semi-circle. Prove that $P\widehat{O}B$ is twice as large as $P\widehat{A}B$.

8 Triangle ABC is isosceles with AB = AC. [CB] is extended to D, and [AB] is extended to E such that BE = DE. Show that [DE] and [AC] are parallel.

9

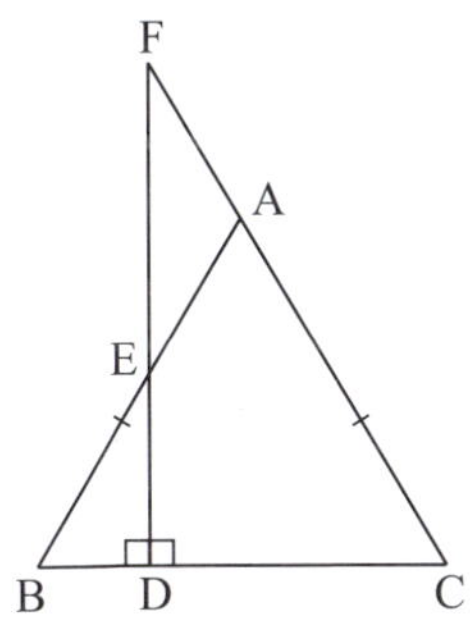

ABC is an isosceles triangle in which AB = AC. [DE] is perpendicular to [BC] and is produced to meet [CA] produced at F.
Show that triangle AEF is also isosceles.

10 Consider the **Opening Problem** on page **434**.
Prove that the five angles of the star add to 180°.

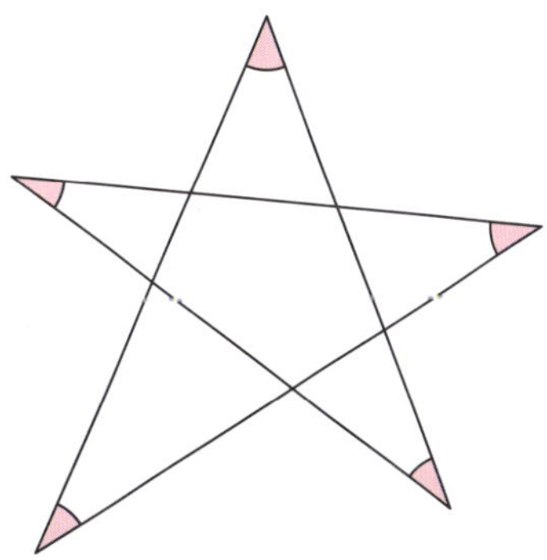

B CIRCLE THEOREMS

Before we can talk about the properties and theorems of circles, we need to learn the appropriate language for describing them.

- A **circle** is the set of all points which are equidistant from a fixed point called the **centre**.
- A **semi-circle** is a half of a circle.
- A **chord** of a circle is a line segment joining any two points on the circle.

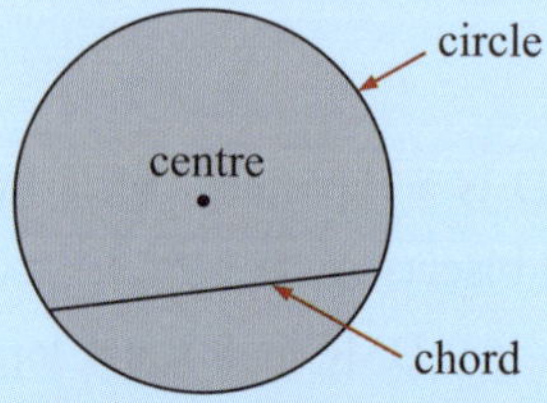

- A **diameter** of a circle is any chord passing through its centre.
- A **radius** of a circle is any line segment joining its centre to any point on the circle.

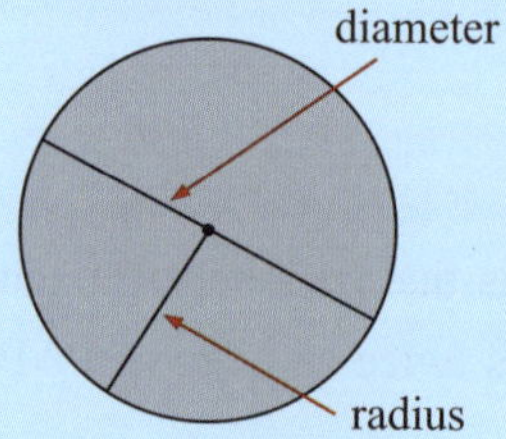

- A **tangent** to a circle is any line which touches the circle in exactly one point.

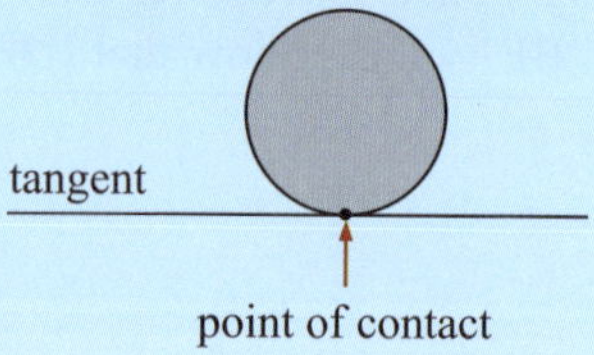

INVESTIGATION 1 CIRCLE THEOREMS

This Investigation is best attempted using the **geometry package**. However, it can also be done using a compass, ruler, and protractor.

Part 1: The angle in a semi-circle

GEOMETRY PACKAGE

1. Draw a circle and construct a diameter. Label it as shown.
2. Mark any point P not at A or B on the circle. Draw [AP] and [PB].
3. Measure angle APB.
4. Repeat for different positions of P and for different circles. What do you notice?
5. Copy and complete:
   The angle in a semi-circle is ......

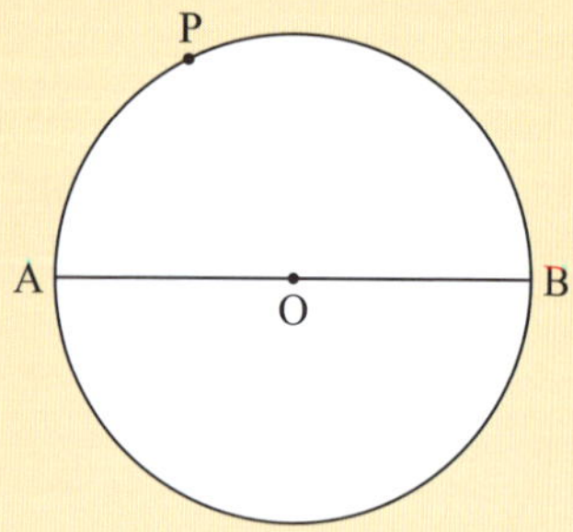

Part 2: Chords of a circle theorem

GEOMETRY PACKAGE

1 Draw a circle with centre O. Construct any chord [AB].

2 Construct the perpendicular from O to [AB] which cuts the chord at M.

3 Measure the lengths of [AM] and [BM]. What do you notice?

4 Repeat this procedure with another circle and chord.

5 Copy and complete:

The perpendicular from the centre of a circle to a chord ......

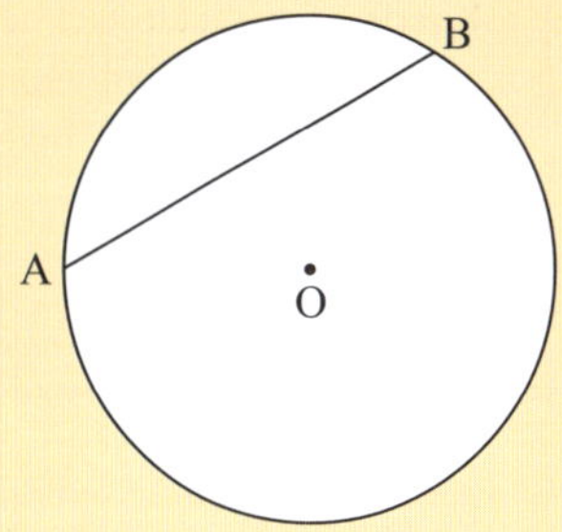

Part 3: Radius-tangent theorem

GEOMETRY PACKAGE

1 Draw a circle with centre O, and mark on it a point A.

2 At A, draw as accurately as possible a tangent [TA].

3 Draw the radius [OA].

4 Measure angle OAT.

5 Repeat this procedure with another circle and tangent.

6 Copy and complete:

The tangent to a circle is ...... to the radius at the point ......

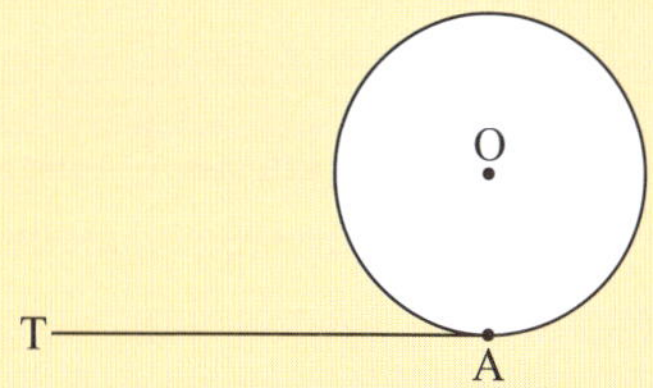

Part 4: Tangents from an external point

GEOMETRY PACKAGE

1 Draw a circle with centre O.

2 From an external point P, draw as accurately as possible the two tangents to the circle. Let the tangents meet the circle at A and B.

3 Measure [AP] and [BP].

4 Repeat this procedure with another circle of different size.

5 Copy and complete:

Tangents from an external point to a circle are ......

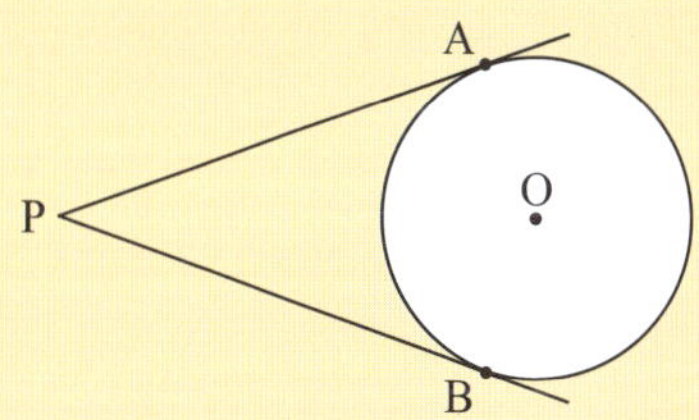

From the **Investigation** you should have discovered the following **circle theorems**.

Name of theorem	*Statement*	*Figure*
Angle in a semi-circle	The angle in a semi-circle is a right angle.	$A\widehat{B}C = 90°$ 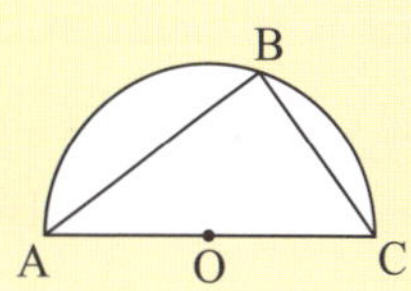

Name of theorem	*Statement*	*Figure*
Chords of a circle	The perpendicular from the centre of a circle to a chord bisects the chord.	AM = BM 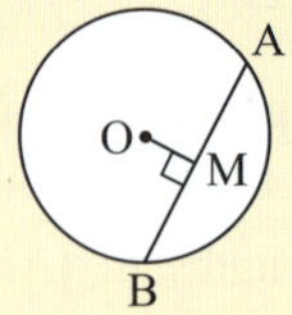
Radius-tangent	The tangent to a circle is perpendicular to the radius at the point of contact.	$\text{O}\widehat{\text{A}}\text{T} = 90^\circ$ 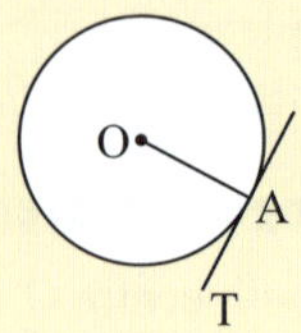
Tangents from an external point	Tangents from an external point are equal in length.	AP = BP 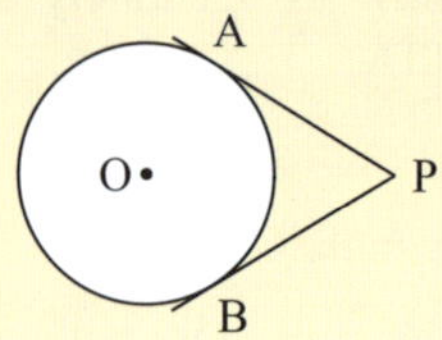

Example 4 — Self Tutor

Find x, giving brief reasons for your answer.

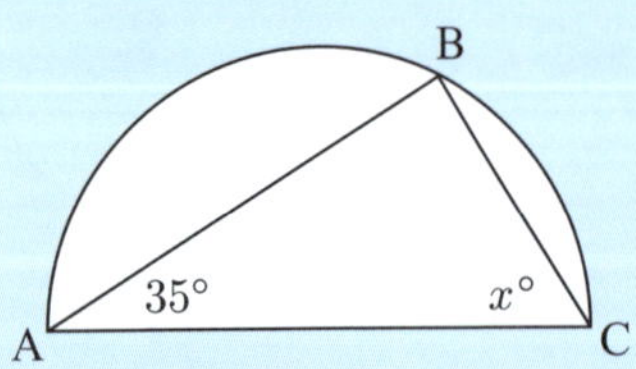

$\text{A}\widehat{\text{B}}\text{C}$ measures 90° {angle in a semi-circle}

$\therefore\ 35 + 90 + x = 180$ {angles of a triangle}

$\therefore\ 125 + x = 180$

$\therefore\ x = 55$

EXERCISE 22B

1 In this question we will prove the *angle in a semi-circle* theorem. Suppose [AC] is the diameter of a circle with centre O, and B is any point on the semi-circle.

Let $\text{O}\widehat{\text{A}}\text{B} = x^\circ$ and $\text{O}\widehat{\text{C}}\text{B} = y^\circ$.

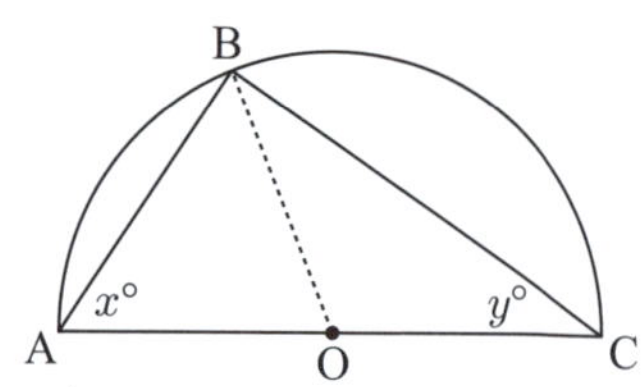

a Find $\text{O}\widehat{\text{B}}\text{A}$ and $\text{O}\widehat{\text{B}}\text{C}$ in terms of x and y.

b Hence, show that $\text{A}\widehat{\text{B}}\text{C} = 90^\circ$.

2 Find the value of x:

a

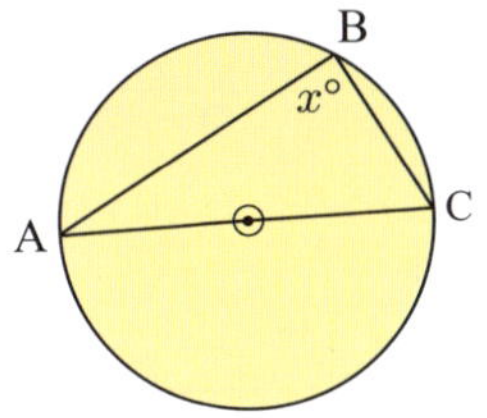

b

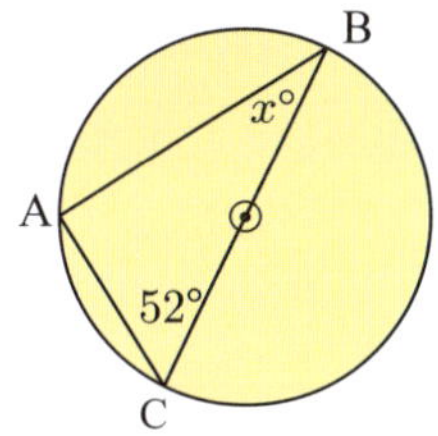

c

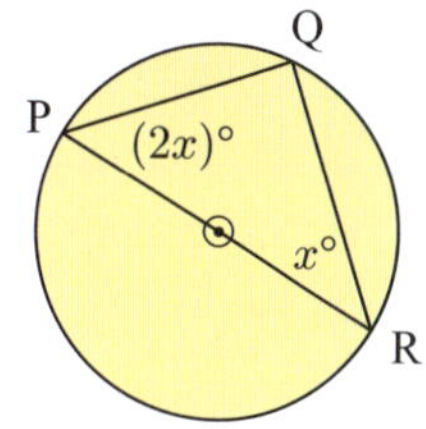

d

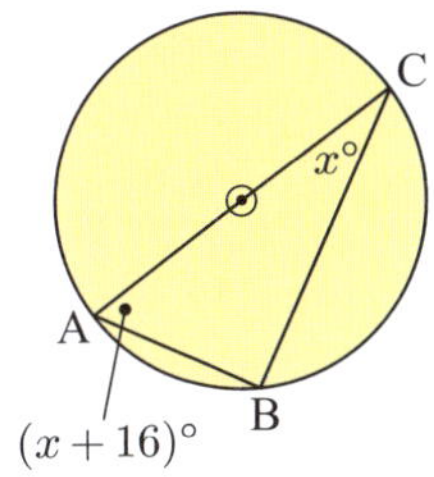

e

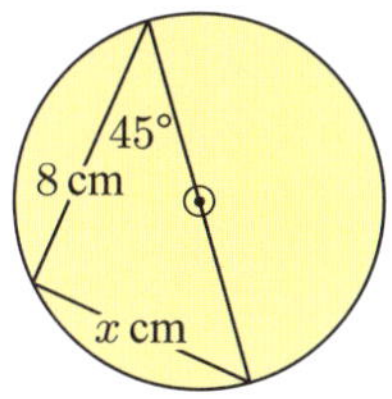

f

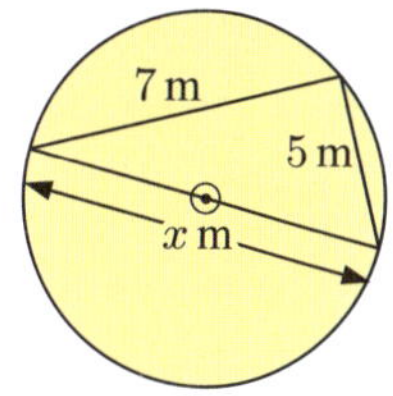

3

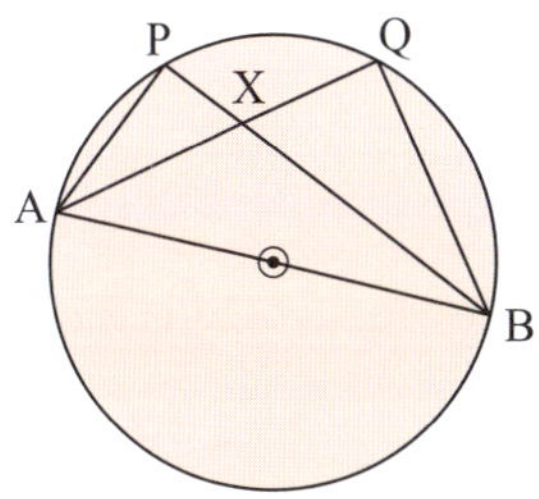

P and Q are any two points on a circle on the same side of diameter [AB].
Show that triangles APX and BQX are similar.

Example 5 **Self Tutor**

A circle has a chord of length 8 cm. The shortest distance from the circle's centre to the chord is 2 cm. Find, to the nearest mm, the length of the circle's radius.

Suppose the radius has length r cm.

The shortest distance from the centre to the chord is the perpendicular distance.

Let M be the foot of the perpendicular from the centre O to the chord [AB].

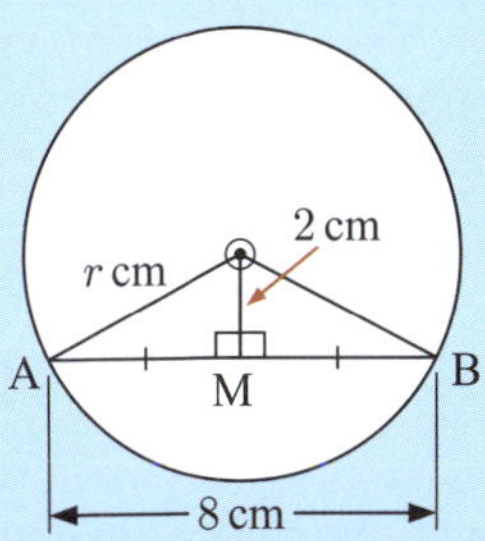

$\therefore$ M is the midpoint of [AB] {chords of a circle}

$\therefore$ AM = 4 cm

In $\triangle$OAM, $r^2 = 2^2 + 4^2$ {Pythagoras}

$\therefore r^2 = 20$

$\therefore r = \sqrt{20} \approx 4.5$ {as $r > 0$}

So, the radius is about 4.5 cm long.

4 In this question we will prove the *chords of a circle* theorem. Suppose [AB] is a chord of a circle with centre O, and [OM] is perpendicular to [AB].

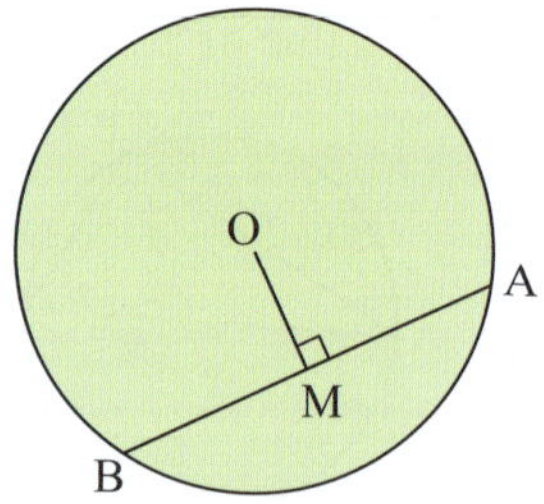

a Show that triangles OAM and OBM are congruent.

b Hence, show that AM = BM.

5 Find x, rounded to 1 decimal place:

a

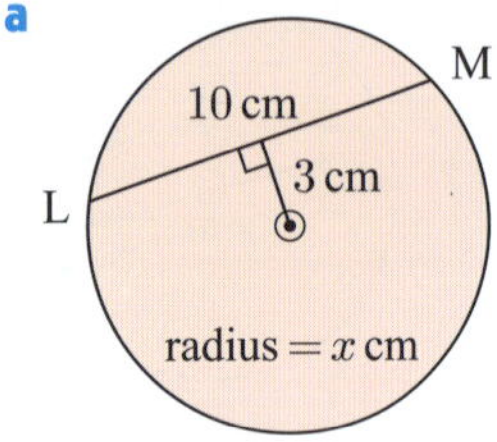

b

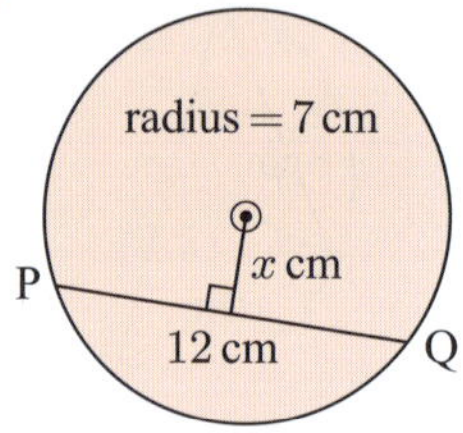

c

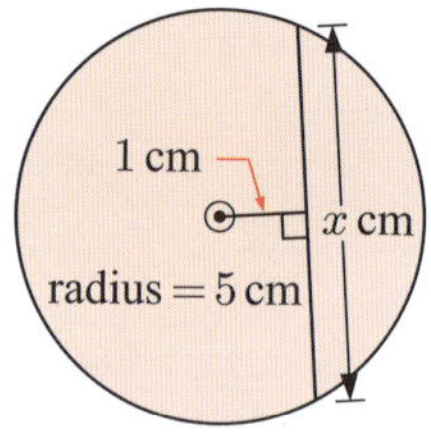

6 A circle has a chord of length 12 cm. The shortest distance from the chord to the circle's centre is 2 cm. Find, to the nearest mm, the length of the circle's radius.

7 A circle has a chord of length 15 cm. The radius of the circle is 9 cm. Find, to the nearest mm, the shortest distance from the chord to the circle's centre.

8 A circle has radius 5 cm. The shortest distance from the circle's centre to a chord is 2 cm. Find, to the nearest mm, the length of this chord.

9 [AB] is a diameter of a circle with centre O, and P is a point on the circle. Perpendiculars are drawn from O to the chords [AP] and [BP].

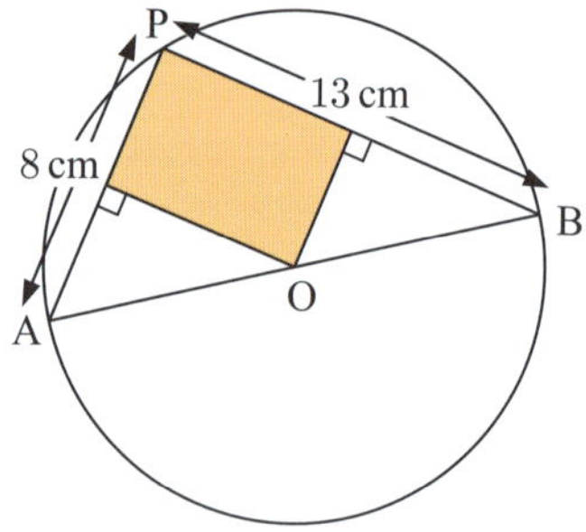

a Find the diameter of the circle.

b Find the area of the shaded region.

Example 6 **Self Tutor**

A circle has radius 6 cm. At point T on the circle a tangent is drawn. The tangent is 10 cm long, and ends at point P. Find the distance from P to the centre of the circle.

Let the distance from P to the centre of the circle be x cm.

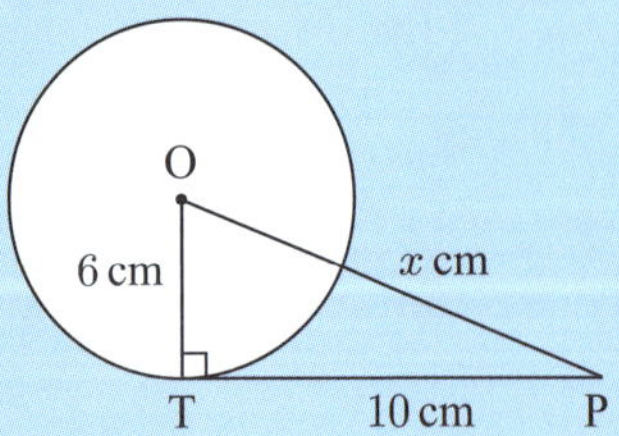

$\widehat{OTP} = 90°$ {radius-tangent}

$\therefore\ x^2 = 6^2 + 10^2$ {Pythagoras}

$\therefore\ x^2 = 136$

$\therefore\ x = \sqrt{136} \approx 11.7$ {as OP > 0}

So, P is about 11.7 cm from the centre of the circle.

10 In the following diagrams, O is the centre of the circle, and [TP] is a tangent. Find the value of x, rounding your answer to 1 decimal place where necessary.

a

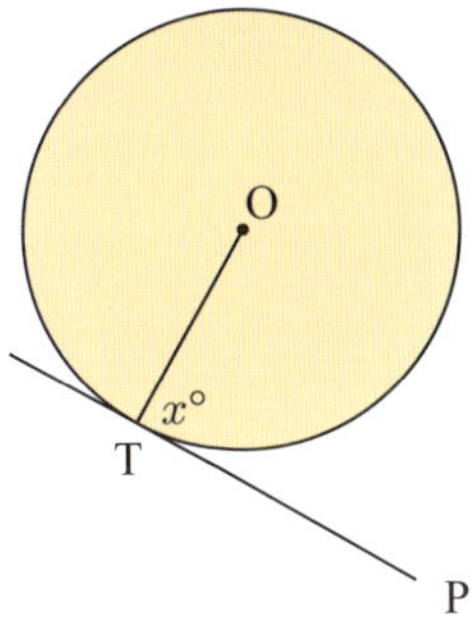

b

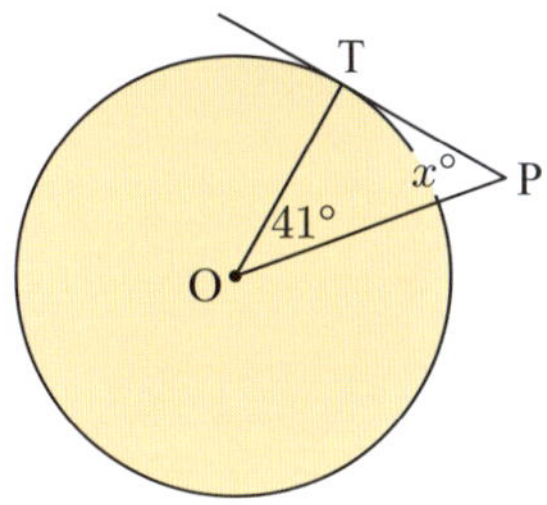

c

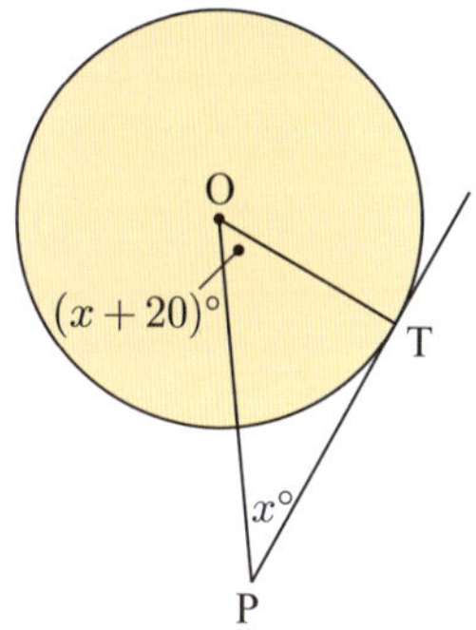

d

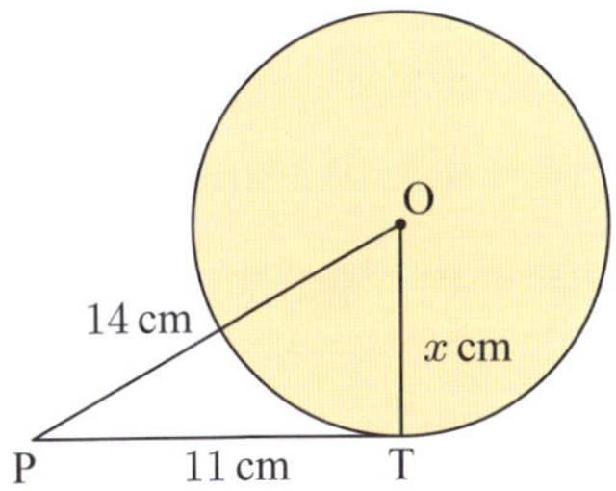

e

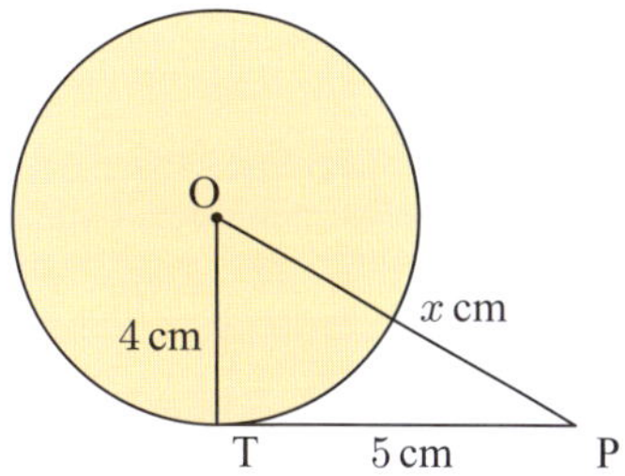

f

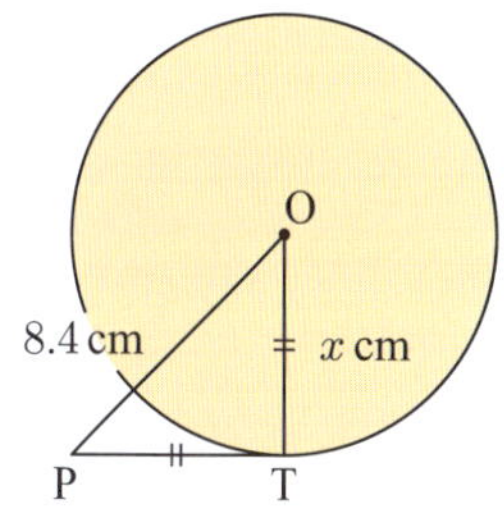

11 Point P is 9 cm from the centre of a circle. A tangent from P to the circle is 5 cm long. Find the radius of the circle.

12 Point Q is 12 cm from the centre of a circle of radius 7 cm. A tangent is drawn from Q to touch the circle at P. Find the length of [PQ].

13 In the figure alongside, the tangent [TP] has the same length as the radius of the circle. What proportion of triangle OTP is inside the circle?

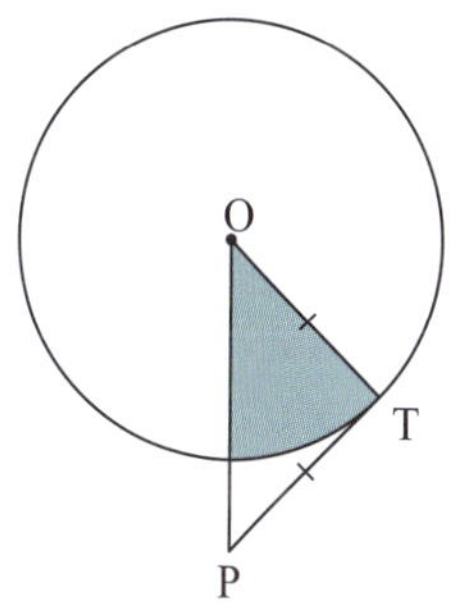

14 In this question we will prove the *tangents from an external point* theorem.

a Show that:

i $AP^2 = OP^2 - OA^2$

ii $BP^2 = OP^2 - OB^2$.

b Hence, show that $AP = BP$.

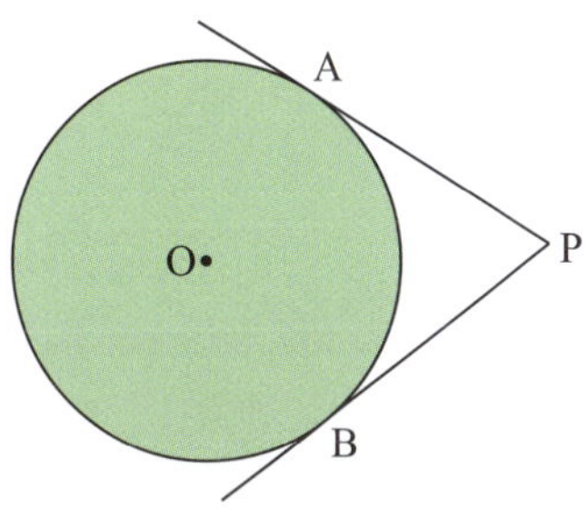

15 In the following diagrams, O is the centre of the circle, and [TP] and [SP] are tangents. Find the values of the unknowns.

a
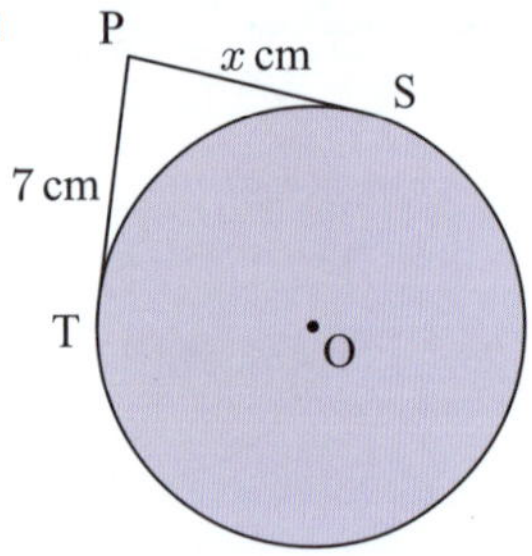

b
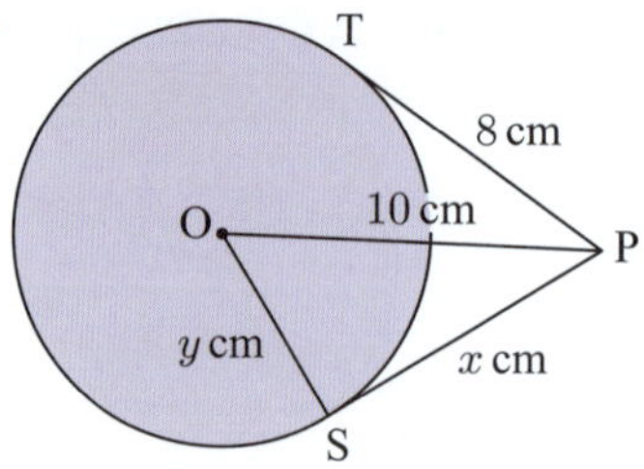

c
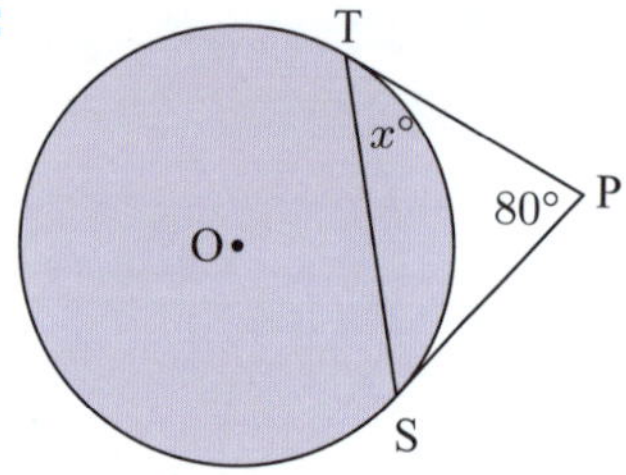

d
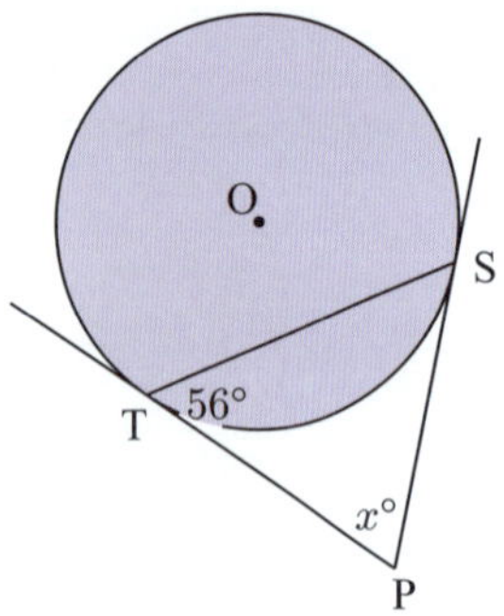

e
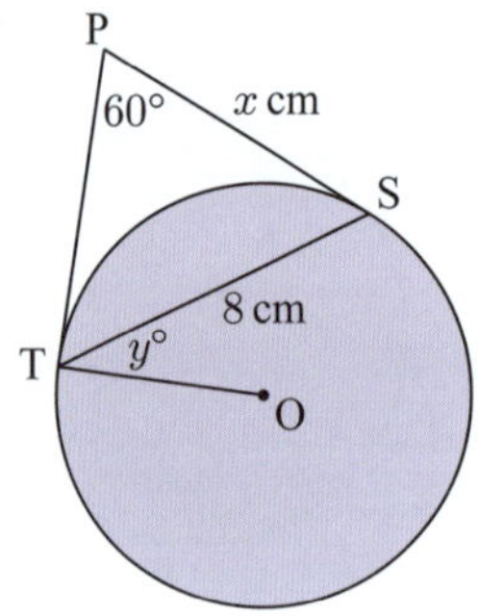

f
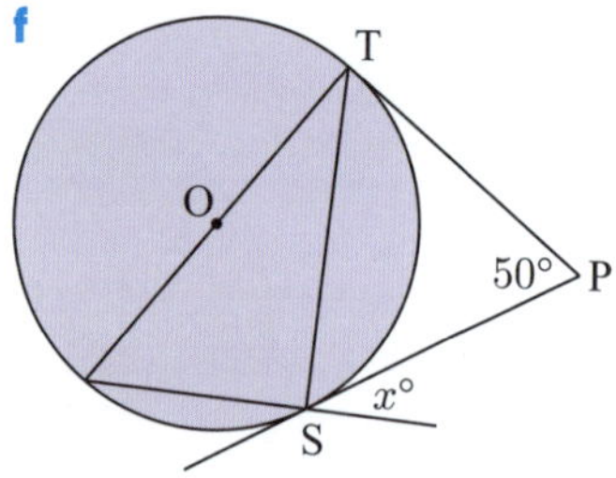

16 In the figure alongside, O is the centre of the circle, and [AP] and [BP] are tangents to the circle.

Show that $A\widehat{O}B$ is twice the size of $B\widehat{A}P$.

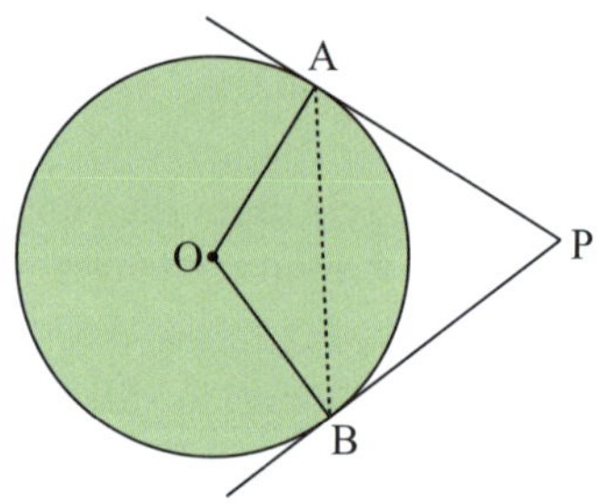

17 [AB], [BC], [CD], and [AD] are tangents to the circle. Find the length of [AD].

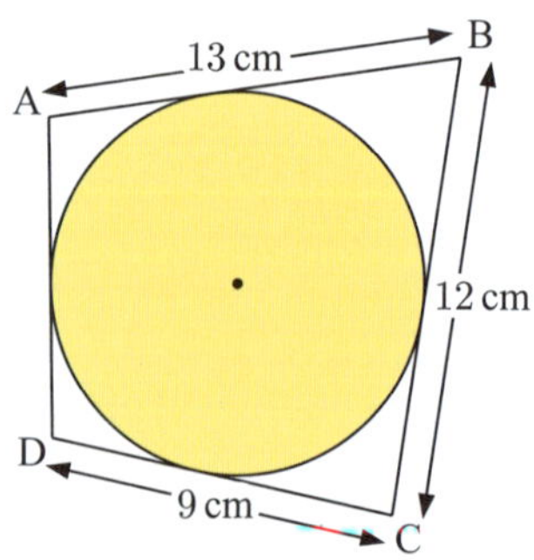

18
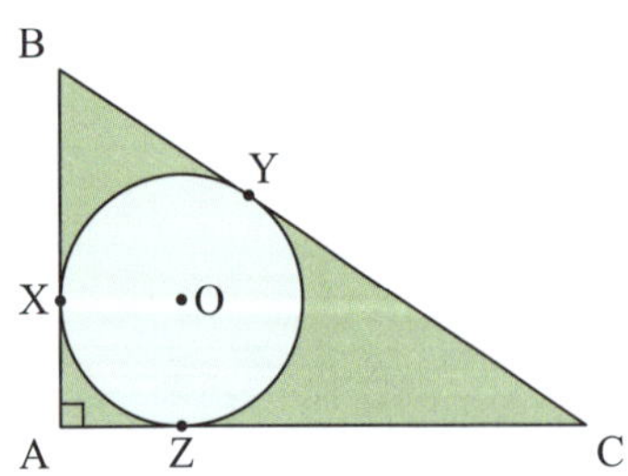

A circle is drawn within the right angled triangle ABC. The circle has radius 3 cm, and $BX = 4$ cm.

a What can be deduced about quadrilateral AXOZ?

b How far is C from Z?

19 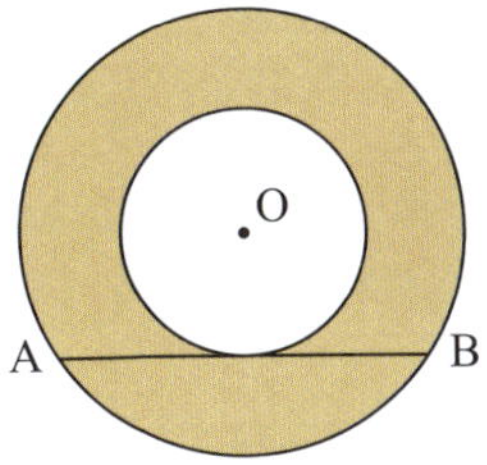

[AB] is a tangent to the inner circle and is a chord of the larger circle. If [AB] is 6 m long, find the area of the shaded annulus.

20 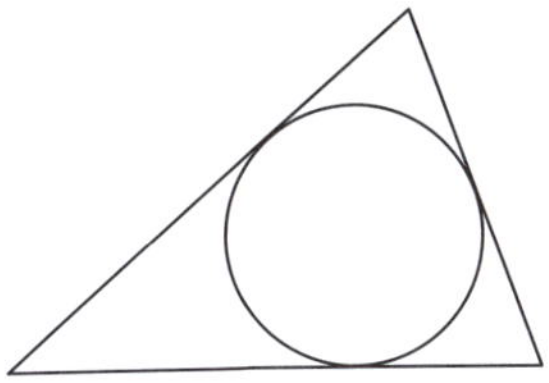

A circle is inscribed in a triangle. The triangle has area A and perimeter P. Show that:

a the radius of the circle is given by $r = \frac{2A}{P}$

b the ratio of perimeters of the circle and triangle is the same as the ratio of their areas.

C THE MIDPOINT THEOREM

The **midpoint theorem** deals with the line segment joining the midpoints of any two sides of a triangle.

INVESTIGATION 2 THE MIDPOINT THEOREM

Use the **geometry package** or a ruler and protractor to complete this Investigation.

GEOMETRY PACKAGE

What to do:

1. Construct a large triangle and label it like the triangle shown.
2. As accurately as you can, mark the midpoints P and Q of [AB] and [AC] respectively.
3. Join [PQ].
4. Measure the lengths of [PQ] and [BC]. What do you notice?
5. Measure $A\widehat{P}Q$ and $A\widehat{B}C$. What conclusion can be drawn about [PQ] and [BC]?
6. Repeat the procedure with different sized triangles.
7. Copy and complete: "*The line joining the midpoints of two sides of a triangle is ...... to the third side and ...... its length.*"

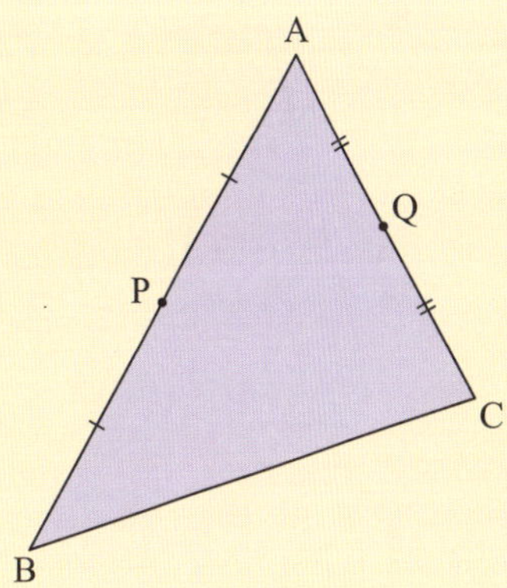

From the **Investigation**, you should have discovered:

THE MIDPOINT THEOREM

The line joining the midpoints of two sides of a triangle is parallel to the third side and half its length.

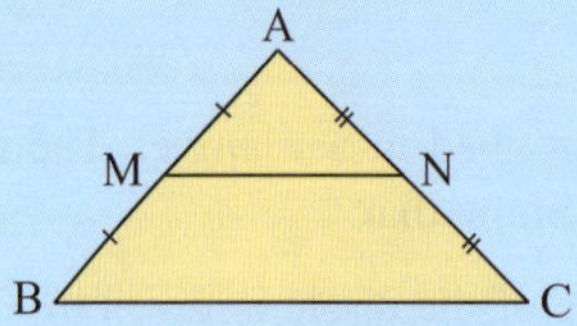

- [MN] is parallel to [BC]
- $MN = \frac{1}{2} \times BC$

CONVERSE OF MIDPOINT THEOREM

The line drawn from the midpoint of one side of a triangle, parallel to a second side, bisects the third side.

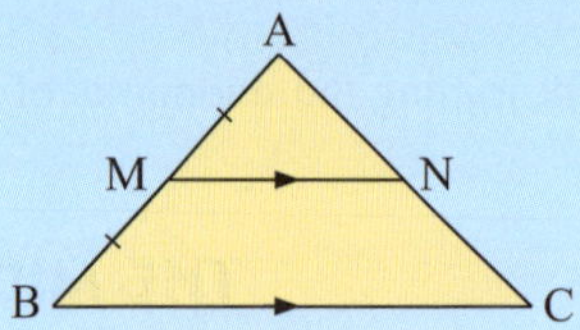

AN = NC

EXERCISE 22C

1 Find the values of the unknowns, giving reasons for your answers.

a

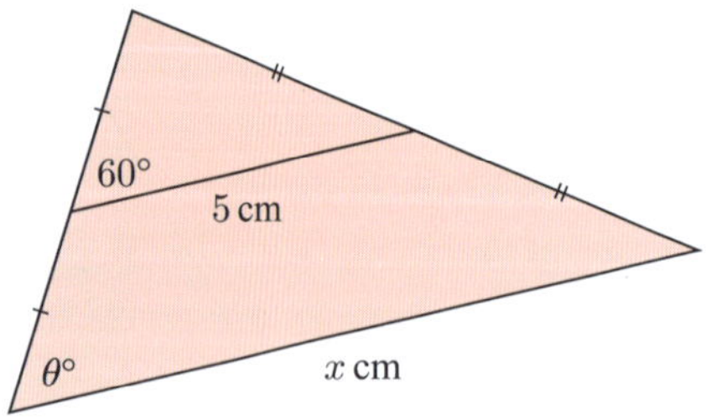

b

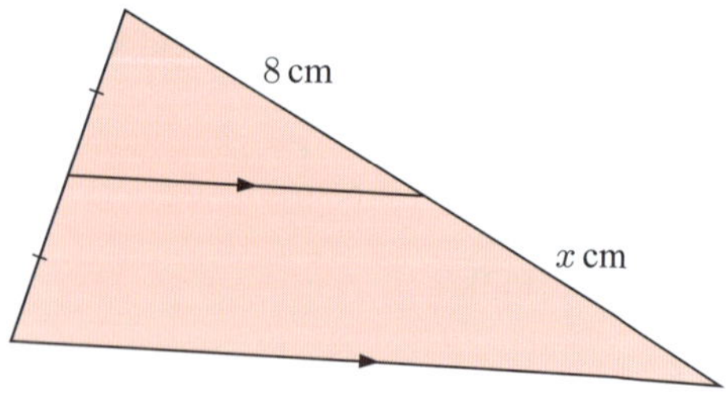

2

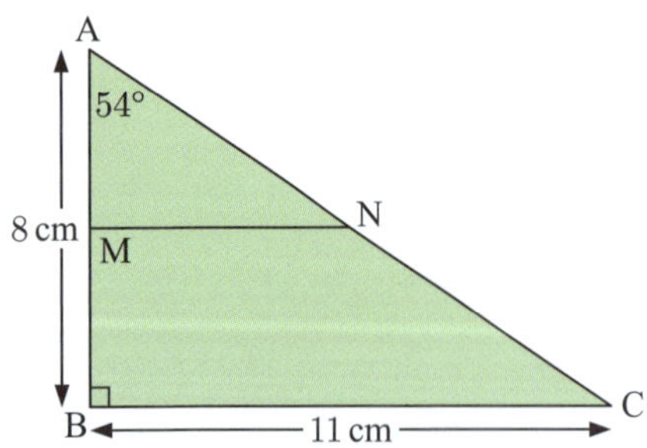

M and N are the midpoints of [AB] and [AC] respectively. Find:

a the length of [MN]

b the size of $A\widehat{N}M$

c the area of $\triangle ANM$.

Example 7 Self Tutor

ABCD is a parallelogram. [AB] is produced to E such that AB = BE.
[AD] is produced to meet [EC] produced at F. Prove that EC = CF.

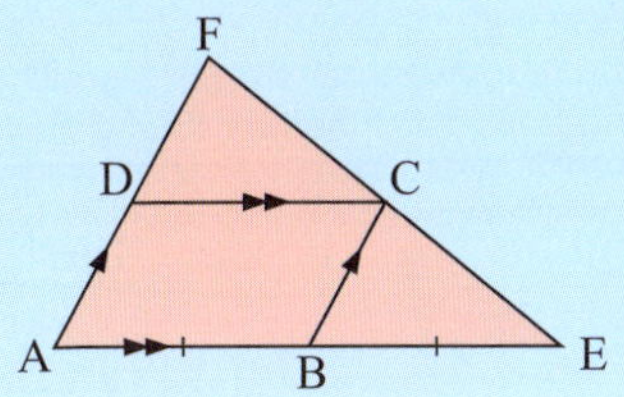

In $\triangle$AEF, [BC] is parallel to [AF].
{as ABCD is a parallelogram}

Now B is the midpoint of [AE],
so C is the midpoint of [EF].
{converse of midpoint theorem}

$\therefore$ EC = CF.

3 ABC is a triangle. D, E, and F are the midpoints of its sides as shown.

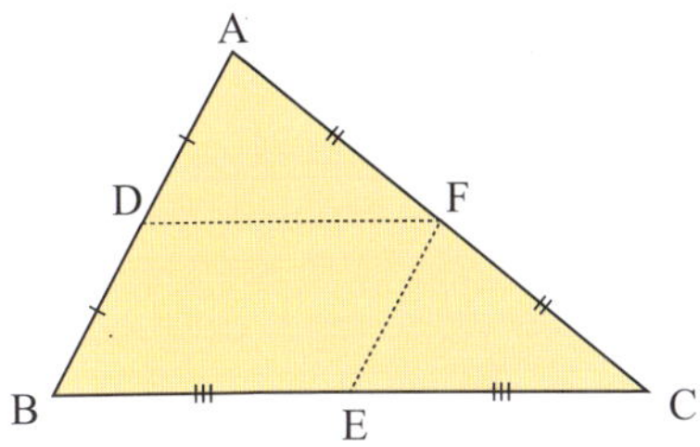

a Show that DFEB is a parallelogram.

b Name two other parallelograms which can be drawn using the given points.

4 ABCD is a parallelogram whose diagonals meet at E. M is the midpoint of [AD]. Show that [ME] is parallel to [AB] and half its length.

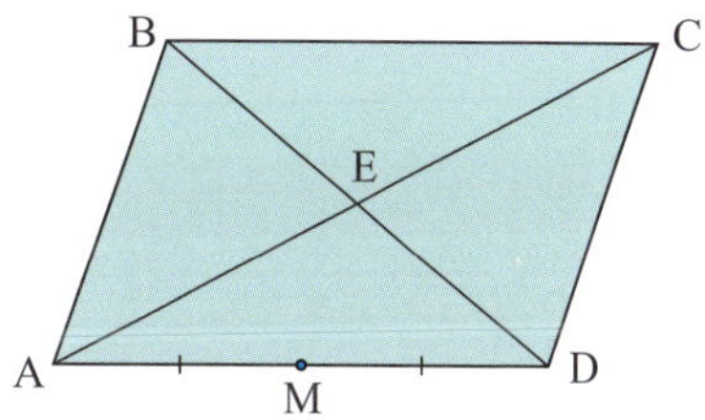

5 A and B are the midpoints of sides [PQ] and [PR]. Y is any point on [QR]. Prove that X is the midpoint of [PY].

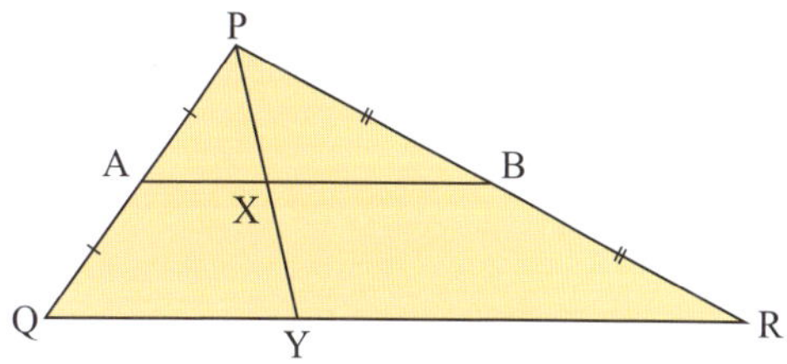

6 Stacey says "I have drawn dozens of different quadrilaterals. When I join the midpoints of adjacent sides of each one of them, the figure formed appears to be a parallelogram."

GEOMETRY PACKAGE

a Draw two quadrilaterals of your own choosing to check Stacey's conjecture.

b By drawing one diagonal of a labelled quadrilateral, prove that Stacey's conjecture is correct.

7 To **prove** the midpoint theorem in $\triangle$BAC, we produce [PQ] to meet a line from C which is parallel to [BA] at R. Show that:

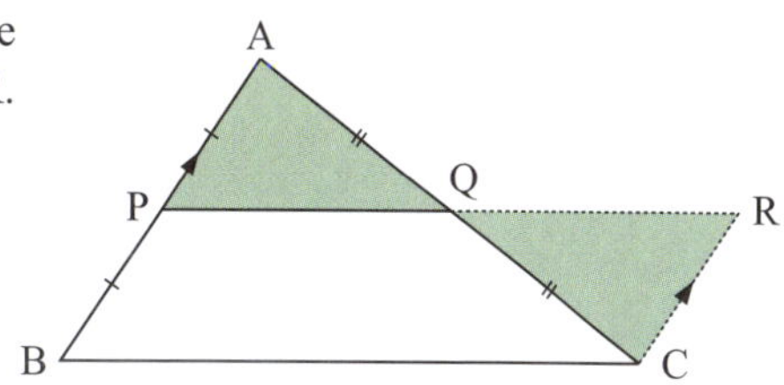

a triangles APQ and CRQ are congruent

b BPRC is a parallelogram

c [PQ] is parallel to [BC] and $PQ = \frac{1}{2}BC$.

8 Use similar triangles to prove the converse of the midpoint theorem.

REVIEW SET 22A

1 Find the value of x:

a

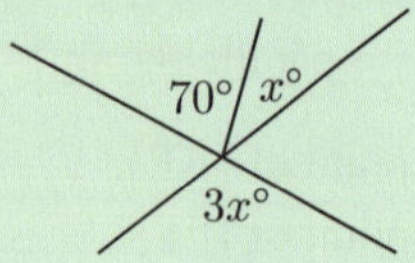

b

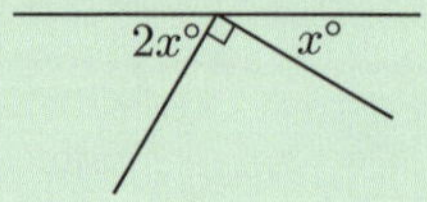

c

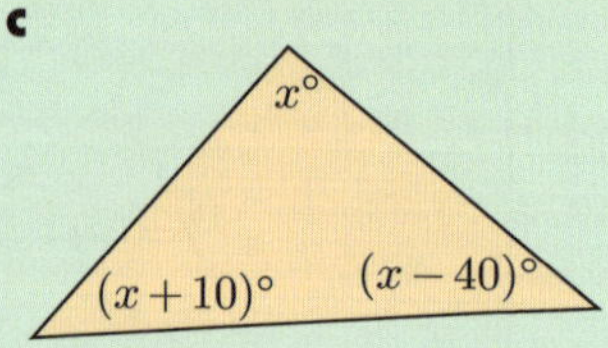

d

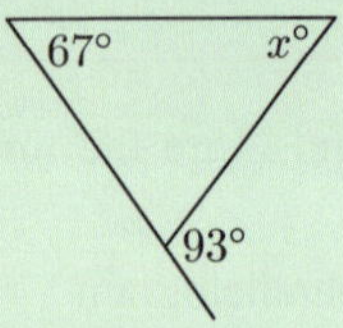

e

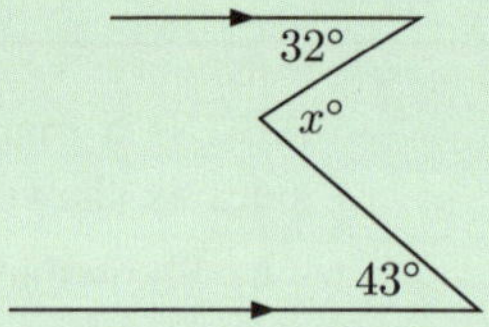

f

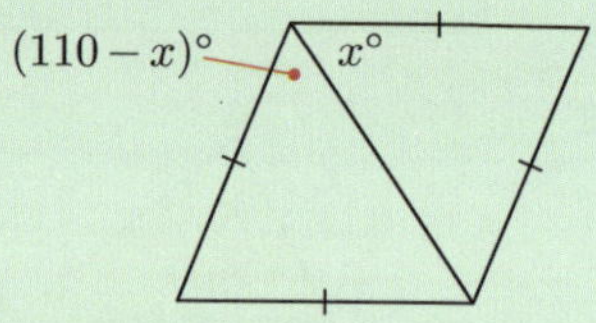

2

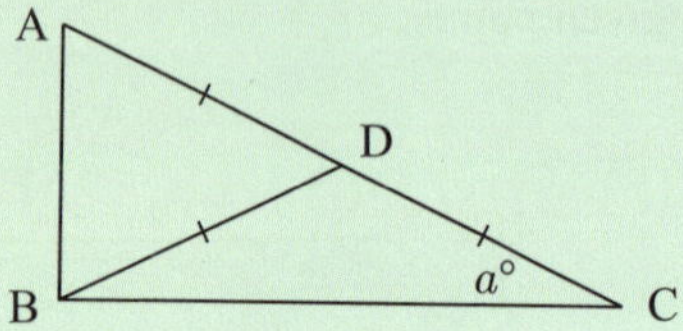

a Find, in terms of a:

i $\widehat{DBC}$ **ii** $\widehat{ADB}$ **iii** $\widehat{DBA}$

b Show that $\widehat{ABC}$ is a right angle.

3

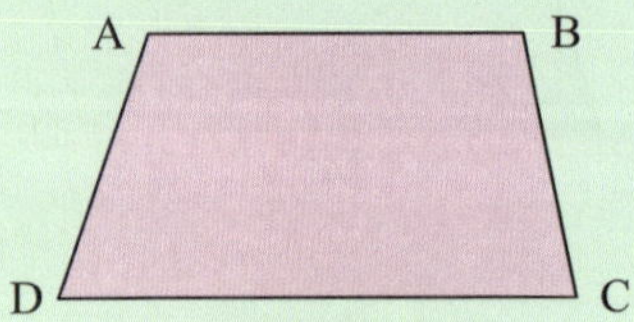

In the given figure $\widehat{DAB} = \widehat{CBA}$ and $\widehat{ADC} = \widehat{BCD}$.
Show that [AB] is parallel to [DC].

4 Find x:

a

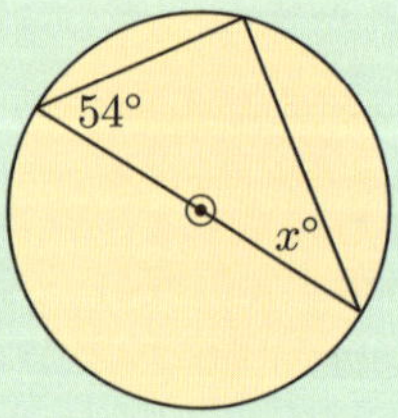

b

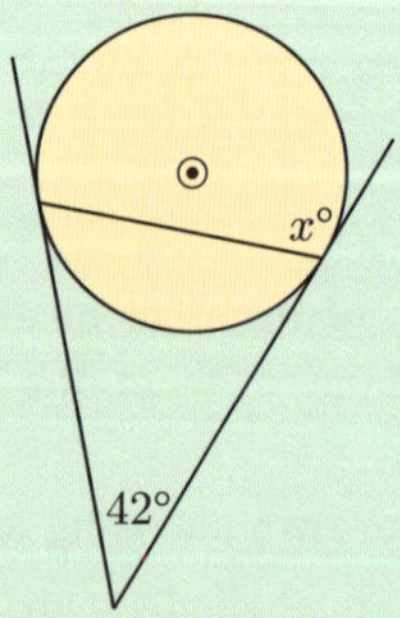

c

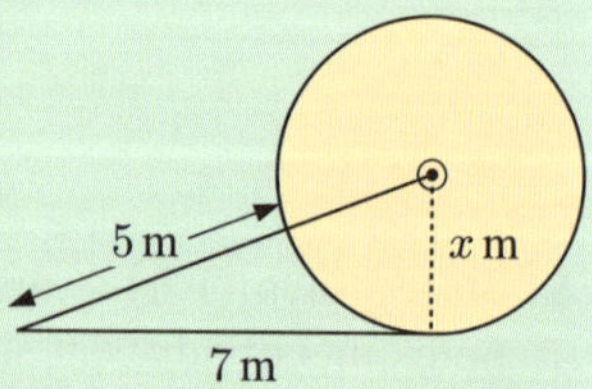

5 O is the centre of the circle, and [TP] is a tangent such that $\triangle$OTP has the same area as the circle. Show that the length of the tangent is equal to the circumference of the circle.

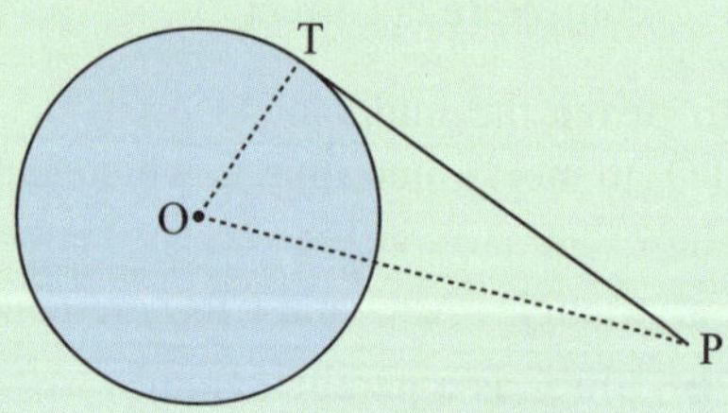

6 A circle has a chord of length 12 cm. The shortest distance from the circle's centre to the chord is 5 cm. Find, to the nearest mm, the length of the circle's radius.

7 Find the values of the unknowns:

a

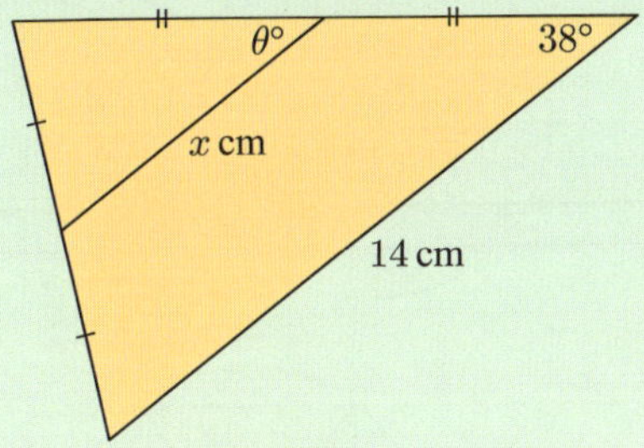

b

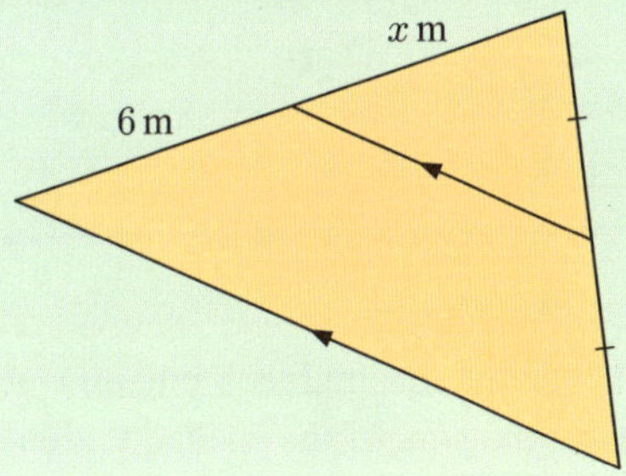

8 ABC is an equilateral triangle. D, E, and F are the midpoints of its sides, as shown.

a Show that DFEB is a rhombus.

b Name two other rhombuses which can be drawn using the given points.

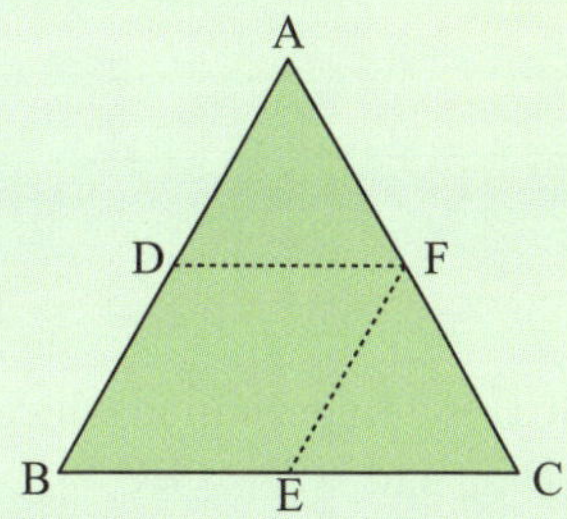

REVIEW SET 22B

1 Find the value of x:

a

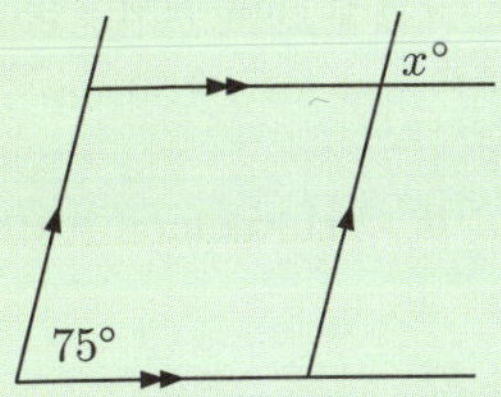

b

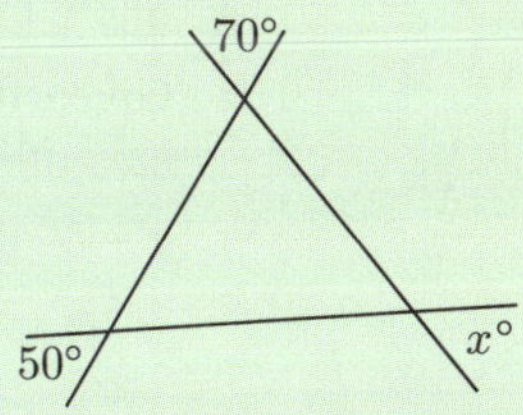

c

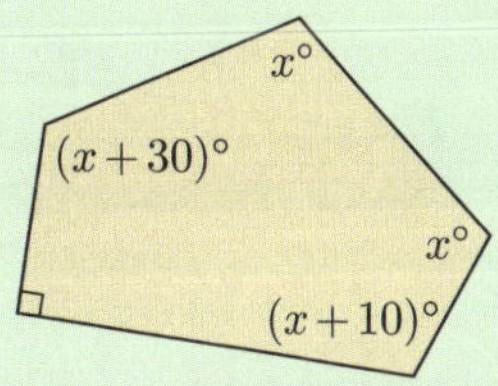

d

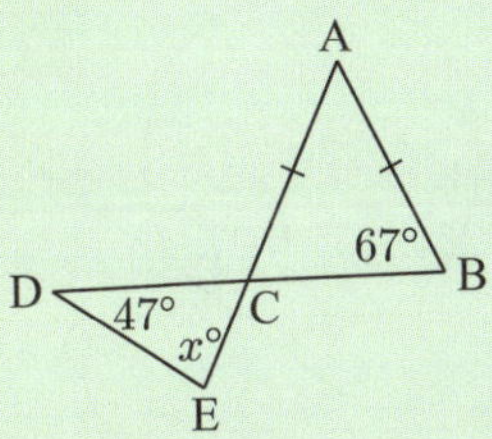

e

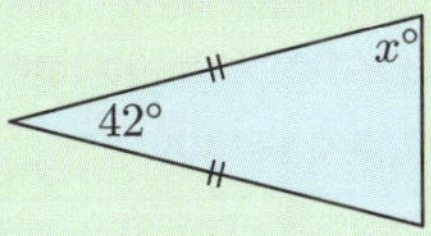

f

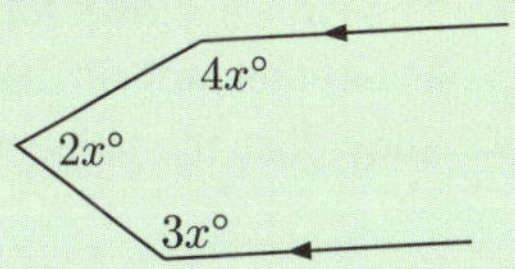

2 ABC is an isosceles triangle with AB = AC.
From a point X on the base [BC], perpendiculars to [AB] and [AC] are drawn, meeting these line segments at Y and Z respectively.
Show that $B\widehat{A}C$ is twice the size of $C\widehat{X}Z$.

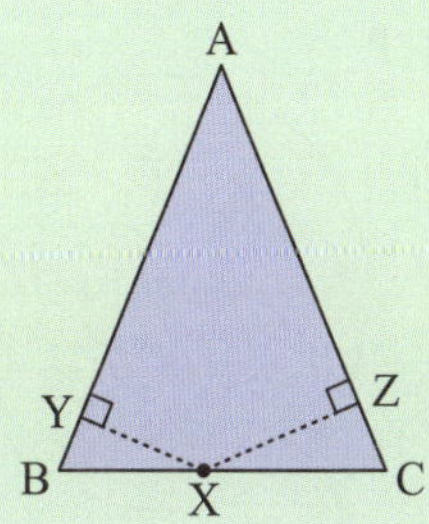

3 Find the values of the unknowns:

a

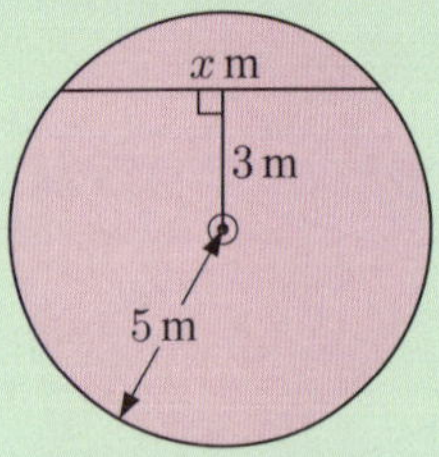

b

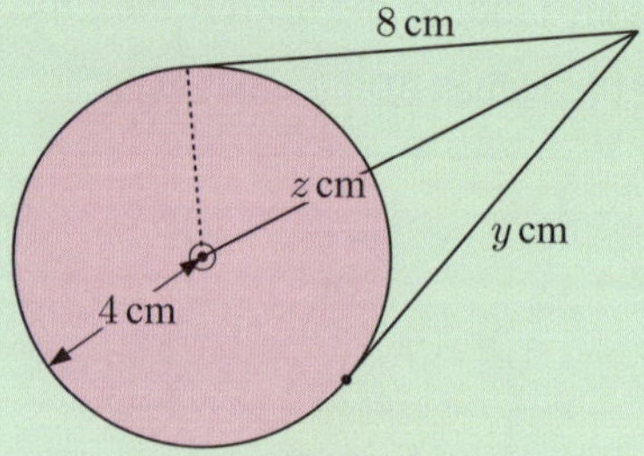

4 This circle has radius 5 cm, and AB = 6 cm. Find the area of the shaded region.

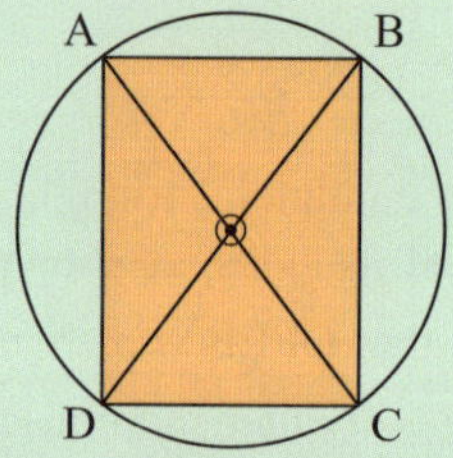

5 ABCD is a trapezium in which [AB] is parallel to [DC], and AD = BC.

Prove that $\widehat{ADC} = \widehat{BCD}$.

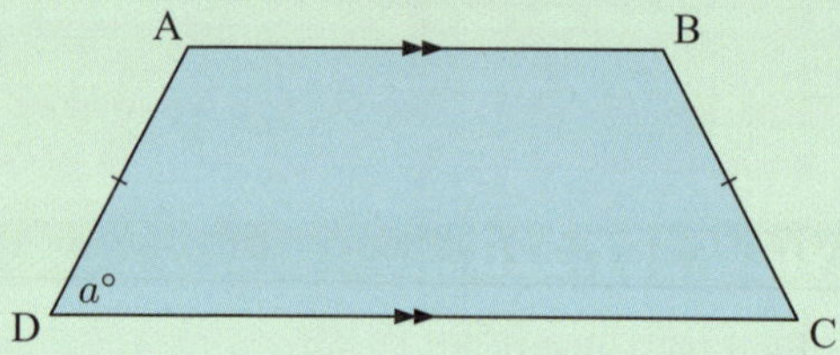

6

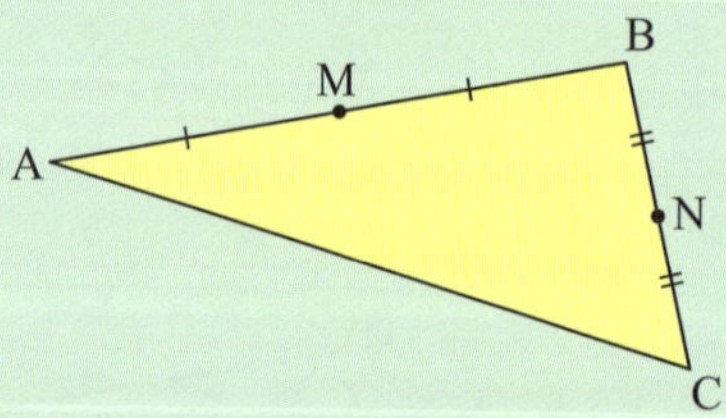

a Show that triangles BAC and BMN are similar.

b Given that $\triangle$BAC has area 100 cm^2, find the area of:

i $\triangle$BMN **ii** quadrilateral AMNC.

7 The circles shown *touch* at P, which means they share a common tangent at this point. XPY is a straight line. Show that the tangents at X and Y are parallel.

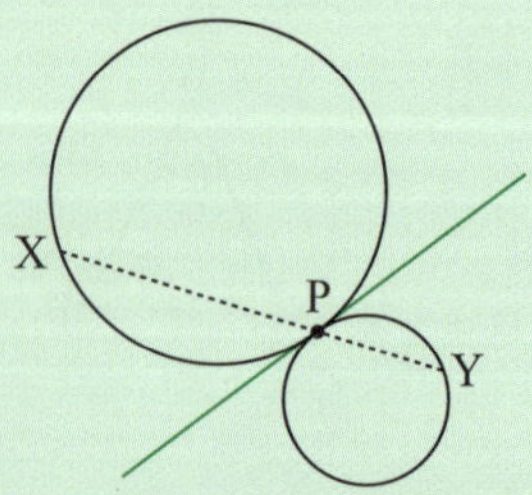

8

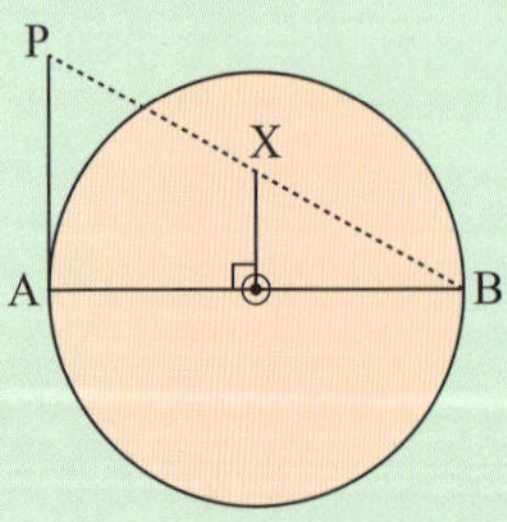

[AB] is a diameter of the circle shown, and [AP] is a tangent.

A line is drawn through the centre of the circle and perpendicular to [AB], meeting [PB] at X.

Show that X is the midpoint of [PB].

Chapter 23

Exponential and rational functions

Contents:

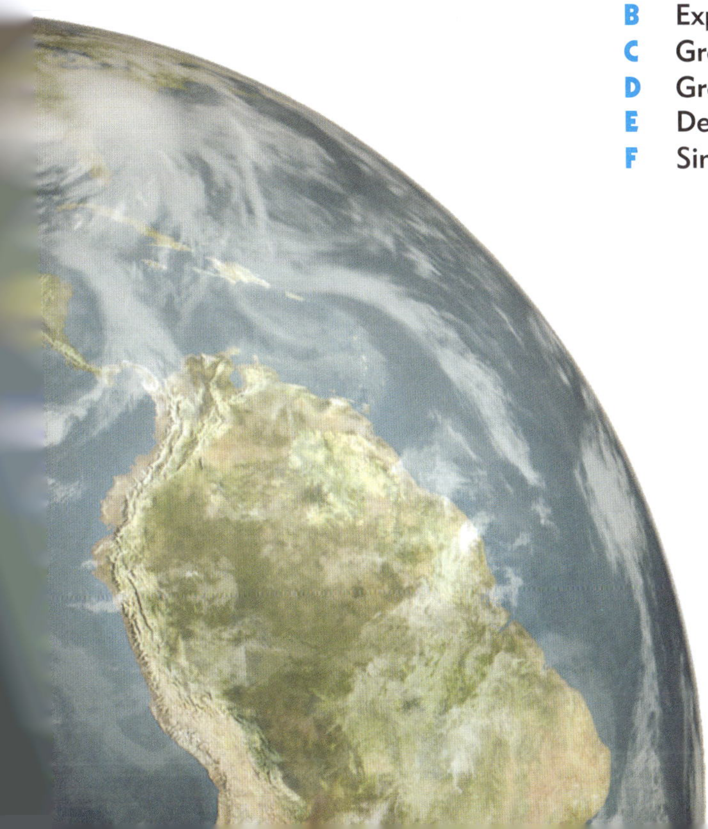

OPENING PROBLEM

An ant nest initially contains 50 ants, and the population of ants doubles each week.

Things to think about:

a How many ants will be in the nest after:

  i 1 week **ii** 2 weeks **iii** 3 weeks?

b What would the graph of the ant population over time look like?

c What will the ant population be after 2.5 weeks?

d Can you write an equation connecting the ant population and the time in weeks?

The population growth in the **Opening Problem** can be described using an **exponential function**.

In this chapter, we will study exponential functions and their graphs. We will also study **exponential equations** and **rational functions**.

A EXPONENTIAL EQUATIONS

An **exponential equation** is an equation in which the unknown occurs as part of the index or exponent.

For example: $3^x = 9$ and $5 \times 4^x = 8$ are both exponential equations.

For the case $3^x = 9$, we notice that $3^x = 3^2$. Thus $x = 2$ is a solution to the equation, and it is in fact the only solution to the equation.

To solve exponential equations, we try to write both sides of the equation with the **same base**. We can then **equate indices**.

If $a^x = a^k$ then $x = k$.

Example 1 Self Tutor

Solve for x:

a $3^x = 27$

b $2^{x+1} = \frac{1}{8}$

a $3^x = 27$

$\therefore\ 3^x = 3^3$

$\therefore\ x = 3$ {equating indices}

b $2^{x+1} = \frac{1}{8}$

$\therefore\ 2^{x+1} = 2^{-3}$

$\therefore\ x + 1 = -3$ {equating indices}

$\therefore\ x = -4$

EXERCISE 23A

1 Solve for x:

a $2^x = 2$ b $2^x = 4$ c $5^x = 25$ d $3^x = 1$

e $2^x = \frac{1}{2}$ f $3^x = \frac{1}{3}$ g $2^x = \frac{1}{8}$ h $7^x = \frac{1}{49}$

2 Solve for x:

a $2^{x+1} = 8$ b $5^{x-2} = 5$ c $3^{x+3} = 81$ d $10^{x-4} = 1000$

e $2^{x-2} = \frac{1}{4}$ f $3^{x+1} = \frac{1}{27}$ g $2^{2x} = 64$ h $11^{2x-3} = \frac{1}{11}$

Example 2 Self Tutor

Solve for x: $5 \times 2^x = 80$

$5 \times 2^x = 80$

$\therefore\ 2^x = 16$

$\therefore\ 2^x = 2^4$

$\therefore\ x = 4$ {equating indices}

3 Solve for x:

a $2 \times 3^x = 18$ b $7 \times 2^x = 56$ c $5 \times 6^x = 5$

d $21 \times 7^x = 3$ e $20 \times 2^{x+1} = 5$ f $100 \times 5^{1-x} = 4$

Example 3 Self Tutor

Solve for x: $4^x = \frac{1}{2}$

$4^x = \frac{1}{2}$

$\therefore\ (2^2)^x = 2^{-1}$

$\therefore\ 2^{2x} = 2^{-1}$

$\therefore\ 2x = -1$ {equating indices}

$\therefore\ x = -\frac{1}{2}$

4 Solve for x:

a $4^x = 32$ b $8^x = 4$ c $9^x = 27$ d $27^x = \frac{1}{9}$

e $4^x = \frac{1}{8}$ f $25^x = \frac{1}{5}$ g $8^{x+2} = 32$ h $8^{1-x} = \frac{1}{4}$

i $4^{x-1} = \frac{1}{16}$ j $9^{x-2} = 3$ k $(\frac{1}{2})^x = 8$ l $(\frac{1}{3})^{x+2} = 9$

B EXPONENTIAL FUNCTIONS

An **exponential function** is a function in which the variable occurs as part of the index or exponent.

Examples of exponential functions are $y = 5^x$, $y = 3^{x+1}$, and $y = 2^{-x} - 6$.

Example 4 — Self Tutor

For the function $y = 2^x - 3$, find:

a y when $x = 4$ **b** x when $y = 5$.

a When $x = 4$,
$y = 2^4 - 3$
$= 16 - 3$
$= 13$

b When $y = 5$,
$2^x - 3 = 5$
$\therefore 2^x = 8$
$\therefore 2^x = 2^3$
$\therefore x = 3$ {equating indices}

EXERCISE 23B

1 Determine whether the following are exponential functions:

a $y = 3^x$ **b** $y = 2 \times 7^{x-1}$ **c** $y = x^5$

d $y = 2^{-x} - 4$ **e** $y = x^3 - x$ **f** $y = 9 \times (\frac{1}{3})^x + 2$

2 Consider the function $y = 2^x + 3$. Find the value of y when:

a $x = 0$ **b** $x = 1$ **c** $x = 2$ **d** $x = -1$ **e** $x = -2$

3 Consider the function $y = 5 \times 3^x$. Find the value of y when:

a $x = 0$ **b** $x = 1$ **c** $x = 2$ **d** $x = -1$ **e** $x = -2$

4 Consider the function $W = 5^{-x} + 1$. Find the value of W when:

a $x = 0$ **b** $x = 1$ **c** $x = 2$ **d** $x = -1$ **e** $x = -2$

5 Consider the function $y = 3^x + 2$.

a Find the value of y when: **i** $x = 1$ **ii** $x = 4$ **iii** $x = -2$

b Find the value of x when: **i** $y = 11$ **ii** $y = 3$

6 Consider the function $P = 6 \times 2^x$.

a Find the value of P when: **i** $x = 0$ **ii** $x = 3$ **iii** $x = -4$

b Find the value of x when: **i** $P = 96$ **ii** $P = 3$

C GRAPHS OF EXPONENTIAL FUNCTIONS

One of the simplest exponential functions is $y = 2^x$. To help graph the function, we can construct a **table of values**.

x	-3	-2	-1	0	1	2	3
y	$\frac{1}{8}$	$\frac{1}{4}$	$\frac{1}{2}$	1	2	4	8

When $x = -3$, $y = 2^{-3} = \frac{1}{8}$

When $x = 0$, $y = 2^0 = 1$

When $x = 3$, $y = 2^3 = 8$

Using these values, we can generate the graph of $y = 2^x$:

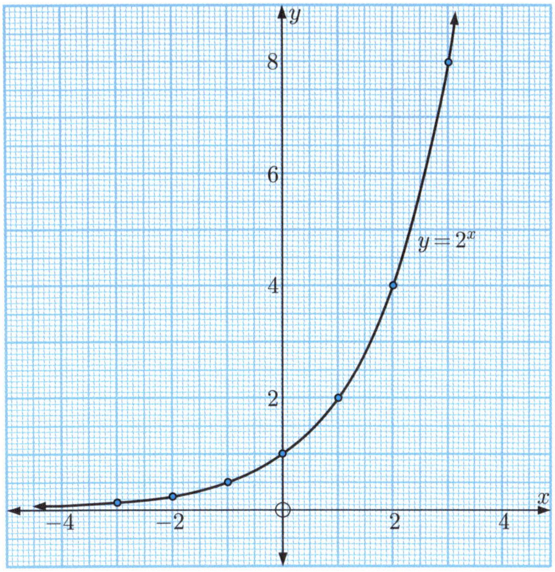

To the left of the y-axis, the graph gets closer to the x-axis but always lies above it. We say that the graph of $y = 2^x$ is **asymptotic** to the x-axis, or the x-axis is a **horizontal asymptote** of the graph.

To the right of the y-axis, the graph becomes very steep as the x values increase.

We can use the graphs of exponential functions to estimate the value of numbers raised to **decimal** powers. We can also use them to help solve exponential equations such as $2^x = 5$, which cannot easily be solved by equating indices.

Example 5 Self Tutor

Use the graph of $y = 2^x$ to estimate, to 1 decimal place:

a the value of $2^{1.3}$

b the solution to $2^x = 5$.

a Reading from the graph, when $x = 1.3$, $y \approx 2.5$.

$\therefore$ $2^{1.3} \approx 2.5$

b Reading from the graph, when $y = 5$, $x \approx 2.3$.

$\therefore$ the solution to $2^x = 5$ is $x \approx 2.3$

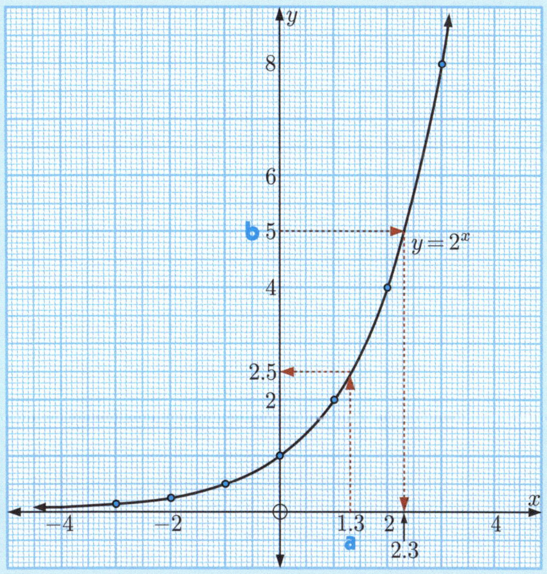

You can use the **graphing package** or a **graphics calculator** to check your answers.

GRAPHING PACKAGE

GRAPHICS CALCULATOR INSTRUCTIONS

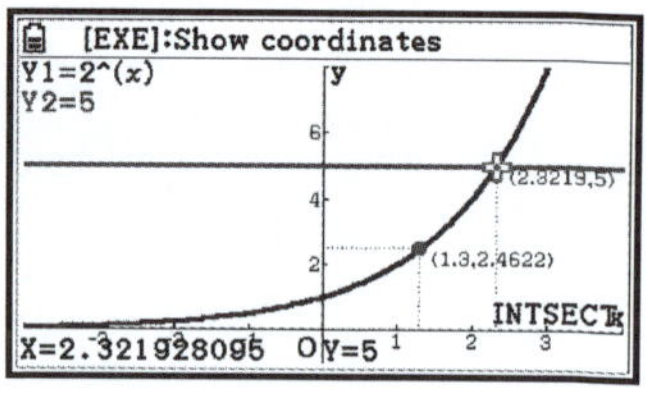

EXERCISE 23C

1 Alongside is the graph of $y = 2^x$.
Use the graph to estimate, to 1 decimal place:

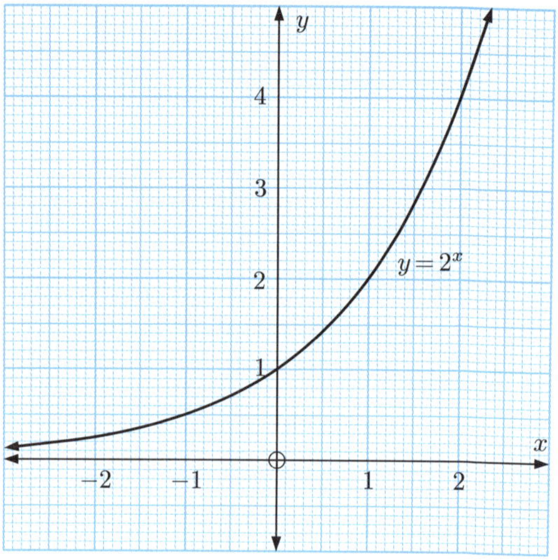

a the value of:
- **i** $2^{0.5}$ **ii** $2^{2.1}$

b the solution to:
- **i** $2^x = 3$ **ii** $2^x = 0.4$

2 Alongside is the graph of $y = 3^x$.
Use the graph to estimate, to 1 decimal place:

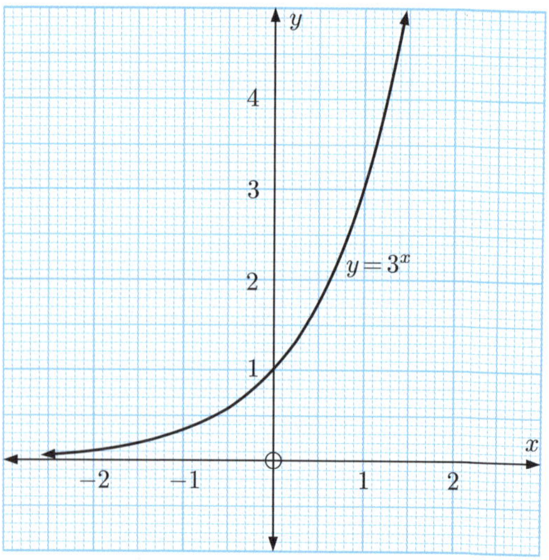

a the value of:
- **i** $3^{0.5}$ **ii** $3^{-1.2}$

b the solution to:
- **i** $3^x = 4.2$ **ii** $3^x = 0.6$

3 Alongside is the graph of $y = 3^{-x}$.
Use the graph to estimate, to 1 decimal place:

a the value of:
i $3^{-0.5}$ **ii** $3^{1.4}$

b the solution to:
i $3^{-x} = 2$ **ii** $3^{-x} = 0.2$

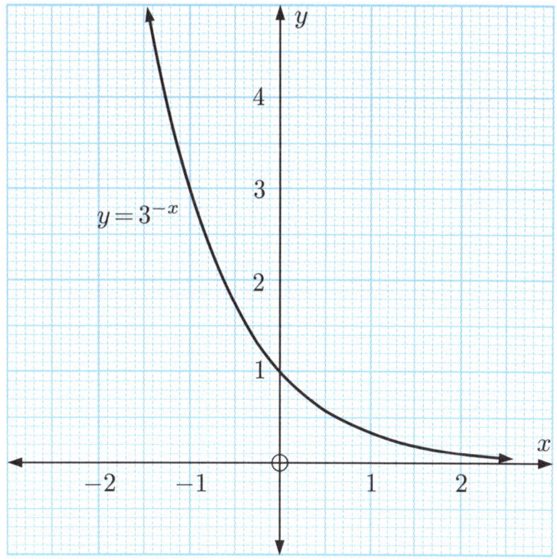

4 Consider the function $y = 2^{-x}$.

a Copy and complete the table of values:

x	-3	-2	-1	0	1	2	3
y							

b Use the table in **a** to sketch the graph of $y = 2^{-x}$.

c Is the x-axis a horizontal asymptote of $y = 2^{-x}$? Explain your answer.

d Use your graph to estimate, to 1 decimal place:
i the value of $2^{-0.5}$ **ii** the solution to $2^{-x} = 3.5$.

D GROWTH

In the **Opening Problem**, an ants nest initially contains 50 ants, and the population of ants doubles each week.

Initially, there are $50 = 50 \times 2^0$ ants.

After 1 week, there are $100 = 50 \times 2^1$ ants.

After 2 weeks, there are $200 = 50 \times 2^2$ ants.

After 3 weeks, there are $400 = 50 \times 2^3$ ants.

So, the population P after t weeks is given by $P = 50 \times 2^t$ ants.

The graph of P against t is shown alongside.

We can use the function to find P for any value of $t \geqslant 0$.

For example, when $t = 2.5$, $P = 50 \times 2^{2.5}$
≈ 283

So, after 2.5 weeks there are 283 ants.

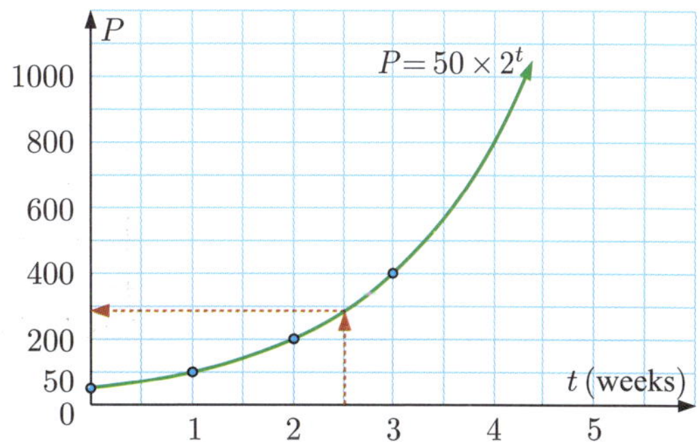

Populations of people, animals, and bacteria are examples of quantities which may show **exponential growth**.

Example 6 — Self Tutor

The population of rabbits on a farm is given by the function $R = 50 \times (1.07)^n$ where n is the number of weeks after the rabbit farm was established.

a What was the original rabbit population?

b How many rabbits were present after 15 weeks?

c How many rabbits were present after 30 weeks?

d Sketch the graph of R against n for $n \geqslant 0$.

e How long will it take for the population to reach 500?

a When $n = 0$, $R = 50 \times (1.07)^0$
$= 50 \times 1$
$= 50$ $\quad\therefore$ there were 50 rabbits originally.

b When $n = 15$, $R = 50 \times (1.07)^{15}$
≈ 137.95 $\quad\therefore$ there were 138 rabbits after 15 weeks.

c When $n = 30$, $R = 50 \times (1.07)^{30}$
≈ 380.61 $\quad\therefore$ there were 381 rabbits after 30 weeks.

d

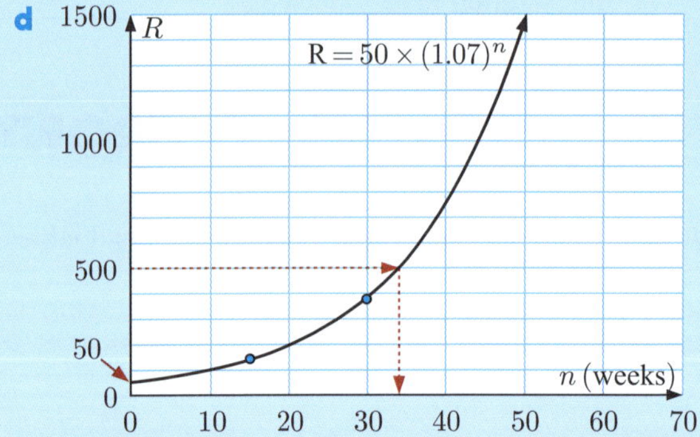

e From the graph, the number of weeks to reach 500 rabbits is approximately 34.
Alternatively, we could use technology to find where the graph cuts the line $R = 500$.

Casio fx-CG20

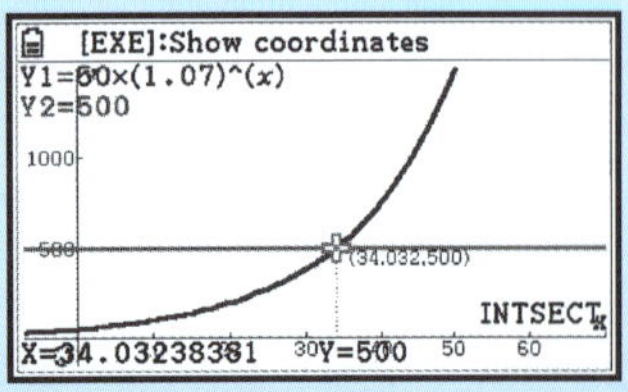

TI-84 Plus

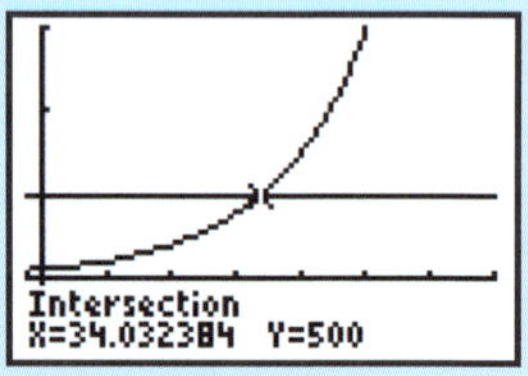

TI-*n*spire

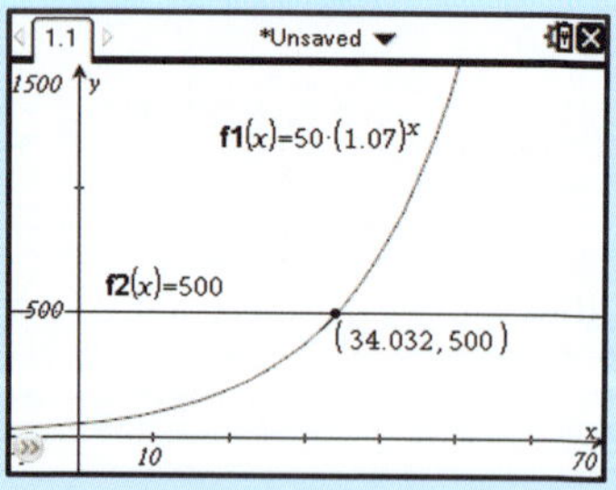

$\therefore$ it will take about 34 weeks.

EXERCISE 23D

You are encouraged to use technology to help answer the following questions.

1 The population of mice in a field after n weeks is given by $P = 500 \times (1.12)^n$.

- **a** How many mice were originally in the field?
- **b** How many mice were in the field after: **i** 10 weeks **ii** 20 weeks?
- **c** Sketch the graph of P against n for $n \geqslant 0$.
- **d** How many weeks will it take for the mouse population to reach 2000?

2 The weight of bacteria in a culture t hours after it has been established is given by the formula $W = 20 \times (1.007)^t$ grams.

- **a** Find the original weight of bacteria in the culture.
- **b** Find the weight of the bacteria after 24 hours.
- **c** Sketch the graph of W against t for $t \geqslant 0$.
- **d** How long will it take for the weight to reach 50 grams?

3 The population of wasps in a nest n days after it was discovered is given by $P = 250 \times (1.06)^n$.

- **a** How many wasps were in the nest originally?
- **b** Find the number of wasps after: **i** 25 days **ii** 8 weeks.
- **c** Sketch the graph of P against n for $n \geqslant 0$.
- **d** How long will it take for the population to double?

4 The population of a city was determined by census at 10 year intervals:

Year	1970	1980	1990	2000	2010
Population (thousands)	23.0	27.6	33.1	39.7	47.7

Suppose x is the number of years since 1970, and P is the population (in thousands).

- **a** Draw the graph of P against x with P on the vertical axis.
- **b** The variables P and x are connected by the exponential function $P = a \times b^x$ where a and b are constants.
  Find the value of:
  i a, using the 1970 population data **ii** b, using the 2010 population data.
- **c** Using the values of a and b found in **b**, check that your exponential formula fits the data from 1980, 1990, and 2000.
- **d** Use your formula to predict the city's population in the year: **i** 2020 **ii** 2060.

E DECAY

When the value of a variable decreases exponentially over time, we call it **exponential decay**.

Examples of quantities which decay exponentially are the mass of radioactive materials, the current in circuits, and the depreciating value of goods.

Example 7 **Self Tutor**

When a diesel-electric generator is switched off, the current decays according to the formula $I = 24 \times (0.25)^t$ amps, where t is the time in seconds.

a Find the value of I when $t = 1$, 2, and 3.

b What current flowed in the generator at the instant when it was switched off?

c Plot the graph of I against t for $t \geqslant 0$ using your results from **a** and **b**.

d How long will it take for the current to fall to 4 amps?

a When $t = 1$,
$I = 24 \times (0.25)^1$
$= 6$ amps

When $t = 2$,
$I = 24 \times (0.25)^2$
$= 1.5$ amps

When $t = 3$,
$I = 24 \times (0.25)^3$
$= 0.375$ amps

b When $t = 0$, $I = 24 \times (0.25)^0 = 24$

$\therefore$ 24 amps of current flowed at the instant the generator was switched off.

c

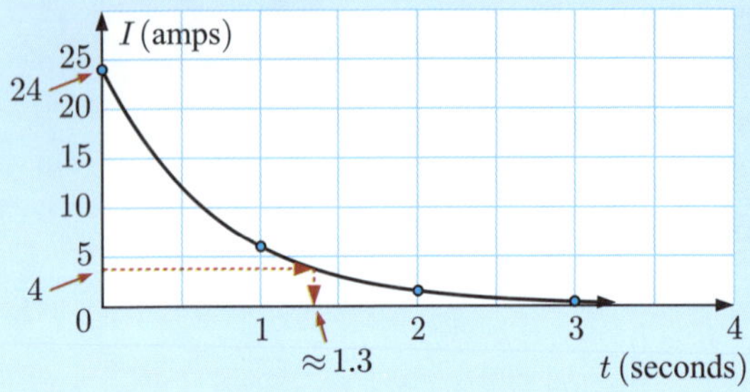

d

Casio fx-CG20

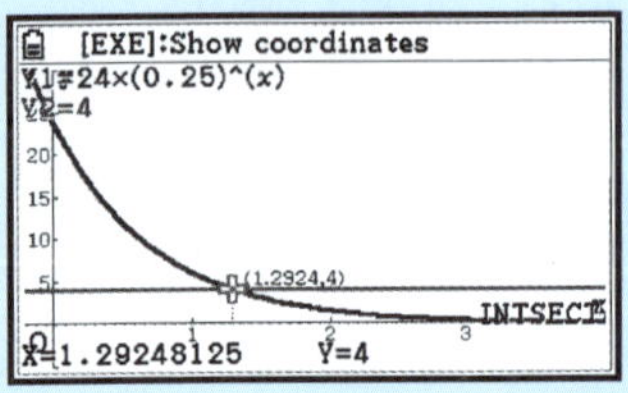

TI-84 Plus

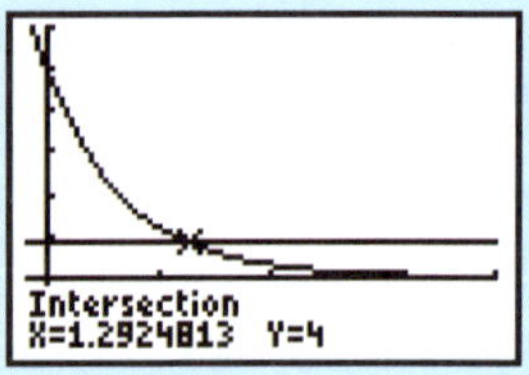

TI-*n*spire

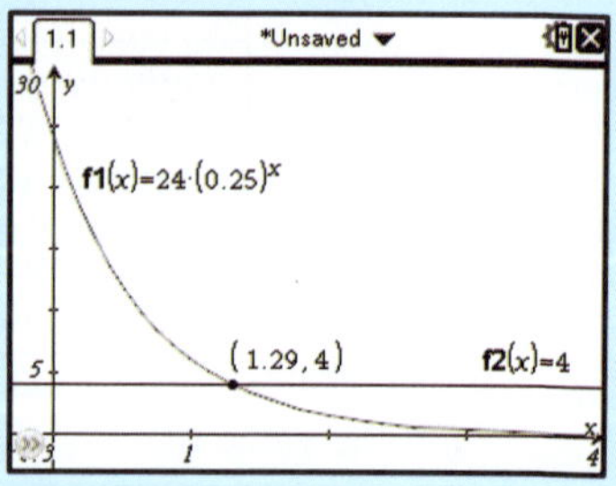

It will take ≈ 1.29 seconds for the current to fall to 4 amps.

EXERCISE 23E

1 When a container of liquid is left to cool, its temperature in °C is given by $T = 100 \times (0.933)^t$, where t is the time in minutes.

a Find the initial temperature of the liquid.

b Find the temperature after:
  i 10 minutes **ii** 20 minutes **iii** 30 minutes.

c Draw the graph of T against t for $t \geqslant 0$, using your results from **a** and **b**.

d How long will it take for the liquid's temperature to fall to:
  i 40°C **ii** 10°C?

2 The weight of a radioactive substance t years after being discovered is given by $W = 150 \times (0.997)^t$ grams.

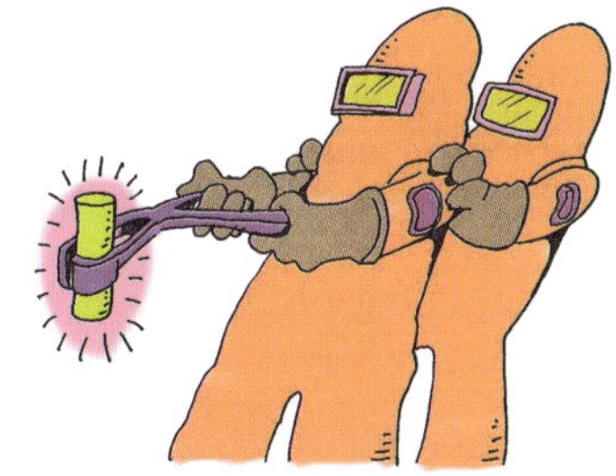

- **a** How much radioactive substance was discovered?
- **b** Determine the weight of the substance after:
  - **i** 100 years **ii** 200 years **iii** 400 years.
- **c** Sketch the graph of W against t for $t \geqslant 0$, using your results from **a** and **b**.
- **d** How long will it take for the substance to decay to 25 grams?

3 The marsupial *Eraticus* is endangered. There is only one colony remaining, and research into its population has shown its decline over the last 25 years:

Year	1985	1990	1995	2000	2005	2010
Population	255	204	164	131	105	84

Let n be the time since 1985 and P be the population size.

- **a** Graph P against n with P on the vertical axis.
- **b** It is believed that P and n are connected by the function $P = a \times b^n$ where a and b are constants. Find:
  - **i** a, using the 1985 population
  - **ii** b, using the 2010 population.
- **c** Using the values of a and b found in **b** check that your exponential function fits the data from 1990, 1995, 2000, and 2005.
- **d** In what year do you expect the population size to be reduced to 50?

ACTIVITY — CARBON DATING

You have probably heard of **carbon dating** as a technique for estimating the age of artefacts or bones. To understand how carbon dating works, we first need to understand the concept of **half-life**.

HALF-LIFE

The **half-life** of a radioactive substance is the time it takes for the substance to decay to half of its original amount.

If we know the half-life of a substance, we can use an exponential function to model the mass of the substance over time.

What to do:

Suppose the masses of substances A, B, and C are modelled by $M_A = 80 \times 2^{-t}$ grams, $M_B = 100 \times 2^{-\frac{t}{2}}$ grams, and $M_C = 60 \times 2^{-\frac{t}{5}}$ grams respectively, where t is in years.

1. Draw the graphs of M_A, M_B, and M_C on the same set of axes. Find the original mass of each substance.
2. From the graphs, determine how long it takes for each substance to decay to half of its original mass.
3. Hence, conjecture a formula for the mass M of a substance with an initial mass of M_0 grams and a half-life of n years.

CARBON DATING

All living organisms contain the element carbon. In nature, carbon exists as two *isotopes* called carbon-12 and carbon-14. The ratio of carbon-12 to carbon-14 is the same in each living organism. For every carbon-14 atom present, there are approximately 10^{12} carbon-12 atoms.

When an organism dies, the amount of carbon-12 in the organism remains constant, but the carbon-14 begins to decay.

It is known that the half-life of carbon-14 is approximately 5700 years. So, by comparing the ratio of carbon-12 to carbon-14 with that of a living organism, we can determine how long the carbon-14 has been decaying for, and hence how long ago the organism died.

What to do:

4 When a piece of wood is analysed, it is found to contain 10^{25} carbon-12 atoms and 4×10^{12} carbon-14 atoms.

- **a** Determine how many carbon-14 atoms were present in the wood when it died.
- **b** Use **a** and the half-life of carbon-14 to obtain a function to model the number N of carbon-14 atoms present in the wood t years after it died.
- **c** Draw a graph of N for $0 \leqslant t \leqslant 10\,000$.
- **d** Use your graph to estimate the age of the wood.

5 An artefact dealer claims to own an object from Ancient Egypt which is 4000 years old. Analysis of the object shows that there are 10^{30} carbon-12 atoms and 8.4×10^{17} carbon-14 atoms present. Test the artefact dealer's claim.

F SIMPLE RATIONAL FUNCTIONS

The simplest rational function has an equation of the form $y = \frac{k}{x}$ where k is a constant.

It has a graph which is called a **rectangular hyperbola**.

There are examples of hyperbolae in the world around us:

- When an aeroplane flies faster than the speed of sound, which is around 1200 km h^{-1}, we say it breaks the sound barrier. It sets up a shock wave in the shape of a **cone**, and when this intersects the ground it does so in the shape of an **hyperbola**. The **sonic boom** formed hits every point on the curve at the same time, so that people in different places along the curve on the ground all hear it at the same time. No sound is heard outside the curve, but the boom eventually covers every place inside it.

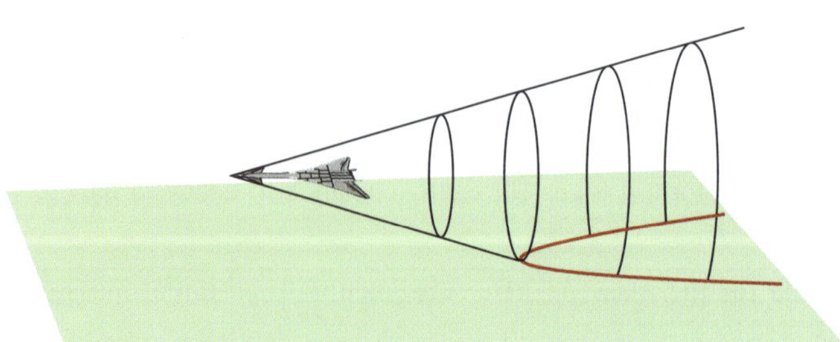

- 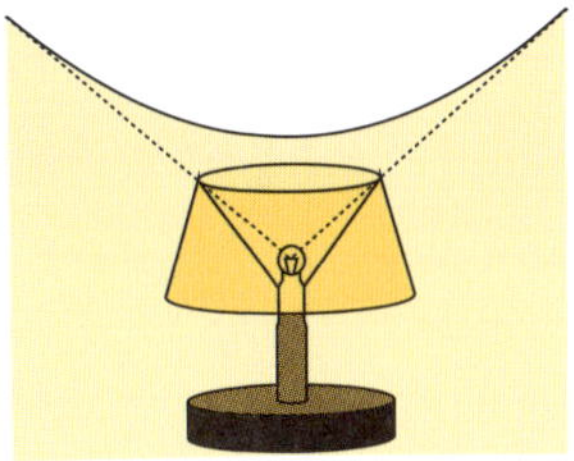

We can also notice an hyperbola in the home when a lamp is close to a wall.

There are many situations in which two quantities vary inversely. They form a **reciprocal relationship** which can be described using a rational function.

For example, the pressure and volume of a gas at room temperature vary inversely according to the equation $P = \frac{77.4}{V}$.

If P is graphed against V, the curve is one branch of an hyperbola.

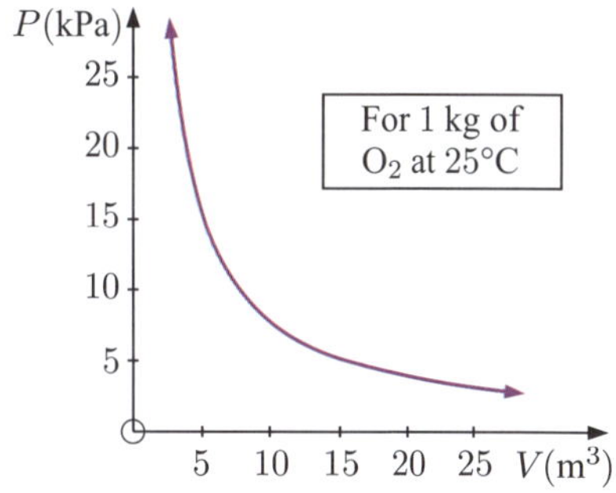

INVESTIGATION THE FAMILY OF CURVES $y = \frac{k}{x}$

In this Investigation, you should use a **graphing package** or **graphics calculator** to draw curves of the form $y = \frac{k}{x}$ where $k \neq 0$.

GRAPHING PACKAGE

What to do:

1 **a** On the same set of axes, draw the graphs of $y = \frac{1}{x}$, $y = \frac{4}{x}$, and $y = \frac{8}{x}$.

b Describe the effect of the value of k on the graph for $k > 0$.

2 **a** On the same set of axes, draw the graphs of $y = \frac{-1}{x}$, $y = \frac{-4}{x}$, and $y = \frac{-8}{x}$.

b Explain how these graphs differ from the graphs in **1**.

3 For any graph of the form $y = \frac{k}{x}$, explain why there is never:

a a y-intercept **b** an x-intercept.

You should have made the following discoveries about the graph of $y = \frac{k}{x}$:

- For $k > 0$, the graph has shape

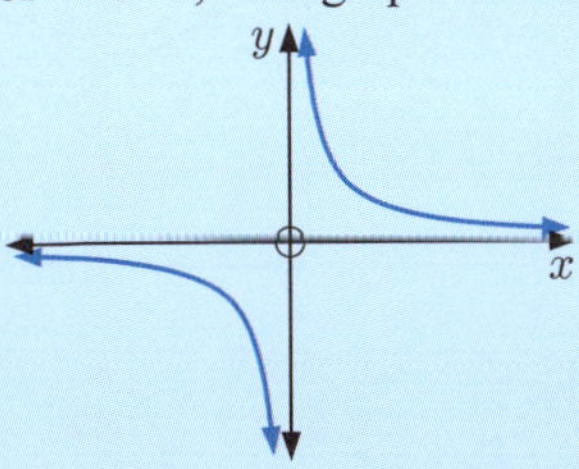

- For $k < 0$, the graph has shape

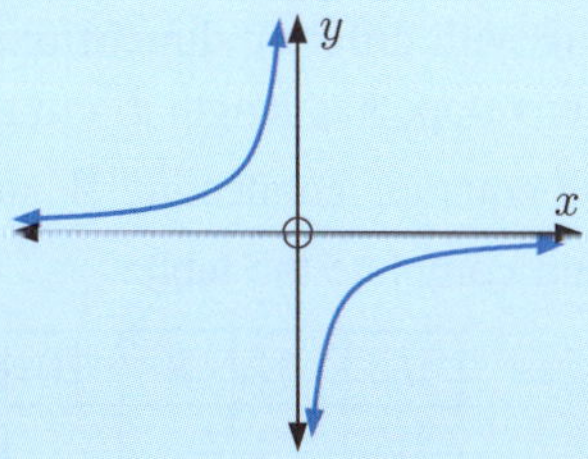

- $x = 0$ is a vertical asymptote.
- $y = 0$ is a horizontal asymptote.

Since there are no axes intercepts, it is useful to determine some other points which can help us draw a graph.

We notice that the curve $y = \frac{k}{x}$ passes through the points $(1,\ k)$, $(k,\ 1)$, $(-1,\ -k)$, and $(-k,\ -1)$.

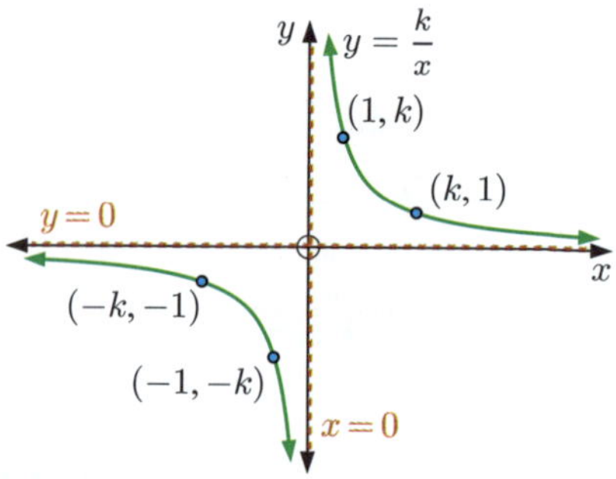

Example 8 — Self Tutor

Sketch the graph of $y = \frac{5}{x}$.

The graph of $y = \frac{5}{x}$ passes through the points $(1,\ 5)$, $(5,\ 1)$, $(-1,\ -5)$, and $(-5,\ -1)$.

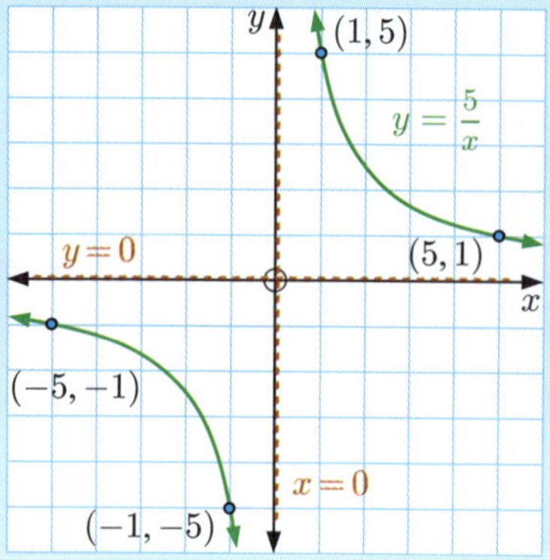

EXERCISE 23F

1 Sketch the following rational functions:

a $y = \frac{2}{x}$ b $y = -\frac{6}{x}$ c $y = \frac{10}{x}$ d $y = -\frac{4}{x}$

2 Determine the equations of the following rectangular hyperbolae:

a
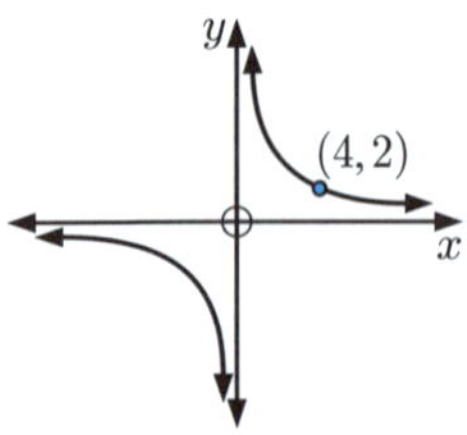

b
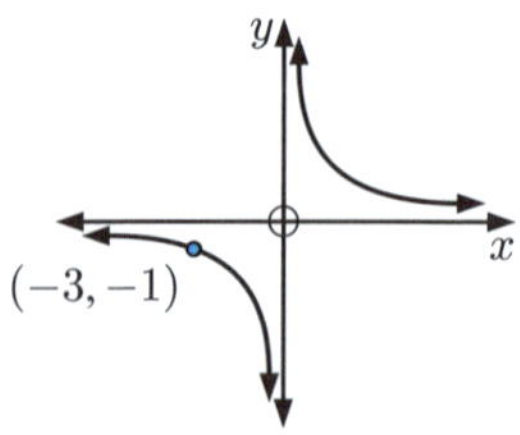

c
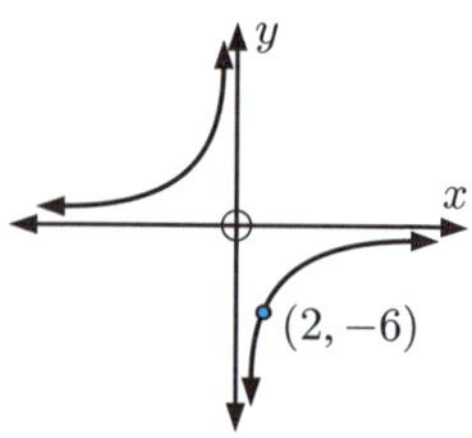

3 Neville wants to choose an envelope with area 120 cm^2, and is experimenting with different dimensions.
Suppose an envelope has length L cm and width W cm.

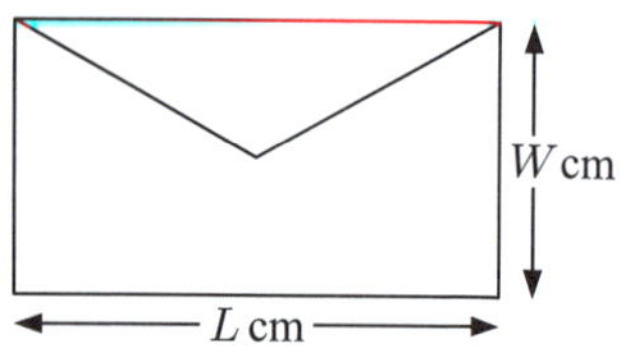

a Write an equation connecting W and L.

b Copy and complete the table:

L	6	8	10	12	15	20
W						

c Use the points in **b** to plot W against L.

4 Kate needs to make 30 costumes for her school play. How fast she can make the costumes will determine how long the job takes her.
Suppose Kate can make c costumes per hour, and the job takes her t hours.

a Write an equation connecting c and t.

b Complete this table of values:

Costumes per hour (c)	3	4	5	6	8	10
Time taken (t)						

c Draw a graph of c versus t, with c on the horizontal axis.

5 Find the axes of symmetry for the rational function $y = \frac{k}{x}$, where k is a constant.

REVIEW SET 23A

1 Solve for x:

a $4^x = 64$ **b** $2^{x-3} = \frac{1}{4}$ **c** $5^{1-x} = \frac{1}{5}$

2 Solve for x:

a $2 \times 3^x = 54$ **b** $8^x = 16$

3 If $y = 3^x - 2$, find the value of y when:

a $x = 0$ **b** $x = 2$ **c** $x = -2$

4 Consider the function $K = 2^x + 5$.

a Find the value of K when: **i** $x = 0$ **ii** $x = 3$

b Find the value of x when: **i** $K = 37$ **ii** $K = \frac{11}{2}$

5 The temperature of a liquid t minutes after it was heated, is given by $T = 80 \times (0.913)^t$ °C.

a Find the initial temperature of the liquid.

b Find the temperature of the liquid after:

i 12 minutes **ii** 24 minutes **iii** 36 minutes.

c Sketch the graph of T against t for $t \geqslant 0$.

d Find the time taken for the temperature to fall to 25°C.

6 A population of termites t weeks after a new nest is founded, is given by $P = 1500 \times (1.087)^t$.

a Find the initial population.

b Find the population of termites after: **i** 3 weeks **ii** 6 months.

c Sketch the graph of P against t.

d Find the time at which the population reaches 50 000.

7 Draw the graph of: **a** $y = \frac{7}{x}$ **b** $y = -\frac{8}{x}$

8 Explain why the graph of $y = \frac{k}{x}$ lies entirely in:

a quadrants 1 and 3 if $k > 0$

b quadrants 2 and 4 if $k < 0$.

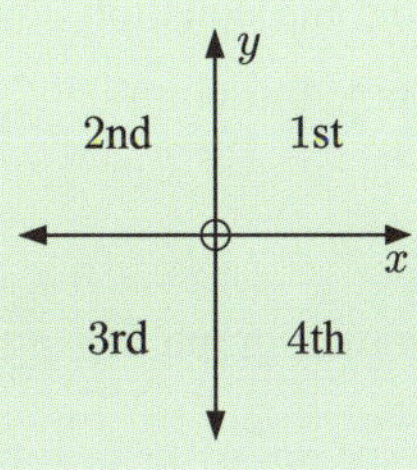

REVIEW SET 23B

1 Solve for x:

a $5^x = 125$ **b** $11^{2x-1} = 1$ **c** $3^{1-x} = \frac{1}{9}$

2 Consider the function $y = 2^{x-1}$. Find the value of y when:

a $x = 0$ **b** $x = 3$ **c** $x = -2$

3 A population P of zebra t years after an initial count is given by $P = 815 \times (1.16)^t$.

a Find the original number of zebras.

b Find the number of zebras after:

i 6 years **ii** 12 years **iii** 18 years.

c Sketch the graph of P against t for $t \geqslant 0$.

d Determine the time taken for the population to reach 8000 zebras.

4 Solve for x:

a $3 \times 2^{x+2} = 48$ **b** $9^{x-1} = \frac{1}{27}$

5 Consider the function $y = 4^x$.

x	-2	-1	0	1	2
y					

a Copy and complete the table of values:

b Use your table to sketch the graph of $y = 4^x$.

c Use your graph to estimate, to 1 decimal place:

i the value of $4^{0.8}$ **ii** the solution to $4^x = 10$.

6 Determine the equation of this hyperbola:

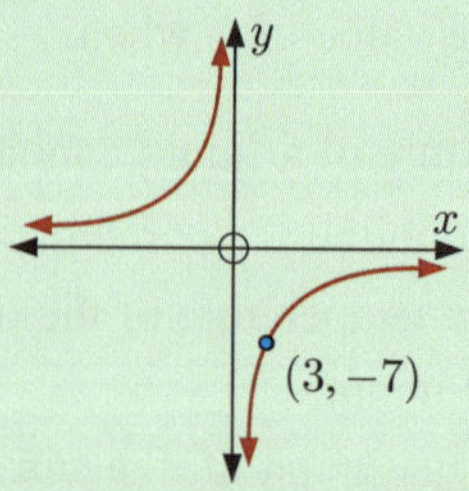

7 The weight of a radioactive substance after t years is given by $W = 1500 \times (0.993)^t$ grams.

a Find the original weight of radioactive material.

b Find the weight of radioactive material remaining after:

i 400 years **ii** 800 years.

c Sketch the graph of W against t for $t \geqslant 0$.

d Find the time taken for the weight to reduce to 200 grams.

8 Gino is travelling from Los Angeles to San Francisco, which is a distance of approximately 600 km. Let s be Gino's average speed in km/h, and t be the time the trip takes in hours.

a Find the equation connecting s and t.

b Copy and complete:

Speed (s km/h)	40	50	60	80	100	120
Time (t hours)						

c Draw a graph of t against s, with s on the horizontal axis.

Chapter 24

Proportion

Contents:

OPENING PROBLEM

A shop sells cans of soft drink for \$2 each. Suppose we buy n cans and the total cost is \$$C$.

To study the relationship between the *number of cans* and the *total cost*, we can use a table of values or a graph.

n	0	1	2	3	4	5
C	0	2	4	6	8	10

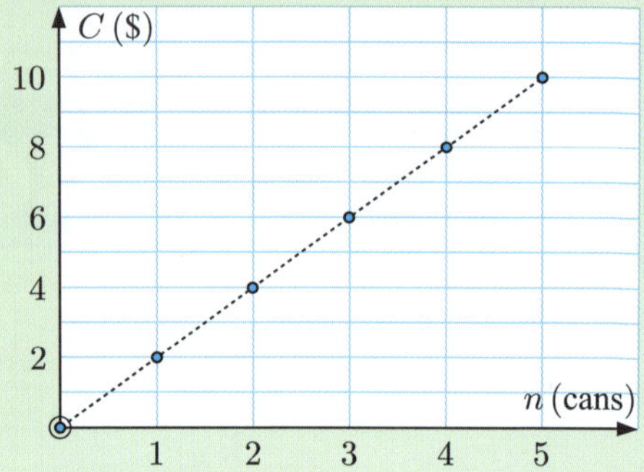

The total cost *depends* on the number of cans bought.

The number of cans is the independent variable, so is placed on the x-axis.
The total cost is the dependent variable, so is placed on the y-axis.

Since we can only buy a whole number of cans, the graph of C against n consists of discrete points. However, an imagined line passing through these points would also pass through the origin.

Things to think about:

a Which of the following are true?

- **i** doubling the number of cans doubles the total cost
- **ii** halving the number of cans halves the total cost
- **iii** increasing the number of cans by 30% increases the total cost by 30%.

b How can we describe the relationship between n and C?

A DIRECT PROPORTION

The rectangles alongside are **similar**. Their sides are in the same **ratio**.

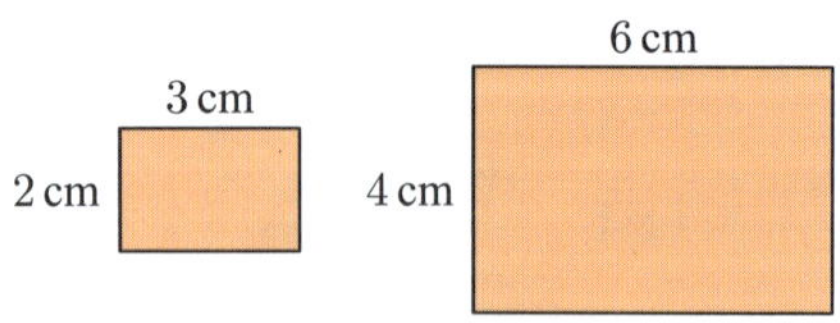

In previous years we have seen that a **proportion** is a statement that two ratios are equal.

For example, the ratios of the side lengths of the rectangles can be written as the proportion $2 : 3 = 4 : 6$ (×2, ×2)

To convert one ratio to the other, we multiply both parts by the same constant 2.

A rectangle with sides x cm and y cm will be similar to the above rectangles if $x : y = 2 : 3$. This occurs when $y = \frac{3}{2}x$, for any x.

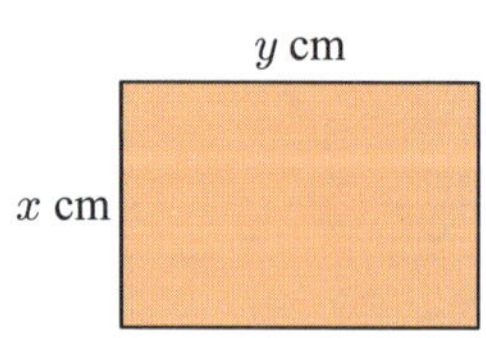

Since the sides are in proportion, we say that y is **directly proportional** to x.

Two variables are **directly proportional** if multiplying one of them by a number results in the other one being multiplied by the same number.

In the **Opening Problem**, C *is directly proportional to* n. We write $C \propto n$. For example, when n is tripled from 1 to 3, C is tripled from 2 to 6.

If we plot C against n, the points lie in a straight line with gradient 2, which passes through the origin. The variables are therefore connected by the formula $C = 2n$.

If two quantities x and y are **directly proportional**, we write $y \propto x$.

If $y \propto x$ then $y = kx$ where k is a constant called the **proportionality constant**.

When y is graphed against x, the graph is a straight line with **gradient** k, which passes through the **origin**.

Example 1 Self Tutor

Fruit buns are sold for 60 cents each.

a Use a graph to show that \$$y$, the total price paid, is directly proportional to x, the number of fruit buns bought.

b Find the law connecting x and y.

a

x	0	1	2	3	4	5
y	0	0.60	1.20	1.80	2.40	3.00

The points lie in a straight line passing through (0, 0), so $y \propto x$.

b The gradient of the line $= \frac{0.60 - 0}{1 - 0} = 0.6$

$\therefore$ the proportionality constant k is 0.6.

$\therefore$ the law connecting x and y is $y = 0.6x$.

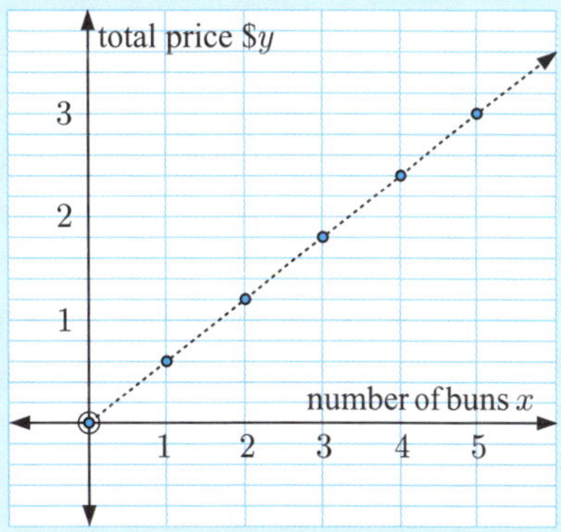

EXERCISE 24A.1

1 Which of the following graphs indicates that y is directly proportional to x?

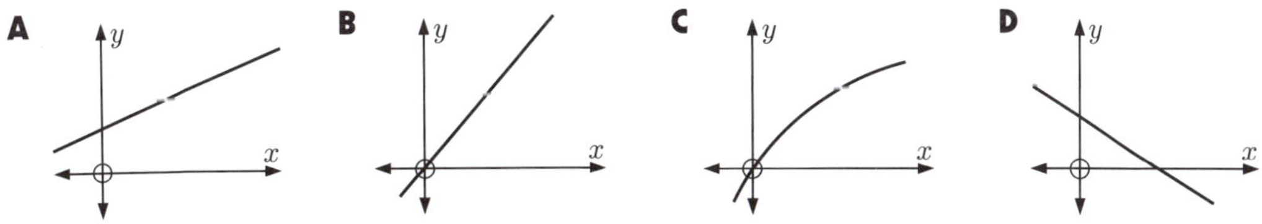

2 For each of the following tables of values:

i Plot the graph of y against x.

ii Determine whether y is directly proportional to x. If it is, find the proportionality constant.

a

x	0	1	2	3	4
y	0	3	6	9	12

b

x	0	1	2	3	4
y	3	5	7	9	11

c

x	0	1	3	5	6
y	0	4	8	12	16

d

x	2	3	5	6	8
y	5	7.5	12.5	15	20

3 The table alongside shows the total wages $\$W$ earned for working h hours.

h	0	1	2	3	4	5
W	0	12.50	25.00	37.50	50.00	62.50

a Draw a graph of W against h.

b Explain why W and h are directly proportional.

c Determine the law connecting W and h.

4 A dripping tap leaks water at a rate of 10 mL per minute.

a Copy and complete the table of values alongside:

Time (t minutes)	0	2	4	6	8	10
Volume of water leaked (V mL)						

b Plot the graph of V against t.

c Are V and t directly proportional? If so, find the proportionality constant.

5 A dry cleaner charges \$5 for cleaning each of the first three items, then \$2 for each extra item.

a Copy and complete the table of values alongside:

Number of items (n)	0	1	2	3	4	5
Cost ($\$C$)						

b Plot C against n.

c Are C and n directly proportional? If so, find the proportionality constant.

6 For each of the following equations:

i draw the graph of y against x

ii determine whether y is directly proportional to x.

a $y = 4x$ **b** $y = x + 2$ **c** $y = \frac{7x}{3}$ **d** $y = 6 - x$

7 Suppose y is directly proportional to x. State what happens to:

a y if x is doubled

b y if x is trebled

c x if y is doubled

d x if y is halved

e y if x is increased by 20%

f x if y is decreased by 30%

g y if 2 is added to x

h y if 3 is subtracted from x.

8 The law connecting the circumference C and radius r of a circle is $C = 2\pi r$.

a Explain why $C \propto r$.

b Find the proportionality constant.

c State what happens to:

i C if r is doubled

ii r if C is increased by 50%.

PROBLEM SOLVING USING DIRECT PROPORTION

If two variables are directly proportional, we can use either a formula or a multiplier to find the value of one variable given the value of the other. In general, using a multiplier is the quicker method.

Example 2 — Self Tutor

Suppose $y \propto n$ and that $y = 40$ when $n = 3$. Find n when $y = 137$.

Method 1:

Since $y \propto n$, $y = kn$ where k is the proportionality constant.

Since $n = 3$ when $y = 40$,

$$40 = k \times 3$$
$$\therefore \frac{40}{3} = k$$
$$\therefore y = \frac{40}{3}n$$

So, when $y = 137$,

$$137 = \frac{40}{3}n$$
$$\therefore 137 \times \frac{3}{40} = n$$
$$\therefore n \approx 10.3$$

Method 2:

n	3	?
y	40	137

$\times \frac{137}{40}$

To change y from 40 to 137, we multiply by $\frac{137}{40}$.

Since $y \propto n$, we must also multiply n by $\frac{137}{40}$

$\therefore n = 3 \times \frac{137}{40} \approx 10.3$

EXERCISE 24A.2

1 Suppose $y \propto x$ and that $y = 35$ when $x = 7$. Find:

- **a** y when $x = 28$
- **b** x when $y = 5$.

2 Suppose $y \propto x$ and that $y = 123$ when $x = 7.1$. Find:

- **a** y when $x = 13.2$
- **b** x when $y = 391$.

3 The amount of petrol Jill's car uses each day is directly proportional to the distance she travels. Yesterday she travelled 81 km, and used 6 litres of petrol.

- **a** If Jill travelled 54 km in one day, how much petrol would she use?
- **b** Jill used 9 litres of petrol in one day. How far did she travel?

4 The resistance R ohms to the flow of electricity in a wire varies in direct proportion to the length l cm of the wire. A 10 cm length of wire has resistance 0.06 ohms.

- **a** Find:
  - **i** the law connecting R and l
  - **ii** the resistance when the wire is 50 cm long
  - **iii** the length of wire which has a resistance of 3 ohms.
- **b** Plot the graph of R against l.

5 A 4 litre can of paint will cover 18 m^2 of wall. Find:

- **a** the area of wall which can be covered using 10 litres of paint
- **b** the amount of paint required to paint a room with wall area 40.5 m^2.

6 The speed of a falling object is directly proportional to the time for which it falls. The speed after 5 seconds is 49 m/s.

a Find the speed of an object which falls for 8 seconds.

b How long will it take a falling object to reach a speed of 100 m/s?

7 The dimensions of a model car are directly proportional to the dimensions of the actual car it was designed from. The actual car is 4 m long, and the model car is 30 cm long.

a The actual car is 1.4 m wide. Find the width of the model car.

b The model car is 9 cm high. Find the height of the actual car.

B POWERS IN DIRECT PROPORTIONS

The formula for finding the area A of a circle of radius r is $A = \pi r^2$.

r	0	1	2	3	4
A	0	3.14	12.57	28.27	50.27

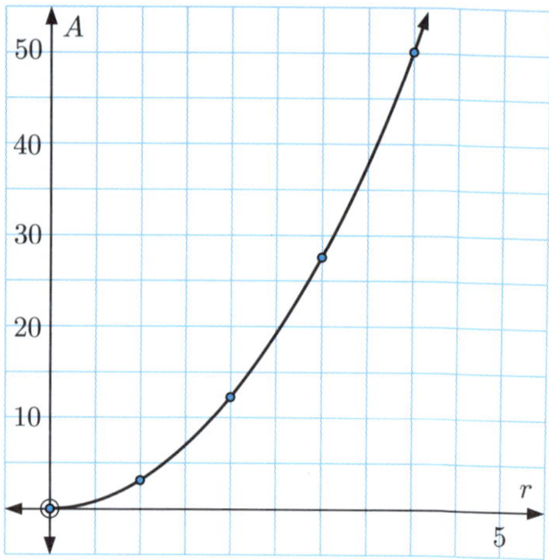

The graph of A against r is shown alongside. The graph is not a straight line, but rather is part of a **parabola**. So, A is *not* directly proportional to r.

However, suppose we include in the table a row for the values of r^2:

r	0	1	2	3	4
r^2	0	1	4	9	16
A	0	3.14	12.57	28.27	50.27

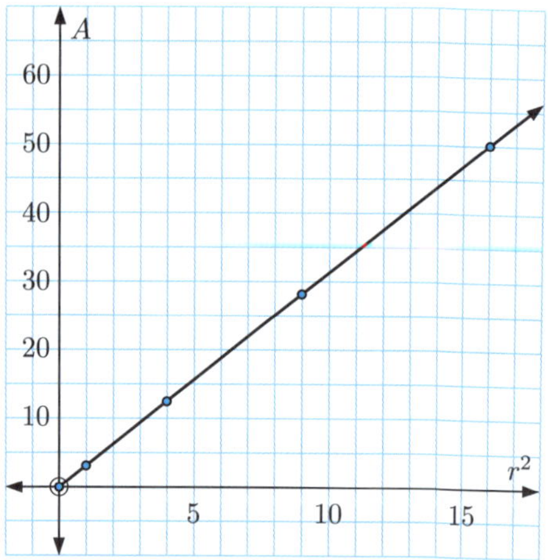

The graph of A against r^2 is a straight line through the origin O. A is directly proportional to r^2, and we write $A \propto r^2$. In this case we know from the formula that the proportionality constant is $k = \pi$.

Notice from the table that if r is doubled from 1 to 2, both r^2 and A are multiplied by 4. If r is trebled from 1 to 3, both r^2 and A are multiplied by 9.

Example 3 Self Tutor

Consider $y = \frac{x^4}{5}$. State which two variables are directly proportional and determine the proportionality constant k.

Since $y = \frac{1}{5}x^4$, so $y \propto x^4$ and $k = \frac{1}{5}$.

y is directly proportional to the fourth power of x, and the proportionality constant $k = \frac{1}{5}$.

EXERCISE 24B

1 State which two variables are directly proportional, and determine the proportionality constant k:

a $A = 4.9t^2$ **b** $K = 2x^3$ **c** $T = \frac{\sqrt{l}}{5}$

d $V = 500t^4$ **e** $P = 4\sqrt[3]{x}$ **f** $V = \frac{4}{3}\pi r^3$

Example 4 Self Tutor

Consider the table of values shown.

a By plotting y against x^2, establish that $y \propto x^2$.

b Hence, write down the rule connecting y and x.

x	2	4	6
y	6	24	54

a

x	2	4	6
x^2	4	16	36
y	6	24	54

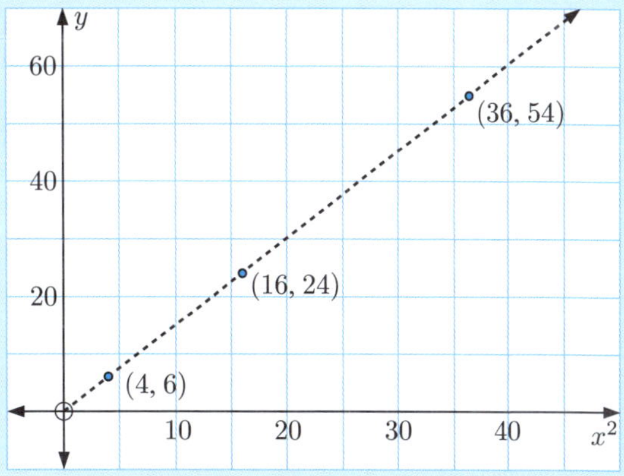

The graph of y against x^2 is a straight line through the origin, so $y \propto x^2$.

b The line has gradient $= \frac{24-6}{16-4}$

$= \frac{18}{12}$

$= 1.5$

$\therefore \quad y = 1.5x^2$

2 By plotting y against x^2, establish that $y \propto x^2$ and state the rule connecting y and x.

a

x	2	3	5
y	16	36	100

b

x	1	3	4	5
y	3	27	48	75

3 By plotting y against x^3, establish that $y \propto x^3$ and find the proportionality constant k.

a

x	1	2	3
y	5	40	135

b

x	2	3	5
y	4	13.5	62.5

Example 5 **Self Tutor**

Suppose T is directly proportional to d^2, and that $T = 100$ when $d = 2$. Find:

a T when $d = 3$ **b** d when $T = 200$, given $d > 0$.

a

$\times\frac{3}{2}$

d	2	3
T	100	

d is multiplied by $\frac{3}{2}$

$\therefore$ d^2 is multiplied by $\left(\frac{3}{2}\right)^2$

$\therefore$ T is multiplied by $\left(\frac{3}{2}\right)^2$ {as $T \propto d^2$}

$\therefore$ $T = 100 \times \left(\frac{3}{2}\right)^2 = 225$

b

d	2	
T	100	200

$\times 2$

T is multiplied by 2

$\therefore$ d^2 is multiplied by 2 {as $T \propto d^2$}

$\therefore$ d is multiplied by $\sqrt{2}$ {as $d > 0$}

$\therefore$ $d = 2 \times \sqrt{2} = 2\sqrt{2}$

4 Suppose P is directly proportional to a^2, where $a > 0$. When $a = 5$, $P = 300$. Find:

a the value of P when $a = 2$ **b** the value of a when $P = 2700$.

5 Suppose M is directly proportional to the cube of x, and that when $x = 2$, $M = 24$. Find:

a the value of M when $x = 3$ **b** the value of x when $M = 120$.

Example 6 **Self Tutor**

The power required to drive a motor boat is directly proportional to the cube of its speed. If an 80 hp motor drives a boat at 20 km/h, what power is required to drive a boat at 30 km/h?

Let P be the power in hp required to produce a speed of s km/h.

$\therefore$ $P \propto s^3$

$\times\frac{3}{2}$

s	20	30
P	80	

s is multiplied by $\frac{3}{2}$

$\therefore$ s^3 is multiplied by $\left(\frac{3}{2}\right)^3$

$\therefore$ P is multiplied by $\left(\frac{3}{2}\right)^3$ {as $P \propto s^3$}

$\therefore$ $P = 80 \times \left(\frac{3}{2}\right)^3 = 270$

So, a 270 hp motor is needed.

6 The value of a gem stone varies in direct proportion to the square of its weight. If a 4 carat stone is valued at \$200, find the value of a 5 carat stone of the same quality.

7 The surface area of a sphere varies in direct proportion to the square of its radius. A sphere of radius 6 cm has a surface area of about 452 cm^2. Find, to the nearest mm, the radius of a sphere with surface area 1000 cm^2.

Example 7

Self Tutor

The *period* or time for one complete swing of a pendulum is directly proportional to the square root of its length. When the length is 25 cm, the period is 1.00 seconds.

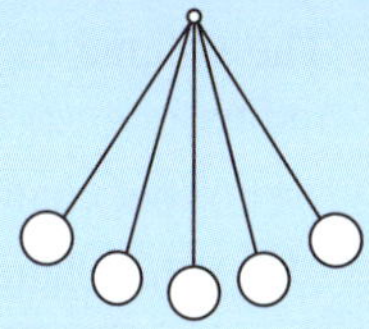

a If the length of the pendulum is 70 cm, find its period. Give your answer to 2 decimal places.

b What would the length be if the period of the pendulum was 2 seconds?

If T is the period and l is the length, then $T \propto \sqrt{l}$.

a ($\times\frac{70}{25}$)

l	25	70
T	1.00	

l is multiplied by $\frac{70}{25}$

$\therefore$ $\sqrt{l}$ is multiplied by $\sqrt{\frac{70}{25}}$

$\therefore$ T is multiplied by $\sqrt{\frac{70}{25}}$ {as $T \propto \sqrt{l}$}

$\therefore$ $T = 1.00 \times \sqrt{\frac{70}{25}} \approx 1.67$

So, the period would be 1.67 seconds.

b

l	25	
T	1.00	2

($\times 2$)

T is multiplied by 2

$\therefore$ $\sqrt{l}$ is multiplied by 2 {as $T \propto \sqrt{l}$}

$\therefore$ l is multiplied by 2^2

$\therefore$ $l = 25 \times 2^2 = 100$

So, the length would be 100 cm.

8 When a stone falls freely, the time taken for it to hit the ground varies in direct proportion to the square root of the distance fallen. If it takes a stone 4 seconds to fall 78.4 m, how long will it take for a stone to fall 500 m down a mine shaft?

9 At sea, the distance in km of the visible horizon is directly proportional to the square root of the height in metres of the observer's eye above sea level.

So, $D \propto \sqrt{h}$.

If the horizon is 9 km when my eye is 5.4 m above sea level, how far can I see from a height of 10 m?

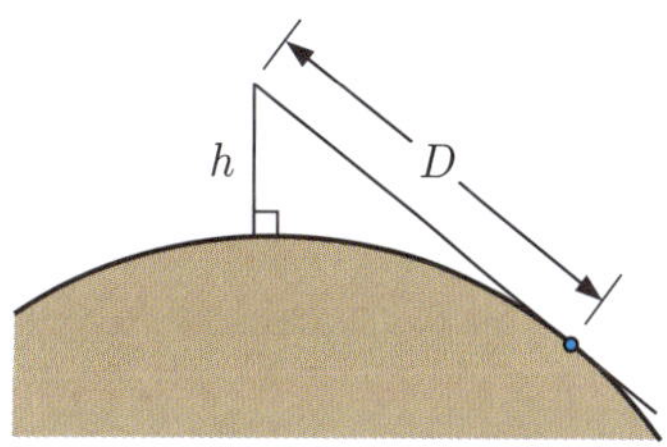

Example 8

Self Tutor

The volume of a cylinder of fixed height varies in direct proportion to the square of the base radius. Find the change in volume when the base radius is increased by 18%.

If the volume is V and the radius is r, then $V \propto r^2$.

When the radius is increased by 18%,

r is multiplied by 118% or 1.18

$\therefore$ r^2 is multiplied by $(1.18)^2$

$\therefore$ V is multiplied by $(1.18)^2 = 1.3924$

$\therefore$ V is multiplied by 139.24%

$\therefore$ the volume is increased by 39.24%.

10 The volume of blood flowing through a blood vessel is directly proportional to the square of the internal diameter. If the diameter is reduced by 20%, what reduction in blood flow would occur?

11 The volume V of a sphere varies in direct proportion to the cube of its radius.

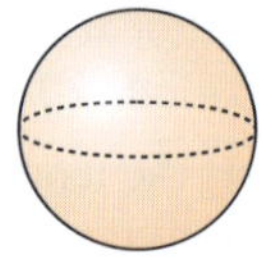

a What change in radius is necessary to double the volume?

b What change in volume reduces the radius by 10%?

12 The volume of a cone is given by $V = \frac{1}{3}\pi r^2 h$, where r is the radius and h is the height.

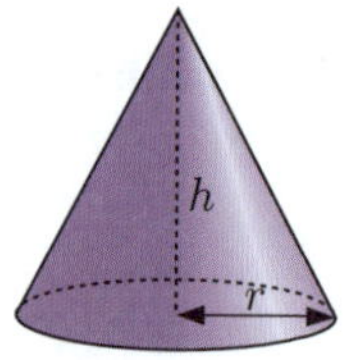

a If we consider cones of fixed radius but variable height, what proportionality exists between V and h?

b If we consider cones of fixed height but variable radius, what proportionality exists between V and r?

c What happens to the volume of a cone if the base radius is increased by 20% and the height is decreased by 15%?

C INVERSE PROPORTION

If two painters can paint a house in three days, how long would it take four painters to paint the house, each working at the same rate?

The four painters will be able to do twice as much work in the same amount of time. So it will therefore take them a half of the time, or $1\frac{1}{2}$ days, to complete the job.

The two variables in this example are the *number of painters* and the *time taken.*

In a case like this where doubling one variable halves the other, we have an **inverse proportion**.

Two variables are **inversely proportional** or **vary inversely** if, when one is *multiplied* by a constant, the other is *divided* by the same constant.

Dividing by k is the same as multiplying by $\frac{1}{k}$, so if one of the variables is multiplied by 2, the other must be multiplied by $\frac{1}{2}$.

Consider again the example of two painters completing a job in three days.

Suppose x is the number of painters and y is the number of days to complete the job. Clearly x must be a positive integer.

The table below shows some possible combinations of x and y.

x	1	2	3	4	6
y	6	3	2	$1\frac{1}{2}$	1

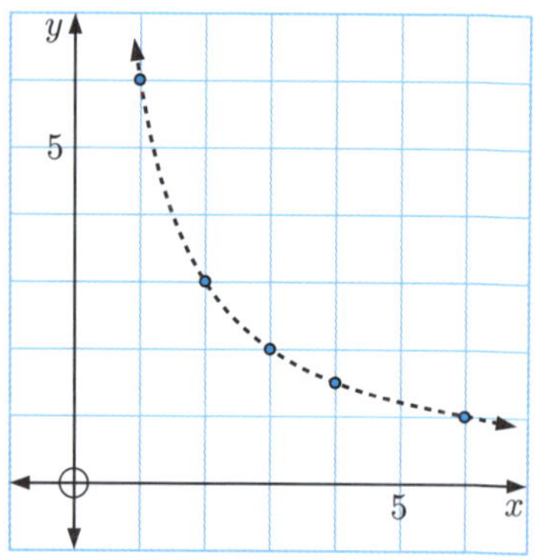

When these points are plotted, they form part of an **hyperbola**.

Now suppose we include in the table a row for the values of $\frac{1}{x}$.

x	1	2	3	4	6
$\frac{1}{x}$	1	$\frac{1}{2}$	$\frac{1}{3}$	$\frac{1}{4}$	$\frac{1}{6}$
y	6	3	2	$1\frac{1}{2}$	1

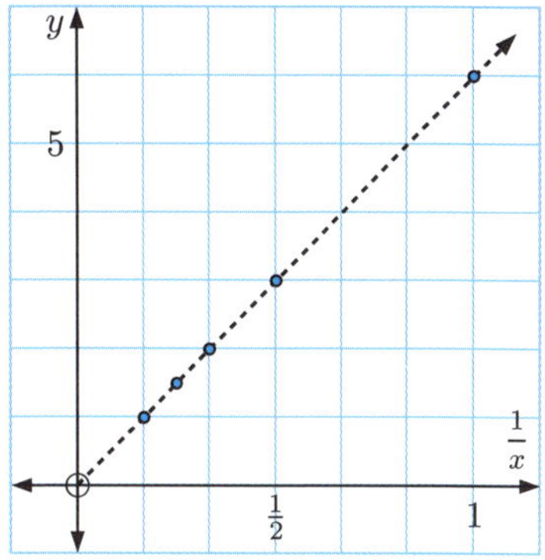

The points on the graph of y against $\frac{1}{x}$ form a straight line which passes through the origin, so y is directly proportional to $\frac{1}{x}$.

If y is **inversely proportional** to x, then y is **directly proportional** to $\frac{1}{x}$.

Consequently, $\boldsymbol{y = \frac{k}{x}}$ or $\boldsymbol{xy = k}$.

In the painters example, notice that $xy = 6$ for all points in the table. So, in this case $k = 6$.

EXERCISE 24C.1

1 Calculate the value of xy for each point in the following tables. Hence, determine for each table whether x and y are inversely proportional. If an inverse proportionality exists, determine the law connecting the variables, and draw the graph of y against x.

a

x	1	2	3	6
y	12	6	4	2

b

x	2	3	6	9
y	12	8	4	3

c

x	7	6	12	1
y	12	14	7	84

2 Consider the table of values alongside. By plotting the graph of y against $\frac{1}{x}$, show that y is inversely proportional to x.

x	1	2	4	8
y	40	20	10	5

3 Which of the following graphs could indicate that y is inversely proportional to x?

A

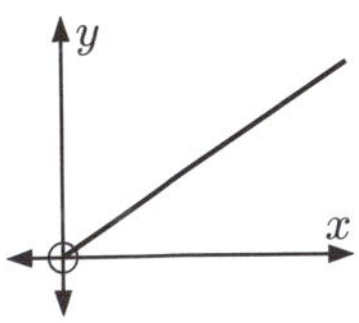

B

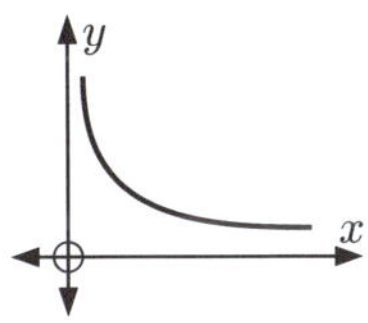

C

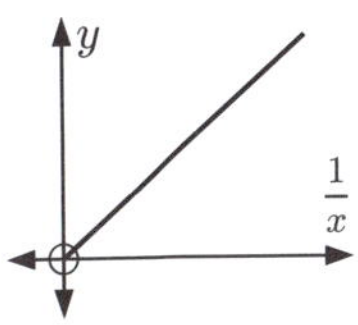

D

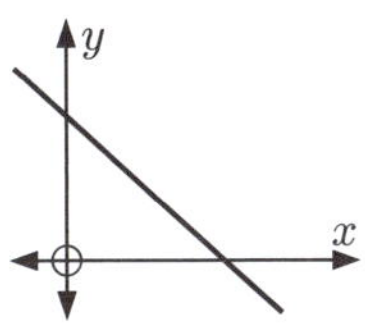

4 The formula connecting average speed s, distance travelled d, and time taken t, is $s = \frac{d}{t}$.
Complete the following statements by inserting 'directly' or 'inversely':

a If d is fixed, s is ...... proportional to t.

b If t is fixed, s is ...... proportional to d.

c If s is fixed, d is ...... proportional to t.

5 Suppose y is inversely proportional to x. Describe what happens to:

a y if x is multiplied by 4

b y if x is multiplied by $\frac{7}{9}$

c x if y is multiplied by $\frac{8}{3}$.

Example 9 — Self Tutor

Suppose M is inversely proportional to t, and that $M = 200$ when $t = 3$. Find:

a M when $t = 5$

b t when $M = 746$

Since M is inversely proportional to t, $M \propto \frac{1}{t}$.

a

$\times\frac{5}{3}$

t	3	5
M	200	

t is multiplied by $\frac{5}{3}$

$\therefore$ $\frac{1}{t}$ is multiplied by $\frac{3}{5}$

$\therefore$ M is multiplied by $\frac{3}{5}$ $\{$as $M \propto \frac{1}{t}\}$

$\therefore$ $M = 200 \times \frac{3}{5} = 120$

b

t	3	
M	200	746

$\times\frac{746}{200}$

M is multiplied by $\frac{746}{200}$

$\therefore$ $\frac{1}{t}$ is multiplied by $\frac{746}{200}$ $\{$as $M \propto \frac{1}{t}\}$

$\therefore$ t is multiplied by $\frac{200}{746}$

$\therefore$ $t = 3 \times \frac{200}{746} \approx 0.80$

6 Suppose y is inversely proportional to x, and that $y = 20$ when $x = 2$. Find:

a the value of y when $x = 5$

b the value of x when $y = 100$.

7 Suppose p is inversely proportional to n, and that $p = 24$ when $n = 6$. Find:

a the value of p when $n = 8$

b the value of n when $p = 12$.

Example 10 — Self Tutor

The velocity V of a body travelling a fixed distance is inversely proportional to the time taken t to complete the journey. When the velocity is 40 cm/s, the time taken is 280 seconds. Find the time taken when the velocity is 50 cm/s.

$V \propto \frac{1}{t}$

t	280	
V	40	50

$\times\frac{50}{40}$

V is multiplied by $\frac{50}{40} = \frac{5}{4}$

$\therefore$ $\frac{1}{t}$ is multiplied by $\frac{5}{4}$ $\{$as $V \propto \frac{1}{t}\}$

$\therefore$ t is multiplied by $\frac{4}{5}$

$\therefore$ $t = 280 \times \frac{4}{5} = 224$

$\therefore$ the time taken is 224 seconds.

8 The time taken to complete a certain job is inversely proportional to the number of workers doing the task. If 20 workers could do the job in 6 days, find how long it would take 15 workers to do the job.

9 The chamber of an airtight cylinder is filled with the yellow gas fluorine. For a given mass of gas kept at a constant temperature, the volume is inversely proportional to the pressure. When $V = 10\text{ cm}^3$, the pressure $P = 40$ units.

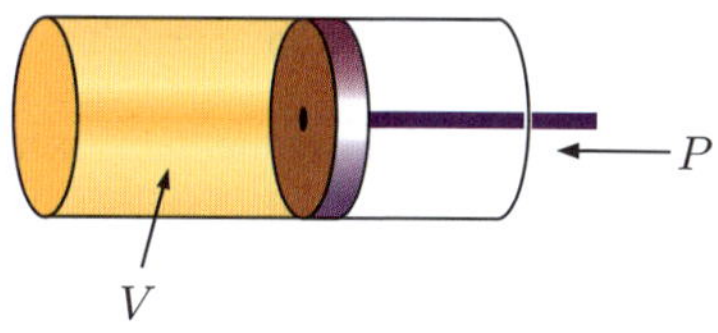

- **a** Find the pressure when the volume is 20 cm^3.
- **b** Find the volume when the pressure is 100 units.
- **c** Plot the graph of P against V.

10

The average court time for each player in a netball team is inversely proportional to the number of players the team has. If the team has 10 players, the average court time for each player is 28 minutes.

- **a** If the team has 12 players, what is the average court time for each player?
- **b** If the average court time is 35 minutes, how many players does the team have?
- **c** Given that 7 players are allowed on court at one time, how long does each game last?

POWERS IN INVERSE PROPORTIONS

Example 11 **Self Tutor**

Suppose y is inversely proportional to the square of x, and that $y = 36$ when $x = 5$.

a Find y when $x = 15$. **b** Given $x > 0$, find x when $y = 49$.

a

$\times 3$

x	5	15
y	36	

x is multiplied by 3

$\therefore$ x^2 is multiplied by $3^2 = 9$

$\therefore$ $\dfrac{1}{x^2}$ is multiplied by $\frac{1}{9}$

$\therefore$ y is multiplied by $\frac{1}{9}$ $\quad \{\text{as } y \propto \dfrac{1}{x^2}\}$

$\therefore$ $y = 36 \times \frac{1}{9} = 4$

b

x	5	
y	36	49

$\times \frac{49}{36}$

y is multiplied by $\frac{49}{36}$

$\therefore$ $\dfrac{1}{x^2}$ is multiplied by $\frac{49}{36}$ $\quad \{\text{as } y \propto \dfrac{1}{x^2}\}$

$\therefore$ x^2 is multiplied by $\frac{36}{49}$

$\therefore$ x is multiplied by $\sqrt{\frac{36}{49}} = \frac{6}{7}$ $\quad \{\text{as } x > 0\}$

$\therefore$ $x = 5 \times \frac{6}{7} = \frac{30}{7}$

EXERCISE 24C.2

1 Suppose y is inversely proportional to the square of x, and that $y = 4$ when $x = 10$.

- **a** Find y when $x = 5$.
- **b** Given $x > 0$, find x when $y = 25$.

2 Suppose F is inversely proportional to the square root of c, and that $F = 20$ when $c = 11$.

  a Find F when $c = 33$. **b** Find c when $F = 30$.

3 A heavy ball is suspended from a spring, and its weight is initially supported. The ball is then dropped so it will bounce up and down. The time between successive bounces is called the *period* of the motion, and varies inversely to the square root of the stiffness of the spring. The period of a spring of stiffness 100 units is 0.2 seconds. Find:

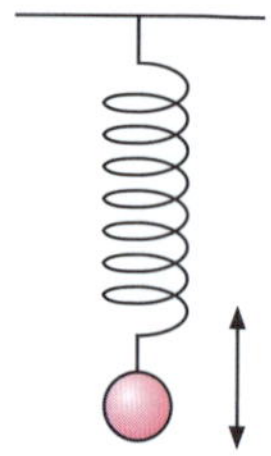

  a the period of a spring of stiffness 300 units
  b the stiffness required for a 0.1 second period.

4

The amount of radiant heat received by a body varies inversely to the square of the distance of the body from the heat source. If the distance from the source is decreased by 20%, what effect does this have on the heat received?

5 The children in a class are each given the same amount of plasticine, and asked to build a solid cylinder. The radius of the cylinder produced is inversely proportional to the square root of the height. Xian's cylinder has radius 2.7 cm and height 13.1 cm.

  a Beth's cylinder has radius 5.4 cm. Find its height.
  b Roger's cylinder is 8 cm high. Find its radius.
  c What volume of plasticine was each child given?

6 The intensity of light on a screen varies inversely to the square of the distance between the screen and the light source.
If a screen is illuminated by a light source 20 m away, the intensity is one fifth of what is required. Where should the light be placed?

REVIEW SET 24A

1 Consider the table of values alongside.

x	2	3	6	11
y	6	9	18	33

  a Plot the graph of y against x.
  b Explain why y is directly proportional to x.
  c Find the law connecting x and y.

2 Suppose x is inversely proportional to y, and that $y = 65$ when $x = 10$. Find:

  a the law connecting x and y **b** y when $x = 12$ **c** x when $y = 150$.

3 The variable M is directly proportional to t. What happens to M if t is trebled?

4 The variables x and y in the table are inversely proportional.
Find a and b.

x	1	3	6	a
y	24	8	b	60

5 The surface area of a sphere is directly proportional to the square of its radius. Find:

a the change in surface area if the radius is decreased by 19%

b the change in radius if the surface area is trebled.

6 Suppose y is inversely proportional to x. What happens to:

a y if x is trebled **b** x if y is increased by 50%?

7 A school's quiz night raises money in direct proportion to the number of people attending. Last year 85 people attended the quiz night, and the event raised \$1275. This year the entry fees were the same, but the school only raised \$1080. How many people attended this year?

8 The kinetic energy E of a solid with mass m and speed v, is given by $E = \frac{1}{2}mv^2$.

a State the proportionality that exists between:

i E and m **ii** E and v^2 **iii** m and v^2.

b Describe the change in the kinetic energy of an object if its speed is reduced by 12%.

9 The Dentists' Association has found that the average number of fillings required at a patient's checkup is inversely proportional to the time the patient spends cleaning their teeth each day.
Over the last 10 years, Thomas has spent 40 seconds brushing his teeth each day, and has needed 3 fillings. If Thomas increases his brushing time to 2 minutes each day for the next 10 years, how many fillings will he need at his next checkup?

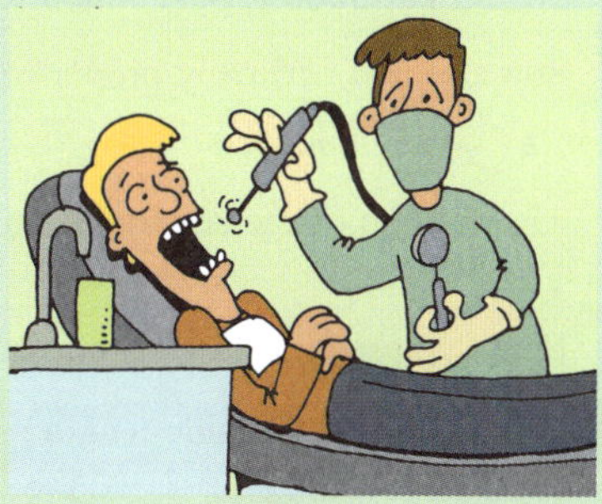

10 Consider the graph alongside.

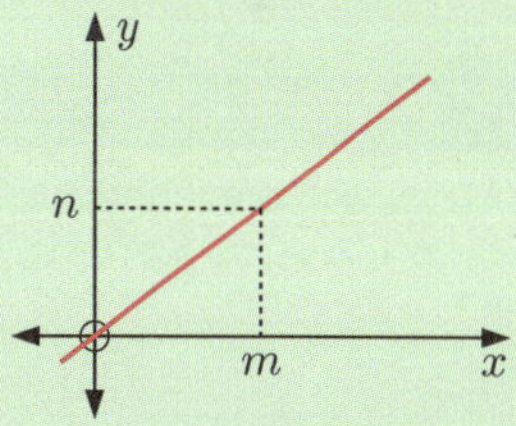

a Describe the relationship between x and y.

b Consider the point where $x = m$ and $y = n$. If the value of x is increased to $(m + 10)$, the value of y is increased to $(n + 30)$. Find the law connecting x and y.

c Find the value of x when $y = 72$.

REVIEW SET 24B

1 The variables p and q in the table are directly proportional. Find a and b.

p	2	4	7	a
q	6	12	b	42

2 If 7 litres of petrol are needed to drive 100 km, how far could you travel on 16 litres of petrol?

3 The table alongside shows the total price \$$P$ of buying n books at \$36 each.

Number of books (n)	1	2	3	4
Total price (\$$P$)	36	72	108	144

a Show that P and n are directly proportional.

b Find the law connecting P and n.

c Plot the graph of P against n.

4 Suppose $y \propto x$ and that $y = 77.9$ when $x = 4.1$.

a Find the proportionality constant. **b** Find the value of y when $x = 6.3$.

5 The volume of a cylinder of fixed length is directly proportional to the square of its radius. Find:

a the change in volume produced by doubling the radius

b the change in radius needed to produce a 60% increase in volume.

6 Catherine's weekly pay is directly proportional to the number of hours she works. Last week she worked 14 hours and received \$357. This week Catherine received \$140.25. How many hours did she work this week?

7 By plotting the graph of y against $\frac{1}{x}$, show that y is inversely proportional to x.

x	2	3	5	10
y	30	20	12	6

8 A small pump empties a dam in 45 minutes at the rate of 1200 L per minute. A larger pump operates at 2000 L per minute. How long would this pump take to empty the dam?

9 Suppose y is inversely proportional to the square of x. Describe what happens to:

a y if x is halved **b** x if y is increased by 44%.

10 The volume of the square-based pyramid illustrated is given by $V = \frac{1}{3}l^2h$ cm^3.

If the volume of the pyramid is 212 cm^3, find the volume of a pyramid with base side length $\frac{l}{2}$ cm and height $3h$ cm.

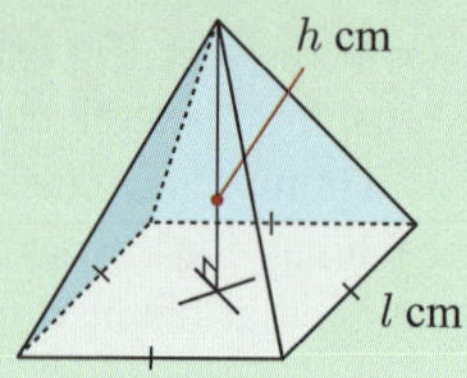

Chapter 25

Non-right angled triangle trigonometry

Contents:

OPENING PROBLEM

Andy is a surveyor who measures the exact lengths of property boundaries. Andy normally uses a laser measuring device, but today he cannot directly measure from A to B because a number of trees block his vision.

He measures the distance from point A to the corner of a shed C to be 124 m. He measures angle ACB to be 130°, and the distance CB to be 105 m.

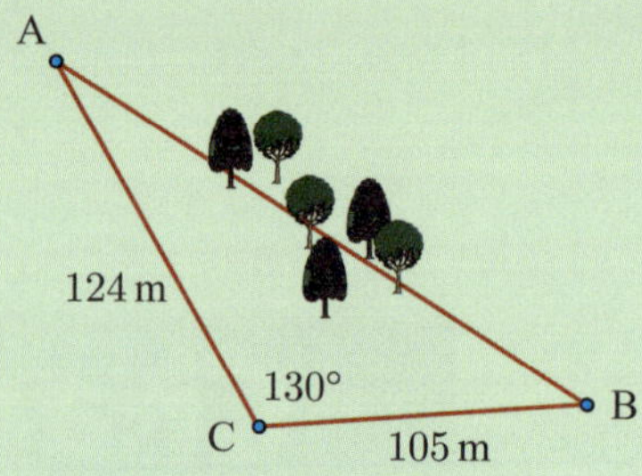

Things to think about:

a The angle ACB is obtuse. Does it make sense to talk about trigonometric ratios such as $\sin 130°$ and $\cos 130°$?

b How can we use trigonometry to find:

i the area of triangle ABC **ii** the length of the boundary [AB]?

If a triangle is right angled, then its remaining angles *must* be acute. However, if the triangle is not right angled, it may contain an obtuse angle.

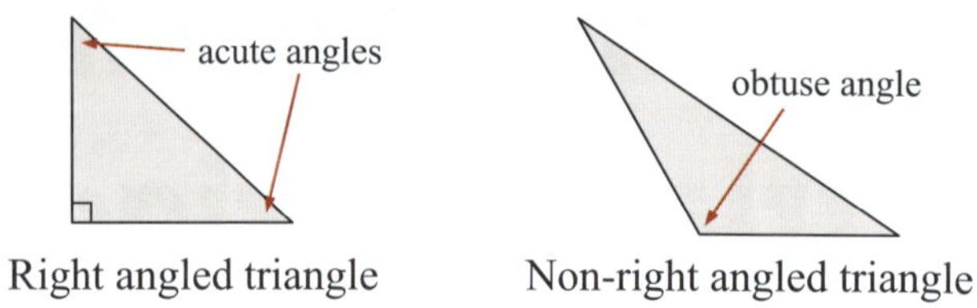

Therefore, before we can use trigonometry in non-right angled triangles, we must first give meaning to the trigonometric ratios of obtuse angles.

A THE UNIT CIRCLE

The circle of radius 1 unit with its centre at the origin O, is called the **unit circle**.

THE UNIT QUARTER CIRCLE

Consider the quarter of the unit circle in the first quadrant. We call this the **unit quarter circle**.

Suppose $P(a, b)$ is a point on the circle so that [OP] makes an angle θ with the x-axis. We see that a and b are both positive, and that

$$\cos\theta = \frac{\text{ADJ}}{\text{HYP}} = \frac{a}{1} = a \quad \text{and} \quad \sin\theta = \frac{\text{OPP}}{\text{HYP}} = \frac{b}{1} = b.$$

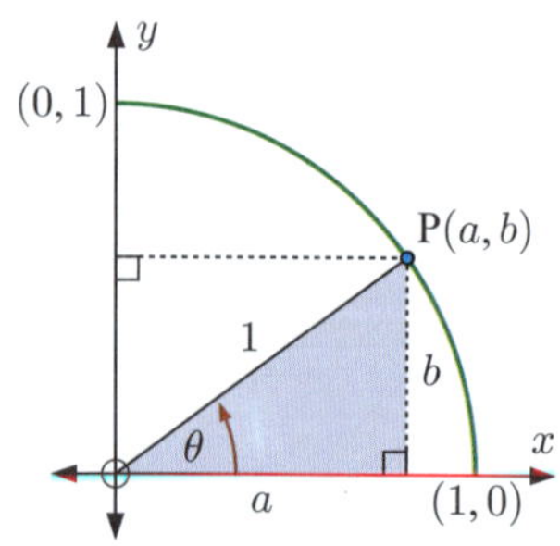

If P is a point on the unit quarter circle such that [OP] makes an angle θ with the positive x-axis, then:

- $\boldsymbol{\cos\theta}$ is the x-coordinate of P
- $\boldsymbol{\sin\theta}$ is the y-coordinate of P.

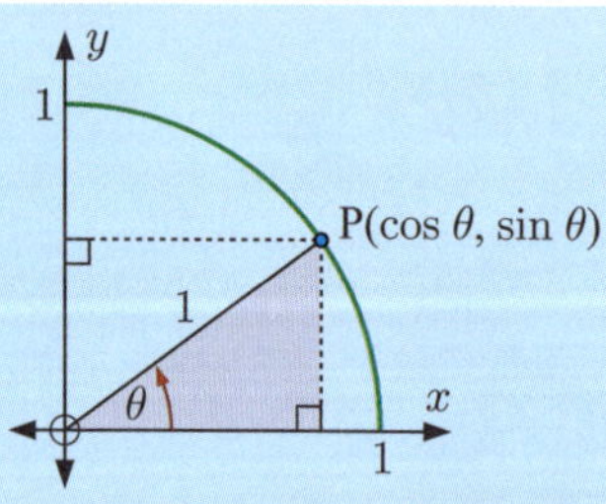

We can use an accurate unit quarter circle diagram to estimate the value of trigonometric ratios.

PRINTABLE DIAGRAM

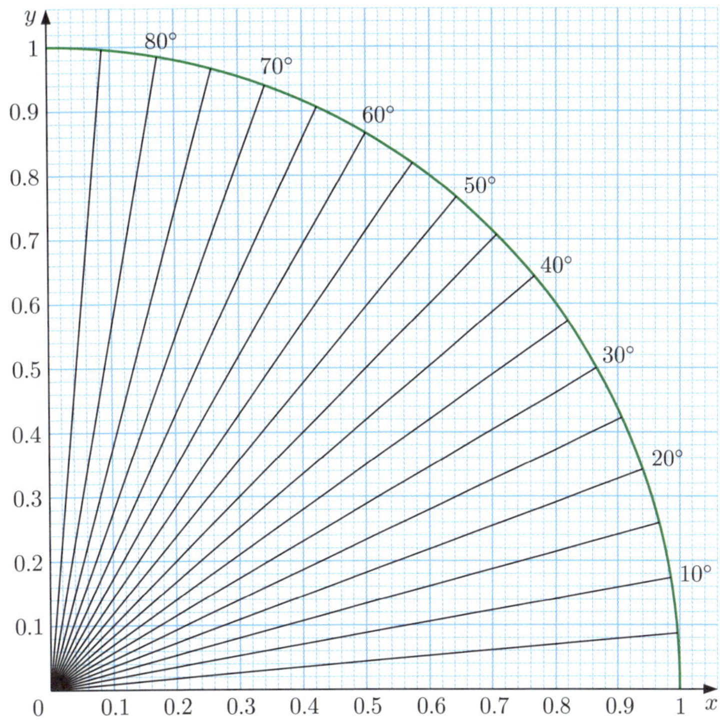

Example 1 Self Tutor

Use the unit quarter circle diagram to estimate the value of:

a $\cos 20°$ **b** $\sin 55°$

a $\cos 20° =$ the x-coordinate at an angle of $20°$
≈ 0.94

b $\sin 55° =$ the y-coordinate at an angle of $55°$
≈ 0.82

EXERCISE 25A.1

1 Use the unit quarter circle diagram to estimate the value of:

a $\cos 0°$ **b** $\cos 15°$ **c** $\cos 25°$ **d** $\cos 30°$
e $\cos 45°$ **f** $\cos 60°$ **g** $\cos 75°$ **h** $\cos 90°$

Check your answers using your calculator.

2 Use the unit quarter circle diagram to estimate the value of:

a $\sin 0°$ **b** $\sin 15°$ **c** $\sin 25°$ **d** $\sin 30°$
e $\sin 45°$ **f** $\sin 60°$ **g** $\sin 75°$ **h** $\sin 90°$

Check your answers using your calculator.

3 Use your calculator to find the coordinates of P and Q, to 3 decimal places.

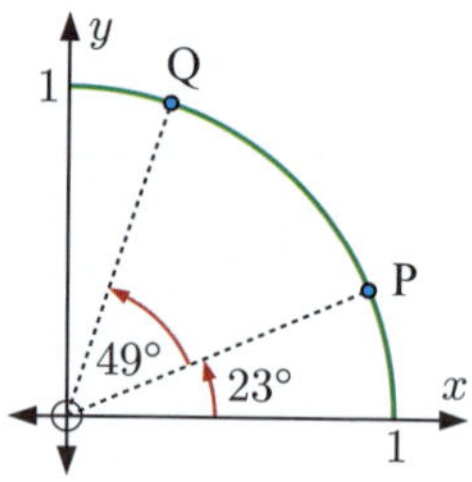

4 Find θ, to 1 decimal place:

a

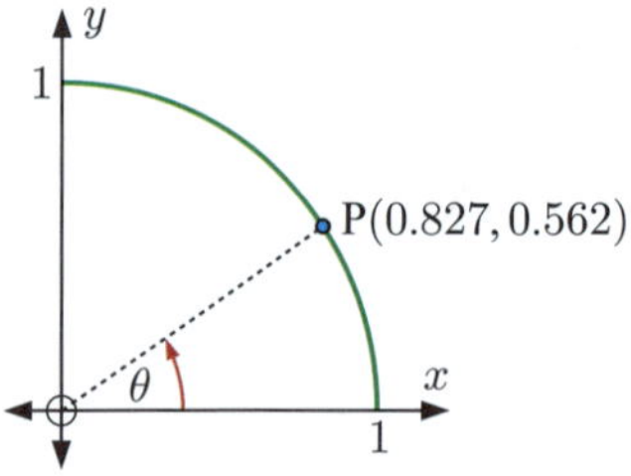

b

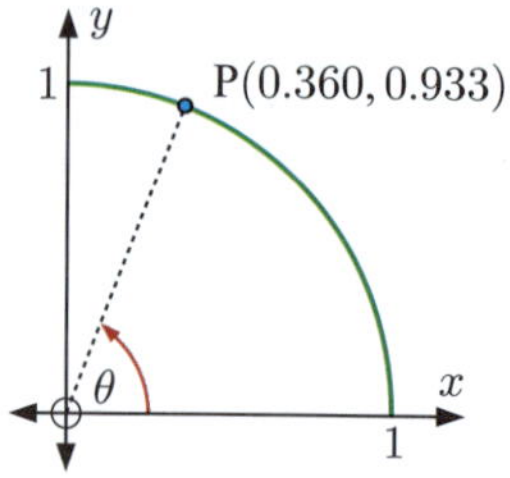

THE UNIT HALF CIRCLE

So far we have only considered angles between $0°$ and $90°$, which are the **acute** angles.

Obtuse angles have measurement between $90°$ and $180°$. In order to display obtuse angles, we extend the unit quarter circle to the **unit half circle** shown below.

PRINTABLE DIAGRAM

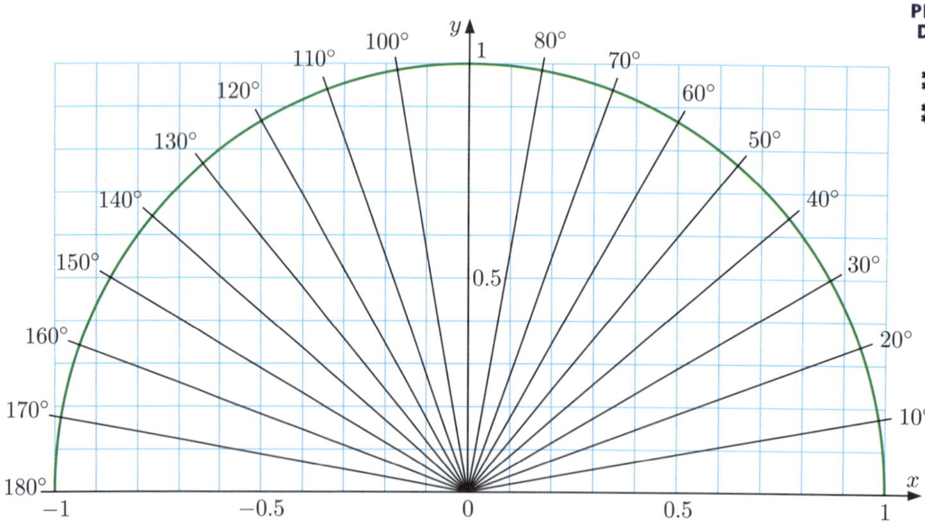

We can apply the same definitions for $\cos\theta$ and $\sin\theta$ as we did before.

If P is a point on the unit circle such that [OP] makes an angle θ with the positive x-axis, then:

- $\cos\theta$ is the x-coordinate of P
- $\sin\theta$ is the y-coordinate of P.

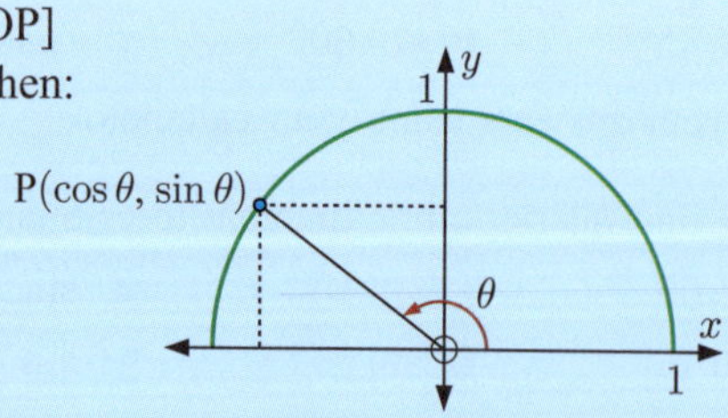

We can see that for an obtuse angle θ, P lies in the second quadrant. This means that $\cos\theta$ will be negative, and $\sin\theta$ will be positive.

Example 2 — Self Tutor

Use the unit half circle diagram to estimate the value of:

a $\cos 140°$ **b** $\sin 100°$

a $\cos 140° =$ the x-coordinate at an angle of $140°$
≈ -0.77

b $\sin 100° =$ the y-coordinate at an angle of $100°$
≈ 0.98

EXERCISE 25A.2

1 Use the unit half circle diagram to estimate the value of:

a $\cos 100°$ **b** $\cos 80°$ **c** $\cos 120°$ **d** $\cos 60°$
e $\cos 130°$ **f** $\cos 50°$ **g** $\cos 180°$ **h** $\cos 0°$

Use your calculator to check your answers.

2 Use the unit half circle diagram to estimate the value of:

a $\sin 110°$ **b** $\sin 70°$ **c** $\sin 120°$ **d** $\sin 60°$
e $\sin 130°$ **f** $\sin 50°$ **g** $\sin 180°$ **h** $\sin 0°$

Use your calculator to check your answers.

3 Use your calculator to find the coordinates of P and Q, to 3 decimal places.

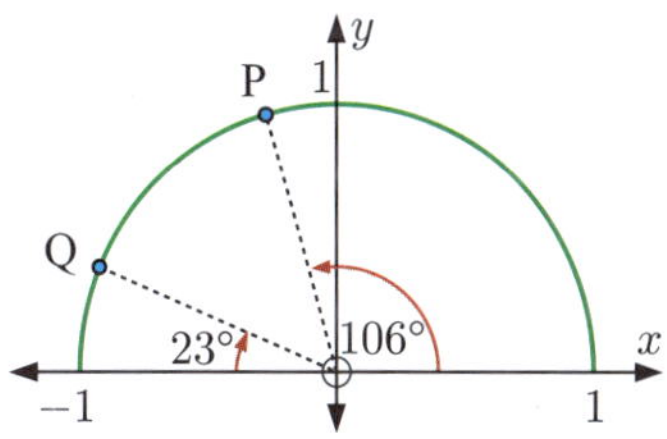

4 **a** Use your results from **1** to copy and complete:
$\cos(180° - \theta) = ...$

b Justify your answer using the diagram alongside.

c Check your answer by using a calculator to find:

i $\cos 63°$ and $\cos 117°$
ii $\cos 26°$ and $\cos 154°$.

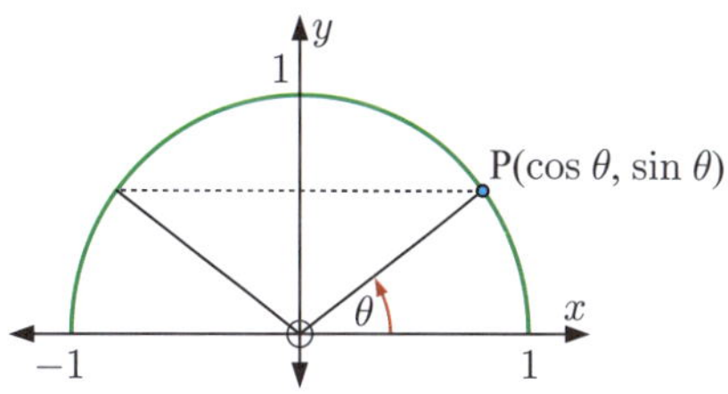

5 **a** Use your results from **2** to copy and complete:
$\sin(180° - \theta) = ...$

b Justify your answer using the diagram alongside.

c Check your answer by using a calculator to find:

i $\sin 71°$ and $\sin 109°$
ii $\sin 8°$ and $\sin 172°$.

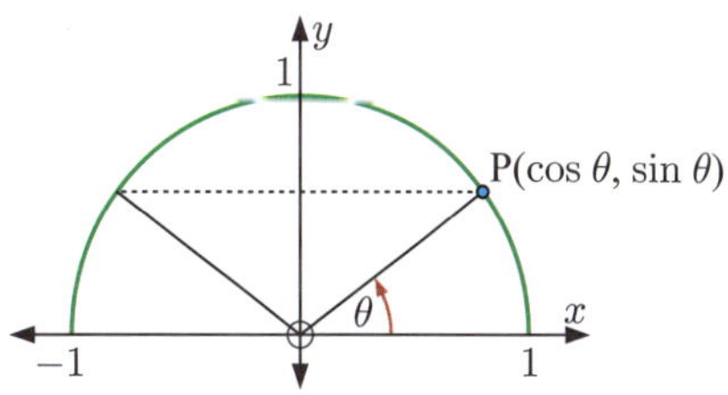

6 Find the obtuse angle which has the same sine as:

a $26°$ **b** $45°$ **c** $69°$ **d** $86°$

7 Find the acute angle which has the same sine as:

a $98°$ **b** $127°$ **c** $156°$ **d** $168°$

8 Without using your calculator, find:

a $\sin 112°$ given $\sin 68° \approx 0.9272$

b $\sin 26°$ given $\sin 154° \approx 0.4384$

c $\cos 168°$ given $\cos 12° \approx 0.9781$

d $\cos 49°$ given $\cos 131° \approx -0.6561$

e $\sin 145°$ given $\sin 35° \approx 0.5736$

f $\cos 98°$ given $\cos 82° \approx 0.1392$

From the previous **Exercise**, you should have found that:

$$\sin(180° - \theta) = \sin\theta$$
$$\cos(180° - \theta) = -\cos\theta$$

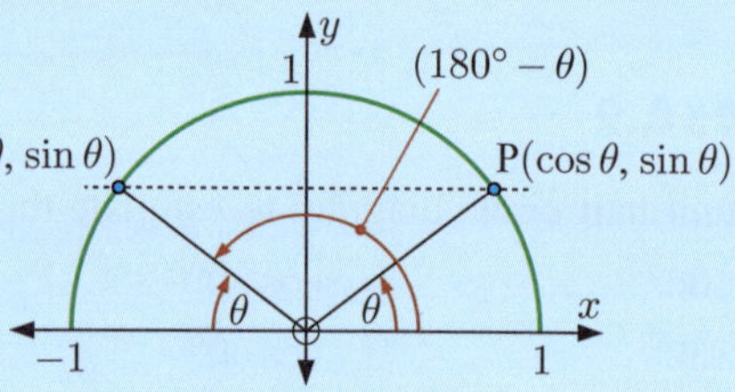

B THE AREA OF A TRIANGLE

So far we have only been able to calculate the area of a triangle using the base and the perpendicular height.

However, we may not know the perpendicular height, especially if the triangle is not right angled.

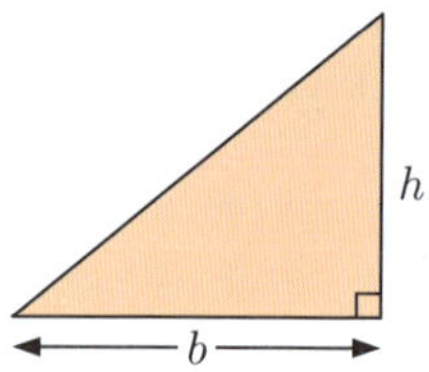

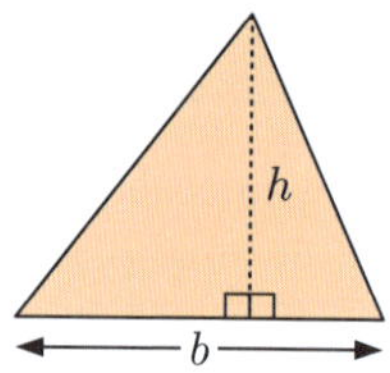

We can use trigonometry to find the area of a triangle if we are given the lengths of **two sides**, as well as the **included angle** between the sides.

The **area of a triangle** is a half of the product of two sides and the sine of the included angle.

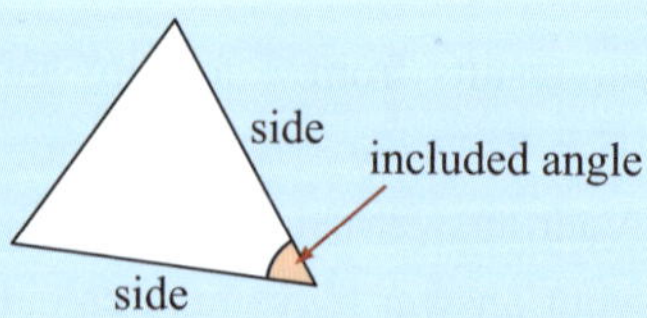

Proof (for the case where A is acute):

Consider triangle ABC in which the sides opposite angles A, B, and C are labelled a, b, and c respectively.

Suppose N lies on [AB] such that [CN] is perpendicular to [AB].

The area of $\triangle$ABC$= \frac{1}{2}\times$AB$\times$CN$= \frac{1}{2}ch$ .... (1)

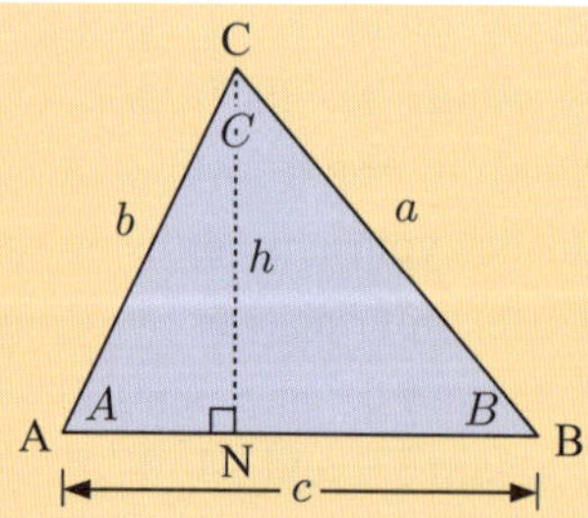

However, in $\triangle$ANC, $\sin A = \dfrac{h}{b}$

$\therefore \quad h = b\sin A \quad .... (2)$

Substituting (2) into (1) gives area $= \frac{1}{2}c(b\sin A) = \frac{1}{2}bc\sin A$

Similarly, if the altitudes from A and B were drawn, it could be shown that area $= \frac{1}{2}ac\sin B = \frac{1}{2}ab\sin C$.

The formula **area $= \frac{1}{2}ab\sin C$** is worth remembering.

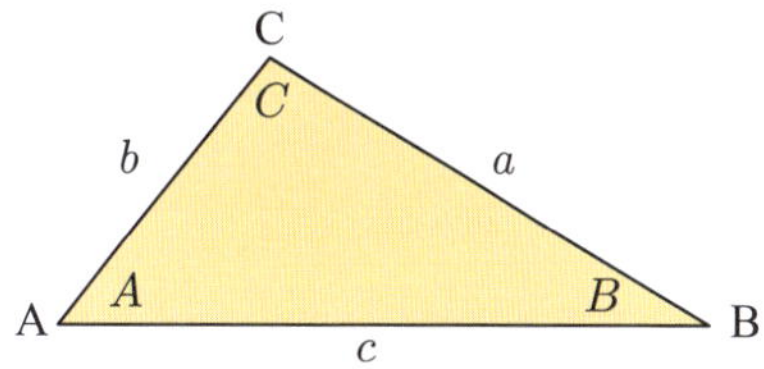

Example 3 Self Tutor

Find the area of $\triangle$ABC:

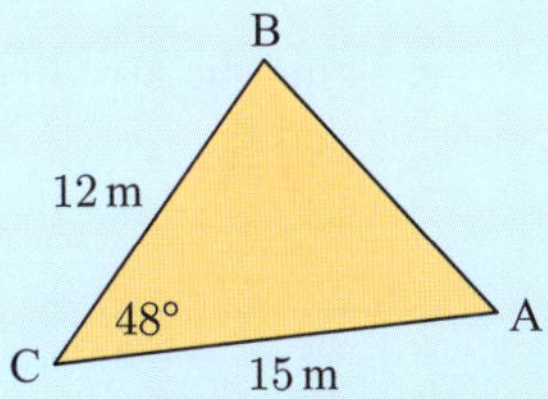

$$\begin{aligned}\text{Area} &= \tfrac{1}{2}ab\sin C\\ &= \tfrac{1}{2}\times 12\times 15\times \sin 48^\circ\\ &\approx 66.9\ \text{m}^2\end{aligned}$$

EXERCISE 25B

1 Find the area of:

a

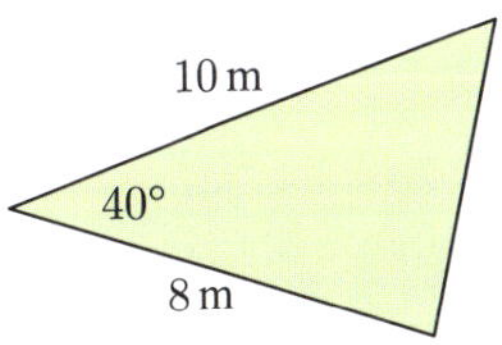

b

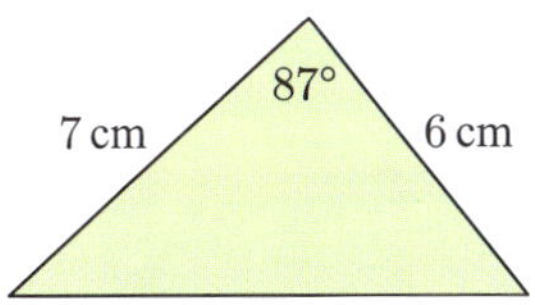

c

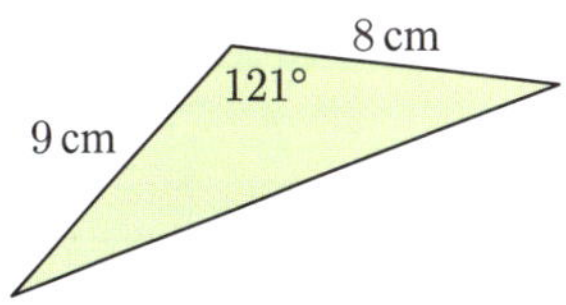

d

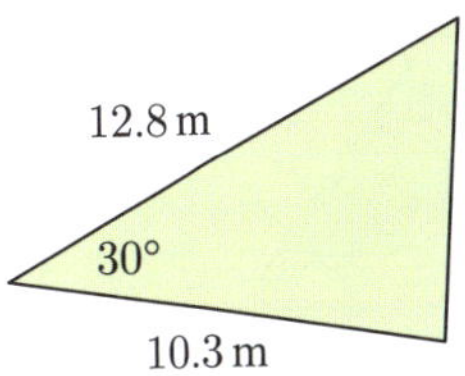

e

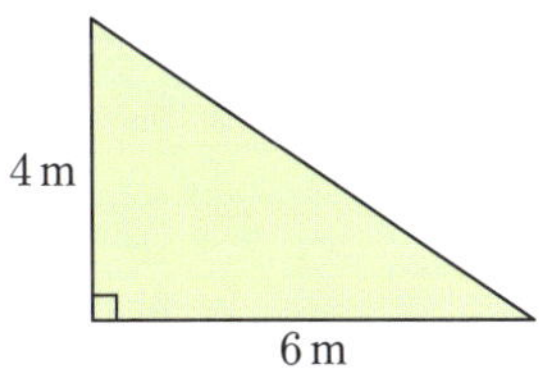

f

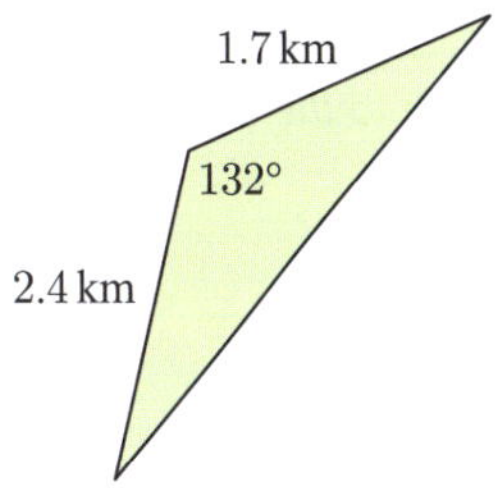

2 Find the area of:

a

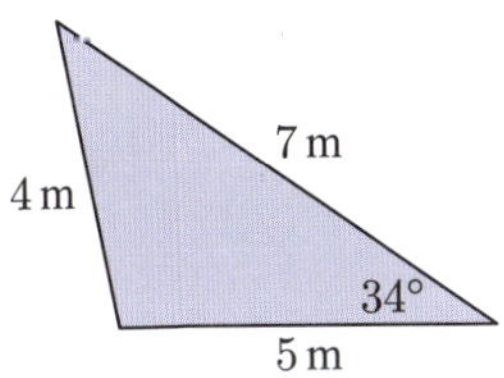

b

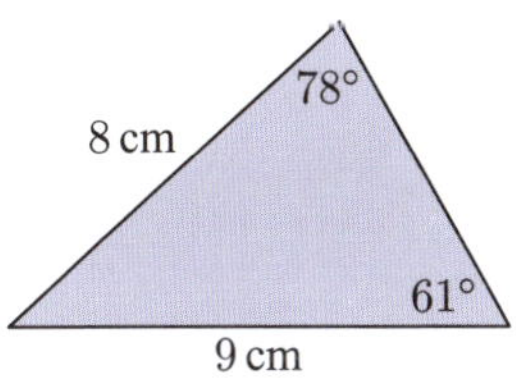

c

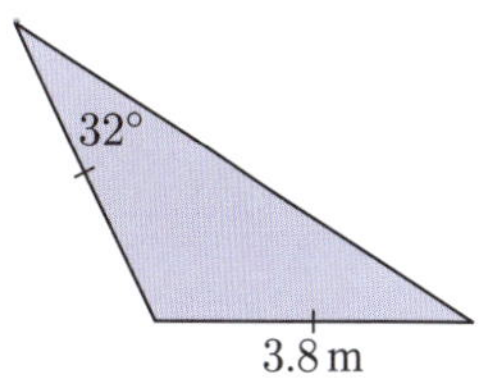

3

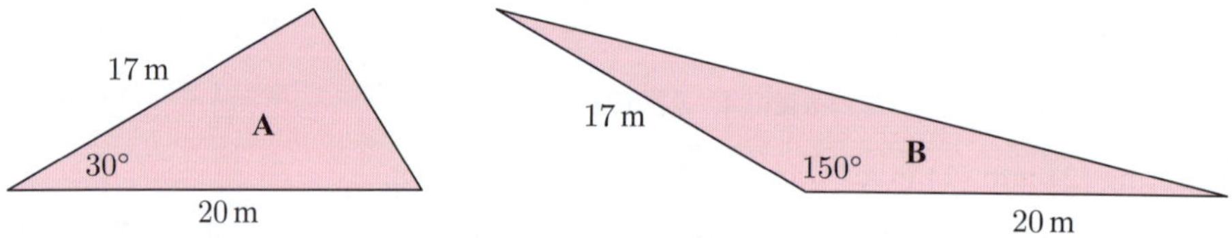

a Find the areas of triangles **A** and **B**.

b Explain why the triangles have the same area.

DEMO

4 The area of this triangle is 127.3 m^2. Find x.

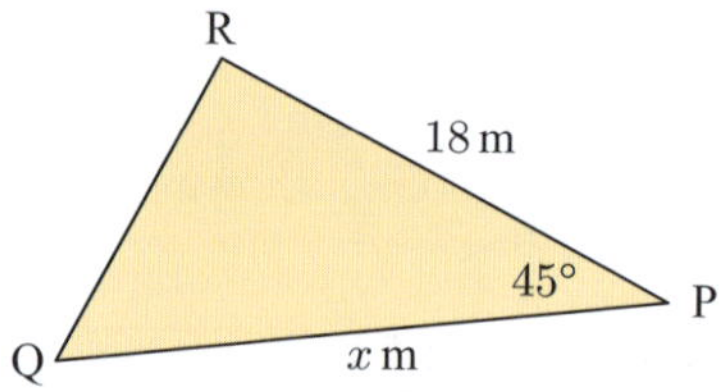

5 Find the area of this parallelogram:

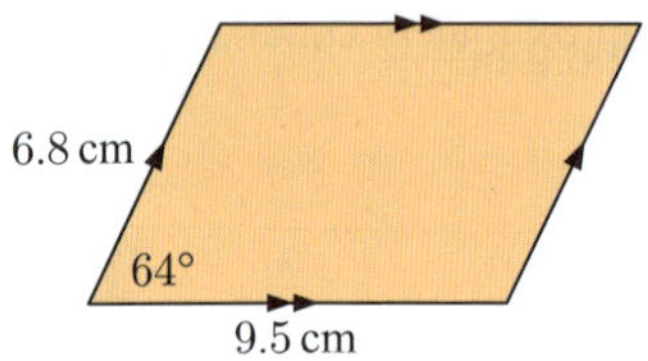

6 Find the area of triangle PQR.

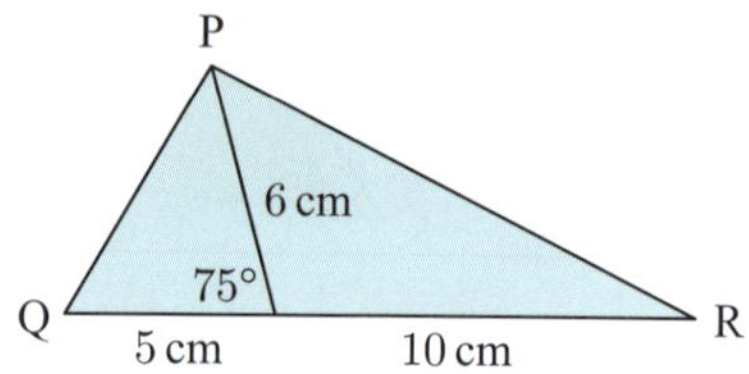

7

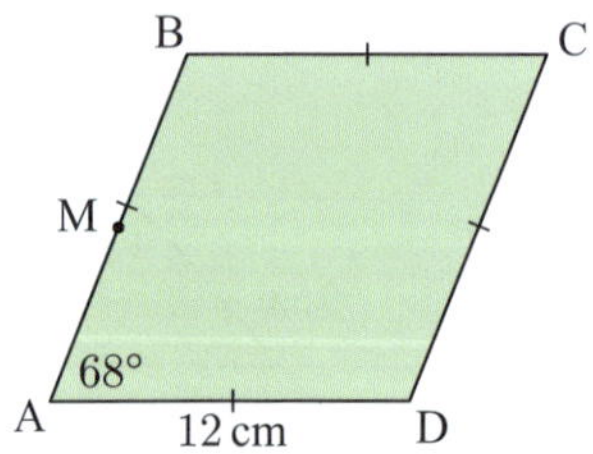

In the rhombus ABCD, M is the midpoint of [AB]. Find the area of $\triangle$BMC.

8 In triangle ABC, M is the midpoint of [AB], and N is the midpoint of [AC]. Show that triangles AMC and ANB have the same area.

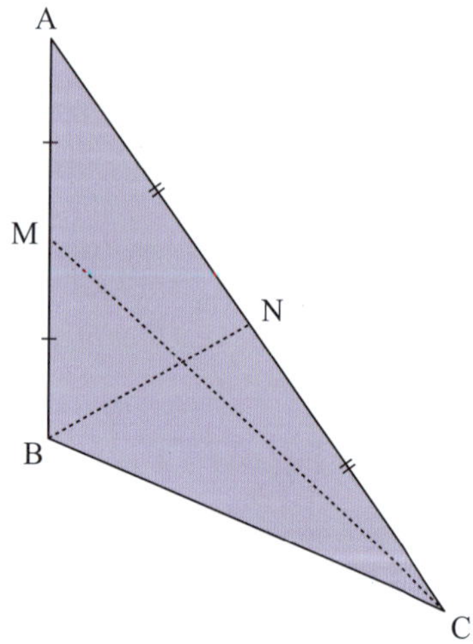

C THE SINE RULE

The **sine rule** is a set of equations which connects the lengths of the sides of any triangle with the sines of the opposite angles.

INVESTIGATION 1 THE SINE RULE

You will need: ruler, protractor, paper

What to do:

1 Use a ruler to draw a large triangle on a sheet of paper.

2 Label the side lengths a, b, and c, and the opposite angles A, B, and C as shown.

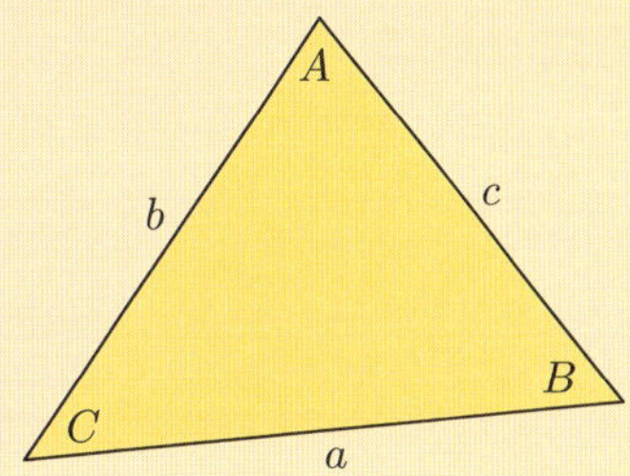

3 Measure a, b, and c, recording your answers to the nearest mm.

4 Use your protractor to measure angles A, B, and C, recording your answers to the nearest degree.

5 Calculate $\dfrac{a}{\sin A}$, $\dfrac{b}{\sin B}$, and $\dfrac{c}{\sin C}$. Are these values approximately equal?

6 Repeat steps **1** to **5** with a different triangle.

7 Copy and complete:
In any triangle, the ratio of the sides and the sines of their opposite angles is ......

GEOMETRY PACKAGE

THE SINE RULE

In any triangle ABC with sides a, b, and c units, and opposite angles A, B, and C respectively,

$$\frac{\sin A}{a}=\frac{\sin B}{b}=\frac{\sin C}{c} \quad \text{or} \quad \frac{a}{\sin A}=\frac{b}{\sin B}=\frac{c}{\sin C}.$$

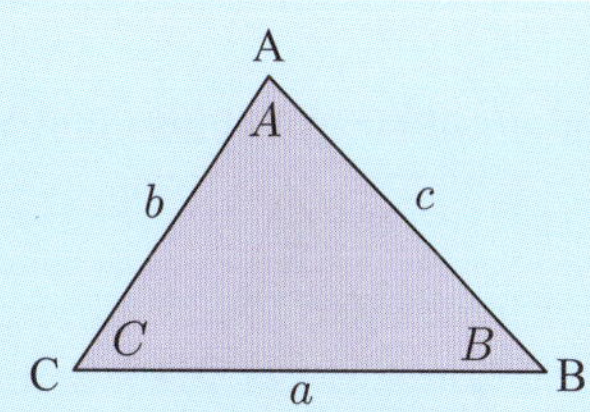

Proof:

The area of any triangle ABC is given by $\frac{1}{2}bc\sin A=\frac{1}{2}ac\sin B=\frac{1}{2}ab\sin C$.

Dividing each expression by $\frac{1}{2}abc$ gives $\dfrac{\sin A}{a}=\dfrac{\sin B}{b}=\dfrac{\sin C}{c}$.

FINDING SIDES

If we are given two angles and one side of a triangle, we can use the sine rule to find another side length.

When finding sides, we use the form $\dfrac{a}{\sin A}=\dfrac{b}{\sin B}$ so that the unknown is in the numerator.

Example 4 — Self Tutor

Find x:

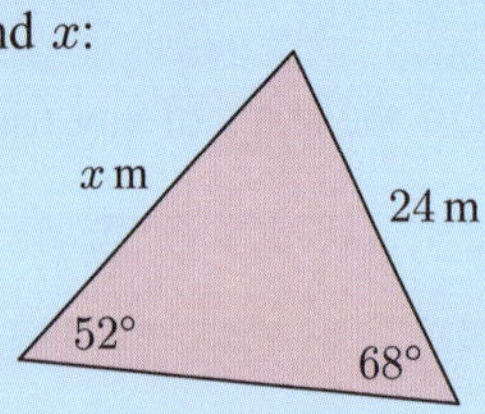

Using the sine rule,

$$\frac{x}{\sin 68^\circ} = \frac{24}{\sin 52^\circ}$$

$$\therefore \quad x = \frac{24 \times \sin 68^\circ}{\sin 52^\circ}$$

$$\therefore \quad x \approx 28.2$$

EXERCISE 25C.1

1 Find the value of x:

a

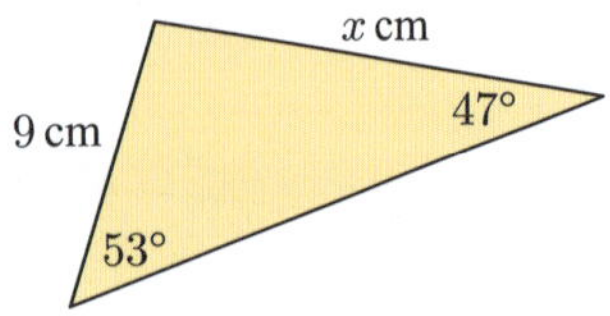

b

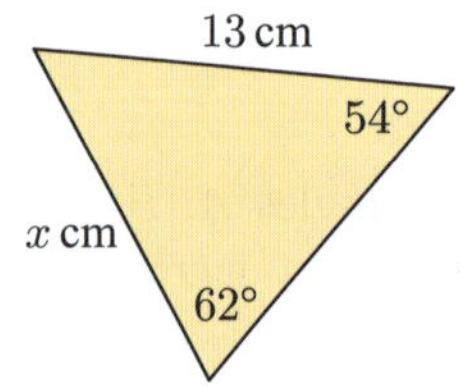

c

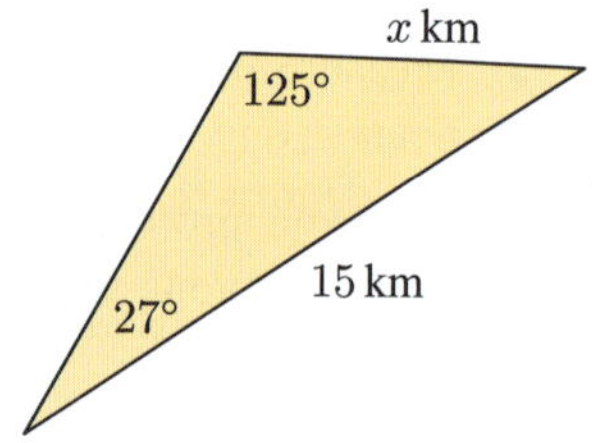

2 Find the value of x:

a

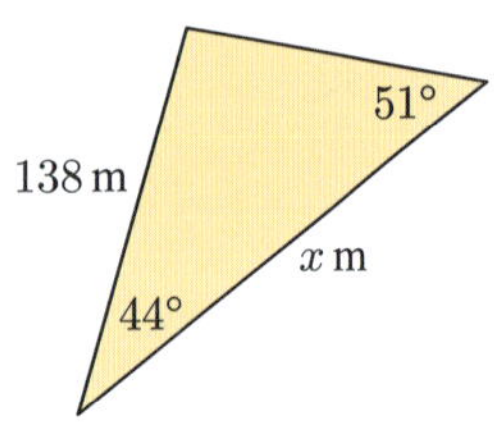

b

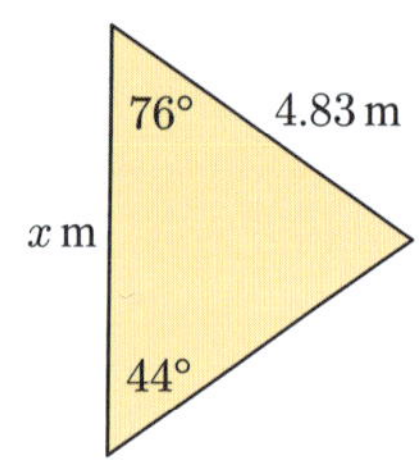

c

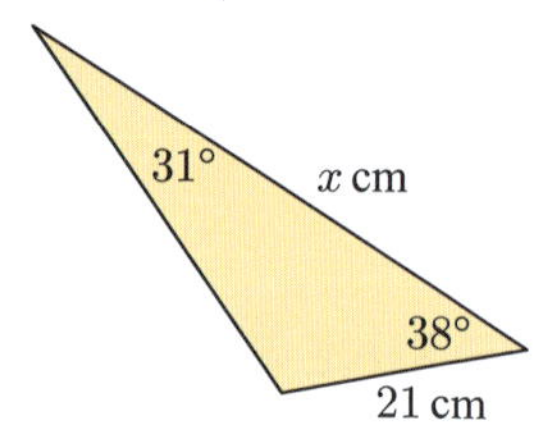

3 Find *all* unknown angles and sides of:

a

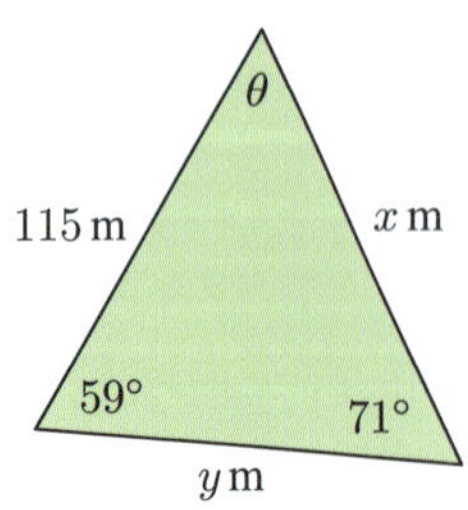

b

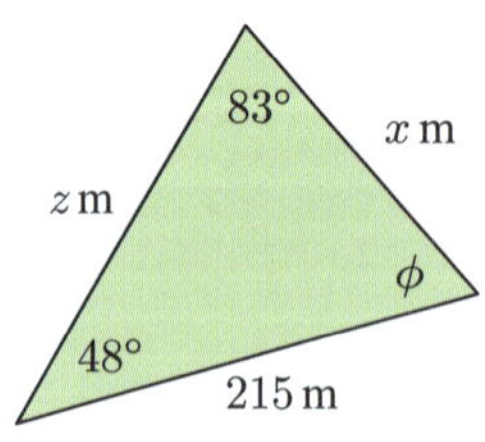

c

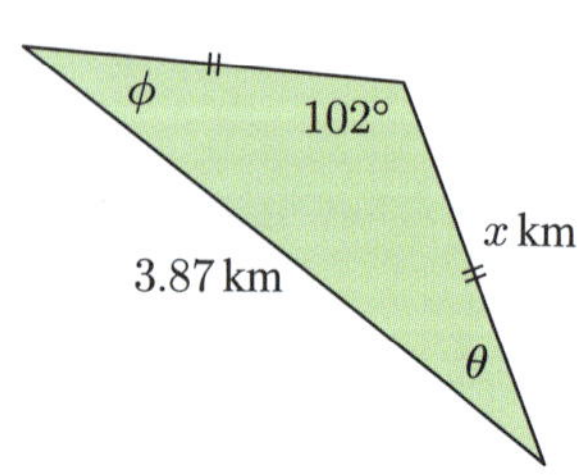

4 Find the unknown side lengths and angles:

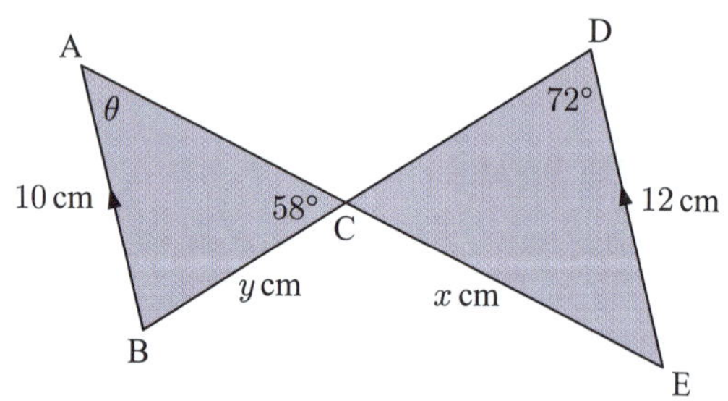

FINDING ANGLES

If we are given two side lengths and a non-included angle of a triangle, we can use the sine rule to find another angle.

Finding angles using the sine rule is more complicated than finding side lengths, because there may be two possible answers.

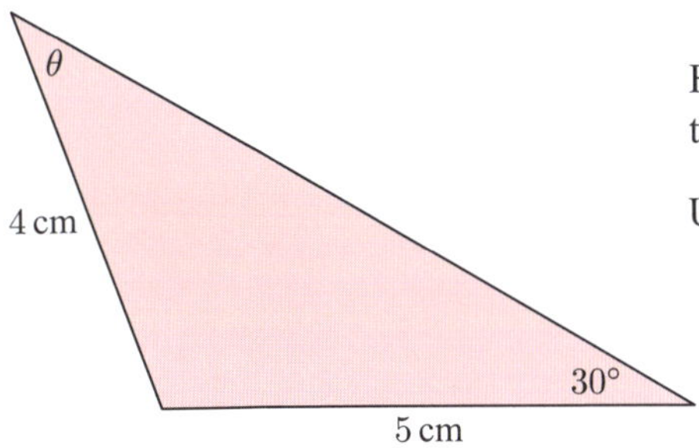

For example, suppose we want to find the angle θ in the triangle alongside.

Using the sine rule, $\dfrac{\sin\theta}{5} = \dfrac{\sin 30^\circ}{4}$

$\therefore \quad \sin\theta = \dfrac{5 \times \sin 30^\circ}{4}$

$\therefore \quad \sin\theta = \frac{5}{8}$

Using a calculator to solve $\sin\theta = \frac{5}{8}$ gives $\theta \approx 38.7^\circ$, since this is the acute angle with a sine of $\frac{5}{8}$.

Casio fx-CG20

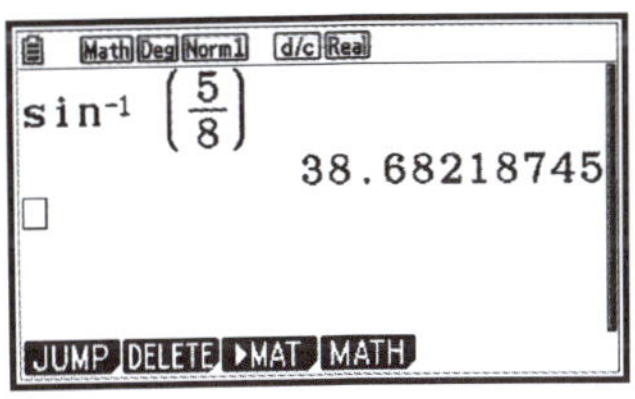

TI-84 Plus

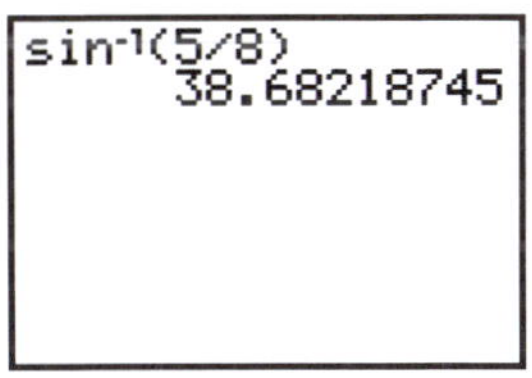

TI-$n$spire

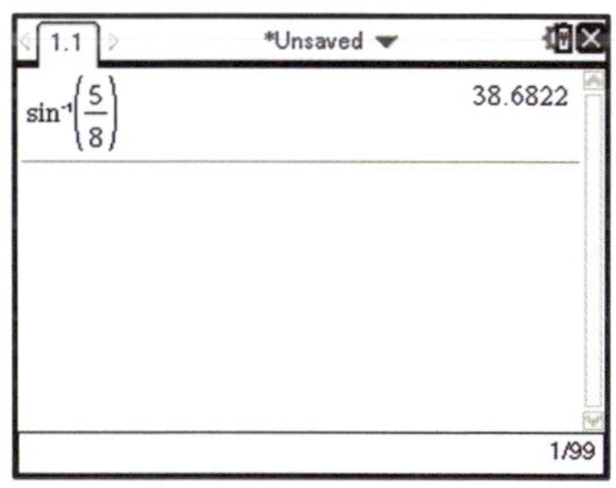

However, we know from **Section A** that there is also an obtuse angle with sine $\frac{5}{8}$. So, the two solutions to $\sin\theta = \frac{5}{8}$ are $\theta \approx 38.7^\circ$ and $180^\circ - 38.7^\circ \approx 141.3^\circ$.

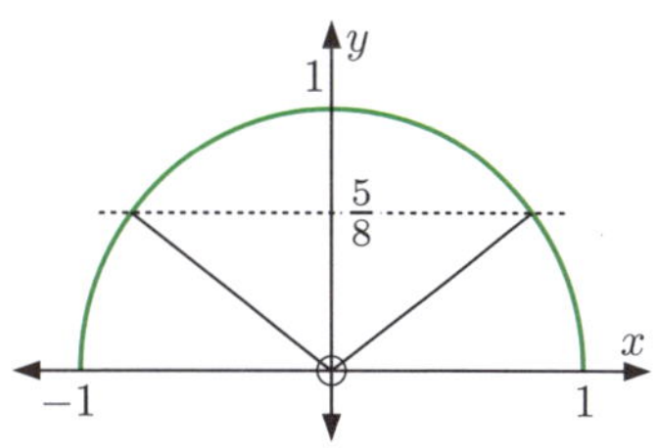

In this section, all of the diagrams are drawn approximately to scale. We can therefore use the diagram to determine whether the unknown angle is acute or obtuse.

For example, in the triangle above, θ is clearly acute, so $\theta \approx 38.7^\circ$.

Example 5 **Self Tutor**

The diagrams below are drawn approximately to scale. Use the sine rule to find θ, rounding your answer to one decimal place.

a

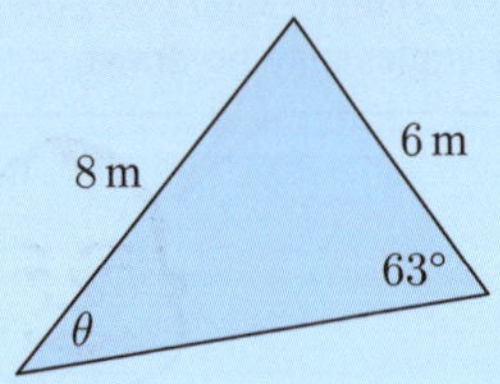

b

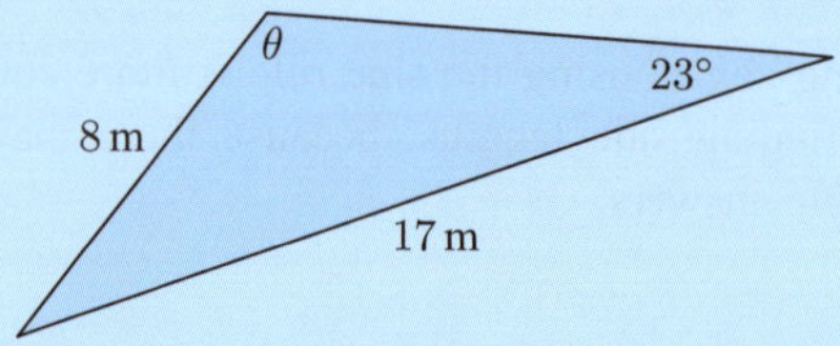

a Using the sine rule,

$$\frac{\sin\theta}{6} = \frac{\sin 63°}{8}$$

$$\therefore \sin\theta = \frac{6 \times \sin 63°}{8}$$

$$\therefore \theta = \sin^{-1}\left(\frac{6 \times \sin 63°}{8}\right)$$

{as θ is clearly acute}

$$\therefore \theta \approx 41.9°$$

b Using the sine rule,

$$\frac{\sin\theta}{17} = \frac{\sin 23°}{8}$$

$$\therefore \sin\theta = \frac{17 \times \sin 23°}{8}$$

$$\therefore \theta = 180° - \sin^{-1}\left(\frac{17 \times \sin 23°}{8}\right)$$

{as θ is clearly obtuse}

$$\therefore \theta \approx 123.9°$$

EXERCISE 25C.2

1 The diagrams below are drawn approximately to scale. Use the sine rule to find θ, rounding your answer to one decimal place.

a

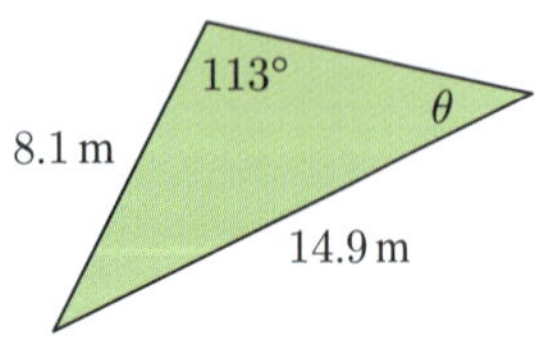

b

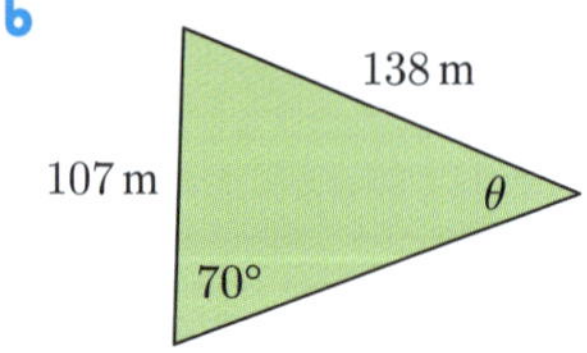

c

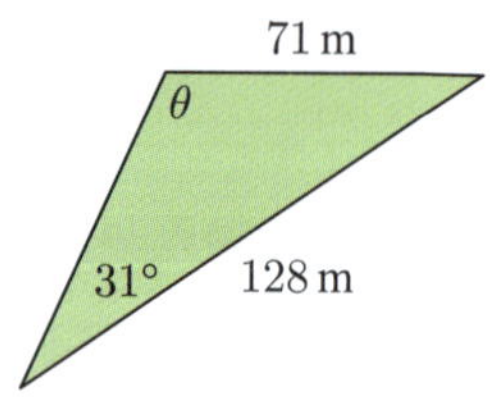

d

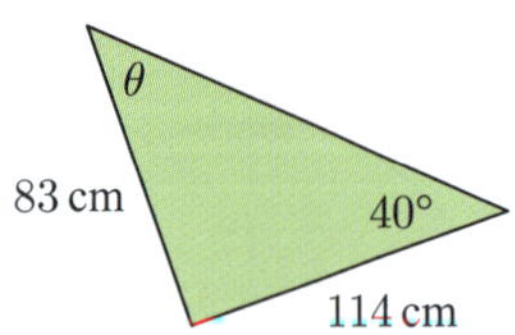

e

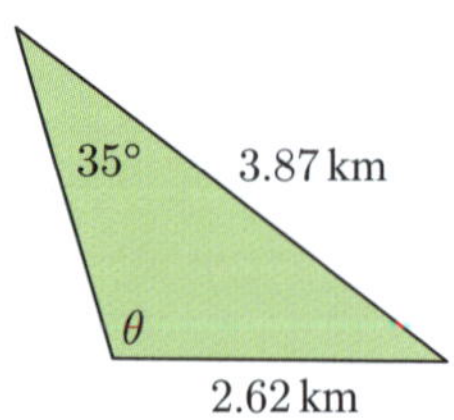

f

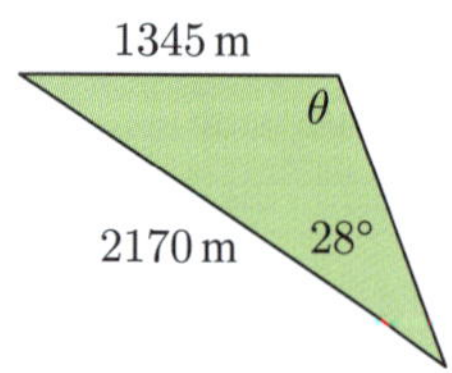

g

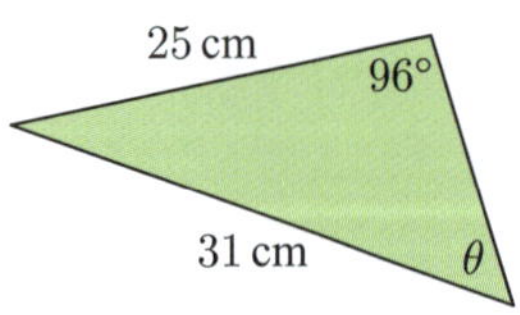

h

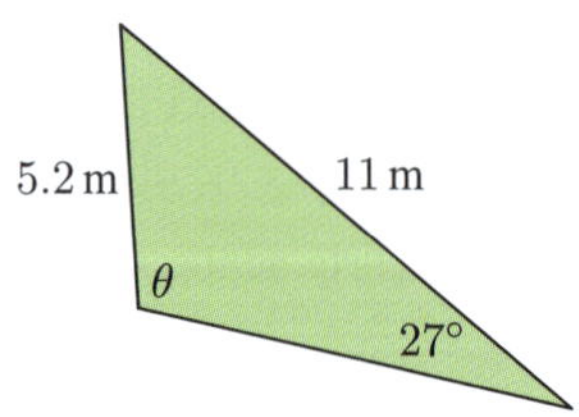

i

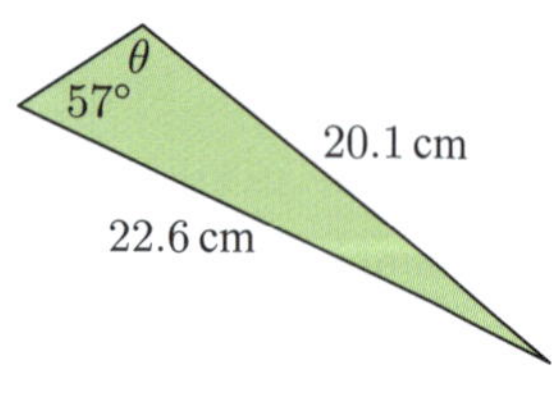

2 The diagrams below are drawn approximately to scale. Find all unknown angles and sides.

a

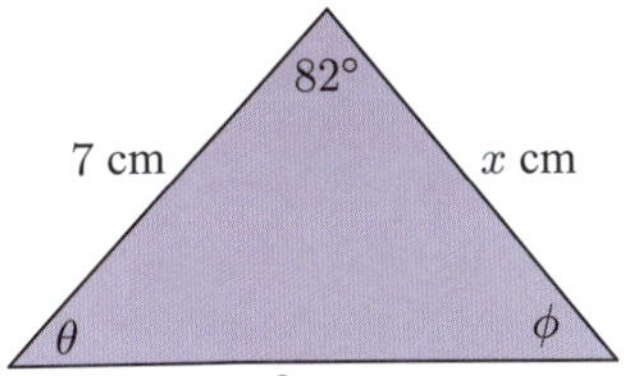

b

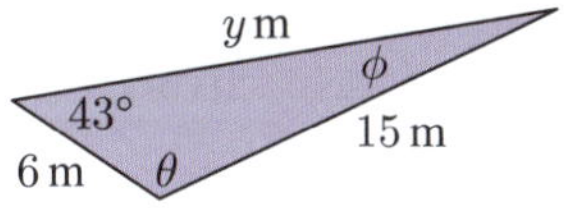

c

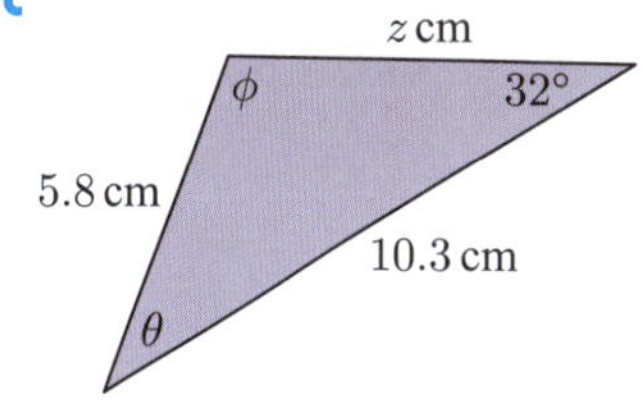

D THE COSINE RULE

The **cosine rule** relates the three sides of a triangle and one of the angles.

INVESTIGATION 2 THE COSINE RULE

In this Investigation you will discover the cosine rule for a triangle which is acute angled.

What to do:

1 For the figure alongside, copy and complete:

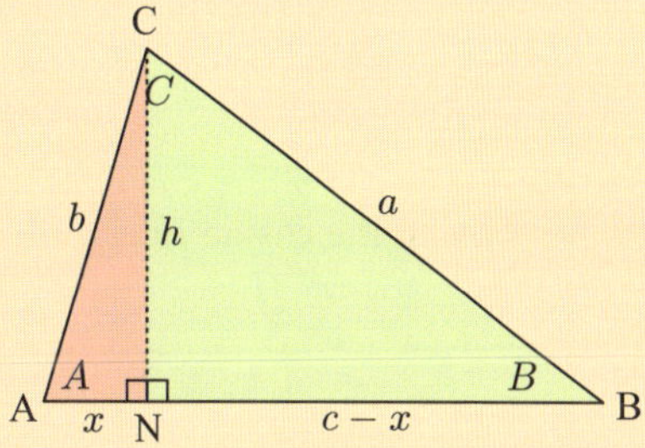

In $\triangle$ANC, $b^2 = ... + ...$ {Pythagoras}

$\therefore\ h^2 = ... - ......$ (1)

In $\triangle$BNC, $a^2 = ... + ...$ {Pythagoras}

$\therefore\ h^2 = ... - ......$ (2)

By equating (1) and (2), $a^2 - ...... = b^2 - ......$

Expanding and collecting terms gives $a^2 = ...$

But in $\triangle$ACN, $\cos A = \frac{....}{....}$

$\therefore\ x = ...$

$\therefore\ a^2 = ...$

$\therefore\ a^2 = b^2 + c^2 - 2bc\cos A.$

2 Copy and complete the following:

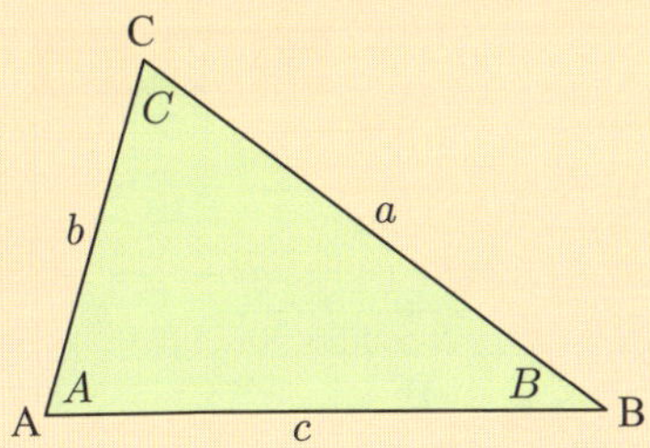

In any triangle with sides a, b, and c units and opposite angles A, B, and C respectively,

$a^2 = b^2 + c^2 - ......$

$b^2 = a^2 + c^2 - ......$

$c^2 = a^2 + b^2 - ......$

These formulae are also valid for triangles which have an obtuse angle.

THE COSINE RULE

In any triangle ABC with sides a, b, and c units, and opposite angles A, B, and C respectively,

$$a^2 = b^2 + c^2 - 2bc\cos A$$
$$b^2 = a^2 + c^2 - 2ac\cos B$$
$$c^2 = a^2 + b^2 - 2ab\cos C.$$

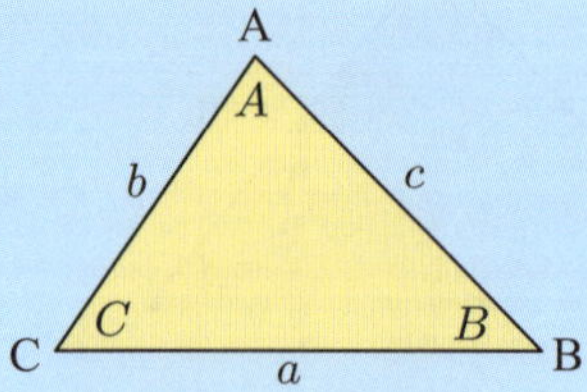

If we are given two sides of a triangle and the included angle, we can use the cosine rule to find the third side.

Example 6 **Self Tutor**

Use the cosine rule to find x, rounding your answer to 1 decimal place.

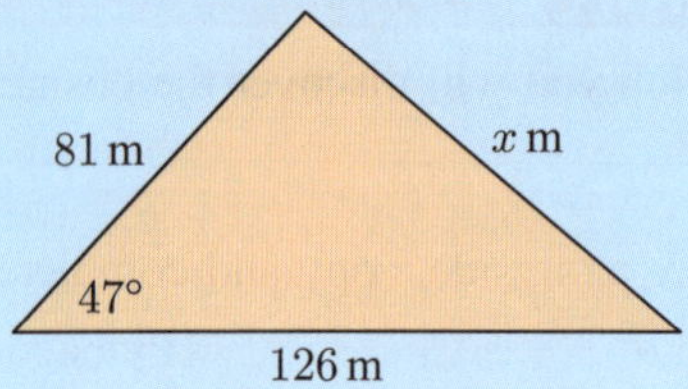

Using the cosine rule, $x^2 = 126^2 + 81^2 - 2 \times 126 \times 81 \times \cos 47°$

$\therefore\ x = \sqrt{126^2 + 81^2 - 2 \times 126 \times 81 \times \cos 47°}$

$\therefore\ x \approx 92.3$

If we know all three side lengths of a triangle, we can use the cosine rule to find any of the angles. To do this, we rearrange the original cosine rule formulae:

$$\cos A = \frac{b^2 + c^2 - a^2}{2bc}, \quad \cos B = \frac{a^2 + c^2 - b^2}{2ac}, \quad \cos C = \frac{a^2 + b^2 - c^2}{2ab}$$

Example 7 **Self Tutor**

Find the measure of angle B in the given figure.

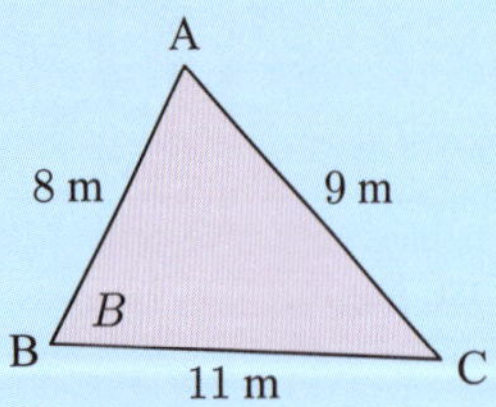

$\cos B = \dfrac{a^2 + c^2 - b^2}{2ac}$

$\therefore\ \cos B = \dfrac{11^2 + 8^2 - 9^2}{2 \times 11 \times 8}$

$\therefore\ \cos B = \dfrac{104}{176}$

$\therefore\ B = \cos^{-1}\left(\dfrac{104}{176}\right)$

$\therefore\ B \approx 53.8°$

EXERCISE 25D

1 Find the length of the remaining side of each triangle:

a
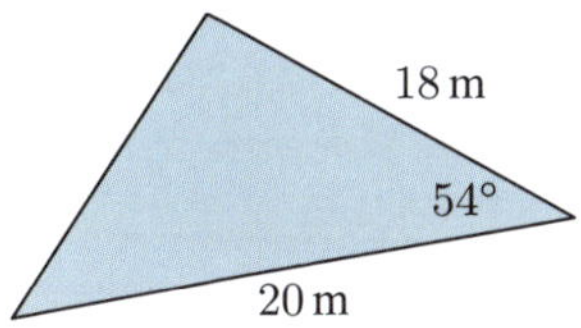

b
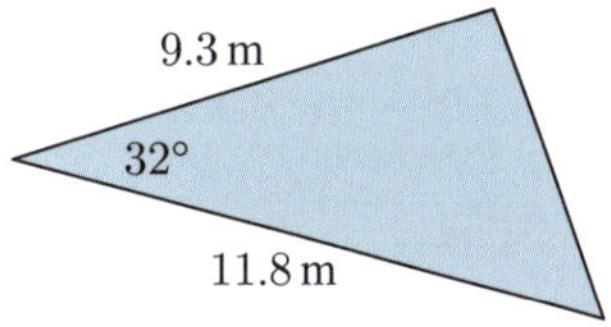

c
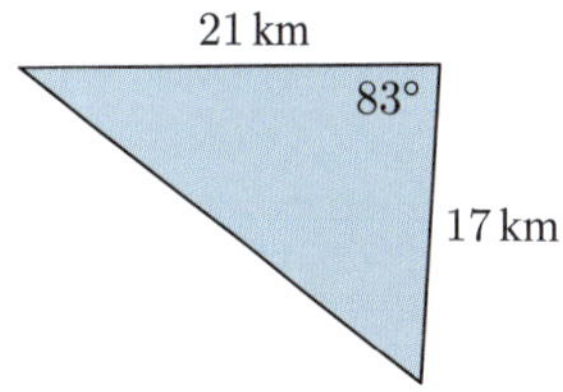

d
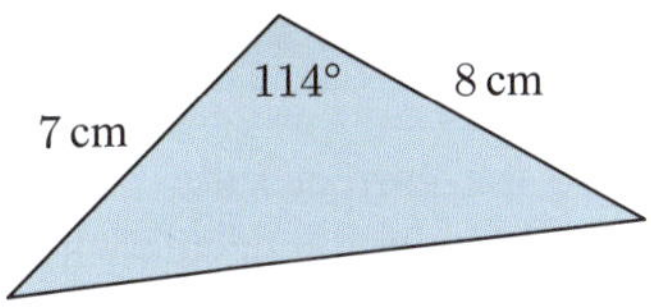

e
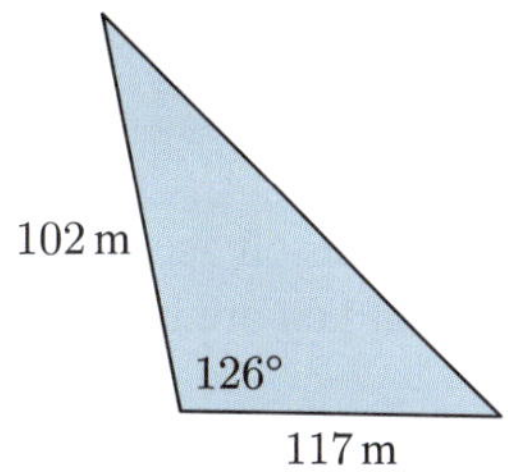

f
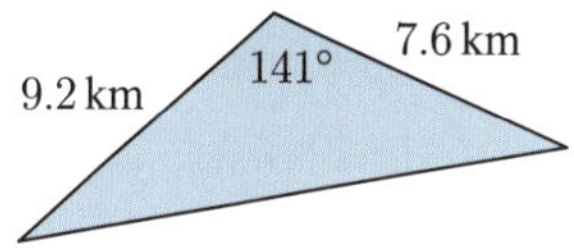

2 Find, to 1 decimal place, the value of the unknown:

a
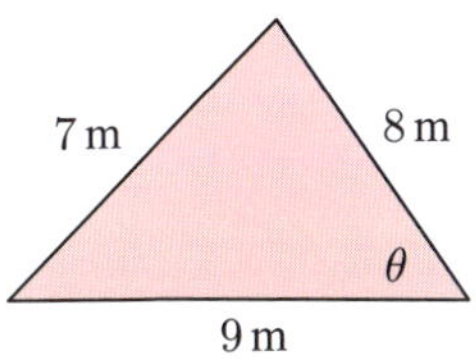

b
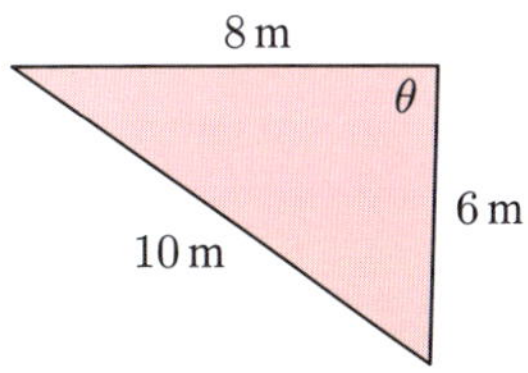

c
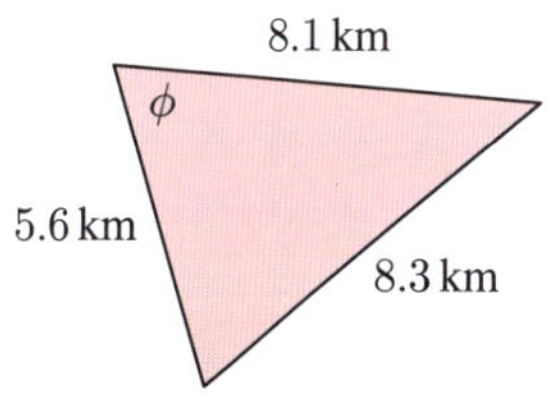

d
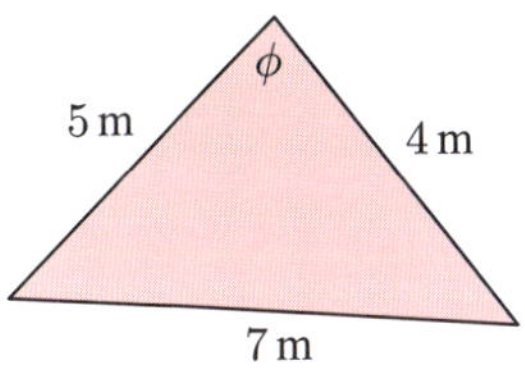

e
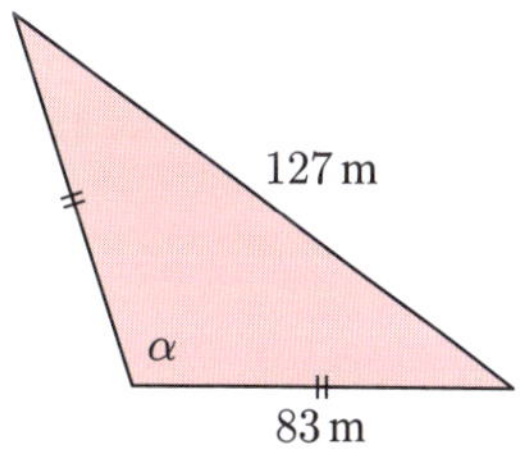

f
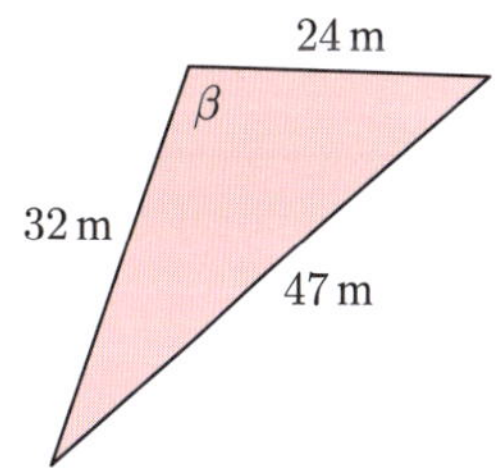

3 Find all of the angles of each triangle:

a
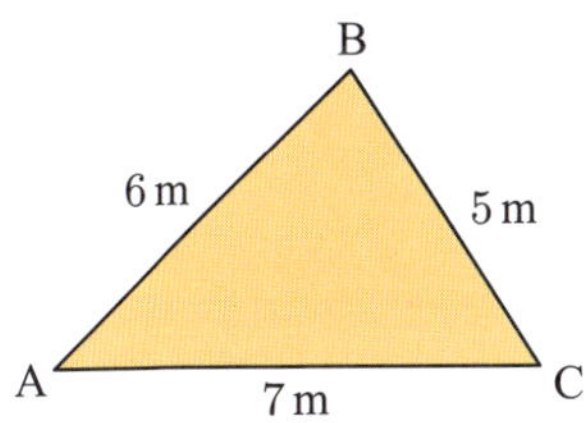

b
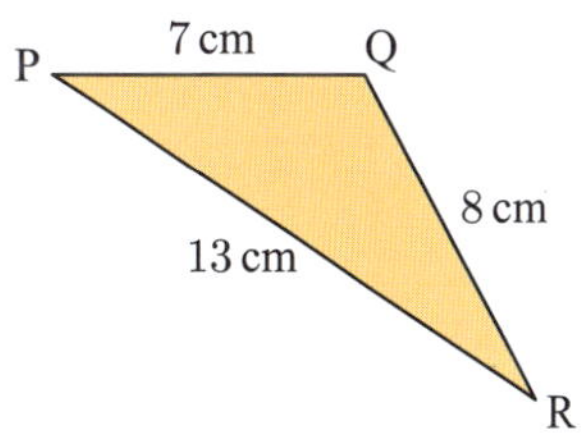

4 Find the measure of all unknown sides and angles:

a
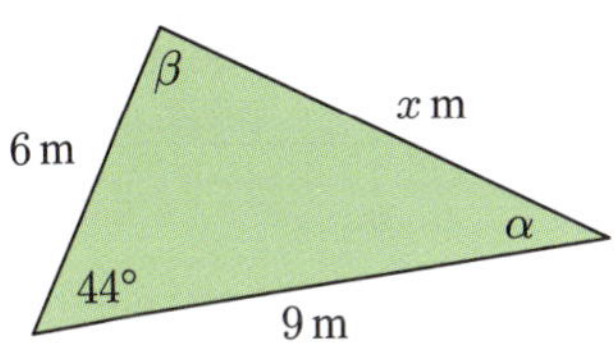

b
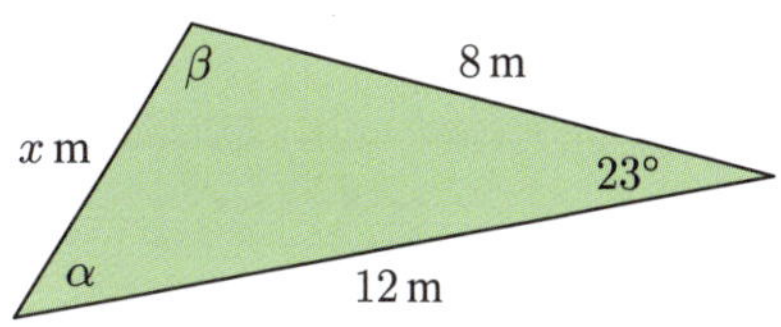

5 Find the area of this triangle:

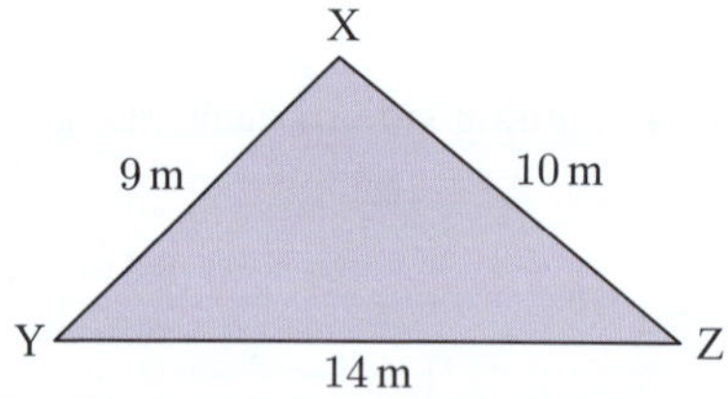

E PROBLEM SOLVING USING THE SINE AND COSINE RULES

When using trigonometry to solve problems, it is important to use the correct rule. Examining the information you are given, as well as what you are trying to find, should help you decide whether to use the sine or cosine rule.

Example 8 Self Tutor

[AD] is a vertical mast and [CE] is a vertical flagpole. Angle ABD is 30° and angle EBC is 50°.

Calculate:

a the length of [DE]

b the size of angle EDB.

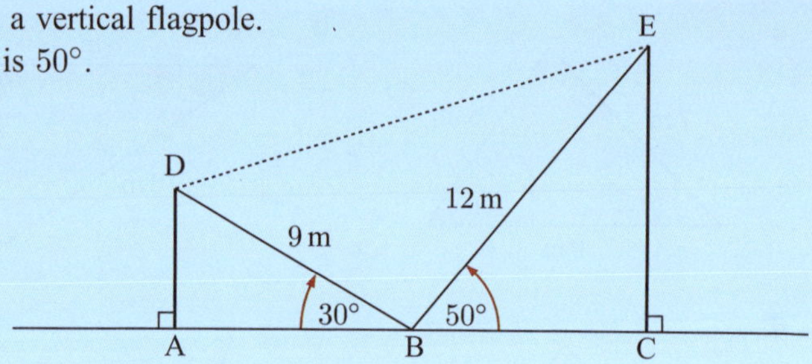

a angle DBE $= 180° - 30° - 50°$
$= 100°$

Using the cosine rule,

$DE^2 = 9^2 + 12^2 - 2 \times 9 \times 12 \times \cos 100°$

$\therefore \; DE = \sqrt{9^2 + 12^2 - 2 \times 9 \times 12 \times \cos 100°}$

$\therefore \; DE \approx 16.2$ m

b

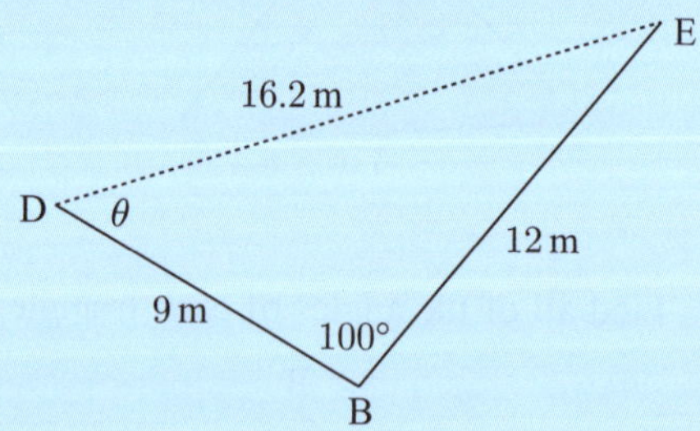

Using the cosine rule:

$\cos\theta \approx \dfrac{9^2 + 16.2^2 - 12^2}{2 \times 9 \times 16.2}$

$\therefore \; \theta \approx \cos^{-1}\left(\dfrac{9^2 + 16.2^2 - 12^2}{2 \times 9 \times 16.2}\right)$

$\therefore \; \theta \approx 46.8°$

$\therefore$ angle EDB is about 46.8°.

EXERCISE 25E

1 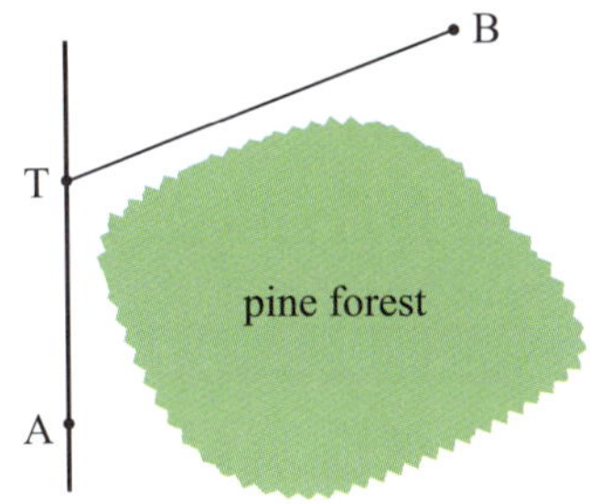

A and B are towns on either side of a pine forest. To get from A to B by car, you need to turn off the main road at T, and then drive along the straight road to B. Point T is 43 km from A and 61 km from B, and $A\widehat{T}B = 116°$.
Find the distance travelled by a helicopter flying directly from A to B.

2 A and B are radar stations 86 km apart along the coast. A ship located at point C is in distress. Radar station A records angle CAB as 47°, while radar station B records angle CBA as 68°.

a Draw a reasonably accurate diagram of the situation, showing all the given information.

b Rescue squads are located at A and B. By how much is B closer to the ship?

3 A field is triangular with sides of length 213 m, 178 m, and 198 m. Find one angle of the field, and hence find the area of the field in hectares.

4 Answer the **Opening Problem** on page **486**.

5 A town has three banks A, B, and C as shown. The post office is halfway between A and B. How far is the post office from bank C?

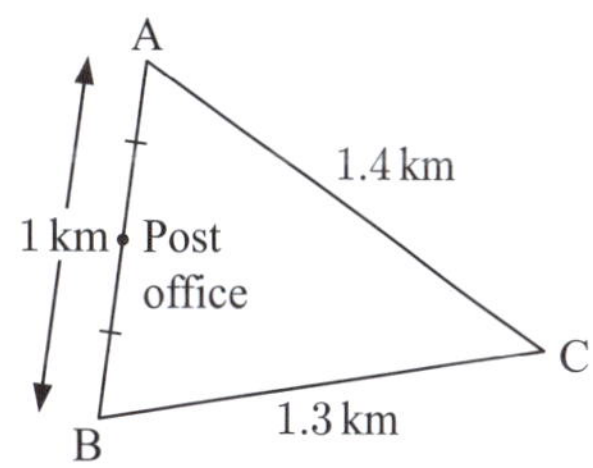

6 [XY] and [RS] are roads which intersect at C at an angle of 77°. There is an explosion at a factory at point D on [RS].

Observers at A and B note that $C\widehat{A}D$ and $C\widehat{B}D$ are 28° and 38° respectively.
A and B are 32.8 km apart.

a Find θ.

b Find the distance [BD].

c How far is D from the intersection C?

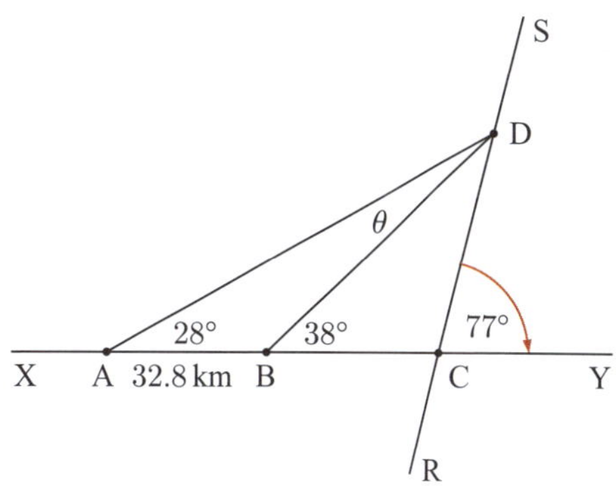

7 In the quadrilateral alongside, the diagonal [PR] is 15 m long. Find the length of the other diagonal [SQ].

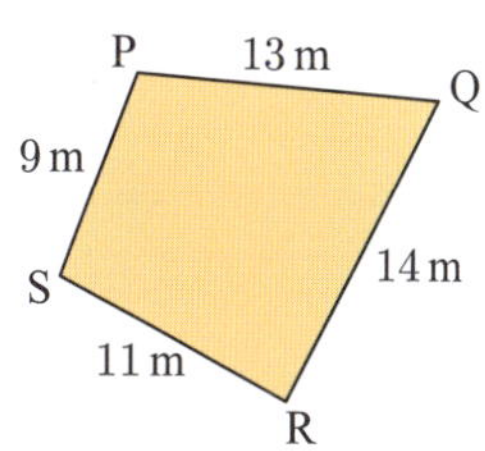

8 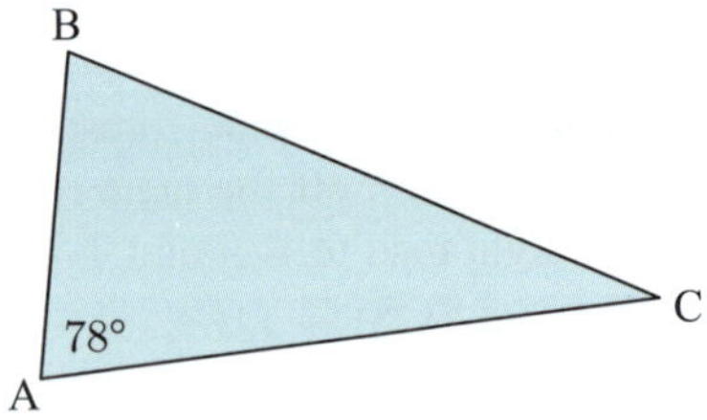

The figure alongside is drawn approximately to scale. [BC] is twice the length of [AB], and $B\widehat{A}C = 78°$.

a Find $B\widehat{C}A$.

b Is A or B closer to C?

c If the triangle has area 10 m^2, find its perimeter.

9 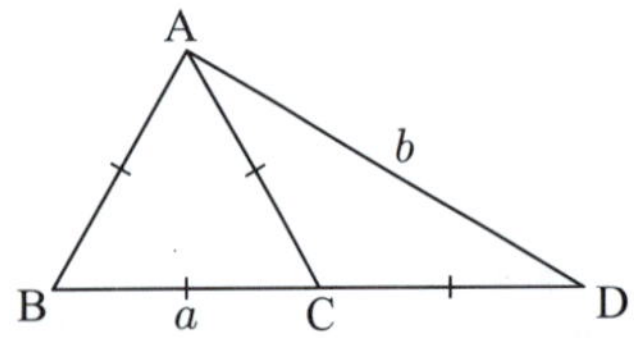

Show that $b = a\sqrt{3}$.

REVIEW SET 25A

1 Use your calculator to find the coordinates of P, to 3 decimal places.

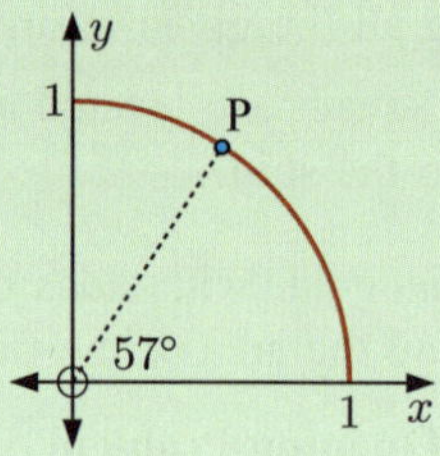

2 Find the obtuse angle which has the same sine as 77°.

3 Find the area of each given figure:

a 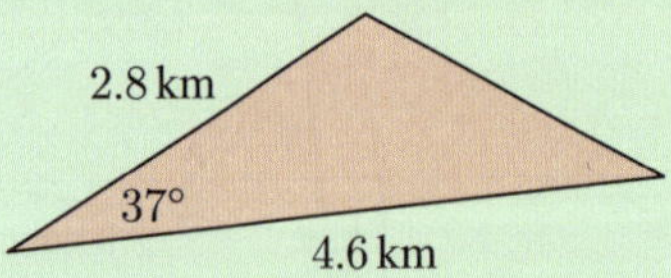

b 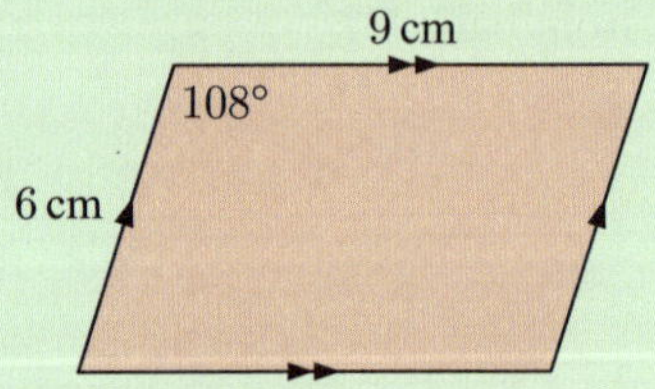

4 Find the value of x:

a 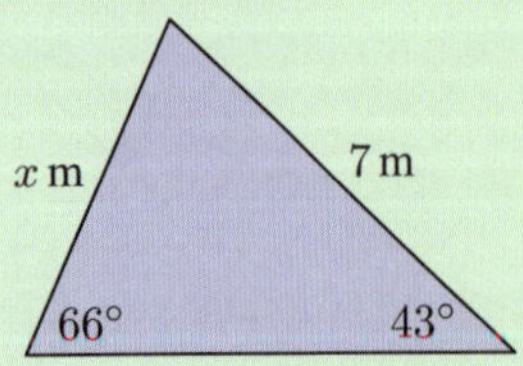

b 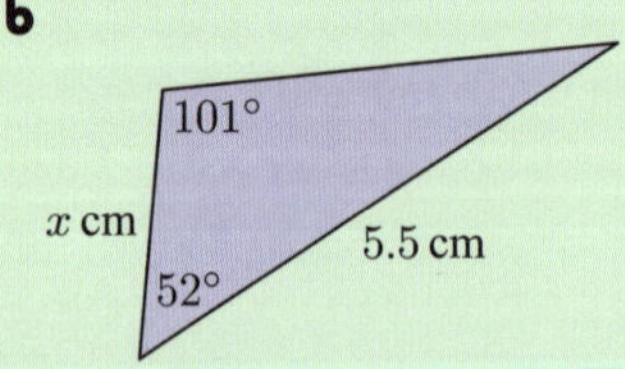

c 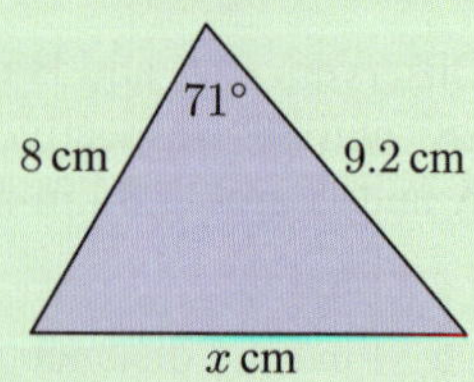

5 Find θ in the diagram alongside, which is drawn approximately to scale.

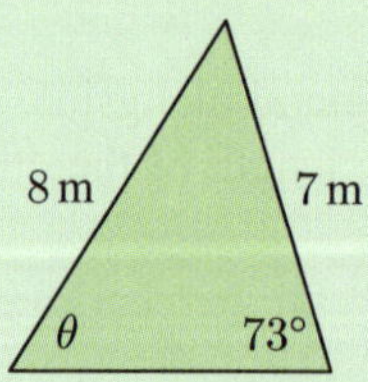

6 Find the length of the remaining side of each triangle:

a

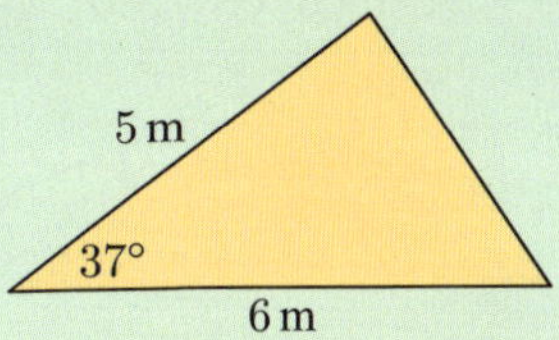

b

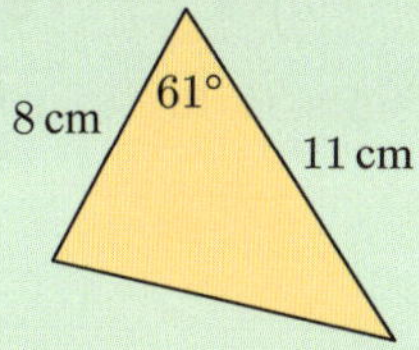

c

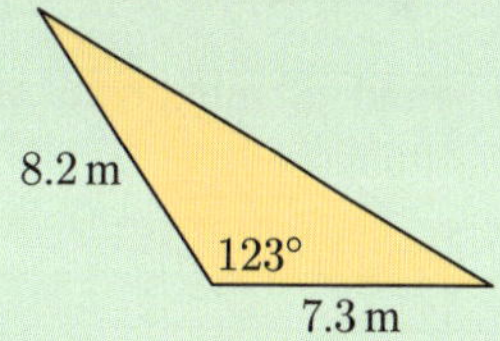

7 Consider the triangle ABC alongside.

a Find the measure of:

i $A\widehat{B}C$ **ii** $B\widehat{C}A$

b Find the area of the triangle using:

i $A\widehat{B}C$ **ii** $B\widehat{C}A$

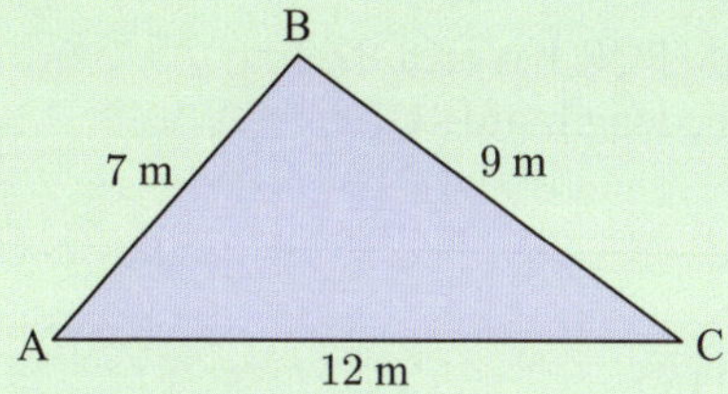

8

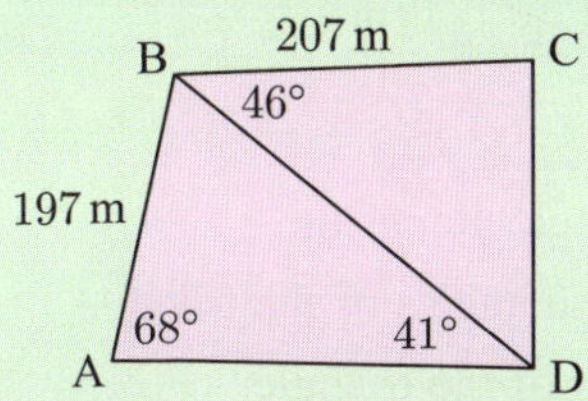

Find:

a the length of [BD]

b the total area of quadrilateral ABCD.

9

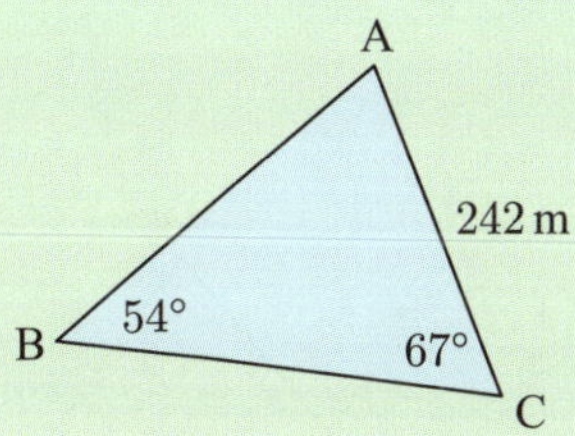

Jason owns a triangular farm with the dimensions shown. Find:

a the length of the fence [AB]

b the area of the farm in hectares.

REVIEW SET 25B

1 Find θ, to 1 decimal place:

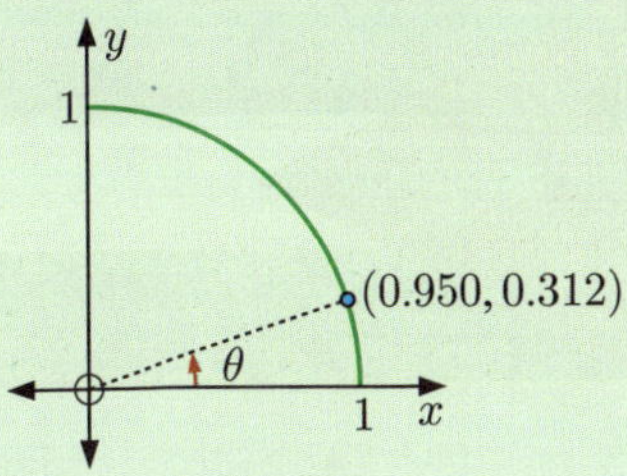

2 Use your calculator to find the coordinates of Q, to 3 decimal places.

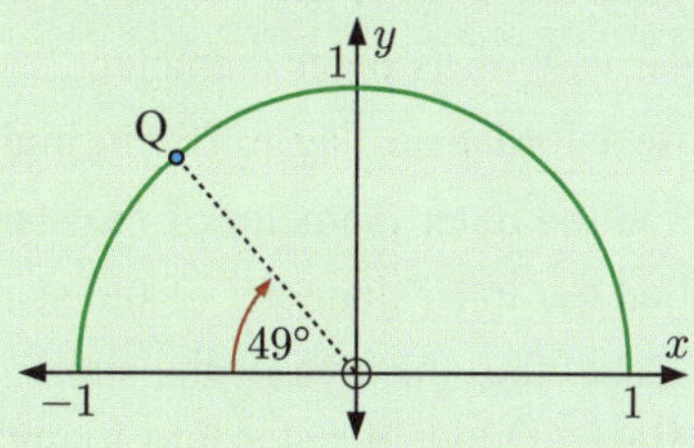

3 Find the area of:

a

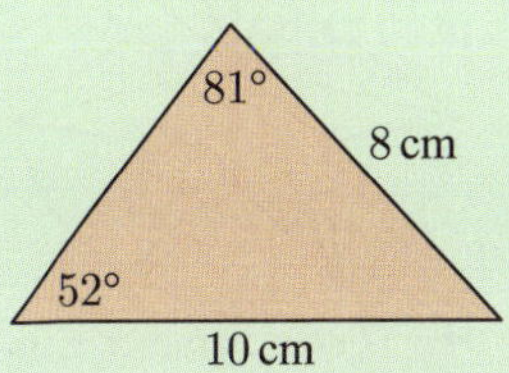

b

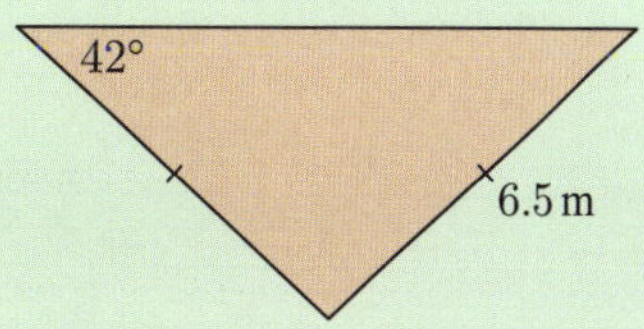

4 Copy and complete:

  a If $\sin\theta = a$, then $\sin(180° - \theta) = ...$ **b** If $\cos\theta = b$, then $\cos(180° - \theta) = ...$

5 Find all unknown angles and sides in the triangle alongside.

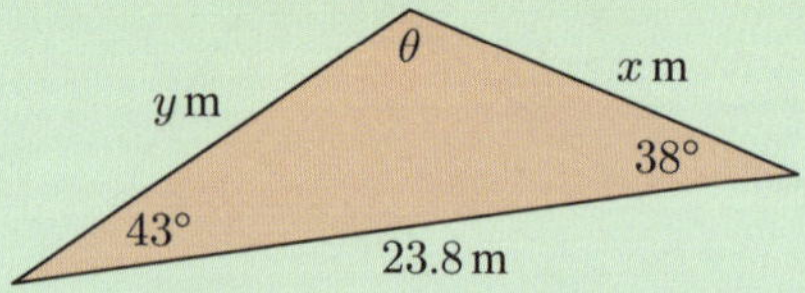

6 Triangle PQR has area 82.2 m^2.
Find the length of [RQ].

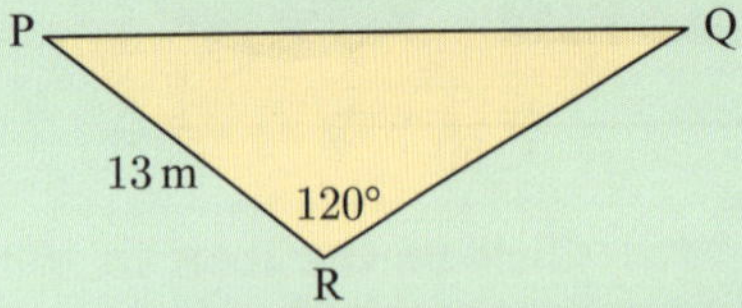

7 The diagram alongside is drawn approximately to scale.
Find all unknown angles and sides.

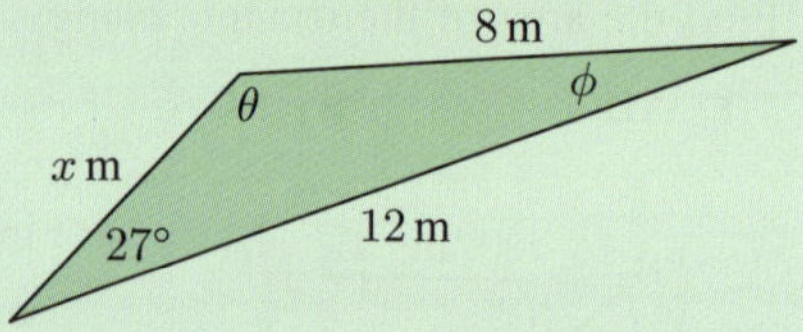

8

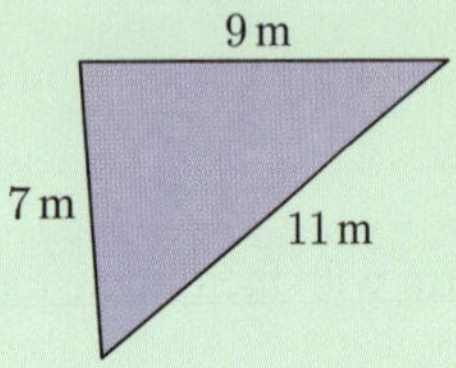

Consider the triangle alongside.

  a Find all of the angles of the triangle.

  b Find the area of the triangle.

9 Triangle PQR has been enlarged with scale factor k to give triangle P′Q′R′.

  a Show that the angle at Q is equal to the angle at Q′.

  b Hence, show that the area of $\triangle$P′Q′R′ $= k^2 \times$ area of $\triangle$PQR.

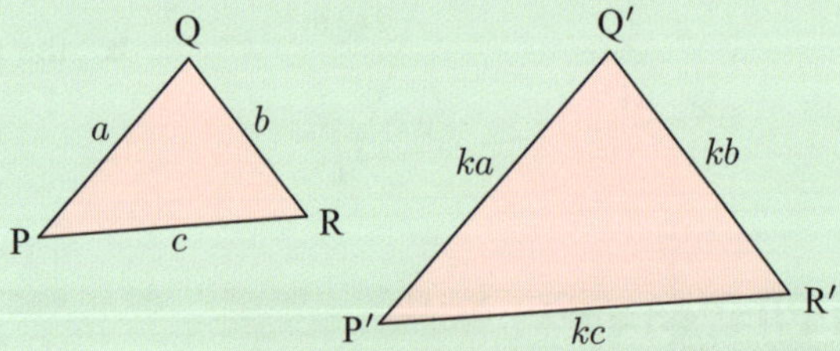

10 The Jordanian flag contains a 7-pointed star. Its outer diameter is $\frac{1}{7}$ of the flag's height, and its inner diameter is $\frac{1}{2}$ of its outer diameter.
Suppose a Jordanian flag is 70 cm high.

  a Find the outer diameter of the star.

  b Find the inner diameter of the star.

  c Hence, find the area of the star.
  Hint: Divide the star into triangles as shown.

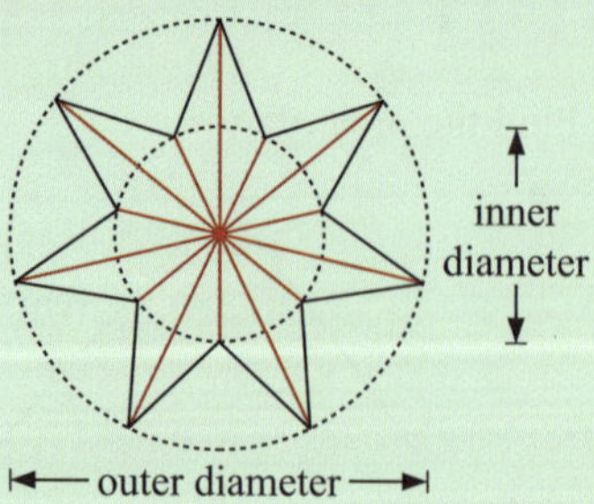

Chapter 26

CHAPTER

Click on the icon to access this chapter

Vectors

Contents:

Chapter 27

CHAPTER

Click on the icon to access this chapter

Logic

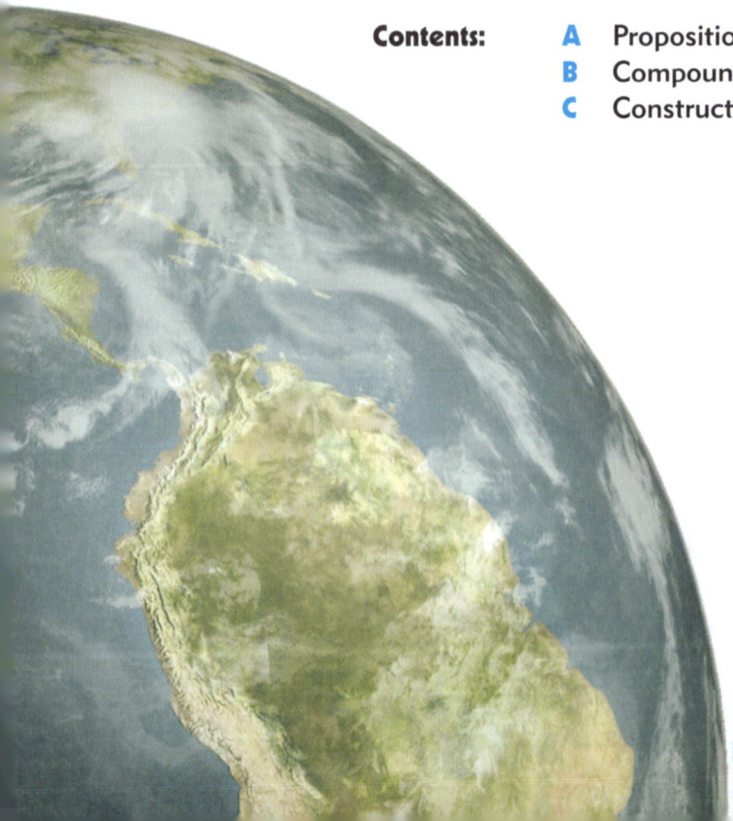

Contents:

ANSWERS

EXERCISE 1A.1

1 a $9x$ b $5f$ c $2a+3b$ d $2a-2b$
e $ac+2b$ f $2x-3y$ g $4ab$ h $2yz$
i $8(a-b)$ j $6(c+d)$ k $xy(z+5)$ l $3lmn$

2 a $4k$ b $2a+3b$ c $3t-2u$ d $3m+5$
e $8-2c$ f $g+3h+9$

3 a b^4 b $4x^2$ c $7y^3$ d $6p^2q^3$
e $2c^3d$ f $5m^2+6n^3$

4 a $y\times y\times y$ b $2\times 2\times x\times x\times x\times x$
c $3\times a\times a\times b$ d $2\times 2\times 3\times 3\times p\times p$
e $2\times g\times g+h\times h\times h$
f $m\times m\times m\times m\times m-7\times n\times n$

5 a $9a-4ab^2$ b t^2+s^2t c $2xy^2z-3z^2$
d a^3-2a e $3x-8y^2$ f m^3n^2-2m
g $3d-6cd$ h $2s+5st^3$ i $4g^3-3g$
j $2x-xy^2+5$ k $3ab^2+ab+b$
l $10v^3w-6vw^2+7w$

EXERCISE 1A.2

1 a double a b the product of p and q
c the square root of m d the square of a
e 3 less than a f the sum of b and c
g the sum of double x and c h the square of twice a
i twice the square of a j a minus the square of c
k the sum of a and the square of b
l the square of the sum of a and b

2 a $a+c$ b $p+q+r$ c ab d $r+s^2$
e $(r+s)^2$ f r^2+s^2 g $2a+b$ h p^3+5
i b^2-a j $\dfrac{a+b}{2}$ k $a+\dfrac{b}{4}$ l $\sqrt{m+n}$
m $x+\dfrac{1}{x}$ n $\frac{1}{4}(a+b)$ or $\dfrac{a+b}{4}$ o $\sqrt{x^2+y^2}$

3 $\sqrt{x^2-13}$ 4 a €$(x+50)$ b €$(x+70)$

5 a \$$2x$ b \$$5(x+30)$ c \$$2x$ + \$$5(x+30)$

EXERCISE 1A.3

1 a L is equal to the sum of a and b.
b K is equal to the average of a and b.
c M is equal to 3 times d.
d N is equal to the product of b and c.
e T is equal to the product of b and the square of c.
f F is equal to the product of m and a.
g K is equal to the square root of the quotient of n and t.
h c is equal to the square root of the sum of the squares of a and b.
i A is equal to the average of a, b, and c.

2 a $S=p+r$ b $D=b-a$ c $A=\dfrac{k+l}{2}$
d $M=a+\dfrac{1}{a}$ e $K=t+s^2$ f $N=gh$
g $y=x+x^2$ h $P=\sqrt{d+e}$

EXERCISE 1B

1 a 12 b 45 c 30 d -30
e -9 f -2 g -11 h -58

2 a $-\frac{4}{3}$ b -2 c $\frac{1}{2}$ d -2
e $\frac{3}{2}$ f -2 g $\frac{3}{2}$ h 3

3 a 1 b 64 c 26 d 36
e -126 f 45 g -1 h 50

4 a 1 b $\sqrt{3}$ c $\sqrt{7}$ d 1
e 5 f 2 g 3 h 4

5 a $\frac{3}{16}$ b $\frac{1}{5}$

6 a i $2h$ m ii $(2h-3)$ m b 10 m c 5 m

7 a $\dfrac{24-x}{2}$ chocolates
b i 10 chocolates ii 8 chocolates iii 5 chocolates

EXERCISE 1C

1 a $a+8$ b $a+6$ c $2b+5$ d $c+3$
e $2x-3$ f $5m$ g $3m$ h $-2t$
i $-4t$ j $2t$ k cannot be simplified
l $2x^2+2x$ m cannot be simplified n $-2a^2+3$
o $7x-4$ p $3x$ q $4ab$ r $8x^2y$

2 a 0 b $5x$ c cannot be simplified
d $3cd$ e $-2pq$ f $2m^2$ g $9b^2-6$
h $4a+6$ i $-2a+2$ j $c+7d$ k $2a+6$
l $3x^2+4x$ m $3xy-1$ n $-4b-3$ o $3a^2+4a$
p cannot be simplified q cannot be simplified
r $2x^2-11x$

3 a $6pq$ b $3m^2-1$ c $4g-8$ d $-6a-2b$
e $2ab-2a+4b$ f cannot be simplified g $3x^3+5x^2$
h cannot be simplified i $4x^3+x^2-3x$

EXERCISE 1D

1 a $10x$ b $12x$ c $-12x$ d $-21x$ e $3x^3$
f $18x^2$ g $-4d^2$ h $18x^3$ i $-2x^4$ j $-15x^3$
k a^2 l $4p^2$ m $3k^4$ n $6x^4$ o $-10t^4$
p $8x^5$

2 a $14x$ b $30a-14b^2$ c $16d^2$ d $10xy$
e x^2 f $-7xy$ g $15ab$ h $-4xy$ i $-7cd$

3 a $5xy-3x^2y$ b $12a^2b^3-a^3b^2$

REVIEW SET 1A

1 a $8t$ b $7ab$ c $3y+2z$

2 a $2y-3y^4$ b a^3b-2b c $5p^3q^2+pq$ d x^2+4xyz^2

3 a $a+\sqrt{b}$ b $\sqrt{a+b}$

4 a 3 more than the square root of a
b the square root of the sum of a and 3

5 -2

6 a T is equal to 4 times s.
b P is equal to the square root of the quotient of x and y.

7 a $\left(\dfrac{100-x}{4}\right)$ dollars b i \$20 ii \$16 iii \$10

8 a $x+8$ b $5x-6y$ c $2a^2+4a$
d cannot be simplified e $5x^3y+2xy^3$ f x^3+2x^2

9 a $-28x$ b $45x^2$ c $9d^2$

10 a $22x$ b $18xy+2x^2y$ c $-a^2$

REVIEW SET 1B

1 **a** half of the sum of a and b **b** the sum of a and half b

2 **a** $5k^3$ **b** $6a^2 - bc^3$ **3** **a** $x^2 + 3$ **b** $(3+x)^2$

4 **a** $(x+30)$ cm **b** $(x+35)$ cm

5 **a** $K = m + n$ **b** $L = \sqrt{pq}$ **6** 3

7 **a** $7 - x^2 + 6x$ **b** $3t^3 + t^2 + t$

8 **a** $50c^2d^3$ **b** $2xy - 5x^2$

9 **a** $7z$ **b** $5c^2d - 2cd^2$ **c** $16m^2 - 6m^2n$

10 **a** $(x-5)^2$ **b** **i** 25 **ii** 1 **iii** 81
c No, as the expression is a perfect square which cannot be negative.

EXERCISE 2A.1

1 **a** 8 **b** 25 **c** 128 **d** 216
e 2250 **f** 1176 **g** 2475 **h** 52 000

2 **a** $2^1 = 2,\ 2^2 = 4,\ 2^3 = 8,\ 2^4 = 16,\ 2^5 = 32,\ 2^6 = 64$
b $3^1 = 3,\ 3^2 = 9,\ 3^3 = 27,\ 3^4 = 81$
c $5^1 = 5,\ 5^2 = 25,\ 5^3 = 125,\ 5^4 = 625$
d $7^1 = 7,\ 7^2 = 49,\ 7^3 = 343$

3 **a** $2^3 \times 5$ **b** $2^2 \times 3 \times 7$ **c** $2^2 \times 5^2$
d $2^3 \times 3^2 \times 5$ **e** $2 \times 3^2 \times 5 \times 11$
f $2^7 \times 3^2$ **g** $2 \times 3^3 \times 7^2$ **h** $3^3 \times 5^2 \times 7$

4 **a** 2^6 **b** 2^9 **c** 2^{11} **5** **a** 5^3 **b** 5^5 **c** 5^6

6 1 **7** $a = 2,\ b = 5$ or $a = 5,\ b = 2$

8 **a** The factors of 30 have been written in the form $2^a3^b5^c$ where $a = 0$ or 1, $b = 0$ or 1, and $c = 0$ or 1. So, there are 2 different values for each of a, b, and c, which gives $2 \times 2 \times 2 = 8$ different values of $2^a3^b5^c$.
b 24 factors **c** $(a+1)(b+1)(c+1)$ factors

EXERCISE 2A.2

1 **a** 1 **b** -1 **c** -1 **d** 1 **e** -16 **f** 16
g -16 **h** -25 **i** 125 **j** -27 **k** 81 **l** -49

EXERCISE 2A.3

1 **a** 4096 **b** $-78\,125$ **c** -81 **d** 823 543
e 32 768 **f** -729 **g** -729 **h** 8.911 650 327
i $-0.653\,121\,490\,1$ **j** $-5.871\,463\,646$

2 **a** **i** $0.\overline{142\,857}$ **ii** $0.\overline{142\,857}$ **iii** $0.\overline{1}$ **iv** $0.\overline{1}$
v 0.015 625 **vi** 0.015 625 **vii** 1 **viii** 1
b **i** It is the reciprocal of the number raised to the positive power.
ii Any non-zero number raised to the power zero is 1.

EXERCISE 2B

1 **a** 5^6 **b** 6^{11} **c** a^6 **d** a^{10}
e b^{14} **f** a^{2+n} **g** b^{m+9} **h** p^7

2 **a** 3^7 **b** 7^4 **c** 5^3 **d** a^5
e b^6 **f** p^{n-2} **g** y^{5-b} **h** b^{2n-1}

3 **a** 2^{18} **b** 3^{21} **c** 5^{28} **d** a^{12}
e q^9 **f** d^{6n} **g** x^{8y} **h** g^{10a}

4 **a** 1 **b** 1 **c** 1 **d** 5 **e** 1 provided $a \neq 0$
f 1 **g** 3 provided $y \neq 0$ **h** 1 provided $y \neq 0$
i d^2 provided $d \neq 0$ **j** 1 provided $c \neq 0$ **k** 3 **l** 7

5 **a** b^3 **b** $15c^5$ **c** a^2 **d** $6d^5$
e $7p^2q^3$ **f** $36s^4t^3$ **g** y^2 **h** q^3
i $2y$ **j** 1 **k** $\frac{1}{b}$ **l** $2p^4$

6 **a** 2^4 **b** 3^3 **c** 5^3 **d** 2^6
e 3^6 **f** 2^{a+2} **g** 7^{t-1} **h** 2^{4k}
i 2^{5-x} **j** 5^3 **k** $7^{4(a-1)}$ **l** 3^2
m 3^{y-2x} **n** 2^{2y-3x} **o** 11^{2x} **p** 5^3

7 **a** p^3q^3 **b** a^5b^5 **c** x^4y^4 **d** $p^5q^5r^5$
e $27a^3$ **f** $25b^2$ **g** $32n^5$ **h** $64a^3b^3$
i $81a^2b^2$ **j** $\frac{x^4}{y^4}$ **k** $\frac{a^7}{b^7}$ **l** $\frac{c^6}{64d^6}$
m $\frac{9x^2}{y^4}$ **n** $\frac{w^4}{4v^2}$ **o** $\frac{8r^3}{27s^6}$ **p** $\frac{81a^4}{b^4}$

8 **a** $9a^4$ **b** $\frac{16}{p^8q^4}$ **c** $27c^3d^{12}$ **d** $\frac{a^8}{9n^{14}}$
e $\frac{8x^6}{y^{18}}$ **f** $32s^{15}t^{20}$ **g** $\frac{16g^{10}}{h^4}$ **h** $\frac{81x^8y^{12}}{256z^4}$

9 **a** $\frac{1}{2}$ **b** $\frac{1}{3}$ **c** $\frac{1}{6}$ **d** $\frac{1}{8}$ **e** 9 **f** $\frac{1}{9}$
g 8 **h** $\frac{1}{8}$ **i** 25 **j** $\frac{1}{25}$ **k** 100 **l** $\frac{1}{100}$

10 **a** $\frac{1}{9}$ **b** $\frac{1}{32}$ **c** $\frac{1}{16}$ **d** $\frac{1}{49}$ **e** 3 **f** $2\frac{1}{2}$
g $\frac{3}{4}$ **h** 12 **i** $\frac{2}{7}$ **j** $\frac{2}{3}$ **k** $1\frac{1}{7}$ **l** $3\frac{1}{2}$
m $1\frac{7}{9}$ **n** $5\frac{1}{25}$ **o** $\frac{4}{25}$ **p** $\frac{9}{49}$

11 **a** $\frac{1}{2x}$ **b** $\frac{2}{x}$ **c** $\frac{3}{b}$ **d** $\frac{1}{3b}$ **e** $\frac{1}{4b^2}$ **f** $\frac{b^2}{4}$
g n^2 **h** $\frac{n^2}{3}$ **i** $\frac{1}{ab}$ **j** $\frac{a}{b}$ **k** $\frac{a}{b^2}$ **l** $\frac{1}{a^2b^2}$
m $\frac{1}{2ab}$ **n** $\frac{2}{ab}$ **o** $\frac{2a}{b}$ **p** a^2b^3

12 **a** 3^{-1} **b** 2^{-1} **c** 5^{-1} **d** 2^{-2} **e** 3^{-3} **f** 5^{-2}
g 2^{-3x} **h** 2^{-4y} **i** 3^{-4a} **j** 3^{-2} **k** 5^{-2} **l** 5^{-3}
m 2^4 **n** $2^0 = 3^0 = 5^0$ **o** $2^{-3} \times 3^{-3}$ **p** $2^4 \times 5^2$

EXERCISE 2C.1

1 C

2 **a** $376 = 3.76 \times 10^2$ **b** $8000 = 8 \times 10^3$
c $0.04 = 4 \times 10^{-2}$ **d** $0.005\,07 = 5.07 \times 10^{-3}$
e $9\,040\,000 = 9.04 \times 10^6$ **f** $0.000\,000\,23 = 2.3 \times 10^{-7}$

3 **a** 4.25×10^2 **b** 4.25×10^5 **c** 4.25×10^0
d 4.25×10^{-1} **e** 4.25×10^{-5} **f** 2.01×10^1
g 2.01×10^4 **h** 2.01×10^{-3} **i** 2.01×10^6
j 2.01×10^9 **k** 2.01×10^{-6} **l** 1×10^1

4 **a** 2.49×10^8 km **b** 5×10^{-4} kg
c 5×10^{-3} mm **d** 1.4162×10^{-7}
e 1.5×10^7 °C **f** 3.444×10^7 hectares

5 **a** 50 000 **b** 3000 **c** 18 000 000 **d** 810
e 650 000 **f** 11 **g** 275 000 000 **h** 8 000 000

6 **a** 0.03 **b** 0.000 09 **c** 0.007
d 0.000 41 **e** 0.000 008 2 **f** 0.761
g 0.000 000 325 **h** 0.000 000 02

7 **a** 0.000 000 475 m **b** 6 850 000 000 people
c 0.000 012 m **d** 40 100 km **e** 0.000 001 5 kg

8 **a** 2.2×10^8 **b** 3.5×10^{-5} **c** 4.9×10^3
d 6.41×10^8 **e** 1.01×10^{-6} **f** 9×10^{-8}

9 **a** 9.11×10^{-28} g **b** 1.67×10^{-24} g **c** proton

10 **a** 6×10^7 **b** 2.4×10^7 **c** 4×10^{12} **d** 8.1×10^{19}
e 9×10^8 **f** 4.9×10^{-5} **g** 4×10^{-1} **h** 2×10^3

EXERCISE 2C.2

1 a 6.00×10^{-10} b 2.39×10^{10} c 2.19×10^{10}
d 4.01×10^{10} e 6.24×10^{-10} f 1.03×10^{-7}

2 a 9.55×10^{9} b 9.62×10^{-7} c 4.90×10^{-14}
d 6.50×10^{-7} e 4.00×10^{-7} f 4.37×10^{14}

3 a 9.60×10^{9} litres b 6.72×10^{10} litres
c 3.51×10^{12} litres

4 a 1.80×10^{10} m b 2.59×10^{13} m c 9.46×10^{15} m

EXERCISE 2D

1 a 2 b $\frac{1}{2}$ c 3 d $\frac{1}{3}$ e 6 f $\frac{1}{6}$
g 3 h $\frac{1}{3}$ i 10 j $\frac{1}{10}$ k 5 l $\frac{1}{5}$

2 a $11^{\frac{1}{2}}$ b $11^{-\frac{1}{2}}$ c $12^{\frac{1}{2}}$ d $12^{-\frac{1}{2}}$
e $26^{\frac{1}{3}}$ f $26^{-\frac{1}{3}}$ g $7^{\frac{1}{4}}$ h $7^{-\frac{1}{5}}$

3 a ≈ 2.11 b ≈ 3.13 c ≈ 0.33 d ≈ 0.66

REVIEW SET 2A

1 a 144 b 100 c 270 2 a -1 b -27 c -4

3 a q^7 b a^2 c t^{4x} d 1

4 a a^3b^3 b $\dfrac{3w}{v}$ c $\dfrac{16a^8}{d^{12}}$

5 a 1 b 5 provided $a \neq 0$ c 5

6 a $\frac{1}{16}$ b $\frac{2}{9}$ c $\frac{5}{6}$ 7 a 2^{-3} b 5^{-4} c 3^{-5t}

8 a 0.007 71 b 40 000 000

9 a 3.06×10^{-4} b 1.69×10^{9} c 4.75×10^{8}

10 1.43×10^{5} km 11 a 9 b $\frac{1}{4}$

12 a $30^{\frac{1}{3}}$ b $11^{-\frac{1}{4}}$

REVIEW SET 2B

1 $2^2 \times 3^3 \times 11$ 2 a c^{12} b 10

3 a $\dfrac{3h^2}{g}$ b x^5 c $8a^6b^3c^9$ 4 a 2^0 b 2^{6k}

5 9 6 a $\frac{1}{81}$ b $\frac{8}{3}$ c $1\frac{7}{8}$ 7 a $\dfrac{b^3}{a^2}$ b $\dfrac{25b^6}{a^4}$

8 0.000 000 000 35 kg 9 a 2 b $\frac{1}{10}$

10 3.9×10^{-4} 11 a ≈ 3.42 b ≈ 0.41

12 a 0.000 000 000 000 035 8 m b 1.405×10^{10} years
c i $\approx 3.853 \times 10^{-22}$ g ii $\approx 2.595 \times 10^{22}$ atoms

EXERCISE 3A

1 a i $C = \{$yellow, black$\}$ ii $D = \{2, 5, 9\}$
iii $L = \{$E, L$\}$
b i 2 ii 3 iii 2

2 a $P = \{1, 3, 5, 9, 15, 45\}$ b $Q = \{$A, U, S, T, R, L, I$\}$
c $n(P) = 6$, $n(Q) = 7$

3 a $4 \in P$ b $9 \notin D$ c $\{2, 3\} \subseteq \{1, 2, 3, 6\}$
d $n(A) = 6$

4 a $\varnothing, \{a\}, \{b\}, \{a, b\}$
b $\varnothing, \{-1\}, \{0\}, \{2\}, \{-1, 0\}, \{-1, 2\}, \{0, 2\}, \{-1, 0, 2\}$

5 a $A \cap B = \{b, d\}$, $A \cup B = \{a, b, c, d, f, h\}$
b $A \cap B = \{\ \}$ or $\varnothing$, $A \cup B = \{1, 2, 3, 4, 5, 6, 7, 8\}$
c $A \cap B = \{$vanilla, chocolate, strawberry$\} = A$
$A \cup B = \{$vanilla, lemon, chocolate, strawberry, coffee$\}$
$= B$

6 a $\{\ \}$ or $\varnothing$ b yes 7 a true b true

8 a yes b no c no d yes

9 a $A = \{2, 3, 5, 7, 11, 13, 17, 19\}$
b $B = \{1, 2, 4, 5, 10, 20\}$ c $n(A) = 8$, $n(B) = 6$
d $A \cup B = \{1, 2, 3, 4, 5, 7, 10, 11, 13, 17, 19, 20\}$,
$A \cap B = \{2, 5\}$
e $n(A \cap B) = 2$

10 a $x = 2, 4, 6,$ or 8 b $x = 4$

EXERCISE 3B

1 a true b false c true d true
e false f true g true h false

2 a $0 \in \mathbb{Z}$ but $0 \notin \mathbb{Z}^- \cup \mathbb{Z}^+$
b $\mathbb{Z}^- \cap \mathbb{Z}^+ = \varnothing$ $\therefore$ $n(\mathbb{Z}^- \cap \mathbb{Z}^+) = 0$

3 a true b true c true d true
e true f true g false h true

4 a $\mathbb{R}$ b $\{\ \}$ or $\varnothing$ c $\mathbb{Q}'$ d $\mathbb{R}$

EXERCISE 3C

1 a $\{x \mid 10 < x < 100,\ x \in \mathbb{Z}\}$
b $\{x \mid x < 200,\ x \in \mathbb{R}\}$ c $\{x \mid -1 \leqslant x \leqslant 1,\ x \in \mathbb{Q}\}$
d $\{x \mid -2 \leqslant x \leqslant 1,\ x \in \mathbb{Z}\}$ e $\{x \mid 2 < x \leqslant 7,\ x \in \mathbb{R}\}$

2 a finite b infinite c infinite d infinite

3 a i The set of all integers x such that x lies between -2 and 5.
ii $B = \{-1, 0, 1, 2, 3, 4\}$ iii $n(B) = 6$
iv

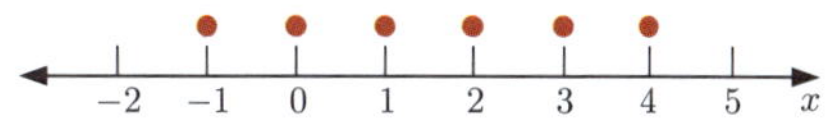

b i The set of all positive integers x such that x lies between -6 and 3 inclusive.
ii $B = \{1, 2, 3\}$ iii $n(B) = 3$
iv

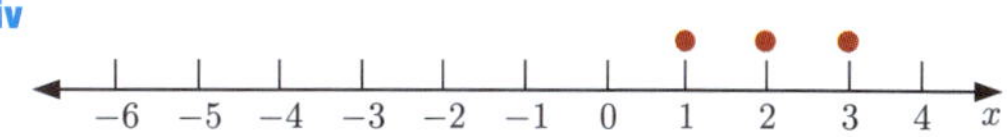

c i The set of all real numbers x such that x lies between 1 and 2.
ii not possible iii $n(B)$ is infinite
iv

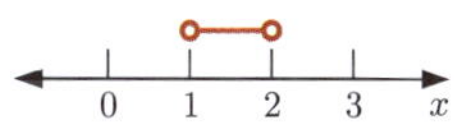

d i The set of all real numbers x such that x lies between 4 and 7, including 4.
ii not possible iii $n(B)$ is infinite
iv

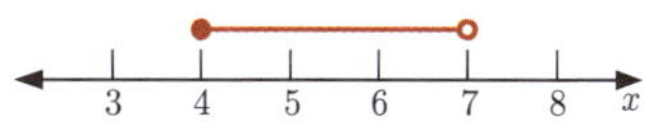

4 a yes b yes c yes d no

5 a i $P \cap Q = \{x \mid 30 \leqslant x \leqslant 80,\ x \in \mathbb{Z}\}$
ii $P \cup Q = \{x \mid 10 \leqslant x \leqslant 120,\ x \in \mathbb{Z}\}$
b i $n(P) = 71$ ii $n(Q) = 91$
iii $n(P \cap Q) = 51$ iv $n(P \cup Q) = 111$

EXERCISE 3D

1 a $B' = \{1, 3, 5, 7\}$

2 a $D' = \mathbb{Z}^-$ b $D' = \{$even positive integers$\}$
c $D' = \{x \mid x > 10,\ x \in \mathbb{Z}\}$
d $D' = \{x \mid 3 < x < 5,\ x \in \mathbb{Q}\}$

3 **a** **i** $\{1, 2, 3, 4, 6, 8, 12, 24\}$ **ii** $\{4, 8, 12, 16, 20, 24\}$
iii $\{0, 5, 7, 9, 10, 11, 13, 14, 15, 16, 17, 18, 19, 20, 21, 22, 23\}$
iv $\{0, 1, 2, 3, 5, 6, 7, 9, 10, 11, 13, 14, 15, 17, 18, 19, 21, 22, 23\}$
b $\{1, 2, 3, 6\}$. These are factors of 24 that are not multiples of 4.

4 **a** **i** $\{12, 13, 14, 15, 16, 17, 18, 19, 20\}$
ii $\{13, 15, 17, 19, 20\}$ **iii** $\{13, 14, 16, 17, 19, 20\}$
iv $\{12, 14, 15, 16, 18\}$ **v** $\{13, 17, 19, 20\}$
b $(A \cup B)' = \{13, 17, 19, 20\}$ which is $A' \cap B'$

5 **a** **i** {January, June, July}
ii {January, March, May, July, August, October, December}
iii {February, March, April, May, August, September, October, November, December}
iv {February, April, June, September, November}
b All months not containing 31 days.
c {February, April, September, November}
All months not beginning with "J" and not containing 31 days.
d {March, May, August, October, December}
All months which do not begin with "J" and have 31 days.

6 **a** **i** $\{1, 2, 3, 4, 5, 6, 7, 8, 9, 10, 11, 12, 13, 14\}$
ii $\{3, 4, 5, 6\}$ **iii** $\{1, 2, 7, 8, 9, 10, 11, 12, 13, 14\}$
iv $\{1, 3, 4, 5, 7, 8, 9, 11, 12, 13\}$
v $\{1, 2, 6, 7, 8, 9, 10, 11, 12, 13, 14\}$ **vi** $\{3, 4, 5\}$
b **i** 11 **ii** 3

EXERCISE 3E.1

1 **a** $\{1, 4, 5, 7\}$ **b** $\{2, 5, 7\}$ **c** $\{2, 3, 6\}$ **d** $\{1, 3, 4, 6\}$
e $\{5, 7\}$ **f** $\{1, 2, 4, 5, 7\}$ **g** $\{1, 2, 3, 4, 5, 6, 7\}$

2 **a** $\{a, c, f, i, j\}$ **b** $\{c, d, g, h, j\}$ **c** $\{b, d, e, g, h, k\}$
d $\{a, b, e, f, i, k\}$ **e** $\{c, j\}$ **f** $\{a, c, d, f, g, h, i, j\}$

3 **a**
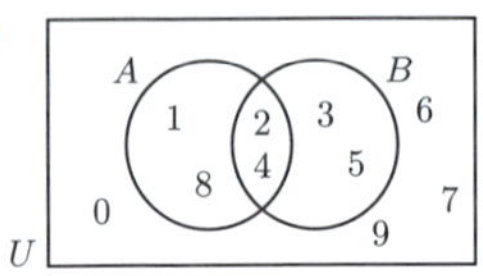

b
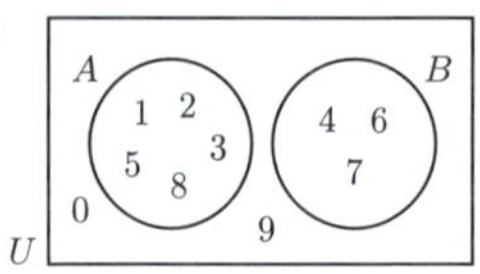

c
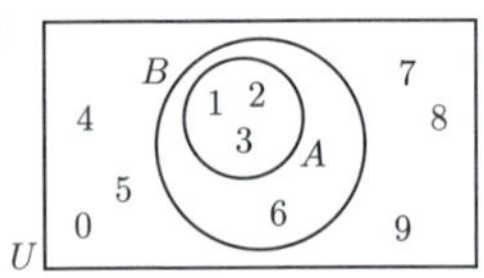

d
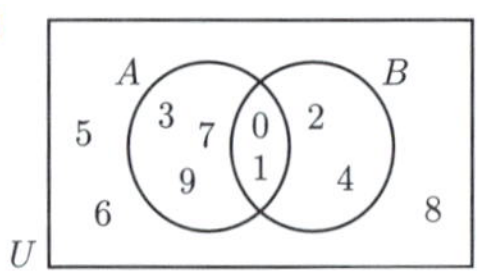

4 **a** $A = \{1, 2, 3, 4, 6, 12\}$, $B = \{2, 3, 5, 7, 11\}$
b $A \cap B = \{2, 3\}$, $A \cup B = \{1, 2, 3, 4, 5, 6, 7, 11, 12\}$
c
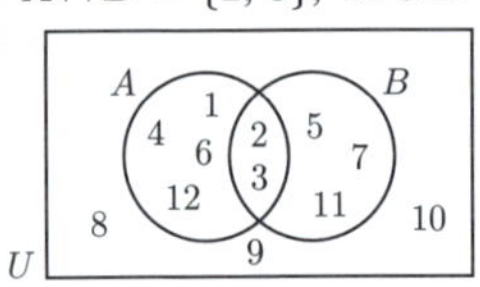

5 **a** $R = \{2, 3, 5, 7, 11, 13, 17, 19\}$,
$S = \{4, 6, 8, 9, 10, 12, 14, 15, 16, 18\}$
b $R \cap S = \{\ \}$ or $\varnothing$,
$R \cup S = \{2, 3, 4, 5, 6, 7, 8, 9, 10, 11, 12, 13, 14, 15, 16, 17, 18, 19\}$

c
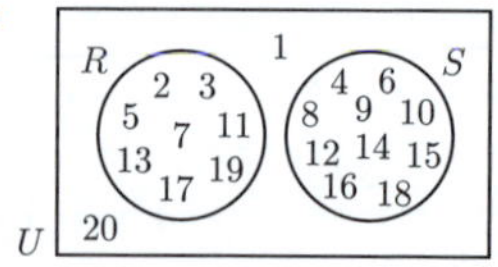

6 **a** M = {Shaun, Ken, Ling, Don, Vivian},
H = {Paul, Ken, Pam, Ling, Stefan, Reg, Don}
b
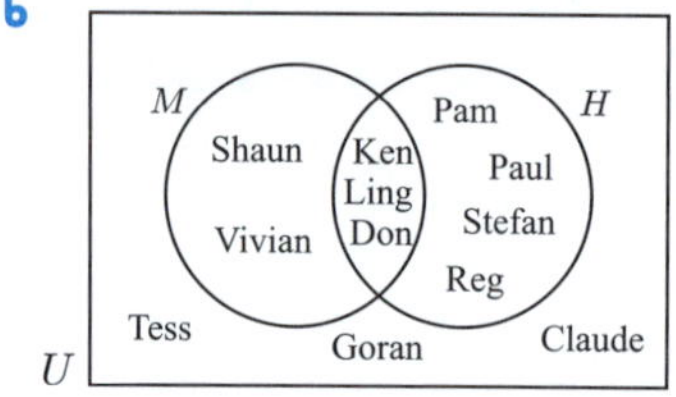

c {Paul, Shaun, Ken, Pam, Ling, Stefan, Reg, Don, Vivian}
All students who scored 15 or more for at least one of the tests.
d {Shaun, Vivian}
All students who scored 15 or more for the Mathematics test, but less than 15 for the History test.

7 **a**
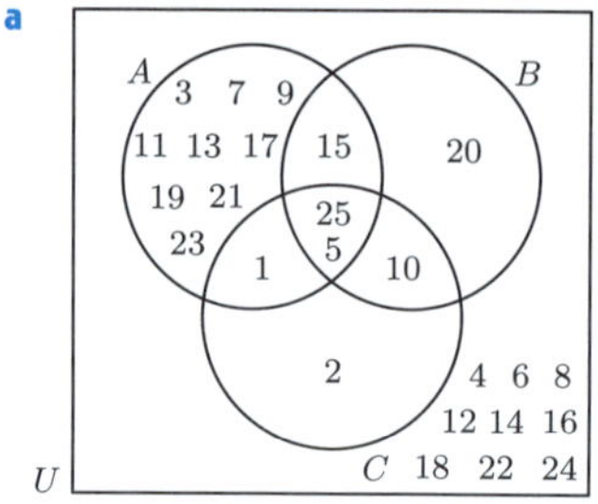

b **i** $\{5, 15, 25\}$ **ii** $\{3, 7, 9, 11, 13, 15, 17, 19, 21, 23\}$
iii $\{10\}$

EXERCISE 3E.2

1 **a**
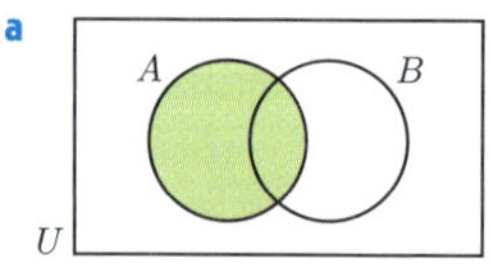

b
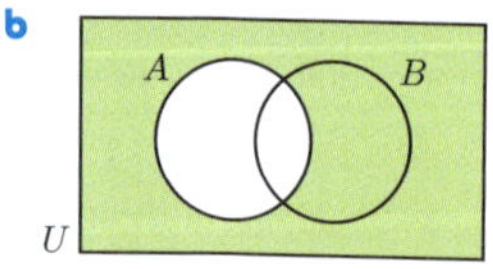

c
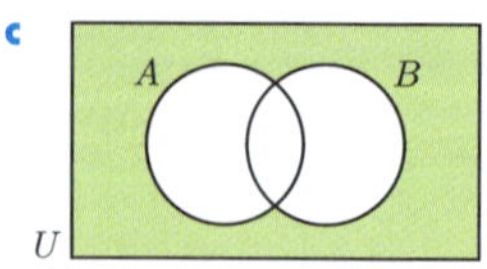

d
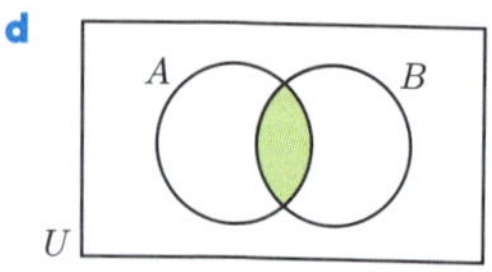

e
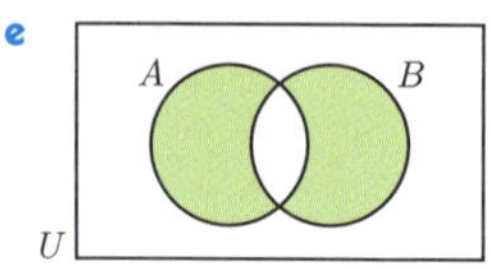

2 **a**
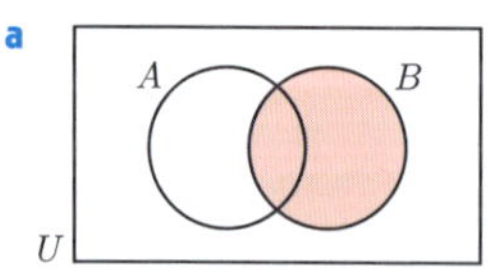

b
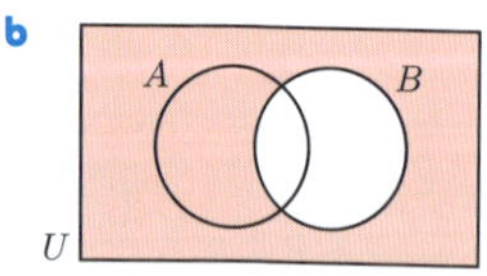

c 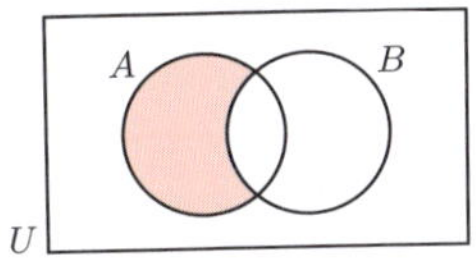

d 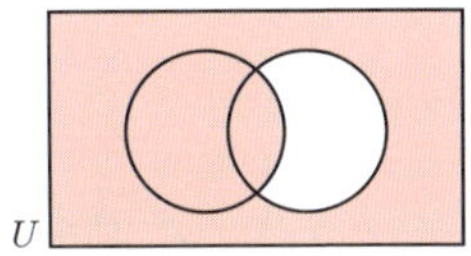

e 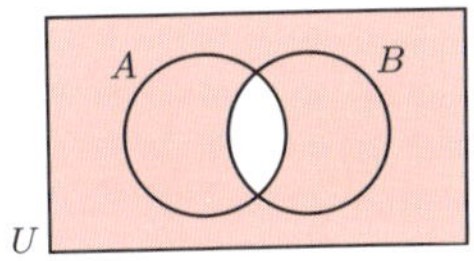

f 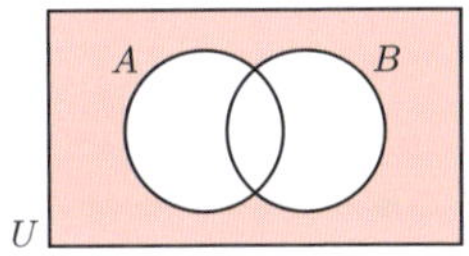

3 a 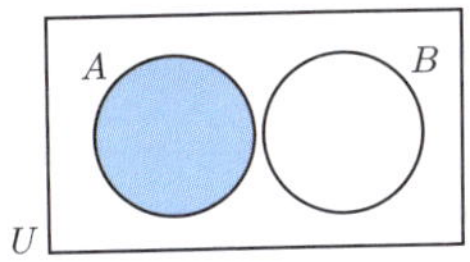

b 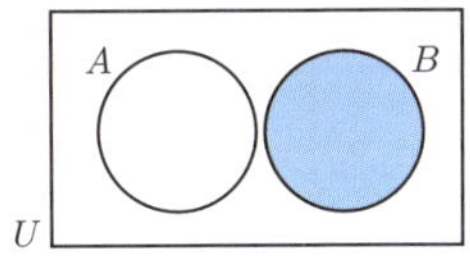

c 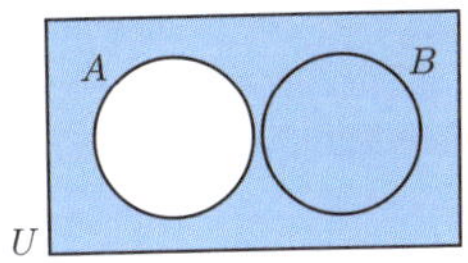

d 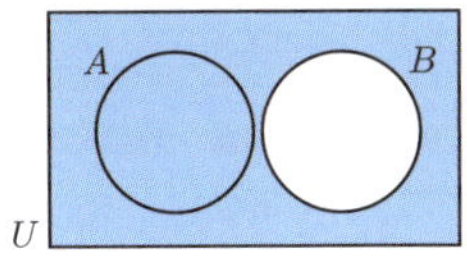

e 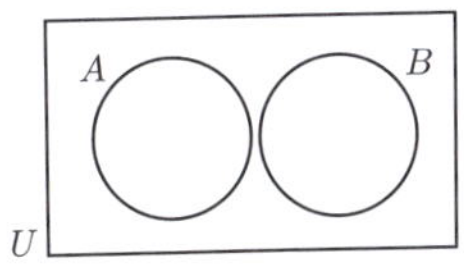

f 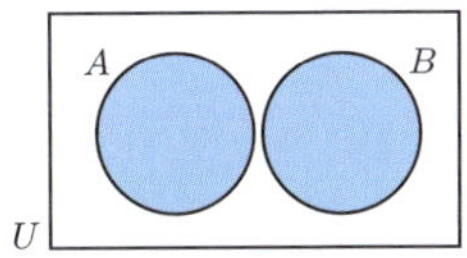

g 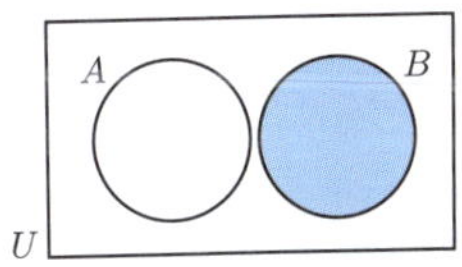

h 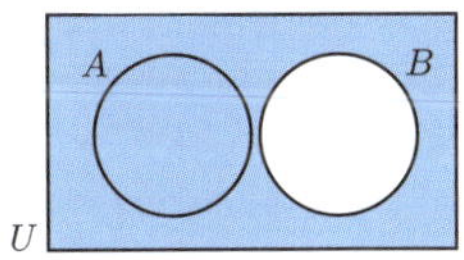

i 

4 a 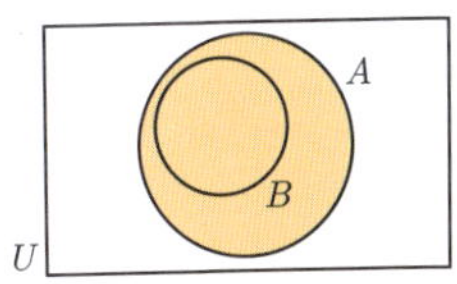

b 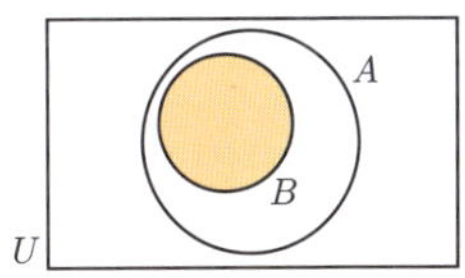

c 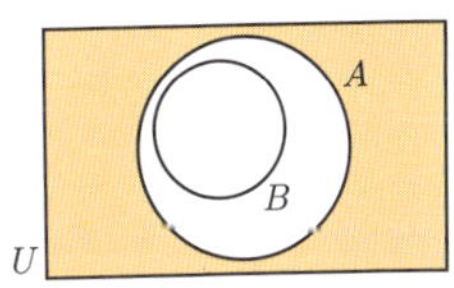

d 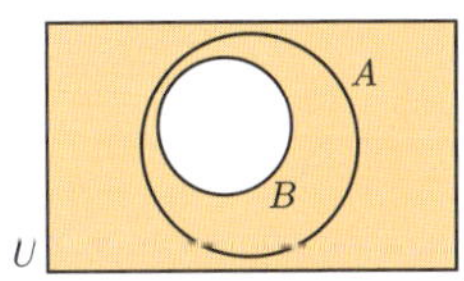

e 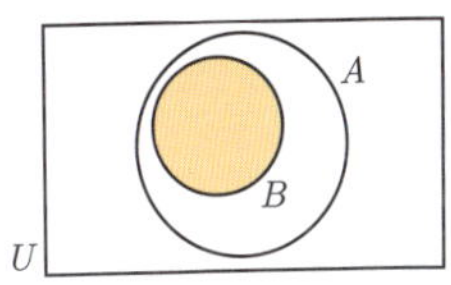

f 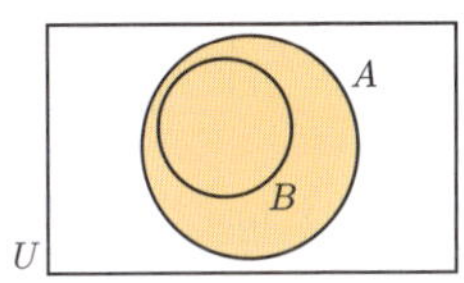

g 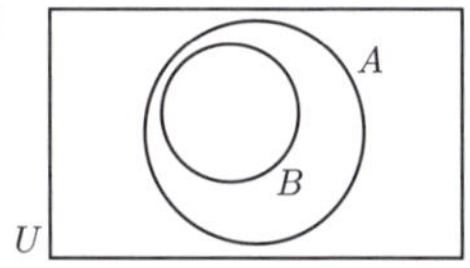

h 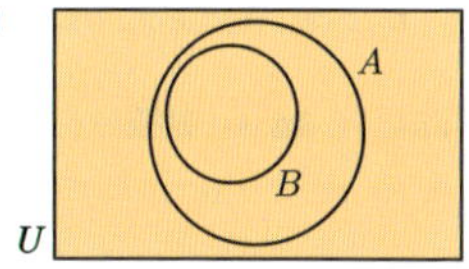

i 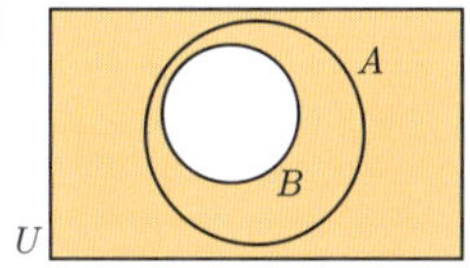

5 a 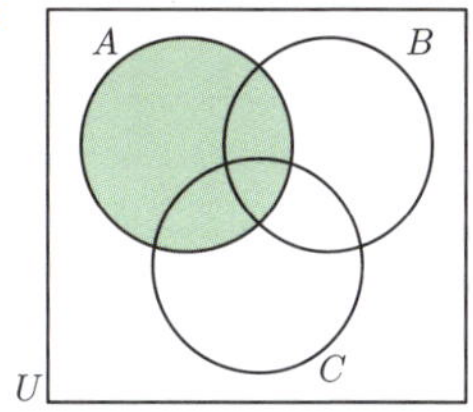

b 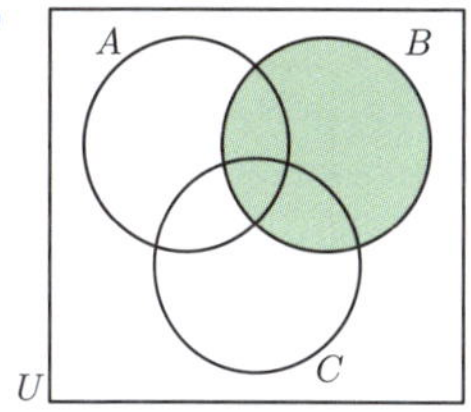

c 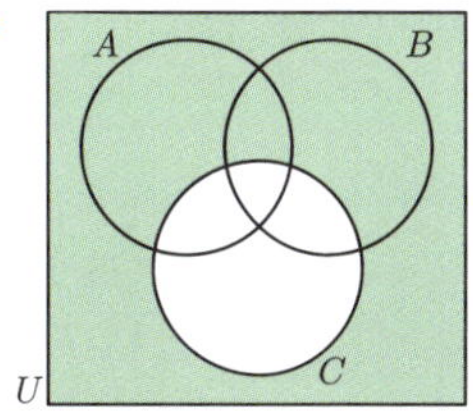

d 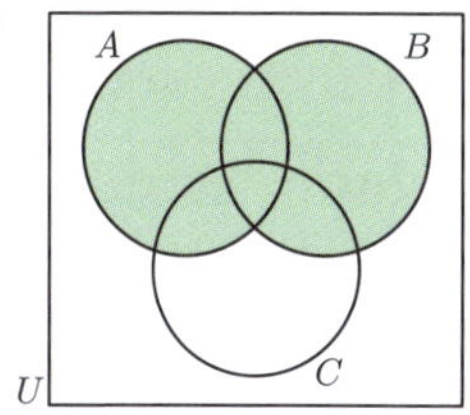

e 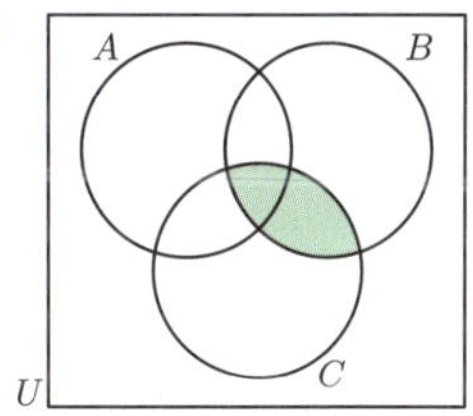

f 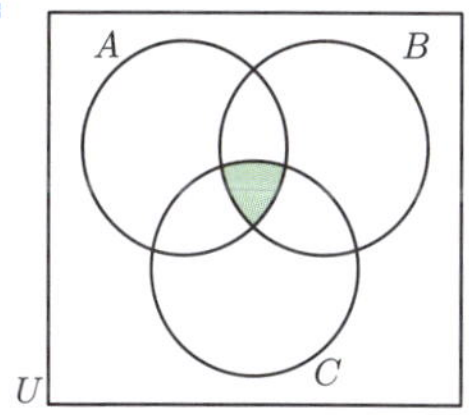

g 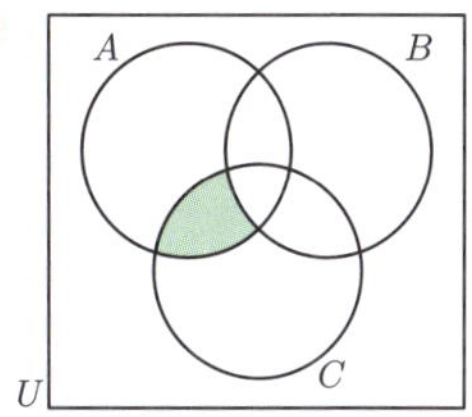

h 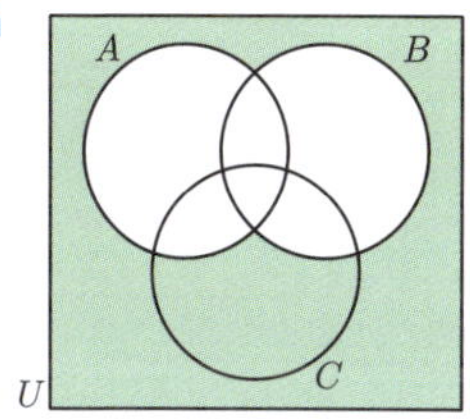

i 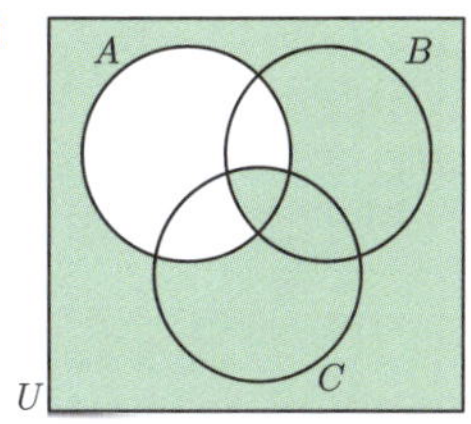

6 a i 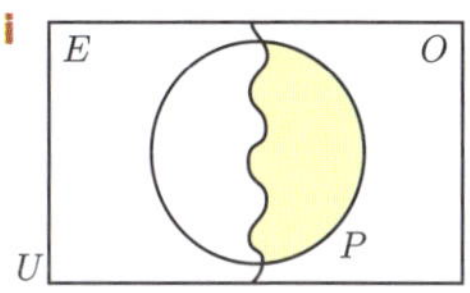

ii 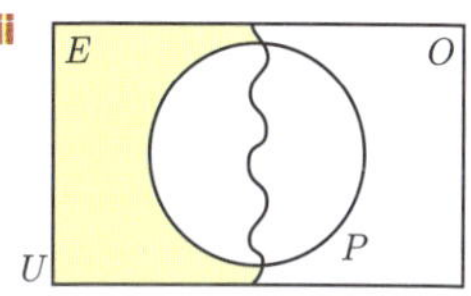

b even prime numbers, $n(E \cap P) = 1$

EXERCISE 3E.3

1 **a** 7 **b** 9 **c** 15 **d** 8 **e** 5 **f** 19

2 **a** 12 **b** 16 **c** 42 **d** 18 **e** 8 **f** 4

3 **a** $b+c$ **b** $c+d$ **c** $a+c$ **d** b
e $a+c+d$ **f** d

4 **a** **i** $2a-4$ **ii** $5a-4$ **iii** $4a+5$ **iv** $6a+1$
b $a=7$

5 **a** **i** $a+b+c$ **ii** $a+b+c$
b $n(A \cup B) = n(A) + n(B) - n(A \cap B)$

6 **a**

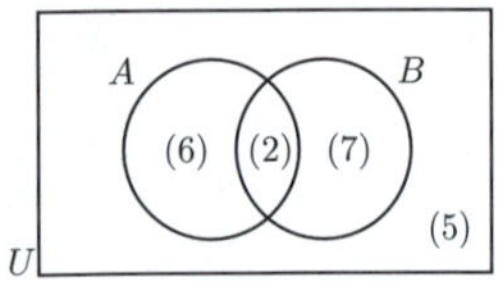

b **i** 15 **ii** 7

7 **a** 19 **b** 4 **c** 15

8 **a** 15 **b** 23 **c** 21 **d** 3 **e** 33
f 2 **g** 3 **h** 6 **i** 43 **j** 7

EXERCISE 3F

1 **a**

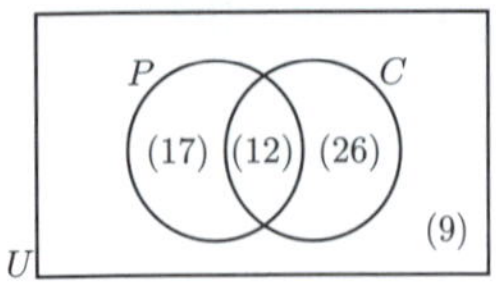

b 64 people
c **i** 17 entrants **ii** 26 entrants **iii** 26 entrants **iv** 35 entrants

2 **a**

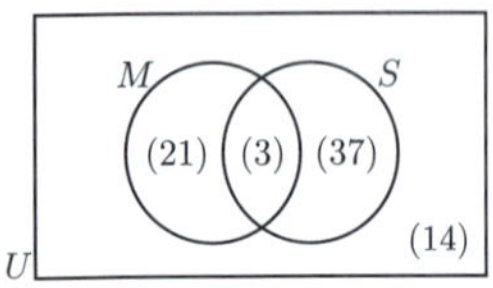

b **i** 21 students **ii** 37 students **iii** 3 students

3 **a**

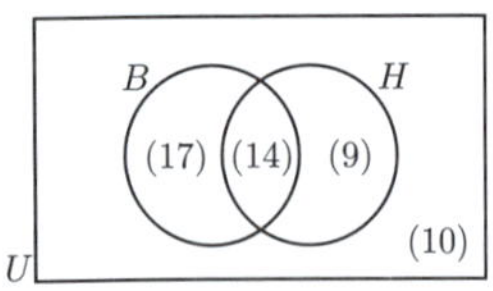

b **i** 17 tourists **ii** 14 tourists

4 **a**

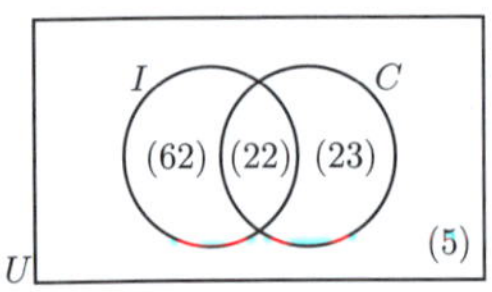

b 62 families
c $\frac{22}{112} = \frac{11}{56}$

5 **a**

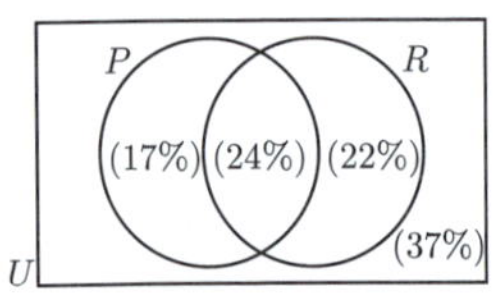

b **i** 24% **ii** 63% **iii** 17% **iv** 39%

6 **a** Let G represent the countries that won gold medals, S represent the countries that won silver medals, and B represent the countries that won bronze medals.

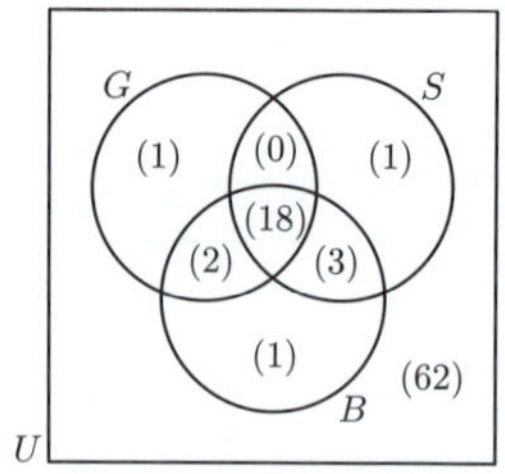

b **i** 62 countries **ii** 3 countries **iii** 3 countries

7 **a**

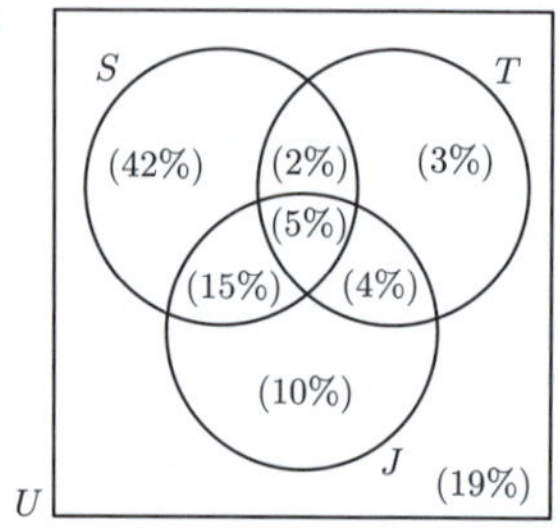

b **i** 42% **ii** 71% **iii** 14%

REVIEW SET 3A

1 **a** **i** true **ii** true **b** 8

2 **a** $\{x \mid 4 \leqslant x \leqslant 8,\ x \in \mathbb{R}\}$ **b** $\{x \mid x < 30,\ x \in \mathbb{Q}\}$

3 **a** $F = \{1, 2, 4, 8, 16, 32\}$, $M = \{4, 8, 12, 16, 20, 24, 28\}$
b no
c **i** $\{4, 8, 16\}$ **ii** $\{1, 2, 4, 8, 12, 16, 20, 24, 28, 32\}$
d ∞

4 **a** **i** $\{c, f, g, h\}$ **ii** $\{a, b\}$ **iii** $\{a, b, d, e, g\}$
iv $\{g\}$ **v** $\{a, b, c, d, e, f, h\}$
b

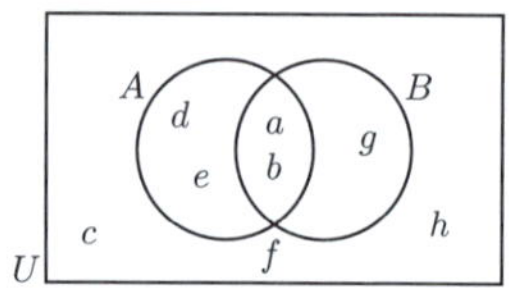

5 **a** **i**

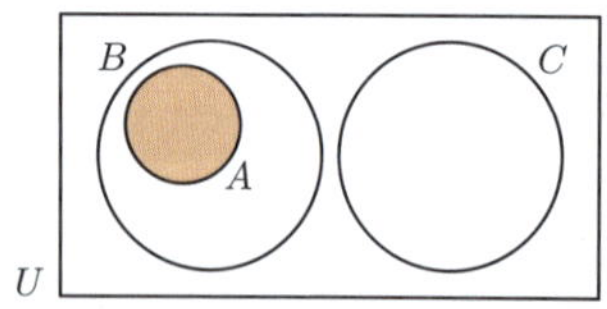

ii

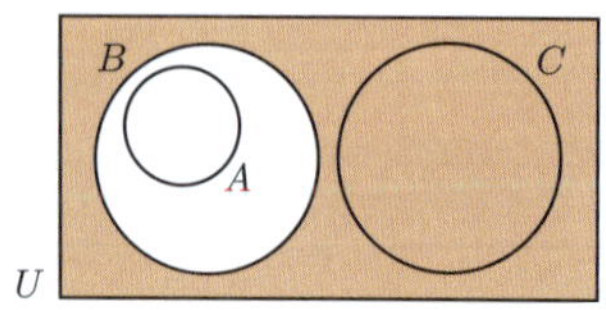

iii

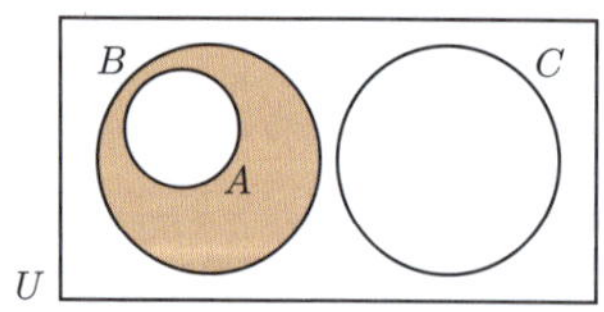

b **i** true **ii** true **iii** false

6 **a** 10 **b** 7 **c** 11

7 **a** 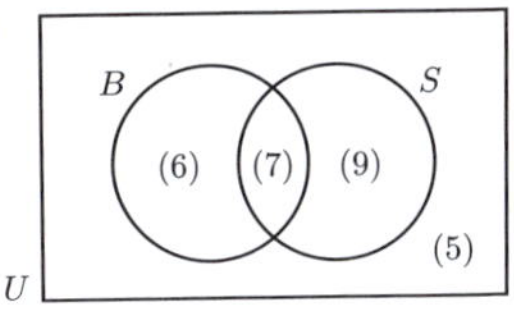

b **i** 7 students **ii** 9 students

8 **a** **i** $2x-2$ **ii** $3x+1$ **b** $x=6$

9 **a** 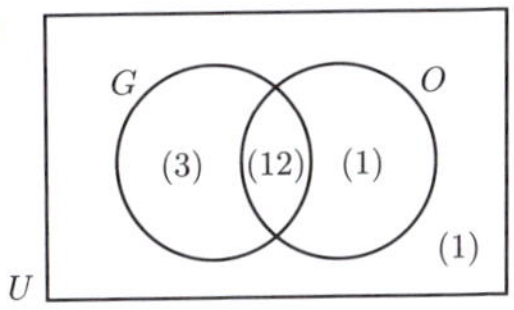

b **i** 3 dishes **ii** 1 dish **iii** 1 dish

10 **a** 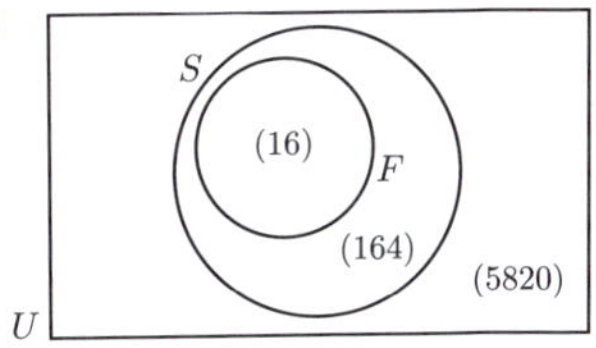

b **i** 164 people **ii** 5820 people **c** 0.267%

11 **a** 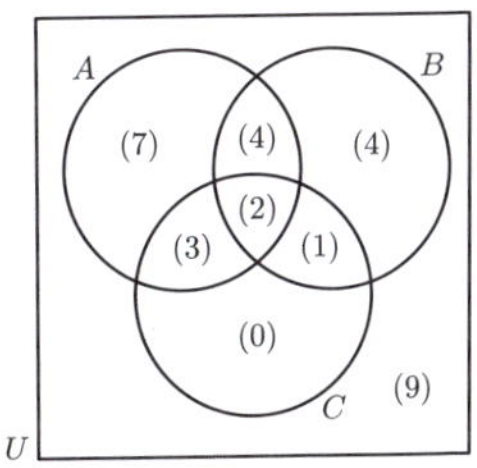

b **i** 17 numbers **ii** 8 numbers

c $35=5\times 7$, but no numbers that are a multiple of 5 but not a multiple of 2 and 3 have been selected, so 35 has not been selected.
$60=2\times 2\times 3\times 5$ and $30=2\times 3\times 5$. These are the only numbers from 1 to 75 that are multiples of 2, 3, and 5. From the information two such numbers have been selected, so 30 and 60 have been selected.

REVIEW SET 3B

1 $C=\{$red, white, green, yellow, black, blue$\}$

2 **a** $\{a, b, c, d, e, i, o, u\}$ **b** $\{a, e\}$

3 **a** $\{4, 16, 36\}$ **b** 3

4 All natural (counting) numbers can be expressed as fractions of integers, so $\mathbb{N}\subseteq\mathbb{Q}$. For example, $7=\frac{7}{1}$.

5 **a** The set of all negative integers greater than -8 inclusive.
b $\{-8, -7, -6, -5, -4, -3, -2, -1\}$
c 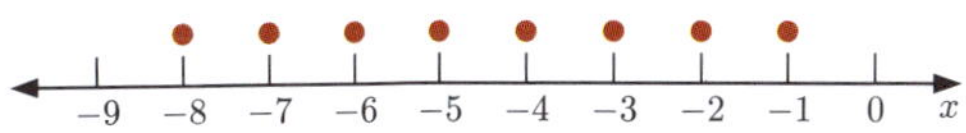

6 **a** $\{5, 10\}$ **b** $\{1, 2, 3, 4, 6, 7, 8, 9, 11, 12, 13, 14, 15\}$
c $\{15\}$

7 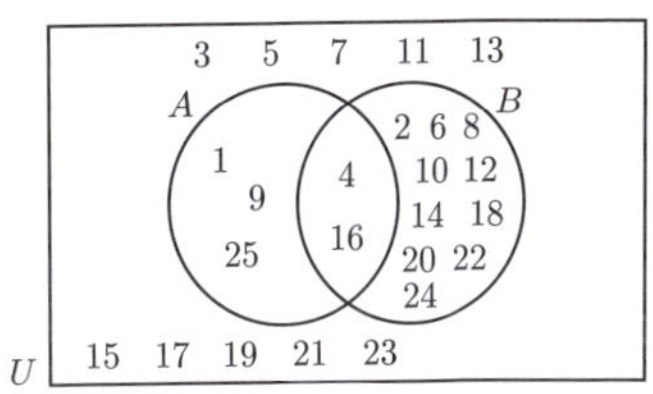

8 **a** 26 **b** 28 **c** 36

9 **a** 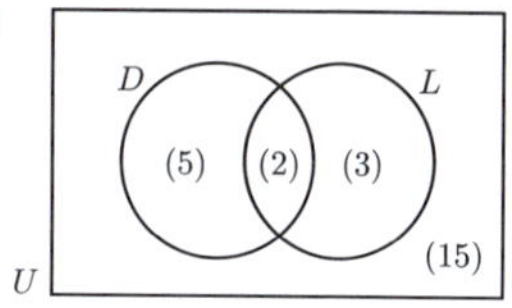

b 15 students

10 **a** 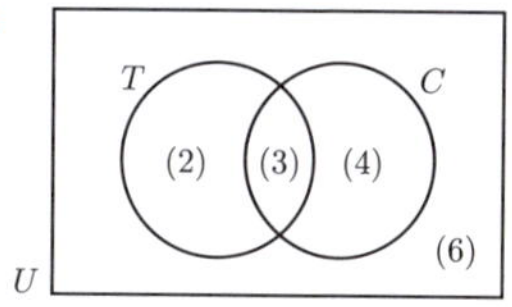

b 3 workers

11 **a** $\{x \mid x\leqslant 30,\ x\in\mathbb{N}\}$
b **i** $\{5, 10, 15, 20, 25, 30\}$ **ii** $\{10, 20, 30\}$
c 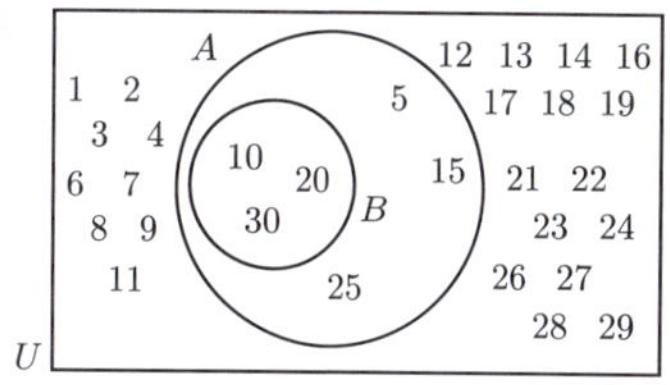

d $\{5, 15, 25\}$
e $3=n(A\cap B')$

12 **a** 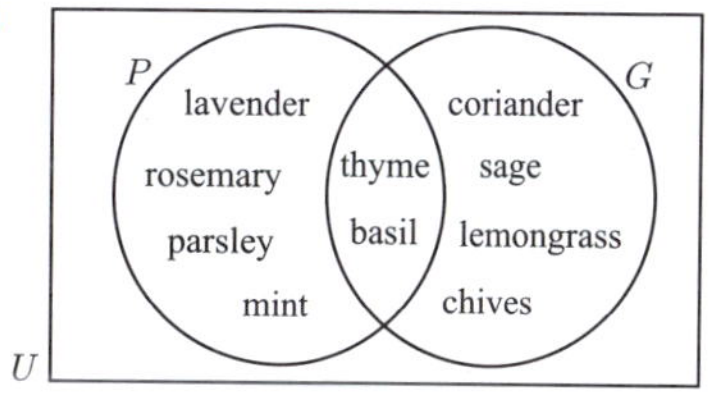

b **i** thyme, basil **ii** lavender, rosemary, parsley, mint
iii lavender, rosemary, parsley, mint, coriander, sage, lemongrass, chives

c 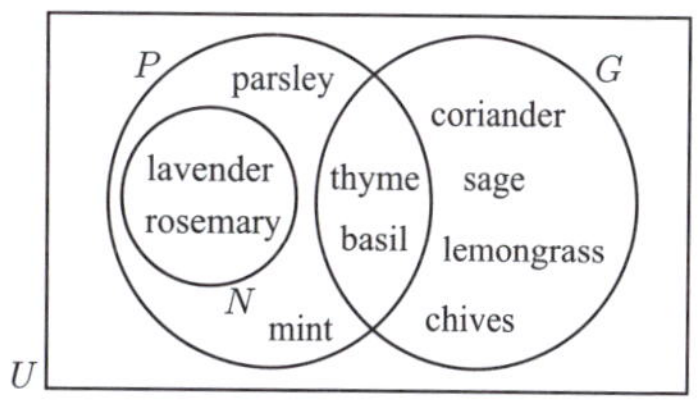

EXERCISE 4A

1 **a** $2x+2$ **b** $5x+10$ **c** $4x-24$ **d** $7x-7$
e $6a+3b$ **f** $6x+18y$ **g** $8a-8b$ **h** $4m^2-4n^2$
i $-3x-6$ **j** $-8a+4$ **k** $-a^2+5a$ **l** $3b^2-12b$
m $-y-2$ **n** $-5x+15$ **o** $-2+b$ **p** $-4a+4b$
q x^2+xy **r** x^2+xy **s** $2a^2-a$ **t** $-21x+12x^2$

2 **a** $4x+4y+20$ **b** $6a-10b+2$
c $3x^2-6xy+9x$ **d** $-2a+6b-10c$
e $6x^3+30x^2-60x$ **f** $4n^3-8n^2-32n$

3 **a** $3x+5$ **b** $-2x+5$ **c** $4x+3$
d $-2x-12$ **e** a^2 **f** $12m-2m^2$
g $3ab+a^2$ **h** $-2b^2+9b$ **i** $3a^2-12a$

4 **a** $5x+6$ **b** $a+3b$ **c** $-a+5b$
d $-2x+26$ **e** -15 **f** $17n-15$

g $x-y$ **h** $a^2+7a-15$ **i** $x^2+2x+12$
j $2a^2+3a$ **k** $-2a^2+a$ **l** x^2-y^2
m $-2x+16$ **n** $9x-9$ **o** $5x^2-16x$

EXERCISE 4B

1 **a** ac **b** ad **c** bc **d** bd
e $ac+ad+bc+bd$ or $(a+b)(c+d)$
$\therefore\ (a+b)(c+d)=ac+ad+bc+bd$

2 **a** $x^2+7x+10$ **b** $x^2-2x-24$ **c** $x^2+4x-21$
d x^2-25 **e** $x^2-6x-16$ **f** $12x^2+11x+2$
g $-6x^2-x+1$ **h** $-2x^2+7x+30$
i $12x^2-5x-3$ **j** $-5x^2+16x+16$
k $-2x^2+11x+40$ **l** $9x^2+12x+4$

3 **a** x^2-4 **b** a^2-16 **c** $36-x^2$
d $9x^2-1$ **e** $25a^2-4$ **f** $49a^2-9$

4 **a** x^2+6x+9 **b** $x^2-8x+16$ **c** $1-2x+x^2$
d $4+4x+x^2$ **e** $4x^2+4x+1$ **f** $4-12x+9x^2$
g $9+30x+25x^2$ **h** $4x^2-4xy+y^2$

5 **a** x^2+8x+1 **b** $x^2+7x-10$ **c** $3a^2-30$
d $5x^2-2x-15$ **e** $3x^2+8x-24$ **f** $-y-31$
g $-k^2+2k-7$ **h** $2x^2-40$

EXERCISE 4C

1 **a** x^2-9 **b** x^2-9 **c** $9-x^2$ **d** $9-x^2$
e x^2-4 **f** $4-x^2$ **g** x^2-36 **h** a^2-16
i b^2-1 **j** p^2-q^2 **k** $25-n^2$ **l** $49-y^2$

2 **a** $4x^2-1$ **b** $25x^2-4$ **c** $16a^2-9$
d $9b^2-25$ **e** $16x^2-1$ **f** $1-16x^2$
g $49-4y^2$ **h** $9-4x^2$ **i** $9x^2-4$

3 **a** $9a^2-b^2$ **b** a^2-9b^2 **c** $36x^2-y^2$
d $25x^2-4y^2$ **e** $16x^2-25y^2$ **f** $4x^2-49y^2$

EXERCISE 4D

1 **a** a^2 **b** ab **c** ab **d** b^2
e $a^2+2ab+b^2$
Conclusion: $(a+b)^2=a^2+2ab+b^2$

2 **a** $(5+3)^2=8^2=64,\quad 5^2+3^2=25+9=34$
b $5^2+2\times5\times3+3^2=25+30+9=64$

3 **a** x^2+4x+4 **b** $x^2+10x+25$ **c** $a^2+8a+16$
d $y^2+12y+36$ **e** $1+2b+b^2$ **f** $4+4x+x^2$

4 **a** $x^2-8x+16$ **b** x^2-4x+4 **c** $a^2-14a+49$
d $b^2-16b+64$ **e** $9-6x+x^2$ **f** $25-10y+y^2$

5 **a** $4x^2+12x+9$ **b** $16a^2-8a+1$ **c** $9y^2+30y+25$
d $9a^2-24a+16$ **e** $4x^2-28x+49$
f $64+48a+9a^2$ **g** $4+20b+25b^2$
h $36-60x+25x^2$ **i** $16-40y+25y^2$

6 **a** x^4+6x^2+9 **b** y^4-4y^2+4
c $4a^4+12a^2+9$ **d** $1-8x^2+16x^4$
e $a^4+2a^2b^2+b^4$ **f** $a^4-2a^2b^2+b^4$

7 **a** $-x^2-2x-3$ **b** $x^2+13x+23$ **c** $2x^2+2x-15$
d $-2x-17$ **e** $10x^2-5x-5$ **f** $8x^2+16x+9$
g $24x^2-4x-8$ **h** $4x^2-9x-13$ **i** $2x^2-2x+13$
j $-2x+5$

EXERCISE 4E

1 **a** x^3+3x^2+3x+2 **b** x^3+4x^2+x-6
c x^3+3x^2-2x+8 **d** x^3+5x^2-7x-6
e $3x^3+4x^2+16x+5$ **f** $5x^3-7x^2+22x-8$
g $2x^3+7x^2+x-6$ **h** $6x^3-5x^2+9x-4$

2 **a** x^3+3x^2+3x+1 **b** $x^3+9x^2+27x+27$
c x^3-3x^2+3x-1 **d** $x^3-6x^2+12x-8$
e $8x^3+12x^2+6x+1$ **f** $8x^3-36x^2+54x-27$
g $x^4+4x^3+6x^2+4x+1$ **h** $x^4-4x^3+6x^2-4x+1$

3 **a** x^3+6x^2+8x **b** x^3-2x^2-15x
c x^3-5x^2+4x **d** $2x^3+2x^2-12x$
e $-2x^3+12x^2+14x$ **f** $-x^3+3x^2+18x$
g $-10x^3-25x^2+15x$ **h** $4x^3-13x^2+3x$
i $3x^4+12x^3+12x^2$

4 **a** $x^3+7x^2+14x+8$ **b** x^3-4x^2+x+6
c $x^3-4x^2-7x+10$ **d** $2x^3-11x^2+13x-4$
e $3x^3+16x^2+23x+6$ **f** $16x^3+64x^2-x-4$
g $3x^3+x^2-8x+4$ **h** $-5x^3+28x^2-13x-10$

5 **a** **i** x^3-1 **ii** x^4-1 **iii** x^7-1 **b** $x^{n+1}-1$

6 6 and 7, *or* -6 and -7

REVIEW SET 4A

1 **a** $3x-12$ **b** $-10x+15$ **c** $x-x^2$
2 **a** $4x-1$ **b** $-2x-5$ **c** $-9x+11$
3 **a** $x^2+11x+28$ **b** $x^2+6x-16$ **c** $-5x^2-x+6$
d $2a^2-21a+27$
4 **a** x^2+8x-6 **b** $n-9$
5 **a** $x^2+8x+16$ **b** $x^2-20x+100$
c $4x^2+20x+25$ **d** $9-24x+16x^2$
6 **a** x^2-81 **b** $9x^2-4$ **c** $16x-x^3$
7 **a** $x^3+5x^2-3x-28$ **b** $4x^3-7x^2-17x+15$
8 **a** **i** $(x+10)$ cm **ii** $(x+6)$ cm
b $(x+10)x=(x^2+10x)\text{ cm}^2$
c $(x+10)x(x+6)=(x^3+16x^2+60x)\text{ cm}^3$

REVIEW SET 4B

1 **a** $-3x-12$ **b** $xy+3x$ **c** $2a^2-10a$
2 **a** $x^2+5x-36$ **b** $2x^2+5x-42$ **c** $2a^2-11a+12$
3 **a** y^2+6y+9 **b** $9x^2+12x+4$ **c** $16a^2-8ab+b^2$
4 **a** $25-x^2$ **b** $9y^2-16$ **c** $36a^2-25b^2$
5 **a** $-x^2-6x-9$ **b** $y^2+5y-40$
6 **a** x^3+6x^2-32 **b** $x^3-4x^2-3x+18$
7 **a** $-x^3+5x^2-10x+8$ **b** $x^3-21x+20$
8 **a** $x^3-9x^2+27x-27$
b $x^4-12x^3+54x^2-108x+81$

EXERCISE 5A

1 **a** $\sqrt{5}$ is a surd, $\sqrt{5}\approx 2.2361$
b $\sqrt{36}=6$, $\sqrt{36}$ is a rational radical.
c $\sqrt{72}$ is a surd, $\sqrt{72}\approx 8.4853$
d $\sqrt{81}=9$, $\sqrt{81}$ is a rational radical.
e $\sqrt{\frac{25}{49}}=\frac{5}{7}$, $\sqrt{\frac{25}{49}}$ is a rational radical.
f $\sqrt{\frac{1}{10}}$ is a surd, $\sqrt{\frac{1}{10}}\approx 0.3162$
g $\sqrt{\frac{1}{64}}=\frac{1}{8}$, $\sqrt{\frac{1}{64}}$ is a rational radical.
h $\sqrt{\frac{3}{14}}$ is a surd, $\sqrt{\frac{3}{14}}\approx 0.4629$

2 $\sqrt{1}, \sqrt{4}, \sqrt{9}, \sqrt{16}, \ldots, \sqrt{961}=31$ are all rational radicals from the given set.
$\therefore\ 1000-31=969$ are surds.

EXERCISE 5B

1 **a** 3 **b** $3\sqrt{3}$ **c** $9\sqrt{3}$ **d** $\frac{1}{3}$ **e** 7 **f** $7\sqrt{7}$ **g** $\frac{1}{7}$ **h** $\frac{9}{7}$ **i** 5 **j** 25 **k** 5 **l** 20

2 **a** 8 **b** 32 **c** 12 **d** 27 **e** 20 **f** 45 **g** 28 **h** 40 **i** 490

3 **a** 24 **b** 30 **c** 70 **d** 32 **e** 147 **f** -6 **g** 30 **h** -42 **i** -22

4 **a** $\sqrt{6}$ **b** $\sqrt{14}$ **c** $\sqrt{34}$ **d** $\sqrt{21}$ **e** $10\sqrt{6}$ **f** 48 **g** $5\sqrt{14}$ **h** $6\sqrt{30}$ **i** $-10\sqrt{14}$ **j** $2\sqrt{21}$ **k** $24\sqrt{5}$ **l** $80\sqrt{6}$

5 **a** 2 **b** $\frac{1}{3}$ **c** $\sqrt{6}$ **d** $\frac{1}{5}$ **e** $\sqrt{15}$ **f** $\sqrt{\frac{1}{15}}$ **g** 3 **h** $\frac{1}{4}$

6 **a** $\frac{1}{4}$ **b** $\frac{3}{4}$ **c** $\frac{7}{8}$ **d** $\frac{11}{3}$ **e** $\frac{5}{2}$ **f** $\frac{4}{3}$ **g** $\frac{8}{5}$ **h** $\frac{8}{3}$

7 **a** $2\sqrt{2}$ **b** $3\sqrt{2}$ **c** $5\sqrt{2}$ **d** $7\sqrt{2}$ **e** $10\sqrt{2}$ **f** $12\sqrt{2}$ **g** $100\sqrt{2}$ **h** $\frac{1}{2}\sqrt{2}$

8 **a** $2\sqrt{3}$ **b** $3\sqrt{3}$ **c** $5\sqrt{3}$ **d** $\frac{1}{3}\sqrt{3}$

9 **a** $2\sqrt{5}$ **b** $3\sqrt{5}$ **c** $5\sqrt{5}$ **d** $\frac{1}{5}\sqrt{5}$

EXERCISE 5C

1 **a** $2\sqrt{3}$ **b** $2\sqrt{7}$ **c** $3\sqrt{6}$ **d** $2\sqrt{15}$ **e** $3\sqrt{11}$ **f** $2\sqrt{13}$ **g** $2\sqrt{10}$ **h** $3\sqrt{7}$ **i** $4\sqrt{3}$ **j** $5\sqrt{5}$ **k** $7\sqrt{3}$ **l** $5\sqrt{7}$ **m** $4\sqrt{11}$ **n** $5\sqrt{6}$ **o** $5\sqrt{11}$ **p** $20\sqrt{5}$

2 **a** $\frac{\sqrt{96}}{\sqrt{6}} = \sqrt{\frac{96}{6}} = \sqrt{16} = 4$

b $\sqrt{96} = \sqrt{16 \times 6} = 4\sqrt{6}$ $\therefore$ $\frac{\sqrt{96}}{\sqrt{6}} = \frac{4\sqrt{6}}{\sqrt{6}} = 4$

EXERCISE 5D

1 **a** $2\sqrt{2}$ **b** 0 **c** $\sqrt{2}$ **d** $\sqrt{3}$ **e** $7\sqrt{7}$ **f** $-3\sqrt{5}$ **g** $-3\sqrt{2}$ **h** $8\sqrt{5}$ **i** $6\sqrt{2}$ **j** $-3\sqrt{2}$ **k** $4\sqrt{3}$ **l** $10\sqrt{5}-10$

2 **a** $2\sqrt{2}+7\sqrt{3}$ **b** $5\sqrt{2}-\sqrt{3}$ **c** $-7\sqrt{2}+4\sqrt{3}$ **d** $11\sqrt{5}-5\sqrt{2}$ **e** $2\sqrt{2}-10\sqrt{7}$ **f** $2\sqrt{2}+3\sqrt{11}$ **g** $1-3\sqrt{2}$ **h** $9\sqrt{3}-13$

3 **a** $\sqrt{12}+\sqrt{27} = 2\sqrt{3}+3\sqrt{3} = 5\sqrt{3} = \sqrt{75}$
b $\sqrt{80}-\sqrt{5} = 4\sqrt{5}-\sqrt{5} = 3\sqrt{5} = \sqrt{45}$

EXERCISE 5E

1 **a** $12+4\sqrt{2}$ **b** $3\sqrt{2}+3\sqrt{3}$ **c** $20-5\sqrt{7}$ **d** $6\sqrt{11}-24$ **e** $\sqrt{2}+2$ **f** $2-5\sqrt{2}$ **g** $2\sqrt{3}+6$ **h** $3-\sqrt{6}$ **i** $6\sqrt{5}-5$ **j** $10-\sqrt{5}$ **k** $10+\sqrt{15}$ **l** $2\sqrt{7}+7+\sqrt{14}$

2 **a** $-4\sqrt{2}-2$ **b** $-3\sqrt{2}+2$ **c** $-2+\sqrt{14}$ **d** $-3\sqrt{3}-3$ **e** $-5\sqrt{3}+3$ **f** $-6-\sqrt{15}$ **g** $2\sqrt{10}+\sqrt{15}$ **h** $-4-2\sqrt{6}$ **i** $-20\sqrt{2}+20$ **j** $-14-4\sqrt{7}$ **k** $-2\sqrt{11}+11$ **l** $-2\sqrt{3}+4\sqrt{6}$

3 **a** $7+6\sqrt{2}$ **b** $11+6\sqrt{2}$ **c** $\sqrt{2}$ **d** $9+\sqrt{3}$ **e** 1 **f** $1-\sqrt{7}$ **g** 9 **h** $2\sqrt{2}+1$ **i** $8+2\sqrt{2}$

4 **a** $4+2\sqrt{3}$ **b** $27+10\sqrt{2}$ **c** $11-6\sqrt{2}$ **d** $8+2\sqrt{7}$ **e** $5-2\sqrt{6}$ **f** $21-8\sqrt{5}$ **g** $8+2\sqrt{15}$ **h** $15-6\sqrt{6}$ **i** $9-6\sqrt{2}$ **j** $17+12\sqrt{2}$ **k** $17-12\sqrt{2}$ **l** $59-30\sqrt{2}$

5 **a** 7 **b** 2 **c** 22 **d** -13 **e** -2 **f** 2 **g** -18 **h** -16 **i** 14 **j** 3 **k** -4 **l** 7

EXERCISE 5F.1

1 **a** $\frac{\sqrt{3}}{3}$ **b** $\sqrt{3}$ **c** $3\sqrt{3}$ **d** $\frac{11\sqrt{3}}{3}$ **e** $\frac{\sqrt{6}}{9}$ **f** $\sqrt{2}$ **g** $3\sqrt{2}$ **h** $6\sqrt{2}$ **i** $\frac{\sqrt{6}}{2}$ **j** $\frac{\sqrt{2}}{8}$ **k** $\sqrt{5}$ **l** $3\sqrt{5}$ **m** $-\frac{3\sqrt{5}}{5}$ **n** $40\sqrt{5}$ **o** $\frac{\sqrt{5}}{15}$ **p** $\sqrt{7}$ **q** $3\sqrt{7}$ **r** $\frac{2\sqrt{11}}{11}$ **s** $2\sqrt{13}$ **t** $\frac{\sqrt{3}}{9}$

2 **a** $\frac{\sqrt{5}}{5}$ **b** $\frac{\sqrt{15}}{3}$ **c** $\frac{\sqrt{14}}{2}$ **d** $\frac{\sqrt{30}}{10}$ **e** $\frac{\sqrt{14}}{4}$

EXERCISE 5F.2

1 **a** $\dfrac{3-\sqrt{2}}{7}$ **b** $\dfrac{6+2\sqrt{2}}{7}$ **c** $-2+\sqrt{5}$ **d** $\sqrt{2}+1$ **e** $2\sqrt{6}+2$ **f** $\dfrac{\sqrt{21}-2\sqrt{3}}{3}$ **g** $-3-2\sqrt{2}$ **h** $\dfrac{4\sqrt{3}+3}{13}$ **i** $2\sqrt{2}+4$ **j** $-7-3\sqrt{5}$ **k** $\dfrac{5+3\sqrt{3}}{2}$ **l** $\dfrac{-38+11\sqrt{10}}{6}$

2 **a** $-\frac{9}{7}-\frac{3}{7}\sqrt{2}$ **b** $4-2\sqrt{2}$ **c** $-\frac{2}{23}-\frac{5}{23}\sqrt{2}$ **d** $-4+2\sqrt{2}$

3 **a** $-2-2\sqrt{3}$ **b** $12-6\sqrt{3}$ **c** $3+2\sqrt{3}$ **d** $-\frac{1}{2}+\frac{5}{6}\sqrt{3}$

4 **a** -4 **b** $-\frac{9}{7}+\frac{9}{7}\sqrt{2}$ **c** $\frac{11}{4}+\frac{7}{4}\sqrt{5}$

5 $2+\sqrt{3}$ **6** **a** 1 **b** 4 **c** 2

REVIEW SET 5A

1 **a** $\sqrt{7} \approx 2.646$ **b** $\sqrt{22} \approx 4.690$ **c** $\sqrt{187} \approx 13.675$

2 **a** 12 **b** 32 **c** $6\sqrt{10}$ **d** $\frac{7}{2}$

3 **a** $\sqrt{5}$ **b** $\sqrt{5}$ **c** $\sqrt{7}$ **d** $\frac{1}{\sqrt{10}} = \frac{\sqrt{10}}{10}$

4 **a** $3\sqrt{11}$ **b** $6\sqrt{7}-3\sqrt{3}$

5 **a** $\sqrt{15} \times \sqrt{15} = 15$ $\therefore$ each side is $\sqrt{15}$ cm long.
b **i** $(15-3\sqrt{15})$ cm^2 **ii** $(30-3\sqrt{15})$ cm^2
c $(10+2\sqrt{15})$ cm

6 $4\sqrt{5}$

7 **a** $2\sqrt{3}+2$ **b** $3\sqrt{2}-2$ **c** $8+2\sqrt{7}$ **d** $9-4\sqrt{5}$ **e** 19 **f** $1-2\sqrt{2}$

8 **a** $2\sqrt{5}$ **b** $\dfrac{1+\sqrt{3}}{2}$ **c** $\dfrac{-4-\sqrt{7}}{3}$

9 **a** $-6-3\sqrt{5}$ **b** $\frac{5}{2}-\frac{1}{2}\sqrt{5}$ **c** $\frac{9}{2}-\frac{3}{2}\sqrt{5}$

10 **a** $9-4\sqrt{2}$
b As $\left(3+\sqrt{2}\right)\left(5-3\sqrt{2}\right) = 9-4\sqrt{2}$

then $\dfrac{9-4\sqrt{2}}{3+\sqrt{2}} = 5-3\sqrt{2}$

c **Hint:** $\dfrac{9-4\sqrt{2}}{3+\sqrt{2}} = \left(\dfrac{9-4\sqrt{2}}{3+\sqrt{2}}\right)\left(\dfrac{3-\sqrt{2}}{3-\sqrt{2}}\right)$

REVIEW SET 5B

1 **a** $\sqrt{10}$ **b** 2 **c** 54 **2** **a** $\sqrt{19}$ **b** ≈ 4.3589

3 **a** $\frac{1}{2\sqrt{2}}$ **b** $\frac{1}{7}$ **c** $\frac{10}{3}$ **4** **a** $2\sqrt{11}$ **b** $3\sqrt{10}$

5 **a** $6\sqrt{2}+2\sqrt{5}$ **b** $-4+3\sqrt{7}$

6 **a** $\sqrt{14}-\sqrt{7}$ **b** $27-10\sqrt{2}$ **7** **a** $5+2\sqrt{6}$ **b** -1

8 **a** $\frac{8\sqrt{3}}{3}$ **b** $\frac{\sqrt{91}}{13}$ **c** $\frac{17-7\sqrt{5}}{4}$

9 **a** $\frac{45}{11}+\frac{9}{11}\sqrt{3}$ **b** $-2-\sqrt{3}$ **c** $\frac{1}{2}-\frac{1}{2}\sqrt{3}$

10 **a** **i** 10 and 2 **ii** 6 and $\frac{1}{2}$

b Let $\sqrt{a}\sqrt{b}=k$ where $a, b \in \mathbb{Z}^+,\ k \in \mathbb{Z}$

then $\dfrac{\sqrt{a}}{\sqrt{b}}=\dfrac{\sqrt{a}}{\sqrt{b}}\times\dfrac{\sqrt{b}}{\sqrt{b}}=\dfrac{\sqrt{a}\sqrt{b}}{b}=\dfrac{k}{b}$

11 $\dfrac{1-10\sqrt{2}}{7}$

EXERCISE 6A.1

1 **a** $x=-4$ **b** $x=7$ **c** $x=5$ **d** $x=-11$

2 **a** $x=3$ **b** $x=2$ **c** $x=6$ **d** $x=2$
e $x=-3$ **f** $x=-2$ **g** $x=\frac{17}{8}$ **h** $x=\frac{1}{3}$

3 **a** $x=100$ **b** $x=27$ **c** $x=-32$ **d** $x=-20$
e $x=-17$ **f** $x=13$ **g** $x=-\frac{2}{3}$ **h** $x=5$
i $x=\frac{19}{2}$ **j** $x=-\frac{8}{3}$ **k** $x=18$ **l** $x=\frac{19}{5}$

EXERCISE 6A.2

1 **a** $x=3$ **b** $x=4$ **c** $x=-5$ **d** $x=-\frac{3}{2}$
e $x=-1$ **f** $x=-3$ **g** $x=4$ **h** $x=1$
i $x=-6$ **j** $x=\frac{1}{2}$

2 **a** $x=5$ **b** $x=2$ **c** $x=\frac{3}{5}$ **d** $x=-\frac{9}{5}$
e $x=\frac{3}{7}$ **f** $x=\frac{3}{2}$ **g** $x=-\frac{5}{4}$ **h** $x=-\frac{1}{8}$
i $x=1$ **j** $x=2$ **k** $x=-7$ **l** $x=5$
m $x=\frac{26}{5}$ **n** $x=-\frac{2}{17}$

3 **a** x can be any real number, that is, $x\in\mathbb{R}$, since LHS $=$ RHS $=6x-5$. This equation is an identity.
b There is no solution for x as it reduces to $-12=3$.

EXERCISE 6B

1 **a** $x=\frac{10}{9}$ **b** $x=2$ **c** $x=-3$
d $x=\frac{13}{11}$ **e** $x=\frac{24}{19}$ **f** $x=-\frac{9}{14}$
g $x=-4$ **h** $x=\frac{25}{11}$ **i** $x=\frac{7}{8}$

2 **a** $x=\frac{21}{2}$ **b** $x=\frac{36}{5}$ **c** $x=\frac{10}{3}$
d $x=\frac{7}{6}$ **e** $x=\frac{6}{5}$ **f** $x=-28$
g $x=-18$ **h** no solution (x cannot be 0. Why?)

3 **a** $x=\frac{1}{7}$ **b** $x=-\frac{13}{10}$ **c** $x=-\frac{1}{4}$ **d** $x=\frac{29}{3}$
e $x=-6$ **f** $x=-\frac{2}{9}$ **g** $x=\frac{5}{12}$ **h** $x=\frac{2}{3}$
i $x=\frac{8}{15}$ **j** $x=2$ **k** $x=\frac{1}{2}$ **l** $x=\frac{13}{4}$

4 **a** $x=12$ **b** $x=-\frac{24}{5}$ **c** $x=-\frac{22}{3}$ **d** $x=-\frac{9}{7}$
e $x=4$ **f** $x=\frac{3}{4}$

5 **a** $x=\frac{36}{11}$ **b** $x=1$ **c** $x=-44$ **d** $x=-\frac{7}{5}$
e $x=-25$ **f** $x=\frac{4}{17}$

EXERCISE 6C.1

1 The number is 7. **2** The number is -4.
3 The number is 12. **4** There are 7 passengers in each row.
5 Carol drank 580 mL of water.
6 There were 4 people at Table A. **7** The number is 11.
8 The number is $\frac{7}{3}$. **9** The fraction is $\frac{8}{12}$.
10 The fraction is $\frac{-2}{-6}$.

EXERCISE 6C.2

1 Ellie's son is 10 years old now.
2 George is 4 years old now. **3** Adrian is 7 years old now.
4 Lana has 11 20-cent coins and 15 50-cent coins.
5 The corner shop has 15 600 mL cartons.
6 2700 €20 tickets, 900 €35 tickets, and 1300 €50 tickets were sold.
7 Olivia used 9 kg of dried fruit and 21 kg of nuts.
8 Tom invested $4500 in Canadian Airways shares.
9 Inka invested €2400 in telecommunications shares and €7200 in mining shares.
10 Bruce has 8 5-cent stamps, 48 10-cent stamps, and 16 20-cent stamps.
11 Andy invested $9250 in Baldish shares, $18 500 in Creweman shares, and $12 250 in Droners shares.

EXERCISE 6D

1 **a** $S\leqslant 40$ **b** $A\geqslant 18$ **c** $a>3$ **d** $b\leqslant -3$
e $d<5$ **f** $-20\geqslant x$ **g** $4<y$ **h** $z\geqslant 0$

2 **a** $x>2$ **b** $b<5$ **c** $c\geqslant 2\frac{1}{2}$ **d** $d\leqslant -7$
e $a<-19$ **f** $p>-3$

3 **a**
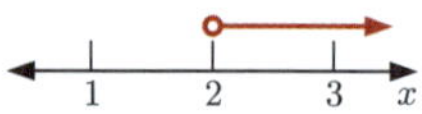

b
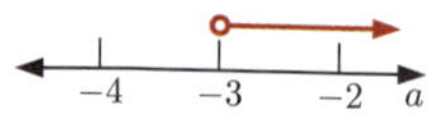

c
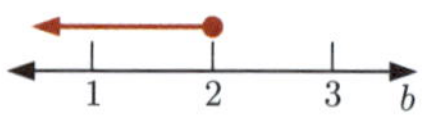

d
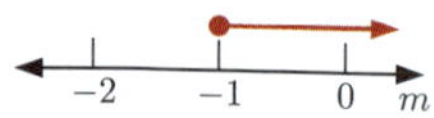

e
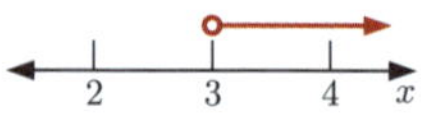

f
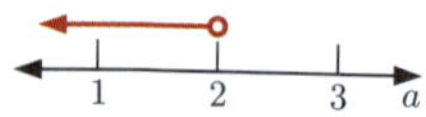

g
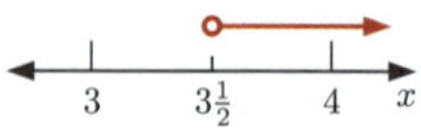

h
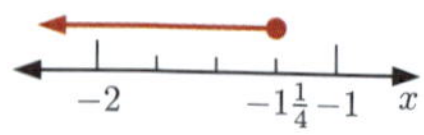

4 **a** true **b** false **c** false **d** false **e** true **f** false

5 **a**
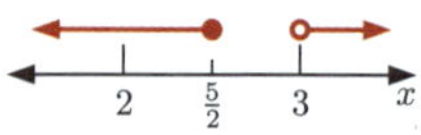

b $x=3$

EXERCISE 6E

1 **a** $a>2$
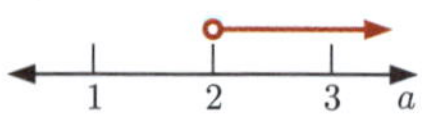

b $b\leqslant -3$
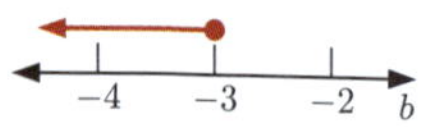

c $s<6$
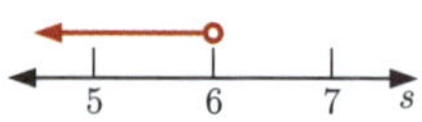

d $c<10$
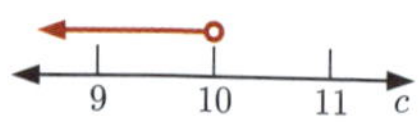

e $x\geqslant 5$
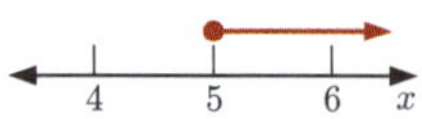

f $b<-4$
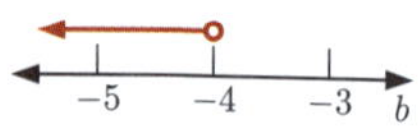

g $t>-5$
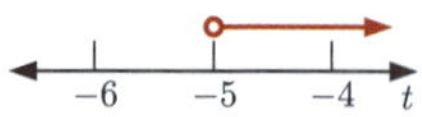

h $k<-6$
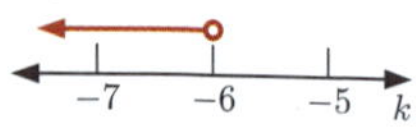

i $m \leqslant -60$

2 **a** $x > 3$ **b** $m \leqslant 4$

c $a \geqslant \frac{10}{3}$ **d** $a < -2$

e $b > 5$ **f** $s > -2$

g $a \leqslant 2$ **h** $b < 3$

i $b \geqslant \frac{1}{3}$ **j** $n < -2$

k $x < 13$ **l** $b \geqslant \frac{7}{4}$

3 **a** $x > 9$ **b** $b \leqslant -5$

c $c \geqslant 16$ **d** $x < 9$

e $x \geqslant -\frac{14}{3}$ **f** $x > -6$

g $x \leqslant 20$ **h** $x > \frac{5}{2}$

i $x < \frac{28}{3}$

4 **a** $a < 15$ **b** $b \geqslant -1$

c $c \leqslant -2$ **d** $a < -6$

e $x \geqslant \frac{17}{3}$ **f** $x \leqslant 3$

5 **a** $x < 1$ **b** $c \geqslant -2$ **c** $b < -2$ **d** $a \geqslant 21$

e $d > 14$ **f** $p \leqslant 0$

6 **a** $x > 2$ **b** $x \leqslant -3$

c $x < -1$ **d** $x < \frac{4}{5}$

e $x \leqslant -3$ **f** $x \geqslant -2$

REVIEW SET 6A

1 **a** $x = -2$ **b** $x = -2$ **2** **a** $x = \frac{47}{5}$ **b** $x = -9$

3 The number is 6. **4** **a** $x = \frac{26}{3}$ **b** $x = \frac{14}{15}$

5 **a** $x = \frac{13}{12}$ **b** $x = -\frac{18}{5}$ **6** The number is $\frac{13}{5}$.

7 Evan is now 7 years old.

8 Elaine has 36 5-cent coins and 52 10-cent coins.

9 Simon invested €2400 in technology shares and €4800 in petroleum shares.

10 **a** **b**

11 **a** $x \leqslant \frac{7}{2}$ **b** $x < -\frac{9}{2}$

c $x \leqslant -\frac{4}{7}$

12 **a** $x > -10$ **b** $x \leqslant \frac{1}{3}$

c $-10 < x \leqslant \frac{1}{3}$

REVIEW SET 6B

1 **a** $x = 5$ **b** $x = \frac{1}{9}$ **2** **a** $x = \frac{13}{12}$ **b** $x = \frac{19}{15}$

3 Dara went on 4 rides. **4** **a** $x = -5$ **b** $x = -\frac{24}{35}$

5 **a** $x = \frac{39}{2}$ **b** $x = -\frac{15}{2}$ **c** $x = -\frac{155}{6}$

6 The fraction is $\frac{20}{24}$. **7** Amanda is 9 years old now.

8 40 90 g tins and 72 200 g tins were sold.

9 Edward mixes 4 kg of jelly fruits and 11 kg of chocolate buttons.

10 **a** **b**

c

11 **a** $x \leqslant \frac{3}{2}$ **b** $x < 10$ **c** $x \geqslant 0$

12 **a** $x = \frac{10}{3}$

b **i** If $x = -2$, $\frac{3}{5}$ is not $< \frac{2}{-2} = -1$.

ii In Henry's method he is really multiplying both sides by $5x$. This cannot be done unless we know for sure that $5x$ is positive.

EXERCISE 7A

1 **a** 21 cm **b** 18 cm **c** 12 m **d** 144 mm

e 12 m **f** 29 cm **g** 22 cm **h** ≈ 18.8 m

i ≈ 44.0 mm **j** ≈ 20.6 cm **k** ≈ 23.0 m **l** ≈ 60.1 cm

2 **a** 36 mm **b** 22 cm **c** ≈ 32.9 m **d** ≈ 451 m **e** ≈ 82.3 cm **f** ≈ 77.1 mm **g** ≈ 94.2 cm **h** 38 m **i** ≈ 37.7 cm

3 **a** $P = 4a + 2b + 2c$ **b** $P = 12a$ **c** $P = 2r + \frac{3}{2}\pi r$ **d** $P = 2\pi x + 2y$

4 **a** 24.4 m **b** 97.6 m **c** \$966.24 **5** 31.1 m

6 **a** **i** ≈ 565.5 m **ii** ≈ 502.7 m **b** ≈ 62.8 m

7 ≈ 133 times **8** $\approx$ €1531.53 **9** ≈ 1 h 35 min

10 61 cm

EXERCISE 7B

1 **a** 3.84 ha **b** 257 000 m^2 **c** 35 000 cm^2 **d** 50 m^2 **e** 1800 mm^2 **f** 4.6 cm^2 **g** 9 000 000 m^2 **h** 3500 ha **i** 5 km^2

2 **a** 40 cm^2 **b** 625 m^2 **c** 42 m^2 **d** 39 cm^2 **e** 15 cm^2 **f** 45 cm^2 **g** 180 m^2 **h** 24 cm^2 **i** 25 cm^2 **j** ≈ 50.3 cm^2 **k** ≈ 63.6 m^2 **l** ≈ 18.2 cm^2 **m** ≈ 19.6 cm^2 **n** ≈ 3.22 cm^2 **o** ≈ 3.67 mm^2 **p** ≈ 70.4 m^2

3 **a** 387 cm^2 **b** 147 cm^2 **c** 65 cm^2 **d** ≈ 66.0 cm^2 **e** 36 cm^2 **f** ≈ 6960 m^2

4 ≈ 2.85 m^2 **5** $\approx 46.2\%$ **6** 23 flights **7** $\approx 25.5\%$

8 **a** **i** ≈ 245.74 cm^2 **ii** ≈ 214.47 cm^2 **iii** ≈ 163.08 cm^2 **iv** ≈ 180.96 cm^2
b ≈ 804.25 cm^2, area of whole pizza ≈ 804.25 cm^2

9 **a** 7500 washers **b** 3050 washers

10 **a** $A = \frac{1}{2}\pi r^2$ units^2 **b** $A = \pi(R^2 - r^2)$ units^2
c $A = (x + 2a)^2 - \pi a^2$ units^2
d $A = 2ab + \pi a^2$ units^2 **e** $A = \frac{1}{3}\pi r^2$ units^2
f $A = 4a^2 - a\sqrt{b^2 - a^2}$ units^2

11 $x = 7$

12 **Hint:** Consider area of $\triangle$PQR and area of 3 semi-circles individually. Use these areas to find an expression for the shaded area.

EXERCISE 7C.1

1 **a** 104 cm^2 **b** 4400 cm^2 **c** 41.5 m^2

2 $A = 2(ab + bc + ac)$ units^2

3 **a** 360 cm^2 **b** 1344 m^2 **c** ≈ 9840 cm^2

4 **a** 576 cm^2 **b** 384 m^2 **c** ≈ 537 m^2 **d** ≈ 823 m^2 **e** ≈ 1380 cm^2 **f** ≈ 313 cm^2

5 22 m^2 **6** $8\sqrt{3}$ cm^2 ≈ 13.9 cm^2 **7** ≈ 26.5 m^2

EXERCISE 7C.2

1 **a** ≈ 207 cm^2 **b** ≈ 339 cm^2 **c** ≈ 196 cm^2 **d** ≈ 56.7 m^2 **e** ≈ 124 m^2 **f** ≈ 79.5 cm^2

2 **a** ≈ 204 cm^2 **b** ≈ 56.5 m^2 **c** ≈ 298 cm^2 **d** ≈ 49.5 mm^2 **e** ≈ 302 cm^2 **f** ≈ 95.1 m^2

3 **a** ≈ 50.3 cm^2 **b** ≈ 707 m^2 **c** ≈ 201 m^2 **d** ≈ 145 km^2 **e** ≈ 84.8 cm^2 **f** ≈ 115 mm^2

4 **a** ≈ 179.1 cm^2 **b** ≈ 238.8 cm^2 **c** ≈ 332.2 m^2 **d** ≈ 216.8 m^2 **e** ≈ 682.3 cm^2 **f** ≈ 640.9 cm^2

5 ≈ 18.1 cm^2 **6** ≈ 33.8 cm^2 **7** 141 spheres

8 \$1100 **9** ≈ 213 cm^2

EXERCISE 7D.1

1 **a** 8650 mm^3 **b** 86 cm^3 **c** 0.3 m^3 **d** 124 000 mm^3 **e** 0.3 cm^3 **f** 3 700 000 cm^3

2 300 000 erasers

EXERCISE 7D.2

1 **a** 72 m^3 **b** 80 cm^3 **c** 525 m^3 **d** ≈ 226 cm^3 **e** ≈ 177 m^3 **f** ≈ 183 cm^3

2 2000 cm^3 **3** ≈ 0.982 cm^3 **4** \$595 **5** ≈ 0.290 m^3

6 **a** $\approx 55\,100$ cm^3 **b** $\approx 25\,600$ cm^3 **7** $\approx 14\,600$ cm^3

8 ≈ 872 cm^3 **9** **a** ≈ 2380 cm^3 **b** ≈ 6 kg

10 $\approx 15\,900$ handles

11 **a** **i** ≈ 3.93 m^3 **ii** ≈ 9.05 m^3 **b** ≈ 13.0 m^3 **c** \$1842.42

EXERCISE 7D.3

1 **a** 32 m^3 **b** ≈ 66.0 cm^3 **c** 120 cm^3 **d** ≈ 233 m^3 **e** ≈ 314 m^3 **f** ≈ 5.33 m^3

2 **a** ≈ 37.7 m^3 **b** ≈ 738 cm^3 **c** 192 cm^3 **d** ≈ 42.7 m^3 **e** ≈ 1.89 m^3 **f** ≈ 310 m^3

3 **a** ≈ 789 cm^3 **b** ≈ 264 m^3 **c** ≈ 2020 m^3

4 4500 m^3 **5** ≈ 2.48 m^3 **6** ≈ 258 cm^3

7 611 garden stakes

EXERCISE 7D.4

1 **a** ≈ 268 m^3 **b** ≈ 905 mm^3 **c** ≈ 102 cm^3 **d** ≈ 16.8 m^3 **e** ≈ 89.8 cm^3 **f** ≈ 1.07 m^3 **g** ≈ 80.4 cm^3 **h** ≈ 101 m^3 **i** ≈ 22.4 m^3

2 ≈ 0.905 m^3 **3** ≈ 0.116 m^3 **4** ≈ 3.85 cm^3

5 **a** ≈ 4.92 cm **b** **i** ≈ 3.63 cm **ii** ≈ 165 cm^2

6 **a** 47 400 sinkers **b** ≈ 1.06 cm **c** ≈ 14.1 cm^2

7 **a** 143 spheres **b** ≈ 1.00 cm^3 **c** ≈ 6.21 mm

EXERCISE 7E

1 **a** 5600 mL **b** 524 000 L **c** 0.28 L **d** 3500 kL **e** 0.743 ML **f** 840 000 mL

2 **a** 3 L **b** ≈ 198 mL **c** 300 mL **3** ≈ 24.4 mL

4 **a** 13 500 m^2 **b** $\approx 23\,000$ kL

5 **a** ≈ 53.8 L **b** ≈ 618 km **6** ≈ 1410 kL

7 265 bottles **8** ≈ 888 mL **9** ≈ 0.707 m

10 **a** ≈ 18.1 kL **b** ≈ 0.177 m

11 **a** **A**: 364 mL, **B**: ≈ 209 mL **b** ≈ 3.07 cm

12 **a** 1000 cm^3 **b** 6.20 cm **c** ≈ 484 cm^2

13 $\frac{91}{108}$ cm

REVIEW SET 7A

1 **a** 2 m^2 **b** 5 000 000 cm^3 **c** 480 cm^3

2 **a** ≈ 158 cm **b** 48 m **c** ≈ 25.1 cm **d** ≈ 209 mm

3 ≈ 565 m

4 **a** ≈ 105 cm^2 **b** 30 cm^2 **c** ≈ 177 cm^2

5 **a** 34 cm^2 **b** 36 cm^2 **c** ≈ 51.8 m^2

6 **a** 400 cm^3 **b** 340 cm^2 **7** ≈ 227 cm^2

8 **a** 5196 cans **b** ≈ 154 m^2 **9** 3 min 36 s

10 **a** ≈ 9557.33 cm^3 **b** ≈ 98.50 cm^3 **c** ≈ 37.70 m^3

11 ≈ 41.8 m^3 **12** ≈ 21.7 m^2

13 **a** **i** 255 cm^2 **ii** 250 mL **b** ≈ 7.07 cm

14 **a** **i** $\frac{x}{2}$ m **ii** $\frac{x}{4}$ m **b** Show both areas are $\frac{\pi x^2}{4}$ m^2.

REVIEW SET 7B

1 a $150\,000\ m^2$ b $3\,600\,000\ mm^3$ c $4.5\ m^3$

2 ≈ 31.9 cm

3 $P = (10x + 2\pi x)$ m, $A = (12x^2 + 2\pi x^2)\ m^2$

4 $\approx 1.21\ m^2$ 5 530 bollards

6 a $\approx 14\,300\ cm^2$ b $\approx 28.3\ m^2$ c $496\ mm^2$

7 ≈ 1.24 cm 8 ≈ 10.9 cm

9 a $\approx 1570\ cm^3$ b $\approx 1190\ cm^3$ c $\approx 0.905\ m^3$

10 $\approx 52.4\%$ 11 a $\approx 754\ cm^3$ b $\approx 415\ cm^2$

12 ≈ 10 min 28 s

13 a $15\ cm^3$ b 15 mL c 133 ice blocks d $\approx 37.4\ cm^2$

14 a $\approx 23\,979.2\ cm^3$ b $5184.2\ cm^2$

EXERCISE 8A.1

1 a 5 units b 4 units c 3 units d $\sqrt{41}$ units e 5 units f $\sqrt{13}$ units

2 a $\sqrt{34}$ units b $\sqrt{29}$ units c $\sqrt{34}$ units d $\sqrt{26}$ units e $\sqrt{17}$ units f $\sqrt{13}$ units

3 a

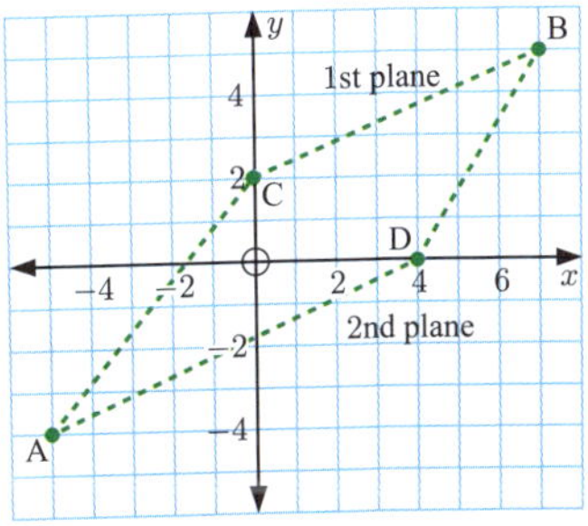

b 150 km

c i ≈ 154.3 km ii ≈ 156.8 km

d the first plane

4 a

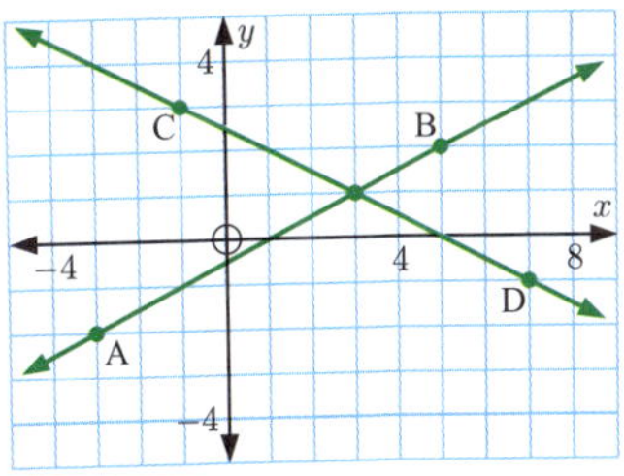

b (3, 1)

c B, $\sqrt{5}$ units

EXERCISE 8A.2

1 a $\sqrt{10}$ units b $\sqrt{13}$ units c $4\sqrt{2}$ units d $\sqrt{41}$ units e $5\sqrt{2}$ units f $\sqrt{13}$ units g $2\sqrt{41}$ units h $2\sqrt{6}$ units

2 a isosceles b scalene c isosceles d equilateral

3 a right angled at A b not right angled c not right angled d right angled at B

4 13.7 units

5 a LM = MN = $\sqrt{20}$ units $\therefore$ $\triangle$ is isosceles

b $LM^2 + MN^2 = LN^2 = 40$ $\therefore$ $\triangle$LMN is right angled at M c $45°$

6 a WX = YZ = $\sqrt{5}$ units, XY = WZ = $3\sqrt{5}$ units

b $5\sqrt{2}$ units

c In $\triangle$WXY, $WX^2 + XY^2 = 5 + 45 = 50$ and $WY^2 = 50$ $\therefore$ $\triangle$WXY is right angled at X.

d rectangle

e

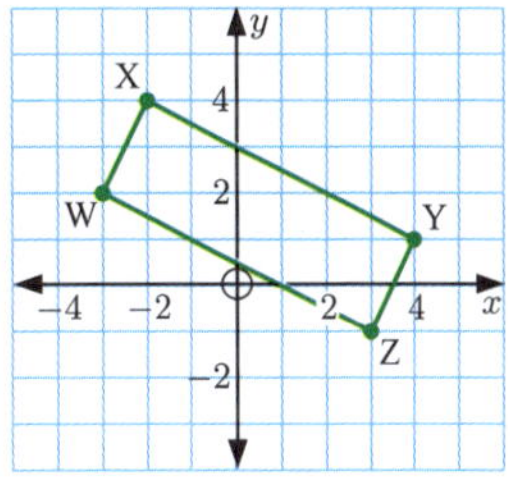

f 15 units^2

EXERCISE 8B

1 a M(−1, 2) b AM = MB = $\sqrt{5}$ units

2 a (3, 5) b (4, 0) c (−1, 1) d $(3, 1\frac{1}{2})$ e $(-1\frac{1}{2}, 2)$ f (1, 1) g $(2\frac{1}{2}, 4)$ h $(-3\frac{1}{2}, 3\frac{1}{2})$ i $(\frac{a+2}{2}, 2)$ j $(0, 2b)$

3 a Q(−1, 11) b Q(1, −10) c Q(6, 0) d $Q(-1\frac{1}{2}, 6)$

4 a A(10, −9) b A(−6, −8)

5 a B(0, 5) b D(−2, 9) 6 a A(−2, 7) b 10 units

7 a $X(3\frac{1}{2}, 4\frac{1}{2})$ b C(6, 7) 8 $N(4, 3\frac{1}{4})$

EXERCISE 8C.1

1 a $\frac{1}{2}$ b $\frac{1}{5}$ c undefined d 0 e $-\frac{3}{2}$ f $-\frac{1}{2}$ g -3 h $-\frac{2}{5}$

2 a **A** and **D** b **B**

3 a

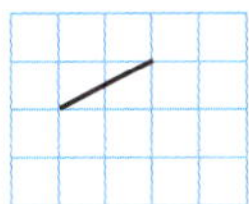

b

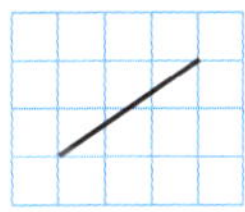

c

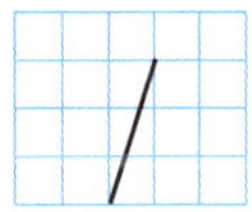

d

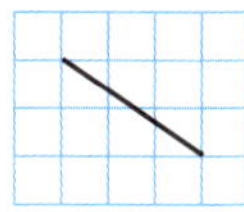

e

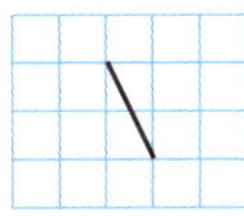

f

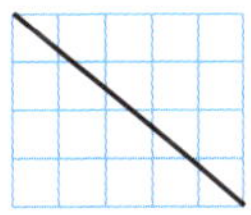

g

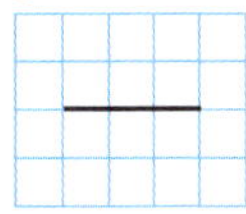

4

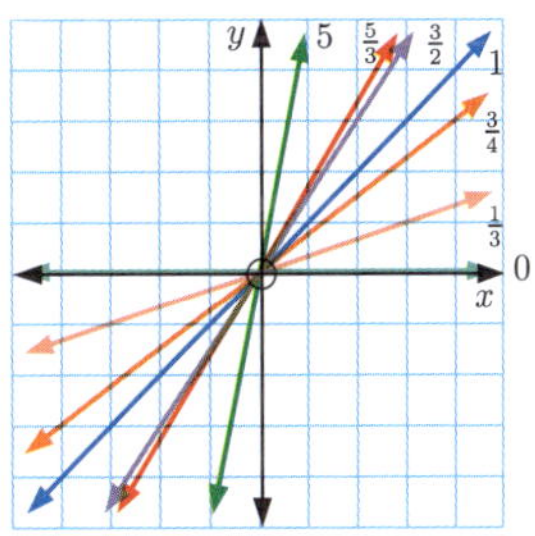

5

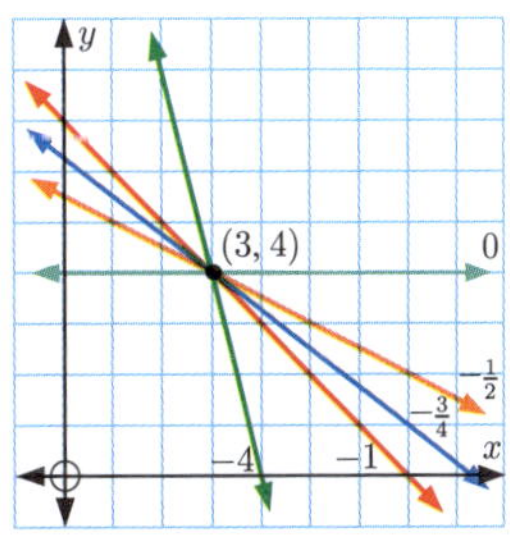

EXERCISE 8C.2

1 **a** $\frac{2}{7}$ **b** -4 **c** $-\frac{2}{3}$

2 **a** 5 **b** 4 **c** $\frac{1}{4}$ **d** $-\frac{3}{4}$ **e** undefined
f 0 **g** $\frac{7}{2}$ **h** $-\frac{7}{2}$ **i** 2

3 **a** $\frac{4}{5}$ **b** $-\frac{9}{2}$ **c** $-\frac{6}{19}$

4 **a** **i** $\frac{2}{3}$ **ii** $\frac{2}{3}$
b [AB] and [BA] are the same line segment and so must have the same gradient.

5 **a** gradient [AB] : $\frac{1}{2}$, gradient [BC] : $-\frac{1}{3}$, gradient [CD] : $\frac{8}{3}$, gradient [AD] : -1
b **i** [AB] and [CD] **ii** [BC] and [AD]
c **i** 2 **ii** $\frac{5}{6}$

6 **a** 8 **b** The sprinter had a constant speed of 8 m/s.
c Yes, as the gradient of the line is constant.

7 **a** 15 **b** Sales assistants are paid \$15/hour.
c **i** \$90 **ii** \$270

8 **a** **i** 70 km/h **ii** ≈ 56.7 km/h
b 72 km/h **c** From 0 h to 2 h and from 5 h to 7 h.

EXERCISE 8D

1 **a** $\frac{3}{5}$ **b** $-\frac{5}{3}$ **c** 6 **d** $-\frac{1}{6}$

2 **a** $-\frac{5}{4}$ **b** -3 **c** $\frac{7}{3}$ **d** $-\frac{1}{2}$ **e** $\frac{1}{11}$
f undefined **g** 0 **h** $\frac{2}{3}$

3 **a** **i** $\frac{2}{3}$ **ii** -1 **iii** $\frac{2}{3}$ **iv** $\frac{11}{9}$ **v** $\frac{1}{9}$ **vi** 1
b **i** [AB] and [CD] **ii** [BC] and [BD]

4 **a** **i** 5 **ii** $-\frac{1}{5}$ **b** perpendicular **c** right angled at B

5 **a** $AB = \sqrt{20}$ units, $BC = \sqrt{8}$ units, $AC = \sqrt{20}$ units
$AB = AC$ so the triangle is isosceles.
b M(2, 3)
c gradient of [AM] = 1, gradient of [BC] = -1
Gradients are $-$ve reciprocals so [AM] $\perp$ [BC]
d

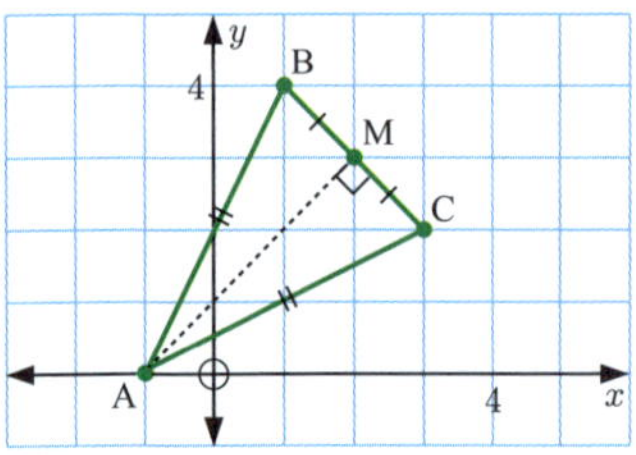

6 **a** M(3, 2), N(4, -1), gradient MN = gradient AC = -3
b $MN = \sqrt{10}$ units, $AC = \sqrt{40} = 2\sqrt{10}$ units

7 **a** **i** gradient [AB] = 0, gradient [DC] = 0
$\therefore$ [AB] $\parallel$ [DC]
ii gradient [AD] = $\frac{3}{5}$, gradient [BC] = $\frac{3}{5}$
$\therefore$ [AD] $\parallel$ [BC]
b a parallelogram

8 **a** $a = 4$ **b** **i** $\frac{1}{2}$ **ii** -2
c Gradients are negative reciprocals so [AP] $\perp$ [BP]
$\therefore$ $\widehat{APB}$ is a right angle.

EXERCISE 8E.1

1 **a** gradient-intercept **b** general **c** general
d gradient-intercept **e** gradient-intercept **f** general

2 **a** gradient = 2, y-intercept = 5
b gradient = -6, y-intercept = -1
c gradient = $\frac{3}{5}$, y-intercept = $-\frac{2}{5}$
d gradient = -1, y-intercept = 10
e gradient = -4, y-intercept = 1
f gradient = $-\frac{7}{4}$, y-intercept = $\frac{5}{4}$
g gradient = $\frac{3}{7}$, y-intercept = $\frac{1}{7}$
h gradient = $\frac{1}{2}$, y-intercept = $-\frac{5}{8}$
i gradient = $-\frac{7}{9}$, y-intercept = $\frac{2}{3}$

3 **a** $2x + y = 11$ **b** $3x - y = 4$ **c** $x + 5y = 6$
d $6x + 7y = 5$ **e** $5x - 8y = 1$ **f** $4x - 9y = -18$

4 **a** **i** $y = -3x + 5$ **ii** -3
b **i** $y = -\frac{2}{5}x + 2$ **ii** $-\frac{2}{5}$
c **i** $y = -\frac{7}{4}x - \frac{9}{4}$ **ii** $-\frac{7}{4}$
d **i** $y = 6x - 1$ **ii** 6
e **i** $y = \frac{5}{11}x - \frac{2}{11}$ **ii** $\frac{5}{11}$
f **i** $y = \frac{9}{2}x + \frac{5}{2}$ **ii** $\frac{9}{2}$

EXERCISE 8E.2

1 **a** yes **b** no **c** yes **d** yes

2 **a** yes **b** no **c** no **d** yes

3 **a** $c = 4$ **b** $c = -2$ **c** $m = \frac{1}{5}$ **d** $a = -2$

4 **a** $k = 11$ **b** $k = 19$ **c** $k = -4$ **d** $k = 10$
e $k = -6$

5 **a** above, (5, -2) is above (5, -7).

EXERCISE 8F.1

1 **a**

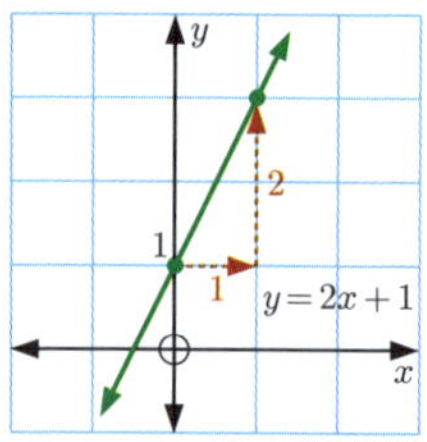

b

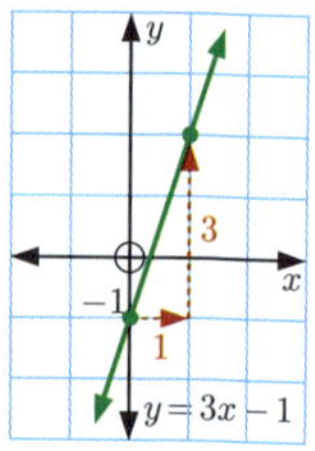

c

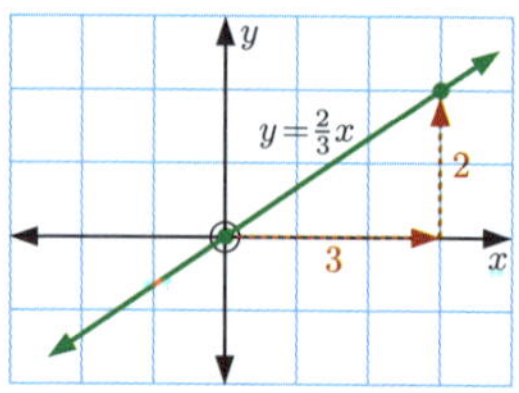

d

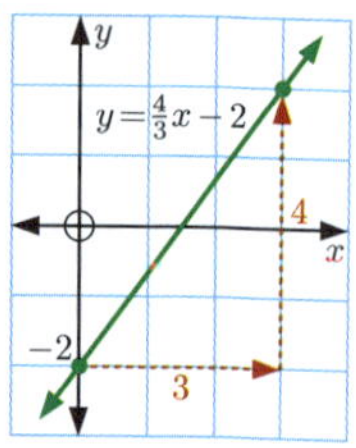

e

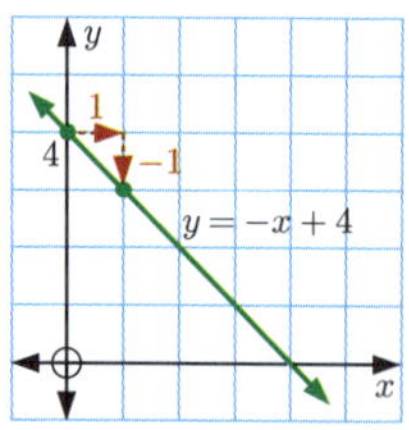

f

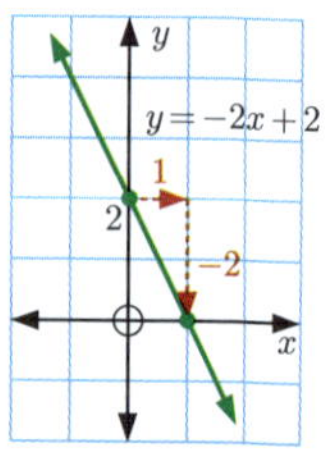

g
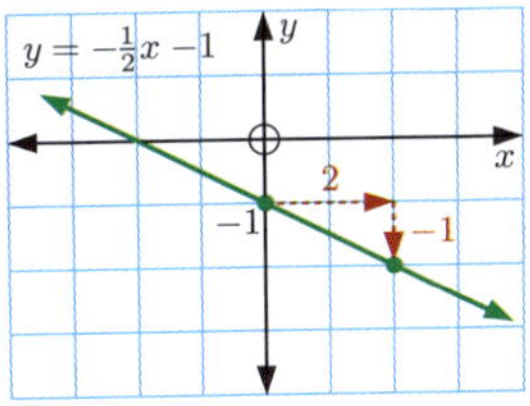

h
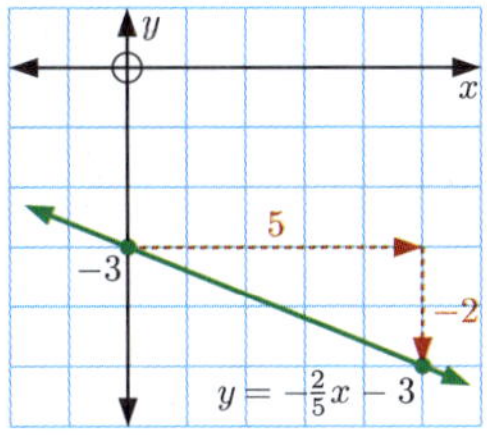

2 **a**
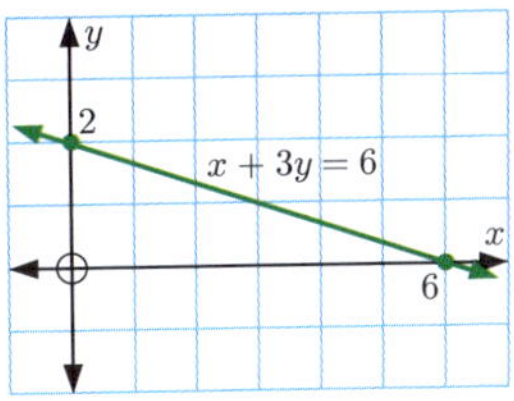

b
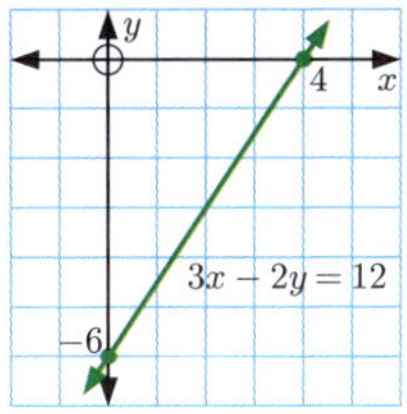

c
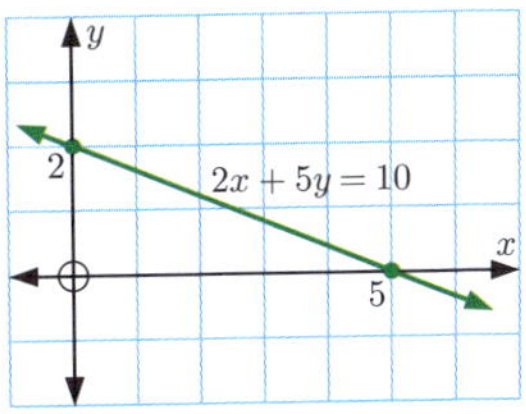

d
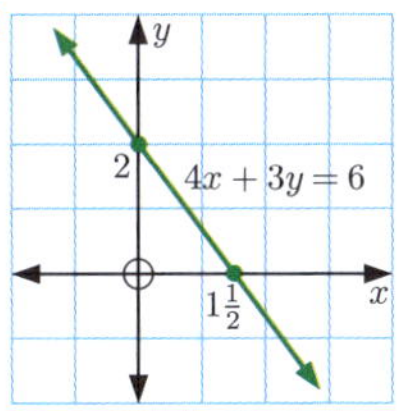

e
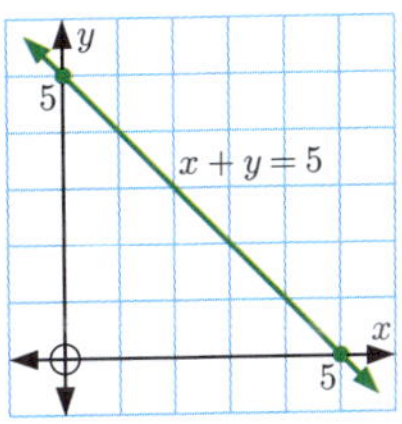

f
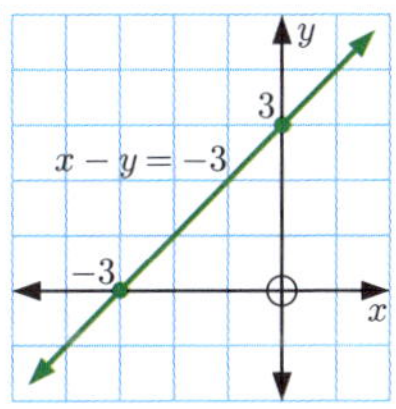

g
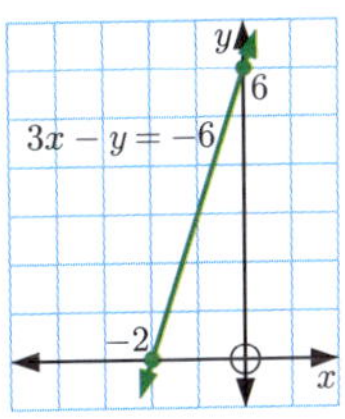

h
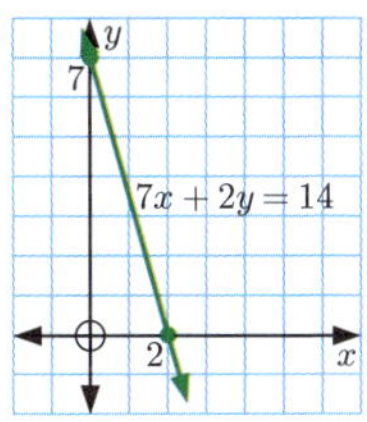

i
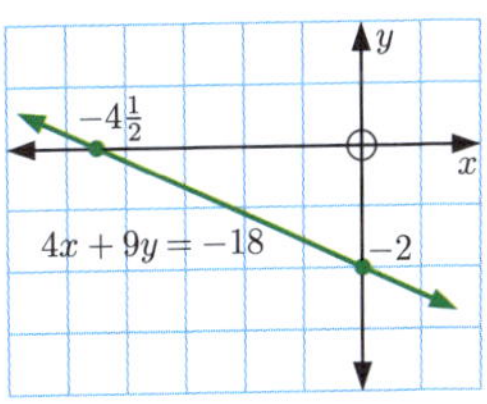

3 **a** **i** $-\frac{2}{3}$
ii 4

b **i** yes
ii no
iii yes

c
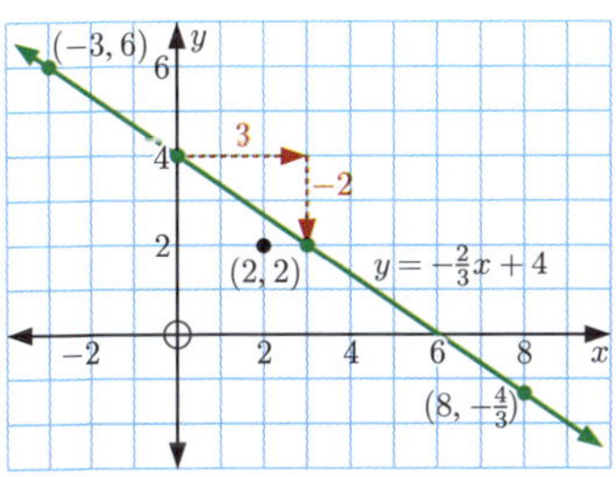

4 **a** **i** 5
ii -3

b **i** yes
ii no

c $c=-\frac{6}{5}$

d
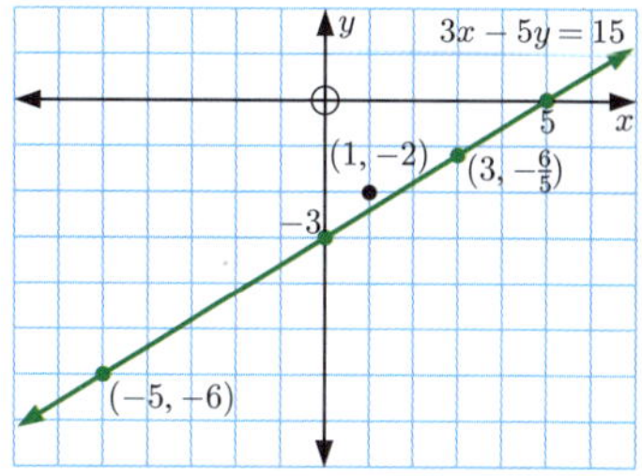

EXERCISE 8F.2

1 **a**
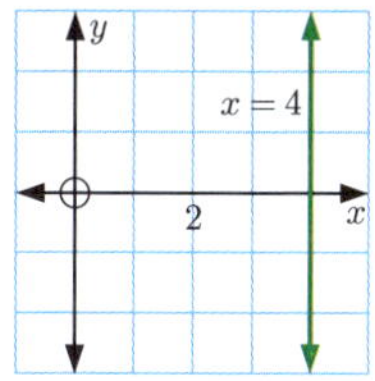

b
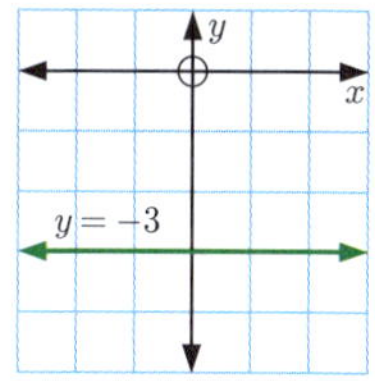

c
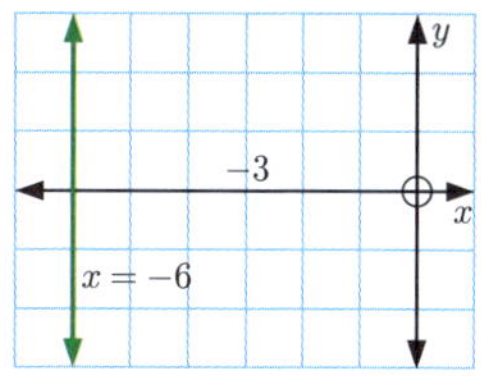

d
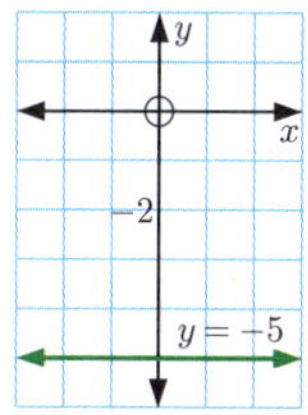

e
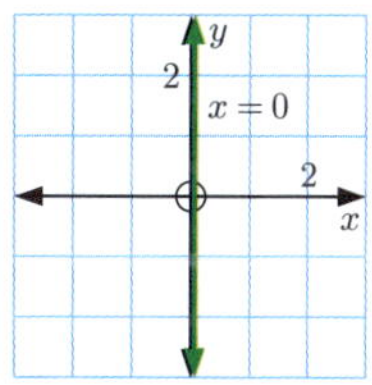

f
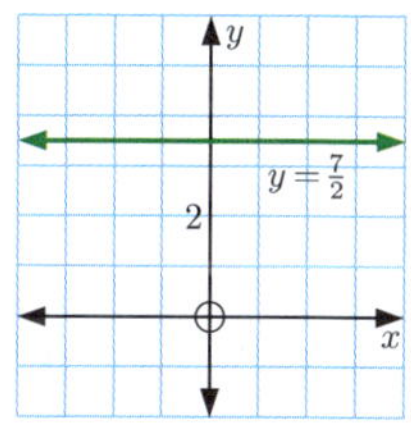

g
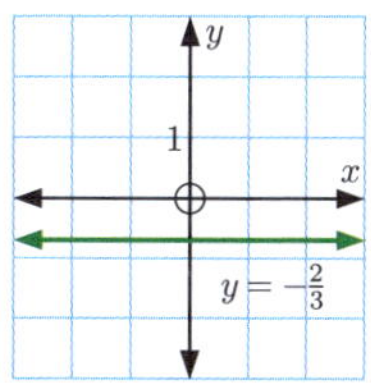

h
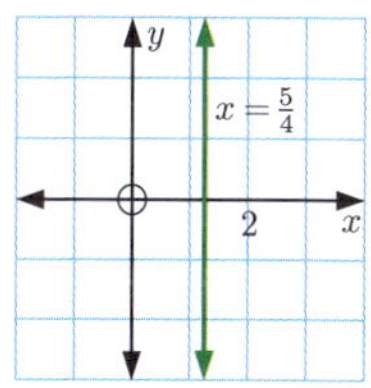

2 $a=5$ **3** b is any real number

EXERCISE 8G

1 **a** $y=3x+4$ **b** $y=7x-1$ **c** $y=-x+\frac{1}{3}$
d $y=\frac{1}{2}x$ **e** $y=5$ **f** $y=-\frac{5}{2}x-\frac{2}{3}$

2 **a** $y=2x+6$ **b** $y=-3x-4$ **c** $y=5x+14$
d $y=\frac{3}{4}x+6$ **e** $y=4$ **f** $y=-\frac{1}{2}x-\frac{13}{2}$

3 **a** $y=2x+2$ **b** $y=\frac{2}{5}x+3$ **c** $y=-x-1$
d $y=-2x+4$ **e** $y=\frac{1}{2}x-\frac{1}{2}$ **f** $y=-\frac{2}{5}x$

4 **a** $y=4x-5$ **b** $y=-3x+7$ **c** $y=\frac{3}{8}x-1$

5 **a** $y=2x+3$ **b** $y=-2x+12$ **c** $y=\frac{2}{3}x-2$
d $y=-3$ **e** $y=\frac{3}{2}x-5$ **f** $y=-\frac{4}{3}x+\frac{1}{3}$

6 **a** $HA=\sqrt{73}$, $HB=\sqrt{72}$, $HC=\sqrt{74}$ $\therefore$ from B
b $y=x+1$

7 **a** $y=3x-2$ **b** $y=-\frac{1}{3}x+4$ **c** $y=7$
d $y=-\frac{2}{5}x-2$ **e** $y=\frac{5}{4}x+\frac{5}{2}$ **f** $x=-4$

8 **a** F **b** G **c** D **d** H **e** C **f** A **g** E **h** B

9 **a** $y=3x+1$ **b** $y=-2x-2$ **c** $y=\frac{1}{3}x-\frac{7}{3}$
d $y=-x+3$ **e** $y=\frac{5}{2}x-\frac{3}{2}$ **f** $y=-3x+3$

10 red line: $y=\frac{2}{3}x-3$
green line: $y=-\frac{3}{2}x+10$
yellow line: $y=-\frac{3}{2}x-6$
blue line: $y=\frac{5}{2}x+10$

REVIEW SET 8A

1 **a** $6\sqrt{5}$ units **b** $(1, 0)$ **2** D$(1, 2)$

3 **a** $\frac{5}{2}$ **b** undefined

4 **a** 40 km/h **b** 20 km/h **c** ≈ 26.7 km/h

5 **a** yes, as $2(-4)-5=-13$ **b** $c=-4$

6 **a** $-\frac{1}{4}$ **b** $\frac{3}{5}$ **c** 4 **d** $-\frac{5}{3}$

7 **a** $3x-y=-7$ **b** $4x+5y=2$

8 **a** isosceles $(\text{PQ}=\sqrt{52},\ \text{PR}=\sqrt{52},\ \text{QR}=\sqrt{104})$
b right angled at P

9

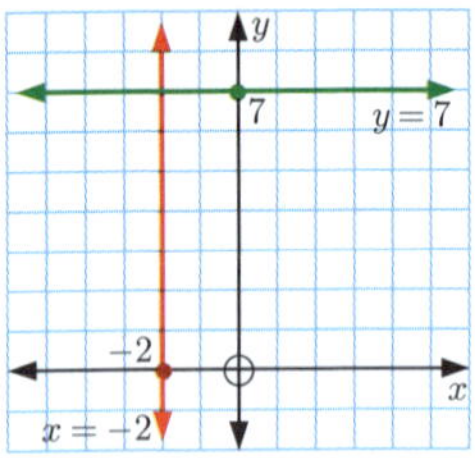

10 **a** $y=2x-3$ **b** $y=-\frac{1}{2}x+2$ **c** $y=-x+2$

11 **a** $(\frac{1}{2}, \frac{3}{2})$ **b** $(2, 0)$ **c** $\sqrt{29}$ units

12 **a**

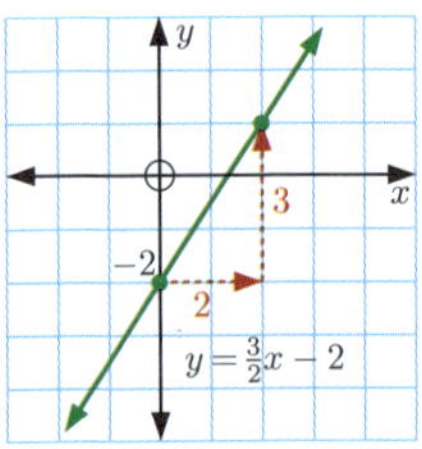

b

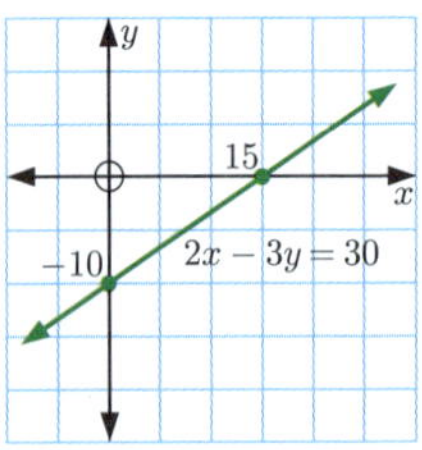

c

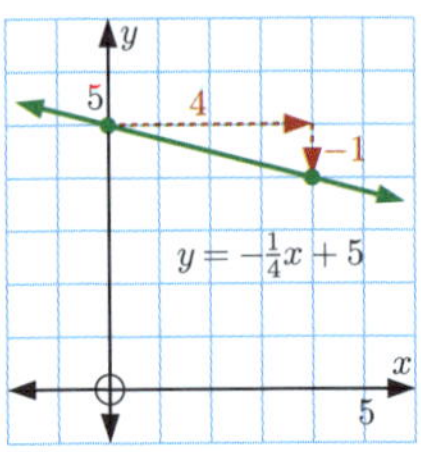

d

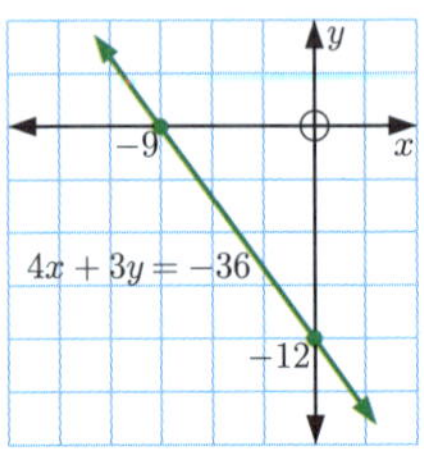

13 $y=-4x-3$

REVIEW SET 8B

1 **a** $(1, -\frac{3}{2})$ **b** $(12, 2)$ **2** $y=\frac{1}{3}x+\frac{23}{3}$ **3** $\frac{3}{2}$

4

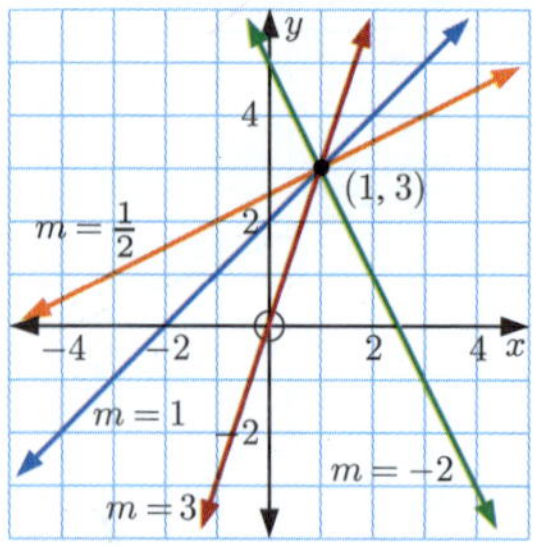

5 $y=-\frac{1}{2}x+3$

6 **a** 36 **b** It is the cost of hiring the hall ($36/hour). **c** $324

7 **a** **i** $\sqrt{52}$ units **ii** $\sqrt{52}$ units **iii** $\sqrt{52}$ units
iv $\sqrt{52}$ units
b [AC] and [BD] are both $\sqrt{104}$ units
c midpoints are both $(4, -3)$ **d** a square

8 **a** $\frac{1}{2}$ **b** $-\frac{4}{3}$ **c** $\frac{3}{10}$ **9** **a** $y=\frac{3}{4}x-6$ **b** $\frac{3}{4}$

10

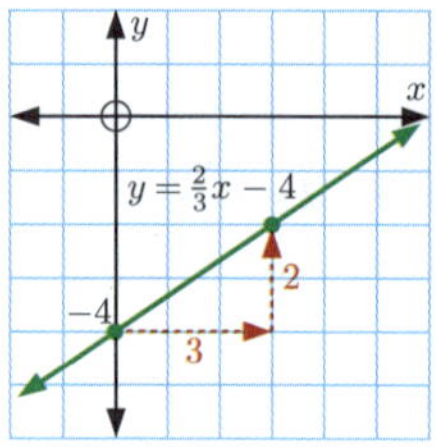

11 $k=-1$

12 **a** **i** $-\frac{3}{5}$ **ii** $\frac{7}{11}$ **iii** $\frac{5}{3}$
b $-\frac{3}{5}\times\frac{5}{3}=-1$ $\therefore$ [PQ] $\perp$ [PR] **c** 34 units^2
$\therefore$ $\widehat{\text{QPR}}$ is a right angle.

13 **a** **i** 4
ii 10
b **i** yes
ii no
c $b=\frac{15}{2}$
d

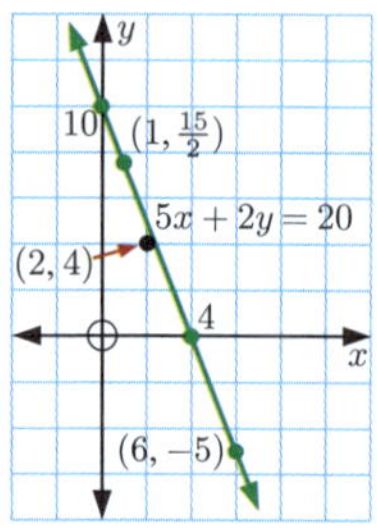

EXERCISE 9A

1 **a** $x(3x+5)$ **b** $x(2x-7)$ **c** $3x(x+2)$
d $4x(x-2)$ **e** $x(9-2x)$ **f** $-3x(x+5)$
g $4x(2x-1)$ **h** $-5x(1+2x)$ **i** $4x(3-x)$
j $x(x^2+x+1)$ **k** $x(2x^2+11x+4)$ **l** $a(b+c+d)$
m $ax(x+2)$ **n** $ab(b+a)$ **o** $ax^2(x+1)$

2 **a** $(x+5)(3+x)$ **b** $(b+3)(a-5)$ **c** $(x+4)(x+1)$
d $(x+2)(2x+5)$ **e** $(c-d)(a+b)$ **f** $(y+2)(y-1)$
g $(x-1)(ab+c)$ **h** $(x+2)(a-1)$

3 **a** $(x+2)(x-3)$ **b** $(x-1)(x-4)$ **c** $(x+1)(x+3)$
d $(x-2)(x+1)$ **e** $(x+3)(x+4)$ **f** $(x+4)(x+6)$
g $(x-3)(x-4)$ **h** $(x+4)(x+2)$ **i** $(x-4)(x-9)$
j $(x+2)(x+5)$ **k** $(x+1)^2(x+2)$
l $(a+b)(a^2+2ab+b^2+1)$ **m** $(x+1)(2x+3)$
n $(x-2)(3x-7)$ **o** $2(a+b)(2a+2b-1)$

EXERCISE 9B

1 **a** $(x+2)(x-2)$ **b** $(2+x)(2-x)$ **c** $(x+9)(x-9)$ **d** $(5+x)(5-x)$ **e** $(2x+1)(2x-1)$ **f** $(3x+4)(3x-4)$ **g** $(2x+3)(2x-3)$ **h** $(6+7x)(6-7x)$

2 **a** $3(x+3)(x-3)$ **b** $2(2+x)(2-x)$ **c** $3(k+5)(k-5)$ **d** $5(1+x)(1-x)$ **e** $2(2t+3)(2t-3)$ **f** $3(5+3x)(5-3x)$ **g** $x(x+7)(x-7)$ **h** $n^2(8+n)(8-n)$ **i** $7x(2x+3)(2x-3)$

3 **a** $(x+\sqrt{3})(x-\sqrt{3})$ **b** no linear factors **c** $(x+\sqrt{15})(x-\sqrt{15})$ **d** $3(x+\sqrt{5})(x-\sqrt{5})$ **e** $(x+1+\sqrt{6})(x+1-\sqrt{6})$ **f** no linear factors **g** $(x-2+\sqrt{7})(x-2-\sqrt{7})$ **h** $(x+3+\sqrt{17})(x+3-\sqrt{17})$ **i** no linear factors

4 **a** $(x+3)(x-1)$ **b** $4(x+2)(x-1)$ **c** $(x-5)(x+3)$ **d** $3(x+1)(3-x)$ **e** $(3x+2)(x-2)$ **f** $(2x+3)(4x-3)$ **g** $(3x-1)(x+3)$ **h** $8x(x-1)$ **i** $3(2x+1)(2x-3)$

EXERCISE 9C

1 **a** $(x+3)^2$ **b** $(x+4)^2$ **c** $(x-3)^2$ **d** $(x-4)^2$ **e** $(x+1)^2$ **f** $(x-5)^2$ **g** $(y+9)^2$ **h** $(m-10)^2$ **i** $(t+6)^2$

2 **a** $(3x+1)^2$ **b** $(2x-1)^2$ **c** $(3x+2)^2$ **d** $(5x-1)^2$ **e** $(4x+3)^2$ **f** $(5x-2)^2$

3 **a** $-(x-1)^2$ **b** $2(x+2)^2$ **c** $-3(x+5)^2$ **d** $2(4x-1)^2$ **e** $4(3x+5)^2$ **f** $-2x(x-7)^2$

EXERCISE 9D

1 **a** $(b+c)(x+y)$ **b** $(p+q)(x+3)$ **c** $(2a+b)(3x+1)$ **d** $(a+b)(m-n)$ **e** $(r-s)(3d+1)$ **f** $(a+b)(2c-5)$

2 **a** $(x+5)(x+7)$ **b** $(x-2)(x-6)$ **c** $(x+3)^2$ **d** $(x+1)(x+8)$ **e** $(x+1)(2x+3)$ **f** $(x-1)(3x+1)$ **g** $(x+5)(2x+3)$ **h** $(3x+2)(2x-1)$ **i** $(x+2)(4x+1)$ **j** $(2x+1)(3x+5)$ **k** $(2x+3)(3x-2)$ **l** $(x-2)(4x+1)$ **m** $(x+11)(3x+4)$ **n** $(3x-2)(6x+1)$ **o** $(2x-7)(5x+2)$

3 **a** $(x-1+a)(x-1-a)$ **b** $(x+a)(x-a+1)$ **c** $-(x+2+b)(x+2-b)$ **d** $-(x-3+c)(x-3-c)$ **e** $(x-y)(x+y-1)$ **f** $2(a-b)(a+2b)$ **g** $(x+2+m)(x+2-m)$ **h** $(x+a+b)(x+a-b)$ **i** $(x+y)(x-y-3)$

EXERCISE 9E

1 **a** 3 and 4 **b** 3 and 5 **c** 2 and 8 **d** 2 and 9 **e** -3 and 12 **f** 3 and -12 **g** 2 and -6 **h** -2 and 15

2 **a** $(x+1)(x+3)$ **b** $(x+3)(x+8)$ **c** $(x+3)(x+7)$ **d** $(x+6)(x+9)$ **e** $(x+4)(x+5)$ **f** $(x+3)(x+5)$ **g** $(x+4)(x+6)$ **h** $(x+2)(x+7)$ **i** $(x+2)(x+4)$

3 **a** $(x-1)(x-2)$ **b** $(x-1)(x-3)$ **c** $(x-2)(x-3)$ **d** $(x-3)(x-11)$ **e** $(x-3)(x-13)$ **f** $(x-3)(x-16)$ **g** $(x-4)(x-7)$ **h** $(x-2)(x-12)$ **i** $(x-2)(x-18)$

4 **a** $(x-8)(x+1)$ **b** $(x+7)(x-3)$ **c** $(x-2)(x+1)$ **d** $(x-4)(x+2)$ **e** $(x+8)(x-3)$ **f** $(x-5)(x+2)$ **g** $(x+9)(x-6)$ **h** $(x+9)(x-8)$ **i** $(x-7)(x+3)$ **j** $(x-3)(x+2)$ **k** $(x-12)(x+5)$ **l** $(x+12)(x-5)$ **m** $(x+6)(x-3)$ **n** $(x+2)(x-9)$ **o** $(x-5)(x-7)$

5 **a** $2(x-4)(x+1)$ **b** $3(x+4)(x-1)$ **c** $5(x+3)(x-1)$ **d** $4(x+5)(x-4)$ **e** $2(x-5)(x+3)$ **f** $3(x+7)(x-3)$ **g** $-2(x-5)(x+4)$ **h** $-3(x-2)^2$ **i** $-7(x+4)(x-1)$ **j** $-(x+2)(x+1)$ **k** $-(x-3)(x-2)$ **l** $-(x-3)(x-6)$ **m** $5(x+5)(x-2)$ **n** $-2(x+7)(x-3)$ **o** $-(x-8)(x+4)$

EXERCISE 9F

1 **a** product $=18$, sum $=11$
Numbers are 9 and 2. $\therefore$ $11x=2x+9x$
b product $=8$, sum $=9$
Numbers are 8 and 1. $\therefore$ $9x=x+8x$
c product $=-60$, sum $=4$
Numbers are 10 and -6. $\therefore$ $4x=-6x+10x$
d product $=-60$, sum $=-11$
Numbers are -15 and 4. $\therefore$ $-11x=-15x+4x$
e product $=-9$, sum $=-8$
Numbers are -9 and 1. $\therefore$ $-8x=x-9x$
f product $=144$, sum $=-25$
Numbers are -16 and -9. $\therefore$ $-25x=-9x-16x$

2 **a** $(2x+3)(x+2)$ **b** $(2x+5)(x+5)$ **c** $(2x+5)(x+3)$ **d** $(3x+1)(x+2)$ **e** $(3x+5)(x+3)$ **f** $(3x+2)(2x+5)$

3 **a** $(2x-1)(x+3)$ **b** $(2x+3)(x-1)$ **c** $(3x-2)(x+2)$ **d** $(3x+1)(x-3)$ **e** $(5x+2)(x-2)$ **f** $(2x+3)(3x-1)$

4 **a** $(2x-1)(x-5)$ **b** $(2x-3)(x-4)$ **c** $(2x-5)(x-2)$ **d** $(3x-2)(x-4)$ **e** $(5x-2)(2x-1)$ **f** $(3x-1)(3x-4)$

5 **a** $(3x+4)(4x-1)$ **b** $(3x+2)(2x+5)$ **c** $(4x+3)(2x-1)$ **d** $(3x-2)(2x-5)$ **e** $(5x+1)(2x-3)$ **f** $(4x+3)(5x+4)$

6 **a** $2(2x-3)(x+2)$ **b** $3(3x+2)(2x+1)$ **c** $2(2x-1)(2x-3)$

REVIEW SET 9A

1 **a** $3x(x+4)$ **b** $3x(5-x)$ **c** $(x+3)(x-1)$ **d** $(2x+3)(2x-3)$ **e** $6(x+2y)(x-2y)$ **f** $(x+\sqrt{13})(x-\sqrt{13})$ **g** $(x+1+\sqrt{6})(x+1-\sqrt{6})$ **h** $(x+4)^2$ **i** $(x-5)^2$

2 **a** $2(x+1)(x+3)$ **b** $5(x-1)^2$ **c** $(x+2)(a+b)$ **d** $(c-d)(2x-1)$ **e** $(x+4)(3x+2)$ **f** $(2x+3)(3x-1)$ **g** $(x-8)(x-3)$ **h** $(x+2+a)(x+2-a)$ **i** $-(x+7)(x-4)$

3 **a** $(2x+1)(x+8)$ **b** $(2x-1)(x+8)$ **c** $(2x-1)(x-8)$ **d** $(2x-5)(3x+2)$ **e** $(3x+2)(4x-1)$ **f** $(2x-3)(6x+5)$

REVIEW SET 9B

1 **a** $4x(x-2)$ **b** $8x(2-x)$ **c** $2x(2x-1)$
d $(3+5x)(3-5x)$ **e** $2(3+a)(3-a)$
f $(x+\sqrt{23})(x-\sqrt{23})$ **g** $(x+2+\sqrt{3})(x+2-\sqrt{3})$
h $(x-6)^2$ **i** $2(x+2)^2$

2 **a** $(x+3)(x-3)$ **b** no linear factors
c $9(x+1)(x-1)$ **d** $3(x+\sqrt{3})(x-\sqrt{3})$
e $3x(x-3)$ **f** $3(\sqrt{3}x+1)(\sqrt{3}x-1)$

3 **a** $3(x-3)(x+1)$ **b** $7(x+2)^2$
c $(m+n)(x-y)$ **d** $(3a+b)(a-2b)$
e $(2x+3)(x+4)$ **f** $(2x+3)(2x-1)$
g $-(x-4)^2$ **h** $-2(x+3)(x-2)$
i $-3(x-4)(x+1)$

4 **a** $(3x+1)(x-6)$ **b** $(3x-1)(x-6)$
c $(3x-1)(x+6)$ **d** $(4x+1)(3x+1)$
e $(x-2)(12x+1)$ **f** $(3x+2)^2$

EXERCISE 10A

1 **a** discrete numerical **b** continuous numerical
c categorical **d** categorical **e** categorical
f continuous numerical **g** continuous numerical
h continuous numerical **i** discrete numerical

2 **a** continuous numerical **b** categorical
c categorical **d** continuous numerical
e continuous numerical **f** discrete numerical
g categorical

3 **a** discrete; 1, 2, 3, 4, 5, 6 **b** continuous; 2 cm - 8 cm
c discrete; 0, 1, 2, 3, 4, 5, 6, 7, 8, 9, 10
d continuous; 9.6 sec - 25 sec
e discrete; 1, 2, 3, 4, ..., 50 say

EXERCISE 10B

1 **a** 20 shops **b** 4 shops **c** **i** 10% **ii** 15%

2 **a** 17 games **b** **i** 76 points **ii** 137 points
c positively skewed **d** 10 times **e** $\approx 11.8\%$

3 **a**

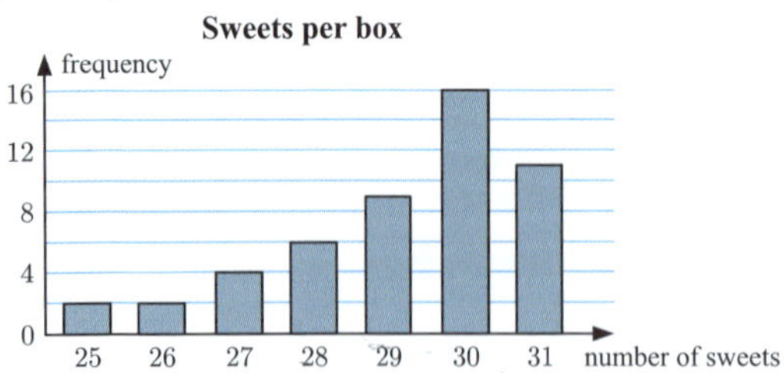

b negatively skewed

4 **a**

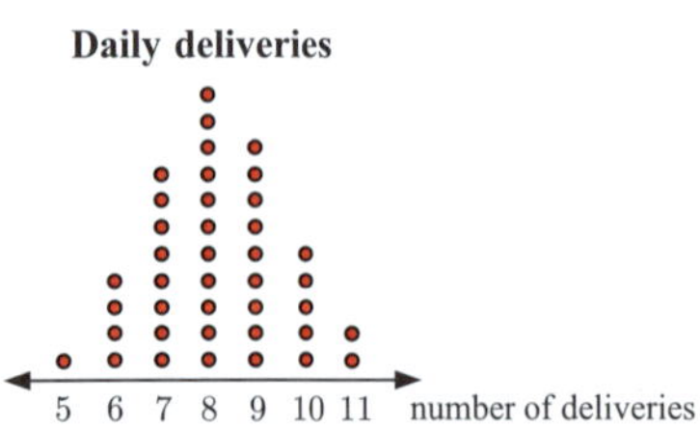

b 7 days **c** approximately symmetric

5 **a**

Number of goals	*Tally*	*Frequency*
0	𝍸	5
1	𝍸 |||	8
2	𝍸 ||	7
3	𝍸	5
4					3
5			1
6		0
7			1
	Total	30

b

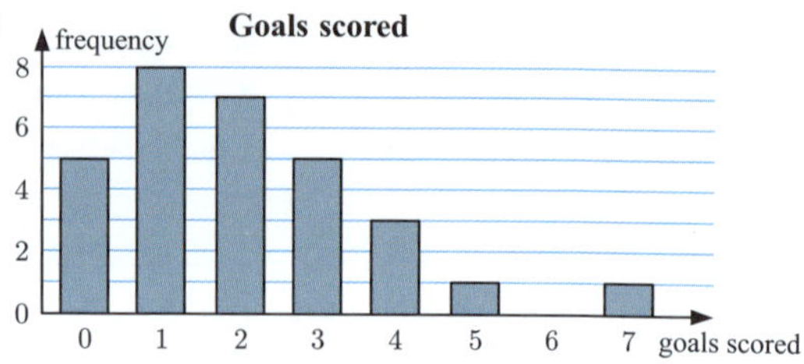

c positively skewed

6 **a** **i** minimum: 27, maximum: 81
ii approximately symmetric
b **i** minimum: 63, maximum: 152
ii positively skewed
c **i** minimum: 1.4, maximum: 8.7
ii negatively skewed

7 **a** **Test results**

Stem	*Leaf*
0	7
1	8
2	7 9
3	2 2 5 6 8 9 9
4	0 3 4 4 5 6 6 7 8

Scale: 1 | 8 = 18

b 9 students **c** 20% **d** 90% **e** negatively skewed

8 **a** **Number of peanuts in a jar**

Stem	*Leaf*
17*	8
18	2
18*	5 5 6 8 9 9
19	1 2 3 3 4 4 4 4
19*	5 5 6 6 7 8 8 9
20	0 1 1 1 1 1 3 3 3 3 4 4 4 4
20*	5 6 7 7 7 8 8 9
21	0 0 3 3 3
21*	5 6 7
22	1 4 4
22*	7 8
23	1

Scale: 18 | 2 = 182
18*| 5 = 185

a 60% **b** $\approx 13.3\%$ **c** approximately symmetric
d no, $\approx 88.3\%$

EXERCISE 10C

1 **a** 700 vehicles **b** $\approx 25.7\%$ **c** $\approx 15.7\%$
d \$22 680

2 **a** Height can take any value within each class.

b **Height of basketballers**

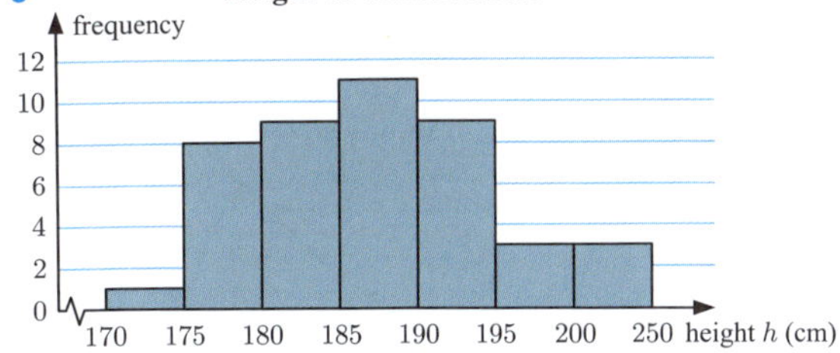

c $185 \leqslant h < 190$ cm. This class has the highest frequency.
d approximately symmetric

3 **a** 60 students **b** 9 students

c **Time to bake a cake**

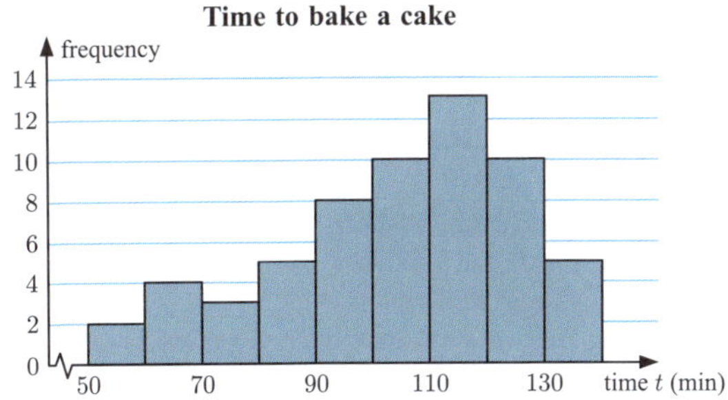

d $110 \leqslant t < 120$ min **e** negatively skewed **f** 25%

4 **a**

Water usage w (kL)	*Frequency*
$7 \leqslant w < 8$	5
$8 \leqslant w < 9$	9
$9 \leqslant w < 10$	10
$10 \leqslant w < 11$	7
$11 \leqslant w < 12$	4
$12 \leqslant w < 13$	2
$13 \leqslant w < 14$	1
Total	38

b **Monthly water usage**

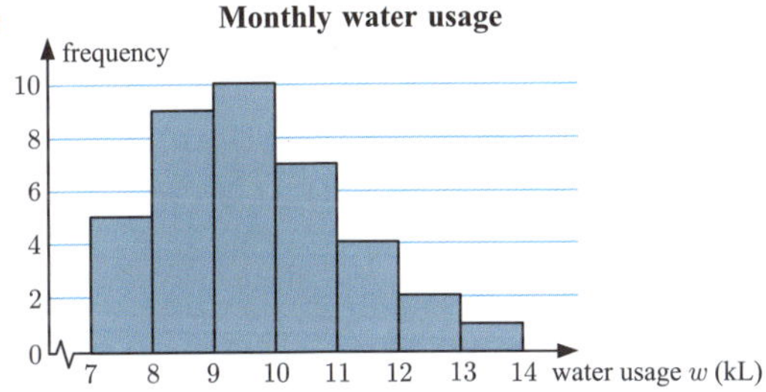

c $9 \leqslant w < 10$ kL **d** positively skewed **e** $\approx 36.8\%$

EXERCISE 10D.1

1 **a** 6 **b** 26.2 **c** ≈ 5.85
2 **a** 29 **b** 108 **c** 149.5
3 **a** mean = £363 770, median = £347 200
The mean is affected by extreme values whereas the median is not. So the mean has been 'pushed up' higher than the median by the higher values.
b **i** mean selling price **ii** median selling price
4 mean = 5.6 presents, median = 6 presents
5 mean $\approx$ 34.6 chocolate almonds, median = 35 chocolate almonds
6 **a** mean = 7.2, median = 7.5 **b** mean = 41, median = 38
7 **a** **A**: positively skewed; **B**: negatively skewed; **C**: approximately symmetric
b **A**: mean $\approx$ 3.29, median = 3;
B: mean = 5.6, median = 6;
C: mean = 4.45, median = 4.5
c **i** "For positively skewed data, the mean is *greater* than the median."
ii "For negatively skewed data, the mean is *less* than the median."
iii "For symmetric data, the mean and median are approximately *equal*."
8 105.6 **9** 1712 km **10** \$2 592 000
11 **a** $x = 9$ **b** $x = 12$ **12** 27 marks out of 30
13 $\approx$ 17.3 goals per match **14** ≈ 17.7

EXERCISE 10D.2

1 **a** **Distance travelled by football**

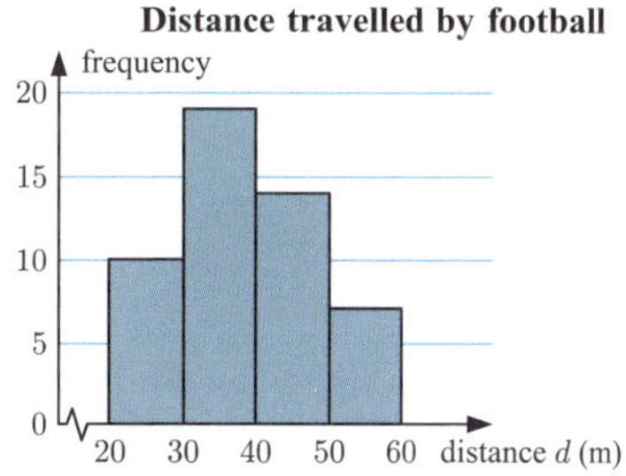

b No; as the actual data values are lost when combining them in classes.

c

Distance d (m)	*Frequency*	*Interval midpoint*	*Product*
$20 \leqslant d < 30$	10	25	250
$30 \leqslant d < 40$	19	35	665
$40 \leqslant d < 50$	14	45	630
$50 \leqslant d < 60$	7	55	385
Total	50		1930

d $\approx$ 38.6 m

2 **a**

Number of times	*Frequency*	*Interval midpoint*	*Product*
0 - 4	19	2	38
5 - 9	24	7	168
10 - 14	10	12	120
15 - 19	5	17	85
20 - 24	2	22	44
Total	60		455

b $\overline{x} \approx 7.58$ times

3 **a** $600 \leqslant A < 700$ m^2 **b** $\overline{x} \approx 661$ m^2

4 **a** positively skewed

b

Weight w (kg)	*Frequency*	*Interval midpoint*	*Product*
$0 \leqslant w < 10$	10	5	50
$10 \leqslant w < 20$	25	15	375
$20 \leqslant w < 30$	20	25	500
$30 \leqslant w < 40$	15	35	525
$40 \leqslant w < 50$	10	45	450
$50 \leqslant w < 60$	10	55	550
$60 \leqslant w < 70$	5	65	325
$70 \leqslant w < 80$	5	75	375
Total	100		3150

c $\overline{x} \approx 31.5$ kg

5 **a**

Time t (min)	*Frequency*	*Interval midpoint*	*Product*
$10 \leqslant t < 15$	3	12.5	37.5
$15 \leqslant t < 20$	1	17.5	17.5
$20 \leqslant t < 25$	6	22.5	135
$25 \leqslant t < 30$	4	27.5	110
$30 \leqslant t < 35$	5	32.5	162.5
$35 \leqslant t < 40$	5	37.5	187.5
Total	24		650

b Estimated mean ≈ 27.1 min

c Exact mean $= 27.0875 \approx 27.1$ min. The estimated mean and exact mean are the same to 1 decimal place.

EXERCISE 10E

1 **a** 60 students **b** **i** ≈ 20 students **ii** ≈ 15 students

2 **a**

Height h (cm)	*Frequency*	*Cumulative frequency*
$120 \leqslant h < 130$	5	5
$130 \leqslant h < 140$	7	12
$140 \leqslant h < 150$	12	24
$150 \leqslant h < 160$	10	34
$160 \leqslant h < 170$	6	40

b 40 dancers

c Height of dancers

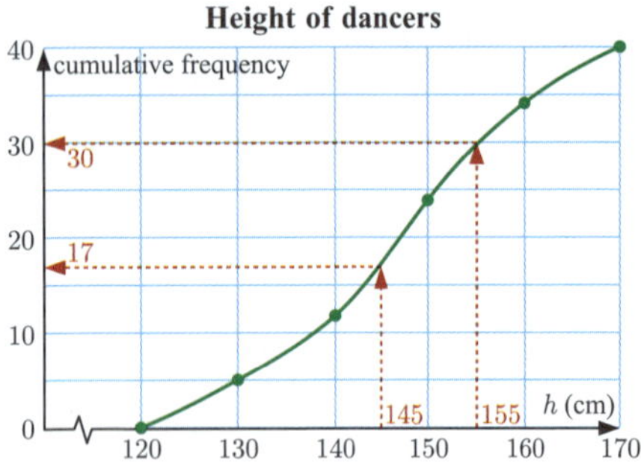

d **i** ≈ 30 dancers **ii** ≈ 23 dancers

3 **a**

Speed v (km/h)	*Frequency*	*Cumulative frequency*
$200 \leqslant v < 220$	12	12
$220 \leqslant v < 240$	16	28
$240 \leqslant v < 260$	21	49
$260 \leqslant v < 280$	18	67
$280 \leqslant v < 300$	13	80

b Speed of planes

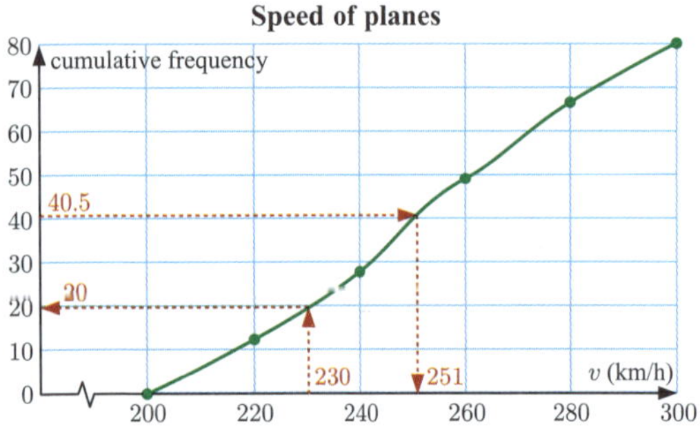

c ≈ 60 planes **d** ≈ 251 km/h

4 **a** $\overline{x} \approx 75.8$ kg

b

Weight w (kg)	*Frequency*	*Cumulative frequency*
$50 \leqslant w < 60$	14	14
$60 \leqslant w < 70$	19	33
$70 \leqslant w < 80$	28	61
$80 \leqslant w < 90$	23	84
$90 \leqslant w < 100$	16	100

c Weight of boxers

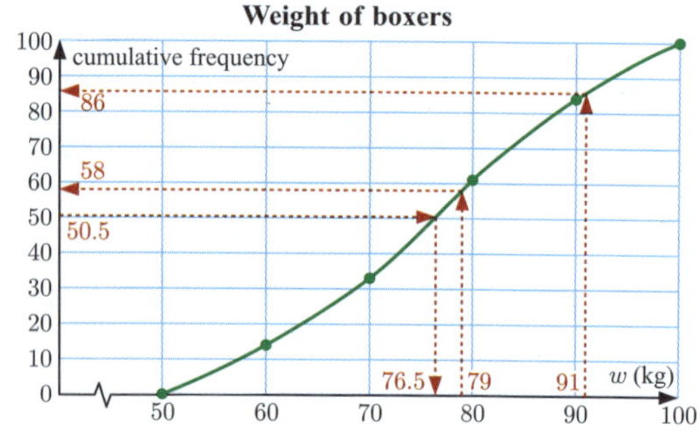

d ≈ 28 boxers **e** ≈ 76.5 kg

EXERCISE 10F

1 **a** 19 **b** 38 **c** 6.9 **2** 12 people

3 **a** 11 **b** $Q_1 = 7$, $Q_3 = 14$ **c** IQR $= 7$

4 **a** \$120 **b** $Q_1 = \$90$ **c** $Q_3 = \$150$ **d** IQR $= \$60$

5 **a** **i** 1.7 min **ii** 1.1 min **iii** 2.4 min

b range $= 2.7$ min, IQR $= 1.3$ min

c **i** "50% of the waiting times were greater than 1.7 minutes."

ii "75% of the waiting times were less than 2.4 minutes."

iii "The minimum waiting time was 0.7 minutes and the maximum waiting time was 3.4 minutes. The waiting times were spread over 2.7 minutes."

6 **a** 6 cm **b** 10.1 cm **c** 8.2 cm **d** 7.3 cm **e** 8.95 cm **f** 4.1 cm **g** 1.65 cm

7 **a** 64 cars **b** 223 people

c range $= 5$ people, IQR $= 2.5$ people

8 **a** range $= 27$, IQR $= 12$ **b** range $= 38$, IQR $= 12$

c range $= 47$, IQR $= 14.5$ **d** range

EXERCISE 10G

1 **a** min $= 3$, $Q_1 = 7.5$, med $= 11$, $Q_3 = 13$, max $= 15$

b 12 **c** 5.5

2 **a** **i** 98 marks **ii** 30 marks **b** 73 marks

c 68 marks **d** 75% **e** 22 marks **f** no

3 **a** min $= 7$, $Q_1 = 9.5$, med $= 12$, $Q_3 = 14.5$, max $= 17$

b

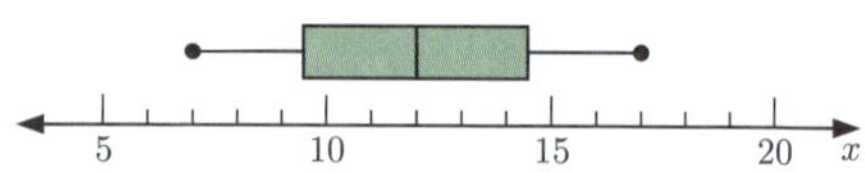

c **i** 10 **ii** 5

4 **a** median $= 6$ beans, $Q_1 = 5$ beans, $Q_3 = 8$ beans

b

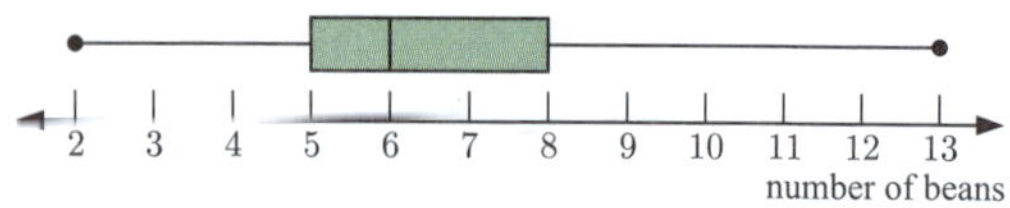

c range $= 11$ beans, IQR $= 3$ beans

5 **a** min $= 13$ tickets, $Q_1 = 17$ tickets, med $= 21$ tickets, $Q_3 = 24$ tickets, max $= 28$ tickets

b

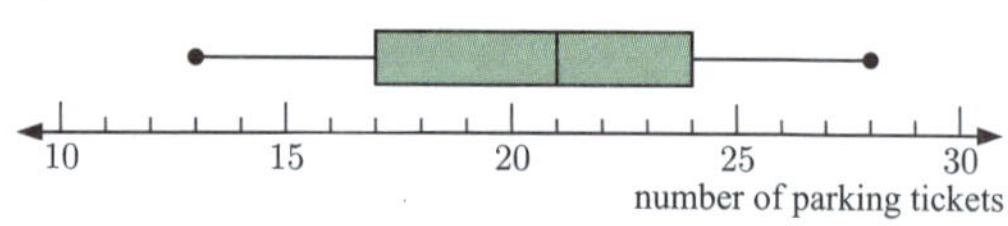

c The inspector handed out more tickets this month. The minimum, Q_1, median, Q_3, and maximum have all increased from last month.

REVIEW SET 10A

1 **a** discrete numerical **b** categorical
c continuous numerical

2 **a** Swim times

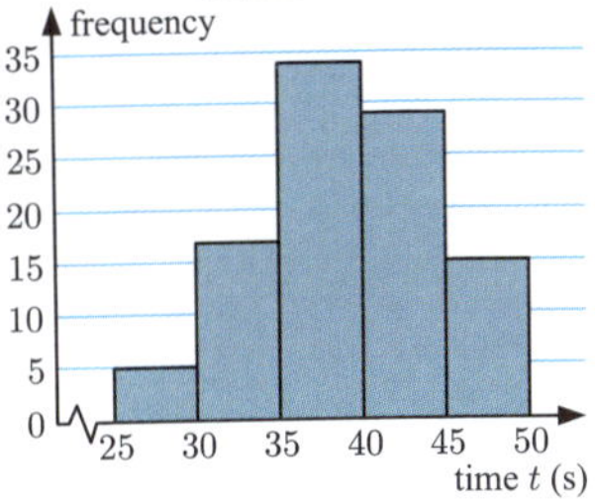

b $35 \leqslant t < 40$ sec **c** 39.1 seconds **d** 22%

3 **a** $\overline{x} \approx 29.6$ points **b** 29 points **c** 45 points
d 19.5 points

4 **a** 49 businesses **b** 15 businesses **c** $\approx 26.5\%$
d positively skewed **e** $\overline{x} \approx 3.51$ employees

5 15

6 **a** 12 **b** 43 **c** 29 **d** 20
e 34.5 **f** 31 **g** 14.5

7 **a** **i** $\overline{x} \approx 8.47$ **ii** 9 **iii** 4
b This class performed better than the national average.
c negatively skewed

8 **a**

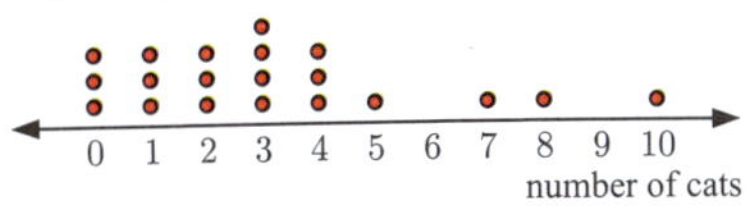

b mean = 3.15 cats, median = 3 cats

9 **a** 4 letters
b min = 6 letters, $Q_1 = 7$ letters, med = 8 letters, $Q_3 = 9$ letters, max = 10 letters
c

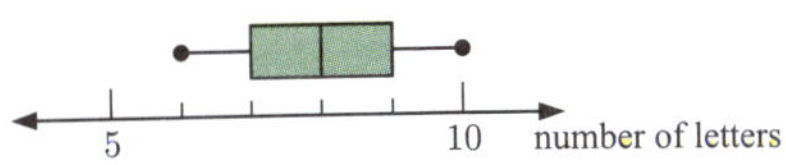

10 **a** **i** ≈ 19 values **ii** ≈ 13 values **b** ≈ 25

REVIEW SET 10B

1 **a** number of children per household
b discrete; the number of children is counted and can only take distinct values.
c

Children	*Frequency*
1	5
2	7
3	4
4	2
5	1
6	0
7	0
8	1

d Number of children per household

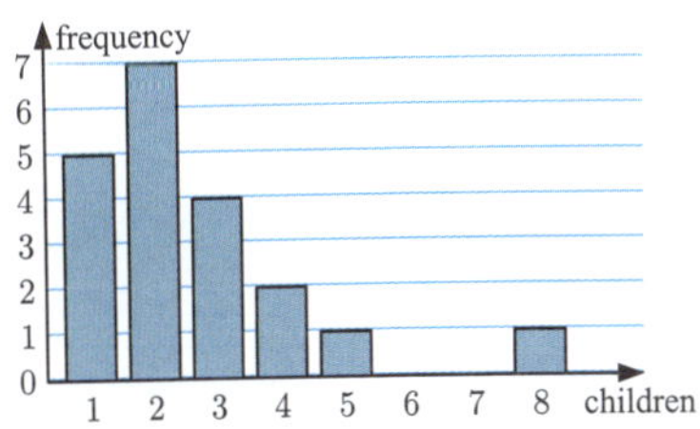

e positively skewed
f 2 children

2 ≈ 8.53 potatoes per plant **3** $x = 7$

4 **a** median = 101.5, $Q_1 = 98$, $Q_3 = 105.5$
b 7.5, this is the range of the middle half of the data.

5 **a**

0	4 8 9
1	3 5 7
2	1 5 6
3	1 4 5 6 7 9
4	1

Scale: $1 \mid 3 = 1.3$ cm

b 62.5%

6 20.7 cm

7 **a** Number of goals scored

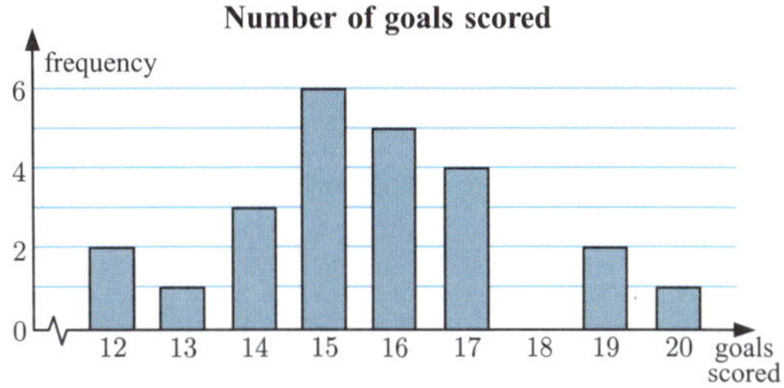

b **i** $\overline{x} \approx 15.6$ goals **ii** 15.5 goals **iii** 8 goals

8 **a** 2 kg
b Mass of domestic cats

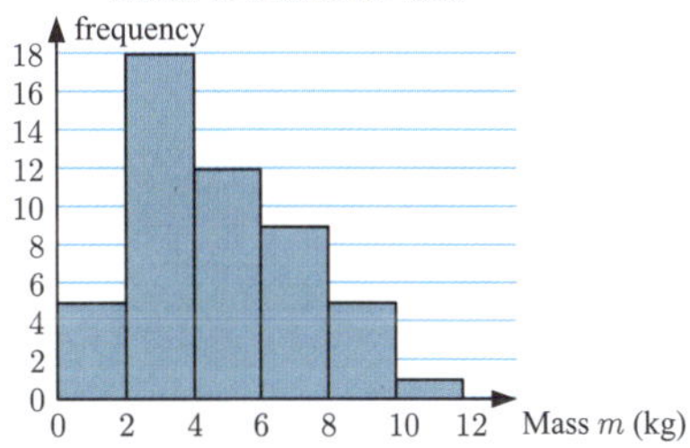

c $2 \leqslant m < 4$ kg **d** ≈ 4.76 kg **e** 12%

9 **a** Ages of children at a scouts camp

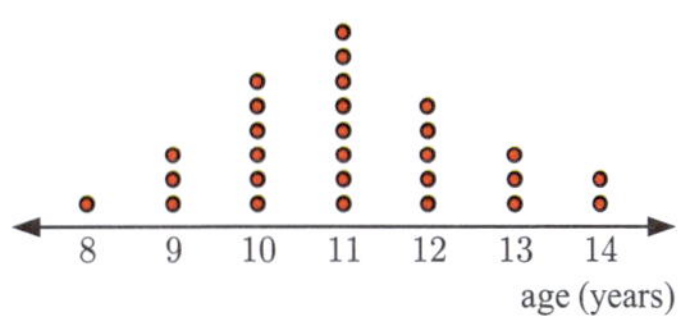

b approximately symmetric
c min = 8 years, $Q_1 = 10$ years, med = 11 years, $Q_3 = 12$ years, max = 14 years
d

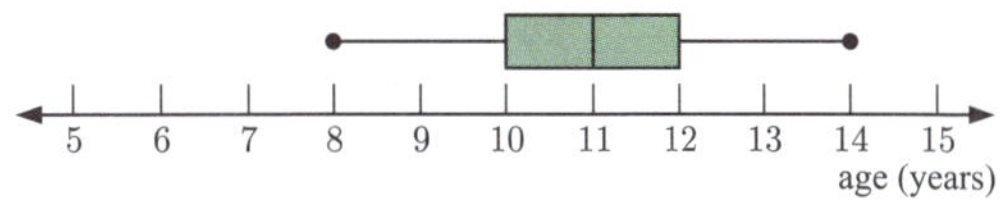

10 **a**

Time t (min)	*Frequency*	*Cumulative frequency*
$175 \leqslant t < 190$	11	11
$190 \leqslant t < 205$	35	46
$205 \leqslant t < 220$	74	120
$220 \leqslant t < 235$	32	152
$235 \leqslant t < 250$	8	160

b Cumulative frequency graph of Ben Nevis climb data

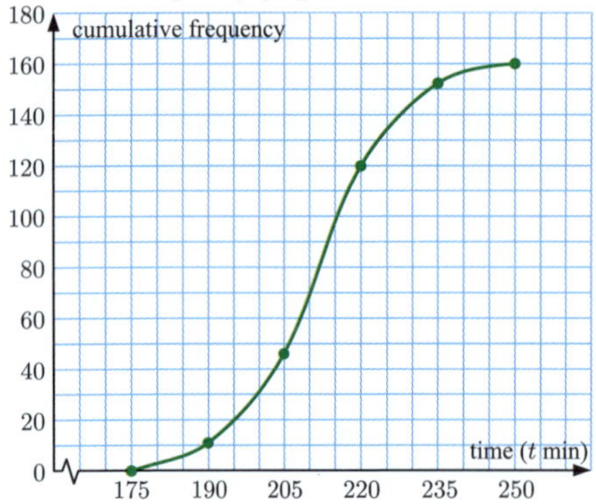

c ≈ 212 min **d** ≈ 20 trampers

EXERCISE 11A

1 **a** 1 **b** $\frac{1}{2}$ **c** 2 **d** $1\frac{3}{5}$ **e** 1 **f** $4\frac{1}{3}$
g 2 **h** -1 **i** 2 **j** 1

2 **a** -2 **b** 3 **c** 1 **d** -1 **e** 5 **f** 1
g $-\frac{1}{2}$ **h** $-\frac{1}{3}$ **i** -1 **j** $-\frac{1}{2}$ **k** $\frac{1}{2}$ **l** 5
m -5 **n** $1\frac{1}{4}$ **o** $-\frac{7}{9}$

EXERCISE 11B.1

1 **a** $\dfrac{10}{a}$ **b** $2x$ **c** 3 **d** $\frac{1}{2}$ **e** c **f** $\dfrac{1}{b}$
g $\dfrac{y}{z}$ **h** a **i** $\dfrac{1}{a}$ **j** 1 **k** $\dfrac{2}{3y^2}$ **l** $\dfrac{4y^2}{x}$

2 **a** $-\dfrac{a}{2}$ **b** $-\dfrac{1}{2a}$ **c** $\dfrac{a^2}{4}$ **d** -4 **e** $\dfrac{2}{t}$ **f** $-2d$
g $-\dfrac{b}{2}$ **h** $-\dfrac{2b}{3a}$

3 **a** $25a$ **b** $5a^2$ **c** $\dfrac{5}{a}$ **d** $-4a$ **e** $\dfrac{3b^2}{2}$ **f** $2c^2$
g 9 **h** e^3

4 **a** cannot be simplified **b** n **c** cannot be simplified
d 1 **e** cannot be simplified

5 **a** $\dfrac{a}{2}+2$ **b** $a+2$ **c** $x+\frac{3}{2}$ **d** $x+4$
e $2x+\frac{3}{2}$ **f** x^2+2x **g** $\dfrac{x^2}{2}+x$ **h** $1-\dfrac{6}{x}$
i $x-6$ **j** $\dfrac{x^2}{6}-x$

6 **a** $5 \neq 14$, $\therefore$ false $\dfrac{3x+8}{4}=\dfrac{3x}{4}+2$
b $3=3$, $5=5$, $7=7$ $\therefore$ likely to be correct
c $7\frac{1}{2} \neq 9$, $\therefore$ false $\dfrac{4x+3}{2}=2x+\frac{3}{2}$
d $3 \neq 7$, $\therefore$ false $\dfrac{2a+b+10}{5}=\dfrac{2a+b}{5}+2$

7 **a** $a+3$ **b** $2(x+2)$ **c** $2(c+3)$
d $4(b+5)$ **e** $\dfrac{d-3}{3}$ **f** $\dfrac{1}{2(z+3)}$
g $\dfrac{2}{x+1}$ **h** $\dfrac{3}{2-x}$ **i** $\dfrac{3}{2(1+x)}$

8 **a** 5 **b** $\frac{1}{3}$ **c** 2
d $\frac{3}{4}$ **e** $\dfrac{x+y}{3}$ **f** $\dfrac{y}{3}$
g $\dfrac{y+2}{3}$ **h** $\dfrac{3(p+q)}{2}$ **i** $\dfrac{x+1}{3}$

EXERCISE 11B.2

1 **a** $x+5$ **b** cannot be simplified **c** cannot be simplified **d** $x-3$
e $6a+2$ **f** cannot be simplified **g** cannot be simplified **h** $b-2$
i $x-5$ **j** cannot be simplified **k** cannot be simplified **l** $2y+1$
m $2x+8$ **n** $x+4$ **o** $\dfrac{x+4}{2}$ **p** $x+2$

2 **a** $\frac{2}{3}$ **b** $\frac{2}{3}$ **c** 2 **d** $1\frac{2}{3}$ **e** 7 **f** $\frac{1}{2}$ **g** x
h a **i** 3 **j** $a+c$ **k** b **l** $a+c+1$

3 **a** $x+3$ **b** $x-2$ **c** $3x$ **d** $5x$ **e** 4 **f** $3\frac{1}{2}$
g $2x$ **h** x **i** x^2-1 **j** x **k** b^2 **l** $\dfrac{y}{2}$

4 **a** -1 **b** cannot be simplified **c** -2 **d** $-1\frac{1}{3}$
e $-\frac{1}{3}$ **f** $-\frac{1}{4}$ **g** $-\dfrac{1}{r}$ **h** $-\dfrac{y}{2}$

5 **a** $x+1$ **b** $a-1$ **c** $\dfrac{1}{2-m}$ **d** $x-y$
e $\dfrac{x-1}{3}$ **f** $-(x+4)$ **g** $\dfrac{a+b}{b}$ **h** $-\dfrac{x}{y}$

6 **a** $x+1$ **b** $-(x+y)$ **c** -1 **d** $x-2$
e $\dfrac{1}{x-5}$ **f** $x+7$ **g** $\dfrac{x+6}{x+4}$ **h** $\dfrac{x+2}{x-4}$
i $\dfrac{x-3}{x}$ **j** $\dfrac{1-x}{x+3}$ **k** $\dfrac{2(x-3)}{x+2}$ **l** $\dfrac{x+4}{x-4}$
m $\dfrac{x+5}{x-8}$ **n** $\dfrac{4-x}{x}$ **o** $\dfrac{x+5}{5-x}$

EXERCISE 11C

1 **a** $\dfrac{x}{4}$ **b** $\dfrac{b}{2}$ **c** $\dfrac{2}{a}$ **d** $\dfrac{x^2}{6}$

2 **a** $\frac{1}{2}$ **b** $3x$ **c** $\dfrac{12}{x}$ **d** $4n$ **e** $\dfrac{m}{7}$ **f** $\dfrac{3a}{4b}$
g 1 **h** 2 **i** $\dfrac{ab}{5}$ **j** $\dfrac{x^2}{9}$ **k** 1 **l** $2x$

3 **a** $\dfrac{m^2}{4}$ **b** $\dfrac{4}{c}$ **c** $\dfrac{8}{x}$ **d** $\dfrac{2}{b^2}$ **e** $\dfrac{a^2}{8}$ **f** $\dfrac{1}{m^3}$
g $\frac{2}{3}$ **h** $\dfrac{6}{a^2}$ **i** 8 **j** 3 **k** $\frac{1}{2}$ **l** 1

EXERCISE 11D

1 **a** a **b** $\dfrac{a}{b^2}$ **c** $\dfrac{b^2}{a}$ **d** a **e** $\frac{4}{3}$ **f** $\dfrac{c^2}{12}$
g $\frac{3}{4}$ **h** $\dfrac{12}{c^2}$ **i** a^2 **j** $\dfrac{1}{3x}$ **k** $5x$ **l** $\dfrac{1}{b^2}$

2 **a** $6a$ **b** $\dfrac{5}{2x}$ **c** $\dfrac{c}{b^2}$ **d** $\dfrac{16}{ab}$
e $\dfrac{x^3}{3}$ **f** $\dfrac{10z^2}{y^2}$ **g** $\dfrac{7a^3}{3b^3}$ **h** $\dfrac{9}{2y^3}$

EXERCISE 11E

1 **a** $\dfrac{2b}{3}$ **b** $\dfrac{a+4}{5}$ **c** $\dfrac{2x+1}{3}$ **d** $\dfrac{5b}{6}$
e $\dfrac{a}{6}$ **f** $\dfrac{2a+x}{12}$ **g** $\dfrac{17a}{24}$ **h** $\dfrac{3a-4b}{12}$

i $\frac{17n}{10}$ **j** $\frac{-11p}{21}$ **k** $\frac{17k}{12}$ **l** $\frac{-s}{8}$

2 **a** $\frac{n+3}{3}$ **b** $\frac{20+y}{4}$ **c** $\frac{14-d}{2}$ **d** $\frac{4n}{3}$
e $\frac{3x}{2}$ **f** $\frac{m-6}{3}$ **g** $\frac{3-2x}{3}$ **h** $\frac{9x}{4}$

3 **a** **i** $\$2x$ **ii** $\$\frac{2x}{3}$ **b** $\$\frac{11x}{3}$
c **i** $11 **ii** $55 **iii** $27.50

4 **a** $\frac{5}{x}$ **b** $\frac{5}{2x}$ **c** $\frac{x+2}{2x}$ **d** $\frac{2-x}{2x}$

5 **a** $\frac{5q+3p}{pq}$ **b** $\frac{7s+2r}{rs}$ **c** $\frac{4q+3p}{pq}$ **d** $\frac{3t}{2s}$
e $\frac{ac+2b}{bc}$ **f** $\frac{9}{2b}$ **g** $\frac{5a+3b}{5b}$ **h** $\frac{3a+c}{3b}$

6 **a** $\frac{3+a}{a}$ **b** $\frac{b-2}{b}$ **c** $\frac{c^2+1}{c}$ **d** $\frac{2-3d}{d}$
e $\frac{4p+3}{p}$ **f** $\frac{7q-2}{q}$ **g** $\frac{24r+1}{3r}$ **h** $\frac{8s-3}{2s}$

7 **a** $\frac{13x}{12}$ **b** $\frac{11y}{6}$ **c** $\frac{4}{p}$ **d** $\frac{13}{6q}$
e $\frac{2y+5x}{xy}$ **f** $-\frac{2}{3x}$ **g** $\frac{19a}{6}$ **h** $\frac{5x}{2}$

8 **a** $\frac{3a+2}{a^2}$ **b** $\frac{5+b}{b^2}$ **c** $\frac{6d-1}{d^2}$ **d** $\frac{5-4x}{x^2}$
e $\frac{2+x^2}{2}$ **f** $\frac{x^2+1}{x^2}$ **g** $\frac{x^2+2}{x}$ **h** $\frac{x^3+4}{x^2}$
i $\frac{20-a^2}{4}$ **j** $\frac{6b^2-2}{b^2}$ **k** $\frac{2x+x^2}{2}$ **l** $\frac{x^3-1}{x^2}$
m $\frac{4x^2+2}{x}$ **n** $\frac{6x^2-1}{x}$ **o** $\frac{3x^2-4}{4x}$ **p** $\frac{1+2x^2}{2x}$

REVIEW SET 11A

1 **a** $-\frac{3}{4}$ **b** -2 **c** 5 **2** **a** $\frac{a^2}{2}$ **b** $\frac{x^2}{2y}$ **c** $-4x^2$

3 **a** $\frac{c}{2}+3$ **b** $3x-\frac{1}{3}$ **c** $\frac{x^2}{2}-x$

4 **a** $x+5$ **b** 3 **c** -1 **5** **a** $\frac{1}{2}$ **b** $\frac{5k}{2}$ **c** $\frac{25}{p^2}$

6 **a** $\frac{5x}{2}$ **b** $x-5$ **c** $\frac{x+5}{2x}$ **7** **a** $\frac{1}{4}$ **b** $\frac{3}{t^2}$ **c** $\frac{a^3}{6}$

8 **a** $\frac{4x}{5}$ **b** $\frac{7x}{10}$ **c** $\frac{n+12}{2}$

9 **a** $\frac{10}{3x}$ **b** $\frac{5b-4a}{ab}$ **c** $\frac{x^3+4}{x}$

10 **a** $\frac{x+6}{3}$ **b** $\frac{6-x}{3}$ **c** $\frac{6}{a}$
d $\frac{3a^3}{50}$ **e** $\frac{1}{2x}$ **f** $\frac{5x+4}{2x^2}$

REVIEW SET 11B

1 **a** -2 **b** -1 **c** $-1\frac{1}{3}$ **2** **a** $4x^3$ **b** $\frac{xy}{5}$ **c** $\frac{4}{x}$

3 **a** 1 **b** cannot be simplified **c** $\frac{1}{b}$ **d** cannot be simplified

4 **a** $\frac{x-3}{2}$ **b** $\frac{3}{5}$ **c** $\frac{9(x+1)}{4}$

5 **a** $\frac{15}{x}$ **b** 12 **c** $\frac{n^3}{8}$ **6** **a** $\frac{3a}{b^2}$ **b** $6x$ **c** $\frac{2c^3}{d^2}$

7 **a** $\frac{x+2}{x-2}$ **b** $\frac{x-1}{x+3}$ **c** $\frac{x-5}{x-3}$

8 **a** $\frac{11a}{28}$ **b** $\frac{k-5}{5}$ **c** $\frac{t^2-4}{t}$

9 **a** $\frac{15}{2x}$ **b** $\frac{24+5a}{6a}$ **c** $\frac{2x^2-3}{3x}$

10 **a** $\frac{9a}{10}$ **b** $4x$ **c** $\frac{6}{ab}$
d $\frac{-2a}{5}$ **e** $\frac{8x+2}{x^2}$ **f** $\frac{3a+8b}{2ab}$

EXERCISE 12A.1

1 **a** 1.1 **b** 0.9 **c** 1.45 **d** 0.55 **e** 0.92
f 1.03 **g** 2 **h** 7 **i** 0

2 **a** 12% increase **b** 23% increase **c** 4% decrease
d 15% decrease **e** 45% increase **f** 33% decrease
g 140% increase **h** 70% decrease

3 **a** $52.50 **b** €28 **c** £624 **d** $78.20
e $3200 **f** €1125

4 **a** 25% increase **b** 20% decrease
c $\approx$ 3.53% decrease **d** 12.5% increase
e $14\frac{2}{3}$% increase **f** $\approx$ 31.4% decrease

5 $24.09 **6** $26 **7** $\approx$ 26.8% increase

8 $\approx$ 11.6% loss

EXERCISE 12A.2

1 **a** $33.60 **b** $77.50 **c** $162 **d** $416 **e** $57.60

2 $1.17 **3** **a** €3.68 per kg **b** €1.44 each **c** €2.04 each

4 £8.25 per kg **5** $594 **6** 80% mark-up

7 $17.59 per kg **8** $1.05

EXERCISE 12A.3

1 **a** $56 **b** €120 **c** $156 **d** ¥26 250

2 $17 **3** $697 **4** $392 **5** 8% discount

6 22.5% discount **7** $29 **8** $80

EXERCISE 12A.4

1 **a** £288 **b** £48 **2** **a** $2294.90 **b** $129.90

3 **a** €420 **b** €70

4 **a** HRK51 229.51 **b** HRK11 270.49

EXERCISE 12A.5

1 $37.51 **2** $504 **3** €42 **4** $560 **5** $1.40

EXERCISE 12B

1 €582.06 **2** **a** $51 392.88 **b** $\approx$ 28.5%

3 £2426.76 **4** $3060 **5** CNY25 056.30

6 $1655.36 **7** $\approx$ 1.38% decrease

8 $\approx$ 1.70% depreciation

9 **a** $267 980.06 **b** $\approx$ 0.748% depreciation

10 $\approx$ 1.32% appreciation

EXERCISE 12C

1 **a** $660 **b** €2062.50 **c** £3881.25 **d** $15 500

2 **a** €21 **b** $170 **c** ¥288 750 **d** $1625

3 $235.60 **4** **a** £529.64 **b** $12 929.64

5 **a** $6800 **b** $6280 **c** $4346.67

6 **a** $8695 **b** $144.92 **7** €268.75 **8** RUB2040

EXERCISE 12D.1

1 **a** \$29 775.40 **b** \$4775.40 **2** **a** £300 **b** £315.25
3 **a** 579.60 pesos **b** \$38.45 **c** \$2370.57

EXERCISE 12D.2

1 **a** \$7920 **b** \$8162.93 **2** **a** €15 750 **b** €21 295.66
3 **a** \$4615.87 **b** \$1615.87 **4** 6045.20 dirhams
5 **a** Option 1: \$10 160, Option 2: \$10 077.70
$\therefore$ choose Option 1
b yes, Option 1: \$11 600, Option 2: \$11 754.62

EXERCISE 12E

1 **a** €329.90 **b** €19 794 **c** €4794
2 **a** £606.50 **b** £21 834 **c** £3834
3 **a** €3792.80 **b** €9800
Gunter should try to pay the loan off in 2 years rather than 5 years so he pays much less interest overall.
4 **A** \$14 382 **B** \$9198
Trina should borrow from **B**, Cash Credit Union.

REVIEW SET 12A

1 \$241.80 **2** €45.22 **3** \$532 168.69 **4** £6273.18
5 \$2643.30 **6** **a** \$23 506.89 **b** \$21 156.20 **7** €1400
8 **a** 755 000 rupees **b** 10 486.11 rupees
9 **a** \$10 099.82 **b** \$2099.82
10 **a** ¥19 643 **b** ¥707 159 **c** ¥107 159

REVIEW SET 12B

1 \$41.65 **2** $\approx 12.7\%$ **3** €53.30 **4** \$80
5 **a** \$104.35 **b** \$15.65
6 **a** \$1824.98 **b** \$2298.95
c Yes, as the final value of the investment after 8 years would have been \$2776.40.
7 \$20 100
8 **a** 35 000 pesos **b** 135 000 pesos
c **i** 60 repayments **ii** 2250 pesos
9 Interest earned with option **A** would be €$0.325x$, whereas interest earned with option **B** would be €$0.288x$. So, choose option **A** for greater return.
10 **a** 300 000 rupees **b** 6448.20 rupees **c** 386 892 rupees
d 86 892 rupees

EXERCISE 13A

1 **a** **i** [BC] **ii** [AC] **iii** [AB]
b **i** [KM] **ii** [KL] **iii** [LM]
c **i** [PR] **ii** [QR] **iii** [PQ]
2 **a** **i** [PR] **ii** [QR] **iii** [PQ] **iv** [PQ] **v** [QR]
b **i** [AC] **ii** [AB] **iii** [BC] **iv** [BC] **v** [AB]
c **i** [LM] **ii** [MN] **iii** [LN] **iv** [LN] **v** [MN]

EXERCISE 13B

1 **a** **i** $\frac{3}{5}$ **ii** $\frac{4}{5}$ **iii** $\frac{3}{4}$ **b** **i** $\frac{5}{13}$ **ii** $\frac{12}{13}$ **iii** $\frac{5}{12}$
c **i** $\frac{15}{17}$ **ii** $\frac{8}{17}$ **iii** $\frac{15}{8}$ **d** **i** $\frac{1}{\sqrt{5}}$ **ii** $\frac{2}{\sqrt{5}}$ **iii** $\frac{1}{2}$
e **i** $\frac{4}{7}$ **ii** $\frac{\sqrt{33}}{7}$ **iii** $\frac{4}{\sqrt{33}}$
f **i** $\frac{8}{\sqrt{73}}$ **ii** $\frac{3}{\sqrt{73}}$ **iii** $\frac{8}{3}$
2 **a** PQ $\approx$ 59 mm, PR $\approx$ 54 mm, QR $\approx$ 24 mm
b **i** ≈ 0.407 **ii** ≈ 0.915 **iii** ≈ 0.444
c **i** ≈ 0.407 **ii** ≈ 0.914 **iii** ≈ 0.445
3 **a** **i** ≈ 0.587 **ii** ≈ 0.630 **iii** ≈ 1.378 **iv** ≈ 0.778
b **i** ≈ 0.588 **ii** ≈ 0.629 **iii** ≈ 1.376 **iv** ≈ 0.777
4 **a** **i** ≈ 0.9240 **ii** ≈ 0.0760 **iii** 1
iv ≈ 0.1786 **v** ≈ 0.8214 **vi** 1
b $\cos^2\theta + \sin^2\theta = 1$
5 **a** **i** 1 **ii** 1 **iii** 1
b $\tan\theta \times \tan(90° - \theta)$ is always 1
6

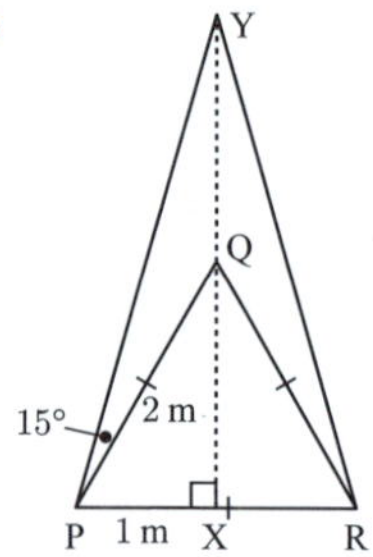

- Show $QX = \sqrt{3}$.
- Consider:

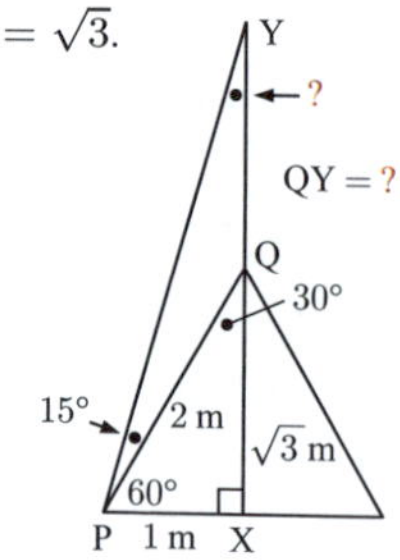

EXERCISE 13C

1 **a** $\sin 65° = \frac{x}{a}$ **b** $\cos 32° = \frac{x}{b}$ **c** $\tan 56° = \frac{x}{c}$
d $\cos 37° = \frac{d}{x}$ **e** $\tan 49° = \frac{e}{x}$ **f** $\tan 73° = \frac{f}{x}$
g $\sin 54° = \frac{g}{x}$ **h** $\tan 27° = \frac{x}{h}$ **i** $\cos 59° = \frac{i}{x}$
2 **a** $x \approx 17.62$ **b** $x \approx 8.91$ **c** $x \approx 6.88$
d $x \approx 16.73$ **e** $x \approx 14.98$ **f** $x \approx 22.94$
g $x \approx 23.47$ **h** $x \approx 36.05$ **i** $x \approx 7.42$
j $x \approx 13.77$ **k** $x \approx 13.07$ **l** $x \approx 20.78$
3 **a** $\theta = 62°$, $a \approx 12.4$, $b \approx 6.6$
b $\theta = 27°$, $a \approx 15.7$, $b \approx 7.1$
c $\theta = 52°$, $a \approx 54.6$, $b \approx 33.6$

EXERCISE 13D

1 **a** $\theta \approx 51.3°$ **b** $\theta \approx 41.8°$ **c** $\theta \approx 39.6°$
d $\theta \approx 38.0°$ **e** $\theta \approx 34.1°$ **f** $\theta \approx 47.6°$
g $\theta \approx 36.5°$ **h** $\theta \approx 35.8°$ **i** $\theta \approx 36.9°$
2 **a** $\theta \approx 38.7°$, $\phi \approx 51.3°$, $x \approx 6.2$
b $\alpha \approx 60.2°$, $\beta \approx 29.8°$, $x \approx 7.0$
c $a \approx 41.5°$, $b \approx 48.5°$, $x \approx 8.4$
3 These three triangles do not exist. The hypotenuse must be longer than the other sides.

EXERCISE 13E

1 ≈ 2.03 m **2** ≈ 26.0 m **3** ≈ 1.93 km **4** ≈ 4.14 m
5 $\approx 44.4°$ **6** $\approx 66.4°$ **7** ≈ 49.0 m **8** ≈ 30.3 m
9 **a** ≈ 10.3 m **b** ≈ 12.2 m **10** $\approx 55.2°$
11 $\approx 23.6°$ **12** **a** ≈ 5.33 m **b** ≈ 41.7 m^2 **13** $\approx 36.0°$
14 **a** **i** ≈ 7.98 m **ii** ≈ 12.6 m **b** $\theta \approx 21.1°$
15 ≈ 412 m **16** $\approx 80.3°$ **17** $\approx 64.0°$
18 **a** ≈ 47.0 cm **b** ≈ 82.9 cm^2 **c** ≈ 8.27 cm
d $\approx 17.1\%$
19 **a** **i** triangular slices each ≈ 41.6 cm^2,
kite slice ≈ 60.9 cm^2
ii no
b $x° \approx 33.7°$, $y° \approx 22.6°$, $z° \approx 33.7°$
20 $\theta \approx 15.6°$

EXERCISE 13F

1 **a**
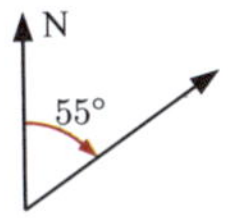

b
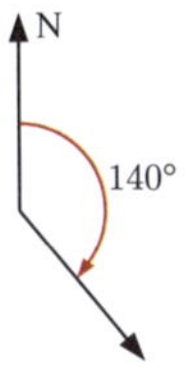

c
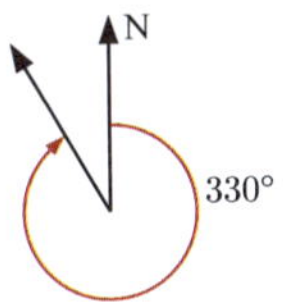

d
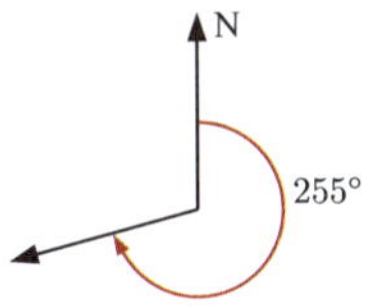

2 **a** 304° **b** 248° **c** 064° **d** 141°

3 **a** **i** 040° **ii** 130° **iii** 310° **iv** 090° **v** 220° **vi** 270°
b **i** 025° **ii** 150° **iii** 330° **iv** 105° **v** 205° **vi** 285°

4 **a** **i** 045° **ii** 225° **iii** 105° **iv** 285° **v** 135° **vi** 315°
b **i** 120° **ii** 300° **iii** 070° **iv** 250° **v** 035° **vi** 215°

5 **a** east, 090° **b** southeast, 135° **c** south, 180°
d west, 270° **e** southwest, 225° **f** northwest, 315°
g north, 0° **h** northeast, 045°

6 **a** ≈ 2.24 km, 063° **b** ≈ 3.16 km, 162°
c ≈ 3.16 km, 072° **d** ≈ 4.47 km, 207°
e ≈ 3.61 km, 304°

7 236° **8** **a** ≈ 55.0 m **b** ≈ 40.0 m

9 ≈ 14.3 km **10** 650 km

EXERCISE 13G

1 **a** ≈ 1.41 cm **b** $\approx 35.3°$

2 **a** 5 cm **b** $\approx 64.9°$ **3** **a** 13 cm **b** $\approx 47.3°$

REVIEW SET 13A

1 **a** [PQ] **b** [PR]

2 **a** **i** $\cos 42° \approx 0.743$ **ii** $\sin 64° \approx 0.899$
iii $\tan 48° \approx 1.114$ **iv** $\sin 26° \approx 0.438$
b **i** $\cos 42° \approx 0.743$ **ii** $\sin 64° \approx 0.899$
iii $\tan 48° \approx 1.111$ **iv** $\sin 26° \approx 0.438$

3 $\sin\theta = \frac{5}{13}$, $\cos\theta = \frac{12}{13}$, $\tan\theta = \frac{5}{12}$

4 **a** $\sin 29° = \dfrac{a}{b}$ **b** $\tan 76° = \dfrac{c}{b}$

5 **a** $\phi \approx 36.1°$ **b** $\phi \approx 48.4°$ **c** $\phi \approx 54.0°$

6 **a** $b \approx 6.17$ **b** $b \approx 3.01$ **c** $b \approx 14.5$

7 ≈ 0.585 m **8** ≈ 71.7 m **9** $\approx 9.21°$

10 $\approx 0.382°$ **11** ≈ 8.22 m **12** ≈ 29.4 km

13 **a**
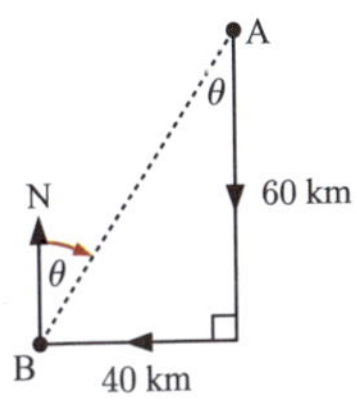

b ≈ 72.1 km
c 034°

14 **a** ≈ 11.2 cm **b** $\approx 32.1°$

REVIEW SET 13B

1 **a** p **b** r **c** r **d** p

2 **a** **i** $\frac{\sqrt{3}}{2}$ **ii** $\frac{\sqrt{3}}{2}$ **iii** $\frac{1}{2}$ **iv** $\frac{1}{2}$
b $\cos\theta = \dfrac{b}{c}$, $\sin(90° - \theta) = \dfrac{b}{c}$ $\therefore$ $\cos\theta = \sin(90° - \theta)$

3 **a** $\frac{5}{\sqrt{74}}$ **b** $\frac{7}{\sqrt{74}}$ **c** $\frac{5}{7}$

4 **a** $\theta \approx 38.7°$ **b** $\theta \approx 63.4°$ **c** $\theta \approx 50.5°$

5 **a** $a \approx 7.39$ **b** $a \approx 42.6$ **c** $a \approx 96.7$

6 **a** $y = \sqrt{12} = 2\sqrt{3}$ **b** [AB]
c **i** $\cos\theta = \frac{1}{2}$ **ii** $\sin\theta = \frac{\sqrt{3}}{2}$ **iii** $\tan\theta = \sqrt{3}$
d $\theta = 60°$, $\widehat{BAC} = 30°$

7 $\theta \approx 25.8°$ **8** ≈ 19.9 cm **9** ≈ 342 m

10 ≈ 77.9 m **11** Yes, by about 11 cm.

12 **a** ≈ 107 km **b** $\approx 306°$

13 **a** ≈ 3.16 km, 018° **b** ≈ 3.61 km, 146°
c ≈ 2.24 km, 063° **d** ≈ 4.12 km, 194°
e ≈ 3.16 km, 288°

14 **a** $\approx 56.3°$ **b** $\sqrt{29} \approx 5.39$ cm **c** $\approx 29.1°$

EXERCISE 14A

1 **a** 0.47 **b** 0.53 **2** **a** ≈ 0.78 **b** ≈ 0.22

3 0.325 **4** **a** ≈ 0.0324 **b** ≈ 0.291

EXERCISE 14B

1 **a** ≈ 0.0895 **b** ≈ 0.126

2 **a** ≈ 0.487 **b** ≈ 0.051 **c** ≈ 0.731

3 **a** ≈ 0.240 **b** ≈ 0.702 **c** ≈ 0.678

4 **a** 0.16 **b** 0.16 **c** 0.6

5 **a** 43 days **b** 127 calls
c **i** ≈ 0.0465 **ii** ≈ 0.186 **iii** ≈ 0.465

EXERCISE 14C

1 **a** {BB, BG, GB, GG}
b {ABC, ACB, BAC, BCA, CAB, CBA}
c {HHHH, HHHT, HHTH, HTHH, THHH, HHTT, HTHT, THHT, HTTH, THTH, TTHH, HTTT, THTT, TTHT, TTTH, TTTT}
d {ABCD, ABDC, ACBD, ACDB, ADBC, ADCB, BACD, BADC, BCAD, BCDA, BDAC, BDCA, CABD, CADB, CBAD, CBDA, CDAB, CDBA, DABC, DACB, DBAC, DBCA, DCAB, DCBA}

2 **a**
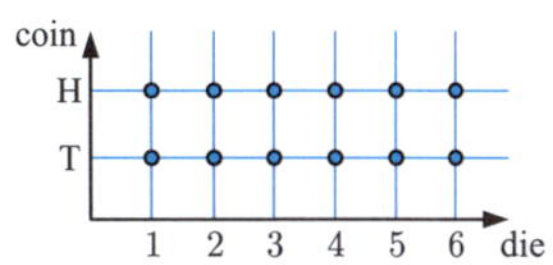

b 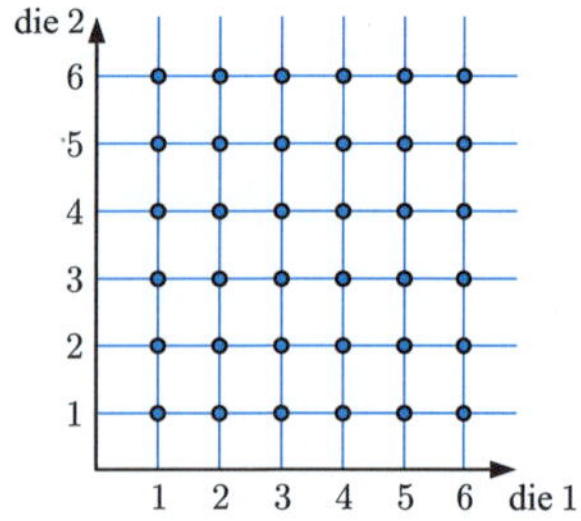

c 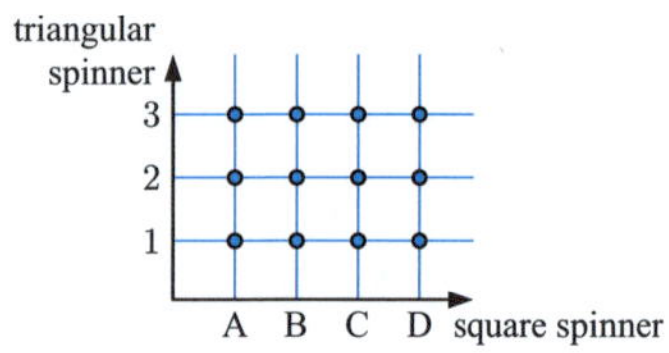

d 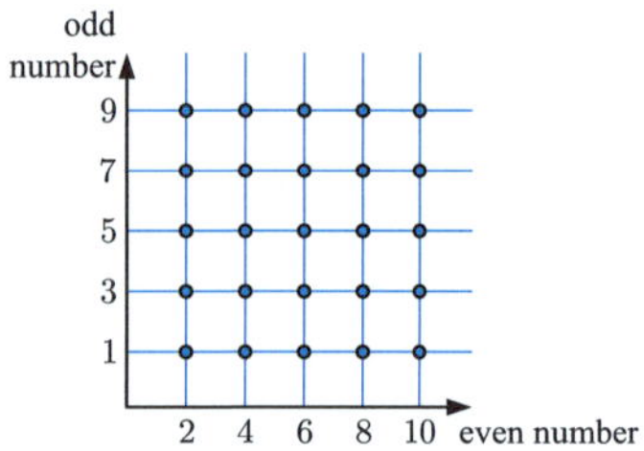

3 **a** 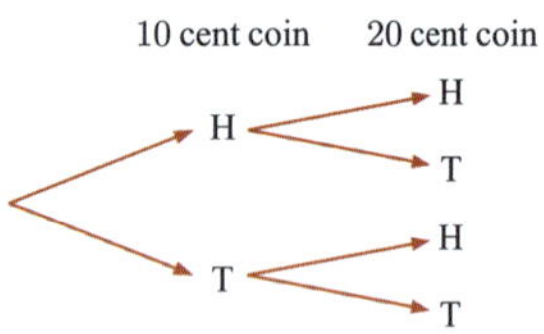

b 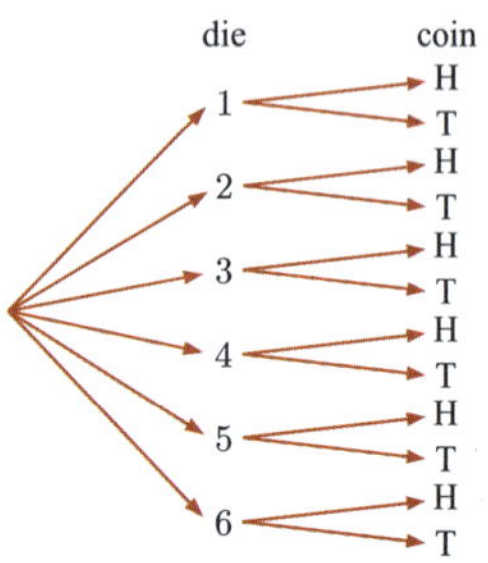

c 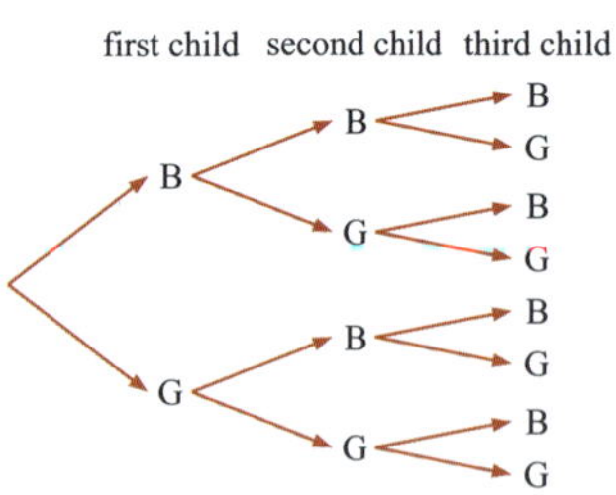

4 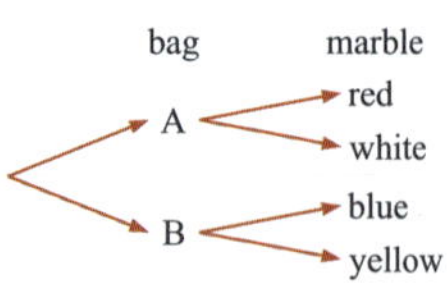

5 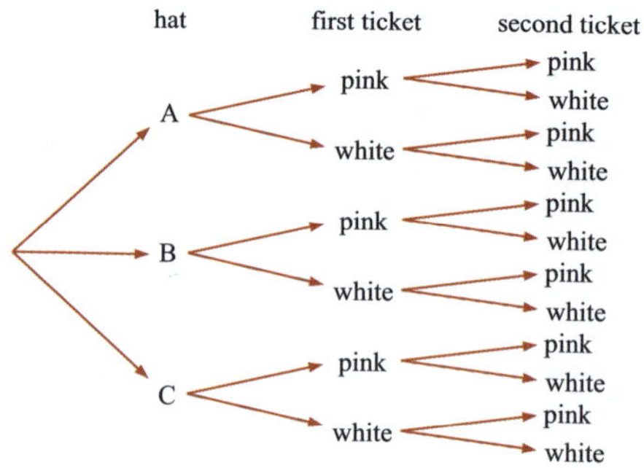

EXERCISE 14D

1 **a** $\frac{1}{5}$ **b** $\frac{2}{5}$ **c** $\frac{3}{5}$ **2** **a** $\frac{1}{8}$ **b** $\frac{1}{2}$ **c** 0

3 **a** $\frac{4}{7}$ **b** $\frac{3}{7}$ **c** 1

4 **a** $\frac{1}{13}$ **b** $\frac{1}{2}$ **c** $\frac{1}{4}$ **d** $\frac{16}{52} = \frac{4}{13}$

5 For a 4-year cycle: **a** $\frac{124}{1461}$ **b** $\frac{113}{1461}$

6 **a** E' is the event that the chosen muffin is not blueberry, that is, it is raspberry.

b $\text{P}(E) = \frac{1}{3}$ **c** $\text{P}(E') = \frac{2}{3}$

7 **a** **i** $\frac{3}{37}$ **ii** $\frac{8}{37}$ **iii** $\frac{26}{37}$

b 1. This is the sum of all possible outcomes. **c** $\frac{3}{11}$

8 **a** **i** $\frac{3}{14}$ **ii** $\frac{4}{7}$ **b** **i** $\frac{3}{13}$ **ii** $\frac{4}{13}$ **9** $\frac{1}{9}$

10 **a** 120 positions

b **i** $\frac{1}{2}$ **ii** $\frac{7}{15}$ **iii** $\frac{1}{24}$ **iv** $\frac{1}{12}$

11 The 11 possible outcomes are **not** equally likely to occur.
$\text{P}(10) = \frac{3}{36}$, $\text{P}(11) = \frac{2}{36}$, $\text{P}(12) = \frac{1}{36}$
$\therefore$ the probability of Sean winning is $\frac{6}{36} = \frac{1}{6}$.

12 $\frac{1}{\sqrt{2}}$ m

EXERCISE 14E

1 **a** 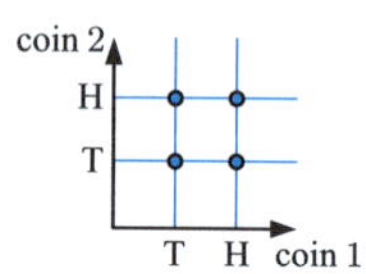

b **i** $\frac{1}{4}$ **ii** $\frac{1}{2}$ **iii** $\frac{3}{4}$

2 **a** 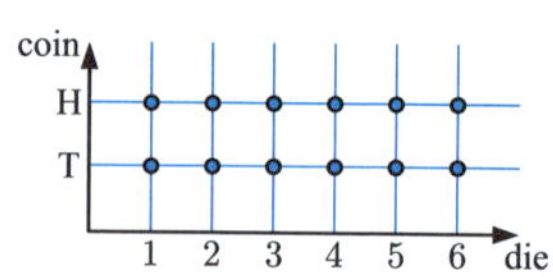

b **i** $\frac{1}{12}$ **ii** $\frac{7}{12}$ **iii** $\frac{2}{3}$ **iv** $\frac{5}{12}$ **v** $\frac{1}{4}$ **vi** $\frac{3}{4}$

3 **a** 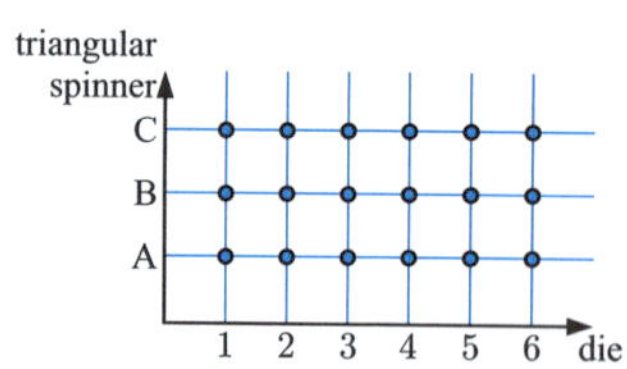

b **i** $\frac{1}{18}$ **ii** $\frac{1}{6}$ **iii** $\frac{2}{9}$

4 **a** $\frac{1}{36}$ **b** $\frac{1}{18}$ **c** $\frac{5}{9}$ **d** $\frac{11}{36}$ **e** $\frac{5}{18}$ **f** $\frac{25}{36}$
g $\frac{1}{6}$ **h** $\frac{2}{9}$ **i** $\frac{5}{18}$ **j** $\frac{13}{18}$

5 **a** $\frac{2}{49}$ **b** $\frac{24}{49}$ **c** $\frac{11}{49}$ **d** $\frac{3}{7}$ **6** $\frac{21}{64}$

EXERCISE 14F.1

1 $\frac{1}{10}$ **2** $\frac{1}{3}$ **3** $\frac{1}{8}$

4 **a** 0.18 **b** 0.28

5 **a** 0.03 **b** 0.68 **c** 0.17

6 **a** 0.343 **b** 0.147 **c** 0.063

EXERCISE 14F.2

1 **a** $\frac{7}{15}$ **b** $\frac{7}{30}$

2 **a** **i** $\frac{2}{15}$ **ii** $\frac{1}{3}$ **iii** $\frac{4}{15}$ **iv** $\frac{4}{15}$

b 1. This is the sum of all possible outcomes.

3 **a** **i** $\frac{2}{15}$ **ii** $\frac{1}{3}$ **b** $\frac{8}{15}$

4 **a** $\frac{1}{4950} \approx 0.000\,202$ **b** $\frac{49}{2475} \approx 0.0198$

c $\frac{4753}{4950} \approx 0.960$

5 **a** 0.56 **b** 0.12 **6** **a** $\frac{14}{55}$ **b** $\frac{1}{55}$

7 **a** $\frac{5}{28}$ **b** $\frac{1}{56}$ **8** **a** $\frac{8}{65}$ **b** $\frac{1}{13}$ **c** $\frac{28}{195}$

EXERCISE 14G

1 **a**

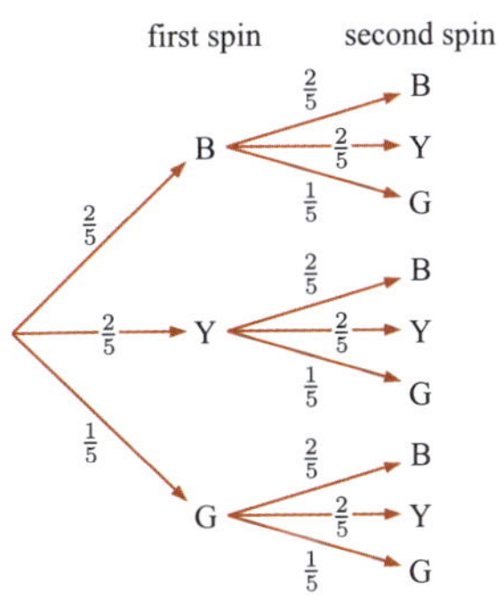

b **i** $\frac{4}{25}$ **ii** $\frac{1}{25}$ **iii** $\frac{16}{25}$ **iv** $\frac{16}{25}$

2 **a**

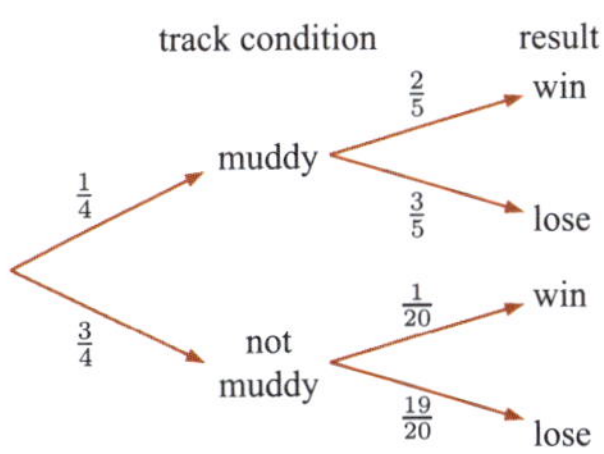

b $\frac{11}{80}$

3 0.752 **4** 0.034 **5** $\frac{23}{60}$

6 **a** $\frac{23}{42}$ **b** $\frac{19}{42}$ **7** $\frac{1}{5}$

8 **a**

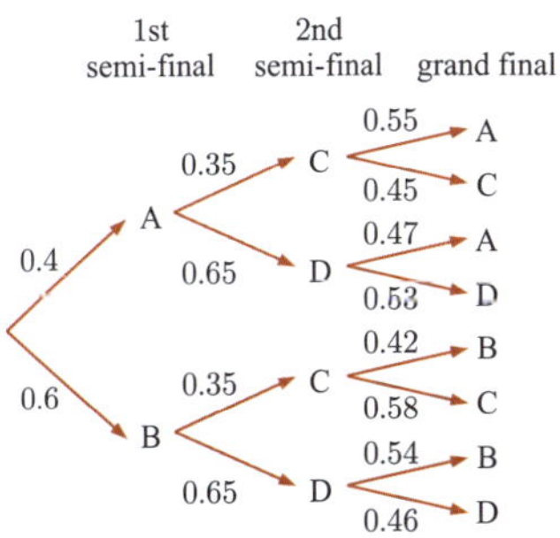

b 0.1992 **c** player D **d** player B

EXERCISE 14H

1 **a**

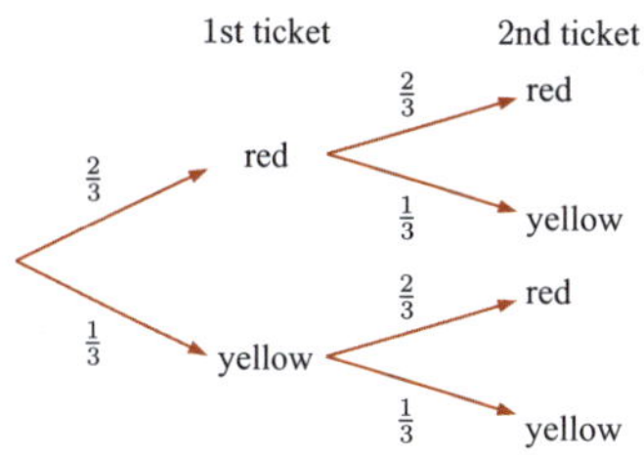

b **i** $\frac{4}{9}$ **ii** $\frac{1}{9}$ **iii** $\frac{2}{9}$ **iv** $\frac{4}{9}$

2 **a** $\frac{2}{7}$ **b** $\frac{1}{7}$ **c** $\frac{2}{7}$ **d** $\frac{4}{7}$

3 **a** **i** $\frac{1}{6}$ **ii** $\frac{5}{18}$ **iii** $\frac{5}{18}$ **iv** $\frac{5}{18}$

b These cases cover all possible outcomes, so their probabilities must add up to 1.

4 **a**

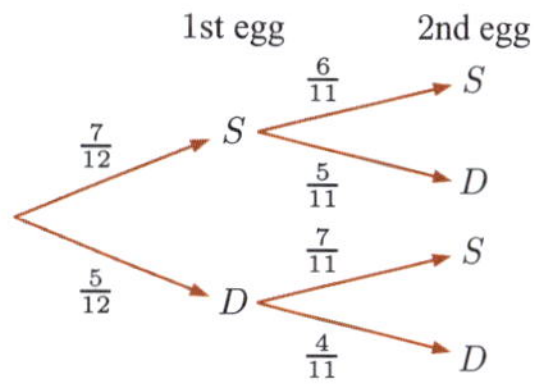

b **i** $\frac{5}{33}$ **ii** $\frac{7}{22}$

5 **a** **i** $\frac{5}{9}$ **ii** $\frac{4}{9}$

b **i** $\frac{25}{81}$ **ii** $\frac{16}{81}$ **iii** $\frac{20}{81}$ **iv** $\frac{40}{81}$

6 **a** $\frac{1}{25}$ **b** $\frac{24}{25}$ **c** $\frac{7}{4975}$ **d** $\frac{4584}{4975}$ **e** $\frac{8}{199}$

7 **a** without replacement

b

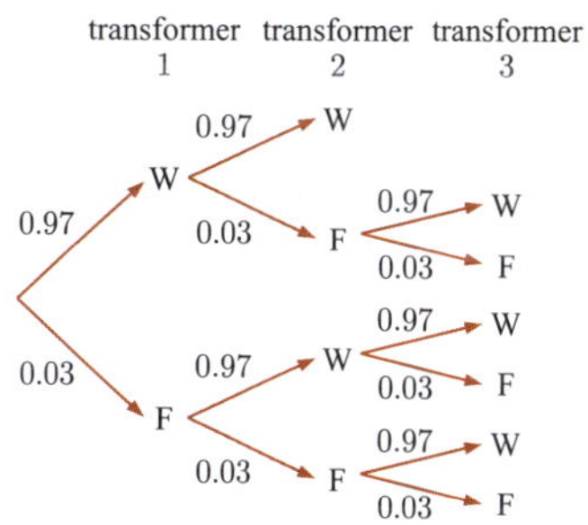

c $\approx 0.002\,65$

EXERCISE 14I

1 **a**

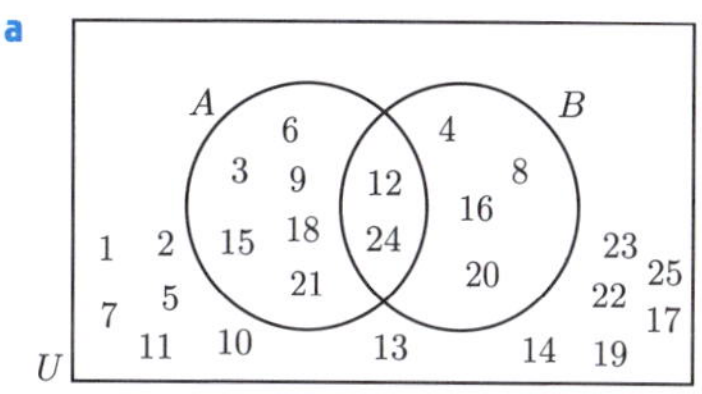

b **i** $\frac{8}{25}$ **ii** $\frac{2}{25}$ **iii** $\frac{4}{25}$ **iv** $\frac{2}{5}$

2 **a** $\frac{3}{26}$ **b** $\frac{1}{13}$ **c** $\frac{11}{26}$ **d** $\frac{15}{26}$

3 **a** $\frac{19}{25}$ **b** $\frac{13}{25}$ **c** $\frac{6}{25}$

4 **a** $\frac{19}{40}$ **b** $\frac{1}{2}$ **c** $\frac{4}{5}$ **d** $\frac{5}{8}$ **e** $\frac{13}{21}$

5 **a** $\frac{7}{15}$ **b** $\frac{1}{15}$ **c** $\frac{2}{15}$

6 **a**

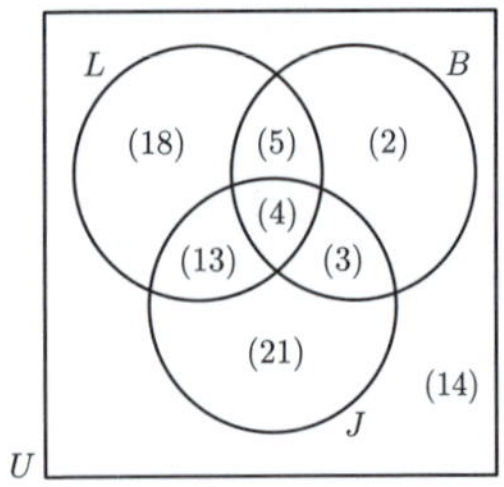

b **i** $\frac{21}{80}$ **ii** $\frac{47}{80}$ **iii** $\frac{1}{16}$ **iv** $\frac{7}{39}$ **v** $\frac{11}{20}$

EXERCISE 14J

1 27 goals **2** 15 days **3** 200 times **4** 30 times

5 **a** 300 people **b** **i** 0.55 **ii** 0.29 **iii** 0.16

c **i** 4125 people **ii** 2175 people **iii** 1200 people

6 150 times **7** **a** 20 people **b** 60 people

8 **a** 16 games **b** 6.8 games **c** $\approx$ 16.3 games

d $\approx$ 17.8 games

REVIEW SET 14A

1 **a** ≈ 0.0571 **b** ≈ 0.571 **c** ≈ 0.943

2 **a** $\frac{1}{13}$ **b** $\frac{7}{26}$ **c** $\frac{7}{26}$

3 **a**

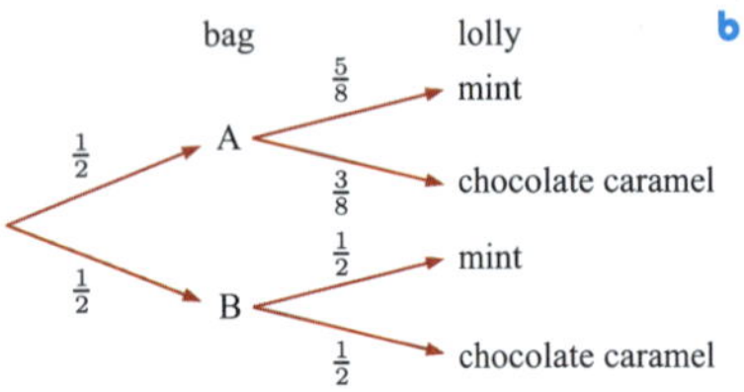

b **i** $\frac{7}{16}$ **ii** $\frac{9}{16}$

4 $\frac{5}{18}$ **5** **a** ≈ 0.238 **b** 0.0375 **c** 0.0125 **d** ≈ 0.713

6 **a**

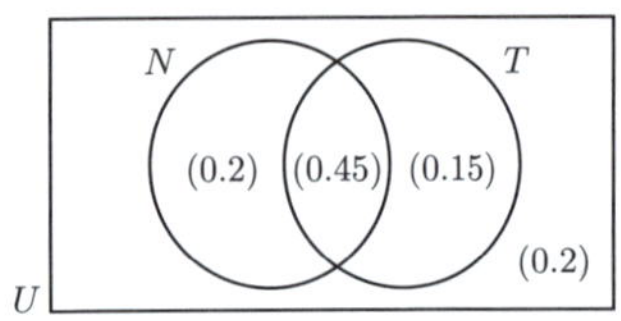

b **i** 0.65 **ii** 0.2 **iii** 0.8 **iv** 0.35

7 **a**

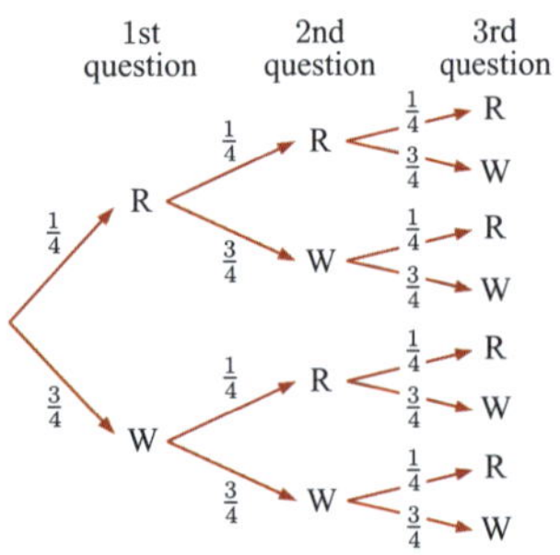

R = right answer
W = wrong answer

b **i** $\frac{9}{64}$ **ii** $\frac{27}{64}$

8 $\approx$ 62 waves **9** $\frac{5}{33}$

10 **a** 0.73 **b** $\approx$ 266 long showers

11 **a**

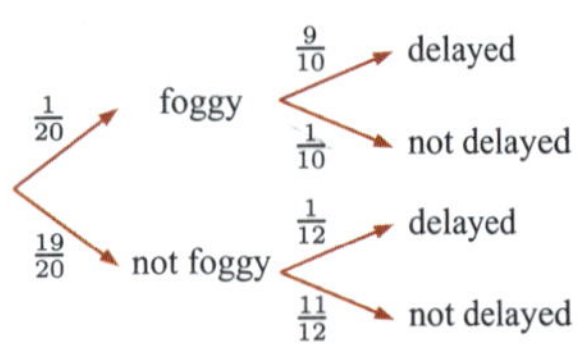

b **i** $\frac{9}{200}$ **ii** $\frac{149}{1200}$ **iii** $\frac{1051}{1200}$

c A flight is either delayed or not delayed, so these cases cover all possibilities and the probabilites sum to 1.

12 **a**

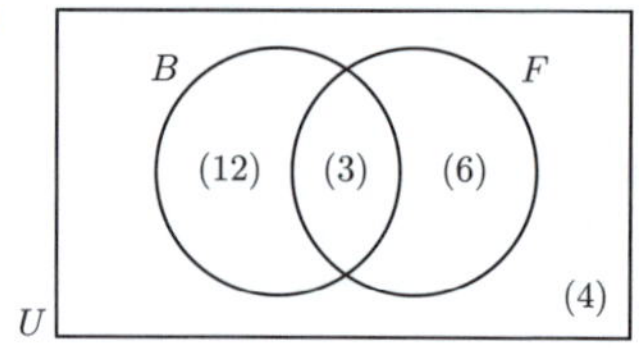

b **i** $\frac{4}{25}$ **ii** $\frac{12}{25}$ **iii** $\frac{1}{5}$ **iv** $\frac{2}{5}$

REVIEW SET 14B

1 If two events are independent then the occurrence of one does not affect the occurrence of the other.

2 **a** ≈ 0.60 **b** ≈ 0.40 **3** **a** ≈ 0.269 **b** ≈ 0.548

4

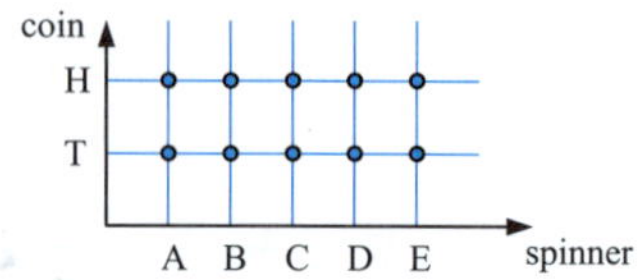

5 **a** $\frac{1}{3}$ **b** $\frac{1}{2}$

6 **a**

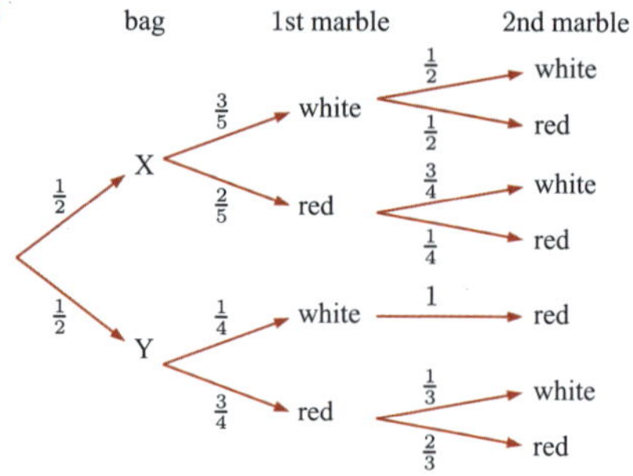

b $\frac{9}{20}$

7 $\frac{1}{6}$

8 **a**

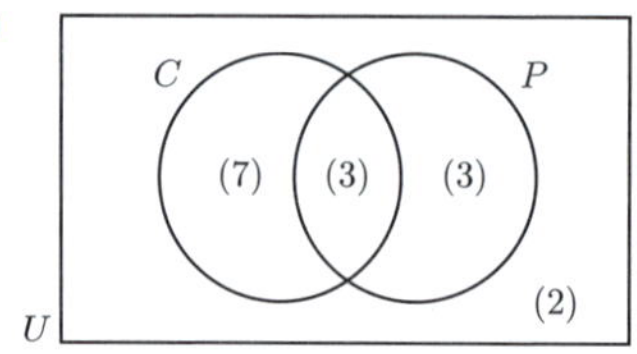

b $\frac{2}{15}$

9 6 goals

10 **a** $\frac{3}{14}$

b

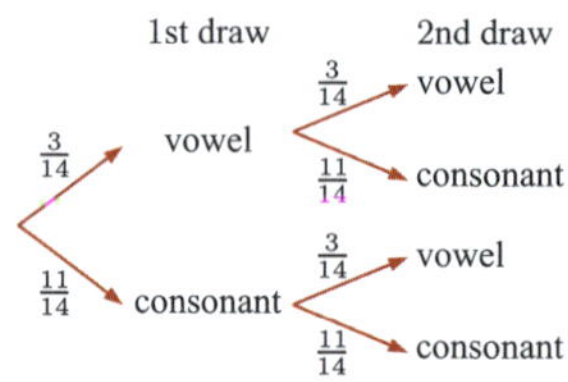

i $\frac{33}{98}$ **ii** $\frac{75}{196}$

11 **a**

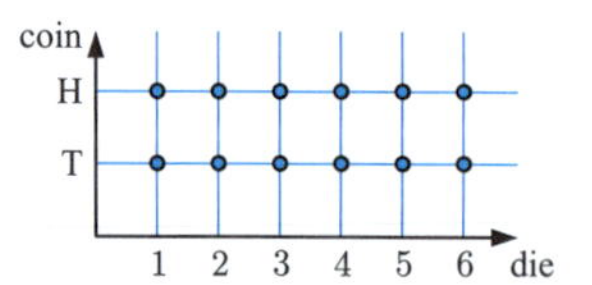

b **i** $\frac{1}{4}$ **ii** $\frac{5}{12}$ **iii** $\frac{1}{3}$ **iv** $\frac{3}{4}$

c $\frac{1}{24}$

12 **a** 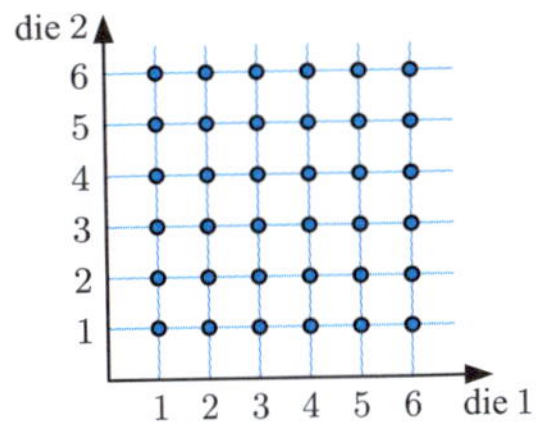

b **i** $\frac{1}{36}$ **ii** $\frac{11}{36}$ **iii** $\frac{1}{6}$ **iv** $\frac{1}{4}$

c 375 times

EXERCISE 15A

1 **a** $F = 30 + (40 \times 2)$ **b** $F = 30 + 40h$

2 **a** $A = 250 \times 6$ **b** $A = 250m$ **c** $A = Dm$

3 **a** $M = 20 + 12 \times 5$ **b** $M = 20 + 5b$ **c** $M = 20 + bn$ **d** $M = w + bn$

4 **a** $A = 100 - 6 \times 12$ **b** $A = 100 - 12b$ **c** $A = 100 - bp$ **d** $A = d - bp$

5 **a** $W = 3500 - 10 \times 200$ **b** $W = 3500 - 200r$ **c** $W = 3500 - mr$ **d** $W = L - mr$

6 **a** $H = 2 \times 5 + 4 \times 8$
b $H = 5r + (7 - r) \times 8 = 56 - 3r$
c $H = 5r + (d - r) \times 8 = 8d - 3r$

7 **a** $G = \dfrac{14 - 3 \times 2}{3 + 1}$ **b** $G = \dfrac{14 - 3w}{3 + 1}$ **c** $G = \dfrac{d - 3w}{3 + 1}$ **d** $G = \dfrac{d - nw}{n + 1}$

8 **a** $P = 20 + (2 \times 50) + 20\pi$ **b** $P = a + (2 \times 50) + 20\pi$ **c** $P = a + 2b + 20\pi$ **d** $P = a + 2b + \frac{\pi}{2}c$

9 **a** $A = 2 \times 3 \times 8 + 3 \times 2$
b $A = 2 \times m \times 8 + m \times (8 - 2m)$
$= 24m - 2m^2$
c $A = 2 \times m \times n + m \times (n - 2m)$
$= 3mn - 2m^2$

10 **a** $V = \pi \times 3^2 \times 2 + \frac{1}{3} \times \pi \times 3^2 \times 4$
b $V = \pi \times 3^2 \times x + \frac{1}{3} \times \pi \times 3^2 \times 4$
c $V = \pi \times 3^2 \times x + \frac{1}{3} \times \pi \times 3^2 \times \sqrt{y^2 - 3^2}$
d $V = \pi \times z^2 x + \frac{1}{3} \times \pi \times z^2 \times \sqrt{y^2 - z^2}$

EXERCISE 15B

1 **a** $F = 17$ **b** $g = 7$ **c** $h = \frac{12}{5} = 2.4$

2 **a** \$4000 **b** 6.6%

3 **a** **i** 22 500 kg m/s **ii** 1.5 kg m/s **b** 40 m/s

4 **a** ≈ 209 cm^3 **b** ≈ 15.6 cm **c** ≈ 2.15 m

5 **a** €7.50 **b** $\approx$ €2.87 **c** ≈ 19.9 cm

6 **a** ≈ 20.3 m^2 **b** $2\sqrt{2}$ m^2 ≈ 2.83 m^2

7 **a** ≈ 40 km **b** ≈ 121 m

EXERCISE 15C.1

1 **a** $x = \dfrac{y}{5}$ **b** $x = y + 3$ **c** $x = \dfrac{y - 2}{3}$ **d** $x = 2y + 2$ **e** $x = 3 - y$ **f** $x = \dfrac{y}{2}$

2 **a** $x = \dfrac{B - n}{7}$ **b** $x = \dfrac{T}{3y}$ **c** $x = G + 5p$ **d** $x = \dfrac{l - K}{9}$ **e** $x = \dfrac{P - yz}{w}$ **f** $x = \dfrac{C + z}{4y^2}$

3 **a** $y = \dfrac{9 - 2x}{3}$ **b** $y = 3x - 4$ **c** $y = \dfrac{3 - 4x}{2}$

4 $T = \dfrac{PV}{nR}$

5 **a** $a = xy$ **b** $a = \dfrac{c}{K}$ **c** $a = \dfrac{Mn}{bc}$ **d** $a = pT - s$ **e** $a = \dfrac{b + d}{2N}$ **f** $a = \dfrac{2Fg + d}{3}$

6 **a** $M = DV$ **b** $V = \dfrac{M}{D}$

7 **a** $a = \dfrac{P - 2b}{3}$ **b** $b = \dfrac{P - 3a}{2}$

8 **a** $x = \sqrt{D - 1}$ **b** $x = \sqrt{\dfrac{K}{5}}$ **c** $x = \sqrt{Fy}$ **d** $x = \sqrt{\dfrac{7z}{N}}$ **e** $x = \sqrt{\dfrac{C - y^2}{2}}$ **f** $x = \sqrt{\dfrac{T + w}{3z}}$

9 **a** $M = BH^2$ **b** $H = \sqrt{\dfrac{M}{B}}$ **10** $r = \sqrt{\dfrac{S}{4\pi}}$

11 $u = \sqrt{v^2 - 2as}$

12 **a** $x = \dfrac{3}{2 - y}$ **b** $x = \dfrac{2y}{y - 3}$ **c** $x = 5 - \dfrac{9}{y^2}$

EXERCISE 15C.2

1 **a** $q = \dfrac{10}{p}$ **b** $y = \dfrac{wz}{x}$ **c** $h = \dfrac{fg}{e}$ **d** $f = \dfrac{eh}{g}$ **e** $r = \dfrac{t^2}{s}$ **f** $m = \dfrac{k^3}{2n}$

2 **a** $H = \dfrac{Bh}{b}$ **b** $B = \dfrac{bH}{h}$ **c** $h = \dfrac{bH}{B}$ **d** $b = \dfrac{Bh}{H}$

3 **a** $x = -\dfrac{19y}{2}$ **b** $x = \pm\dfrac{y}{\sqrt{2}}, \quad y \neq 0$ **4** $u = \dfrac{fv}{v - f}$

EXERCISE 15D

1 **a** $t = \dfrac{Ms^3}{4}$ **b** **i** $t = 81$ **ii** $t = 30$ **iii** $t = 62.5$

2 **a** $r = \dfrac{C}{2\pi}$ **b** **i** ≈ 3.18 cm **ii** ≈ 7.16 cm **iii** ≈ 9.92 cm

3 **a** $h = \dfrac{2A}{b}$ **b** **i** 10 cm **ii** 8 m **iii** 7.8 mm

4 **a** $d = st$ **i** 180 km **ii** 120 km **iii** ≈ 127 km
b $t = \dfrac{d}{s}$ **i** 3 h **ii** 1 h 40 min **iii** 2 h 12 min

5 **a** $a = \sqrt{c^2 - b^2}$ **b** **i** $a = 6$ **ii** $a = 2\sqrt{6} \approx 4.90$ **iii** $a \approx 2.24$

6 **a** $h = \dfrac{V}{x^2}$ **i** 10 cm **ii** 128 m
b $x = \sqrt{\dfrac{V}{h}}$ **i** 8 cm **ii** ≈ 9.26 cm

7 **a** $P = 6r + \pi r$ **b** $r = \dfrac{P}{6 + \pi}$ **c** **i** $r \approx 0.547$ **ii** $r \approx 0.875$ **iii** $r \approx 1.09$

8 **a** $A = 10x + 25$ **b** $x = \dfrac{A - 25}{10}$ **c** **i** $x = 5$ **ii** $x = 10$ **iii** $x = 17.5$

REVIEW SET 15A

1 **a** $V = 8 \times 6$ **b** $V = 8n$ **c** $V = ln$

2 **a** $C = 2 + 1.2 \times 1$ **b** $C = 2 + 1.2 \times 5$ **c** $C = p + 1.2b$ **d** $C = p + bx$

3 **a** $M = A + 50V$ **b** $M = A + hV$

4 **a** $M = -27$ **b** $b \approx 44.3$ **5** **a** 50 m/s **b** 4.8 s

6 **a** $x = \dfrac{y-c}{m}$ **b** $x = \pm\sqrt{\dfrac{2}{a}}$

7 **a** $V = \frac{1}{3}\pi d^3$ **b** 9π m^3 or ≈ 28.3 m^3 **c** $d \approx 2.67$ m

8 **a** $y = \dfrac{10-3x}{4}$ **b** $y = \dfrac{5x-8}{3}$ **c** $y = \dfrac{3b}{a}$

9 **a** $A = 2lr - \dfrac{\pi r^2}{2}$ **b** $l = \dfrac{2A + \pi r^2}{4r}$ **c** $l \approx 9.02$

10 **a** ≈ 207 cm^2 **b** $h = \dfrac{S - 2\pi r^2}{2\pi r}$

c It is useful if we want to find h a number of times, without having to rearrange the formula each time.

11 **a** r m **b** $S = 2\pi r^2$ m^2 **c** $A = 2\pi rh$ m^2 **d** $S = (2\pi r^2 + 2\pi rh)$ m^2

REVIEW SET 15B

1 **a** $C = 40 + 30 \times 3$ **b** $C = 40 + 30n$ **c** $C = F + 30n$ **d** $C = F + an$

2 ≈ 1.72 seconds **3** $c = 7$

4 **a** $A = ab + \dfrac{ac}{2}$ **b** $P = a + 2b + c + \sqrt{a^2 + c^2}$

5 **a** $y = 2x - 4$ **b** $m = 2$ **c** $c = -4$

6 **a** $x = \dfrac{3q - y}{2}$ **b** $x = \dfrac{4}{3y}$

7 **a** ≈ 7230 cm^3 **b** ≈ 0.943 m

8 **a** $b = \dfrac{c-2a}{3}$ **b** $b = \dfrac{4}{a-c}$ **c** $b = \pm\sqrt{3e}$ **d** $b = \dfrac{4c}{a}$

9 $a = 12$ **10** **a** $a = \dfrac{2d}{t^2}$ **b** $a = 7$

11 **a** ≈ 7854 m^2 **b** $r = \sqrt{\dfrac{A}{4\pi}}$ **c** ≈ 1.99 m

EXERCISE 16A

1 **a** $\begin{pmatrix}3\\-2\end{pmatrix}$

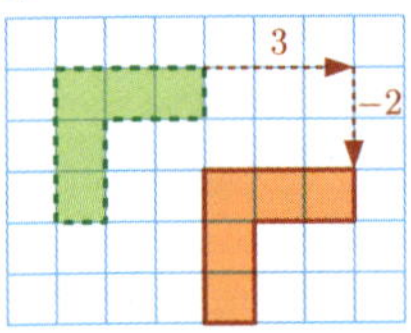

b $\begin{pmatrix}4\\0\end{pmatrix}$

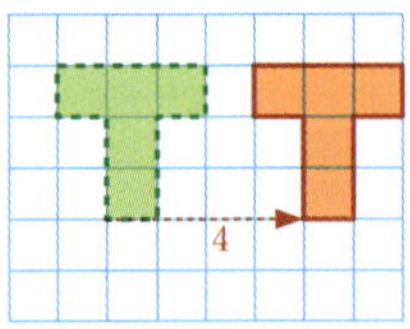

c $\begin{pmatrix}0\\-3\end{pmatrix}$

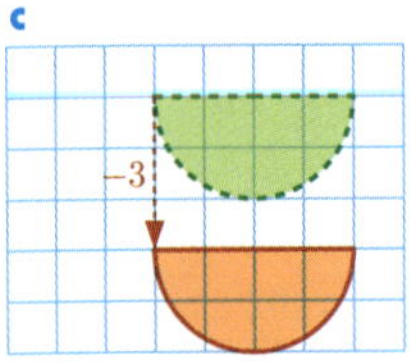

d $\begin{pmatrix}-3\\1\end{pmatrix}$

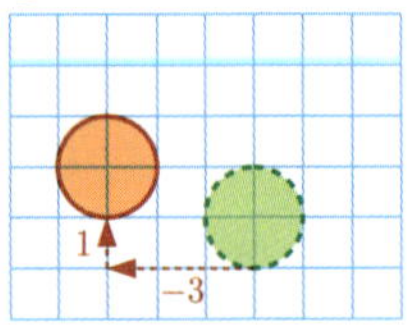

e $\begin{pmatrix}-1\\-4\end{pmatrix}$

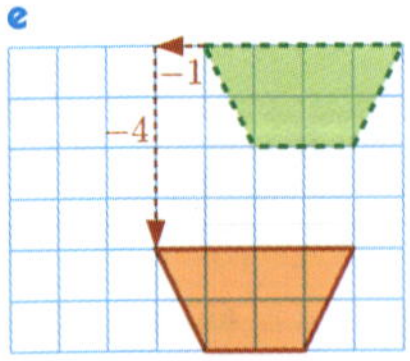

f $\begin{pmatrix}-3\\3\end{pmatrix}$

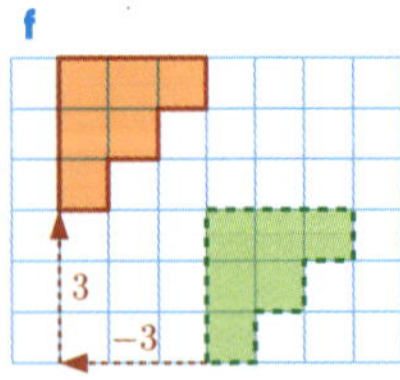

2 **a** 5 units right, $\begin{pmatrix}5\\0\end{pmatrix}$ **b** 3 units down, $\begin{pmatrix}0\\-3\end{pmatrix}$

c 2 units right, 2 units down, $\begin{pmatrix}2\\-2\end{pmatrix}$

d 4 units left, 1 unit up, $\begin{pmatrix}-4\\1\end{pmatrix}$

e 3 units right, 3 units down, $\begin{pmatrix}3\\-3\end{pmatrix}$

f 3 units left, 3 units down, $\begin{pmatrix}-3\\-3\end{pmatrix}$

3 **a** $\sqrt{34}$ units **b** 5 units left, 3 units up, $\begin{pmatrix}-5\\3\end{pmatrix}$

4 **a**

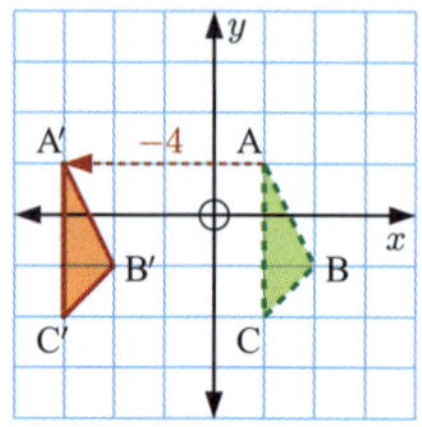

A$'(-3, 1)$
B$'(-2, -1)$
C$'(-3, -2)$

b

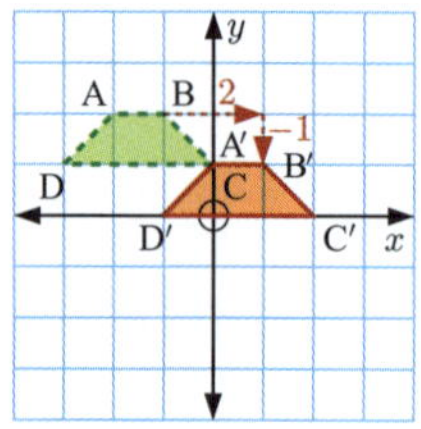

A$'(0, 1)$
B$'(1, 1)$
C$'(2, 0)$
D$'(-1, 0)$

c

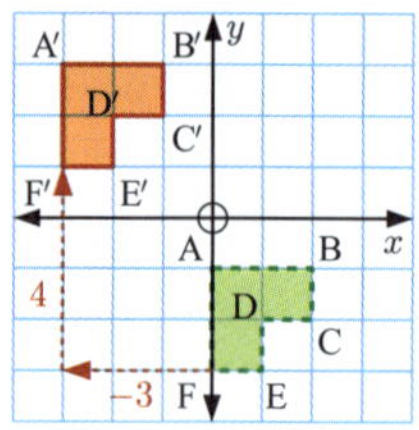

A$'(-3, 3)$
B$'(-1, 3)$
C$'(-1, 2)$
D$'(-2, 2)$
E$'(-2, 1)$
F$'(-3, 1)$

5 **a** $\begin{pmatrix}-4\\2\end{pmatrix}$ **b** $\begin{pmatrix}6\\1\end{pmatrix}$ **c** $\begin{pmatrix}2\\3\end{pmatrix}$

6 **a** No **b** Yes, $\begin{pmatrix}2\\-3\end{pmatrix}$ **c** No

7 **a** I, $\begin{pmatrix}-5\\-6\end{pmatrix}$ **b** D, $\begin{pmatrix}7\\5\end{pmatrix}$ **c** H

8 **a**

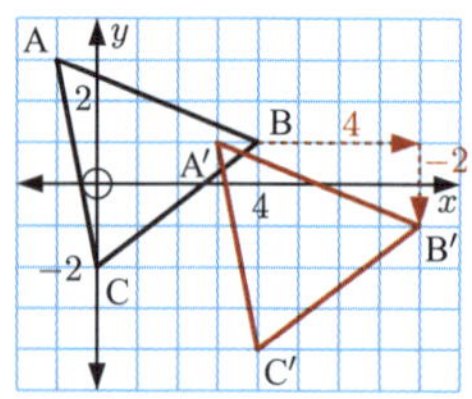

b A$'(3, 1)$, B$'(8, -1)$, C$'(4, -4)$ **c** $-\frac{1}{2}, -\frac{1}{2}, -\frac{1}{2}$

d $2\sqrt{5}$ units

EXERCISE 16B.1

1 **a**

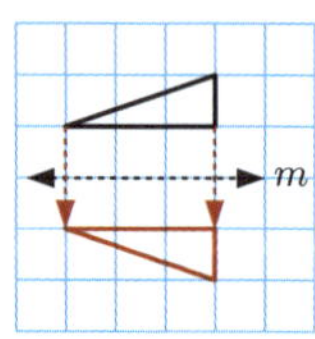

b

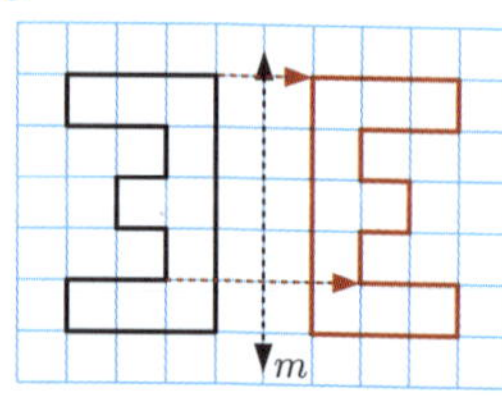

c

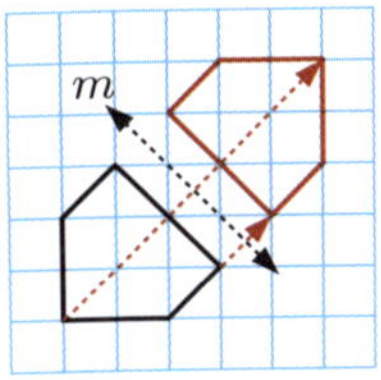

d

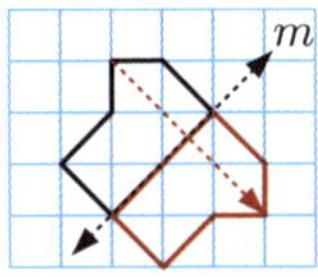

e

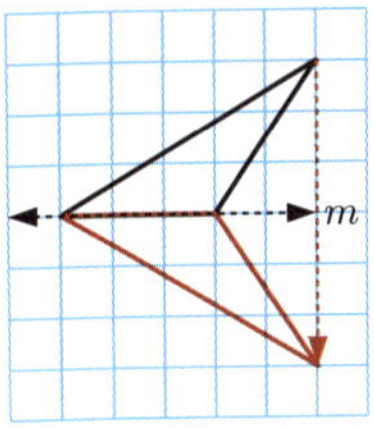

f

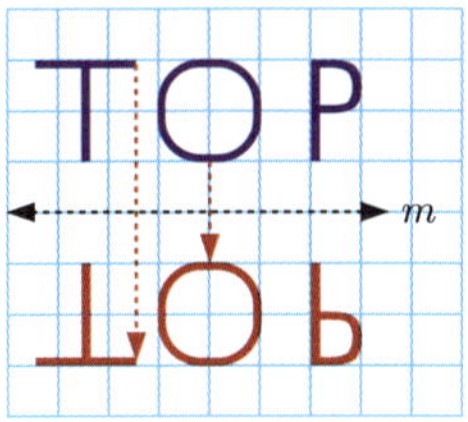

2 **a** and **c** represent reflections.

3 **a**

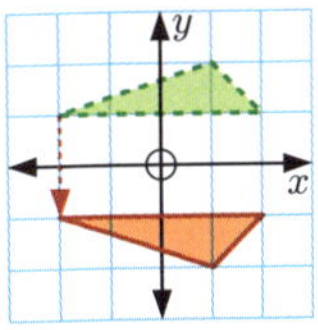

b

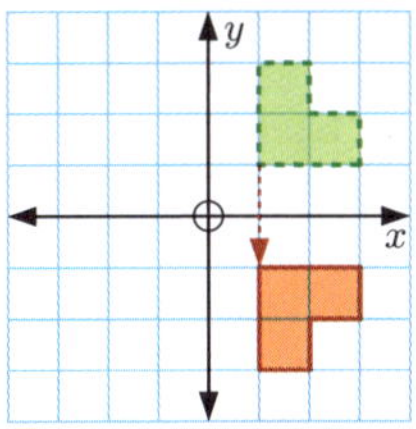

c

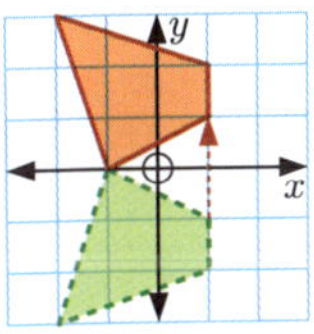

4 **a**

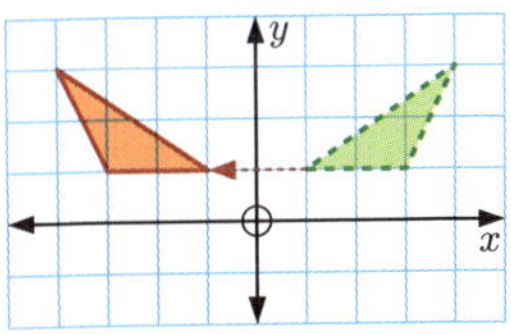

b

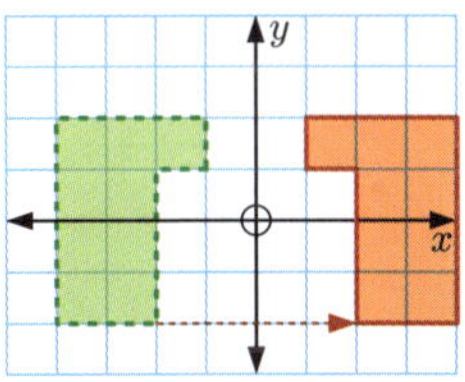

c

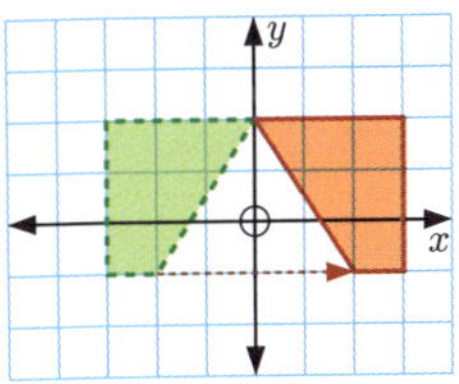

5 **a** **i**

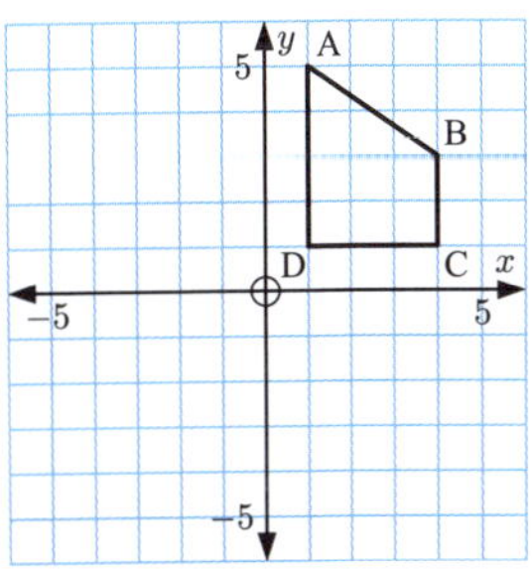

ii

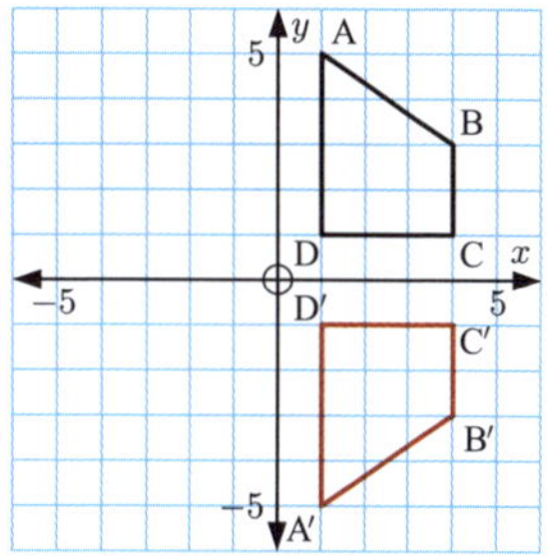

A′(1, −5)
B′(4, −3)
C′(4, −1)
D′(1, −1)

iii

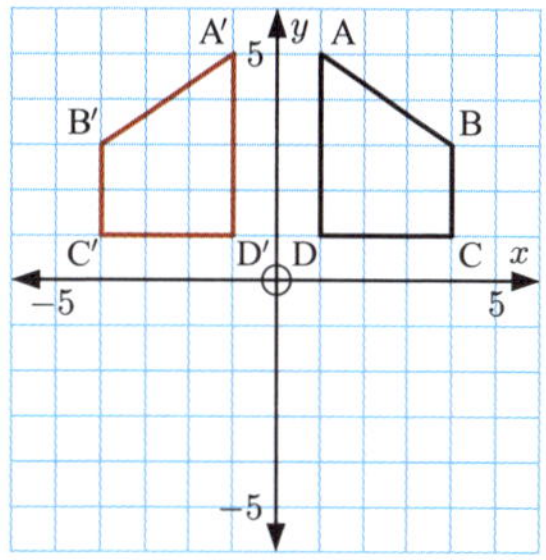

A′(−1, 5)
B′(−4, 3)
C′(−4, 1)
D′(−1, 1)

iv

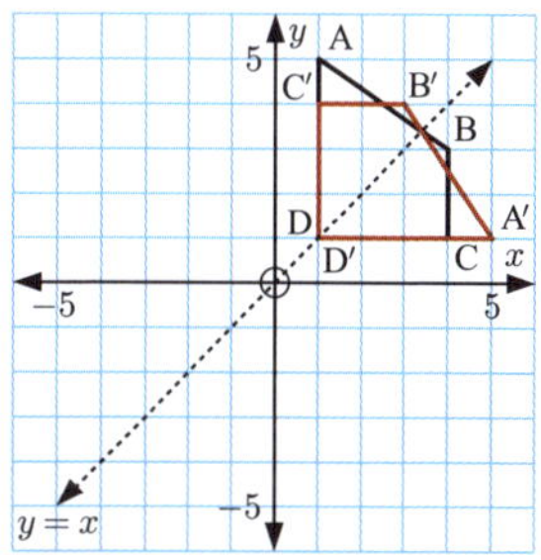

A′(5, 1)
B′(3, 4)
C′(1, 4)
D′(1, 1)

b **i** When the point (a, b) is reflected in the x-axis, the image has coordinates $(a, -b)$.

ii When the point (a, b) is reflected in the y-axis, the image has coordinates $(-a, b)$.

iii When the point (a, b) is reflected in the line $y = x$, the image has coordinates (b, a).

6 **a** reflection **b** not a reflection **c** not a reflection

7 **a** A and C **b** x-axis **c** A and D

8 **a** $\triangle$BXB′ is isosceles since the pipeline perpendicularly bisects [BB′].

b The shortest distance between A and B′ is through X, and since BX = B′X, AX + XB = AX + XB′ must also be the shortest distance from A to B via the pipeline.

EXERCISE 16B.2

1 **a**

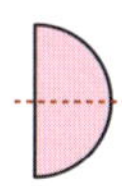

b

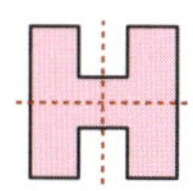

c

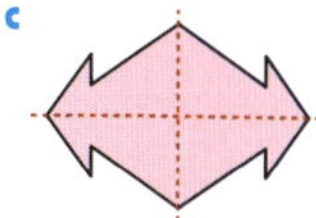

d

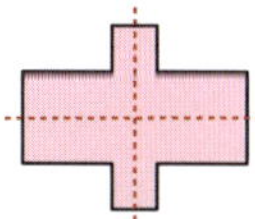

e

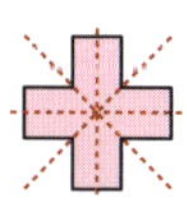

f

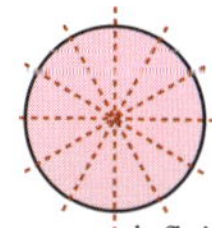

infinite

2 **a** 4

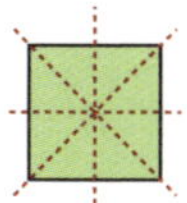

b 3

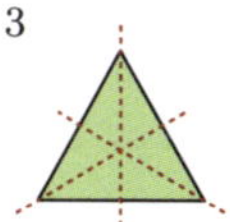

c 0

d 2

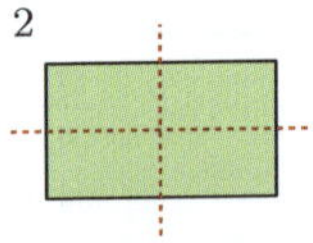

e 2

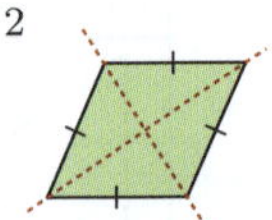

f 5

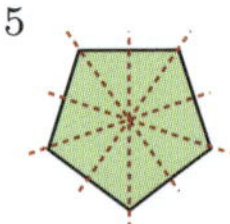

g infinite

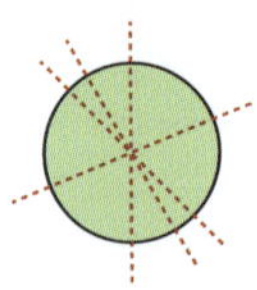

h 1

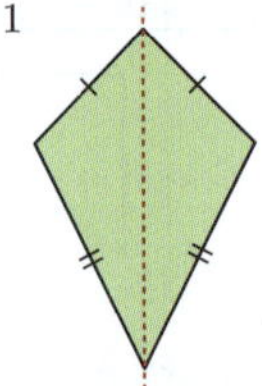

i 1

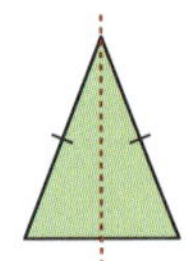

EXERCISE 16C

1 **a**

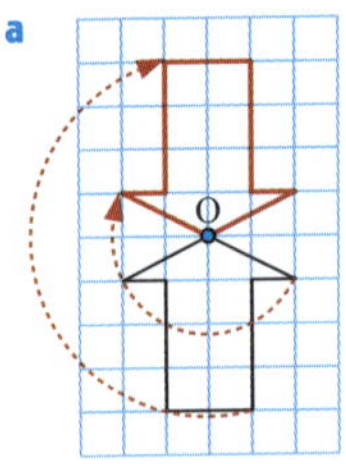

b

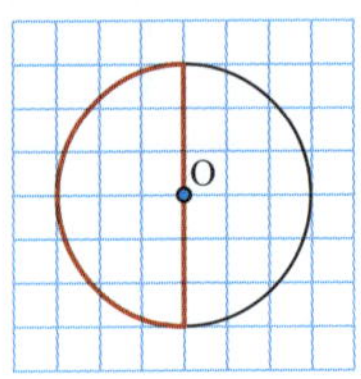

c

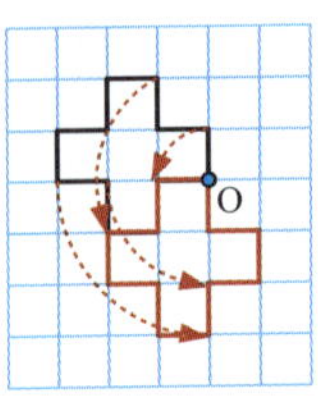

d

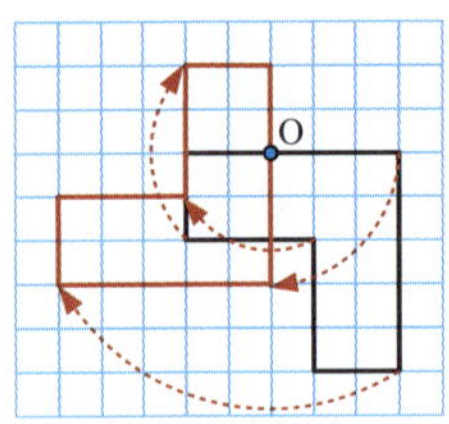

e

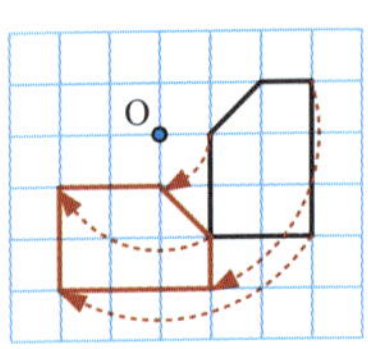

f

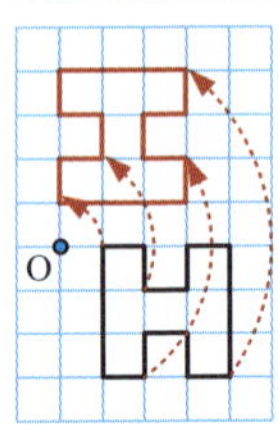

2 H, I, M, N, O, S, W, X, and Z either remain the same or become a different letter under a rotation.

3 **a** D **b** B **c** C

4 Rotate 75° anticlockwise about O. **5** **a** C **b** 180°

6 **a** P(0, −2)
Q(2, −1)
R(2, −4)

b

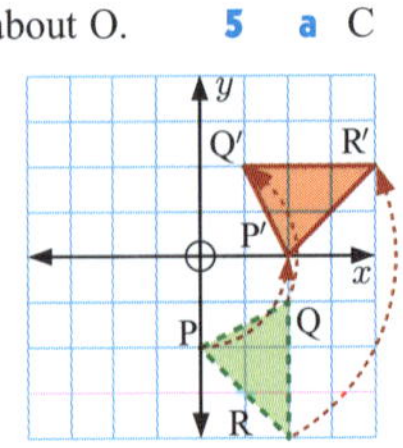

c P′(2, 0)
Q′(1, 2)
R′(4, 2)

7 **a, b**

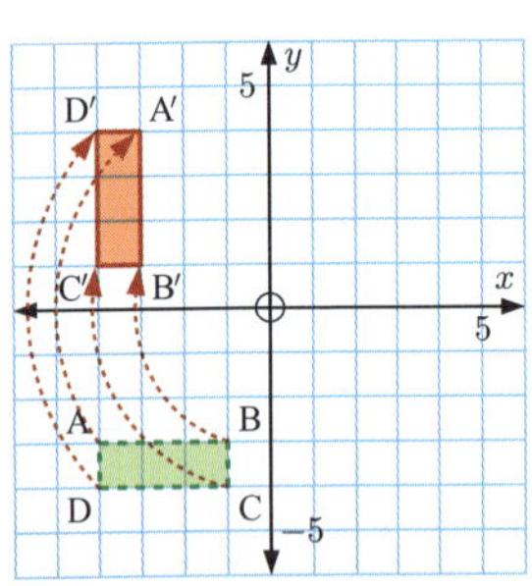

c A′(−3, 4)
B′(−3, 1)
C′(−4, 1)
D′(−4, 4)

8 **a** **i** (−1, 4) **ii** (2, 5) **iii** (5, 0) **iv** (5, −1)
v (−4, −3)
b $(-y, x)$

9 **a** **i** (3, −5) **ii** (−3, 0) **iii** (−1, −4) **iv** (3, 2)
v (−4, 2)
b $(-x, -y)$

10 **a** **i** (−2, 0) **ii** (2, −4) **iii** (−5, 1) **iv** (−2, −4)
v (2, 3)
b $(y, -x)$

11

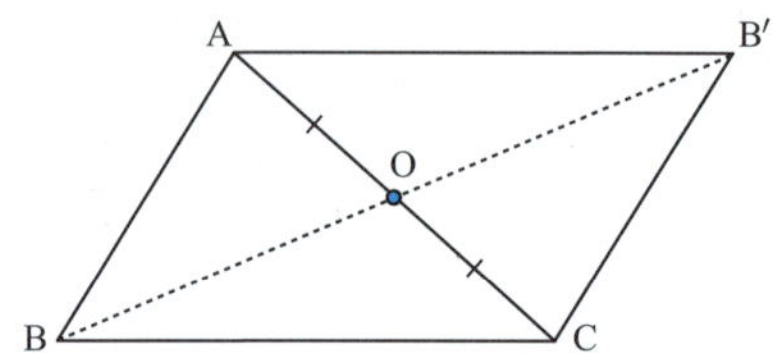

a ABCB′ is a parallelogram.
b opposite sides are equal and parallel, diagonals bisect each other, opposite angles are equal
c When $A\widehat{B}C$ is a right angle, ABCB′ is a rectangle.
opposite sides are equal and parallel, all angles are right angles, diagonals are equal and bisect each other

EXERCISE 16D.1

1 **a** 2 **b** 3 **c** $\frac{3}{2}$ **2** **a** $\frac{1}{2}$ **b** $\frac{1}{6}$ **c** $\frac{3}{4}$

3 **a** enlargement with scale factor 2
b reduction with scale factor $\frac{1}{2}$
c enlargement with scale factor 3
d reduction with scale factor $\frac{1}{3}$
e enlargement with scale factor $\frac{3}{2}$
f reduction with scale factor $\frac{2}{3}$

EXERCISE 16D.2

1 **a**

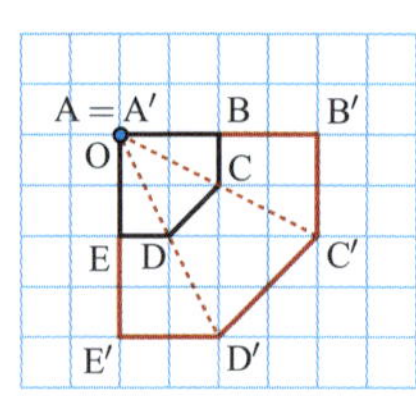

b

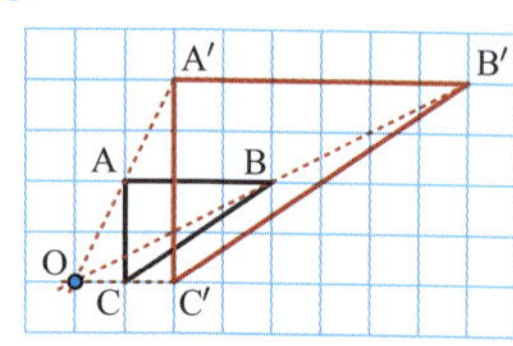

c

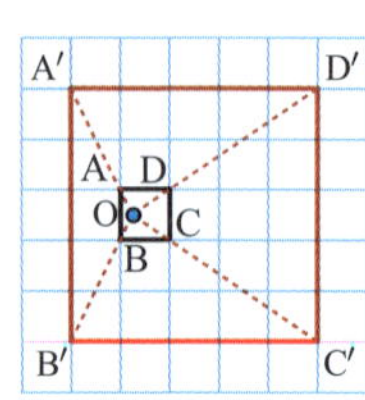

2 **a**

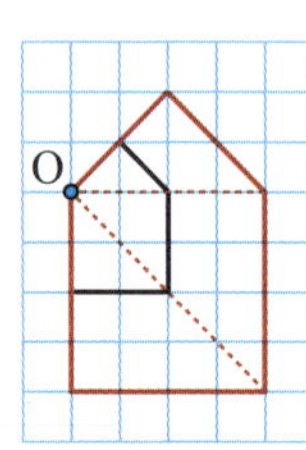

b

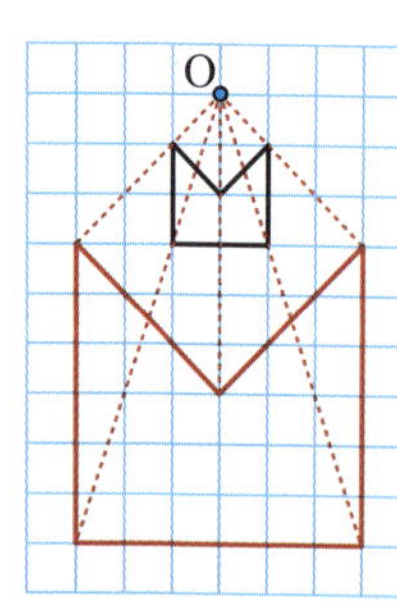

c

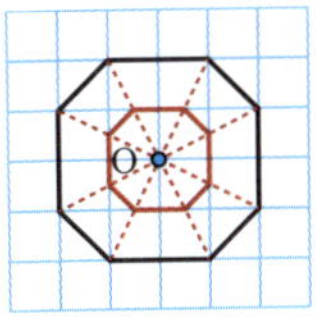

d

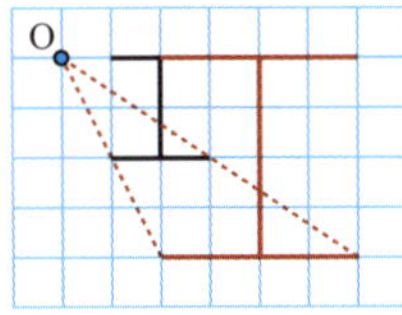

e

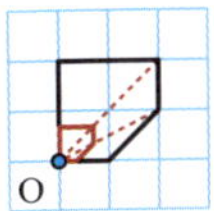

f

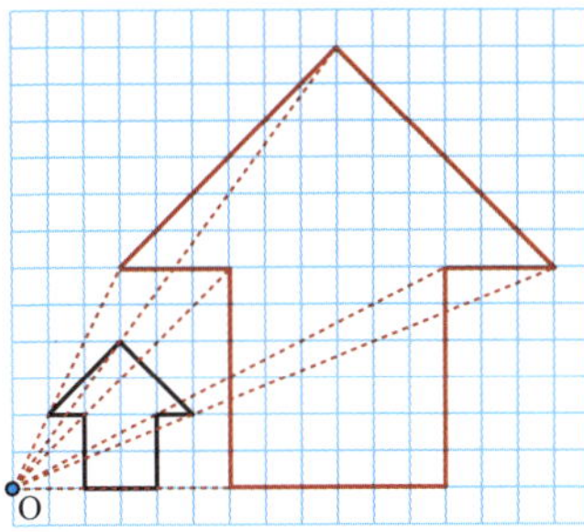

3 **a** **i**

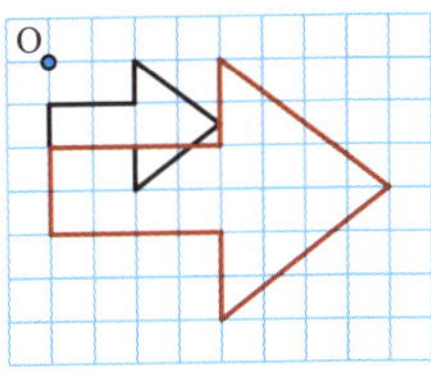

ii

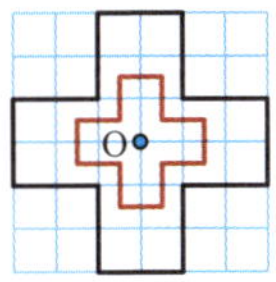

iii

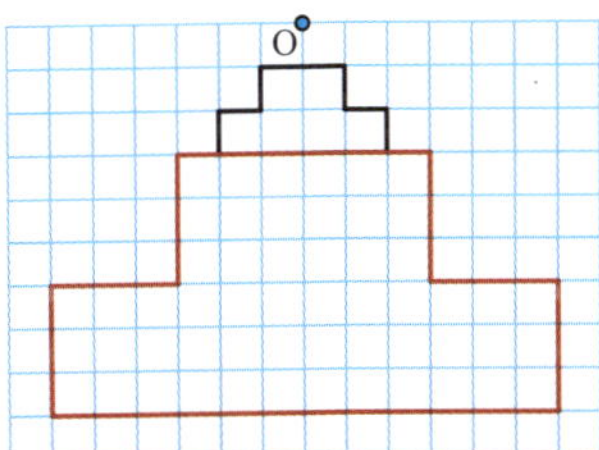

b

	Area of object	*Area of image*	$\dfrac{\textit{Area of image}}{\textit{Area of object}}$	k	k^2
i	5 units^2	20 units^2	4	2	4
ii	20 units^2	5 units^2	$\frac{1}{4}$	$\frac{1}{2}$	$\frac{1}{4}$
iii	6 units^2	54 units^2	9	3	9

c $\dfrac{\text{area of image}}{\text{area of object}} = k^2$ or

area of image $= k^2 \times$ area of object

4 56 m^2 **5** **a** $k = 2.5$ **b** 20 cm **c** \$18.75

6 $5\sqrt{2}$ m $\times$ $8\sqrt{2}$ m ≈ 7.07 m $\times$ 11.31 m

REVIEW SET 16A

1 **a**

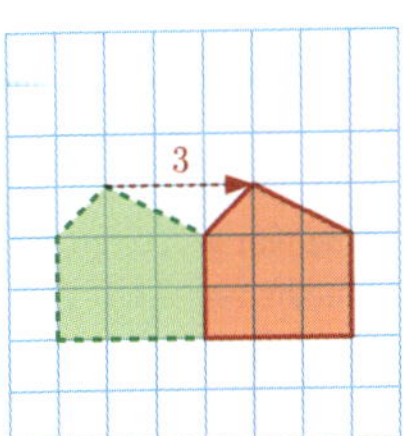

b

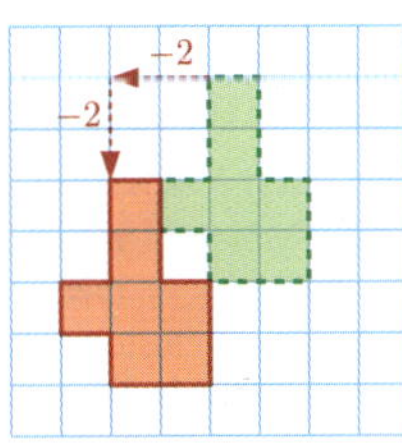

2 **a** $\begin{pmatrix} -4 \\ -2 \end{pmatrix}$ **b** $\begin{pmatrix} 6 \\ -2 \end{pmatrix}$ **c** $\begin{pmatrix} 2 \\ -4 \end{pmatrix}$

3

	P	P'
a	(1, 5)	(−1, 5)
b	(−2, 4)	(2, 4)
c	(−4, −3)	(4, −3)
d	(3, 2)	(−3, 2)

4

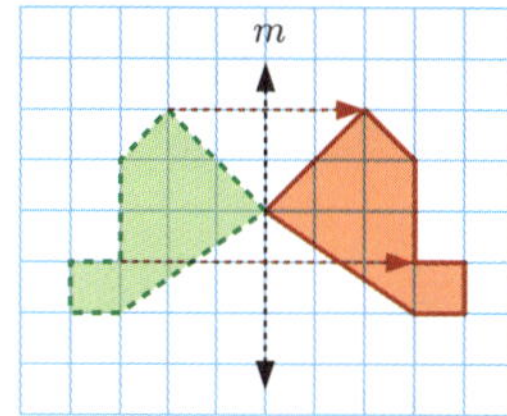

5 **a** A translation of 4 units right and 3 units up.
b a reflection

6

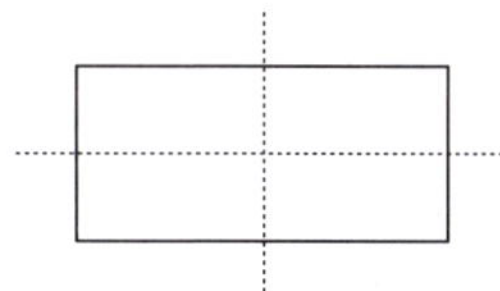

7 **a**

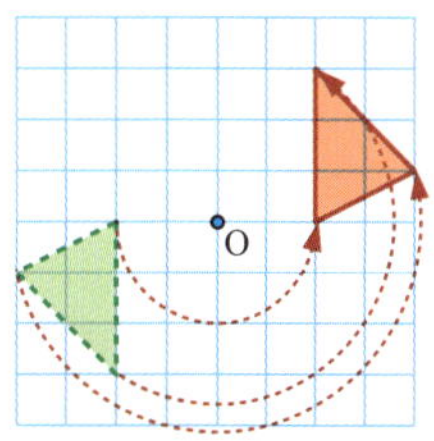

b

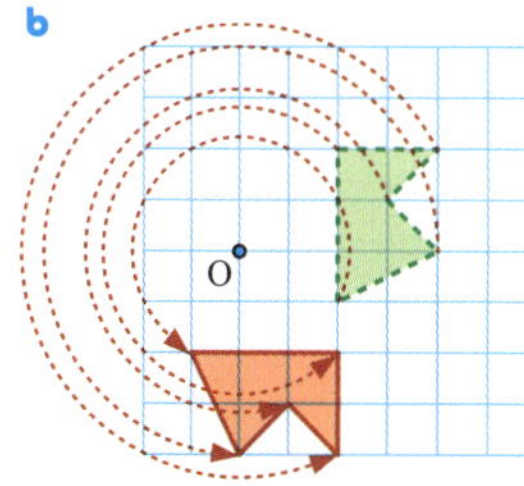

8 **a** D **b** 90° clockwise

9 Area is enlarged with a scale factor of 4.

10 **a**

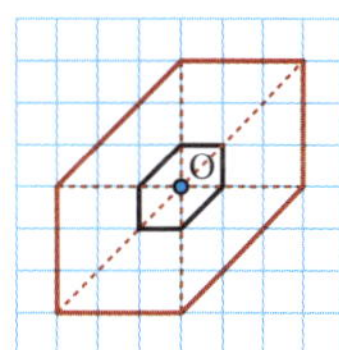

b

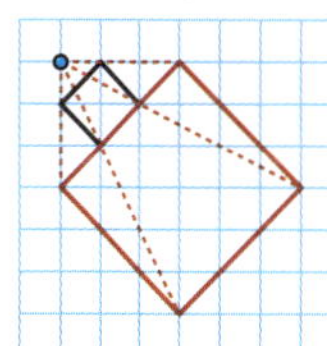

11 **a**

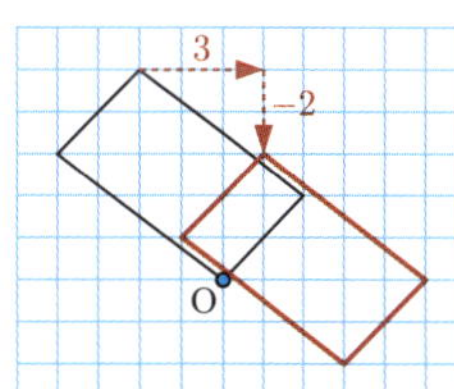

b

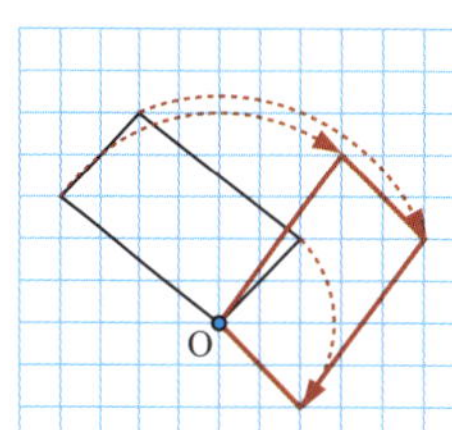

c

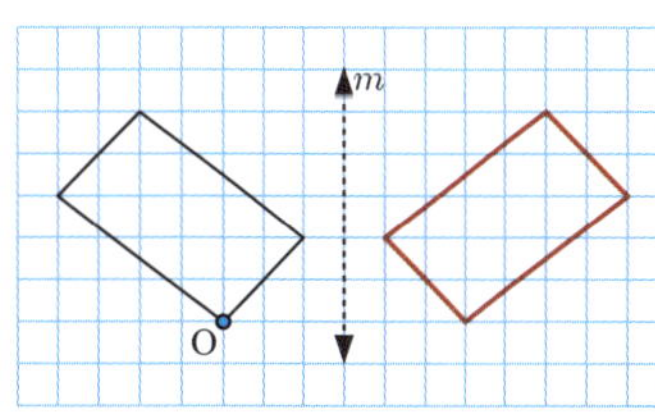

d 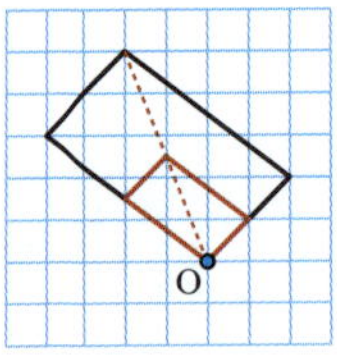

12 **a** A and C **b** B and D **c** the x-axis

REVIEW SET 16B

1 **a** $\begin{pmatrix} 3 \\ -2 \end{pmatrix}$ **b** $\sqrt{13}$ units **c** $\begin{pmatrix} -3 \\ 2 \end{pmatrix}$

2 **a** 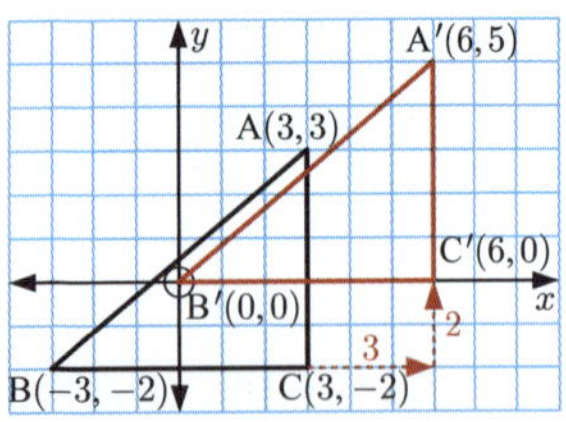

b A′(6, 5), B′(0, 0), C′(6, 0)

3

	P	P'
a	(1, 4)	(1, −4)
b	(−2, 3)	(−2, −3)
c	(−3, −1)	(−3, 1)
d	(4, 2)	(4, −2)

4 **a** 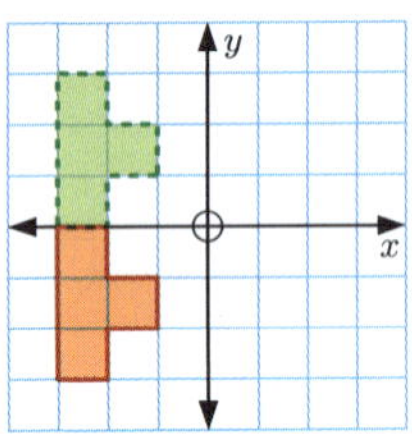

b 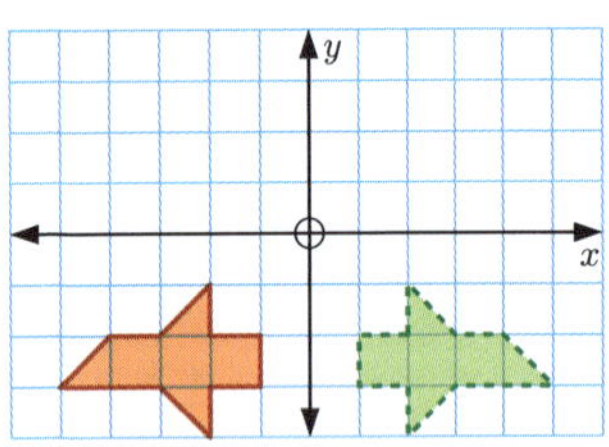

5 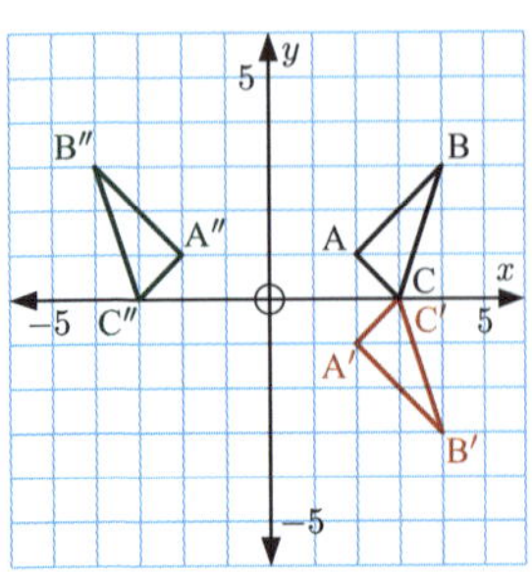

b A′(2, −1) B′(4, −3) C′(3, 0)

c A″(−2, 1) B″(−4, 3) C″(−3, 0)

6 **a** (−2, 0) **b** (0, −5) **c** (3, 1) **d** (4, −2) **e** (−2, −3)

7 **a** 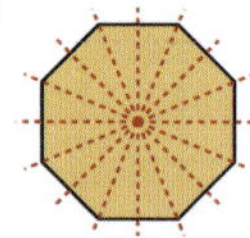

b 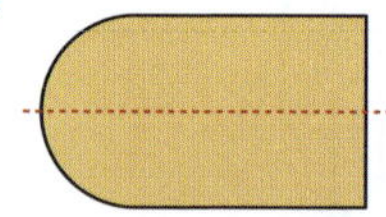

c no lines of symmetry exist

8 **a** P(−4, −3) Q(−2, 0) R(−1, −2) S(−1, −3)

b 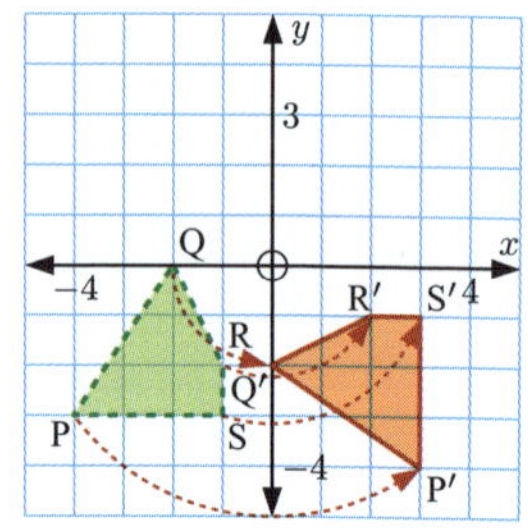

c P′(3, −4) Q′(0, −2) R′(2, −1) S′(3, −1)

9 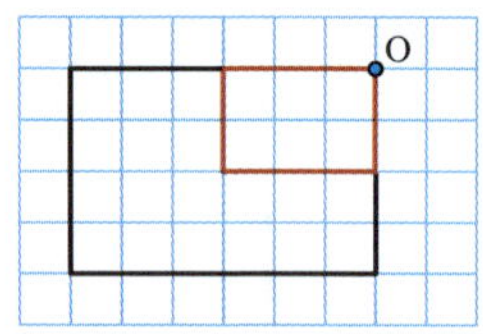

10 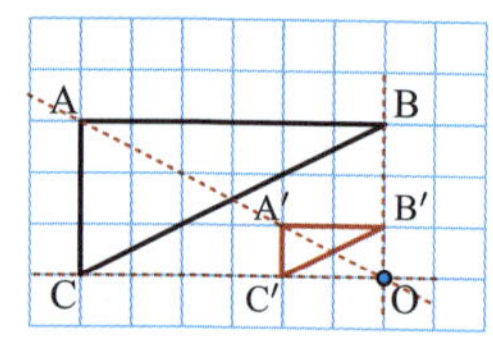

11 **a** 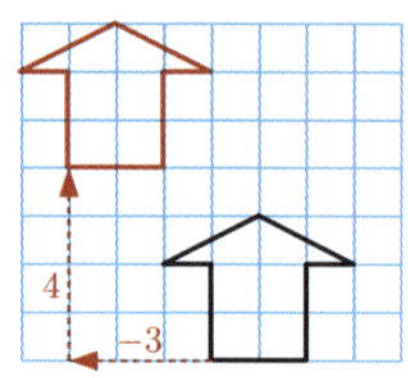

b 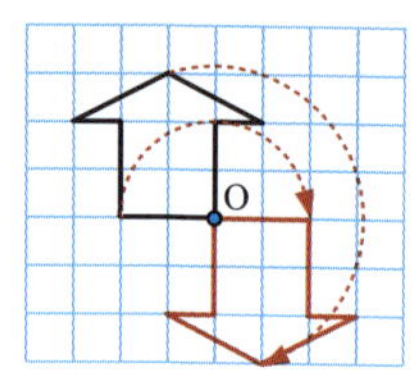

c 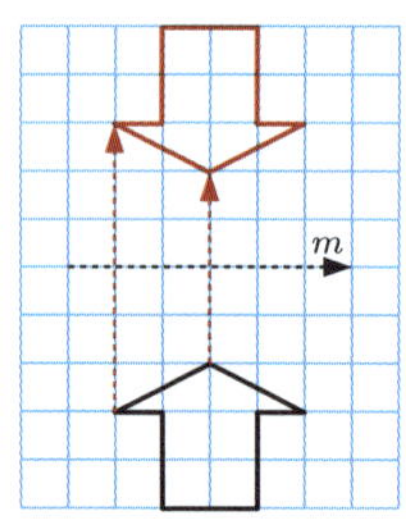

d 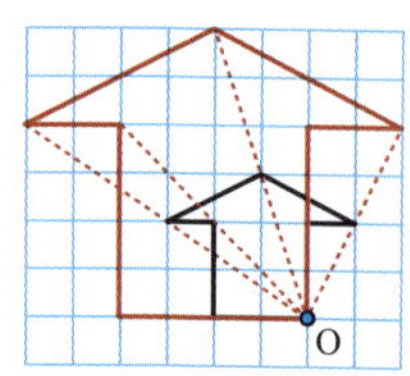

12 **a** B **b** D **c** C **d** A reflection in the y-axis.

EXERCISE 17A

1 **a** High Street **b** Park Street

2 Team B (Team A: ≈ 91.3, Team B: ≈ 91.8)

3 **a** Class 9A: 4 students; Class 9B: 2 students
b Class 9A: 4 sick days; Class 9B: 8 sick days
c Class 9A: 5 sick days; Class 9B: 7 sick days
d Class 9B **e** Class 9B

4 **a** The Rookies: mean = 1189 points, median = 1180 points;
Best Mates: mean = 1137 points, median = 1120 points
b The Rookies: range = 282 points, IQR = 131 points;
Best Mates: range = 438 points, IQR = 155 points
c The Rookies (mean and median values are higher)
d Best Mates (range and IQR are higher)

5 **a** Sydney: mean ≈ 17.4°C, median = 17.3°C;
Paris: mean ≈ 23.1°C, median = 23.1°C
b A conclusion cannot be made. Any conclusion is very unreliable because July is the middle of winter in Sydney whereas it is in summer in Paris.

EXERCISE 17B

1 **a** California: 30 classes; Oregon: 11 classes

b California: negatively skewed;
Oregon: approximately symmetric

c "The selected classes from *California* generally have larger class sizes."

2 **a** Group A: 30; Group B: 20

b Group A: negatively skewed;
Group B: approximately symmetric

c Group A **d** Group A

3 **a** **Weight-loss results**

Before program		After program
	6	8 8
9	7	4 8 9 9
6 3 2	8	3 3 4 5 9
9 8 7 5 3	9	2 3 5 6 9 9
4 4 1 0	10	2 6 6 7
9 7 4 1 1	11	0 3
5 3 2 1 0	12	2 5
5 2	13	

Scale: 7 | 8 = 78 kg

b Before program: mean ≈ 107 kg, median = 104 kg;
After program: mean = 93.4 kg, median = 93 kg

c Yes; both measures of centre indicate that the weight of the men has decreased. (mean dec by ≈ 12.7%, median dec by ≈ 10.6%)

4 **a**

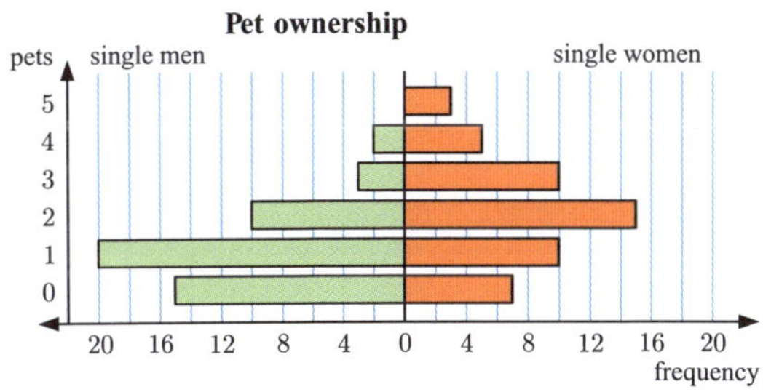

b Single men: mean = 1.14 pets, median = 1 pet;
Single women: mean = 2.1 pets, median = 2 pets

c In general, single women have more pets than single men.

5 **a** **Vending machine sales**

Drinks		*Snacks*
8 7 5 4 3 2	2	3 6 6 8
9 9 7 6 6 3 2 2 1 0	3	1 1 5 7 7 8 9
9 8 8 7 4 2 1 0	4	0 2 2 3 4 6 7 8 9
7 6 5 2 2 0	5	0 0 1 2 3 3 6 8
8 7 4 1 0	6	0 1 2 6 7 7 8
5 4 2	7	0 3 3 4 7
4 1	8	

Scale: 2 | 3 = 23

b Drinks: positively skewed;
Snacks: approximately symmetric

c Drinks: mean ≈ 47.0 items, median = 45.5 items;
Snacks: mean ≈ 49.8 items, median = 49.5 items

d Drinks: range = 62 items, IQR = 26 items;
Snacks: range = 54 items, IQR = 23 items

e snacks **f** drinks

6 **a**

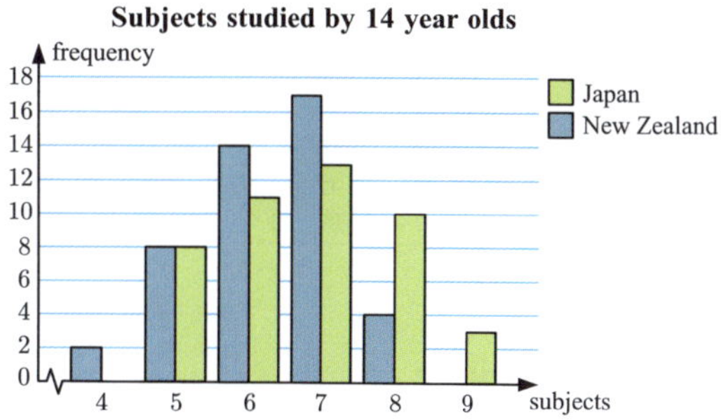

b New Zealand: mean ≈ 6.29 subjects, median = 6 subjects;
Japan: mean ≈ 6.76 subjects, median = 7 subjects

c New Zealand: 4 subjects; Japan: 4 subjects

d "The selected students from *Japan* generally studied more subjects."

7 **a** Tigers: 13 players; Panthers: 10 players

b Tigers: $175 \leqslant h < 180$ cm;
Panthers: $170 \leqslant h < 175$ cm

c Tigers: approximately symmetric;
Panthers: approximately symmetric

d Tigers: ≈ 177 cm; Panthers: ≈ 175 cm

e The Tigers appear to have a slightly greater average height. This conclusion may be unreliable as we can only estimate the means.

EXERCISE 17C

1 **a** **i** Set B, as the box-and-whisker plot is further to the right.
ii Set A, as the box-and-whisker plot covers a greater spread of values.

b **i** Set B, as the box-and-whisker plot is further to the right.
ii Set B, as the box-and-whisker plot covers a greater spread of values.

2 **a** **i** 50% **ii** 75% **b** bus B

3 **a**

	Maria	Sophie
min.	0	0
Q_1	2	1
med.	3	2.5
Q_3	4	4
max.	6	8

b Maria: range = 6 runs, IQR = 2 runs;
Sophie: range = 8 runs, IQR = 3 runs

c Maria was more consistent (her scores had less spread) and had a higher median than Sophie. Sophie had a higher maximum but her scores were positively skewed.

4 **a**

	Zimbabwe	India
min.	2	3
Q_1	4	6
med.	11	17
Q_3	19	38
max.	25	90

b

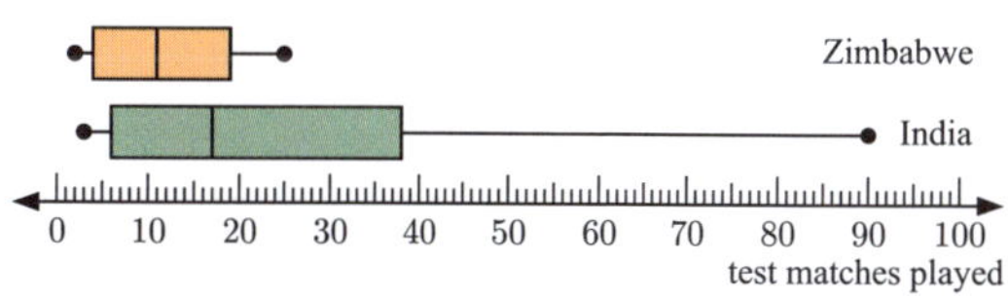

c India **d** India has a far greater spread than Zimbabwe.

5 **a**

	Frank	Competitor
min.	128	76
Q_1	136	87
med.	142.5	89
Q_3	146	92
max.	163	97

b

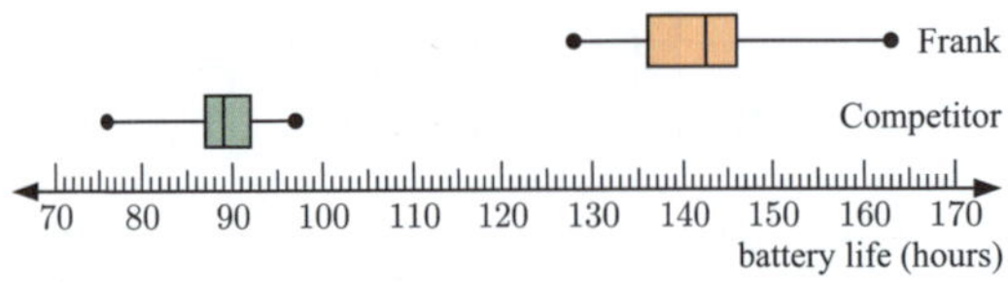

c Frank: $\approx$ 142 hours; Competitor: $\approx$ 89.2 hours

d No. The values in the five-number summary for Frank's batteries are not double their corresponding value for the competitor's batteries. The mean for Frank's batteries is not double the mean of the competitor's batteries.

6 **a** Which machine produces fewer faulty screw caps? (We suspect the answer is Machine A.)

b Machine A: 5.98 caps; Machine B: 7.56 caps

c Machine A: range = 5 caps, IQR = 1 cap
Machine B: range = 8 caps, IQR = 2 caps

d

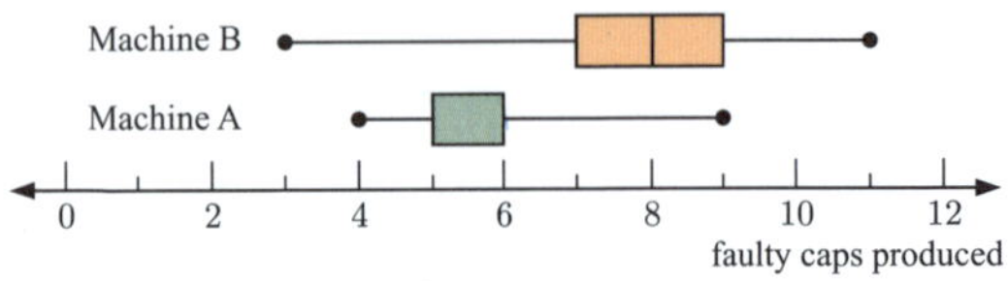

e Machine A does cause fewer faulty caps than machine B (lower mean and box-and-whisker plot is more to the left).

Machine A works with better consistency than machine B (lower range and IQR).

REVIEW SET 17A

1 Team B

2 **a** Monkey bars: positively skewed;
Swings: approximately symmetric

b **i** 4 weeks **ii** 1 week

c The swings appear to be slightly more dangerous (the swing data has a higher mean and median).

3 **a**

Peter		John
9 5 3	9	7
	10	
8 6	11	0 0 6
	12	7 9
2	13	9 9
7 0	14	2 7
5 4 3	15	2 3 8 9
8	16	0 1 1 2 3
	17	4
8 3 2	18	0
6 4 1 0	19	
8 5 5	20	
	21	5
3	22	3
3	23	
	24	1
	25	
	26	0
0	27	

Scale: 12 | 7 = $127

b Peter: mean $\approx$ \$169.52, median = \$182;
John: mean $\approx$ \$159.12, median = \$158

c Peter: range = \$177, IQR = \$64.50;
John: range = \$163, IQR = \$34.50

d Peter; he has higher mean and median fare totals.

4 **a** **i** Set B generally has larger values. It has a higher median and the middle half of the data is more to the right on the diagram.

ii Set A has a greater spread of values. There is a greater distance between the minimum and maximum values (the ends of the 'whiskers'), and also between Q_1 and Q_3 (the 'box' part of the diagram).

b **i** Set A generally has larger values. The entire box-and-whisker plot for set A is more to the right than that for set B.

ii Set B has a greater spread of values. There is a greater distance between the minimum and maximum values, and also between Q_1 and Q_3.

5 **a**

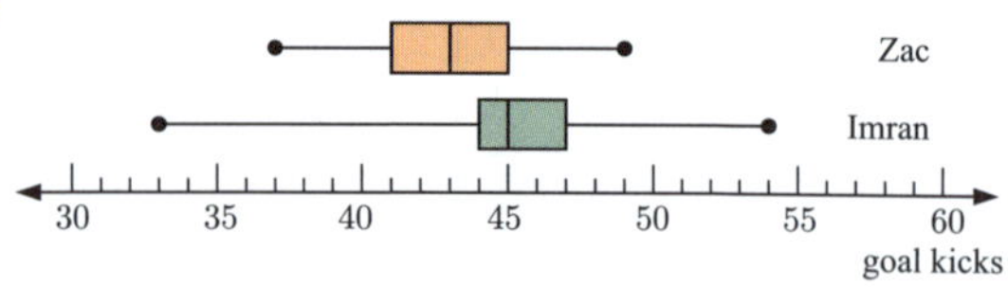

b Imran, because the box-and-whisker plot is more to the right.

c Imran, as indicated by the far greater range.

REVIEW SET 17B

1 **a** **i** mean: Before $\approx$ 28.3 s, After $\approx$ 28.0 s

ii median: Before $\approx$ 28.4 s, After $\approx$ 28.4 s

iii IQR: Before $\approx$ 10.6 s, After = 9.95 s

b No, the program has not been effective. There has been no significant decrease in the measures of centre.

2 **a**

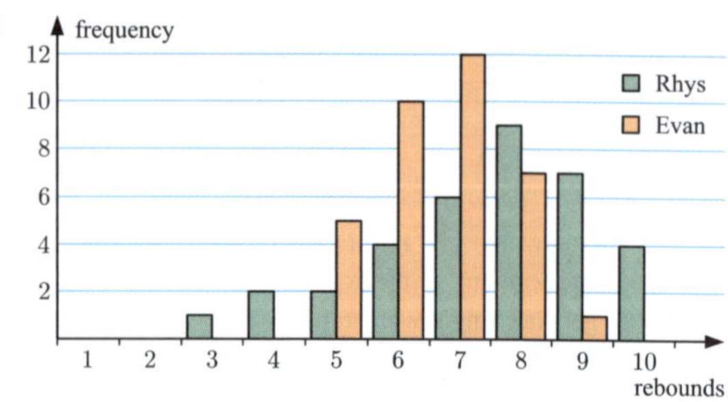

b Rhys, as he has more values at the higher end of the graph.

3 **a**

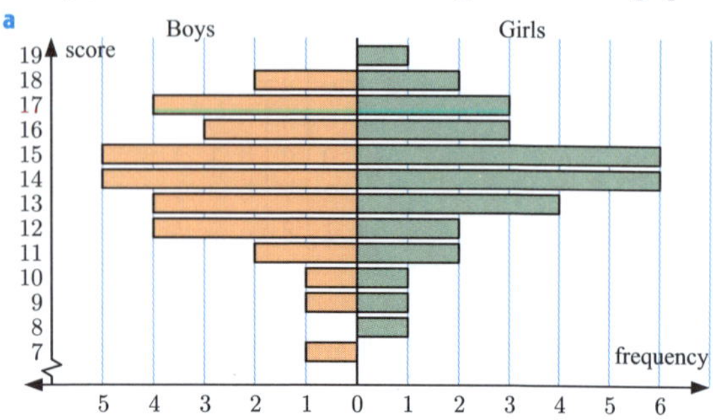

b Boys: mean $\approx$ 13.9 marks, median = 14 marks,
range = 11 marks, IQR = 4 marks
Girls: mean $\approx$ 14.2 marks, median = 14 marks,
range = 11 marks, IQR = 3 marks

c The boys had a slightly greater spread of scores as their IQR is higher. However, the ranges are the same.

d The medians are the same, but the girls' mean is slightly higher meaning they perfomed better than the boys.

4 **a**

		start	*end*
i	fastest	142 s	140 s
ii	slowest	180 s	170 s
iii	median	160 s	153 s
iv	range	38 s	30 s
v	IQR	20 s	13 s

b **i** 50% **ii** 25%

c Yes, there was an improvement. The median time improved by 7 seconds. The slowest and fastest times improved also.

d The 'start of the season' times show greater variability as indicated by greater range and IQR values.

5 **a**

	Merry-go-round	Bouncy Castle
min.	59	32
Q_1	62	63
med.	67.5	75.5
Q_3	77	85
max.	82	98

b

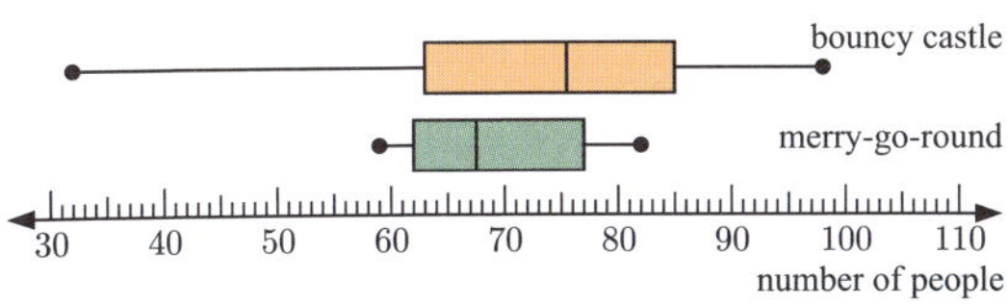

c The bouncy castle appears to be more popular. While the minimum is smaller for the bouncy castle data, all other values in its five-number summary are higher than their corresponding values for the merry-go-round data.

d The bouncy castle data has a greater spread, as indicated by the greater range and IQR.

EXERCISE 18A

1 **a** $x = \pm 4$ **b** $x = \pm\sqrt{19}$ **c** $x = \pm 3$
d $x = \pm\sqrt{6}$ **e** no real solutions **f** $x = 0$
g $x = \pm 5$ **h** no real solutions **i** $x = \pm\sqrt{3}$

2 **a** $x = 5$ or -1 **b** $x = 1$ or -9
c no real solutions **d** $x = 4 \pm \sqrt{2}$
e no real solutions **f** $x = 2$ **g** $x = -5 \pm \sqrt{3}$
h $x = 6 \pm \sqrt{2}$ **i** $x = -8 \pm \sqrt{13}$ **j** $x = -\frac{5}{3}$
k $x = \frac{1}{5}$ or $-\frac{3}{5}$ **l** $x = \frac{1+\sqrt{2}}{2}$ or $\frac{1-\sqrt{2}}{2}$

EXERCISE 18B

1 **a** $p = 0$ or $q = 0$ **b** $a = 0$ **c** $p = 0$
d $x = 0$ or $y = 0$ **e** $m = 0$ or $n = 0$
f $a = 0$ or $c = 0$ **g** $w = 0$ or $x = 0$
h $w = 0$ or $x = 0$ or $y = 0$ or $z = 0$
{at least one of them is 0}

2 **a** $x = 0$ or -3 **b** $x = 0$ or 5 **c** $x = 1$ or 3
d $x = 0$ or 2 **e** $x = 0$ or $-\frac{1}{2}$ **f** $x = -2$ or $\frac{1}{2}$
g $x = -\frac{3}{2}$ or $-\frac{1}{2}$ **h** $x = -2$ or 7 **i** $x = 5$ or $-\frac{2}{3}$
j $x = 3$ **k** $x = -\frac{1}{2}$ **l** $x = 0$
m $x = 0$ or $\frac{5}{3}$ **n** $x = 0$ or $\frac{6}{5}$ **o** $x = -\frac{1}{3}$ or $\frac{5}{2}$
p $x = -\frac{1}{3}$ **q** $x = \frac{3}{4}$ **r** $x = 0$ or $\frac{4}{3}$

EXERCISE 18C

1 **a** $x = 0$ or -12 **b** $x = 0$ or -3 **c** $x = 0$ or 4
d $x = 0$ or 7 **e** $x = 0$ or $-\frac{7}{2}$ **f** $x = 0$ or 9
g $x = 0$ or $\frac{7}{3}$ **h** $x = 0$ or $\frac{9}{4}$ **i** $x = 0$ or $-\frac{8}{3}$

2 **a** $x = -2$ or -4 **b** $x = -3$ or -8 **c** $x = -2$
d $x = \pm 7$ **e** $x = -8$ or 6 **f** $x = 5$
g $x = 5$ or -3 **h** $x = 7$ or -2 **i** $x = 3$ or 4
j $x = 5$ or 12 **k** $x = 11$ or -2 **l** $x = 6$ or -3

3 **a** $x = -8$ or 3 **b** $x = -2$ or -6 **c** $x = 5$ or -4
d $x = -7$ or 3 **e** $x = 6$ or -1 **f** $x = 1$
g $x = 3$ or 6 **h** $x = -3$ **i** $x = -15$ or 2

4 **a** $x = -\frac{1}{2}$ or -5 **b** $x = \frac{1}{5}$ or 4 **c** $x = \frac{1}{3}$ or 2
d $x = \frac{2}{3}$ or -4 **e** $x = -\frac{1}{2}$ or 7 **f** $x = -\frac{3}{7}$ or 2
g $x = \frac{2}{3}$ or 5 **h** $x = \frac{5}{2}$ or 4 **i** $x = \frac{5}{3}$ or -2
j $x = \frac{15}{2}$ or -1 **k** $x = -\frac{3}{2}$ or 5 **l** $x = -\frac{1}{3}$ or 4

5 **a** $x = -7$ or 2 **b** $x = -7$ or 4 **c** $x = 14$ or -1
d $x = -3$ or 1 **e** $x = -\frac{1}{2}$ or 3 **f** $x = -\frac{5}{3}$ or -2

6 **a** $x = 0$ or $-\frac{8}{5}$ **b** $x = 13$ or -3 **c** $x = \frac{1}{5}$ or $-\frac{4}{3}$
d $x = 4$ **e** $x = 11$ or -3 **f** $x = 0$ or $\frac{11}{2}$
g $x = -11$ or 4 **h** $x = 2$ or $\frac{2}{5}$ **i** $x = 2$ or -9
j $x = 5$ or -1 **k** $x = -\frac{3}{4}$ or $\frac{5}{2}$ **l** $x = \frac{2}{5}$ or -3

EXERCISE 18D

1 **a** 4 **b** 4 **c** 9 **d** 25 **e** 25 **f** 64
g 81 **h** 100 **i** $\frac{9}{4}$ **j** $\frac{9}{4}$ **k** $\frac{25}{4}$ **l** $\frac{25}{4}$

2 **a** $(x+1)^2 = 3$ **b** $(x-1)^2 = 5$ **c** $(x+3)^2 = 5$
d $(x-3)^2 = 16$ **e** $(x+5)^2 = 2$ **f** $(x+4)^2 = 3$
g $(x+3)^2 = 11$ **h** $(x+\frac{3}{2})^2 = \frac{5}{4}$ **i** $(x-\frac{7}{2})^2 = \frac{69}{4}$

3 **a** $x = 2 \pm \sqrt{6}$ **b** $x = -1 \pm \sqrt{7}$ **c** $x = 1 \pm \sqrt{3}$
d $x = -2 \pm \sqrt{5}$ **e** $x = -3 \pm 2\sqrt{5}$ **f** $x = 3 \pm 2\sqrt{2}$
g $x = -4 \pm \sqrt{11}$ **h** $x = \dfrac{-1 \pm \sqrt{13}}{2}$ **i** $x = \dfrac{3 \pm \sqrt{29}}{2}$

4 **a** no real solutions **b** $x = -2 \pm \sqrt{3}$ **c** no real solutions
d $x = -3 \pm \sqrt{11}$ **e** $x = \dfrac{1 \pm \sqrt{17}}{2}$ **f** no real solutions

5 **a** We need the coefficient of x^2 to be 1.
b **i** $x = \dfrac{-4 \pm \sqrt{14}}{2}$ **ii** $x = \dfrac{3 \pm \sqrt{3}}{3}$
iii $x = \dfrac{-5 \pm 2\sqrt{5}}{5}$

EXERCISE 18E

1 The original number was 3 or 5.

2 The numbers are -2 and 11 or 2 and -11.

3 The original number was $-\frac{1}{3}$ or $\frac{1}{2}$.

4 **a** $x = 4$ or 12, but $x > 7$, $\therefore$ $x = 12$
b $x = 5 + \sqrt{70} \approx 13.4$
c $x = \frac{3}{2}$ or 8, but $x > \frac{7}{3}$, $\therefore$ $x = 8$

5 **a** $(9 + 2\sqrt{7})$ cm ($x = 9 - 2\sqrt{7}$ is < 7, so we reject it.)
b ≈ 14.3 cm

6 It is 180 m by 120 m.

7 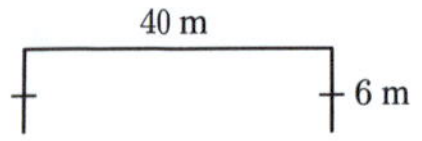

Case A

or

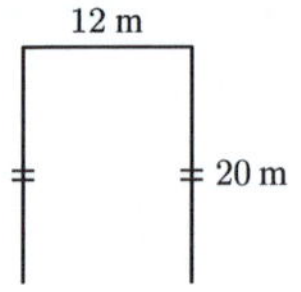

Case B

8 9 m × 12 m or 6 m × 18 m **9** 6 rows

10 **a** No **b** $(6-x)(20-x) = (120 - 26x + x^2)$ cm^2
c 4 cm × 4 cm

11 39 cm

REVIEW SET 18A

1 **a** $x = \pm 4$ **b** $x = \pm 5$

2 **a** $x = 5$ or -7 **b** no real solutions **c** $x = \dfrac{1 \pm \sqrt{5}}{3}$

3 **a** $m = 0$ or $n = 0$ **b** $p = 0$ **c** $a = 0$ or $b = 0$

4 **a** $x = 0$ or -4 **b** $x = 6$ or -11 **c** $x = 4$ or 9

5 **a** $x = -3$ or 4 **b** $x = -1$ or 5 **c** $x = \frac{3}{2}$ or -3

6 **a** 16 **b** 36 **c** $\frac{49}{4}$

7 **a** $x = -7$ or 2 **b** $x = \frac{2}{3}$ or 4

8 **a** $x = 1 \pm \sqrt{11}$ **b** $x = \dfrac{-5 \pm \sqrt{33}}{2}$

9 2 or 5 **10** 2 m

REVIEW SET 18B

1 **a** $x = \pm\sqrt{11}$ **b** no real solutions

2 **a** $x = -2 \pm \sqrt{17}$ **b** $x = -\frac{1}{2}$ **c** no real solutions

3 **a** $x = 0$ or -3 **b** $x = 1$ or 3 **c** $x = \frac{1}{3}$ or $-\frac{7}{2}$

4 **a** $x = 0$ or $\frac{9}{2}$ **b** $x = 2$ or 9 **c** $x = -3$ or 7

5 $(x-3)^2 = 19$

6 **a** $x = -6$ or 5 **b** $x = 5$ or 10 **c** $x = -\frac{3}{4}$ or 5

7 $x = 7$ **8** **a** $x = 5 \pm \sqrt{27}$ **b** no real solutions

9 **a** $x = 9$ or -12 **b** $x = -\frac{3}{2}$ or $\frac{7}{2}$ **10** 4 m

EXERCISE 19A.1

1 **a**

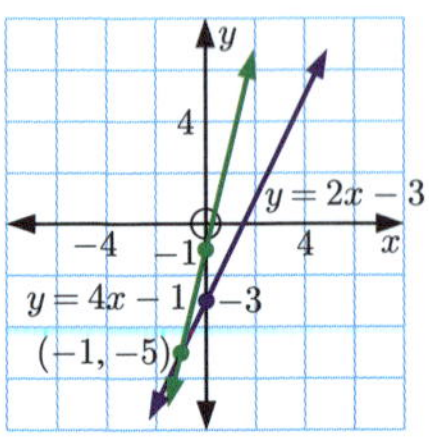

$x = -1, \quad y = -5$

b

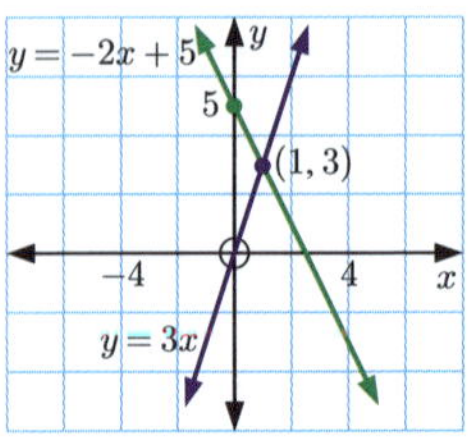

$x = 1, \quad y = 3$

c

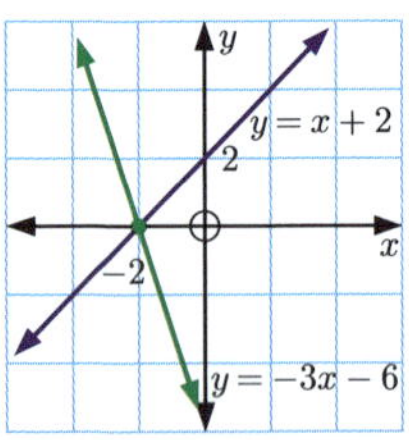

$x = -2, \quad y = 0$

2 **a**

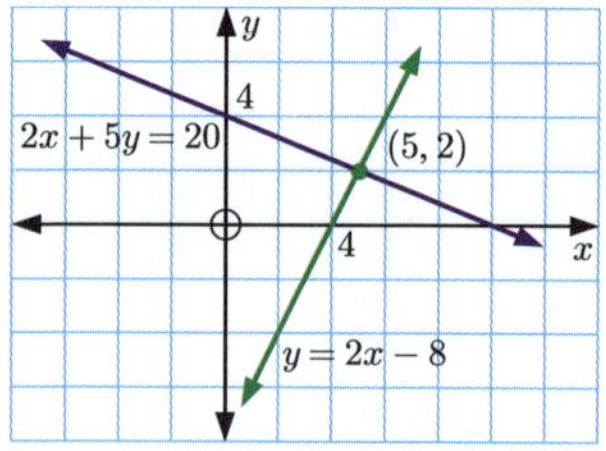

$x = 5, \quad y = 2$

b

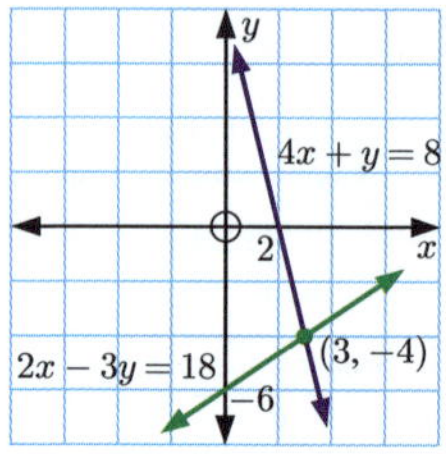

$x = 3, \quad y = -4$

c

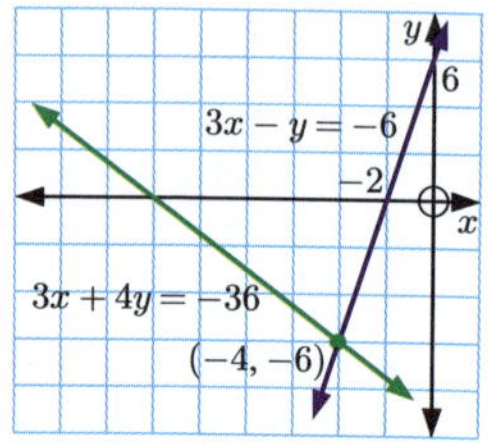

$x = -4, \quad y = -6$

3 **a**

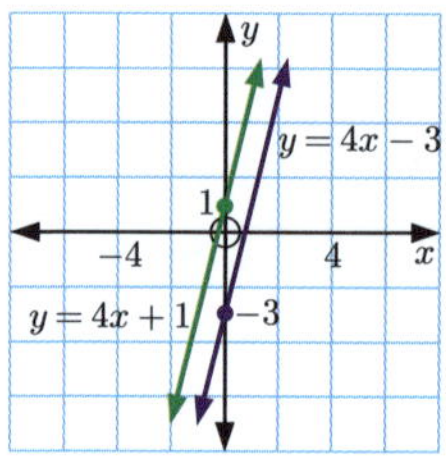

No solutions, the lines are parallel.

b

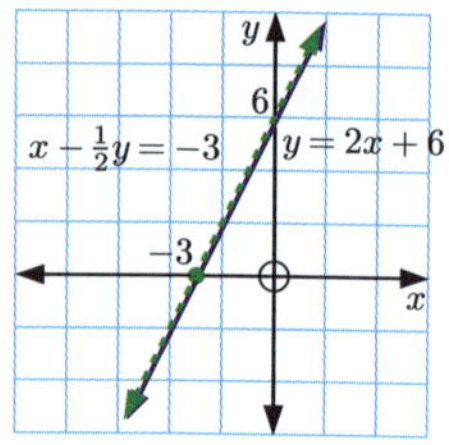

Infinitely many solutions, the lines are coincident.

c

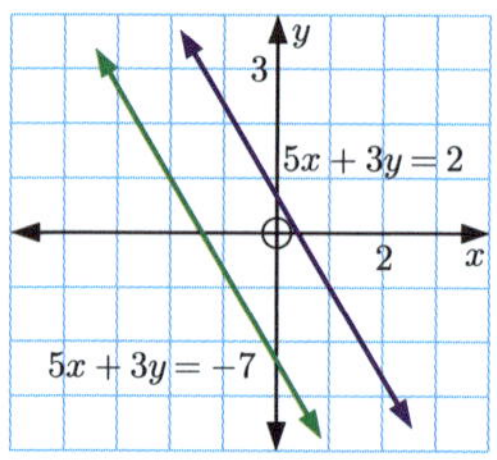

No solutions, the lines are parallel.

EXERCISE 19A.2

1 **a** $x = 0.75, \quad y = 1.75$ **b** $x = 1.4, \quad y = -1.8$
c $x = -0.25, \quad y = -4.75$ **d** $x = 1.8, \quad y = 4.2$
e $x = 1.\overline{27}, \quad y = -4.\overline{63}$ **f** $x \approx -2.71, \quad y \approx -1.53$

2 **a** infinitely many solutions **b** no solutions
c no solutions

EXERCISE 19B

1 **a** $x = 7, \quad y = 4$ **b** $x = -1, \quad y = -5$
c $x = 3, \quad y = -7$ **d** $x = 5, \quad y = -2$
e $x = -\frac{1}{2}, \quad y = \frac{5}{2}$ **f** $x = \frac{20}{47}, \quad y = \frac{1}{47}$

2 **a** $x = 4, \quad y = -5$ **b** $x = -2, \quad y = -3$
c $x = -2, \quad y = \frac{5}{2}$ **d** $x = \frac{1}{3}, \quad y = -\frac{4}{3}$
e $x = -\frac{3}{4}, \quad y = \frac{5}{8}$ **f** $x = \frac{23}{31}, \quad y = -\frac{29}{31}$

3 **a** $x = 5, \quad y = \frac{1}{2}$ **b** $x = -\frac{4}{3}, \quad y = 2$
c $x = 3, \quad y = -\frac{1}{4}$ **d** $x = \frac{2}{5}, \quad y = -\frac{6}{5}$
e $x = \frac{3}{7}, \quad y = -\frac{5}{7}$ **f** $x = \frac{7}{2}, \quad y = -\frac{8}{5}$

4 **b** no solutions

5 **b** infinitely many solutions

6 **a** $y = \dfrac{-4x-4}{5}$ **b** $x = -\frac{17}{7}, \quad y = \frac{8}{7}$

EXERCISE 19C

1 **a** $11x = 11$ **b** $4y = 12$ **c** $9x = 9$ **d** $9x = 6$
e $-y = 11$ **f** $-11y = -11$

2 **a** $x = 2, \quad y = 6$ **b** $x = 1, \quad y = -2$
c $x = -1, \quad y = -3$ **d** $x = 1, \quad y = -5$
e $x = 2, \quad y = -2$ **f** $x = -\frac{17}{4}, \quad y = \frac{5}{2}$

3 **a** $10x + 25y = 5$ **b** $-3x + y = -4$
c $3x - 21y = 24$ **d** $-10x - 8y = -18$
e $-18x - 12y = 12$ **f** $-16x + 8y = -12$

4 **a** $x = 5, \quad y = -2$ **b** $x = 1, \quad y = 2$
c $x = 1, \quad y = -6$ **d** $x = \frac{1}{2}, \quad y = -3$
e $x = 1, \quad y = -\frac{2}{3}$ **f** $x = -\frac{3}{4}, \quad y = 3$

5 **a** $x = 5, \quad y = 1$ **b** $x = 1, \quad y = 1$
c $x = -3, \quad y = 4$ **d** $x = \frac{8}{5}, \quad y = \frac{22}{5}$
e $x = \frac{15}{2}, \quad y = -\frac{5}{2}$ **f** $x = -\frac{10}{3}, \quad y = -\frac{11}{3}$

6 **a** infinitely many solutions, the lines are coincident
b no solutions, the lines are parallel

EXERCISE 19D

1 56 and 16 **2** 22 and 8 **3** 17 and 51
4 fish costs £2.20 each, chips cost £0.75 per serve
5 coffee costs €2.90 per cup, muffins cost €1.25 each
6 30 fifty cent coins; 26 ten cent coins
7 Anthony has €26.10, Michelle has €39.15
8 33, 600 mL cartons; 60, 1 L cartons **9** $a = 3, \quad b = 5$
10 length 18 cm, width 10 cm
11 before 3 pm: $2.50/h, after 3 pm: $4/h
12 6 small sets and 5 large sets

EXERCISE 19E

1 **a** $x = -3, \ y = 20$ and $x = 2, \ y = 5$
b $x = 2, \ y = 13$ and $x = -9, \ y = 35$
c $x = -3, \ y = 24$ and $x = 1, \ y = 4$
d $x = -2, \ y = 18$ and $x = 5, \ y = 46$
e $x = -2, \ y = 1$ and $x = 6, \ y = 73$
f $x = -3, \ y = 19$

2 **a** $x = -5, \ y = -4$ and $x = 4, \ y = 5$
b $x = 4, \ y = 8$
c $x = -\frac{5}{3}, \ y = -3$ and $x = \frac{1}{2}, \ y = 10$

3 **a** no real solutions
b $x = -2 - \sqrt{10}, \ y = 2 - \sqrt{10}$ and
$x = -2 + \sqrt{10}, \ y = 2 + \sqrt{10}$
c no real solutions

4 **Hint:** Try to solve $\begin{cases} x + y = 10 \\ xy = 30 \end{cases}$.

5 **a** $x = 0, \ y = 0$ and $x = 1, \ y = 1$
b $x = 1, \ y = 1$
c $x = -1, \ y = -2$ and $x = 0, \ y = 0$ and
$x = 1, \ y = 2$

REVIEW SET 19A

1 **a**

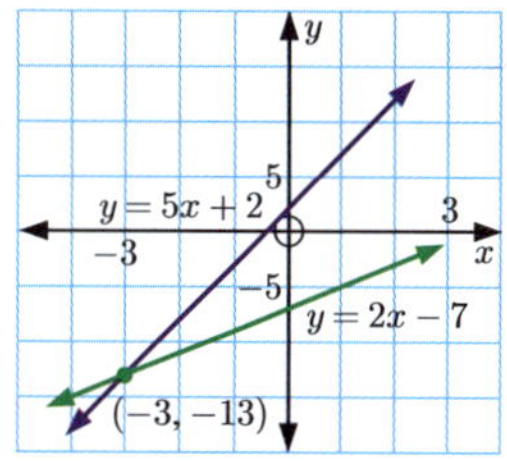

$x = -3, \quad y = -13$

b

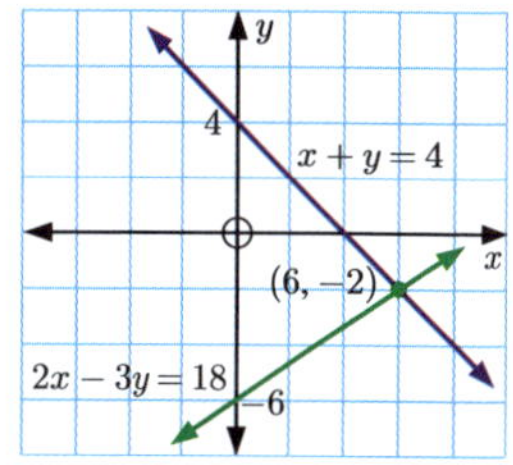

$x = 6, \quad y = -2$

2 **a** $x = 3.25, \quad y = 5.75$ **b** $x = 0.2, \quad y = -0.6$

3 **a** $x = 2, \quad y = 2$ **b** $x = -\frac{2}{9}, \quad y = \frac{17}{9}$

4 $10 \neq 3 \quad \therefore$ no real solutions. The lines are parallel.

5 **a** $6x = -6$ **b** $-4y = 40$

6 **a** $x = -3, \quad y = 1$ **b** $x = \frac{3}{2}, \quad y = -\frac{5}{2}$

7 **a** $x = 4, \quad y = -3$ **b** $x = \frac{4}{3}, \quad y = -\frac{2}{3}$

8 adult $18, student $12

9 **a** **i, ii, iii**

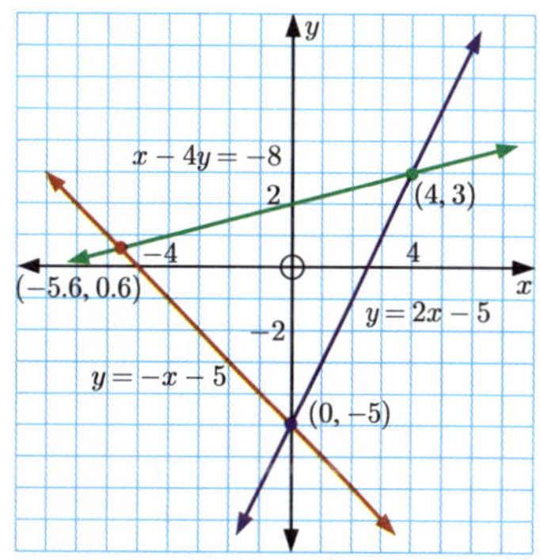

b **i** $x = 4, \quad y = 3$ **ii** $x = 0, \quad y = -5$
iii $x = -5.6, \quad y = 0.6$

10 14 fifty cent coins; 7 twenty cent coins

11 $x = -3, \ y = 16$ and $x = 5, \ y = 0$

12 $x = -1, \ y = 3$ and $x = 2, \ y = 0$

REVIEW SET 19B

1

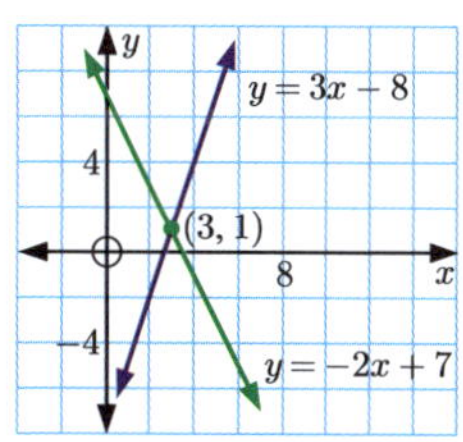

$x = 3, \quad y = 1$

2 $x = \frac{23}{9}, \quad y = \frac{29}{9}$

3 The lines are coincident. There are infinitely many solutions.

4 $-6x + 9y = 12$ **5** $x = -\frac{13}{5}, \quad y = \frac{29}{5}$

6 **a** $x = -\frac{4}{3}, \quad y = -2$ **b** $x = -43, \quad y = -31$

7 **a** $4 = 4$ which is true for all values of x.
b infinitely many solutions

8 7 points

9 **a** **i** $8 **ii** $5 **b** **i** $5 **ii** $2

10 11 packs **11** $x=-6,\ y=-6$ and $x=3,\ y=12$

12 **a** **i** 3 minutes **ii** 5 minutes
b 38 minutes

EXERCISE 20A

1 **A** and **D**, **B** and **E** **2** **A**, **D**, and **M**; **C** and **O**; **I** and **L**

3 **a** **i** [QR] **ii** $\widehat{QRS}$ **b** **i** [NO] **ii** $\widehat{NOM}$
c **i** [QR] **ii** $\widehat{QRP}$ **d** **i** [HE] **ii** $\widehat{HEF}$

EXERCISE 20B

1 **a** congruent {SAS}
b not congruent {could be different sizes}
c congruent {SSS}
d not congruent {equal sides not in corresponding positions}
e congruent {RHS} **f** congruent {AAcorS}

2 **a** A and C {SAS} **b** B and C {AAcorS}
c B and C {SSS} **d** A and C {RHS}

EXERCISE 20C

1 **a** **Hint:** Use AAcorS. **b** 5 cm

2 **Hint:** Show that $\triangle AOC \cong \triangle BOD$.

3 **Hint:** Show that $\triangle PQR \cong \triangle PTS$.

4 **a** **Hint:** Use RHS.

5 **Hint:** Let P be any point on the perpendicular bisector, and M be the midpoint of [AB].
Show that $\triangle AMP \cong \triangle BMP$.
Consider also the case where P = M.

6 **a** **Hint:** Show that $\triangle AED \cong \triangle BFC$.
b **Hint:** Show that $\triangle ACD \cong \triangle BDC$.

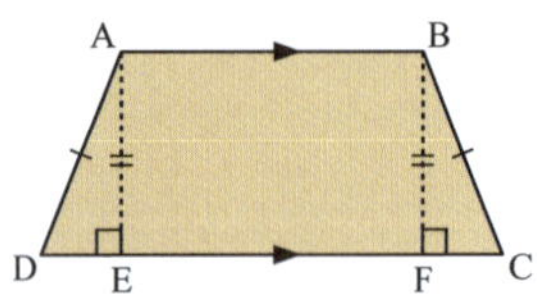

EXERCISE 20D.1

1 **a** similar **b** not similar **c** similar
d not similar **e** similar **f** similar

2 **Hint:** Show that all pairs are equiangular.

3 **Hint:** Show that [AB] is parallel to [CD].

EXERCISE 20D.2

1 **a** $x=8$ **b** $x=6$ **c** $x\approx 11.3$ **d** $x=9.75$

2 **a** $x=8$ **b** $x=3.75$ **c** $x=4.8$
d $x\approx 7.71$ **e** $x=4.8$ **f** $x\approx 13.3$
g $x=17.5$ **h** $x=12$ **i** $x=8.75$

3 $\frac{50}{3}$ m **4** $x=\frac{12}{7}$

EXERCISE 20E

1 9.22 m **2** **a** 1650 m **b** 1677 m

3 43.75 m **4** 6.5 m **5** **b** 13.3 m

6 **a** 25 m **b** 19.2 m **7** PR $\approx$ 3.67 m

8 **a** $\frac{9}{20}$ **b** 9 : 2

EXERCISE 20F

1 **a** $x=32$ **b** $x=16$ **c** $x=5$ **d** $x=4.8$

2 **a** $k=4$ **b** 20 cm and 24 cm **c** 16

3 **b** $k=\frac{5}{2}$ **c** **i** 100 cm^2 **ii** 84 cm^2

4 ≈ 1.45 cm **5** 4 units^2

EXERCISE 20G

1 **a** $V=270$ **b** $V=31.25$ **2** **a** $x=8$ **b** $x\approx 5.26$

3 12 cm **4** $k\approx 1.19$ **5** **a** 4.116 m^3 **b** 3.75 m^2

6 No, comparing lengths $k=1.6$, comparing capacities $k\approx 1.37$.

REVIEW SET 20A

1 **a** [GH] **b** $\widehat{FEH}$

2 **a** not congruent {need another pair of equal angles or sides}
b congruent {SSS}

3 **a** **Hint:** Use SAS. **4** **a** not similar **b** similar

5 **a** $x=3$ **b** $x=4$ **6** 16 cm **7** **c** 6.4 cm

8 **a** $x=4$ **b** $x\approx 42.7$ **9** **a** $x=15$ **b** $y=32$

10 648 cm^3

REVIEW SET 20B

1 **a** not congruent
{equal sides are not in corresponding positions}
b not congruent {the triangles may be different sizes}

2 $x=9$ **3** **B** and **C** {SAS} **4** **b** 7.2 cm

5 **Hint:** Draw lines from P perpendicular to [AB] and [CB], meeting [AB] and [CB] at X and Y respectively. Show that PX = PY.

6 **a** $A=7$ **b** $x\approx 8.14$

7 **Hint:** Let $\widehat{DAX}=\alpha$, find other angles in terms of α.

8 5.4 cm^2 **9** **a** $x=4.5$ **b** 8 : 27

10 **a** SU = 5.5 m, BC = 8.2 m
b No, on BC, ≈ 8.31 m from B.

EXERCISE 21A.1

1 **a** quadratic function; $a=2,\ b=1,\ c=4$
b not a quadratic function; no x^2 term
c quadratic function; $a=-1,\ b=-1,\ c=-1$
d quadratic function; $a=1,\ b=0,\ c=0$
e quadratic function; $a=\frac{1}{2},\ b=0,\ c=-6$
f not a quadratic function; contains x^3 term
g quadratic function; $a=-3,\ b=0,\ c=0$
h not quadratic function; contains x^{-2} term
i quadratic function; $a=-2,\ b=0,\ c=-\frac{5}{2}$

2 **a** $y=10$ **b** $y=-5$ **c** $y=-5$

3 **a** $y=4$ **b** $y=14$ **c** $y=34$

4 **a** $y=1$ **b** $y=1$ **c** $y=\frac{9}{2}$

EXERCISE 21A.2

1 **a**

x	-3	-2	-1	0	1	2	3
y	9	4	1	0	1	4	9

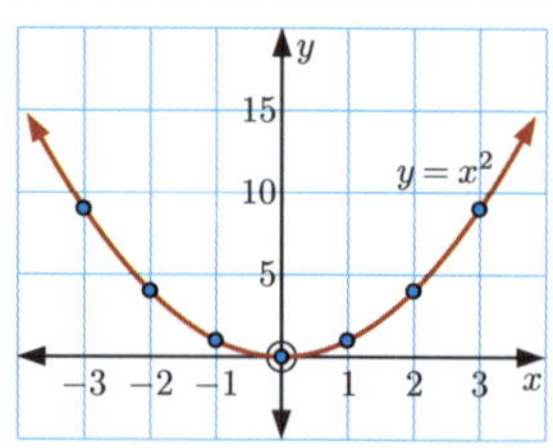

b

x	-3	-2	-1	0	1	2	3
y	15	8	3	0	-1	0	3

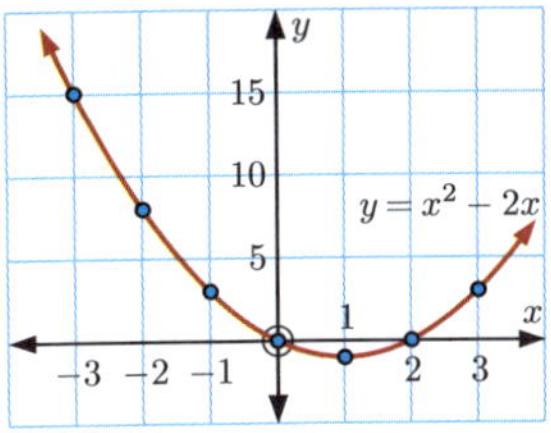

c

x	-3	-2	-1	0	1	2	3
y	13	6	1	-2	-3	-2	1

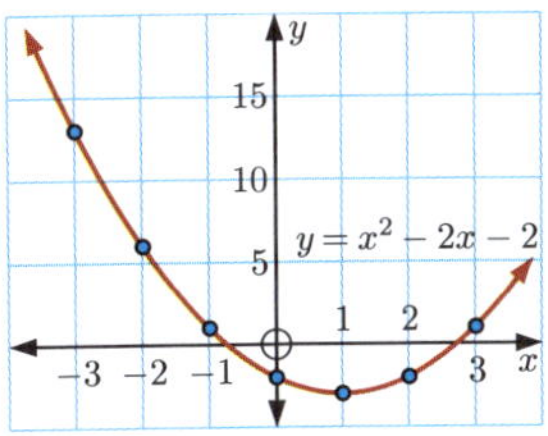

d

x	-3	-2	-1	0	1	2	3
y	18	8	2	0	2	8	18

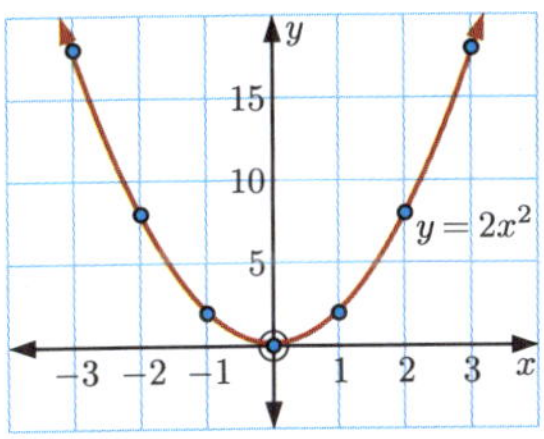

e

x	-3	-2	-1	0	1	2	3
y	19	9	3	1	3	9	19

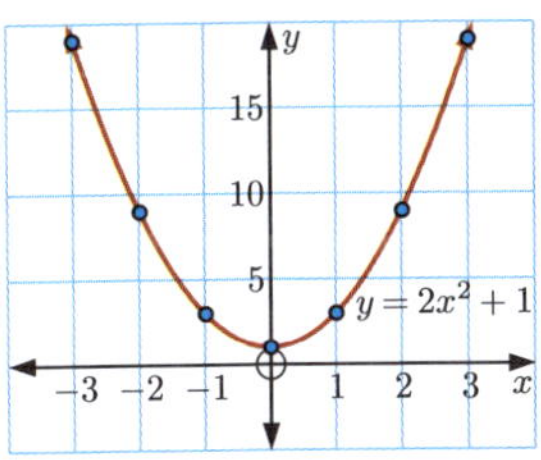

f

x	-3	-2	-1	0	1	2	3
y	$\frac{5}{2}$	0	$-\frac{3}{2}$	-2	$-\frac{3}{2}$	0	$\frac{5}{2}$

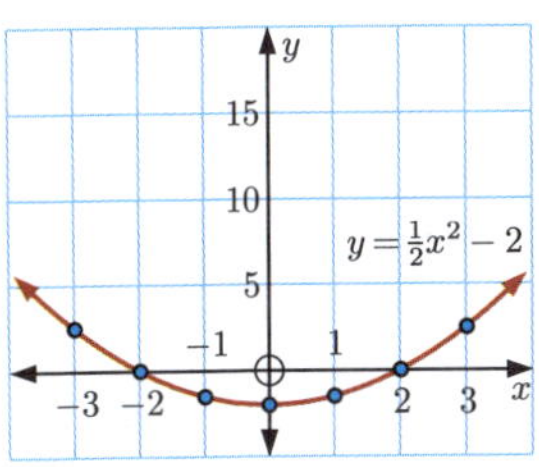

g

x	-3	-2	-1	0	1	2	3
y	-9	-4	-1	0	-1	-4	-9

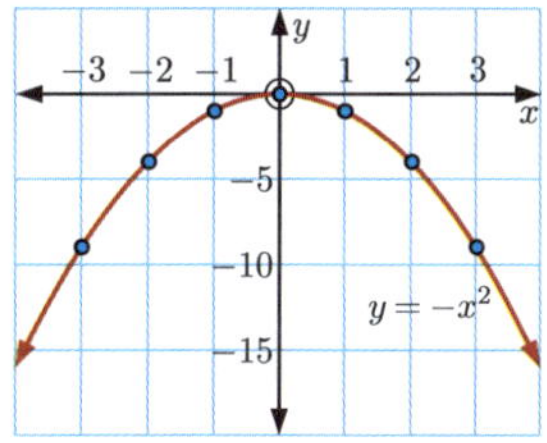

h

x	-3	-2	-1	0	1	2	3
y	-12	-6	-2	0	0	-2	-6

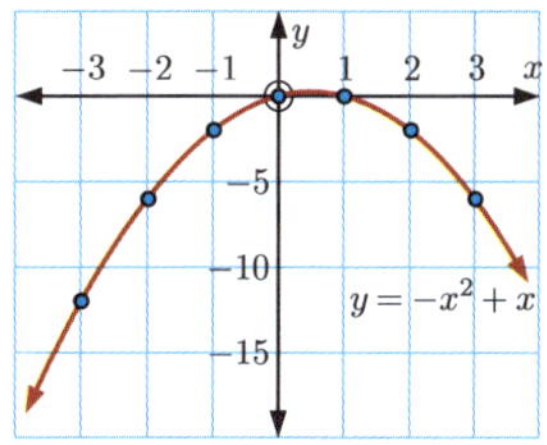

i

x	-3	-2	-1	0	1	2	3
y	-14	-7	-2	1	2	1	-2

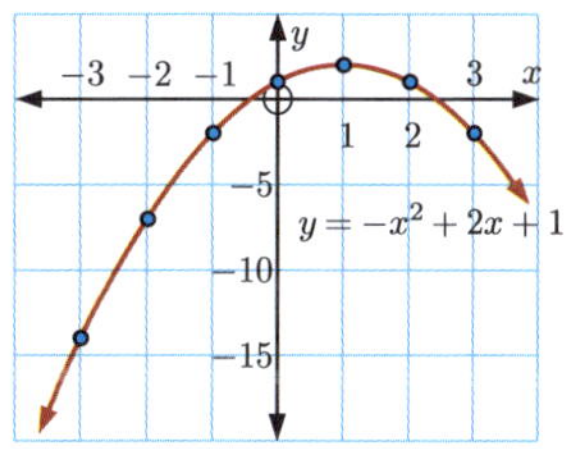

2 A does not, B does

3 P does, Q does not, R does

4 **a** the sign of a, if $a > 0$  , if $a < 0$ 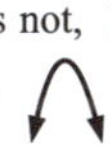

b the magnitude of a; if $a < -1$ or $a > 1$ the graph is thinner than $y = x^2$, if $-1 < a < 0$ or $0 < a < 1$ the graph is wider than $y = x^2$

c the y-intercept is c

5 **a** $k = 3$ **b** $d = -39$ **c** $m = -7$ **d** $n = 4$

6 **a** $(2, 1)$ does, $(-2, -13)$ does not

b $k = 1$, $m = -5$

c

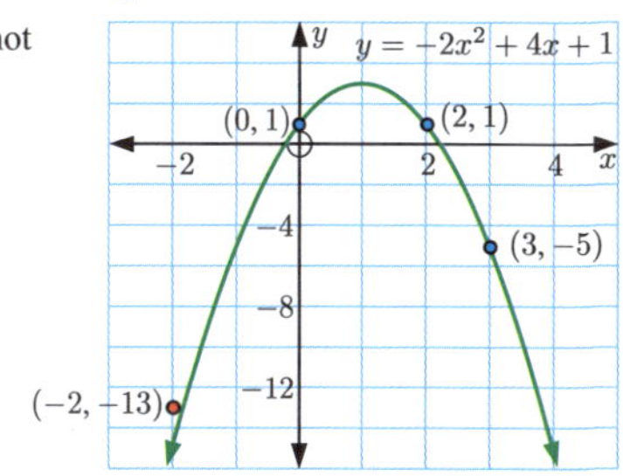

EXERCISE 21A.3

1 **a** $x = 0$ or -2 **b** $x = -1$ **c** $x = 3$ or -5

2 **a** $x = -1$ or 5 **b** $x = 0$ or 4 **c** $x = 2$

3 **a** **i** $y = 70$ **ii** $y = -20$

b **i** $x = -2$ or -3 **ii** $x = 2$ or -7

4 **a** $m = 2$ or -5 **b** $n = 1$ or -11

5 **a** **i** 55 m **ii** 135 m
b **i** 2 sec and 10 sec **ii** 4 sec and 8 sec **c** 12 sec

6 **a** **i** \$320 loss **ii** \$640 **iii** \$880
b 3 or 13 cabinets

7 **a** a quadratic **b** **i** 6 m **ii** 28.4 m
c yes, by 1.1 m

8 **a** **i** two points **ii** one point **iii** no points
b **i** $k > -4$ **ii** $k = -4$ **iii** $k < -4$

EXERCISE 21B.1

1 **a**
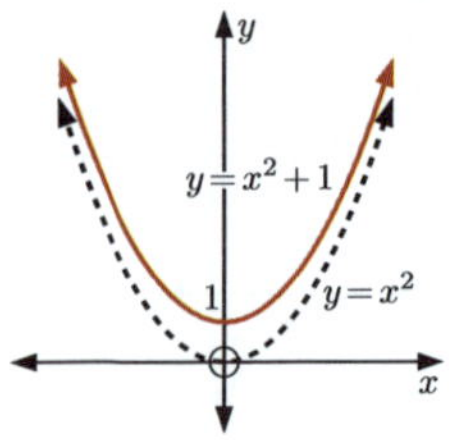

b
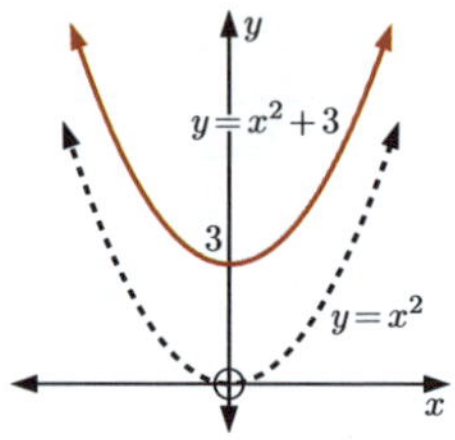

c
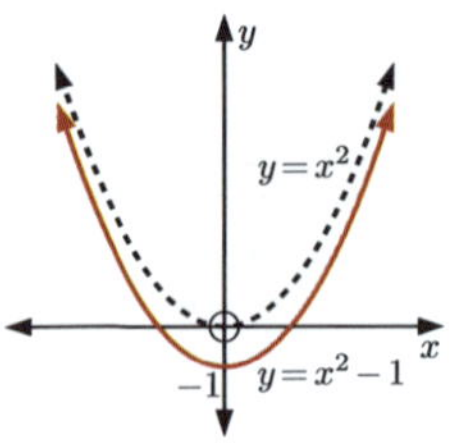

d
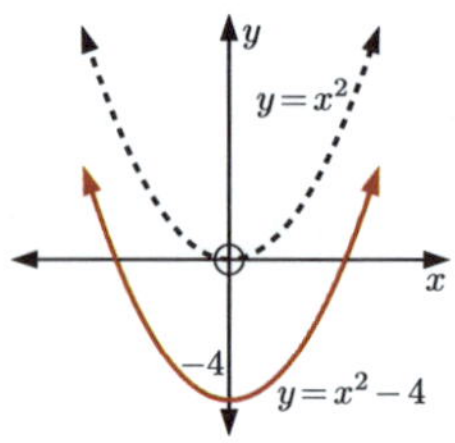

e
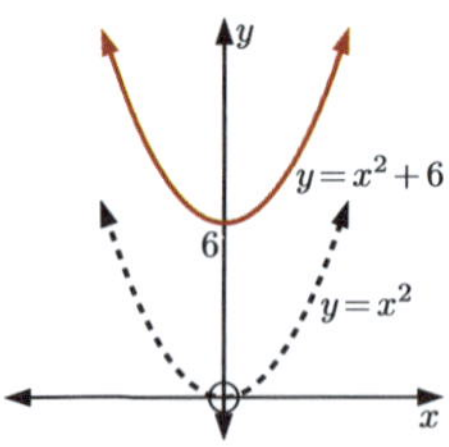

f
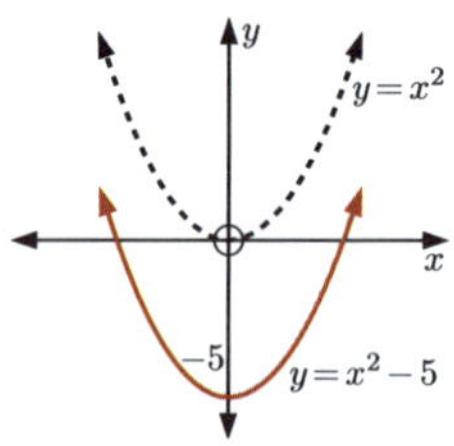

2 **a**
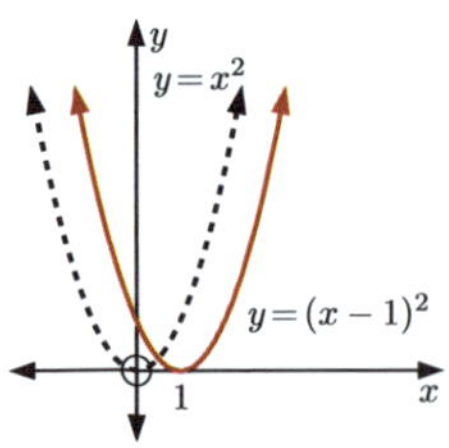

b
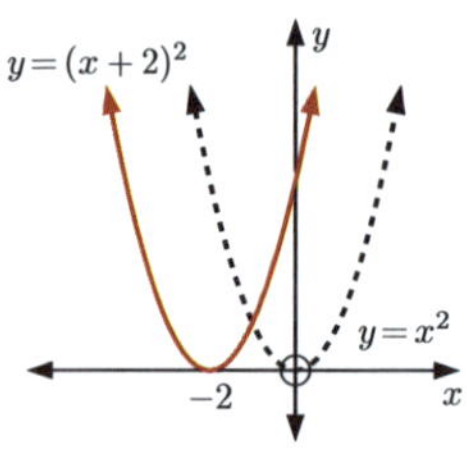

c
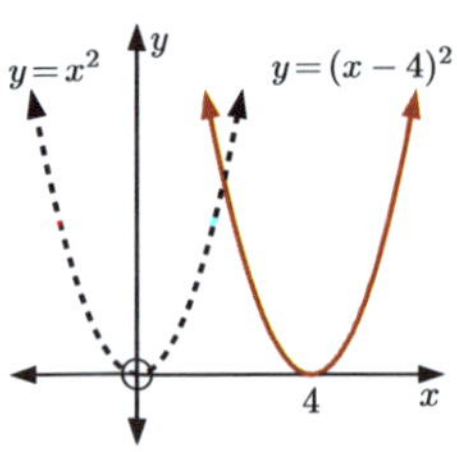

d
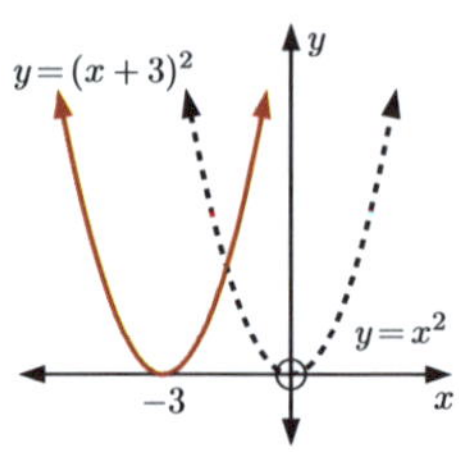

e
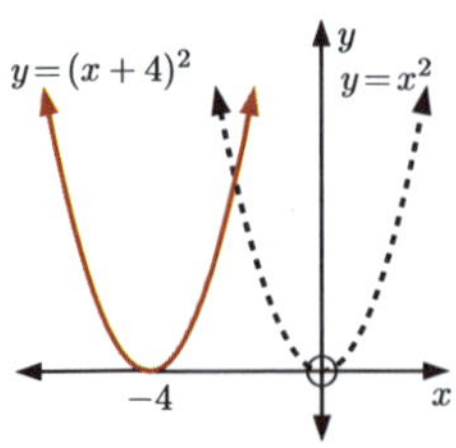

f
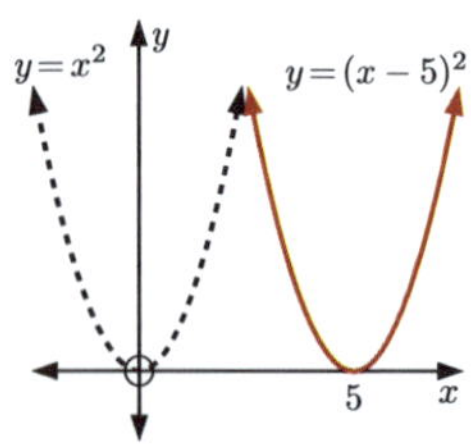

3 **a** $(0, 2)$ **b** $(-1, 0)$ **c** $(0, -3)$ **d** $(6, 0)$

4 **a**
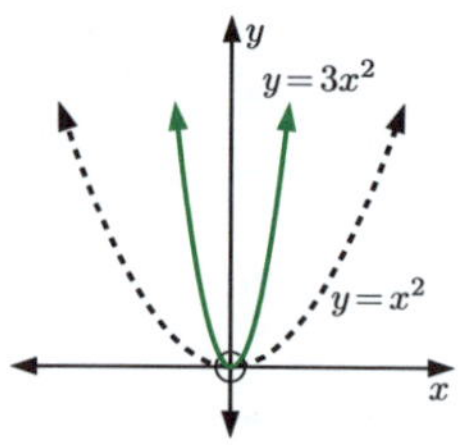

$y = 3x^2$ is 'thinner' than $y = x^2$, and the graph opens upwards.

b
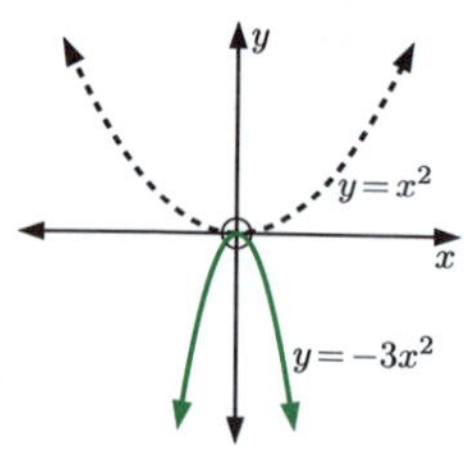

$y = -3x^2$ is 'thinner' than $y = x^2$, and the graph opens downwards.

c
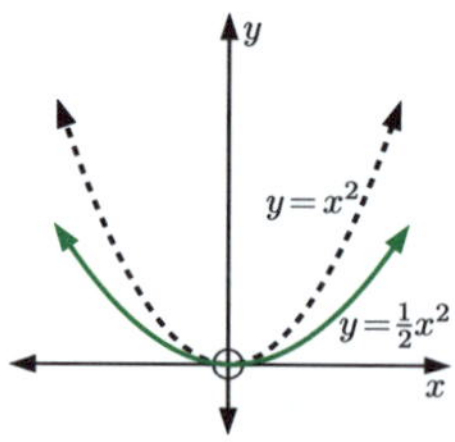

$y = \frac{1}{2}x^2$ is 'wider' than $y = x^2$, and the graph opens upwards.

d
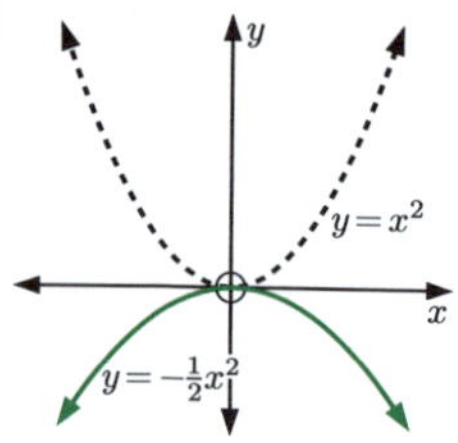

$y = -\frac{1}{2}x^2$ is 'wider' than $y = x^2$, and the graph opens downwards.

e
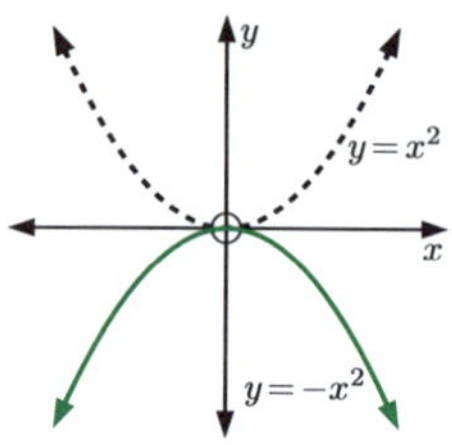

$y = -x^2$ is the same width as $y = x^2$, and the graph opens downwards.

f
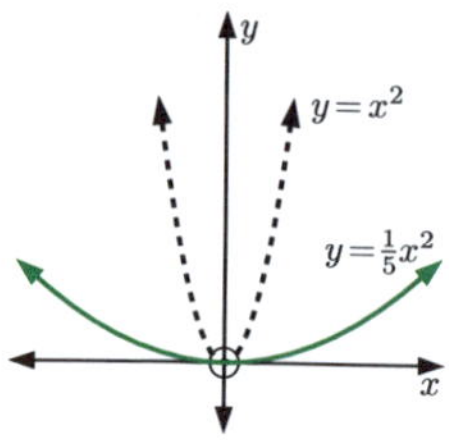

$y = \frac{1}{5}x^2$ is 'wider' than $y = x^2$, and the graph opens upwards.

5 **a** $y = (x-2)^2 + 1$
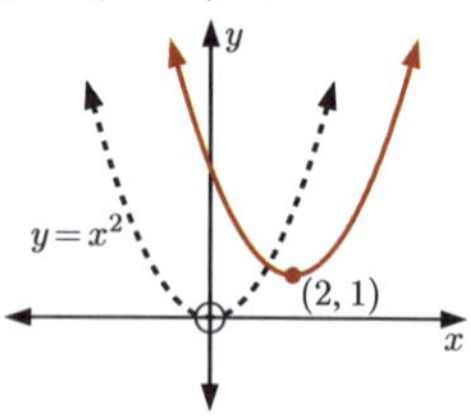

vertex is at $(2, 1)$

b $y = (x+3)^2 + 4$
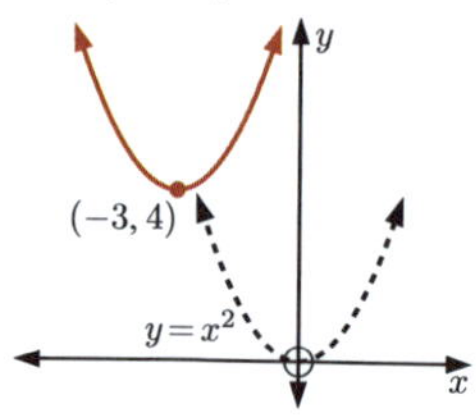

vertex is at $(-3, 4)$

c $y = (x-1)^2 - 3$
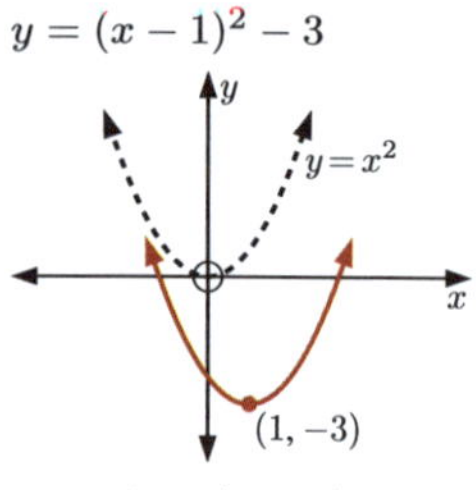

vertex is at $(1, -3)$

d $y = (x+4)^2 - 2$
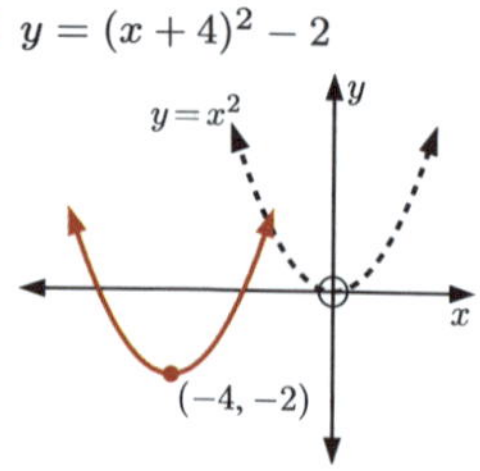

vertex is at $(-4, -2)$

e $y=(x-4)^2-4$

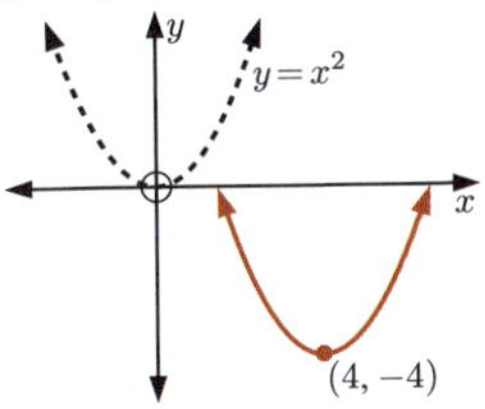

vertex is at $(4,\ -4)$

f $y=(x+5)^2+1$

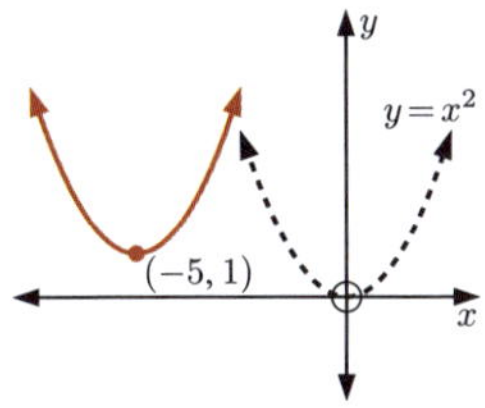

vertex is at $(-5,\ 1)$

EXERCISE 21B.2

1 **a** $y=(x+1)^2+1$

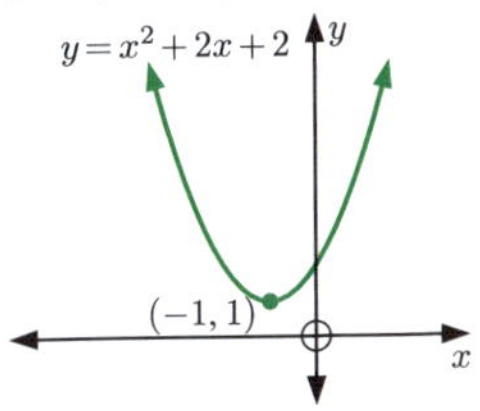

vertex is at $(-1,\ 1)$

b $y=(x-2)^2-2$

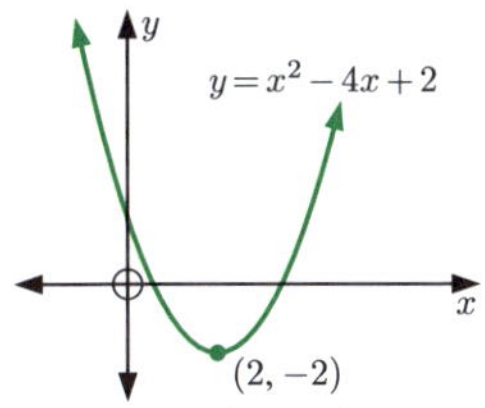

vertex is at $(2,\ -2)$

c $y=(x+1)^2-4$

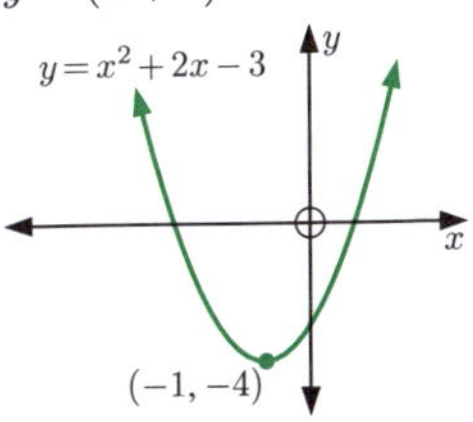

vertex is at $(-1,\ -4)$

d $y=(x-3)^2-8$

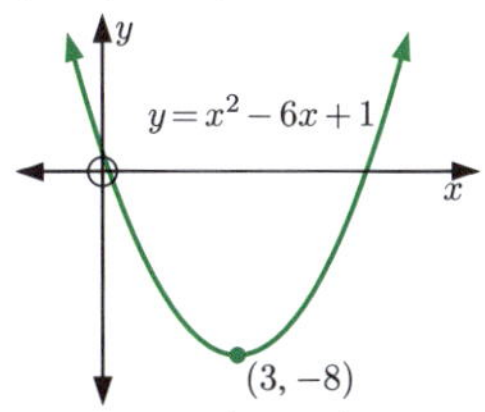

vertex is at $(3,\ -8)$

e $y=(x+1)^2-1$

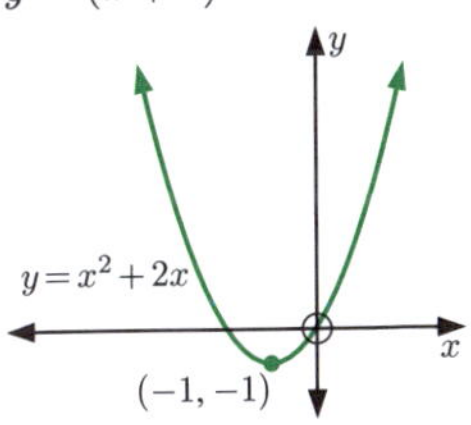

vertex is at $(-1,\ -1)$

f $y=(x-2)^2-4$

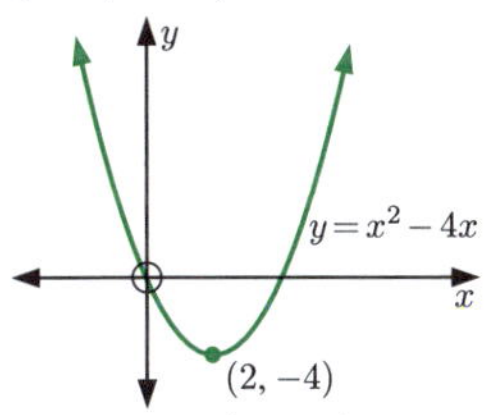

vertex is at $(2,\ -4)$

g $y=(x+\frac{1}{2})^2-\frac{5}{4}$

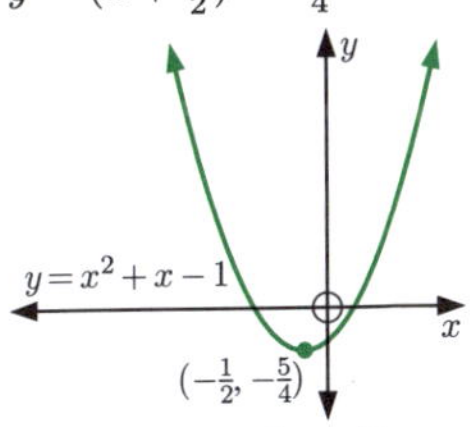

vertex is at $(-\frac{1}{2},\ -\frac{5}{4})$

h $y=(x-\frac{3}{2})^2-\frac{1}{4}$

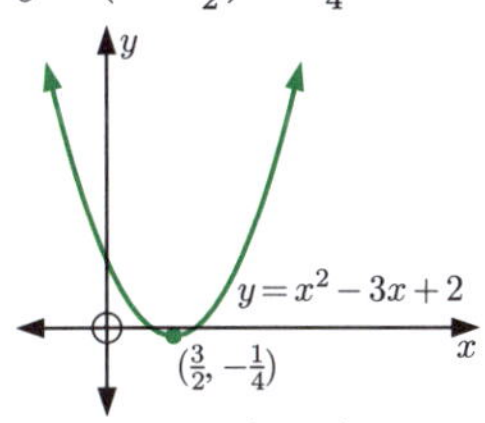

vertex is at $(\frac{3}{2},\ -\frac{1}{4})$

i $y=(x+\frac{5}{2})^2-\frac{45}{4}$

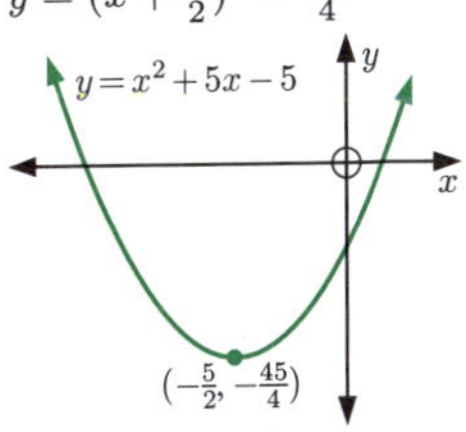

vertex is at $(-\frac{5}{2},\ -\frac{45}{4})$

EXERCISE 21C

1 **a** 1 **b** 3 **c** -2 **d** 4 **e** 6 **f** 1
g 4 **h** 3 **i** 7

2 **a** -2 **b** -3 **c** 2 **d** 0 **e** 4 **f** -10

3 **a** 1 and -3 **b** -2 and 3 **c** 3 and -7
d 2 and 5 **e** -5 **f** 7

4 **a** 2 and -2 **b** 3 and -3 **c** 3 and 4
d 3 and -2 **e** 0 and 3 **f** 2 and 5
g 3 **h** -1 **i** 1 and -6
j 1 and -5 **k** -2 and 8 **l** 3 and -7
m $\frac{1}{2}$ and -4 **n** $-\frac{1}{3}$ and 2 **o** $-\frac{1}{2}$ and $\frac{5}{2}$

EXERCISE 21D

1 **a**

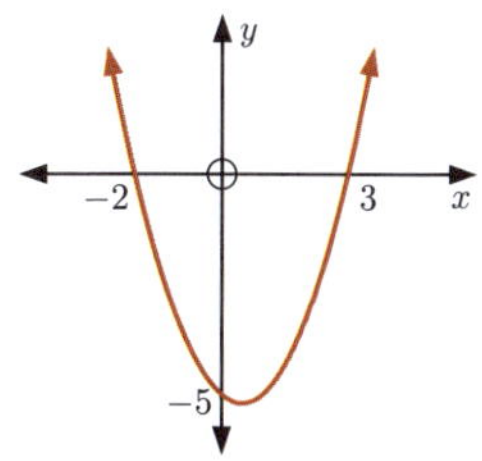

b

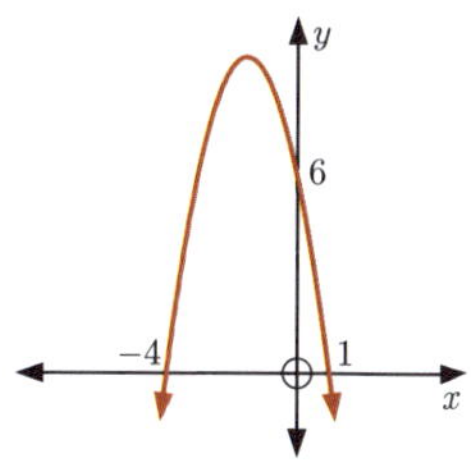

c

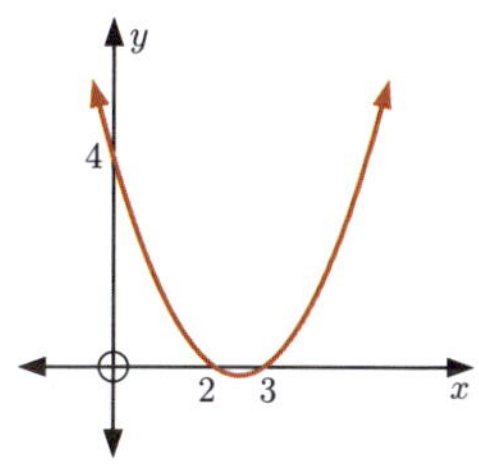

d

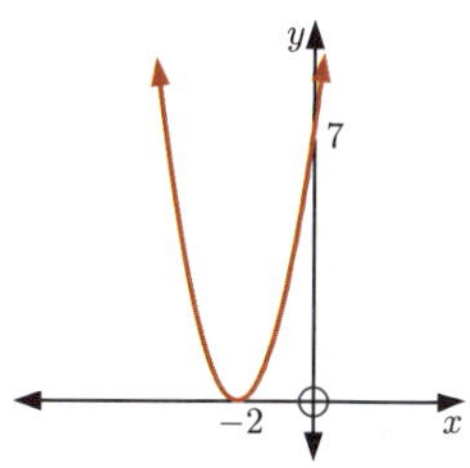

e

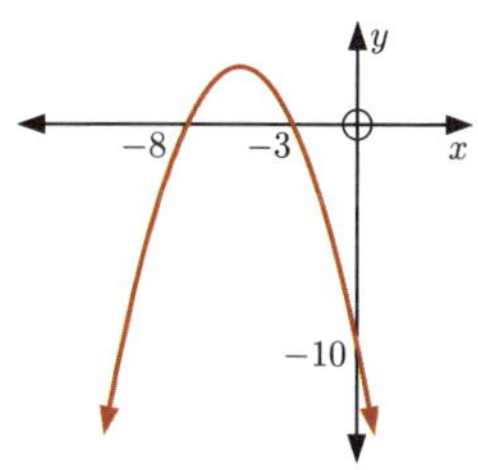

f

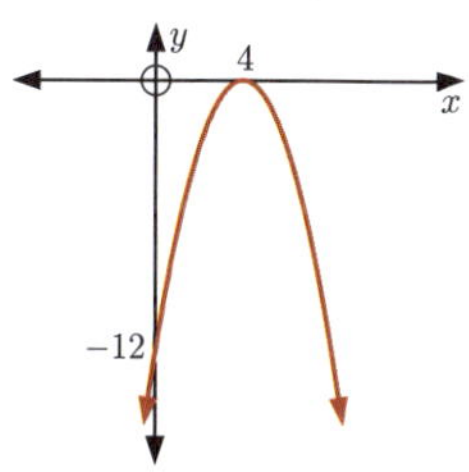

2 **a** **i** 1 and 3 **ii** 3
iii

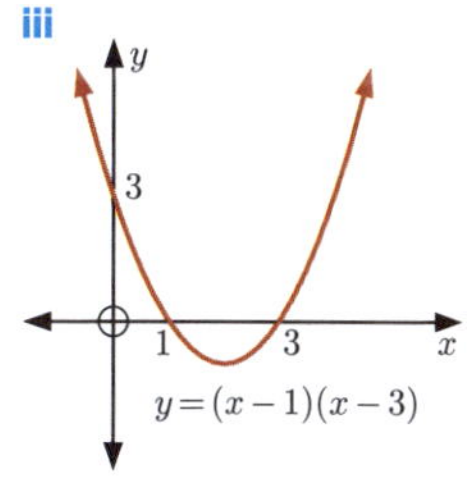

b **i** 1 and 5 **ii** -5
iii

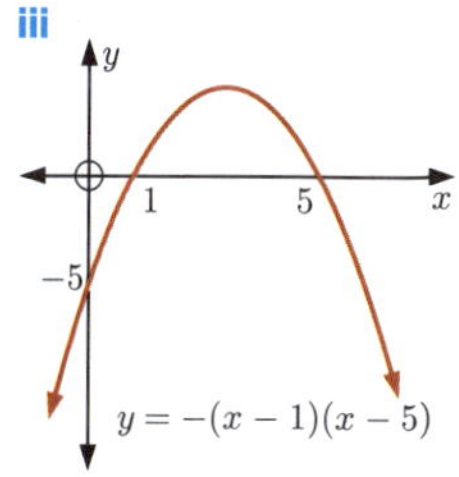

c **i** -2 and 4 **ii** -8
iii

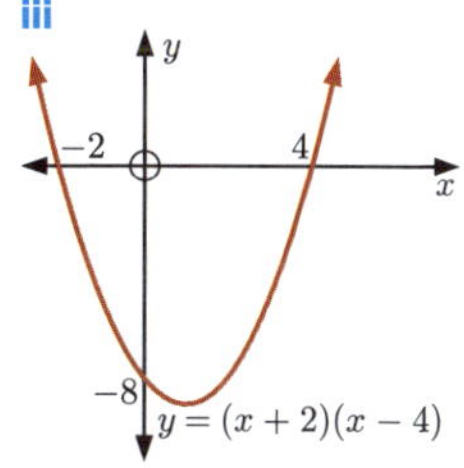

d **i** 1 **ii** 1
iii

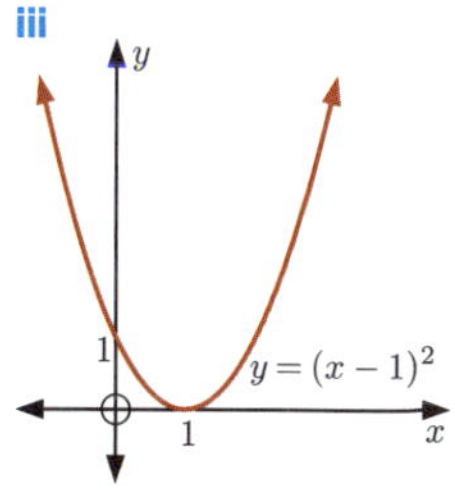

e **i** -2 and 3 **ii** -18
iii

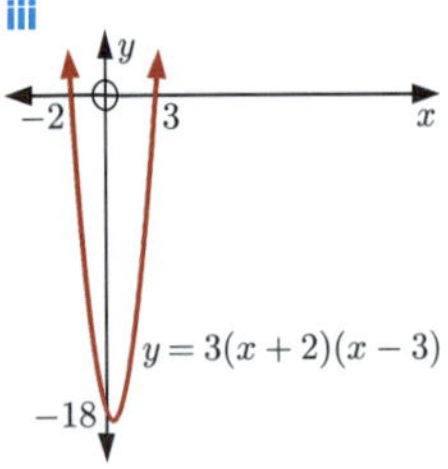

f **i** -4 and 2 **ii** 16
iii

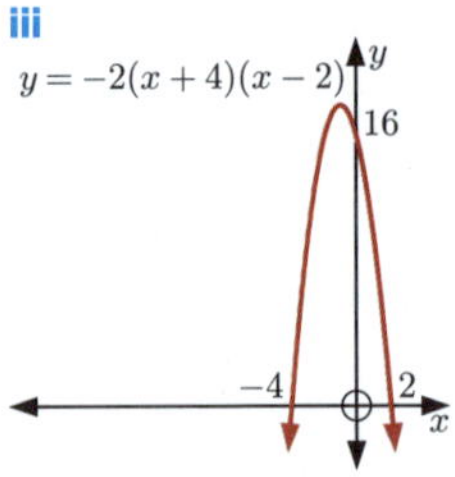

g **i** -3 **ii** -3
iii

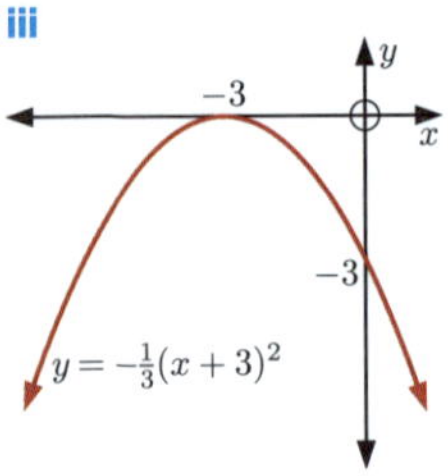

h **i** -5 and 2 **ii** -10
iii

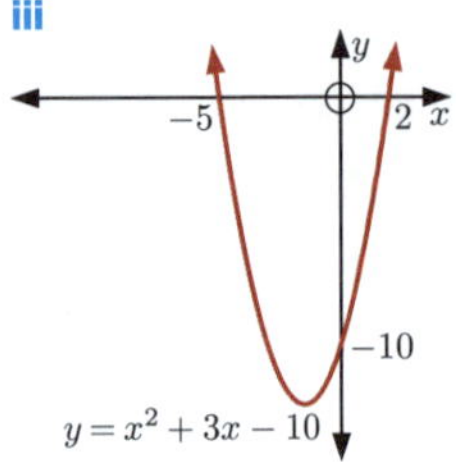

i **i** -2 **ii** -4
iii

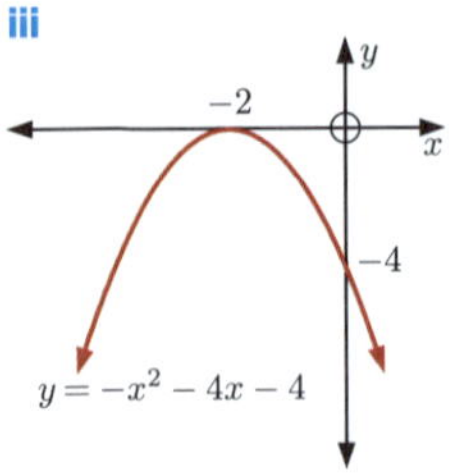

j **i** -6 and -1 **ii** 6
iii

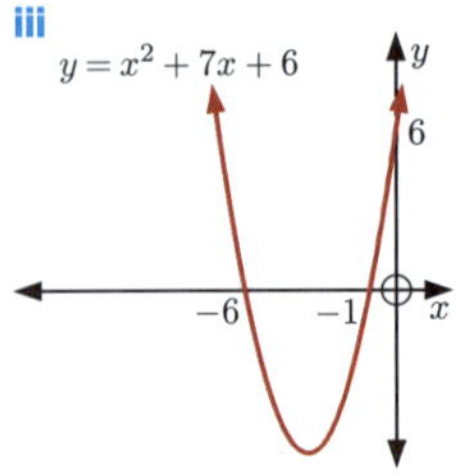

k **i** 5 **ii** 50
iii

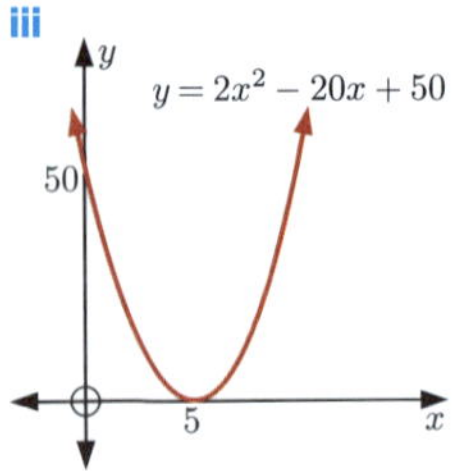

l **i** -2 and $\frac{2}{3}$ **ii** -4
iii

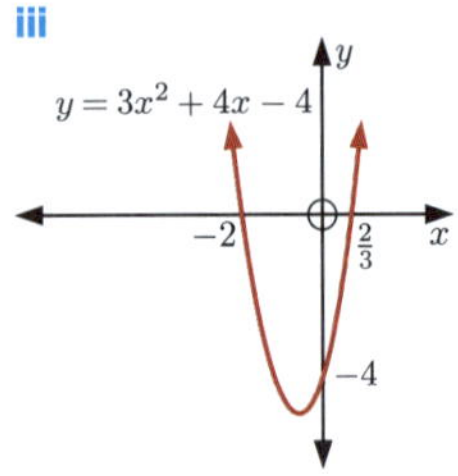

3 **a** B **b** C **c** E **d** A **e** F **f** D

REVIEW SET 21A

1 $t = 0$ or 3 **2** **a** $y = 24$ **b** $y = 3$ **c** $y = 29$

3 **a** **i** $644 **ii** $800
b **i** 0 or 30 bats **ii** 5 or 25 bats

4

x	-3	-2	-1	0	1	2	3
y	18	11	6	3	2	3	6

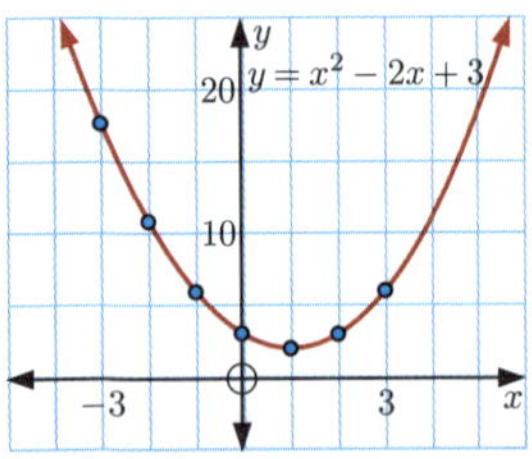

5 **a**

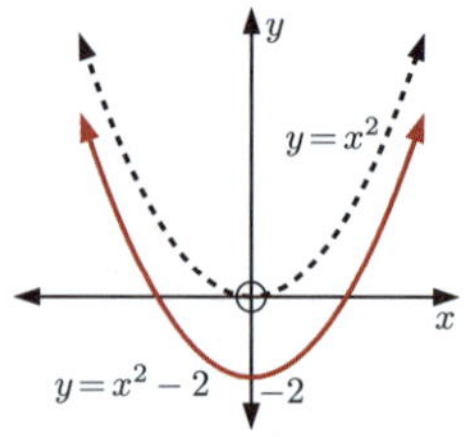

vertex is at $(0, -2)$

b

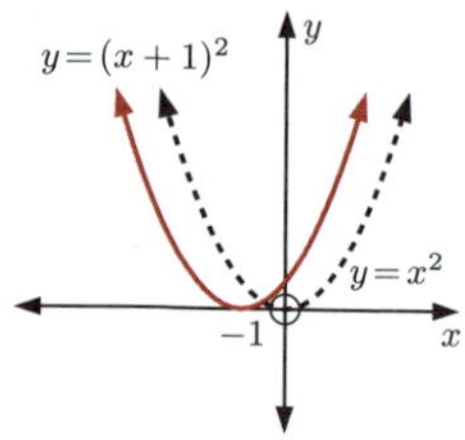

vertex is at $(-1, 0)$

6 **a**

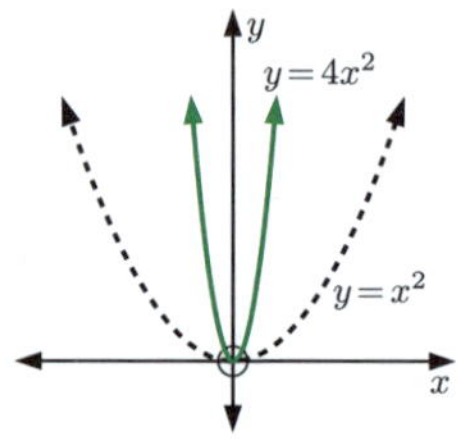

$y = 4x^2$ is 'thinner' than $y = x^2$, and the graph opens upwards.

b

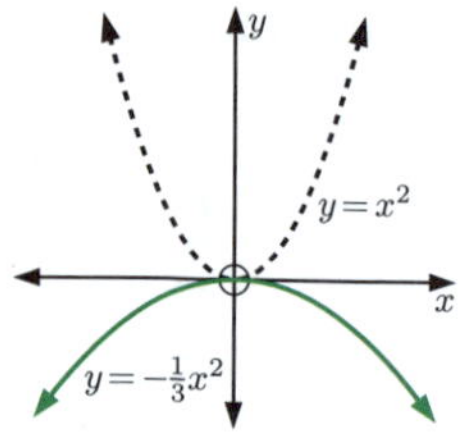

$y = -\frac{1}{3}x^2$ is 'wider' than $y = x^2$, and the graph opens downwards.

7 **a** $y = (x-2)^2 - 1$

b

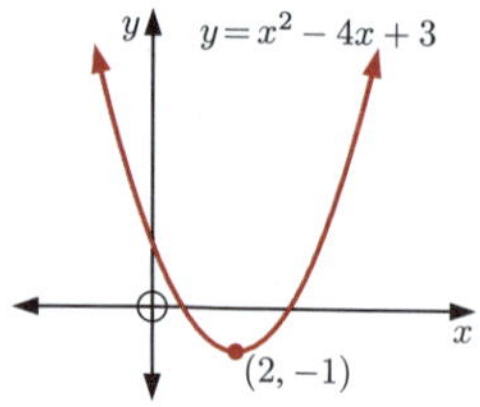

vertex is at $(2, -1)$

8 **a** -2 and 3 **b** -3 and 4 **9** **a** 10 **b** 9

10 **a** -3 and 2
b -6
c

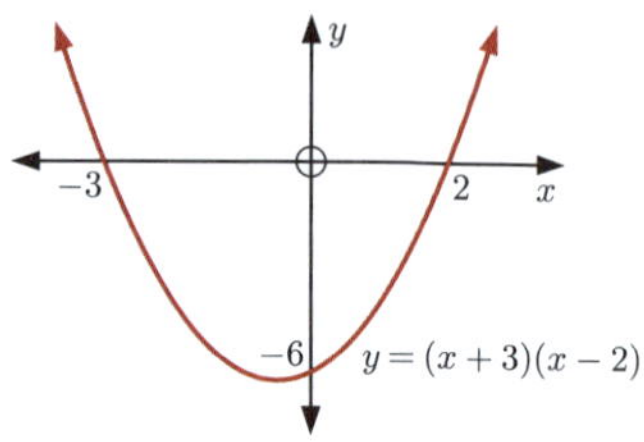

REVIEW SET 21B

1 $a = -1$ or -2 **2** **a** $y = 11$ **b** $x = -1$ or 5

3

x	-3	-2	-1	0	1	2	3
y	-4	0	2	2	0	-4	-10

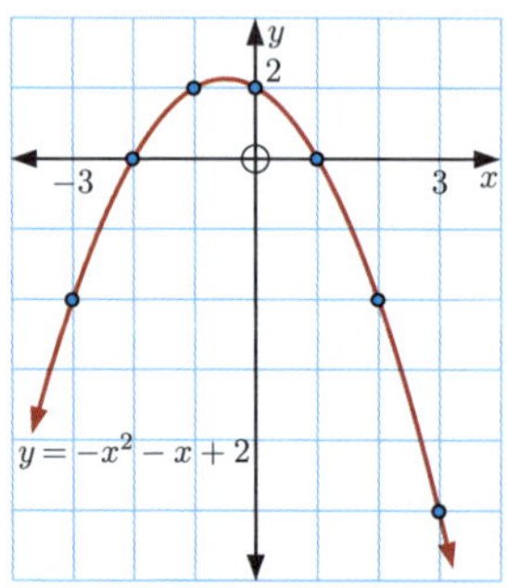

4 **a**

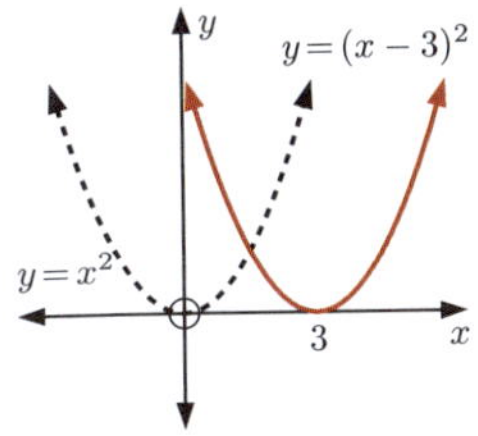

vertex is at (3, 0)

b

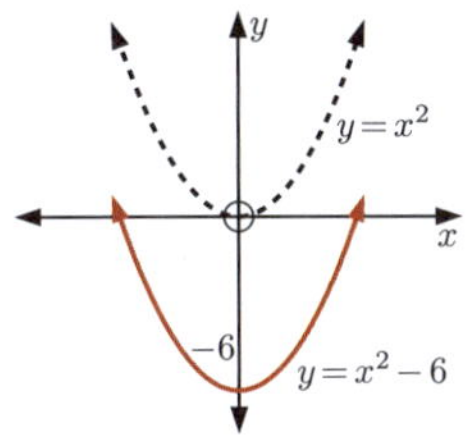

vertex is at (0, −6)

c

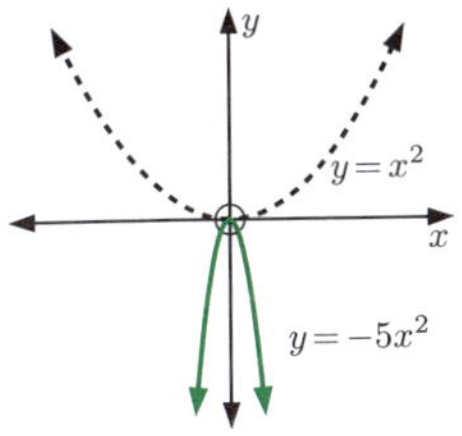

vertex is at (0, 0)

5 **a** 1 or 3 sec **b** 2 sec

6 **a** $y = (x+2)^2 + 5$

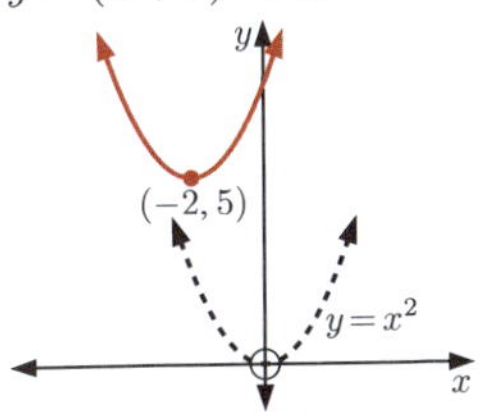

vertex is at (−2, 5)

b $y = (x-5)^2 - 3$

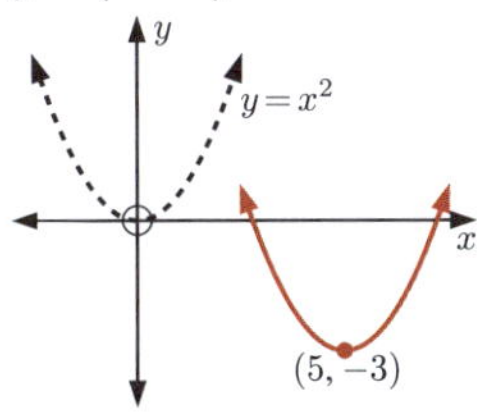

vertex is at (5, −3)

7 **a** 0 **b** −21

8 **a** $y = (x - \frac{1}{2})^2 + \frac{11}{4}$

b

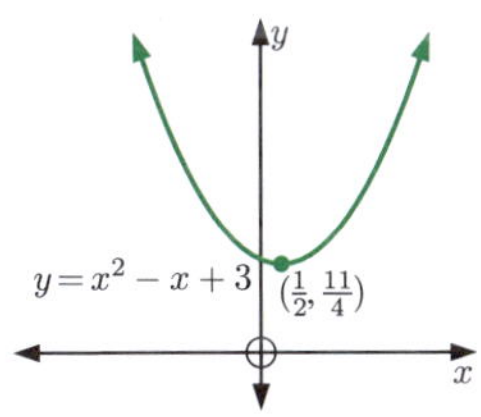

vertex is at $(\frac{1}{2}, \frac{11}{4})$

9 **a** −3 and 6 **b** $\frac{3}{2}$ and −2

10 **a** −10

b $-\frac{2}{3}$ and $\frac{5}{2}$

c

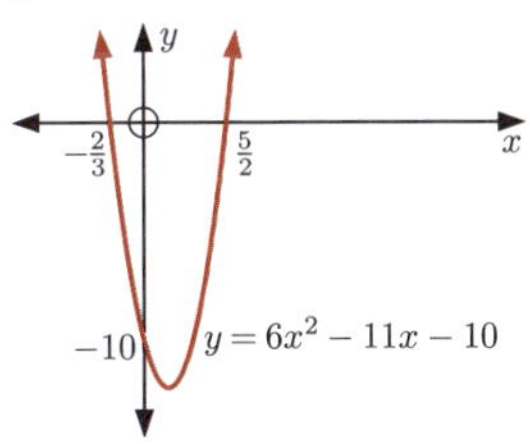

EXERCISE 22A

1 **a** $x = 100$ **b** $x = 40$ **c** $x = 70$ **d** $x = 35$
e $x = 36$ **f** $x = 90,\ y = 110$
g $a = 35,\ b = 70,\ c = 70$ **h** $x = 33\frac{1}{3}$ **i** $x = 107\frac{1}{2}$

2 **a** $a = 85$ **b** $a = 61$

3 **a** $2\alpha + 2\beta = 360°$ {sum of interior angles of a quadrilateral}
b $\alpha + \beta = 180°$ **c** co-interior angles are supplementary
d the quadrilateral is a parallelogram.

4 **a** **i** $\widehat{ACB} = x°$ **ii** $\widehat{CAE} = x°$ **iii** $\widehat{DAE} = x°$
b The line segments [AE] *bisects* $\widehat{CAD}$.

5 **Hint:** Extend the line [DC] to meet [AB].

6 **Hint:** △ADC is isosceles.

7 **Hint:** △AOP is isoceles.

8 **Hint:** Show that $\widehat{BDE} = \widehat{BCA}$.

9 **Hint:** Let $\widehat{ABC} = \alpha$. Find $\widehat{AEF}$ and $\widehat{AFE}$ in terms of α.

10 **Hint:** Label the angles inside the central pentagon $a°$, $b°$, $c°$, $d°$, and $e°$.

EXERCISE 22B

1 **a** $\widehat{OBA} = x°$, $\widehat{OBC} = y°$ **b** **Hint:** $\widehat{ABC} = (x+y)°$

2 **a** $x = 90$ **b** $x = 38$ **c** $x = 30$ **d** $x = 37$
e $x = 8$ **f** $x \approx 8.60$

3 **Hint:** Show that the triangles are equiangular.

4 **a** **Hint:** Use RHS.

5 **a** $x \approx 5.8$ **b** $x \approx 3.6$ **c** $x \approx 9.8$

6 6.3 cm **7** 5.0 cm **8** 9.2 cm

9 **a** 15.3 cm **b** 26 cm²

10 **a** $x = 90$ **b** $x = 49$ **c** $x = 35$ **d** $x \approx 8.7$
e $x \approx 6.4$ **f** $x \approx 5.9$

11 7.48 cm **12** 9.75 cm **13** $\frac{\pi}{4} \approx 78.5\%$

14 **a** **Hint:** $\widehat{OAP} = \widehat{OBP} = 90°$ **b** **Hint:** OA = OB

15 **a** $x = 7$ **b** $x = 8,\ y = 6$ **c** $x = 50$
d $x = 68$ **e** $x = 8,\ y = 30$ **f** $x = 25$

16 **Hint:** Let $\widehat{BAP} = \alpha$. **17** 10 cm

18 **a** AXOZ is a square. **b** 21 cm **19** 9π m²

20 **a** **Hint:** Show that
$A = r(a+b+c)$.

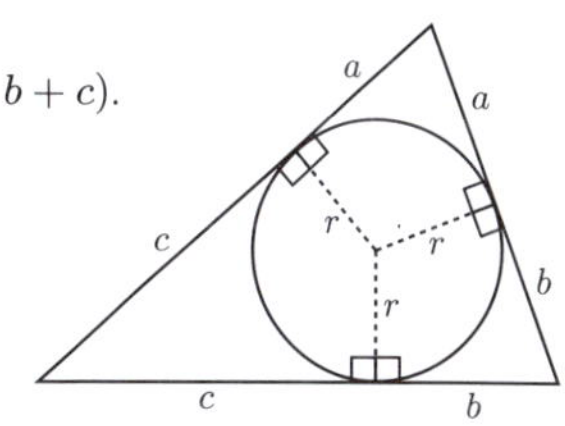

EXERCISE 22C

1 **a** $x = 10,\ \theta = 60$ {midpoint theorem, corresponding angles}
b $x = 8$ {converse of midpoint theorem}

2 **a** 5.5 cm **b** $\widehat{ANM} = 36°$ **c** 11 cm²

3 **a** **Hint:** Use the midpoint theorem. **b** ADEF, DECF

4 **Hint:** The diagonals of a parallelogram bisect each other.

5 **Hint:** Use the midpoint theorem and the converse.

6 **a**

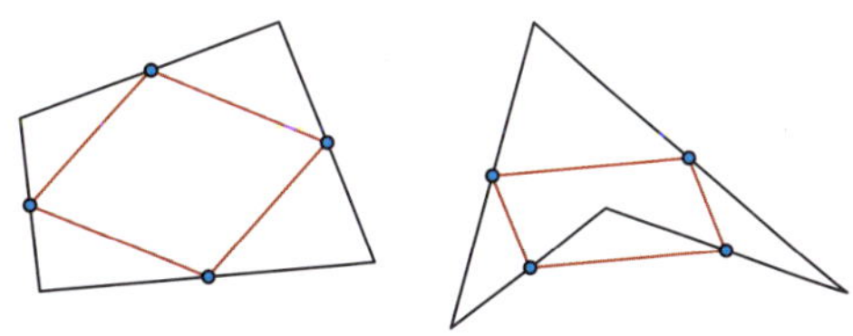

Note: Other quadrilaterals are possible.

7 **a** **Hint:** Use AAcorS **b** **Hint:** Show that BP = CR.

8 **Hint:** Show that $\triangle$s ABE and ACD are similar.

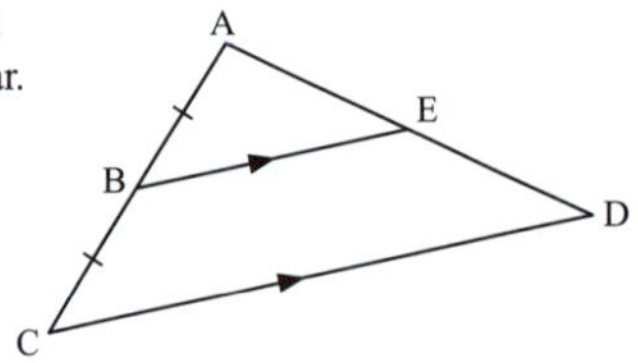

REVIEW SET 22A

1 **a** $x = 35$ **b** $x = 30$ **c** $x = 70$ **d** $x = 26$ **e** $x = 75$ **f** $x = 55$

2 **a** **i** $D\widehat{B}C = a°$ **ii** $A\widehat{D}B = (2a)°$ **iii** $D\widehat{B}A = (90 - a)°$

3 **Hint:** Show that co-interior angles are supplementary.

4 **a** $x = 36$ **b** $x = 111$ **c** $x = 2.4$

5 **Hint:** $O\widehat{T}P = 90°$ **6** 7.8 cm

7 **a** $x = 7,\ \theta = 38$ **b** $x = 6$

8 **a** **Hint:** Use the midpoint theorem. **b** ADEF, DECF

REVIEW SET 22B

1 **a** $x = 75$ **b** $x = 60$ **c** $x = 102\frac{1}{2}$ **d** $x = 66$ **e** $x = 69$ **f** $x = 40$

2 **Hint:** $A\widehat{B}C = A\widehat{C}B$

3 **a** $x = 8$ **b** $y = 8,\ z \approx 8.94$ **4** 48 cm^2

5 **Hint:** Draw a line from B to [DC], parallel to [AD].

6 **a** **Hint:** Use the midpoint theorem.

b **i** 25 cm^2 **ii** 75 cm^2

7 **Hint:** Use the *tangents from an external point* theorem.

8 **Hint:** Show that [AP] and [OX] are parallel.

EXERCISE 23A

1 **a** $x = 1$ **b** $x = 2$ **c** $x = 2$ **d** $x = 0$ **e** $x = -1$ **f** $x = -1$ **g** $x = -3$ **h** $x = -2$

2 **a** $x = 2$ **b** $x = 3$ **c** $x = 1$ **d** $x = 7$ **e** $x = 0$ **f** $x = -4$ **g** $x = 3$ **h** $x = 1$

3 **a** $x = 2$ **b** $x = 3$ **c** $x = 0$ **d** $x = -1$ **e** $x = -3$ **f** $x = 3$

4 **a** $x = \frac{5}{2}$ **b** $x = \frac{2}{3}$ **c** $x = \frac{3}{2}$ **d** $x = -\frac{2}{3}$ **e** $x = -\frac{3}{2}$ **f** $x = -\frac{1}{2}$ **g** $x = -\frac{1}{3}$ **h** $x = \frac{5}{3}$ **i** $x = -1$ **j** $x = \frac{5}{2}$ **k** $x = -3$ **l** $x = -4$

EXERCISE 23B

1 **a** exponential function **b** exponential function **c** not an exponential function **d** exponential function **e** not an exponential function **f** exponential function

2 **a** $y = 4$ **b** $y = 5$ **c** $y = 7$ **d** $y = 3\frac{1}{2}$ **e** $y = 3\frac{1}{4}$

3 **a** $y = 5$ **b** $y = 15$ **c** $y = 45$ **d** $y = 1\frac{2}{3}$ **e** $y = \frac{5}{9}$

4 **a** $W = 2$ **b** $W = 1\frac{1}{5}$ **c** $W = 1\frac{1}{25}$ **d** $W = 6$ **e** $W = 26$

5 **a** **i** $y = 5$ **ii** $y = 83$ **iii** $y = 2\frac{1}{9}$

b **i** $x = 2$ **ii** $x = 0$

6 **a** **i** $P = 6$ **ii** $P = 48$ **iii** $P = \frac{3}{8}$

b **i** $x = 4$ **ii** $x = -1$

EXERCISE 23C

1 **a** **i** 1.4 **ii** 4.3 **b** **i** $x \approx 1.6$ **ii** $x \approx -1.3$

2 **a** **i** 1.7 **ii** 0.3 **b** **i** $x \approx 1.3$ **ii** $x \approx -0.5$

3 **a** **i** 0.6 **ii** 4.7 **b** **i** $x \approx -0.6$ **ii** $x \approx 1.5$

4 **a**

x	-3	-2	-1	0	1	2	3
y	8	4	2	1	$\frac{1}{2}$	$\frac{1}{4}$	$\frac{1}{8}$

b

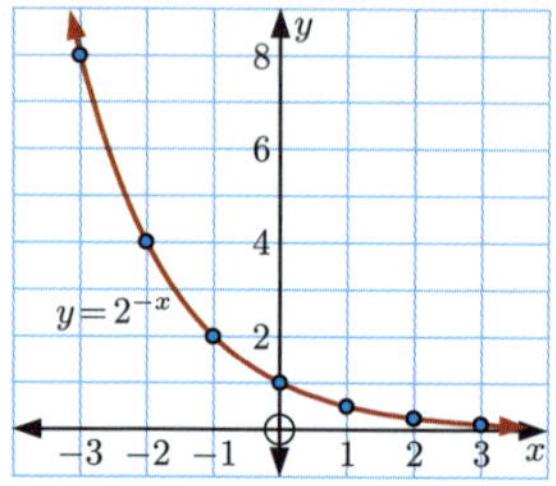

c Yes. As x increases, the graph gets closer to the x-axis but always lies above it.

d **i** 0.7 **ii** $x \approx -1.8$

EXERCISE 23D

1 **a** 500 mice **b** **i** 1550 mice **ii** 4820 mice

c

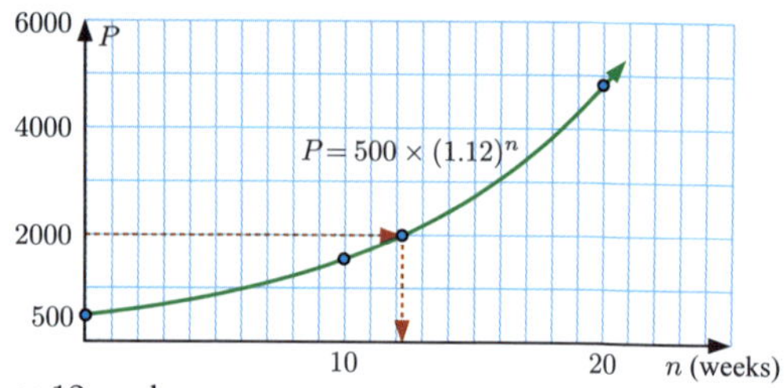

d $\approx$ 12 weeks

2 **a** 20 g **b** $\approx$ 23.6 g

c

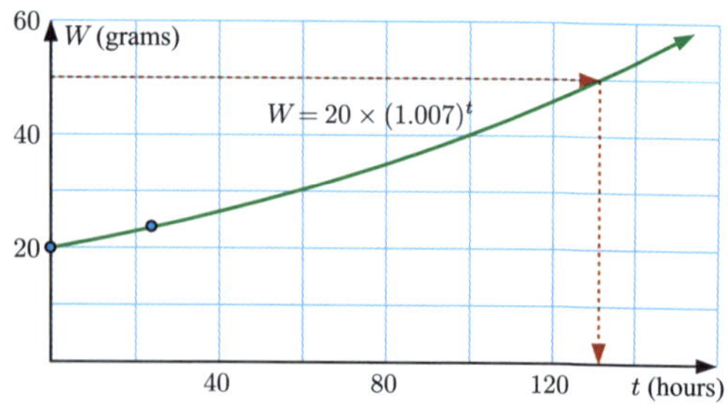

d $\approx$ 131 hours

3 **a** 250 wasps **b** **i** 1070 wasps **ii** 6530 wasps

c

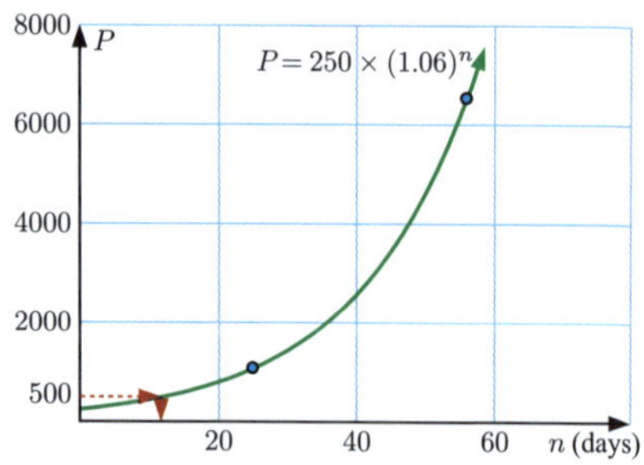

d $\approx$ 12 days

4 **a**

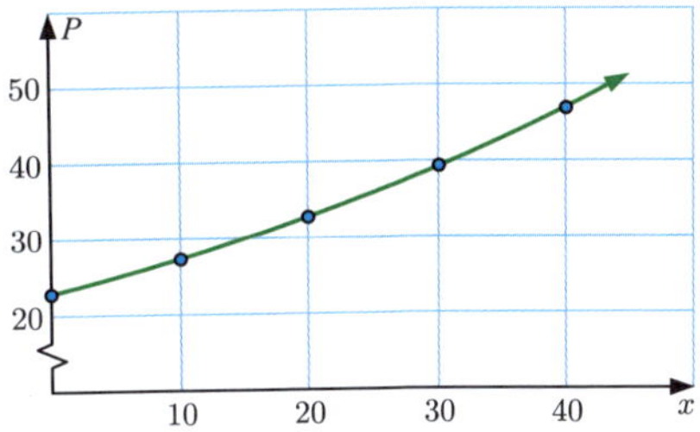

b **i** $a = 23$ **ii** $b \approx 1.018$

c 1980, $P \approx 27.6$; 1990, $P \approx 33.1$; 2000, $P \approx 39.7$

d **i** $\approx 57\,300$ **ii** $\approx 119\,000$

EXERCISE 23E

1 **a** $100°C$

b **i** $50.0°C$ **ii** $25.0°C$ **iii** $12.5°C$

c

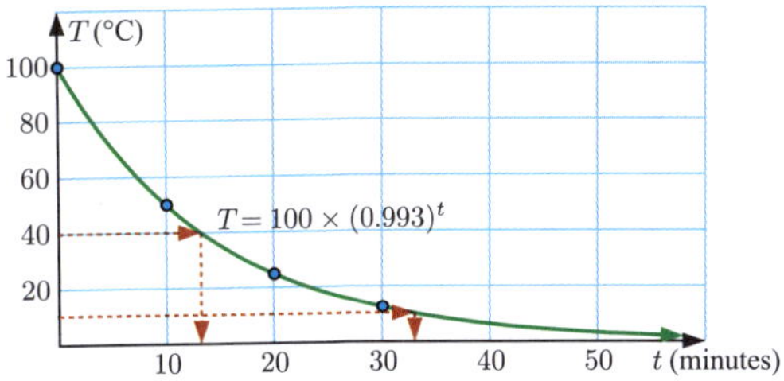

d **i** ≈ 13 mins **ii** ≈ 33 mins

2 **a** 150 g

b **i** ≈ 111 g **ii** ≈ 82.2 g **iii** ≈ 45.1 g

c

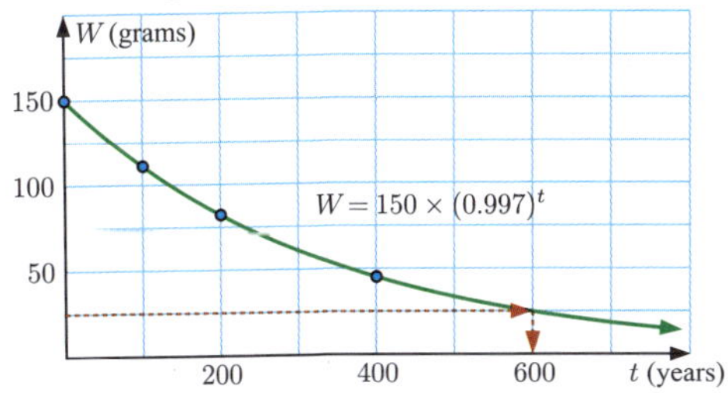

d ≈ 596 years

3 **a**

n	0	5	10	15	20	25
P	225	204	163	131	104	84

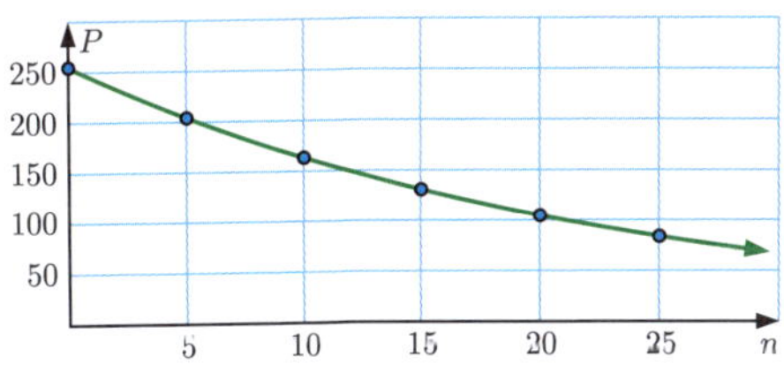

b **i** $a = 255$ **ii** $b \approx 0.957$

c So, $P = 255 \times (0.9566)^n$
1990, $P \approx 204$; 1995, $P \approx 164$; 2000, $P \approx 131$;
2005, $P \approx 105$

d 2021

EXERCISE 23F

1 **a**

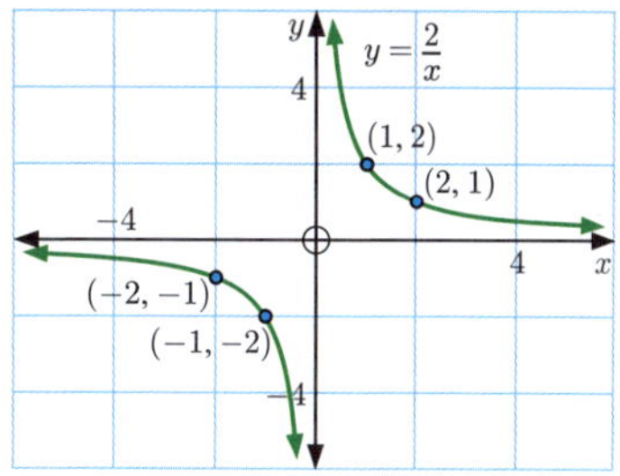

b

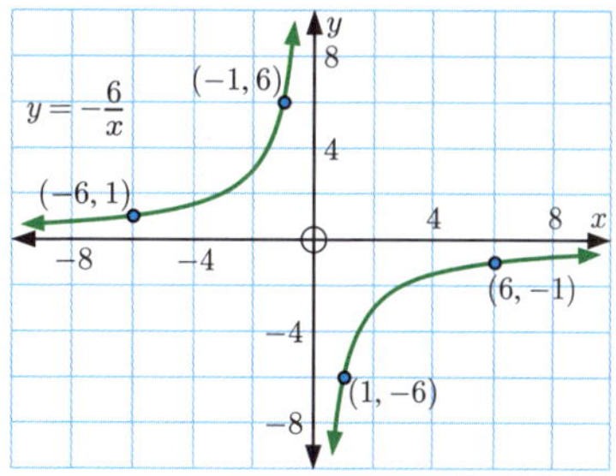

c

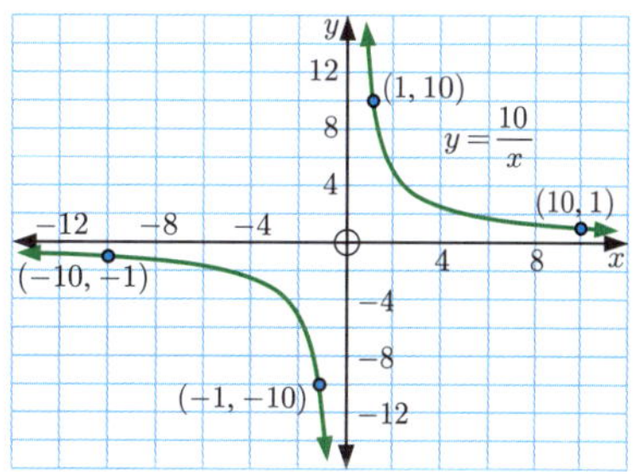

d

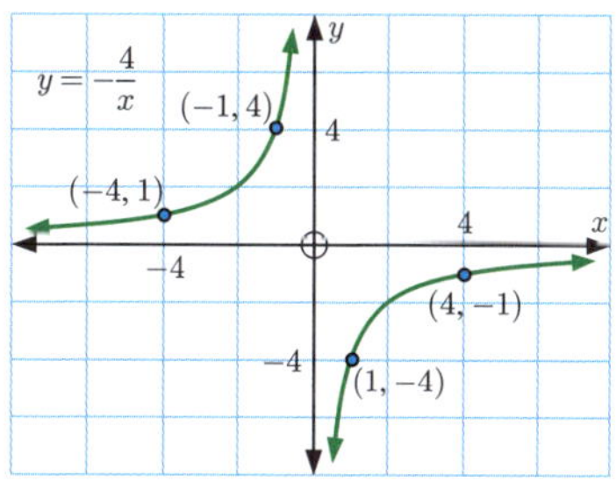

2 **a** $y = \dfrac{8}{x}$ **b** $y = \dfrac{3}{x}$ **c** $y = -\dfrac{12}{x}$

3 **a** $W = \dfrac{120}{L}$ (or $LW = 120$), $L > 0$

b

L	6	8	10	12	15	20
W	20	15	12	10	8	6

c

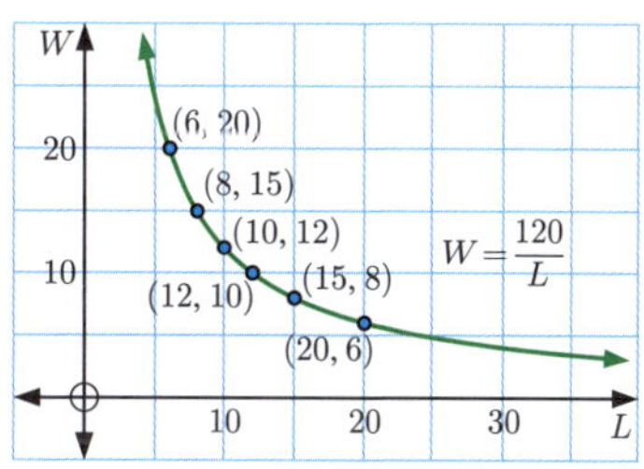

4 **a** $t = \dfrac{30}{c}$, $c > 0$ **b**

c	3	4	5	6	8	10
t	10	$7\frac{1}{2}$	6	5	$3\frac{3}{4}$	3

c

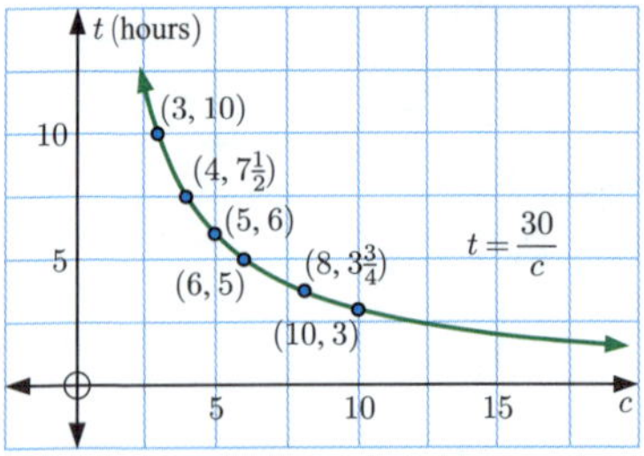

5 $y = x$, $y = -x$

REVIEW SET 23A

1 **a** $x = 3$ **b** $x = 1$ **c** $x = 2$

2 **a** $x = 3$ **b** $x = 1\frac{1}{3}$

3 **a** $y = -1$ **b** $y = 7$ **c** $y = -1\frac{8}{9}$

4 **a** **i** $K = 6$ **ii** $K = 13$ **b** **i** $x = 5$ **ii** $x = -1$

5 **a** 80°C **b** **i** 26.8°C **ii** 9.00°C **iii** 3.02°C

c

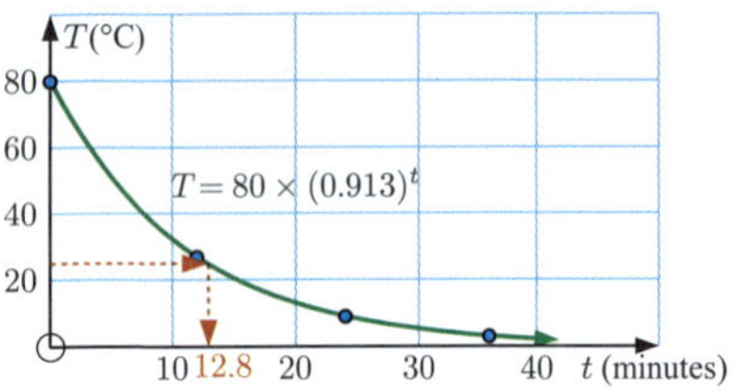

d $\approx$ 12.8 minutes

6 **a** 1500 termites **b** **i** 1930 termites **ii** 13 100 termites

c

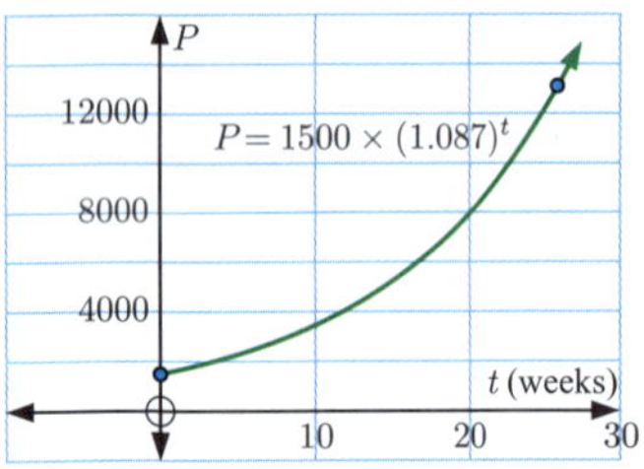

7 **a**

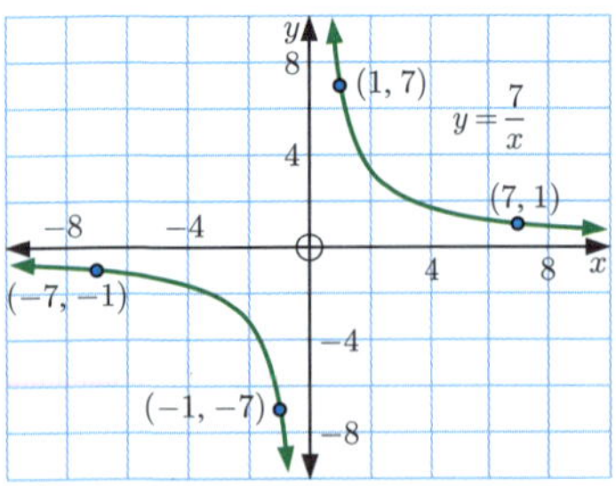

b

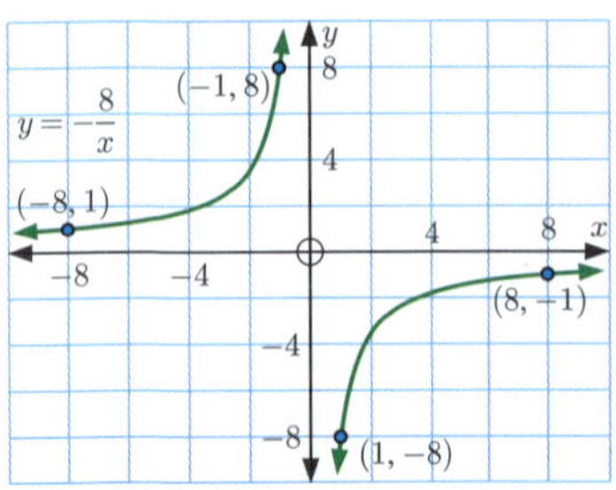

8 **a** If $k > 0$ then $xy > 0$, so x and y have the same sign, $\therefore$ lie entirely in Q_1 or Q_3.

b If $k < 0$ then $xy < 0$, so x and y have opposite signs, $\therefore$ lie entirely in Q_2 or Q_4.

REVIEW SET 23B

1 **a** $x = 3$ **b** $x = \frac{1}{2}$ **c** $x = 3$

2 **a** $y = \frac{1}{2}$ **b** $y = 4$ **c** $y = \frac{1}{8}$

3 **a** 815 zebras

b **i** $\approx$ 1990 zebras

ii $\approx$ 4840 zebras

iii $\approx$ 11 800 zebras

c

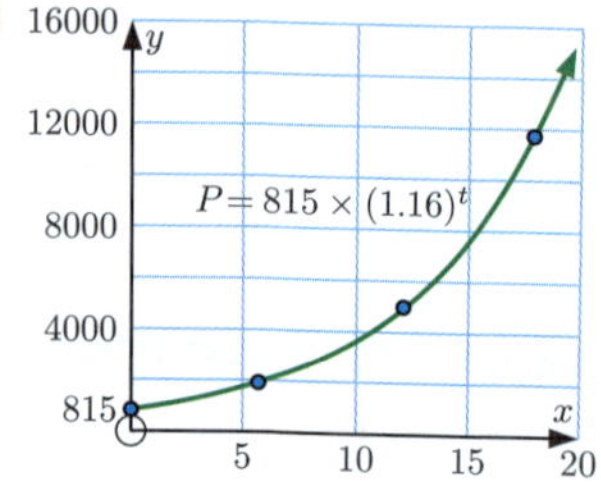

d $\approx$ 15.4 years

4 **a** $x = 2$ **b** $x = -\frac{1}{2}$

5 **a**

x	-2	-1	0	1	2
y	$\frac{1}{16}$	$\frac{1}{4}$	1	4	16

b

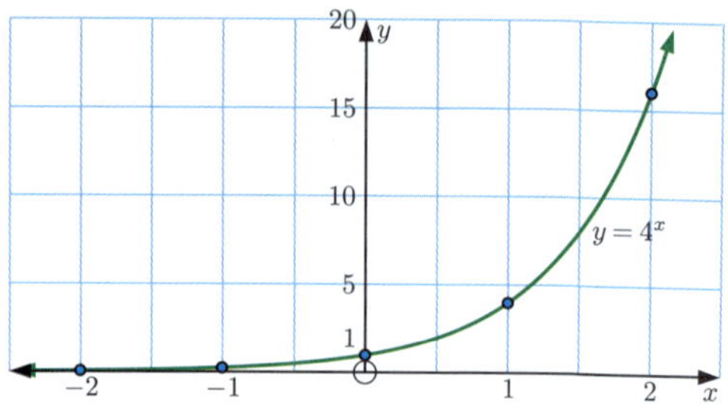

c **i** ≈ 3.0 **ii** $x \approx 1.7$

6 $y = -\dfrac{21}{x}$

7 **a** 1500 g **b** **i** 90.3 g **ii** 5.44 g

c

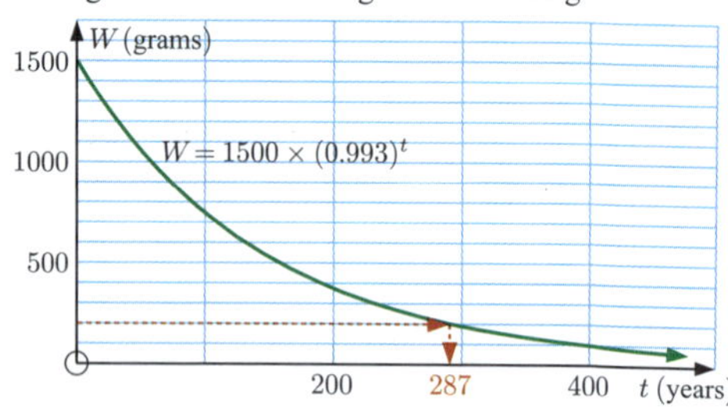

d $\approx$ 287 years

8 **a** $t = \dfrac{600}{s}$, $s > 0$

b

Speed (s km/h)	40	50	60	80	100	120
Time (t hours)	15	12	10	7.5	6	5

c

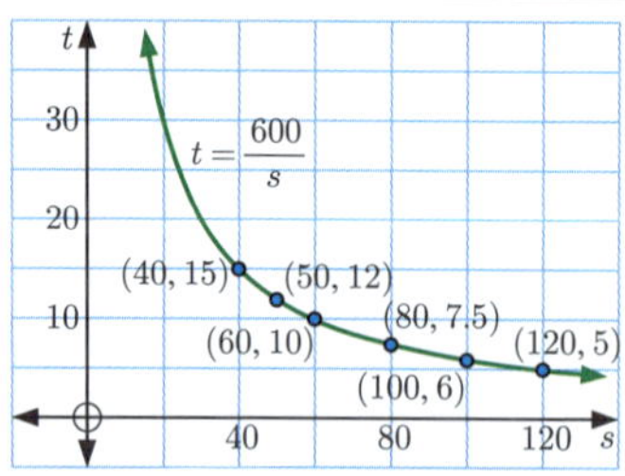

EXERCISE 24A.1

1 **B** (a straight line through O)

2 **a** **i**

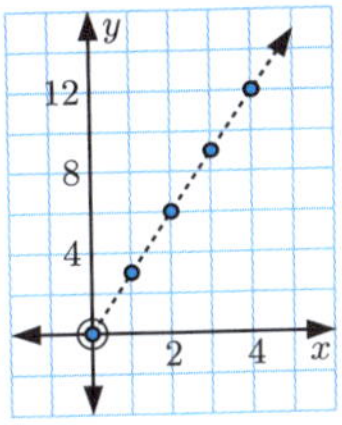

ii directly proportional $k = 3$

b **i**

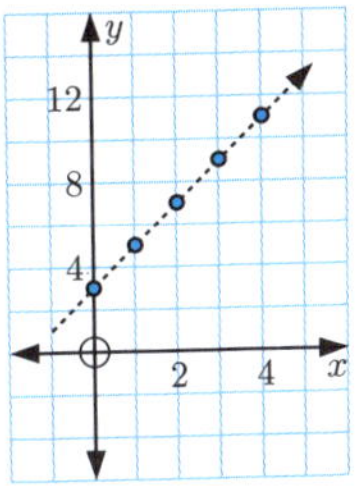

ii not directly proportional

c **i**

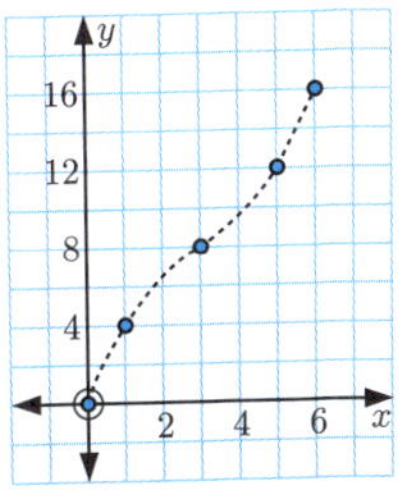

ii not directly proportional

d **i**

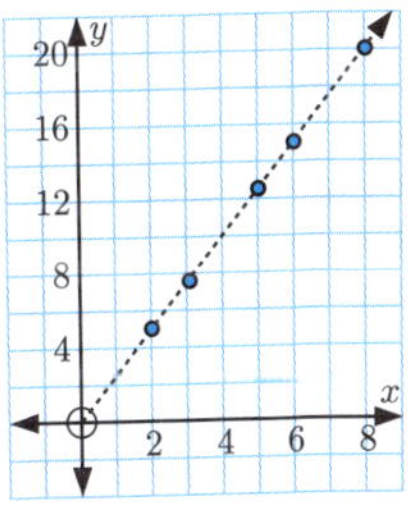

ii directly proportional $k = 2.5$

3 **a**

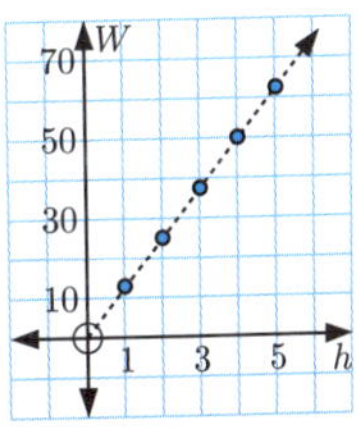

b The graph of W against h is a straight line through the origin.

c $W = 12.5h$

4 **a**

Time (t *minutes*)	0	2	4	6	8	10
Volume of water leaked (V mL)	0	20	40	60	80	100

b

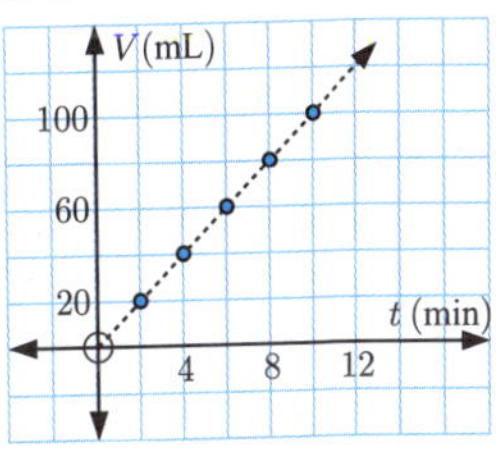

c yes, $k = 10$

5 **a**

Number of items (n)	0	1	2	3	4	5
Cost ($\$C$)	0	5	10	15	17	19

b

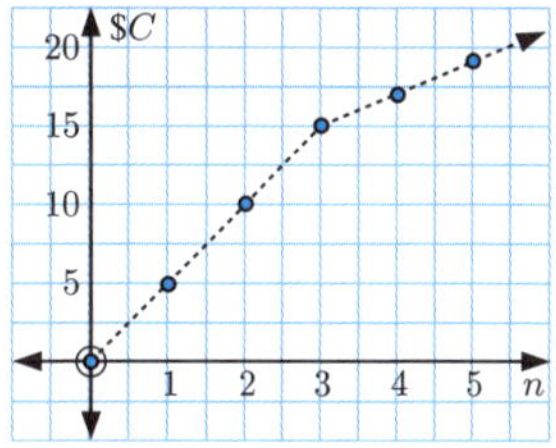

c no

6 **a** **i**

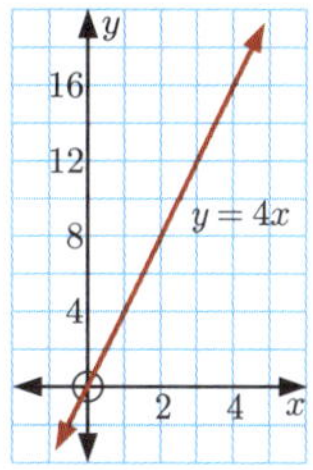

ii directly proportional, $k = 4$

b **i**

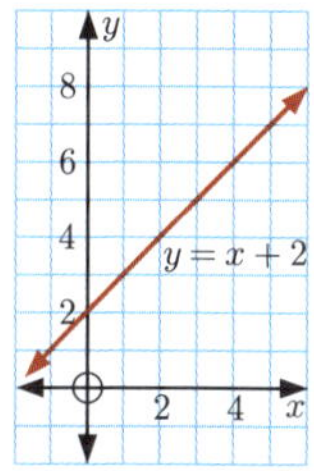

ii not directly proportional

c **i**

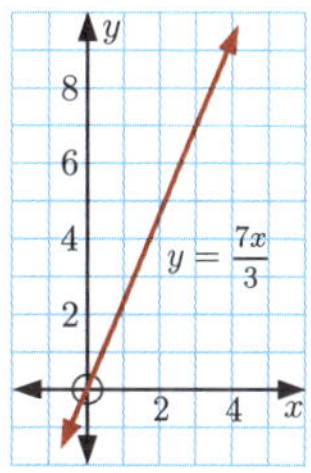

ii directly proportional, $k = \frac{7}{3}$

d **i**

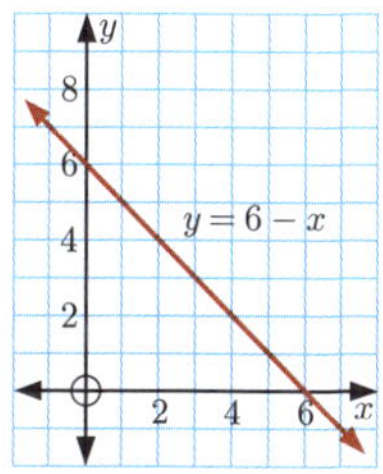

ii not directly proportional

7 **a** y is doubled **b** y is trebled
c x is doubled **d** x is halved
e y is increased by 20% **f** x is decreased by 30%
g cannot be determined **h** cannot be determined

8 **a** 2π is a constant so the graph of C against r is a straight line through the origin.
b $k = 2\pi$
c **i** C is doubled **ii** r is increased by 50%

EXERCISE 24A.2

1 **a** $y = 140$ **b** $x = 1$ **2** **a** $y \approx 229$ **b** $x \approx 22.6$

3 **a** 4 litres **b** 121.5 km

4 **a** **i** $R = 0.006l$ **ii** 0.3 ohms **iii** 500 cm

b

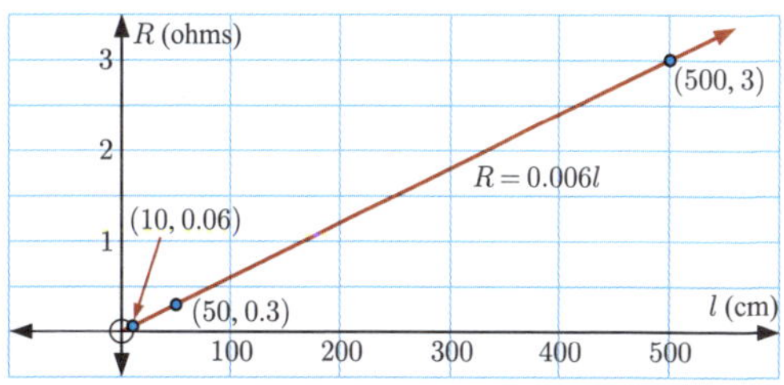

5 **a** 45 m^2 **b** 9 litres

6 **a** 78.4 m/s **b** ≈ 10.2 seconds

7 **a** 10.5 cm **b** 1.2 m

EXERCISE 24B

1 a $A \propto t^2$, $k = 4.9$ b $K \propto x^3$, $k = 2$
c $T \propto \sqrt{l}$, $k = \frac{1}{5}$ d $V \propto t^4$, $k = 500$
e $P \propto \sqrt[3]{x}$, $k = 4$ f $V \propto r^3$, $k = \frac{4}{3}\pi$

2 a

x^2	4	9	25
y	16	36	100

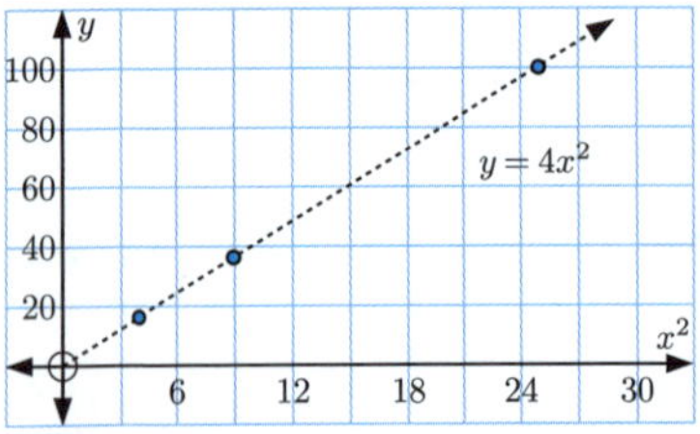

The graph is a straight line through the origin.
$\therefore\ y \propto x^2$
Rule is $y = 4x^2$

b

x^2	1	9	16	25
y	3	27	48	75

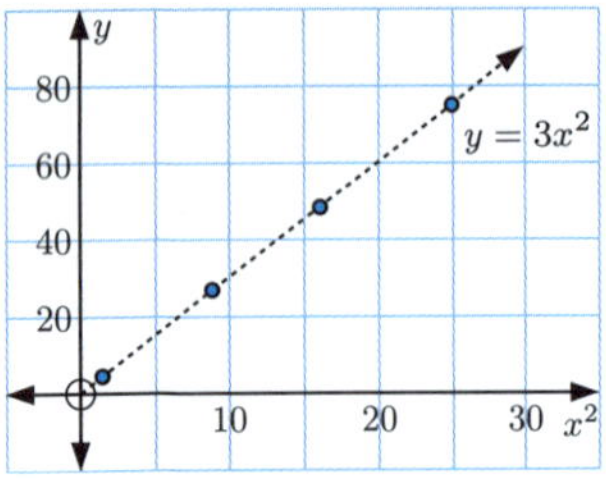

The graph is a straight line through the origin.
$\therefore\ y \propto x^2$
Rule is $y = 3x^2$

3 a

x^3	1	8	27
y	5	40	135

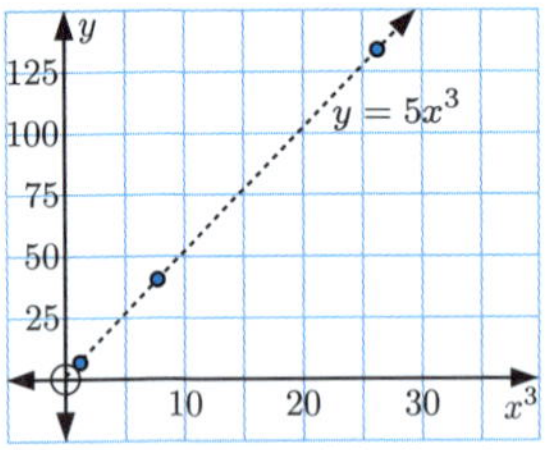

The graph is a straight line through the origin.
$\therefore\ y \propto x^3$
Rule is $y = 5x^3$
$\therefore\ k = 5$

b

x^3	8	27	125
y	4	13.5	62.5

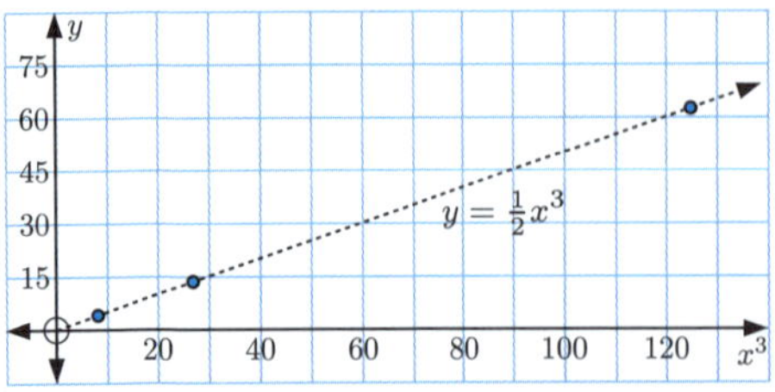

The graph is a straight line through the origin.
$\therefore\ y \propto x^3$
Rule is $y = \frac{1}{2}x^3$
$\therefore\ k = \frac{1}{2}$

4 a $P = 48$ b $a = 15$
5 a $M = 81$ b $x \approx 3.42$
6 \$312.50 7 ≈ 8.9 cm 8 ≈ 10.1 seconds
9 ≈ 12.2 km 10 36% reduction
11 a multiplied by $\sqrt[3]{2}$ ($\approx 26.0\%$ increase)
b 27.1% reduction
12 a $V \propto h$ b $V \propto r^2$ c 22.4% increase

EXERCISE 24C.1

1 a $xy = 12$ b not inversely proportional

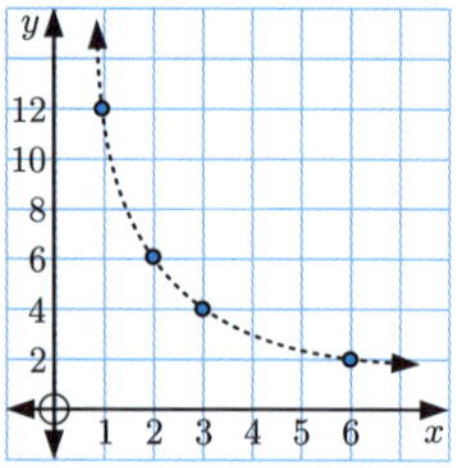

c $xy = 84$

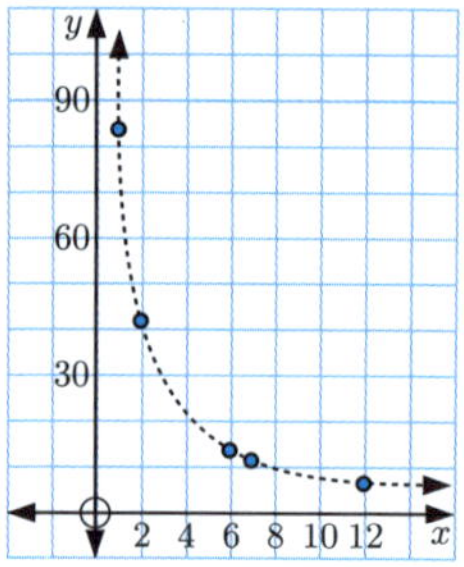

2

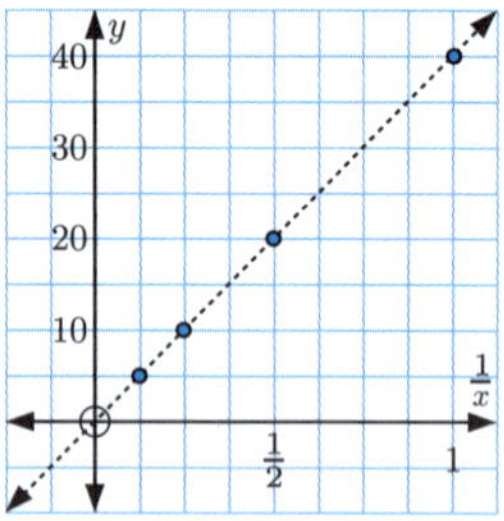

The points on the graph of $y = \dfrac{1}{x}$ form a straight line through the origin, $\therefore\ y \propto \dfrac{1}{x}$.
$\therefore\ y$ is inversely proportional to x.

3 **B** and **C**

4 a If d is fixed, s is **inversely** proportional to t.
b If t is fixed, s is **directly** proportional to d.
c If s is fixed, d is **directly** proportional to t.

5 a y is multiplied by $\frac{1}{4}$ b y is multiplied by $\frac{9}{7}$
c x is multiplied by $\frac{3}{8}$

6 a $y = 8$ b $x = \frac{2}{5}$
7 a $p = 18$ b $n = 12$ 8 8 days

9 **a** 20 units **b** 4 cm^3

c

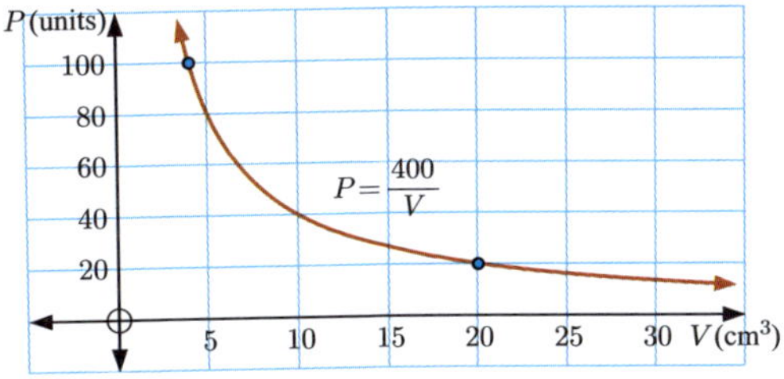

10 **a** 23 min 20 s **b** 8 players **c** 40 min

EXERCISE 24C.2

1 **a** $y = 16$ **b** $x = 4$

2 **a** $F = \frac{20}{\sqrt{3}} \approx 11.5$ **b** $c = \frac{44}{9} \approx 4.89$

3 **a** ≈ 0.115 seconds **b** 400 units

4 56.25% increase

5 **a** ≈ 3.28 cm **b** ≈ 3.46 cm **c** ≈ 300 cm^3

6 ≈ 8.94 m away

REVIEW SET 24A

1 **a**

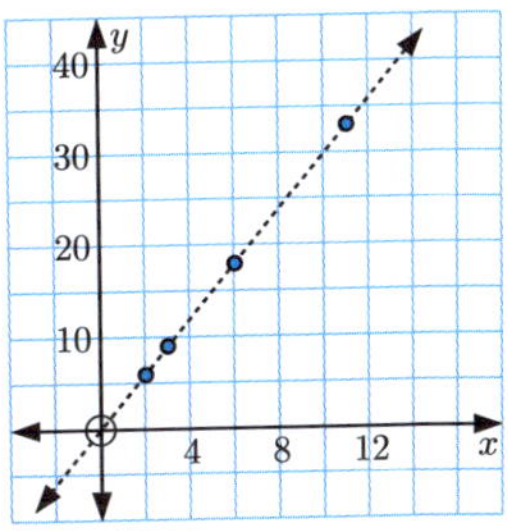

b The graph of y against x is a straight line through the origin. $\therefore\ y \propto x$

c $y = 3x$

2 **a** $xy = 650$ **b** $y \approx 54.2$ **c** $x \approx 4.33$

3 M is trebled **4** $a = 0.4,\ b = 4$

5 **a** $\approx 34.4\%$ reduction **b** $\approx 73.2\%$ increase

6 **a** y is divided by 3 **b** x is decreased by 33.3%

7 72 people

8 **a** **i** $E \propto m$ if v is constant

ii $E \propto v^2$ if m is constant

iii $m \propto \frac{1}{v^2}$ if E is constant

b If m is constant, $E \propto v^2$.

$\therefore$ E has a $\approx 22.6\%$ reduction.

9 1 filling **10** **a** $y \propto x$ **b** $y = 3x$ **c** $x = 24$

REVIEW SET 24B

1 $a = 14,\ b = 21$ **2** ≈ 229 km

3 **a** Each P value is 36 times each n value. $\therefore\ P \propto n$

b $P = 36n$

c

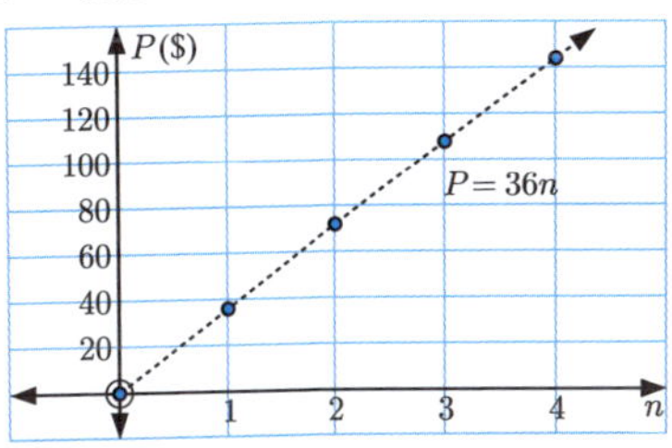

4 **a** $k = 19$ **b** $y = 119.7$

5 **a** The volume is multiplied by 4.

b The radius is increased by $\approx 26.5\%$.

6 5.5 hours

7

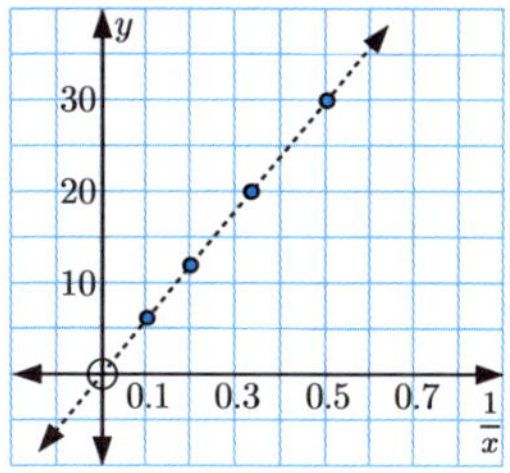

The graph of y against $\frac{1}{x}$ is a straight line through O.

$\therefore\ y \propto \frac{1}{x}$

8 27 minutes

9 **a** y is multiplied by 4. **b** x is reduced by $\approx 16.7\%$.

10 159 cm^3

EXERCISE 25A.1

1 **a** 1 **b** ≈ 0.97 **c** ≈ 0.91 **d** ≈ 0.87
e ≈ 0.71 **f** 0.50 **g** ≈ 0.26 **h** 0

2 **a** 0 **b** ≈ 0.26 **c** ≈ 0.42 **d** 0.50
e ≈ 0.71 **f** ≈ 0.87 **g** ≈ 0.97 **h** 1

3 P(0.921, 0.391), Q(0.309, 0.951)

4 **a** $\theta \approx 34.2°$ **b** $\theta \approx 68.9°$

EXERCISE 25A.2

1 **a** ≈ -0.17 **b** ≈ 0.17 **c** -0.5 **d** 0.5
e ≈ -0.64 **f** ≈ 0.64 **g** -1 **h** 1

2 **a** ≈ 0.94 **b** ≈ 0.94 **c** ≈ 0.87 **d** ≈ 0.87
e ≈ 0.77 **f** ≈ 0.77 **g** 0 **h** 0

3 P(-0.276, 0.961), Q(-0.921, 0.391)

4 **a** $\cos(180° - \theta) = -\cos\theta$

b

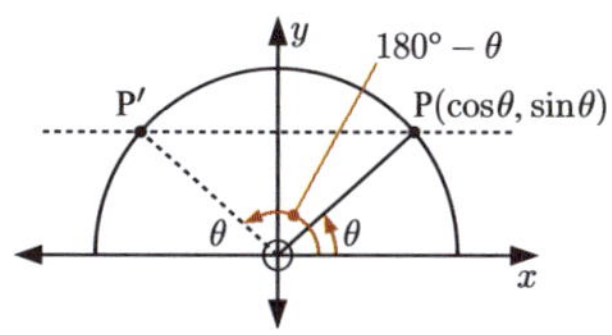

P′ has x-coordinate $\cos(180° - \theta)$. But P′ is a reflection of P in the y-axis.

$\therefore$ P′ has x-coordinate $-\cos\theta$.

$\therefore$ $\cos(180° - \theta) = -\cos\theta$

c **i** $\cos 63° \approx 0.454$, $\cos 117° \approx -0.454$

ii $\cos 26° \approx 0.899$, $\cos 154° \approx -0.899$

5 **a** $\sin(180° - \theta) = \sin\theta$

b Comparing y-coordinates using the diagram in **4 b**,
P′ has y-coordinate $\sin\theta$ {reflection in y-axis}
But P′ has y-coordinate $\sin(180° - \theta)$
$\therefore$ $\sin(180° - \theta) = \sin\theta$

c **i** $\sin 71° \approx 0.946$, $\sin 109° \approx 0.946$

ii $\sin 8° \approx 0.139$, $\sin 172° \approx 0.139$

6 **a** 154° **b** 135° **c** 111° **d** 94°

7 **a** 82° **b** 53° **c** 24° **d** 12°

8 **a** ≈ 0.9272 **b** ≈ 0.4384 **c** ≈ -0.9781
d ≈ 0.6561 **e** ≈ 0.5736 **f** ≈ -0.1392

EXERCISE 25B

1 **a** ≈ 25.7 m^2 **b** ≈ 21.0 cm^2 **c** ≈ 30.9 cm^2
d ≈ 33.0 m^2 **e** 12 m^2 **f** ≈ 1.52 km^2

2 **a** ≈ 9.79 m^2 **b** ≈ 23.6 cm^2 **c** ≈ 6.49 m^2

3 **a** Area **A**: 85 m^2, Area **B**: 85 m^2
b Triangles have the same area because the sides used are the same length and $\sin 30° = \sin 150°$.

4 $x \approx 20.0$ m **5** ≈ 58.1 cm^2 **6** ≈ 43.5 cm^2

7 ≈ 33.4 cm^2

8 **Hint:** Let AM $= a$, AN $= b$, and B$\widehat{\text{A}}$C $= \theta$.
Show that both triangles have area $ab\sin\theta$.

EXERCISE 25C.1

1 **a** $x \approx 9.83$ **b** $x \approx 11.9$ **c** $x \approx 8.31$

2 **a** $x \approx 177$ **b** $x \approx 6.02$ **c** $x \approx 38.1$

3 **a** $\theta = 50°$, $x \approx 104$, $y \approx 93.2$
b $\phi = 49°$, $x \approx 161$, $z \approx 163$
c $\phi = \theta = 39°$, $x \approx 2.49$

4 $\theta = 50°$, $x \approx 13.5$, $y \approx 9.03$

EXERCISE 25C.2

1 **a** $\theta \approx 30.0°$ **b** $\theta \approx 46.8°$ **c** $\theta \approx 112°$
d $\theta \approx 62.0°$ **e** $\theta \approx 122°$ **f** $\theta \approx 131°$
g $\theta \approx 53.3°$ **h** $\theta \approx 106°$ **i** $\theta \approx 109°$

2 **a** $\phi \approx 50.4°$, $\theta \approx 47.6°$, unknown side ≈ 6.71 cm
b $\phi \approx 15.8°$, $\theta \approx 121°$, unknown side ≈ 18.8 m
c $\phi \approx 110°$, $\theta \approx 38.2°$, unknown side ≈ 6.77 cm

EXERCISE 25D

1 **a** ≈ 17.3 m **b** ≈ 6.29 m **c** ≈ 25.4 km
d ≈ 12.6 cm **e** ≈ 195 m **f** ≈ 15.8 km

2 **a** $\theta \approx 48.2°$ **b** $\theta = 90°$ **c** $\phi \approx 72.0°$
d $\phi \approx 101.5°$ **e** $\alpha \approx 99.8°$ **f** $\beta \approx 113.4°$

3 **a** $A \approx 44.4°$, $B \approx 78.5°$, $C \approx 57.1°$
b $P \approx 32.2°$, $Q = 120°$, $R \approx 27.8°$

4 **a** $x \approx 6.27$, $\alpha \approx 41.7°$, $\beta \approx 94.3°$
b $x \approx 5.59$, $\alpha \approx 34.0°$, $\beta \approx 123.0°$

5 ≈ 44.8 m^2

EXERCISE 25E

1 ≈ 88.7 km

2 **a**

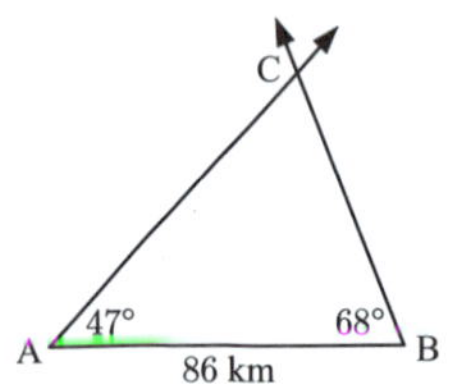

b ≈ 18.6 km

3 ≈ 1.64 hectares

4 **a** Yes; $\sin 130° = \sin 50°$ and $\cos 130° = -\cos 50°$ (consider the unit half-circle).
b **i** ≈ 4990 m^2 **ii** ≈ 208 m

5 ≈ 1.25 km

6 **a** $\theta = 10°$ **b** ≈ 88.7 km **c** ≈ 56.0 km

7 ≈ 17.8 m

8 **a** B$\widehat{\text{C}}$A $\approx 29.3°$ **b** A is closer to C **c** ≈ 16.0 m

9 **Hint:** Show that A$\widehat{\text{C}}$D $= 120°$.

REVIEW SET 25A

1 P(0.545, 0.839) **2** 103°

3 **a** ≈ 3.88 km^2 **b** ≈ 51.4 cm^2

4 **a** $x \approx 5.23$ **b** $x \approx 2.54$ **c** $x \approx 10.0$ **5** $\theta \approx 56.8°$

6 **a** ≈ 3.62 m **b** ≈ 9.98 cm **c** ≈ 13.6 m

7 **a** **i** A$\widehat{\text{B}}$C $\approx 96.4°$ **ii** B$\widehat{\text{C}}$A $\approx 35.4°$
b **i** Area ≈ 31.3 m^2 **ii** Area ≈ 31.3 m^2

8 **a** BD ≈ 278 m **b** $\approx 46\,700$ m^2

9 **a** ≈ 275 m **b** 2.86 hectares

REVIEW SET 25B

1 $\theta \approx 18.2°$ **2** Q is $(-0.656, 0.755)$

3 **a** ≈ 29.3 cm^2 **b** ≈ 21.0 m^2

4 **a** If $\sin\theta = a$, then $\sin(180° - \theta) = a$.
b If $\cos\theta = b$, then $\cos(180° - \theta) = -b$.

5 $\theta = 99°$, $x \approx 16.4$, $y \approx 14.8$ **6** RQ ≈ 14.6 m

7 $\theta \approx 137°$, $\phi \approx 15.9°$, $x \approx 4.83$

8 **a** $\approx 85.9°$, $\approx 39.4°$, $\approx 54.7°$ **b** ≈ 31.4 m^2

9 **a** **Hint:** $\triangle$PQR and $\triangle$P$'$Q$'$R$'$ are similar
b **Hint:** Find area $\triangle$PQR and area $\triangle$P$'$Q$'$R$'$ using sine Q.

10 **a** 10 cm **b** 5 cm **c** ≈ 152 cm^2

INDEX